From Space to Place

2nd International Conference on Remote Sensing in Archaeology

Proceedings of the
2nd International Workshop,
CNR, Rome, Italy, December 4-7, 2006

Edited by
Stefano Campana
Maurizio Forte

BAR International Series 1568
2006

Published in 2016 by
BAR Publishing, Oxford

BAR International Series 1568

From Space to Place: 2nd International Conference on Remote Sensing in Archaeology

ISBN 978 1 84171 998 6

BAR Publishing is the trading name of British Archaeological Reports (Oxford) Ltd. British Archaeological Reports was first incorporated in 1974 to publish the BAR Series, International and British. In 1992 Hadrian Books Ltd became part of the BAR group. This volume was originally published by Archaeopress in conjunction with British Archaeological Reports (Oxford) Ltd / Hadrian Books Ltd, the Series principal publisher, in 2006. This present volume is published by BAR Publishing, 2016.

Printed in England

BAR
PUBLISHING

BAR titles are available from:

BAR Publishing
122 Banbury Rd, Oxford, OX2 7BP, UK
EMAIL info@barpublishing.com
PHONE +44 (0)1865 310431
FAX +44 (0)1865 316916
www.barpublishing.com

Contents

SESSION 7
Environment Analysis for Remote Sensing Archaeology

SESSION 8
3D Visualization of Place and Landscapes

SESSION 9
Virtual Archaeological Reconstruction

SESSION 10
Landscapes, CRM and Ethics

POSTER SESSIONS

Introduction

In 2001, UNESCO and the European Space Agency (ESA) launched the 'Open Initiative on the Use of Space Technologies to Monitor Natural and Cultural Heritage of UNESCO Sites'. The 'Open Initiative' is a framework of cooperation to assist countries to improve the observation, monitoring and management of natural and cultural sites as well as of their surroundings, through space technologies. In this field of operations a group of experts, called International Working Group of Space Technologies for World Heritage, was created under the coordination of UNESCO, the present membership including representatives of CNR-ITABC (Italy), GORS (Syria), the Chinese Academy of Sciences (China), NASA (US), ETH (Switzerland) and other European research centres and institutions.

In accordance with the Beijing Declaration issued at the International Conference on Remote Sensing Archaeology in October 2004, the International Working Group of Space Technologies for World Heritage has already started its work. At the Beijing conference the topics discussed demonstrated clearly that the concept of Remote Sensing was significantly wider than in the past and involved the integration of numerous different technologies and fields of application: photogrammetry, air photography, air-photo mapping, airborne multi-spectral and thermal imagery, satellite imagery, geophysics, GIS but also, laser scanning, visualization displays, space models virtual reality.

This conference at Rome in December 2006, building on these ideas, will aim to continue in this direction, promoting the use of integrated methodologies in remote sensing archaeology so as to help in the creation of new and sustainable policies in the monitoring, interpretation, fruition and communication of the cultural heritage.

The event has been organized by representatives of CNR-ITABC and the Laboratory of Landscape Archaeology and Remote Sensing at the University of Siena. During the four days of the conference about 100 papers will be presented and discussed by over 200 authors and an estimated 150 active participants. The speakers will come from 25 countries and represent four continents.

Archaeology and Remote Sensing

For a long time remote sensing consisted almost entirely of aerial archaeology. As is well known, the first application was related to the documentation of archaeological excavations. Italy and Italians have played prominent parts in the history of aerial archaeology, starting with the famous shots of Foro Romano taken in 1899 by Giacomo Boni and followed some years later in 1910 by photographs of Pompei. By this time photographs from balloons were already being used for mapping purposes, along the Tiber near Rome in 1902-3 and 1908, and then in and around Venice, Ostia and other locations. The Great War gave a huge impetus to the development of aircraft, cameras and films, and to their use in photo intelligence by all of the combatants. The war had, however, introduced a number of pilots and observers to the archaeological potential of air photography. In Britain one of these was O G S Crawford while in the Middle East Antoine Poidebard, the two of them considered worldwide as the fathers of aerial archaeology and its application to landscape studies. In 1928 O G S Crawford published (with Alexander Keiller) *Wessex from the Air*, which demonstrated the vast potential of exploratory air photography and established the main principles of the technique.

The Second World War brought about enormous technical developments in both aerial cameras and new sensors. In the 1950s new platforms became available through the use of satellites and high-altitude aircraft, and new sensors were introduced in the form of near-, medium- and thermal infrared imagery, along with microwave and multispectral data. To take account of the new perspectives introduced by these new sensors and the early satellites, scientists coined a new term: *remote sensing*, adding a new term to the technical lexicon.

Remote Sensing is now commonly used to describe the science of identifying, observing, interpreting and measuring objects or surfaces without coming into direct contact with them. This process involves the detection and measurement of radiation of different wavelengths reflected or emitted from distant objects or materials.

But what happened in the archaeological scientific community? For a long period the technologies involved in Remote Sensing were considered to be 'leading edge' and it was fairly unusual for archaeologists to access them. It must be acknowledged, too, that the relationship between archaeologists and technology has not always been free of pain and frustration. Despite this, in the last twenty years technology – and particularly computer science – has become more and more common in the archaeological laboratory and also – although with some delay – in the field.

Remote Sensing began to become more widespread in the field of archaeology in the later 1980s and early 1990s. In

this period we can point to occasional examples of the application of innovative techniques, including the analysis of satellite imagery, the acquisition through airborne sensors of multispectral, hyperspectral and radar data, along with geophysical methods such as magnetometry, electrical resistivity tomography (ERT) and ground-penetrating radar (GPR).
Experience in the following decades, along with technological progress and an increasing understanding of the extraordinary complexity of the archaeological record, led to the inescapable conclusion that only through the integration of remote sensing techniques, of archaeometry and of traditional methods could we possibly achieve the quantum leap in quality that everybody hoped for and expected. It is our belief that one instrument in particular played a central role in the maturing of Remote Sensing in archaeology. This is synonymous with the integration and management of the subject's inherent complexity: we are thinking of GIS.

While engineers, physicists and computer scientists improved the quality or effectiveness of individual systems, sensors and techniques, or designed entirely new ones, archaeologists through the application of GIS started thinking *beyond* the individual image so as to produce and map broader integration and therefore interpretations, bringing together a wide variety of data. The stratification and overlaying of the information in a single 'container' provided an essential tool in the search for and development of an integrated approach.

'Integration' has been the most term in scientific meetings of the last ten years and many thousands of words have been spent on its discussion. In the meeting at Rome we would like to add only one thought. In recent years we have sometimes gained the impression that there has been a tendency to seek the integration mainly of higher-level technological applications. Caution is needed here not to ignore more 'traditional' techniques. We need to integrate whatever is useful to reach our stated archaeological goals. For instance, aerial survey does not in any real sense represent 'new technology' but it has in the past provided, and will in the future still provide, an enormous amount of information about archaeological sites and landscapes. The same, of course, applies in a rather different way to excavation, environmental archaeology and documentary research.

Finally we believe that the contribution of Remote Sensing, and in a broader sense of all kinds, of new technology are indispensable in the field of the archaeological research. Remote Sensing technologies will help archaeologists to see more and so to achieve better archaeology. Another consequence should be recognized in the increasing of the complexity of archaeological research. New instruments bring new knowledge and hitherto unexpected career profiles. In conclusion, we would like to highlight one further point – the need to achieve or maintain a balance between archaeologists and technicians, as well as archaeology and technology. In our experience we believe that the relationship, for instance, with computer scientists, engineers and physicist is extremely important but we also believe that everything should spring in the first instance from the desire to address historical and archaeological questions and to create sustainable policies for landscape monitoring and conservation.

The landscape

The landscape belongs to the science of complexity and, in this sense, we think that the most adequate method of reconstruction, or the best cybernetic map, is provided by Virtual Reality, intended as an approach to 'map' or 're-map' the landscape. Today we are facing serious problems of inadequacy in our digital tools if these are not supported by an adequate cybernetic and epistemological approach. We are not dealing any more with a simple lack of technology but with a need for the wholesale re-thinking of our cultural processes.

The goal of the process we are dealing with is a 'multi-layered' visualisation of the archaeological landscape. By 'multi-layered' we mean the visualisation of different categories of complex data within the same georeferenced space. With Virtual Reality applications we can manage, in the same metric environment *acquired data*, such as survey results, geological and geodetic prospections, GPS data, laser scanner point clouds, air photography, satellite imagery, archives, meta-data or other textual sources, but also *processed data,* such as remotely sensed data, or *interpreted data*, such as thematic layers elaborated through GIS analysis, etc. Another important advantage is that it is even possible to integrate data at different scale of accuracy (metric – GPS, centimetric – DGPS, millimetric – DGPS, scanner laser). The concept that allows this kind of integration is the *multiresolution* approach, through the depiction of different levels of detail (LOD) depending on the depth of the user's penetration into the VR application and the type of 'behaviour' being used at the time.

A 'behaviour' is an immersive digital action aimed at creating a feedback in a 3D space, inter-connecting all the data in the same real-time visualization. This capacity to integrate different information allows us, as in a bi-dimensional and static GIS, to overlay for instance remotely sensed data processed from different sources: laser scanner, geophysics analysis, lidar, radar data, aerial imagery etc, and to compare them in the same space. Their interpretation becomes easier and new elements or interpretations may arise unexpectedly.

This approach leads us to re-think even the traditional interpretative process of archaeology. In many cases an archaeologist uses his/her mental patterns in order to use a *top-down* interpretation: e.g. by mental comparison of archaeological models, structures, landscapes, artefacts, settlements, etc. Typically, the entire work of the archaeologist results in a scientific publication, often printed many years after an excavation or a survey. The Remote Sensing approach, in terms of digital observation of the landscape, is different, since it starts from digital and remotely sensed data and ends with their digital analysis, in a *bottom-up* method.
Top-down and bottom-up aspects are rarely integrated, with the result that often some information is lost. The innovative aspect of Virtual Reality applications is not the technological impact but its use even as a scientific tool, not only as a visualisation system. The visual and interactive integration of the bottom-up and top-down approaches in the same space increases the facility of interpretation in a holistic sense: the scientific and cognitive impact of the data is very high because of the relationships created through exploration within the VR system.

From space to place

Sense of place is also sense of time, the difference between space and place, between the 'global' and the 'local'. The world process of globalization is removing places and multiplying spaces, reducing cultural differences. In particular the dissemination of not-places, stations, hypermarkets, hotels, etc carries a risk of creating a uniformity in our perceptions, reducing what we perceive of the world to a few mental maps. For this reason the title of the conference aims to draw attention to one the fundamental tasks of remote sensing archaeology, namely the capacity to use spatial technologies for recovering and identifying places and the sense of place in collaboration with local communities.
Destruction and obliteration of the collective processes of memory transmission risks the exclusion of local communities from the environment they live in, rescinding relationships/inter-relationships or constructing false relationships. This shows the importance of re-creating virtual and mind landscapes in order to have a collective memory of the past.

Indeed the self-organization, the *autopoiesis* of the landscape, or better of the environment, can create additional places where the spaces seem out of control, re-creating the 'local' perception. This re-created Local will create unplanned maps and feedback activities on the basis of the anthropological needs of the territory. The theory of 'mindscape' shows that the use of virtual reality is a key factor in the reconstruction of mental maps of the past because it involves the ways through which we perceive information in time and space. In this scenario digital technologies, archaeology and anthropology can have a social role, very important in reading the territory and in catalysing the diachronic perception of the landscape. In this direction remote sensing technologies, if supported by an epistemological analysis of information, can attempt to reconstruct our mental maps of the past while preserving the transparency of data and meta-data used.

Understanding of the landscape will have a social impact on local people, on tourists and on visitors who, without 'maps', cannot gain access to the mental code of environmental interaction. Finally, each process of sustainable development cannot stand aside from a correct perception of the archaeological and ancient landscape. This trend towards spatial anthropology and remote sensing, supported by digital immersive technologies, should help local communities to re-possess power and sense of place, and to guarantee an adequate cultural transmission.

Maurizio Forte (Chair – ITABC-CNR, Virtual Heritage Laboratory)

Stefano Campana (Co-Chair – University of Siena, Landscape Archaeology, LAP&T)

Acknowledgments

The organisation and support for this conference has depended on the creation and maintenance of successful links between people and institutions. Below we want to thank all of those whose help has permitted us to bring this international even to a successful conclusion.

We are grateful to all those who helped in the publication of this book especially to EPOCH for providing a grant for this.

EPOCH

People

We are grateful to everyone responsible for all the organization and arrangements: Dr. Robert Bewley, Cristina Felici, prof. Riccardo Francovich, Mariaelena Ghisleni, prof. Eugenio La Rocca, Lorenzo Marasco, Alessia Moro, Chris Musson, Francesco Pericci, Sofia Pescarin, Lucrezia Ungano.

Special thanks, however, are due to Barbara Frezza, Augusto Palombini whose enormous effort helped the set up of the conference.

Companies and Institutions

The bulk of the funding was provided by UNESCO, SEAT, ESRI Italia, LEICA GEOSYSTEMS.

Funding and logistical support were also provided by City of Rome, CNR, Soprintendenza ai Beni Culturali Comune di Roma, University of Siena, Aerial Archaeology Research Group, Culture 2000 project *European Landscapes: Past, Present, Future*.

Education and Culture

Culture 2000

aarg

Aerial Archaeoloy

Research Group

Abstracts

RECORDING AND MODELING OF CULTURAL HERITAGE OBJECTS WITH STRUCTURED LIGHT PROJECTION SYSTEMS

D. Akca, F. Remondino, A. Gruen, D. Novak

Institute of Geodesy and Photogrammetry, ETH Zurich, Switzerland
E-mail: <devrim><fabio>@geod.baug.ethz.ch
http://www.photogrammetry.ethz.ch

Active sensors, i.e. laser scanners and structured light systems, are used for many kinds of 3D reconstruction tasks and recently very much for the recording and 3D documentation of cultural heritage objects. They have become a very common source of documentation data in recent years, in particular for non-expert users, as they provide range data of surfaces in high resolution and with high accuracy. Compared to image-based approaches, active sensors provide directly and quickly 3D information of the surveyed object in form of range data (pointclouds), without requiring external software with a mathematical model to convert the 2D image information into 3D data. In active sensor systems the conversion is integrated and not visible to the user. Active sensors are suitable for different scales and objects. While the recording devices are still relatively expensive, important progress has been made in recent years towards an efficient processing and analysis of range data.

Structured light systems consist of one (or two) camera(s) and an active light source, which illuminates the object with a known pattern of light sequence. Based on the triangulation principle, the 3D object coordinates are generally recovered in ca 2-3 seconds with a potential accuracy of 50 microns or even better.

The coded structured light technique, also called topometric technique, is based on a unique codification of each light token projected onto the object. When a token is detected in the image, the correspondence is directly solved by the de-codification. There exist many codification methods. The time-multiplexing, also called temporal codification, with a combined gray code and phase shifting, is the most commonly employed technique. The volume of measurement is strictly related to the baseline distance, focal length of the camera and the light projector set-up. Typically, projects conducted with a stripe projection system involve volumes of approximately 50 dm^3. Coded structured light projection systems provide for highly precise and detailed results in short time. They have some distinctive advantages over the triangulation-based laser systems. The use of incoherent light reduces speckle noise and provides better surface smoothness. Furthermore, they do not penetrate so much into the object surface, unlike laser light whose penetration property is well known, e.g. for marble. All these reasons make structured light projection systems a suitable choice for Cultural Heritage applications.

This paper reports about two case studies where a structured light system (Breuckmann optoTOP-HE) is used for the precise 3D digitization and documentation of Cultural Heritage objects. It includes all necessary steps of the 3D object modeling pipeline from data acquisition to 3D visualization.

The first study is the 3D modeling of a part of a Herakles statue, named "Weary Herakles". This is a marble statue of the Greek demigod Herakles, which dates back to the 2nd century AD (Fig. 1a). It is a copy of an original bronze statue of Herakles sculptured about 330-320 BC by the Greek master Lysippos of Sikyon. Many artisans devoted their skills to replicating this original around that period. This particular example was probably carved in the Hadrianic or Antonine (Roman) period. It is identified as the "Herakles Farnese" type on the basis of its similarity to a more complete copy in the Museo Archeologico Nazionale di Napoli (Naples National Archaeological Museum, Italy).

The statue was broken in two pieces. The upper half is currently displayed at the Boston Museum of Fine Arts, USA. The lower part, found by Prof. Dr. Jale Inan during an excavation in Perge (Antalya, Turkey) in 1980, is now conserved in the Antalya Museum, together with a photograph of the upper half part of the statue. Our goal was the digitization of both parts to recreate a virtual 3D model of the entire statue. With the help of the Turkish authorities and the Antalya Museum we were able to complete our work on the lower part, but access to the Boston Museum was denied.

Therefore, so far, we have digitized the lower part of the statue, which is approximately 1.1 meters in high. A total of 67 scans were performed and each scan has approximately 1.3M points. Each individual point cloud was registered into a common coordinate system using an in-house developed method, called Least Squares 3D matching. The average accuracy of the pairwise registrations is between 50 and 100 microns. Afterwards a block adjustment by an independent models procedure was carried out for the global registration, reaching an accuracy of 47 microns. After the data registration, a surface model was generated and textured by means of digital images, which were externally taken with a 4M pixel CCD Leica Digilux 1 camera (Fig. 1b). The surface modeling was performed in Geomagic Studio, a commercial 3D modeling software. The final 3D model contains approximately 5 million triangles.

In the second case study we report the 3D digitization and documentation of a Khmer sandstone stone head (Fig. 1c), originally found in Cambodia and now available in the collection of the Rietberg Museum in Zurich, Switzerland. The head spans a diameter of ca. 30 cm and contains small details and ornaments. It is from the late 12th or early 13th century, and carved in Bayon style.

The processing workflow consists of measurement, registration, surface wrapping, editing, texture mapping and visualization. A total of 18 scans were performed and after the global registration, an accuracy of 29 microns was achieved. The model consists of approximately 1.4 million triangles (Fig. 1d). The geometric model was afterwards textured and it is now used for visualization purposes and for the generation of physical replicas.

3D documentation and visualization of Cultural Heritage objects is an expanding application area. The selection of the right technology for these kinds of applications is very important and strictly related to the project requirements, budget and user's experience. In this contribution we report our experience in the 3D digitization of two objects. We cover all the necessary steps of the 3D object modeling pipeline with structured light technique and we discuss the capabilities of the used technology.

HYPERSPECTRAL REMOTE SENSING APPLICATION IN THE CELTIBERIAN CITY OF SEGEDA

Rejas Ayuga J. G.[1], Burillo Mozota F.[2], López R.[2], Farjas Abadía M.[3]
rejasaj@inta.es
[1]Instituto Nacional de Técnica Aeroespacial (INTA), España
[2]Facultad de Humanidades y Ciencias Sociales de Teruel, Universidad de Zaragoza (UZ), España
[3]E.T.S.I. en Topografía, Geodesia y Cartografía, Universidad Politécnica de Madrid (UPM), España

Key Words: hyperspectral remote sensing , archaeology, SAR, field radiometry, GPS/IMU.

From a point of view of remote sensing investigation, the incipient state of excavation of the archaeological site of Segeda's Celtiberian city supposes a special advantage for the development and validation of methodologies. To this level of study, in which one tries to confirm indications of buried structures and manmade remains, the surfaces response in reflective (VIS-SWIR) and thermal (TIR) spectral regions can provide a relevant information as support in archaeological exploration.
This paper describes the hyperspectral campaign of data acquisition in Segeda area and the strategies of integration with airborne SAR (*Synthetic Aperture Radar*) data in a multisource study. The digital methodology processing focused in the thermal bands and the first results with hyperspectral images and aerial photos fused are presented. The main points for the analysis have been studied and checked with field radiometry information.
During the last thirty years the Archaeological Institute (today, Dipartimento di Beni Culturali) of the University of Palermo has carried out archaeological survey in the hinterland of Himera and Alesa, ancient towns close to the northern coast of Sicily, whose hinterland is a very suitable area for landscape archaeology. *ArcView* 8.3 and *Manifold* 7.0 has been employed to carry out thematic maps (visibility, hillshade, geological, morphological and hydrographic layers). The finds have been positioned by a handheld computer (*Recon*) with *ArcPad* software 7.0.
Survey of S. Leonardo valley is another step of Himera project, suitable to investigate economic and political relationship between Punic, Greek and native population in archaic/classical/hellenistic times, and the evolution of landscape from Prehistory to Roman and Medieval ages.
Alesa, founded on 403 B.C., is located about 40 km East of Himera, at the foot of Nebrodi mountains, about 1 km from the sea. Around Alesa we have explored part (about 45 kmq) of the Tusa stream valley, between the modern town of Tusa and, South, towards the Nebrodi, today, and maybe since the past, covered by woods.
An *Access* database (linked to the GIS) allows to query the data of the survey, and to manage a graphic and photographic archive. GIS project is also going to be on line by *Web Mapping*, an *open source* technology, which allows a global data sharing. Our Web GIS is a flexible tool, built up on a common language, available to students and researchers for downloading and viewing. First results of our research will be shown by a *beta* version of the forthcoming web digital archive.

SEMI-AUTOMATIC SEARCH FOR CULTURAL HERITAGE SITES IN SATELLITE IMAGES

Lars Aurdal1,[a], Line Eikvila, Hans Korena, Anke Loska[b]
[a]Norwegian Computing Center, Oslo, Norway
[b]Norwegian Directorate for Cultural Heritage, Oslo, Norway

Introduction
It is generally recognized that the increasingly intensive use and modification of the landscape resulting from modern demands for efficient infrastructure and land use (agricultural production, mining, energy sources, leisure/tourism facilities etc.) exerts growing pressure on cultural heritage in the landscape.
In recognition of this, the Norwegian Directorate for Cultural Heritage (Riksantikvaren), the Norwegian Space Center and the Norwegian Computing Center in collaboration with several municipal administrations have started a project with the overall aim of developing a cost-effective method for locating, surveying and monitoring cultural heritage sites on a regional and national scale. Given the enormous costs of surveying the tracts in question by traditional field work, alternatives must be sought. One possible approach is through the use of satellite images. Current experience with manual analysis of these images indicates that cultural heritage sites with no apparent above ground structures still may be observable in such images.
The working assumption in this project is that cultural heritage sites with no visually apparent structures above ground will be detectable in satellite images due to alterations in the spectral signature of the bare soil or that of smaller plants growing out of the soil. The project is currently limited to sites located in agricultural fields, detecting such sites in forests is expected to be a much more complicated problem. In satellite images of agricultural fields, cultural heritage sites may show up as patches having particular shapes and different spectral properties compared to their surroundings.
The costs of acquiring and analyzing such images is not negligible, and it is not reasonable to assume that such an approach can entirely replace traditional field work. Nevertheless, it is clear that if the potential of satellite images for detecting and monitoring such sites could be clearly established, this would be an invaluable addition to the traditional methods that would be very useful in monitoring this important resource on a regional and national scale.
Methods
We have started to develop methods and software for assisting archaeologists in the process of scanning satellite images for the presence of potential cultural heritage sites. The system is operated through a simple graphical user interface (GUI) that provides the user with two different methods for running the system; one fully automatic method and one stepwise method providing some more user control. In both approaches care has been taken to keep the necessary knowledge of technical details to a minimum. A user with little knowledge of image processing and remote sensing should still be able to run the system without too much training.
The operation of the system is based on two central processing steps, *segmentation* to detect potential cultural heritage sites and *classification* in order to provide an initial type for each detection. The segmentation step is based on the use of adaptive thresholding. The classification step extracts a number of features from the different detections and performs a minimum distance

classification based on these features.
Segmentation is the process of dividing the areas of the satellite image into different categories based mainly on their spectral characteristics. The current segmentation method works on panchromatic images, it operates by identifying areas that are darker or brighter than the surroundings. This is obtained using Niblack's method for threshold selection. The method is used in two passes, first to find dark regions and then to find bright regions.
The method works by estimating the mean and the variance in a local area. A local threshold is found by multiplying the local variance with a weight factor and *adding* it to the local mean. This threshold can then be used to find regions brighter than their surroundings. In order to find regions that are darker than their local surroundings, another threshold is found by multiplying the local variance by a (possibly different) weight factor and *subtracting* it from the local mean.
The reason for using a locally adaptive algorithm is that the local brightness varies a lot from field to field and what we are looking for are actually regions that stand apart from the local surroundings. The performance of this algorithm heavily depends on the size of the window over which we calculate the local means and variances. The size of this window as well as the different weights can be adjusted.
Once dark and bright regions have been detected like this, we postprocess them by removing regions consisting of too few pixels. We also verify that the regions have a reasonable difference from their surroundings by looking at the variance in an area slightly larger than each region.
Classification is performed on the regions resulting from the segmentation. In this process the spatial characteristics of the segmented areas are taken into consideration in addition to the spectral characteristics of the area and it's surroundings, to determine whether this could be a potential cultural heritage site.
During classification, features are extracted from the segmented regions of unknown class and based on the statistical class descriptions, a minimum distance classifier is used to determine the most probable class for each region.
The classes that have been defined currently consist of two classes representing potentially interesting objects, and four classes of different types of noise objects. The features that are currently used to characterize regions are the following:

1. Ratio between perimeter and area
2. Gradient along boundary of region
3. Relation between height and width (major and minor axes)
4. Standard deviation of grey levels within region
5. Difference in local grey levels inside and outside the region

Results
The developed methods have been tested on two satellite images of the agricultural areas surrounding the Lågen river in Vestfold County as well as those surrounding the Rygge Community in Østfold County in Norway. Initial results are promising. The process used to detect candidate regions is sensitive and is capable of detecting poorly defined sites. Subsequent processing of these initial detections using classification approaches removes a large number of erroneous detections and provides an initial interpretation of the retained detections.
Conclusions and future work
We have started to develop methods and software for use in detecting possible cultural heritage sites in satellite images. Such sites presenting no above ground structure will typically manifest themselves as spots having particular shapes and different spectral properties compared to their surroundings. Our approach operates in two steps, the first step being a segmentation in order to detect candidate sites followed by a classification to provide an initial sorting of the sites. The methods have been tested on two satellite images providing promising results. The chosen approach can be trained to meet classification criteria established by archaeological specialists. Work in the near future will be focused in the following areas:

1) Drawing on information from archaeological specialists the system will be more extensively trained.
2) Efficient testing by specialists requires a graphical user interface; the current version will be refined.
3) The current focus has been on panchromatic satellite data; in the near future we will include multispectral data in the analysis.
4) The current segmentation and classification methods are designed to produce a large number of false positives. This is beneficial in the sense that it reduces the risk for missing interesting sites, however, if the number of false positives becomes too large the user will spend too much time going through the detections. We will seek to reduce the number of false positives in future work.

The final aim of this work is to produce software that will be used by Norwegian county administrations in their cultural heritage conservation programs.
Acknowledgements
This work was financed jointly by the Norwegian Directorate for Cultural Heritage (www.ra.no) and the Norwegian Space Center (www.spacecenter.no).

WEBGIS AND TOURIST PERSONALIZED ITINERARIES FOR EXPLOITATION OF CALABRIAN CULTURAL AND ARCHAEOLOGICAL HERITAGE

P. A. Bertacchini[1], A. Dell'Accio[2], S. Giambò[3], G. Naccarato[2], P. Pantano[2]

[1]Department of Linguistic, University of Calabria pa.bertacchini@unical.it,
[2]Department of Mathematics, University of Calabria dellaccio@fis.unical.it, piepa@unical.it, giuseppenaccarato@alice.it
[3]Department of Mathematics, University of Messina giambo@mat520unime.it

Calabrian History and Culture have very ancient roots, so that Southern Italy could be considered the first territory to weave a net of relationships with the Mediterranean populations. The region Calabria offers a wide variety of its cultural heritage, witness of its many rulers (Greek-Hellenistic, Byzantine, Norman, and so on). In fact, these hegemonies marked in a determining way the Calabrian culture, enriching the land of architectural and artistic works. Moreover, important archaeological finds can be admired in several museums and in the archaeological sites of Sibari, Locri, Roccelletta di Borgia and Capo Colonna.

In order to promote the protection, the preservation and the exploitation of these cultural resources, there were many projects, like for example the "Virtual Museum System of Magna Graecia" project, financed by the POR Calabria (annuity 2000 – 2003). Its objective was the offering of original procedures for the fruition of Calabrian cultural heritage, in particular of Magna Graecia period, using the most advanced technologies. In fact, the system foresees the access and the visit of a remote site: in this way the Web becomes a way to spread knowledge and promote historical places and a specific culture, with obvious development of tourist and economical aspects of the region. A central role is given to the multimedia technologies: modules of virtual visits, three-dimensional reconstructions of archaeological finds, documentary films are the frame of the traditional information, enriching its own semantic content.

A WebGIS module, a system of navigation through a dynamical interactive map, was specifically designed by open source systems, for an on line fruition of the cultural heritage (a monument as well as a church, a find or an archaeological site), that was not separated from the territory determining the localization. The variations of scale allow to pass from a panoramic sight connecting the relationships among the objects present in the territory to a detailed perspective, in which the specific object is important. Moreover, the user has the possibility to access to different levels of the informative content in an immediate manner.

The use of the above-written tools and technologies allowed the building of real world models and to create bi-dimensional scenarios, but they also allow the simulation of three-dimensionality: the aim was to offer to the user the opportunity to test virtually a possible future tourist visit.

In this work we present the development activity of a tool that can create tourist personalized itineraries for the user of the "Virtual Museum System of Magna Graecia". It is a relatively complex optimization problem involving several parameters which are not well defined in some cases; the solution involves the utilization of AI (Artificial Intelligence) and the support of tools like the GIS (Geographical Information System), because of the geographical data.

Itinerary is the central part of any tourist visit and it is an important strategic element for the exploitation of the territory. Its planning includes defined objective elements, for example the distance between two places or the limit of speed on a road, as well as subjective elements, for example the importance of a monument for a specific tourist or the necessary time to visit it. In this case, the definition of these elements is more articulated and complex. The possibility for a computer to develop the typical reasoning of the human mind is an application of the Artificial Intelligence.

We can suppose that a tourist interested in the archaeological sites of Calabria wants to visit the most important ones in a specific number of days. The words "most important" create a problem of interpretation: in fact, the concept of importance is not only based on recognizable objective pointers, but it is linked to the sensibility of the users, that could have different opinions about the design of the itinerary. In general terms, we assign to every cultural place, a possible visit in the itinerary, a degree of importance or "value" that we call "*v*".

From a mathematical point of view v is a function of different variables $x_1, \ldots, x_p$, that is:

$v = f(x_1, \ldots, x_p)$ with $x \in X$

where X is the set of pointers that describe the features of a cultural good. Particularly, v depends from the age of construction, the state of preservation, the degree of accessibility of the site, the artistic quality, and so on. Besides v, other considerable parameters for the definition of the problem are the optimal duration of the fruition, t_f, that could be determined for example from the visit time obtained by a sample of users, and the cost c when the fruition requires a ticket.

The information needed to fill the indicators, result of a consistent research effort, has been stored in a database, whose optimal management is insured by the use of a DBMS (Data Base Management System) that guarantees reliability and security. In particular, these stored data concern the immovable goods that belong to the calabrian cultural heritage, including architectural monuments, archaeological sites, monumental centers, as well as buildings where the movable cultural goods are exposed and/or preserved, such as museums, libraries, archives. Every cultural immovable good has a descriptive format that defines its own characteristics and location.

A tourist itinerary consists of a set of steps, coinciding with the places of interest, and of the possible connections among them. Therefore, it is convenient to represent it through a graph-based formalization. A graph G is defined as an ordered pair, $G = (V,E)$, where V is a set of vertices (the steps) and E is a set of links between pairs of vertices, called edges (communication channels).

Initially, the parameter tij, that represents the time needed to cover a link $(i,j) \in E$, has been assigned to each edge of G. The time t_{ij} depends on a certain number of factors, including distance, road type, speed limits, mean of transport. The value v_i and the fruition time t_{fi} of the cultural good has been assigned to each vertex. Another important parameter for travel planning is the user time T, that is not necessarily boundless.

Our purpose is to find the itinerary that maximizes the travel performance in terms of importance of the visited goods with respect to the time limits or to any other constraint defined earlier (i.e. cost of the trip, existence of forced layovers, etc.), taking into account a certain number of preferences expressed by the user, such as category of the cultural good to be visited or geographical zone of interest.

Once the relevant parameters have been chosen, we need to formulate a model able to describe the problem. If one forgets about the connections between steps, the problem can easily be solved using the known Knapsack model. This model allows to find the set of objects that maximize the profit without overcoming the maximum capacity of the knapsack, for a given collection of objects characterized by a value v_i and a volume c_i, and for a given knapsack of fixed capacity C.

The Knapsack problem belongs to the class of NP-hard optimization problems whose solution can be found through an heuristic approach.

Including connections between steps complicates the analysis, moving the problem to the application field of Network Optimization. Indeed, in formulating the constraints on the time span that user can employ, one has to take into account the time to cross the connections between different steps.

So, it is possible to propose a first formulation of the itinerary personalized problem in the following way:

$$max\ z(x) = \sum_{i=1}^{N} v_i x_i$$

s. a.

$$\sum_{i=1}^{N} tf_i x_i + \sum_{(i,j)\in E} t_{ij} y_{ij} \leq T$$
$$x_i \in \{0,1\} \quad \forall i = 1,\ldots,N$$
$$y_{ij} \in \{0,1\} \quad \forall (i,j) \in E$$

This formulation can be, obviously, generalized by the addition of further constraints or objectives, in such a way to guarantee a high level of personalization from the user.
Once the mathematical model has been defined, one can develop appropriate resolution algorithms for the particular problem under study. From an introductive survey of the state of the art, it is easily seen that the problem can be faced in different ways.
A traditional approach consists in using the optimization techniques. In this case, the solution for a given optimization problem P is based on research, use and modification of the approximate algorithm A, present in literature. For a given instance x of P, this algorithm gives an approximate solution $A(x)$, in a time that is a polynomial function of $|x|$.
Using the Fuzzy Linear Programming techniques allows to take in account the not well defined ("fuzzy") nature of many involved parameters. The advantage of this method is that it does not force the decision maker to define the problem in precise terms, as it would be needed for a not fuzzy approach.
An alternative approach of which it is also interesting to appraise the applicability consists in using the Disjunctive Logic Programming (DLP), that is also able to simply and naturally represent way of thinking, incomplete knowledge and, in general, "difficult" problems. It is currently applied in different fields, such as the expert systems, the artificial intelligence and the evolved systems for database management.
The inclusion of a tool for tourist personalized planning in the Virtual Museum System of Magna Graecia constitutes an innovation in the field of exploitation of Calabrian cultural and archaeological heritage. In this context, WebGIS will constitute the natural interface through which the consumer can express his/her preferences, also in terms of geographical chooses, regarding the itinerary and by which this can be returned and visualized.

THE FLAMINIAN WAY IN UMBRIA: AN INTEGRATED SURVEY PROJECT FOR THE STUDY AND CONSERVATION OF THE HISTORICAL, ARCHITECTURAL AND ARCHAEOLOGICAL FEATURES

Stefano Bertocci, Sandro Parrinello

University of Florence, Architectural Planning Department Faculty of Architecture
Viale A. Gramsci 42, Firenze

The Umbrian territory crossed by the ancient Roman road, the Flaminian Way, which goes through the municipalities of Massamartana, Acquasparta e Sangemini, presents monuments and art works connected with the old road network. During the centuries, the Roman road track, which was marked in 223 a.C., has coagulate an amount of settlements and monuments, like the Massamartana catacombs, rural churches, monasteries and castles. All these products prove the remarkable anthropologic activities of the last two millenniums and they embody therefore a great interest for the archaeology, but also for the history, the architecture and the environment.
The conservation and the accessibility of all the sites lay down problems not only concerning merely the management of the survey, study and conservation compaign, but also, if the aim is to create a museum network, those related on one side with the improvement of a correct use of the sites, on the other with the realisation of an executive program and planning of the works in order to guarantee the best conservation of the heritage itself. These considerations must be taken especially regarding the open-air archaeological fields, which are constantly exposed to stress caused by tourism and by the necessity of keeping pace with the economical and social development of the whole region.
The two methodologies of the remot sensing and the photo-interpretation with the integration of site surveys and documents are used during the research in order to gain the definition and the structure of the connective principles of the revealing material and the digital data bases, in order to get an aimed use for the project purposes.
The air photo-interpretation, the study and the creation of virtual models with the support of the laser scanning are determined steps for the generation of a connecting system, which supports the drawing records, the cognitive research and the conservation of the heritage.
The GIS (Geografic Informatic System) with its characteristic of flexible use, also on different levels, becomes a basic support for the development of an open-air museum network which tries not only to preserve the image and the survey data of the single finds and the monuments related to the Flaminian Way, but also to build a procedure for the management, the accessibility, the protection and the conservation of the heritage itself.

LE STRADE SU ARGINE DELLE VALLI GRANDI VERONESI MERIDIONALI TRA L'AEROFOTOINTERPRETAZIONE ED IL CONTROLLO MIRATO A TERRA DELLA STRATIGRAFIA SEPOLTA: UN *FEEDBACK* POSITIVO.

A. Betto, G. De Angeli, F. Sartor

Il presente contributo riporta i risultati di una ricerca svolta nelle Valli Grandi Veronesi Meridionali nell'ambito del più ampio progetto Alto Medio Polesine Basso Veronese, coordinato da Claudio Balista, Armando De Guio e Alessandro Vanzetti. L'areale di indagine è caratterizzato dalla presenza di insediamenti di tipo terramaricolo di cui i principali (Castello del Tartaro, Fondo Paviani, Fabbrica dei Soci) sono già ampiamente noti in letteratura (BALISTA, DE GUIO 1997) e per la sua storia geomorfologica-un estesa terra di paludi redenta solo dalle bonifiche '800-risulta particolarmente conservativo di tracce antropiche e naturali sepolte a limitate

profondità, che lo rendono un caso straordinario di *fossil landscape*, palinsestico e multifase (BALISTA ET ALII 1998).
In particolare questo paesaggio è segnato dalle tracce di due grandi costrutti viari su argine:
-un percorso "alto" noto come Strada del Cavariolo, che è stato indagato tramite una serie di carotaggi ed analisi di transetti pedo-alluvionali nel tratto prossimo a Castello Del Tartaro (BALISTA 1996), ricavandone anche una proposta di datazione al periodo che vede la formazione del paesaggio centuriato di età romana;
-ed un percorso "basso", denominato Strada Meridionale su Argine, ancora di incerta attribuzione cronologica assoluta, ma in forza dei suoi rapporti di cronologia relativa ascrivibile con ogni probabilità al paesaggio antropico dell'età del bronzo.
La ricerca da noi condotta si è concentrata su quest'ultima evidenza, con l'obiettivo di fornirne una datazione univoca ed una proposta di modello formativo all'interno del paleopaesaggio naturale ed antropico in cui risulta pienamente inserita.
L'analisi aerofotointerpretativa è stata realizzata su serie diacroniche di fotogrammi e ha interessato il territorio compreso tra le località di Corte Cavariolo e di Ponte Moro (Legnago – Verona).
Diversi sono stati i processi applicati alle immagini (*image processing* ed *image enhancement*) che hanno permesso di mettere in luce diverse tipologie di tracce, definite come unità minime osservabili e poste in rapporto stratigrafico tra loro. Verifiche sul campo mediante l'analisi di sequenze portate in esposizione sulle pareti di scoline attuali, integrate con carotaggi manuali, hanno permesso di sottoporre a verifica (*ground truth*) la veridicità delle ipotesi sviluppate tramite l'osservazione sul supporto remoto: si è dovuto necessariamente passare dalla scala microregionale di individuazione geomorfologica dei grossi complessi deposizionali coinvolti, fino al dettaglio dell'analisi stratigrafica fra le interfacce dei brani dei singoli corpi deposizionali individuati a scala locale, passando attraverso la ricomposizione delle varie unità corrispondenti alle singole facies pedo-alluvionali individuate sui transetti campionari..
L'indagine si è conclusa con un inquadramento della Strada Meridionale su Argine nell'ambito del paesaggio antropico dei vicini villaggi terramaricoli dell'età bronzo medio-recente, di cui viene a costituire un'importante asse di comunicazione.
La ricerca è stata l'occasione per sviluppare alcune note di metodo, qui proposte e che costituiscono il nucleo dell'intervento, in merito alle dinamiche di *feedback* positivo che esistono fra trattamento ed analisi dell'aereofoto ed il controllo a terra, mirato, della stratigrafia sepolta tramite carotaggi e transetti campionari effettuati su sezioni esposte, con una notevole economia di mezzi e di tempo impiegato rispetto a campagne più estese, senza sacrificare tuttavia la complessità dei problemi offerti dagli scenari sepolti.
Non da ultimo è stato possibile verificare come talora il tentativo di creare un costrutto harrisiano delle tracce da foto aerea possa portare ad interpretazioni talora ambivalenti, che solo il controllo a terra ha potuto dipanare.
Bibliografia citata all'interno dell'abstract:

BALISTA, DE GUIO 1997 = C. BALISTA, A. DE GUIO, *Ambiente ed insediamenti dell'età del bronzo nelle Valli Grandi Veronesi*, in *Le terramare la più antica civiltà padana,* catalogo della mostra, a cura di M. Bernabò Brea, A. Cardarelli, M. Cremaschi, Modena 1997, pp. 137-165.
BALISTA 1996 = C. Balista, *Geoarcheologia delle formazioni superficiali: linee guida e casi studio dal progetto AMPBV. I risultati di una ricerca volta al definitivo inquadramento stratigrafico delle strade su argine delle Valli Grandi Veronesi,* in La Ricerca Archeologica di superficie in area padana, a cura di E. Maragno, Stanghella, pp. 319-349.
BALISTA ET ALII 1998 =C. Balista, M. Bagolan, F. Cafiero, A. De Guio, S. Levi , A. Vanzetti, R. Whitehouse, J. Wilkins, *Bronze-Age "Fossil Landscapes" in the Po Plain, Northern Italy*, a cura di Hansel B., in Man and Environment in European Bronze Age, Oetker-Voges Verlag, Kiel, pp.493- 499.

"EMBANKED ROADS" OF THE SOUTHERN VALLI GRANDI VERONESI BETWEEN AIRPHOTO-INTERPRETATION AND GROUND EVALUATION OF BURIED STRATIGRAPHY: A POSITIVE FEEDBACK.

A. Betto, G. De Angeli, F. Sartor

This note reports the results of a study carried out in the southern area of the Valli Grandi Veronesi as a part of a wider project, called "AMPBV – Alto Medio Polesine Basso Veronese", co-ordinated by C. Balista, A. De Guio and A. Vanzetti. This area is characterized by the presence of some *terramara*-type settlements, the most important of which (Castello del Tartaro, Fondo Paviani, Fabbrica dei Soci) have been widely investigated and published (BALISTA, DE GUIO 1997). Thanks to its geo-morphological history – a widespread marshland, reclaimed only in the 19th century - it is an extraordinary example of palimpsestical and multiphase fossil landscape, particularly rich of well preserved buried natural and anthropic marks (BALISTA ET ALII 1998).
Specifically, this landscape is marked by the tracks of two large embanked roads:

- a northern road, named "Strada del Cavariolo", investigated by core borings and pedo-alluvial trenches, realized near the site of Castello del Tartaro (BALISTA 1996), and dated up to the Romanisation period.
- a southern road, named "Strada Meridionale su Argine", uncertainly dated, but probably related to the bronze-age anthropic landscape because of its relative chronology relations.

Our research concerned the latter structure, in order to fix a strict dating and to make a proposal for a formative model inside the ancient natural and anthropic landscape, in which it is perfectly inserted.
The Remote Sensing Analysis (*image processing* and *image enhancement*) of the territory between Corte Cavariolo and Ponte Moro (Legnago, VR), realized on diachronic series of aerial photographs, brought a several types of features, called "minimum-visible units", to light, and put them in stratigraphic relationship.
We checked our suppositions (*ground truth*) through stratigraphic sections exposed on already existing drains, integrated with manual core borings. Necessarily, we passed from a micro-regional scale of geomorphologic individuation of the big depositional wholes involved, to the detail of the stratigraphic analysis of the single depositional bodies picked out at a local scale, passing through the reconstruction of the many units related to the single pedo-alluvial facies differentiated on the stratigraphic sample sections.
The investigation ended up with a chronological setting of the "Strada Meridionale su Argine" within the late bronze-age framework of the nearside *terramara*-type settlements landscape, of which this road turns out to be an important connection axis.
This research has also given us the chance to develop some methodological remarks, hereafter pointed out, and which are the core of the present report, concerning the positive feedback dynamics occurring between aerial photographs processing and analysis phase and the *ground truth* phase, with a noteworthy economy of means and time spent, when comparing with more extensive excavation

fields, nevertheless without disregarding the complexity of the buried landscapes.
Last but not least, it has been possible to verify as the attempt to create a Harrisian matrix out of aerial photographs may sometimes bring to ambivalent interpretations, cleared only by ground checks.
Bibliography cited in the abstract:
BALISTA, DE GUIO 1997 = C. BALISTA, A. DE GUIO, *Ambiente ed insediamenti dell'età del bronzo nelle Valli Grandi Veronesi*, in *Le terramare la più antica civiltà padana,* exhibition catalogue, M. Bernabò Brea, A. Cardarelli, M. Cremaschi ed., Modena 1997, pp. 137-165.
BALISTA 1996 = C. Balista, *Geoarcheologia delle formazioni superficiali: linee guida e casi studio dal progetto AMPBV. I risultati di una ricerca volta al definitivo inquadramento stratigrafico delle strade su argine delle Valli Grandi Veronesi,* in La Ricerca Archeologica di superficie in area padana, E. Maragno ed., Stanghella, pp. 319-349.
BALISTA ET ALII 1998 =C. Balista, M. Bagolan, F. Cafiero, A. De Guio, S. Levi , A. Vanzetti, R. Whitehouse, J. Wilkins, *Bronze-Age "Fossil Landscapes" in the Po Plain, Northern Italy*, Hansel B. ed., in Man and Environment in European Bronze Age, Oetker-Voges Verlag, Kiel, pp.493- 499.

'EUROPEAN LANDSCAPES – PAST, PRESENT AND FUTURE'
A CULTURE 2000 PROJECT: REALITIES, OBJECTIVES AND FUTURE PROSPECTS

R. Bewley, C. Musson

The talk will review the very mixed picture of provision in various parts of Europe for the acquisition and use of new and existing air-photo data relating to the historic environment and landscape studies. The Culture 2000 project, which started in September 2004 and will run until the end of August 2007, has brought together 15 organisations (museums, universities and national institutions) from 11 countries throughout Europe. The aim is to share ideas and expertise in the use of aerial photography and other remote sensing techniques in ways that will contribute to conservation by throwing light on common threads and regional diversities in the archaeological and landscape heritage across Europe. The ideas of public understanding and landscape conservation are central themes in the project, which involves institutions and countries where the integrated use of remote-sensing and ground-based techniques is already well advanced, along with others where these techniques and methodologies are in their relative infancy. The emphasis will be very much on practicalities in relation to the differing 'structural' situations (academic, political, financial) in the countries and institutions concerned, and the effects that these may have on attempts at promoting and combining aerial survey and other landscape-recording techniques. Arrangements which bring success in one country or institution cannot of course be transferred automatically to different national or institutional contexts. But there may be common themes or useful lessons that could help to reconcile the sometimes conflicting but often mutually supportive interests of research, conservation and public awareness in different parts of Europe. Whatever success the Culture 2000 project may achieve within its lifetime, there is also a need for it to have a longer-term effect. What might these long-term effects be? Will approaches or solutions differ radically from country to country? How fast can things change, and what factors come into play in achieving change? Above all, can we envisage new initiatives in the coming years to help promote the more effective deployment of air photo and other remote-sensing techniques in the study, conservation and presentation of the archaeological and landscape heritage of Europe?

SURFACE MODELLING OF COMPLEX ARCHAELOGICAL STRUCTURES AND WALLS BY DIGITAL CLOSE-RANGE PHOTOGRAMMETRY

Gabriele Bitelli[1], Valentina Alena Girelli[1], Fabio Remondino[2], Luca Vittuari[1]

[1]DISTART – University of Bologna, Italy E-mail: <valentina.girelli><gabriele.bitelli><luca.vittuari>@mail.ing.unibo.it
[2]Institute of Geodesy and Photogrammetry – ETH Zurich, Switzerland E-mail: <fabio>@geod.baug.ethz.ch

3D modeling of an object can be seen as the complete process that starts from the data acquisition and ends with a virtual model in three dimensions visible interactively on a computer. Different applications and fields require 3D models, from the traditional industrial inspections and robotics to the recent interests for visualization, documentation and preservation of Cultural Heritages. In fact the documentation and visualization of Cultural Heritage sites has become a very interesting and challenging research field, in particular if performed with non-metric or amateur digital cameras. Nowadays the generation of computer 3D models is mainly achieved with active sensors (laser scanners and structural light systems) or image data while in some cases other information like CAD, surveying or GPS data are also integrated in the project. Active sensors provide directly and quickly 3D information of the surveyed object, without requiring a mathematical model to convert the measurements into 3D data. Active sensors are suitable for different scales and objects, are frequently used also by non-expert due to their speed and easiness, but they are still costly and not often able to provide texture information. On the other hand, images can be acquired with inexpensive systems and contain all the information for the generation of detailed and photo-realistic 3D models.
Photogrammetry, the science of obtaining reliable and accurate measurements from images, is nowadays much more frequently used to generate visualization and animation products, documentation material for conservation, restoration and future planning as well as traditional maps of the investigated areas. This is mainly due to two facts. On one side Unesco and other major agencies and authorities are paying more attention to the digital documentation of cultural and natural sites. On the other side a wide array of new technologies is available which greatly support the efficient generation, administration and analysis of such models.
But to recover a complete, accurate and realistic 3D model from images is still a difficult task, in particular if images acquired with non-conventional geometrical configuration or uncalibrated are used, or in case of complex architectures like ruins of temples, with lots of details and often not well defined edges. In addition, automation of the whole process and in particular of corresponding point transfer in image triangulation and 3D surface generation via matching is in most cases not possible due to the complex objects and the convergent images but also due to the poor performance of commercial systems.

In this article we report about the detailed photogrammetric reconstruction and documentation of some structures of the cultural heritage site of Tilmen Höyük, in Turkey.
The Archaeological Mission of the University of Bologna started working at this town in 2004; the area of investigation is an ancient settlement located 10 km. east of Islahiye town within Gaziantep province in the South Eastern Turkey and dating back to 3400 BC. The palace complex with a temple, inner and outer strong defence walls surrounding the city bring Tilmen Höyük to an important position from archaeological point of view not only in the region, but in the whole country.
The work consists in the image-based modelling of different areas of the archaeological site (walls, rooms, stairs, etc.), using various type of images, mainly acquired with off-the-shelf digital cameras. Amateur digital cameras are very practical for site documentation, but need very precise calibration procedures to retrieve accurate 3D data. In practical situations (e.g. during field campaigns), the image network for the object reconstruction is not appropriate for precise camera calibration. Therefore, rather than simultaneously calibrate the internal camera parameters and reconstruct the object, it may be better first to calibrate the camera at a given setting using the most appropriate network design and afterwards recover the object geometry using the calibration parameters at the same camera setting.
After the image calibration and orientation, automated or semi-automated measurement procedures are employed to retrieve digital surface model of the area of interest. These 3D data are then used to derive different kinds of representation and documentation products, to perform stability and structural analysis or simply for visualization of the archaeological site.
The major contribution reported in this work is the high-accuracy and automated generation of a dense surface model of complex and detailed objects. This is achieved using an in-house surface measurement program developed to match close-range images and based on multi-photo geometrically constrained least squares matching. Although with such complex objects full automation still has a long way to go, we demonstrate that a certain degree of automation can reduce production time and improve the results. Comparisons with results obtained by commercial software are also reported and commented.

3D DIGITAL SURVEY AND VISUALIZATION OF THE ETRUSCAN "IPOGEO DEI VOLUMNI" AT PERUGIA AND THE SIMULATED TIMBER STRUCTURES OF THE CARPENTRY.

Daniel Blersch, Marcello Balzani

University of Ferrara, Italy, Department for Development of Integrated Automatic Procedures for Restoration of Monuments D.I.A.P.Re.M.

The impressive Hypogeum of the Volumnis (Ipogeo dei Volumni), prominent family from Perugia, was built in the Hellenistic period (probably III - II B.C.) of the Etruscan civilization.
The tomb was discovered (1840) in original conditions with incineration *urnae,* still preserved and now exposed *in situ* which allowed Scholars to propose a genealogical tree.
The monument is excavated in the soft sandstone with an anthropomorphic plan, articulated in a large *atrium* with a final *tablinum*, two *alae* and six lateral *cellae*; the *atrium* (3,60 x 7,30 m; h. 4,40) has a two-slopes ceiling made in a way to resemble the intrados of a wooden hut roof: therefore the entire simulated carpentry – ridge beam, rafters, joists, boards etc. are carefully carved in the rock, in a true scale; the same for the a*lae* and *tablinum* which have square corbelled "wooden" coverings. The arrangement is made to give to the death's afterlife the same comfortable setting of the house of the living people.
There are a few more examples of this kind of apparatus in Italy (e.g. some Cerveteri Tombs) and in Sardinia (Puttu Codinu and others) but none made in such a realistic way.
The survey was worked out by means of 3D laser scanner technology combined with topographical data obtained by means of a total station. The scan data has been acquired in difficult conditions of humidity and illumination in very short time without interfering with the affluence of tourist visits on site.
In laboratory the acquired point cloud was meshed and, by means of a sequence of repeated intelligent data reduction, an organic meshed true scale model has been worked out. The research project deals with the different experimentations and approaches to obtain dimensional and geometric information from an archaeological survey worked out by means of 3D laser scanner technology. During the research process there is shown a logical path of elaboration, visualization and comparison between three different kinds of models: departing from the point cloud model, passing to the organic meshed model and finally arriving at a summarizing 3D surface model of the main ceilings.
Using 3D-Laser Scanner technology combined with industrial deviation calculation procedures required a new approach in experimentation of advanced 3D modelling with extremely large input data by developing a methodology for processing colour scale images for the recognition of sensible global deformations on local elements of an ancient deteriorated structure.
The analysis of deformation morphology shows surprising results: carpentry members simulate flections in stress conditions of true wooden beams in similar conditions of support, span, and loads.
Many Scholars had studied the Monument and made careful surveys but none of them faced the problem in a comprehensive way and none noticed this realistic peculiarity. The deformations, in fact, are very slight and the traditional survey methods, even the optical ones, were not suitable to detect this kind of connotations of the pretended wooden ceiling. In order to fully document the whole monument with complete account of the features, researchers of the University of Ferrara together with Prof. Gennaro Tampone from the Department of "Restauro e Conservazione dei Beni Architettonici" D.I.R.E.S., University of Florence, decided to use the laser scanner methodology integrated with geological observations and survey.
Detailed geometrical analysis of the scan model, for the first time, makes suspecting the presence of optical corrections in sense of enhanced perspectivity in late Etruscan funeral architecture, and allows dimension handling between repeated inaccessible points, in order to evaluate numerical series by Gaussian statistical results obtaining important information about the use of ancient dimension units.
The paper focuses on the visualization of the spatial distribution, geometry and morphology of the archaeological monument and concludes with the typological study of the simulated timber structure.
The investigations and the survey lead to very interesting results and conclusions such as:

- Visualization of a subterranean ancient architecture with full three-dimensional control about morphology and dimension;
- Updating and comparison with existing traditional surveys of the object with evidence of their limits and the new possibilities given by 3D laser scanner technology;
- Handling of the 3D model of an architecture excavated by subtraction like as it was a constructed architecture;
- Reconstruction of a typical carpentry of the time;
- Account on the structural planning concepts and building processes of the time;
- Advanced 3D-modelling methodology for deviation check of ancient buildings and the possibility to extend the used appraisal methodology to similar cases.

ANCIENT MAPS – MODERN DATA SETS: DIFFERENT INVESTIGATIVE TECHNIQUES IN THE LANDSCAPE OF THE EARLY IRON AGE PRINCELY HILL FORT HEUNEBURG, BADEN-WÜRTTEMBERG, GERMANY

Dr. Jörg Bofinger*, Dr. Siegfried Kurz, Sascha Schmidt M. A*.**

*Regierungspräsidium Stuttgart, Landesamt für Denkmalpflege
** University of Tübingen

Since 2004 new research activities around the princely hill fort Heuneburg near the upper Danube are in progress. The Culture 2000 program "European Landscapes – past, present, future" is closely related to this project which is a part of a priority research program of the DFG (Deutsche Forschungsgemeinschaft).
Archaeological fieldwork (survey and excavations) in the area of the Heuneburg near the Danube valley/Upper Svabia precinct and the so called Außensiedlung (peripheral settlement) and supplementary investigative techniques (air photos, LIDAR, geophysics) are employed side by side. This combination offers a large bandwidth of information about archaeological structures and development of the landscape surrounding the Heuneburg. The main objective is to extend knowledge of the different archaeological sites and their environment in the complex system of hill fort, precinct, peripheral settlement, fortification systems of ramparts as well as necropolis and burial mounds. This would offer a sound basis for well directed site preservation.
The excavations of the year 2004 in the area of the fortified precinct of the Heuneburg pointed to an extended settlement during the late Hallstatt period. Wooden construction elements, located in the fortifications' ditches, gave us a clear idea of the age of these trenches. The rampart-system of the Heuneburg was built in the early Iron Age, as indicated by the dendrochronological analyses (dates ca. 580 BC) of these wood findings. The whole defence-system is very clearly to recognise in shaded relief images based on LIDAR data. Of special interest is the comparison of maps dating of the 1820s on the one hand side and results of analysis from LIDAR data.
In the peripheral settlement there are systems of different ditches known too. The analysis of LIDAR data sets offers a good possibility to locate these systems even in areas covered by woods and forests. Therefore in cultivated zones, these ditches could be identified by the use of geophysical investigation methods in combination with air-photo mapping and LIDAR data.
Another possibility of visualisation are three dimension-block models. The digital terrain-model of the Danube valley near the Heuneburg, based also on these LIDAR data could be combined with simulation of high tide and floodings of the river and its impact on the surface. This method offers new insights in to possible settlement areas and traffic routes. For example, there is a Hallstatt settlement located near the Bettelbühl necropolis, both situated on this ridge. Furthermore the roman road leading through the Danube valley is exactly located on a non flooded ridge. It's very likely that this the traffic route was used already in prehistoric times and acted as an important connection of the Heuneburg to long haul channels of trade.
By the way, studying these sensible structures offers good possibilities in comparison of different resolution of LIDAR data. A detailed study of the accuracy of two LIDAR data sets with different grid width was undertaken in the region of the Heuneburg:

- DTM project Baden-Württemberg (grid 1m point spacing, elevation accuracy +/- 0,15m), costs ca. Euro 70/km^2
- Special data-collection provided by TOPOSYS Ltd (grid 0.5m point spacing, elevation accuracy +/- 0,05m); costs: ca. Euro 500/km^2

The analysis of the two models demonstrates, that the cheaper solution offers a good possibility for visualisation of archaeological remains. This means profit on the economy of future LIDAR work. Only for very detailed and fable structures the very high resolution has to be employed.

THE PATH OF THE VIA ANNIA ROMAN ROAD ALONG NORTHERN ADRIATIC SEA FROM REMOTE SENSING, HISTORICAL CARTOGRAPHY AND GEOMORPHOLOGIC SURVEY

Bondesan A.(*), Favaretto S. (*), Ferrarese F. (*), Fontana A. (*), Furlanetto P. (*), Levorato C. (*), Meneghel M. (*), Miola A. (**), Mozzi P., (*) Primon S. (*), Sostizzo I. (*), Valentini G. (*P)

*Università di Padova, Dipartimento di Geografia
**Università di Padova, Dipartimento di Biologia

Via Annia is a roman road following the coastal rim of Northern Adriatic Sea from Adria to Aquileia, passing through main Roman cities of North-Eastern Italy.
Recent investigation allowed to define most of the path of the road. These investigation were performed by aerial photo interpretation on multi-year photo collection, dating from 1944 to 2005, including recent Google Earth™ scenes found in the web; also satellite imagery were used, mostly Landsat and Ikonos.. A very detailed geomorphological survey published by the Authors in 2004 give also the chance to correlate coastal track and geomorphological evolution occurring during the life of the road which is still nowadays used along some tracts of its path. Some further studies allowed to define the geomorphologic constrains on the direction of the road and the correlation with the geomorphological framework gave the chance to discover new settlements and bridges that

were lately archeologically excavated. Some topical sites were also studied in their sedimentological, mineralogical and paleobotanical content: these studies permitted to get paleogeographical reconstruction of different sites from late Pleistocene to the present.

"GIS AND WEB MAPPING OF HALAESA HINTERLAND"

Aurelio Burgio, Maria Assunta Papa, Angelo Ceraulo

Università di Palermo - Dipartimento di Beni Culturali – Laboratorio di Topografia Antica

The hinterland of the ancient town of Halaesa, close to the northern coast of Sicily, is one of the most suitable areas for landscape archaeology. In the 16th century, among the ruins, the "Tables of Alesa" were found, a big marble-slab inscribed in Greek language (*IG* XIV, 352). The "Tables" describe in detail the landed property, the distribution and destination of the fields around Halaesa, rented by the city to the farmers. Some scholars have tried hypothetical reconstructions of the agrarian landscape, based on topographical and morphological data offered by the "Tables"; these assumptions, illustrated by sketches, have been used to restore the distinctive peculiarities of the so-called "mediterranean garden", marked by a specialized arboriculture connected with a woody and pastoral economy.
During the last years, an *équipe* of University of Palermo has carried out an archaeological survey around the ancient town, exploring the areas (about 45 kmq) between Alesa (W), the modern towns of Castel di Tusa (N), Motta d'Affermo and Pettineo (E), and, South, towards the Nebrodi mountains, today, and maybe since the past, covered by woods. The archaeological survey has been supported by remote-sensing analysis (MIVIS module, Multispectral Infrared and Visible Imaging Spectrometer). About one hundred archaeological sites (*Unità Topografiche* – UT – graves, farms, pipe-line) have been identified on the ground, and results are showed by historic thematic maps. The UT have been positioned by a handheld computer (*Recon*) with *ArcPad* software 7.0; *ArcView* 8.3 has been employed to carry out thematic maps (visibility, hillshade, geological, morphological and hydrographic layers).
An *Access* database (linked to the GIS) allows to query the data of the survey, and to manage a graphic and photographic archive. GIS project is also going to be on line by *Web Mapping*, an *open source* technology, which allows a global data sharing. Our Web GIS is a flexible tool, built up on a common language, available to students and researchers for downloading and viewing. First results of our research will be shown by a *beta* version of the forthcoming web digital archive.

SHARING INTERPRETATION: THE CHALLENGE OF OPEN SOURCE WEB APPROACH

Luigi Calori[1], Carlo Camporesi[2], Augusto Palombini[2], Sofia Pescarin [2]

[1] CINECA Visit Lab, Bologna, Italy , l.calori@cineca.it
[2] CNR ITABC, VHLab, Rome, Italy, [carlo.camporesi].[augusto.palombini],[sofia.pescarin]@itabc.cnr.it

The impact of digital technologies is going to cause a drastic change both in the working process and in theoretical and epistemological settings of archaeological science and cultural heritage management, forcing everybody to rethink the traditional ways of information validation and diffusion.
Today, digital technologies let a large amount of information to be available and transmittable in short time. Almost everybody can potentially create and diffuse data to the whole world population, or at least to that who can reach the web, setting up relevant problems in terms of scientific validation. Science, in this sense, seems to have to change its tools: from a narrow scientific community who controlled the access to publication, to an *open* approach in which everybody can be an actor, and information reliability relies on the *transparency* of its sources and of the development process (e.g. Camporesi et al. 2006, Feller and Fitzgerald 2002; Forte, 2005).
According to this approach, an *open process* can not only be focussed on the adoption of Open Source software, but it implies also the application of an *open approach* in the use of interpretation and validation structure.
A famous example that explains the aspects of *Open Source* is the one described by Eric Raymond in his famous book "the cathedral and the bazaar", where he describes *Commercial* and *Open* software development philosophies as *Cathedral* and *Bazaar* models (Raymond 1999). In the *Cathedral* model, software projects are developed by small groups of experts, while in the *Bazaar* model everybody can look at the people working, participate, give suggestions and critics.
It is interesting to think that this metaphor could be also applied to a general scientific interpretation process, distinguishing a traditional *Cathedral* way from a different *Bazaar* process, open to public or at least to a wider evaluation by the transparency of its creation process.
In this latter approach, the open dimension of information and the use of open source software are two elements of the same process, that recall each other.

REMOTE SENSING AND GROUND-THRUTING OF A MEDIEVAL MOUND (TUSCANY - ITALY)

S. Campana[1], R. Francovich[2], L. Marasco[3]

[1] University of Siena, Department of Archaeology and History of Arts, Landscape Archaeology, campana@unisi.it
[2] University of Siena, Department of Archaeology and History of Arts, Medieval Archaeology, francovich@unisi.it
[3] University of Siena, Department of Archaeology and History of Arts, PhD Student, marasco@unisi.it

The site we present in this paper has been "discovered" from the air during the XIVth International Summer School in Archaeology in summer 2005. The evidence is close to the castle of Scarlino along the seaside on the west coast in the province of Grosseto.
This landscape has been studied from the Department of Archaeology of the University of Siena since 1979, when Riccardo

Francovich started the archaeological excavation of the castel of Scarlino (Francovich 1985). Almost at the same time, in relationship with the excavation, followed a field walking survey that contributed to make this landscape one of the more studied in the province of Grosseto (Cucini 1985).
So, our site has been already pointed out in the past by air photo interpretation of vertical historical coverage. Unfortunately the absence of a mapping project of the features and the unavailability of GPS device for navigation made the identification of the evidences on the field more and more difficult. The field walking survey and the ground-thruting ended up always positioning itself on a side or on the other of the site, without never gathering the position and the entity of it (Chapman 2001; Campana 2005). Despite the analysis of the aerial photo allowed already the interpretation of the site as an triple enclosure, the lack of correspondence on the ground with any material evidence avoid us to understand the chronology of the site and also any contribution to the development of new settlement pattern. About that we may consider the difference of knowledge on this kind of site between Continental Europe or United Kingdom and Italy. I mean that in Germany for instance it should be quite easy to interpret this kind of feature without any need of ground-thruting but in Tuscany this kind of evidence don't find any comparison.
In this paper we will show how the quality and the quantity of the information grow up in the last study we did and starting from the good results obtained we will discuss the new perspective for the archaeological research in the landscape around the site.

A PURPOSE FOR THE DIGITAL STORAGE AND SHARING OF REMOTELY SENSED ARCHAEOLOGICAL DATA

Stefano Campana[1], Barbara Frezza[2]

[1] University of Siena, Department of Archaeology and History of Arts, Landscape Archaeology campana@unisi.it
[1] University of Siena, Department of Archaeology and History of Arts, Landscape Archaeology & Remote Sensing LAB, frezza@lapetlab.it

The Department of Medieval Archaeology at the University of Siena has been actively engaged in programmes of landscape archaeology and remote sensing for over twenty-five years. The research activity produced a large collection of data and particularly a huge amount of photographs with related information. For instance twenty years of aerial reconnaissance produced 6000 thousand vertical aerial photograph or again in the last six years in the aerial survey research programme we collected an archive of about 35000 thousand obliques. Obviously digital data base represent the solution. We started our experience working on multimedia data base in the 1999 with Canto Cumuls software. This product showed immediately many problems of flexibility and of data sharing.
In fact digital technology, in addition to offering powerful tools for the cataloguing, storage, and analysis of archaeological data, become a common and quickly language for the communication of information. Some years later we had the opportunity to move to a data base that could share information through the internet. A first attempt has been realized between 2004 and 2005, building a web data base. The result was excellent for the management and the research of the data but unfortunately was not possible to use this system everyday for the storage of the data (http://www.paesaggimedievali.it/atlante/index.html). That system forced us to catalogue the data using Canto Cumulus Software, after that we need to export the whole data base to update the web DB. Now the project of the new data base project have to satisfy the following tasks:

- Real time update.
- Easy and quickly communication of the data within the laboratory as well as with the general public, private individuals and other institutions, etc.
- First step of a web GIS that could manage the data integration: graphical, geographical and alphanumeric data.

FROM SPACE TO PLACE: THE AIALI PROJECT

S. Campana[1], S. Piro[2], C. Felici[3], M. Ghisleni[4]

[1] Landscape Archaeology, Department of Archaeology and History of Arts, (LAP&T), Università di Siena a Grosseto campana@unisi.it
[2] ITABC–CNR, P.O. Box 10–00016 Monterotondo Sc. (Rome), ITALY. Salvatore.Piro@itabc.cnr.it
[3] Ph.D.student, Department of Archaeology and History of Arts, (LAP&T), Università di Siena felicicri@unisi.it
[4] Ph.D. student, Department of Archaeology and History of Arts, (LAP&T), Università di Siena at Grosseto ghisleni@lapetlab.it

Aiali, is a place name sited on the lowland between the medieval town of Grosseto and the Roman town of Roselle in central Italy. The archaeological area has been detected in 2001 during the Aerial Archaeology Research School we organized in Siena (Campana et al. 2006). Aerial survey allowed us to recognize an area within which the growth of the wheat varies in such a way as to reveal an articulated group of traces that make up the plan of a complex of structures interpreted as a Roman villa that cover an extent of 4 hectares. Year after year Aiali become the most important test site of the Laboratory of Landscape Archaeology and Remote Sensing. From 2001 we collected, processed and interpreted many different kind of data: Quickbird-2 satellite imagery, historical and recent vertical coverage (starting from 1954 to 2001), oblique photographs in different years, seasons and light condition, field walking survey and geophysical survey (magnetometry, GPR, EM, ERT).
Aiali project means from one side to apply the highest available level of intensity of the archaeological prospection methods on a large, complex and stratified site, from Etruscan period, through the Roman and the Medieval age. From the other side Aiali is the starting point to enlarge this approach to the landscape between Grosseto and Roselle. This is an original project in Italy but it's clear that we are trying to emulate the strategy applied for more than thirty years from Dominc Powlesland in his Vale of Pickering (Tipper 2004; Powlesland 2001; Powlesland 2006).
Certainly should be easy to understand that it will never be possible to extend this approach to larger areas and with particular regard to the scale of the archaeological mapping projects we are used. As an example, the archaeological map of Grosseto province work on an extent of 4030 sqkm and that of Tuscan region of about 22990 sqkm!
As Chris Musson argued - during one of the long discussion I had with him on these topics - the archaeological meaning of the Aiali

project has be recognized mainly on the critical effect on the kinds of information that are available for recording: "in assessing the potential or interpretation of a landscape it is at least as important to know what may *not* be visible as to appreciate what *is* visible".

MOBILE COMPUTING IN ARCHAEOLOGICAL PROSPECTION

Stefano Campana, Matteo Sordini

Landscape Archaeology, LAP&T, Dept. of Archaeology and History of Arts, University of Siena at Grosseto campana@unisi.it
LAP&T, Dept. of Archaeology and History of Arts, University of Siena at Grosseto Sordini@lapetlab.it

We should emphasize that from the 1990s, when in Italy we began to use Geographic Information System for the management and analysis of archaeological data, we have felt a progressive schism between work in the laboratory and work in the field. While the availability of advanced technologies has been rapidly growing, activities in the field have continued to make use of instruments and methodologies developed in the 1970s. The problems arising from this situation are mainly inherent in the collection of data that lack the accuracy required by GIS, or which rely work to a different kind of rational logic and the unavailability in the field of the large amount of data stored in the desktop GIS. We would like to emphasize in this paper is that the technological merging between PDA and GPS devices or better a mobile GIS goes far beyond the level of increased fieldwork and aerial work efficiency, in at last making possible the systematic application of strategies and methodologies developed in the past but applied only rarely up till now because of the excessive amount of time involved in their use. In the last year we developed a mobile GIS system for PDA computer aimed to the field walking survey and a mobile GIS system for Tablet PC designed specifically for aerial survey. Both systems achieve a real improvement in the acquisition of new data for our GIS systems and consequently in the type and quality of the analyses which we can then carry out. Moreover, in our experience this new development goes a long way towards restoring the link between active work in the field and the management and analysis of heritage data in the laboratory.

AERIAL SURVEY PROJECT: SEVEN YEARS OF FLIGHTS OVER TUSCANY

S. Campana, R. Francovich, M. Corsi, F. Pericci

In year 2000 the Laboratory of Landscape Archaeology and Remote Sensing at the University of Siena began an annual programme of aerial survey over Tuscany. In the first seven years over 400 hours have been flown, producing more than 35000 oblique aerial photographs and documenting around 2200 archaeological sites. From the experience gained so far it is clear that aerial survey in Tuscany, like all the other methods of remote sensing, will not produce significant results if applied in isolation. For this reason we wish to emphasise in this contribution the way in which aerial survey forms just one part of a wider strategy developed in the last ten years to counteract the low level of archaeological visibility in Tuscany, mainly due to heavy clay soils and woodland areas. Source-integration now represents the prime focus of our research. Without this approach we foresee little possibility in an area like Tuscany of obtaining results which will have a real effect on our understanding of the development of the landscape across time.

3D WATERMARKING

Vito Cappellini, Francesca Uccheddu

University of Florence, Department of Electronic and Telecommunications

Watermarking of 3D meshes has received up to now a limited attention due to the difficulties encountered in extending the algorithms developed for 1D and 2D signals to topological complex objects such as meshes. Other difficulties arise from the wide variety of attacks and manipulations 3D watermarks should be robust to. For this reason, most of the 3D watermarking algorithms proposed so far adopt a non-blind detection. In this work we present a new blind watermarking algorithm for 3D meshes. In order to simultaneously achieve watermark imperceptibility and robustness a multiresolution framework is adopted.
To do so we assume that host meshes are semi-regular ones, a property that permits to first perform a wavelet decomposition and then to embed the watermark at a suitable resolution level. One of the more examined issues is the possibility of hiding the highest amount of information without affecting the visual quality of the host data. The goal of our research is to provide a perceptually-based method for the efficient watermarking of 3D models. Watermark detection is accomplished by computing the correlation between the watermark signal and the to-be-inspected mesh features. Robustness against geometric distortions such as rotation, translation and uniform scaling is achieved by embedding the watermark in a normalized version of the host mesh, obtained by means of Principal Component Analysis. Experimental results show the validity of the proposed algorithm both in terms of imperceptibility and robustness against a wide class of attacks including noise addition, geometric transformations, and smoothing.

AIR PHOTO INTERPRETATION AND THE STUDY OF MEDIEVAL SETTLEMENTS IN MAREMMA (CENTRAL ITALY)

Anna Caprasecca

Università di Siena a Grosseto, Area Archeologia Medievale, Laboratorio di Archeologia dei Paesaggi e Telerilevamento
caprasecca.a@unisi.it

The use of 'early' air photographs in the historical and archaeological analysis of landscapes has a long and well-documented tradition. A fundamental consideration in planning the photo-interpretation of an area is the quality and ease (or difficulty) of access of the material available for examination. The photographs are often held in public collections where failure to understand their documentary value has resulted in poor archival maintenance or even the complete loss of material which constitutes the only direct testimony for the landscape of the time. The most frequent problem for 'historical' photographs of this kind in under-funded or isolated archives is the absence of any system of cataloguing or classification, and hence the effective 'loss' of potentially valuable information. For a photo interpreter the best opportunity arises with the discovery of historical photographs of good quality from before the period of heavy agricultural ploughing. A search of out-of-the-way archives is worthwhile for this very reason. The survival of air photographs from the 1930s and 1940s has in the case reported here strongly influenced the choice of study area, notably through the chance discovery of a collection of military air photographs taken between 1934 and 1938. The photographs, covering the northern part of Viterbo include that part of the Maremma which lies on the border between Tuscany and Lazio. It was immediately obvious that restitution of these three-dimensional images could help us the evaluation of anomalies and potential archaeological sites, including the erosion of the archaeological levels by subsequent heavy ploughing. These images from many years ago will enable us to understand the degree of erosion and morphological deterioration suffered by individual sites. To obtain a better understanding of the transformation of the landscape we are now planning a renewed programme of oblique aerial photography, covering the most interesting sites from the photo interpretation of the early military photographs. The work is being carried in association with the Culture 2000 project *European Landscapes: Past, Present and Future*, an important element of which is the use of aerial photographs in the understanding and conservation of the landscapes of the past.

FROM TOWN TO COUNTRY: AN INTEGRATED APPROACH FOR REMOTE SENSING AT TILMEN HÖYÜK IN SOUTH-EAST TURKEY

Barbara Cerasetti, Bernardo Rondelli

Through their application on the case study of Tilmen Höyük (a 2nd mill. BC town in south-eastern Turkey excavated by Bologna University – in cooperation with Istanbul University and Gaziantep Museum – under the direction of prof. Nicolò Marchetti), this work stresses the close connection of different methods, effective and of limited cost. They are all needed at the same time if one aims at fully extracting the historical information carried by ancient monuments and landscape. The setting of the site within its landscape is done, at survey level, through diversified hight technologies and remote sensing methodologies, working on high and medium resolution satellite images (multispectral and radar), allowing for a close GIS archive.

In addition to the historical and archaeological data, panchromatic and multispectral satellite imagery of the region enabled the use of new analytical methods. This type of supports has been used on large scale in Near and Middle East archaeological projects to define ancient landscapes and we therefore directed our research towards the interpretation of the territorial changes in relation of one of the main archaeological sites as Tilmen. The results obtained on the basis of SRTM mosaic, integrated with the other vertical platforms as CORONA and QuickBird images, allowed us to investigate with a higher degree of confidence the environmental, geomorphological and archaeological history of the area around Tilmen Höyük. The integration into the GIS of the data from different archives, collected during three years of intensive work by several specialists from different scientific disciplines since 2003, has been used to reconstruct the settlement distribution pattern in a virtual archaeological landscape (building upon a survey carried out in the sixties by a Turkish team directed by prof. Bahadır Alkım). By using the GIS as a tool for the systematic testing of alternative explanations and simulation models, we may direct archaeological data to enhance historical interpretations for the formative stages of Near Eastern civilisations.

ANALYSIS OF THE EFFECTIVENESS OF AIRBORNE LIDAR INTENSITY FOR PREDICTING ORGANIC PRESERVATION POTENTIAL OF WATERLOGGED DEPOSITS

Keith Challis, Andy J Howard,[2] Ben Gearey,[3] David Smith[2]

[1]HP Vista Centre, Birmingham Archaeology, University of Birmingham, k.challis@bham.ac.uk
[2]Institute of Archaeology and Antiquity,University of Birmingham, Edgbaston, Birmingham a.j.howard@bham.ac.uk
Birmingham Archaeology, University of Birmingham, Edgbaston, Birmingham

Airborne remote sensing techniques have traditionally been employed to great effect in mapping cultural archaeology and to a lesser extent the geomorphology of valley floor aggregate bearing landscapes. While the two dimensional record of air-photography provides an approach to mapping the quantity of archaeological material (both cultural and geoarchaeological) within aggregate bearing landscapes, it provides no indication of the state of preservation of that material or the associated cultural evidence. In particular, aerial photography provides no indication of the moisture level and the potential for the preservation of organic preservation of sediments. This is a significant shortfall in the usefulness of the data, since the presence of moist, wet, organically - rich sediments may greatly increase the archaeological value of deposits.

This paper describes the preliminary results of the authors' on-going research to systematically investigate the potential of backscattered intensity data from airborne laser altimetry to remotely determine soil properties, including organic content and moisture levels. The research aims to create a tool to allow the rapid assessment of valley floor corridors to provide baseline information on the environmental potential of organic sediments. Analysis of lidar intensity potentially provides a tool for remotely predicting organic preservation of waterlogged deposits in aggregate bearing environments. Such rapid assessment may save aggregate companies the expense of prospection within areas that have high potential for organic preservation, or paying for expensive mitigation strategies in areas where the organic potential of organic deposits is low.

LANDSCAPE ARCHAEOLOGY AND GIS FOR THE ARCHAEOLOGICAL HERITAGE MANAGEMENT OF THE AKSUM REGION, ETHIOPIA

Rossano Ciampalini , Andrea Manzo*, Cinzia Perlingieri*, Luisa Sernicola*

* Università degli Studi di Napoli "l'Orientale"

A Landscape Archaeology approach was adopted to reconstruct the diachronic changes in the landscape in the area of Aksum, Ethiopia, the capital city of the Aksumite kingdom in N Ethiopia (ca. 400 BC-AD 800). The present landscape can be considered as a stratified palimpsest of all past evidences of ecosystemic interaction that is studied to reconstruct the different aspects of the Man-Environment relationships. For this reason a special attention is given not only to the distribution and chronological and functional classification of archaeological sites but also to the study of the off-site man-made features (ancient fields, roads, terraces, facilities for water management, etc.) and of their environmental settings.

The dynamics of human exploitation, manipulation, and management of the territory, also related to environmental stress and to social pressures in the whole N Ethiopian plateau, is heavily affected by human modification, environmental deterioration and soil erosion.

The survey and the excavations conducted over the last years gave important insights into the typology of the structures which characterized the ancient landscape of the Aksum Region; particular interest was concentrated on the symbolic and power landscape, and to the economic landscape. The study of the components of the territory and their relationships greatly contributed also to the archaeological heritage management of the whole Aksum region.

INCALLAJTA, PERFORMANCE CENTER OF THE INKAS: A DIGITAL RECONSTRUCTION AND VIRTUAL REALITY ANALYSIS

Larry Coben

Prior analyses of the monumental Inka site of Incallajta, Bolivia have argued that it was an administrative, political, ceremonial, or military center. Most recently, I have argued that it was a Peircean replica of the Inka capital Cuzco, designed at least in part as the setting for the performance of the most important and sacred Inka ceremonies that had been restricted to Cuzco itself. These Cuzcos, and the spectacles and rituals within them, were an important strategy in Inka imperial expansion, as the empire grew in a fifty year period from a single valley in Peru to encompass Peru and significant portions of Bolivia, Ecuador, Chile and Argentina.

According to the historical chronicles, the Inka constructed their capital as a physical representation of their worldview (Rowe 1944, 1946, Zuidema 1990). Cuzco was a sacral theater, the site of an elaborate series of massive structured spectacles performed according to a ritual calendar (Betanzos 1996 [1551]: Book 1, Ch. XV, Cobo1990: Ch. 25-30, D'Altroy 2002:152-153). Special ceremonies and rituals were also performed in Cuzco in times of war or crisis (Cobo 1990[1653]:112, Ch.31). The idols of newly conquered peoples were frequently transported to Cuzco, and local elites were required to worship their idols there in temples built for such purpose (Cobo 1990[1653]). The Inka later built a limited number of "new or other Cuzcos" in specified regions of their empire, including one in a region now defined approximately by modern day Bolivia (Guaman Poma 1980[1603], Cieza 1959[1552], Garcilaso 1987[1609]). Employing primarily an analytical framework drawing upon theories of semiotics and pragmatics which have been successfully utilized in the study of ritual in many parts of the world, (Parmentier 1994; Preucel and Bauer 2001), I have argued that new or other Cuzcos contain buildings, landscapes and other features arranged in a particular pattern, and that the site of Incallajta is one such Cuzco. ,serving as a theatrical space and locus of performance as discussed above. More recently, I have suggested that new or other Cuzcos were strategically placed only in areas of war and rebellion where the utilization of ritual performance to maintain, reinforce, and manipulate Inka ideology and identity was a critical element of imperial strategy as the polity expanded from a single valley in highland Peru to encompass most of the Andes.

Integral to my analysis of the performance and theatrical qualities of Incallajta is a digital reconstruction of the site, utilizing the latest in computer technology. I n this paper, I will briefly describe the methods and processes utilized for such reconstruction. Through a series of digitally reconstructed "animated walk-through" simulations, I analyze how access to the site's core was orchestrated through narrow and controlled entry points and winding passages. Similarly, I explore the architecture of the kallanka, the largest single room structure in the Inka empire, and demonstrate how its construction, physical scale, and relationship to adjacent areas, particularly large plazas, would have been shaped both a performer's and a viewer's the experience of the space and the performances within it.

OPERATIVE ACTION FOR THE CONSERVATION OF THE ARCHAEOLOGICAL COMPLEX OF CHAN CHAN, PERÙ

F. Colosi*, G. Fangi, R. Gabrielli*, R. Orazi***, D. Peloso***

* CNR- Istituto per le Tecnologie Applicate ai Beni Culturali (ITABC), V. Salaria Km 29,300 – 00016 – Monterotondo Sc. (Roma). (francesca.colosi@itabc.cnr.it, roberto.gabrielli@itabc.cnr.it, daniela.peloso@itabc.cnr.it)

** Università Politecnica delle Marche – DARDUS. (g.fangi@univpm.it)

***Direttore della Missione Italiana in Perù. (roberto.orazi@gmail.com)

The town of Chan Chan represents the most important witness of the Chimù culture, one of the many civilizations that arose in the valleys and along the northern coast of Perù since the beginning of our Era.

The town, entirely built with *adobe* (850 AD – 1470 AD), has a nuclear part of about 6 hectares, where 9 *ciudadelas*, 5 *huacas* (stepped pyramids) and intermediate residences are located.

Though inscribed in the UNESCO "List of Word Heritage" and "List of in danger Word Heritage", Chan Chan is suffering a dramatic process of material and urban degrade due to the building technique, the extension of the settlement and the luck of funds.

From 2001 the MIPE (Missione Italiana in Perù) is operating in Chan Chan carrying on a wide action of documentation, conservation

and exploitation. The most important purpose of the work is represented by the restoration of Palacio Rivero, the smallest of the Chan Chan palaces, and by the protection of the site from the aggressive and uncontrolled growth of the near town of Trujillo.
Therefore the MIPE project is mainly pointed to the creation of a wide Archaeological Park, a Documentation Centre and a Museum in order to save the original landscape and to produce an organic and controlled development of the socio – economic structure of the region and of the tourist sector.
Different techniques have been taken into consideration for the documentation of the site so that the actual shape of the buildings has been represented by mean of total station, Differential GPS and aerial photogrammetry.
For what is regarding the study of such a wide territory an important and recent investigation tool has been represented by high resolution satellite images. We have been using the Quickbird Satellite images, kindly provided by Eurimage SpA, which are available in real colours, infrared and panchromatic, this last with the highest resolution.
The most interesting characteristic of these images is represented by the possibility of obtaining very strong enlargements and therefore by the great advantage to work at different scales (architecture and territory on the same image).
With the aim to improve the quality of the information a fusion between the real colour and the panchromatic (architectural and archaeological structures) and between the infrared and the panchromatic (soil and vegetation) images has been performed. The new images keep all the information of the real colour and the infrared shots but achieve the high resolution of the panchromatic one.
The new images, beside the visible archaeological structures, show also architectural alignments which are not visible on the ground, providing therefore an important tool for the identification of buried structures.
The satellite images are particularly useful for the project of the "Chan Chan Archaeological Park". In order to face the uncontrolled development of Trujillo, three different areas have been planned: the archaeological area (archaeological investigation), the protection area (no constructions allowed) and the area with inspected development (structures for handicraft, agricultural activities and tourists welcome).
For what is regarding the restoration of Palacio Rivero we have started the three-dimensional documentation of the structures using the Laser Scanner (Callidus of Trimble SpA). The laser technology, that comes out as the most appropriate for the geometrical survey of structures with undefined outlines as those of Chan Chan, has been integrated with procedures of digital photogrammetry in those parts of the palace which were particularly damaged.
The 3D documentation will be particularly useful for the study of the monument and for the restoration project since it contains any type of graphic drawing (maps, profiles, sections, elevations). Furthermore the laser survey will allow a virtual reconstruction of the original shape of the palace and which becomes every day more important in consideration of the slow loss of the material witness and which will be visible at the Documentation Centre. As a matter of fact an important element of the project is to provide the scientific researcher and the tourist with a first multi media approach to the archaeological complex.

REFLECTION TRANSFORMATION IMAGING ON LARGER OBJECTS: AN ALTERNATIVE METHOD FOR VIRTUAL REPRESENTATIONS

Massimiliano Corsini, Matteo Dellepiane, Marco Callieri, Roberto Scopigno

ISTI-CNR, Via Moruzzi 1, Pisa, Italy, firstname.lastname@isti.cnr.it

Virtual representations of cultural heritage artifacts are becoming more important, as the technology behind them improves. Their field of application includes a range that goes from the simple visualization to the support of the analysis and restoration. 3D scanning techniques have reached a very good level in terms of precision, but there are also other interesting technologies: one of these is Polynomial Texture Mapping. Using a mathematical model presented in 2001, this set of image processing methods describes the luminance information for each pixel in an image, in terms of a function representing the direction of incident illumination. The per-pixel information is able to record properties including surface inter-reflection, subsurface scattering, and self-shadowing. Moreover, PTMs can communicate useful shape information using purely image based transformations, without any use of 3D reconstruction techniques. PTMs are calculated from a set of photos taken, from a fixed point of view, under several different positions of a single point wise light.
This technique has been already used for visualization and study of cultural heritage artifacts: cuneiform epigraphy, fossil specimens, stone tools, ancient coins. Many features of these objects are clearly readable only under certain light conditions. The use of PTM, in addiction with methods like specular enhancement and diffuse gain, gave scientist the possibility to discover details which were not visible during physical inspection.
The size of all the objects considered by previous studies was quite small, also because the usual capture systems are based on a small light dome (e.g. 50 cm radius), with a camera mounted on the apex and a variable (24 to 50) number of lights.
In the present paper we analyzed the application of the PTM technique to bigger artifacts. The aim was to demonstrate that Polynomial Texture Maps can be an alternative (low cost) method to visualize and explore cultural heritage objects.
We adapted the complete PTM "pipeline" (capture system, acquisition planning, acquisition, PTM building and visualizing) to produce high-resolution PTMs in a simple and fast way. We also made a preliminary study of the quality of the illumination model, comparing it to the corresponding 3D scanning product.
First step of the pipeline was the definition of the capture system, since it was not possible to create a dome with a radius of more than two meters and dozens of lights. The set of photographs was created using an 8 Megapixel digital camera, fixed perpendicularly to the center of the object, and a single 1000W flood light, whose position was changed manually in order to create all the light directions needed.
Since bigger object can't often be moved from their place, we created a simple software which helped us to plan the acquisition. The software calculates the position and height of the light for each photo, taking into account: the dimensions of the object, its height from the floor and the available space around it.
Since we wanted to create and visualize high-resolution PTMs, we developed a new builder and a new viewer, which preserve all the detail of the acquisition. Due to the large amount of data involved, the images are divided in several parts and stored separately. In order to preserve a real time navigation of the model, the detailed subparts of the image are loaded only when needed.

We applied the pipeline to some examples: we show here a medieval capital and a particular of a XIVth Century Sarcophagus. On the second one we performed also a comparison in terms of quality with the corresponding 3D scanned model.
Finally, a hi-resolution PTM of a part of a Roman Sarcophagus in Camposanto Monumentale of Pisa was produced.
The results of the acquisitions showed that, in some cases, PTM can be a good solution for the visualization and exploration of artifacts. Moreover, the capture system is low-cost, and after an accurate planning, the acquisition and production of the illumination model can be easy and fast.
Therefore, PTM can be a valid alternative to the classic acquisition techniques, not only for small objects, but also for bigger artifacts, like basrelieves and statues.

THE EMERGING LANDSCAPE OF EAST LOTHIAN, SCOTLAND – MAPPING, ANALYSIS AND PRESENTATION

Dave Cowley, RCAHMS

Email: dave.cowley@rcahms.gov.uk

The coastal plain in East Lothian, in south-east Scotland, is a predominately arable area where the majority of known archaeological monuments are plough-levelled sites recorded as cropmarks on aerial photographs. This dataset has emerged from 50 years of aerial reconnaissance, but especially since 1976 with the aerial survey programme of the Royal Commission on the Ancient and Historical Monuments of Scotland (RCAHMS). Much of this dataset has been added to the archive of RCAHMS with little analysis or mapping and it is therefore relatively poorly understood. This material is now being mapped in detail across the entirety of the coastal plain, providing an interpreted digital map base, which underpins more detailed analysis and presentation, as well as providing accurate locations and extents of sites for planning and conservation purposes. This paper presents some preliminary results from this project.

USE OF REMOTE SENSING AND GIS IN THE MANAGEMENT AND CONSERVATION OF HERITAGE PROPERTIES AT AGRA

D.Dayalan

In the recent years the necessary has been felt for utilization of remote sensing, GIS and related geo-information technologies for the conservation and maintenance of the cultural and natural properties. The entire surface of the earth, including the most remote and inaccessible places can be observed by satellites The new development in the technology like, microwave remote sensing, Global Positioning System (GPS), etc are also being utilized for various studies. The activities wherein the remote sensing technology can be utilized for the management and conservation of cultural and natural properties is:

1. Creation of data-base and maps of monuments and their environments, ancient settlements, etc.
2. Evaluation and management of the monuments and sites
3. Study of chronological changes in the nearby area of the monument/site.
4. Monitoring of encroachment in the surrounding of the properties
5. Preparing base maps for drawing master plan for integrated development in and around the heritage properties
6. Survey and monitoring of natural properties
7. Development of Natural Disaster Management System and monitoring of river pollution.

Satellites images can be obtained in digital form (so directly workable by computer processing) and provide reliable, repetitive, non-invasive, rapid and cost-effective information. The remote sensing technology is a valuable tool to assist conservation activities. All information is exactly localized and gathered in one tool. Information can be continuously updated. The direct extraction of topographic and thematic maps for terrain use is also possible with this technology.
Agra (270 10' N ;780 1'E), the prime destination of tourism in India, is not only an important urban centre in the medieval period but also blessed with three world heritage monuments including the world famous Taj Mahal. Spread to an area of 22.44 hectare, Taj Mahal has to be visualized along with the river bank, river Yamuna and adjoining monuments, gardens and landscape as they are the integral part of this monument. The remote sensing and GIS techniques are applied for creating data-base, maps and various other purposes. These techniques are also used for the management of the buffer zone, preparation of the integrated scheme in and around the monument, conservation activities, etc. The satellite images are useful to visualize the overall urban settlement pattern during the medieval period. Fatehpur Sikri, another world heritage monument near Agra is the planned city of the Mughal Emperior Akbar (1556-1605) constructed on the bank of a large natural lake. The ancient city wall, which in its 9 kms run is pierced by nine gates holds a large area with undisturbed archaeological remains in the form of palaces, mosques, sarais, hammams, water works and many other secular, religious and defense edifices. The department is the process of applying the remote sensing and GIS techniques for mapping the various structures, monitoring the monumental and the periphery areas and for the conservation activities. The detail paper will provide the vivid information about the use of remote sensing and GIS in these world heritage monuments.

GEOARCHAEOLOGICAL SITE LOCATION MODELLING BASED IN THE TERRITORY OF A CLASSICAL TOWN – CASE STUDY SAGALASSOS (SOUTHWEST TURKEY)

V. De Laet[1], E. Paulissen[1], H. Vanhaverbeke[2], M. Waelkens[2]

[1]Physical and Regional Geography, Katholieke Universiteit Leuven, Celestijnenlaan 200E, 3001 Heverlee, Belgium
[2]Eastern Mediterranean Archaeology, Katholieke Universiteit Leuven, Blijde Inkomststraat 21 3000 Leuven, Belgium

Since the eighties archaeologists are studying the distribution of archaeological sites, but often modelling studies are limited to the Prehistoric period. This research attempts to create a methodological framework for the prediction of locations likely to contain

archaeological settlements in the territory of the antique city of Sagalassos according to various archaeological time units: Late Bronze Age – Early Iron Age; Hellenistic times; Early Roman – Mid Roman times and Late Roman – Early Byzantine times. Instead of using regular logistic regression methods, rare event logistic regression analysis in order to overcome the problems related to rare event data (much more non-site locations than site locations). Further distances were calculated taking into account friction or the difficulty to travel trough a region. The environmental variables used to accomplish such kind of analysis were based on a high amount of field data and GIS information.
The analysis showed that during the turbulent periods in the history of the territory, settlement strategies were followed that were in sharp contrast to settlement preferences during the high days of the region. All results correspond well with literature observations for the region.

VIA FLAMINIA PROJECT: RELIEF AND POST PROCESSING DATA TECHNIQUES

Nicolò Dell'Unto[1], Fabrizio Galeazzi[2], Marco Di Ioia[2]

[1] IMT Insititute of Advanced Studies, Lucca, Italy
[2] CNR-ITABC, Istituto per le Tecnologie Applicate ai Beni Culturali, VHLab, Montelibretti, Roma, Italy

In autumn 2005, CNR-ITABC signed a research agreement with ARCUS (Art & Culture Society), in collaboration with the Archaeological Superintendence of Rome, for the "reconstruction of the archaeological landscape of the ancient road Flaminia with virtual reality systems". The aim is to build a digital ecosystem that individualizes two levels of perception: the first one represents an holistic vision of the road from Rome to Rimini, based on historical maps, information on the archaeological excavation, technical cartography and aerial and satellite photos able to support the realization of a web gis; the second one represents and contextualizes the local entities within the "ancient landscape", going into a micro-space vision that focuses the attention on four sites: Villa di Livia, Malborghetto-Tor di Quinto, Ponte Milvio, Grottarossa.
This paper describes the methodologies and the problems faced during the acquisition and the elaboration of the elements which compose the virtual scene of the Villa di Livia case. These processes were studied to produce a virtual reality application.
The instruments used to acquire and generate the data are: laser scanner, photogrammetry, computer vision, Total Station and differential GPS. All the data acquired with these tools have been elaborated to exist in the same system.
The different nature of all the archaeological sites studied leads to the problematic relationship between optimization and details of every single structure. These two elements are strongly connected together: the greater is the detail of an object (in number of polygons), the more difficult it will be to use it in a virtual reality environment.
Due to the complexity of the archaeological structures, we have developed different strategies of data acquisition, i.e. the integration of different techniques and the use of many softwares.

UN SISTEMA INFORMATIVO ARCHEOLOGICO: APPLICAZIONE PRESSO LA SCAVO DI FERENTO (VITERBO)

De Minicis E.*, Gabrielli R., Peloso D.****

* Dipartimento di Scienze del Mondo Antico,University of Tuscia, Viterbo, Italy
** Institute for Technologies Applied to Cultural Heritage – CNR, Rome, Italy

This paper relates the specific experience for the collaboration between the Institute for Technologies Applied to Cultural Heritage and the University of Tuscia in managing archaeological excavations on a GIS platform. The basic idea was the reproduce on a graphic level the exact situation we find in the field. We therefore organised our objects according to an overall composite plan representing all the excavated layers, as well as the necessary landscape features, related only spatial terms; detailed alphanumerical data and interpreted information were derived from the DBMS using specific identifiers.

ARCHAEOLOGICAL REMOTE SENSING IN YEMEN, THE JABALI TEST SITE. FROM LARGE-SCALE SURVEY TO FIELD INVESTIGATION

Jean-Paul Deroin[1], Florian Tereygeol[2], Paul Benoit[3], Mohammed Al-Thari[4], Ismail N. Al-Ganad[5], Jürgen Heckes[6], Audrey Peli[7], Sophie Pillault[8] and Nicolas Florsch[9]

1 EA 2455, Observation de la Terre et Information Géographique, Université de Marne-la-Vallée, 5 boulevard Descartes Cité Descartes Champs sur Marne 77454 Marne-la-Vallée cedex 2, France
2 UMR 5060, Institut de Recherche sur les Archéomatériaux (Belfort) et UMR 9956, laboratoire Pierre Süe, CEA Saclay, France
3 UMR 8589, LAMOP, Université Paris 1-CNRS, Paris, France
4 ZincOx Resources plc, Hadda, Sana'a, Republic of Yemen
5 Yemen Geological Survey and Mineral Resources Board (GSMRB), Al-Tahrir, Sana'a, Republic of Yemen
6 Deutsches Bergbau-Museum, Bochum, Germany
7 UMR 8167, Orient et Méditerranée, Université Paris 1-CNRS, Paris, France
8 UMR 5060, Institut de Recherche sur les Archéomatériaux (Belfort)
9 UMR 7619, SISYPHE, Université Pierre et Marie Curie-CNRS, Paris, France

The Jabali area, about 70km as the crow flies north-east from Sana'a, the capital city of the Republic of Yemen (Sana'a and Ma'rib districts), is currently explored for zinc-ore. The mineral rights are held through an exploration licence which is leaded by ZincOx. Jabali contains a geological resource of 12.6 million tonnes of oxide ore, grading about 9% zinc, 1.2% lead and 68g/t silver. The mineralization comprises oxides and carbonates (smithonite, hydrozincite, etc.). Silver is mainly related to sulfides, which are

represented, even less than oxides: sphalerite (relatively rare), galena, etc
During the pre-islamic period and until the IXth century AC, the Jabali area has been mined for silver. The activity ceased at the end of the IXth century AC, according to the description given in the Xth century AC by al-Hamdani, the famous Yemenite geographer who named the site, al-Radrad, one of the most important silver mines of the Abbassid world. In the early 1980's, the site has been rediscovered by the French Geological Survey (BRGM). From 2004 onward, archaeological investigations have been carried out. The last mission has been done in March 2006, including archaeological field investigations (topography, mapping of archaeological remains, slags and ceramics sampling, etc.), geomorphological and geological mapping and sampling, and field geophysics (magnetometry), supported by GPS measurements and satellite imagery as well.
Using Jabali as a test area, we have evaluated different types of remote sensing data. At a large-scale, remote sensing data have been used for drawing new geological maps on the area. Landsat ETM+, Landsat TM and Spot panchromatic data have been widely interpreted for distinguishing lithologies either in the Proterozoic substrate or in the sedimentary cover, mainly Jurassic to Cretaceous in age. Cenozoic events are represented by two phases of volcanism, first at about 30Ma, constituting large traps in the Sana'a region, second during the Quaternary, forming basaltic volcanoes aligned along faults linked with the opening of the Red Sea and the Afar tectonics. Recent deposits are also identified such as deflation reg, alluvial deposits, colluvions, sand dunes, etc. The mineralization of Jabali is put in light when using ETM+ band no 3, *i.e.*, the red range of the electromagnetic spectrum. This is clearly explained by the iron oxides concentrated in palaeokarst and linked with (de)dolomitization phenomena. Dolomites could be easily identified by their spectral response and their texture as well.
At a more detailed-scale, SPOT 5 stereoscopic data (pixel-size: 2.5m) have been used for extracting a detailed Digital Elevation Model (DEM) on the test area. This allows a 3D-vizualisation of the site, which is about 6km long from north to south. From the top (about 1800m a.s.l.) to the bottom (about 1400m a.s.l.), the following subsites could be shortly described. (1) First, the mine area in the higher part, which includes a number of ancient mines. It corresponds to carbonate layers deeply weathered and karstified. (2) In the sector of Jabali itself, palaeometallurgy is revealed in the field by slags, palaeochannels, alluvial deposits linked to washery, etc. (3) 5km to the north, along Wadi Al-Khaniq, slag heaps are encountered on about 2km long. Different terraces have been identified. A detailed mapping of these deposits and terraces is in progress. On the right bank of Wadi Al-Khaniq, a carbonate plateau shows a lot of slags associated with archaeological remains (walls, buildings, graves, etc.). This plateau is about 15ha-wide. Geophysical (magnetic) measurements have been tested on that plateau. An other place with slags and remains has been identified close to a right bank tributary of Wadi Al-Khaniq. Very high resolution Quickbird data (pixel-size: 0.61m) have been also used by the colleagues from DBM for a detailed study of the sector of mines.
The ancient silver mine of Jabali is located within the southernmost outcrops of the Jurassic limestones of the Amran Group, close to the contact between the Proterozoic substrate and the sedimentary cover, an area deeply affected by volcanism and tectonics. It perfectly illustrates the interest of an integrated study in the topics of archaeological remote sensing, using image processing, 3D visualization, geophysics, etc. The landscape has to be studied at three time scales represented by different morphoclimatic conditions. The current setting is characterized by an arid mountainous area with temporary watercourses (wadis) and rare human settlements. The mediaeval setting (VIIth-IXth century AC) evidences more temperate conditions with relatively abundant water and wood used for metallurgy and villages scattered from the mine to Wadi Al-Khaniq. The Quaternary setting could be rebuilt using surficial deposits, palaeoterraces, structural studies especially those of the Quaternary volcanism, and neotectonics. Imagery and GPS constitute powerful tools for such a work.
Acknowledgements. GSMRB and ZincOx facilitated the study. We specially thank Dr Brett Grist from ZincOx, Sana'a. We are also indebted to Major Khaled Al Naphani and his group from the National Guard for assistance with field work. CEFAS (French-Yemenite Archaeological Centre, Sana'a) provided facilities. The French Ministry of Foreign Affairs, CNRS and University Paris 1 financed field investigations for the French part.

THE FOURTH DIMENSION OF PLACES: LANDSCAPE AS AN ENVIRONMENTAL AND CULTURAL DYNAMIC PROCESS IN THE MAREMMA REGIONAL PARK

Michele De Silva
Università degli Studi di Siena mdesilva@unifi.it

The landscape should be considered the result of a historical sedimentation process where different natural events and human activities have all left traces in forming an interwoven and complex whole. In order to understand, preserve and appreciate a landscape it is not sufficient to study places as they are, or as they were, in all three dimensions, but it is necessary to diachronically explore, in the fourth dimension of time, the natural and human processes which constitute its identity from a historical and environmental point of view.
Through the analysis of aerial photographs and historical sources - in particular medium to large scale historical cartographies - correlated with survey data, it is possible to reconstruct the territorial settings of the past and subsequently identify the dynamics and transformation processes of the landscape. The diachronic approach to the Landscape Archaeology allows us to recognize the role of natural events and human intervention in the transformation processes that constitute the matrix of the present territorial setting.
In the case study presented, the comparison between land use derived from the cadastre of the beginning of the 19th century and more recent land use data derived from former and current aerial photographs, made possible through GIS technology, is employed as an indicator of changes and continuity in the landscape of the Alberese area (Maremma Regional Park, Southern Tuscany, Italy) over the last two centuries. Although today the Maremma Regional Park may appear an example of natural wilderness as opposed to other manmade areas of Tuscany, we should not undervalue the role of human activity in shaping this landscape. A further comparative evaluation of land use distribution throughout the entire mainland of the Grand Duchy of Tuscany in 1830 underlines the distinctive nature of the study area.
In the Alberese area at the beginning of the 19th settlements are rare whereas woodlands and pasture occupy a large portion of the territory as opposed to sowable lands. Furthermore, wetlands are still present in the area despite the fact that several reclamation works had already been undertaken at that time. Pinewoods are present only in a restricted area and olive groves seem to be altogether absent. A few small areas classified as 'oleasters' may represent relicts of former degraded olive groves.

In recent times sowable lands extend throughout the northern and eastern areas whereas marshes and wetlands have virtually disappeared and pinewoods cover the flat coastal belt. Woodlands are considerably reduced, pastures are limited to small areas near woodlands, and olive groves reach a significant coverage. New rural settlements and small parcels of vineyard and olive groves are scattered throughout the inland plain.
Analysis of the evolution of the land use in the study area over the last two centuries, obtained through a cross-tabulation analysis, highlights that changes have mainly occurred on the plains. The main areas of continuity in land use are represented by the hilly woodlands, certain sowable lands and a small area of pinewood. Sowable lands and pinewoods have replaced pastures and woodlands in the plains, and olive groves have covered the inland lower hills.
From an ecological point of view, the most relevant changes concern the disappearance of the plain woodlands and wetlands, the substitution of pastures by sowable land and the expansion of pinewoods. In addition, the coastline has regressed by approximately 1 kilometre near to the Ombrone river mouth, whereas it has advanced up to 400 meters in other areas
The Alberese Farm at the beginning of the 19th century is characterized by the prevalence of woodlands and pastures, bearing witness to the existence of an economy based mainly on breeding and woodland exploitation. The role of grain farming seems to be marginal considering the limited extent of sowable land.
We would, furthermore, like to stress the role that human activity has played in shaping the landscape. Economic and social choices such as reclamation works and farming activities, or in other words historical events, are undoubtedly the main causes of landscape transformations.
In due consideration, study of the current landscape is insufficient for competent territorial management and must be coupled with a diachronic investigation, in other words, an analysis of the historical dynamic processes of transformation that constitute the fourth dimension of landscape. Only in this way can more convincing and tenable choices be made relating to the planning of activities and interventions that are compatible with the preservation and appreciation of our natural and cultural heritage.

PROGETTO S. BENEDETTO PO "LETTURA STRATIGRAFICA" DEL TERRITORIO DELLE BONIFICHE: REALIZZAZIONE DI UN SISTEMA INFORMATIVO GEOGRAFICO (GIS).

Carla di Francesco*, Cristina Ambrosini*, Marina De Marchi*, Maurizio Boriani, Susanna Bortolotto**, Sergio Alifano**.**

* Ministero per i Beni e le Attività Culturali, Direzione Regionale per i Beni Culturali e Paesaggistici della Lombardia.
** Politecnico di Milano, Dipartimento di Progettazione dell'Architettura, Laboratorio di Diagnostica per la Conservazione e il Riuso del Costruito.

Il progetto nasce dalla volontà della Direzione Regionale della Lombardia di applicare ad un caso concreto quanto espresso nel Codice dei Beni Culturali e del Paesaggio (D.Lgs. 22.01.2004 n.42) intendendo quest'ultimo come una parte omogenea di territorio i cui caratteri derivano dalla natura, dalla storia umana o dalle reciproche interrelazioni.
Il progetto, si pone l'obiettivo di fornire una conoscenza il più possibile completa ed articolata di un bacino territoriale che in area lombarda presenta specifiche ambientali derivate dalla presenza del grande fiume Po e dai processi di trasformazione indotti dagli interventi di bonifica da sempre considerati necessari allo sviluppo produttivo ed al miglioramento delle condizioni di vita dei suoi abitanti. Le bonifiche hanno disegnato sul territorio un lungo processo iniziato in età preromana e prolungatosi nel corso dei secoli fino ai primi decenni del Novecento, quando i Consorzi diedero corso alle opere di bonifica più estese ed incisive. Siamo di fronte ad un paesaggio modificato dall'operare umano dai tempi più antichi, che permette di cogliere la complessa stratigrafia che ha mutato nel corso del tempo le modalità insediative, la rete viaria e i suoi snodi, il paesaggio rurale e la maglia dei canali irrigui e dei manufatti di bonifica idraulica.
Il progetto vuole essere un'esperienza innovativa e di eccellenza in grado di collegare i beni paesaggistici a quelli culturali per ricostruire con studi ed analisi approfonditi, un sistema territoriale, e restituire alla popolazione lo spessore storico del territorio, cioè il suo più profondo significato ai fini di una pianificazione consapevole.
Il GIS realizzato consente di far dialogare i dati delle sezioni tematiche per singoli periodi storici e contemporaneamente leggere il succedersi degli eventi nel tempo.
Per la georeferenziazione si è rivelato idoneo alle esigenze del progetto il programma ArcMap 8 della Esri, collegato a *database* realizzati con programmi software Access ed Excel della Microsoft.
Le "sezioni" attivate attualmente sono: Area di Studio, Basi cartografiche, Geologia, Idrografia, Toponomastica, *Archeologia (georerefenziazione siti "puntuali " ed "areali" con riferimento a schede censimento survey e carta archeologica),* Edifici di culto e monastici, Viabilità storica, Destinazione uso suolo: ricostruzione, sulle basi cartografiche utilizzate (Carta del Regno Lombardo-Veneto 1833/56; IGM 1888-89 e 1933-35; CTR 1994), dell'evoluzione colturale del paesaggio agrario nel contesto territoriale di studio; Censimento beni archeologici, architettonici e manufatti idraulici.

ARCHITECTURAL LECTURES TROUGH THREE DIMENSIONAL STRATIGRAPHY INTERPRETATION: VILLA ADRIANA IN TIVOLI.

Sergio Di Tondo, Filippo Fantini

University of Florence, Facoltà di Architettura

The wide use of laser scanning technology for site and archaeological monuments survey is nowadays recognized as a good and fast solution for contemporary data collection and to produce new effective readings and analysis on monuments with a range of solutions that go from documentation to 3d data sharing and database referencing to 3d interface.
This is a presentation of same case studies conducted by the survey lab of the *Dipartimento di Progettazione dell'Architettura* of the Florence University on some archaeological monuments. The main objective is to underline the use of laser scan point clouds for viewing and managing stratigraphy in a point based 3D Model.

The main subject of the presentation will be the digital surveys of some area from Villa Adriana. Starting from the digital survey based on 3d laser scanning of those monuments (conducted using Leica HDS systems) we will show our procedure to manage and archive the data of those survey together with some evolved strategies for transferring traditional two-dimensional solutions for reading and drawing into the three-dimensional datascape. This solution improve the possibility to investigate masonry in 3D spatial model, and allow to create a relationship between the stratigraphy interpretation and the guideline of lying plane. All the survey presented come from our latest experience in laser scan survey, while our laboratory has work, during the last two years, on almost forty survey of monumental architecture, some of them with main archaeological features. Finally: some considerations about how innovative could be the three-dimensional stratigraphy solution for a better understanding, documentation and diffusion of excavation results while working on the site as when the site is open for tourist frequentation.
The production of well suited product of digital survey with an appropriate language towards archaeology, based on the understanding of the monument characteristics instead of mass-gathering information and unshaped documentation. The development of easy to use procedure for stratigraphy building in 3d data models, with a workflow system suitable for technical operators as well for naïve users.
The procedure for managing and building this specialized structure of 3d space can be used by entry level operator, but if used from advanced user show a great solution for reading behind the consolidated perception of the monument and allows to investigate deeper the project characteristics and the compositive pattern.
In this way, the subdivision of the pointcloud, formed during the survey process, allow to compare single parts of the monuments and to verify with effective results the organization of masonry with direct spatial access to the shape of the architecture. In this way is possible to obtain an architectural lectures of the building in relationship with the characteristics of the masonry.
In order to Villa Adriana, the research try to investigate the compositive pattern that regulate the position and the guideline of the single buildings erected during Adriano's empire. So a corrected lectures of masonry is necessary to separated the parts of buildings that don't belong to this period (117-131 d.C.) from the original Adriano's parts of Villa. In order to answer this question during September 2004 Dipartimento di Progettazione dell'Architettura made a 3D Laser scanner survey on Grandi Terme and its environment (Grande Vestibolo, Pretorio), and during the September 2005, Paletsra and the main front of Cento Camerelle. Both surveys allow to have enough information to create a complete model of the landscape front of Rocca Bruna's valley.
Original survey missions were conducted during the International project competition Premio Piranesi 2004 and Premio Piranesi 2005 DARC. Scientific coordination: prof. Marco Bini, Prof. Pierfederico Caliari; survey coordination: Dott. Giorgio Verdiani; topographical survey cordination: Arch Mauro Giannini, Arch. Francesco Tioli. survey squad: Dott. Carlo Battini, Dott. Sergio Di Tondo, Dott Filippo Fantini, Dott Michele Cornieti, Dott Laura Angelini.

LIDAR-SUPPORTED ARCHAEOLOGICAL PROSPECTION IN FORESTED AREAS

Michael Doneus, Martin Fera, Martin Janner

University of Vienna, Department for Pre- and Early History, Franz-Kleingasse 1., A-1190 Wien

Systematic detection of archaeological sites in woodland is one of the unsolved problems in archaeological prospection resulting - not only in Austria – in a big deficit of archaeological knowledge of forested areas. Over the past few years airborne laser scanning, also known as LIDAR (Light Direction and Ranging), has been used to produce high precision terrain models. Its applications in archaeology are successful but still rare, and it turned out to be a possible tool that can help to solve the problems with the recognition and measurement of preserved sites in forested areas.
To explore the potential of LIDAR for Archaeological Prospection in a densely forested area, a project was launched beginning of 2006 at the Institute for Prehistory and Early History in Vienna. Within the project, which is funded by the Austrian Science Fund (P18674-G02), we try to evaluate an approx. 190 sq km forest area within the Leitha mountain range, 40 km southeast of Vienna.
The scans are realized using the latest generation of Airborne Laser Scanners, which digitally samples and stores the entire echo waveform of the reflected laser pulses. These full-wave scanners have the potential to overcome the problem of conventional systems, where the height difference of the scanned objects must be 1.5 m to be able to distinguish the returns, which makes it difficult to get the surface height in areas e.g. covered with bushes. Therefore, we expect a more reliable classification of the laser points and a higher accuracy of the terrain points from full wave in comparison with the conventional LiDAR data.
During the first phase of the project, a test scan covering two 4 sq km large areas was performed. The areas were carefully chosen and represent different kind of canopy (bushes, trees with and without brushwood) above already known archaeological sites (earthwork, tumuli, ruined buildings, stone quarry and single walls).
The analysis of the data is under development. Both analysis and georeferencing of the full-wave data is done in cooperation with the Institute of Photogrammetry and Remote Sensing of the Technical University of Vienna. For the filtering of the data we use the software package SCOP++ where robust interpolation with an eccentric and unsymmetrical weight function is used.
Several of the covered archaeological features were surveyed terrestrially using a total station and a terrestrial laser scanner. The results of the ground survey will be compared with the results of LiDAR to assess its potential as well as its limitations for archaeological prospection. During the lecture, the results of the test scan and the comparison will be demonstrated and discussed.

ACTIVE AND PASSIVE 3D SURVEY MERGING. THE CASE STUDY OF THE WATER CHANEL SYSTEM IN AL HABIS CASTLE, JORDAN

Drap P.*, Franchi R., Gabrielli R.***, Peloso D.*** and Angelini A.*****

* UMR 694 MAP – CNRS- Marseilles School of Architecture, France
**Istituto di Geodinamica e Sedimentologia & Centro Studi Archeologici "CESAR", University of Urbino, Italy
*** Institute for Technologies Applied to Cultural Heritage – CNR, Rome, Italy

This research project is funded by the Italian Ministry of Education, University and Research and the Ministry of Foreign Affairs and has developed as a joint project between the University of Urbino, CNRS, UMR 694-MAP of Marseille and the CNR, ITABC of Rome. The aim of the work is the development of an integrated system of automated topography applied to the remains of the Crusaders' castle of al-Habis at the western boundary of the archaeological area of Petra.
In particular, the monumental site of Petra is located on the left rim of the Rift Valley, in central-southern Jordan. The only visible part of the architectonic heritage consists of tombs and temples dated to the Nabataean period and sculptures within the rock walls.
Past inspections carried out in the field have highlighted the key role played by rainwater flowing down the facades of the monuments and causing their advanced deterioration. Today, this drainage system is no longer operational, as it has been blocked by both the accumulation of debris and collapses caused by landslides and earthquakes.
One of the main objectives of the architectural and patrimonial survey is to provide a precise documentation of the *status quo* of the surveyed objects (monuments, buildings, archaeological objects and sites) in order to preserve and protect them, to study and restore the monuments and to present them to the people. Complex objects, not planar or with ornaments and decoration, require the high-density and the high-resolution of the spatial data.
The laser scanning techniques and close range photogrammetry can offer two complementary sets of instruments and technologies able to answer to the specific requirements of architectural and archaeological survey.
The results of this operation is an extremely detailed and measurable 3-D model of the channels that can be used both for classification and study purposes as well as for virtual tests and simulations on the waters flow. The creation of this detailed model allowed us to extract information on sections along the course of a channel and revealed the degree of deterioration of the side walls of the water channel.

AIR PHOTO APPLICATIONS IN WALES, UK. EXPLORATION, LANDSCAPE ANALYSIS, CONSERVATION AND PUBLIC PRESENTATION

Toby Driver, Tom Pert and Chris Musson

The object of this paper is to describe (briefly) some aspects of air-photographic work in Wales, as illustrated in the work of the Royal Commission on the Ancient and Historical Monuments of Wales (RCAHMW) and related bodies and individuals. The authors are all current or former members of staff at the Royal Commission. The emphasis will be on practical lessons from the use of active aerial survey and the air photo mapping to assist research, field survey, conservation and public presentation of the archaeological and landscape resource in Wales. For the most part technical or methodological details will be omitted in the spoken version of the paper, though they will be presented at least in outline in the published version.
Four aspects have been chosen as illustrative of the potentialities and difficulties of applying 'aerial' techniques in ways which will be economical and effective. Questions of organisation, finance and 'who does what' are seen as major influences on the effectiveness of attempts to study, conserve and present the historic environment in the particular conditions which apply in Wales. It is not claimed that the examples are in themselves remarkable or outstandingly innovative in a technological sense. Rather, the aim is to show how differing objectives, organisations and individuals, working together over a considerable period of time, can have an important influence on the integration of 'aerial' techniques into the everyday conduct of research, conservation and public presentation. It is hoped that other participants in the meeting may see parallels or contrasts with their own situations, and be able to contribute to discussion of ways for promoting similar or contrasting initiatives in the differing structural, financial or cultural situations which apply in their own countries.
Four examples have been chosen for illustration.

- The regular monitoring, through air photography, of nationally-protected archaeological sites, and the effect that such 'routine' work can have on the creation and use of a national archive of air photo images.
- The contribution that exploratory air survey has made in three areas of Wales and its Borderland, along with subsequent attempts at analysis and integration of the aerial information alongside ground-based evidence.
- The use of air-photo mapping prior to or in partnership with ground-based survey in the creation of local, regional and national archaeological records.
- A recent initiative in the development of an interactive method for encouraging children and adults to develop and understanding of the historic environment of their own areas.

3D LASERSCAN ARCHIVING AND DOCUMENTATION OF SELECTED WORLD HERITAGE SITES

Gerhard Ehgartner, Hubert Schöndorfer, Harald Suitner

(a) LASCADO
GEOID Geographic Information and Data Management, www.geoid.at,
(in co-operation with RIEGL Measurement Systems, Schildwächter Ingenieure, ZGIS-Univ. Salzburg)

The objectives of LASCADO are to capture data of World Heritage sites by means of the latest 3D laserscan technology, to process

this data for further scientific archiving and documentation, to build up a digital global data archive of our World Heritage, and to transform the captured data into information which can be disseminated to the interested public via various media (Atlas, DVD, web). Depending on the objective and the related level of data detail (LoD) various target groups can be identified such as (i) conservation authorities (primary data, LoD1), (ii) UNESCO and/or scientific research (primary/secondary data – LoD1, 2), (iii) public, private users, industry (tertiary data – LoD3).
Technical solutions – Data capture: Campaigns are executed with LMS-Z420i in combination with Nikon D10 and appropriate software. 3D documentation including digital terrain models (DTM) of World Cultural Heritage sites are generated (panorama scans, detailed scans, image acquisition). Resulting point clouds with millimeter resolution are be triangulated and visualised. Additional data sources such as satellite data, aerial images, archaeological maps are integrated into a GIS. Criteria and technical requirements for a (global) database and archiving solution have been and will be further developed.
Technical solutions in visualisation comprise (i) virtual walks and fly through, (ii) reconstruction of historic scenarios, (iii) large scale simulations (overviews), (iv) virtual exposees (ancillary information, maps, plans, etc.). Languages such as VRML (*Virtual Reality Modeling Language*) or X3D, GML and Java3D have made individual navigation in a computer-generated world a new experience for the user. The use and potential of these languages shall be further examined in the course of the project. Other VR-elements to be investigated are 3D game engines which offer real time 3D rendering combined with simple but logic navigation tools. User requirements (a.o.) are (i) compatibility of data interfaces and export possibilities, (ii) possibilities to ex post editing (in construction, level of detail, data volume and size), (iii) general usability, and (iv) variety of possible use cases.
Pedagogically valuable content is a major part in LASCADO. Apart from high tech solutions in captureing and archiving, World Cultural Heritage shall be disseminated to a wide public in an interesting and multilingual way.

HIGH RESOLUTION DTM FOR THE GEOMORPHOLOGICAL AND GEOARCHEOLOGICAL ANALYSIS OF THE CITY OF PADUA (ITALY)

Ferrarese F.[1], Cervo F.[1], Mozzi P.[1], Veronese F.[2]

[1]Dipartimento di Geografia, Università degli Studi di Padova, via del Santo 26, 35100 Padova (Italy); e-mail: paolo.mozzi@unipd.it, francesco.ferrarese@unipd.it, pippo156@hotmail.com
[2]Laboratorio di Archeologia, Dipartimento di Scienze del Mondo Antico, Università degli Studi di Padova, P.zza Capitaniato 7, 35139 Padova (Italy); e-mail: archeolabor@tiscali.it

The first known archaeological evidence in the city of Padua are of the Bronze Age. Since then, the city centre records a continuous occupation through the Iron Age, the Roman period and the Middle Ages until modern times. The history of Padua is strictly linked to the evolution of the hydrographic network, as the inner and oldest part of the city developed on the banks of a former meander of the Brenta river, later occupied by a minor river, the Bacchiglione. The archeology and geomorphology of Padua are quite well known, with a long research history of collaboration between archaeologists and environmental scientists but, still, there are a lot of significant details to be investigated. The georeferenced database of spot heights of the Technical Maps at scale 1:1000 of the Municipality of Padua, made available by the Cartographic Office, has allowed the creation of a high resolution DTM of this territory. About 90,000 spot heights exist on an area of 9.3 km^2; only Lidar detection could achieve better resolution. The resulting DTM shows very interesting natural and archaeological features. The analysis of the relief, integrated with other data such as topographic maps from the 18th century onwards, geomorphological maps, aerial photographs, satellite images and archeological stratigraphies, has been of great importance for improving the knowledge on the geomorphological and geoarchaeological evolution of the area. To the NW and S of the city centre some meanders of the Brenta River are evident. They probably are of middle Holocene age and have been only partially mapped in previous studies. Another important landform recognized in the DTM is the fluvial ridge built by the Brenta river since the 2nd millennium B.C. to present day. The structure of the large mound that corresponds to city centre, about 10 m higher than the surrounding plain, is also well defined. This upbuilding is entirely artificial, due to the superimposition of archaeological layers. Two prominent elevations are present on the banks of the meander, where the early settlements were located, while some linear features seem to reflect the Roman urban structure.

FROM REMOTE SENSING TO VIRTUAL REALITY IN ARCHAEOLOGY: AN INTEGRATED APPROACH

Maurizio Forte, Sofia Pescarin, Eva Pietroni

The long process of information from the archaeological fieldwork to interpretation and communication is based on a complex system of micro and macro relations: topology, accuracy, communication, representation, perception, ecology of mind. At standard level of the archaeological research this process is constituted of 2D maps with many manual activities for collecting and representing the data: a key question is: how much information do we lose gathering the data ? Are we able to reconstruct all the path of data capture and archaeological interpretation step by step ? Is there a direct link between topology, archaeological spatial relations, and representation and communication of information ?
We interpret what we perceive, therefore it is important to increase the factor of interaction/feedback inside a virtual reality system able to show all the phases of the digital processing in a 3D domain. This dynamic interaction in a dedicated VR system can multiply the faculty of interpreting archaeological data from the remote sensing applications to the final virtual reconstruction, monitoring all the digital ontologies of observable and unobservable processing of research. In this paper we will show how integrated technologies and spatial information (DGPS, GIS, Remote Sensing, 3D reliefs) can be input in a VR domain so that to have an increased knowledge of the landscape. For this reason a dynamic interaction is the best approach for uncovering relations and feedbacks, comparing present and past, mental maps and mindscapes. "The simulation of tours through man made or natural landscape such as cities, buildings, campuses, gardens, or imaginary geographies [...] offers a dynamic experience of space that contrasts with the static representation of the map [...] the tour temporalizes the experience of space by revealing it one visual frame at a time. Whereas the

map is an abstract model of space, the walk-through is a lived experience." Two projects concerning Rome, "The Appia antica project" and "Via Flaminia Project", will be presented as case studies.

TRANSPARENCY, INTERACTION, COMMUNICATION AND OPEN SOURCE IN VIRTUAL ARCHAEOLOGY

M. Forte[1], S. Pescarin[1], E. Pietroni[1]

[1]CNR ITABC, Virtual Heritage Lab, Rome, Italy, [maurizio.forte],[sofia.pescarin],[eva.pietroni]@itabc.cnr.it

While 3-dimensional visualization methods are now employed in a wide range of contexts to assist in the research and communication of cultural heritage, it is now recognized that, to ensure that such work is intellectually and technically rigorous, and for its potential in this domain to be realized, there is a need both to establish standards responsive to the particular properties of 3d visualization, and to identify those that it should share with other methods (London Charter, March, 2006).
In the last years indeed, the widespread of powerful software, hardware, devices and the exponential growing of digital models built by a wide community, forced researchers to get to a new need, before rarely perceived. If scientific publications have always been a reference instrument in the humanities branch, and not only, what kind of reference can be pointed out in the digital domain? The paper will face up, first of all, into some general observations on what we have to intend with "scientific" in this field, and some possible solutions will be then taken into account, such as transparency, interaction, communication, dynamic modeling, epistemology and "open" approach. The "Flaminia" project (the ancient and famous Roman road), and its VR network, will be proposed as case study and example for the discussed thesis.

GEOPHYSICS AND THE ARCHAEOLOGY OF AN ETHNIC CLEANSING: THE CASE OF GRAND-PRÉ NATIONAL HISTORIC SITE OF CANADA

Jonathan Fowler

Doctoral Candidate, University of Oxford fowler@ns.sympatico.ca

During the 17th and early 18th centuries, French colonists settled in what are now Canada's Maritime Provinces. They developed a unique culture based on marshland agriculture, relative political autonomy, and positive relations with the region's aboriginal inhabitants, known as the Mi'kmaq. Life in the forests and fields of North America transformed these immigrants' identities, and within a few decades they became known as Acadians.
The British conquest of Acadia in 1710 created a series of challenging dynamics between the Acadian, the Mi'kmaq, and the British, and the following years present a complex story of tension and compromise, occasionally punctuated by violence. As France and Britain edged toward an international war in 1755, Britain's colonial government, fearing an Acadian uprising, took drastic measures. The Acadians were rounded up and deported to Britain's colonies along the Atlantic seaboard of North America, where it was hoped they would be assimilated, and their villages were systematically burned. In the centuries since, these communities have been virtually erased from the map as English-speaking colonists gradually moved in and reorganized the landscape. Today, practically all that remains of the Acadian communities are the occasional place names, or withered old trees said to have been planted by Acadian hands. Even the cellars of the homes themselves have been backfilled and ploughed over by farmers.
Grand-Pré National Historic Site of Canada commemorates the settlement of Grand-Pré, founded in 1680, and its destruction by New England soldiers in 1755. Occupying a slight rise overlooking the vast expanse of dykeland that gave this community its name, it is one of the country's best-known national historic sites, and a place of great significance to the Acadian people, who remain a distinct ethnic group to this day. Originally created in the early 20th century, partly in an effort to prevent treasure hunters from disturbing Acadian graves in the area, the national historic site is now an ornamental garden with a memorial church, supposedly marking the ruins of the old Acadian parish church. Surprisingly little archaeological work has been done on this site, and nothing of the destroyed Acadian community is visible on the surface. It follows that little tangible evidence exists to corroborate the traditional knowledge suggesting that this place was once at the center of one of the largest and most prosperous of the Acadian villages.
Since 2000, Saint Mary's University, Parks Canada, and the Société Promotion Grand-Pré have sponsored archaeological excavations at Grand-Pré National Historic Site in an effort to explore the remains of the destroyed Acadian community. The Grand-Pré Archaeological Field School Project has run every summer since 2001, providing a hands-on learning experience for students as well as a rare opportunity for the visiting public to see archaeology up close. The project's research design includes archaeological geophysics, employing the Em38b, a ground conductivity meter designed and marketed by Geonics Ltd. The Em38b measures both apparent conductivity and magnetic susceptibility, and it is the latter channel that provided critical archaeological insights at Grand-Pré, owing to the special nature of the buildings stone employed by the Acadians. Thanks to the application of geophysical survey, this project has located at least three colonial structures, including a small house that was destroyed by fire some time after 1732.

This paper offers a brief introduction to the Acadian colonial settlement of Grand-Pré, its destruction, and subsequent landscape history. It describes the application of geophysical survey to a set of basic historical archaeological research questions, and reviews the results of six seasons of ongoing excavations, which have offered an excellent opportunity to refine or understandings of the relationship between geophysical data and the archaeological resources they describe.

AERIAL ARCHAEOLOGY IN DAUNIA (MORTHERN PUGLIA, ITALY). NEW RESEARCH AND DEVELOPMENTS

Roberto Goffredo

Since 2002 a sistematic project of aerial survey has been carried out in Daunia (northern Puglia,Italy) by Foggia University. It aims at studying the landscape changes and its relative settling patterns from the Prehistory to the Middle Ages. Working on the twin track of

recovering earlier documentation and acquiring new data, it was also possible to carry out a programme of exploration and monitoring of the archaeological heritage in northern Puglia, whether published or not.
This paper illustrates some preliminary results and demonstrates how important is the contribution of aerial survey for the study of the archaeological landscapes in Daunia; it's focused on the data which have enable us to increase our understanding, in particular, of the Roman period and Late Antiquity, especially regarding field systems and different site types.

RECONSTRUCTING THE ANCIENT REPUBLICS (*JANAPADAS)* OF THE INDIAN SUB-CONTINENT

Pallavee Gokhale, Shreenand Bapat

Archaeology is an integrated study of the ancient landscape, resources and human society; landscape forming the 'foundation', resources establishing the 'energy sources' and society forming the 'way of life'. Understanding the ancient signs of human society is crucial as it touches human psychology, relationships, and day-today life of the people. Since the society is evolving continuously, the landscape and resources play a major role in this evolution. They seemed to regulate the political influence of a society or a human-group over others because of limited modes of communication and transportation in the past.
Identifying the geographical boundaries of such groups in the past and mapping them on the present site provides a different perspective for correlating such sites. Moreover these boundaries often seem to change with the upcoming dynasties, their acceptance in the society, wars or conflicts etc. These geographical locations are not only important from the excavation point of view but also for recognising the political significance of the sites during ancient era. These are the actual footprints of the historic times. Hence defining these frequently changing boundaries becomes a prerequisite for the historical/ archaeological studies. This paper discusses modern geospatial techniques to undertake this task.
The aim of this study is to delineate the administrative boundaries of ancient Indian states, known as 'Janapadas' using SRTM data. These republics or kingdoms existed in the Indian sub-continent between 800 BC to 300 BC approximately. The literal meaning of the term 'Janapada' is 'Foothold of a tribe'. In the Pre-Buddhist time, the NW part of Indian subcontinent was divided in several 'Janapadas'. These were named after the tribes or the 'Janas' who had settled in them. By 600 BC, many of these Janapadas had further evolved into larger political entities, known in Buddhist traditions as the Mahajanapadas or the great nations. The continuous process of land-grabbing eventually led to the formation of these kingdoms. Buddhist and other scripts incidentally refer to 16 Mahajanapadas. (Sanskrit: Maha = great, Janapada = country)
The historical references about the locations of these units are in relation with geographical features eg. 'Kosala' kingdom was divided in North and South by the river 'Sharayu', the cities of 'Saket' or 'Ayodhya' were in 'Kosal', etc. Such broad descriptions make it hard to narrow down the search-window for micro-level studies. The main objective of the present research work is to define these 'broadly described' boundaries using river basins as a 'geographical unit'. The drainage basins are later merged based on the order of the river, size of the basin, other prominent geographical features. Besides these, the historic references of wars, inscriptions on the walls of the caves, linguistic difference contribute to form possible accurate units of Mahajanapadas. These boundaries of Mahajanapadas generate the ancient political map of Indian subcontinent. The historic milestones identified during this period are considered as cross-sections of time-line to create the maps. These maps are overlaid on the present political map of India to correlate the archaeological studies in various parts of the countries.

HYPERSPECTRAL, MULTISPECTRAL AND GEO-CHEMICAL PERSPECTIVES ON THE PREHISTORIC CULTURAL LANDSCAPE – OR - REVEALING THE CHEMICAL TRUTH ABOUT PREHISTORIC SETTLEMENT AND INFRASTRUCTURE

Ole Grøn,[1] Pietro Orlando, Finn Christensen, Ivar Baarstad

[1]Institute of Archaeology, University College London

There is an increasing focus on geo-chemistry in archaeology and archaeological remote-sensing. The preservation of some of the chemical patterns created by the prehistoric and historic inhabitants of our landscapes seems to represent a window for extraction of important information about the prehistoric societies from the small-scale level and upwards.
The use of multi-spectral and hyperspectral images for mapping of dwellings, mounds, roads and other small-scale features that are no longer visible in the landscape's topographies and therefore difficult to distinguish by means of traditional archaeological survey has produced promising results. Chemical signatures for the different types have begun to appear in the local study areas, and an important next step will be to study the over-regional similarities.
The paper discusses results obtained so far and outlines a strategy for further development.

RECONSTRUCTING AN IRON AGE AND ROMAN LANDSCAPE NEW RESEARCH IN THE FOULNESS VALLEY, EAST YORKSHIRE, ENGLAND

Peter Halkon

Department of History, University of Hull

1. Introduction

The analysis of the whole of the catchment of the River Foulness in East Yorkshire, England, has enabled the construction of a new model of Iron Age and Roman landscape development within this region, which contains one of the largest Iron Age iron industries yet found in Britain. In an award-winning community-based research project undertaken since 1980, one of the longest-running and most detailed carried out in the North of England, over 1,448 ha have been field walked, large-scale geophysical surveys carried out and 13 seasons of excavation undertaken. Within this 20x30 km area, over 1245 crop mark sites from the English Heritage National

Mapping Programmes of the Yorkshire Wolds and Vale of York, and our own aerial survey, have been studied. Examination of Sites and Monuments Registers and museum records, together with field survey, has also revealed over 600 Iron Age and over 525 Roman sites within this area. The aim of this paper is to illustrate how the application of GIS has recently enabled this rich archaeological dataset to be collated and analysed against a background of landscape change, with significant results.

2. Landscape background

Through study of topography, drainage, soils, vegetation, climate, sea-level rise and coastal change, three distinct landscape zones have been identified within the Foulness Valley, each with a different trajectory of development. The southern landscape sector contained a tidal estuarine inlet which replaced a forested fresh-water landscape at the beginning of the Iron Age, opening up the wider region to the Humber and the North Sea beyond. It was through this route that the tradition of the square barrow and chariot burial, characteristic of the Iron Age Arras Culture, and perhaps iron production itself, may have arrived into this region. Certainly, like the burial traditions, there are parallels between the iron technology of northern France and that of the Foulness Valley. Within this region iron technology may have reached its zenith, exemplified by the internationally important cache of five swords in highly decorated sheaths, and a bundle of 35 spears, found recently near South Cave, within the study area. These were buried at the time of the Roman conquest of what is now Yorkshire.

Recent research suggests that the tidal inlet with *Petuaria* (Brough-on-Humber), the *civitas* capital of the Parisi, on its eastern shore, may be identified with "*opportunus sinus*" or the safe-haven bay referred to by Ptolemy. The inlet continued to be influential in the Roman period, as it dictated the route of the main Roman roads running through the area. The importance of this inlet and related waterways has been underestimated, as they appear to have been a factor in the conquest of the region and its development. It is often forgotten that this river system provided a route to the legionary fortress and provincial capital of *Eboracum* (York).

3. Iron Age and Roman Landscape development

The main focus of this paper will be the characterisation and analysis of the crop mark sites of the Foulness Valley, which differ from each other in both density and form, within the three landscape zones, possibly due to environmental factors such as drainage. Through examination of crop mark morphology, with field walking and comparison with excavated sites, it was possible to propose a sequence of development of settlement patterns within the landscape.

In the Later Bronze Age, a network of long distance linear earthworks was constructed, developing through the Iron Age. Ranging in complexity from single banks and ditches to complex systems, these features appear to be associated with hill-forts and other curvilinear enclosures, and may represent deliberate division of the landscape. Many linear earthworks corresponded closely with landscape features, such as springs and streams; others controlled routes through the landscape or acted as barriers.

Several curvilinear enclosures, similar in form to Staple Howe, the Early Iron Age type site, were identified in the lowland areas, as well as the Yorkshire Wolds uplands. Some of these enclosures may have been "marsh forts" and others may have related to ritual activity associated with water.

Clear contrasts could be determined in the distribution of the single rectilinear enclosures of the Middle and later Iron Age, which are found mainly in the lowland zones, and the linear enclosure complexes known as "ladder settlements", which are mainly located on the uplands. The development of complex enclosure systems suggests intensification of activity in the later Iron Age and Roman periods, though this was not constant. Whereas the interrelationship of features in some areas may illustrate evolution over a long time-span, there was also evidence for major landscape re-alignment especially in the Roman period, particularly where the Roman roads cut through pre-existing crop mark systems.

Field walking proved to be an effective method for helping to determine the character and chronology of crop mark sites. Settlement shift was also identified by the relative presence/absence of artefacts dating to each period and distinct differences in the types of sites within and between the Landscape Zones, the quantity and variety of surface material collected from the settlements nearest the Roman roads, demonstrating firmer integration into the Roman system than those in the rural hinterland.

During the Roman period, as in the Iron Age, there appears to have been a strong correlation between site location and landscape constraints, for example the position of the Roman forts closely relates to soils, terrain and watercourses. In the later Roman period there was a contrast between the lower Foulness Valley, where a pottery industry was situated in a landscape of wetland and woodland, continuing a tradition of furnace-based production, and the more open, elevated and fertile Wold margins, where Roman villas developed. During the fieldwork project "new" villa sites were identified, confirming that East Yorkshire has by far the densest concentration of villas in the North of England. The ability of villas to flourish within the study area, especially at the boundary of Zones 1 and 2, may represent a combination of an ideal landscape for agriculture, and convenient access to the main Roman roads. The roads themselves run along the same alignment as the Zone 1 and 2 boundaries, their position being partly determined by geographical factors such as drainage.

A close relationship between cult and countryside has also been identified, confirmed by a recent survey of a possible Roman rural sanctuary on the Yorkshire Wolds in the north of the study area. Rites associated with water seem to have been especially important throughout the period of study, and the positioning of burials also appears to be closely related to the landscape.

As well as enabling the better understanding of human-landscape interaction within this dynamic past landscape, GIS has provided the platform on which to build virtual worlds to promote its better understanding, through a website which documents the results of this research (www.ironmasters.hull.ac.uk) and the development of a 3D immersive landscape allowing real-time exploration, topics which will be addressed in another paper.

TEANO, *TEANUM SIDICINUM*. A GEOPHYSICAL URBAN LANDSCAPE SURVEY.

Sophie Hay, Simon Keay, Martin Millett

Under the aegis of the British School at Rome, The University of Southampton and The University of Cambridge, large-scale geophysical and topographical surveys have examined a series of Roman urban centres in Italy, focusing on sites in the Tiber Valley. In view of the success of the work at sites such as Falerii Novi, Ocriculum and Portus, it was decided to explore extending the survey to examine sites elsewhere in Italy with a particular emphasis on investigating sites with defined chronologies which might contribute to our understanding of patterns in the development of urban planning in Roman Italy. The current ongoing project at Teano (*Teanum Sidicinum*) was commissioned by the Soprintendenza per i beni Archeologici delle Province di Napoli e Caserta.

Teano is located some 30km to the north west of Caserta and over looks the impressive gorge cut by the River Savone. The town is situated on a rock outcrop on the south eastern slopes of the extinct volcano of Roccamonfina, at the point where the mountain range drops away into the Campanian plain. The origins of the town lie in a series of sanctuaries and settlements of the 6th– 4th centuries BC. These were complemented by the construction of a new planned centre from the fourth century BC, the focus of which was a hilltop enclosure beneath the modern town. An orthogonally planned settlement spread down the hill alongside the via Latina to the south-east of the town centre during the later Republic and early imperial periods and a series of major public buildings are known (including a theatre, a bath complex, an amphitheatre and a macellum).
The initial geophysical survey of Teano focused on the terraces of the sanctuary of Loreto and revealed the complex sacred landscape and potential archaic burials surrounding the discovery of a series of temples dating to the 4th century BC made by Johannowsky during his excavation campaign in the 1960s. In an attempt to understand the broader development and extent of the urban landscape the magnetometer survey was extended and clear indications of the form and layout of the settlement were detected including the discovery of a second, possibly earlier, amphitheatre and a dual street grid system. Evidence of a probable urban villa was also identified. The magnetometer survey has successfully identified a broad chronological and functionally diverse urban landscape and in conjunction with a detailed topographic survey will aid the understanding of the settlement pattern and development of Teano.

DIGITAL RECONSTRUCTION OF ARCHAEOLOGICAL OBJECTS USING HYBRID SENSING TECHNIQUES -THE EXAMPLE PORTA NIGRA AT RIER

Hoffmann A., Huxhagen U., Welter D., Boochs F.

I3mainz, Institute for Spatial Information and Surveying Technology Holzstr. 36, D-55116 Mainz, Germany
huxhagen@geoinform.fh-mainz.de

Digital reconstructions of archaeological objects may serve for different purposes and accordingly vary in the information provided. Simple web-based visualisations primarily need textural data combined with more or less accurate spatial geometry, classical documentations require complete and precise 3D-models allowing to eventually rebuild the object in case of damage, whereas basic research to the object itself, the history of its edification, the origin of the material used and the analysis of exiting damages demands a maximum of textural and geometrical information. In correspondence to these aims techniques for the data collection have to be selected, which will result in higher effort and higher demands to performance and versatility when more complex reconstructions are needed. Looking at the actual technological progress, 3d-scanning, both structured light and laser based, and digital photogrammetry , for example, are providing possibilities to collect maximal information at a large spectrum. On the other hand, the potential of these new technologies sometimes will be overestimated, whereas effort and need for an adopted processing of data and a customised use of instruments often is underestimated. It therefore seems necessary to give a look into an interdisciplinary project of historians and surveyors dedicated to evaluate the value of these new sensing technologies. Main focus of the project is the combined use of basic surveying methods together with laser scanning and high resolution digital photogrammetry for a detailed textural and geometrical documentation. The collected data shall serve as base for a research to the history of the building done by an interactive analysis and manual interpretation of geometry and appearance of the object. In addition to this research with help of a digital reconstruction of the object this will be done conventionally by manual inspection and evaluation at the objects surface. The comparison of traditional proceeding and a readapted, surely gainful strategy then gives an impression of the value of the high resolution data for these research purposes. In the framework of this paper the project will be explained, principal considerations when using such hybrid data sources will be noticed, insight into practical aspects will be given and, finally, the value of high resolution hybrid data sets for research purposes in the context of art and building history will be commented.

APPLICATIONS OF REMOTE SENSING ARCHAEOLOGY TECHNOLOGIES IN CHINA

Guo Huadong[1], Wang Changlin[1] Nie Yueping[1], Fan Xiangtao[1], Yang Lin[2]

[1]Institute of Remote Sensing Applications, Chinese Academy of Sciences
[2]China National Museum

Remote sensing archaeology, which is associated to multi-disciplines combining social sciences with natural sciences, is an important branch of remote sensing, and is showing the increased demands worldwide. With the increasing spatial resolution of earth observation data, the remote sensing data can not only be applied to landscape and environmental analysis for the archaeological sites, but also be applied to detailed mapping of the archaeological sites. In recent years, attentions have been paid greatly to remote sensing technologies in China in the research of heritage monitoring, investigation and conservation in the effort of promoting the applications of space technologies for natural and cultural heritage conservation, shown as 1) establishing the Joint Laboratory of remote Sensing Archaeology affiliated to Chinese Academy of Sciences, Ministry of Education, and State Administration of Cultural Heritage in 2001, 2) Held the first National Symposium on Remote Sensing Archaeology in 2002, 3) Held the Xiangshan Science Forum on "Understanding the Cultural Heritage from Space" in 2003, and 4) Held the First International Conference on Remote Sensing Archaeology. In this paper, we will introduce the capacity building of China's national earth observing systems, some research work on aerial and satellite data for archaeological work, virtual reality reconstruction for the Great Wall, and other sites, and will discuss some issues related to better understanding the cultural heritages with space technology.

3D BLOG - A NEW WAY FOR SUPPORTING COMMUNICATION OVER CULTURAL HERITAGE -

Rieko Kadobayashi

National Institute of Information and Communications Technology, Japan rieko@nict.go.jp

The number of 3D digital recording and modeling of cultural heritage and 3D visualization of them has been rapidly increasing. These 3D models and visualization are utilized for wide variety of purpose such as research, preservation, and presentation. Although there are a lot of case studies and applications, most of them provides only limited interactivity and communicative functions. From the interactivity viewpoint, experts may interact with the 3D models using proprietary software or expensive commercial software while the general public are allowed only to view them passively, e.g. to see still images or animation in most cases. From the communicative point of view, the most dominant communication is from experts to the general public like in presentation use. The experts convey their interpretation of cultural heritage via 3D visualization while the general public does not have such means. As a result, 3D models and visualization have been basically for experts. However, we should take the general public or local community into account since they are also one of stakeholders involved in cultural heritage and it is very important to promote participation of the general public or local community in utilization of 3D models and visualization. We have been developing "3D Blog" which allows a user to interact with a 3D model, annotate anywhere on the model, and publish the annotations as blog so that the user can easily communicate with other people and share interests, ideas, questions and so on about the 3D models the user saw on the user's browser. One of the main features of 3D Blog is that each annotation is saved as a blog entry with 3D information of the user's viewpoint, point of interest, and the gaze direction. Readers of someone's 3D Blog can reconstruct the exactly same 3D scene as that of the blogger when reading the blog entry and therefore the readers may understand it well. The another one is that the 3D Blog system detects similar entries based on the distance of the viewpoint, point of interest and the similarity of gaze direction and send trackback link automatically in order to let users to find easily people who may have similar interests and to promote emergence of communities consisting of people who shares their interests. In order to make 3D Blog widely available, standards for 3D model format and 3D model metadata are needed and standards for 3D Blog entries and RSS will be needed too. Moreover we should develop means to find 3D models effectively. These issues are also discussed in this paper.

TEACHING AND USING REMOTE SENSING IN ARGENTINE ARCHAEOLOGY: EVALUATING THE UNIVERSITY CURRICULA AND THE SCIENTIFIC PRODUCTION OF THE LAST DECADE

Débora M. Kligmann

CONICET (Consejo Nacional de Investigaciones Científicas y Técnicas), UBA (Universidad de Buenos Aires)

INTRODUCTION

Remote sensing in archaeology can be addressed from three different angles:

1) creation of technologies (developers)
2) transmission of knowledge (teachers)
3) use of technologies (researchers)

In this paper, I will mainly deal with education and research because these areas are still underdeveloped in argentine archaeology. In other words, before we can use the knowledge we need to acquire it.

ADVANTAGES AND DISADVANTAGES OF REMOTE SENSING TECHNIQUES AS APPLIED TO ARCHAEOLOGY

The principal advantage of remotely sensed imagery in archaeology is that it provides general information pertaining to very large areas, revealing widespread spatial distributions and patterns of vegetation and landforms.

Satellite images and aerial photographs can be very helpful when, as a result of scale, pedestrian surveys are ineffective at discerning subtle patterns produced by archaeological remains and where visibility is low because of heavy vegetation.

Archaeologists can obtain valuable data about where sites might be found, in addition to getting images of the sites themselves.

Images can also be used to locate outcrops of raw materials (both minerals and rocks) available in the study area and potentially used in the past to manufacture artifacts (e.g. ceramic vessels and projectile points).

Geologic analysis of satellite images can offer useful information about areas of erosion and deposition, thus allowing predictive models on archaeological site location to be generated.

Among the limitations, both the spatial and temporal scales have to be mentioned.

As far as the *spatial scale* is concerned, remote sensing instruments often cannot detect anything as small as individual artifacts, unless they are deposited in rather dense clusters. Besides, archaeological sites are usually too small to be seen in "low resolution" images.

Regarding the *temporal scale*, the challenge is how to use present information to study areas that have been occupied several hundred or even several thousand years ago. We cannot assume that what we see today was exactly the same in the past.

To sum up, remote sensing techniques were not developed for archaeological purposes and thus have some limitations. If used carefully, however, they can be very useful. The key issue is knowing what kind of archaeological questions these techniques can answer.

When working in archaeology, we should take into account:

- the characteristics of the archaeological remains (e.g. artifact raw material, artifact density structure size, height and shape, and structure location),
- the characteristics of the landscape where these remains are located (e.g. ground cover and vegetation cover) and
- the characteristics of the remote sensing techniques used (e.g. spatial and spectral resolution).

REMOTE SENSING IN ARGENTINA: CONAE AND THE EARTH OBSERVATION SATELLITES

Argentina is one of the few countries in the world that has a National Space Agency. CONAE (National Commission on Space Activities), created in 1991, is in charge of the national satellites (entirely designed and built in the country) as well as of a ground station, located in the province of Córdoba. In addition to data from one national satellite, data from 12 foreign orbiting satellites (including Landsat 5 and 7, Radarsat, SPOT, the NOAA series, EROS, Orbview 2, Terra, Aqua and ERS) are downloaded, transmitted and stored.

CONAE's satellite missions are devoted to fill vacant areas of space information at the local and regional levels, thus fulfilling the needs of argentine users.

The SAC-C (Satellite of Scientific Applications), launched in 2000, is fully operational and two more satellites called SAOCOM 1A and 1B will be launched in the near future.

SAC-C (with sensors in the visible and infrared range of the electromagnetic spectrum) is the first argentine Earth observation satellite and is part of the AM Constellation of Earth observation together with Landsat 7, EO-1 and Terra US satellites.

SAOCOM 1A and 1B (with sensors in the active microwave range) integrate the SIASGE Constellation (Italian-Argentine Satellite System for Emergency Management) together with four satellites of the COSMO-SkyMed Italian Constellation. The aim of this cooperation program is to help in emergency situations by preventing, detecting and monitoring natural and anthropogenic disasters such as fires, floods, hurricanes, earthquakes, volcanic eruptions, landslides, draughts, oil spills and agricultural plagues. The information generated can be used by archaeologists, agronomists, biologists, geologists and other scientists interested in the environment.

It is important to mention that CONAE cooperates with argentine researchers by providing training, technical support and images free of charge for scientific projects.

ARGENTINA AND UNESCO'S "OPEN INITIATIVE"

In 2001, UNESCO and ESA (European Space Agency) launched the "Open Initiative on the Use of Space Technologies to Monitor Natural and Cultural Heritage of UNESCO Sites". The "Open Initiative" is a framework of cooperation to assist countries to improve the observation, monitoring and management of natural and cultural sites as well as of their surroundings, through space technologies.

Argentina adhered to the World Heritage Convention in 1978. Of the eight argentine properties inscribed on the list, four are cultural: Cueva de las Manos (Patagonia), Jesuit Block and Estancias of Córdoba (Córdoba Province), Jesuit Missions of the Guaranis (Northeast Argentina, property shared with Brazil) and Quebrada de Humahuaca (northwest Argentina). Some more cultural properties (such as the Qhapaq Ñan or Main Andean Road) are currently being considered to be included in this list.

One of the main goals of the "Open Initiative" is that of capacity building. This is carried out through the implementation of workshops and the development of joint research projects with local people in order to provide training.

REMOTE SENSING IN ARGENTINE ARCHAEOLOGY: EDUCATION AND RESEARCH

The huge technological development of the last decade in the area of remote sensing has not been followed by an intensification in the archaeological applications of these technologies. This is happening worldwide and is also happening in Argentina.

In this paper, I present an analysis of the archaeological university curricula and evaluate the archaeological scientific production of the last decade in the remote sensing arena.

A detailed analyses of the curricula reveals that in none of the universities where archaeology is taught remote sensing is included as a class, or even as part of one. When looking at the list of thesis and dissertations an interesting issue emerges: very few of them deal with remote sensing. I believe this is a consequence of the lack of proper education on the topic.

The same situation can be noticed while reviewing the papers presented at national scientific meetings or published in peer-reviewed journals in Argentina: very few of these papers use remote sensing technologies.

In summary, we have a National Space Agency that is willing to cooperate with argentine researchers, we have free satellite images, we have lots of sites to protect and we have lots of questions that can be solved through remote sensing technologies. What we do not have is the knowledge. The question that follows is why argentine archaeologists are not getting the proper training so that we can take advantage of what we do have?

At the University of Buenos Aires, the absence of remote sensing classes can be explained by the following issues:

- lack of trained archaeologists in remote sensing technologies and
- lack of adequate infrastructure, including hardware, software and scanners.

This probably happens in all other universities of the country as well.

Even when students can take remote sensing classes outside the archaeology department (in other departments of the same university or even in other universities), these classes are not applied to archaeology. This means that although students can learn theoretical and methodological issues, they still need to read and discuss specific applications in order to understand how archaeology can benefit from remote sensing.

This can be partially solved by teaching a remote sensing class for archaeologists at the undergraduate or graduate level. Thus future argentine archaeologists can be trained to use remote sensing technologies both for their own research as well as for conservation of the National Heritage.

A NEW TOOL FOR 3D ARCHAEOLOGICAL AND EPIGRAPHIC RECORDING

Dimitri Laboury

Regarding the technological and epistemological problems raised by the necessity for Archaeology to record the precise shape of the objects of its study, the OSIRIS (Optical Systems for Interferometric Relief Investigation and Scanning) project aimed, under the auspices of the European Center for Archaeometry at the University of Liège (Belgium), to develop a device that allows by optoelectronic processes an accurate, quick and easy to use recording, dedicated to the specific and very demanding needs of archaeological and epigraphic researches. The presentation will explain how this new innovative device takes advantage of monochromatic light projection, polarization states splitting technique and interferometry principles to reach that goal. Some examples of its application to in situ archaeological recording problems (i.e. in Museums and on archaeological and open-air sites) will also be given, in order to demonstrate the advantages of the new technique, especially when naked-eye inspection fails.

CULTURAL ATLASES AS DATA REPOSITORIES: INNOVATION AND INFORMATION MANAGEMENT

Lewis Lancaster

University of California, Berkeley

The Electronic Cultural Atlas Initiative (ECAI) , for nearly a decade, has been involved in the process of exploring ways of using GIS software for the purpose of archiving and retrieving cultural data. In this extended effort, new ways of approaching digital data have been explored and adopted by researchers in several hundred projects.

ECAI has followed the recommendations of its affiliates in looking for more efficient ways of display and use of data. The work has been focused on the digital library models of how to preserve, index, and acquire data in the electronic formats. The membership has explored and discussed the issues in 20 conferences and workshops and from these sources developed an expanded view of the GIS approach. It is obvious for the humanities and many of the social sciences as well as art history and anthropology that we need not only "where" but also "when," "who," and "what." This had led the ECAI teams to work on GIS, gazetteers, named time periods (as opposed to year dates), biographies, and markup for events. Each of the four "Ws" has exhibited complexities that challenge us.

In this paper, I will explore the ways in which we have: used digital library catalogues to find named time periods; created historical place name thesauri for expanding existing gazetteers; drawn up biographical data into useable formats; expanded the idea of "who" with "who else" to show social networks within a GIS structure; and finally the research on how to index events through some form of automation.

The work that is going forward in ECAI is in part based on the idea that while digital libraries has begun to store large amount of data and to provide search and retrieval methods, these processes do not exhaust the traditional role of the librarian. The reference rooms of the past and the reference librarians were essential parts of the functions of the library. The role of the reference library has not yet been fully developed within the digital domain. ECAI hopes to use the four "Ws" as the beginning of researching avenues for how references can be established within the new technology.

As reference technology expands and improves, it can help to provide links between the "silos" of knowledge that scholars are producing in digital sites. It is a reality that data will not be centrally housed and that the future will be one in which we has widely distributed information. At the heart of this ECAI approach is the idea that "where" i.e. space/place, remains a key component for control and management of information.

PERFORMANCE EVALUATION OF DATA FUSION AND EDGE DETECTION ALGORITHMS FOR THE DETECTION OF ARCHAEOLOGICAL FEATURES BY USING SATELLITE QUICKBIRD DATA

Rosa Lasaponara[1], Nicola Masini[2]

[1] Istituto di Metodologie per l'Analisi Ambientale, IMAA-CNR, C.da S.Loja – 85050 Tito Sc. (PZ), Italy, Tel. ++39 0971 427 214 – fax ++39 0971 427 217 – e_mail: lasaponara@imaa.cnr.it

[2] Istituto per i Beni Archeologici e Monumentali, IBAM-CNR, C.da S. Loja, 85050 Tito Scalo (PZ), Italy, Tel. ++39 0971 427.321 – fax ++39 0971 427.222 – e_mail: n.masini@ibam.cnr.it

The recent availability of Very High Resolution (VHR) satellite imagery, such as IKONOS (1999) and QuickBird (2001), may be able to open new perspectives in the field of archaeological remote sensing. In particular, QuickBird offers panchromatic and multispectral imagery with the highest spatial resolution currently available within the satellite sensors. It has panchromatic and multispectral sensors with spatial resolutions (SR) of 61-72cm and 2.44-2.88m, respectively, depending upon the off-nadir viewing angle (0-25 degrees).

The high spatial resolution and spectral capability can make the VHR satellite images a valuable data source for archaeological investigation ranging from synoptic view (i.e. identification of high probability locations of ancient buried sites) to small details (i.e. single subsurface building).

Nevertheless, the satellite-based detection of archaeological marks faces several challenges, particularly in the case of buried remains. In fact, the presence of underlying structures produces weak signals, that can be easily covered by noise. Responses from true features and those from noise can not be distinguishable. This kind of problem could be reduced using data fusion and edge detection algorithms. The data fusion enables the integration of the high spatial resolution of the panchromatic image with the spectral capability of multispectral images. Thus, allowing to achieve improved capabilities that are not possible using solely the individual datasets. The edge detection allows the enhancement of spatial features, so, facilitating their identification.

The most common data fusion and edge detection algorithms were tested in this paper in order to evaluate their performance for the detection of archaeological features. The evaluation is performed by using satellite QuickBird imagery acquired for some study cases located in the South of Italy.

DEVELOPMENTS IN AERIAL PHOTOGRAPHIC RECTIFICATION AND MAPPING AT RCAHMS

Kevin H.J. Macleod

RCAHMS Email: Kevin.MacLeod@rcahms.gov.uk

The Royal Commision on the Ancient and Historic Monuments of Scotland (RCAHMS) has for some 30 years run a programme of aerial survey across lowland areas to identify plough-levelled archaeological sites appearing as cropmarks (also recording upstanding monuments, townscapes, industrial sites and landscapes).

The resulting oblique aerial photographs of cropmark sites are held in the archive of RCAHMS. In order to locate these sites to the UK National Grid (BNGR) RCAHMS adopted the *Aerial* software programme in the late 1980's, which has been written and maintained by John Haigh of Bradford University. This software is used routinely to produce a BNGR digital transcription of the plough-levelled archaeology. RCAHMS is currently working on a large area mapping of the coastal plain in East Lothian, dealing with the totality of the cropmark record. Other work includes thematic projects such as Roman sites along the Antonine Wall and specific requests. This paper outlines recent developments in this programme of work.

TRANSCRIPTION PROCEDURES

The suite of aerial photographs for each site (which may be taken over a number of years) is assessed for those images with the best view of the cropmarks, as well as the quality of the ground control points on the Ordnance Survey (OS) map, which will be used together to create the transcription. Three datasets are assembled before transcription:

1. A digital OS Landline map extract covering the site, extracted from the RCAHMS GIS.
2. The relevant OS Profile Digital Terrain Model (5m interval).
3. The selected images are desktop scanned at high resolution from a print.

The transcription process involves geo-rectifying the chosen aerial photograph(s) employing user-selected map control points, which are on both the image, and the OS map. *Aerial* provides dynamic control refinement functionality, resulting in high levels of metrical accuracy for each point (commonly <1m accuracy). Reference may be made to the extensive RCAHMS vertical photographic collection, where control may be awkward and for additional sites not represented in the oblique collection.

The rectified image is saved (***ML3156re.tif***), along with an ArcGIS "worldfile" (***ML3156re.tfw***) that allows the image to be viewed in the GIS in the correct spatial location. The archaeology is then digitised on-screen in *Aerial 5* and each 3D vector is tagged with attributes that will be carried forward into the GIS (e.g. ***ML3156_SM-ENCL-RECTILINEAR***). This attribute data begins with the negative number of the image used (e.g. ML3156) and includes both a broad archaeological classification (e.g. in this case SM for Settlement) and more detailed interpretation (e.g. ENCL-RECTILINEAR for a rectilinear enclosure).

When the digitisation is completed and the interpretation discussed and peer reviewed, the data is migrated through *Autocad* for any minor editing, then imported into a master Aerial Photographic Transcriptions GIS dataset (shapefile), which is accessible through the RCAHMS GIS (ArcGIS software).

ARCHIVING AND USING IMAGERY

The current archiving system in RCAHMS is based on 5x5 km units of the OS 1:10,000 mapsheet e.g. NT57SW. The individual rectified, high resolution (300dpi greyscale/colour) aerial photograph for each site is archived in the GIS archive structure again indexed by mapsheet.

All the rectified aerial photographs for every site are then assembled onto a geo-referenced raster image (a low resolution JPEG) covering the whole OS 1:10,000 mapsheet. *Aerial* and *Photoshop* are used to control this stage. This is also archived in the GIS archive structure by mapsheet. File naming conventions indicate that the sheet is completed, or incomplete, awaiting upgrade. Edge matching issues are addressed at this stage. The mapsheets themselves are then compiled into a single GIS "Image Catalogue" covering a 100 km square (e.g. the file for NT is '***nt index of images transcribed.dbf***').

RCAHMS TRANSCRIPTION PRODUCTS

Therefore, the final metrical depiction of the cropmark archaeology for a site comprises:

1. Interpreted 3D vector data, coloured by broad classification, and supported by detailed attribute data (e.g. negative number, archaeological interpretation, database unique identifier).
2. Vector data tags accessible by on-screen selection.
3. Individual high-resolution, geo-referenced aerial photograph (TIF).
4. Lower resolution (JPEG) composite showing all rectified images per 1:10,000 mapsheet.
5. Access to RCAHMS *Oracle* database lookup table for each site, showing the unique identifier for the archaeological description of the site and its associated collection of imagery and other information (including administrative).

OUTPUTS

All the digital data can be presented in a variety of publication formats, with *Illustrator*, *Photoshop* and *Autocad* used to prepare illustrations. 3D animations can be created and recorded in *ArcScene*, as the vector data is transcribed in 3D, and geo-referenced raster images can be lifted up onto the relevant DTM surface.

FUTURE WORK

The RCAHMS aerial transcription programme has its origins in the mapping of information from oblique photography, but we are now making routine use of our extensive collection of vertical photographs, both to assist with control and for the information they hold, which may not be on the obliques. Aerial photography from survey at RCAHMS is now entirely digital and an ongoing project will look at the use of these digital images and any implications of the change in media.

AN APPLICATION OF GOOGLE EARTH FOR ARCHAEOLOGICAL SITE SURVEY AND SITE DISCOVERY

Scott Madry

The University of North Carolina at Chapel Hill madrys@email.unc.edu

This paper will present the results of an investigation into the utility of Google Earth for regional archaeological survey and site prospection. The author has worked for over 25 years in the Burgundy region of France, and has conducted several years of low level aerial survey and the analysis of aerial photography over the area. The development of Google Earth has provided a new tool which has great potential for supporting these activities. The author has conducted an extensive manual analysis of color .7 meter Digital Globe imagery in the area of Dijon, France. This analysis has located over 100 new and previously recorded archaeological sites, dating from the Iron Age, Gallo-Roman, and Medieval periods. Field visits have been made in 2006 to validate the site data. The paper will discuss the methods developed for this analysis, the results of the survey, and will present an evaluation of the strengths and weaknesses of the Google Earth environment for professional archaeological survey and analysis. The large number of sites that have been located from the author's lab in North Carolina, ease of use, possibility of incorporating existing GIS data, and ease of sharing the data with French colleagues demonstrate that this technique has excellent potential for use by the professional archaeological community.

THE NORTH CAROLINA DEPARTMENT OF TRANSPORTATION PREDICTIVE MODEL DECISION SUPPORT SYSTEM-FINAL RESULTS

Scott Madry

A team of researchers has recently completed a three-year, million dollar archaeological predictive model and Decision Support System for the North Carolina Department of Transportation. The final predictive model, the infrastructure of GIS data, easy user interfaces, and updated cultural resource database have been integrated into a Decision Support System (DSS) that will create a tool to analyze and rank proposed highway corridor alternatives. The long range project tasks included: 1) development of digitized environmental and cultural information for the State of North Carolina, including computerizing all existing archaeological site files and site maps at the North Carolina Office of State Archaeology (OSA); 2) development of GIS archaeological predictive models using the digitized information for the three provinces of North Carolina and verifying their accuracy; 3) creation of an internet-based DSS and GUI for use by NCDOT staff; 4) application of the GIS archaeological predictive models to multiple transportation projects to aid in the selection of a preferred corridor/ alternative; 5) field testing of the models in actual NCDOT projects using intense archaeological survey and GPS mapping to quantitatively test and refine the models. This paper will provide an overview of the final results of this project and will discuss problems that were encountered as well as lessons learned for future projects.

DETECTING MYCENAE: SYSTEMATIC REMOTE SENSING SURVEY IN THE 'LOWER CITY' – TOWARDS THE DISCOVERY OF THE MYCENAEAN SETTLEMENT OUTSIDE THE CITADEL

Christofilis Maggidis, Antonia Stamos

In 2003 a long-term systematic geoprospection and remote-sensing survey was launched at Mycenae by Dickinson College on behalf of the Athens Archaeological Society and in collaboration with INSTAP and the Univ. of Patras. Several methods and techniques are being combined (including Ground Penetrating Radar, Electrical Resistivity, Magnetometry, Geoacoustics, Global Positioning System, Total Station, and Aerial Photography) to establish the size, extent, and limits of the settlement of Mycenae, to locate and map all visible and buried structures in the area around the citadel (buildings, sectors and roads), and ultimately to create a 3-D digital model of the buried city of Mycenae.

The first targeted areas, located immediately to the south of the citadel on the terraced fields flanking the dry riverbed gorge, were selected after careful observation of the geomorphology and consideration of several variables, study of the spatial distribution and orientation of visible ruins, analysis of high-resolution aerial photographs showing "ghost traces" of walls as crop/soil marks, and use of a site predictive program (ERDAS) which determines plausible locations by comparing visible, recognizable features to possible features that match a set of parameters. Those areas which produced ghost traces or crop/soil marks in the aerial photographs, and were red-flagged as high-probability areas by ERDAS, were targeted as the best candidates for the first phase of the geoprospection survey. As of today, a total of 32 grids have been surveyed with GPR and/or electrometer on six separate terraces.

The geoelectric survey, which is conducted by the Patras Univ. team, employs a digital electrometer (GEOSCAN RM4), special cables and four electrodes in a twin lay-out, which allow precise measurement of soil electrical resistivity (450-900 Ohms) and detection of subsurface wall ruins to a depth of 1.5 m. It appears that there are definite geometric features located in the northern half of several aligned grids, all of which appear to have the same orientation and may possibly be part of a similar structure or structures. The most interesting feature is a series of parallel oblong rooms with a N–S orientation, reminiscent of palatial storerooms.

In the GPR remote-sensing survey, which is conducted by the INSTAP team and a Dickinson student team headed by Antonia Stamos, data is collected using a Subsurface Interface Radar or (SIR)-2000 system with a 400MHz antenna and is processed using GSSI's RADAN NT software. A total of 13 grids on five different terraces have been surveyed and processed so far, several of which overlap with the grids of the geoelectric survey. The grids were collected in 25cm transects in order to produce more detailed results; furthermore, all of the grids were collected in two datasets: one of E–W orientation overlaid by one of N–S orientation, resulting in an orthogonally cohesive grid.

The preliminary results reveal wall corners, structures, features, gates and roads at two different depths (0.40–1m and below 1.20–1.40m) which define two superimposed occupation horizons, most likely a Hellenistic level and an underlying Mycenaean level, accordingly. Detected structures include series of rooms with the same orientation, isolated rooms, and building complexes. The most interesting discovery was made on the grids of the SWB-V terrace which are parallel to the modern asphalt road leading up to the citadel, and roughly aligned with the Oil Merchant's House to the north: what appears to be a road surface (3m wide) running N–S (roughly parallel to the modern road) was detected at a depth of –0.68m, which may very likely be the Hellenistic or the Mycenaean highway leading up to the citadel through the lower city. If this is the case, then the two Γ-shaped, solid features to the north on either side of the "road" may be gates or towers, considering their diagnostic shape and large size (at least 3.5m wide, preserved to a total height of approx. 1m). The discovery and excavation of this important landmark will shed new light on the topography and road system of the settlement of Mycenae around the citadel.

As the subsurface mapping gradually expands, it will be integrated as an interface in a specially designed GIS program. The final results of this geoprospection survey will be published annually in scientific journals and collectively as a second volume to the *Atlas of Mycenae*, and will form the basis for better understanding the topography of Mycenae, interpreting settlement pattern changes diachronically, and for preparing the systematic excavation of the "lower city" in the very near future.

DIGITAL FIELD DOCUMENTATION: DO WE STILL NEED PAPER? THE TELL LEILAN APPROACH

Stacey Maples,[1] Cristiano Putzolu

YALE University Map Collection, USA

Post-excavation synthesis of archaeological excavation data is one of the greatest challenges to the archaeologist. Highly personal recording vocabulary, mapping styles and thoroughness all contribute to the difficulty of moving from the excavation notebook,

through the analysis phase to the archaeological monograph. The archaeological laboratory, its walls lined with years, if not decades, worth of paper-based excavation notes waiting to be transferred into some standardized and useable form is as familiar a phenomenon in archaeology as diagnostic ceramic sequences.
During the 2006 fieldwork campaign of the Tell Leilan Project (Prof. Harvey Weiss, Yale University, USA) a digital system of excavation data recording was tested.
The goal is to create a paper-less digital documentation of field notes and ensure a good level of accuracy in the daily registration of data.
The tools used consist of a pair of Itronix Duo-Touch Tablet PCs running ESRI ArcPad 7.0 and a Sokkia SET 3130R3 Total Station.
From the previously designed Access forms an ArcPad application was designed: when for example a new Lot (any defined context or stratigraphical unit) has to be documented the excavator draws a sketch map of its perimeter in ArcPad environment and immediately the Lot sheet, divided in 4 pages, pops up.
Many fields have a drop down menu making the data entry faster. Prior to excavation, careful thought must therefore be given to the conceivable types of deposits, soil types, feature types and other kinds of archaeological data one might encounter. This first season has been useful in expanding the vocabulary of drop down categories, which will speed recording in subsequent seasons.
A few memo fields allow the excavators to register their impressions and other non-standarizable data. This has been particularly helpful for recording individual excavators' personal impressions of the archaeology, as well as describing the techniques employed to uncover the remains in prose form. The Itronix Tablet PC is capable of handwriting recognition from a touch-pen, which makes the entry of such additional descriptions as simple as writing in a field notebook. The handwritten notes are converted into digital format, which can be easily transferred to any document software. With the addition of voice-recognition software in future seasons, even this already simple procedure can be sped up.
The topographical survey of the excavation is carried out using a Total Station and a digital camera.
Both the coordinates from the Total Station and the images from the camera (orthorectified and georeferenced with Rolleimetric MSR 4.0) are used in ESRI ArcGIS environment to draw day by day the map of the excavation.
This methodology of survey combining total station and monoscopic photogrammetry is used now by many surveyors almost everywhere and there is no need to add anything about its procedures.
What we think could be considered new (at least for Near Eastern Archaeology) is the survey of Lots (in other words pits and floors) and Walls as surfaces and volumes not just as perimeters.
The complexity of an excavation is mostly on spatial relationships between Stratigraphical Units.
We decided to make them clearly visible at least in the 3D environment of ArcScene.
Each Lot is surveyed with a number of points defining its perimeter and with many points inside its perimeter.
The density of internal points changes according to the surface to be surveyed (the flatter the surface, the fewer the points) allowing us to reach a good compromise between time consumption and data quality (of course a Laser Scanner would ensure the best quality in a quicker time but with enormously higher costs!).
The idea is to process each dataset as a DTM, interpolating the surface with the Natural Neighbour Algorithm.
Adding the different DTM in ArcScene we can easily visualize the phase we are interested in from infinite points of view, we can check which units were just above or belove it, we can calculate volumes (pits) or thicknesses (floors), using the information that a surface store and a polygon can't have.
The next steps will be to implement a Bluetooth connection between the Tablet PCs and between them and the Total Station, so that the sketch map of Lots could be more precisely drawn following the points just surveyed, rather than dangerous approximations based on excavators' impressions; and the addition of a voice recognition system that can speed up the entry of memo fields and the daily log.

"VALLE D'AGREDO": A PALEOENVIRONMENTAL AND GEOARCHEOLOGICAL RECONSTRUCTION BASED ON REMOTE SENSING IMAGES ANALYSIS

B. Marcolongo,[1] A. Ninfo[1] and M. Simone[1]

[1] Consiglio Nazionale delle Ricerche, Ist Ricerca Protezione Idrogeologica Corso Stati Uniti, 4. 35127 PADOVA
[2]Dipartimento di Geografia Università di Padova, Via del Santo, 26 35126 PADOVA
[3] Collaboratore presso Consiglio Nazionale delle Ricerche, Ist Ricerca Protezione Idrogeologica

"Valle d'Agredo", donated on Christmas 926 A.D. by Ugo di Provenza, king of Italy, to Adalberto, bishop of Treviso, is historically remembered like a vast plain area that include the present towns of Castelfranco, Camposampiero, Noale e Mirano.
The area of interest is completely included in the morphological units of the Brenta river mega-fan, deposited in the late Pleistocene (mainly between 25 and 18 ka BP, and with decreasing intensity until 14 ka BP).
Nowadays the boundaries of this area, through the provinces of Padova and Treviso, are not well defined, with not evident morphological features and a homogeneous flat surface giving the appearance, at the first glance, of a monotonous context.
The present human pressure is related to heavy urbanization, dense industrial plants growth, gravel and sand open air quarries expansion particularly across the "groundwater blowout band".
These activities mask and complicate identification and reconstruction of the ancient "Valle d'Agredo", which surely represented for long time a solid physiographic reality and influenced so much the local culture to be often named in popular tradition and literature.
The analysis of remote sensing images, chosen with suitable spatial and temporal resolution, brings a synthetic understanding of the "object" and its problematic with a great efficiency.
The numerous multi-spectral sensors carried by various satellite platforms furnish useful tools to describe the tight relationships between natural resources and ancient settlement model.
This information is complementary to the data gathered through direct field recognition, bibliographic research, historical cartography, aerial photography and DEM.
The first image analyzed has been a cosmic spectro-zonal photo Sojuz KFA 1000, acquired in June 1990 from a spatial Russian

platform, placed on an orbit of 270 km of altitude, having geometrical resolution of 5 m.
On it there is a good evidence of a wide stripe of terrain with a darker tonality, elongated NW-SE, where are lying all the four towns mentioned in the donation of Ugo di Provenza. Most probably is a band rich in soil moisture content, containing a preferential groundwater flow direction, which corresponds to a palaeo-river bed of Brenta.
The successive interpretation of ASTER images acquired on December 2001 and February 2002 (geometric resolution of 15 m) has confirmed the presence of a landscape unit with high humidity and homogeneous grains size soil, in coincidence with the ancient way Piombino Dese – Trebaseleghe – Noale – Mirano (NW-SE). The town of Trebaseleghe is lying on the sandy ridge (expression of an important Brenta's paleo-course in the upper Venetian plain) extended for more than 20 km and now clearly understood as the major physiographic axis of the ancient "Valle d'Agredo".
To go deeper around the central area of Trebaseleghe, IKONOS images acquired in the July 2002 (geometrical resolution of 1m for panchromatic and 4m for the multi-spectral bands) have been elaborated (geometric and radiometric corrections, resampling to 1m pixel, etc.) and interpreted.
The principal contribution of IKONOS images analysis is the recognition and detailed description of the agrarian parcels mosaic (agricultural pattern). This shows with his geometrical grid of canals and roads (direction and module) the survival of an older agricultural subdivision, interpreted as the westernmost portion of the Altino's roman graticule ("centuriazione").
Moreover in the territory of Trebaseleghe, 3 sub-circular anomalies inserted in this regular pattern have been identified, all located on sandy natural levees along an ideal direction perpendicular to the physiographic axis of "Valle d'Agredo". These anomalies are interpreted as expression and relicts of artificial mounds ("motte"), most likely sites of Bronze-Iron age related to the strategic control of trade and communications between the Venetian lagoon and the "high plain".
The central band of "Valle d'Agredo", overlaying the late-glacial Palaeo-Brenta, represented during proto-historic and historic periods the selected way connecting Venetian perilagoon area and Valsugana valley, with the activation in late roman age and Middle Ages time of an important way joining Adriatic Sea to the pre-alpine region and the further alpine continental area.
The complex spatial organization, developed through millennia around this way oriented NW-SE, and its present consistence shown by the many physiographic and geomorphological relicts primarily recognized on satellite images, testifies how in the modern landscape the youngest morphology often is guided by more ancient forms and present matches past.

THE KARGALY PROJECT: MODELLING BRONZE AGE LANDSCAPES IN THE STEPPE.

M. Isabel Martínez Navarrete[1] Santiago Ormeño[2] Juan Vicent[3]

[1]Institute of History, C.S.I.C., Madrid, Spain
[2]Universidad Politécnica de Madrid, Spain
[3]Institute of History, C.S.I.C., Madrid, Spain

The Kargaly region is located in the *oblast* of Orenburg, in the Southeastern limit of the Russian Federation. During the Bronze Age its extensive copper deposits were object of an intensive mining and metallurgical activity, that reaches its phase of maximal intensity during the Late Bronze Age. Thanks to its huge copper production and its strategic position, in the limits of Volga and Ural basins, the Kargaly complex plays an outstanding role in the process of early development of metallurgy in steppes. Since 1995, a Russian-Spanish research team developes archaeological, archeo-metallurgical and paleo-environmental investigations in Kargaly.
The historical understanding of the Kargaly complex poses relevant questions on the reconstruction of Bronze Age landscapes, among them the evolution of forest resources and subsistence practices during Bronze Age. A specific program of Landscape Archaeology within the Kargaly project focuses those questions. As we seen in this paper, the analysis of satellite images, as source of data for landscape modelling, plays an outstanding role in this program. Its objective is to create a suitable comparative frame for the contextual interpretation of paleo-environmental data, specially palynological ones, produced by a complementary paleo-environmental research program. Data are collected in archaeological and natural deposits within the work area.
Landscape Archaeology research design is based in the use of the actual landscape as a "present analogue" of the different landscape states of the past. This "analogue" takes the form of a statistical model of structural correlations between (1) the variability of actual pollen dispersal patterns, established on the basis of a representative sampling of present day pollen rain in the work area, and (2) the local and regional variability in the patterns of distribution of vegetation and land uses, determined after the analysis and classification of satellite images. Thanks to G.I.S. techniques, the qualitative and quantitative variability of pollen signatures may be analyzed in relation to the spatial variability of vegetation, land uses and other geographical factors, like topographical ones. This model is used to test hypothesis on the significance of the Bronze Age palynological evidence and its variations through time, established by C14 dating of pollen sequences.
Preliminary results of this program, based on the analysis of a single Landsat 5 TM image, were presented in the previous International Conference on Remote Sensing Archaeology (Beijing, 2004). Now, we present new advances in the research. The interpretation of the most recently obtained paleo-palynologic evidence, related with the problem of the early spread of agriculture in the steppes, demands to improve the quality, complexity and precision of basic thematic cartography and terrain models, beyond the ambiguities and limitations of the previous model, based in a single set of TM data. A multi-temporary and multi-sensor approach to the problem of land use modelling is proposed, consisting in the analysis of a series of ASTER TERRA images, and its integration with the TM data previously used. This approach is presented and discussed in the paper, and its results are evaluated in relation with the objectives of the program and the Kargaly project.

EVALUATION OF THE SPECTRAL CAPABILITY OF QUICKBIRD IMAGERY FOR THE DETECTION OF ARCHAEOLOGICAL BURIED REMAINS.

Nicola Masini[1], Rosa Lasaponara[2]

[1] Istituto per i Beni Archeologici e Monumentali, IBAM-CNR, e_mail: n.masini@ibam.cnr.it
[2] Istituto di Metodologie per l'Analisi Ambientale, IMAA-CNR, e_mail: lasaponara@imaa.cnr.it

The spectral capability of satellite QuickBird imagery for the identification of surface anomalies linked to the presence of archaeological buried remains is here discussed for five archaeological sites located in the South of Italy. The selected study cases are related to sites buried under surfaces with different soil characteristics and vegetation covers. This allowed the evaluation of the QuickBird spectral ability in detecting the different types of archaeological marks (crop, soil and shadow-marks) in relation with different surface characteristics. The spectral features obtained for the five archaeological showed that:

1) Red channel was able to better enhance soil marks observed for bare surfaces;
2) NIR channel was able to better enhance crop marks observed for vegetated surfaces;
3) the use of vegetation indices, such as NDVI (Normalized Difference Vegetation Index) or EVI (Enhanced Vegetation Index) allowed to better emphasize crop marks observed for surfaces covered by green and healthy herbaceous plants;
4) Panchromatic imagery generally provide good results for the detection of archaeological features for most of surface and soil characteristics. In particular, they were found to be more capable than the spectral channels to enhance shadow marks.

USING VEGETATION INDICES TO STUDY ARCHAEOLOGICAL AREAS

P. Merola, A. Allegrini, D. Guglietta; S. Sampieri

CNR Atmospheric Pollution Institute, Rome-Italy

INTRODUCTION

In the recent years there has been a growing interest to use of remote sensing in archaeology.

In past years archaeologists used air pictures to investigate the archaeological areas. Since the launch of the first earth resources satellite in 1972, the development of digital remote sensing has provided an additional investigation instrument, such as digital multispectral imaging sensors, thermal imaging radiometers and Radar images. These new instruments have increased the wavelength range at which images can be acquired into the short wave, near-middle e thermal infra-red and microwave range.

Until very recently the archaeological research activities by digital remote sensing have been limited because the optical satellite images have been of low spatial resolution.

In Italy since 1994 the CNR-LARA (Airborne Laboratory for Environmental Research) has carried out research activities involving both remote sensing and archaeology, by using Daedalus AA5000 MIVIS (Multispectral Infrared and Visible Imaging Spectrometer) airborne hyperspectral scanner to recognize the soil discontinuities connected to buried anthropic elements.

The MIVIS spectrometer is a passive remote sensing instrument which collect the earth surface radiation from an airborne platform. The detector separates incoming radiation into four optical ports covering the Visible (0.43 - 0.83 μm), the Near-InfraRed (1.15 - 1.55 μm), the Mid-InfraRed (2.0 - 2.5 μm), and the Thermal-InfraRed (8.2 - 12.7 μm) region, total 102 spectral bands. It operates with an Instantaneous Field Of View (IFOV) of 2 mrad. The air platform aboard of which MIVIS is installed is a CASA 212/200.

The MIVIS data, that can be enhanced, rectified and reclassified using several algorithms and specific software, permit the identification of buried archaeological structures because that any buried ruins, either of human or natural origin, affects over time, soil surface characteristics creating anomalies.

These anomalies are due to different factors, such as physical and chemical soil features and vegetation cover status. The above factors are strictly connected and are responsible of surface spectral responses.

In particular the spectral response of vegetated areas presents a complex mixture of vegetation, soil brightness, environmental effects, shadow, soil colour and moisture. The scientists have developed vegetation indices for qualitatively and quantitatively evaluating vegetative cover using spectral measurements.

MIVIS remote sensing permits to analyse simultaneously a wide range of different wavelengths: these research have shown that the use of the red and near-infrared channels of the MIVIS sensor are particularly well useful for the study of cover vegetation.

Spectral Vegetation Indices are important products in observing spatial and temporal variations of vegetation biophysical properties and photosynthetic activities, by which is possible to analyze the effects of buried ruins presence on vegetation cover status.

VEGETATION INDICES

Vegetation indices are quantitative measurements indicating the vigor of vegetation. These indices are interesting because they constitute notably a method for the detection of land use changes, the evaluation of vegetative cover density.

The different vegetative covers can be distinguished according to their unique spectral behaviour in relation to overall ground elements: visible radiation in the red is absorbed by chlorophyll while radiation in the near-infrared is reflected by leaf cellular structures.

The deviation observed between the red and near-infrared constitutes a variable sensitive to the presence of green vegetation.

Vegetation Indices have been classified in two categories: ratio indices and orthogonal indices. Ratio vegetation indices, combining two bands, are simpler to use than those derived from more complex combinations: RVI, VIN and NDVI ect.

Orthogonal vegetation indices include two-channels perpendicular vegetation index (PVI), the four-channel green vegetation index (GVI) the six-channel and n-dimension spectral index.

METHODOLOGY AND STUDY AREA

Many scientists have focused on the factors affecting vegetation indices. These factors belong two different domains: a biological domains and physical domains. A vegetation area can be characterized by: water content, age, mineral deficiencies, cover geometry, row spacing, and orientation and the presence of buried archaeological structures (archaeological traces).

An archaeological trace is meant to be the effect on an image of the interaction of different elements which encompass or hide the archaeological object; these elements, of chemical-physical, biological and morphological nature, are humidity, humus, vegetation and relief .
The archaeological traces can be classified as follows:

- Humidity traces: moist soil can be assimilated to an imperfect reservoir in which the presence of buried archaeological elements produces anomalies in water distribution, giving origin to areas with differentiated humidity; in the image these areas exhibits a different colour shade respect the surrounding soil, as they are lighter in case of positive archaeological elements (walls) and darker in case of negative elements (ditches, canals, etc.). The perception of this phenomenon depend on the depth of the burial, the firmness of the structures and the climatic and environmental conditions.
- Vegetation traces: even though similar in behavior to the foregoing class, they differ as occurring on surfaces where vegetation grows more luxuriant and green, due to abundance of humus and humidity.
- Soil composition alteration traces: they depend on the presence of fragmented material (remains of ancient destroyed structures, large amount of fictile elements, etc.) in some part of the soil, highlighted by a colour changes due to grain size differences with respect to the soil nearby.
- Survival traces: some examples are the Roman centuriae or current roads precisely corresponding to ancient routes.
- Microrielief traces.
- Underwater remain traces.
- Anomaly traces: not included within the previous ones.

The aim of this work is to assess the usefulness of vegetation indices in order to identify archaeological traces and to verify the quantitative estimates of presence of buried archaeological structures in every type of elaboration.
Statistical analysis were conducted on several Italian archaeological test sites processing by hyperspectral MIVIS (Multispectral Infrared and Visible Imaging Spectrometer) various typologies of vegetation cover.

STUDYING VISUALIZATION OF 3D DATA ON A SMALL ISLAND IN CHINA

Yu Ming,[1,2] Ai Ting Hua[2]

[1]School of Geographical SciencesFujian Normal University Fuzhou, 350007, Fujian, China;
[2]School of Resource and Environmental Science, Wuhan University, Wuhan, 430079, Hubei, China
Email myu@fjnu.edu.cn or mingfj@hotmail.com

Most of the display facilities in GIS and mapping packages are concerned with data represented in two dimensions. To some extent this reflects the fact that most digital geographically referenced data are indeed 2D. There is however increasing interesting 3D data display. As usage of GIS technology has become more widespread so has an awareness of the possibilities of exploiting the technology. Thus it is now possible to create simulations of 3D scenes for purposes of geovisualization.
Taking the Nantai island located in the Minjiang estuary as the study area. It is the maximal island in the Minjiang valley and an district of Fuzhou whose area is about 118.2km^2. Most of the physiognomy is low mountain and hill. Fuzhou city is the capital of Fujian province, and also the politics, economy, technology, education and culture center of the whole province. The vegetation and soil of the island have the coherent distributing characteristic to the climate here. Red-soil (changing to yellow-soil in mountain) and evergreen broad leaved forest in East China lie in middle subtropical ocean monsoon climate region; latosol and south subtropical ocean monsoon tropical forests lie in south subtropical ocean monsoon climate region. The soil of the investigation region consists mostly of red-soil, latosol, paddy slurries formed by manpower. It is a type of complex bayou wetland ecosystem.
Space information visualization is an effective application of computer's technology. Depending on changing geographic environment, we cognize and analyze them through seeing landscape. Taking a small island located at city of Fuzhou in Fujian province in China as research area the author discuss some approaches which come true visualization of three dimension data and can created with ease based on ERDAS and ARCGIS9.0.

THE ARCHAEOLOGICAL AND NATURALISTIC PARK OF 100 RURAL ROMAN FARMS IN LUCCA, ITALY

Alessandro Mrakic – Giovanni Pancani

Facoltà di archtettura di Firenze, Dipartimento di Progettazione dell'Architettura

The "100 Rural Roman Farms in Lucca" project has been developed during the last years by Forum UNESCO – University and Heritage (ten universities from the Mediterrenean area and five from the Americas) and through the financial aid of the Tuscany Region, the Province of Lucca and the local Municipalities of Porcari and Capannori. The project has been carrying out archaeological research in the area of 100 rural Roman farms and their environment in Lucca's plain where stratified sites and landscapes were found. Its history goes from the middle Bronze Age (approximately 3500 years ago) through the Etruscan, the late Imperial Age (approximately 1500 years ago), the Middle Age to the 19th century when a draining project was realized.
In this archaeological sounding and preserving context, new technologies have been applied to obtain, process and interpret data: Remote sensing, Aerial photo, Scanner laser, GPS, Topographic and Photogrammetry survey, Geological prospection, GIS analysis, 3D recostruction and Web-database GIS.
The complete survey and analysis of the area was broached with the intent of finding out new archaeological sites and studying in depth the management, restoration and conservation of them. A particular attention of our study was therefore paid to research opportunities enabled by new technologies to the surveyor, with specific reference to applications derived from the use of a laser scanner and to the possibilities of integrating data obtained from various different surveying methods.
The survey produced a considerable amount of information, considering this large quantity of data obtained and its complexity, we questioned ourselves about the necessity of having adequate surveying protocols and appropriate systems of restitution for surveys

carried out by traditional methods integrated with new technologies.
The laser scanning the shots were taken with an amplitude of 40 degrees on the abscissa plane and 40 degrees on the ordinate plane.
The realization of this archive has proved to be particularly useful for the following work of restitution and for the integration of the numerical survey. This has consented us to achieve some remarkable and unsuspected results: the integration of the cloud of points, generated by the laser scanner, with the importation inside itself of the three-dimensional model of areas surveyed with other methodologies (3D recostruction of a roman farm and its environment); straightening with absolute geometric accuracy and the study of the surrounding landscape.
An important result that let understand and program in 3D-VR the restoration and management project on the archaeological site.

THE FLYING PATCHWORK QUILT: AERIAL RESPONSES ACROSS EUROPE

Chris Musson
for the Aerial Archaeology Research Group and Culture 2000

Introduction
It would be nice to think that the aerial viewpoint could provide a flying carpet to archaeological enlightenment. Aerial work has many merits – speed of operation, the 'landscape' view, the recovery of information not attainable by any other means. But it also leaves uncertainties about date and function and it responds unevenly to differing geologies and land-use patterns. There are also what we might call 'structural' problems, the extent to which it is implemented or integrated into archaeological studies in individual countries and regions.
As the title of my talk implies, we see in Europe not a magic carpet but a sort of 'patchwork quilt', of difficulties and differing provision, of partial solutions and significant achievements. The quilt sometimes looks quite pretty but it often hides significant shortcomings underneath. Nevertheless, there may be some lessons that we can learn from the very *variety* of responses that we have seen across Europe in the past decade of international work by members of the Aerial Archaeology Research Group (AARG).
The Group's aim is to provide a forum for the exchange of ideas and information for all those actively involved in aerial photography, photo interpretation, field archaeology and landscape history, including the formulation of conservation strategies.
From small beginnings at Cambridge we have now expanded into Europe and beyond. Our annual conferences are held alternately in Britain and on Continental Europe, we publish a twice-yearly informal journal, *AARGnews*, and we have nearly 200 members, about half of them from outside Britain (but only one, so far, from Ireland).
In the past five years we have organised or taken part in ground-based seminars or workshops in several Continental countries as well as the UK. But one of our main preoccupations over the past decade – with help from grant-giving bodies and the EC's Culture 2000 programme – has been a series of week-long ground-and-air training schools, in Hungary, Poland, Italy three times, Germany, and (in the coming year) Italy again, Germany again and the UK.
Our aim in these schools has been to help countries, institutions and individuals in their advancement – or first-time use – of aerial archaeology, in particular exploratory air-survey by archaeologists themselves. But the schools always pay equal attention to ground-based work, in mapping, interpretation and use of the resulting information, including that provided by pre-existing or newly-commissioned vertical photography.
Our travels across Europe have shown us a wide variety of responses to the initiation or application of aerial studies. I will highlight just a few before trying to draw some general advice from our own experience in Britain and from what we have seen in Europe.
The European situation: mainland Britain
I will start with mainland Britain, which to most of our European colleagues looks idyllic. From the days of Crawford, Allen, StJoseph and others our aerial work has always had a strong 'exploratory' element and to an extent a 'landscape' view of the past. In the '70s and '80s we perhaps became over-concerned with cropmark discoveries, not least in the partly-funded local flying that has complemented the national coverage started by the Royal Commissions, in England in the mid-1960s and in Scotland and Wales one and two decades later.
Importantly, however, our flying now responds to a much wider 'portfolio' of projects or targets. Individual flights might involve a mixture of pre-development survey, landscape characterisation, the search for cropmark, soilmark or earthwork evidence, the recording of buildings, townscapes and industrial areas, the monitoring of protected monuments, the recording of new and already-known sites for mapping and interpretation, and the continuing enhancement of local and national photographic records to assist conservation, education and the fostering of public awareness.
In Britain we have never had any significant bureaucratic, legal or military restrictions on getting into the air, nor any problems in gaining access to Ordnance Survey maps (at a variety of scales) or to nationally-funded vertical air photographs, literally millions of which, from RAF and then Ordnance Survey sorties, are now housed in the open archives of English Heritage and the Royal Commissions in Scotland and Wales. Many other archives, including that at Cambridge, are also open for public consultation. I mention these points about 'open skies', maps and photo-archives because it is easy to assume that things will be the same in other countries. They are often not.
An important point about the UK in the past decade or more relates to photo-interpretation, mapping and record creation. English Heritage has since the early 1990s had a comprehensive National Mapping Programme, to record at 1:10,000 scale, now in digital format, all the defunct features that can be seen on readily-accessible vertical or oblique air photographs. About a third of England has been covered so far. The resulting records (as both maps and text) are passed on to the conservation sections of county and municipal councils and are regularly used in development control and conservation work. Air photo analyses are used widely, but not always with sufficient skill, in pre-development assessments, and aerial views bulk large in both specialist and popular publications. In Wales and Scotland we do some of these things slightly differently, but not significantly so.
So, through the public service at national and local level we have a reasonably effective framework of policy and provision. With a few notable exceptions, however, we have not effectively penetrated the University scene. The veritable flood of new information from reconnaissance and air-photo mapping has not been fully incorporated into academic perspectives, nor always into a more 'landscape-oriented' view of the past.

We have a few published landscape studies based largely on air photo evidence, with others pending. There are several books on 'something' or 'somewhere from the air'. There is also David Wilson's *Air Photo Interpretation for Archaeologists* and a recent set of essays on practical aspects of aerial survey, *From the Air*. Unlike Italy, we no longer have a specialist journal, other than *AARGnews*. Nor a manual covering the whole field from initial survey, through mapping to archaeological interpretation (though paradoxically, Italy does, as a result of our Culture 2000 work).

So not everything in the British garden is rosy. But this still represents the most comprehensive provision that we have seen in our travels round Europe.

The rest of Europe

Much of Europe, indeed, has scanty provision. Around the Baltic Sea (and even in Iceland), aerial work is in its early stages, with a concentration on educational work and the traces of relatively recent events. There is currently no activity in Norway and Sweden. Sweden, like Holland, had some small provision in the past, but lost it through staff retirements. It is dangerous to depend on so few people (or rather posts). Denmark is currently trying to revive aerial survey there, partly by planning a training school for 2007.

Germany enjoys some fine exploratory work through Otto Braasch and a few others. But this depends on the individual regional museums or conservation services, with no national co-ordination and very little mapping beyond 'points on a map'. In Belgium the Universities of Leuven and Ghent, in a semi-competitive/semi-cooperative sort of way, have built up a reasonably comprehensive coverage for reconnaissance survey. Of France I can say little because the French speak only to themselves, and then only sometimes, and always in French.

The former Soviet States of Bulgaria and the Ukraine are still shrouded in bureaucratic darkness but Romania is making a start through exploratory work by Glasgow University and – within the Culture 2000 scheme – through the commitment of individual staff in the Institute for Cultural Memory. But there are difficulties, hopefully soon to be overcome, in getting access to large-scale maps and existing vertical photographs – a problem, still, in some countries of the former Soviet bloc.

University archaeologists in the Czech and Slovak Republics rapidly took to the air after the political changes of the early '90s, in one case buying their own aircraft. But here as elsewhere on the Continent there is a certain reluctance to trust air-photo sites until a hole has been dug in them to show that they exist. In Slovenia Darja Grosman, of the University of Ljubljana, has developed ways of coping with aerially-difficult terrain, taking part in interdisciplinary projects across the national boundary into Croatia. Greece and Turkey are a closed book, notionally at war with one another, and Spain and Portugal have as yet no real aerial provision.

Austria, through Michael Doneus, has developed highly-productive (but expensive) computer and fieldworking methods for integrating remote-sensing and field data. He now flies *just* enough to identify when cropmarks in his study-area west of Vienna are at their best. Then he calls in the military, who carry out total vertical survey as 'training exercises'. What bliss!

Hungary, for the most part, has relied on flights by Otto Braasch since he started going there in the early '90s. But in a small way Culture 2000 is now helping in the development of 'home-grown' work through a project with the Baranya County Museum.

Poland, another 'stamping ground' for Otto Braasch, has found a champion in Wlodek Raczkowski, of Poznam University. He and a few supporters have done good work on very limited funds, but they are struggling against two structural, or rather intellectual, problems. In the 1930s exploratory air photography was 'tried' by Polish archaeologists, but with seemingly negative results. They were just flying at the wrong time of year. But the apparent failure has coloured Polish academic attitudes ever since. As has the dominance of field-walking survey, the subject of a comprehensive national programme started in the 1970s and now about 75% complete. This has become the sole non-destructive source of survey information in Poland, making it difficult, because of ingrained academic and bureaucratic attitudes, to mobilise any significant resources for aerial work. In reality, of course, the two should complement one another, each acting as a check on the weaknesses of the other. An opportunity was also lost when the then head of the state Inspectorate, one of our students at the first training school in 1996, tried to create some kind of general policy for aerial work amongst his otherwise self-willed inspectors. Unfortunately political changes removed him from office and the initiative died.

In Italy, something similar applies. The Soprintendenza is a dispersed body in which local policy lies at the whim of individual inspectors. Few see any value in aerial archaeology. They feel they have enough on their plates. Not everybody in Italy agrees, of course. One who did not was Professor Riccardo Francovich, of the University of Siena. But for 60 years from 1939 a domestic law so restricted the conduct of photography from light aircraft that Italian archaeologists virtually lost all idea of doing it themselves, concentrating instead on sophisticated photogrammetric mapping, mainly of individual sites and local areas rather than the broader landscape. Francovich, however, never believed the Italian mantra that 'it won't work here and anyway half of the country is covered in trees'. It is true that Italy (like other countries) has many areas, such as the Po Valley and the clay-lands of Tuscany, which will always prove difficult, but others parts are ideal for exploration from the air, as anyone who has flown over central or southern Italy will testify.

When the 1939 law was at last repealed Francovich prompted the first Italian training school, at Siena in 2001, and others have followed. But it seems clear that there will never be a concerted national policy in Italy, nor a co-ordinated programme of aerial reconnaissance or air-photo mapping. Instead, progress will have to come from the initiative and inventiveness of individual university researchers, co-operating and competing with one another to integrate this (for them) 'new' technique alongside their earlier expertise. Fortunately, they begin with a clear concept (in some cases at least) of the 'landscape' potential of active aerial survey, and of mapping through simple desktop computers. They also start with a balanced approach to combining remote sensing and field-based techniques.

There is a clear contrast here with mainland Britain, where most aerial work takes place in isolation from fieldwork in the 'plannable' public service, rather than in the quixotic but perhaps more inventive ambience of university research. If only we could combine the two! But one of the lessons from Europe is that you cannot pick and choose – you must work within the constraints and possibilities of your own country.

Conclusions

So, can we offer some general advice from our experience in mainland Britain and other parts of Europe? I hope so, and if I simply reinforce the words of other speakers take it as confirmation that the advice is good.

- Build on the experience and mistakes of others. Join AARG and become part of the European network of advice and experience. And possibly set up your own all-Ireland interest group.
- There is no magic formula for success. Each country, and sometimes each region, must find its own way of making aerial studies 'work' within its own geography, environment, academic and organisational framework.

- Seek the participation of your principal official bodies, in the academic world as well as in public service. If you can develop some kind of national policy, or just some 'community of approach', you will benefit greatly. That said, push on doors that are already half-open, not on those which will remain obstinately shut.
- Aerial archaeology is not a 'subject in its own right'. Start from the assumption that air survey analysis, whether from new flying or existing photography, is just *one* of the techniques that you can deploy in a balanced programme of interdisciplinary investigation.
- Do not assume that aerial archaeology takes place *only* in the air, or that it can be done *wholly* on the ground. New exploratory work is essential but you must also make full use of existing vertical coverage, especially that which shows the landscape before the recent decades of rural, urban and industrial development.
- Treat record creation, mapping and landscape interpretation as a standard part of your practice from the outset.
- When you do fly, do so with maps marked up with all the known archaeology, and carry with you a 'portfolio' of targets or subject-areas. Then you will nearly always come down with something useful to show.
- Fly at the right time of day and year for the kind of information that you are trying to gather. Good timing is crucial, and this comes from planning and experience.
- Do not expect or demand instant results. In some cases you may achieve them. In others you may simply produce absence of evidence rather than evidence of absence. Where conditions are at all favourable it may be decades rather than years before new discoveries fall below a significant proportion of the total sites photographed on each year's 'exploratory' flights.
- For any targeted air photography allow a realistic 'lead-in' time: good visibility and low sunlight cannot be conjured up in a matter of weeks, nor cropmarks at the flick of a finger.
- Do not leap into the air without initial training. If you try to do aerial work without basic skills and equipment, or without understanding the *limitations* of the art, you will risk failure and will prejudice rather than advance the progress of aerial studies.
- Think about long-term training programmes. Try to have the subject taught in universities. Send students and practitioners to training schools. Perhaps set up training experiences here in Ireland (but don't ask AARG or Culture 2000 for the money, or not much of it anyway – it's already committed elsewhere).
- If you need help with standards for such things as photographic or computer equipment, cataloguing and record creation, mapping scales and conventions, and image-capture, processing and archiving, use the AARG network to access a wealth of advice, based on years of sometimes painful experience
- Go digital from the outset, but be aware of the problems that this decision brings, as well as the many benefits.
- And finally, conform to the national stereotype and enjoy the luck of the Irish.

THE USE OF DMC SMALL SATELLITE DATA IN THE INVESTIGATION OF CHINESE GREAT WALL

Yueping Nie, Jingjing Dai, Xianfeng Zhang

Institute of Remote Sensing Applications Chinese Academy of Sciences

The Great Wall of China, one of the greatest wonders of the world, was enlisted in the World Cultural Heritage by UNESCO in 1987 owing to its architectural grandeur and historical significance. Just like a gigantic dragon, the Great Wall winds up and down across deserts, grasslands, mountains and plateaus stretching approximately 6,700 kilometers (4,163 miles) from east to west of China. It starts from Shanhaiguan Pass in the east to Jiayuguan Pass in the west traversing provinces of Liaoning, Hebei, Beijing, Tianjin, Shanxi, Inner Mongolia, Ningxia, Shaanxi and Gansu. It was originally built in the Warring States Periods as a defensive fortification, and it went through constant extensions and repairs in later dynasties with a history of more than 2000 years.

Until now it is still a problem to definitely study its distribution and dating situation owing to historical and technical factors, what's more, some of the sections of the great wall are now in ruins or even entirely disappeared due to natural or human damage. So it is a very urgent task to effectively investigate and protect our precious cultural heritage.

Chinese government has paid great attention to these problems, and has launched a project called "The Great Wall Project", aiming at investigating its distribution and conservation in detail all over China. Because of huge spatial span of the Great Wall, field archaeological methods can be very time-consuming, expensive and ineffective. Besides, some segments are difficult to access. So spatial technology based on remote sensing, geographic information system, global positioning system and visual reality may play an important role in the project.

For detailed study of the Great Wall, high spatial resolution satellite data is very important. As we all know, high spatial resolution data such as IKONOS and QUICKBIRD data is very expensive. Thus, we are very expecting to low-cost satellite data with high spatial resolution. Beijing NO.1 small satellite was launched by China in November 2005, which acquire three multi-spectral bands with the spatial resolution of 32 meters and one panchromatic band with the spatial resolution of 4 meters. And it has some other advantages such as large swath (600km*600km) and high temporal resolution (less than 72 hours). So it can be a good data source for archaeological study such as the great wall investigation.

To examine the utility of the DMC data for the great wall investigation, we conducted an experimental study in Beijing area. We also check the archaeological application potential of DMC (Beijing NO.1) data. The objectives of the study specifically include:

1) DMC (Beijing NO.1) data processing for remote sensing archaeology
2) Interpretation signatures for the Great Wall
3) Experiment with automated extraction of the walls
4) Comparison with SPOT data
5) Field verification

Though our initial study, we find that DMC (Beijing NO.1) data can be used for interpreting the Great Wall features in Beijing area. However, some technical problems in automated extraction of the wall features need to be further studied.

MODELING PARTHIAN SITE DISTRIBUTION PATTERNS IN THE CENTRAL ZAGROS, WESTERN IRAN

Kamal Aldin Niknami[1], Ya'koub Mohammadifar[2]

[1]Department of Archaeology, University of Tehran, Tehran, Iran. E-mail: kniknami@ut.ac.ir
Department of Archaeology, University of Bo Ali Sina, Hamadan, Iran. E-mail: yamohamadi@yahoo.com

The project study area lies in the Central Zagros archaeological region, Western Iran. Archaeological survey has been conducted in the study area, and approximately 340 Parthian (ca. 250 B.C.-225 A.D.) archaeological sites have been discovered and recorded. The goal of this project is to develop a set of simple processes that can be widely used to build basic models of site locations and settlement patterns. Two main issues were concerned in this paper, (1) technical aspects regarding basic methods for data acquiring, and GIS based processing and analysis, (2) modeling archaeological site distribution patterns related to the available archaeological and environmental data. In order to explore further the location of archaeological sites, a Geographical Information System (GIS) was developed combining all existing information produced from the recent survey of the area. The combination of this information into a GIS has resulted in a digital archive of the surveyed area that allows the user to simultaneously visualize and analyze all data within their original spatial contexts allowing a more comprehensive investigation of the sites. This digital database has been developed to synthesize information on site type, site size, site dating and site functions, which are used to address the issues regarding the changing settlement patterns in the study area. We found that settlement patterns are, to a large extent, directly shaped by widely held cultural needs, they offer a strategic point for the functional interpretation of Parthian archaeological culture.

A (GIS)-BASED PREDICTIVE MAPPING TO LOCATE PREHISTORIC SITE LOCATIONS IN THE GAMASIAB RIVER BASIN, WESTERN IRAN

Kamal Aldin Niknami, Mohammad Reza Saeedi

Use of geographic information system (GIS)-based predictive mapping to locate areas of high potential for prehistoric archaeological sites is becoming increasingly popular among archeologists. Generally, these techniques assume that the selection of sites by original inhabitants was at least partially based on a set of favorable environmental factors, such as distance to water or topographic setting. Another assumption is that modern day GIS layers consistently characterize changes from the prehistoric condition of the region sufficiently well that they can be used to help discover additional sites. Knowledge of the environmental variables influencing activities of original inhabitants is used to produce GIS layers representing the spatial distribution of those variables. The GIS layers are then analyzed to identify locations where combinations of environmental variables match patterns observed at known prehistoric sites. Presented here are the results of a study to locate high-potential areas for prehistoric sites in a largely unsurveyed area of 11000 square km in the Gamasiab River Basin, Western Iran, including details of the analysis process, based on the multivariate statistics. The project used environmental data from many known sites in other parts of the region and the results corresponded well with known sites in the study area. The predictive model, developed over a two-year period. The results of model will be incorporated into the earliest phases of project concept planning, and aware of the possible locations of archaeological sites. The aim of this paper is to review the conceptual framework of predictive modeling as well as the model formulation and model applications.

GPR TESTS FOR 3D IMAGING OF THE HISTORIC WALL OF PONTEVEDRA (SPAIN)

Alexandre Novo, Fernando I. Rial, Manuel Pereira, Henrique Lorenzo, Pedro Arias

EUET Forestal. University of Vigo. Campus A Xunqueira s/n. 36005-Pontevedra (Spain). alexnovo@uvigo.es

Ground-penetrating radar (GPR) surveys can provide reliable information about buried archaeological remains. Thus, GPR archaeological surveys are being more and more utilized in the last years, especially related to 3D visualisation due to the capability of getting high resolution three-dimensional images of buried archaeological remains.

This paper shows some results about the use of GPR in an urban environment in order to locate the historic wall of Pontevedra city (Spain), which has Roman origin although with a lot of subsequent modifications. The historic wall of Pontevedra was pulled down in the nineteenth century and its remains are big irregular stones of granite. Nowadays the wall shows some preserved stretches and other ones partially excavated while in some places its precise position is still unknown. The current survey has been carried out in a place where the presence of the wall is perfectly documented (it has been excavated, documented and buried again), the remains of the wall are buried in the shallow subsurface, this information is a well-known fact thanks to the archaeological excavations directed by D Juan Carlos Castro Carrera, archaeologist in charge of the excavations in this stretch of the wall. So this place is a perfect area in order to test different survey methodologies.

The main aim is to assess the most proper kind of antennas and the optimum surveying methodology so as to extend the survey to the detection of other unknown position stretches. The surveys have been carried out using a RAMAC/GPR system with one 500 MHz shielded antenna and two 800 MHz shielded antennas combining different polarities as well as paying special attention in the assessment of the most appropriate space between profiles in order to built 3D models that reliably show remains in the subsurface, their dimensions, position and depth.

A multichannel unit control, a laptop with special transflective display for real interpretation and the antennas were placed in a non-metallic trolley with an odometer in order to obtain the position of the profiles. In this survey, medium-high frequency antennas were used so as to achieve a good balance between depth of penetration and high resolution. The multichannel unit control of the system with the two 800 MHz bistatic antennas (with independent transducers) let us obtain a profile with different configurations of the transmitting and receiving dipoles simultaneously.

In order to get a high resolution 3D image of the remains of the wall, a very dense grid was designed for the data acquisition. The selected area (108 m^2) was surveyed along parallel 16 cm spaced profiles using two 800 MHz antennas and along parallel 50 cm spaced profiles using a 500 MHz antenna.

Another aim of this survey was to get conclusions about the proper real-time processing procedure. During data acquisition different filters were applied so as to remove the DC component from the data, gain filters in order to highlight the data and also a Band-Pass filter. This data processing does not have influence in the post-processing data because the data are stored as raw data.
All profiles registered are processed with a specific software which is capable to create 3D cubes of the surveyed area in order to achieve a global interpretation much more understandable than B-scans. Then it is possible to extract plane or volumetric C-scans at different depths from those 3D images.
The more suitable antenna for this place, where the remains of the wall are situated at a shallow depht, was the 800 MHz one. Nevertheless, the 500 MHz antenna will perform better in areas where the remains of the wall are situated deeper than in the surveyed area.
Moreover, it has been checked that if the distance between parallel profiles is more than four times a recommended value and the antenna does not have enough resolution, the 3D visualisation of the archaeological remains is not evident. Hence the determination of the precise situation and dimensions of the remains is very hard.
The results show the possibility to reach high resolution three-dimensional information of the shallow subsurface. The data interpretation is perfectly compatible with the results of the previous archaeological excavations that have been carried out in the surveyed area. However, it is arguable whether in other GPR surveys, where the situation of the wall is still unknown, it is worth to multiply the fieldwork so as to obtain detailed 3D information or it is sufficient a traditional survey by means of a non-dense grid.

APPLICATION OF REMOTE SENSING IN THE DETECTION OF MAYA ARCHAEOLOGICAL SITES IN SOUTH-EASTERN CAMPECHE, MEXICO

Krištof Oštir, Žiga Kokalj and Ivan Šprajc

Scientific Research Centre of the Slovenian Academy of Sciences and Arts, Ljubljana, Slovenia
kristof@zrc-sazu.si

Remote sensing is a fast and efficient mean to obtain data over larger and difficultly accessible areas. Archaeological applications of air and space borne earth observation can provide characteristics of sites and their indirect or direct observation. In the paper possibilities of contemporary medium resolution imaging systems, image processing methodology and problems, are described. For selected environmental parameters the difference between archaeological and random points is shown.
To detect archaeological sites in the area of central Yucatan in the Mexican state of Campeche, both aerial photography and satellite remote sensing data has been applied. Aerial photographs have been taken in 1995 and 1996 to support the mapping of the Calakmul Biosphere Reserve (Conabio). Approximately 400 photographs have already been scanned, observed visually, using a stereoscope, and photogrametrically processed to provide high resolution imagery and corresponding digital elevation models of the area. Another set of images will follow in near future.
Aerial photographs have been acquired in the scale 1:20.000 with Wild RC 30 analogue camera and scanned with the resolution of 0.6 m. They have been processed to produce othophotos with the same resolution m and a digital elevation model with the resolution of 6 m. Due to lack of characteristic (man made) objects and mostly out-of-date maps it was difficult to collect ground control points. The final images are therefore not positioned well absolutely. Nevertheless, the results enable visual stereo interpretation and detection of pyramid like structures. The digital elevation model produced is better than any existing model in the area and enables observation of "unusual" features in the relief. The major problem with DEM analysis is however the fact that the elevations represents canopy tops and not real surface.
Beside aerial data a set of satellite images from different optical (SPOT and Landsat) and radar (ERS) sensors have been used. Two radar images from the ERS (European Remote Sensing Satellite) have been interferometrically processed. Both images have been collected by the ERS 1 satellite on 1996-02-11 and 1996-04-22. The time difference was selected to observe the differences in the humidity and try to relate them to anomalies. However, due to decorrelation in forested areas interferometric processing was almost impossible. The images, combined with coherence (which is a measure of similarity between them), have been thus observed visually only.
Better results were obtained from SPOT and Landsat satellite data. Two panchromatic SPOT images, acquired on 2002-11-23 and 2001-03-31 covering different parts of the study area have been tested. The 10 m resolution of earlier SPOT systems is to coarse to detect structures directly, but enables observation of "unusual" features that can be related to sites. Simple and advanced edge detection has been used, including filtering in the direction of the Sun. Again some of the sites were visible on original or processed images, but the amount of non-relevant information was relatively high.
Despite the fact that Landsat has lower spatial resolution then SPOT, its spectral characteristics are far superior. In the study one Landsat 7 Enhanced Thematic Mapper Plus (ETM+) image, acquired in on 2000-03-27 has been used. Before performing further site analysis responses in the individual bands and different band ratios have been studied and we used a combination of bands 4, 5, and 7 to compute the ratios 5/4, 7/5, and 7/4 and combined them into colour composite image (5/4 showin in red, 7/5 in green, and 7/4 in blue).
Detection of archaeological sites is possible by observing contrasts between archaeological features and the natural background. When archaeological features possess physical properties different from those of the surrounding matrix, the distinction may be noticed. The analysis of the environmental characteristics was performed for the sites colleted in recent field work. The archaeological database included 63 sites with appropriate attributes – coordinates, type, and size. In most of the analyses only larger sites were considered, i.e. major and medium centres (33 in total). The main reasons for this selection is to eliminate the influence of unknown lower rank sites and to use only most important (believably also characteristic) sites. Additional limiting factor were spectral features of the locations. Sites that are close to settlements, roads or have been cleared or cultivated (secondary vegetation) have not been used in the analysis of satellite data, leaving 20 sites in the study set. In this way uncharacteristic spectral responses have been eliminated.
Environmental characteristics of the sites have been studied to find site indicators regarding Landsat ETM+ image data and digital

elevation model. A buffer zone that includes an important surrounding of each site has been produced. Different radiuses (50, 250, and 1000 m) have been tested, to see the properties of the immediate vicinity and larger influence area. Indicators have been compared with the background; 60 points distributed randomly in the area. These random points could also exhibit archaeological sites, since the area has been surveyed only partly and not all sites have been found.
The digital values (reflectance) in individual bands of the Landsat ETM+ image, acquired on 2000-03-27, show difference between random points and larger archaeological sites. However the deviations are very small and only noticeable in bands 4, 5, and 7. Small differences can be enhanced by computing band ratios, i.e. normalized differential vegetation index and band ratios 5/4, 7/5, and 7/4. The archaeological sites have slightly smaller vegetation index, meaning that the vegetation is less intensive or disturbed in growth. Studies of the relief determined that it is not possible to distinguish sites from random points based on elevation and orientation (aspect). However, slope and curvature show a much clearer picture. All sites, and especially larger ones, have been found in steeper and more agitated areas. Statistical evaluation even shows evident and consistent relation of the site size and selected curvature parameters.
A significant number of detected Maya sites are found close to bajo areas. Since the vegetation in bajos is characteristic (typically less intensive) these areas could influence the lower vegetation index found within the archaeological sites. The bajo areas have been therefore delineated by reclassifying the vegetation classification data provided by Conabio. None of the sites in our database lies in bajo and only one random point was positioned in it; the results of satellite or relief data are therefore not biased. In contrast, interesting issues can be observed by analysing the distance of sites from bajos and the surface of bajos in their proximity. Especially larger centres are closer to bajos as one could expect from random points. The analysis should be further improved with better bajo definition (exclusion of smaller ones, clearer delineation), by using elevation data and additional selection of (representative) sites.
Although the results presented here are based on preliminary studies, both GIS and remote sensing proved to be very efficient tools for archaeological research in Maya regions. We have been able to define common environmental characteristics of archaeological sites and observe their relations with various factors. While some of the answers have been found, there are still many questions open, and therefore the research will be continued in several directions. First, it is necessary to obtain additional data: high resolution multispectral (or even hyperspectral) satellite images, multitemporal Landsat data to cover seasonal changes, multipolarization or multifrequency radar images, and better elevation data (e.g. from lidar).
The results obtained so far have shown that only advanced, sophisticated processing methodology is able to provide indicators of archaeological sites under dense vegetation canopy. However, even preliminary results enabled us to understand the location and characteristics of the sites. It has been determined that a predictive model for (large) sites should be based on slope and curvature of the relief, supported by the (subtle) vegetation status indicators.

PALEORELIEF DETECTION AND PALEODEM MODELLING A CASE OF STUDY IN EASTERN LANGUEDOC (FRANCE)

Krištof Oštir *, Laure Nuninger **

* Institute of Anthropological and Spatial Studies, ZRC SAZU (Slovenia)
** Chrono-Ecology, UMR 6565 CNRS/University of Franche-Comté (France)

The main focus of the paper is paleoDEM (paleo digital elevation model) detection and modelling, a multidisciplinary approach involving geodesy, archaeology, and paleoenvironmental studies.
Why paleoDEM production is necessary? To understand settlement pattern and socioenvironmental contexts and evolution, archaeologists are using GIS analysis based on digital elevation model. For example, DEM help to analyze viewsheds, to define access time between places, to simulate ancient paths, etc. Considering a period of three millennia, some topographical areas can be considered without fundamental change. Within littoral zones or fluvial areas the situation is totally different, however. Sedimentation and erosion, due to alluvial and sea level dynamic, model the relief permanently. A flat area nowadays could be hilly in prehistoric times. Usage of present DEM for the analysis is therefore not possible, since it would distort the expected result completely.
Our case of study is situated in Languedoc (France), on the littoral plain of Mauguio between the cities of Montpellier to the South and Nîmes to the North. This area covers around 250 square kilometres. From an archaeological point, the zone is very well known zone with almost 350 sites inhabited from Bronze Age to Middle Ages, showing also parcellar evidences obtained from aerial photo interpretation. In addition, from 1990s several geo-archaeological surveys were done in the main deltas of Lez and Vidourle rivers. Unfortunately the geo-archaeological coring has been performed to understand paleochannel profiles in the Vidourle delta and the cores are concentrated in three zones. Considering a regional scale, only three average points could be use. The geo-archaeological approach in itself brings indices about paleorelief but since the measurement are represented as points they are not enough to produce generalisation and compute a proper past DEM. Additionally the approach is slow and relatively expensive. We have therefore used remote sensing, optical image processing and radar interferometry, in combination with spatial analysis to detect the paleorelief. The first step was to create an accurate present DEM that had a resolution suitable for analyses. Then, this DEM as well as SPOT and Landsat satellite images were processed with several filtering to detect some "evidences" of paleorelief, called "paleofeatures". Features were digitalised and their spatial distribution was statistically tested according to archaeological data distribution. At last, we compared and tried to qualify the paleofeatures with the results obtained by photo-interpretation, classification and simulation.
The digital elevation model has been made as a combination (weighted sum) of all available sources. First, a DEM has been produced with radar interferometry (InSAR) from two ERS radar images, taken on 1996-04-04 and 1996-04-05. The resolution of this partial model is 25 m and the vertical accuracy is approximately 8 m and the positional accuracy is almost 5 m (one fifth of pixel size). Additionally the following DEMs were used: IGN DEM (50 m, provided by IGN), Aster DEM (30 m, made from Aster stereo satellite images), SRTM DEM (90 m, produced with radar interferometry from the Shuttle Radar Topography Mission data). These models have different resolution and different coverage of the area: from ~60% for Aster to almost 100% for InSAR and SRTM and complete for IGN. All standard deviations are less than 10 m in the area; Aster and InSAR are systematically higher because of vegetation cover. According to the elevation accuracy they can be ranked as IGN (the lowest standard deviation from control points),

SRTM, InSAR and Aster. The final (combined) model takes the rough shape of IGN DEM and adds details of the others, particularly InSAR. It has a resolution of 25 m and is very close to the control points (-0.2 m, standard deviation 3.7 m). The model is rather homogenous and could be used for further analysis.

Optical images (Landsat and SPOT) were enhanced prior interpretation. We have been trying to observe paleofeatures indirectly through detection of edges, particularly related to humidity and vegetation anomalies. Landsat with its seven bands gives much better spectral information than SPOT while SPOT has a better spatial resolution (SPOT 4 pan with 10 m resolution has been used only). Simple and advanced edge detection has been performed, including filtering in the direction of the Sun. To increase the spatial and preserve spectral resolution of Landsat resolution merge of panchromatic and selected multispectral channels has been performed (pan sharpening). A similar procedure has been repeated with panchromatic SPOT and multispectral Landsat data. It has to be mentioned the huge difference between 15 m resolution of panchromatic Landsat and 10 m of SPOT. The latter is sharper, what is probably the influence of different sensor types, and simplifies the observation of anomalies.

The pre-processed DEM and satellite imagery enabled paleo feature delineation. Manual on screen digitalization has been performed and all possible features have been included. The following data has been used to digitalize potential features:

- panchromatic and edge filtered SPOT,
- multispectral and pan-sharpened Landsat, edge filtered panchromatic Landsat, and
- shaded and edge filtered combined DEM.

The features have been over-digitalized and later cleaned. Maps in the scale 1:25000 were used to inspect the features by eliminating existent relief characteristics (roads, buildings ...) and water courses (rivers, channels ...). SPOT and shaded DEM helped the most in feature detection and all other layers were used only to update or check the results.

To test if the detected paleofeatures can be linked to real paleorelief an analysis of archaeological sites in the area has been performed. The distribution of sites in relation to the features (buffer zones, site proximity analysis) has been observed for Prehistory, Roman period and medieval period. The sites were compared to random point distribution. For all known sites the distance from digitalized paleofeatures has been determined. Only sites that are less than certain distance (2.5 km) from the features have been considered. It has been statistically proven that the average distance of sites for all periods, especially for Prehistory, is smaller than for random points. For Prehistory the average distance is only ~450 m, while for random points it is more than 800 m, almost twice as large. One can also observe, that the majority of the sites lie less then 500 m from the detected features.

In order to define the different type of paleofeatures as paleochanel, edge of natural paleo-depression, ancient road, terrace (etc.), the digitized layer was compared with geology, pedology, cadastre with contour lines at 1 m interval, aerial photography from allied missions in 1944. First, the map interpretation give attribute information to perform an automatic typology based on classification method. In addition, since especially IGN DEM contained artefacts of previous processing (contour lines), an automatic processing based on altitude analysis of each points of palaefeature is in progress to recognize artificial stage.

Finally, to test the reliability of the longer paleofeature which seems to be an ancient coastline, sea level and river deposit modelling has been performed. The sea level parameter plays as an equalizer regarding the general sedimentation from small rivers and silting-up of the Etang de Mauguio. River deposit modelling based on geo-archaeological data is very basic and sea level was integrated without true references but as a parameter of simulation, this process gave a first visualisation of paleoDEM for Iron Age, Roman and medieval periods. It offered an additional opportunity to test the validity of our assumption regarding paleocoast.

With the case study we have proven that remote sensing can help in the detection of paleorelief. It is not possible to detect past environment directly, but several data layers (in our case most successfully SPOT and DEM) can be used to observe paleofeatures and to relate them to relief. The results obtained present a first step and should be improved with further processing and data inclusion (lidar, high resolution satellite images).

SATELLITE SURVEY AND CORING IN RECONSTRUCTING ANCIENT LANDSCAPES IN THE NILE FLOODPLAIN AND AT TELL-EL AMARNA (MIDDLE EGYPT): A MULTIFACETED APPROACH

Sarah Parcak

University of Alabama-Birmingham, Department of Anthropology

Introduction

One of the crucial issues within the field of archaeology is how we reconstruct what used to exist, from 100 years to 1 million years ago. Landscape reconstruction is another crucial issue: so many modern landscapes continues to influence how we perceive overall ancient landscapes. It is difficult to step back and let go of preconceived notions of land levels, flood patterns and ancient settlement locations. Ancient history is, after all, embedded in landscapes, and how we have interpreted key datasets is ultimately bound within our understanding of forests, deserts, rivers, valleys and floodplains in ancient times. Ultimately, we must rely on a variety of archaeological, technological and geomorphological tools in reconstructing past landscapes.

Project Overview

It is with this in mind that the author has undertaken work from 2004 to the present through the Middle Egypt Survey Project across from the well-known ancient Egyptian site of Tell el-Amarna. The project developed a methodology for dealing with landscape evolution in the ancient Near East, utilizing a combination of satellite remote sensing, ground survey and coring to locate 70 previously unknown archaeological sites, finding evidence for ancient Nile river courses, and determining landscape evolution over the past 3300 years.

In general, archaeologists working across the ancient Near East have focused on the visual capabilities of satellites, and have not developed set methodologies for mutispectral analysis. Through work in Sinai, the East Delta and Middle Egypt with Landsat, SPOT, Corona and Quickbird satellites, the author has developed a remote sensing methodology with between a 93-98% success rate for ancient site location, and has successfully located 114 previously unknown sites through various classification methods such as Principal Components Analysis and Unsupervised Classification. Multiple survey seasons allowed for ground-truthing of the sites, collecting ancient material culture and overall settlement pattern analyses.

Using a combination of satellite images and coring, the Middle Egypt Survey Project aims to reconstruct past settlement patterns and geomorphological developments in the Nile Valley. The 2004/5 seasons saw the discovery of 70 primarily unknown archaeological sites using satellite remote sensing (Landsat, Corona and Quickbird satellite images) in Middle Egypt, followed by ground survey and coring work. In 2005, the project included coring, and revealed additional insights into the geomorphology of the region, especially regarding annual Nile silt deposition.

Detecting archaeological site destruction: another valuable use for satellite remote sensing in Egypt, the Middle East and the world

Landscape geomorphology is only one issue at hand within the parameters of the Middle Egypt Survey project. As in many areas across the globe, Egypt is facing archaeological site loss at tremendous rates from an imploding population, massive urban expansion and rising water tables. How to track this site loss is another way that satellite remote sensing can contribute to the field of archaeology. The 2006 focused on an examination of overall site loss at Tell el-Amarna using a combination of old maps, aerial photographs and satellite images (in particular Quickbird). Over the last 200 years, the cultivation surrounding Tell el-Amarna has increased significantly. This project sought to document the exact rate of site loss through remote sensing and ground survey.

It is difficult to say how much the site was cut into over time. We can chart earlier overall destruction of Amana based on early maps from 1820- 1930, where it is easy to see expanding villages, and the agriculture cutting into the site. Overall, the amount of destruction in the past 200 years is widespread. It is most concerning that 90 hectares of the site have gone in the last 25 years, showing a sped up rate of destruction. If this rate continues, then in 50-60 years, very little of Egypt's most well-known sites will remain.

This methodology for detecting site loss using satellite remote sensing, old maps and aerial photographs could be applied across the globe, and would help to address many issues facing governments, cultural heritage agencies such as UNESCO and numerous archaeological projects.

Satellite remote sensing and Landscape geomorphology

A crucial aspect of this work is coring, used in cultivated areas along the East Bank to attempt to determine the original extents of the ancient city, as well as the original location of the Nile during the Amarna Period (ca. 1300 BC). The purpose of the work was to look at just how much of Tell el-Amarna, if any, remained beneath the fields, and to get a better idea of the geomorphology of the area, as no coring had even been done within the East Bank cultivation in conjunction with the satellite remote sensing results.

This project utilized an Eijelkamp bailer boring set, complete with core bits for use in sandy, silty and dense clay soil levels. Treating each core as an archaeological "spit" level permitted the comparison between different cores taken along similar points, with each core taken recorded as an archaeological locus, later sieved. Through the use of a GPS, the project recorded the Easting and Northings of each coring point for placement on a Quickbird satellite image of the entire Tell el-Amarna region for comparative purposes. Taking cores at set intervals within the floodplain running north to south meant that a full range of results could be obtained. The project took thirty cores, ranging from 70cm to 7m in depth.

The geomorphology of the floodplain next to Tell el-Amarna can now be reconstructed. One can clearly see from a cross section a definite sloping layer of the desert gebel beneath the ground surface, on top of which can be seen a fairly consistent number of sloping clay, sand and silt layers. They suggest a consistent deposition of silts since ca. 400 AD, which appears to be the time when the silt and clay layers starting building up on the East Bank, and seems to be the main point in time when a large section of Tell el-Amarna was simply washed away.

It seems that the Nile river course has shifted in the past 3700 years, in some areas more dramatically than others. It is virtually impossible to determine exactly when this shift took place, but satellite images assisted in the location of ancient cities in the Late Roman Period which would have been located along the edge of the Nile ca. 400-800 BC. Some of these towns are located some 600-1000m to the West of the current Nile edge. Considering that there is a distinct bend to the Nile next to Tell el-Amarna, and that the aforementioned towns would have likely been located along the edge of the Nile during the Late Roman Period, one can suggest an ancient Nile river course a minimum of 600-1000m to the West of the current location in antiquity. With a shift of some 600+m in 1600 years, this would suggest an annual shift of .3-.65m. Does this mean that the Nile would have been located even further west during the Amarna Period?

Little of the ancient city remains beneath the current floodplain level, which appears to have developed in the past 2000 years based on the concentration of Late Roman Period sherds in the silt layers some 2-5m below the ground. "Modern" (i.e. post Late Roman Period) cultivation and the Nile flood levels appear to have combined to obliterate any remaining traces of the city, save a few areas where deep foundations preserved Amarna Period material culture. The exact overall size of Tell el-Amarna is uncertain due to this cutting into the site, but initial coring and remote sensing work suggest it could have been significantly larger.

Quickbird imagery: a multispectral approach.

Quickbird's multispectal capabilities have not yet been explored within any archaeological arena. Using cutting edge technology can reveal a landscape hidden to our naked eyes using Quickbird satellite imagery, with a 62cm pixel resolution. Like any new technology, a number of scholars find them useful for conference presentations. Given that Quickbird images can cost many thousands of dollars, it is a tremendous waste of resources as aerial, balloon, or kite photographs can do the same thing for a fraction of the price. One must recognize the advantages of satellite images and have the background to process them correctly. Using Quickbird's multispectral capabilities, we can gain a clearer picture of what lies beneath the soil. The multispectral bands of Quickbird allow for subsurface mud brick to be differentiated from the surrounding sands, due to their higher moisture content, which allowed for the majority of Tell-el Amarna to be mapped, in particular the subsurface remains. The implications for other sites around Egypt and the Middle East are numerous.

Conclusion

We must combine satellite remote sensing results with the geomorphological data to give a complete landscape history, and we must think beyond the landscape in front of us today in Egypt, which is different in so many ways from what existed 3300 years ago and more.

LA CARTA DELLE RISORSE ARCHEOLOGICHE DELLA PROVINCIA AUTONOMA DI BOLZANO – ALTO ADIGE: ARCHAEOBROWSER

Parnigotto

La carta delle risorse archeologiche, così è sottotitolato l'Archaeobrowser, è un potente motore di ricerca basato su una enorme mole di dati geografici digitali.
Una carta archeologica on line, alla quale tutti gli interessati potranno avere comodamente accesso da casa propria digitando l'indirizzo: http://www.provincia.bz.it/beni-culturali/1302/archaeobrowser_long_i.htm
ArchaeoBrowser è un servizio di cartografia tematica, che consente a tutti i cittadini di consultare dati sulle risorse archeologiche presenti nel territorio della provincia di Bolzano direttamente su Internet utilizzando un semplice browser. Archaeobrowser è un servizio che Ufficio Beni Archeologici, Ripartizione 13 Beni Culturali, ha voluto istituire non solo per esperti o per altri enti pubblici, ma principalmente per i cittadini, per agevolare e incrementare la conoscenza e il rispetto per il patrimonio archeologico presente nel territorio provinciale.
La grande diffusione degli strumenti informatici e la consapevolezza delle potenzialità di diffusione degli applicativi internet hanno suggerito l'impiego di questo strumento di informazione per divulgare i dati relativi al patrimonio archeologico.
L'archivio di ArchaeoBrowser contiene l'elenco delle particelle catastali alle quali è applicato, sulla base dell'art. 10 del Codice dei beni culturali e del paesaggio, il vincolo archeologico e inoltre indicazioni sulla natura, sulla cronologia e sulla distribuzione dei giacimenti archeologici provinciali già individuati, parzialmente indagati o completamente esauriti, compresi quelli che non ricadono sotto le misure di vincolo.

PALAEOHYDROGRAPHY AND ANCIENT SETTLEMENTS IN THE ADIGE RIVER PLAIN, BETWEEN ROVIGO AND ADRIA (ITALY).

Silvia Piovan[1], Raffaele Peretto[2], Paolo Mozzi[1]

[1]Dipartimento di Geografia, Università degli Studi di Padova, via del Santo 26, 35100 Padova (Italy); e-mail: paolo.mozzi@unipd.it, silvia.piovan@unipd.it
[2]Comune di Rovigo-Museo dei Grandi Fiumi, piazzale S. Bartolomeo 18, 45100 Rovigo (Italy); e-mail: raffaele.peretto@comune.rovigo.it

The Adige river is the second Italian river by length (410 km) and the third by catchment area (12,200 km²). It enters the Venetian plain near the city of Verona, after having followed a long, north-south alpine valley cut in carbonatic and porfiric rocks. During the Last Glacial Maximum the Adige valley was occupied by a glacier down to the valley mouth. In the so-called "high plain", near the Alpine piedmont, the river is of the braided type and the deposits are manly gravels and sands. Downstream, in the "low plain", it becomes meandering and the sediments are finer. Here, the sandy-silty-clay Adige alluvial plain interfingers with the adjacent sedimentary systems of the Po, Tartaro and Brenta rivers. Since the 16th century the Venetian Republic started a land reclamation program with embankments and artificial cut-offs. Due to river management, the recurrent floods, avulsions and crevasses which chararcterized the Adige have greatly diminished in the last centuries. The study area is delimited respectively to the north and to the south by the Adige and Po rivers, between the cities of Rovigo and Adria. This stretch of low plain consists of late Holocene fine sediments. Some portions belong to the Adige system while others to the Po one. From the archaeological viewpoint, the area is of great interest due to the presence of important Bronze and Iron Age sites and the existence of an exceptionally well-preserved roman *centuria*, which extend for 250 km^2 from the city of Rovigo almost until the Lagoon of Venice. This research aims at understanding how the geomorphic and palaeoenvironmental evolution of the area influenced the pattern and typology of the ancient settlements and infrastructures. Prominent cases are the areas of Villadose and Pettorazza Grimani, where the investigations are respectively focussed on a crevasse splay and a palaeo-meander.
The investigation takes advantage of several methods: microrelief analysis, remote sensing and boreholes down to the depth of 4-6 m with the Edelman hand auger. Microrelief analysis is one of the main tools for a geomorfological study in a flat land. A digital elevation model of the investigated area has been developed, based on contour lines with spacing of 0.5 m. For the construction of the contour lines, a manual interpolation of spot heights of the "Carta Tecnica Regionale del Veneto" at scale 1:5000 has been adopted. This method allows to discard the points that are related to present-day artificial structures and to reconstruct the natural swale-and-ridge systems. Landsat TM and Aster images have been used for the analysis of major landforms while photo-aerial interpretation has been essential for a finer detection of natural morphologies and anthropic structures. Both oblique and vertical panchromatic aerial photographs were used to map the palaeochannels and crevasse splays, and to analyse their interaction with ancient roads, ditches and field patterns.
In the Villadose case, aerial photos and microrelief indicated the convenience of drilling a set of boreholes across a crevasse splay placed north of the *decumanus maximus,* nowadays called "*Via per Villadose*". The cross section shows the interactions between the crevasse splay and the roman road; it also clarifies the environmental conditions before local centuriation took place. Pettorazza Grimani meander site was chosen because of the absence of present-day dikes (the meander cut-off dates back to 1783) and the presence of an abandoned road of unknown age, about 30 m wide, with side ditches, which follows the palaeochannel. The cross section of the meander shows a complex sedimentary history characterized by different depositional environments. Periods of rapid aggradation due to the upbuilding of fluvial ridges and crevasse splays alternated with periods when overbank clays and peats accumulated with low depositional rates. The radiocarbon dating of the backswamp peats, currently under way, will help to date the major avulsion which, during the Middle Ages, led the terminal stretch of the Adige river to follow the present direction.

EXPLORING THE ARCHAEOLOGICAL LANDSCAPE THROUGH A LOCAL PERSPECTIVE: SPACES AND PLACES IN THE PREHISTORY OF THE FLORENTINE PLAIN

Giovanna Pizziolo and Lucia Sarti

Dipartimento di Archeologia e Storia delle Arti – Università di Siena

A large number of rescue excavation carried out since 1982 in the Florentine plain, and in particular in the Sesto Fiorentino area, by the Universities of Siena and Firenze provided a huge quantity of archaeological data which are currently being input into a GIS. The setting up of the system, beyond the importance of building up a geographical and archaeological framework to be commonly shared by laboratories of university and local community administrations, is the fundamental step to perform landscape archaeology analysis. Different kind of sources are nowadays available for our study area: in particular we are focusing our attention on stratigraphic information (derived by archaeological excavations and a large number of preliminary test pits), aerial photographs (historical, present day colour and infrared photos) and historical cartography. Aim of the project is to relate these sources throughout an integrated approach getting further the simple juxtaposition of data. The combination of archaeological, geomorphological and stratigraphical information is oriented towards the definition of prehistoric landscape. Since the fifth millennium B.C. a progressive peopling of the area took place according to different modalities. In the course of this settling trend the plain was densely frequented especially from the third to the second millennium B.C. and during the Bell Beaker phases. The research is actually focused on exploring the relationship between the changing dynamics of landscape and settlement strategies occurred in the plain environment. In order to perform this analysis, a local perspective needs to be undertaken concerning the interpretation of possible peculiarities of such a system both at a micro and macro scale. Since prehistoric time the study area exerted a continuous attraction as a place for settling activities which is widely testified by numerous archaeological sites often characterised by multiperiod evidence. Was the Florentine plain an attractive "space of resources" to be exploited and/or a "space of living and meeting" to be crossed over? We would propose to examine this continuous settlement activities starting from the local characteristics which may have played a role in the process of this appropriation of landscape.

INTEGRATED TECHNOLOGIES FOR THE RECONSTRUCTION OF THE ANCIENT LANDSCAPE OF POSEIDONIA-PAESTUM

A.Pontrandolfo, A.Santoriello, L.Sirangelo, F.U.Scelza, M.Marinaro, L.Pugliese, S.Scarpetta A.Esposito- Centro

This work has been developed within the Innova project (Regional Centre for the Development and Transfer of the Innovation applied to Cultural and Environmental Goods) whose leading subject is the University of Salerno which has stipulated an agreement for the purpose with the "Soprintendenza Archeologica delle Province di Salerno, Avellino e Benevento". Its main goal was the definition of mentioned patterns relevant to the ancient landscape of Poseidon-Paestum and of its territory, and to the forms and ways of employ of the human groups. The present work also provides monitoring features for the tutelage and valorisation of the cultural resources. The attention was particularly focused on environmental analyses carried out through the use of integrated GIS test. Among the several features the so-structured system uses multi-spectral satellite images on which some algorithms have been applied in order to confirm cognizance and to point out the environments which humanity faced in the course of time. This work aims to illustrate some techniques of analyses applied to Ikonos II imagery.

The interested area is the territory of the city Poseidon-Paestum and especially the southern part of the alluvial plain of Salerno gulf. This area is bordered northerly by the Sele river, easterly by hilly reliefs – such as Calpazio, Sottano, Soprano – and southerly by Agropoli hills and the Capodifiume river – which laps the city on its southern part – and the Solofrone and Testene rivers. The urban centre of Poseidon, which arose around 600 b.C. corresponded to the calcareous platform slightly lifted above the sea level and situated in the middle of the southern plain (Sele left bank side). The boundaries of this area are defined by the necropolis, which, from their foundation until Roman colonization (273 b.C), surrounded the city both in the North (Arcioni-Gaudio and Andriuolo) and in the South (Santa venera-Spinazzo on SE and Licinella on SW) side.

All the information derived from the excavations of this site - which have been carried out from the beginning of '900 onwards - led to the creation of an Informatics System. This system was able to elaborate – on a regional and territorial scale – all data coming from several sources and their relevant analyses' methods and techniques which led to the creation of synthetic products, such as thematic geo-referential maps, both territorial and detailed, with both recognized and interpreted elements. The numerous and different used tools were managed within the GIS (Geographical Information System), by using a unique cartographic projection system (Gauss-Boaga, East Area) and by realizing an informative topographical basis through several multi-temporal and multi-scale tools and the use of different informatics treatments. A list of them is traced here, departing from widely detailed (great scale) data to small scale data:

- planimetries of excavations from the Archive of the Archaeological Supervising Body;
- vectorial reproductions of the ancient inhabited Poseidon-Paestum area, resulted from the excavations, plotting and restoration from past century up to nowadays;
- cadastral maps of the historical records, on a 1:2000 scale;
- detailed photogrametries on 1:5000 scale provided by Capaccio municipality;
- Italy geological maps on 1:100.000 scale, whose information were used as a departure point for defining the geo-environmental analysed territory;
- air photographs from past and present flights. We especially focused on photograms and films relevant to the flights made in 2002 and 1944;
- Landsat satellite imagery with a soil resolution of 30X30m, Spot with soil resolution 5X5m, Ikonos 2 with soil resolution 1-4m

As stated before, the main objective of this paper is the discussion of some results obtained in the satellite environmental analysis part of the project.

FROM SPACE TO PLACE

While the value of satellite imagery to archaeology is increasingly apparent, most current applications involve its use for environmental reconstruction.
Landsat Thematic Mapper imagery has been used to aid the identification of environmental factors, particularly useful in landscape archaeology approaches. Because of the relatively low spatial resolution of Landsat multispectral data (e.g. 30 m for Landsat multispectral data), these are of limited applicability for the analysis in sites where human activity has radically transformed (and in most cases fragmented) the aspect of natural landscape.
However the continuing improvement in the resolving power of the last generation of satellites is changing the possible uses of satellite imagery so that in the right circumstances the information drawn from, for example, Ikonos-2 imagery is beginning to stand comparison with that of vertical air photography. In the present research, in order to take advantage of both regional information driven from Landsat support and territorial details contained in Ikonos and air photograph and having to bridge the two different scales, we acquired also a Spot 5 panchromatic scene.
One of the useful approaches to the analysis of satellite multispectral images involves the grouping of image data into a finite number of discrete clusters or classes that identify the distribution, over the land, of environmental factors such as soils, vegetation types , waters, and artificial surfaces. However classification of intricate, high-dimensional spectral signatures has turned out to be far from trivial (both in the environment analysis and in the archaeological field). Discrimination among many surface cover classes, discovery of spatially small, interesting spectral species proved to be an insurmountable challenge to many traditional clustering and classification methods. By customary measures (for example by means of linear transformation such as the Principal Component Analysis, PCA) the spectral dimensionality of multispectral images can be reduced. Yet dimensionality reduction to such low numbers has not been often successful in terms of preservation of important class distinctions. A faithful topological mapping is necessary to preserve the information contained in the multispectral image.
The spectral bands, many of which are highly correlated, may lie on a low-dimensional but non-linear manifold, which is a scenario that push research into advanced and novel approaches for remote sensing image analysis and, in particular, neural networks.
That's the reason why, having to face an environmental analysis on a territory, the Paestum Area described above, which appears strongly fragmented nowadays, the neural classifier issue, and in particular the Self Organzing Map (SOM) one seemed to be the right attempt.
The Kohonen's Self-Organizing Map (SOM) is an unsupervised neural network algorithm that has turned out to be an efficient tool, in various applications, for data exploration tasks. Recently it has been used also for the analyses of remote sensing multispectral images with quite successful results. It carries out a nonlinear mapping of input data onto a two-dimensional grid of neurons or nodes. The mapping preserves the most important topological and metric relationships of the data.
In this paper we employ the Self Organizing Map (SOM) as a strategy for an unsupervised analysis of IKONOS MultiSpectral (MS) images over the archaeological area. The segmentation goal was to identify both environmental factors (soils, vegetation types, water,...) and manmade structures.
In order to ease the successive photo-interpretation of the image segmentation, the Sammon Algorithm was employed to project the SOM prototypes into a three dimensional colour space, thus defining a Look Up Table (LUT) that preserves in its colour composition the topological information exploited by the SOM strategy
A IKONOS 2 Multispectral (MS) image from January 2004 was processed for this study using IDRISI Andes ® (version 15, released April 2006) geo-analytic and image processing system (http://www.clarklabs.org), which includes among its New Features, a SOM module conceived to perform unsupervised classification tasks. A successive analysis- interpretation step was performed by importing the segmentation results in a PCI Geomatica © (version 10) geo-processing environment (http://www.pcigeomatics.com).
Comparing the results of the SOM Classification with the prior knowledge of the territory we found that the thesis that the final distribution of colours over the segmented scene is such that similar hues identify objects that are homogeneous in a physical sense (for example different types of vegetation, or different types of rocks) is fairly confirmed. For instance, a group of only two shades of yellow isolate all artificial surfaces (temples, buildings, greenhouses, streets, and rails). In the same manner, the sea is represented with three shades of Blue, and these different shades are probably due to different sea levels.
Some few mixed classes still remain. We believe that with the insertion of additional information such as textural information it will be possible to better separate these mixed classes.
Indeed the insertion of textural information is one of the future step of our project, also in view of the ability of the SOM network to extract information from high-dimensional dataset.
With the same aim of refining the outcomes of the Ikonos 2 segmentation step , the macro classes obtained have been compared with the other satellite analysis products realized from Landsat and Spot and also with the results of photo-interpretation conducted on air photos at different scales and taken in different dates. This refining task also takes advantage of the above analysis results with information extracted from literature and archives.
As an illustration of the procedures and methodologies herein adopted, an interpretation is given of the distribution of Necropolis in the northern part of the town-walls . This is a discontinuous distribution along an axis that goes from the ancient Town to the Heraion Temple situated at the mouth of River Sele.
The choice to realize a specific product for the necropolis came from the fact that the actual amount of funerary documents seems to be the most relevant amongst the existing ones for a study of the Paestum city. Secondly, the analyses of the funerary areas was important in understanding the organization of the entire Greek polis.

3D VISUALIZATION Of ARCHAEOLOGICAL PLACE Of CORZANO). DOCUMENTATION, SEARCH, COMMUNICATION MEDIUM

Paola Puma, Carlo Battini, Lorenzo Bianchini, Francesca Concas, Michele Cornieti, Francesco Tioli

Extents, purposes and context of the research
The report we are to present is the result of a former research, which was commissioned in order to constitute the documental basis for the material and signification recovery of Corzano –a fortified settlement on mountains between Toscana and Emilia-Romagna.

This research was conducted through the study and the documentation of surviving fortifications in a place, which, during the centuries, has represented an important presence in the life of the community and of the surrounding landscape.

The ranging and the proliferation of fortified settlement and garrisons along the main roads interests the communities on the Tosco-Romagna mountains especially between the XIIth and the XIVth centuries. The seigniorial powers with military origins and the same monastic orders rise and take roots (until the Florentine conquest, which is completed at the beginning of the XVth century) in peripheral territories of the imperial and ecclesiastical jurisdictional seats, faraway from big urban centres, but with an important strategic role for the viability connecting the Po valley to central Italy.

Here, the exercise of power is executed controlling the ancient *romee* roads, which pilgrims and armies from the north ride to reach the central Italy through the passes of the *Alpes appenninae*: in fact, along these roads are settled many *castra* and *rocchae*, as well documented by the contemporary sources and the current remains of material structures.

At the end of 2005, the Municipality of Bagno di Romagna, in the province of Forlì-Cesena, has commissioned a research to the Department of *Progettazione dell'Architettura* of the University of Florence for the creation of documents and the formulation of hypothesis on the recovery of constructional works in Corzano, an outstanding example of a garrison with a military eminent character that controlled important trans-Apennine roads conducting to the Val di Bagno.

Articulation of the research

The surviving constructional works in this place, the ruins of the Fortress (which is datable to XIIth) and the Sanctuary (which is documented since XIII th), have been surveyed at the beginning of 2006 during three thematically articulated campaigns, with following base restitutions and thematic elaborations aiming to the critical evaluation of a range of aspects in connection with the possible individuation of remains, which can conduct to the typical settling fortified typologies of that territorial context, the specific individual settling and territorial control modalities, morphology, building techniques, the possible original configuration of the castle.

The plan of work has been drawn considering different steps their own methodology, tools and proceedings: but these steps are characterised by strong integration, aiming to the production of descriptive elaborations of the elements complexity, though they are intentionally simplified and friendly designed for not specialists final users.

The whole work has been articulated in four steps:

1. Data collection: creation of a data base, documentation and integrated survey;
2. Thematic restitution: morphological, metrical and material data;
3. Critical elaborations: structural, building, geometric and preservative consistence of the constructional works;
4. Planning and restoration proposals.

Data collection

During the first organizing step, the structure of the data base has been created: at first the historical documentation had to be collected in this data base, in order to be systematically ordered, configured and referred to for the following creation of links among historical, surveyed, reinstated and elaborated data.

Integrated survey

The planning of the integrated survey campaign has been necessary to the collection of data from:

- Digital survey of the fortress, by means of Leica HSD 2500 laser scan with ToF (time of flight) technology;
- Photogrammetric survey of the Sanctuary;
- Topographic survey, by means of a total station Leica 706 no-prism, for the necessary references;
- Direct survey, supporting the photogrammetric survey.

Thematic restitutions and critical elaborations

In this case, the particular morphological and environmental conditions of the place, with the intrinsic characteristic of the constructional work and its grade of preservation, asked for and indispensable integration among the different tools, surveying and data interpretation procedures. In fact, every methodology, from the most traditional to the most technically evolved one, has been chosen because it was the most suitable to investigate specific aspects. The surveying campaign and the following interpretation of the Corzano Castle's structures have put in evidence the necessity of a systematic confrontation during the operations between the data that were collected by means of the laser scan technology and those collected with the traditional investigating tools for architecture and archaeology, such as the analysis of geometrical configuration and stratigraphy of the fronts. The undifferentiated and massive collection of the metric data through the laser scan technology on one hand gives a complete, precise and fast survey as unimaginable before, on the other hand shows the lack in the surveying action of the fundamental intellectual operation: the critical discrimination of the constructional work. During this necessarily autopsic step, the surveyor is immediately bound to read and to formulate interpretations on the investigated object, while with the laser scan technology the operator risks to remain at a superficial understanding of the object: in fact, the cloud of points model does not supply an exhaustive picture of the possible information of the constructional work, but its efficiency appears as it is supported by other investigative tools, so that it can clarify complex situations through the three dimensional view.

In the investigation on the Castle of Corzano, the interpretation of the complete three-dimensional cloud of points model has put in evidence discontinuities and anomalies in the plans and in the walls alignment, which could not be perceived with traditional tools, due to the characteristics of the place and the inaccessibility of some structures. Their congruence has been verified through the information of the traditional reading of wall disposition.

This has helped in the individuation of late or restoration interventions, which sometimes have modified the perimeter of the original plan in the fortified settlement. In those contiguous walls portions, which have been realised in different times, the different superficial degradation of the material do not seem to affect the view of the cloud of points with false colours, due to the material reflectivity (a local sandstone).

From the drawing to the 3d model: three-dimensional view as analysing, verifying and communication tool

The collected data that have been collected during the steps 1 and 2 have been elaborated and divided into different database, which have been related under finalized rules.

The first use of the collected information has allowed the production of a wide range of graphical elaborations, constituting the necessary documental base for the understanding and the interpretation of data. More, they represented synthetic documents of the analysis that have been conducted in a direct contact with the place, such as the stratigraphic interpretation of walls, the geometric

and measure relations among the parts, interpretative keys which have contributed to clarify the historical and constructive facts of the structures that have determined its current configuration.
The second use of these collected data has involved the regulation of tools allowing the analysis of the examined object by means of interactive systems that is manageable by final user too. These tools have been defined in a plan that foresees the creation of three-dimensional digital models with different scale of detail: textured VRML model, average detail definition model, high detail definition model.
An important goal of the working group has constantly been the will of using an interpreting and representing system of surveyed data where the three dimensional visualization is not a simple representation of both the current state and the hypothesis of reconstruction, but a tool for investigation, analysis and scientific control.
Under this profile the study of the place has been conducted creating an easy manageable general model, which is indispensable for the morphologic and geometric analysis of the constructional work. Later, the detail definition has been improved to investigate the building, stratigraphic and structural characteristics: this has allowed the formulation of a former hypothesis on the genesis and the development of the building, which has converged on the final model. The final model has been set on an average scale of representation, in order to better suit for a fast visualization within a web browser.
The third use of the surveys has been the production of a virtual model of Corzano place, as a synthesis of the study and allowing the proposal of the castle's original configuration.
The purpose of the different scaled three dimensional representations rises from the necessity of producing elaborations that can be used not only on dedicated systems with suitable software and hardware, but also within a wider project that indicates the web as source of information exchanges and cultural growth.
In fact, the planned research included the creation of informative web pages.
The elaborated three-dimensional models have been used as a tool for immediate visual communication, with links to other pages with more definition and information. This simple and intuitive tree-structure allows both the simple visitor and the expert to concentrate his attention also on detailed portions of the constructional work, without loosing the starting general configuration.
The plan foresees the possibility of interaction with data banks for admitted people, who can insert annotations and results of researches on analogous themes.

AIRBORNE LASER SCANNING OF CULTURAL REMAINS IN A NORWEGIAN FOREST

Ole Risbøl

The Norwegian Institute for Cultural Heritage research

This paper will present some results achieved in an ongoing Norwegian research project using airborne laser scanning to map archaeological remains in forest areas. Forest covers approximately 38 % of the Norwegian land area. We know that these forest areas hold a great number of cultural remains of which only a fraction is surveyed and mapped. Some major archaeological projects carried out in Norwegian forests and outfield areas during the last couple of decades have demonstrated a long term and versatile human use of forest resources. The poor situation with regard to records of cultural remains in forests is critical especially with a view to the threat of destruction that these remains are exposed to primarily due to forestry.

To carry out systematical surveys of these large forest areas using the common raise a posse (or search party) method is not considered a realistic solution to the problem because of the expenses connected with having several search parties walking in the forests. Thus remote sensing is an alternative method and especially airborne laser scanning has a large potential as a method for improving the survey situation. As opposed to satellite remote sensing and ordinary aerial archaeology using vertical or oblique photos, laser scanning offers an opportunity to reveal remains below the forest canopy. This is because of the ability of laser beams to penetrate vegetation which makes it possible to map small height variations on the ground surface even in areas covered with vegetation. The laser data collected in this manner can be linked with digital image-processing techniques in order to create relief maps and digital terrain models (DTM), showing all kinds of anomalies on the ground surface. This remote sensing approach makes it possible to scan large areas which can be analysed digitally in order to point out for instance cultural remains. It is an assumption that the remains are visible on the surface and it is a challenge to make a distinction between cultural remains and natural features. To overcome this problem different digital techniques can be used to manipulate the data in order to optimize the interpretation of the anomalies shown on the maps and models.

In our project a research area was chosen in SE-Norway. The area which comprises 10 square kilometres was laser scanned and the data collected constitutes the basis of our analyses combined with the results of an ordinary raise a posse (search party) survey done a few years ago covering approximately 20 % of the area. The dominant human impact in this area consists of remains of pre-industrial iron production dating from the late Viking Age to the early Middle Ages (app. AD 950-1250). This production, based on bog ore and charcoal as the main resources, has left slag heaps at the production sites and charcoal pits in the surrounding areas. The charcoal pits are rather uniform as far as size and form is concerned while the slag heaps are less uniform and less distinct. Tentatively our analyses will involve different approaches in order to sort out the advantages and disadvantages connected with different visual and digital analytic methods used in interpreting the models. The project and the preliminary results of these analyses will be presented in the paper.

AN EXAMPLE OF USE OF THE HISTORICAL AEROFOTOGRAFIES: THE CASE OF THE INHABITED AREA AND THE TERRITORY OF PONTECAGNANO (SALERNO-ITALY)

A: Rossi, A. Santoriello

Università degli Studi di Salerno - Dipartimento di Beni Culturali

Pontecagnano is a large Etrusco-campanian settlement in southern Italy (8 km South of Salerno) where there are more than 8000 graves from the 9th century b.C. to the Roman age. The Lerici Foundation drew the ancient town, with necropolis on three sides, by geoarchaeological prospectings. The archaeological excavations interested till now two preroman sanctuaries and the rests of *Picentia*, founded by the Romans in the 268 b.C.
The rests of houses delimited from two roads point out that the ancient town was according to a regular scheme.
New excavations for the Salerno-Reggio C. highway (about 2 km) are bringing to light houses and elements of the landscape from the 6th to the 1st century b.C.
The first study about the use of the landscape was led by M. Guy and A. S. Stefan on 1990 for the Archaeological Superintendence of Salerno.
This study, using Landsat TM, Spot Image and aerophotographies of different levels, pointed out two kinds of agrarian organization: one North 31°-33° East; the other North 45°Est.
To interact the new archaeological data with the ancient topographical context, it need to study the historical aerophotographical documents related to the flights: RAF, scale 1:10000, on August 1943 (ftg. 4166-4170); WWS, scale 1:38000 ca., on July of 1955 (ftg. 11337); IGM, scale 1:15000 ca. on 1956. With this documentary basis the analysis was both analogical (reading the stereoscopic pairs, identification of marks etc.) both digital with the aid of a WS Stereographic and of a scanner Vexcel 5000 and the realization of a DTM.
First, the study pointed to valorize all the anomalies (actual system of land use, the agrarian works, public structures), that changed the agrarian landscape with the disappearing of small ownerships.
Then, level by level (individualization, identification, classification, deduction), it was possible defining every mark corresponding directly and/or indirectly to the ancient agrarian use.
For that, we applied metrological rules to correctly understanding the areal development of the settlement, avoiding of drawing cadastres on photos or maps.

AIRBORNE MULTI SENSOR REMOTE SENSING OF EXPOSED AND SUBSURFACE ARCHAEOLOGICAL REMAINS AT ITANOS AND ROUSSOLAKKOS, CRETE

Aled Rowlands[1] and Apostolos Sarris[2]

[1]Institute of Geography and Earth Sciences, University of Wales Aberystwyth

A suite of airborne remote sensing techniques have been applied to the Itanos and Palaikastro (Rousollakkos) archaeological sites in eastern Crete, Greece, to assess their potential for locating known and unknown buried remains, and to delineate subsurface remains beyond the spatial limits of ground geophysical data in order to permit future targeted geophysical surveys and archaeological excavations. The sites have previously been the focus of intensive geophysical surveys. For Itanos, transects and grids of geophysical data consist of soil resistance mapping, 2D and 3D electrical tomography, magnetic mapping, electromagnetic mapping, seismic refraction and ground penetrating radar. These surveys have shown that archaeological remains at Itanos exist from the surface to 1-3 meters below the surface. Lately, large sections around the Minoan coastal site of Palaikastro (Rousollakkos) have been also surveyed by magnetic and electrical resistance techniques.
Airborne imagery was acquired over a two day period in April 2004 as part of the NERC Mediterranean Flight Campaign. Data from three primary sensors, namely the Airborne Thematic Mapper (ATM), the Compact Airborne Spectrographic Imager (CASI) and LiDAR will be considered. The CASI sensor is limited to the visible and near infrared (NIR) and was programmed to operate in spatial mode acquiring data in 15 channels from 449 nm to 940 nm at a spatial resolution of 2.0 m, whilst the ATM sensor acquired data in 11 broader bands, with a spatial resolution of 2.5m. The ATM sensor provides important information in the Short Wave Infrared (SWIR) and the thermal (TIR) portions of the spectrum with respect to soil properties and heat capacity. The LiDAR data were acquired using an Optech ALTM 3033 high-resolution airborne laser scanner with a point density of 1.0 m. Soil temperature profiles were acquired with a series of Skye Instruments sensors coupled to a data logger in an area adjacent to the archaeological site in order to calibrate and interpret the thermal data.
A range of processing techniques (e.g., Reed-Xiaoli anomaly detection) have been applied to the reflectance data derived from the CASI and ATM sensors as well as the intensity and elevation data from the LiDAR instrument. The fundamental premise is that buried remains are likely to alter the physical and chemical characteristics of the soil compared with those of the surroundings due to variations in soil depth and moisture, which may be manifested within the reflectance and intensity data.
Through a combination of CASI, ATM and LiDAR data, surface remains have been classified and mapped effectively using an object oriented approach. Whilst surface archaeology has been well mapped by the various sensors, the delineation of subsurface remains based on remote sensing remains problematic, although tentatively it is claimed that certain features have been identified. Difficulties stem from the presence of outcropping rock, vegetation and mixed pixels in key archaeological areas. The study has shown that multi- and hyper- spectral imagery are best employed within a multidisciplinary approach incorporating other remote sensing tools. A key consideration in being able to determine the presence of both surface and shallow subsurface features is that the remote sensing takes place at a scale appropriate to the spatial resolution of the objects. The advantage of opting for the high spatial resolution configuration is that data are acquired across the spectral range at a resolution close to the actual spatial patterns of structures such that the likelihood of unmixed pixels is greater. Even so, in this particular case study, a spatial resolution of 2.0m for CASI and 2.5m for the ATM was still slightly too coarse to detect known subsurface remains. Remote sensing can therefore complement geophysical and surface surveys and has the potential of delineating areas for further detailed field based investigations.

ANCIENT MAN IN PALMYRA –SYRIA AND THE PLEOGEOGRAPHICAL SITUATION IN QUATERNARY
Mohamed Rukieh

Nobody can doubt about the importance of studying ancient man and its effect in serving better understanding to the development of human being civilizations. We can't make a comprehensive study for this subject without studying the ancient man in Syria, where it is known as the "cradle of civilization". In this study, we will focus on the ancient man in Palmyra.
An important pleogeographical study has been carried out using images and field works to Palmyra area in the Quaternary age, where its deposits spread out in many sites since lower Pleistocene till recent age. We carried out this study for discovering and analyzing the existence of ancient man in Palmyra; determining climatic changes, which affect deeply man's existence, the role of Al-Moh Lake formation, in addition to determining the living sources which were common at that time and helped man in continued living for thousand hundred years.
The research showed that ancient man has been lived in this area since more than 750 thousand years. His existence essentially, associated with the rotation of the five Ice Ages during Pleistocene continued till the last ten thousand years. The formation of the fresh Al-Moh lake in Tadmor - during Pleistocene and its expansion during middle Pleistocene, where it reached 600km2- help the availability of suitable circumstances for living,existing of natural sources (vegetable, animal, water, rock) on which ancient man depended in his living. This matter has developed later in Neolite, at the beginning of agrarian villages formation since 8 thousand years B.c. using rocky quarries since more than 2 thousand years in North Tadmor within upper Cretaceous rocks and in its South within Neogene for raising developed cities, especially Tadmor where its remains still until now. The climate began to change towards dryness during the last thousand years till arrives its recent shape. Fresh water of Al-Moh Lake changed into salt sabkha, and it is one of the salty sources in Syria now.

LASER SCANNER, QUICK STEREO-PHOTOGRAMMETRIC SYSTEM, 3D MODELING: NEW TOOLS FOR THE ANALYSIS AND THE DOCUMENTATION OF CULTURAL ARCHAEOLOGICAL HERITAGE
Paolo Salonia, Serena Scolastico, Valentina Bellucci

CNR - Istituto per le Tecnologie Applicate ai Beni Culturali paolo.salonia@itabc.cnr.it

The critical process of ancient monuments analysis and conservation has been facilitated by the introduction of Computer Science in the field of documentation and survey of architectural heritage. Indeed, computer-based techniques have strongly modified the acquisition phase as well as the subsequent operations such as computation and management of all the information necessary to describe usefully the features and peculiarities of buildings and environments.
New survey technologies (quick stereo-photogrammetric systems, laser scanner) allow to acquire and elaborate a great deal of geometric data; processing data software guarantee the control and the utilization of the same geometric data in 3D visualization environments both raster and vectorial. The set of all these innovative tools lets administrative bodies, entrusted to preservation, safeguard and valorisation of cultural heritage, to avail themselves with great levels of documentation and then of knowledge.
How much said becomes more important in the case of archaeological artefacts reconstruction, where the difficulties of survey are diversified and the complexity of representation, transfer and communication of acquired information, directed or not to a specialist user, is of particular value to the goals of *sharing* knowledge.
A quick non-conventional stereo-photogrammetric system was first of all applied for the complete stereoscopic coverage of different ancient archaeological artefacts. It is a digital mono-camera system which has the same advantages of a bi-camera based system and allows stereoscopic acquisition that can be used directly for restitution. This system guarantees the same accuracy of geometric data acquisition as a traditional stereoscopic method, which is fundamental in the process of diagnosing the state of conservation. Although the system does not need topographical support, the latter was nevertheless provided in order to verify the reliability of the test system, the accuracy of the results obtained and to set the stereoscopic models properly in real space. All the photographic stereoscopic pairs have been processed using dedicated software that allow control points to be added (natural and targets deriving from topographic survey) in order to minimize errors in overlapping stereoscopic models.
The final result is composed of metric raster stereoscopic model, explorable in a stereoscopic environment, where it is possible to perform three-dimensional measures and to survey conservation status. In this way is possible to avoid the vectorial phase which often represents a subjective abstraction from the truth: the raster models provide a corpus of information deriving from the geometrically controlled stereoscopic pairs, which are qualitatively and quantitatively richer than a traditional survey. This tools make it possible to investigate the whole construction with a very high degree of accuracy and with certain geometric rigour and, at the same time with levels of subsequent zooming to "approach the object" made possible by the high definition of the photographic apparatus which would be inconceivable in a visual inspection of the real object.
Moreover these stereoscopic model can be further on processed in order to obtain orthophotos of the artefact elevations, vectorial restitution of conventional graphics representations (plans, front figures, sections, etc.) and at last geometrically controlled three-dimensional models that could be spatially inserted in the archaeological landscape, also virtually reconstructed and explorable in real-time, with the possibility to integrate, to the geometric models, descriptive information coming from different disciplinary fields.
Laser scanner technology has also been applied for the documentation of ancient artefacts that had a complex and difficult to survey geometry. Indeed this technology allows to obtain points cloud that represent an objective and discrete representation of the artefacts and that can easily and quickly describes complex geometry surfaces. A punctual three-dimensional survey, as that you can acquire by the laser scanner, allows to take the x-y-z spatial coordinates of a lot of points of the studied surface, whose reciprocal distance can be fixed by the operator in relation to the need for precision and quality of the survey product. The processing of this great number of points and the subsequent definition of the surveyed surfaces permits to describe, analyze, and "communicate" the studied monuments through digital models from whom it's possible to obtain measurements, projections, sections that can be utilized for historical-architectural analysis and for other specific studies of the artefacts themselves.
For the survey activities a Leica HDS2500 laser scanner (former Cyrax 2500) has been used. It is a time of flight laser scanner, able

to scan objects up to 100 meters with a single point accuracy of 4 millimeters (although maximum recommended range is 50 meters). Moreover Cyclone software of Cyra Technologies has been utilized for the acquisition, the registration and the alignment of single points clouds captured from different scanner position and necessary for the complete coverage of the different part of a monuments. Through the use of this laser scanner the complex geometry of ancient artefacts, which remains are represented only by the internal construction in *opus caementitium*, has been realized. The use of this advanced techniques has guaranteed optimized working times and thus cost optimization and, above all, more rigour and precision of the geometric data acquired then that one that would have been obtained by the use of traditional topographic survey systems. The acquired points clouds were in fact imported in an AutoCAD environments, where, by the use of a special plug-in furnished with the software Cyclone, it was possible to obtain a great number of horizontal and vertical sections of the artefacts, necessary for geometry object understanding.
From the acquired data it would be also possible to realize three-dimensional virtual wireframe model and then to make render of them, applying materials information. In this sense in order to obtain photorealistic model, i.e. an as realistic as possible affinity between real and virtual materials, it would be interesting to texture the model with the photos obtained from the stereoscopic acquisition. In this way it would be possible to realize a complete description of the whole geometric, dimensional, chromatic and materical information of an architectural building.
Different ancient artefacts have been chosen for being surveyed both by photogrammetry and laser scanner in order to compare the results of these measurement techniques, to define which method would be the best for specific kind of monuments and at last to experiment the possibility of combination of these two technologies.
This paper focuses on multimodal approaches, through the presentation of experimental results deriving from the application of innovative instrumental survey systems to the documentation of ancient monuments of the Appia Antica archaeological park, in the framework of a joint research project carried out by CNR Institute for Technology Applied to Cultural Heritage (CNR-ITABC) and the Municipal Archaeological Superintendence of Rome. The Public Administration has required a reliable scientific documentation finalised to the realisation of an Informative System and 3D explorable and navigable environments of the Appia Antica archaeological heritage. The necessity of this scientifically reliable documentation, leads to the experimentation of the integration of different survey technologies.

AERIAL ARCHAEOLOGY IN SPAIN: HISTORIOGRAPHY AND EXPECTATIONS.

José Carlos Sánchez Pardo, Iván Fumadó Ortega

Escuela Española de Historia y Arqueología en Roma-CSIC)

The work has two main objectives: First, we would like to carry out a critical revision of the underdevelopment of the Spanish Aerial Archaeology, with a general overview of its characteristics and their current uses, and remind some basic ideas about the character and objectives of these studies. In second place, we would want to analyze the special possibilities of the Aerial Archaeology in our country, studying the different types of grounds and terrains in Spain and its specific potential for Aerial Archaeology studies. Through these two main lines we would like to create a basic and necessary departure point for the later development of the Spanish Aerial Archaeology.
We want to stress the great potentiality the Aerial Archaeology has in the Iberian Peninsula. On one hand and looking at physical questions, the country has large areas without heavily built-up urban environments, vast cultivated fields and scarce vegetation on the ground. The advantages of the wheather conditions aren't worthless, specially appropriate to fly the most part of the year. Paradoxically, the tentatives in this performative field had been rather shy. On the other hand and looking at already available documents, Spain also possesses an interesting collection of files of aerial pictures. Besides diverse series of historical pictures taken previously to the 60s, stands out the called *American Flight* (1955-1957), that for the first time photographed the whole country. That's obviously a privileged file because was filmed before the deep transformations and uncontrolled growth of the 60s and 70s (in strong contrast with other European countries, where changes had begun before). This collection of aerial pictures, although the flight height is a bit elevated, can contribute with a lot of historical information. However, there is not been thoroughly exploited in a systematic program of archaeological studies. Spain has also a wide number of images of modern flights, carried out periodically in the last decades by the *Instituto Geográfico Nacional* and for the *Servicio Cartográfico del Ejército* as well. It can be added the Internet available files served both by the *Ministerio de Agricultura, Pesca y Alimentación* and the *Ministerio de Fomento* of diverse recent aerial views and satellite images. It has to take on account that Spanish laws have not involved obstacles to the free use of the aerial pictures, to the contrary to other countries like Italy, where laws were restrictive.
As we have already pointed out, this great potentiality did not yet become visible for the most part of the Spanish Archaeology. The first impression after a brief overview of the bibliography is the limited quantity of publications on this topic. Specially dramatic is the absence of a manual about Aerial Archaeology in Spanish. Moreover, only a few foreigners works are little diffused, diametrically opposite to other countries where exist a large number of specialize reports.
In general, works made in Spain concerning Aerial Archaeology can be classified in three categories. In first place can talk of some few works of theoretical considerations to general concepts. Among these kind we can highlight the contributions of Sánchez Palencia and Orejas Saco del Valle (1991) and Orejas Saco del Valle (1995). In second place there are some specific projects that had had the aerial surveying and aerial photographic interpretation like main working method. The most interesting efforts in this class are those of Del Olmo Martín (1990, 1993, 1996, 1999 y 2001) in Castilla-León, as well as the aerial surveying campaigns of *Casa Velázquez* in different areas of the Iberian Peninsula between 1978 and 1982 (Bazzana and Humbert 1983) and the *pioneer* work of Roselló Bordoy (1970) in Mallorca. There are so many studies in the third group, which use some aerial photographies only in a complementary way (it can be found from a beautiful but simple illustration of the archaeological site until serious tentatives of aerial surveying without a realistic and systematic character). From this kind of approach we will mention only few examples (Ariño Gil and Rodríguez Hernández 1997, Urbina Martínez 1999). Thus far anything like the English programs (standing out those carried out by the *Aerial Archaeological Research Group*) or the Italien ones (with the commendable results of the *Area di Archeologia Medievale dell'Università degli Studi di Siena*). The only areas that had been quite intensive analyzed are the provinces of Zamora, León, Valladolid, Mallorca and Valencia. The country has another 46 provinces that remains unexploited from our point of view.

Therefore, it would be very interesting and productive to develop a group of works of revision of the wide photographic material existent combined with systematic aerial surveys in diverse areas of Spain. This is not simple expectation: there are multiple and concrete historiographic problems where Aerial Archaeology could contribute to its clarification in a decisive or, at least, in a very important way. We can mention some examples of diverse chronological moments, as the study of the fortified settlement in Iron Age, the search and localization of roman villages, the analysis of the early medieval settlement, the reconstruction of the network of medieval and modern castles... and much more topics.

But even more urgent than the development of this series of works, it would be necesary for the spanish archaeology to rethink and meditate on some background problems in this topic. At first place it would be convenient to clarify and to distinguish diverse concepts that frequently spanish archaeologists confuse: *arqueología aérea, fotografía aérea, prospección aérea, aerofotografía...* On the other hand it would be necessary to enter in another series of problems with a big "charge" of reflection in this whole topic: is Aerial Archaeology an specific discipline inside the archaeology or is simply another working technique?, which is the influence level of subjectivity in the interpretation of the archeological marks?... And linked with this, it would be also interesting to develop a bigger reflection about the possibilities that Aerial Archaeology can offer in the management and diffusion of the cultural and environmental patrimony, topic that in Spain and in general in all Europe recently has had an special development. Anyway, in this work we only can outline these topics and invite to reflect and dialogue about them, because, on one hand, these pages, as we have already said, only tries to be an starting point for new developments in Aerial Archaeology, and on the other hand, we know that the participants in this conference know all these questions perfectly.

In all these problems that we had outlined we consider that Aerial Archaeology should not be understood like a simple method for searching archaeological vestiges but rather as a type of scientific focus that can -and should- help us to understand and to recreate historical landscape, its internal structure and the relationships between its components. In this way, Aerial Archaeology is very near to the bases of Landscape Archaeology. Therefore, following this idea, Aerial Archaeology should also be diachronic, as Landscape Archaeology is, or should be: we should not looking for an specific type of vestige from one historical moment, neither we should not surveying or studying photographies with a preconceived idea in our mind, but rather it is necessary to analyze all the whole landscape and trying to extract all the information that it contains about the past, like a board where are printed diverse marks of different historical moments and the relationships between them. It`s in this point where we think that Aerial Archaeology can offer its biggest potential.

Nevertheless, our main goal in this report is to set where and when aerial surveying would be more efficient in the search for sites of antiquity in Spanish landscapes. We would like to make the most of the experiences that had been accomplished since several yeas ago in Europe, and that would be quickly apply thanks to GIS technology and available georeferenced files with accurate land cover data (CORINE Land Cover). In this way and according to Jones and Evans (1975) we can reasonably estimate an index depending on kind of culture (deepest roots and plough often are the best ones) geological data (soils with high reflectivity and arid composition in its first 30/50 cm are optimal) and also meteorological factors (calculating the moths with more potential transpiration average for each area). GIS software allow us to carry out geostatistical and spatial analysis, bringing us distribution maps of our index. This would show us how to planificate the future aerial surveys improving the efficiency all over the country.

MULTI-RESOLUTION IMAGE-BASED VISUALIZATION OF ARCHAEOLOGICAL LANDSCAPES IN PALPA, PERU

Martin Sauerbier,[1] Gerhard Schrotter,[1] Henri Eisenbeiss,[1] Karsten Lambers[2]

ETH Zurich, Institute of Geodesy and Photogrammetry, email: (martin.sauerbier, gerhard.schrotter, henri.eisenbeiss)@geod.baug.ethz.ch
[2]German Archaeological Institute, KAAK Bonn, email: lambers.kaak@gmx.de

In this paper we describe the visualization of a combined 3D model composed of three textured 3D models with different resolution and coverage of the archaeological region of Palpa, Peru, using a view and distance dependent mult-resolution visualization approach, and the use of the combined model for archaeological research at different scales.

The Nasca-Palpa Archaeological Project has been conducted since 1997 by the German Archaeological Institute (DAI, KAAK Bonn), the Institute of Geodesy and Photogrammetry of ETH Zurich, and further project partners from Peru, Switzerland, Germany, and Austria. The interdisciplinary project aims at investigating the prehispanic cultural history of Palpa in the Nasca region on the south coast of Peru, applying a wide range of methods and techniques from the natural and engineering sciences, among them photogrammetry and remote sensing. Archaeological research has been undertaken simultaneously at different scales, ranging from partial excavations of settlements to the study of sites in their local context, and finally to the investigation of changing settlement patterns at a regional scale. In the course of these investigations, various textured 3D models of parts of our study area were produced for different purposes by photogrammetrically analyzing imagery from different sensors. Eventually it was decided to visualize these different models as a single one, without actually integrating them, for purposes of visual archaeological analysis as well as presentation.

Imagery on which the three original 3D models were based ranged from digital aerial images at a scale of 1:4,000 taken by a digital CMOS camera with a resolution of 6 megapixels carried by a mini UAV (Unmanned Aerial Vehicle), to analog aerial images at a scale of 1:7,000 acquired from an airplane with an aerial camera during a traditional photogrammetric photoflight, and to ASTER satellite imagery with a footprint of 15 m. From each of these image datasets, digital terrain and surface models (DTM/DSM) as well as orthoimages were derived photogrammetrically in order to produce stand-alone 3D models, each of them textured using the corresponding orthoimages. The three individual 3D models can be characterized in terms of mesh size of the DTM/DSM, the area covered, texture resolution, and radiometric properties of the orthoimages.

The mesh size of the DTMs and DSMs derived from the different image datasets differs widely, as does the size of the covered area. The DSM based on the UAV images covers an area of 200 x 300 m^2 at a mesh size of 10 cm. The DTM derived from the aerial images covers 89 km^2 at 2 m mesh size, while the DSM generated from ASTER imagery covers several thousand km^2 with a mesh size of 30 m. The resolution of texture ranges from 3 cm for the orthoimage produced from UAV imagery, to 15 cm for the orthoimage derived from the scanned aerial images, and finally to 15 m for the orthoimage generated from spaceborne ASTER images. Image radiometry is another important factor when combining multi-sensor imagery due to its impact on the quality of visualization. The UAV images were acquired in RGB color space and therefore cover the whole visible spectrum. The analog aerial images and the corresponding orthoimage are available in greyscale mode, whereas the ASTER orthoimage results from a combination of the red, green and blue channels of the ASTER sensor, with the blue channel not covering the whole visible spectrum. This combination of the three channels is required in order to achieve a more realistic radiometry than the original RGB ASTER channels that optically overemphasize the red and green part.

The area covered by the DSM based on UAV images falls within the area of the DTM derived from analog aerial images, which in turn is part of the area covered by the ASTER images. As all three datasets were georeferenced in the UTM system (Zone 18S, horizontal and vertical datum WGS 84), no coordinate transformation was required. Nevertheless, care had to be taken to ensure that the different DEMs fitted together to avoid visible steps in Z direction as well as horizontal misalignments in X and Y direction along the borders of different datasets that may be caused by different accuracy of image orientations and DEMs. For this purpose, least squares 3D surface matching was applied such that the dataset with the highest accuracy, in this case the DSM derived from UAV images, served as a reference while the other two surfaces were registered to the reference by means of three translations, three rotations and one scale factor.

Due to inefficient memory management, most available commercial software for terrain visualization cannot handle large datasets consisting of multi-resolution DTM/DSMs and multi-resolution texture imagery. In order to overcome the lack of such a software tool for the generation of virtual flyovers, which are usually stored as image sequences in various file formats such as avi or mpeg, we developed a set of software modules based on a view and distance dependent multi-resolution visualization approach. These tools are suited to produce a high quality flyover in batch mode based on digital terrain data, represented as regular rasters, and orthoimages. Terrain and image data are distributed to several computers in a network, which allows memory constraints of single machines to be avoided. Furthermore, processing speed is increased by distribution of calculations, such as the determination of the field of view, the distance to the object of interest, visibility, triangulation of visible points, texture mapping and others more. The final rendering including anti-aliasing runs as well as a distributed process on different computers. Following this approach, it is possible to generate a smooth flight path, to merge multi-resolution DTM/DSMs and imagery online in order to obtain smooth transitions, and to generate a high resolution image sequence. The final result then is an image sequence where image size can be adapted to the resolution of the presentation device.

While the three original 3D models had already proven to be helpful tools for archaeological research at different scales (single sites, core study area, region), the combined multi-resolution model provides a broad view of the setting of single sites in their environment, of the study region in general, and of the relationships between natural and cultural features within the archaeological landscape. It is furthermore highly useful for the purpose of presenting and explaining the results of the archaeological investigations in Palpa. Due to the flexibility of the modeling and visualization approach, new stand-alone 3D models can be visualized as part of the combined model as they become available.

INTERACTIVE SEGMENTATION IN LARGE TERRAIN MODELS

M. Schneider, R. Klein

University of Bonn, Department or Computer Science II, ms@cs.uni-bonn.de, rk@cs.uni-bonn.de

We present aninteractivesegmentationtool thatpermitssegmentationdirectly withinlarge3d terrain environments created on the basis of air photography and digital elevation models. Since fully automatic segmentation is still an unsolved problem we propose a semi-automatic method that allows the combination of expert knowledge through user input and low-level segmentation techniques. The integration of the method into a 3d terrain visualization engine allows the exploration of the environment and segmentation of objects of interests at the same time. The visualization of the landscape in its natural 3d representation during the segmentation process makes it easier for the user to locate and identify objects of interest compared to the segmentation on 2d air photos as it is commonly done. The segmentation algorithm is based on Intelligent Scissors which formulates the boundary detection problem in an image as an optimal path search in a graph. The goal is to find the optimal path from a start node to a goal node where pixels in the image represent nodes and edges exist between neighboringpixels. Anoptimalpathis definedby theminimum cost path from a start pixel to a goal pixel using a weighted sum of gradient edge features as edge costs. The user canstartthesegmentationprocessby planting an initial seed point for which a path map is constructed. The path map provides the minimum cost path from the seed node to every other node. By moving the mouse a path from the seed point to the current cursor position is interactively displayed snapping to nearby object boundaries. If the path starts to digress from the desired object boundary a new seed point can be established by the user. This causes a new boundary segment to be extended from the new seed while fixing the previous boundary segment. In order to use this algorithm to map objects in a 3d environment we project the user input from 3dto the underlying air photo on which the algorithm is then performed. The resulting object boundary is projected back onthe3d terrainsurface providingimmediate feedback to the user.

The main problem that arises when applying the algorithm to terrain data is the sheer size of the air photos often comprising several gigabytes of data. This prohibits the application of the original Intelligent Scissors algorithm because it does not permit computation of the image gradient features and the path map in real-time for images of this size not to speak of memory requirements. We present an algorithmthat usesaquadtreerepresentationof thedata to drastically reduce computational costs and memory requirements. The segmentation starts at a certain level-of-detail (LOD) of the quadtree that can either be specified by the user or is chosen automatically. Within this level we make use of the tiling that is imposed by the quadtree. Instead of applying the algorithm on the complete image we only run it on the quadtree tiles sourroundingtheactual mouseposition. This domain is incrementally extended and the path map is updated accordingly when the mouse is moved across tiles not considered so far. An e□cient incremental update of the path map is achieved by utilizing the D* algorithm, a dynamic version of theA* algorithm. D* does not need to recompute the whole map but in most cases only a small subset of it. The computation of image gradient features is accelerated by performing the convolution with the appropriate kernels entirely on the GPU using framebu□er objects and fragment shader. After the computation of the object boundary on the start level we exploit the hierarchical structure of the quadtree. We propagate the solution from the coarse leveltothe nexthigherresolutionlevel. Hereweperform the segmentation onlywithin a narrow band surrounding the projected result from the coarser level. Since the algorithm is only run on the subgraph that comprises the narrow band, the additional computation at the fine resolution is significantly less than running it on the full graph. Additionally, since only thesubgraphhastobe establishedmemory consumption is also drastically reduced.

3-D VISUALIZATION OF ANCIENT PROCESSIONAL LANDSCAPES IN COSTA RICA

Payson Sheets,[1] Tom Sever[2]

[1]Anthropology, University of Colorado, Boulder CO USA
[2] NASA Huntsville, AL, USA

For decades one of us (PS) has been collecting cases of ancient Middle American societies affected by explosive volcanic eruptions. Sheets began with cases in southern Mesoamerica because he was doing his doctoral research in El Salvador. The sample of cases expanded to cover much of Mesoamerica because there were so many examples of eruptions affecting ancient societies in Central Mexico and the Gulf Coast area. He found that complex societies were often quite vulnerable to the unanticipated great stresses. Because complex society emerged early in Mesoamerica, compared to the surrounding culture areas, the sample was weak in egalitarian societies. Sheets therefore turned to Costa Rica to try to correct that imbalance, by adding cases involving less complex societies.

We began the "Proyecto Prehistorico Arenal" in the 1980s with the objective of exploring the reactions of egalitarian societies to the sudden massive stresses from the ten explosive eruptions of Arenal volcano during the past 4000 years. We were surprised to find these "simple" but clearly very sophisticated societies were exceptionally resilient to the disruptions caused by the big eruptions. We have expended considerable efforts to try to understand why and how these egalitarian societies could have developed such resiliency. Based on our efforts to date, we believe the key factors were (1) low regional population densities and availability of uncontested refuge areas, (2) alliances from exogamy and from multiple community feasting, (3) reliance on wild foods rather than a domesticated dietary staple, and (4) the household or sometimes the village was the decision-making unit rather than having centralized authority. Sedentary villages have been dated to as early as 3000 BC, with sophisticated ceramics and at least some maize agriculture. Secondary burials were placed with the households. A major cultural change occurred around 500 BC when villagers began burying their dead in cemeteries that were located at a distance from the village. Ethnographic accounts by traditional Cuna and Bribri Indians of lower Central America describe the spirits of the deceased needing separation from the noise of villages. That may provide a reason for that separation. And the spirits of the dead are powerful, and are better kept at a distance from the functioning community. The archaeological evidence is clear that people walked single-file in their processions to and from the cemeteries, for reasons that are unknown. The routes they followed were strikingly straight, as we found from mapping the ancient paths, and traditional ethnographic accounts may shed light on the reasons for that. Bribri accounts mention stretching a string to help guide the spirit from the community to the grave. An unanticipated consequence of generations of people following the same path is entrenchment, often to 2 to 4 meters below the surrounding ground surface. Using entrenced entrances into special places must have become a cultural standard. And such entrenched paths certainly focus one's attention on the objective, as the surrounding countryside disappears on both sides of the path. When one enters the cemetery it opens up to view, very much like emerging from a tunnel. All Arenal area societies apparently were egalitarian from the earliest (Clovis horizon, about 11,500 years ago) to the Spanish conquest about 500 years ago.

More complex societies did emerge to the east of the Arenal area, by about AD 500, as exemplified by the Cutris chiefdom. What could be called the "mentality of monumentality" develops when egalitarian principles no longer govern society, and the chiefly elite seeks ways to physically express their centralized authority. The chiefs at Cutris apparently chose the entrenched straight entryways that became so important in the Arenal area as something to magnify immensely in their search for monumentality. Chiefs mobilized large workforces to construct and maintain deep and wide entryways that connected satellite communities some 6-10 km distant with their site center. And we can only imagine the processions of people using the huge sunken entryways into the site center. The entryways are almost 50 meters wide, and many meters deep, as they disgorge their participants into the site center. Ring roads allow processional participants to connect with the radial entryways without being seen from the site center. The radial principal entryways and the ring roads are almost exclusively of earthen construction. As this is a tropical wet rainforest environment, with some 3000 mm of annual mean precipitation, it is unknown how erosion and water management were achieved.

The emergence of complex society at Cutris was followed by other chiefdoms emerging, including Fortuna, Las Mercedes, and Guayabo de Turrialba. These largely date from AD 1000 to 1500. The favored construction material of the entryways changes to stone, perhaps because of possible erosional problems of the earlier earthen construction. This later monumentality was achieved by slightly different means, but the net effect was quite similar to Cutris.

The paths were discovered in remote sensing imagery provided by Tom Sever of NASA. Particularly effective are color infrared aerial photography and IKONOS satellite imagery for discovery of the ancient paths. Other imagery that contributed include LIDAR, Radar, color and black-and-white aerial photography, and the Thermal Infrared Multispectral Scanner (TIMS). We have developed a field and lab methodology for confirming, and disconfirming, anomalies in the remote sensing imagery as ancient footpaths. The detailed tephrochronology of the ten large explosive eruptions of Arenal volcano assists in our dating the initiation and the cessation of use of a particular path or path segment.

Three-dimensional interactive visualizations have been constructed of the Arenal landscape, including the sunken pathways linking cemeteries with villages. The paths also link cemeteries with the sources of stone used in their construction. And the paths link cemeteries with springs, as people needed large amounts of water to support themselves in construction efforts and particularly in feasting and ritual. The 3-D visualization allows one to fly along or over a pathway and experience the topography in ways that cannot be done by any other technology.

The Arenal 3-D visualization was part of a broader program including Mesoamerica and Central America. To create the terrain file for Mesoamerica and Central America the research team selected 15-meter resolution imagery from Landsat 7 that covered the entire

region from southern Mexico to northern Colombia, plus selected IKONOS scenes, at one-meter resolution, for special areas of interest. The imagery was draped over 90 meter resolution elevation data from the Shuttle Radar Topography Mission (SRTM).
In order to better understand the context, function and relationship of ancient footpaths in the Arenal Region of northern Costa Rica and their evolution to the natural terrain and connectivity to ancient villages and sites, a three dimensional model of the terrain with overlaying features was developed The resultant visualization allows the investigator to "fly over" the footpaths at various altitudes, directions, and tilt angles and zoom in upon features of interest upon the landscape. The aerial vantage point leads to the realization that everything observable in the ancient landscape has meaning and interconnection, factors that sometimes cannot be appreciated from the ground. This gives the archeologist the ability to understand the complex evolution of human factors for settlement patterns, subsistence strategies, use of resources, and the social, political, religious, and economic relationships that connect the study of the past with the present.
The terrain model was produced from several data sources. High resolution, pan-sharpened IKONOS data were processed using ERDAS software to produce both natural and color infrared images at one-meter resolution. These high resolution IKONOS scenes and 30-meter resolution LANDSAT imagery were draped over a Digital Elevation Model (DEM) derived from digitized contour lines of the 1:50,000 scale topographic maps of Costa Rica using TerraBuilder - a 3D terrain construction application from Skyline Software Corporation.
TerraBuilder provides a three-dimensional spherical Earth with latitude/longitude coordinates as a base for holding imagery, elevation and vector data types. Because the IKONOS data were formatted using a UTM coordinate system, Terra Builder had to convert the files into latitude and longitude using built in conversion procedures. (Terra Builder is quite versatile in it's acceptance of a wide range of data formats including GeoTiff, .img files output by ERDAS Imagine, and MrSID compressed files to name a few.) The multi-resolution terrain file produced by TerraBuilder combines the imagery and elevation data with built-in imagery of the entire globe.
Using the resultant multi-resolution terrain file another Skyline application – Terra Explorer Pro (TEP) - was used to interact with the file, add and remove vector data, and add or remove other assets such as labels, images, and icons. For instance, the location of the ancient footpaths, determined from previous analysis of NASA airborne scanner data and color infrared photography, and ground verified using Global Positioning System (GPS) measurements, were input as vectors onto the terrain file. In addition, the location of ancient sites and cemeteries were also incorporated as vector layers. Terra Explored Pro was also used to create and record animated "flights" over the terrain and support analyses of the terrain through a series of tools that measure distance, produce contour maps, and locate best path to name a few. Due to the flexibility of TEP other GIS data can be added to the mosaic and any of the existing data layers can be toggled on or off.

THE POTENTIAL OF LIDAR IN ASSESSING ELEMENTS OF CULTURAL HERITAGE HIDDEN UNDER WOODLAND CANOPIES. POSSIBILITIES AND LIMITS IN DETECTING MICRORELIEF STRUCTURES FOR ARCHEOLOGICAL SURVEYS

Benoît Sittler

Institute for Landscape Management University of Freiburg (D)

Knowledge of the historic landscape is far stronger in agricultural areas than in woodlands because conventional aerial photography that has been the mainstay of airborne archeological reconnaissance is unable to detect finer features hidden under forest canopies.
As Lidar (Light Detection and Ranging) remote sensing enjoys the advantages of excellent vertical and horizontal resolution range and has the capacity to record the topography even under vegetation, it now offers news prospects for heritage assessment purposes. This technique has been tested in various projects for obtaining high resolution assessments of earthworks that survived in woodlands, such as ridge and furrow or burial mounds near Rastatt in South West Germany.
These examples show to what extent data obtained from airborne laser scanning missions carried out by the state survey agency of Baden-Württemberg for the purposes of upgrading the state's altimetric database can also be of use for heritage assessment and preservation purposes. Examples presented here and showing also the detection range for fine scale structures provided by this method may be helpful for optimizing any other airborne laser scanning approach (requested point density, use of method in mountainous areas, etc.).

REALITY-BASED 3D MODELING OF ANGKOR TEMPLES USING AERIAL IMAGES

Till Sonnemann, Martin Sauerbier, Fabio Remondino

Institute of Geodesy and Photogrammetry – ETH Zurich, Switzerland
E-mail: <till><msb><fabio>@geod.baug.ethz.ch
Web: http://www.photogrammetry.ethz.ch

The documentation and visualization of large Cultural Heritage sites has become a very interesting and challenging research field. Photogrammetry, the science of obtaining reliable and accurate 3D-measurements from images, is nowadays much more frequently used to generate visualization and animation products, documentation material for conservation, restoration and future planning as well as traditional maps of the investigated areas. This is mainly due to two facts. First of all UNESCO is paying more attention to large cultural and natural sites. On the other hand a wide array of new technologies is available which greatly support the efficient generation, administration and analysis of such models. Data from high-resolution satellite sensors and digital aerial cameras are increasingly becoming available. The same applies to laser scanners, digital panoramic and still video cameras. Moreover, the automation of the photogrammetric processing chain has made great progress recently. Geographic Information System (GIS) technology provides great tools for data administration and analysis. Finally, visualization and animation software is also becoming affordable at better functionality and lower costs. In this contribution we report about the photogrammetric reconstruction and

documentation of the Angkor area by means of aerial images. The UNESCO world heritage site of Angkor in Cambodia is said to be the largest religious structure in the world. More than 1000 structures have been found within an area of 200 km^2 that the Khmer empire had constructed between 800 and 1327 A.D. in the lowlands north of the Tonle Sap Lake. Abandoned after the fall of the Khmer kingdom and covered by forest, a team of French archaeologists started in the 19th century to excavate the ruins. The most impressive Khmer buildings are the temples of Angkor Wat and further north the temple of Bayon, which is located in the centre of the former capital city Angkor Thom. Today all the temples are surrounded by farmland and only a slim belt of green remains as a buffer area. Enclosed by water basins, the sandstone structures of the protected area surmount the surrounding jungle only outplayed in height by the hill of Phnom Bakheng that houses another large temple. The goal of our work is to check the possibility to derive detailed and reality-based 3D models of complex historic buildings from aerial images. The aerial images were acquired in 1997 within the 'Topographic Mapping for Angkor Archaeological Area' campaign to produce detailed maps (1:5000). We received from JICA (Japan International Cooperation Agency) a block of 55 mostly black and white copies of an original set of 357 aerial photos. The images cover an area of approximately 6x7 km and include the temples of Angkor Wat, Bayon and Phnom Bakheng. The average scale of the photos is 1:5100 and those displaying the main temples are in colour. For photogrammetric processing, the photos were digitized with a resolution of 15 μm.
For the image triangulation, geo-referencing, digital elevation model and ortho-photos generation, commercial and in-house programs were used. Because only 3 of the available 22 ground control points were within the processed area, additional height and control points were derived from GIS maps produced during the 1997 mapping campaign. During the photogrammetric processing, problems raised due to low overlap between adjacent images and the dense vegetation, which made difficult the identification of corresponding and well distributed points between the images. The final accuracy achieved after the photogrammetric triangulation was $\sigma = 7.3$ μm. After the image orientation, the produced Digital Terrain Model (DTM) displays a completely flat area, only interrupted by the Phnom Bakheng hill and the artificial banks surrounding the moats of the temples. For the 3D reconstruction of the temple structures, stereo measurements were performed. The resolution of the available images is not high enough for a detailed modeling of the entire area, therefore mostly horizontal surfaces (such as roofs) were measured and the walls extruded to the ground. For a more detailed reconstruction of the temples, terrestrial acquisition (image or range data) or high resolution aerial images with larger overlap would be required. Nevertheless we produced one of the first 3D model of the Angkor area based on aerial images (see figures below). In addition to four large temples, a variety of about 30 other structures, such as walls and modern houses, were reconstructed. The data are now imported in a 3D GIS for documentation, preservation and future restoration of the UNESCO site.

TIDGROVE WARREN FARM ARCHAEOLOGICAL PROJECT: AN INTEGRATED APPROACH TO THE STUDY OF AN ARCHAEOLOGICAL LANDSCAPE IN HAMPSHIRE, UK

Kristian Strutt

This paper presents results from a collaborative archaeological community and research project, run by the University of Southampton and the Kingsclere Heritage Association, at Tidgrove Warren Farm in Hampshire, in the UK. The work is focusing on an area of landscape situated at the head of the Test valley that has ample evidence of settlement from the Neolithic onwards. The project has focused in particular on the environment and archaeology of the Iron Age and Roman period, and the medieval use of the area.
The project has utilized a series of different non-destructive methods, including geophysical survey, fieldwalking, geochemical analysis and study of aerial photography, and combined this with the excavation of the two settlements in the immediate area of Tidgrove Warren Farm. This strategy has enabled the material from the excavation to be placed in the context of the surrounding landscape, in particular the location of a rural Iron Age and Roman settlement relating to extensive Celtic field systems visible on air photographs, and a hunting lodge or house associated with the medieval deer park, possibly the predecessor of the royal hunting lodge at Freemantle Park some 2km to the north.

ARCHAEOLOGICAL USABILITY OF HYPERSPECTRAL IMAGES: SUCCESSES AND FAILURES OF IMAGE PROCESSING TECHNIQUES

Arianna Traviglia

Università Ca' Foscari di Venezia Dipartimento di Scienze dell'Antichità e del Vicino Oriente E-mail: arianna.traviglia@unive.it

In the decade, particular interest in the use of airborne hyperspectral sensors applied to archaeological research has occurred in Italy, where the MIVIS sensor has found large application. The different projects involving its use have tried to demonstrate that MIVIS imagery represents an important complementary source of information for archaeological goals. However in most of the cases the limits and the problems generated by using hyperspectral images have never been stressed enough, while the spotlight has been focused only on the positive outcomes. In this paper, through the discussion of the results of research based in NE Italy, in the suburban area of Aquileia, together with the presentation of positive results, an overview will be presented of the main problems encountered when dealing with hyperspectral images. Problems such as resolution, atmospheric corrections, noise reduction along with problems such as redundancy of data and the large number of images to handle, that becomes exponential while testing different processing using different bands, will be presented. Focus will be placed also on the fact that standard procedures borrowed from multispectral analysis can not be largely applied without consideration of the landscape characteristics and soil coverage and, over everything else, of the goals of the archaeological research. With this approach in mind, based on a personal experience, the paper will expose what can be reasonably expected to be obtained exploiting the spectral informative content of the MIVIS images and how it can be used to give prominence to the presence of certain ancient buried sites and structures. Starting with the most common technical issues encountered in pre-processing the raw images (lack of specific algorithms for the atmospheric correction, high geometrical distortion etc.) the paper will go through the enhancement techniques that allow to increase and improve the optic distinction between traces recorded in the scene by generating a new image where the useful information is more easily identifiable. Various treatments and processing of the images that have been applied will be critically discussed and their results compared in order to define the ones that fit better the different research goals. Among them, Vegetation Indexes find prominence being one of the most used applications in the archaeological research, since studies about the quality of the vegetation, monitoring variations in its vigor, can enable the identification of archaeological deposits that allow or, in opposition, limit the growth of the vegetation. As well known, heterogeneity of the texture of the subsoil has a strong impact on the growth of the vegetation, creating the appearance of the so called "cropmarks". Despite the large use of NDVI in analogous archaeological studies with no regard to the environmental conditions, it will be shown through the application of different arithmetic operations that other Vegetation Indexes can work better for archaeological research, depending on the soil coverage and the density of the canopy. These two factors must be kept in mind when selecting the appropriate process to evaluate the biomass of a certain area, since the alteration due to the disturbance of background can strongly affect the results of the process.

Having gained support in the identification of canopy anomalies through the Vegetation Indexes and recognizing the need for an analogous tool for bare soil areas, a specific Soil Line Index has been developed to support the investigation over non-vegetated soil. The possibilities offered by studying the tonal variations of the bare soil (damp-marks) are well known in archaeology: their existence in fact can be an indication of the presence in the subsoil of archaeological structures or ditches that inhibit or facilitate the absorption of rainwater and of the rising of humidity. The studying of the different degrees of water absorption of a soil becomes therefore particularly useful for the identification of buried archaeological sites. The Soil Index created for hyperspectral MIVIS data aims to constitute a support for the identification of traces over non-vegetated soil, emphasizing the wetness or the dryness of a portion of the ground. Starting from the well-known concept of the linear relationship between the near-infrared and red reflectance, the S.L. Index produces an image where the driest areas are represented in brighter and lighter colors compared to humid areas.

Different processes, typical of the multispectral analysis, have also been applied to the hyperspectral shots, like classifications, P.C.A. and Tasseled Cap Transformation. While classification techniques, both supervised and unsupervised, have shown their limits and were unable to provide any original or unique results that cold not be reached through other processes,

P.C.A. and T.C.T., with their reduction of the redundancy in data by transforming a set of correlated variables into a new set of uncorrelated ones, have produced more valid results. In the case of the P.C., it will be demonstrated how the increasing of the number of principal components combined into color composites can be successful. When handling multispectral images, the most common procedure of visual inspection consists of generating a R.G.B. false color composites with the first 3 principal components of a scene (P.C.1, P.C. 2, P.C.3): this creates 6 different combinations. In the case of MIVIS data, applying the same procedure would mean excluding from the composite many of the lower order principal components that have shown to carry interesting information. For this reason, bringing up to 4 or 6 the number of principal components to combine in R.G.B. composites generates a total of 24 to 120 different combinations and minimizes the loss of information. Despite the advantages obtainable applying P.C.A., many details that are visible analyzing singularly the original bands cannot always be recognized in the P.C. composites because in there they are covered by the overlaying information from other bands. The best results in terms of recognizable features can be reached through the composite of spectral subsets (S.P.C.A., Selective Principal Components Analysis), using a dedicated correlation matrix in order to identify the best usable subset able to ensure no loss of the information stored in the original bands.

The amount of processes tested over the hyperspectral shots generates clearly also an issue with the number of produced images. We are dealing here with a 102 bands image; even considering to realistically use 1/10 of the bands for the multiple processes, the number of produced images that must be subjected to visual analysis and interpretation is too large. For this reason it is necessary to define a set of procedure to reduce the number of final images to be used, selecting the ones that do not carry redundant information. Consequently, comparison tables, cross process coverage tables, correlation and selection matrixes have been applied in order to reduce the images used in the next step of the research to a small number and able to insure the complete coverage as regards to the detected traces.

Finally, it will be discuss how the use of G.I.S. technology provides and manages all the archaeological and topographical data necessary to eventually recognize the detected surface anomalies as ancient origin traces, attributing to each of them a value of visibility and archaeological reliability as result of the interpretation process.

A SIMULATION OF THE MEDIEVAL ENVIRONMENT AND ITS CHANGE AROUND MEDIEVAL CASTLES – SPECIAL CASE IN FINLAND

Kari Uotila, Isto Huvila, Anna-Maria Vilkuna, Terttu Lempiäinen, Elisabeth Grönlund, Pentti Zetterberg

1. General Overview of the Subjects of Research

The research project will be multidisciplinary, combining the latest results of the humanities, the natural sciences, social sciences, information sciences and teleinformatics. The multivariate utilization of the different source types will provide an excellent basis for the study and understanding of both long-term and short-term phenomena and processes.

The subjects of research will be the Kuusisto Castle in Varsinais-Suomi (Finland Proper) and the Häme Castle located in the Heart of Häme.

Kuusisto Castle was, during the Middle Ages, the castle of Turku's Catholic bishop. The earliest document of Kuusisto is from the year 1295.

The castle was in possession of the Church continuously to the beginning of the 1520s, when it was transferred in connection with the Reformation from the Church to the Crown. Kuusisto became an agricultural estate under Turku Castle, and agriculture directed toward the castle's needs was practiced thereat.

Häme Castle was established as a base for the Swedish royal power in the midst of a prosperous Häme settlement at the turn of the 13th and 14th centuries. The castle retained its position as administrative center of the Häme Province until the 1630s, when it was transferred to provincial management under governors and the national authority moved from the castle to the newly-established town of Hämeenlinna.

Earlier source and research data which are available from the Häme and Kuusisto Castles provide a good point of departure for study of the habitat. The multiple ways in which the settlements, castle construction and everyday activities affected the environment will be illustrated by the aid of the research results. In addition to the castles' role in representing the state with their physical being, they were also significant economic centers where the affairs of several dozen persons were managed on a daily basis and large-scale agriculture was practiced.

The Middle Ages in this connection will be defined as the period from the 11th century to the beginning of the 17th century, covering the time period from the early Middle Ages to the beginning of the early modern period. The permanent settlement that sprang up in the research areas of Varsinais-Suomi and the Heart of Häme already in the Iron Age continued to become consolidated. The new secular and spiritual power gradually got a foothold, utilizing the natural resources of the environment and affecting people's physical as well as mental world. The cultural characteristics to be studied retained a very strong consistency for the entire research period.

The objective of the project's natural science section is to clarify the area of, and the changes in, the environment of the selected subjects of research from the Middle Ages to the beginning of the modern period making use of pollen and plant-remains (macrofossil) research and tree-ring analysis. During the last decades, an increasing number of paleobotanical studies have been applied in the field of environmental history.

Paleobotanical methods, mainly pollen and macrofossil analyses, have been successfully utilized in reconstructions of past vegetation changes and patterns. Subjects of research will be, in addition to useful plant species, different time periods of environmental change. Botanical results will be particularly important, as knowledge of the vegetation in different time periods will be crucial for the computer models produced by modern technology. Very uneven models have been produced, especially during the last decade. These still, however, almost completely lack "supplements" corresponding to the research results reflecting the environment and nature. Research data obtained by botanical methods will be needed here.

Pollen analysis accompanied by with macrofossil analyses of geological deposits allows for comparison of vegetation characteristics and dynamics before extensive human activity with periods of intensive anthropogenic influence. These studies have often focused on dating of earliest cultivation phases locally and regionally. However, in the present project, the idea will be to concentrate on the archaeologically and historically well-documented central regions around the castles of Hämeenlinna and Kuusisto.

Pine-tree annual-ring data starts in the Turku region in the year 1012 and in the Hämeenlinna region in the year 1094, so they cover the research period, 1100 to 1600. The data collected from both research areas are, however, used very seldom in other than ordinary dating. Thus, there is great research potential in the data. To this can be added, as entirely new material, graph data being prepared for oak. The dated wood material also indirectly reveals variations in human activity according to the date of felling: for example, in the 1350s and 1360s a lot of trees were felled, while again, during the 1340s, for example, hardly any trees were felled.

The senior researchers of the project are leading researchers of their own field. Doc. Kari Uotila ja Ph.D. Anna-Maria Vilkuna are the leading castle researchers of the young generation in Finland. The modeling group for the research project has for many years made models of the past world different objectives under the direction of Doc. Kari Uotila. Doc. Terttu Lempiäinen´s interests are investigations of the vegetation history, palaeoethnobotany and plant taxonomy and systematics. At the moment she is the only scientist in Finland teaching palaeoethnobotanical and macrofossil analysis methods. Ph.D. Elisabeth Grönlund has worked for years with utilisation of pollen analysis, other sediment microfossils and physico-chemical analyses in environmental reconstructions. Ph.Lic. Pentti Zetterberg is the leading scientist in Finland in the field of dendrochronological dating. One of his main achievements has been the construction of a 7 600-year continuous tree-ring master chronology, which is one of the longest in the world and has proven to have outstanding applications in climatic studies.

GRIDJAM. A NETWORKED 3D IMMERSIVE PERFORMANCE

Judith van der Elst

http://jackox.net/pages/Gridjam1.html

Overview

GridJjam is a real-time, geographically distributed, networked multimedia event. It is an experimental project that brings together visual artists, composers, musicians and computer scientists, while using the new high speed international LambdaRail network. *GridJam* will demonstrate real-time, low latency, interactive, distance computing through the complexity of the live, partly improvised, 3D visualized, musical performance, being both a world-class work of art and a research project into high performance collaborative network computing.

GridJam will utilize international artist Jack Ox and programmer David Britton's Virtual Color Organ™,visualizing distinguished composer Alvin Curran's electronic music performed by musicians located indistant locales but connected via next generation networking technologies. The Virtual Color Organ™(VCO) is a 3D immersive environment in which music is visually realized in colored and image-textured shapes as it is heard. The visualization remains as a 3D graphical sculpture after the performance. Thecolors, images, shapes and even the motions and placement of the visualized musical shapes are governedby artist-defined metaphoric relationships, created by hand as aesthetic and symbolic qualities rather thanalgorithmically. The VCO visually illustrates the information contained in the music's score, the composer's instructions to the musicians, and the musicians contributions to the score as they improvise in reaction to each other's performances and to the immersive visual experience. Illustrative of synesthesia andintermedia, the VCO displays the emergent properties within the meaning of music, both as informationand as art.

GridJam is significant both as an artistic work and as a demonstration of the capabilities of next generationarts and entertainment technology venues. *GridJam* is based on key observations regarding current and future media technologies:

Cinematic presentation has yet to incorporate paradigmatic technological shifts. Current cinematic venues are 2-D (visual) and minimal 3-D (audio) rear projection environments that have remained essentially unchanged since the emergence of color, sound, and relatively low-resolution projection standards. Given the convergence of gaming, computing, cinematic, telecommunications and networking technologies, new next generation venues for art, entertainment and telecommunications will surely emerge. Grigjam will propose, explore, document and demonstrate some answers to some important questions about next-generation venues and environments.

How will new immersive and networking technologies impact the evolution of public arts,entertainment and presentation venues? Ten, twenty years in the future, will people go tomovies, popular and fine arts performances in the same way they do now or will

computingtechnology produce fundamental changes in these aspects of life, as they have in many otheraspects of contemporary life?
What comes after cinema? How will cinematic arts evolve in a world of high resolutionhome theatres, VR, interactive networked gaming and a generation raised with on-demand high definition video, computing and gaming in the home?
What roles will interactivity play, and how will it change the fundamental experience ofcinema? What about the social aspects of movies, concerts and shows; will it be lost in favorof chat rooms and tele-presence? Can distance learning and entertainment capture the essence of the human experience of physical presence?
What are possible conventions and specifications for new, networked presentation,performance and entertainment spaces that will enable new immersive media venues to provide a high degree of intimacy and immediacy, suggestive of physical rather than virtualpresence? What materials, knowledge and practices from telecommunications, computerscience, military applications, gaming and simulation technologies, as well as computergraphics and audio will be used in new media venues?
How will next-generation public media venues scale to home and personal entertainmentsystems, and how will they integrate with personal computing?
What kinds of content and new experiences can artists offer as a demonstration of the powerof new media arts and entertainment technologies, given ample time and access, and absent aprofit motive?
Participants
Internationally recognized artist Jack Ox, a specialist in visualization of music, and recently appointedAdjunct Professor in the department of Fine Arts at UNM, connected to the Art/Technology Center andnewly created ARTSLab, provides the graphical and overall design. David Britton, a pioneer in virtualreality art and expert multimedia software developer provides the core software development. Internationalcomposer and Mills College faculty member, Alvin Curran provides the music's compositional structure,composed score and electronic transformations. UCSD's CRCA music software pioneer Peter Otto provides sound spatialization software and supervises the electronic music programming. Tim Thomas,Deputy Director of High Performance Computing at UNM will be our networking, LambdaRail specialist.Pierre Boulanger from the U. of Aberta will be providing 3D avatars in the form of the musicians and their instruments, actually sitting in the VR desert environment.
Figure 1. View across *Strings, Piano, Woodwinds*, and *Vocal*landscapes in the desert organ stop of the Virtual Color Organ™
The 21st C. Virtual Color Organ™
The VCO is a computational system for translating musical compositions into visual performance. This interactive instrument consists of three basic parts:
A set of systems or syntax that provides transformations from a musical vocabulary to a visual one.
A 3D visual environment that serves as a performance space and the visual vocabulary from which the 3D environment was modeled. This visual vocabulary consists of landscape and/or architectural images and provides the objects on which the syntax acts.
A software environment that serves as the engine of interaction for the first two parts.
The VCO is capable of having multiple visual organ stops. An organ stop on a traditional organ is a voice that affects the entire keyboard of notes. An organ stop in the VCO is the 3D immersive environment in which the visualized music will exist and also the visual vocabulary applied to the musical objects. *GridJam* will take place in the black and white, hand drawn desert sand and rock structures coming from real deserts in California and Arizona. The original drawings by Ox, from which the 3D modeling was made, serve as the basic texture maps for all of the musical object.
The Music by Alvin Curran
I have come up with a plan for the musical part of this project which is inspired both by the purefundamental "synesthetic" goals of Jack's visual structures as by the technical and theatrical nature of the domed projection spaces, the local acustics and "global"nature of the work itself.
The music itself will be composed using a fluid mix of structured indeterminacy, syncronized composition, spontaneous music structures, and live electronic processing. But the essence of the music -whose goal is to become interchangeable with the images is to create another set of "images" real and sonic, which is the music-theater itself. The L-Rail communicatioins will enable the complete syncronization of the dislocated music groups and their ability to react as if in the same room together.
The choice of instruments, which represent and embrace the main musical cultures of the world -will become another prime visual element imbedded in the rich texts of Jacks desert and VR sound-objects. The trios of instruments (played by a group of selected artists both local and non local) are:
Disklavvier (laptop-sampler) ContraAlto Clarinet Violoncello using the original source samples
Giant Taiko Drum Tuba Electric guitar
Cembalon Accordion Violin/Soprano Voice
Shofar (rams horn) Giant Gong (TamTam) performance Glove (laptop) processed (bowed scratched and struck)
In each locality there will be one fulltime Laptop player assigned only to the electronic processing of thesounds of the trio in situ. These processing devices -may be a combination of Spectral Filters (by TomErbe/Sound Hack), granulators, or personlized set of processing tools created locally. The live processingwill be yet another "timbral" component to the music which will be present (as major soundtransformations) discreet and sparse moments.
The music (gestures and typologies) itself will range from almost inaudible sparse noises to discreetisolated tones to pulsing synchronous rhythms to clouds and walls of sound. characterizing the music throughout will be the presence of recorded (natural) sounds from animals, people and machines andphenomena from around the world (which however embedded in the texture, are at the source of the visualmaterials generated)... Midi information from the processing devices will also serve as real-time transformers of some aspects of the visual objects.
GridJam: The Visualization
The original building blocks of *GridJam* come from a list of almost 200 collected sounds belonging toCurran. The sound files include John Cage reciting short phrases, Maria Calas singing a high note, animalsounds, objects like coins being tossed, and various musical sounds etc. Ox put them through a Maxprogram creating graphs of melody and dynamics. She sorted them into 8 groups based on these simplevisualizations and made 3D models reflecting melody from the front and dynamics on the top of the object. There are eight groups because there are eight landscape texture maps. Each object in a group has as thebase of the texture map the same hand drawn landscape picture. The colors and materials are based on thetimbre of each specific sound sample. These colors can be rather complicated gradients, often in layers. Ifthe sound sample has a sequence of vowel sounds then the colors in the gradient

will show these based onOx's color system, which defines how and where vowels are made in the vocal track. Other timbre colors come from the extensive list of timbre created by Ox.
The 3D objects made by Ox from Curran's collected sound files will be able to appear in small bits of thewhole, based on the duration of the played sound. If the Disklavier key remains depressed for longer thanthe actual sound file, the object will begin to appear for a second time, moving along the time line. Thetime line follows a path from the center of the virtual desert out, until it curves up and back towards thecenter. .Before reaching the center it will curve up again and move outward once more, repeating thispattern for as many times as necessary in order to accommodate the playing. Because of this path, time willmove both horizontally and vertically.
At the end of the performance the visualization will remain and can be more thoroughly inspected as aplace with moving objects in a black and white desert environment. We will also have recorded the entiremultimedia experience so that it can be played back in digital environments, such as the ever more ubiquitous digital planetarium theaters.

SPACE AND TIME: VIRTUAL REALITY AND THE ART OF EXPERIENCE

Judith van der Elst, Jack Ox

Department of Anthropology, University of New Mexico
Department of Fine Arts

We often study cultures or communities that have/had a different perception of space and time than the typical Western researcher. Yet, we employ our Western frame of reference where space and time is linear, to understand all of human behavior.
Visualization and Virtual reality offers possibilities to explore different perspectives (of space and time) within a virtual landscape. Within this paper the authors explore the link between visual culture, different cultural perspectives, and a new art project by Jack Ox and David Britton in order to expand the way in which virtual reality can be used in cultural heritage, cultural understanding and communication.
The purported usefulness of collaborations between artists and scientists has prompted many questions and discussions. However, the authors agree in large part with other researchers that science can learn from art and vice versa (Arends and Thackara 2003, Casti and Karlqvist 2003). Both artists and scientist are driven by curiosity, an urge to find out, sharing their methods and practices can lead to new insights into human experience. What we will present here is an example of this kind of collaboration and how this research is mutually beneficial. For instance, the benefit of drawing, a method belonging to the arts, has been successfully used to better understand and solve complex problems in science (Hjort 2003).
Several researchers from diverse backgrounds have recently addressed the issue of forgotten, ignored, warped, or eliminated histories. Ideological justification for this can be traced back to Rousseau and later Social Darwinism. Confronted by alternative conceptions of other societies, Western reality became reified as representing something 'better' – the last stage on the way to civilization, reflecting 'higher orders' of thinking, (Smith 2005: 48).
Ideas such as the dualism of Mind and Body are concepts belonging to the Western knowledge system. Similarly, within a Western framework the concept of Space is differentiated from the concept of Time, each can be individually assessed. In addition, space is often viewed as being static or divorced from time; space can be controlled. According to Smith what makes ideas 'real' is the system of knowledge, formations of culture and the power relations in which these ideas are located.
Within the discipline of archaeology (as in other scientific disciplines) the notion of Space is 'universally' understood as being 'appropriated by Mathematics', it can be better understood and measured if compartmentalized. However, in many indigenous languages there is no distinction between space and time, others do not even have a word for space nor time. If within the western framework lines can measure space, this dominant view has limited what we 'see'. Archaeologists and other researchers working from a different viewpoint or epistemology have criticized the fact that intellectual Non-Western contributions toward cultural understanding and perception are largely ignored (Schmidt and Patterson 1995). For instance, oral histories are often neglected as sources of knowledge, even though they may provide us with a richer and different account of history. Sometimes these stories are accepted within a scientific discourse, but only it they are supported by physical evidence, such as in the case of the Mahican Indian Tribe's story (Handsman and Richmond 1995). Other examples inform us of cases elimination, the practice of renaming the land, "newly named land became disconnected from songs and chants used by indigenous peoples to trace their histories, to bring forth spiritual elements or to carry out the simplest of ceremonies" (Smith 2005: 51).
(Hi)stories and (Virtual) Reality
Critical theory seeks to "to liberate human beings from the circumstances that enslave them" (Horkheimer 1982, 244). Researchers have begun to integrate critical theory into geographical information systems applications, exploring the use and experience of space and time by different gender, ethnic groups etc. Critical GIS, as it has come to be known crosses many disciplinary boundaries and it concerned with the power embedded in the use and production of technology (*Cartographica,* vol 40, no 4, 2005).
According to Critical theory the physical world is as much a construction of mind as the world constructed from Virtual Reality cues. Virtual reality raises the issue of differences between artificial and real physical space, such as: how do we verify reality? (http://digitalarts.lcc.gatech.edu/unesco/vr/home.html). This brings us back to the notion of reality discussed earlier, the reality belonging to a specific knowledge system. Within a virtual environment we can create different realities as experienced or created by different individuals or groups of people. Furthermore, as is stated on the UNESCO website on digital art, "the history of art further suggests that communication is not simply just a sending of sensual cues as many engineers would like to imagine. An artist is practiced in how subtleties and nuances within a given medium may achieve a richer sense of presence than one created by someone who lacks this specific type of knowledge" (ibidem).
Within a virtual environment we can explore space and time so that we can begin to understand the people, societies and cultures, whose material remains we like to study as archaeologists, but whose walks of life we can not understand if we do not open our mind to other worlds of perception? The world is not flat; there is a spot where we can all meet.
ART PROJECT: *Gridjam*
Gridjam is a real-time, geographically distributed, networked multimedia event. It is an experimental project that brings together

visual artists, composers, musicians, and (computer) scientists, while using the new high-speed international LambdaRail network. *Gridjam* will demonstrate real-time, low latency, interactive, distance computing through the complexity of the live, partly improvised, 3D visualized, musical performance, being both a world-class work of art and a research project into high performance collaborative network computing.

Gridjam will utilize international artist Jack Ox and programmer David Britton's Virtual Color Organ™, visualizing distinguished composer Alvin Curran's electronic music performed by musicians located in distant locales but connected via next generation networking technologies. The Virtual Color Organ™ (VCO) is a 3D immersive environment in which music is visually realized in colored and image-textured shapes as it is heard. The visualization remains as a 3D graphical sculpture after the performance. The colors, images, shapes and even the motions and placement of the visualized musical shapes are governed by artist-defined metaphorical relationships, created by hand as aesthetic and symbolic qualities rather than algorithmically. The VCO visually illustrates the information contained in the music's score as they improvise in reaction to each other's performances and to the immersive visual experience. Illustrative of synesthesia and intermedia, the VCO displays the emergent properties within the meaning of music, both as information and as art.

The VCO is a computational system for translating musical compositions into visual performance. This interactive instrument consists of three basic parts: 1) A set of systems or syntax that provides transformations from a musical vocabulary to a visual one, 2) A 3D visual environment that serves as a performance space and the visual vocabulary form which the 3D environment was modeled. This visual vocabulary consists of landscape and/or architectural images and provides the objects on which the syntax acts, 3)) A software environment that serves as the engine of interaction for the first two parts.

The VCO is capable of having multiple visual organ stops. AN organ stop on a traditional organ is a voice that affects the entire keyboard of notes. An organ stop in the VCO is the 3D immersive environment in which the visualized music will exist and also the visual vocabulary applied to the musical objects. *Gridjam* will take place in the black and white, hand drawn desert sand and rock structures coming from real deserts in California and Arizona. The original drawings by Ox, from which the 3D modeling was made, serve as the basic texture maps for all of the musical object.

LINKING ART AND SCIENCE TO EXPLORE DIFFERENT REALITIES

During his inspirational speech at iGRID2005, Larry Smarr told the audience that we have already gone past Thomas Friedman's notion of the world as flat.

With the emergence of this new technology the Earth has the possibility to be one spot. So far one can only access this "spot" in a limited number of places, however as pioneering artists devote themselves to designing complex performances for the LambdaRail, they can help to develop the paths that increasingly more people will have access to in the future, while also forging methods to put their own communities back together again.

The purpose of this paper is to show the applicability of this technology for archaeology/anthropology to investigate and explore issues that have been discussed on a theoretical level, but have so far had only limited practical application. It is our task to seek out these different, culturally diverse, experiences and realities, and bring these together within a virtual environment, rethinking notions of space and time. The cross-fertilization of art and science can help us better understand different (cultural) perspectives in space and time and in doing so broaden our toolkit to understand past perspectives of space and time, recreating places of (past) human experience.

THE IMPORTANCE OF THE RELIEF AND THE SOURCES TO INTERPRET AND COMMUNICATE THE CULTURAL HERITAGE

The cultural heritage must communicate with its addressees, therefore it's impossible not to take into consideration its expressive and emotional ability, or to forget that scientific knowledge is at the base of a reliable and transparent communication.

The present study is proposed, only after having obtained accessible data, in order to reach the knowledge of the cultural heritage and to spread it throughout new technologies.

The plan in issue, carried out at the Virtual Heritage Lab of the Itabc at the CNR (commissioned from ARCUS, in collaboration with the Archaeological Superintendence of Rome and the participation of the University of Barcelona), looks at the "reconstruction of the archaeological landscape of the ancient Flaminia with virtual reality systems", building a digital ecosystem that individualizes two levels of perception: the first one represents an holistic vision of the road from Rome to Rimini, based on technical cartography and aerial and satellite photos able to support the realization of a web gis; the second one, which dips into a micro-space vision that focuses the attention on four monographies, with the aim to represent and contextualize the local entities within the "ancient landscape" entity.

The areas of monographic interest are Grottarossa, Ponte Milvio, Malborghetto-Tor di Quinto and Villa of Livia.

A firm knowledge of the cultural heritage is at the base of all the successive processes of interpretation and communication.

For the restoration of a reliable and verifiable plan of reconstruction in all its parts it is necessary to begin from a first phase collection of high level data, both the cognitive aspect of the state of the places ("bottom up") and the collection of texts, instances, archives etc. (in close collaboration with the Superintendence of Cultural Heritage) relative to actual places and on study cases morphologically and temporarily comparable ("top down").

The first category of information about the cognition of the actual state of the sites is based on the possibility of using advanced technologies for the data acquisition that is intimately connected to the typology and the particularities of the entities to examine.

In the case of Malborghetto, for the simplicity of the geometry, the technique of photo-modelling was chosen, and, together with the relief of strategic points and the total station and the Gps, has concurred to refer the site geographically and reconstruct the landscape.

The importance of starting from objective and accessible data, in the case of the Villa of Livia has turned out even more necessary, due to the complexity of the architectonic system and the insufficient and partial credibility of the planimetry in public archives and of scarce information caused by incomplete excavations.

In fact, the great archaeological complex of Villa Livia has made use of advanced technology in the laser scanner to obtain 3d reliefs with a resolution of 6mm. This technology represents an enormous potential of publishable, up-to-date, available information and, above all, georeferenceble, fundamental for an archaeological contextualized knowledge of the territory and in a position to transmit a strong evocative and communicative component.

Through the 3D reconstruction of the architectonic heritage aimed at a real time fruition, we approach an effective and sustainable interpretation that involves the public and transmits the importance of the conservation and communication of the cultural heritage.

They decode data and information that often belong to the scientific ambit that are not accessible or comprehensible to the majority of the unspecialized public.
The interpretation increases the experience and vice versa: the possibility to interact with the cultural heritage increases the respect and the understanding of the public, transmitting the importance of the conservation, knowledge and the sharing and exchange of data and contents.
The acquisition of 3D data with the laser scanner generates a high resolution model with questionable and exact data, with generated and interpolated faces. The same technique is used to detect both the architectonic compound of the Villa of Livia and the hill on which it exists.
The acquired and elaborate models are further optimized within other three and bidimensional graphic softwares (3D Studio Max, Photoshop CS2), used for texturing, decimation and the improvement of the graphical quality of the rendering (techniques of multires, normal maps, UVWrap, render to textures).
Moreover, this complicated model gives the possibility to refer to 2D reliefs that enable an accurate and deeper analysis.
In fact, by the cloud of points method, through the cloudwork tool, it is possible to elaborate horizontal and vertical sections in CAD format that constitute an analyzable documentation from the architectonic point of view as a base of the typological and structural study of the building.
The second category of data has put into evidence the problem of the interpretation of the ancient architecture which, connected with the limitations of the literary sources, make the architecture become the main way to understand the roman architecture, placing the problem of how the public interpret it. The history of the buildings is not a collection of irremovable facts, but is a continuous process, a set of various inclinations and interpretations which transform the object, as always, everyone transforms the past adapting it to its own experiences.
The observation of its geometry and materials, the study of typological parallels, the static verifications and the arithmetical calculations become the fundamental instrument in order to understand a building. The only historical surviving treatise of the roman age is *De architectura Libri Decem* of Vitruvio. Unfortunately it is not sufficient to reconstruct the architecture. The careful observation of architecture and design remains the only solid foundation.
Design is a process, a dynamic interaction between concept and contingency, the generic and the specific, that progressively evolves in a variety of decisions that are assimilated in total.
Through the formal and material analysis of the architecture, the original morphology, its operation and history can be understood, realizing re-constructive hypotheses that are based on constructive logic and cultural background. The structural verification through the graphical static's calculation is a fundamental instrument in order to ratify the re- constructive hypotheses. The study of materials supplies another important aspect: we may learn of its mechanical abilities which condition architectonic aspects such as the height of walls, cover systems, etc.
In the passage from acquisition to communication it is fundamental to have a method which renders transparent the processes through which we arrive at the conclusions translatable in 3D images.
The information loses neutrality from the moment in which it is elaborated: since acquisition is chosen the final result is determined. A re-constructive process must be visible and transparent; the re-constructive processes and the analysis of the information can be illustrated, for example, through the spatial superposition of maps and 3D models, sights of the 3d reconstruction that leave space to fixed images, monochromatic and mapped models etc.
The cultural wealth at our disposal and the application of advanced manageable technologies give the possibility to communicate and share the common heritage rendering it completely available to everyone in its historical, artistic and scientific valences.

A REMOTE CONTROLLED INFRARED EYE IN THE SKY

Geert Verhoeven, Jo Loenders

Since more than 100 years, oblique aerial photography has proven to be an extremely valuable tool for (landscape) archaeologists. However, the vast majority of the pictures taken only records the reflected visible portion of the electromagnetic spectrum produced by the sun. In this lecture, a newly build apparatus will be presented which allows archaeologists from Ghent University to take both "regular" and infrared pictures of archaeological interesting landscapes and loci in Italy (the Potenza Valley Survey in the Regione Marche) as well as in Flanders (Belgium).
Rather than taking the usual false-colour infrared approach by capturing also a portion of the visible light, this application enables only the near-infrared to be registered. As this approach demands longer shutter speeds, the aerial platform needs to be very steady. To satisfy the requirements, the application makes use of a unique concept – a helikite – to hold a remotely controlled camera in the air. By joining kite wings with a helium balloon, the helikite combines the advantages of both appliances, hereby reducing the drawbacks to a minimum. The balloon enables the helikite to take off in windless weather conditions (which a kite is not able to do), whereas the kite components play two important roles in case there is wind: they lift the helikite higher up in the air, while additionally stabilizing this lighter-than-air construction. Consequently, they overcome the knocking down of the helikite in windy conditions, a major drawback of all normal balloons.
The camera-supporting construction, attached to the helikite line by a picavet construction, is characterized by a fully carbon-aluminium design and can be remotely controlled in all directions from the ground.
In order to frame the shot correctly, a video live link is applied, making the archaeologists on the ground aware of what will be captured in the air.
More than focusing on the technical details of this construction alone, this lecture also wants to point to the extra archaeological and financial advantages of this HAP (Helikite Aerial Photography). Besides offering an infrared eye in the sky – hereby hopefully revealing more information than the standard photography – HAP will in the near future also be employed to take stereopairs from excavations, which will afterwards be photogrammetrically processed into detailed orthophotomaps. Thirdly, it is also a means for taking archaeologically large-scale photographs from a low altitude.
As the whole creation is still a prototype and its use in an initial phase, the presentation will conclude with possible further improvements as well as the shortcomings of HAP in general.

INTEGRATING AERIAL PHOTOGRAPHY FOR THE STUDY OF ROMAN TOWNS IN ITALY

Frank Vermeulen

Ghent University, Belgium

In this paper will be argued that an integrated non-destructive survey-approach, with a large contribution of aerial photography and remote sensing operations, is one of the best ways to achieve diachronic size/density reconstructions of Roman towns during some of their stages, without having to fall back on large scale and expensive excavation work. Thanks to such an approach, preferably applied on (partially or completely) deserted sites with a high potential for surface reconnaissance, we are in the position to look not only at such important matters as size and form but also at function and functional zoning. Interesting aspects about functional zoning that can be studied thanks to this particular intra-site approach are: discontinuity in the use of urban space, fluctuations in the size of the built-up area and gradual shifts in the location of the nuclei of the town sites. Short of total excavation of the urban areas of Roman Italy, we are convinced that intensive overall artefact survey combined with geophysical prospections, and checked by aerial imagery and (if available) some stratigraphic data, is one of the best approaches to detect all three of these dimensions of change. In this presentation the achievements of active oblique aerial photography, combined with fine examination of vertical aerial imagery will be especially emphasized.

To demonstrate the possibilities of such an approach a case study from Central-Adriatic Italy is chosen. We will present some of the most striking results of new archaeological survey work, carried out since 2000, in and around the coastal Roman town of *Potentia* (Porto Recanati). The intensive surveys carried out by a team from Ghent University in and around this Roman colony at the mouth of the river Potenza include low altitude aerial photography, systematic archaeological field walking, artefact studies, re-study of excavated evidence, detailed geomorphologic field mapping, geophysical surveys and fine topographic mapping. This urban site has an excellent potential for accumulation of new evidence by way of surface survey techniques, because of the small scale of excavations previous to our work in the area. Equally important is the current state of the terrain. Like many Roman towns in the Marche region *Potentia* in some way became 'unsuccessful' from Late Antiquity onwards, and was completely abandoned. As almost no present-day habitation covers the former town area and because agricultural practice still dominates the zones where its ruins lie buried, the ancient topography of the city can still be well-studied with the help of intensive survey methods. In this paper we will present some of the possibilities elaborated during the on-going *Potentia*-surveys for a different approach towards functional zoning and changes in the use of space in an average 'new' Roman town. This multi-disciplinary approach builds further on the recent development of a wide range of surface-survey methodologies which have enhanced the potential for investigating ancient urban sites.

RECONSTRUCTING COMMUNITY STRUCTURE AT CERRO BAÚL, PERU WITH SPATIAL DISTRIBUTIONS AND REMOTE SENSING

Benjamin Vining

There has been some criticism of environmentally-based archaeological models, due to a presumed inherent determinism that underemphasizes communal and individual cultural behavior in favor of behavior that is economically or materially-focused. As they interact primarily with physical environmental parameters, digital landscape methods, including satellite remote sensing, particularly are exposed to such criticism; the reliance on environmental data is seen as a reliance on environmental factors as causative explanations (e.g., Gaffney and van Luesen1995). These criticisms pertain more to analytic paradigms than limitations inherent in remote sensing as a method.

Remote sensing data, which are environmentally responsive, can easily be integrated with archaeological data that are socially sensitive. This case study presents an analysis of lithic artifact materials within domestic contexts at Cerro Baúl, Peru, in light of a high-resolution (15 m) lithological map derived from the spectral classification of ASTER data. The domestic context, as a basic social unit of interaction and production (Bermann 1994) is used as the fundamental analytic unit and conventional artifact analysis is integrated with environmental and spatial data through image processing and GIS. Vector data representing archaeological entities are integrated with environmental data in raster format, where the categorical raster data form the basis for statistically rigorous one-sample tests of correspondence. In turn, the raster data derived from image classification can be integrated with conventional statistical measures of diversity and correspondence to evaluate the relationship between space and landscape use among households.

Cerro Baúl is situated in a geologically active environment at the lower edge of the Western Andean Escarpment. The region is highly fractured and is the contact between depositional, sedimentary environments to the west and uplifting igneous and volcanic environments to the east. This results in a lithologically highly heterogeneous landscape and there are significant differences in the type and quality of exploitable resources with distance from the site. The transport of raw material is a sensitive indicator of the extent and intensity of behavior across the landscape, but high-resolution mapping of lithological resources is a requisite. This condition is met by ASTER. ASTER's high resolution is on a scale commensurate with archaeological phenomena and this analysis integrates these data with artifact databases through a site-based GIS.

This analysis reveals social divisions in the way in which lithic raw material available within the local landscape was exploited. The use of raw material followed social divisions which were economically as well as ethnically negotiated and characterized the Cerro Baúl community (Vining 2005, Williams, Isla and Nash 2001). There were scalar differences in the degree of interaction that individual social units at Cerro Baúl had with their surrounding space. These scalar interactions were realized either directly, with the engagement on the individual with their physical surrounding, indirectly through an interpersonal network, or both. Such differences reflect the notion of an ideational landscape which is created through the behavior of community which acts out its social order and cultural dispositions during their engagement with space and the elements in it (Knapp and Ashmore 2000: 12 – 13).

This case study illustrates some of the potential for analyzing remotely sensed data according to archaeological behavioral criteria and the crystallization of ideational landscapes. With such approaches, digital landscape reconstructions can include cultural behavior and move beyond environmental analysis.

SATELLITES, SURVEY, AND SETTLEMENT: THE LATE CLASSIC MAYA UTILIZATION OF BAJOS (SEASONAL SWAMPS) AT TIKAL AND YAXHA, GUATEMALA

Errin T. Weller

Anthropology, University of Colorado, Boulder CO, USA.

The past decade has seen a resurgence of interest in the low-lying *bajos* (seasonal swamps) comprising 40% of the ancient Maya landscape of Guatemala. *Bajo* research is an outgrowth of earlier research on wetlands, permanently moist areas as opposed to seasonally moist *bajos*. In some documented cases, wetlands were transformed into *bajos* due to the environmental impact of the ancient Maya on the landscape. The dominating *bajo* feature of inland Mesoamerica has fascinated explorers and archaeologists since the early 20th century because of the imposing archaeological sites located on *bajo* margins. This has led some scholars to suggest the *bajos* were an important resource providing available land for habitation, water, or agriculture to support the large populations of urban Late Classic (AD 550-850) Maya centers. Others, however, find insufficient evidence to support *bajo* use and suggest they were a marginal resource at best.

Advances in remote sensing technology, especially the advent of high-resolution (with sub-meter pixel size) multi-spectral imagery, allow for re-analysis of these *bajo* questions. In 2006, I began the Bajo Regional Archaeological Project with the objective of exploring the utilization of bajos surrounding the sites of Tikal and Yaxha in the northeast Peten of Guatemala. I chose two bajos: the 10km x 10km Bajo la Justa, Yaxha and the larger Bajo de Santa Fe, Tikal. This project is integrating acquired Global Positional System (GPS) data from previous *bajo* projects and survey maps with satellite imagery (Landsat and IKONOS), RADAR (AIRSAR and SRTM), Geographic Information System (GIS)-based spatial analysis and future ground survey to investigate the purpose and extent of landscape modifications within and surrounding these *bajos*. These results will be directly applicable to questions of ancient Maya settlement, subsistence, and population.

Conducting research in the humid rainforest environments of Mesoamerica is challenging, and in the dense, noxious thorn forests of the bajos even more so. Only limited ground survey of the *bajos* has been undertaken, leading to the prevalent view that the *bajos* were uninhabited. To obtain a more coherent understanding of settlement, my research is studying the landscape for evidence of the following: 1) agriculture (seasonal or continuous), including canals, planting platforms, terraces, and dams to control water flow; 2) habitation (on natural or raised areas); 3) water procurement and/or storage including reservoirs and aguadas (natural clay-laden sunken areas); and 4) causeways (roadways).

The dominant settlement view in Maya archaeology is the Concentric Zonation Model (CZM) with the highest density of monumental architecture in the center of the site with structure size and density decreasing as distance from the monumental center increases. A Maya center is organized in concentric rings with a ceremonial core surrounded by a zone of elite residences and in turn by dispersed non-elite houses. This model has failed to consider *bajos* as part of the landscape when constructing this view of ancient Maya settlement. When scholars are conducting on-the-ground research to test CZM, they likewise do not incorporate *bajos* into their research.

The Bajo Communities Project documented a range of sites within and surrounding the Bajo la Justa, with sites ranging from isolated mounds to minor centers to Yaxha itself. Recent models are also suggesting not only "bajo communities" with habitation where settlement focused on the *bajos* as a resource but also "aguada communities" to exploit water and "terrace communities" to exploit agriculture.

While these models provide innovative new ways to consider the settlement of the ancient Maya, they provide essentially top-down views that should be complemented with another approach. I would suggest a heterarchical approach. Carole Crumley defines heterarchy "as the relation of elements to one another when they are unranked or when they possess the potential for being ranked in a number of different way, depending on conditions." This approach can be seen in the work of Dr. Marshall Becker at Tikal who critiques archaeologists' concern with structure size as a descriptive and limiting approach that should be replaced with a focus on the form of the group. He identifies 10 plaza plans (PP), group plans, that occur in consistent percentages across the site and range in size from very small to very large within the same plaza plan. Of particular interest to this study is that it is an emic approach and, to date, Becker has identified an anomalous cluster of a residential cluster of plaza plan number 2 occurring on a peninsula in the Bajo de Santa Fe. This anomaly will be further studied by remote sensing.

Access to the dataset of multispectral satellite imagery including Landsat 7 and high-resolution IKONOS imagery combined with RADAR from the Shuttle RADAR Topography Mission (SRTM) and Airborne Synthetic Aperture RADAR (AIRSAR) was provided by Dr. Tom Sever of the National Aeronautics and Space Administration (NASA). An additional acquired data source is approximately 23km of survey transects conducted by the now completed Bajo Communities Project in the Bajo la Justa. The transects recorded vegetation changes every 25m as well as any archaeological features. These GPS points were used in supervised classifications of the vegetation of the Bajo la Justa on IKONOS imagery. Differences in vegetation are essential in identifying causeways (due to their elevated nature), habitation on raised areas (known as "bajo islands") and agriculture (palm *bajo* retains moisture and is favorable for agriculture). Additionally, these GPS points form vectors for known sites and vegetation sites.

Wide ranges of analysis techniques are currently being employed on the data in RSI-ENVI. This research is ongoing to develop and refine a methodology for confirming or denying anomalies identified in the remotely-sensed data as ancient Maya modifications. Previous bajo research conducted by T. Patrick Culbert, Vilma Fialko, Tom Sever, and also Dr. William Saturno have successfully employed differences in vegetations to differentiate *bajos* from the surrounding uplands on both Landsat and IKONOS imagery, particularly false color composites. Using ENVI's 3D SurfaceView to drape Landsat 7 imagery over 90 meter SRTM resolution elevation data as well as IKONOS data over the AIRSAR Digital Elevation Model (DEM) is also effective. Saturno and Sever's recent identification of a vegetation signature for identifying archaeological sites, particularly within *bajos*, has also proved useful for identifying target sites for ground-truthing in the upcoming field season.

ROADS SCHOLARSHIP: A METHODOLOGICAL APPROACH TO THE ROLE OF TRADE IN THE TIWANAKU STATE OF THE SOUTH AMERICAN ANDES

Patrick Ryan Williams,[1] Charles Stanish,[2] Michael Moseley,[3] Benjamin Vining,[4]

[1]The Field Museum
[2]UCLA
[3]University of Florida
[4]Boston University

The role of trade in the development of early states throughout the prehistoric world stands as one of the major theoretical questions in anthropological archaeology. As one of a handful of areas where archaic states developed, the Central Andes is one of the few areas where such theoretical issues can be addressed in a comprehensive manner. In the last two decades, many archaeologists have argued that extensive, state-supported interregional trade was a significant factor in the development of two early Andean states: the Middle Horizon (ca. AD 600-1100) polities of Tiwanaku and Wari. While much indirect evidence exists to support this proposition, it has never been directly tested in the southern part of the Central Andes, the only area where Tiwanaku and Wari settlements are currently known to overlap.

The methodology elaborated in this paper is designed to assess whether formal exchange routes existed between the core territory of one of these states (Tiwanaku) and a known area where both states had interests (the Moquegua Valley). Existing data indicate that there very clearly was some kind of exchange between the altiplano and this coastal valley. This research will therefore assess: 1) if formal state-maintained trade routes existed between the Tiwanaku homeland and the largest known Tiwanaku colony in the west. Two other alternatives will be tested: 2) nonstate-sponsored exchange (down-the-line and/or autonomous caravan exchange) between these areas and, 3), the lack of trade links in the optimal travel areas and the existence of intermediary exchange routes elsewhere between the Moquegua and Lake Titicaca regions. The research results will inform our models of Tiwanaku state formation in particular, and the role of interregional exchange in the development of archaic states in general.

The archaeological correlates of formal and nonstate exchange have been worked out through analysis of ethnographic and ethnohistorical data as well as decades of survey and excavations in the area by scores of archaeologists. The archaeological research design to test this model will use full regional coverage and large-site reconnaissance survey in key areas between the Moquegua valley and the Tiwanaku heartland. These areas include the key "corridor" between the central place of Mazocruz and the last major water source of above Moquegua around Lake Suches.

The model elaborated here is based on the definition of 2500 km of footpaths from historical maps and photographs that would be associated with both state and non-state exchange. The research area is a 250 km x 50 km rectangle between Moquegua and Tiwanaku defined by 1) GIS analysis of the least-cost travel route between the sites, 2) remote sensing evidence for water and pasture resources linking the two regions to each other, 3) Early Colonial period and ethnographic sources that define traditional llama caravan routes and, 4) the use of historical analogy of roads in the historic and modern periods.

This chapter thus lays out a methodology for predicting the locales of state sponsored road infrastructure and way stations over a vast survey region comprising some 12500 square kilometers. A full coverage survey of the area at 2 km^2 per day would require over 20 years of continuous survey with a crew of 4. Using remotely sensed data to define historical routes, water resources, and GIS-based probability indexes, we propose a model that can be tested with 6 months of survey fieldwork.

NEW AERIAL AND SATELLITE REMOTE SENSING METHODS FOR HERITAGE CONSERVATION

Ray A. Williamson

The Space Policy Institute, The George Washington University, Washington, DC 20052, rayw@gwu.edu

New methods of collecting and analyzing Earth observations data from satellites and aerial platforms are developed and introduced to the marketplace every year. Taken as a whole, these data sources and analytical methods have steadily improved heritage conservation. In particular, they have enhanced the ability of heritage conservation professionals to improve their analysis of heritage values and, where needed, make the case for conserving heritage sites. Whether for "natural" or built environments, this information also enhances the ability of heritage conservation professionals to educate the general public about the historical and cultural importance of heritage conservation. Geospatial data and analysis also augments the ability of local communities to explore their own heritage, enhancing their sense of place and creating the foundation for sharing this understanding with others.

This paper surveys some of the sources of data and methods of analysis and display that have become available within the past decade, and examines some of the trends in future development. It also explores some of the issues faced by heritage conservationists as they attempt to bring these methods into broader use within their fields of expertise. Finally, it explores development trends in these powerful technologies that promise to make them even more powerful and useful for heritage conservation.

THE APPLICATION OF REMOTE SENSING AND AERO-PHOTOGRAPHIC ARCHAEOLOGY IN THE WEST OF INNER MONGOLIA AUTONOMOUS REGION

lbyanglin@263.net

Centre for Remote Sensing and Aero-photographic Archaeology, National Museum of China
The Archaeology academe, Inner Mongolia Autonomous Region, China

The working group took color photos for the archaeological sites and integrated a light digital aircraft remote sensing system in the west of Inner Mongolia Autonomous Region by MF-11 ultra-light airplane with camera specification and communication system.

Unexpectedly, the working group discovered a new temple site, and the group did its best to find the key of the secret. It was a landmark of the consummation of the research in the academia.

MULTI-TEMPORAL MONITORING OF LANDSLIDES IN ARCHAEOLOGICAL MOUNTANIOUS ENVINRONMENTS USING OPTICAL REMOTE SENSING IMAGERY

Freddy Yugsi1, Henri Eisenbeiss[2], Fabio Remondino[2], Wilfried Winkler[1]

[1]Geological Institute, ETH Zurich, CH-8092 Zurich, Switzerland
[2]Institute of Geodesy and Photogrammetry, ETH Zurich, CH-8093 Zurich, Switzerland

Ground-based techniques for the characterization and analysis of landslides, including geological mapping, geophysical surveys, and field based geotechnical and geodetic monitoring, are not always cost-effective or practical, especially in developing countries where both land use planning and risk policies are under-valued, and therefore few funds are assigned for natural hazard investigations. Photogrammetry and remote sensing (we usually talk about remote sensing when land use and vegetation studies are also involved) are image-based techniques for the extraction of metric and semantic information from images. Traditionally terrestrial photogrammetry, aerial photogrammetry and satellite remote sensing were developed along separate lines, both in terms of types of sensors used, processing methodology, tools and products. Nowadays, within an almost totally digital environment, a strong convergence is appearing with a more common integration of these technologies, leading to formidable results. Moreover, technological advances have resulted in a vast array of relevant and efficient image acquisition tools including: high, medium and low-resolution satellite images, large format digital aerial cameras, hyperspectral sensors with several hundred channels, model helicopters (UAV) with off-the-shelf digital cameras, panoramic cameras and a large number of diverse consumer still video cameras and camcorders. These products are augmented by GPS/INS systems for precise navigation and positioning. With these advances digital photogrammetry provides a powerful tool for 3D data determination and scene recording. Between the possible applications, photogrammetry can be used for landslide detection and monitoring using aerial photographs and/or satellite imagery. Moreover, as there are archives containing multi-temporal image data, they can be employed for monitoring, analysis and prediction. Mountainous regions have always been considered as zones prone to the occurrence of landslides. Therefore archaeological sites located in mountainous areas should be evaluated and monitored with the recent technologies to provide relevant information for planning preservation and protection strategies. In this contribution, we report on the results of multi-temporal monitoring of the archaeological area surrounding El Tambo (Ecuador). Ingapirca, Ecuador's major Inca settlement, is located approximately 5 kilometers away from the village of El Tambo. This site is on the northern flank of a N-S valley between two cordilleras forming the Andes in the zone. Due to its mountainous location and geological condition, different large-size landslides (Figure 1) have occurred in the area, most of them active for a long period of time. These landslides currently affect around 25000 people living in the two principal towns (El Tambo and Cañar) in an area of approximately 300 km^2.

Multiple sets of aerial images dating from 1963 to 1995 were processed with our newly developed image matching tools resulting in a set of four high resolution Digital Terrain Models (DTMs) and the corresponding orthophotos (Figure 2). For the photogrammetric modeling pipeline, four different packages were used. For the ground control and tie points collection as well as for the orthophoto generation, we employed the commercial software LPS (Leica Photogrammetry Suite). Image triangulation was done using both a commercial software (ORIMA) and an in-house software (BUN) developed by the Institute of Geodesy and Photogrammetry of ETH, Zurich. For the DTM extraction, another in-house package called SAT-PP (Satellite Imagery Precision Processing) was used. With the recovered 3D data we could identify five different areas where active landslides are currently present and quantify the long term rate and the direction of the larger displacements. The obtained results can be imported into GIS software, for planning and preservation of the archaeological area and to create new land use policies or to select places to install early warning systems. We demonstrate how modern image-based technologies can be reliably employed for monitoring, studying and planning of archaeological sites located in critical and mountainous areas. The methodology reported here can be applied to other places exposed to landslide hazards, and in particular areas where cultural heritage sites are at risk.

REMOTE SENSING ARCHAEOLOGY FOR ANCIENT CITIES STRUCTURE OF PINGYAO AND LIANGZHU, YUHANG CITY, ZHEJIANG PROVINCE, CHINA

ZHANG Li, WU Jianping

Joint Laboratory of Remote sensing Archaeology, Chinese Academy of Sciences, Ministry of Education and Cultural Relics Bureau, Shanghai, China Lzhang@geo.ecnu.edu.cn
Geography Information Science Laboratory, Ministry of Education, East China Normal University, Shanghai, China Jpwu@ geo.ecnu.edu.cn

Liangzhu culture is neolithic culture in Taihu area, China. Initially it was discovered at Liangzhu area, Zhejiang Province, China. Since 1936 a lot of exquisite culture relics were discovered one after another. Archaeologists locally unearthed many times. But because the area is very large and the shape of the earth's surface is reduced to fragment, Liangzhu's buildings layout structure is still a puzzle to be resolved. Today we processed aerial photographs and satellite images with different temporal, scale and band□discovered four important regions of Liangzhu period in the investigated area.

1. Central urban districts

The Building area is near Mojiajiao. The whole area is surrounded by river system. West part is high and southeast and north part are low. The building is divided into 3-floor structure. East, south and north building base are built on alluvial plain 3~4 meter above sea level, the building base on the south end of west part is built on the hill 9-10 meters above sea level. The building bases on west end and north end of west part are built on upland 5-6 meters above sea level.

Mojiajiao city is located in the west by north, artificial flat, is based on a natural tableland. Flattop is higher than east plain around 7 meters. Mojiajiao city is 600 meters long from east to west, 450 meters wide from south to north. The border is city wall around the city. East and south city wall are slightly wider than west and north city wall, . There are city gates on all sides□

There are 3 artificial rectangle soil platforms in the city of Mojiajiao: Big Mojiajiao-Shan, Small Mojiajiao-Shan, Wugui-Shan. Relative altitude of Mojiajiao -Shan platform base is 3~4 meters, 162 meters from east to west, 73 meters from south to north. Building base is divided into east part and west part. On the west the building pattern appears Chimes traditional rectangular courtyard. East building pattern appears right angle with broken line. On the north of Big Mojiajiao-Shan platform base there are regularly built steps. Around Big Mojiajiao-Shan platform base there is Protection Rivers intermittently.

Small Mojiajiao-Shan is located in the right west of Big Mojiajiao-Shan, 73 meters away. Relative attitude of platform base is 3 meters, 85 meters long from east to west, 25 meters wide from south to north.

Wugui-Shan is located on the right south of Small Mojiajiao-Shan 85 meters away. Relative altitude of platform base is 3 meters, 109 meter long from east to west, 65 meters wide from south to north. There are 3 bases from east to west. Around platform bases there are marks of Protection River.

There is a round two floors soil altar in the center of Mojiajiao city. The soil altar's height is 2 meters. The diameter of lower alter is 25 meters. The diameter of top alter is 12 meters.

There is a long step from south to north up near to west wall of Mojiajiao City. The step may go up to west gate of Mojiajiao City. In the middle of journey we can enter into Western City. The Western City is a rectangle expanding from south to north. North half part base is lower slightly than the ground surface in Mojiajiao City. South half part base is slightly lower than north half part.

The east and northeast Mojiajiao Cities are built on the lower city of 3-4 meters above sea level plain; there are many wall bases and building platform lands. The platform land is 5-6 meters high. The platform base in right east is more regular. Because the terrain is lower in the west, these regions have been broken and unendurable.

Near southeast wall of Mojiajiao City there is a little river. The river is narrow. But in the image it was a 20 meters wide beheaded river. The head of river was located at east gate of Mojiajiao City. River water went into main river course in the south Mianchang gang.

Near west end of north wall of Mojia,. there is a protruding square platform land. Platform's height is 7 meters, 154 meters long from east to west, 90 meters wide from south to north, there is no wall.

In the south of Mojiajiao City there is a naturally formed earth hummock and appears arc. There are many building contours cut into block on top. It is obvious on the top of tableland in middle.

In summary, Total area of the center city region is around 5 square meters. The volume of earthwork of main building base is around 2, 400,000 cubic meters

2.Large scale settlement region

It is around Xunshan. Area is around 2.1 square kilometers. In the middle Xunshan is table hillock ,higher than plain 30 meters, pyramid shape the slope of 4 slopes is 30°. The top of Xunshan had been built artificially to rectangular platform. There is a barricade on the north of platform. There is a square soil platform on the middle of top. There is a spiral pathway going up in the south slope of Xunshan. In half way up the table hillock there are two square platforms. There are three-built pyramid shape table hillock In the south of the Xunshan, there are dry lands which are 5-6 meters above sea level. Ground surface is broken. There is a trace that people resided here from time to time.

3. Small scale settlement region

Small scale settlement region are mainly along roots of hills on northeast in investigated area. They are located 4-5 meter above sea level. These sites are not very large and scattered. The largest one is in Yaojiadun. There are 5 platform lands with scope 600 m×600 m.

4. Irrigation works area

Irrigation works area is mainly located in northwest of the investigated area.There are many dams in the southeast and north of investigated area by hills. The dams in southeast are uncontinuous and shortdams. The dams in the north are many and long. A dam among them is a continuous dam extending several kilometers, the head of dam is located in Laojiatou, Fengjiatou, Xujiatou, Wengjiatou., the end of dam is at southwestern Chiba. The total length (including the part formed by natural massif) is 7.7 kilometers. Among them artificial part is 3.5 kilometers. Including single dam, double dam and quadruple dam, the whole northern artificial dam length is 4.2 kilometers. Volume of earthwork is around 1,680,000 cubic meters.

Besides dams in north of the investigated area there are a gathering water region based on natural arc depression.On the bottom of depression there are artificial polders located different height. There is different volume of water in different polder under different water situation. In addition there are many interception facilities on the bottom of depression to reduce water speed. Among them there is a string of "回" shape dam. "回" goes bigger from north to south. "回" dam body is 8 meter wide. The clearest dam is in the middle of string of dams. The dam is square with 40 meters side.

Finally we reconstruction Mojiajiao City by means of computer. It is found that the investigated area is the largest one among embryo cities discovered at present in China. It is the outstanding representative of South China's city origin.

Semi-automatic search for cultural heritage sites in satellite images

Lars Aurdal,[1] Line Eikvil,[1] Hans Koren,[1] Anke Loska[2]

[1]Norwegian Computing Center, Oslo, Norway
[2]Norwegian Directorate for Cultural Heritage, Oslo, Norway

1 Introduction

It is generally recognized that the increasingly intensive use and modification of the landscape resulting from modern demands for efficient infrastructure and land use (agricultural production, mining, energy sources, leisure/tourism facilities etc.) exerts growing pressure on cultural heritage sites in the landscape.

In recognition of this, the Norwegian Directorate for Cultural Heritage (Riksantikvaren), the Norwegian Institute for Cultural Heritage Research, the Norwegian Space Center and the Norwegian Computing Center in collaboration with several county administrations have started a project with the overall aim of developing a cost-effective method for locating, surveying and monitoring potential cultural heritage sites on a regional and national scale. Given the enormous costs of surveying the tracts in question by traditional field work, alternatives must be sought. One possible approach is through the use of satellite images. Current experience with manual analysis of these images indicates that cultural heritage sites with no apparent above ground structures still may be observable in such images.

The working assumption in this project is that such cultural heritage sites will be detectable in satellite images due to alterations in the spectral signature of the bare soil or that of smaller plants growing out of the soil. The project is currently limited to sites located in agricultural fields, detecting such sites in forests is expected to be a much more complicated problem. In satellite images of agricultural fields, cultural heritage sites may show up as patches in the images having particular shapes and different spectral properties compared to their surroundings.

The costs of acquiring and analyzing such images are not negligible, and it is not reasonable to assume that such an approach can entirely replace traditional field work. Nevertheless, it is clear that if the potential of satellite images for locating, surveying and monitoring such sites could be clearly established, this would be a valuable addition to the traditional methods that would be very useful in the management of this important resource on a regional and national scale (Grøn 2002).

2 Methods

We have started to develop methods and software for assisting archaeologists in the process of scanning satellite images for the presence of potential cultural heritage sites. The aim in developing the system is not to replace the archaeologist in the process of searching for potential cultural heritage sites, but rather to assist the archaeologist in the process of detecting *potentially interesting* sites. Once such a site has been detected, the final decision as to the interpretation of the site and possible follow-up actions are left to the operator. The system is therefore intentionally made to produce numerous false positives.

The operation of the system is based on two central processing steps, *segmentation* to detect potential cultural heritage sites and *classification* in order to provide an initial type for each detected region. The segmentation step is based on the use of adaptive thresholding. The classification step extracts a number of features from the different segmented regions and performs a minimum distance classification based on these features.

It is important to bear in mind in the following discussion that the system is designed to detect **candidate sites** and that no claim is made that these candidates are **true cultural heritage sites**. Even human specialists can not make such an assertion based only on satellite imagery. The verification of a potential site always depends on some kind of ground intervention (Grøn 2003, Aurdal 2005 and Aurdal 2006).

Segmentation

Segmentation is the process of dividing the areas of the satellite image into different regions based mainly on their spectral characteristics. The current segmentation method works on panchromatic images, it operates by identifying areas that are darker or brighter than the surroundings. This is obtained using Niblack's method (Niblack 1986) for threshold selection. The method is used in two passes, first to find dark regions and then to find bright regions.

Niblack's method works by estimating the mean and variance in a local window W. A local threshold $t_{W,bright}$ is found by multiplying the local standard deviation σ_W with

[1] Corresponding author, aurdal@nr.no.

a weight factor ω and adding it to the local mean μ_W. The overall formula for threshold selection is thus given by:

$$t_{W,bright} = \mu_W + \omega\sigma_W$$

This threshold can be used to find regions that are brighter than their local surroundings. In order to find regions that are darker than their local surroundings, another threshold $t_{w,dark}$ is found by multiplying the local variance by a (possibly different) weight factor and subtracting it from the local mean, thus:

$$t_{W,dark} = \mu_W - \omega\sigma_W$$

The reason for using a locally adaptive algorithm is that the local brightness varies from field to field and what we are looking for are actually regions that stand apart from their local surroundings. The performance of this algorithm depends on the size of the window W over which we calculate the local means and variances as well as on the choice of value for the weight factor ω.

Once dark and bright regions have been detected in this manner, we postprocess them by removing regions consisting of too few pixels. We also check if the physical extent of the region is too large, regions having widths or heights that are above a certain threshold are excluded from further processing. Finally, we verify that the regions have a reasonable contrast to their surroundings by looking at the variance in an area defined by each region and a band of pixels surrounding each region.

Classification

Classification is performed on the regions resulting from the segmentation. In this process, the spatial and spectral characteristics of the segmented regions are taken into consideration to determine whether they are potential cultural heritage sites or not.

During classification, features are extracted from the segmented regions of unknown class. Based on the statistical class descriptions, a minimum distance classifier is used to determine the most probable class for each region. The classes that have been defined currently consist of two classes representing potentially interesting regions, and four classes of different types of "noise" regions. More specifically, we use the following classes:

1. Potential cultural heritage site, bright.
2. Potential cultural heritage site, dark.
3. Plow furrow, bright.
4. Other noise, bright.
5. Other noise, dark.
6. Object, dark (typically this will consist of buildings).

The features that are currently used to characterize regions are the following:

1. *Complexity:* This is calculated as the ratio between a regions perimeter and area. When the perimeter of the region is long in comparison to the area enclosed by the perimeter, the region has a complex shape. We expect regions corresponding to cultural heritage sites to have a low complexity compared to several other types of regions such as plow furrows.
2. *Border quality:* This is calculated by summing the gradient information along the border of the region (and normalizing by the border length). This provides a measure of the quality of the border of the region. If this measure is high it means that the entire border is well defined, if it is low, at least parts of the border are poorly defined. We expect cultural heritage sites to have well defined borders.
3. *Aspect ratio:* This is calculated as the ratio of the lengths of the major and minor axes of an ellipsis adapted the region. Long, narrow regions will have a large ratio between these axes. More compact objects will have a smaller such ratio. We expect cultural heritage sites to have a relatively low aspect ratio.
4. *Uniformity:* This is the standard deviation of grey levels within the region. Relatively homogeneous regions will have a low standard deviation of the pixels in the region. Similarly, if the pixels have highly different spectral values, then standard deviation will be high. Cultural heritage sites will typically be relatively uniform in comparison to for instance plow furrows.
5. *Contrast:* This is the difference in mean grey levels inside the region and in an area surrounding the region. This is essentially a measure of contrast and provides an indicator of how well the region stands apart from its surrounding. Cultural heritage sites have higher contrast than certain regions but lower contrast than for instance buildings.

The system was trained by inspecting detected regions from the areas surrounding the Lågen River in Vestfold County. These were then labeled with a label corresponding to the classes listed above. Based on this manual inspection, we derived the features described above from both the interesting and uninteresting regions. It is then possible to derive a statistical description of both the interesting and the uninteresting classes. This statistical description is then input to the classifier for use in classifying the segmented regions.

Overall system

The system is operated through a simple graphical user interface (GUI) that provides the user with two different methods for running the system; one fully automatic method and one stepwise method providing some more user control. In both approaches care has been taken to keep the necessary knowledge of technical details to a minimum. A user with little knowledge of image processing and remote sensing should still be able to run the system without too much training. In addition to letting the user initiate segmentation and classification of the input images, the interface lets the user generate masks to limit the processing to agricultural fields. These masks are derived based on GIS data giving the boundaries of these fields. The interface also lets the user load a mask and a satellite image into a common reference frame. Finally, the interface provides functionality for manually deleting regions that are deemed to be of no interest.

3 Results

The developed methods have been tested on two satellite images of the agricultural areas surrounding the Lågen River in Vestfold County as well as those surrounding the Rygge Community in Østfold County in Norway. Initial results are promising. The following figures exemplify the results obtained on part of the image of the Lågen River. This area is rich in known cultural heritage sites and is also expected to contain a large number of unknown such sites.

Figure 1 shows the subsection of the Lågen image that has been considered. This region contains an excavation site (Bronze Age grave mounds) as well as several known cultural heritage site locations. The area is expected to contain numerous unknown sites. For this location, map data giving the border of the agricultural fields is readily available. This has been transformed into masks delimiting the processing only to agricultural areas.

Figure 2 shows a subsection of this satellite image (from the lower central part of the image in Figure 1. Here a number of potential sites are indicated with red arrows. In the segmentation step we seek to detect these and other possible sites based primarily on their contrast to the local background.

The result of the segmentation step is illustrated in Figure 3. In this image, regions that are darker than their local background are labeled red, those that are brighter than their local background are labeled green. Notice that the three dark spots marked with arrows in Figure 2 have all been detected. Also notice that a large number of other regions are detected. Comparison of Figures 2 and 3 indicate that some of these detected regions are obvious errors. Notice for instance that part of the road crossing the lower right part of the image in Figure 2 has been considered a bright region. Plow furrows, field borders and small groups of trees have also given rise to detections of regions.

There are several reasons for this large number of detections. First of all, it is essential to remember that the segmentation step will consider local pixel intensity values only, thus no contextual information is used at this stage. Secondly, the parameters of the segmentation process are deliberately set to values that make this process very sensitive. This is done in order to avoid loosing possible detections in the segmentation step.

Figure 1: *Part of the Lågen River in Vestfold County, Norway. The agricultural areas surrounding the river are known to contain large numbers of cultural heritage sites and are also expected to contain many unknown such sites.*

Figure 4 shows the result of applying the preclassification step to the segmentation result shown in Figure 3. Notice particularly that parts of the road network originally detected by the segmentation step are now excluded from further processing. The image in Figure 4 is now input to the classification step for further analysis of the detections.

In the classification process features are extracted from each region and the regions are then analyzed based on these features.

The final class of a region is determined by finding the statistically most likely class given the features. The result of the classification step is shown in Figure 5. This figure shows only the regions that belong to either class 1 or 2, that is, the two classes corresponding to potential cultural heritage sites.

Figure 2: *Zoom of part of the imagePotential cultural heritage sites will typically appear as indicated here (red arrows), either as spots that are darker or brighter than their surroundings.*

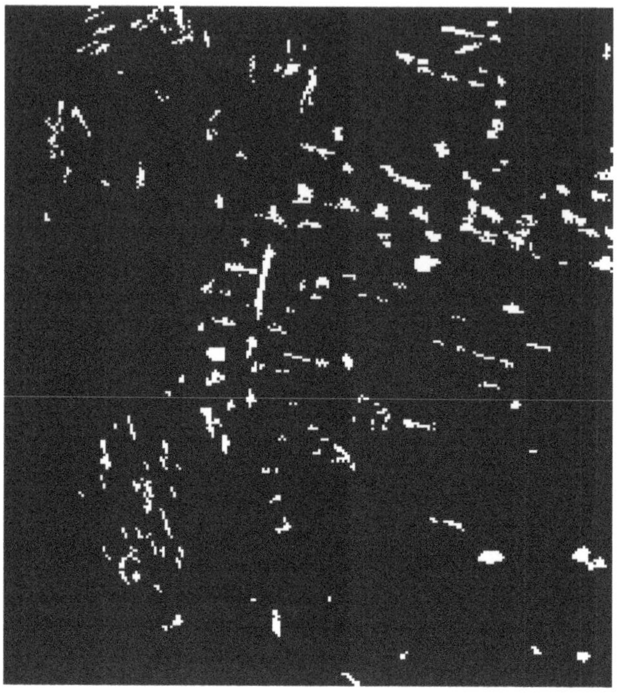

Figure 4: *Result of preclassification applied to the image shown in Figure 3. Notice that the long structures associated with field boundaries etc. have been suppressed.*

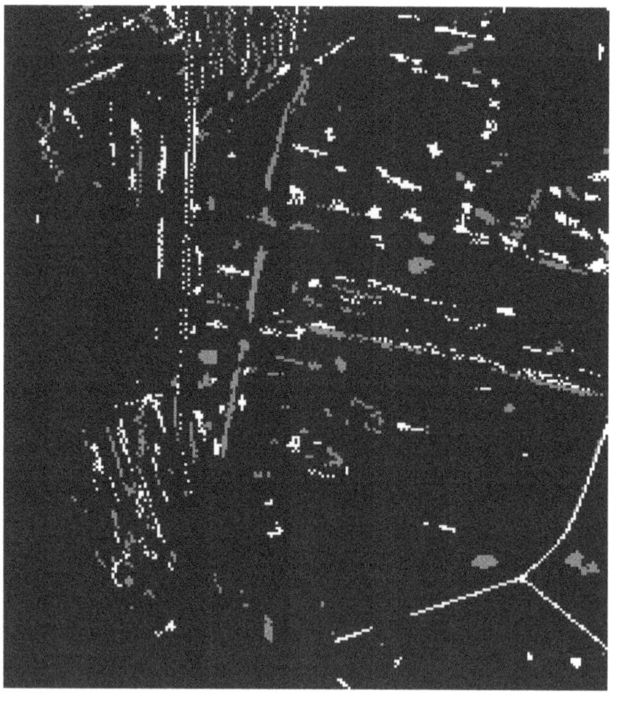

Figure 3: *Result of segmenting the image shown in Figure 2. Dark detections are marked in grey, bright detections are marked in white.*

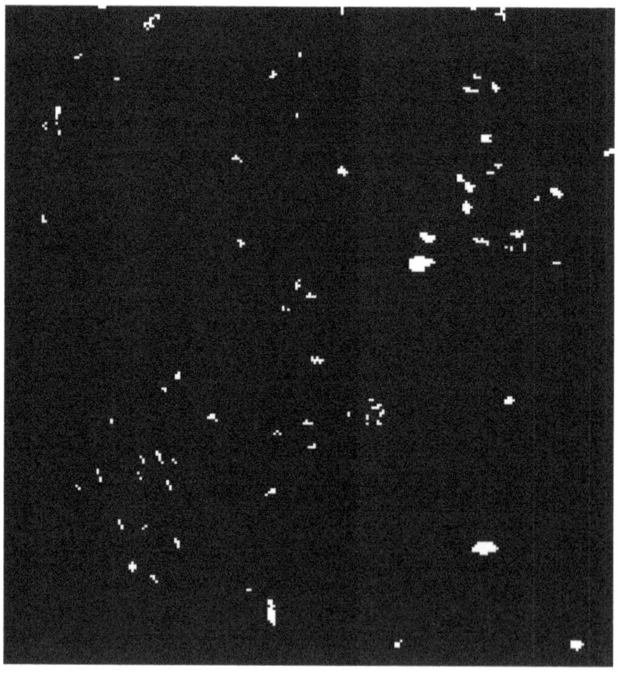

Figure 5: *Result of classification applied to the image shown in Figure 4.*

Notice the large decrease in the total number of regions. Since the classifier performs a much more detailed analysis of the regions compared with the previous steps, it will be much more selective.

The classification result shown in Figure 5 is also shown on top of the original image in Figure 6. Notice that in order to facilitate interpretation the contrast has been artificially enhanced in this image. Apart from the three potential sites indicated with arrows in Figure 2, the system has also detected numerous other sites as being potential cultural heritage sites. Some of the regions that have been classified as potential cultural heritage sites are clearly errors. Notice for instance the small detections surrounding the small group of trees in the central part of the image. In this region, the mask for agricultural fields has not been sufficiently precise, and trees have been present in the regions that were processed. This will often lead to erroneous detections. Also, notice that some plow furrow structures have been classified as cultural heritage sites in the lower left field of the image.

4 Conclusions and future work

We have started to develop methods and software for use in detecting possible cultural heritage sites in satellite images. Sites presenting no above ground structure manifest themselves as spots having particular shapes and different spectral properties compared to their surroundings. Our approach operates in two steps, the first step being a segmentation in order to detect candidate regions followed by a classification in the second step. The methods have been tested on two satellite images providing promising results. The chosen approach can be trained to meet classification criteria established by archaeological specialists.

Training of the system has so far been done by operators who are not specialists in archaeology. In the near future, the training will be extended to include input from archaeologists. Since our aim is to make a system that will be used by archaeologists in different county and municipality administrations, we will involve these in extensive testing of the system. This is important in order to detect potential methodological weaknesses of the system, it will also provide valuable feedback on the functionality of the graphical user interface.

One methodological improvement that is foreseen in the near future is the inclusion of multispectral data in the classification process. Currently, only features derived from the panchromatic data are used in the classification step. Inclusion of multispectral data in this step might improve the performance.

The current segmentation and classification methods are designed to produce a large number of false positives. This is beneficial in the sense that it reduces the risk of missing interesting sites. However, if the number of false positives becomes too large the user will spend too much time going through the detected sites. Hence, in future work we will seek to reduce the number of false positives. There are numerous ways of achieving this. In collaboration with archaeologists, we will try to tune the system better so that the interesting sites are still detected while uninteresting sites to a larger degree are rejected. The current system is expected to be overly prone to detecting false positives. Better preprocessing, for instance in order to detect and reduce the effects of plow furrows might be effective. We have so far only experimented with the use of Niblack's method for segmentation, it is possible that other methods are better adapted. This will be investigated in the near future.

Figure 6: *Result of the final classification overlaid on part of the original image.*

In the future we hope that this system will be used in Norwegian county and municipality administrations in order to assist archaeologists in gaining an overview of potential cultural heritage sites in their areas of responsibility. This may allow for better inventorying of this very valuable resource which in turn may lead to a better management and protection of cultural heritage sites. It will also allow for a higher degree of predictability in modern infrastructure construction processes, currently such processes are routinely halted due to discoveries of unknown cultural heritage sites.

5 Acknowledgements

This work was financed jointly by the Norwegian Directorate for Cultural Heritage (www.ra.no) and the Norwegian Space Center (www.spacecenter.no).

6 References

AURDAL, L., SOLBERG, R. & LOUS, J. 2005 Computer assisted detection and classification of possible cultural heritage sites: User requirements and conceptual design. Norwegian Computing Center.

AURDAL, L., EIKVIL, L. & KOREN, H. 2006 Cultural Heritage Search - Software Prototype Description. Norwegian Computing Center.

GRØN, O. & LOSKA, A. 2002 Development of methods for satellite monitoring of cultural heritage sites. NIKU, NSC and RA.

GRØN, O., AURDAL, L., CHRISTENSEN, F., TØMMERVIK, H. & LOSKA, A. 2004 Locating Invisible Cultural Heritage Sites in Agricultural Fields: Evaluation of Methods for Satellite Monitoring of Cultural Heritage Sites-Results 2003. NIKU, NSC and RA.

NIBLACK, W. 1986 An Introduction to Digital Image Processing, Prentice Hall.

Archaeological remote sensing in Yemen, the Jabali test site From large-scale survey to field investigation

Jean-Paul Deroin,[1] Florian Téreygeol,[2] Paul Benoit,[3] Mohammed Al-Thari,[4] Ismail N. Al-Ganad,[5] Jürgen Heckes,[6] Annette Hornschuch,[6] Audrey Péli,[7] Sophie Pillault,[8] Nicolas Florsch[9]

[1] Université de Marne-la-Vallée, EA 2455, *Observation de la Terre et Information Géographique*, Champs-sur-Marne, France
[2] UMR 5060, *Institut de Recherche sur les Archéomatériaux*, Belfort, and UMR 9956, CEA Saclay, France
[3] UMR 8589, LAMOP, Université Paris 1-CNRS, Paris, France
[4] ZincOx Resources plc, Hadda, Sana'a, Republic of Yemen
[5] Yemen Geological Survey and Mineral Resources Board, Sana'a, Republic of Yemen
[6] Deutsches Bergbau-Museum, Bochum, Germany
[7] UMR 8167, *Orient et Méditerranée*, Université Paris 1-CNRS, Paris, France
[8] UMR 5060, *Institut de Recherche sur les Archéomatériaux*, Belfort, France
[9] UMR 7619, SISYPHE, Université Pierre et Marie Curie-CNRS, Paris, France

1 Introduction

During the pre-Islamic period and until the 9th century AD, the Jabali area was mined for silver. This intense activity ceased at the end of the 9th century AD, according to the description given in the 10th century AD by al-Hamdani, the famous Yemenite geographer who named the site, al-Radrad, one of the most important silver mines of the Abbassid world (Al-Hamdani, published in 1957, 1968 and 2003). In the early 1980's, the site was rediscovered by the French Geological Survey (BRGM) using also data from the archaeologists. From 2004 onwards, archaeological investigations have been carried out by both French (Benoit *et al.*, 2003) and German (*DBM, Deutsches Bergbau-Museum*) teams. The last mission carried out by the French research group has been done in March 2006, including observation of ancient objects, which are visible on the surface and in the mines as far as they are accessible. Archaeological field investigations focused on topography, mapping of archaeological remains, slags and ceramics sampling. Geological mapping and sampling have been also carried out, with emphasis on lithologies from the one hand, and geomorphology of alluvial deposits in the lower part of the area from the other hand. Some geophysical measurements have been acquired using magnetometry. All these investigations have been supported by GPS measurements and satellite imagery as well. DBM has deeply explored the cartographic potential of the QuickBird data for detailed topography.

The aim of the present paper is to summarise the methodology used for processing and analysing remote-sensing data at different scales and their respective contribution to the archaeological, geological and environmental knowledge of the area under investigation. The particular interest for topographical and geological mapping of the very high resolution QuickBird data is put into light.

2 The Study Area

The Jabali area, about 70km as the crow flies north-east from Sana'a, the capital city of the Republic of Yemen (Sana'a and Ma'rib districts), is a Zn-Pb-Ag deposit currently explored for zinc-ore (Figure 1). The mineral rights are held through an exploration licence which is led by ZincOx. Jabali contains a geological resource of 12.6 million tonnes of oxide ore, grading about 9% zinc, 1.2% lead and 68g/t silver. The mineralization comprises carbonates and oxides (smithsonite, hydrozincite, etc.). Silver is mainly related to sulfides, which are represented, even less than oxides: sphalerite (relatively rare), galena, etc. ZincOx planned the development of the Jabali deposit at the rate of 800,000 tonnes per annum of ore. A small amount of lead is recovered and sold but lead is generally not of an economic grade. The silver, however, passes to the tailings and is recoverable. Mining will be from an open pit, with a 2:1 ratio of waste to ore. The ore will be treated by means of a proprietary hydro-metallurgical process, test work on which has indicated a 77% zinc recovery. Production is expected to commence during the second half of 2008. The planned facilities will have a destructive impact on the old pits and processing heaps in the immediate mine area. Archaeological remains and slag heaps on the plateau below will be largely unaffected.

The ancient silver mine of Jabali is located within the southernmost outcrops of the Jurassic carbonates of the Arabian platform, the so-called Amran Group (Table I), close to the contact between the Proterozoic basement (Whitehouse *et al.*, 2001) and the sedimentary cover (Al-Ganad, 1991). The area is deeply affected by volcanism and tectonics. Cenozoic events are represented by two phases of volcanism, first at about 30Ma, constituting large traps in the Sana'a region and the Jabal Amrah (andesites), a volcano in the Jabali area, second during the Quaternary, forming basaltic volcanoes aligned along faults (for example near Ma'rib).

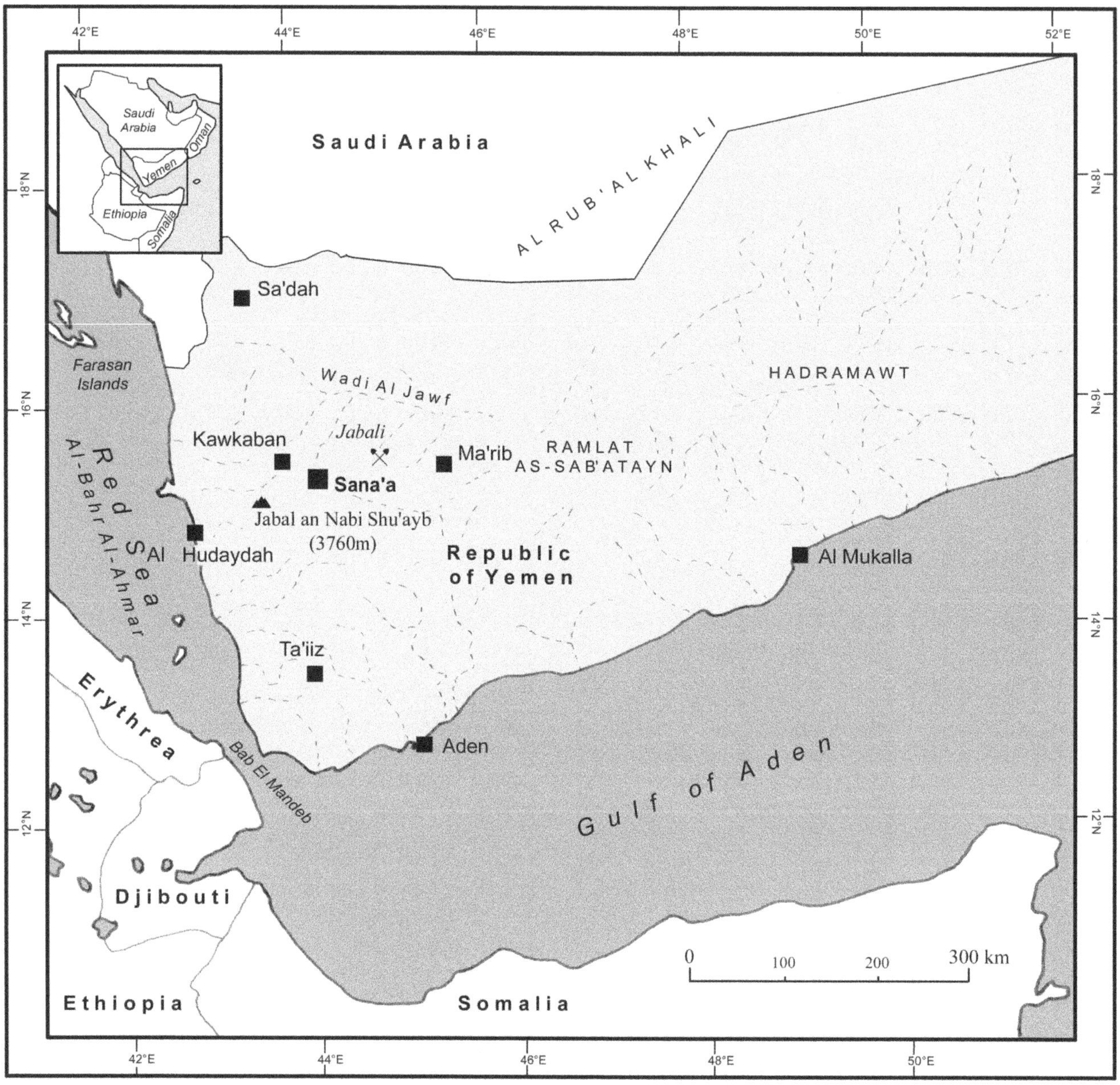

Figure 1 *The Republic of Yemen and the Jabali site.*

Table I. *Simplified geology of the Jabali area (for details see SHARLAND* et al.*, 2004)*

AGE	NAME	LITHOLOGIES	PLACES
Quaternary (≤1 Ma)	Second magmatic episod	basalts, isolated volcanoes and lava flows	Sana'a and Ma'rib
Mid Oligocene (30-31 Ma)	First magmatic episod	basalts, rhyolites, gabbros, granites	Jabal Amrah
Cretaceous to Palaeocene?	Tawilah Group	mainly sandstones (≤ 300 m)	Kawkaban, etc.
Callovian to Tithonian	Amran Group	limestones, dolomites, marls (Al Jawf)	Zn-Pb-Ag deposit
Permian? to Dogger	Kohlan Formation	conglomerates, sandstones, shales	Wadi Lahman
Permian	Akbarah Series	argilites (≤ 130 m)	North Yemen only
Cambrian to Ordovician	Wajif Formation	sandstones (≤ 300 m)	North Yemen only
Proterozoic	Volcano-sedimentary belts, gneiss and migmatites		Basement of Amran G.

3 Remote-sensing data Used

A short history

Close to Ramlat-As-Sab'Atayn, geoarcheological investigations of the Qatabanian irrigation system of Wadi Bayhan were carried out by means of Soyouz Kate 200 data (Marcolongo and Morandi Bonacossi, 1997). More recently, different remote-sensing techniques have been compared in order to identify ancient caravan routes near Ma'rib (Andreeff, 2001). For geological mapping of Yemen, Landsat-MSS data (Grolier and Overstreet, 1978) and Landsat-TM data (Kruck and Schäffer, 1991) were exploited. Christmann and Scanvic (1987) made preliminary observations on the Jabali area using Spot-panchromatic data. The geological tele-analysis has been deeply evaluated using multisource data including Landsat and QuickBird imageries (Deroin *et al.* 2006). The more recent geological map on the area under investigation is that drawn by NRP (2000). It is a synthesis of both field works and previous remote-sensing analyses.

Imagery

We have evaluated different types of remote-sensing data (Table II). At a large-scale, Landsat ETM+ (*Enhanced Thematic Mapper*), Landsat TM (*Thematic Mapper*) and Spot panchromatic data have been used for drawing new geological maps of the area. In the near future, it is expected to evaluate other sources such as ASTER and ALOS imageries.

QuickBird data represent the new generation of very high resolution imageries, which are very useful for archaeological investigations (Lasaponara and Masini, 2005, 2006a, 2006b). QuickBird data have been used for a detailed study of the sector of mines. The Jabali area is covered by the QuickBird scenes 02MAR30073205-P2AS (panchromatic, 0.7 m ground resolution at nadir) and 02MAR30073205-M2AS (multispectral, 2.44 m ground resolution at nadir), both as standard imagery with a position accuracy of 23 m. At first, both scenes were allocated by Erdas resolution merge to a pan sharpened RGB color model.

Some GPS ground control points carried out during a first visit in Spring 2004 by DBM showed differences between 30 and 50 m. Further GPS ground control measurements of obviously identifiable natural points and GPS-tracks along roads - in both directions - were carried out during the campaign in Spring 2005 (Garmin MAP 76S, accuracy 1-3 m by a registration time of two minutes per point, Kleibrink and Joemann 2004).

Digital Elevation Model, DEM

With an about 90m-accuracy, the SRTM DEM is worldwide available (Figure 2). A DEM has been also generated using Spot 5 data (supermode). The QuickBird scene should be processed as a 3D model further. Unsatisfactory results yielded the transformation on the two SRTM and ASTER terrain models. The best adaptation of the scene to the area could be calculated by the transformation on a digital terrain model from digitized elevation lines of the topographic map sheet 1:50 000 'Jabal Bahrah' (Series YAR 50(E), Sheet 1544 B4, 1-OSD 1988). In comparison with the GPS ground control there is still a residual error of approximately 10 m in the horizontal plane.

4 Some Results

Landsat and Spot imageries are particularly suitable for geological mapping at medium scale (DEROIN *et al.*, 1998 and 2005). On Figure 3, the mineralization of Jabali is put in light when using ETM+ band no 3, *i.e.*, the red range of the electromagnetic spectrum. This is clearly explained by the alteration halo of iron oxides concentrated around the deposit and linked with (de)dolomitization phenomena. Dolomites can be easily identified by their spectral response and their texture as well. The link between the Proterozoic basement and the mineralization is clear. It seems that Jabali is situated just above a major Proterozoic shear zone separating to terranes, the Asir/Afif terrane to the north-west and the Abas Gneiss terrane to the south-east. Cenozoic thermal events contributed to concentrate in Jabali elements initially present in the basement.

Table II. *Synthesis of the data used on the Sana'a-Jabali-Ma'rib region (spectral range, V: visible, NIR: near infrared, SWIR: short wave infrared)*

Date	References	Sensor	Captor	Spectrum	Resolution	Cover	Area
1975 10 06	178-49	Landsat 2	MSS	V-NIR	80 m	185 x 185 km	Sana'a-Ma'rib
1987 02 04	165-49	Landsat 5	TM	V-NIR-SWIR	30 m	185 x 185 km	Jabali-Ma'rib
1995 04 21	Nimadata	Spot 3	P	V	10 m	env 50 x 30 km	Sana'a-Ma'rib
2000 02 11-22	S 45-09	STS - Shuttle	SAR-SRTM	Radar band X	90 m	5° x 5°	Regional
2000 03 19	165-49	Landsat 7	ETM+	V-NIR-SWIR	15 - 30 m	185 x 185 km	Jabali-Ma'rib
2000 05 13	166-49	Landsat 7	ETM+	V-NIR-SWIR	15 - 30 m	185 x 185 km	Sana'a
2002 03 30	30073205	QuickBird	MS/P	V	0.70 – 2.44 m	env 10 x 6 km	Jabali area only
2002		Spot 5	P	V	5 m – 2.5 m	6,2 x 18,6 km	Jabali area only

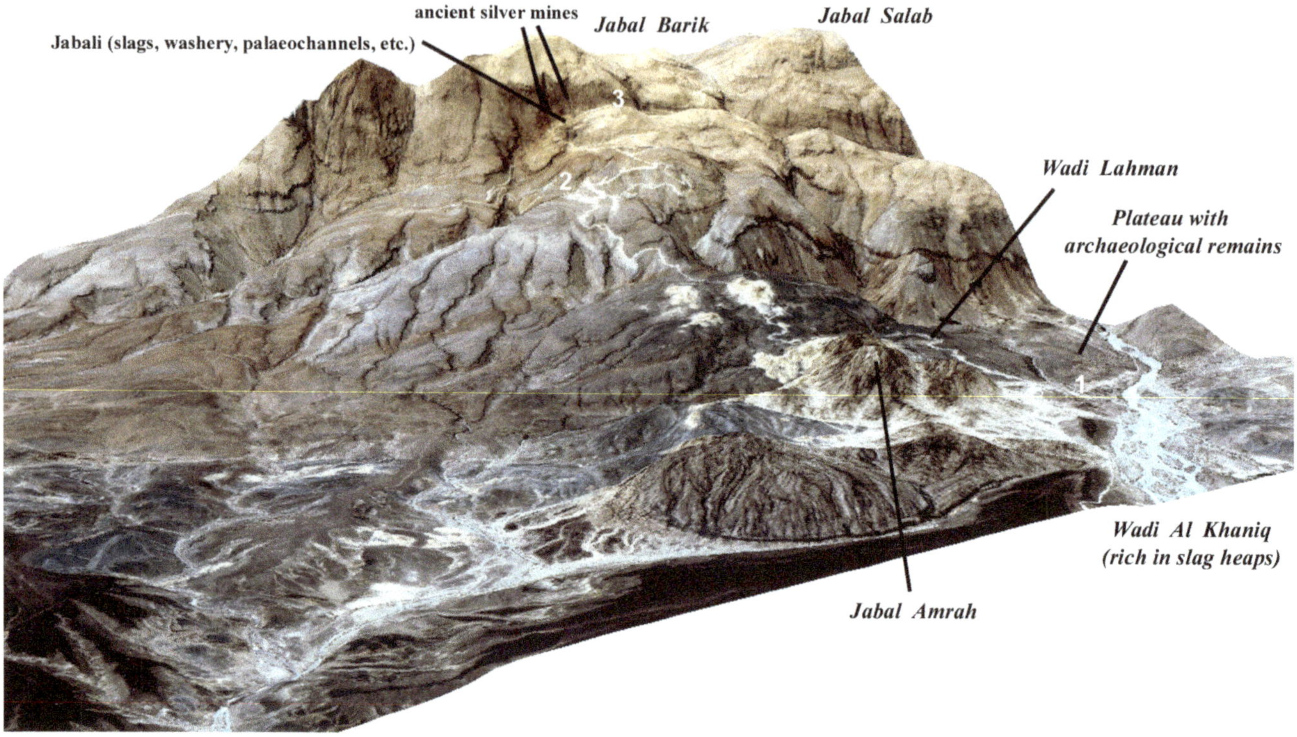

Figure 2 *The QuickBird data draped over the DEM generated by the Shuttle Radar Topography Mission. View from the North-West to the South-East. Size of the area: about 4 km from East to West, about 6 km from North to South. White numbers refer to three military camps (vertical exaggeration: ×4)*

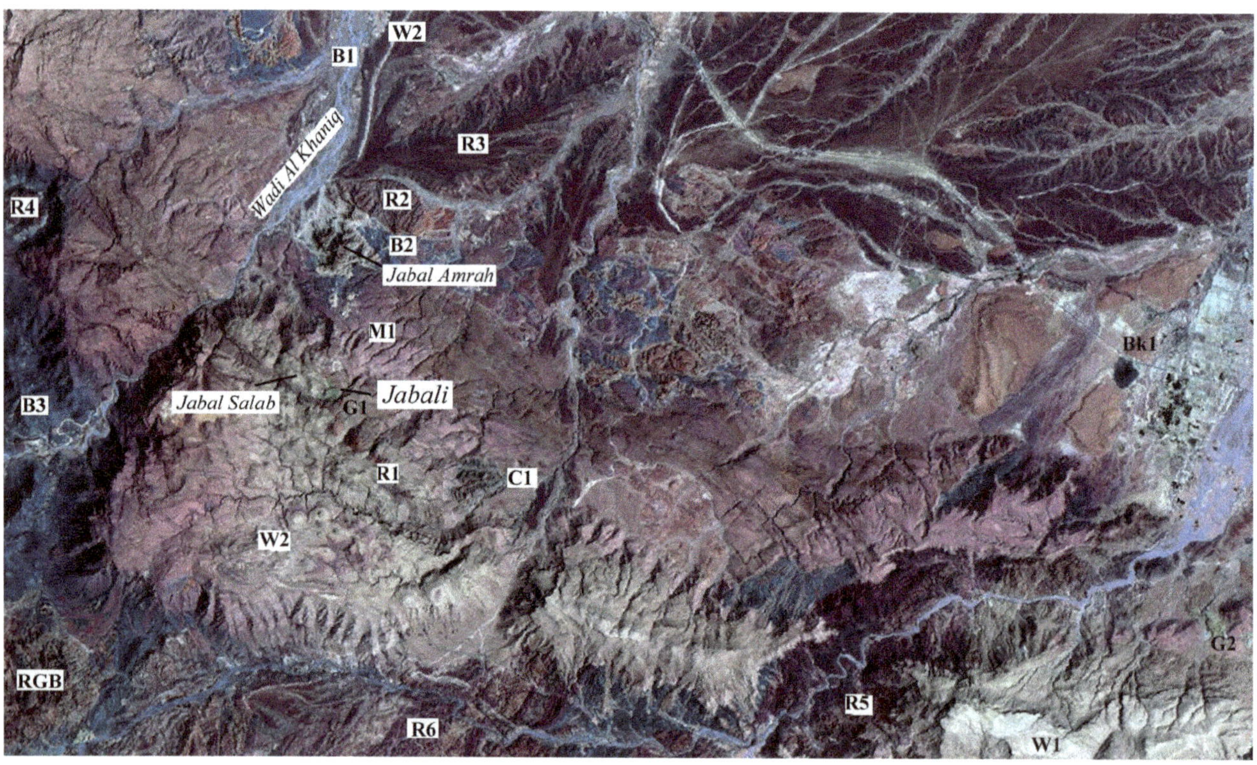

Figure 3 *Landsat ETM+ data, 165-49 acquired on March 19th, 2000 (RGB=Band 7-Band 3-Band 1). About 22 km×15 km*
Blue: B1 Alluvials, B2 Contact aureole of Jabal Amrah laccolith, B3 Contact metamorphism. Black: Bk1 Quaternary basalts. Cyan: C1 Alluvials. Green: G1 Jabali oxidized cap, G2 gossan? Magenta: M1 platy limestones (Dogger-Malm). Red: R1 secondary dolomitization, R2 andesites North Jabal Amrah, R3 Calcareous marls (Dogger), R4 small granitic intrusion, R5 Paragneiss (Proterozoic), R6 Metavolcanites (Proterozoic). RGB Cenozoic granitic pluton. White: W1 Carbonates Jabal Ailan (Malm), W2 Dolomites Jabal Barik (Malm)

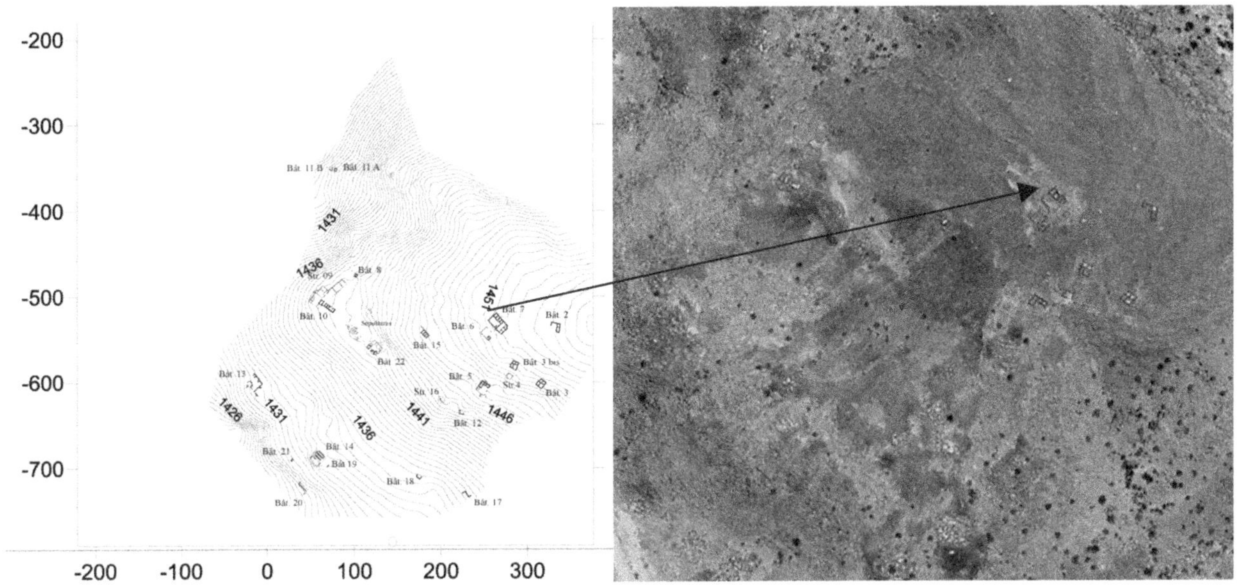

Figure 4 *Comparison between topography (left) carried out using field techniques (elevation calibrated using GPS data) and the QuickBird data (© Digitalglobe, 2003), slighly enlarged. Buildings are concentrated incompact grey tones areas characterizing reg surfaces, whereas circles correspond to rare shrubs growing along fractures within the substrate.*

On the right bank of Wadi Al-Khaniq, a carbonate plateau shows a lot of slags associated with archaeological remains (walls, buildings, graves, etc.). This plateau is about 15ha-wide (Figure 4). Geophysical (magnetic) measurements have been tested on that plateau. A 50 m x 50 m surface has been surveyed. Results show magnetic anomalies, which could be associated to old furnaces remains. Future excavations are now necessary to validate this result. It should be emphasised the potential of geophysics to reduce archaeological fieldwork. From a 2,500 m²-area investigated, excavation could be limited to a 100 m².

An other place with slags and remains has been identified close to a right bank tributary of Wadi Al-Khaniq.

5 Discussion

The Jabali area perfectly illustrates the interest of an integrated study in the topics of archaeological remote sensing, using image processing, 3D visualization, and field geophysics.
From the top (about 2000m a.s.l.) to the bottom (about 1400m a.s.l.), the following subsites can be shortly described.

(1) First, the mine area in the higher part, which includes a number of ancient mines. It corresponds to carbonate layers deeply weathered and karstified.

(2) In the sector of Jabali itself, palaeometallurgy is revealed in the field by slags, palaeochannels, alluvial deposits linked to washery, etc.

(3) 5km to the north, along Wadi Al-Khaniq, slag heaps are encountered on about 2km long.

(4) Different terraces have been identified. A detailed mapping of these deposits and terraces is in progress. One of the terrace, the so-called plateau with archaeological remains indicated on Figure 2, is rich in building remains. The cartography of these building is facilitated by the very high resolution QuickBird imagery (Figure 4).

The description of the site has been given by Al-Hamdani thousand years ago. It mentions a great quantity of water, a lot of palm trees and an important human settlement (Benoit *et al.*, 2003). When visiting the current setting, all these aspects have changed. Climatic conditions were milder and moister.

The landscape has to be studied at three time scales represented by different morphoclimatic conditions. The current setting is characterised by an arid mountainous area with temporary watercourses (wadis) and rare human settlements. The mediaeval setting (7th-9th century AD) evidences more temperate conditions with relatively abundant water and wood necessary for metallurgy and villages scattered from the mine to the Wadi Al-Khaniq. The Quaternary setting could be rebuilt using surficial deposits, palaeoterraces, structural studies especially those of the Quaternary volcanism, and neotectonics.

Imagery and GPS, complementary to field investigations, constitute powerful tools for such a work.

Acknowledgements. GSMRB and ZincOx facilitated the study. We specially thank Mr Brett Grist from ZincOx, Sana'a office. We are also indebted to Major Khaled Al Naphani and his group from the National Guard for assistance with field work. CEFAS (French-Yemenite Archaeological Centre, Sana'a) provided facilities and particularly Dr Mounir Arbach. The French Ministry of Foreign Affairs, CNRS and University Paris 1-Sorbonne financed field investigations for the French part.

References

AL GANAD, I. 1991 *Etude géologique et métallogénique du gisement Zn-Pb-Ag de Jabali (bordure Sud du bassin de Wadi Al Jawf, République du Yémen).* Thèse doctorat Université d'Orléans, 181 p.

AL-HAMDÂNÎ, *Kitâb al-jawharatayn al-`atîqatayn al-mâ'i`atayn min al-safrâ' wa-l-baydâ'.* 1. Ed. Toll, C., translation in german, *Die beiden Edelmetalle Gold und Silber,* Upsala, 1968. 2. Ed. `Abd Allâh (Yûsuf M.), Sanaa, 1424/2003. 3. Partly translated, Dunlop, D.M. 1957 Sources of gold and silver in Islam according to al-Hamdânî, *Studia Islamica,* 8:29-49.

ANDREEFF, W. 2001 *A comparison of remote sensing analysis techniques in the search for ancient caravan routes near Marib, Yemen.* BSc. Degree Memoir, University of Calgary, Canada, 16 p. (unpublished).

BENOIT, P., J. FÉRAUD, F. MICHAUD & F. TÉREYGEOL 2003 Nouvelles recherches sur la mine d'al-Jabalî. *Chroniques Yéménites.* 11:47-66.

CHRISTMANN, P. and J.-Y. SCANVIC 1987 Apport de l'imagerie SPOT à la cartographie géologique en pays aride. Exemple de la République Arabe du Yémen. *Bulletin de la Société française de Photogrammétrie et Télédétection.* 106: 23-32.

DEROIN, J.-P., I. N. AL-GANAD, M. AL-THARI, P. BENOIT, F. TEREYGEOL, J. HECKES & A. HORNSCHUCH 2006 The effect of spatial resolution for teleanalytical geological mapping, example of the Kawkaban-Sana'a-Ma'rib region, Republic of Yemen. *Photo-Interprétation,* 3, in press.

DEROIN, J.-P., E. MOTTI and A.SIMONIN 1998 Comparison of potential for using optical data and synthetic aperture radar for geological mapping in arid regions (Atar, Western Sahara, Mauritania). *International Journal of Remote Sensing.* 19, 6 :1115-1132.

DEROIN, J.-P., V. RABEMANANA and P. ANDRIEUX 2005 Contribution of Landsat's Short Waves Infrared (SWIR) for geological mapping in arid regions, the Androy example, Southern Madagascar. *Photo-Interprétation.* 1: 9-19 and 43-47.

GROLIER, M.J. and W.C. OVERSTREET 1978 *Geologic map of the Yemen Arab Republic (San'a).* U.S.G.S. Misc. Inv. Series, map I-1143B, map at 1:500,000.

HUCHON, P. and K. KHANBARI 2003 Rotation of the syn-rift stress field of the northern Gulf of Aden margin, Yemen. *Tectonophysics.* 364: 147-166.

KLEIBRINK, M. and B. JOEMANN 2004 *Genauigkeits-analyse von GPS Receivern aus dem Low-Cost-Bereich.* University of Applied Sciences Georg Agricola, Bochum, diploma thesis.

KRUCK, W. and U. SCHÄFFER 1991 *Geological Map of the Republic of Yemen, Sheet San'a'* 1 : 250 000, Ministry of Oil and Mineral Resources (Republic of Yemen) and Federal Institute of Geosciences and Natural Resources, Hannover.

LASAPONARA, R. and N. MASINI 2005 QuickBird-based analysis for the spatial characterization of archaeological sites: Case studies of the Monte Serico medieval village. *Geophysical Research Letters.* 32, 12:L12313.

LASAPONARA, R. and N. MASINI 2006a Detection of archaeological crop marks by using satellite QuickBird multispectral images. *Journal of Archaeological Sciences.* 1-8.

LASAPONARA, R. and N. MASINI 2006b On the potential of QuickBird data for archaeological prospection. *International Journal of Remote Sensing.* 27, 6:3607-3614.

MARCOLONGO, M. and D. MORANDI BONACOSSI 1997 L'abandon du système d'irrigation qatabanite dans la vallée du wadi Bayhan (Yémen). *Comptes Rendus de l'Académie des Sciences, Paris.* 325:79-86.

NRP, Natural Resources Project, 2000. Geological Map San'aa, Sheet 15G, 1 : 250,000, Robertson Group plc.

SHARLAND, P.R., D.M. CASEY, R.B. DAVIES, M.D. SIMMONS and O.E. SUTCLIFFE 2004 Chrono-sequence stratigraphy of the Arabian plate. *GeoArabia.* 9, 1:199-214, 2 posters.

THURMOND, A.K., M.G. ABDELSALAM, J.B. THURMOND 2006 Optical-radar-DEM remote sensing data integration for geological mapping in the Afar Depression, Ethiopia. *Journal of African Earth Sciences.* 44: 119-134.

WHITEHOUSE, M.J., B.F. WINDLEY, D.B. STOESER, S. AL-KHIRBASH, M.A.O. BA-BTTAT and A. Haider 2001 Precambrian basement character of Yemen and correlations with Saudi Arabia and Somalia. *Precambrian Research.* 105: 357-369.

Performance evaluation of data fusion algorithms for the detection of archaeological features by using satellite QuickBird data

Rosa Lasaponara,[1] Nicola Masini [2]

1 Istituto di Metodologie per l'Analisi Ambientale, IMAA-CNR, C.da S.Loja – 85050 Tito Sc. (PZ), Italy, Tel. ++39 0971 427 214 – fax ++39 0971 427 217 – e_mail: lasaponara@imaa.cnr.it
2 Istituto Beni Archeologici e Monumentali, IBAM-CNR, , C.da S.Loja – 85050 Tito Sc. (PZ), Italy, Tel. ++39 0971 427 321 – fax ++39 0971 427 222 – e_mail: n.masini@ibam.cnr.it

1 Introduction

The satellite-based identification of spatial features linked to the presence of buried archaeological remains is one of the most complex and challenging tasks faced by computer vision and photogrammetry communities. This is due to the fact that archaeological deposits tend to induce small spatial and spectral anomalies, that can be characterized by various kinds of marks, such as, soil, crop and shadow marks (Dassie 1978; Wilson 1982). These marks are generally not evident on the ground, but, they could be recognized from air. The visibility of such marks strongly depend on vegetation cover, soil types, illumination conditions and view geometry. For example, soil-marks can appear as changes of color or texture in ploughed fields. They can be found before the germinating crops mask the surface of the soil. The crop-marks (or weed-marks) are induced by the presence of vegetation over buried archaeological deposits, that could be ditches or walls. In the first case, the re-filling of archaeological features stores more water and nutrients. For this reason, the plants on the top of them will grow higher ("positive marks"), stay green for a longer period and will have a darker green than the surrounding plants. The opposite situation will take place for plants growing over buried walls.

On the basis of remotely sensed multispectral data, these different archaeological marks could be detected by using specific spectral channels and their spectral combinations. This fact, in the last 25 years, has caused the increased exploitation of satellite imagery, such as, Landsat Temathic Mapper (30 m) or Spot (10 m) in the field of archaeological prospection. Nevertheless, the spatial resolution of early satellite sensors did not allow the identification and mapping of small features which represent a large amount of the archaeological heritage.

The current availability of Very High Resolution (VHR) satellite images, can provide new capabilities in the field of archaeological prospection. The main advantages of VHR satellite images compared to aerial photos, are the synoptic view and, above all, the possibility of performing a combined analysis with multispectral images by using data fusion algorithms whose benefits have been assessed in different field of applications (see, for example, Ranchin et al. 2003).

Recent investigations pointed out the high spectral capability of satellite Quickbird images for the identification of ancient buried structures (Lasaponara and Masini 2005, 2006).

This paper deals with the use of data fusion algorithms applied to QuickBird images for the enhancement of archaeological marks. The most common data fusion algorithms were tested for a study base in order to evaluate their performance for the detection of archaeological features. The study case is a deserted medieval village named Monte Irsi, which is located on the top of a hill in the Northeast of Basilicata, already studied by using QuickBird imagery (Lasaponara and Masini, 2006)

The evaluation is performed by using both visual inspection and an indicatorset up to assess the spectral separability between the detected archaeological features (Lasaponara and Masini, 2006) and their surroundings areas.

2 QuickBird data description

QuickBird satellite has panchromatic and multispectral sensors with resolutions of 61-72cm and 2.44-2.88m, respectively, depending upon the off-nadir viewing angle (0-25 degrees). It provides a good stereo geometry and a high revisit frequency of 1-3.5 days depending on latitude. The reader is referred to technical notes available on line at the web site:
http://www.digitalglobe.com/product/product_docs.shtml
The panchromatic sensor collects information at the visible and near infrared wavelengths and has a bandwidth of 450 – 900 nm. The multispectral sensor acquires data in four spectral bands from blue to near infrared (NIR). Both panchromatic and multispectral sensors offer 11 bit (2,048 grey levels) resolution. The QuickBird imagery products are available at different processing levels (basic, standard, ortho) serving the needs of different users.

One of the main advantage of VHR satellite is the possibility of combining multispectral and panchromatic data to achieve improved accuracies and better inference about the surface characteristics of the single sensor. In particular, the data fusion allows the enhancement of spatial anomalies linked to the presence of archaeological marks; thus, making their recognition and extraction easier.

Compared to aerial photo the QuickBird imagery can be promptly geo-referenced and offer a very large coverage. This makes them ideal for investigations on regional scale as well as for researches performed in areas where aerial photography is restricted because of military or political reasons.
Nevertheless, there are also some considerable drawbacks. Although, the use of data fusion allowed the identification of features with the spatial detail of panchromatic (0.6 m) instead of multispectral (2.40m) this spatial resolution is still lower than those obtained from aerial photo. Thus could be a limitation for the identification of small features.

3 Satellite data processing: methodological approach

The idea underlying the approach suggested for satellite data analysis is that archaeological features could be enhanced and efficiently studied by exploiting the high spatial resolution of satellite QuickBird panchromatic imagery and the multispectral properties of the four spectral channels. The methodological approach adopted for the enhancement and extraction of archaeological marks is mainly based on the use of data fusion.

The data fusion enables the integration of the high spatial resolution of the panchromatic image with the spectral capability of multispectral images. Thus, allowing to achieve improved capabilities that are not possible using solely the individual datasets.

Over the years, a number of algorithms have been developed for data fusion. The question is to assess which of the available algorithms can perform better when applied to satellite QuickBird imagery for the detection of archaeological marks. The evaluation of performance of the existing algorithm is crucial in order to address this question. Section 4 summarizes a brief discussion of advantages and disadvantages obtained from the most commonly used algorithms developed for data fusion.

4 Data fusion methods

Data fusion refers the process of combining multiple images of a scene to obtain a single composite image. The different images to be fused can come from different sensors of the same basic type or they may come from different types of sensors. The composite image should contain a more useful description of the scene than provided by any of the individual source images.

Due to the objective of this paper, to investigate the influence of image fusion on archaeological feature detection, different fusion techniques were evaluated. From the widely variety of data fusion algorithms, a selection of the most commonly used techniques was performed.

4.1 Fusion method based on Principal Component Analysis (PCA)

Principal Component Analysis (PCA) can be adopted as a data transform to enhance features and anomalies that can not be easily distinguishable in the original spectral variables. Over the years, PCA has been widely adopted for processing satellite data for archaeological investigations. That is the main reason why, in this study, it was selected to perform data fusion.

The PCA (see, for example, Richards, 1986) is a linear transformation which decorrelates multivariate data by translating and/ or rotating the axes of the original feature space, so that the data can be represented without correlation in a new component space.

In order to do this, the process firstly computes the covariance matrix (S) among all input time series dates (each element of S is calculated by using formula 1), then eigenvalues and eigenvectors of S are calculated in order to obtain the new feature components.

$$cov_{k1,k_2} = 1/nm \sum_{i=1}^{n} \sum_{j=1}^{m} (SB_{i,j,k1} - \mu_{k2})(SB_{i,j,k2} - {}_{k2}) \quad [1]$$

where k_1, k_2 are two input spectral channels, SB i,j, spectral value of the given channel in row i and column j, n number of row, m number of columns, μ mean of all pixel SB values in the subscripted input channels.

The percent of total dataset variance explained by each components is obtained by formula 2:

$$\% i = 100 * \lambda_i / \sum \lambda_i \quad [2] \quad i=1$$

where λ_i are eigenvalues of S

Finally, a series of new image layers (called eigenchannels or components) are computed (formula 3) by multiplying, for each pixel, the eigenvector of S for the original value of a given pixel in the input bands

$$P_i = \sum_{i=1} P_k \, u_{k,i} \quad [3]$$

where P_i indicates a spectral channel in component i, u k,i eigenvector element for component i in input band k, Pk spectral value for channel k, number of input band.

A loading, or correlation R, of each component i with each input date k can be calculated by using formula 4.

$$R_{k,i} = u_{k,i} (\lambda_i)^{1/2} (var_k)^{1/2} \quad [4]$$

where var_k is the variance of input date k (obtained by reading the kth diagonal of the covariance matrix)

The application of PCA should allow a transformation of the input multispectral bands able to make the identification of distinct surface types or discontinuity easier. This is a direct result of the high correlation that exists among channels for homogeneous areas that do not change significantly over the space and the relatively low correlation associated with areas characterized by the presence of surface anomalies and/or discontinuity. The major portion of the variance in a multi-spectral data set is associated with homogeneous areas, whereas localised surface anomalies will be enhanced in later components. In particular, each successive component contains less of the total dataset variance. In other words, the first component contains the major portion of the variance, whereas, later components contain a very low proportion of the total dataset variance. Therefore, they may represent information variance for a small area or can be essentially noise and, in this case, it must be disregarded. Some problems can arise from the fact that eigenvectors can not have general and universal meaning since they are extracted from the series itself.

The data fusion method based on the use of PCA, the first principal component is replaced by high resolution image after PCA transformation has been achieved. Then, the inverse transformation is carried out from PCA back to multispectral data.

4.2 Fusion method based on Tasseled Cap Transformation TCT

Tasseled Cap Transformation (TCT) has been widely adopted for extracting archaeological features linked to ancient human transformations of the landscape (REF). For this reason, in this study it was selected to perform data fusion. The TCT, also known as Kauth-Thomas technique, was devised for enhancing spectral information content of satellite data. The reader is referred to Crist and Kauth (1986), for an introduction. The TCT is a linear affine transformation substantially based on the conversion of given input channel data set in a new data set of composite values. The transformation depends on the considered sensor. The original TCT was derived (Kauth and Thomas, 1976) for the four bands of the Landsat MSS sensor. Later, the TCT was extended to the Landsat TM (Crist and Cicone, 1984), ETM (as available, for example, in a routine of geomatica, PCI software) and IKONOS sensor (Horne 2003). All the existing TCTs are performed on a pixel basis to best show the underlying structure of the image by using weighted sums of the input channels. Among the available weighted sums, the processing of QuickBird spectral channels was performed using the weighted specifically developed for IKONOS data (see equations 4 to 7). Note, that usually there are just three composite variables tasseled cap transform bands that are often used:

- TCT-band 1
- TCT-band 2
- TCT-band 3

The weighted sums developed by Horne (2003) for IKONOS input channels:

TCT_{IKONOS}-band 1 = 0.326 BLUE + 0.509 GREEN + 0.560 RED + 0.567 NIR [5]

TCT_{IKONOS}-band 2 = -0.311 BLUE - 0.356 GREEN - 0.325 RED + 0.819 NIR [6]

TCT_{IKONOS}-band 3 = -0.612 BLUE - 0.312 GREEN + 0.722 RED -0.081 NIR [7]

TCT_{IKONOS}-band 4 = -0.650 BLUE + 0.719 GREEN - 0.243 RED -0.031 NIR [8]

To perform data fusion, the TCT-band 1 is replaced by high resolution image after TCT transformation has been achieved. Then, the inverse transformation is carried out from TCT back to multispectral data. Note that the TCT is herein proposed as data fusion technique for the fist time.

4.3 IHS fusion method

Over the years, RGB has been widely used for emphasizing archaeological features in satellite images. That is the main reason why, in this study, IHS fusion method was selected to be tested in our investigations.

The IHS fusion method perform the data fusion by converting a color image from the RGB (Red, Green, Blue) space into the IHS color space. Because the intensity (I) band resembles a Pan image, it is replaced by a high-resolution Pan image in the fusion. A reverse IHS transform is then performed on the Pan together with the hue (H) and saturation (S) bands, resulting in an IHS fused image. AS in the previous cases the fusion approach substitutes the value (or intensity component with the panchromatic band. Although the HIS methods are very old fusion approaches they are still used and improved (see, for example, Ehlers, 2004)

4.4 Fusion method based on multiplicative approach

Multiplicative approach is based on the following simple arithmetic integration of the two raster data sets:

$$DN_{Bi\text{-}new} = DN_{Bi} \text{ x } DN_{\text{hig res. Image}} \quad [9]$$

where DN denotes digital number, $DN_{Bi\text{-}new}$ denotes the fused product obtained for the given band i, DN_{Bi} denotes the original spectral band, and $DN_{\text{hig res. Image}}$ denotes the highest resolution panchromatic image.

4.5 Fusion method based on Brovery transformation

Brovery transformation is a simple numerical method to merge different source of data (Vrabel 1996, Liu 2000). Brovery equation is designed on the basis of the assumption that spectral range of the panchromatic image is the same as that covered by multispecral data. The transformation is defined by equation 10.

$$Y_k(i, j) = X_k(i, j)\, X_p(m, n) / \sum_{K=2}^{n} X_k(i, j) \quad [10]$$

where Y_k (i, j) and X_k (i, j) are the kth fused multispectral band and the original multispectral band respectively.; I and j denote the pixel and line number respectively. X_p (m, n) is the original panchromatic band., and m, and n denote the pixel and line number.

In the current case, the image fusion for QuickBird data is carried as follows:

(i) selection of spectral bands

(ii) resample of them to panchromatic spatial resolution

(iii) perform Brovery transformation for the resampled new image data

The resulting image consist of a combination of the n multispectral bands and panchromatic image.

4.6 Fusion method based on Zhang's approach.

Zhang (2004) suggested that all the existing data fusion approaches were not adequate for VHR satellite data being strongly affected by (i) the colour distortion, and (ii) dataset dependency. For this reason he developed a data fusion algorithm specifically devised for VHR satellite images. This algorithm was adopted by Digital Globe http://www.pcigeomatics.com/support_center/tech_papers/techpapers_main.php] and it is also available in a PCI-Geomatica routine (PANSHARP). In the PANSHARP routine, if the original MS and Pan images are geo-referenced, the resampling process can also be accomplished together with the fusion within one step. All the MS bands can be fused at one time. The fusion can also be performed solely on user-specified MS bands. The statistics-based fusion technique devised by Zhang (2004) is able to solve the two major problems in image fusion – color distortion and operator (or dataset) dependency. Such a data fusion algorithm is quite different from the other techniques in the following two principle ways:

(i) Firstly, to reduce the color distortion, it utilizes the least squares technique to find the best fit between the grey values of the image bands being fused and to adjust the contribution of individual bands to the fusion result;

(ii) Secondly, to eliminate the problem of dataset dependency, it employs a set of statistic approaches to estimate the grey value relationship between all the input bands.

5 Discussion and results

5.1. Performance evaluation criteria of datafusion methods

In this study, the assessment of data fusion techniques was performed using both visual inspection and an indicator specifically adopted to this aim. Image fusion performance is generally assessed using three properties: (i) each fused band is reduced to its original spatial resolution and then compared to its reference, (ii) each band of the reference image is resampled to a reduced resolution prior to fusion to simulate a coarser resolution input data which is compared to the reference, (iii) band – based comparison in order to account for multispectral image information changes. Each of these properties can be critical for the availability of the actual information. For this reason we adopted a simple index based on bandwise statistics, namely mean and standard deviation. The rationale for adopting this indicator is mainly based on the fact that we intend to assess if the higher spatial resolution can allow to better identify the spatial features linked to archaeological buried remains. Thus, we focused on the spectral separability between archaeological features and their surroundings areas.

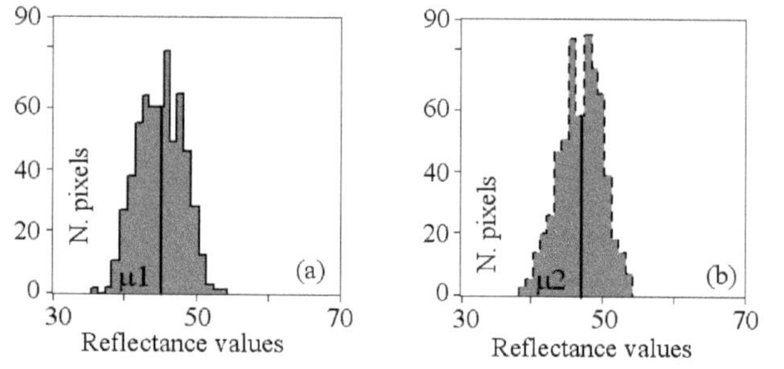

Figure 1. *Histograms of spectral values of archaeological features (a) and their surroundings*

Such a spectral separability (SSI) is evaluated using the following formula:

$$SSI = \Delta\mu/(\sigma 1+\sigma 2) \qquad [11]$$

where $\Delta\mu$, $\sigma 1+\sigma 2$ are respectively the difference of spectral values and standard deviation observed in a given data fusion band between archaeological features and their surroundings.

5.2. Study area

The investigation was performed for Monte Irsi, a medieval deserted village located close to the border between Basilicata and Apulia, in the South of Italy. This site is a well documented from the archaeological point of view (see, for example, Small et al. 1998) as well as well studied using QuickBird imagery (Lasaponara and Masini, 2006). Monte Irsi is a hilly plateau , near the confluence of the Bradano and Basentello rivers, characterized by the presence of vegetation made up of dense herbaceous plants. The strategic location of Monte Irsi favoured a long and intensive human activity from Paleolithic to Middle Ages, as testified by archaeological remains (Small *et al.* 1998). As regards to the Middle Ages, documentary sources state the existence of a village and a monastery. The village achieved its maximum expansion between the 12th and 13th centuries and was abandoned in the 15th century, whereas the monastery was destroyed around 1370. Today, the only building preserved is the church of the monastery. Archaeological remains near the same church are related to a Roman villa. Finally, the medieval age is proved by some remains as well as by archaeological features at the highest part of the hill, detected by using QuickBird imagery (Lasaponara and Masini, 2006).

Just the buried remains related to medieval settlement were adopted as "archaeological features" to perform the comparison of performances obtained by applying the datafusion methods above described.

5.3 Results

The figures 2 and 3 show the results obtained in the previous study carried out on Monte Irsi (Lasaponara and Masini, 2006). In particular, the integration from panchromatic and NIR datafusion product, performed by using Zhang (2004) algorithm, allowed to detect several crop-marks as well as to obtain a detailed physical and geometrical characterization of the archaeological site.

Table 1 shows the scarce capability of the other bands (green and red). The different spectral performance, is due to vegetation cover which characterizes the site surface. In particular, NIR is more sensitive than the other channels to different vegetational growth and status. For this reason the NIR band is regarded as a reliable indicator of the presence of buried structures and ditches. This is proved by the value of SSI of NIR (0,359) that is higher than the SSI values computed for the other bands (SSI 0,155 and SSI 0,039 respectively obtained for the green and red datafusion products).

Imagery			min	max	μ	σ	Δμ	SSI=Δμ/(σ1+σ2)
Panchromatic		mask1	47	76	58,99	4,41	2,65	**0,308**
		mask2	48	73	56,34	4,18		
Zhang algorithm	blue	mask1	32	42	36,37	1,49	0,15	0,057
		mask2	33	40	36,22	1,15		
	green	mask1	44	57	49,1	2,54	0,73	0,155
		mask2	42	57	48,37	2,17		
	red	mask1	22	47	35,7	3,98	0,27	0,039
		mask2	27	46	35,43	3,01		
	NIR	mask1	59	131	80,57	8,91	5,73	**0,359**
		mask2	59	99	74,84	7,03		
PCA	green	mask1	196,25	212,76	203,25	3,33	1,39	0,218
		mask2	195,17	212,76	201,86	3,05		
	red	mask1	10,68	33,27	10,68	4,39	2,23	0,269
		mask2	8,45	31,67	8,45	3,91		
	NIR	mask1	-9,75	12,47	3,42	3,27	-1,65	-0,272
		mask2	-6,95	11,47	5,07	2,8		
TCT	green	mask1	10,4	30,25	21,59	2,57	0,22	0,046
		mask2	14,57	30,39	21,37	2,23		
	red	mask1	14,64	33,58	25,17	2,64	0,39	0,079
		mask2	18,16	34,71	24,78	2,3		
	NIR	mask1	38,11	75,66	53,16	6,97	4,22	**0,335**
		mask2	36,79	66,18	48,94	5,63		
Multiplicative approach	pan*green	mask1	2208	3960	2919,8	320,37	162,77	0,258
		mask2	2156	3960	2757,03	310,44		
	pan*red	mask1	1512	3264	2141,85	318,59	114,75	0,193
		mask2	1470	3105	2027,1	274,53		
	pan*NIR	mask1	3149	7070	4727,8	723,68	467,8	**0,352**
		mask2	3168	6552	4260	606,22		
Brovey transform	green	mask1	11,66	18,16	14,4	1,02	0,42	0,209
		mask2	11,78	18,25	13,98	0,99		
	red	mask1	7,1	14,51	10,56	1,19	0,29	0,131
		mask2	8,03	14,31	10,27	1,03		
	NIR	mask1	16,84	33,04	23,29	2,68	1,71	**0,348**
		mask2	16,85	29,78	21,58	2,23		
IHS	I: Intensity	mask1	49	68	57,68	3,86	2,48	**0,340**
		mask2	49	66	55,2	3,44		
	H: Hue	mask1	216	242	231,98	3,65	-0,26	-0,044
		mask2	215	242	232,24	2,2		
	S: Saturation	mask1	24	66	37,41	8,11	3,6	0,254
		mask2	23	52	33,81	6,07		

Table 1 *Data fusion results evaluation. The SSI (see equation 11) bold values indicate the more relevant products obtained using the different algorithms of data fusion. The SSI values for panchromatic image is computed in order to compare it with the SSI of data fusion products. Note that archaeological features are indicated as mask 1 and their surroundings as mask 2.*

The comparative test between the considered datafusion methods (paragraphs 4.1, 4.2, 4.3, 4.4, 4.5 and 4.6), confirms the high performance of NIR by using TCT, multiplicative approach, Brovey transform (see figures 4 and 5) and IHS transformation, as it is showed in table 1. In particular, by applying the multiplicative approach the SSI values of the detected archaeological features is almost equal to that calculated by using Zhang (SSI 0,348 and SSI 0,359 obtained respectively from the multiplicative approach and Zhang algorithm). Whereas, the SSI values of NIR by using TCT, Brovey transform and IHS transformation are slightly lower than the values obtained by applying Zhang algorithm. Finally, SSI value by applying PCA is considerably lower (0,272) than those obtained by the other datafusion methods.

If we enlarge the attention to the performance of the considered data fusion methods related to the other channels we discover a surprising behavior.

The SSI values related to the other channels (green and red) are very low by using Zhang algorithm and TCT (respectively 0,155 and 0,046 for green, 0,039 and 0,079 for red). Whereas, PCA, multiplicative approach and Brovey transform provide higher values of SSI (respectively 0,218, 0,258 and 0,209 for green, and 0,269, 0,193 and 0,131 for red) than those obtained by the above mentioned data fusion methods.

6 Conclusions

The potentiality of VHR satellite data in the field of archaeological can considerably be improved by using data fusion methods. They enables the integration of the high spatial resolution of the panchromatic image with the spectral capability of multispectral images. Thus, allowing to achieve improved capabilities that are not possible using solely the individual datasets.

In order to select the best method, it is fundamental to assess the performance of the method itself by using a quantitative indicator.

In this paper an indicator to evaluate the spectral separability between archaeological features and their surroundings areas was proposed.

The indicator was applied to the considered datafusion methods. The main findings of the application can be summarized as follows:

1) Zhang algorithm and TCT method provided the highest values of spectral separability for the NIR channel. This last one emphasizes more than the other channels the archaeological features for vegetated surfaces, as in the considered study case.
2) Good results were obtained also by IHS transform, multiplicative approach and Brovey transformation for the same channel.
3) A different behaviour was observed for the other channels.
4) The SSI values related to green and red channels are very low by using Zhang algorithm and TCT; whereas, higher values of SSI were obtained by PCA, multiplicative approach and Brovey transform.

The fact that NIR channel performs better than the other spectral images is mainly linked to the surface characteristics of the study areas that was covered by herbaceous plants when the satellite data were acquired In this conditions it is expected that NIR could give the better performance for archaeological analysis being that NIR is more sensitive than other spectral bands to different vegetation growth and status.

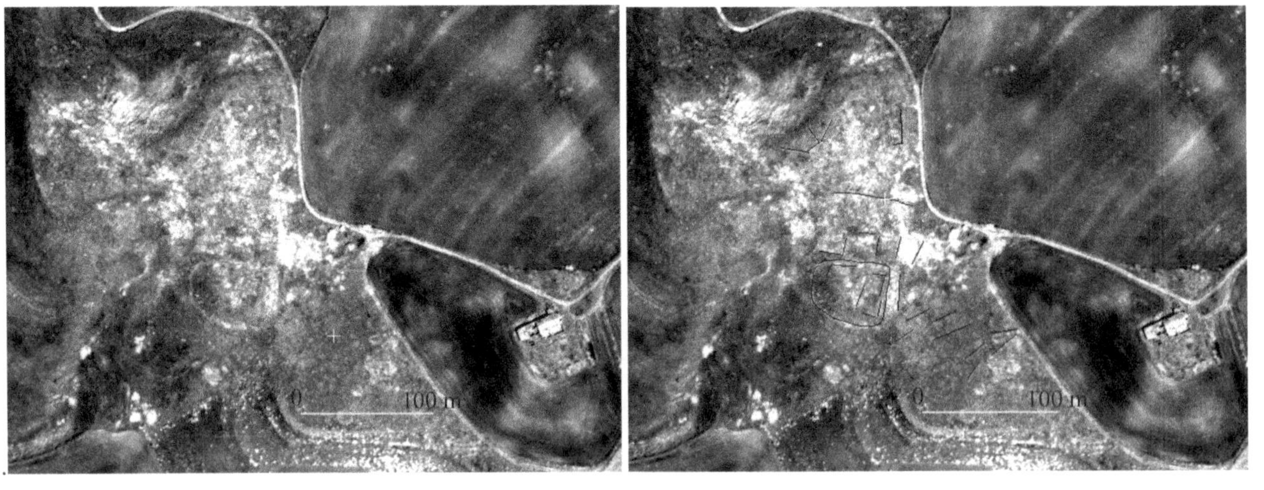

Figure 2 *(a) Irsi panchromatic image at 0.70 m resolution ; (b) the same image as on the left with detected archaeological features (Lasaponara and Masini, 2006)*

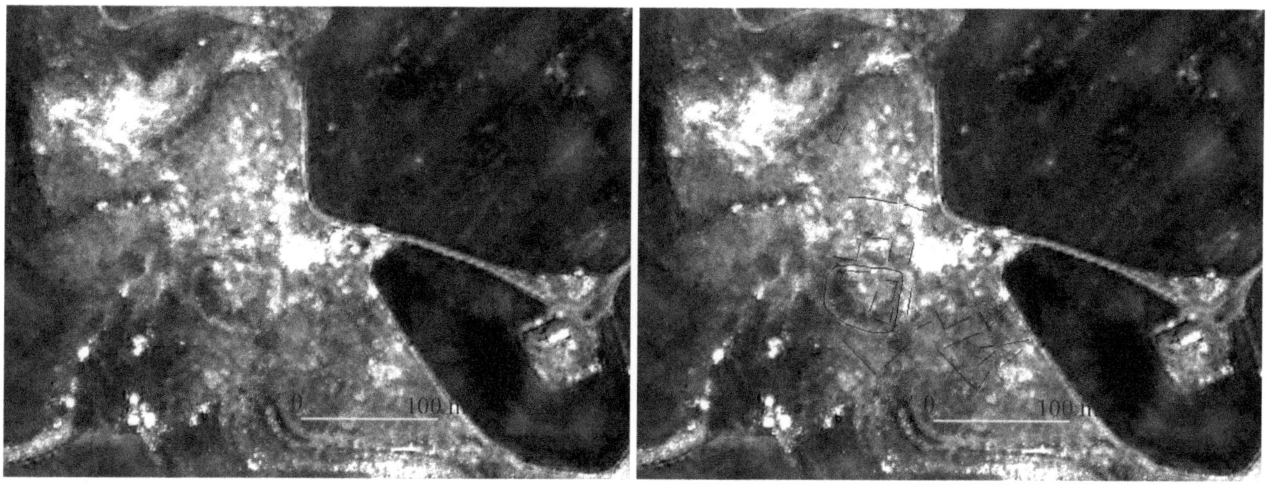

Figure 3 *(on the left) Irsi NIR datafusion products (algorithm by Zhang, 2004); (on the right) the same image as on the left with detected archaeological features (Lasaponara and Masini, 2006)*

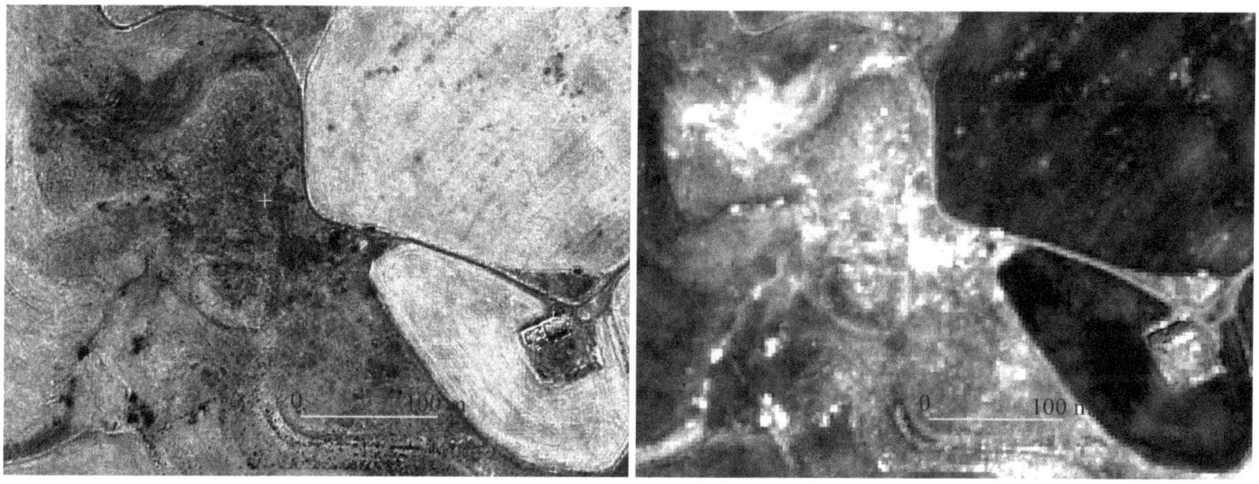

Figure 4 *Irsi: data fusion obtained for NIR channel using PCA (on the left) and TCT (on the right).*

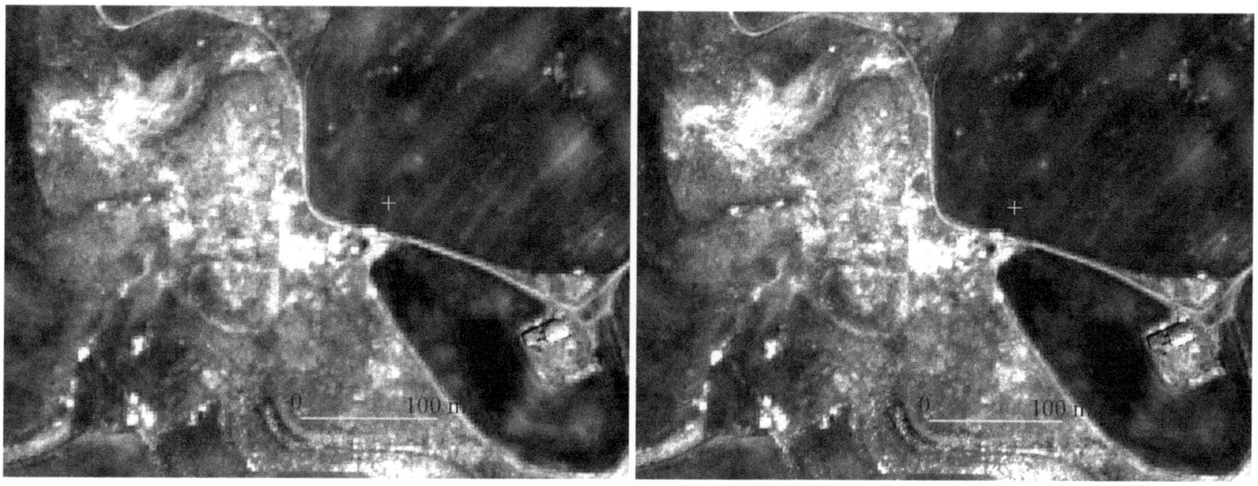

Figure 5 *Irsi: data fusion obtained for NIR channel using (on the left) multiplicative approach and (on the right) Brovey transform.*

References

CRIST, E.P. & R.C. CICONE 1984. A Physically-Based Transformation of Thematic Mapper Data – The TM Tasseled Cap. *IEEE Transactions of Geoscience and Remote Sensing*, 22 (3): 256-263.

CRIST, E.P. & R.J. KAUTH 1986. The Tasseled Cap De-Mystified. *Photogrammetric Engineering and Remote Sensing*, 52 (1): 81-86.

DASSIE, J. 1978. *Manuel d'archeologie aerienne*, Paris, Editions Technip.

EHLERS, M. 2004. Spectral characteristic preserving image fusion based on Fourier domain filtering. In Proceedings of SPIE vol. 5574, 13-16 September 2004, Maspalomas, Spain.

HORNE, J.2003. A Tasseled Cap Transformation for IKONOS Images. In: Proceedings of the *ASPRS: 2003 Annual Conference and Technology Exhibition.*

KAUTH, R.J. & G.S THOMAS 1976. The Tasseled Cap – a graphical description of the spectral-temporal development of agricultural crops as seen by Landsat. In Proceedings of the *Symposium on Machine Processing of Remotely Sensed Data*, Purdue University, West Lafayette, Indiana, pp. 4B41-4B51.

LASAPONARA, R., & N. MASINI 2005. QuickBird-based analysis for the spatial characterization of archaeological sites: case study of the Monte Serico Medioeval village, *Geophysical Research Letter*, 32, (12): L12313.

LASAPONARA, R., & N. MASINI 2006. On the potential of panchromatic and multispectral Quickbird data for archaeological prospection. *Int. J. Remote Sensing*, 27: 3607-3614.

LIU, J. G. 2000. Evaluation of Landsat -7 ETM + panchromatic band for image fusion with multispectral bands. *Natural Resources Research*, 9: 269-305.

RANCHIN, T., B. AIAZZI, L. ALPARONE, S. BARONTI & L. WALD 2003. Image fusion the ARSIS concept and some successful implementation schemes. *ISPRS Journal of Photogrammetry and Remote Sensing, 58* (1-2): 4-18.

RICHARDS, J. A. 1986. *Remote Sensing Digital Image Analysis.*, Springer-Verlag.

SMALL, A., C. SMALL, I. CAMPBELL, M. MACKINNON, T. PROWSE & C. SIPE, 1998. Field survey in the Basentello valley on the Basilicata-Puglia border, *Echos du Monde Classique. Classical views*, XLII, 337-371.

VRABEL, J. 1996. Multispectral imagery band sharpening study. *Photogrammetric Engineering and Remote Sensing* , 62: 1075-1083.

WILSON, D. R. 1982. *Air photo interpretation for archaeologists*, London, St. Martin's Press.

ZHANG, Y. 2004. Understanding Image Fusion. *Photogrammetric Engineering & Remote Sensing*, 70: 657-661.

Evaluation of the spectral capability of quickbird imagery for the detection of archaeological buried remains

Nicola Masini,[1] Rosa Lasaponara[2]

[1] Istituto Beni Archeologici e Monumentali, IBAM-CNR, C.da S.Loja – 85050 Tito Scalo (PZ), Italy, Tel. ++39 0971 427 321 – fax ++39 0971 427 222 – e_mail: n.masini@ibam.cnr.it.

[2] Istituto di Metodologie per l'Analisi Ambientale, IMAA-CNR, C.da S.Loja – 85050 Tito Scalo (PZ), Italy, Tel. ++39 0971 427 214 – fax ++39 0971 427 217 – e_mail: lasaponara@imaa.cnr.it

1 Introduction

The application of space technology to cultural heritage and archaeological research has been paid great attention worldwide, mainly because very high resolution (VHR) satellite data such as, IKONOS (1999) and QuickBird (2001), are able to match with aerial photogrammetric images. During the last century, aerial photos have been the most common remote sensing data source used in the field of archaeological investigations. The low spatial resolution of early satellite sensors, such as Landsat Temathic Mapper (30 m) or Spot (10 m) did not provide sufficient precision for the identification and inspection of archaeological sites, thus, restricting the application of satellite imagery to paleogeographic environment studies (Parry, 1992, Drake, 1997) or to human ecology and landscape archaeology investigations (Clark et al. 1998, Sever, 1998). Up to now QuickBird is the commercial satellite that provides images with the highest spatial resolution from both panchromatic (61-72cm) and multispectral (2.44-2.88m) sensors (for additional information, the reader is referred to technical notes available on line at the web site http://www.digitalglobe.com/product/product_docs.shtml).

The availability of satellite VHR data impones new challenges in the field of archaeological remote sensing. The main issues that should be addressed can be summarized as follows. Could the VHR satellite be able to furnish a valuable data source for archaeological investigation ranging from synoptic view (Landscape archaeology) to small details (i.e. single subsurface building)? Could the satellite QuickBird data detect buried archaeological remains and provide an effective spatial characterization of them?

In this study, the spectral capability of satellite QuickBird imagery for the identification archaeological marks is presented for some archaeological sites located in the South of Italy. The study cases were selected mainly because they are buried under surfaces with different soil characteristics and vegetation covers. This allowed the evaluation of the QuickBird spectral ability in detecting the different types of archaeological marks (crop, soil and shadow-marks) in relation with different surface characteristics.

2 QuickBird Satellite data

The QuickBird satellite has panchromatic and multispectral sensors with resolutions of 61-72cm and 2.44-2.88m, respectively, depending upon the off-nadir viewing angle (0-25 degrees). The sensor has a coverage of 16.5-19km in the across-track direction. In addition, the along-track and across-track capabilities provide a good stereo geometry and a high revisit frequency of 1-3.5 days. The panchromatic sensor collects information at the visible and near infrared wavelengths and has a bandwidth of 450 – 900 nm. The multispectral sensor acquires data in four spectral bands from blue to near infrared (NIR). Both panchromatic and multispectral sensors offer 11 bit (2,048 grey levels) resolution. The QuickBird imagery products are available at different processing levels (basic, standard, ortho) serving the needs of different users.

3 Rational basis: Quickbird spectral capability for archaeological application

From a rational point of view, the use of QuickBird panchromatic image should be able to detect the typical marks expected in the presence of buried structures as in the case of aerial photos (Wilson 1982). The ability to correctly identify archaeological remains by remote sensing is largely determined by the image spatial resolution and their spectral capability.

Recently, some evaluations concerning the capability of satellite Quickbird images for archaeological applications were performed by Lasaponara and Masini (2005, 2006a,b). These investigations pointed out that the use of the red channel can allow better enhancement of shadow and soil marks. This is due to the fact that the presence of archaeological deposits under bare surface, lacking in vegetation, can strongly influence the soil constituents and, above all, moisture content. This causes a different spectral response that can be used for the identification and spatial characterization of buried structures.

Lasaponara and Masini (2006a) suggested that QuickBird Near InfraRed (NIR) channel can be used to enhance crop-marks, as already verified for other satellite data (M.J.F. Fowler and Y. Fowler 2005). This is because the presence of archaeological remains buried under

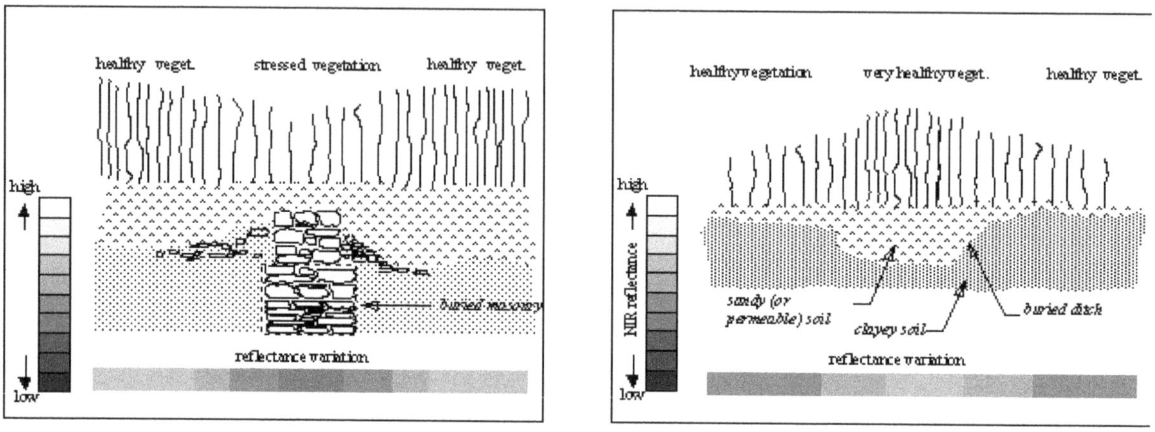

Figure 1a *Qualitative spectral characterization of crop marks*

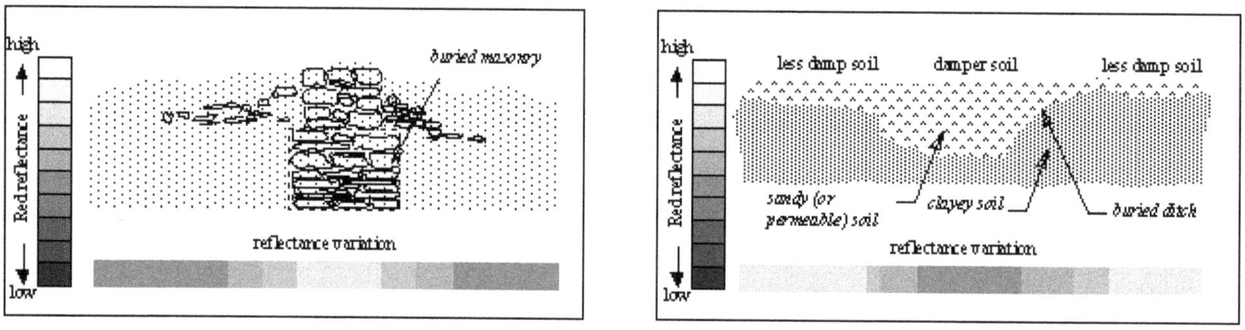

Figure 1b *Qualitative spectral characterization of soil-marks*

vegetated surface strongly influences the state of plants; enhancing or reducing their growth, by increasing or reducing soil nutrient and moisture availability. These subtle effects can often be evident using solely an aerial view. The small differences in the plant growth and/or color can be characterized by different spectral responses that, on the basis of QuickBird spectral channels, can be enhanced in the NIR channel. Green and healthy plants tend to exhibit high NIR reflectance values; whereas, vegetation under stress, due to lack of water or nutrient deficiencies, is characterized by low NIR reflectance values depending on the stress levels.

Additionally, on the basis of multispectral remotely sensed data, vegetation-marks (crop or weed) can be successfully identified by exploiting vegetation indices that are spectral combinations of different bands.

Such indices are quantitative measures, based on the vegetation spectral properties, that attempt to measure biomass or vegetative vigor. Vegetation indices are mainly derived from reflectance data from the Red and NIR bands. They operate by contrasting intense chlorophyll pigment absorption in the red against the high reflectance of leaf mesophyll in the near infrared. The simplest form of vegetation index is the ratio between two digital values from the red and near infrared spectral bands. The most widely used index is the well-known Normalized Difference Vegetation index (NDVI) obtained by using the following formula:NDVI = [NIR-Red]/ [NIR+Red].

The normalization of the NDVI reduces the effects of variations caused by atmospheric contamination. NDVI is indicative of plant photosynthetic activity and has been found to be related to the green leaf area index and the fraction of photosynthetically active radiation absorbed by vegetation. High values of the vegetation index identify pixels covered by substantial proportions of healthy vegetation, whilst disease or stressed vegetation exhibits low values. It is expected that crop-marks created by vegetation patterns can be emphasized by using NDVI map.

4 Results: reconnaissance and spectral characterization of archaeological marks

The investigations were performed for some archaeological sites located in the South of Italy (see Figure 2). These areas were selected because they are well documented and above all are characterized by different features from the geological, pedological and land-use point of view. These different characteristics allowed the analysis of performance obtained from satellite data for different features and surface characteristics. The spectral characterization of different archaeological marks such as soil, crop and shadow marks was carried out by using the analysis of the histograms observed for both archaeological features and their surroundings. The identification of archaeological marks was performed for widely investigated test sites (see Lasaponara and Masini 2005, 2006a,b; Masini and Lasaponara 2006).

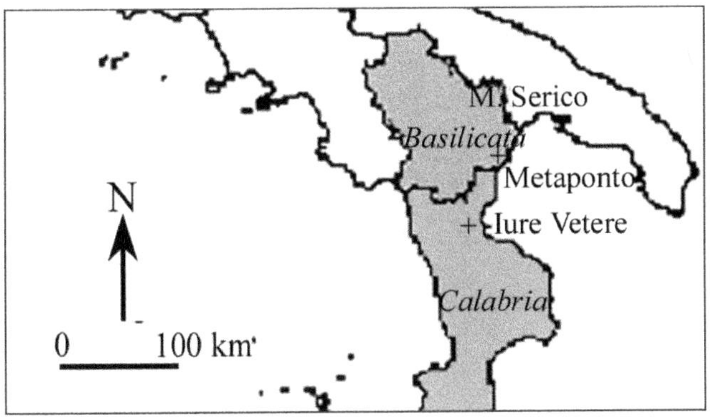

Figure 2 *Location of study areas*

4.1 Monte Serico test case

4.1.1 Study area

The site is found on Monte Serico, a hill (520 above sea level) located in Southern Italy (see Figures 1 and 2). From a geological point of view, the stratigraphic sequence is composed of Subappennine Clays, Monte Marano sands and Irsina conglomerates. Sporadic herbaceous plants grow over the investigated area. Historical sources state that around the 11th century, a castle was built on the hill (Masini, 1985); whereas a village was founded in the 13th century and gradually abandoned between the end of the 14th and the first half of 15th century. Today the only buildings remaining are the castle and a church. On the southern side of the hill, the presence of earthenware, pottery and crumbling building materials, indicates the existence of a buried settlement.

4.1.2 Satellite based analysis

In the Monte Serico site, erosion caused a chaotic distribution of building materials that coupled with the presence of surface rocks have made difficult the detection of marks linked to the buried medieval village. Results from the QuickBird single channel inspection showed that panchromatic image and the blue, green and NIR channels do not show significant marks; whereas, the red channel puts more rectilinear marks that are clearly referable to human activity into evidence (Lasaponara and Masini, 2005). The red channel is capable of better emphasizing the presence of buried structures mainly because the archaeological remains are buried under a bare surface.

Results from data fusion processing confirm the fact that the best results were obtained by the panchromatic and red channel data fusion. The urban plan reconstructed on the basis of satellite QuickBird data analysis agrees with results of independent studies performed on the basis of ground survey and aerial photointerpretation (Masini, 1995), and some excavation tests. Figures 3.1 show some details of the buried medieval village.

Figures 3.1 (a to c) show the QuickBird panchromatic image and (3.1.a) and red data fusion product (3.1.b). Figures 3.2 and 3.3 show the histograms obtained from the archaeological features (denoted as mask 1 in Table 1) and their surroundings (denoted as mask 2 in Table 1). A visual inspection of figures 3.1.a) and histograms of figure 3.2 clearly show that the archaeological features can not be discriminated on the basis of the QuickBird panchromatic image due to the very low detectable contrast. This is confirmed by Table 1 that shows the difference in grey values computed for archaeological features and their surroundings. Such differences are at around 1,31 and these value ranges are really too low to be estimated by human eyes and even by statistical processing of digital images.

Table 1 also shows the spectral difference values observed in red data fusion product between archaeological features and their surroundings.

The difference in reflectance values computed in this case is at around 3,48. Although such a value is still low it is actually higher than the difference calculated for the panchromatic image. The spectral separability between archaeological features and their surroundings can be evaluated using the following indicator $\Delta\mu/(\sigma1+\sigma2)$ where $\Delta\mu$, $\sigma1+\sigma2$ are respectively the average spectral differences and standard deviation observed for the two mask (mask 1 and mask 2). This indicator allows to take also the influence of noise into account. Table 1 shows that $\Delta\mu/(\sigma1+\sigma2)$ is at 0,188 and 0,358 for pan and red respectively. Thus, strongly confirms the fact that for the investigated area red channel provided the best results.

The spectral contrast between underlying structures and their surrounding areas can be identified by a visual inspection (compare Figures 3.1 a and 3.1b) and it can be more enhanced by applying a. statistical approach that can enable the extraction of spatial discontinuity and make the reconnaissance of geometric pattern distribution easier.

Monte Serico reflectance values								
Mask data	PAN				RED			
	μ	σ	$\Delta\mu$	$\Delta\mu/(\sigma1+\sigma2)$	μ	σ	$\Delta\mu$	$\Delta\mu/(\sigma1+\sigma2)$
Mask1	56,86	3,46	1,31	0,188	64,18	4,93	3,48	0,358
Mask2	55,55	3,49			60,7	4,8		

Table 1 *Monte Serico reflectance values observed for the archaeological features (mask 1) and their surroundings (mask 2)*

The capability of QuickBird Red spectral channel to better enhance spatial anomalies for bare or sparsely vegetated surfaces is due to many reasons. Firstly, the presence of near-surface archaeological deposits (soil marks) strongly influences the soil constituents and especially moisture content. Soils reflectance is dependant on several factors such as the colour, constituents and especially the moisture content. Water is a relatively strong absorber of all wavelengths. The overall reflectance of soil tends to decrease as a soil moisture content increases. Thus, coupled with the presence of building materials, creating different spectral response linked to the different superficial patterns that, for the given soil type, are strongly enhanced by red reflectance.

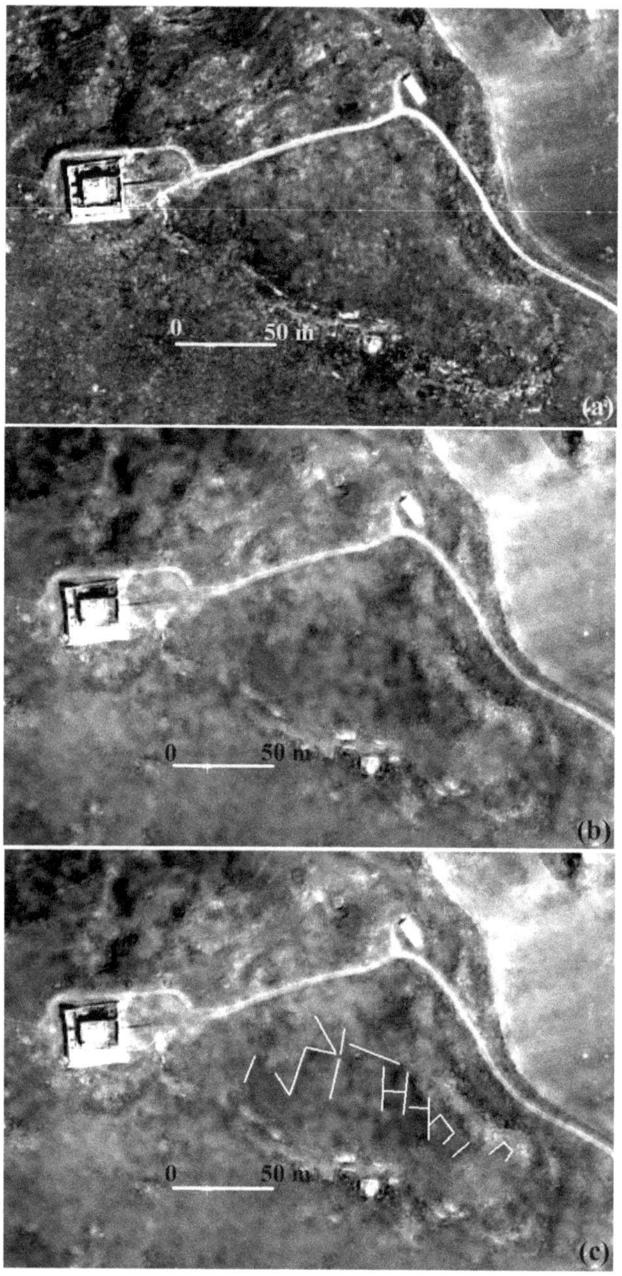

Figure 3.1 *Monte Serico: (a) panchromatic image; (b) red data fusion product; (c) red data fusion product with marks related to buried structures of the medieval village.*

Secondly, the sunlight angle conditions of the QuickBird data acquisition made visible the presence of micro-topographic relieves (shadow marks) by shadowing. Such micro-topographic relieves are remains of ancient buildings that crop out of the ground. This causes small differences in height level between buried structures and neighbouring areas strongly affected by erosion processes due to the lack of vegetation.

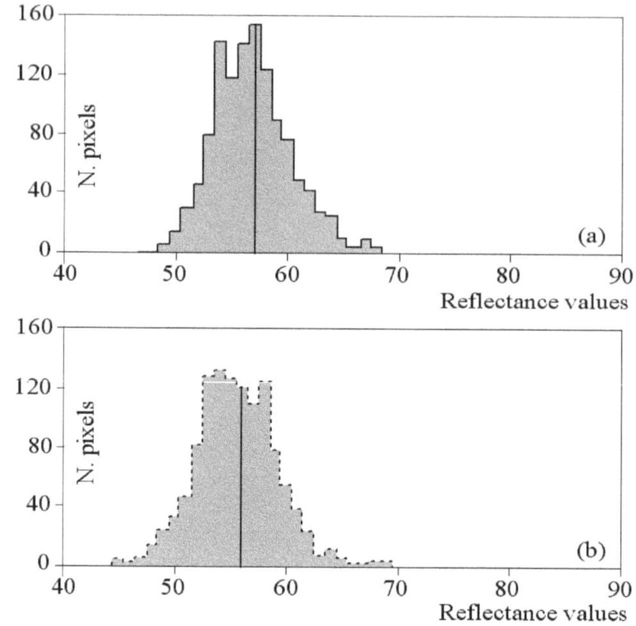

Figure 3.2 *Monte Serico panchromatic image: histograms of reflectance values of archaeological features (a) and their surroundings (b)*

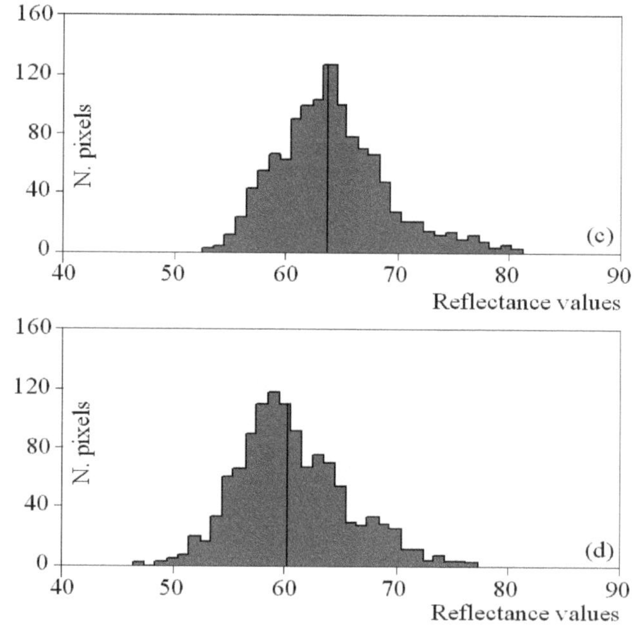

Figure 3.3 *Monte Serico red channel: histograms of reflectance values of archaeological features (a) and their surroundings (b)*

4.2 Metaponto test site

4.2.1 Study area

Metaponto is one of the most important archaeological sites in the South of Italy. It is located in the Basilicata Region (see Figure 2). The several archaeological campaigns performed by Adamesteanu (1974) stated human presence since the half of the 8th century B.C. when Metaponto was founded by Greeks coming from Acaia region. Between the Greek colonization (700 BC-

200 BC) and the Roman age (200 BC – 400 AD) the territory was characterized by an intense human presence, as revealed by the several rural sites brought to light by excavations, and also by the presence of an extensive system of parallel land divisions (Adamesteanu, 1973). In particular, these lines are thought to have been a network of country lanes or drainage canals (Carter 1983, 1990). The drainage canals were built because of rise of water table that occurred in the 5th century. The satellite-based detection of these land divisions is the main objective of the Metaponto study case.

The site investigated (see Figures 2 and 4), named San Salvatore, is near to a burial site dating back between the 5th century and the middle of the 3th century. The investigated area was scarcely vegetated when the satellite images were acquired.

4.2.2 Satellite based analysis

The satellite analysis was carried out by using QuickBird images) acquired on August 22nd 2004 with an off nadir view angle of 2 degrees (see table 1).

The best results (see Masini and Lasaponara, 2006) were obtained from panchromatic, NIR channel images. However, the Blue and Green channels did not show any regular features of archaeological interest. Figures 4.1 (a to c) show the QuickBird panchromatic image and (4.1.a), NIR (4.1.b) and red (4.1b) channels. Figure 4. 2 shows the map of land division obtained from satellite imagery. Figures 4.3, 4.4, and 4.5 show the histograms obtained from the archaeological features (denoted as mask 1 in Table 1) and their surroundings (denoted as mask 2 in Table 1).

A visual inspection of histogram related to panchromatic that is shown in figures 4.3 clearly show that the archaeological features can not be discriminated due to the very low detectable observed in red channel between archaeological features and their surroundings. The difference in reflectance values computed in this case is at around -2.1. As in the previously considered case (section 4.11) such a value is still low, but, it is actually higher than the difference calculated for the panchromatic image. Finally, the NIR channel exhibit the highest differences in reflectance values between pixels with underlying archaeological structures and their surroundings.

The spectral separability between archaeological features and their surroundings can be evaluated as in the previous case by using the figure metric $\Delta\mu/(\sigma1+\sigma2)$ where $\Delta\mu$, $\sigma1+\sigma2$ are respectively the average spectral differences and standard deviation observed for the two mask (mask 1 and mask 2).

This indicator allows to take also the influence of noise into account. Table 2 shows that $\Delta\mu/(\sigma1+\sigma2)$ is at 0,052 , -0,347 and 0,387 for pan, red and NIR. Thus, strongly confirms the fact that for the investigated area red and NIR channels provided the best results.

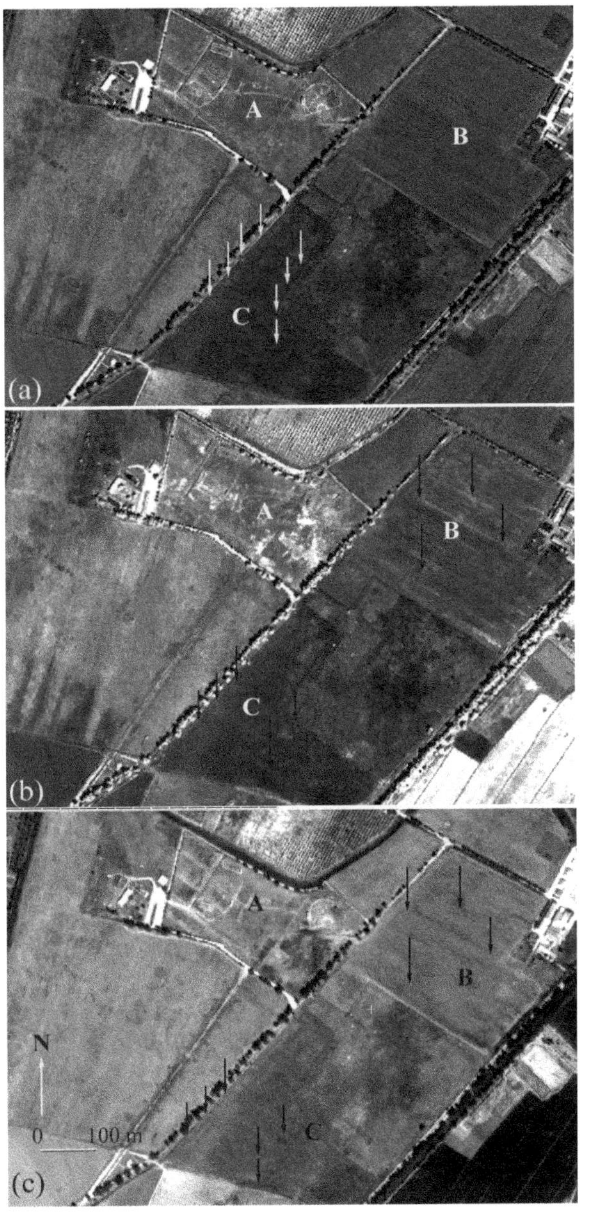

Figure 4.1 *Metaponto: site investigated near the archaeological park of Apollo Licio. (a) Panchromatic image; (b) NIR channel image; (c) red channel image. Letter A denotes the archaeological park, whereas letters B and C indicate the areas in which land divisions were detected. The arrows put in evidence the weed-marks related to ancient land divisions.*

The figure metric could also been considered in its absolute value, but, we suggest to take the sign into account to better understand the nature of the marks, namely the type of structures that are underlying. The spectral contrast between underlying structures and their surroundings could be further enhanced by applying a statistical approach that can enable the extraction of spatial discontinuity and make the reconnaissance of geometric pattern distribution easier.

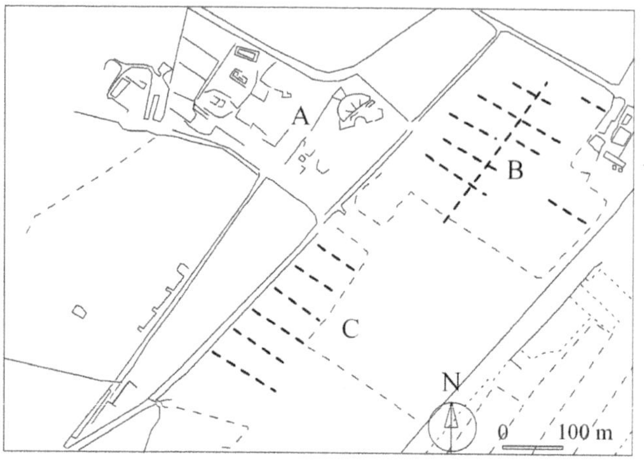

Figure 4.2 *Metaponto: map of land divisions (dashed lines) drawn by the QuickBird imagery.*

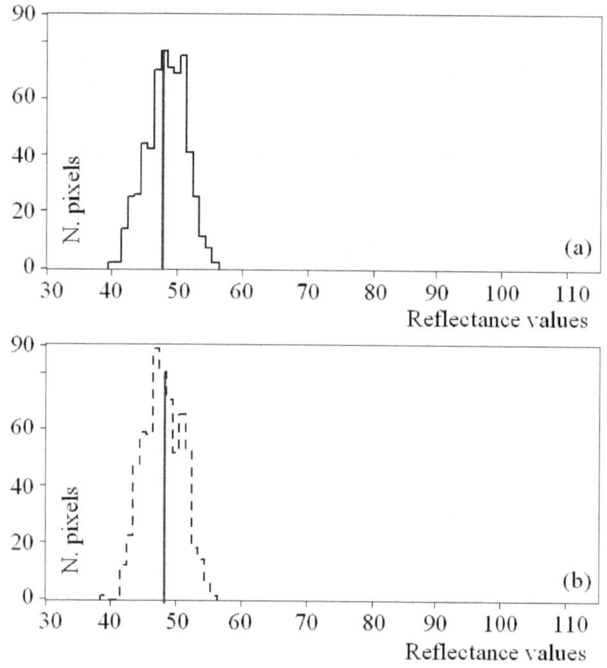

Figure 4.3 *Metaponto, Archaeological Park: site B. Panchromatic image: histograms of reflectance values of archaeological features (a) and their surroundings (b)*

The detected anomalies are the typical marks induced by the presence of vegetation that, in this case, is mainly composed by weeds. Such anomalies, indicated respectively as B and C in figure 4.1 and 4.2, are evident in the east and southeast parts of the Archaeological Park. In particular, for the B area, the weed-marks related to land divisions are not visible by using the panchromatic image.

They are only delineated by using NIR channel images (both single channel and NIR data fusion product). The use of NDVI map makes such marks more evident. There are seven rectilinear features: six of these features have an orientation of 55°-60° West of North (that are distanced from each other by about 35 to 54 m) and one is orthogonal to them.

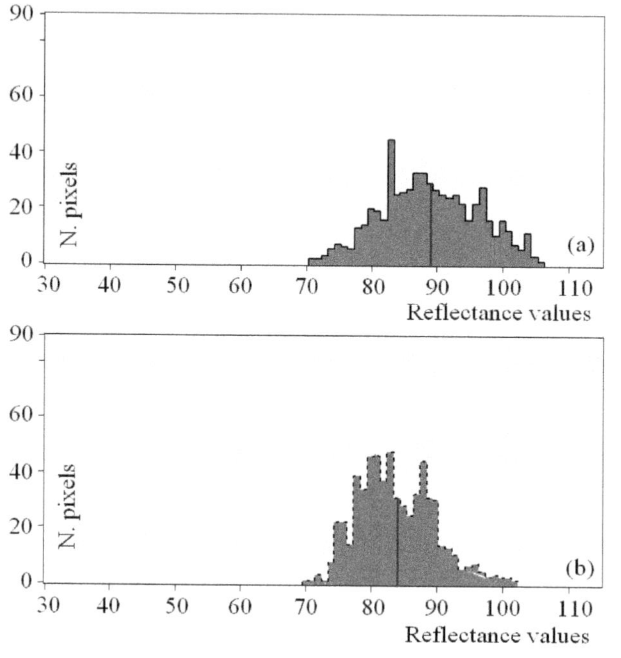

Figure 4.4 *Metaponto, Archaeological Park: site B. NIR channel: histograms of reflectance values of archaeological features (a) and their surroundings (b)*

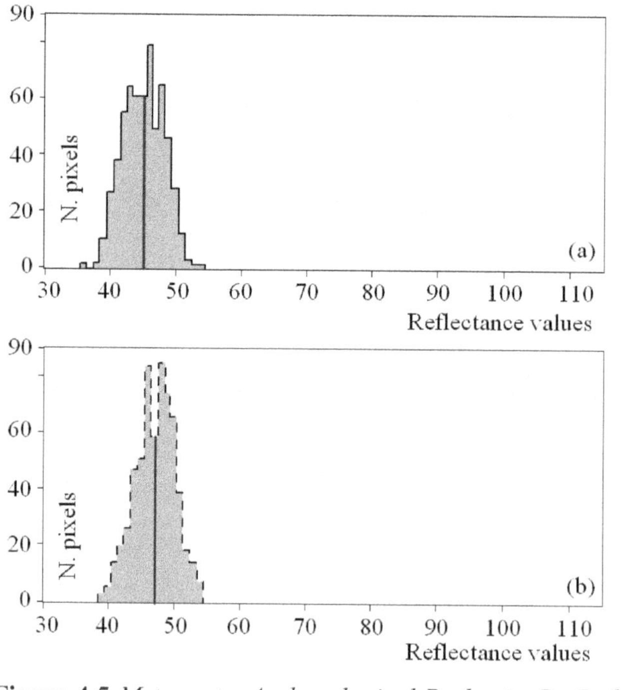

Figure 4.5 *Metaponto, Archaeological Park: site B. Red channel: histograms of reflectance values of of archaeological features (a) and their surroundings (b)*

For the C area, the weed-marks are visible using only panchromatic and NIR images. Here, there are 6 rectilinear features with an orientation of 50°-55° West of North that are distant each other approximately 46 to 48m. The fact that the same kind of archaeological mark

Metaponto reflectance values												
Mask data	PAN				RED				NIR			
	μ	σ	Δμ	Δμ/(σ1+σ2)	μ	σ	Δμ	Δμ/(σ1+σ2)	μ	σ	Δμ	Δμ/(σ1+σ2)
Mask1	48,37	3,02	0,31	0,052	45,14	3,06	-2,1	-0,347	89,02	7,46	5,13	0,387
Mask2	48,06	2,96			47,24	3			83,89	5,78		

Table 2 *Metaponto reflectance values.*

exhibits different spectral features, can be linked to the different surface characteristics observed for the two investigated areas. Both of them were covered by weeds that were dry at the end of summer when the satellite images were acquired (August, 22nd 2004).

However, area B (see Figure 4.1) had a more packed weed cover compared to that present over area C. Therefore, for area B the reflectance variations induced by the presence of buried archaeological deposits are stronger than those observed for area C. These spectral characteristics are evident by a visual inspection of the histograms observed for the archaeological features (see mask 1) and their neighbouring (see mask 2) as shown in Figures 4 and Table 2.

4.3 Jure Vetere test case

4.3.1 Study area

Jure Vetere is the first monastery of Florense Orden was founded by Gioacchino da Fiore at the end of the 12th century. It is located in the Calabria Region (see Figure 2).

The archaeological area is located on the ridge of a hilly plateau characterized by the presence of vegetation made up of dense herbaceous plants that were green and blooming when the satellite images were acquired. The site faces the watershed between the basins of the Arco river and its affluent. From a geological point of view Jure Vetere is within the Monte Garigliano unit made up of high grade metamorphic rocks, such as graniodiorites and magmatites. The soils are mainly composed of siliceous sands, products of the chemical and mechanical alteration of acid igneous rocks, with a minor part of clays made up of kaolins and vermiculites (Roubis et al. 2003).

Before starting archaeological excavations (2002-04) preliminary investigations based on aerial photos (1998-1999) and geophysical prospections were performed in order to identify the location of buried settlements. Results obtained from both aerial photos and geophysical prospections identified the highest probable location of ancient buried remains. Such results were confirmed by the excavation campaigns that unearthed a large structure related to the church of the monastery. Actually, the location of the rest of the medieval monastery is still unknown. For this reason, the archaeologists enlarged the study area to the whole plateau. The enlargement of the study area and its physical characteristics suggested the integration of the remote sensing data set with satellite QuickBird imagery.

4.3.2 Satellite based analysis

In the case of Jure Vetere, the most interesting outputs (Lasaponara and Masini, 2006b) were obtained from the processing of NDVI map (see Figure 5), whereas panchromatic and multispectral channels processed at their own spatial resolution did not provide additional information to identify surface anomalies of archaeological interest. Figure 5 shows panchromatic, NIR and NDVI map

The processing of NDVI map showed the presence of crop-marks having a rectilinear morphology (Figure 5). Geophysical prospections (Bavusi and Chianese, 2006) and archaeological trials showed the nature of these features which are related to drainage canals, maybe coeval with monastery. The capability of QuickBird NDVI to better enhance spatial anomalies for surfaces covered by green and blooming herbaceous plants is due to the fact that the NDVI is indicative of plant photosynthetic activity and related to the green leaf area index. In fact, traditionally, vegetation monitoring by remotely sensed data has been carried out by using NDVI, that is a quantitative measure of biomass and vegetative vigor. The NDVI operates by contrasting intense chlorophyll pigment absorption in the red against the high reflectance of leaf mesophyll in the near infrared. NDVI is indicative of plant photosynthetic activity and related to the green leaf area index. A growing plant lacking certain nutrients may produce less chlorophyll. Thus spectral reflectance variations are a function of plant state which in return depends on growth conditions. The relationships existing between growth conditions and plant state as well as those existing between plant state and reflectance ability allows the identification of vegetation abnormal status caused by underlying archaeological deposits.

5 Final remarks

Satellite Quickbird imagery were used in order to assess their capability to detect archaeological marks, such as soil, crop and shadow marks. The analyses were performed for some archaeological sites buried under surfaces characterized by different characteristics. In particular, also the status of vegetation (dry/green) was different when the satellite data were acquired. Results from our investigations showed that the multispectral VHR satellite data provide valuable information for

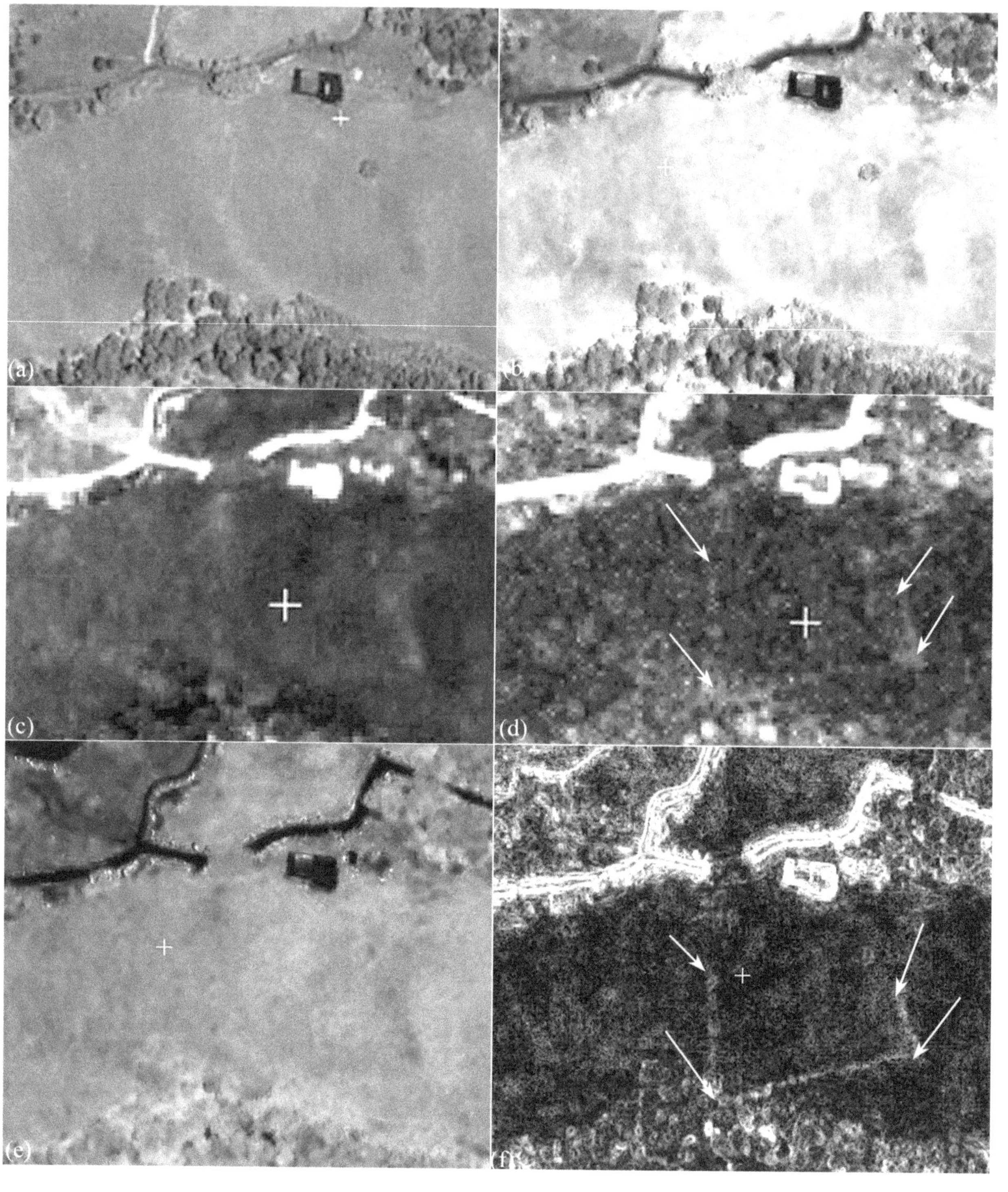

Figure 5 *Jure Vetere QuickBird images: (a) panchromatic image; (b) NIR channel data fusion product; (c) NDVI map (2,80 m); (d) results from NDVI map processed according to Lasaponara and Masini (2006b); (e) NDVI data fusion map; (f) results from NDVI data-fusion map denote the location of marks of archaeological interest.*

vegetated areas, where the use of aerial photos can be limited since the visibility of crop marks strongly depends on many factors such as, vegetation type and status, soil conditions, sun-sensor geometry and film sensitivity.

In particular, the main findings of the performed investigation can be summarized as follows:

1. QuickBird red channel was able to better enhance soil marks observed for bare surfaces;
2. NIR channel was able to better enhance crop marks observed for vegetated surfaces;
3. the use of vegetation indices, such as NDVI (Normalized Difference Vegetation Index allowed to better emphasize crop marks observed for surfaces covered by green and healthy herbaceous plants.

4. Panchromatic imagery were generally found good for the detection of archaeological features for most of surface and soil characteristics.
5. the spatial resolution of VHR satellite imagery is still lower than those obtained from aerial photo, thus should limit the identification of small features.

References

ADAMESTEANU, D. 1973. Le suddivisioni di terra nel metapontino. In *Problems de la terre en Grece Ancienne* ed. M I Finley, Paris, Mouton, pp. 49-61.

ADAMESTEANU, D. 1974. *La Basilicata Antica. Storia e Monumenti*, Cava dei Tirreni.

BAVUSI, M. and D. CHIANESE 2006. Indagini geofisiche a Iure Vetere, in press.

CARTER, J.C. 1983. *The territory of Metaponto, 1981-82*. Austin: Institute of Classical.

CARTER, J.C. 1990. *Between the Bradano and Basento: Archaeology of an Ancient Landscape* in Kelso W and Most R (Eds.) *Earth Patterns. Essays in Landscape Archaeology*. Charlottesville : University of Virginia Press, pp. 227-243.

CLARK, C.D., S.M. GARROD, & M. PARKER PEARSON 1998. Landscape archaeology and Remote Sensing in southern Madagascar. *International Journal of Remote Sensing*, vol. 19, No 8, 1461-1477.

DRAKE, N.A. 1997. Recent aeolian origin of superficial gypsum crusts in Southern Tunisia: geomorphological, archaeological and remote sensing evidence, *Earth surface processes and landforms*. 22 : 641-656.

M.J.F. FOWLER, Y. FOWLER, Detection of Archaeological Crop Marks on Declassified CORONAKH-4B Intelligence Satellite Photography of Southern England, Archaeological Prospection. 11(4) (2005) 237-246.

LASAPONARA, R., & N. MASINI 2005. QuickBird-based analysis for the spatial characterization of archaeological sites: case study of the Monte Serico Medioeval village, *Geophysical Research Letter*, 32, (12): L12313.

LASAPONARA, R., & N. MASINI 2006a. On the potential of panchromatic and multispectral Quickbird data for archaeological prospection. *Int. J. Remote Sensing*, 27: 3607-3614.

LASAPONARA, R., & N. MASINI 2006b. Identification of archaeological buried remains based on Normalized Difference Vegetation Index (NDVI) from Quickbird satellite data IEEE *Geoscience and remote sensing Letters*, 3: 325-328.

MASINI, N. 1995. Note storico-topografiche e fotointerpretazione aerea per la ricostruzione della ""forma urbis" del sito medievale di Monte Serico. *Tarsia*, 16: 45-64.

MASINI, N., & R. LASAPONARA 2006. Satellite-based recognition of landscape archaeological features related to ancient human transformation. *Journal of Geophysics and Engineering*, 3: 230-235.

ROUBIS D., F. SOGLIANI., & M. LAZZARI 2003, Ricerche archeologiche a Jure Vetere (S. Giovanni in Fiore): la campagna di scavo del 2002 e le indagini geopedologiche, in *Siris.*, 4 : 99-118.

SEVER T.L. 1998. Validating prehistoric and current social phenomena upon the landscape of Peten, Guatemal", in Liverman, D., Moran, E. F., Rinfuss R. R., and Stern, P. C., (eds.), *People and pixels: Linking Remote Sensing and Social Science*, Washington D.C., National Academy Press.

WILSON, D. R. 1982. *Air photo interpretation for archaeologists*, London, St. Martin's Press.

Satellites, Survey, and Settlement: the Late Classic Maya Utilization of Bajos (Seasonal Swamps) at Tikal and Yaxha, Guatemala

Errin T. Weller

Department of Anthropology
University of Colorado at Boulder, USA

Introduction

Ancient Maya civilization is best known for stunning monumental pyramids set in lush rainforest. This research, however, has been biased towards the high-canopied uplands leaving our understanding of past landscape incomplete. The last decades have seen a resurgence of interest in the low-lying *bajos* (seasonal swamps) comprising 40% of the ancient Maya landscape of lowland Guatemala. This prominent *bajo* feature has fascinated explorers and archaeologists since the early 20th century because of the imposing archaeological sites located on *bajo* margins. *Bajo* research is an outgrowth of earlier research on wetlands, permanently moist areas as opposed to seasonally moist *bajos*. Some scholars suggest the *bajos* were an important resource providing available land for habitation, water, or agriculture to support the large populations of urban Late Classic (AD 550-850) Maya centers. The prevailing belief, however, has been that the *bajos* are, and were, uninhabitable. This belief has permeated research to such an extent that one of the most intensive intersite surveys, between the two archaeological sites of Tikal and Yaxha, to test a settlement model, completely skipped over *bajo* environments.

The sheer magnitude of *bajos* has been daunting to researchers using traditional methods of survey transects. The Maya lowlands present many challenges to data acquisition because of the rainforest environment. Rich biomass, thick vegetation, heavy cloud cover, and rainfall characterize the area. For example, the rainy season typically begins in June and extends through December (encompassing the summer months when many academics are able to conduct field research) with yearly averages of 2000 mm to 4500 mm of precipitation. The dry season, peaking in April, has limited rainfall but can experience temperatures in excess of 30 degrees Celsius with the overall average temperature being 24 to 26 degrees Celsius.

Survey is expensive, laborious, and in tangled overgrowth it is easy to pass by archaeological remains. Therefore, this method is not going to yield sufficient answers to research questions. Advances in remote sensing technology, especially the advent of high-resolution (with sub-meter pixel size) multi-spectral imagery, allow for re-analysis of *bajo* questions. It provides a method to analyze large regions and aid in prioritizing and selecting optimum research areas in which to ground-truth. In 2006, I began the Bajo Regional Archaeological Project with the objective of exploring the utilization of *bajos* surrounding the sites of Tikal and Yaxha in the northeast Peten of Guatemala. I chose two *bajos*: the approximately 10km x 10km Bajo la Justa, by Yaxha and the larger Bajo de Santa Fe, by Tikal. Both areas are located in the protected Maya Biosphere Reserve meaning there is limited modern settlement, although illegal encroachment with slash-and-burn agriculture is an increasing problem.

This project is integrating acquired Global Positional System (GPS) data from previous *bajo* projects and survey maps with satellite imagery (Landsat and IKONOS), RADAR (AIRSAR and SRTM), spatial analysis and future ground survey to investigate the purpose and extent of landscape modifications within and surrounding these *bajos*. These results will be directly applicable to questions of ancient Maya settlement, subsistence, and population. To obtain a more coherent understanding of settlement, my research is studying the *bajo* landscape for evidence of the following: 1) agriculture (seasonal or continuous), including canals, planting platforms, terraces, and dams to control water flow; 2) habitation (on natural or raised areas); 3) water procurement and/or storage including reservoirs and aguadas (natural clay-laden sunken areas); and 4) causeways (roadways).

In this paper, I will discuss previous research conducted in the Bajo la Justa and the Bajo de Santa Fe, the preliminary results of my analysis, interpretation, and suggest the impacts this may have on future research.

Background

Bajo La Justa

The work conducted here builds upon that of the now completed Bajo Communities Project in the Bajo la Justa directed by T. Patrick Culbert and Vilma Fialko. The aim of this project, also conducted in cooperation with Dr. Tom Sever and Daniel Irwin of the National Aeronautics and Space Administration (NASA), was to determine how settlement was affected by the *bajo* through documenting settlement patterns, settlement history, and the presence of agriculture. The project documented a range of settlement complexity from isolated mounds on raised area ("*bajo* islands") to minor centers both within the *bajo* itself and the immediate surroundings. In the Bajo la Justa, the data suggest a hierarchy of sites with long occupation. Their final goal, the detection of raised agriculture, remains equivocal (Sever and Irwin 2003).

Bajo de Sante Fe

Only very limited research has been undertaken within the Bajo de Santa Fe, documenting some habitation, a cluster of linear anomalies of unknown function and an area of pine savanna. Previous soil cores taken in the *bajo* suggests it was once a permanent wetland or even a shallow lake that was transformed by the heavy deforestation undertaken by the ancient Maya. This activity possibly transformed the landscape into a seasonal swamp by the Late Classic and the one we observe today.

Additionally, Fialko (2005) has documented heavy settlement on the bajo margins. From the Late Preclassic (350 BC to AD 150) through the Late Classic (AD 800), the period of maximum population growth, sites exploded along the *bajo* margins growing in both number and size. Within the *bajo* proper, only one *bajo* island has been studied. She and her colleagues suggest that early settlement was attracted by the permanent water source. As population increased and the ancient Maya damaged their environment, they may have transformed a permanent wetland into a seasonal wetland.

Imagery Analysis in the Bajo la Justa

The dataset over the Bajo la Justa includes Landsat 7 ETM+ from 2003, IKONOS satellite imagery covering a majority of the *bajo* acquired in January 2001, complete coverage from the Shuttle RADAR Topography Mission (SRTM), and a small portion of Airborne Synthetic Aperture RADAR (AIRSAR) acquired in 2004. Dr. Culbert of the Bajo Communities Project has provided an additional resource: the Global Positioning System (GPS) data from approximately 27km of survey transects from the Bajo la Justa. The researchers recorded vegetation changes every 25m as well as any archaeological features.

General Analysis

All analysis to date has been conducted in the computer program RSI-ENVI, an industry standard. A dark object subtraction was performed on the IKONOS and Landsat imagery to reduce atmospheric haze. The images were each pan-sharpened, creating 1 meter and 15-meter pseudo-pixels, respectively. The Brovey transformation was particularly useful in false color images because it highlights vegetation differences. HSV and PC Spectral sharpening were also employed.

Habitation

Due to the seasonal inundation of the *bajos*, habitation needs to be raised. Previous research has identified a number of *bajo* islands of habitation. To locate more of these locales, I employed a number of analysis techniques. The first was straight visual analysis in color infrared imagery of the IKONOS (4, 3, 1 and 4, 2, 1). The islands contain vegetation that distinguishes them from their surroundings due to their elevated nature.

Drs. Tom Sever and William Saturno have discovered a vegetation signature in color infrared imagery that indicates the presence of ancient Maya sites. The vegetation covering the sites appears more yellow than their surroundings and they hypothesize that this is due to the decay of limestone, the primary building material available to the Maya. The exact mechanism by which the limestone affects the vegetation is still unknown. The vegetation signature also holds true in true color imagery where there is a correlation between habitation and lighter colored vegetation.

Principal component analysis successfully distinguished known habitation sites from the surrounding *bajo* vegetation. It also aided in identifying new *bajo* islands.

Band ratios can further refine this vegetation signature. Based on the spectral signature of vegetation, soil, and water, I assigned the following ratios: vegetation (4/3), soil (3/1), and water (2/4). I then assigned each ratio a color and combined them into a single image. Differences between archaeological sites emerged, some were clearly covered with more vegetation while others appear to have more visible soil or rock (possible limestone, the main component of site construction) on the surface.

The combination of IKONOS and Landsat imagery and the Digital Elevation Model (DEM) from AIRSAR was helpful in distinguishing larger *bajo* islands. Using ENVI's 3D SurfaceView to drape Landsat 7 imagery as well as IKONOS data over the AIRSAR Digital Elevation Model (DEM) is also effective. However, the 5m-pixel size of the DEM was inadequate for smaller scale features. The large pixel size of the SRTM data was not helpful in delineating *bajo* islands that were not already known features. The *bajo* is characterized by particularly low slope and STRM is not of sufficient resolution.

I have identified 26 new possible *bajo* islands, ranging in size from 60 to 460 meters in length, on the color infrared IKONOS imagery. This is a conservative count and does not include new sites located on the *bajo* margins. On the larger pan-sharpened Landsat imagery, I have identified 27 new *bajo* islands.

By comparing the vegetation signatures of known sites with possible *bajo* islands, I can say with relative certainty that a number of islands remain to be field verified.

Causeways: roadways of monumental construction

Causeways have a high reflectivity in the infrared band and linearity. There is a difference in the moisture of vegetation on causeways versus the surrounding vegetation because vegetation prefers to grow on the better drained construction versus the poorer drained swamp. Color infrared combinations of IKONOS and Landsat as well as band ratios (4/3, 2/4, 3/1) revealed multiple linear anomalies throughout the *bajo*. Principal component analysis was successful in highlighting vegetation differences in the *bajo* vegetation and revealing multiple linear anomalies. The fact that the same anomalies are present in multiple analysis techniques is encouraging and suggests they may be real on-the-ground features. They are also present and enhanced in a decorrelation stretch. There is one anomaly in particular that I believe has a high likelihood of being a causeway. The probability that it is cultural is increased because it is in direct association with one confirmed and one possible archaeological site. In theory, combining the multispectral image with a digital elevation model obtained from AIRSAR should delineate if the anomaly has height consistent with a causeway. In practice, however, it does not. While this anomaly does have height, I have also observed that denser areas of vegetation along a small tributary (the Rio Yaxha) also appear to have height when the multispectral image is draped over the DEM.

Multiple filtering techniques, including edge detection, did not prove useful in detecting these anomalies. As the previous transects did not cross these areas, they will be ground-truthed during the 2007 field season.

Canals

Canals within *bajos* can be detected due to high reflectivity because of increased lush vegetation in a manner similar to causeways. *Bajos* have low slope and water features are naturally very curvy, with a high sinuosity, to flow through the almost flat landscape. Therefore, features that are linear (and have low to no sinuosity) are likely cultural and thus not naturally occurring. The two primary waterways running through the *bajo* are the Rio Holmul and its tributary, the Rio Yaxha. One canal has been confirmed through excavation close to Yaxha (Kunen et al 2000).

Color infrared compositing of IKONOS and Landsat imagery both reveal unusual features in the Rio Yaxha. The tributary makes unnaturally sharp turns through the *bajo* suggesting cultural modification. These angles are further enhanced in the AIRSAR data by running texture filters, specifically various co-occurrence measures, found in ENVI. Another very straight anomaly leads from the Rio Holmul to a *bajo* island, a probable indicator of a canal. This suggests that the tributaries running through the *bajo* were likely modified. Again, previous transects have not crossed this section of the *bajo*.

Agriculture

In a *bajo* environment, agricultural landscape modification would most likely involve cutting irrigation channels into the silt and mounding the dirt beside these cuts to form raised areas suitable for cultivation or recessional agricultural where the areas are cultivated as the water levels recede. Early techniques of detecting agriculture involved visible observation of a lattice vegetation pattern in Southern Quintana Roo, Mexico and northern Belize later confirmed as raised field agriculture (Turner and Harrison 1983). I have not observed such patterns in the visible imagery or in the RADAR. Researchers have also looked for evidence of agriculture, without success, by subjecting Landsat TM (30m pixel size) imagery to false color compositing, contrast stretching and edge enhancement (Pope and Dahlin 1993). These techniques on the Landsat 7 have also proved unsuccessful.

Detailed vegetation classification can indirectly indicate agriculture. In the methodology of Culbert's project in the Bajo la Justa, vegetation mapping was undertaken. They were specifically interested in palm *bajo* (consisting of palms which have heights varying from 15 to 20 meters), because moisture probes showed it retains enough moisture in the dry season to be favorable for agriculture. Guatemalan informants also agreed with this assertion, however, other *bajo* vegetation types such as the common scrub *bajo* of dense, short vegetation (6 to 12 meters in height) varied widely in their moisture retention (Kunen et al. 2000). Therefore, the researchers correlated vegetation observed during survey with Landsat imagery.

With finer-grained IKONOS imagery, I undertook supervised vegetation analysis correlating vegetation identified on-the-ground with the imagery. I identified 12 classes including vegetation (scrub bajo, palm bajo, etc) and other features (such as the large lake). Using various classifications (parallelepiped, minimum distance, and so forth) it was determined that maximum likelihood classified the most pixels. The *bajo* is dominated by palm and scrub *bajo*, findings similar to those obtained by previous researchers suggesting agriculture is a viable possibility. A visual classification obtained similar results suggesting vegetation differences are visible. The imagery can also be classified in true color although the differences are subtler.

Both the probable canal and causeway features discussed above (linear and/or strong angular features) were enhanced with vegetation classification, further suggesting they are distinguished from their surroundings by vegetation.

Unsupervised classifications (IsoData and K-means) are not useful in distinguishing the often mixed and subtle *bajo* vegetation. They could only identify the most general of features in the imagery such as large lakes, clouds, and roads. The only archaeological feature unsupervised classification was able to successfully identify was the large Aguada Maya (discussed in further detail below). These techniques could also separate the lowest *bajo* vegetation from the highest upland vegetation.

To date, no raised agricultural fields have been verified. Landsat imagery analysis that I conducted has revealed two areas of high vegetation reflectance in the *bajo* that resemble *bajo* islands. A singular difference, however, is that the *bajo* islands identified to date have been round in form. The two raised areas are rectangular, indicating possible construction. They are also both in close proximity to the Rio Holmul. This suggests they may be agricultural fields, an assertion in need of further study. Agricultural fields could have been buried by sedimentation.

Water Features

The largest water feature confirmed to date is the 13, 200 sq. m. Aguada Maya. It is a rectangular reservoir that began as a natural depression and was modified by the ancient Maya. It is believed that it could have irrigated nearby land although no agriculture has yet been positively identified. In various combinations of AIRSAR band frequencies, I identified square and circular features encircling the Aguada Maya.

Water features are identifiable because they appear black on infrared imagery. In addition, *bajos* have been documented to have small pockets of perennial wetlands called civales: "treeless, wet areas of herbaceous vegetation" that can be wet in the dry season (Dunning et al. 2002.) As a potential source of water, the cival's location would have been an important resource in this area that does not receive significant rainfall for four to six months. Civales can be identified on imagery because of the contrast with surrounding vegetation and they appear dark in infrared imagery. Aguadas, naturally depressed areas that accumulate and hold water, are also important to identify because that they may attract settlement communities to exploit water.

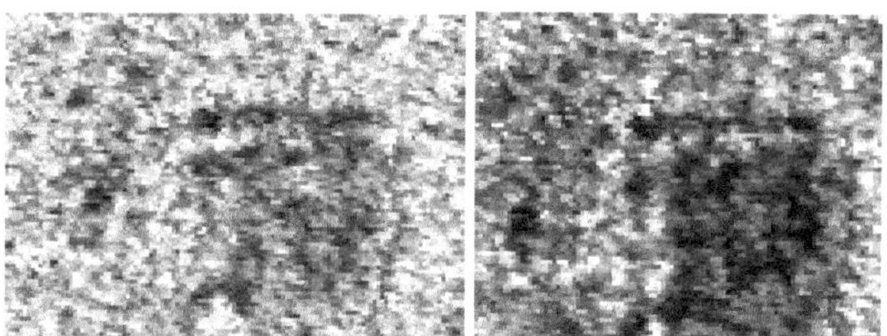

Figure 1 *Aguada Maya in AIRSAR imagery. Note the linear and round lines.*

In addition to water sources at confirmed archaeological sites and the rivers, small pockets of water (no more than 11 meters in width) are visible throughout the *bajo*. I have also identified eight larger aguadas on the Landsat imagery, ranging in size from 35 to 85 meters across.

Imagery Analysis in the Bajo de Santa Fe

The dataset for the Bajo de Santa Fe includes Landsat 7 ETM+ acquired in 2003 that provides complete coverage of the *bajo* as does SRTM. Two IKONOS images subset the *bajo* as well as AIRSAR. The IKONOS images were both acquired in March 2001 and AIRSAR in 2004. GPS data are not available as in the Bajo La Justa because of the limited ground survey.

General Analysis

The same methodologies established for identification of features in the Bajo la Justa were employed in analysis of the Bajo de Santa Fe including pan-sharpened imagery, visual analysis, band ratios, color infrared imagery, and vegetation classification.

Habitation

In the IKONOS imagery (4, 3, 1) there are two clear vegetation signatures: one predominately red and blue and the other predominately red. In the 4, 2, 1 band combination these differentiations are also apparent with one set of *bajo* islands appearing in a combination of red/orange and the other white/blue. There area approximately 30 of the first and approximately 97 of the second. The *bajo* islands of the first type are larger in size (0.5 to 1.0 km) while those of the second type tend to be only 100 meters with few exceptions. These vegetation differences are also observable in true color images, although they are subtler.

Analysis of the false and true color Landsat imagery also identified numerous *bajo* islands across the Bajo de Santa Fe. Within the *bajo*, I have identified 87 islands ranging in size from 100m to 2.5km in size.

In the second IKONOS imagery closest to the archaeological site of Tikal, the habitation in the *bajo* is different. There are fewer discrete areas of *bajo* islands and more ridges, making a habitation count difficult. There is also a large sprinkling of white pixels (in both the true color and CIR images) throughout the *bajo* and along the margins. To date, it is unclear what they are but it is highly likely they are associated with ancient Maya construction and settlement.

Agriculture

As discussed above in the habitation section, there is a type of small vegetation separate from the surrounding *bajo*. Also significant is the location of these small clusters. The majority of them are within the curve of the Rio Holmul, in an area of numerous drainages and are connected to the river by these drainages. Thus, the likelihood is very high that these vegetation clusters are connected to the river itself. Of all the locations within the *bajo*, this locale has the highest probability of having raised field agriculture. These vegetation differences are also evident in various band combinations on the AIRSAR data.

Water

Water is easy to identify in the Bajo de Santa Fe. In total, 204 water features in the IKONOS satellite imagery of various sizes, from a few meters to 50 meters in width. A square water feature, a probable reservoir, of 50 meters in width, has also been identified in the middle of the *bajo* in direct association with the Rio Holmul. The Landsat imagery also revealed a 41 aguadas, 25 to 150 meters in size. Water is apparent throughout the *bajo* even in March, at the height of the dry season.

Linear Anomalies

In the IKONOS imagery near Tikal there are numerous linear anomalies connecting the margins of the *bajo* to likely settlement within the *bajo*. The function of these anomalies remains unclear because they are not highly reflective in the color infrared imagery. They are primarily white. They may be causeways but require ground-truthing.

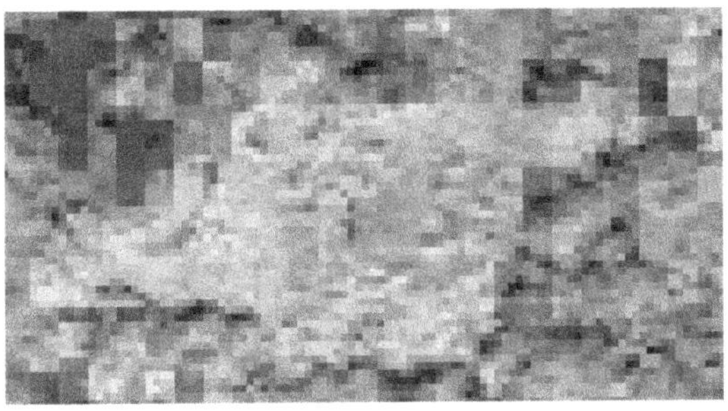

Figure 2 *Square feature visible in IKONOS brovey pan-sharpened imagery. Feature is adjacent to the river.*

Previous research done by Sever and his colleagues has identified a system of linear anomalies in the northwest corner of the *bajo* near Tikal and field verification was unable to determine their function. They were raised in sections and depression in other sections. The ground survey confirmed they are ancient in origin but not if they are irrigation channels, causeways, or of some other function. Further analysis of this area suggests part of this system may in fact form a rectangle. The availability of AIRSAR, not available when the original researchers conducted their field verification, further indicates another anomaly attached to this system of anomalies.

It may be purely coincidental, but this section of the *bajo* is 1.5 – 2 km due east of where the northern Tikal wall (a large embankment adjacent to a ditch) terminates. Currently, the Tikal earthwork is 12.8 km long and averages 4 meters in width.

Unlike a causeway or a canal, it is not straight and meanders across the landscape and often has gaps, making identification difficult. The suite of analysis techniques conducted on the imagery has not rendered the known section of the wall readily identifiable on either the IKONOS or AIRSAR imagery. A somewhat linear anomaly does stretch from shortly after the wall to the system of anomalies identified by Sever in the AIRSAR data. This is highly speculative if it is cultural in natural or a drainage as the resolution of the AIRSAR data would make identification of the earthwork difficult.

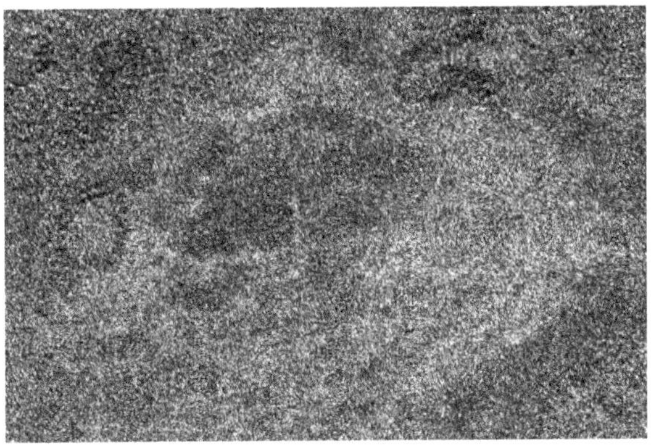

Figure 3 *Linear anomalies and vegetation differences seen in AIRSAR image.*

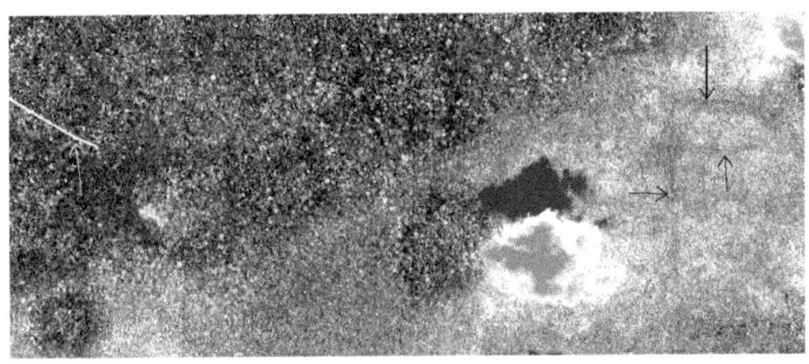

Figure 4 *IKONOS image showing location of Tikal wall (vector overlay, right) and linear anomalies (left).*

Interpretation and Significance

By identifying where archaeological features – including architecture, agricultural systems, causeways, and water -- are located, settlement patterning can be examined. The dominant settlement model in Maya archaeology is the Concentric Zonation Model (CZM) that posits the highest density of monumental architecture is in the center of the site with structure size and density decreasing as distance from the monumental center increases. A Maya center is organized in concentric rings with a ceremonial core surrounded by a zone of elite residences and in turn by dispersed non-elite houses. This model has failed to consider *bajos* as part of the landscape when constructing this view of ancient Maya settlement. When scholars are conducting on-the-ground research to test CZM, they likewise do not incorporate *bajos* into their research.

The CZM has been tested as Tikal, without definitive results. Scholars have argued that it holds true while the researchers I discussed in my introduction tested the model by cutting survey transects but neglected to include *bajo* environments. The presence of archaeological remains on *bajo* islands has suggested to other scholars the presence of *bajo* communities. Scarborough and Valdez (2003) draw attention to settlement in the Three Rivers Region in Northwestern Belize. They propose "resource-specialized communities" where settlement focused on an environmental resource. Their three examples of communities are *bajo* communities, aguada communities to exploit water, and terrace communities to exploit agriculture. They envision a large generalized community connected to these more specialized resource communities. They are beginning to address *bajos* in settlement.

However, research and contentious debate to date has focused on whether *bajos* were suitable for agriculture, habitation, or water storage. No one has yet considered the wider significance of the *bajo* to the ancient Maya; for example, in terms of experiential movement across the landscape or sacred places within the *bajo*. *Bajos* are huge bodies of water and water is sacred to the ancient Maya because both it is a portal to the underworld and it represents the primordial ocean of creation. The Bajo de Sante Fe and the Bajo la Justa both fill with water in the rainy season and desiccate in the dry season.

Tikal Settlement

While these models provide innovative new ways to consider the settlement of the ancient Maya, they provide essentially top-down views that should be complemented with another approach. This approach can be seen in the work of Dr. Marshall Becker at Tikal who critiques archaeologists' concern with structure size as a descriptive and limiting approach that should be replaced with a focus on the form of the group. He identifies 10 plaza plans (PP), group plans, that occur in consistent percentages across the site and range in size from very small to very large within the same plaza plan. Of particular interest to this study is that it is an emic approach and, to date, Becker has identified an anomalous cluster of a residential cluster of plaza plan number 2 occurring on a peninsula in the Bajo de Santa Fe. I have digitized and georeferenced all 9 square km of the Tikal settlement maps and overlain them onto the IKONOS and AIRSAR imagery to determine if individual plaza groups can be identified. I estimate I have an accuracy of 5 meters. While I am still refining my techniques for identifying individual structures, I believe with the vegetation signature identified by Sever and Saturno, it is possible to identify clusters of structures.

We should also consider that the ancient Maya may not have one form of settlement that applies to the entire Maya world because there is great environmental and temporal diversity.

Causeways

Numerous linear anomalies are present in the *bajos*. The presence of monumental causeways raises the following question: why would the Maya have built such impressive construction in the *bajo*? Smaller pathways, 1-2 meters in width, would have been much more practical for avoiding the floodwaters and muck of the *bajo*. Why chose monumentality that requires vastly more time and resources when there was a viable alternative? Furthermore, the probable causeways in the *bajos* connect confirmed and possible archaeological sites throughout the *bajo*. Considering monumentality in combination with the sites the causeways connects raises the possibility that they are not only secular in function but also ceremonial; for moving multitudes of people to pilgrimage sites or sacred places.

Conclusions

Due to the dense vegetation of the ancient Maya lowlands, only the tops of large temples are easily visible above the lush rainforest. Satellite imagery analysis is enabling archaeologists to obtain a more cohesive view of the ancient landscape, filling in an overlooked segment: *bajos*. The work conducted here has contributed to correcting this oversight by identifying numerous areas of the landscape that were modified or likely modified and will be confirmed with future research. This research has opened up numerous areas for further excavation.

Acknowledgements

I would like to thank Dr. Tom Sever, Burgess Howell, and Dan Irwin of NASA; Dr. Payson Sheets; Dr. T. Patrick Culbert; Space Imaging; National Science Foundation; and the University of Colorado.

References

BECKER, MARSHALL J. 2004, Maya Heterarchy as Inferred from Classic-Period Plaza Plans. *Ancient Mesoamerica* 15:127-138.

FIALKO, V. 2005, Diez Anos de Investigaciones Arqueologicas en la Cuenca del Rio Holmul, Region Noreste de Peten. In *XVIII Simposio de Investigaciones Arqueologicas en Guatemala*. Ed. Juan Pedro LaPorte. Instituto de Antropologia y Historia.

KUNEN, JULIE L., T. PATRICK CULBERT, VILMA FIALKO, BRIAN R. MCKEE, LIWY GRAZIOSO 2000, Bajo Communities: a Case Study From the Central Peten. *Culture and Agriculture* 22 (3): 15-31.

POPE, KEVIN O. AND BRUCE H. DAHLIN 1993, Radar Detection and Ecology of Ancient Maya Canal Systems – Reply to Adams et al. *Journal of Field Archaeology* 20(3): 379 – 383.

SEVER, T. AND D. IRWIN 2003, Landscape Archaeology: Remote-sensing Investigation of the Ancient Maya in the Peten Rainforest of Northern Guatemala. *Ancient Mesoamerica* 14: 113-122.

TURNER, B. L. AND PETER HARRISON (editors) 1983, *Pulltrouser Swamp: Ancient Maya Habitat, Agriculture, and Settlement in Northern Belize.* Austin: University of Texas Press.

Culture 2000 Project
'European Landscapes: past, present and future'

Robert Bewley,[1] Chris Musson[2]

[1] English Heritage, 29 Queen Square, Bristol, United Kingdom, BS1 4ND.
[2] Tanyffordd, Pisgah, Aberystwyth, United Kingdom, SY23 4NE.

Through satellite imagery, airborne survey, fieldwork, geophysics and excavation, the aim of the project is to promote the exploration, public appreciation and conservation of heritage sites and landscapes across Europe.

This short presentation summarises the aims, content and progress of the Culture 2000 project *European Landscapes: past, present and future* (Project CH-A2-UK-2077). At the time of writing, in September 2006, the project is entering its third and final year before completion in August 2007. The project is sponsored by English Heritage in the UK, along with six co-organisers in Belgium, Germany (2), Hungary and Italy (2). There are ten co-partners in seven other countries, the Czech Republic, Estonia, Finland, Lithuania, Poland, Romania, Slovakia, plus an additional co-partner in Italy. A full list of the participants is given at the end of this report. Over its three-year life the project will cost just under 900,000 euro, 47% from the European Union and the rest from the participants themselves. Several members of the present partnership also worked on a Culture 2000 project in 2000/2001, entitled *Conservation through Aerial Archaeology*. This built on earlier training schools and exchanges organised by the Aerial Archaeology Research Group (AARG), whose international contacts were crucial in the conception and framing of both of the Culture 2000 projects.

Aims

The aim of the present project is to increase awareness and conservation of the cultural heritage within European archaeological landscapes, especially through non-destructive techniques of investigation such as aerial reconnaissance, air-photo mapping, field survey and the innovative use of satellite and airborne remote sensing. There are nine inter-linked fields of activity, or Actions.

Action 1 The promotion of training schools, workshops and seminars
Action 2 Aerial and ground-based surveys of threatened landscapes
Action 3 Air-photo interpretation and mapping of cultural landscapes
Action 4 Innovative surveys using air-photos, lidar and satellite imagery
Action 5 The search for under-exploited air photo sources from World War II
Action 6 Networking and the exchange of skills across Europe
Action 7 The creation of a European Centre for Aerial Survey and Conservation
Action 8 Public outreach through websites, TV/radio, films and exhibitions
Action 9 Centrally-funded activities, student exchanges and support for meetings etc

Progress to September 2006

The project does not have a single unified programme. Instead, the participants respond to their local situations, contributing to one or more of the Actions according to their individual capacities and needs. The examples given later in this presentation have been chosen to illustrate the variety of programmes and objectives in a few of the participating countries. Good progress has been made in almost all of the intended Actions. The highlights have been several productive training schools, conferences and seminars, focusing activity on individual countries or subjects and helping to spread expertise across Europe. Despite poor weather in some parts of the Continent there have also been successful programmes of aerial survey in both of the last summers, along with exploratory work on the use of laser-scanning technology (lidar) in landscape and archaeological recording. The project's central website (www.e-landscapes.com) has been established and work is in hand on teaching packs and exhibitions, both real and 'virtual'. World War II photo-sources have so far proved elusive, except in Poland, but work will continue throughout the rest of the project. Discussions are well advanced for the eventual creation of the planned European Centre for Aerial Survey and Conservation.

Action 1 Promotion of training schools, workshops and seminars

Successful training schools, with both ground-based and in-air experience for the students, have been held at Grosseto in Italy, Barth in northern Germany (twice) and Cirencester in the United Kingdom. There will be similar events during 2007 at Foggia in southern Italy and again at Barth in Germany. Ground-based workshops and/or seminars have been held in Finland and Romania and others are in hand in Belgium, Estonia, Germany and Italy. Assistance of one kind or another has been given for meetings in Germany, Italy, Belgium and the Czech Republic.

Action 2 Aerial and ground-based surveys of threatened landscapes
Substantial programmes of aerial reconnaissance, mainly concentrating on plough-threatened landscapes, have been conducted in Italy (Foggia and Siena). Exploratory flights directed at specific aspects of the landscape have been undertaken in Belgium, the Czech Republic, Estonia, Germany, Hungary, Lithuania, Poland and Slovakia. In all of these countries important archaeological discoveries have been made and photographs have also been taken of heritage landscapes, townscapes and buildings for the benefit of public presentation and future conservation.

Action 3 Air-photo interpretation and mapping of cultural landscapes
Photo-interpretation and mapping work has been undertaken by many of the project's members, in particular for previously unknown sites or groups of sites discovered from the air during exploratory flights in 2005 and 2006. More intensive 'landscape' mapping will be undertaken in southern Italy and elsewhere in 2006/2007.

Action 4 Innovative surveys using air-photo, laser-scanning and satellite imagery
The use and testing of lidar and other remote-sensing methods has been initiated or continued in Germany and Italy. Co-partners here and in the Czech Republic, Hungary and Slovakia have been experimenting with the integration of aerial techniques alongside more established ground-based methods such as geophysical prospection, field survey and sample excavation.

Action 5 Under-exploited air photo sources from World War II
The search for further WWII photo-maps in Poland has proved frustrating but some important war-time and pre-war air photographs have been identified and the search continues. In Belgium the project is revealing the importance of historical and present-day air photographs in the understanding and conservation of Great War landscapes and military remains.

Action 6 Networking and the exchange of skills across Europe
The European network of specialists involved with air photography and other forms of remote sensing continues to grow, through meetings and exchanges of various kinds and through a wide range of email contacts. New areas of contact have been established in Iceland, Scandinavia and around the Baltic Sea.

Action 7 European Centre for Aerial Survey and Conservation
Discussions are in train with the University of Siena to establish the proposed centre within a wider Centre of Excellence in Landscape Archaeology and Remote Sensing, at Grosseto in Tuscany. The search for funds and support of other kinds will now begin.

Action 8 Public outreach: websites, TV/radio, films and exhibitions
The project's central website went on-line in November 2005 and has been progressively enhanced and updated since. Several partners have added Culture 2000 pages to their own websites. Discussions are in hand with film/TV companies. The collection of material for real and/or 'virtual' exhibitions has been started by most of the project's participants. There will be major exhibitions at the end of the project in Prague and Siena.

Action 9 Centrally-funded activities, student exchanges, support for meetings etc
Funds for pan-European activities have been used mainly to assist in arrangements for or attendance at training schools, workshops, seminars and meetings. Student exchanges have been funded and more are planned in the final year of the project.

Spin-off activities
Through the prestige of the Culture 2000 project co-partners have been able to initiate or assist externally-funded activities relevant to the Culture 2000 objectives. These have included, for example, an aerial archaeology workshop in Rome in November 2005 (funded by the Aerial Archaeology Research Group, ACE and the British Academy) and a conference on GPS and laser scanning in archaeology at Grosseto, Italy, in June 2005 (sponsored by Leica Geosystems and others). Exciting initiatives after the end of the project will be aerial archaeology training schools in Denmark and Holland in 2007/2008 – neither country having had any significant provision for aerial archaeology in the past two decades. Through the project's Hungarian co-organiser the first moves have been made for future exploratory air survey in Syria. The participation of an American student in the UK training school in 2006 may eventually lead to a workshop or training school in the USA, where archaeological air survey from light aircraft is still virtually unknown.

Highlights to September 2006

The following section illustrates some examples from various parts of Europe, to show the range of work and outcomes across the project as a whole.

Belgium: WWI landscapes in Flanders
The Belgian element of the project, organised by the Department of Archaeology and Ancient History of Europe at the University of Ghent, has been focused on the study and use of air-photo and related ground-based evidence for WWI remains in Flanders. The aims have been to locate and examine previously un-exploited military air photographs, to map the wartime features and examine them on the ground, to undertake new flights to assemble comparative photography from the present day, and to organise a workshop on the results towards the end of the project. From archives containing over 25,000 WW1 air photographs 950 were selected for scanning at 400dpi and were then incorporated into a database,

located on the map and geo-rectified. Ninety years after the events it proved possible to gain additional information on the topography of military features and the surrounding terrain and to discover important features that would otherwise have remained hidden. Nine digital stereoscopic views were created from overlapping archive photographs, using specialist software. The geo-referenced pictures were used to create a detailed inventory of all visible military and other archaeological traces in the study area. Approximately 3000 separate features were identified and documented in an area of 49 km^2, most of them belonging to the trench systems of the German second line, which were traced over a distance of more than 10 km. A significant number of medieval moated sites were also recorded, along with various other features revealed by temporary flooding. Subsequent work has aimed at gaining a better understanding of the military remains and formulating a strategy for selective conservation for features which survive above the ground or which might retain valuable below-ground evidence. The planned workshop, on *Military Aerial Photography and Archaeology*, organised in collaboration with the In Flanders Fields Museum, will take place at Ypres in October 2006.

Italy: survey, training schools, workshops, lidar and conservation

The project has supplemented a growing amount of exploratory survey and related work carried out from year 2000 onwards by the University of Siena in Tuscany and the University of Foggia in southern Italy. The project has also involved co-operation with the University of Lecce, in the far south, where aerial studies based on pre-existing vertical photographs have reached high levels of sophistication over the past two decades.

The Department of Medieval Archaeology at the University of Siena, along with its newly-formed Laboratorio di Archeologia dei Paesaggi e Telerilevamento (LAP&T, Laboratory for Landscape Archaeology through Remote Sensing) at Grosseto, has continued to work on active aerial reconnaissance, interpretation and mapping of the resulting information. The University also organised a highly successful aerial archaeology training school at Grosseto in 2006, where for the first time training was provided for pilots as well as for prospective air photographers and air photo interpreters. A special feature of the University's activity, however, has been the way in which it has combined limited Culture 2000 funding with research and sponsorship funds from outside the project, thereby creating a positive impact beyond the project's – and the University's – own restricted resources.

Further south in Italy the Department of Human Sciences at the University of Foggia has continued its programme of oblique air photography in co-operation with Dr Otto Braasch from Germany and the Aero Club di Foggia. Every year of flying over the rich air-photo landscapes of the Tavoliere delle Puglie brings new insights and sometimes startling discoveries. A particular feature of the work in 2005 and 2006 was the initiation of a comparative study of oblique air photography in relation to lidar and mult-spectral imagery. This is focused on the ancient city of *Arpi*, close to the present-day town of Foggia. It became possible through co-operation with a project of airborne survey already in hand by the UK Natural Environment Research Council, the University of Durham and CNR Potenza. Another promising innovation in 2006 was the targeting of some of the annual survey programme on parts of the Tavoliere and the surrounding hills which are under threat from large-scale wind-farm developments. Provision of the resulting information and photographs to planning authorities and the wind-farm companies will hopefully contribute to better conservation of archaeological and landscape features in the future.

Lithuania: air photography and teaching packs

Over the past 50 years the territory of Lithuania has been photographed five times through vertical photography and the resulting photographs are reasonably easily accessible. But no oblique photography for archaeological purposes was undertaken during the Soviet period. In 1996, however, the Department of Lithuanian Heritage Protection began a limited programme of aerial photography for conservation purposes. The Department now holds over 2000 oblique air-photographs, covering over 300 heritage features of various kinds. The Department's contribution to the Culture 2000 project, now carried out in co-operation with the University of Klaipeda in western Lithuania, aims to supplement the annual flying programme, to develop a digital database of the photographs, to engage in the European network of Culture 2000 activities and to organise an international workshop in Lithuania (held at Klaipeda in November 2006, with a resulting publication in preparation).

Another element, of particular relevance to the 'outreach' aspirations of the Culture 2000 Programme, has been the preparation of education pack for schools, using aerial photographs as well as other heritage sources. Some 20 towns and townships with origins going back to prehistoric times have been chosen for the preparation of these packs. Related archaeological and historical information has been selected and detailed preparation of the packs is in progress, with several already completed. The dating evidence lies in prehistoric features both within the towns/townships and in their immediate surroundings – Iron Age hillforts, burial mounds and settlement sites, for example. A castle, a church or an estate attached to a market place are the main components from which the towns and townships developed in the Middle Ages and post-medieval period. The packs each contain historical and recent aerial photographs, supported by other archaeological data, historical information, cartographic material, ancient paintings and ground-based photographs. The information is being specially adapted for schoolchildren so as to show features of the local cultural heritage and to emphasise their importance for society, both historically and in the present day.

Poland: old and new resources for Poland's landscape heritage

Poland and much of the rest of Europe possesses a largely un-recognised landscape resource in a coverage of 1:25.000-scale photo-maps made for German military and intelligence purposes in the late 1930s and early 1940s. These show the landscape as it was before the radical changes of the post-war years and their full historical potential has yet to be assessed. One of the project's objectives in Poland has been to identify the location, number and quality of the photo-maps and to test ways of making them more widely available, for instance as scanned images. During the search for the photo-maps two important air-photo archives from before and during WWII have been identified, though one of them is as yet un-catalogued. Work is in hand on a database for the photo-maps and mosaics, held both in Polish institutions and at the University of Keele in the UK, the resulting information being made available in instalments through the project's central website.

Until now little money has been allocated in Poland to exploratory air photography, which has had to compete for funds with Poland's long-standing programme of ground survey, artefact collection and recording, known as the Polish Archaeological Record (or AZP). This now covers the greater part of Poland through over 500,000 entries. The AZP provides an unrivalled opportunity for Polish aerial survey to be even more effective than in other countries, adding 'shape' to sites known only from surface scatters of artefacts and revealing archaeological and landscape features where no evidence at all is visible at ground level. One of the aims of the Culture 2000 work has been to carry out air photography in five heavily-ploughed areas, along with mapping of the results and the creation of a database to link the aerial evidence to that obtained from field-walking survey. Although limited in scope the flying programmes have produced valuable records of known and previously unrecognised sites, including spectacular traces of Early Neolithic trapezoidal buildings and (for the first time in Poland) pit-alignments of the kind known from other parts of Europe. For specialists and the general public alike these photographs provide striking evidence of the nature and value of the half-hidden traces of the cultural heritage which Poland shares with Europe as a whole.

Another major development has been the design and testing of a database application (*APh_Max*) for storage and analysis of the air-photo archives in various institutions. This is being used initially for mass data entry of the 3000 photographs of the Wielkopolska region held at the Adam Mickiewicz University. This is the third module of a heritage management system under development now for several years at the Poznań Archaeological Museum. It will enhance the management of air photo collections through the creation of unique files for each photograph. These can then be compared with other data-sets such as text descriptions of individual sites, GIS data etc. The database has been designed for linking with those of the AZP so as to assist the protection and management of monuments, whether known from ground-survey or from aerial evidence.

Romania: a future for the past

Aerial survey is as yet virtually unused by Romanian archaeologists for exploration or landscape study, though aerial research by the University of Glasgow has been in hand for several years in the north-west of the country. Through the Culture 2000 project the Institute for Cultural Memory (CIMEC), in Bucharest, is now promoting use of the technique elsewhere in Romania, initially through the examination of historical air photo sources but hopefully in the final year of the project also through the initiation of a small flying programme.

The geographical focus of CIMEC's work is in part of the Lower Danube Valley, archaeologically one of richest areas in Romania from early prehistory to modern times. Many ancient sites there have been destroyed or are now at risk of damage or destruction by modern development and intensive arable cultivation. Many more await discovery through the aerial techniques which have proved so fruitful elsewhere in Europe. By drawing these sites to the attention of officials and cultural resource managers, the project will contribute to their long-term conservation and future enjoyment by schoolchildren, young adults and the public at large. Work on a selected area in the Mostiştea Valley and its environs in the Lower Danube area is now under way, concentrating on a pilot study of the important Malu Roşu site and its surroundings through the analysis of landscape change recorded on air photographs taken between 1972 and 2003. The fortified settlement stands on a high terrace above the Mostiştea Lake, and is now under re-excavation following earlier (un-published) site-work in the 1920s and in 1975.

In addition, training and information exchange have formed key parts of the project so far. A week-long training course in aerial archaeology and the uses of GIS in heritage work was organised at Buşteni in May 2005 through Culture 2000 and the EPOCH Network of Excellence. There were 34 participants from six countries, including six from CIMEC itself. In September 2005, during a week-long course in Bucharest, Rog Palmer from the UK provided five Romanian archaeologists with intensive training in the interpretation of aerial photographs and a day was also spent in targeted field-walking in the Mostiştea Valley. Culture 2000 funding has subsequently enabled two Romanian archaeologists to attend the 2006 annual conference of the Aerial Archaeology Research Group at Bath in the UK, one of them then staying on for a further week of training by Rog Palmer in photo-interpretation and mapping.

Prospects for the future

There have inevitably been some disappointments but by and large the participants are reasonably content with progress so far, both in specific projects and in raising awareness of the role of aerial archaeology and related techniques. The project's ultimate value, however, will lie in its *long-term* effects in the participating countries and institutions once the Culture 2000 funding ends. For the moment, the final year of the project will concentrate on a number of important objectives:

- Intensive 'outreach' work – website, TV/film, possibly a book-of-the-project and final meetings, open to the general public, at various venues.
- Design and mounting of actual and virtual exhibitions in various countries.
- Sustaining initiatives in the Baltic and Scandinavian states and helping to prepare for future projects in Denmark and Holland. Also making the first moves in Spain and Portugal, where active aerial survey is still in its infancy.
- Communicating with heritage and planning agencies throughout Europe to secure better archaeological and landscape conservation in the future.
- Developing ideas for further initiatives and their funding after the end of the current project in August 2007, and especially for the proposed European Centre.
- Further enhancing the European and world-wide network of contacts for sharing advice and expertise about aerial archaeology and landscape studies.

Acknowledgements

This summary draws heavily on accounts prepared by project co-partners for the project's Interim Report to the European Union in February 2006. Particular thanks are due the following for specific sections of the text. For Belgium Jean Bourgeois and Birger Stichelbaut of the University of Ghent. For central Italy Stefano Campana of the University of Siena. For southern Italy Giuliano Volpe, Roberto Goffredo and Valentino Romano of the University of Foggia. For Lithuania Romas Jarockis of the Department of Lithuanian Heritage Protection and the University of Klaipeda. For Poland Włodek Rączkowski of the Institute of Prehistory at Adam Mickiewicz University, Poznań, Andrzej Prinke of the Poznań Archaeological Museum and Dariusz Krasnodębski of the Institute of Archaeology and Ethnology, Polish Academy of Sciences, Warsaw. And for Romania Irina Oberländer-Târnoveanu of the Institute for Cultural Memory in Bucharest.

Participants in the project

The seven co-organisers are: English Heritage, UK, as Project Sponsor; the University of Ghent, Belgium; the Landesdenkmalamt Baden-Württemberg, Germany; the Archaeological State Museum of Mecklenburg-Vorpommern, Germany; the Museums of Baranya County, Pécs, Hungary; the University of Foggia, Italy; and the University of Siena, Italy. The ten co-partners are: the University of West Bohemia, Czech Republic; the National Heritage Board of Estonia; the University of Helsinki, Finland; the Agenzia per il Patrimonio Culturale Euromediterraneo, in partnership with the University of Lecce, Italy; the Department of Lithuanian Heritage Protection; the Adam Mickiewicz University, Poznań, the Poznań Archaeological Museum and the Polish Academy of Sciences, Poland; the Institute for Cultural Memory (CIMEC), Romania; and the Slovak Academy of Sciences, Slovakia.

Education and Culture

Culture 2000

ENGLISH HERITAGE

A historic flight and the study of archeological landscape in Maremma (Central italy)

A. Caprasecca

Università of Siena at Grosseto, Medieval Archaeology Section,
Laboratory of Landscape Archaeology and Remote Sensing

1 The project

This work presents the principal objective of our research[1] project: evaluation of potential archaeological information from a photographic document. To reach this goal it is necessary to employ a strategy of investigation that uses all tools essential for a good methodological formulation. A systematic and essential photo interpretation has been effected through the study of the landscape and through direct comparison between historical cartography, and aerial photos. Then we followed the natural prosecution in the integration of the technologies G.I.S and Remote Sensing. All the material available has been taken in consideration: from the vertical historical flights realized during the two world conflicts 1930 - 40, to digital photographs realized in the last years, keeping in mind all the information, above the historical and archaeological character essential for a correct interpretation of signs left on our territory. We have selected study areas located between two regions: Lazio and Tuscany, which show a very interesting morphological problem. We analyzed cities of Sorano, Pitigliano in Tuscany; Valentano, Latera and Piansano in Lazio. In other words, this is the area in which all the St. Savior's abbey possessions were distributed in the middle ages.[2] The analysis of historical aerial photos has been in partnership with the reading of cadastral maps and comparison with documents St. Savior's abbey file. Our area is located in one of the larger stretches of land which was in possession of St. Savior abbey, displaced among Tuscany and Lazio. We have also had the opportunity of comparison with other historical sources such as "Cronichon Farfense or Leonis IV Privilegium".[3]

After this phase the ground approach followed, based on archaeological surveys, with the aim to verify the information found in documents. This has allowed a new, correct topographical positioning for all St. Savior's abbey settlements with a very small error.

The photo interpretation was following two lines of investigations. First one was a purely traditional formulation, with an aerial photo approach organized on classical system: reading and interpretation of traces and visible anomalies through stereoscopy and comparison on other more recent flights. The second was the "air photos survey," developed in systematic way on all the territory.

This phase has seen the employment of G.I.S. with an initial role of simple support to acquisition and cataloguing of data by pc, through the geo referencing of all anomalies and graphic restitution of traces. The survey, as we already said, concluded the analytical phase. The second phase, of synthesis, concerns the employment of G.I.S. for aerial photos elaboration, images treatment and three-dimensional models construction through digital photogrammetry. The opportunity to get a comparison among three-dimensional models realized through photogrammetry restitution of vertical photos, performed to distance of years, will allow a best understanding of erosion and transformations phenomenon of territory up to the disappearance of some archaeological sites.

Figure 1 *The study area*

Before conclusion of search we hope to increase the research method by using measurements of magnetic gradient on more interesting sites individualized during

[1] Doctorate in History and Archaeology of the Middle Ages Institutions and Files. Area of medieval archaeology. Tutor prof. Riccardo Francovich. Title thesis: The historical aerial photo in the study of medieval landscape. Digital elaboration of images, G.I.S., photogrammetry, three-dimensional modelling: the support of the computer technology to the traditional photo interpretation.

[2] Kurze, Citter 1994, pp. 159 – 186

[3] Some places mentioned in medieval documents are also brought in parchments of Farfa abbey; Cfr. Callise 1894; Raspi Serra, Laganara Fabiano 1987, Del Lungo 2001

survey. The analysis of distribution of archaeological material, through vectorial grate, as well as test-excavation, will help the understanding of specific archaeological hypotheses to the interpretation of individualized traces on vertical, oblique and satellite photos.

2 The aerial flight "Fondo Nistri" and its potentiality in archaeological field

The surveys to edit cadastral maps using aerial photogrammetry method, was used for the first time in 1934 by S.A.R.A the Society Anonymous Shops Galileo and made official in 1941. In that time the provinces of Terni and Viterbo and portions of those of Sondrio and Rome were noticed both in plant and height. The Society Autonomous Surveys Aerial Photogrammery founded by brothers Umberto and Amedeo Nistri in the years 30 - 40, were considered the greatest industrial organization deprived for the execution of the aerial photogrammeric reliefs.[4] Then the flight was totally lost. To say the truth it had not disappeared but it was only forgotten in a secondary archive of the Government Revenue Office, after the period of Fascism. The flight, recently given by Territory Agency[5] to the Government Historic File of Viterbo, is currently in progress of restauration. On back of photos it is the perfectly legible concession: Wing Littoria S.A. Air Photographic Service, Roma;
concession R.A. n° AF00201/1940, for limited use. (R. Decreto Legge 1732 del 22/7/1939/XVII - Art, 9). On October 28th 1934, the Mediterranean Aerial Society became "Ala Littoria S.A." The new company, reassumed by itself the totality of the Italian aerial services. In some cases the photographic format and the better quality of paper, allows us to consider further reprints in more recent periods. The material is composed of photographic plates 12 x 18 and two formats on photographic paper: the first has two different scale factors between 1:10000 and 1:13000; the second format 23 x 29, is in scale factor 1: 4500[6] and has been used for cartographic restitution. The aerial coverage is partial and relative to a limited area of Viterbo province, in north of Lazio. Frames are organized in irregular strips with a double numeration. Unfortunately a graph of flight is not available, except for a limited zone in the area of search. The graph, brought a scale factor 1:7000, and the partial coverage of the territory is clearly visible. It is also possible to see a transversal orientation in comparison to the Cartesian aces: nord/est - sud/ovest. Photos are organized in stereoscopic couples that introduce in some cases a longitudinal cover of over 60% and not always a regular sequence of release. The photographic paper (in particular cartographic restitution format 23 x 29) is rich of annotations related to owners, use of the ground, name of places and possible ruins or underground hollow, such as manors, fountains, abandoned churches, etc.... The descriptive aspect of these documents represents an enormous resource, not only for traditional photo interpretation, but also for historical and archaeological documentation. These photos through interpretation of the "signs", represent the only testimony of landscape that still preserves traces of territorial organizations and settlements of different periods. In some cases the landscape shows complete change to the present day. Enormous transformations have happened in the exploitation of soils. In the last years of records, a bigger and bigger presence of pastures and abandoned areas covered by Mediterranean wood that damaged of cultivated wheat fields can be observed. In the area of Tuscany the vineyards and the olive woods show more diffused plantation with the consequent growth of deep destruction of soil made by ploughing, principal cause of destruction of superficial archaeological layers. The transformation of this territory has its beginning with the works in Maremma in 1950, with the reorganization and subdivision of the great land ownerships, to improve the extension of zones for plants or crops. There was numerous deep ploughing made on the volcanic surface and, the scarce consistence of ground, composed primarily from volcanic deposits of pumices, has brought year by year a real phenomenon of landslide and deep erosion. The landscape change is possible to see by comparison between an aerial photo realized in 1934, and an actual. The recent territory urbanization has involved a modified environment with the consequent disappearance of precedents systems land organizations.

Figure 2 *Different photo formats of Fondo Nistri 1934 Flight*

[4] The first experiments of aerial photogrammetry were made during '20 by the Military Geographical Institute. On '30 the S.A.R.A realized the first cartographic commercial jobs on S. Paolo of Brazil territory and Palermo city. Cfr. Piccarreta, Ceraudo 2000, pp. 49 - 52

[5] We thank the Territory Agency Central Direction for cartography, inventory and publicity, for consultation concession of material: prot. Num. 93536 - Rome 6/12/2004

[6] In lack of the calibration certificate, the calculation of the scale factor has been effected through the formula: Sf = Dc*Sc/Df. We have to consider therefore a greater degree of approximation to determinate photos scale factor. Cfr. Piccarreta, Ceraudo 2002, pp. 46 - 47

3 A practical example - The case of Paterno - S. Lucia in Paterno settlements

The area is situated on the slant of an hill, a relief of border volcanic caldera of Latera. The site could be connected with the documents related to Paterno. In the first document (823) *Paterno* place is called "hamlet". In the Ludovico II (853) precept we have: *"cella of S. Severo in Paterno, cum ipsa curticella de Bisentio"*. The place continues to be present in other imperial precepts (937 - 962 - 964) up to 1210 with the last Ottone IV confirmation. In the Bolla of Urbano V (1369) for the diocese of Montefiascone, S. Lucia is called *"in Paterno"*, reduced to small rural church.[7] The actual hamlet is dated to 1837 and built on a Roman cistern ruins. The structure is articulated in two environments realized in Roman masonry. The whole surrounding zone is sprinkled by ceramic fragments. The dating of the settlements confirms that already brought in the preceding publications, that is the fifth century AD. After this Lombard occupation happens, testified by the recovery of two saxes now preserved at Valentano museum.[8] The most careful examination of the 1934 flight and the fortunate condition of visibility during the survey on the field, have allowed elements unknown in precedence. The flight 1934 underlines a series of irregular traces, positive and negative evidences of probable underground walls, to the east of the hamlet. An excessive erosion during winter rains, brought to the light some structures of the rustic villa, inside a canaletta of drainage.

A floor is well evident in *opus spicatum*, which supported a structure made of square stones.

Figure 3 *S. Lucia in Fondo Nistri 1934 Flight*

[7] Raspi - Serra pp.120; Del Lungo, 2001
[8] Luzi, Mancini 1985

Figure 4 *Result of survey on traces revealed during photo interpretation*

4 A practical example – "Vicus Mariano" a medieval village on an important roman road

On the cadastral map of 1934, the presence of a tower is noticed in correspondence of a place called Marano,[9] also visible in oblique photos of 2002. The archaeological evidences individualized not very far from the late medieval tower, have ceramic that suggest an occupation from the third cent. B.C. to the sixth – seventh centuries A.D. The purified ceramics painted on red and brown strips, discovered a few hundred meters north - west in a raised little hill, would move the dating of site to the ninth to the first half of the twelfth centuries A.D. The interest for this zone is linked to the probable identification with the "vicus Mariano" on whose position has been made numerous hypotheses.[10] Mariano was a village on an important Roman road: the Clodia. During the survey some plates of road surface have been recovered.

[9] Codex Diplomaticus Amiatinus. 12, 28, 34, 39, 55, 91, 94; Prv. Leonis IV 1240 C
[10] The vicus Mariano (VIII – IX on documents) has allowed a lot of hypotheses of location not completely acceptable at moment. Scneider identifies it in the zone between Piansano and Arlena of Castro. Del Lungo seems to identificate it with Poggio Marano in Cellere territory, that our survey has noticed to be an Etruscan installation. At moment it is not even possible to confirm the identification of the vicus with our site, although the ceramics would seem to point out a chronological continuity of place, from the VII to the XII century.

Figure 5 *Medieval tower in 1934 flight*

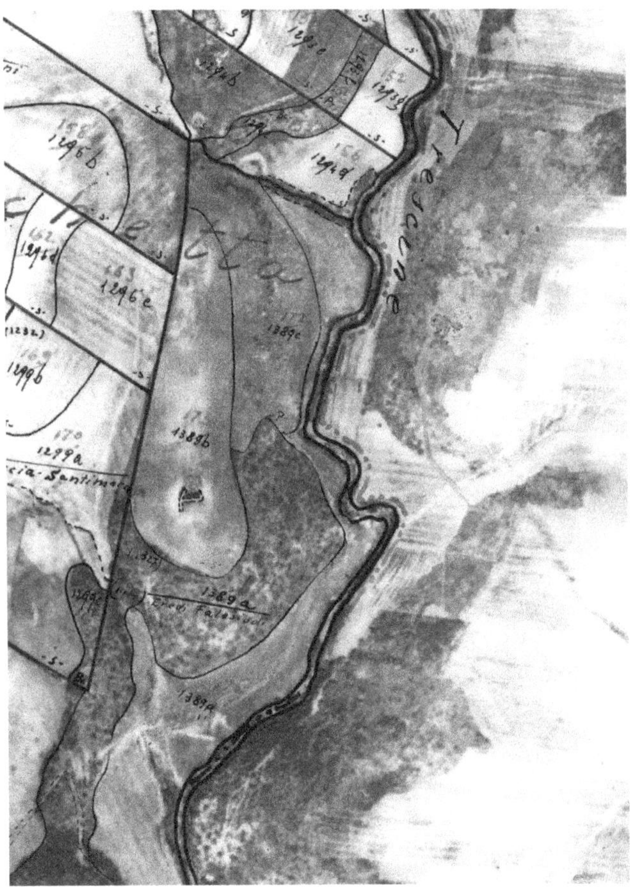

Figure 6 *Medieval tower in oblique photo 2002*

References

CALLISE C. 1894 *Documenti del Monastero di S. Salvatore sul Monte Amiata riguardanti il territorio romano* (secoli VIII – XII). A cura della R. Società Romana, Roma

COSCI M. 1988 *Foto interpretazione archeologica: guida pratica per gli studenti, note introduttive.* Quaderni del Dipartimento di Archeologia e Storia dell'Arte

DEL LUNGO S. 2001 *Presenze abbaziali nell'alto Lazio. San Salvatore al Monte Amiata e le sue relazioni con l'Abbazia di Farfa* (Secoli VIII - XII).

GAZZETTI G. 1985 in Carandini A., (a cura di), *La romanizzazione dell'Etruria. Il territorio di Vulci*, Milano.

GUAITOLI M. 2003 (a cura di), *Lo sguardo di Icaro. Le collezioni dell'Aerofototeca Nazionale per la conoscenza del territorio*, Roma.

LUZI R., MANCINI B. 1985 *Luoghi e tempi del sacro.* Gruppo Archeologico Verentum, Valentano

KURZE W., CITTER C. 1994 *L'occupazione della Maremma toscana da parte dei Longobardi*, in Brogiolo G.P. (a cura di) "*Città, castelli e campagne nei territori di frontiera.*"Atti del 5° seminario di Monte Barro, Monte Barro 1994, Documenti di Archeologia, 6: 159 – 186

MUSSON C., CAMPANA S., PALMER R. 2005 (a cura di) *In volo nel Passato. Aerofotografia e cartografia archeologica:.* 172 - 229

PICCARRETA F., CERAUDO G., 2000 *Manuale di aerofotografia archeologica. Metodologie tecniche e applicazioni.* Bari: 46 – 47; 51 – 52

RASPI SERRA J., LAGANARA FABIANO C. 1987 (a cura di), *Economia e territorio. Il patrimonium Beati Petri nella Tuscia*, Napoli

SILVESTRELLI G. 1970, *Città castelli e terre della regione romana*

The emerging landscape of East Lothian, Scotland Mapping, analysis and presentation

Dave Cowley

RCAHMS, dave.cowley@rcahms.gov.uk

The coastal plain in East Lothian, in south-east Scotland, is a predominately arable area where the majority of known archaeological monuments are plough-levelled sites recorded as cropmarks on aerial photographs. This dataset has emerged from 50 years of aerial reconnaissance, but especially since 1976 with the aerial survey programme of the Royal Commission on the Ancient and Historical Monuments of Scotland (RCAHMS). Much of this material has been added to the archive of RCAHMS with little analysis or mapping and it is therefore relatively poorly understood. This material is now being mapped in detail across the entirety of the coastal plain, providing an interpreted digital map base, which underpins more detailed analysis and presentation, as well as providing accurate locations and extents of sites for planning and conservation purposes. This paper presents some preliminary results from this project.

Mapping a Lowland landscape

The lowland landscape of East Lothian has dense distributions of cropmark archaeological sites and these have emerged from decades of patient survey. The mapping of this landscape is now progressing on an area basis, taking in both oblique observed-directed photography and block-coverage verticals (see Kevin Macleod paper). Alongside this mapping, aerial reconnaissance is targeted to explore particular questions such as gaps in the distributions of sites to establish if they are likely to be an artefact of survival and visibility or if they may represent real gaps in the pattern of past activity. This process provides a solid basis for detailed analysis of settlement and landuse patterns. The mapping project is also beginning to engage with the issues of reconstructing a landscape that has changed enormously in the last 200 years, with an intensification of agriculture that has obscured many landscape features, with, for example, the canalisation or burial of many small watercourses. The vertical photography, in particular several runs from the very dry summer of 1977, have been invaluable in this, providing a broad view across large areas that is not afforded by the obliques.

Space and Place in the later prehistoric landscape

The archaeological map of East Lothian is not producing an even or representative spread across the chronology of settlement. There is an apparent concentration of known sites into the later prehistoric period, spanning the first millennium BC and the first three centuries AD. This is a time when enclosures appear to be one of the main foci of settlement and this makes the period highly visible by contrast to earlier periods when settlements may have been unenclosed, relatively ephemeral and therefore not very responsive to aerial survey. In addition there are systems of land-division that also appear to date to this period. Dating is an ongoing problem, with the vast majority of sites assigned to a broad chronological period by analogy with a few excavated and scientifically dated sites. There is, however, a broad framework to assign some confidence to the broad dating, built out of intermittent campaigns of excavation from the 1970s to the present and including recent excavations by Professor Colin Haselgrove on a range of sites.

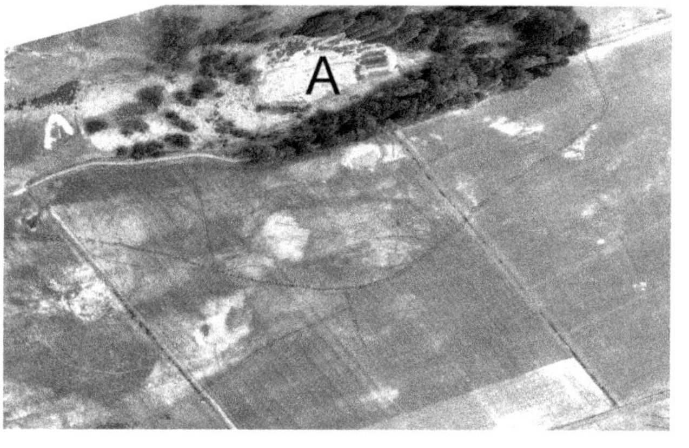

Figure 1 *The later prehistoric fort at Kae Heughs (A) lies within a boundary system that may have been for stock control. Such systems may point to a specialisation in the organisation of the landscape (Crown copyright: RCAHMS).*

In exploring the structure of the later prehistoric landscape on the coastal plain, morphology of major monument types and the spatial patterning of these sites at both a local and regional level has been carried out. These patterns will be interpreted in the framework of an emerging political and economic landscape from the 1st millennium BC and early centuries AD.

The material obtained is challenging in presenting it to a variety of audiences, from the web-mapping applications of RCAHMS to the difficulties of engaging with a landscape and sites, where by definition there is usually nothing to see.

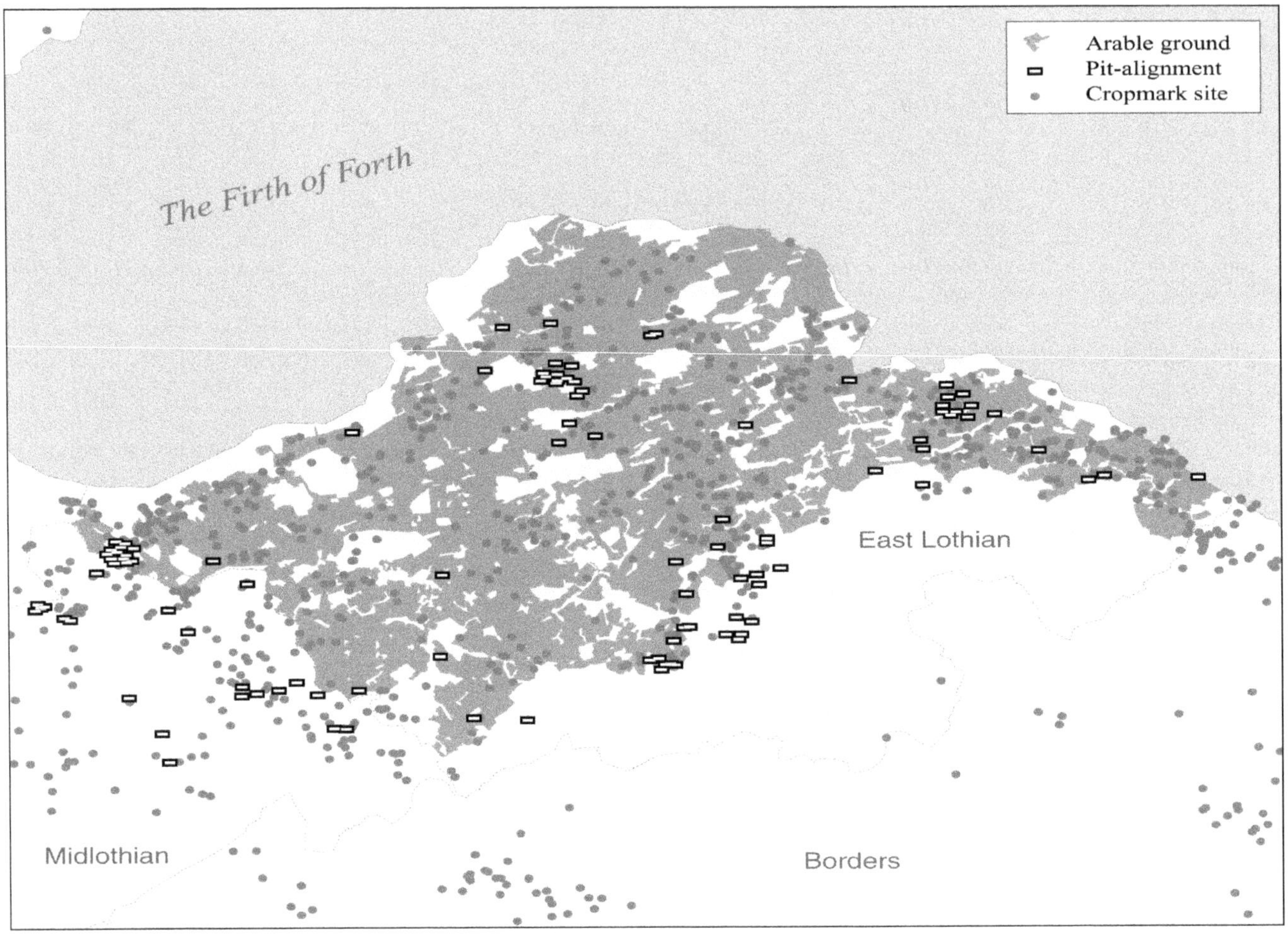

Figure 2 *The distribution of pit-defined boundaries in East Lothian shown against the generalised extent of arable ground and all sites recorded as cropmarks (Crown copyright: RCAHMS; arable ground derived from MLURI mapping based on 1988 aerial photography). This distribution shows that the pattern of pit-defined boundaries is not an artefact of survey bias, but reflects specialisation in the organisation of the landscape.*

Development in aerial photographic rectification and mapping at RCAHMS

Kevin H.J. Macleod

RCAHMS, Kevin.MacLeod@rcahms.gov.uk

For some 30 years the Royal Commision on the Ancient and Historic Monuments of Scotland (RCAHMS) has run a programme of aerial survey across lowland areas to identify plough-levelled archaeological sites appearing as cropmarks (also recording upstanding monuments, townscapes, industrial sites and landscapes).

The resulting collection of oblique aerial photographs are held in the archive of RCAHMS. In order to accurately locate these sites to the UK National Grid (BNGR), RCAHMS adopted the Aerial 5 software programme in the late 1980's, which has been written and maintained by John Haigh (formerly of Bradford University). This software is used routinely to produce a BNGR digital transcription of the plough-levelled archaeology. RCAHMS is currently working on a large area mapping of the coastal plain in East Lothian, dealing with the totality of the cropmark record.

Other work includes thematic projects such as Roman monuments along the Antonine Wall and specific requests from external bodies. This paper outlines recent developments in this programme of work.

RCAHMS Transcription process

A suite of aerial photographs, taken over a number of years, is assembled for each site and assessed to identify those images with the best representation of the archaeological features. In addition, there is a requirement to identify good quality control points visible on the aerial photographs and represented on the Ordnance Survey (OS) map.

Three datasets are assembled before transcription: A digital OS Landline map extract covering the site, extracted from the RCAHMS GIS in DXF format.
The relevant OS Profile Digital Terrain Model (5m interval) in DXF format.

The selected images digitally scanned from a print at high resolution (400dpi greyscale or colour) in TIF format (or derived from a digitally born image for more recent photography).

The transcription process involves the geo-rectification of the chosen aerial photograph. A selection of control

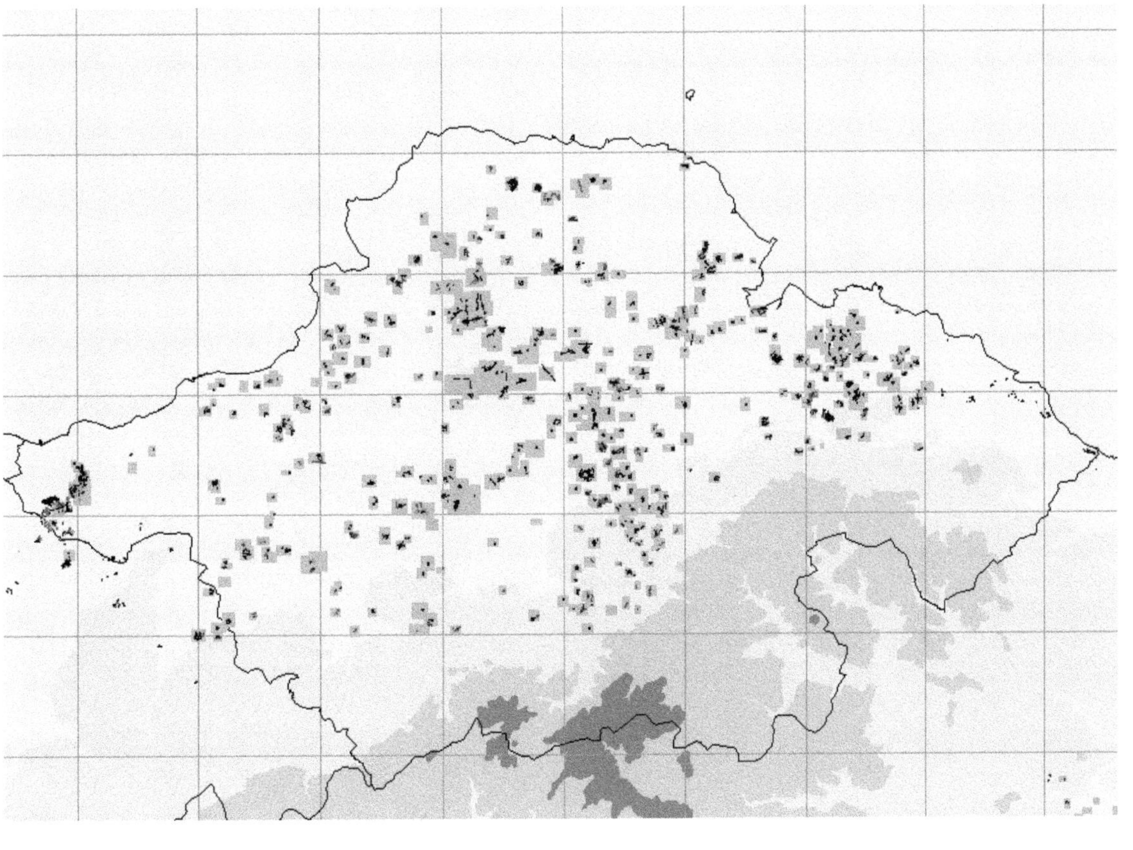

Figure 1 *The extent of cropmark transcription East Lothian, September 2006*

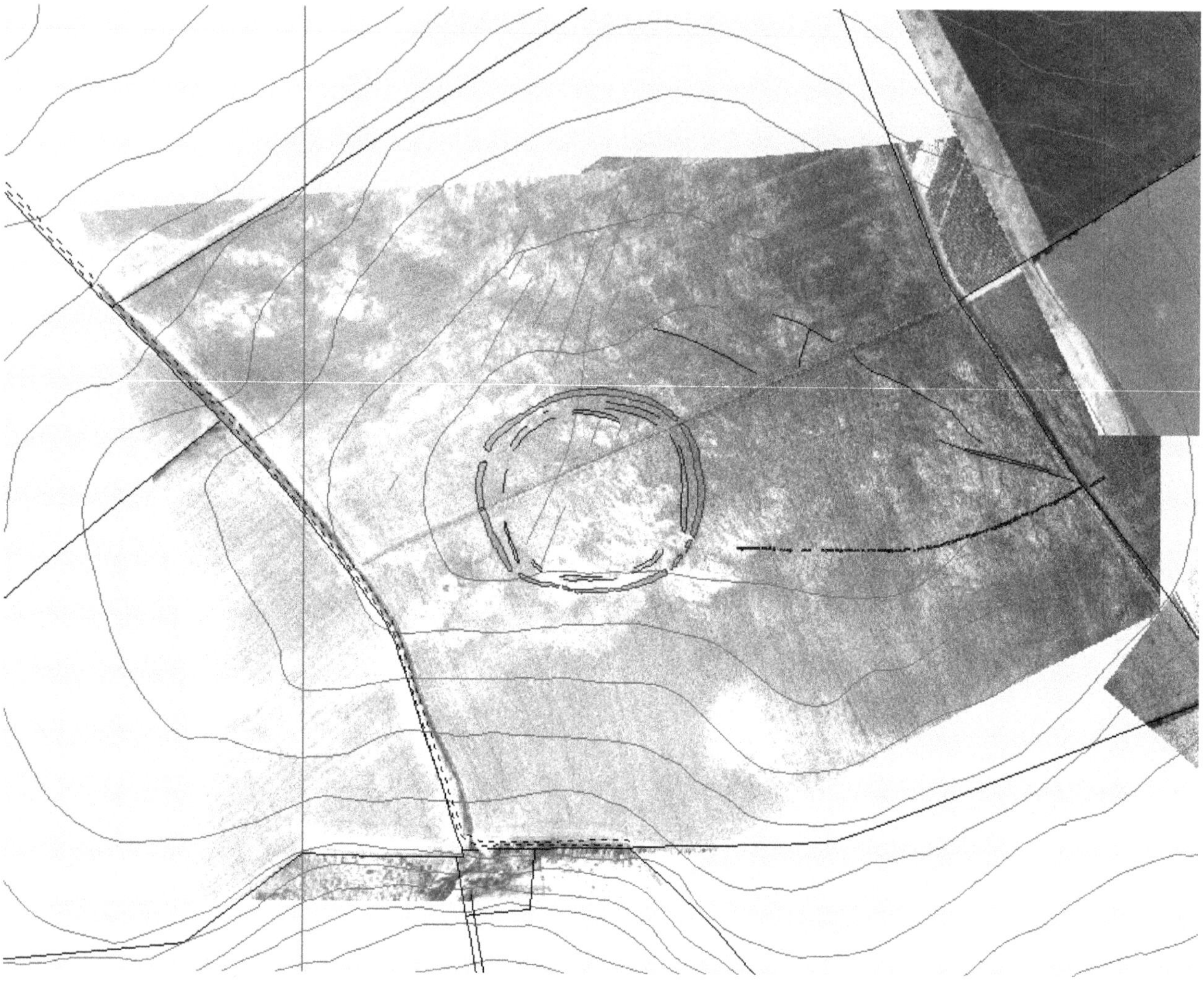

Figure 2 *NT57SE 17, Whitelaw Hill fort, East Lothian – geo-referenced rectified oblique aerial photograph and digitised interpretation of cropmark archaeological features.*

points, which are identified on both the image and the OS map are input into the programme. Aerial 5 can be used to dynamically refine the accuracy of the selected control point, resulting in high levels of metrical accuracy for each point (commonly <1m). Reference may also be made to the extensive RCAHMS vertical photographic collection, where control may be problematic and for additional sites not represented in the oblique collection.

Once a satisfactory fit is obtained, the control point information is saved in a RDA data file. The Aerial 5 format RDA file contains all the chosen information required to rectify the oblique photograph. This is archived should there be a need to revisit this photograph of the site. The oblique photograph is then rectified using this control file data. At this stage the output scale, resolution and size of the image is determined, and the DTM is used to improve the fit of the control and create 3D output vector data. The resulting rectified image is thus geo-referenced to the OS map, and is fully OS referenced whilst "live" in the software.

The archaeology is then digitised on-screen in Aerial 5 and the vector depiction is captured in OS coordinates, and is also recorded with a Z value in metres.

Each 3D vector line is tagged with attributes that will be carried forward into the GIS (e.g. ML3156_SM-ENCL-RECTILINEAR). This attribute data begins with the negative number of the image used (e.g. ML3156) and includes both a broad archaeological classification (e.g. in this case SM for Settlement) and more detailed interpretation (e.g. ENCL-RECTILINEAR for a rectilinear enclosure). In recent years, RCAHMS has developed an in-house classification system for cropmark identification which will allow for intelligent searching within the GIS.

The resulting depiction of the archaeology is then saved in DXF format (ML3156.dxf)

The rectified image is saved (ML3156re.tif), along with an ArcGIS worldfile (ML3156re.tfw) that allows the

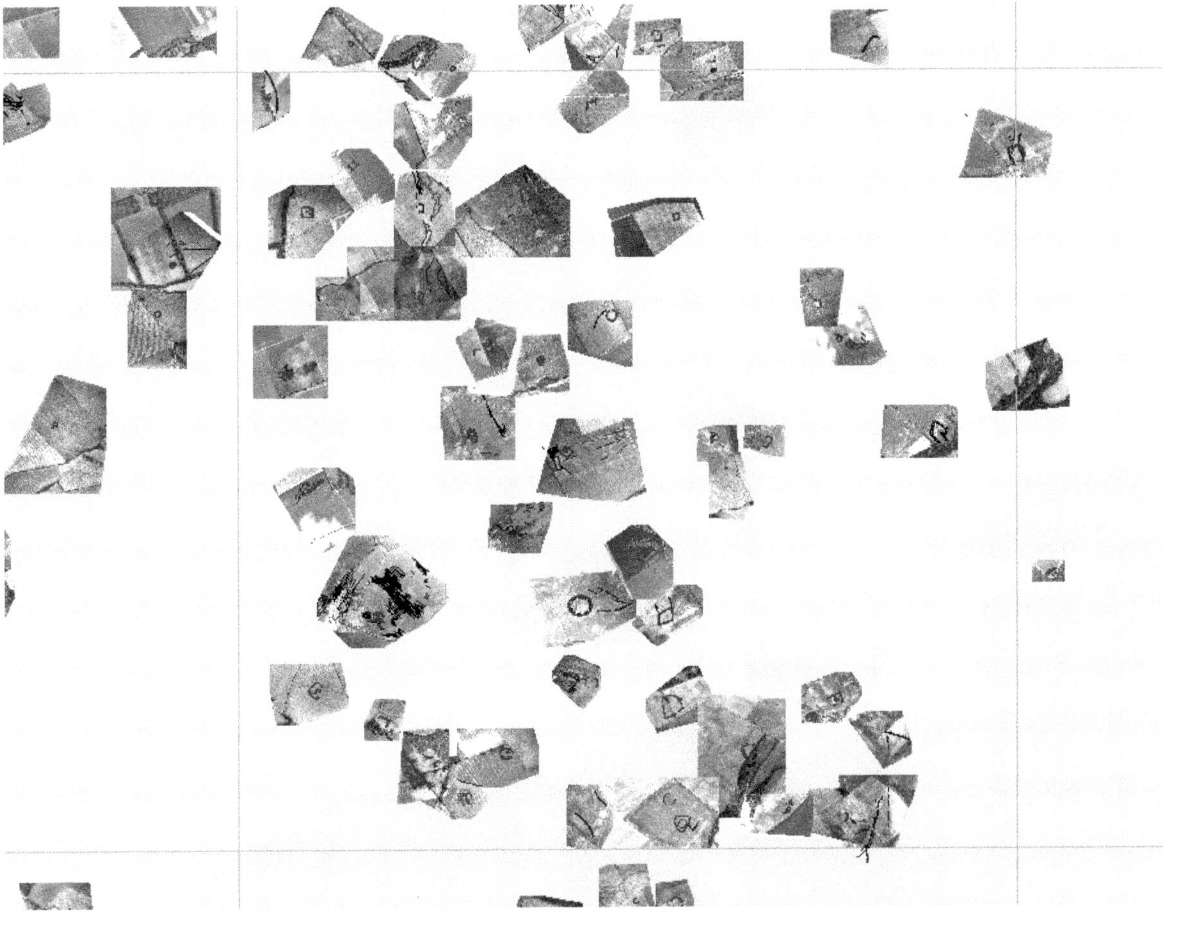

Figure 3 *1:10,000 mapsheet NT57SE, East Lothian – the completed suite of geo-referenced rectified oblique aerial photographs and digitised interpretation of all cropmark archaeological features.*

image to be viewed in the GIS in the correct spatial location.

When the digitisation is completed and the interpretation discussed and peer reviewed, the vector data is migrated through Autocad for any minor editing, and then imported into a master Aerial Photographic Transcriptions GIS dataset (shapefile), which is accessible through the RCAHMS GIS (ArcGIS software).

Archiving and using imagery

The current archiving system in RCAHMS is based on 5x5 km units of the OS 1:10,000 mapsheet e.g. NT57SW. The individual rectified, high resolution (300dpi greyscale/colour) aerial photograph for each site is archived in the GIS archive structure, again indexed by mapsheet.
All the rectified aerial photographs for every site are then assembled onto a geo-referenced raster image (in this case a low resolution JPEG) covering the whole OS 1:10,000 mapsheet. Aerial and Photoshop are used to control this stage.

This "image catalogue" is also archived in the GIS archive structure by mapsheet. File naming conventions indicate that the sheet is either completed, or incomplete and awaiting upgrade. Edge matching issues between overlapping photographs are also addressed at this stage, so that every rectified image is seen in its entirety in the GIS. The mapsheets themselves are then compiled into a further single GIS "image catalogue" covering a 100 km square (e.g. the file for NT is 'nt index of images transcribed.dbf').

RCAHMS Transcription products

Therefore, the final metrical depiction of the cropmark archaeology for a site comprises:

Interpreted 3D vector data, displayed by broad classification (e.g. settlement), and supported by detailed attribute data (e.g. negative number, archaeological interpretation, database unique identifier).

Vector data tags accessible by on-screen selection.

Individual high-resolution, geo-referenced aerial photograph (TIF).

Lower resolution (JPEG) composite, showing all rectified oblique images per 1:10,000 mapsheet.

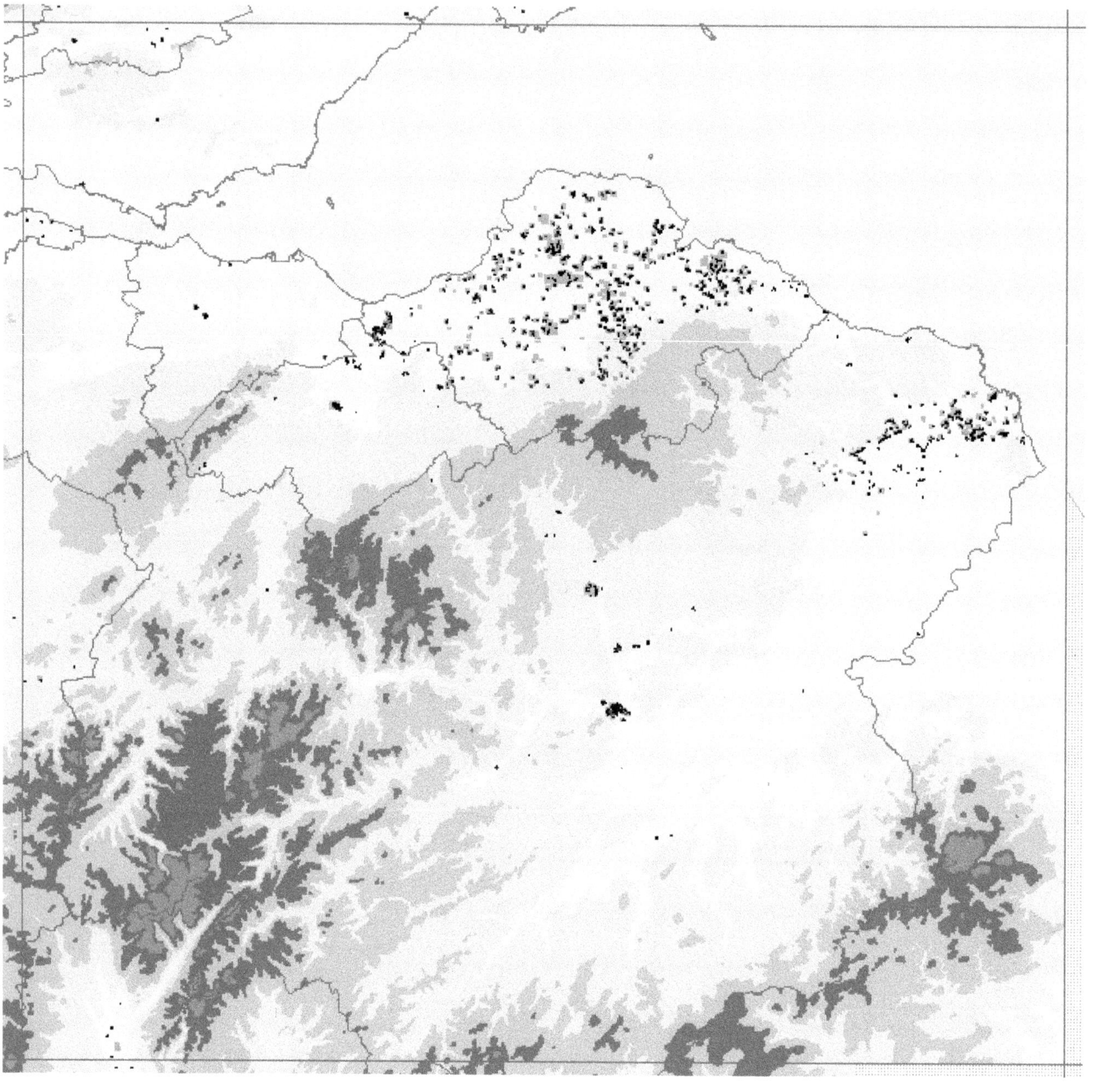

Figure 4 *100 square km, NT, Eastern Scotland – all geo-referenced rectified oblique aerial photographs and digitised interpretation of cropmark archaeological features.*

Access to RCAHMS Oracle database lookup table for each site, showing the unique identifier for the archaeological description of the site and its associated collection of imagery and other information (including administrative data).

Outputs

All the digital data can be presented in a variety of publication formats, with Illustrator, Photoshop and Autocad utilised to prepare illustrations as required. 3D animations can be created and recorded in ArcScene, as the vector data is transcribed in 3D, and geo-referenced raster images can be lifted up onto the relevant DTM surface.

Future work

The RCAHMS aerial transcription programme has its origins in the mapping of information from oblique photography, but we are now making routine use of our extensive collection of vertical photographs, both to assist with control and for the information they hold, which may not be on the obliques. Aerial photography from aerial survey at RCAHMS is now entirely digital and an ongoing project will look at the use of these fully digital images and any implications of the change in media.
To date, some 7000 plough-levelled monuments have been recorded as cropmarks and mapping will progress by area across Scotland's lowland landscapes, recording this hidden archaeological landscape to a consistent

standard that will underpin further analysis and understanding of these important areas. The mapping work will also inform the ongoing annual survey of lowland areas, enabling reconnaissance and photography to be targeted effectively.

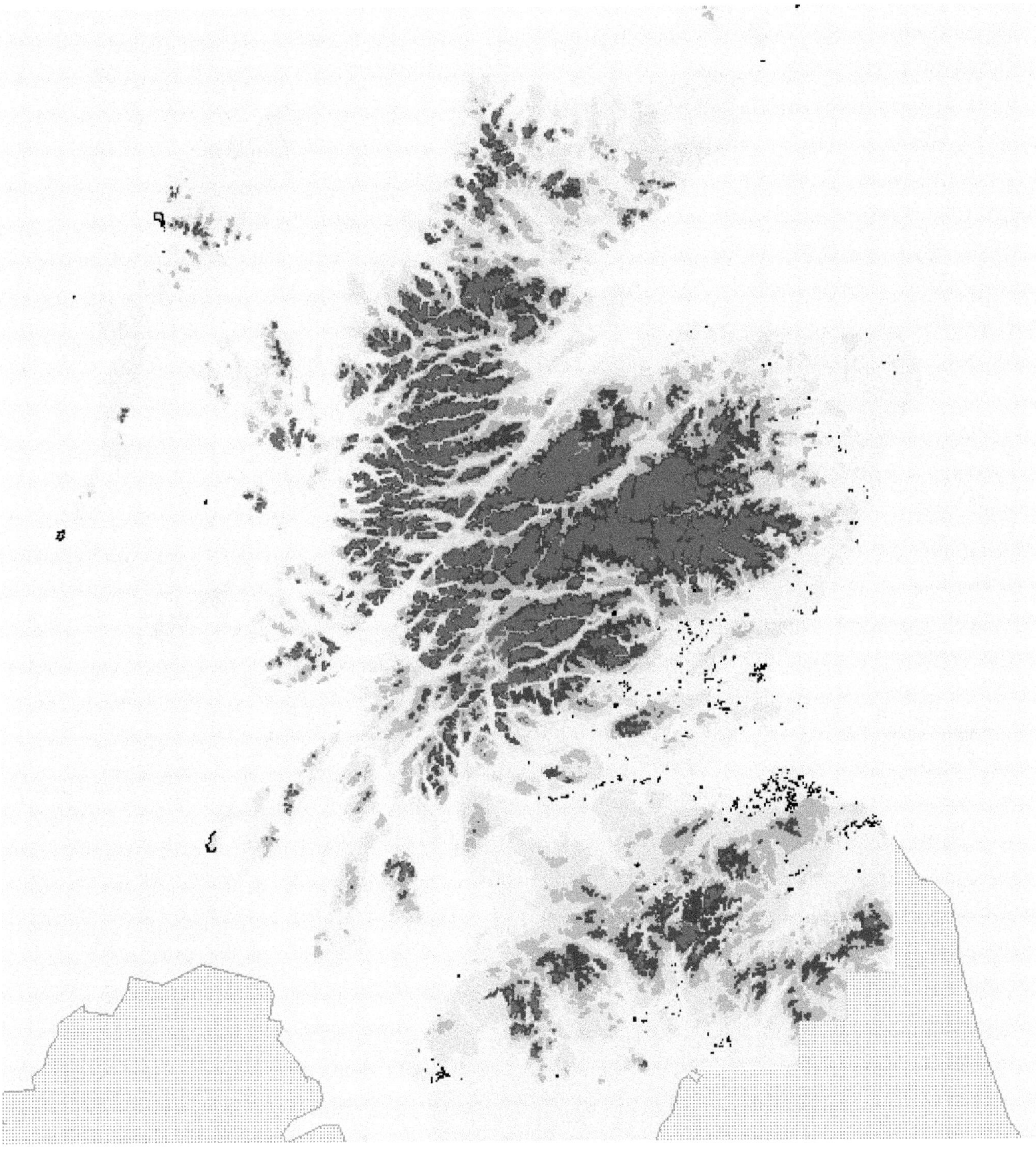

Figure 5 *Scotland – the current extent of geo-referenced rectified oblique aerial photographs and digitised interpretation of cropmark archaeological features.*

Air photo applications in Wales, UK: exploration, landscape analysis, conservation and public presentation

Chris Musson,[1] Toby Driver,[2] Tom Pert[2]

[1] Tanyffordd, Pisgah, Aberystwyth, United Kingdom, SY23 4NE.
[2] Royal Commission on the Ancient and Historical Monuments of Wales, Plas Crug, Aberystwyth, SY23 1NJ.

The object of this paper is to discuss air-photographic work by official, academic and private-sector bodies in Wales over the past 30 years (Wales is the rocky western part of the United Kingdom to which the gentler landscapes of central England are attached). Our aim is to show how differing objectives, organisations and individuals, working together over a considerable period of time, can have an important influence on the integration of aerial techniques into the everyday conduct of archaeological and landscape research, conservation and public presentation. Our emphasis is on the inter-relationship between organisations and work-programmes over time, rather than the technical means through which the work is carried out. We see the latter, essentially, as progressively better mechanisms for meeting broad-ranging objectives which for one reason or another we are incapable of achieving through single institutions alone.

First, however, we must introduce the 'players' in what at times has seemed like a rather stately dance. They are:

- ***Cadw: Welsh Historic Monuments***. A section of the partly-devolved national government of Wales which deals with the protection and presentation of the country's most valued archaeological and historical resources. (We have also used the name for earlier bodies serving broadly similar purposes.)
- ***The Royal Commission on the Ancient and Historical Monuments of Wales***, known for short as 'the Royal Commission' or 'the Commission'. This is a publicly funded body, notionally at one remove from government but almost entirely dependent on public funds. It is the official body of archaeological survey and record in Wales, matching broadly similar bodies in England and Scotland.
- ***Four relatively small University Departments*** in various parts of Wales, carrying out independent research as well as teaching. Their research, however, has often been focused outside rather than inside Wales.
- ***Four independent Archaeological Trusts***, providing a broad 'regional' service for archaeological records, development-control, conservation and project work of various kinds. The Trusts draw on both public and commercial funding.
- Contributions of various kinds are also made by Local Authorities, national and local museums, commercial archaeological firms, amenity societies, voluntary groups and private individuals.

The renewal of aerial survey in Wales, after early forays by Cambridge University, mostly in the 1950s and 1960s, depended very much on the enthusiasm of a few individuals. But their work-time connections with Cadw, the Royal Commission and two of the archaeological Trusts were crucial in the initiation of renewed aerial exploration from the mid-1970s onwards. This culminated in the establishment in 1986 of a dedicated air-photography post in the Welsh Royal Commission, a decade after a similar move in Scotland and twenty years after the formation of a small air photography unit in the Royal Commission in England.

The gradual development of aerial work in Wales, through the interaction of various bodies with differing official remits, forms the focus for the rest of this paper. Questions of organisation, finance and 'who does what' are seen as major influences on attempts to study, conserve and present the historic environment in the particular conditions which pertain in Wales. Parallel but inevitably different relationships will apply in other regions and other countries. Our point is that these relationships, whatever they may be, must be mobilised in a flexible way, sometimes over a substantial period of time, to achieve objectives which all of us tacitly support but which none of us can bring to fruition on our own.

Example 1: exploration, analysis and follow-up research

Our first example concerns aerial exploration by archaeologists themselves through oblique air photography, using industry-standard cameras and readily available light aircraft. No sophisticated technology is involved, though GPS navigation and digital cameras have increased the efficiency of the work in recent years. Improvements in technology have also aided subsequent photo-interpretation, mapping and record-creation from the aerial images, though the emphasis is again on simple desk-top computers and inexpensive programs designed specifically for transcribing archaeological information from both vertical and oblique air photographs.

On the central Borderland of Wales this kind of exploratory air survey, carried out mainly by the local archaeological Trust in the 1970s and 1980s, transformed our understanding of the pre-Roman landscape. What appeared, from surviving earthworks, to be a pattern of defended hilltop settlements with relatively little in the valleys below, became within a decade and a half a densely populated landscape of hilltop, hill-slope and lowland settlements, not very different in distribution and density from the rural landscape of the present day. All this could be mapped and documented in individual site records for the local archive and for onward transmission to the national archive in the Royal Commission. Newly-discovered sites could be quickly incorporated into the local planning process which (now, but less so then) requires all construction projects to be subjected to prior archaeological and environmental assessments. But cropmarks, of themselves, tell us little about the function and dating of the sites, apart from suppositions based on limited excavation or artefactual evidence from elsewhere. Of course this is not a problem that applies only to 'aerial' evidence – archaeology is by its very nature a study which relies on partial rather than total survival of physical evidence from the past.

In the early 1980s one of the more complex settlement on the Welsh Borderland, however, came under threat of destruction by ploughing (over which there is virtually no legal control in Britain). Total excavation of the site was placed in hand through the local archaeological Trust, using funding from Cadw in its role as protector of the historic environment ('rescue' work is now largely funded by the developers themselves, following the establishment in Britain of the principle that 'the polluter pays'). The excavation revealed much about the dating, function and development of the settlement, allowing more informed speculation about other sites and their inter-relationships in the emerging landscape round about. The archaeological Trust then took the initiative in proposing a structured programme of surface collection and small-scale excavation on a *number* of sites so as to provide basic dating – and hopefully functional evidence – across a representative range of sites. After much discussion the project was accepted by Cadw for limited funding and was put into effect in the mid-1980s.

The results, however, were disappointing, for a variety of reasons. Cadw, for instance, required the work to be essentially non-destructive, simply examining the *surface* of the intact archaeological deposits below the disturbed plough-soil. Limited excavations across enclosure ditches were allowed, and a few potentially helpful radiocarbon samples were obtained. But virtually no pottery or other datable artefacts came to light, save for Roman material from later occupation of some of the sites. This part of Wales, in the pre-Roman Iron Age, is virtually a-ceramic, a fact which severely limits the effectiveness of surface collection. At that time, unfortunately, geophysical prospection was also considered an expensive, time-consuming and very specialist means of investigation, and was not used in the fieldwork programme. Overall, the results were felt to be disappointing, perhaps predictably so in the known or suspected conditions of the area, and funding for a continuation of the project was not forthcoming.

Further to the south-west in Wales, from the 1980s until the present day, a previously unsuspected landscape of ditched enclosures, of various shapes and sizes, has been progressively revealed through aerial survey on a coastal plateau north of Cardigan. The aerial exploration was carried out first by the local archaeological Trust and then by the Royal Commission. Again questions of dating and function became matters of increasing concern, in the attempt to understand the increasingly distinctive landscape, or landscapes, that the enclosures represented. For several years the local Trust and others urged Cadw to fund a programme of research on the area, despite the fact that 'research', as such, is not mentioned in the legal framework through which Cadw operates. So for a considerable period of time nothing happened. But there was a growing consciousness of the need to provide *some* kind of protection for this newly discovered landscape, and perhaps for cropmark sites in general. This eventually unlocked the door for a Cadw-funded programme of integrated field survey, sample excavation and geophysical prospection to gain information which might allow a small number of sites to be selected for legal protection as being of 'national importance'. The resulting four-year programme of fieldwork and assessment by the local archaeological Trust is now entering its final year.

By 2004, when the programme began, geophysical survey, using a variety of instruments, had become an almost routine form of non-destructive investigation in Britain. Each season of work has therefore involved both sample excavation and geophysical survey on and around selected sites, with positive results which we do not need to elaborate here. The point is that, following the initial discoveries from the air, the impetus for the next stage of landscape analysis came not from the university sphere, nor indeed form any 'research' imperative as such, but from the need to understand the character and state of preservation of the sites so as to justify their selective preservation through the legal process (paradoxically, there has never been a problem in providing legal protection for Neolithic or Roman sites known only from cropmark evidence).We should mention in passing that an English university has been involved in this project, principally by providing students to assist in the excavation work and geophysical survey. Through their involvement in the project the university students, along with others from around the world, have come to appreciate the 'aerial' nature of the original discoveries, along with the complementary contributions of excavation, geophysical prospection, traditional field survey and legal procedures in an integrated process of research and conservation.

Example 2: aerial monitoring of legally protected sites in Wales

In the mid-1980s Cadw initiated an innovative programme of aerial monitoring of legally protected sites in Wales (about 2400 of them then, around 3500 now). Systematic oblique air photography was carried out in the first instance by the archaeological Trusts, whose staff thereby became familiar with the usefulness of aerial photography in various aspects of their work. More recently the bulk of the monitoring has been carried out by the Royal Commission through its own aerial archaeologist (Toby Driver, who took over the post on Chris Musson's retirement in 1997). Approximately half of the Commission's 60-hour annual flying programme is now spent on monitoring work, the photographs progressively enriching the pictorial and interpretative record for the country's most important heritage sites, from the Neolithic to the remains of industry, war and urban life.

The photographs, critically for Cadw, provide its staff with readily intelligible images for discussions with landowners, tenant farmers and others about the sympathetic management and conservation of the sites in their care. The detailed photographic record, and hence the characterisation of the sites and their possibly changing land-use, is supplemented year by year. The photographs are widely used in Cadw's highly-praised guide-books and analytical works, as well as in a host of other publications for specialists and (more importantly perhaps) for the general public. The monitoring work, which can form part of ***any*** flight throughout the year, provides a framework of assured returns from each individual flight. At any time during a flight, however, attention can be turned from monument monitoring to other kinds of targets such as exploratory survey, excavations in progress, industrial or townscape recording or landscape characterisation. We describe this as a 'portfolio' approach, which makes our flying far more productive, hour-for-hour, than any form of single-subject air photography. Similar monitoring work is now done selectively in England, where there have also been experiments with detailed low-level photography to identify potential damage to individual sites from ploughing or other agricultural activities.

Example 3: air photo interpretation, mapping and field survey

In England a programme of air-photo mapping at 1:10,000 scale, first by high-quality manual transcription and now by computer-aided methods, has been in progress since the early 1990s. Over this time a team of around twelve photo-interpreters and mappers has so far covered just over a third of England, adding vast numbers of previously unrecorded sites to the national archive. The resulting information is passed as digital maps and text-records to Local Authority archaeologists so that it can be used in development-control and conservation work.

In Wales, with far smaller resources, air-photo mapping has been directed at more specific targets, initially through grants from the Royal Commission to the archaeological Trusts, for mapping cropmark evidence in particular areas or of particular site types. More recently the Royal Commission itself, through the part-time work of a single member of staff (Tom Pert, who later worked on the PDA project noted later in this discussion) has been mapping ***upland*** areas prior to their rapid survey by teams of field surveyors provided by the regional Trusts or commercial archaeological firms. The upland mapping projects are variable in the number of previously unrecognised sites which they reveal but they have proved highly effective in providing a skeleton framework of archaeological features which the field surveyors can then check, elaborate and add to in the field (see Example 4 below).

The down-side of the choice is that some ***lowland*** areas, where cropmarks provide the only evidence of otherwise invisible sites or landscapes, remain only partially mapped and analysed. But difficult choices have to be made when financial resources are limited. The decision to concentrate on upland areas has a certain logic since systematic walk-over survey of these areas, under the title of the Uplands Initiative, has been in progress through Cadw and more recently the Royal Commission since the late 1980s. The problem of the lowland areas, however, remains unresolved. The computer-aided mapping of individual cropmark or earthwork discoveries is now routine in the both the Royal Commission and the archaeological Trusts, but the comprehensive examination of *vertical* photography for lowland areas from 1940s onwards still remains to be tackled. There may be a role here for university students, at both undergraduate and postgraduate levels, but little has yet been achieved along these lines.

Example 4: public presentation

In recent decades in Britain there has grown up a general acceptance that archaeological information gathered at public expense belongs ultimately to the public whose taxes (or military and mapping authorities) facilitated its collection. The information, including – with rare exceptions – the location and character of the sites, should therefore be made available to the general public and not treated as the private domain of specialists, whether archaeologists or otherwise. Hence the readily accessible air-photo and other archives of English Heritage and the Royal Commissions in Scotland and Wales, as well as the parallel records created and made available at regional level in Wales by the archaeological Trusts. Since the partial devolution of home affairs to Wales' own government in 1999 two keynotes of public policy have been social inclusion and open access for all. Over time these have added moral support and limited financial assistance to outreach initiatives by official bodies and local groups, aimed at making the physical traces of the past more intelligible and accessible to the whole community, both intellectually and physically. Some archaeologists, of course, have seen this as an alien 'dumbing down' of their esoteric art. Others have seized on the opportunity, arguing that if we do not carry the public with us we can hardly complain if, through their elected politicians, they refuse to give us the increasing amounts of money that we so often seek (but so often fail to get!).

Much of Cadw's work is concerned with public presentation of the monuments in its care, but paradoxically it does not (yet) have a dedicated education officer. The archaeological Trusts have long encouraged and co-operated with local groups whose aim is to increase the understanding, presentation and conservation of heritage features in their areas. There has been a strong push, too, to make the accumulated archaeological records, at both national and regional level, directly accessible to the general public through the Internet. This builds on the early ambition of the four archaeological Trusts to make their combined records available as an Extended National Database (or END). In the late 1970s and early 1980s this proved beyond the technical resources and manpower available. But the concept has remained alive and all of the Trusts, Cadw, the national and local museums, the Royal Commission and other bodies are now engaged in an initiative to make their differing kinds of records seamlessly available over the internet through a single 'historic environment portal'.

Another initiative has been the development by the Royal Commissions in Scotland and Wales of a system for internal web-based access to their ever-growing heritage archives (the system is called SWISH, or Shared Web Information Services for the Heritage). For the general public this partnership manifests itself in Wales as a free, web-based, application called Coflein, meaning *memory-line* in Welsh. Through text-based or map-based searching this provides access to information on more than 72,000 archaeological sites, monuments, buildings and maritime sites, along with an index of drawings, manuscripts and photographs held by the Royal Commission. Thousands of digital images are also available, along with detailed reports in the form of pdf files. Through this system it may soon be possible to make available all newly acquired digital air photographs, along with their related text-records, very shortly after their acquisition.

The Commission has also been developing what might be called 'remote sensing in reverse', delivering heritage information in the field through the use of mobile handsets. These moves began with experiments to see how mobile GIS software, deployed on PDA devices, could benefit the Commission's own field-workers. The combination of GPS-enabled handsets and air photo mapping data removed the need for wind-swept or rain-soaked field-workers to struggle with paper maps and hard-copy air-photographs. Instead, the system guided them accurately to already-mapped archaeological features that might otherwise have been very difficult to find. Use of the PDA also allowed far more information to be accessed in the field, including digital versions of databases, maps of various scales and dates, and both oblique and ortho-rectified air photographs.

These 'internal' developments suggested the possibility of making similar information available to the general public as they walked about town or countryside. The Royal Commission, through a partnership with outside groups and a local high school, was able to access grants for the development of a so-called 'e-trail' around Ruthin, a small historic town in north-east Wales. The aims were:

- To deliver heritage information in a new and innovative way to a wider audience.
- To introduce students to the fundamentals of GIS, GPS and mobile technology.
- To enable students to develop measurable skills for specific educational targets.
- To develop understanding on the delivery of heritage information to mobile devices.

At Ruthin, members of the public are loaned PDAs by the local Record Office for use as they wander the streets. The PDA displays the user's position in real-time and highlights buildings for which historical information is available. The user simply taps on the PDA screen with a stylus and the device links through to a Coflein web page giving information about the chosen building. From here there are links to related images, from historic engravings to 20th-century postcards and recent architectural records created by the Royal Commission. Many of the images, including air photographs, show 'hidden' parts of the town, so that even local residents can gain a new perspective on its buildings, layout and development over time.

Students at the high school were involved in the selection of suitable images and were trained in the use of GIS and photo editing, along with the creation of web pages which could then be incorporated into the e-trail. They were highly enthusiastic about the use of both the PDAs and GIS. The school was able to buy additional PDAs and GIS software at an educational discount, and the availability of national mapping databases has prompted its teachers to think about wider applications for this equipment and software.

The Royal Commission is now working on the next phase of development, to provide a wider range of information, including audio and video playback, for the contrasting historical, archaeological and physical landscapes of a wide-spreading World Heritage Site and a Roman legionary fortress. These initiatives, under active development with the Royal Commission in Scotland, will hopefully advance the long-term aim of achieving wireless delivery of the Royal Commission's accumulated heritage data, including air photographs and related records, to mobile phones and PDAs without the need for expensive specialist software.

Final thoughts

This last example shows how technological developments – especially those focused on our own particular objectives – can sometimes help us to do things which we could only dream of doing in the past. But in a real sense the key element in achieving any kind of progress is co-operation between organisations whose individual remits may be circumscribed in one way or another. The process requires goodwill, time and perseverance. It involves the 'teasing out' of ways in which work or funding from one organisation or source can be broadened or elaborated by another working to a different brief. Progress will almost always be slow and will not always take us precisely where we want to go (if only because, individually, we often want to arrive at slightly different places).

Have we been successful in Wales? And perhaps in Britain generally? In the public service, for information-provision, conservation work and an irreducible residue of rescue archaeology we are inclined to give a qualified 'yes'. But our university teaching of aerial and remote-sensing archaeology, with a few notable exceptions, is weak. Moreover, in Britain as a whole the meagre engagement of university staff and students in free-ranging research contrasts strongly with the active (in some cases almost exclusive) involvement of universities in remote-sensing archaeology and landscape research in some other parts of Europe.

Where we can get to of course depends on where we are starting from, where we *think* we want to go, what means we have of getting there, and how much help we receive from fellow travellers along the way. There is rarely a single answer. The capacity to explore different possibilities, over an extended period of time, must surely be a keyword in any part of Europe, as it has been in Wales over the past three decades.

The flying patchwork quilt: aerial responses across Europe

Chris Musson[1]
On behalf of the Aerial Archaeology Research Group[2] and Culture 2000[3]

[1] Tanyffordd, Pisgah, Aberystwyth, United Kingdom, SY23 4NE.
[2] http://aarg.univie.ac.at/
[3] *European Landscapes: past, present and future*. www.e-landscapes.com

This discussion is based on a brief review of aerial provision across Europe, prepared for a seminar in the Irish Republic in December 2005. Strangely, bearing in mind the Republic's proximity of Britain, aerial archaeology has in recent years been little used in Ireland, though sorties from 'across the water' at various times have shown how much aerial work could contribute to Irish archaeology and landscape studies as an exploratory and illustrative technique. In the 1970s and 1980s there was some exploratory flying by private individuals, and some pioneering work was done on the use of high-level vertical photographs for preparing national monument records. But relatively little active aerial work, or indeed mapping and photo-interpretation, has been done in recent years. Ireland, in fact, shows how aerial studies will not really 'take off' until home-grown exploration, air-photo mapping and landscape analysis find a proper place in public institutions and/or university departments. With good fortune that omission will be remedied in Ireland over the coming years.

Flying carpet or patchwork quilt?

It would be nice to think that the aerial viewpoint could provide a flying carpet to archaeological enlightenment. Aerial work has many merits – speed of operation, the 'landscape' view and the recovery of information that is not attainable by any other means. But it also leaves uncertainties about date and function and it responds unevenly to differing geologies, meteorological conditions and land-use patterns. There are also 'structural' problems – the extent to which the technique is implemented or integrated into archaeological studies in individual countries, regions and institutions.

As the title of this review implies, the pattern of provision across Europe resembles not so much a magic carpet as a patchwork quilt, of difficulties and differing provision, of partial solutions and significant achievements. Like any such quilt it looks quite pretty on the surface but sometimes hides significant shortcomings underneath. Nevertheless, it may be instructive to look at the ***variety*** of responses that we have seen across Europe in the past decade of international work by the Aerial Archaeology Research Group, or AARG. But first a word or two about the Group itself.

The aim of the Group is to act as a forum for the exchange of ideas and information between all those who are involved in aerial photography, photo interpretation, field archaeology and landscape history, including the formulation of conservation strategies. It also concerns itself with other forms of airborne and ground-based remote sensing. From small beginnings at Cambridge early in the 1980s the Group has now expanded into Europe and beyond. Its annual conferences are held alternately in Britain and on the Continent, it publishes a twice-yearly newsletter, *AARGnews*, and has around 180 members across Europe and beyond, almost half of them from outside Britain.

In the past five years the Group has organised or assisted ground-based seminars or workshops in several Continental countries as well as the UK. But one of its main preoccupations over the past decade – with help from grant-giving bodies and the EU's Culture 2000 programme – has been the mounting of a series of week-long ground-and-air training schools, in Hungary, Poland, Italy three times, Germany twice, the UK and (in the coming year) Italy and Germany again.

The aim of these schools has been to help countries, institutions and individuals in the advancement – or first-time use – of aerial archaeology, in particular exploratory air-survey by archaeologists themselves. But the schools always pay equal attention to ground-based work, in photo-interpretation, mapping and use of the resulting information, including that provided by pre-existing or newly commissioned vertical photography.

The Group's travels across Europe have shown a wide variety of responses to the initiation or application of aerial studies. This discussion highlights just a few of these from our experience in Britain and our more fragmentary understanding of the situation elsewhere in Europe.

Mainland Britain

To our Continental colleagues the situation in mainland Britain may look idyllic. From the 1930s onwards our aerial work has always had a strong exploratory element and to an extent a 'landscape' view of the past. In the 1970s and 1980s we perhaps became over-obsessed with cropmark discoveries, not least in the local or regional flying programmes that were by then complementing the national coverage begun by the Royal Commission in England in the mid-1960s and in Scotland and Wales ten and twenty years later.

Importantly, however, our flying in the UK now responds to a much wider 'portfolio' of projects or targets. Individual flights might involve a mixture of pre-development survey, landscape characterisation, the search for cropmark, soilmark or earthwork evidence, the recording of buildings, townscapes and industrial areas, the monitoring of protected monuments, the recording of new and already-known sites for mapping and interpretation, and the continuing enhancement of local and national photographic records to assist conservation, education and the fostering of public awareness.

In Britain we have never had any significant bureaucratic, legal or military restrictions on getting into the air. Nor have we had problems in gaining access to national maps, at a variety of scales, or to pre-existing vertical air photographs, literally millions of which, from RAF and Ordnance Survey sorties, are now housed in the open archives of English Heritage and the Royal Commissions in Scotland and Wales. Many other archives, including that at Cambridge University, are also open for public consultation. It is worth mentioning these points about open skies and access to maps and photo-archives since it is all too easy to assume that things will be the same elsewhere in Europe. This is not always the case.

Another important point about the situation in Britain relates to photo-interpretation, mapping and record creation. English Heritage, through the now-amalgamated Royal Commission, has since the early 1990s had a comprehensive National Mapping Programme. This aims to record at 1:10,000 scale, now in digital format, all the no-longer-used features that can be seen on readily accessible vertical or oblique air photographs. About a third of England has been covered so far. The resulting records, as both maps and text, are passed to the conservation sections of county and municipal councils and are regularly used in development control and conservation work. Air photo analyses are used widely, but not always with sufficient skill, in pre-development assessments, and archaeologically informed aerial views bulk large in both specialist and popular publications. In Wales and Scotland we do some of these things slightly differently, but not significantly so.

So, through the public service at national and local level we have a reasonably effective framework of policy and provision. With a few notable exceptions, however, we have not effectively penetrated the university scene. The veritable flood of new information from aerial reconnaissance and air-photo mapping has not been fully incorporated into academic perspectives, nor always into a more landscape-oriented view of the past. For the most part, too, there is only limited teaching about aerial archaeology.

We have some rather ageing manuals and conference proceedings dealing with air photography and its uses, along with a few pioneering landscape studies based largely on air photo evidence. Others are pending or in press. There are several books on 'something' or 'somewhere from the air' and a recent set of essays about various practical aspects of aerial survey. There is also the late David Wilson's magisterial *Air Photo Interpretation for Archaeologists*. Unlike Italy, however, we no longer have a specialist aerial archaeology journal, other than the informal (but informative) *AARGnews*, nor do we have a manual covering the whole field from initial survey, through mapping to archaeological interpretation (though paradoxically, Italy does, as a result of an earlier Culture 2000 project).

So not everything in the British garden is rosy, though this still represents the most comprehensive provision that we have seen in our travels around Europe.

Continental Europe

Much of Europe, indeed, has scanty provision. In Iceland, Scandinavia and around the Baltic Sea aerial work is in its early stages, often concentrating on traces of relatively recent events (and why not?). There is currently no aerial activity in Norway or Sweden. Both countries, like Holland, had some small provision in the past but lost it through staff retirements. It is dangerous to depend on only one or two people (or rather posts) to provide any kind of national coverage. Denmark is currently trying to revive aerial survey, by hosting the AARG annual conference in 2007 and organising an aerial archaeology training school there in 2008.

Germany enjoys some fine exploratory work through Dr Otto Braasch, Klaus Leidorf and a few others. But this depends on the individual regional administrations (of which there are sixteen) and on their museums or conservation services, with no national co-ordination and very little analysis in most cases beyond points-on-the-map. In Belgium the Universities of Leuven and Ghent, in a semi-competitive/semi-co-operative way, have built up a reasonably comprehensive coverage for exploratory survey in significant parts of the country. Of France we know little, since its aerial archaeologists have never participated in the activities or membership of the Group.

The former Soviet states of Bulgaria and the Ukraine are still shrouded in bureaucratic darkness but Romania is making a start through exploratory work by Glasgow University and – within the Culture 2000 project – through the staff of the Institute for Cultural Memory (CIMEC) in Bucharest. But there are difficulties, hopefully soon to be overcome, in getting access to large-scale maps and existing vertical photographs – a problem, still, in some countries of the former Soviet bloc.

In the Czech and Slovak Republics archaeologists in universities and research institutes rapidly took to the air after the political changes of the early 1990s, in one case buying their own aircraft. But here as elsewhere on the Continent there still seems to be a certain reluctance to trust air-photo evidence in its own right, until a hole has been dug into the detected sites to show that they are really there. In Slovenia Darja Grosman, of the University of Ljubljana, has developed ways of coping with aerially difficult terrain, and has been promoting and taking part in interdisciplinary projects across the border in Croatia. Greece and Turkey are a closed book, still in some senses 'at war' with one another. Spain and Portugal seem to offer ideal ground for aerial archaeology but appear to have little real provision at the moment.

Austria, through Michael Doneus, has developed highly-productive (but expensive) computer and field-working methods for integrating remote-sensing and field data. He now flies just enough to identify when cropmarks in his study-area west of Vienna are at their best. Then he calls in the military, who carry out total vertical survey as training exercises. More recently he and his colleagues have been integrating purpose-made lidar coverage into their studies, and have been carrying out innovative aerial surveys in mountainous areas. But in Austria, as elsewhere, only a small part of the country enjoys significant aerial coverage.

Hungary, for the most part, has relied on flying visits by Otto Braasch since he first started going there in the early 1990s. But in a small way the Culture 2000 project is now fostering home-grown work through a combined programme of aerial survey, photo-interpretation, mapping and geophysical prospection by the Baranya County Museums in the south of the country. But at the moment there is uncertainty about the continuation of this initiative after the Culture 2000 money runs out.

Poland, another country which has benefited from regular visits by Otto Braasch in the past decade and more, has found aerial archaeology champions in Wlodek Rączkowski, of the Adam Mickiewicz University in Poznań and Zbigniew Kobyliński and Darek Krasnodębski of the Polish Academy of Sciences in Warsaw. In recent years they and a few colleagues elsewhere in Poland (Toruń, Olsztyn, Łódź, Leszno) have done excellent exploratory work on very limited funds. But they are struggling against two structural, or rather intellectual, problems. In the 1930s exploratory air photography was 'tried' by Polish archaeologists, but with seemingly negative results so far as cropmarks were concerned. They were simply flying at the wrong time of year. But the apparent failure has coloured Polish academic attitudes ever since, as has the dominance of field-walking survey, the subject of a comprehensive national programme (AZP), now about 85% complete. This has become almost the sole non-destructive source of survey information in Poland, making it difficult, because of ingrained academic and bureaucratic attitudes, to mobilise any significant resources for aerial work. In reality, of course, the two should complement one another, each acting as a check on the shortcomings of the other. An opportunity was also lost a decade or so ago when the then head of the state archaeological service, Zbigniew Kobyliński, one of our students at the first training school in 1996, tried to create some kind of general policy for aerial work amongst his otherwise unconvinced staff. Unfortunately political changes removed him from office soon afterwards and the initiative died. Sadly, something similar has happened again in the present year, despite important aerial discoveries during the last two summers. Fortunately aerial survey has been attracting growing interest in the public service at regional level, so aerial work can be expected to continue in at least parts of the country.

In Italy, in some respects there is a similar situation. The Soprintendenza is a dispersed body in which local policy lies at the command of individual Soprintendenti. Not many see great value in aerial archaeology – they have enough on their plates already. Not everybody in Italy agrees, of course. Until December 2000 there was also another impediment, in the form of a domestic law which since 1939 had so restricted the conduct of photography from light aircraft that Italian archaeologists felt virtually incapable of doing it themselves. Not surprisingly, they concentrated instead on sophisticated photogrammetric mapping and analytical interpretation, both of individual sites and the broader landscapes and communication systems. The often-startling results are illustrated in the magnificent *Sguardo di Icaro*, published in 2003 under the care of Professor Marcello Guaitoli.

When the 1939 law was at last repealed Professor Riccardo Francovich, of the University of Siena, prompted the first Italian training school in 2001 and others have since followed at Foggia, Grosseto and (next year) at Foggia again. So exploratory flying is now taking place in various parts of Italy, with a clear understanding of the 'landscape' potential of the technique, especially when combined with field-walking survey, targeted geophysical prospection and other forms of remote sensing. Air photo mapping through simple desktop computers has also become fairly firmly established in some Italian universities.

But it seems difficult to envisage a concerted national policy of publicly funded aerial work by the Soprintendenza, nor a nationally co-ordinated programme of aerial reconnaissance and air-photo mapping. Instead, progress in Italy will perhaps come from the initiative and inventiveness of university departments and individual research-workers, co-operating and competing with one another in the search for outside funding and in developing their own ways of dealing with the peculiarities of the Italian landscape and its rich fund of heritage information.

There is a clear contrast here with the situation in Britain, where most aerial work takes place, in isolation from active fieldwork, in the public service rather than in the quixotic but perhaps more inventive ambience of university research. If only we could combine the two! But one of the lessons from Europe is that you cannot pick and choose – you must work within the constraints and possibilities of your own country.

Final thoughts

So, can we point to any common threads from across Europe? Not many, I suspect, though we can look at each other's situations and try to draw lessons which might help us to make the best of things in our own countries. We can extend and deepen the network of advice and support that is now available through the Aerial Archaeology Research Group, the Culture 2000 network and co-operative endeavours in our own lands.

Whatever we do, however, there is no magic formula – each country, and sometimes each region, must find its own way of making aerial studies 'work' within its particular geography, climate, land-use pattern and academic and organisational framework. Perhaps we could come to Rome again in ten years' time and look with satisfaction at what is for now an uncertain but challenging future? But only if wheelchair access is provided for some of us!

Aerial Archaeology in Spain: historiography and expectations*

José Carlos Sánchez Pardo,[1] Iván Fumadó Ortega[2]

[1] Escuela Española de Historia y Arqueología en Roma-CSIC, Universidad de Santiago de Compostela
[2] Escuela Española de Historia y Arqueología en Roma-CSIC, Universidad de Valencia

This paper has a double objective: to critically revise the Spanish historiography of Aerial Archaeology, marking its late development, and, on the other hand, to establish a base for future national, regional or even local investigations in this field. Both objectives will be brief, but we hope they will motivate those who manage the necessary resources, and those who could acquire the necessary technical capacities to carry out archaeological projects that are science-based, or at least, projects helped for this indispensable methodology. Spanish investigators and foreigners should work together to put an end to the paradox that supposes the aerial sureying contempt in a country that offers so excellent weather conditions, vast cultivated fields,[1] and large areas without heavily built-up urban environments.

1 A brief revision of the Spanish Aerial Archaeology

As we know, World War I supposed a point without return for the aerial photographs technical and practical development, even with archaeological ends. We want to remember that previously, and especially with military ends, were applied photogrametry techniques for mapping and cadastre management. Spain was not away of these pioneer movements, as demonstrate some publications that go back to 1862,[2] or the fact that since 1902 the officials carrying out aerostatics ascensions must have had in their equipment the photographic camera. The first aerial photographic recognition carried out by the *Ejército del Aire* took place in the north of Morocco, in the context of the 1913's colonial wars (Fernández García 2000, 17). The civil applications began in 1923 with the first attempt from the *Ministerio de Hacienda* and from the *Instituto Geográfico Nacional* (IGN) of using the aerial pictures for making the cadastre. These works were interrupted by Primo de Ribera's dictatorship, recaptured during the Second Republic and again off with the Civil War, although this tragic period meant a development of new methodological development of the aerial photos and their applications. The Spanish files deserve a deep revision of its archaeological documentation during these initial phases, although at the moment it seems that we can consider the first flight with archaeological purpose to be the one made by Montalbán in 1929 around the hinterland of the ancient site of Lixus (Larache, Morocco).[3]

The material and human destructions caused by the Civil War and the absence of quick reconstruction, forced Franco's dictatorship to appeal to the United States Army to obtain the covering of the national territory. The Americans made it in two times using vertical pictures: the first named *serie A* (1945-46) and the second, *serie B* (1956-57). These files are obviously privileged ones because were filmed before the uncontrolled growth's deep transformations in the 60s and 70s (in strong contrast to other European countries where agricultural changes had begun before) but they have hardly been studied for archaeological purposes. One century later, according to the article by the general Terrero, several political changes had reduced the innovative will of the pioneer minority. In fact, we can say that until the decade of the 70s, the activity of the Spanish Aerial Archaeology grew up not much more than two notices, for a total volume of five pages.[4] As we will see shortly, this panorama has varied in the last years, although it is still far behind the guideslines of other European countries.

A new period started thanks to an initiative from *Casa Velázquez* (Madrid) where it was able to congregate different archaeologists and other specialists, Spanish as French as well, in a multidisciplinary project dedicated to the study of the Spanish territory from different perspectives, and with the goal of ending with the traditional "*méconnaissance des services que pouvait rendre ce type de documentation, qui avait arrêté les chercheurs et les organismes*" (Bazzana y Humbert 1983, 5). This volume summarizes the result of four campaigns of aerial surveys in different Spanish landscapes, using the possibilities of the oblique and low height pictures. The quality of the results, the clear theoretical positions

* To Xavier Dupré Raventós, *in memoriam*

[1] 6.8 millions ha. just of cereal fields in 2005 (the area of the entire country of Switzerland is 4.1 million ha.). *Ministerio de Agricultura, Pesca y Alimentación, Subdirección General de Estadística agroalimentaria.* Internet site http://www.mapa.es

[2] Antonio Terrero (1862): Fototopografía, es decir, aplicaciones de la fotografía al levantamiento de planos topográficos. *Asamblea del Ejército y la Armada*, year V, 2nd time, III, p. 31-46.

[3] The aerial survey review and its photographs have been recently found by Carlos Cañete Jiménez. The research grup managed by Carmen Aranegui Gascó works since several years on this site, and it's preparing its study and publication.

[4] Almagro Basch, M. (1943): La colaboración de la aviación española en el campo de la Arqueología. *Ampurias*, 5, p. 247-49. Martínez Santa-Olalla, J. (1945): Aviación y Arqueología. *Boletín Arqueológico del Sureste Español*, 1, p. 45-48.

and the important source of experiences make this an indispensable reading for any future research project in the Iberian Peninsula.

After this publication other studies using aerial photographs were easier to find, although non was a complete survey project. Frequently the research groups agree with an illustrative or too sporadic use of beautiful pics, highlighting nevertheless those examples by Sánchez-Palencia and Orejas Saco del Valle (1991), Sánchez-Palencia and Fernández-Posse (1992), Ibáñez González and Polo Cutando (1993) or Ariño Gil and Rodríguez Hernández (1997). Anyway, in all these studies the use of vertical photography is more frequent, thus the potential of oblique photography has hardly been explored. We can also find some reviews of the existent bibliography and the history of the discipline, with the addition of theoretical views. The one by Orejas Saco del Valle (1995) should be highlighted. In their analysis of the Spanish delay, they stress the study strategies used by the researchers to the present. These local and even simple approaches make unnecessary data obtained through aerial pictures. We agree with this CSIC investigator when she points out the need to overcome the survey (aerial and terrestrial) just as the way of finding new digging locations (op. cit. 20). The curiosity for understanding the landscape as the result of the complex dialogue among the forces of traditions and the desires of intervening in the world to modify it, will be the road that takes us to feel the aerial prospecting approach as a basic need. The Spanish archaeology also requires forgetting the methodological hierarchy that considers that surveying is just the first phase (and subordinate) to the later excavation. In this way, Aerial Archaeology is very near to the Landscape Archaeology and like this one, it should adopt a diachronic perspective.

It is true that the Spanish archaeologists have some manuals in Castilian talking about the aerial photograph techniques. The most complete reference in this field would be that of Fernández García (2000). Unfortunately, it does not directly answer archaeological questions, as English, French or Italian publications do. The limited number of scholars on the topic and the little diffused foreigners' works make the absence of a manual on Aerial Archaeology in Spanish dramatic. Maybe a separate mention deserves the work from an archaeologist not straight inside the academic institutions, but that surely overcomes to all other in flight hours and by his important *know how*. Del Olmo Martín[5] has been carrying flights in the area of Castilla-León regularly for almost two decades. Even when obtaining good results, his work has not had enough institutional support.

In definitive, techniques shyly used and terminology frequently confused that hides conceptual weakness. Therefore we think it is necessary to build the awareness of the great potential of Aerial Archaeology in Spain, and to begin the revision of the wide already existent photographic material and complement it with systematic aerial survey in diverse areas of Spain, that remains without studying in a global way from the perspectives that offers Aerial Archaeology. There is so much work to do in the Spanish horizon to reach the high levels of the English *Aerial Archaeological Research Group* or the firm beginnings of the Italian *Laboratorio di Archeologia dei Paesaggi e Telerilevamento dell'Università degli Studi di Siena*.

To conclude this bibliographical review we will mention the available files for those interested to begin using the rich available material in our photographic archives, before undertaking new flights. Besides the already mentioned North American flights, there are another two general flights: the *serie C* (1967-68) and the flight of 1981-84 (scale 1:30000). The photographic archive of the *Centro Nacional de Información Geográfica,*[6] that of the *IGN,*[7] and that of the *Centro Cartográfico y Fotográfico del Ejército del Aire* (*CECAF*),[8] have several partial flights, that have become more and more frequent since the 80s. The oldest file with this type of aerial pictures was that of the *Compañía Española de Trabajos Fotogramétricos Aéreos* (*CEFTA*) which began to operate in 1927. Unfortunately, the company was closed in the 90s and their files have been sold in fragments to different administrative offices.[9] Several companies have

[5] (1990): Fotografía aérea y prospección arqueológica. Revista de Arqueología, 107, p. 60; (1993): Arqueología Aérea en Valladolid. *Actas de la Reunión sobre Inventarios y Cartas Arqueológicas*: homenaje a Blas Taracena: 50 aniversario de la primera Carta arqueológica de España, Soria 1941-1991. Soria, p. 235-237; (1993): Arqueología Aérea en Castilla y León. Revista de Arqueología, 142, p. 6-7; (1993): Arqueología Aérea en Asentamientos Vacceos. *Arqueología Vaccea: Estudios sobre el mundo prerromano en la cuenca media del Duero*. Valladolid. p. 507-528; (1994): Arqueología Aérea de Emplazamientos de Defensa Medievales en la Provincia de Valladolid. *I Congreso de Castillología Ibérica*. Aguilar de Campoo. p 593-609; (1995): La Villa Romana de los Casares (Armuña, Segovia). *Congreso Internacional La Hispania de Teodosio*. Segovia. p. 675-686; (1996): Arqueología aérea en tres ciudades indígenas romanizadas. *Congreso Internacional Los Orígenes de la ciudad en el Noroeste Hispánico*. Lugo. p. 409-428; (1999): Arqueología Aérea en Castilla y León. *Revista de Arqueología*, 215, p. 44-49; (2001): Arqueología Aérea en Clunia. *Revista de Arqueología*, 244, p. 6-9. Internet site http://www.geocities.com/archeoa/aerea/portada.html

[6] Internet site http://www.cnig.ign.es

[7] Internet site http://www.ign.es

[8] Internet site http://www.ejercitodelaire.mde.es

[9] Following Fernández García (2000, 20 note 7), the pictures corresponding to Asturias were bought by the *Departamento de Geografía de la Universidad de Oviedo*; those from Catalonia by the *Institut Geogràfic de Catalunya*; those from the Vasc Country by the *Departamento de Ordenación del Territorio, Vivienda y Medio Ambiente del Gobierno Autonómico Vasco*; those from Valencia by the *Consellería de Cultura, Educació i Ciència de la Generalitat Valenciana*; those from Murcia by the *Consejería de Política Territorial Obras Públicas del Gobierno Autonómico de Murcia*; those from the Canarias by the *Consejería de Política Territorial del Gobierno Insular*; those from Extremadura by the *Consejería de Medio Ambiente, Urbanismo y Turismo de la Junta de Extremadura*.

carried out similar works throughout the 20th century although many of them have disappeared.

2 Analysis of potential for Aerial Archaeology in Spain

As a little complement and contribution to this reflection about Aerial Archaeology in Spain, we will try now to offer some preliminary data about the potentiality of this discipline in this country that can be used as a first reference when beginning a project of this kind. Our proposed objective would be to obtain a first general map of visibility or potentitality for Aerial Archaeology in Spain that allows us to know which are the most susceptible areas for evidencing soilmarks and cropmarks based on the characteristics of the land and on the current land uses; something that has not been done until now. In this way and according to Jones and Evans (1975) and Musson, Palmer and Campana (2005) we can reasonably estimate an index depending on types of cultivations (deepest roots and plough like cereals often are the best ones, and so on). GIS software allows us to carry out geostatistical and spatial analysis, bringing us distribution maps of our index.

For doing this, our working method will be the GIS-based integration and analysis of the available information about the different factors that influence the practice of Aerial Archaeology. However, given our limitations of time and material, we can only access a limited series of data and documents, so we must underline the basic, general and preliminary character of this analysis that is only a first step for the development of future and deeper works of Aerial Archaeology in Spain. We also want to remember that we are not looking for a predictive or archaeological map, but a map of visibility, or rather, of suitability for the practice of the Aerial Archaeology; it is therefore a merely methodological study, in order to guide future investigations.

The basic documents on which we will work are in the first place the maps of land uses of the Corine Land Cover Project produced by the *European Environmental Agency* (EEA)[10] that describe the type of current land use in all the countries of the European Union, based on the analysis of satellite images of the year 2000. On the other hand we use the statistics of the year 2005 of the *Ministerio de Agricultura, Pesca y Alimentación* of Spain (MAPA) about surfaces and yields of cultivations in each Autonomous Community. Finally we also use a Digital Model of Elevations of the Iberian Peninsula with a precision of 90 meters, that allows us to know the general characteristics of the relief and the slopes of each area. In future and within more complete studies other conditions for the development of the Aerial Archaeology should also be considered, such as geology, the rate of soil moisture deficit (Jones and Evans 1975, 3-8), climate and weather...

[10] Internet site: http://www.eea.europa.eu/

The method of the analysis is conditioned by the varied nature of documentation: the information of the Corine Land Cover has the advantage of being cartographic but on the other hand it is quite generic and it hardly details the cultivation types. On the other hand, the statistical information of the MAPA for Communities describes very thoroughly the types of cultivations but it is not cartographic, they are only generic figures. Therefore, they are complementary documentation but of diverse nature, something that will force us to restructure the information in charts for Autonomous Communities, in order to be finallly able to add it. So in this analysis we will create two types of maps, one general and graduated based on the Corine Land Cover and the DEM, and another one with the statistics for communities, based on the sum of the previous map and the data of the MAPA.

The working process has been the following: after clipping from the Corine Land Cover European raster map the part that corresponds to the Iberian Peninsula, we have classified the types of land uses according to its higher or lower capacity for reflecting archaeological marks that can be seen from the air. For doing this, we have created 6 categories that goe from 0 (null visibility) to 5 (best conditions for evidencing soilmarks and cropmarks) based on data from bibliography and, if possible, in our own experience[11] (Fig. 1). This way, after reclasifying the raster map based on these values, we have already created a first map of archaeological aerial visibility of Spain.

After this, working with the Digital Model of Elevations of the Iberian Peninsula we have created a map of slopes and have reclassified the slopes according to the level of difficulty that they add to the aerial survey and to the observation of archaeological evidences (for example creating shades or limiting the minimum height of flight) (Bazzana & Humbert 1983, 10). We have opted for a very generic classification in 4 groups of slopes with its 4 levels of difficulty: slopes from 0 to 15 degrees (in general we can consider that they are not a problem for the practice of Aerial Archaeology), from 15 to 30 degrees (they offer certain problems), from 30 to 45 degrees (they imply important problems) and slopes higher than 45 degrees (we consider that they prevent the effective practice of Aerial Archaeology). Then we have subtracted the map of potentiality obtained from the Corine Land Cover raster and this map of slopes reclassified in 4 groups, so areas with slopes smaller than 15 degrees don't change their level of potentiality for Aerial Archaeology, those from 15 to 30 degrees lose 1 point in this classification, those from 30 to 45 degrees lose 2 points and areas with slopes bigger than 45 degrees

[11] We are aware that this classification, like the following one based on the data of the MAPA contains, as all the interpretations, a certain subjective component, but it is unavoidable and necessary for doing analysis and comparisons, and on the other hand, as we have just sai, we have tried to minimize this subjectivity based on the data from the available bibliography.

CORINE LAND COVER 2000 (Label Level 3) BY EEA	POTENTIAL 0-5
Continuous urban fabric	0
Discontinuous urban fabric	0
Industrial or commercial units	0
Road and rail networks and associated land	0
Port areas	0
Airports	0
Mineral extraction sites	0
Dump sites	0
Construction sites	0
Green urban areas	0
Sport and leisure facilities	0
Non-irrigated arable land	5
Permanently irrigated land	3
Rice fields	1
Vineyards	2
Fruit trees and berry plantations	1
Olive groves	2
Pastures	5
Annual crops associated with permanent crops	3
Complex cultivation patterns	3
Land principally occupied by agriculture, with significant areas of natural vegetation	5
Agro-forestry areas	3
Broad-leaved forest	2
Coniferous forest	1
Mixed forest	2
Natural grasslands	5
Moors and heathland	5
Sclerophyllous vegetation	2
Transitional woodland-shrub	2
Beaches, dunes, sands	1
Bare rocks	0
Sparsely vegetated areas	4
Burnt areas	1
Glaciers and perpetual snow	0
Inland marshes	0
Peat bogs	0
Salt marshes	0
Salines	0
Intertidal flats	0
Water courses	0
Water bodies	0
Coastal lagoons	0
Estuaries	0
Sea and ocean	0

LAND USE IN 2005 IN SPAIN BY M.A.P.A.	Surface (km2)	Percent of total surface	Potential (0-5)
Cereals grain	6.841.004	13,55%	5
Leguminous grain	410.763	0,81%	3
Tubers	72.438	0,14%	5
Industrial cultivations	728.904	1,44%	3
Fodder crops	852.675	1,69%	3
Vegetables and flowers	236.403	0,47%	2
Fallows	3.319.192	6,57%	3
Fruit-bearing citric	306.557	0,61%	1
Fruit-bearing not citric	1.065.513	2,11%	1
Vineyard	1.149.894	2,28%	2
Olive grove	2.456.719	4,87%	2
Other woody cultivations	59.940	0,12%	1
Nurseries	17.596	0,03%	0
Empty hothouses	25.026	0,05%	0
Family orchards	101.054	0,20%	1
Natural grasslands (irrigable)	869.929	1,72%	5
Natural grasslands (unirrigated land)	70.980	0,14%	5
Pasture in high mountain	323.696	0,64%	3
Pasture	4.657.412	9,22%	4
Pasture scrub	1.407.318	2,79%	2
Poplar	132.167	0,26%	2
Coniferous	6.173.958	12,23%	1
Leafy slow growth	3.057.285	6,06%	1
Leafy quick growth	705.068	1,40%	1
Coniferous and leafy	1.477.798	2,93%	1
Scrub	7.505.176	14,87%	2
Barren	1.597.621	3,16%	2
Esparto	225.380	0,45%	2
Waste	911.180	1,80%	3
Unproductive	1.039.334	2,06%	3
Not agricultural	2.152.809	4,26%	0
Inland waters	537.021	1,06%	0
Total surface of Spain	50.487.812	100,00%	

Figure 1

Figure 2

Figure 3

take value 0 (null visibility). This way we already obtained a first general map of potentiality for Aerial Archaeology in Spain (Figs. 2 and 3).

However to be able to do a comparative analysis, we need to study the specific characteristics of each Community. For doing this, the map of potential has been added to a raster map of Autonomous Communities of Spain in order to obtain the number of pixels of each type –from 0 to 5– that contains each Community. By doing this it has been possible to create a chart of mean values and percentages of the best lands (those with values 4 and 5) for Aerial Archaeology in each Community.

On the other hand, we have summarized and classified the data on surfaces of cultivations of the MAPA, again based on a scale from 0 to 5, and we have calculated what percentage they make in regard to the total surface of the Autonomous Community (Fig. 1). This information, more detailed than the Corine Land Cover one, allows us to know, for example, the surfaces used in grasslands, cereals, potatoes and sugar beet, which, as Jones and Evans indicate (Jones and Evans 1975, 4), are the cultivations that allow a better visibility of archaeological marks in the underground. This information helps us further to value an entire wide series of cultivations, kinds of vegetation and land uses. This way we obtained a second chart of mean values and extension of better areas for Communities.

Finally, we calcultated the mean of the values from the two charts: the first one which contains data from the Corine Land Cover and from the slopes map, which would be a first and more general working surface, and the second one, which contains exact surfaces of cultivations. So, the two charts complement each other, despite the limitations that have already been indicated. This way we finally obtained, as is reflected in the charts, graphics and vectorial maps that represent those values in each Community (Figs. 4 and 5).

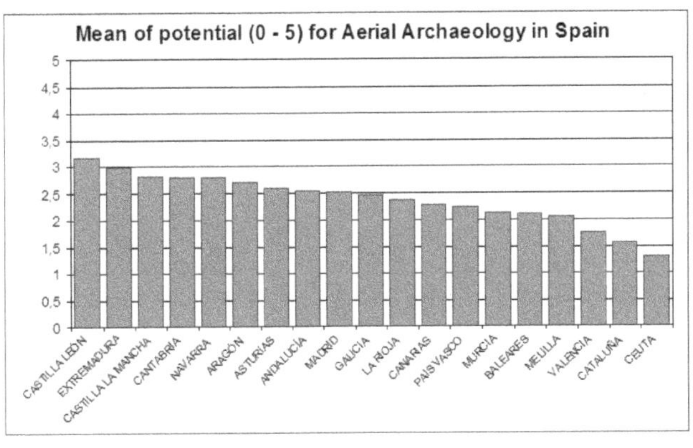

Figure 4

We will now briefly comment on some ideas that emerge from the data and the obtained maps. In first place we have to stress the high potential for Aerial Archaeology for some Autonomous Communities of the inland of Spain such as Extremadura, Castilla La Mancha and mainly Castilla León –where the work by Julio Del Olmo confirmed this great potential–, as much in general potential as in percentage of lands with more aptitude. This result was partly expected, given the conditions of relative plain, dryness[12] and extension of cultivations of cereal in these communities, that converts the inland of Spain into one of the most privileged areas for the practice of Aerial Archaeology in all of Europe. On the

[12] In this analysis it has not been possible to evaluate too many conditions that influence in the emerging of soilmarks. However they would occur more frequently just in this castilian area due to its aridity and the characteristic of the land.

other hand, the high potential that show some communities of the Cantabrian North such as Asturias, Cantabria or Navarra, especially for their high percentage of lands with more aptitude, is more surprising. This may be explained by the fact that this communities, in spite of wide mountainous areas which, as we know, have also been calibrated in this calculation, and that is why their potential mean is not so high, also present a great extension of grasslands and prairies, with a high grade of archaeological visibility from the air. We also find interesting the fact that Communities in the South of Spain such as Andalucia or Murcia don't show a potential mean as high as it could be expected initially, although as we can see from the map, there are some zones with great potentiality. This is due to the presence of mountains such as the Andalusian mountain range, and because of the great extension of cultivations of olive groves that do not allow a good archaeological visibility from air. On the other hand, in general the whole Spanish Mediterranean coast (except areas such as Almería) does not present very favorable conditions for Aerial Archaeology. Concretely Catalonia, mainly in their coastal provinces, and the whole of Spanish Levant, show very low indices of aptitude, probably because of the wide fruit-bearing, coniferous and scrublands extensions. Finally, although we cannot enter the topic here, we point out that almost the whole oriental area of Portugal presents high potentiality for practicing Aerial Archaeology.

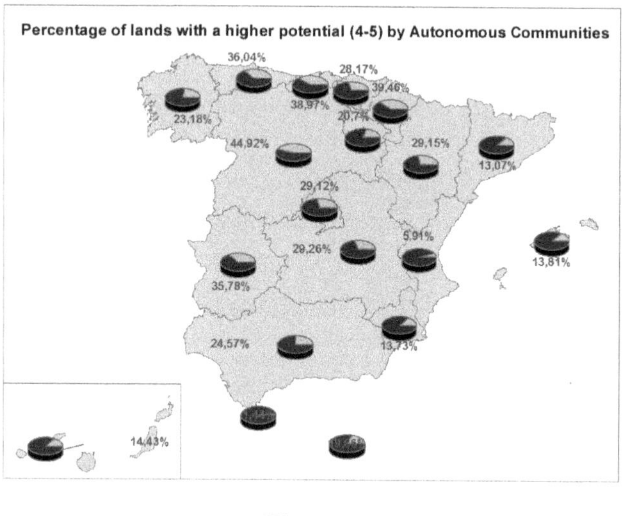

Figure 5

In general the northren half of the Iberian Peninsula seems to present better conditions for the development of Aerial Archaeology, especially the provinces of Álava, Asturias, Burgos, Cantabria, Lugo, Navarra, Palencia, Salamanca, Segovia, Soria, Teruel, Valladolid, Zamora and Zaragoza. Nevertheless, in the south half of the Iberian Peninsula there are provinces with important extensions of good aptitudes lands, such as Albacete, Almería, Badajoz, Cádiz, Ciudad Real, Cuenca, Sevilla and Toledo. In general, we observe that the 24,7% of the surface of Spain, that is, a fourth of the country, has a very good preliminary conditions for the development of Aerial Archaeology, and this is a very significant index for the great potential that our country offers for the development of this discipline.

Of course, all these data and maps are only a first and very general step within the topic, and new and deeper studies that include other variables are necessary. However we think that they can already be used like an initial reference to outline future works. And most importantly we want them to be a stimulus for the necessary development of this discipline in Spain, because we consider that it can have very positive results in a country with good geographical conditions, both physically and in its land uses (we remember, for example that Spain produces almost 18% of cereal of the European Union) and with an important archaeological heritage that in its majority has not been studied through the areal explorations.

Bibliography

ARIÑO GIL, E. and RODRÍGUEZ HERNÁNDEZ, F. (1997): El poblamiento romano y visigodo en el territorio de Salamanca: Datos de una prospección intensiva. *Zephyrus*, 50, p. 225-245.

BAZZANA, A. and HUMBERT, A. (1983): *Prospections aeriennes, les paysages et leur Histoire*. Madrid.

FERNÁNDEZ GARCÍA, F. (2000): *Introducción a la fotointerpretación*. Ariel, Barcelona.

FRANCO, J.; EGEA, J.; MONTORO, I.; OSUNA, M.; CARBONERO, M.; CRESSIER, P. (1992): Aportación de la fotografía aérea a la reconstrucción de los paisajes agrarios medievales: Caso de Andalucía oriental. *Jornadas sobre teledetección y geofísica aplicadas a la arqueologia, Madrid 7-10 de mayo de 1986, Mérida 1-3 de octubre de 1987*. Madrid. p. 137-148.

IBÁÑEZ GONZÁLEZ, E.J. and POLO CUTANDO, C. (1993): La fotointerpretación en la prospección arqueológica: su aplicación en la Carta Arqueológica de Aragón. *Reunión sobre Inventarios y cartas arqueológicas: actas: homenaje a Blas Taracena: 50 aniversario de la primera Carta arqueológica de España, Soria 1941-1991*. Soria. p. 225-234.

JONES, R.J.A. and EVANS, R. (1975): Soil and crop marks in the recognition of archaeological sites by aerial photography. *Aerial reconnaissance for Archaeology*, XII, p. 1-11.

MAGALLÓN BOTAYA, M.A. (1986): Algunos problemas sobre utilización de la fotografía aérea en las investigaciones sobre las vías romanas en Aragón. *Metodología de la investigación científica sobre fuentes aragonesas: actas de las I Jornadas celebradas en Monzón, del 8 al 20 de diciembre de 1985*. Zaragoza. p. 229-234.

MUSSON, C., PALMER, R., and CAMPANA, S. (et al.) (2005): *In volo nel passato*. All'insegna del Giglio, Siena.

OREJAS SACO DEL VALLE, A. (1995): *Del marco geográfico a la Arqueología del paisaje: la aportación de la fotografía aérea*. CSIC, Madrid.

OREJAS SACO DEL VALLE, A. (1996): *Estructura social y territorio: el impacto romano en la cuenca noroccidental del Duero*. Madrid.

PEÑA MONNE, J.L.; RODANES VICENTE, J.M.; MAZO PÉREZ, C.; MONTES RAMÍREZ, M.L. (1986): La fotografía aérea vertical en blanco y negro y su aplicación a la prospección arqueológica y geoarqueológica. *Metodología de la investigación científica sobre fuentes aragonesas: actas de las I Jornadas celebradas en Monzón, del 8 al 20 de diciembre de 1985*. Zaragoza. p. 219-228.

ROSSELLÓ BORDOY, G. (1970): Prospecciones arqueológicas en Mallorca mediante fotografía aérea. *Pyrenae*, 6, p. 59-64.

SÁNCHEZ PALENCIA, F.J. and OREJAS SACO DEL VALLE, A. (1991): Fotointerpretación arqueológica: el estudio del territorio. *Arqueología*. Madrid.

SÁNCHEZ-PALENCIA, F.J. and FERNÁNDEZ-POSSE, D. (1992): Fotointerpretación aplicada a la prospección arqueológica. *Jornadas sobre teledetección y geofísica aplicadas a la Arqueología*. Mérida, 1987. p. 175-87.

Looking through Black-Tinted Glasses – A Remotely Controlled Infrared Eye in the Sky

Geert Verhoeven,[1] Jo Loenders

[1] Research Assistant of the Fund for Scientific Research - Flanders (FWO), Department of Archaeology and Ancient History of Europe, Blandijnberg 2, B-9000 Ghent, Belgium

Introduction

Since more than one hundred years, oblique aerial photography has proven to be an extremely valuable tool for (landscape) archaeologists. However, the vast majority of the pictures taken only record the reflected visible portion of the electromagnetic spectrum produced by the sun, whereas an infrared view can often reveal more information when compared to this "traditional" imagery. Both the technique and use of infrared photography are explored within the multidisciplinary survey project entitled Potenza Valley Survey (PVS – regione Marche, Italy) by archaeologists of Ghent University (Belgium). To acquire data, a newly build apparatus allows the researchers from the PVS to take both "regular" and infrared pictures of archaeological interesting landscapes and *loci* in Italy as well as in Belgium. Moreover, this construction will also prove useful in mapping excavations.

Potenza Valley Survey

In January 2000, the Departments of Archaeology and Geography at Ghent University (Belgium) initiated a geo-archaeological survey project in Central Italy, titled "The Potenza Valley Survey. From Acculturation to Social Complexity in Antiquity: A Regional Geo-Archaeological and Historical Approach". The main aim of the project was (and still is) to reconstruct both the archaeological and geographical landscapes along the Potenza river, which flows over its 80 km long course from the Umbria-Marches Apennines through a wide and fertile Apennine foothill landscape, before it ultimately runs into the generally flat Adriatic coastline zone (figure 1). Although it was already important in Prehistory, the Potenza river valley became one of the most commercial routes of the Central Italian Protohistory and even in later periods the valley remained an important corridor for political, economical and cultural contacts between the

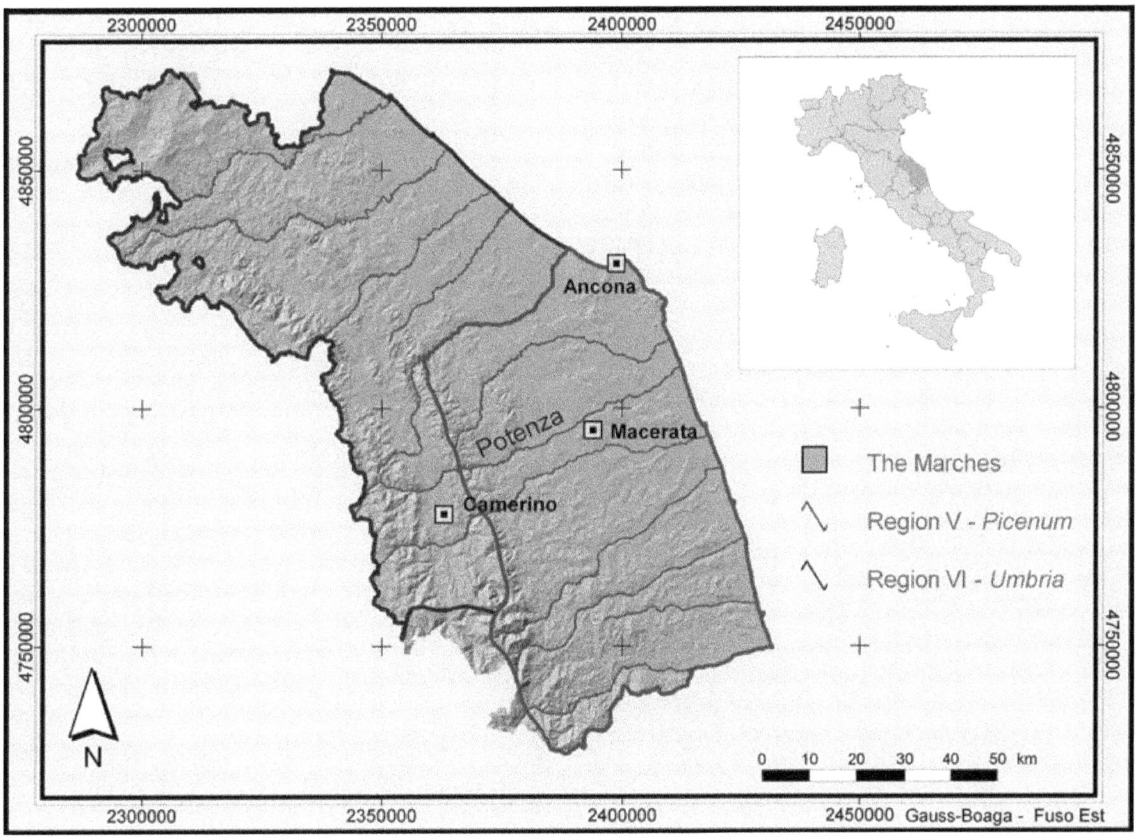

Figure 1 *Location of the Potenza valley*

Tyrrhenian and the Adriatic coast. In Roman times, several cities developed in or near the valley floor and even a southern branch of the *Via Flaminia* passed through it. In the subsequent period, the area remained of importance as it formed the contact zone between Longobards and Byzantines.

To fulfil the initially outlined goals, three specific sample zones, systematically spaced around the Potenza river, are (or were) subjected to major survey techniques: aerial, field, geomorphological, geophysical and a historical survey. The aerial survey comprises the collecting of all (if possible) remotely taken imagery ever taken above the valley. Besides this passive part, taking oblique photographs from a low flying aircraft is also executed. Till now, this has been done in a very thorough way: on the one hand, different flights were made in different seasons; on the other hand, pictures were taken of the whole valley with a specific interest in the three sample zones.

Although started with an analogue small format Single Lens Reflex (SLR) camera, a digital SLR camera (Canon EOS 300D) was bought in 2004 to speed up the workflow. Both systems yielded photographs which captured the visible portion of the electromagnetic spectrum. In 2005, the idea was born to also partly incorporate the infrared radiation.

Electromagnetic Spectrum

Centuries ago, the highly-gifted Englishman Isaac Newton (1642–1727) discovered that white sunlight could be divided into the colours of the rainbow (red, orange, yellow, green, blue, indigo and violet) using a simple prism. Such a range of colours is called a *spectrum* (the Latin word for "apparition"). Newton also put forward that light consisted of tiny particles, every individual particle too small to be seen. This vision was claimed to be wrong by the Dutch physicist Christiaan Huygens (1629–1695), who declared light travelled in the form of waves. Experiments in the early nineteenth century confirmed this wavelike nature of light.

A major next step was taken by the Scottish mathematician and physicist James Clerk Maxwell (1831-1879), who mathematically accomplished to describe all basic properties of electricity and magnetism in only four equations. Both electrical and magnetic fields were discovered to have an intimate relationship, being part of one phenomenon called electromagnetism. Maxwell stated that visible light was in fact an electromagnetic wave. It comprises two oscillating electric and magnetic fields perpendicular to each other as well as perpendicular to the direction of propagation (figure 2), travelling at about 300 000 km/s (more specifically 299 792,458 km/s or 186 282,397 mi/s).

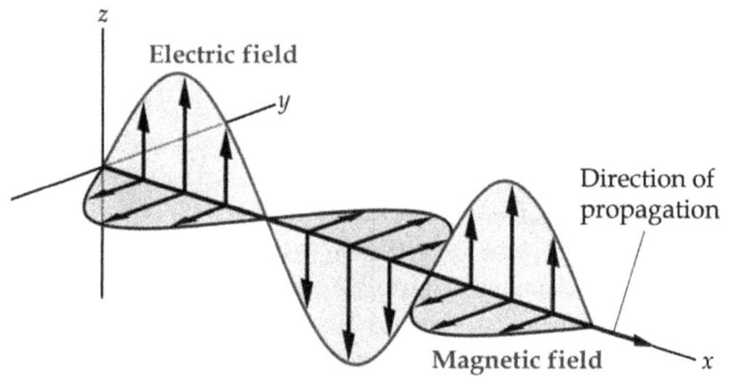

Figure 2 *An electromagnetic wave (adapted from Walker 2004, fig. 25-3)*

When light travels in air, glass, water or other transparent substances, this speed decreases. Being a wave phenomenon, light can be characterised be the length of its waves, called the wavelength (λ). Because wavelength is mostly expressed in nanometre (nm), visible electromagnetic radiation is defined as all light with wavelengths between approximately 380 nm and 730 nm (figure 3).

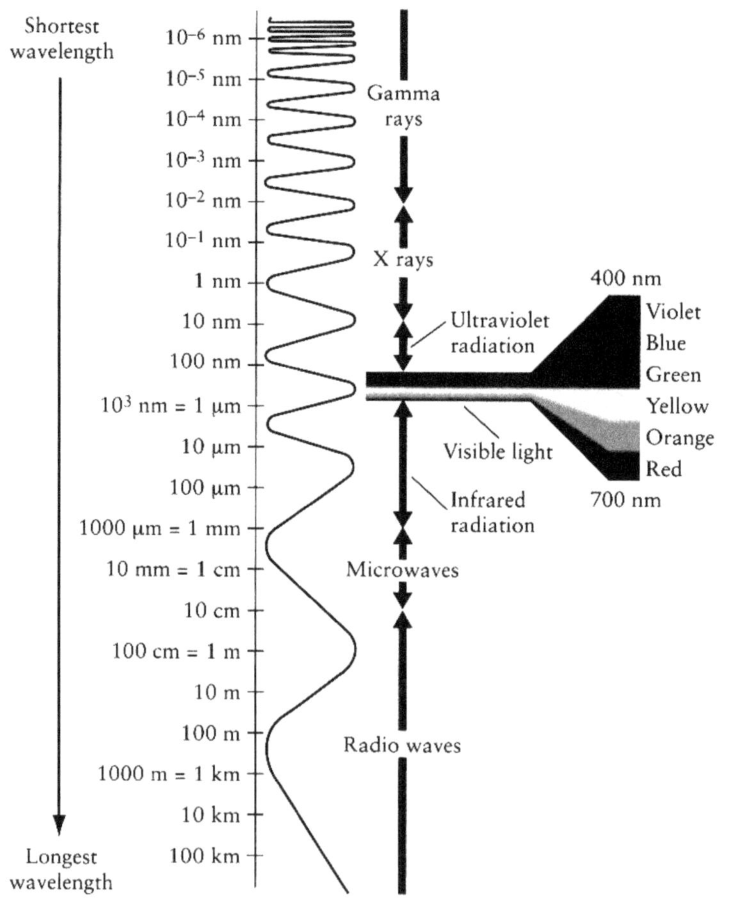

Figure 3 *The electromagnetic spectrum (after Freedman & Kaufmann 2005, fig. 5-7)*

In addition to this visual part of electromagnetic radiation, electromagnetic waves also come in forms human eyes are not sensitive to. In fact, the visible light is only one narrow range out of all electromagnetic waves (called the electromagnetic spectrum – see figure 3).

On both sides of the visible spectrum reside bands which do not produce a visual sensation. They all have smaller or larger wavelengths and are given names as gamma rays, X rays, ultraviolet rays, infrared rays, microwaves and radio waves.

A third attribute – besides wavelength and speed – to characterise electromagnetic waves is frequency, which denotes the number of complete waves passing a certain point in one second. Short wavelengths therefore yield high frequencies. Frequency is always expressed in hertz or cycles per second (abbreviated Hz and denoted by ν).

Today, it is known Newton was not wrong after all. One of the key concepts in quantum mechanics is the *wave-particle duality*, meaning light behaves both as waves and as particles. This particle theory was raked up by the German physicist Max Planck (1858-1947), who in 1900 suggested that light must be emitted and absorbed in certain fixed amounts called quanta, rather than continuous quantities. Five years later, Albert Einstein (1879-1955) writes a paper in which he draws upon Planck's quantum idea to explain the photoelectric effect (i.e. the phenomenon in which electrons are ejected from a metal surface when light shines on the metal). Einstein proposed that these light quanta or discrete energy packets were in fact particle-like entities (now called *photons*). Depending on the wavelength of the electromagnetic radiation, the energy of the photons was different. Due to this dual behaviour, electromagnetic radiation can also be seen as a travelling bundle of photons.

When light (no matter if it is visualised as waves or photons) falls onto an object, the object can do three things: transmit, absorb or reflect the light. The latter is the main idea of passive remote sensing: capturing that specific portion of the electromagnetic radiation which is produced by the sun, partly transmitted by the atmosphere and reflected by objects (e.g. plants, humans, bare soil, buildings).

Infrared Photography

- What?

Besides the small range of the visible spectrum, other wavelengths can be recorded by either electronic sensors or photographic film. However, only a limited portion of electromagnetic radiation can be captured using film. Radio waves for instance will not produce a latent image on the photographic emulsion, but infrared and ultraviolet waves can. The complete range of wavelengths that is able to expose film is denoted the *photographic spectrum*. On both sides of the visible spectrum, wavelengths do exist that still fall in this photographic spectrum. Wavelengths shorter – and hence more energetic photons – than blue visible light are called ultraviolet (UV) rays, while infrared (IR) radiation resides on the other side of the visible spectrum. Therefore, these waves have longer wavelengths (lower frequencies) than light from the visual spectrum. Here, only infrared radiation is considered.

The width of the infrared band is exceptionally wide, spanning three orders of magnitude and comprising wavelengths between 730 nm (the edge of the visible spectrum) and 1 mm. Although the limits are often differently used, the infrared radiation can be divided into five different zones.

1. Near InfraRed (NIR) from 0,73 µm – 1,4 µm (micrometer);
2. Short Wavelength InfraRed (SWIR) from 1,4 µm – 3 µm;
3. Mid Wavelength InfraRed (MWIR) from 3 µm – 8 µm;
4. Long Wavelength InfraRed (LWIR) from 8 µm – 15 µm;
5. Far InfraRed (FIR) from 15 µm to 1000 µm (= 1 mm).

Often, only near (0,73 µm – 5 µm), mid (5 µm – 30 µm) and long (30 µm – 1000 µm) infrared are used. The mid and far IR bands are known as the thermal bands. It is even often said that beyond 1350 nm, the infrared radiation is thermal radiation. Imaging these long wavelengths (called *thermography* or thermal imaging) is not possible with conventional IR film and digital cameras, but only with special non-photographic equipment.

Consequently, just the spectral range from 730 nm to 1350 nm is important for photographic purposes (although the region most accessible for conventional films is limited to around 925 nm, while digital cameras often react to wavelengths of up to 1000 nm and even more). Therefore, photographic emulsions are unable to capture so-called "heat-images" (i.e. infrared radiation emitted by objects). What they record is the portion of near infrared radiation that is produced by the sun and reflected by objects in specific amounts.

- Why?

The reason for capturing this radiation is rather easy to explain. For the human visual system, real-world objects have a certain colour which corresponds to the visible light that is reflected by the object. If light falls on a ripe tomato, the red part of the visible spectrum will be reflected most; hence, the tomato will be perceived red. This principle also holds for infrared recording. Objects reflect or absorb infrared radiation in certain amounts.
As an example, consider green vegetation (e.g. grass, a tree, foliage) which reflects the green part of the visible spectrum due to the chlorophyll – a green pigment –

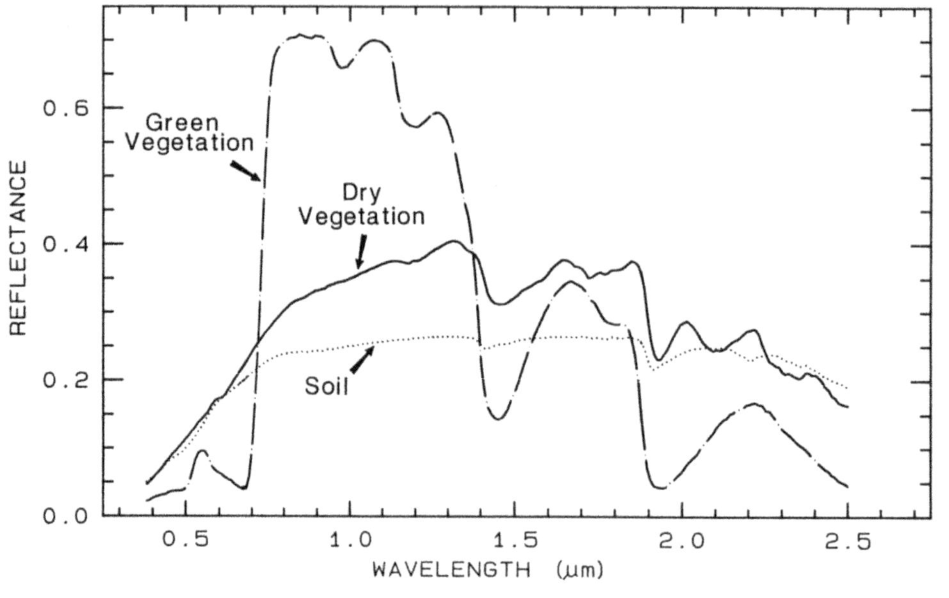

Figure 4 *Reflectance spectra of photosynthetic (green) vegetation, non-photosynthetic (dry) vegetation and a soil (after Clark 1999, fig. 1.18)*

inside. Additionally, chlorophyll also strongly reflects invisible near infrared light. When imaged on colour-infrared film, healthy growing vegetation will appear bright red due to the high amounts of chlorophyll. On the other hand, chlorophyll quantities in stressed, unhealthy or dead vegetation are much lower, leading to less near infrared light being reflected. Hence, these plants will appear light red or darker in the photograph. Due to the fact that chlorophyll reflects around 50% of the near infrared radiation as opposed to around 15% of the visible green component of light, plant stress (e.g. grass growing on a buried wall) is much easier to discern in these non-visible wavelengths (clearly illustrated by the spectral reflectance curve in figure 4). This principle allows aerial survey to monitor the condition of vegetation.

- Equipment

To obtain the photographs, Kodak Ektachrome EIR Infrared film is used in this project. The emulsion, which is sensitive to wavelengths between 380 nm and 900 nm, is a so-called false-colour slide film, meaning the images are created from a wider range spectrum than the visible light and finally remapped into the visual spectrum range. This way, a false-colour image is created because the invisible near infrared wavelengths are displayed as red by the film, whereas red and green objects are depicted as green and blue respectively.

The film is manually set to ISO 200, which is necessitated by the fact that exposure meters are not sensitive to infrared (and hence infrared film does not have a fixed ISO value). Therefore, shooting and evaluating a lot of test strips is the only way to determine the right exposure when photographing infrared.

Besides film, filters can be used to only capture a specific portion of the visible and infrared radiation. Whereas most infrared photography is performed using a visible light filter (often a yellow or red filter), this project applies an infrared filter (B&W 093), delivering a pure infrared picture (as the glass of the filter is completely opaque to all visible wavelengths). By this approach, it is hoped to detect stress in plants growing on archaeological sites much better, as the image will not be "disturbed" by the visible wavelengths. On the other hand, using an infrared filter also has a major drawback: the photographer can not see through them, meaning framing and focusing must be done before screwing the filter on the lens. Moreover, shutter speed can become rather long, meaning the camera must be mounted onto a stable platform to not end up with fuzzy images.

Helikite Aerial Photography

To deal with all previously mentioned difficulties, a stable, easily maintainable and remotely controllable construction had to be created. Moreover, as the archaeological crew from Ghent University goes to Italy only for some limited periods, the apparatus must work in most weather circumstances and be easily transportable. In this section, all components which are part of the finally assembled system will be described separately.

- Helikite

This camera-lifting construction, maybe the most important piece of the whole system, joins a helium balloon with a kite (figure 5). This unique design, patented by Allsopp Helikites Ltd combines the best properties of both objects. The balloon enables a helikite to take of in windless weather conditions – something a kite is unable to do –, whereas the kite components play two important roles in case there is wind: they lift the helikite higher up in the air, while additionally stabilising this lighter-than-air construction. Consequently, they overcome the knocking down of the helikite in windy conditions, a major drawback of all normal balloons. Moreover, the more wind, the stronger the lift (with a maximum threshold depending on the helikite's size).

Figure 5 *Inflating the helikite*

In this respect, the Allsopp helikite is both smaller and more versatile than comparable lighter-than-air constructions, operates in stronger winds than traditional blimps and – by the additional lift through its wings – supports more payload for its size when compared to ordinary aerostats.

Due to the fact that it can be used in adverse weather, it is also more flexible than a remotely controlled (R/C) helicopter. Additionally, R/C helicopters can only hang stable for a while when controlled by an experienced user (and there still is the vibration problem). Although the helikite is not cheap (£ 2100 or around € 3100 for a 7 m³ model), its maintenance cost is low and very expensive crashes are out of the question (which is the reason why R/C helicopters ultimately cost twice the amount of money they are sold for).

The helikite used is – as mentioned – a 7 m³ model, enabling a lift of around 3,5 kg in windless conditions, increasing to even 10 kg in 25 km/h wind (for all prices and facts about the different products, please check www.allsopphelikites.com). By analogy with KAP – which means Aerial Photography by means of a Kite – this type of photography is inaugurated HAP or Helikite Aerial Photography.

- Tether

To securely fly the helikite, a Dyneema tether is used (as is also recommended by Allsopp Helikites Ltd). These tethers combine a small mass with a large breaking strain (in this case 270 kg).

- Picavet system

Named after its French inventor (Pierre L. Picavet) who first described this suspension in 1912, the Picavet system is a construction which allows for dampened suspension (figure 6). It is hanging below the dyneema line by means of two pulleys (which are several meters apart) and must stop the camera from spinning around when manoeuvring with the helikite on the one hand, as well as absorbing vibrations created by the wind on the other. As can be seen from figure 6, the Picavet suspension involves a rigid cross with each of the four ends connected to two attachments on the kite line. The small line which provides these connections is a continuous looping string. Lastly, the crossing two innermost cables are constrained with a ring. The result is a simple, very lightweight system for self-levelling and securing the camera-supporting frame or cradle. (Both the cradle and a suspension system are denoted as camera rig.)

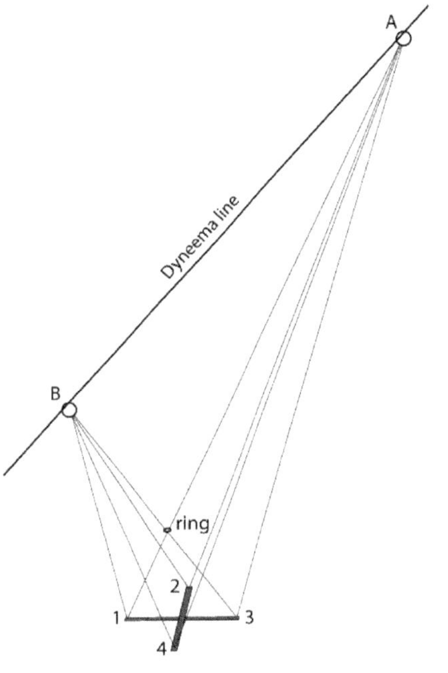

Figure 6 *The Picavet suspension*

- Cradle

The sturdy cradle, devised and built by J. Loenders within the framework of his master's thesis, was specifically designed to allow camera rotation around all three X-, Y- and Z-axes (figure 7). Its four carbon legs allow the construction to stand and take-off independently and protect the camera in case of a tough landing. Moreover, precisely lasered and bended solid aluminium frame profiles allow for extremely fluid rotations, the latter all directed by a 35 Mhz remote control unit which commands three small servo motors. On the frame, an analogue 35 mm SLR camera (Nikon F70) is mounted. By adapting the camera's remote shutter release, completely automatic focusing is made possible as well as ultimately releasing the shutter to take the photograph (both actions are also controlled remotely).

Figure 7 *The cradle with mounted camera*

- Monitor – video live link

Framing the shot correctly is enabled by the Pro X2 (www.hicam.com.au/pro_x2.htm), a very handy video live link system. Using a tiny 4,8 V Hi Cam EO5-380 CCD camera mounted onto the camera's eyepiece, the image that will be captured by the camera is remotely sent to a ground-based monitor, where an archaeologist can decide whether or not to take the image. Thanks to its maximum compactness and minimum weight, the Pro X2 system is ideal for R/C aerial photography. Moreover, most commercial equipment is not capable of sending images further than 100 m line-of-sight. Due to the 200 mW of output power, this 2,4 GHz transmitter enables to send a signal over about 300 m line-of-sight.

Although humans are incapable of seeing something through the camera when the infrared filter is mounted, the CCD of the little camera allows to 'see through' the infrared filter because digital sensors are sensitive to near infrared. This way, the picture that the analogue film will capture is instantaneously shown on the small (7 in) TFT monitor (figure 8), making the archaeologists on the ground aware of the captured features.

- Fishing reel and rod

One can imagine the tractive force of the helikite when flying even in moderate winds. Using a big game fishing reel and accompanying sea-fishing rod, these forces can be managed reasonably well and, on the other hand, allow the helikite's "pilot" to freely walk around. Fixing a large and solid winch to the ground could be more convenient to pull down the helikite, but severely restricts steering and risks the penetration of archaeological layers (e.g. floors).

Figure 8 *Steering the camera by assistance of the video live link*

Problems and Terra Incognita

When combining all these elements, a complete HAP system is created (figure 9) which almost completely fulfils the initial aims.

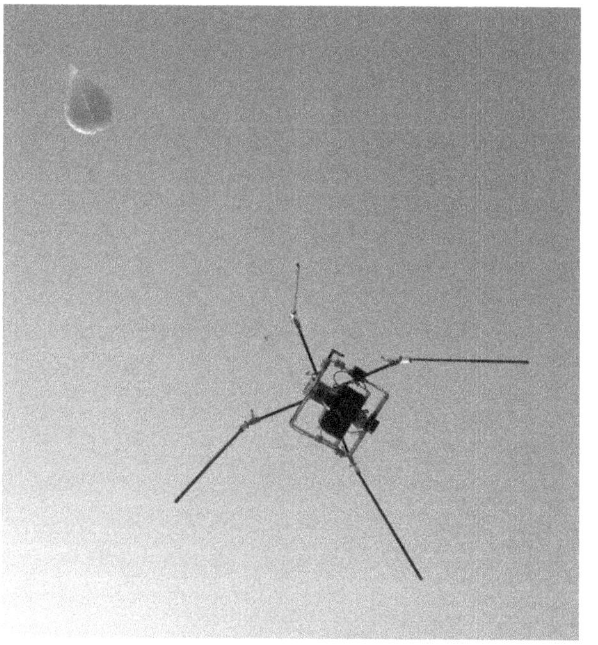

Figure 9 *HAP in action*

However, as this is still a prototype and the first tests were only executed in the summer of 2006, it is likely that changes and improvements will have to be made. One of the first issues already encountered was the communication between the helikite pilot and the person taking the photographs. Bad communication will certainly yield bad photographs. Furthermore, as all equipment is battery-driven, spare (and fully charged!) batteries are a necessity when working in remote areas. Furthermore, a good pair of gloves is essential as well, because fast running tether can burn or severely cut exposed skin when performing KAP or HAP. It also still remains unsure whether or not the carbon legs will have a major influence on the rig's stability when flying in high winds. Finally, it must be admitted that the cost of helium was underestimated. As helium comes with a rather high price tag, filling a helikite of 7 m³ time after time is very detrimental for the project's budget. Consequently, a large trailer which can store the whole system – partly or completely inflated – will be bought.

Archaeological HAP Applications

It must be obvious that HAP will not only be used to take infrared photographs.[1] For instance, the system will also be used as a research instrument to look at the information embedded in the ultraviolet spectrum. Moreover, it allows for detailed photography of archaeological *loci* – something which is more (or too) difficult to achieve when photographing from an aircraft at 300 m altitude. In close relation with the previous application, it is hoped to put HAP into action on excavations (figure 10). By taking two overlapping images from a different point of view (i.e. a stereopair) of the excavation area, photogrammetric software allows to extract an orthophoto (i.e. a completely distortion free and true-to-scale photograph) on which accurate measurements can be carried out. It is hoped that this approach will turn out to be a possible solution for the always occurring accuracy, precision, cost and time issues concerning archaeological excavation mapping.

Figure 10 *HAP on an excavation*

To even further increase the flexibility of acquiring aerial imagery, a digital camera will be implemented into the system. Removing its internal IR cut filter (the so-called *hot mirror*) to make the sensor capture infrared radiation, will allow to directly take near infrared photographs, hence removing the buying, processing and scanning stages as well as the cost of purchasing and processing the film. When taking two individual overlapping images will prove to be unpractical, a stereo rig using two cameras can also be considered. This way, stereopairs would be created automatically without having to manoeuvre the helikite.

What the future will bring for HAP may be uncertain, but it can be stated that from now on HAP will allow the Ghent University researchers to look at archaeology through black-tinted glasses.

Acknowledgements

This paper arises from the first author's ongoing PhD which studies the application of remote sensing in archaeological surveys. The research is conducted with permission and financial support of the Fund for Scientific Research - Flanders (FWO) and supervised by Professor dr. Frank Vermeulen (Department of Archaeology and Ancient History of Europe - Ghent University). Finally, the author's friends Wouter Gheyle and Wouter Van Hecke are acknowledged for taking pictures when testing the helikite and proofreading the article respectively. All errors and misconceptions remain, of course, the author's own responsibility.

Bibliography

ALLSOPP HELIKITES Ltd, 2006, *Allsopp Helikites Limited.*
www.allsopphelikites.com; consulted on 18/02/2006.

BEUTNAGEL, R., WOLFGang, B. & BÖHNKE, O. 1995, Picavet – Past & Present, *The Aerial Eye* 1(4), pp. 6-7 & 18-19.

CLARK, R. 1999, Spectroscopy of Rocks and Minerals, and Principles of Spectroscopy, in: Rencz, A. (ed.), *Remote Sensing for the Earth Sciences*, pp. 3-58, New York. (= Manual of Remote Sensing 3)

FREEDMAN, R. & KAUFMANN, W. 2005, *Universe*, New York.

Hi Cam, 2005, *Pro X2 Systems.*
www.hicam.com.au/pro_x2.htm; consulted on 03/12/2005.

LOENDERS, J. 2006, *Module voor luchtfotografie*, Ghent. (Unpublished dissertation KAHO Sint-Lieven)

WALKER, J. 2004, *Physics*, Upper Saddle River.

[1] Due to the fact that the film development was not finished at the moment of writing, no infrared pictures could be included in this article.

Integrating aerial photography for the study of Roman towns in Italy: a case study from the Adriatic area

Frank Vermeulen[1]

Ghent University, Belgium

1 Introduction

It is widely accepted that the study of towns is a key to any understanding of the Roman World, since the urban centres were the nodes through which the administration, economy and dominating culture were negotiated (Keay et al. 2004). Among Roman archaeologists involved in intensive urban research it is agreed that there is a firm need for the collection of more data about the internal organization and chronology of the full range of urban sites of that era. As much of our knowledge about Roman urbanism in Italy and other parts of the Empire is based on an uncertain sample, of well-preserved but atypical towns such as Pompeii and Ostia, there is a need for more systematically collected data, different from the large body of earlier work on Roman towns, which far too often focused upon individual buildings decontextualized from the rest of their urban environment. Furthermore, a major part of Roman urban archaeology is constrained by medieval and modern developments in those towns that remained occupied today.

One of the aims of this paper is to focus attention upon the potential of the carefully defined strategic use of non-destructive field techniques and especially systematic aerial reconnaissance in investigating such ancient city sites. This implies also careful selection of Roman urban sites to study, selecting within the still available range of abandoned sites with well preserved subsoil structures that allow a fully integrated geo-archaeological approach. To sustain this choice and thus to contribute to the broader historical-archaeological debate on Roman urbanization, we present the most recent results of a project that examines urban settlement patterns in the Potenza valley. Our work aims to complement earlier town-based research in a region of Italy, the Marche, which has a long and distinguished tradition of historical and archaeological work on the development of Roman towns (Luni 2003). This research provided important and detailed information about the development of parts of individual towns and town buildings, involving both the painstaking excavation and publication of particular structures, and the analysis of material evidence, such as that of pottery, inscriptions and sculpture. Our approach, by contrast, provides broader information for a series of urban settlements located in a central area of Marche. We are interested in the patterns of urban development, distribution, size and form and believe that new information can be collected from intensive and integrated surface survey and fieldwork rather than excavation or detailed artifact studies. If detailed excavation evidence is available it must, however, be fully used. Integrating data correctly, can advance our knowledge of the topography and internal organization of the ancient towns and contribute to a wider vision of urban development in the Roman era. Our project complements previous work by providing a broader context for the interpretation and comparison of individual sites.

Elsewhere it has been argued that systematic surface survey and large-scale geophysics contribute to our understanding of Roman urban topography, as was recently demonstrated in several town sites in the Tiber valley (Keay/Millett *et al.* 2000; Keay et al. 2004). Both techniques are easy to use and allow archaeologists to cover substantial areas in a short time span. As we can now demonstrate, this is also true for a detailed investigation of available aerial images and especially for an intensive monitoring of the urban site with the help of low-altitude aerial photography during different seasons and years. It is our firm belief that a simultaneous application of these non-destructive techniques with feedback from applied geomorphological research creates a method ideally suited to characterize the extent, organization, and chronology of urban sites typical of central Italy and other parts of the western Mediterranean. By contrast, the traditional strategy of attempting to excavate urban landscapes is expensive and not always of great scientific value. The Potenza Valley Survey project builds upon these experiences and attempts to refine the non-destructive exploration of Roman urban landscapes.

2 An Adriatic case study: *Potentia*

Between 2002 and 2005 intensive archaeological field survey work has been carried out at the site of the Roman colonial city of *Potentia*, founded in 184 B.C. in the central part of coastal Adriatic Italy, just south of Ancona. This town survey is part of the long term Potenza Valley Survey Project (PVS) of Ghent University (Department of Archaeology). Apart from rural surveys, which were the main focus of the first phase (2000-2003) of the project (Vermeulen/Boullart 2001; Vermeulen/Monsieur/Boullart 2002; Vermeulen *et al.* 2003; Vermeulen *et al.* 2005, Percossi/Pignocchi /Vermeulen 2006), intensive intra-site fieldwork is now also being carried out on the main protohistoric centres (Picene culture) in the valley, as well as on the Roman towns of the Potenza corridor, from West to East: *Septempeda* (San Severino Marche), *Trea* (Treia), *Ricina* (Villa Potenza) and *Potentia* (Porto Recanati). All of these sites have reasonable to excellent potential for surface survey and because of the small scale of excavation work previous to our investigation of the area, much can be learned from an integrated survey approach. Our surveys are designed to improve general understanding of the topography of proto-urban and urban settlement in this part of Italy and to deepen our knowledge of the processes of early (mostly Roman) urbanisation in the Marche. They are intended to integrate the results of non-destructive survey methods, such as oblique aerial photography and airphoto-interpretation, geophysical surveys, geomorphological surveys, detailed topographic measurements and intensive field walking. The results of small-scale, previous or on-going, excavation work on these cities and artefact studies of survey materials and excavation finds are being integrated in this picture, as they progressively become available.

As the results of intra-site survey of *Potentia*, situated in the sample area of the lower valley, have now reached a good level of detail and precision, this coastal city is the first to be presented in a more comprehensive manner. Like almost half of all Roman towns in this part of *Picenum*, from Late Antiquity onwards *Potentia* became 'unsuccessful' in some way and was completely abandoned (Alfieri 1981 and 1985). As almost no present-day habitation covers the former town area and because agriculture still dominates the zones where its ruins lie buried, the ancient topography of the city can effectively be studied by means of our survey methodology. Up till now most survey efforts were concentrated on the all-important intra-mural part of this Roman city, where the conditions for such an integrated analysis are now fulfilled, and on the impact of aerial photography on our results.

From the beginning of the project a program of intensive flying and oblique aerial photography has been developed, with yearly campaigns of systematic recording between the months of April and October. Although the methodology applied here has been explained elsewhere (Vermeulen 2004 and Vermeulen/Verhoeven/Semey 2005), a few practical aspects of our use of this active survey method can be underlined. Flying with a small aircraft piloted by members of the Aeroclub of Ancona is undertaken from several airbases in the area. Most photographs are taken with the help of standard reflex cameras (35 mm) for colour slides and a digital camera (since August 2004). The area of *Potentia* has been continuously monitored from the air, and especially intense flying during the dry spring of 2003 has produced remarkable images, locating and visualizing many aspects of the urban topography and of the suburban landscape, including the discovery of 2 cemeteries, 4 roads connected to the town and several extra-mural settlement areas. Some flights also made it possible to monitor the many geomorphological traces connected with the changing course of the river and the precise location of the beach ridges and lagoons. The flights over the original intra-mural area of *Potentia* resulted in some excellent aerial views of crop and soil marks. The latter are mostly the result of ploughed up larger stone structures, such as the city streets and remains of the circuit wall, but also of a combination of ploughed up occupation layers, zones with different quantities of organic substance in the upper layers and humidity traces caused by differential drying of the soil in some parts of the town. Good results derived from detecting crop marks are connected again with the colony's street grid and circuit wall, but in several areas also with parts of building structures, especially in the southern sector of the city.

The registration and further manipulation of the aerial data has been done in a systematic way, with the use of an MS Access database linked to an original GIS that was elaborated within the project. Important efforts are being undertaken by G. Verhoeven to enhance, rectify and map the most relevant archaeological features. For this mapping and interpretation work we also make use of some existing vertical photography of the area, such as pictures from a 1947 RAF flight over the area and good Italian vertical imagery (made by the Istituto Geografico Militare) of 1972 and 1974. More recent vertical photographs, such as those from a 1992 flight organised by the Regione Marche, provide much less information, because of obvious changes in the landscape and surface situation. The integrated GIS mapping of all the aerial data is facilitated by the use of *Airphoto 3.x*, software specifically developed to rectify archaeological images made with handheld non-calibrated cameras (Scollar 2002, 167).

Although *Potentia* was first identified in the 1940's, and some punctual excavations have since revealed some of its topography and stratigraphy, this new phase of remote sensing allows now the elaboration of a detailed plan of the urban pattern. Our results identify the location of the town defenses, including the gates, the street grid, the forum, several other monumental complexes and many elements of city housing. Although all newly discovered structures seem at first sight to be only "superficial" evidence, their contextualization as a result of fine grid walking and subsequent artifact studies and their integration with partly re-studied excavation data and with the very valuable historical information about the early years of *Potentia*, allow us to make major progress in the archaeological and historical research of the topography, urbanization and settlement history of this Adriatic town, building further upon the excellent recent synthesis of Percossi Serenelli (2001), who focuses on the excavation data. The impact of the remote sensing information is most visible in the following aspects of the site (Fig. 1)

Location and landscape context

The image analysis, in conjunction with field data, demonstrates that the narrow ancient longitudinal beach ridge, lying parallel with the coastline and just north of the Roman mouth of the Potenza stream, partly determined the exact location and size of the colony. Its morphology also affected the results of the aerial

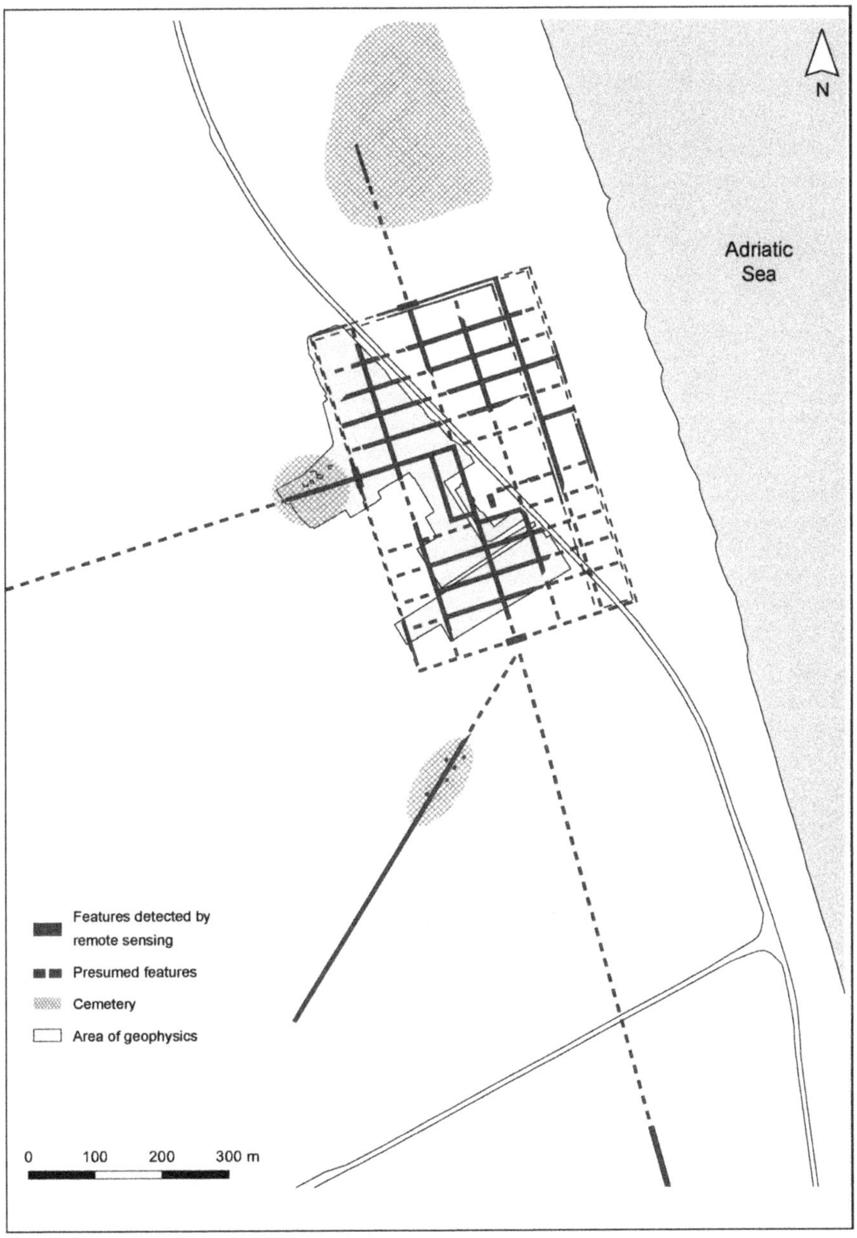

Figure 1

photography: the streets and other structures immediately near the town borders were not identified from the air, because they lie deeply buried under alluvial layers of clay on the edges of the beach ridge. Their remains cease to affect the growth pattern of the grass seen on the surface. This phenomenon is particularly obvious in the still lower lying southern and south-western part of town. Here the arable fields are much eroded due to post-Roman flooding of the river Potenza in this area, an observation again well attested by way of remote sensing and confirmed by geomorphological and geophysical survey and further underlined by the sharp reduction in surface artifacts towards the south.

City defenses

As already partly identified by Moscatelli (1987) on the basis of older vertical photographs, the town defenses of *Potentia* can now be distinguished better due to the many new data. Close re-examination of these images in GIS-overlay with our excellent 2003 oblique aerial photographs and with the information from the magnetic surveys, demonstrates that the situation is quite complex. Although we must remain very careful when considering the diachronic aspects of these defenses, as no regular stratigraphic excavations have touched upon them, we would like to put forward the idea of three major phases in the development of the *Potentia* circumvallation. During the second phase of the town development, possibly initiated ten years after the colony's foundation, the site was fully urbanized and adorned with a circuit wall. The observed structure of an ashlar wall seems to delimit a very regular, rectangular area of some 525 by 343 m or almost 18 ha (measured *intra muros*). In the ancient Roman system this represents some 15 by 10 *actus*. When we include the walls in the total city surface we obtain almost 19 ha. Our aerial information suggests that later (from the reign of Augustus onwards?) this intra mural area was enlarged some 50 m to the east. The initial eastern side of the city wall was then replaced by a street – evidenced by a larger and more pronounced trace in the aerial photographs than most other streets – and the new wall (and ditch?) was built on the edge of what is now a local coastal road bordering the current beach area. This hypothesis, to be verified in the field, would certainly explain well why the sector in the north-eastern corner of the city, excavated by Mercando (Mercando 1979), was only inhabited from the Augustan era (late 1st century B.C.) onwards. It would also explain well why in our aerial photos the E-W streets of the housing blocks in the eastern periphery of the town have a different look than the other city streets.

City gates

Whatever the exact phasing of the defenses may be, it is likely that the town wall knew three major gates, more or less centrally placed in their respective town sides, namely North, West and South. Thanks to our surveys the northern and western gates are now well located in the field, and only await confirmation by excavation. For the northern gate there is not only evidence from several aerial photographs, displaying a clear breach in the city wall and ditch traces, but the grid walking of the plowed field in this area revealed here a dense surface concentration (15x10 m large) of building materials, consisting of many tiles, some pieces of white limestone and several large river pebbles. Some aerial photography traces and especially geophysical evidence hinted at the location and quite complex phasing of the second entrance to town, the western gate. Combining both magnetometer and electric resistivity approaches in this area has revealed a 15m long E-W oriented zone full with anomalies which would seem to indicate the presence of high concentrations of building material and of a complex gate system.

The third gate, in the southern part of the wall, can also be quite well positioned, despite the fact that it is located in an area which is now much eroded by post-Roman stream movements. This southern town exit links up neatly with a road leaving the town area towards the SW to reach the main Roman bridge over the river Potenza (Vermeulen/Verhoeven 2004; Lilli 1999). This road, and also the major roads leaving the northern and western gates, have all been discovered and well mapped thanks to the study of vertical photographs and especially with the help of our low altitude flying. Thanks to these aerial surveys and connected ground control we now know that all three roads leaving the city gates were bordered by cemeteries with funeral monuments facing the roadway.

Local street network

The located gates link up well with the regular NNS-SSE oriented street system. Several tracts of this orthogonal street pattern, mostly seen on our aerial photographs as some 5 to 6 m wide, pale and linear crop marks and soil marks, were well identified on the ground and were complemented by the results of the geophysical surveys in parts of town. Most tracts are characterized in the plowed surface as clear concentrations of river pebbles mixed with some artifacts, mostly pottery and tile. According to the evidence from the excavations near a temple complex (Percossi Serenelli 2001) most streets are probably some 5 m wide and normally consist of a surface of battered river pebbles. The digs also indicate that at least some of them had one or two sidewalks and brick sewers, the latter lying underneath the axis of the road-deck. According to our aerial photography interpretation the E-W oriented somewhat broader (7-8 m) street segments in the younger (?) peripheral eastern part of town seem to be even simpler, with a road-deck of battered earth and a stone lining. The two dominant main N-S and E-W street axes of the planned Roman town, respectively called *cardo maximus* and *decumanus maximus*, were probably constructed with much more care. It is likely that both streets, which link up nicely with the town gates, were paved. This can not only be deduced from their different aspect in the crop marks seen from above, but also from the stronger linear positive anomalies seen in the magnetometer survey. As the nature of their traces is not continuous we may presume that the pavement of these main city roads did not completely survive post-Roman spoliation of the building materials of the town.

The very regular parallel city-streets, as we can now map them with some confidence, subdivide the urban space into housing blocks *(insulae)* of different size. It is difficult to calculate the exact number of blocks of *Potentia* as we still lack details about some areas. It is also not clear yet whether we should interpret some streets as being fully part of the major grid, or just as local subdivisions of *insulae*. For the moment we hypothesize the presence of a total number of 51 to 57 *insulae*.

The forum

It is the great merit of the 2004 geophysical surveys in the large zone adjoining the excavated temple area (Percossi 2001) to have located the forum and to have evidenced that the central *insulae*, around the junction of the main *cardo* and *decumanus*, represent the civic center of *Potentia*. The magnetometer results are very explicit in situating the forum square, a long rectangular and almost N-S oriented open area of some 120 x 30 m, immediately SW of this junction. It seems to have been bordered on both long sides by rows of shops fronted by a portico. The magnetic anomalies testify that several large public buildings surrounded the forum, but their precise nature still has to be established. As this part of the site is taken in by grassland, aerial photography has proven to be of little help here.

Housing

Thanks to the magnetometer surveys and intensive site monitoring from the air, fragments of buildings are visible all over town, as a rule neatly lined according to the street grid. For this type of evidence, which is generally less pronounced than city walls, broad streets and large public buildings, it was necessary to constantly confront and integrate the data from these two approaches. Thanks to this effort, and despite the ascertainment that the magnetic surveys did not produce very clear evidence of housing structures, a good number of partial plans of houses is now discernable. In the most prominent groupings of traces, some of which were clearly linked to the main N-S street axis, we can distinguish remains of large buildings, probably single family residences (*domus* type), with several rooms grouped around what seems to be a central courtyard. In a few cases where the overall plan of the buildings remains obscure, a series of rectangular patches in the crops, visible from the air, could be interpreted as individual floors. As these particular plots often coincide with concentrations of mosaic *tesserae*, a certain quality of architecture is indicated. In several areas simple rectangular structures, facing the street with one of the short sides, suggest the presence of shops and modest houses of the so-called *taberna*-type.

3 Conclusion

Despite the success of the magnetometer and resistivity surveys, the aerial photography from low altitude, the manipulation of vertical images and the analyses of these data in a GISmodel, provide a series of images of the buried remains and it must be stressed that they can reveal little with regards the chronology or specific function of buildings detected (Keay et al. 2000: 9; Vermeulen/Verhoeven/Semey 2005). Viable hypotheses about functions and even general chronologies of Roman urban patterns remain problematic. This apparent weakness of the fieldwork is partly solved by our systematic fieldwalking efforts in grids over large sectors of town, contextualizing the imagery of the remote sensing. Together with further integration of the valuable stratigraphic data from different excavations in the town area, this work has proven to be of crucial importance for our understanding of the city topography and part of its evolution. We are convinced that this integrated methodology can be widely applied beyond the geographical and chronological limits of Roman archaeology and should be particularly useful on all types of large open archaeological sites with some form of regular or understandable patterning. It is particularly clear from the research in *Potentia* that the remote sensing information is crucial in this process. We are therefore convinced that many 'urbanistic studies' elsewhere in Italy and beyond still undervalue the potential of intensive active flying over the ancient urban sites.

References

ALFIERI, N. 1981 Le Marche e la fine del mondo antico. *Atti e Memorie della Deputazione di Storia Patria delle Marche.* LXXXVI: 9-34.

ALFIERI, N. 1985 Le Marche e la fine del mondo antico. *Atti e Memorie della Deputazione di Storia Patria delle Marche.* LXXXVI: 109-24.

KEAY, S., MILLETT, M., POPPY, S., ROBINSON, J., TAYLOR, J. & TERRENATO, N. 2000 Falerii Novi: A New Survey of the Walled Area. *Papers of the BSR.* 68: 1- 93.

KEAY, S., M. MILLETT, S. POPPY, J. ROBINSON, J. TAYLOR & N. TERRENATO 2004 New Approaches to Roman Urbanism in the Tiber Valley, in: PATTERSON, H. (ed.), *Bridging the Tiber. Approaches to Regional Archaeology in the Middle Tiber Valley. Archaeological Monographs of the British School at Rome.* 13. London: 224--251.

LILLI, M. 1999 Caratteristiche tecniche e strutturali di alcuni ponti tra V e VI regio. *Picus.* XIX: 107-38.

LUNI, M 2003 *Archeologia nelle Marche dalla preistoria all'età tardoantica.* Firenze: Nardini Editore.

MERCANDO, L. 1979 Marche. Rinvenimenti di insediamenti rurali. *Notizie degli scavi di Antichità.* 33: 89-296.

MOSCATELLI, U. 1987 Materiali per la topografia storica di Potentia. In: PACI, G., *Materiali di Studi Marchegiani Onori di Fabo Allevi.* Agugliano: 429-38.

PERCOSSI SERENELLI, E. (ed.) 2001 *Potentia. Quando poi scese il silenzio... Rito e società in una colonia romana del Piceno fra Repubblica et tardo Impero.* Porto Recanati, Federico Motta Editore.

PERCOSSI, E., PIGNOCCHI, G. & VERMEULEN, F (eds.) 2006 *I siti archeologici della Vallata del Potenza*, Ancona.

SCOLLAR, I. 2002 Making Things Look Vertical. In: BEWLEY, R. & RACZKOWSKI, W. (eds.) *Aerial Archaeology. Developing Future Practice, NATO Science Series.* Series I: Life and Behavioural Sciences, 337: Amsterdam ,166-72.

VERMEULEN, F. *et al.* 2003 The Potenza Valley Survey: preliminary report on field campaign 2002. *Babesch*, 78: 71-106.

VERMEULEN, F. 2004 Fotografia aerea finalizzata nelle Marche Centrali: Une progetto integrato. *Archeologia Aerea. Studi di Aerotopografia Archeologica.* Libreria dello Stato. Roma. 91-118.

VERMEULEN, F. & BOULLART, C. 2001 The Potenza Valley Survey: preliminary report on field campaign 2000. *Babesch,* 76: 1-18.

VERMEULEN, F., MONSIEUR, P. & BOULLART, C. 2002 The Potenza Valley Survey: preliminary report on field campaign 2001. *Babesch*, 77: 49-71.

VERMEULEN, F., MONSIEUR, P., BOULLART, C., VERREYKE, H., VERHOEVEN, G., DE DAPPER, M., GOETHALS, T., GOOSSENS, R. & DE VLIEGHER, B.M. 2005 The Potenza Valley Survey:

Preliminary Report on Field Campaign 2003, *Babesch* 80: 33-64.

VERMEULEN, F. & VERHOEVEN, G. 2004 The contribution of aerial photography and field survey to the study of urbanization in the Potenza valley (Picenum). *Journal of Roman Archaeology*, 17: 57 – 82.

VERMEULEN, F., VERHOEVEN, G. & SEMEY, J. 2005 The integration of Aerial Photography and GIS in the Potenza Valley Survey. In: Bourgeois, J. and Meganck, M. (eds.), *AP&A 2003. A Century of Information*, Gent, 371-382.

Ancient Maps – modern data sets: different investigative techniques in the landscape of the Early Iron Age princely hill fort Heuneburg, Baden-Württemberg

Jörg Bofinger,[1] Siegfried Kurz,[2] Sascha Schmidt[1]

[1] Landesamt für Denkmalpflege, Regierungspräsidium Stuttgart, Berliner Str. 12, D-73728 Esslingen, Germany
joerg.bofinger / sascha.schmidt@rps.bwl.de

[2] Institut für Ur- und Frühgeschichte, Universität Tübingen, D-72070 Tübingen, Germany
siegfried.kurz@gmx.net

1 Introduction

The Heuneburg is an important Early Iron Age (Late Hallstatt culture, Ha D1-D3) hill fort in the Danube Valley, Upper Swabia, Germany. In the archaeology of the European Iron Age this site is known for a detailed stratigraphy (several phases of settlement and fortifications) and the famous mud brick wall. Links with the Mediterranean during the 6th century BC are proofed by attic pottery and imported amphorae from the Greek colony of Massilia (Marseille).

Several princely burials in the surroundings nearby are dating to the same period. The Iron Age fort was abandoned during the early 5th century BC. Excavations of a great part of the Heuneburg plateau since 1950 - between 1963 and 1979 under the direction of W. Kimmig and E. Gersbach from the University of Tübingen – and the discovery of the famous mud brick-wall on a stone foundation made this site eminent and important in European prehistory (Kimmig 1989; Gersbach 1989).

2 New Research activities at the Heuneburg

During the 1980ths and 1990ths there was only very little archaeological fieldwork in the wide area of the "Heuneburg-Außensiedlung" (peripheral settlement). But the excavated zone under the burial mounds of the Giessübel-Talhau necropolis (Kurz 2000), intense survey activities and some test excavations demonstrated that the settlement structures in the areas surrounding the Heuneburg are surprisingly complex and vast (Kurz 2001). Furthermore, the excavation of the new parking place for the archaeological park, situated northwest of the Heuneburg, showed clearly that settlement structures of Early Iron Age habitat are also located immediately in front of the fortified hill (Reim 1999). Another important result of these activities was the insight that the defence system of the Heuneburg precinct, until these days considered to be of medieval in origin, dates probably into the early Iron Age phases of the Heuneburg (Reim 2003).

In the year 2004 a national priority research program on the late Hallstatt princely hill forts in Middle Europe was established (Krausse 2004). In this program, financed by the German Research Foundation (Deutsche Forschungsgemeinschaft - DFG), projects started on the peripheral settlement of the Heuneburg, the Heuneburg precinct and its relation to interregional long trade systems. Intense archaeological fieldwork was initiated. Supplementary archaeological research techniques for a detailed study of the micro topography of the landscape in the upper Danube valley and its environments today and in former times are closely related to these excavations. Therefore, different methods and techniques are applied side by side: for example the study of LIDAR-scans of a large area around the princely hill fort or aerial photography research in combination with geophysical prospecting methods give new insights into the understanding of complex settlement activities during the late Hallstatt period (7th – 5th cent. BC). The combination of these investigative techniques and excavation offers a large variety of information on archaeological structures and development of the landscape surrounding the Heuneburg.

3 LIDAR – detailed mapping of the Danube valley and the archaeological remains in the landscape around the Heuneburg

Preceding the excavations which started in spring 2004, a good opportunity was seized to collect new data on the landscape surrounding the Early Iron Age hill fort: After the very hot and dry summer of 2003 the old branches of the river Danube could be seen very clearly in the valley and even small elevations and structures in the relief of the landscape became apparent because of low ground water level. Therefore, a high-resolution airborne LIDAR scan of an area of 20 km^2 (Fig. 1) was applied for a very detailed acquisition of data of the known archaeological sites themselves. The second aim was to get a solid basis of well directed prospecting and mapping. LIDAR – acronym for LIght Detection And Ranging is an application with multifarious functions. It can be based on satellite, airplane or terrestrial for various purposes (http://www.ghcc.msfc.nasa.gov/sparcle/sparcle_tutorial.html). The archaeology in Baden-Württemberg does use terrestrial 3D-Laserscanning and ALTM (Airborne Laser Terrain Mapper). We use LIDAR exclusively to create high resolution DEMs as those used to make flooding simulations (http://www.fht-stuttgart.de/fbv/fbvweb/veranstaltungen/ESRI/Vortraege_2004/ESRI_2004_Panczak.pdf).

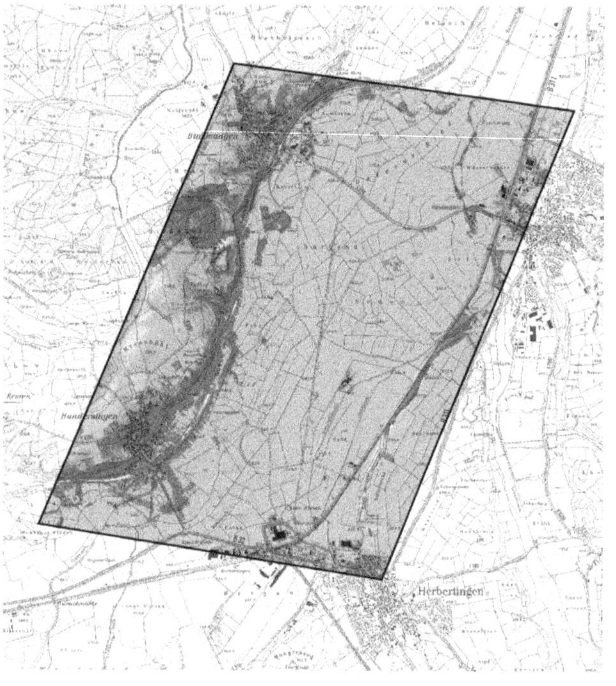

Figure 1 *Part of the topographic map 1: 100 000 with entry of the LIDAR mapped area (20 km^2). (with kind permission Landesvermessungsamt Baden-Württemberg).*

The elevation model is based on the last-pulse data. It is assumed that the last-pulse gives us the biggest chance to depict the surface. There are methodological problems in calculating a correct DEM in bushy areas and on craggy slopes. These problems are not sufficiently solved at the moment. The surface DEM is also biased by modern structures. They are mathematically removed and replaced by an interpolated plain. Especially big structures like factory buildings and sports halls remain visible because of the uncharacteristic smooth surface.

The modern impact on the surface by brickyards and quarries could be demonstrated at different places in the Rhineland (Gerlach et al. 2005). Therefore the modern DEM does not represent the historical surface, even in areas with low "Reliefenergie" (measuring unit for the potential erosion degree). It is difficult to identify historical structures of different periods. For example, modern clay pits show high similarity to the historical pit mining on iron. Comparable problems are known from the aerial archaeology and need not to be discussed here in detail, but it should be mentioned that the methodological limitations of the aerial archaeology have to be considered for LIDAR as well.

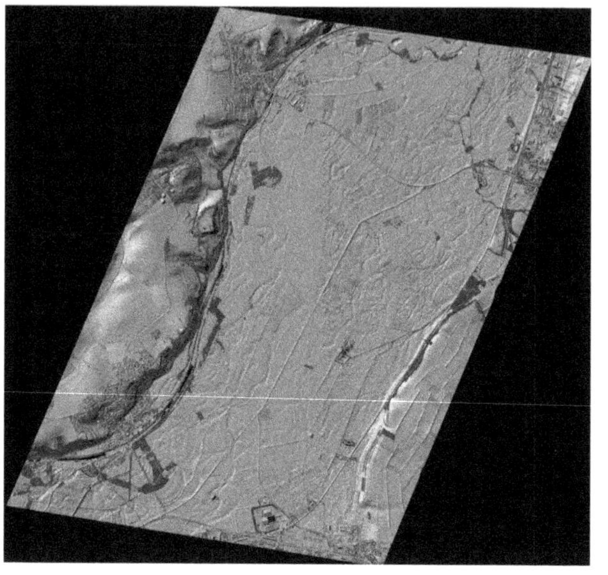

Figure 2 *High resolution shaded relief map of the Danube valley near the Heuneburg.*

Figure 3 *Aerial photograph of the hook-shaped rampart system in the fields northwest of the Heuneburg.*

The LIDAR data set of the selected area around the Heuneburg, collected by TOPOSYS Ltd., Biberach, in September 2003 was recorded in a 0.5 m point spacing grid from 500 m altitude. The result after a four–hour flight was a high resolution DEM (Fig. 2). Even very inconspicuous details such as the little elevation of an ancient rampart system in the fields northwest of the Heuneburg are appearing very distinctly in the shaded relief. This hook-shaped rampart is hardly discernible in the landscape anymore; only the ditch in front of the rampart is clearly visible on aerial photographs (Fig. 3). For the identification of archaeological remains in the fields aerial photography is still very important as shown by the example above. But in the woodlands aerial archaeology is confronted with limiting factors. In this case, the LIDAR airborne scanning is a good and helpful replenishment as shown in Fig. 4. There are various ways to remove the vegetation cover automatically by special algorithms. Even under canopies not only the burial mounds are clearly visible, but also the Late Celtic

"Viereckschanze" next to the "Hohmichele", one of the largest burial mounds in Middle Europe.

While the area of 20 km^2 around the Heuneburg was scanned in very high resolution (point spacing 0.5 m) on special demand, the participation in the Culture 2000 program "European Landscapes – past, present, future" offered the possibility to enlarge the field of work. A LIDAR data set provided by the national topography authorities of Baden-Württemberg of a wider area east of the hill fort has been purchased. These LIDAR data – part of the complete Baden-Württemberg record – are available in 1 m gridwidth. On this basis a good possibility for comparison of resolution of both data sets is given.

	DTM project Baden-Württemberg	Data collection on demand
point spacing	1.5 m	0.5 m
elevation accuracy	± 0.15 m	± 0.05 m
coordinate system	ETRS 89/UTM	WGS 84
costs	1 : 7	

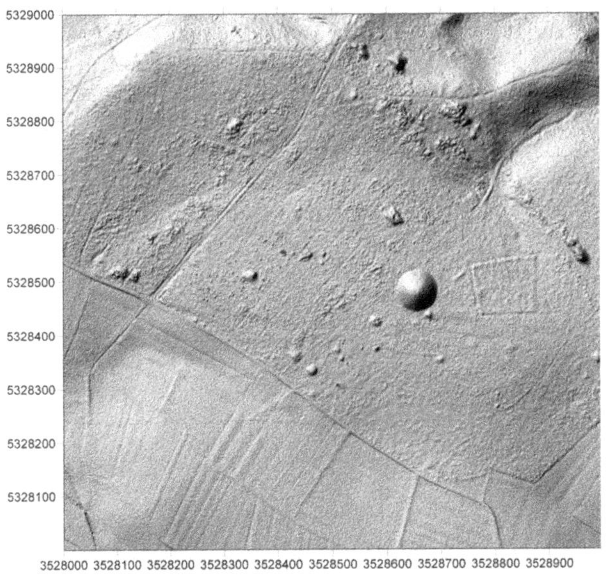

Figure 4 *LIDAR scan of the necropolis "Hohmichele" northwest of the Heuneburg.*

A detailed study of the accuracy of these two LIDAR data with different gridwidth resulted in the insight, that for the identification of archaeological structures the cheaper solution would normally be sufficient. For visualizations of archaeological remains this is a good and cheap possibility. For further investigations, it means to make daily work with LIDAR in archaeology economically more profitable. It is especially useful to employ the high resolution LIDAR data in the case of extremely detailed and sensitive structures such as small walls or wooden elements only. Exceptions to these are very prominent impacts on the landscape due to present changes, as for example flooding or landslides. Then also, a very contemporary recording of "updated" data by special LIDAR flights is recommended.

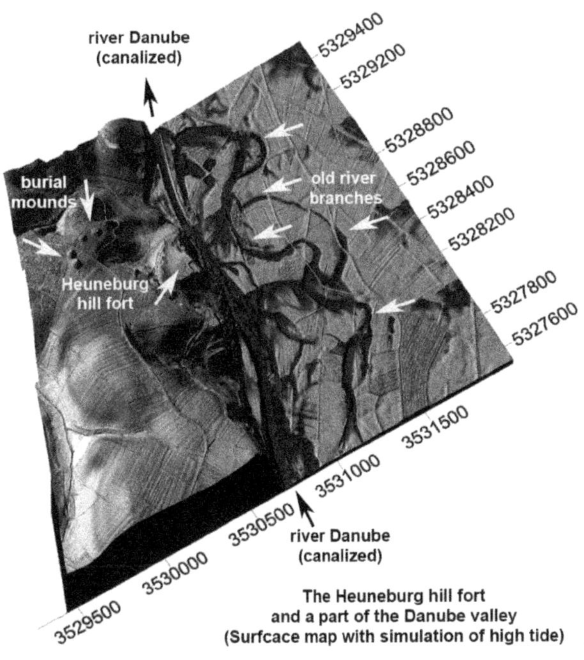

Figure 5 *LIDAR based block model of the Danube valley near the Heuneburg with simulation of flooding.*

As concerns the Heuneburg and the situation near the Danube valley, incidentally, these LIDAR data offer good possibilities for scenic reconstruction and determination of flood areas in the river valley (Fig. 5). Three-dimensional terrain block models based on the LIDAR data could be combined with simulations of high tide and flooding of the river and its impact on the surface. On the one hand this method offers very impressive images of different natural situations. On the other hand the possibility to recognize presumed settlement areas and traffic routes in the valley itself is very important for the evaluation of prehistoric settlement patterns. For example, there is a Hallstatt settlement located near the "Bettelbühl" necropolis (Kurz/Schiek 2002, 91). Both sites are situated on a gravel ridge. This ridge is raised above the flood area during high level of the Danube. Furthermore, the known course of the roman road leading through the Danube valley is also situated on this non flooded ridge. As it is very likely that this roman road is established on a much older long distance trading route, used already in prehistoric times, this zone has to be regarded as a high potential area for further settlements situated near this connection of the Heuneburg to long haul channels of trade.

4 Identification of ditched enclosures in the "peripheral settlement" of the Heuneburg

In the area of the front end of the hill fort and the Heuneburg peripheral settlement different systems of ditches left their marks on the landscape. They are partly known since the excavations of the high prestige burial mounds of the Giessübel-Talhau necropolis. Now, recent field activities and an intense prospecting program have revealed a very complex system of these ditches, still visible in part as landmarks north and south of the Giessübel-forest (Kurz 2006). In particular the wood covered areas preserved the visible structures, for example, in form of uneven grounds.

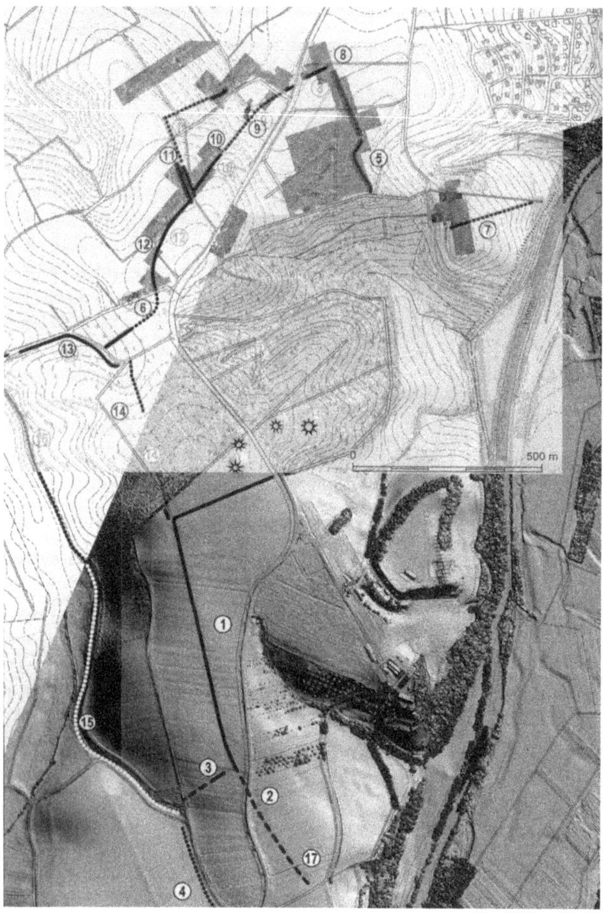

Figure 6 *Map of the ditches of the peripheral settlement of the Heuneburg and the areas prospected by geophysics.*

And yet, as great sectors of this system are not visible in the landscape, geomagnetic prospecting has been initiated since autumn 2003. The aim was to verify the courses of the different parts and to get an impression whether they have formed an integrated system. Even if the soil in this region is not very favourable in some places for this method, it was nevertheless possible to follow the lead of some ditches. As one result, there is now a complete map of theses structures (Fig. 6) which shows, that the system of ditches extends over an area of 2,4 km length (north-south).

Furthermore, a look at the official cadastral maps of the area, dating to the 1920s, is very helpful: These maps are confirming the course of some ditches, because the borders of plots of land and limits of woodlands in some cases are oriented along these trenches. This would mean that in the beginning of the 20th century parts of the ditch systems – nowadays flattened – must still have been visible and used as landmarks.

The localization of such traces indicating remnants of ancient walls and ditches is based - last but not least - on the analysis of LIDAR data and the interpretation of the shaded relief maps. Especially gently sloped embankments could be recognized very clearly. For example the channel-shaped parts of a ditch visible in the DEM south of the Heuneburg are impressive proofs for the accuracy in mapping archaeological structures in the landscape.

5 The Heuneburg precinct: fortification systems in ancient maps, LIDAR and excavation

The remains of the ramparts, still visible in the landscape, have been mapped for the first time in a detailed way by Major Julius Steiner during the 1890s (1893). The LIDAR data now offer good possibilities to complete this first map according to the present state of conservation (Kurz 2006).

Figure 7 *The test excavations in the Heuneburg precinct fortification system. The digital recorded coordinates of the excavations area are mapped directly in the LIDAR image which guarantees high precision mapping of archaeological activities.*

Getting new information about the age of the fortification was the main objective of the recent prospecting activities and excavations in the area of the Heuneburg precinct. Until the 1990s, these ramparts and mighty ditches were dated to the Middle Ages (Gersbach 1989). The first indications for an Iron Age dating emerged from the excavations of parts of the razed wall near the present parking place (Reim 2003). Therefore, three further test excavations at the defence systems have been made since 2004 (Fig. 7).

Figure 8 *Wooden construction elements and first dendrochronological results.*

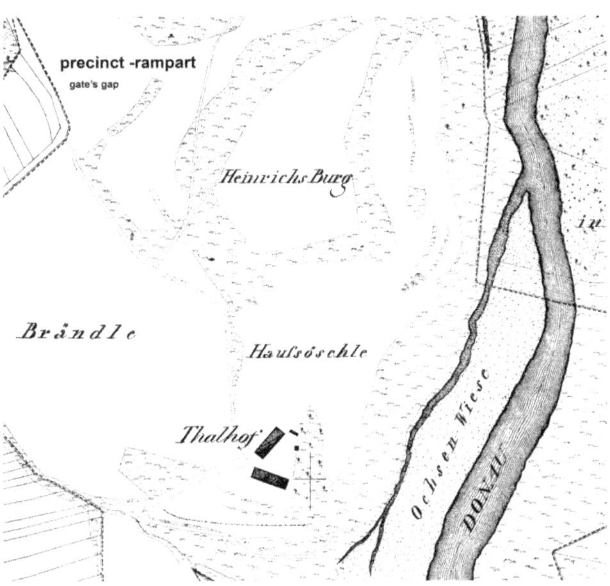

Figure 9 *Historic cadastral map with the rampart and location of the gate.*

Interpreting the data is the major problem. In 2005, A. de Boer (2005) presented a software solution to automatically identify burial mounds in the Netherlands. As the data do not offer chronological information to a considerable chronological depth, archaeological work remains necessary. As we had the benefit of already available LIDAR-data for that area, we concentrated our research aim on the identification of structures. By using LIDAR one can regain "invisible" structures which have been formally denoted on old maps. For example, this has been shown recently at the Fasanengarten of Waldmannshofen (Hahn et al. 2006). Unambiguous archaeological evidence can be identified, if the structure is well-defined like a burial mound. To identify ditches and trenches, ramparts of fortifications and others unspecific archaeological structures, additional information is needed. A short review of the test excavations will show the importance of the archaeological investigation.

Figure 10 *The stone gate of the Heuneburg precinct during the excavations in summer 2005.*

Special geological circumstances are responsible for extraordinary conservation conditions in the ditch on the north-western flank of the Heuneburg. In permanently humid clay layers, organic material persisted over centuries. Therefore, in the ditch at a depth of 6 m an extensive accumulation of carved wood was discovered in summer 2004. Posts and boards in combination with postholes, located on both sides of the embankments, may be seen as remains of a wooden construction, probably a bridge crossing the ditch.

Dendrochronological dating of oak, beech and pine woods was successful: during the first quarter of the 6th century BC most of the trees were cut and used for the wooden construction (Fig. 8). This means, that already during the 6th century BC impressive defence systems protected the Early Iron Age hill fort. Not only a military background but also a demonstration of power and rule might be considered.

Next to the rampart, near the parking place, an interesting discovery was made in summer 2005. During the excavation of settlement areas inside the precinct, two walls obviously belonging to a gate appeared only 30 cm underneath the modern surface. The earliest cadastral maps of this area, dating back to the early 19th century show the entire rampart in the north west of the Heuneburg shortly before it was levelled by the local farmer in the year 1837/38. At this very place the gate has recently been discovered, the map shows a gap in the rampart course which marks the gate's location (Fig. 9). The gate itself consists of two main walls constructed out of well worked ashlars of limestone. The parallel running walls are about 12 m long and 8 m apart (Fig. 10). This gate in its dimensions and architectural construction, built out of limestone is unique for Hallstatt settlements and fortification. Nevertheless the plinth of the mud brick wall is very similar. Since the levelling of the western parts of the rampart, traces of the gate's construction are visible neither in aerial photography nor in LIDAR imaging. Only the study of old maps indicated the existence of interesting archaeological structures.

6 Conclusions

The main objective of the various research activities around the Heuneburg is to extend knowledge of the different archaeological sites and their environment in a very complex system of hill fort, precinct, peripheral settlement, fortification systems and different burial places. This would offer a sound basis for well directed site preservation. The combination of excavation, aerial archaeology, geophysical prospecting and the analysis of LIDAR data together with the study of historic maps increased the information which is necessary to understand use and development of the landscape around the Heuneburg in the Early Iron Age period.

References

BOFINGER, J. 2004 Archäologische Untersuchungen in der Vorburg der Heuneburg - Siedlung und Befestigungssysteme am frühkeltischen Fürstensitz an der Oberen Donau, Gemeinde Herbertingen-Hundersingen, Kreis Sigmaringen. *Arch. Ausgrabungen Baden-Württemberg* 82 – 86.

BOFINGER, J. 2005 Stein für Stein... - Überraschende Befunde im Bereich der Befestigungssysteme der Heuneburg-Vorburg an der oberen Donau, Gemeinde Herbertingen-Hundersingen, Kreis Sigmaringen. *Arch. Ausgrabungen Baden-Württemberg,* 73 – 78.

BOFINGER, J. 2006 Wall und Graben, Haus und Tor – Archäologische Forschungen im Bereich der Befestigung und der Siedlung der Heuneburg-Vorburg an der Oberen Donau, Gemeinde Herbertingen-Hundersingen, Kreis Sigmaringen. In: Bayer-Niemeier E. /Hagmann S. /Lauermann E. (ed.), *Donau, Fürsten und Druiden. Kelten entlang der Donau.* Cat. Asparn an der Zaya/Herbertingen/Künzing. (Haugsdorf) 33 - 41.

DE BOER, A., 2005 Using pattern recognition to search LIDAR data for archeological sites. Paper *CAA* 2005, Tomar (in press).

GERLACH, R./HERZOG, I./VON KOBLINSKI, J. 2005 "Eine Kulturlandschaft mit Löchern - Erfassung, Ausmaß und Bedeutung neuzeitlicher Oberflächenstörungen". Kulturlandschaft digital. *Beiträge zur Landesentwicklung* 58, 2005

GERSBACH, E. 1989 *Ausgrabungsmethodik und Stratigraphie der Heuneburg. Heuneburgstudien VI.* Röm.-Germ. Forsch. 45 (Mainz).

HAHN, M. /OBMANN, J./PFUNDT-TITTELBACH, B. 2006 Der Fasanengarten von Waldmannshofen. Eine Gartenanlage des 17. Jahrhunderts im Dornröschenschlaf. *Denkmalpflege in Baden-Württemberg* 35, 2, 129-133.

KIMMIG, W. 1989 Forschungsgeschichte. In: GERSBACH, E., *Ausgrabungsmethodik und Stratigraphie der Heuneburg. Heuneburgstudien VI.* Röm.-Germ. Forsch. 45 (Mainz), 89-112.

KRAUSSE, D. 2004 Frühe Zentralisierungs- und Urbanisierungsprozesse. Zur Genese und Entwicklung frühkeltischer Fürstensitze und ihres territorialen Umlandes. Ein Schwerpunktprogramm der Deutschen Forschungsgemeinschaft. *Arch. Nachrichtenblatt* 9, Heft 4, 359-374.

KURZ, S. 2000 *Die Heuneburg-Außensiedlung. Befunde und Funde.* Forsch. u. Ber. Vor- u. Frühgesch. Baden-Württemberg 72 (Stuttgart).

KURZ, S. 2001 Siedlungsforschungen im Umland der Heuneburg. Fragestellung und erste Ergebnisse. In: P. Schauer (ed.), DFG-Graduiertenkolleg 462: *„Paläoökosystemforschung und Geschichte". Beiträge zur Siedlungsarchäologie und zum Landschaftswandel.* Regensburger Beitr. Prähist. Arch. 7 (Regensburg) 187 - 226.

KURZ, S./SCHIEK, S. 2002 *Bestattungsplätze im Umfeld der Heuneburg.* Forsch. u. Ber. Vor- u. Frühgesch. Baden-Württemberg 87 (Stuttgart).

KURZ, S. 2006 Befestigungsanlagen im Vorfeld der Heuneburg. In: WOTZKA, H.-P (ed.) *Grundlegungen. Beiträge zur europäischen und afrikanischen Archäologie für Manfred K. H. Eggert.* (Tübingen) 563-579.

REIM, H. 1999 Siedlungsgrabungen im Vorfeld der Heuneburg bei Hundersingen, Gde. Herbertingen, Kreis Sigmaringen. *Arch. Ausgr. Baden-Württemberg* 53-57.

REIM, H. 2003 Eine befestigte Siedlung der jüngeren Späthallstattzeit im Vorfeld der Heuneburg bei Hundersingen, Gde. Herbertingen, Kr. Sigmaringen. *Arch. Ausgr. Baden-Württemberg* 56-61.

Using airborne LiDAR intensity to predict the organic preservation of waterlogged deposits

Keith Challis,[1] Andy J Howard,[1] Derek Moscrop,[1] Ben Gearey,[1] David Smith,[1] Chris Carey,[2] Alan Thompson[3]

[1] Institute of Archaeology and Antiquity, University of Birmingham
[2] School of Geography, University of Exeter
[3] Cuesta Consulting Ltd

1 Introduction

Airborne remote sensing techniques have traditionally been employed to great effect in mapping the cultural archaeology and to a lesser extent the geomorphology of valley floor landscapes. Archaeologists have largely focused their attention on the comprehensive mapping of cropmarks and other features of the archaeological landscape revealed from aerial photographs (eg. Riley 1980; Whimster 1989), and large areas of England have been comprehensively mapped as part of the National Mapping Programme undertaken by English Heritage (Bewley 2003). Aerial photographs have also been employed in mapping geomorphology in alluvial landscapes, for example in extensive studies of the valleys of the Rivers Trent (Baker 2003; Garton and Malone 1998) and Thames (Lambrick 1992; Robinson and Lambrick 1994). Such mapping of fluvial geomorphology provides a context for past cultural landscapes and assists in identifying topographical features of high archaeological potential (for example relict river channels) and isolating areas of past river erosion where little in the way of archaeological material might be expected to survive (*cf* Brown 1997). The systematic reconnaissance, mapping and classification of aggregate bearing valley floor landscapes in this way has played a significant role in the strategic management of the geoarchaeological resource and intimately associated archaeological remains (e.g. fishweirs, bridges, platforms, trackways, etc.) in the face of growing impacts from aggregate extraction and other development pressures (Bishop 2003). More recently, information relating to former fluvial regimes is also being increasingly used in understanding wider issues of floodplain management, particularly with respect to assessing the impacts of changing flood frequency and magnitude with respect to future climate.

While the two dimensional record of air-photography provides an approach to mapping the quantity of archaeological material (both cultural and geoarchaeological) within aggregate bearing landscapes, it provides no indication of the state of preservation of that material or the associated cultural evidence. In particular, aerial photography provides no indication of the moisture level and the potential for the preservation of organic sediments. This is a significant shortfall in the usefulness of the data, since the presence of moist, organic-rich sediments may greatly increase the archaeological value of deposits.

In contrast to photography, airborne lidar provides access to high resolution, high accuracy terrain information and as a secondary output a laser "image" of the land surface derived from measurements of the intensity of reflection of each laser pulse. A typical airborne lidar is able to record multiple returns for each laser pulse (Figure 1) enabling the recording the top of the vegetation canopy and the underlying ground surface. In most systems intensity is taken as being the measurement of the amplitude of the returned pulse, and is usually recorded on an 8-bit digital scale. To date, archaeological applications of Lidar have focused largely on its ability to provide a high resolution record of terrain variation, allowing the detection and mapping of subtle archaeological features (Bewley *et al*, 2005), mapping of fluvial geomorphology (Challis 2005 and 2006) and due to its unique ability to penetrate vegetation cover to map archaeological earthworks under vegetation (Devereux, *et al*, 2005). Laser intensity measurements have largely escaped attention, and indeed do not form a part of the standard data product supplied by Environment Agency (EA: Brown et al, 2003) the chief instigator of lidar survey in the UK (although intensity data is collected on each EA flight and can be accessed by reprocessing original flight data).

The lidar system used by EA, NERC and many UK-based commercial lidar providers (an Optech Airborne Laser Terrain Mapper) utilises a laser that operates in the near infra-red (NIR: 1047nm) and so backscattered intensity is in-effect a record of the reflectance of earth surface materials at this wavelength. While it is recognised that other airborne remote sensing techniques (ATM, CASI, etc.) may be equally or more effective than lidar at detecting geoarchaeological and anthropogenic features, our present study focuses on lidar for a number of reasons. Principally lidar has been chosen as the basis of this study because it is an emerging technology, still poorly understood within the archaeological community. There is a growing demand for and supply of commercial lidar providers. National agencies such as EA already operate lidar, while Ordnance Survey are actively investigating a move into ownership or a lidar system for data acquisition to assist national mapping. Some good work has been done of understanding the parameters of lidar terrain mapping required for archaeological purposes (Challis 2005, Crutchley, pers comm) and so archaeologists are equipped to specify such work when commissioning survey. However, little or no work has been undertaken to investigate intensity, which until now has been largely ignored both by archaeologists and by the lidar industry as a whole.

2 Researching Lidar Intensity

Initial examination of backscattered laser intensity data by the applicants (Challis, 2006) suggests that a fall-off in the intensity of the reflected light corresponds with the position of sediment filled landform features such as palaeochannels (Figure 2). Variations in the reflectivity of various earth surface materials to laser light are well-documented (Wehr and Lohr 1999, 74) and damp soil conditions are known to reduce reflectivity. It is possible that the increased soil-moisture associated with palaeochannels, and perhaps other associated variations in soil and vegetation properties, are responsible for the reduced reflection of the laser pulse. This phenomenon is largely untested but demands further examination as if reflected laser intensity is closely related to soil moisture it may provide a useful means of identifying areas of preferential organic preservation. The rigorous examination of lidar intensity data, both to identify geoarchaeological and anthropogenic features and to remotely determine wetness and other soil and sediment properties forms the core of the research described here.

In most aggregate-rich areas the *greatest* potential for organic preservation is usually associated with palaeochannels, whose position within the landscape is already known from the analysis of aerial photographs and historic maps. Information gained from analysing lidar data is unlikely to identify new areas of organic preservation, but may help in the rapid assessment of valley floor corridors, preventing the need for costly investigations prior to the submission of planning proposals and provide an indication of the preservation potential and degree of waterlogging.

Our on-going research has two components: firstly aimed at determining the soil proprieties characterized by backscattered laser intensity through use of terrestrial laser scanning; secondly at systematic investigation of the backscattered laser intensity component of airborne laser scanning.

3 Terrestrial Lidar and Intensity

The initial stage of the research investigated the land cover and properties influencing laser intensity. Since mobilisation of an airborne system is both costly and time consuming, with no guarantee that data can be collected to suite specific soil and vegetation conditions, work on this stage of the project made use of a NIR terrestrial lidar with appropriate deployment and data manipulation to simulate airborne lidar. Work was undertaken both to investigate the extent to which intensity data might allow identification of buried archaeological remains through their physical expression at the earth surface as crop or soil marks (including those not otherwise evident in the visible portion of the spectrum), and to examine and quantify the aspects of land cover, soil and sediment properties affecting intensity values.

The project investigated the impact of ground conditions and vegetation on lidar intensity through examination of a variety of ground cover types. Briefly, work was undertaken to collect intensity data for sites with bare earth such as ploughed fields, equating to conditions where archaeological soilmarks may be evident, growing crops, where light crop cover is unlikely to give any indication of buried features using conventional techniques, and areas of mature crop where cropmarks might be expected to form. Areas of permanent pasture were similarly investigated to determine to what extent archaeological remains may be detected and analysed through lidar backscatter in such conditions.

Information on weather conditions prior to each survey were collected from the nearest Metrological Office ground station so that some attempt at gauging impact of prevailing conditions on soil properties may be made. Fieldwork collected laser backscatter data using a NIR terrestrial lidar to scan the ground surface at selected test sites (Figure 3). The use of a ground-based lidar allowed collection of dense data (100s points / m2) that were resampled to reflect typical point densities for airborne lidar at varying resolutions. Use of a highly mobile ground-based lidar allowed rapid collection of multiple data on each survey visit to allow for example, the investigation of factors such as scan direction, angle of incidence, and short term environmental conditions on intensity values. Intensity data were compared with simultaneously collected information on land cover, soil and sediment properties. This included analysis of the physical properties of soils through the analysis of sediment samples from each test site in order to assess organic content, moisture, geochemical signature and particle size. Initial results, which take into consideration measurements of volumetric soil moisture only, show a weak inverse relationship between soil moisture and intensity (Figure 4). This suggests that areas of low laser intensity may act as a reliable indicator of damp soil conditions. Further analysis will determine the extent to which other soil properties affect intensity, and the contribution of each recorded property to the overall intensity reading.

4 Airborne Lidar and Intensity

Subsequent stages of the research aim to determine to what extent the properties of lidar intensity predicted though use of a terrestrial laser system hold true for airborne lidar. Airborne data for test sites will be collected and similar calibration with land cover, soil and sediment properties undertaken as for stage one. In order to avoid the cost of new commissioned flights the team will liase with the EA airborne remote sensing facility and with commercial lidar providers to identify areas that will be flown in the future and co-ordinate field data collection with over flights in appropriate areas.

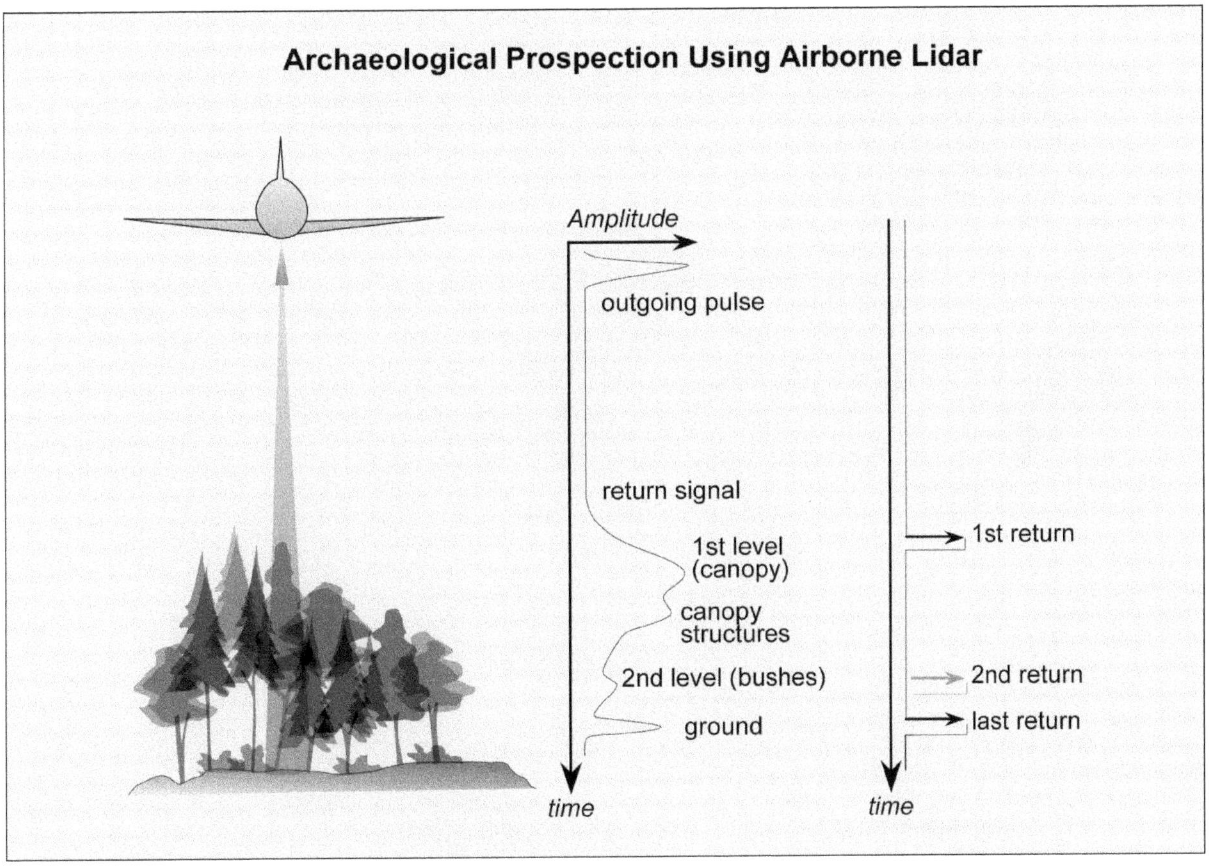

Figure 1 *Diagram showing a typical airborne lidar. Typically, each pulse is reflected from a number of objects on the earth's surface, generating multiple returns that are captured digitally as discrete pulses by the data logging system. The amplitude of each return is logged as the intensity value.*

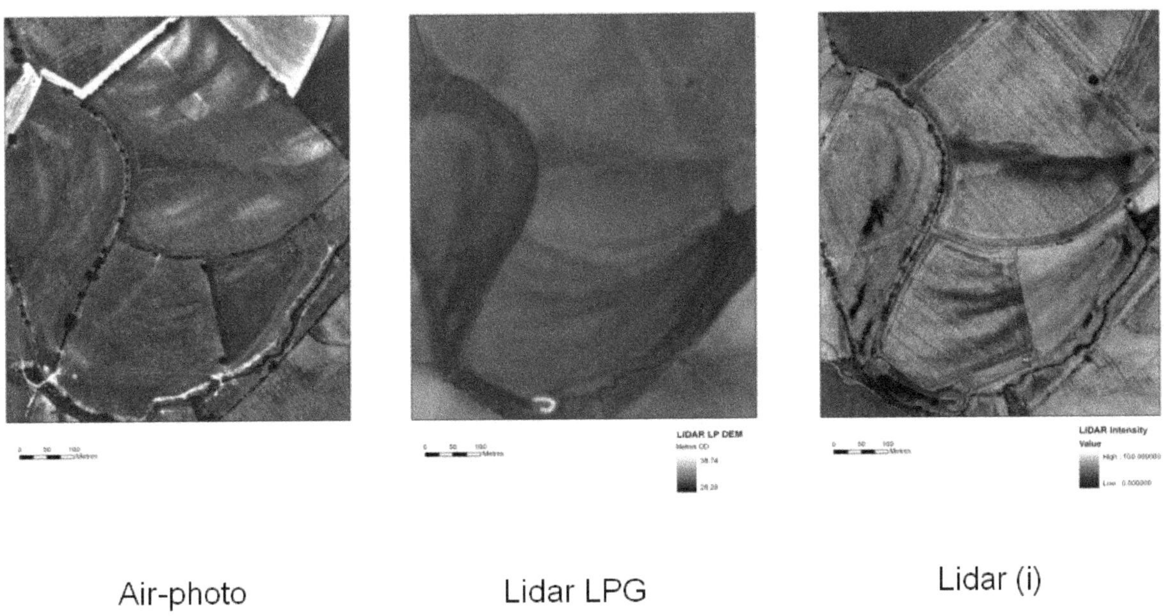

Figure 2 *Comparison of air-photographic evidence, lidar last pulse surface model and last pulse intensity for a part of the Lockington test site. Note that the same palaeochannels are evident to different degrees in each data source.*

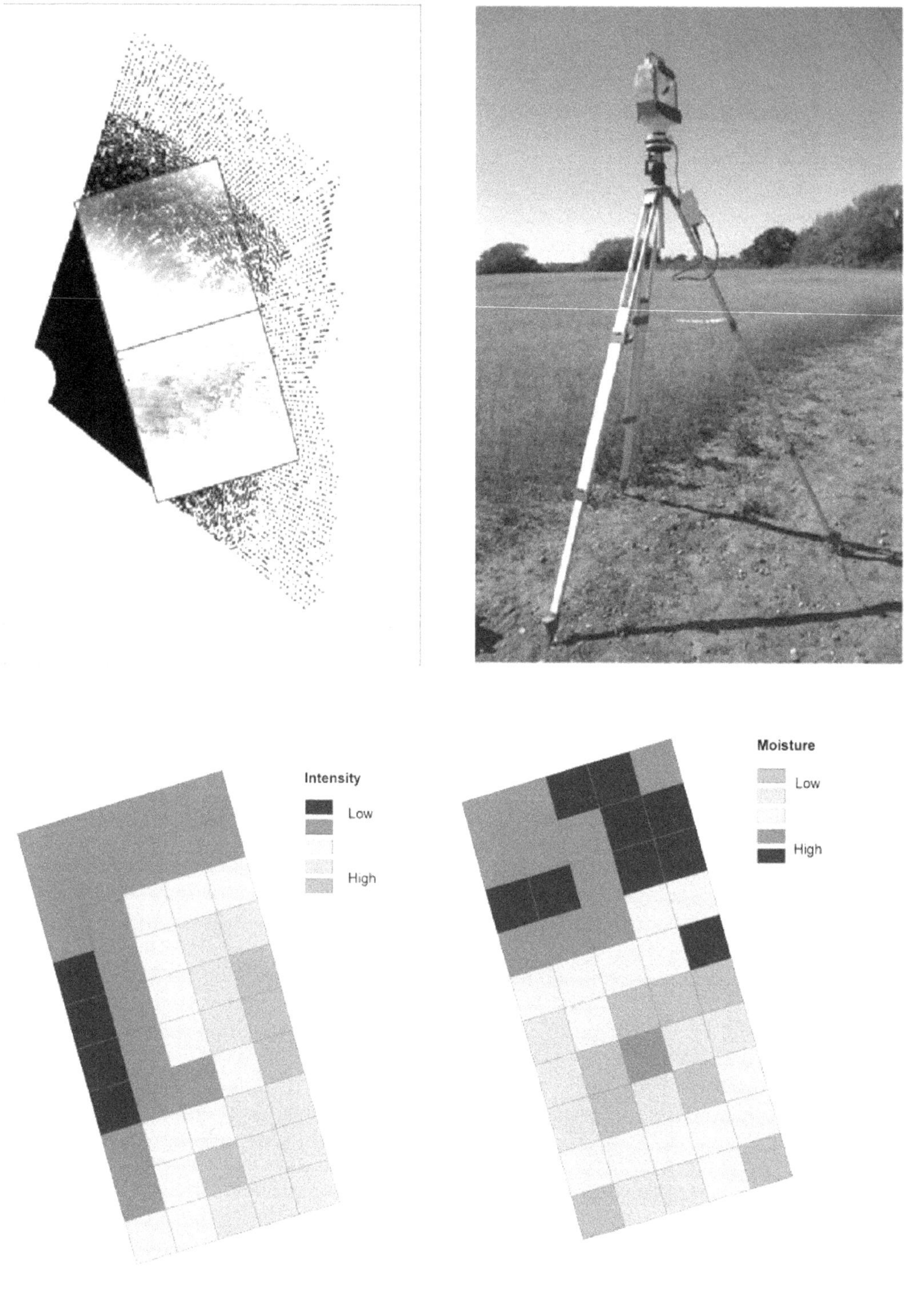

Figure 3 *Terrestrial laser scanning in use to simulate airborne lidar, showing: (top) MDL Laser Ace laser scanner mounted on 4m tripod and scanned point-cloud data with two 20x20m sample areas extracted (bottom) intensity values (aggregated by 4m square, extracted from the scanned point-cloud data) and field collected volumetric soil moisture for the two 20x20 sample areas.*

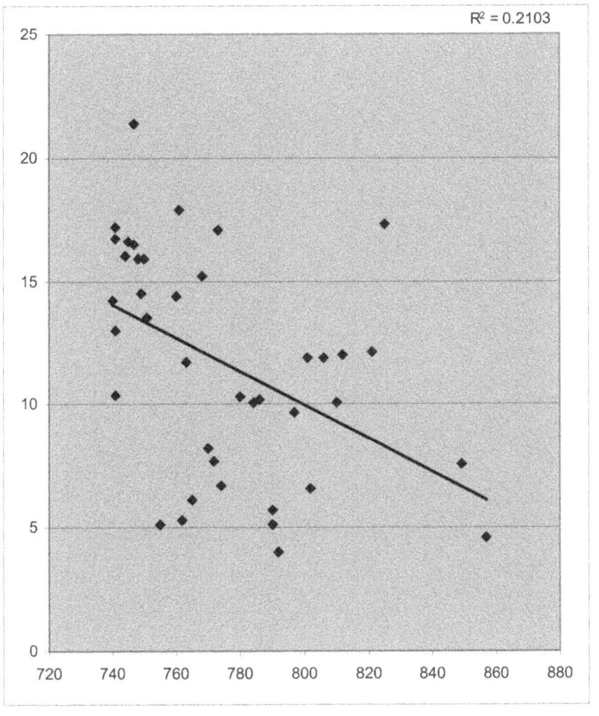

Figure 4 *Scatter plot showing the relationship between intensity and volumetric soil moisture in the two 20x20m sample areas show in figure 3. The R^2 value of -0.21 suggests a weak negative correlation between intensity and soil moisture.*

Investigation of newly collected lidar data will be supplemented by examination of intensity data from archive lidar surveys held by EA and others. A number of problems are anticipated in examining airborne intensity data. Most notably, intensity data from Optech airborne systems has a number of recognised but undocumented issues each of which will require careful investigation and consideration. Intensity values are not constant as sensors may be adjusted from flight to flight to provide different data collection parameters. Within any single flight, intensity values vary markedly depending upon the direction of the flight path so that for example, adjacent flight paths may exhibit different intensity values for similar earth surface materials. Since this means that intensity values are not recorded on an absolute scale and vary across or even within flights analysis will focus on change of intensity value within individual survey flights, and within flights on analysis of intensity values within individual flight lines.

5 Conclusions

The on-going research described in this paper is intended to culminate in a good practice document relating to airborne lidar intensity aimed at those in the archaeological and development control community who might be called upon to commission or interpret lidar surveys. It is hoped that the research will inform the commissioning of future lidar surveys to take into account the potential of intensity data to provide a qualitative assessment of the archaeological resource and encourage the reassessment of existing, archive, intensity data.

Acknowledgements

English Heritage has generously funded the work described in this paper through the Aggregates Levy Sustainability Fund (Grant No 4782) Thanks are also extended to Nick Holden and Bob Davison of the Environment Agency and to Nick Hanmer of Network Mapping Ltd for their assistance.

References

BAKER, S. 2003 *The Trent Valley: Palaeochannel Mapping from Aerial Photographs*. Trent Valley Geoarchaeology Research Report. Nottingham: Trent & Peak Archaeological Unit.

BEWLEY, R.H. CRUTCHLEY, S.P and SHELL, C.A. 2005 New light on an ancient landscape: lidar survey in the Stonehenge World Heritage Site. *Antiquity* Volume: 79 Number 305: 636–647

BEWLEY, R. H. 2003 Aerial survey for archaeology. *The Photogrammetric Record* 18 (104): 273-292.

BISHOP, M. 2003 Issues and agenda in archaeological research and management: a case study from the Trent Valley, UK. In Howard, A.J., Macklin, M.G. and Passmore, D.G. (eds) *Alluvial Archaeology in Europe.* Lisse: Balkema: 123-131.

BROWN, A.G. 1997 *Alluvial Geoarchaeology. Floodplain archaeology and environmental change.* Cambridge: Cambridge University Press.

BROWN, K., DUNCAN, A., O'DWYER, C., DAVISON, B., HOGARTH, P,. BUTLER, D. & SAMPSON, E. 2003 Integrated airborne data collection by the Environment Agency. In Aplin P. and Mather, P.M. (eds) *Proceedings of RSPSoc 2003: Scales and Dynamics in Observing the Environment, Nottingham, 10-12 September 2003*. Nottingham: The Remote Sensing and Photogrammetry Society.

CHALLIS, K. 2006 Airborne laser altimetry in alluviated landscapes. *Archaeological Prospection* **13**: 103-127.

CHALLIS, K. 2005 Airborne Lidar: A Tool for Geoarchaeological Prospection in Riverine Landscapes. in Stoepker, H. (ed) *Archaeological Heritage Management in Riverine Landscapes.* Rapporten Archeologische Monumentenzorg, 126: 11-24

DEVEREUX, B.J. AMABLE, G.S. CROW, P. and CLIFF, A.D. 2005 The potential of airborne lidar for detection of archaeological features under woodland canopies. *Antiquity*, Volume: 79 Number: 305: 648–660

GARTON, D. & MALONE, S. 1998 Geomorphology from aerial photographs in the Trent Valley. In Challis, K. (ed) Fieldwork by Trent & Peak Archaeological Trust in Nottinghamshire, 1996-7. *Transactions of the Thoroton Society of Nottinghamshire* 102: 139-141.

KENWARD, H.K. and LARGE, F. 1998 Recording the preservation condition of archaeological insect fossils. *Environmental Archaeology* 2, 49-60.

LAMBRICK, G., 1992 Alluvial archaeology of the Holocene in the Upper Thames basin 1971-1991: a review, in Needham, S., and Macklin, M.G., eds., *Alluvial Archaeology in Britain*: London, The Society of Antiquaries of London, 209-225

RILEY, D.N. 1980 *Early Landscape from the Air. Studies of cropmark sites in South Yorkshire and North Nottinghamshire.* Sheffield: Department of Archaeology and Prehistory, University of Sheffield.

ROBINSON, M., and LAMBRICK, G., 1984 Holocene alluviation and hydrology in the Upper Thames basin: *Nature* (London), v. 308, 809-814.

WEHR, A. and LOHR, U. 1999 Airborne laser scanning – an introduction and overview. *ISPRS Journal of Photogrammetry & Remote Sensing* 54: 68-82.

WHIMSTER, R. 1989 *The Emerging Past: Air-Photography and the buried Landscape.* London, RCHME.

Full-waveform airborne laser scanning as a tool for archaeological reconnaissance

Michael Doneus,[1] Christian Briese[2]

[1] Department for Prehistory and Early History of the University of Vienna, Austria
[2] Christian Doppler Laboratory "Spatial Data from Laser Scanning and Remote Sensing", Institute of Photogrammetry and Remote Sensing of the Vienna University of Technology, Austria

1 Introduction

Airborne Laser Scanning (ALS), also referred to as LiDAR, is a rather new technique to produce dense and high precision measurements of the topography of the Earth's surface (Ackermann, 1999; Kraus, 2004, 449-470; Wehr and Lohr, 1999). The scanning device is typically mounted at the bottom or below an aeroplane or helicopter. For the determination of object points, the laserscanner emits short infrared pulses into different directions across the flight path towards the earth's surface and a photodiode records the backscattered echo and determines the distance to the reflecting object by the determination of the travel time.

ALS finds application in archaeology because working with archaeological sites and landscapes involves also the description of topography. So far, ALS was used as a tool to produce detailed maps of archaeological sites showing in relief. But it has also a high potential to be used for detection of archaeological sites, especially in densely vegetated areas.

This paper therefore deals with the reconnaissance aspects of ALS. In April 2006, the project "LiDAR-Supported Archaeological Prospection in Woodland" was launched, which is funded by the Austrian Science Fund (FWF P18674-G02). The goal of the project is to explore the potential of ALS for archaeological reconnaissance in a densely forested area; specifically, to evaluate an approx. 190 km² forest area within an archaeological case study. Using the data of a test scan, some issues are presented that have to be decided upon during its workflow, in order to use the output for the detection of archaeological sites.

2 ALS and archaeological surveying

The archaeological application of ALS so far showed that it is extremely useful to measure extensive archaeological sites showing in relief (Bewley, 2003; Challis, 2006; Holden et al., 2002; Motkin, 2001; Shell and Roughley, 2004). The short time needed for the production of data and the detail and high accuracy of the derived DTM made it also a tool to monitor sites and map the occurrence and extent of damage or destruction (Barnes, 2003). Various applications demonstrated that when using ALS it was possible to depict subtle features in relief, which are hardly detectable and comprehensible when surveying terrestrially. In several cases, previously unknown features were detected in ALS data. This expands the potential of ALS from a plain measurement technique to a tool for archaeological reconnaissance.

Recent publications examined its possibility to survey sites in relief even under dense vegetation. Already known archaeological sites, as for example an Iron Age hillfort (Devereux et al., 2005), ridge and furrow (Sittler, 2004), or round barrows (Doneus and Briese, 2006) were scanned using conventional and full-waveform recording systems. The results were more than promising and will have a substantial impact on future archaeological investigations in woodland. Even under difficult canopy, as deciduous trees with dense understorey, or conifers, the archaeological features could be measured in detail. Especially the latest generation of full-waveform recording scanners proved to be extremely useful. The consequent step forward will be to test ALS as a method of archaeological reconnaissance in forests, i.e. to evaluate its potential to detect previously unknown sites.

3 ALS as an archaeological prospection technique

Currently, archaeologists use only two prospection methods for the systematic (i.e. not coincidental) discovery of archaeological sites: field survey and aerial archaeology. Both methods come to their limit in woodland, which results in the fact that we usually know only large and well preserved sites in these areas.

As a prospection technique, ALS takes an intermediate position between aerial archaeology and topographical surveying. With aerial archaeology it has in common that it is an airborne method recording sites in relief in a way, which is similar to vertical aerial photography (where sites are showing as shadow-, flood-, or snow-marks). Both the aerial photograph and the derived DTM from an ALS scan need *interpretation* to identify potential archaeological sites and features. The main difference lies in the fact that aerial photography is a passive remote sensing technique recording the reflected part of the visible and near infrared range of the electromagnetic spectrum on film or a digital sensor. A photo-pixel is a mixture of all radiation within a certain back-scattering area. ALS, on the other hand, is an active technique measuring a dense network of 3d-points of the Earth's surface. A laser pulse can penetrate vegetation to a certain degree and offers the possibility to discriminate different objects within the footprint.

During aerial archaeological reconnaissance flights, only a few parameters, as date and time of flight, flying height, direction of view, and type of sensor can be influenced to improve the detectability of sites. Once an aerial photograph is made, there is only a limited variety of possibilities to enhance it for easier interpretation.

The raw data of an ALS scan are composed of both terrain and off-terrain measurements, which can be modelled as a digital surface model (DSM) and digital terrain model (DTM). The high-resolution models of an ALS scan have a wider range of possibilities for interpretation. It is for example possible to virtually influence factors of shadow characteristic (preserved height and alignment of the structures to the angle of insolation; position of the sun) that we cannot influence in real life, or to derive secondary models like slope, aspect, and curvature in a desk-based environment later on.

The common point of ALS with surveying is that both methods take measurements and both methods should result in an *interpretation* map. The important difference lies in the fact that the archaeological surveyor is literally in touch with an *already known site* and for reasons of time usually only records what is interpreted on site as archaeologically relevant. During an ALS scan, the total area of interest will be documented with the same high point density (3 to 8 points per m^2), regardless if it contains archaeological features or not.

Throughout the process of terrestrially recording a site, an interpretation is already taking place. The interpretation of an ALS scan is a desk-based analysis of the data later on in a remote place. Being in touch with the site has many advantages and it will be no difficulty for the surveyor to distinguish a pile of wood from a barrow. This can be a very difficult task when working with ALS data. On the other hand, what the surveyor doesn't see during the recording procedure at the site (e.g. because of the faint relief or the dense vegetation cover) will be lost in the final record. In an ALS point cloud, all features within the range discrimination of the instrument and which are sampled within the given point density are documented, regardless whether they are observed during the interpretation process or not. Anybody, who afterwards (re)interprets the data, will see the terrain in the same condition, regardless, if this happens the day after or years later. The data can be interpreted time and again and, as experience and knowledge increases and perception changes, more and more information will be probably found.

Re-interpretation of a site (i.e. re-measuring it) is theoretically also possible when surveying it terrestrially, but it is certainly not common practice. More often than wanted, sites cannot be re-measured because they were subject to damage or destruction. In these cases, previously not recorded features are lost for ever.

4 Using ALS for archaeological reconnaissance in forests

With human occupation, the landscape usually gets shaped: pits and ditches are dug, banks and mounds erected, slopes are terraced, houses built, etc. When a site is abandoned, the structures will start to deteriorate. Without human interaction (especially agriculture), the process of decay can be stabilised by vegetation and the archaeological structures will form part of the ground's surface. Therefore, in a forest, we can expect sites, which are still surviving in relief and can be consequently detected in a DTM derived from ALS data.

ALS works in forested areas because it is possible to differ between the first and the last light echo that follow a single measurement pulse. The first echo (= first pulse) is reflected by the earth's surface (also from treetops, high voltage transmission lines, or roof edges), the last echo (= last pulse) usually from the ground beneath. For the time being, the application of ALS seems to be the only possibility to gain elevation information at an acceptable cost to systematically search for unknown archaeological traces in woodland.

To evaluate the potential of ALS to detect previously unknown sites in forests, the project "LiDAR-supported archaeological prospection in woodland" tries to evaluate an approx. 190 km^2 forest area within the Leitha mountain range, 50 km southeast of Vienna, in an archaeological case study. The area is almost entirely covered with mixed woodland of mainly oak and beech trees with a varying range of understorey.

To test the configuration and parameters of the scanning system and the methods of data processing, filtering, and visualisation, a test scan was conducted during the pilot phase of the project. During the workflow of the test scan from data acquisition to the visualisation for final interpretation, several considerations and decisions had to be made, which had a considerable impact on the quality of the result. In the following some of these issues will be discussed.

4.1 Choice of sensor

At the moment, there are two different types of sensor systems available: so called conventional scanners and full-waveform recording systems.

Conventional systems record typically up to four distinct echoes from multiple targets (e.g. treetop, second level of vegetation, surface) touched by a single laser pulse using analogue detectors in real time during the acquisition process. As a result, these systems provide "only" an irregular 3D point cloud containing coordinates of scattering targets and usually the respective un-calibrated intensity information. For the classification of the point cloud into terrain and off-terrain points, only the location information of the individual points is typically used. Therefore, the developed filtering techniques are all based on the local spatial relationships of the 3D points (Wagner et al., 2004, 101). Usually, the height difference of the scanned objects must be 1.5 m to be able to distinguish the returns, which makes it difficult to get the surface height in areas e.g. covered with dense low bushes (Kraus, 2004).

The latest generation of commercially available ALS systems, the so called full-waveform recording scanners, sample the entire echo waveform for each emitted laser beam (typically with an interval of 1 ns) and convert the signal in a digital data stream, which can be post-processed later (Wagner et al., 2004, 105).

The echo-waveform allows gaining further physical observations of the reflecting surface elements, which can be useful for subsequent object classification. As a result, in comparison with the conventional ALS data, a more reliable classification of the laser points and a higher accuracy of the terrain points can be expected.

4.2 Data acquisition

In a forest, only part of the laser pulses will penetrate through to the actual ground. To get a high density of ground points, a slow moving platform with a high frequency scanner was used. The data acquisition took place in the dormant period beginning of April 2006 using a RIEGL Airborne Laser Scanner LMS-Q560 operated by the company *Mi*lan Flug GmbH. Its multi-target range discrimination is 0.6 m.

Flight altitude was about 600 m above ground, which resulted in a laser footprint size of 30 cm on ground. A total area of 9 km^2 was covered with a scan angle of 45 degrees by 26 parallel flight tracks, which had a width of approximately 500 m and an overlap of 50%. The real scan rate was 66 kHz that resulted in an overall mean point density of eight measurements per m^2.

Although it is often argued that scanning a forest demands a narrow opening angle of the scanner to receive only nadir or near-nadir returns, the test scan was done using the maximum scan angle of 45 degrees with an overlap of 50%. This resulted in a large overlapping area of two neighbouring strips, which can be an advantage when georeferencing. Due to the large overlap, it was also ascertained that every object on the ground would be hit at least twice from two sides and there is a good chance, that some of the oblique laser pulses will hit the ground below conifers where the almost vertical laser pulses of a system operating with a narrow scan angle would not get through. The drawback of a wide opening angle with a large overlapping area is the huge amount of data that have to be processed. In this case the entire point cloud had a density of 8 points per m^2 resulting in 22.5 GB of raw data for 9km^2. The remaining point density when considering only the last pulses was still 5 points per m^2. Due to the wide opening angle, especially near the borders of the scanning strip a high number of last pulses were returned from tree trunks and consequently had to be filtered out later on.

4.3 Georeferencing

Georeferencing of the laser data is usually done by the data providers, where a calibration over asphalt or areas with sparse, very low vegetation is usually performed (Pfeifer et al., 2004, 2). This was also the case with the test scan.

The georeferencing process still causes a few problems (Kager, 2004), which usually introduce inaccuracies into the resulting point cloud. These inaccuracies will lead to discrepancies between overlapping laser scanner strips, which can result in noise, and formation of non existing structures (sinusoidal curves, but also edges), which consequently can irritate during the process of interpretation.

These inaccuracies can be improved doing a simultaneous 3D adjustment by least squares. For this, tying features between the overlapping stripes are used analogously to tie points in aerotriangulation. As tying features, homologous planes and straight lines with low noise from covering vegetation are used (e.g. roofs, meadows, roads).

In order to have a reliable description of the plane, there should be a sufficient number of last pulse ALS points (at least 15 points). Furthermore, the plane should be of minimal steepness. It should have not too many outliers (chimneys, trees etc.) (Kager, 2004). In a forest, it can be more difficult to find appropriate tying features. Therefore, a large overlapping area between scan strips is an advantage.

At the time of writing, a simultaneous 3D adjustment by least squares is being performed with the data of the test scan. The results will be presented in a future publication.

4.4 Classification and Filtering

To be able to identify archaeological structures, archaeologists have to *interpret* the resulting topographical data of an ALS scan. This does not pose major problems with large structures. Smaller features, however, are much more difficult to identify. Their appearance in an ALS point cloud can be very similar to natural and recent features, as for example dense brushwood, or piles of twigs or wood, which are – other than archaeological structures – actually off-terrain points.

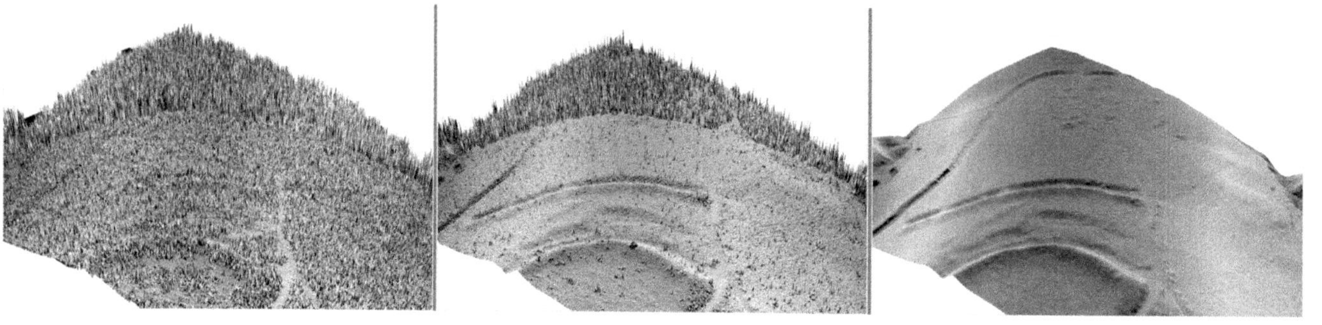

Figure 1 *Left: DSM of the first pulse data showing the canopy of the scanned area. In the foreground, the vegetation consists of dense bushes. In the background there is a dense forest with understorey. Middle: DSM resulting from the unfiltered last echo point-cloud. There are many points, which represent, tree-trunks, very dense vegetation or narrow vegetation, which do not represent the actual ground. Right: filtered DTM showing even faint archaeological traces, as e.g. round barrows with shallow depressions from looting.*

Therefore, to eliminate potential sources of error, a high quality separation of terrain and off-terrain points while maintaining a high point density of the ALS data is essential as a first step before an archaeological interpretation can take place. This can be quite a challenge in densely vegetated (e.g. forested) areas.

The main value of using a full-waveform recording scanner is the availability of additional physical observations of the reflecting surface elements, which can be useful for object classification (see also chapter 4.1).

By modelling the full waveform as a series of Gaussian distribution functions, individual scatterers can be distinguished (Hofton et al., 2000; Wagner et al., 2006). The results are estimates of the location and scattering properties of the individual targets: for each returning echo of a single laser pulse, the estimated coordinates of the scatterer, the echo width, and the amplitude is determined. Using amplitude and echo width, it is possible to investigate the return signal and extract additional ground characteristics. Consequently, much more information is available when classifying the point cloud into solid ground and vegetation cover. In a recent paper, the usage of a simple threshold operation in order to pre-exclude points situated within low vegetation or other structures as for example clearance piles (consisting of twigs and branches) is demonstrated (Doneus and Briese, 2006).

In that way, one has a means to distinguish for example a clearance pile from a round barrow already in the unfiltered data. After eliminating these points, a more reliable DTM will be the result, where most local topographic features are in fact local topographic elevations.

In the data of the test scan, the resulting point-cloud, which was classified as last pulse with narrow echo width, still had a density of 3.7 points per m^2. Many of these points were still off the surface due to random errors (depending on the measurement system), gross errors (due to multi-path effects or buildings, tree-trunks, etc.) and a few systematic errors (vegetation) (Briese et al., 2002). Therefore, for the determination of the DTM additionally advanced filter methods are necessary in order to eliminate remaining off-terrain points from the derived last echo point cloud (Sithole and Vosselman, 2003).

For these tasks the theory of robust interpolation with an eccentric and unsymmetrical weight function (Kraus and Pfeifer, 1998) used within a hierarchic framework, which is implemented into the software package SCOP++ (Kraus and Otepka, 2005), was used (Briese et al., 2002). A brief description of the technique is given in a recent paper (Doneus and Briese, 2006). The whole process of filtering is also demonstrated at the EuroSDR Distance Learning Course "Filtering and Classification of Laser Scanner Data", available under http://www.ipf.tuwien.ac.at/eurosdr/index.htm.

For archaeological purposes, the filter strategies are not allowed to be too rigid. Otherwise, the archaeological objects could be removed, too. Usually, vegetation should be filtered out, while walls, banks, ditches, and smaller archaeological structures should be kept for the resulting DTM. The result of the robust interpolation can be seen in Figure 1. The resulting point cloud had still a density of 2.5 points per m^2, which was sufficient for a grid width of 0.5 m of the final visualisations after DTM interpolation (Figures 2-4).

4.5 Visualisation and Interpretation

The filtering procedure resulted in a DTM, which shows a detailed map of the topography with even fain archaeological structures under the forest canopy. The 260 by 100 m large graveyard with at least 50 round barrows was recorded in detail. The size of the barrows ranges between 5 and 15 m in diameter. The preserved heights vary between 20 cm and 2 m. Many of the barrows even show traces of looting, where shallow depressions (7 - 20 cm deep) on top of the tumuli are interpreted as relics of former treasure hunters.

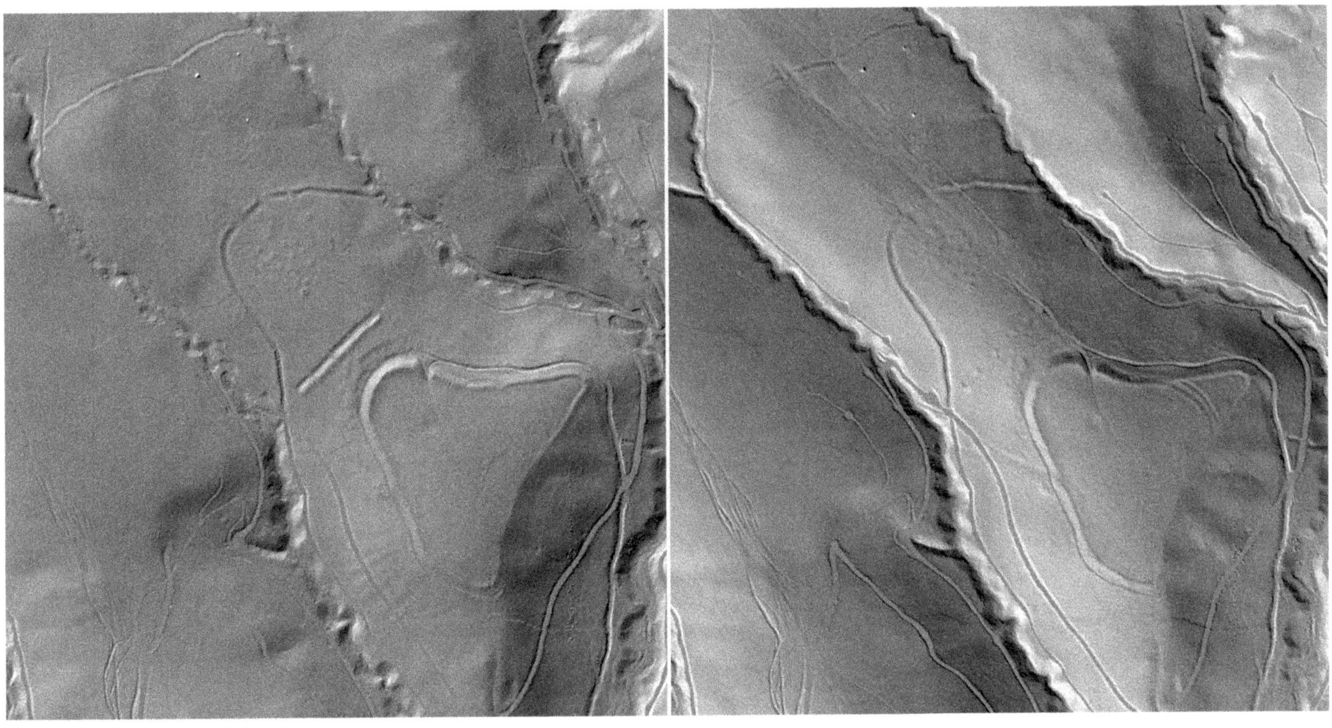

Figure 2 *Shaded DTM after filtering. Note that even large linear features can be missed if they align with the sun and do not cast shadows. Left: illumination from northwest. Right: illumination from southwest.*

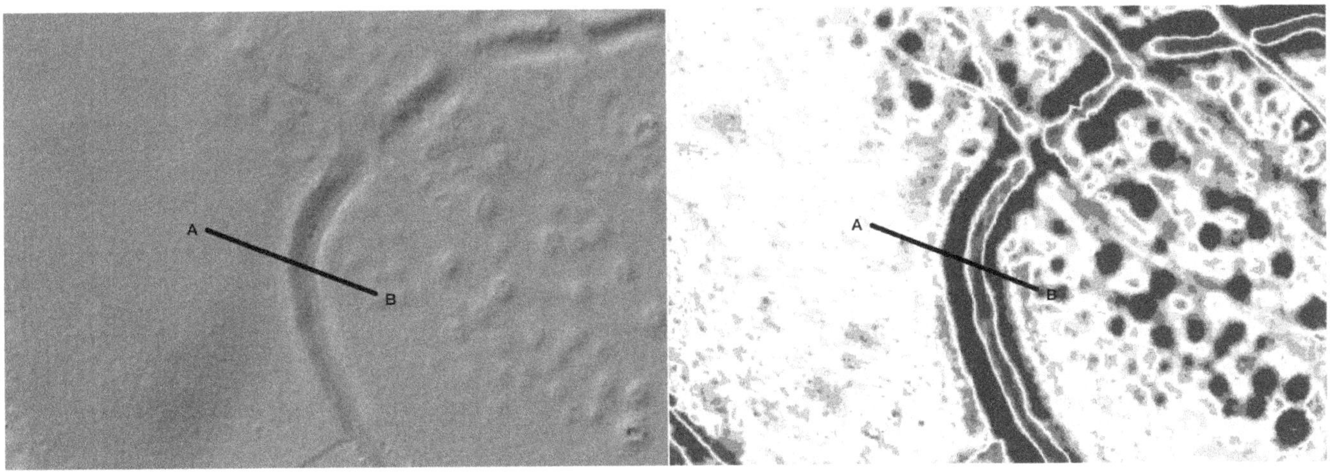

Figure 3 *Left: Shaded DTM (illumination from northwest). Right: Difference model after subtraction of a resampled DTM. The line A-B refers to the cross section in Figure 5. In this case, a point of the 5 m DTM is lying in the ditch of the archaeological fortification. Because of the larger distances between points, the local slope in the resampled terrain is therefore lower than in the original DTM derived from the ALS point-cloud. The resampled DTM will therefore be too low which will result in an "artificial" bank after subtraction from the ALS-derived DTM.*

Although shading of the DTM results in a comprehensible representation of the DTM, there is a certain danger that features will be missed because they align with the sun and do not cast shadows (Figure 2). This problem can be overcome by creating different hillshades with different directions and heights of the illumination source. Another possibility is to subtract a resampled DTM with a crude resolution (in this case: 5 m) from the high resolution DTM with its subtle variations representing banks, ditches, or barrows. The resulting difference model consequently shows only the terrain variations below or above the main topography of the surface, which was flattened by the subtraction. After colour-coding, the difference model is easier to interpret with all features visible regardless of any illumination source (Figure 3).

In Figure 3, the dark grey indicate upstanding features in relief, while lighter grey resent hollows. This visualisation clearly has advantages with smaller features, as the round barrows of Figure 3. To work also proper with large structures, as for example wide banks and ditches, the resampled point cloud of the coarse DTM should not contain points lying on these features.

Otherwise, non existing features as for example an additional bank in Figure 3 can be introduced.

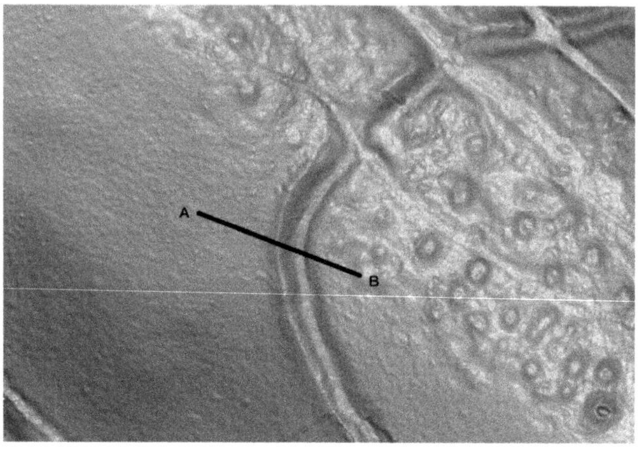

Figure 4 *Slope-map of the area of Figure 3. Dark values indicate steep slopes, while flat areas are represented by bright values. The line A-B refers to the cross section in Figure 5.*

Figure 5 *Cross section of Figures 3 and 4.*

Another possibility is to use a slope mapping, where all features in relief can be clearly distinguished regardless of any illumination source by the change in slope (Figure 4). One problem of using slope mapping is that without additional information it is not possible to tell whether a slope is raising or falling, i.e. whether one sees a bank or a ditch. Additionally, one has to get used to interpret slope maps, because they are similar to shaded images and one therefore tends to "see" shadows.

There are a variety of other possibilities to produce appropriate visualisations for subsequent interpretation and it will be an aim of the current project to develop and investigate the topic of visualisation and interpretation of ALS data.

The interpretation of the visualisations will take place in a GIS-based environment. 'Potential' archaeological sites and features will be identified and drawn. Interpretation is a subjective process and the resulting confidence will vary from site to site. There are many natural and recent features that could affect the interpretation of ALS data. Hence, we maybe identify non-sites as potential archaeological sites or on the other hand give a natural meaning to features, which are in fact of archaeological relevance.

Therefore, to evaluate the accuracy of this identification process, these 'potential' sites will need visiting to ascertain whether they are archaeological sites or not and if not what has caused the response. These visits are essential to the production of an image interpretation key, which should help to improve future interpretation.

5 Conclusion

The project "LiDAR-supported archaeological prospection in woodland" tries to evaluate an approx. 190 km^2 forest area within the Leitha mountain range, 50 km southeast of Vienna, in an archaeological case study. The area is almost entirely covered with forest.

In the course of a test scan over known archaeological sites, which was performed to evaluate the configuration and parameters of the scanning system, several decisions had to be made. Choosing a full-waveform recording scanner for data acquisition and the high point density achieved had the most dramatic effect on the result. In combination with a still rather simple point classification and a sophisticated filtering technique, non-surface points could be reliably eliminated. The potential of ALS to measure even subtle structures in a forested area with dense understorey could be clearly demonstrated. The results were more than promising and will hopefully have a substantial impact on future archaeological investigations in woodland.

The next step of the project will be to test ALS as a method of archaeological reconnaissance in forests, i.e. to evaluate its potential to detect previously unknown sites.

Acknowledgements

This research has been supported by the Austrian Science Fund (FWF) under project no. P18674-G02.

The authors thank Thomas Melzer from the Christian Doppler Laboratory for Spatial Data from Laser Scanning and Remote Sensing, Vienna University of Technology, for the processing of the Gaussian decomposition of the full-waveform ALS data.

References

ACKERMANN, F., 1999. Airborne laser scanning: present status and future expectations. *ISPRS Journal of Photogrammetry and Remote Sensing*, 54(2-3): 64-67.

BARNES, I., 2003. Aerial remote-sensing techniques used in the management of archaeological monuments on the British Army's Salisbury Plain Training Area, Wiltshire, UK. *Archaeological Prospection,* 10: 83-91.

BEWLEY, R., 2003. Aerial survey for archaeology. *The Photogrammetric Record,* 18(104): 273-292.

BRIESE, C., PFEIFER, N. and DORNINGER, P., 2002. Applications of the robust interpolation for DTM determination. In: R. Kalliany, F. Leberl and F. Fraundorfer (Editors), *Photogrammetric Computer Vision. International Archives of Photogrammetry and Remote Sensing,* Vol. XXXIV, 3A, Graz, Austria, pp. 55–61.

CHALLIS, K., 2006. Airborne Laser Altimetry in Alluviated Landscapes. *Archaeological Prospection,* 13(2): 103-127.

DEVEREUX, B.J., AMABLE, G.S., CROW, P. and CLIFF, A.D., 2005. The potential of airborne lidar for detection of archaeological features under woodland canopies. *Antiquity*, 79: 648-660.

DONEUS, M. and BRIESE, C., 2006. Digital terrain modelling for archaeological interpretation within forested areas using full-waveform laserscanning. In: M. Ioannides, D. Arnold, F. Niccolucci and K. Mania (Editors), *The 7th International Symposium on Virtual Reality, Archaeology and Cultural Heritage VAST* (2006), pp. in Press.

HOFTON, M.A., MINSTER, J.B. and BLAIR, J.B., 2000. Decomposition of Laser Altimeter Waveforms. IEEE *Transactions on Geoscience and Remote Sensing*, 38: 1989-1996.

HOLDEN, N., HORNE, P. and BEWLEY, R., 2002. High-resolution digital airborne mapping and archaeology. In: R. Bewley and W. Raczkowski (Editors), *Aerial Archaeology Developing Future Practice. Nato Science Series, Series I: Life and Behavioural Sciences,* Vol. 337, pp. 173-180.

KAGER, H., 2004. Discrepancies Between Overlapping Laser Scanning Strips- Simultaneous Fitting of Aerial Laser Scanner Strips. In: O. Altan (Editor), *Geo-Imagery Bridging Continents. XXth ISPRS Congress, 12-23 July 2004 Istanbul, Turkey International* Archives of Photogrammetry and Remote Sensing, Volume XXXV, Part B/3, Istanbul, pp. 555-560.

KRAUS, K., 2004. *Photogrammetrie. Band1. Geometrische Informationen aus Photographien und Laserscanneraufnahmen.* 7., vollständig bearbeitete und erweiterte Auflage. Walter de Gruyter, Berlin, New York.

KRAUS, K. and OTEPKA, J., 2005. DTM Modelling and *Visualization - The SCOP Approach, Photogrammetric Week.* Wichmann Verlag, Heidelberg, pp. 241-252.

KRAUS, K. and PFEIFER, N., 1998. Determination of terrain models in wooded areas with airborne laser scanner data. ISPRS *Journal of Photogrammetry and Remote Sensing,* 53(4): 193-203.

MOTKIN, D., 2001. The assessment of LIDAR for archaeological use. *AARGNews*, 22: 24-25.

PFEIFER, N., GORTE, B. and ELBERINK, S.O., 2004. Influences of Vegetation on Laser Altimetry - Analysis and Correction Approaches. In: M. Thies, B. Koch, H. Spiecker and H. Weinacker (Editors), *Proceedings of Natscan, Laser-Scanners for Forest and Landscape Assessment - Instruments, Processing Methods and Applications. International Archives of Photogrammetry and Remote Sensing,* Volume XXXVI, Part 8/W2, pp. 283-287.

SHELL, C. and ROUGHLEY, C., 2004. Exploring the Loughcrew landscape: a new airborne approach. *Archaeology Ireland*, 18/2(68): 22-25.

SITHOLE, G. and VOSSELMAN, G., 2003. Comparison of filtering algorithms. In: H.-G. Maas, G. Vosselman and A. Streilein (Editors), *Proceedings of the ISPRS working group III/3 workshop "3-D reconstruction from airborne laserscanner and InSAR data". International Archives of Photogrammetry and Remote Sensing,* Vol. XXXIV, 3/W13, Dresden, Germany, pp. 71–78.

SITTLER, B., 2004. Revealing Historical Landscapes by Using Airborne Laser Scanning. A 3-D Modell of Ridge and Furrow in Forests near Rastatt (Germany). In: M. Thies, B. Koch, H. Spiecker and H. Weinacker (Editors), *Proceedings of Natscan, Laser-Scanners for Forest and Landscape Assessment - Instruments, Processing Methods and Applications. International Archives of Photogrammetry and Remote Sensing,* Volume XXXVI, Part 8/W2, pp. 258-261.

WAGNER, W., ULLRICH, A., DUCIC, V., MELZER, T. and STUDNICKA, N., 2006. Gaussian decomposition and calibration of a novel small-footprint full-waveform digitising airborne laser scanner. *ISPRS Journal of Photogrammetry and Remote Sensing*, 60(2): 100-112.

WAGNER, W., ULLRICH, A., MELZER, T., BRIESE, C. and KRAUS, K., 2004. From Single-pulse to Full-waveform Airborne Laser Scanners: Potential and Practical Challenges. In: O. Altan (Editor), *Geo-Imagery Bridging Continents. XXth ISPRS Congress, 12-23 July 2004 Istanbul, Turkey International Archives of Photogrammetry and Remote Sensing,* Volume XXXV, Part B/3, Istanbul, pp. 201-206.

WEHR, A. and LOHR, U., 1999. Airborne laser scanning - an introduction and overview. *ISPRS Journal of Photogrammetry and Remote Sensing*, 54(2-3): 68-82.

Airborne laser scanning of cultural remains in forests: some preliminary results from a Norwegian project

Ole Risbøl,[1] Arnt Kristian Gjertsen,[2] Kjetil Skare[3]

[1] Norwegian Institute for Cultural Heritage Research,
[2] Norwegian Forest and Landscape Institute,
[3] County Municipality of Hedmark

1 Introduction

This paper will present some results achieved in an ongoing Norwegian research project using airborne laser scanning (or lidar - Light Detection and Ranging) to map archaeological remains in forest areas[1]. Forests cover approximately 38 % of the Norwegian land area. We know that these forest areas hold a great number of cultural remains of which only a fraction is surveyed and mapped. Some major archaeological projects carried out in Norwegian forests and outfield areas during the last couple of decades have demonstrated a long term and versatile human use of forest resources. The poor situation with regard to records of cultural remains in forests is critical especially with a view to the threat of destruction that these remains are exposed to primarily due to forestry (Risbøl 2005).

To carry out systematic surveys of these large forest areas using a common search party method is not considered a realistic solution to the problem because of the expenses connected with having several search parties walking in forests all over the country. Thus remote sensing is an alternative method and especially airborne laser scanning which has a large potential as a method for improving the survey situation. As opposed to satellite remote sensing and ordinary aerial archaeology using vertical or oblique photos, airborne laser scanning offers an opportunity to detect remains below the forest canopy (Devereux et al. 2005). This is because of the ability of laser beams to penetrate vegetation which makes it possible to map small height variations on the ground surface even in areas covered with vegetation (Sittler 2004). The laser data collected in this manner can be linked with digital image-processing techniques in order to create relief maps and digital terrain models (DTM), showing all kinds of anomalies on the ground surface. This remote sensing approach makes it possible to scan large areas which can be analysed digitally in order to identify for instance cultural remains. It is an assumption that the remains are visible on the surface. To make a distinction between cultural remains and natural features as well as identifying insignificant remains is a challenge. To overcome these problems different digital techniques can be used to manipulate the data in order to optimize the interpretation of the anomalies shown on the models.

2 The Study Area

In our project a research area was chosen in Elverum municipality, SE-Norway. The area which comprises 10 square kilometres was laser scanned and the data collected constitute the basis of our analyses combined with the results of an ordinary search party survey done a few years ago covering approximately 20 % of the area. The dominant human impact in this area consists of remains of pre-industrial iron production dating from the late Viking Age to the early Middle Ages (app. AD 950-1300). This production, based on bog ore and charcoal as the main resources, has left slag heaps at the production sites and charcoal pits in the surrounding areas. The charcoal pits are rather uniform as far as size and shape is concerned while the slag heaps are less uniform and less distinct. Charcoal pits are the most common prehistoric cultural feature in Norwegian forests. In these pits which are situated in the area around the iron production sites charcoal used in iron production was produced. The average charcoal pit in this area has a diameter of 5 to 6 meters and is between 0,5 and 1 meter deep and has a dike encircling it.

The visible parts of the iron production sites usually consist of one to four slag heaps at each site. Some of the heaps are round and some are oblong and where there are more than one heap they lie at a distance of 7 to 8 meters from each other. Usually the height is less than 1 meter. The round ones have a diameter of approximately 5 meters and the oblong heaps have a length of approximately 7 to 8 meters and a width of around 4 to 5 meters. In some cases smaller heaps containing charcoal or roasted bog meant for use in the furnace are visible on the production site as well.

Cultural remains representing early tar production are also found in the area. Tar production was carried out either in kilns or oblong ditches situated in sloping terrain. In the test area only oblong ditches were found. They are approximately 10 meters long with a dike on each side and with a total width of 2 or 3 meters. At the lower end of the ditch a pit is normally found where a bucket was placed to collect the tar. Tar was extracted from resinous pinewood and was used for preserving for instance wooden buildings, boats, ropes and fishing nets.

[1] The project is partially financed by the Norwegian Directorate of Cultural Heritage.

3 The Scanning

Laser technology used for mapping terrain and vegetation from aircrafts has been evolving rapidly for the last 10 to 15 years (Næsset 1997). The lidar instrument consists of a powerful laser which emits infrared light and a receiver that records the reflected laser pulses. The receiver measures very precisely the running time back and forth for a laser pulse and makes it possible to calculate the distance to the reflected object by converting time to range using the velocity of light (3 x 10^8 m/s). The range measurements would be useless if it had not been possible to georefer the points. The exact position of the laser and information about the orientation of the airplane is required in order to be able to georefer every single laser point. Airborne lidar instruments can measure an extremely dense net of points on the ground with a pulse frequency up to 100 kHz. In addition to the pulse frequency the density of points is dependent on the altitude of the airplane, its speed and scanning angle. Flying both crosswise and lengthwise is also a way to increase the point density. A new pulse cannot be launched before the previous pulse has been received and in that respect the velocity of light is a limiting factor to the pulse launching frequency. A laser pulse is a pencil of rays that widens in accordance with the range from the laser. The diameter of the pulse is called "footprint". For airborne lidar the footprint will be from 15 to 60 centimeter depending on the altitude of the plane (Sapeta 2000). Most likely the laser ray will hit more objects on its way towards the ground and it will therefore be reflected several times. The first return might be from the top of a tree, the next from a branch of a tree and the last from the ground beneath the tree. The last returns are used to create terrain models. Even in an area with a dense canopy, a part of a laser ray often reach the ground an results in a returning signal.

The lidar survey in our project was carried out by Blom Geomatics AS using the lidar instrument Optech ALTM3100. The data collecting parameters are described in the table below.

The lidar data was delivered to us as two different sets of point data, one representing returns from the ground (the fourth return) and one representing all other points (e.g. from vegetation).

Based on the data points from the ground returns, a terrain model on raster format (DTM) was created. From the DTM, a relief model was created by illuminating it with a light source from respectively north-east and north-west and with a 25 degrees elevation. The relief model enhanced the possibility of gaining information from the DTM. Orthophotos were taken simultaneously with the laser survey in order to be able to bring in forest density as a parameter in our analyses (Figs. 1 and 2).

Technical specifications of the laser scanning in Elverum, Norway

Altitude	1000 m
PRF	70.000 Hz
Scanner frequency	47 Hz
Half angle survey	17°
Aircraft speed	75 m/s
Points per square m.	4 pt.
Date	28.06.2005

4 Results

The relief models give a precise representation of the terrain free of vegetation Divergent structures on the models can be observed due to differences in elevation (the z-values). These anomalies appear as either convex or concave forms and constitute the basis of our archaeological interpretation analysis. A visually based analysis was carried out in order to sort out recognizable structures on the ground surface. It soon turned out to be a great challenge to distinguish cultural features from natural ones, but with our knowledge about the usual location of different types of cultural remains in the landscape, as well as their shape and size, a combined total of 29 convex and concave anomalies were singled out and examined in field (fig. 3 and 4). The selection was done with a view to gaining experience in the interpretation of lidar maps. It was considered important to single out a broad spectrum of formations to work with, and anomalies that were expected to be caused both by cultural remains and natural features were selected.

As many as 17 of the chosen structures were assumed to be iron extraction sites (with one or more slag heaps). The interpretations were considered to be reliable in the case of 13 of the anomalies while four were held to be more uncertain (table 2). The succeeding field work proved that only five of the 17 actually were slag heaps, while ten turned out to be natural features such as a boulder, a wind fallen tree, a pile of rubble and even large anthills. On two locations, with quite large convex structures visible on the relief models, the terrain turned out to be absolutely flat and no structures at all were found . These anomalies might represent errors in the data processing action or could represent a moving object such as an elk.

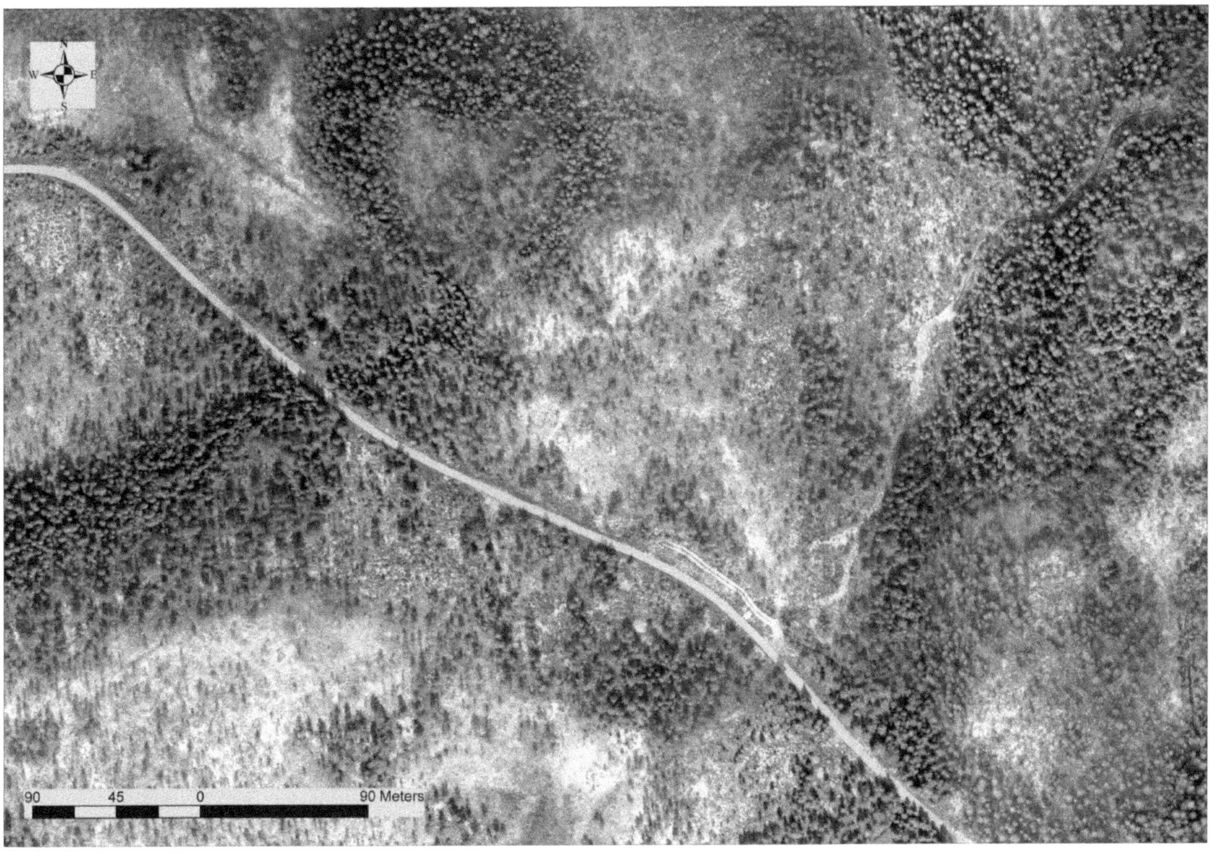

Figure 1 *Orthophoto of a section of our test area.*

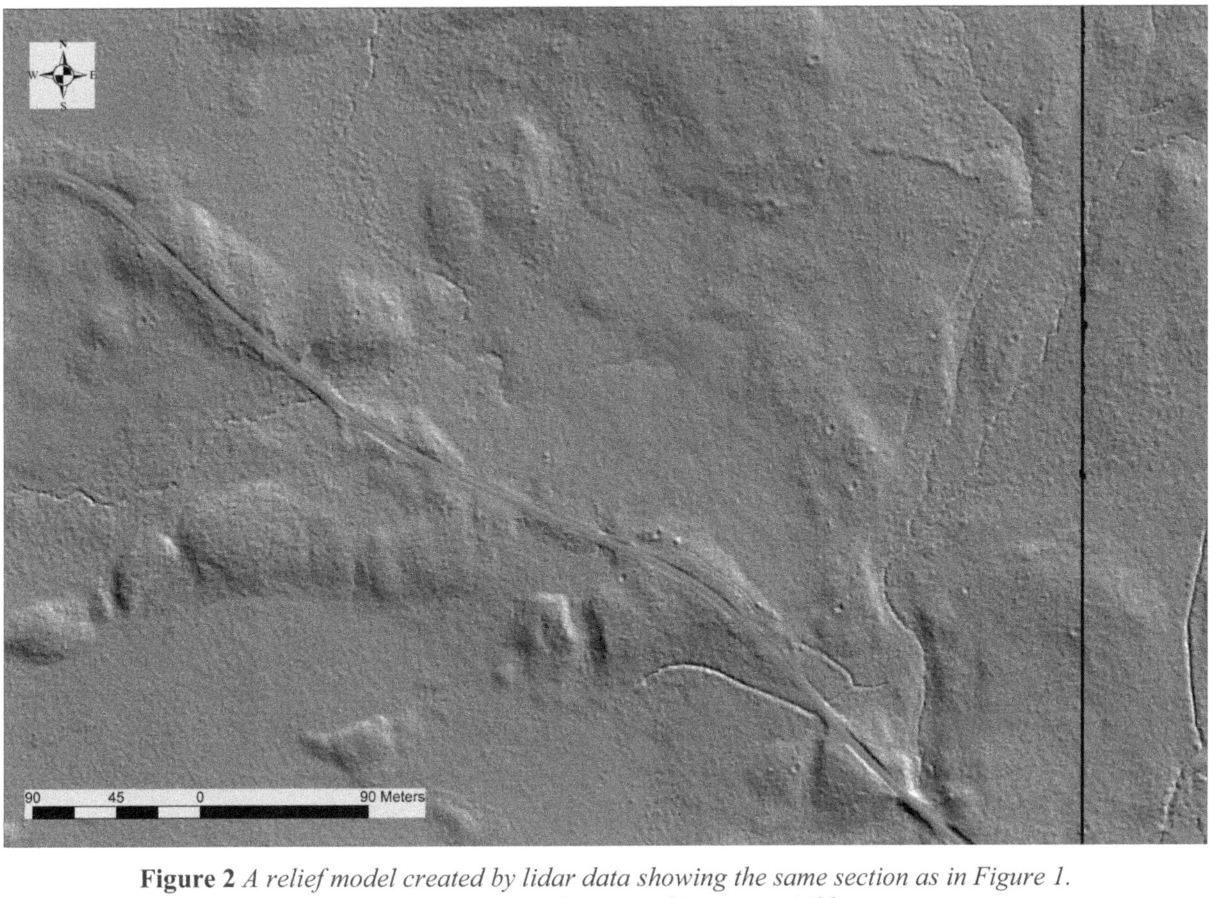

Figure 2 *A relief model created by lidar data showing the same section as in Figure 1. More concave and convex objects are visible.*

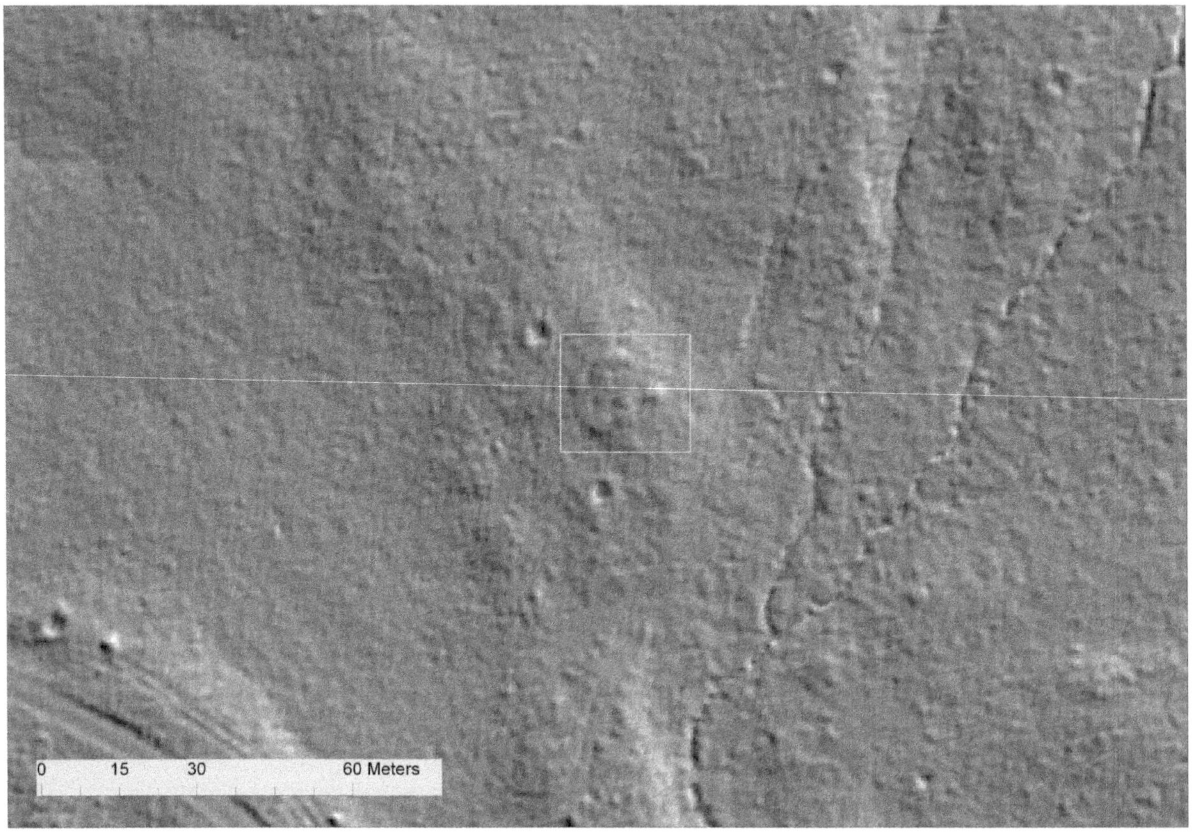

Figure 3 *A section of the relief model showing an iron extraction site with two parallel slag heaps (marked with a rectangle) and with more charcoal pits in the surrounding area.*

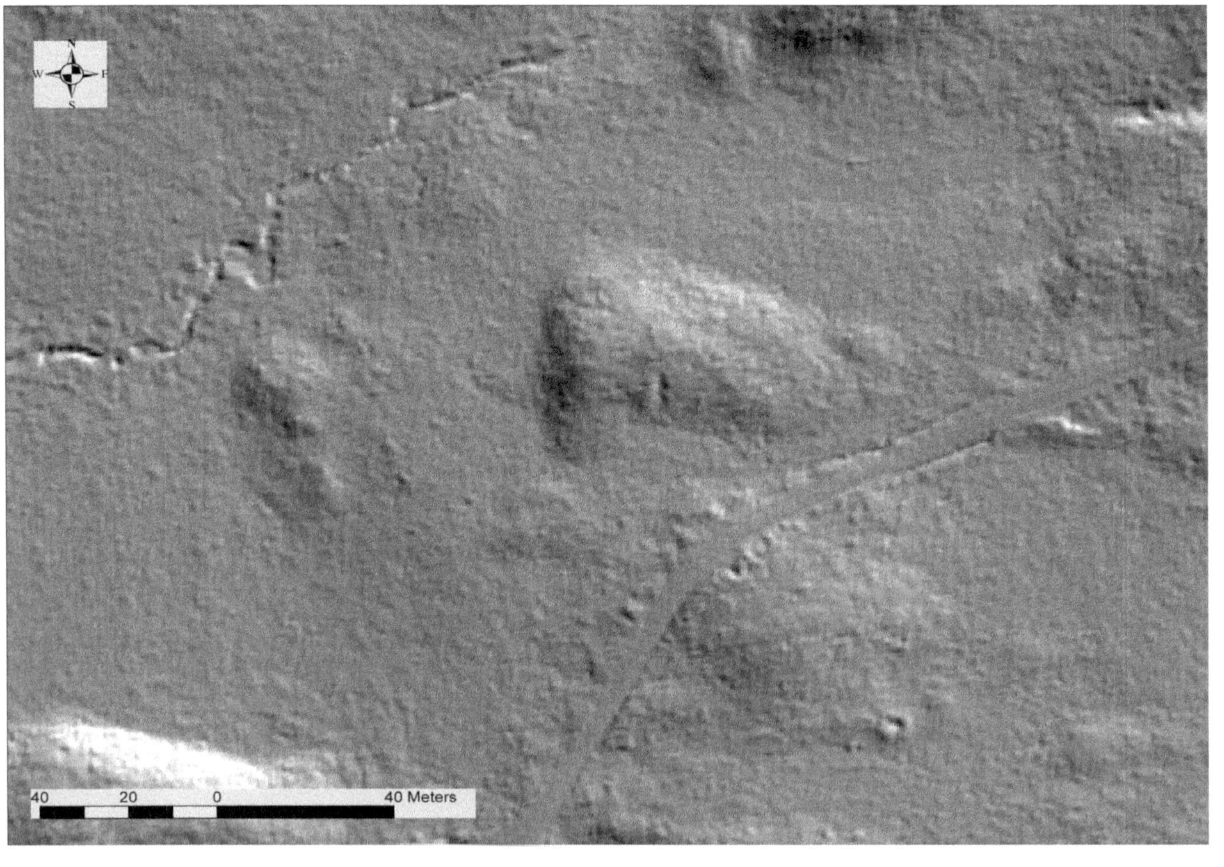

Figure 4 *In this section a ditch for tar production is visible as a crack in the middle of the illustration and there is a charcoal pit SSW of the ditch.*

The table shows the monitor interpretation of 29 structures (in parentheses) compared with the results of the field work.

Field work / Monitor int.	Iron production sites	Charcoal pits	Ditch for tar production	Natural feature	Nothing observed
Iron production sites (17)	5			10	2
Charcoal pit (3)		3			
Ditch for tar production (2)			1	1	
Field system (1)					1
Undef. cultural remain (3)		2	1		
Undef. natural feature (3)	1			2	

As the table shows the interpretation of three structures as charcoal pits turned out to be correct and in addition two of the undefined cultural remains also turned out to be charcoal pits. One of the formations interpreted as a ditch for tar production was correct while the other was created by a mixture of tufts of grass and overgrown tree stumps. So more anomalies expected to be cultural remains turned out to be natural features, but on the other hand one of the three anomalies defined as natural features was in fact an iron extraction site with two parallel slagheaps.

This plain study shows that it is a particular challenge to distinguish between man made structures and natural features. This is especially true with regard to slag heaps which are difficult to distinguish from natural heaps in the terrain. On the other hand, the charcoal pits with their regularity in shape and size are easier objects to define on the relief models. But it is also important to point out that some of the charcoal pits that were slightly distinguishable on the model in fact were very visible in reality. This brings us to the question of vegetation density as a critical factor in the use of lidar generally and in forest areas particularly. There seems to be a connection between the density of vegetation above a cultural remain and how clearly it appears on the model. It is obvious that the number of laser returns from the ground is decisive for the clarity of small terrain details. In order to illustrate this, a comparison was made of an indistinct and a distinct charcoal pit. The density of ground returns or ground points was calculated to be 2,17 p/m2 for the indistinct while the corresponding figures for the distinct pit was 3,00 p/m2.

The distinctiveness of cultural remains on relief models seems to depend on two important circumstances: The size (and regularity) of the structures and the density of the vegetation covering the remains. An obvious solution to the last hindrance is to increase the number of pulses launched from the aircraft, a matter obtained either by reducing the altitude when collecting the data or by flying over the area both lengthwise and crosswise. There is reason to believe that it will be easier to interpret many of the structures on the ground with more points resulting in a better resolution of the lidar derived models. An increased density of ground returns is however a matter both of financial resources and the challenge of handling even larger amounts of data.

A simple analysis was made in order to find out if an increased density of ground returns would improve the interpretation possibilities. The data in our test area were collected by flying east-west, but in order to get a better georeference of the plot, two lines were flown north-south. Along these two lines the traces theoretically have twice as many points. Four almost identical charcoal pits lie on a line in this area – two outside the twice flown trace and two inside. Photographs of these pits show that the vegetation is quite similar around and above the pits, with the exception of the less visible one having a small and bushy birch (Betula) standing in it. This agrees with the contention stated above that the density of the vegetation determines the visibility of the objects and might indicate that increasing the number of points does not necessarily result in noticeable better models. Another thing to learn from this analysis is that the kind of vegetation creating the canopy above the structures is of some importance, where deciduous trees might be a more critical factor than coniferous trees - the latter being easier to penetrate with laser beams. To avoid this problem the best time of the year to carry out the lidar survey is late autumn or early spring where the trees are without leaves and the ground free of snow.

So far we can conclude that it is possible to point out and make a designation of type with regard to a number of the cultural remains situated in forests by visually analyzing the relief maps. Building up experience in analysing the models is expected gradually to improve the accuracy of the interpretations. Still, there are more challenges in using lidar data concerning both the ability to point out structures and to make a designation of these that claim adopting the right analytical tools. As mentioned a couple of times this seems to be especially relevant concerning areas with dense vegetation. But the size of the remains is also of the outmost importance in determining to what extent one can map cultural remains from the air.

As mentioned above the relief models are illuminated by a light source placed with a certain direction and with a certain angle above the horizon in order to make the differences in the elevation visible. We got our data set with light from the north-east and at a 25° angle. To what extent the structures are visible on the models is strongly dependent on the direction of the light. If the orientation of a structure coincides with the direction of the light the visibility will be rather reduced compared with the instances where the light hits the anomaly crosswise.

Geograpic information system software and other kinds of visualization software (e.g. QT-Reader) make it possible to move the light around 360 degrees at the same time as the elevation of the light source can be changed. Another way to improve visibility is to do a digital manipulation of the z-value and in that way increase the size (the height) of the elevation in the relief model. In this way less distinctive remains stand out more clearly and facilitates the possibility of interpretation. The use of these tools while examining a 3-D model will improve the analytical approach to the data set and is expected to lead to better results than a pure visual study of the models. Some of the cultural remains in our test area are quite uniform with regard to shape and size. This also makes it interesting to try out the usefulness of a pattern recognition program. The idea is to feed the program with data describing the most typical features of a cultural remain and carry out a digital scanning of the relief model in order to point out similar structures. In our case the most obvious category to use in a test like this will be the charcoal pits. It is expected that the ditches for tar production also appear so homogeneous and clear that they will be useable while it is more doubtful when it comes to the slag heaps on iron extraction sites.

5 Concluding remarks

Studies of relief models based on laser data sets make it possible to point out and define some types of cultural remains in forest areas. A pure visual approach is not an adequate method if the aim is to show also less visible structures. It is necessary to try out and develop different kinds of analytical tools in order to detect and define as many cultural remains as possible. An optimized use of lidar with the purpose of mapping cultural remains in forest areas requires focus on these three main challenges: 1. How to cope with dense vegetation (and especially deciduous forest), 2. How to distinguish between cultural remains and natural features and 3. How to intercept small and insignificant cultural remains. It is also necessary to work with a larger range of cultural remains in different kinds of forest and landscape in order to extend the frame of reference. Cost-benefit analyses, where expenses are compared between a lidar approach and traditional survey methods, are needed. The experience so far is that lidar is a well suited method in our efforts to map all kinds of visible cultural remains in forest areas, so far in a rather large-meshed manner but indubitably with a very large potential.

6 References

DEVEREUX, B.J., G.S. AMABLE, P. CROWE & A.D. CLIFF 2005: The potential of Airborne Lidar for Detection of Archaeological Features under Woodland Canopies. *Antiquity* 79 (305): 648-660.

NÆSSET 1997: Determination of mean tree height of forest stands using airborne laser scanner data. *ISPRS Journal of Photogrammetry and Remote Sensing* 52: 49-56.

RISBØL, O. 2005: Pre-industrial Technology in Norwegian Forest Areas – a Vulnerable Cultural Heritage. *Heritage of Technology – Gdansk Outlook 4. Proceedings from an international conference.* University of Gdansk, Polen: 269-277.

SAPETA 2000: Have You Seen the Light? LIDAR Technology is Creating Believers. GEOWorld.

SITTLER, B. 2004: Revealing historical landscapes by using airborne laser scanning. A 3-D model of ridge and furrow in forest near rastatt (Germany). *Proceedings of the ISPRS working group VIII/2: Laser-Scanners for Forest and Landscape Assessment.* International Archives of photogrammetry, remote sensing and spatial information sciences, vol. XXXVI, part 8/W2: 258-261. Freiburg 2004.

Airborne multi sensor remote sensing of exposed and subsurface archaeological remains at Itanos and Roussolakkos, Crete

Aled Rowlands,[1] Apostolos Sarris,[2] James Bell[1]

[1]University of Wales Aberystwyth, UK
[2] Institute for Mediterranean Studies, Greece

1 Introduction

A suite of airborne remote sensing images and techniques have been acquired and applied to two archaeological sites in eastern Crete, Greece, to assess their potential for locating known buried remains and to delineate subsurface remains beyond the spatial limits of ground geophysical data in order to permit future targeted geophysical surveys and archaeological excavations. The sites have previously been the focus of intensive geophysical surveys through different ground-based techniques and as such they provide good test areas for the application of airborne remotely sensed data.

2 Study Sites

Data for the areas of Itanos and Rousollakkos in eastern Crete are presented. Itanos is located 10 km north of Palaikastro, in Lasithi prefecture, close to the Vai Palm Forest (Figure 1). Its harbour was of major importance for trade between Crete and the eastern Mediterranean, and its economic growth resulted in it being the first Cretan town to have a currency. Itanos is marked mainly from three periods: Geometric, Roman and Early Byzantine, while the exact periods of original occupation and abandonment are not well established. During the Roman conquest and the first Christian times, the town continued to flourish, until it was gradually destroyed, probably by the Arabs (Spyridakis *et. al.* 1986 and Sanders *et. al.* 1982). The archaeological site covers approximately 16,000 m^2, of which only 1% has been excavated. A defence wall is partially obvious on sloping terraces that separate Itanos from Vai. Archaeological remains at the site include two First Byzantine basilicas, a large Hellenistic cemetery, an observation tower, isolated tombs etc. The harbour and the coastal installations seem to have been covered by alluvial deposits, a process which has been attenuated by the rise in sea level due to local tectonic activity.

The Bronze Age town of Palaikastro is situated at Rousolakkos two kilometres from the village of Palaikastro. It is estimated that the town may have covered 50,000m^2 (MacGillivray *et. al.* 1984). Much of the earliest excavations have revealed houses and streets, with a long main street running east-west, though no palace/temple has yet been found. Geophysical investigations have been carried out in the area by the British School of Athens.

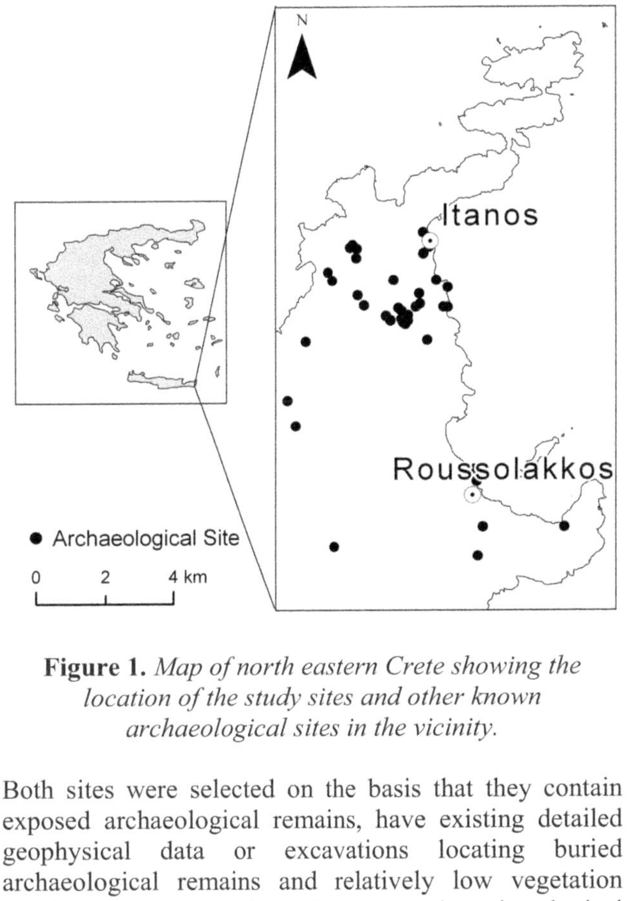

Figure 1. *Map of north eastern Crete showing the location of the study sites and other known archaeological sites in the vicinity.*

Both sites were selected on the basis that they contain exposed archaeological remains, have existing detailed geophysical data or excavations locating buried archaeological remains and relatively low vegetation densities. At both sites the exposed archaeological remains are well conserved and are mainly constructed from local stone. This fact poses a potential problem for remote sensing since their reflectance characteristics are likely to be similar to the surrounding material. At Itanos, transects and grids of geophysical data consist of soil resistance mapping, 2D and 3D electrical tomography, magnetic mapping, electromagnetic mapping, seismic refraction and ground penetrating radar. These surveys have shown that archaeological remains at Itanos exist from the surface to 1-3 meters below the surface (Sarris *et. al.* 1998, Vafidis *et. al.* 2003). Large sections around the Minoan coastal site of Rousollakkos have also been surveyed by magnetic and electrical resistance techniques with promising results.

3 Data Acquisition and Processing

Airborne imagery was acquired over a two day period in April 2004 as part of the NERC Mediterranean Flight Campaign. Data from three primary sensors, namely the Airborne Thematic Mapper (ATM), the Compact

Airborne Spectrographic Imager (CASI) and lidar, will be considered. Day and night time (thermal) image pairs were captured for the Itanos site whilst only day time imagery is available for Rousolakkos. The CASI sensor is limited to the visible and near infrared (NIR) and was programmed to operate in spatial mode acquiring data in 15 channels from 449 nm to 940 nm at a spatial resolution of 2.0 m, whilst the ATM sensor acquired data in 11 broader bands, with a spatial resolution of 2.5m. The ATM sensor provides important information in the Short Wave Infrared (SWIR) and the thermal (TIR) portions of the spectrum with respect to soil properties and heat capacity. The lidar data were acquired using an Optech ALTM 3033 high-resolution airborne laser scanner with a point density of 1.0 m. Soil temperature profiles were acquired with a series of Skye Instruments sensors coupled to a data logger in an area adjacent to the archaeological site in order to calibrate and interpret the thermal data.

A range of processing techniques have been applied to the reflectance data derived from the CASI and ATM sensors as well as the intensity and elevation data from the lidar instrument (Rowlands and Sarris *in press*). The fundamental premise is that buried remains are likely to alter the physical and chemical characteristics of the soil compared with those of the surroundings due to variations in soil depth and moisture, which may be manifested within the reflectance data, whilst minor topographic variations are potentially detectable using the lidar data.

The data have been processed with anomaly detection and edge detection algorithms in order to highlight anomalies and/or linear features that can be related to surface or subsurface remains (Mavrantza and Argialas 2003, Masini and Lasaponara 2006). The latter algorithms are collectively known as convolution filters in which a neighbourhood operation is performed such that each output pixel is the weighted sum of the neighbouring input signals. A suite of different edge detection algorithms have been applied to the data including Laplacian, Sobel Filters and Roberts filters. For the purposes of demonstration in this paper all the images are derived from the application of a directional filter, which is a first derivative edge enhancement filter that selectively enhances the image features having specific direction components.

4 Results

Through a combination of CASI, ATM and lidar data, surface remains have been classified and mapped effectively at both sites using an object oriented approach. Whilst surface archaeology has been well mapped by the various sensors (Figure 2), the delineation of subsurface remains at Itanos based on remote sensing remains problematic, although tentatively it is claimed that certain features have been identified.

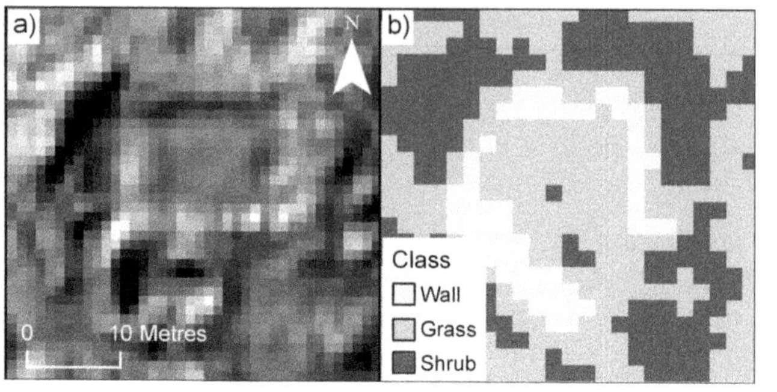

Figure 2 *a) Shaded relief image of the southern most basilica at Itanos derived from the lidar data and b), a classified image of the same area based on the CASI and lidar data derived from an object-oriented approach using the eCognition software.*

At Itanos, construction slabs of surface buildings such as the basilica, consisting mainly of sandstone were easily distinguished from the surrounding soil matrix. Similar was the situation of the surrounding wall of the site, whose trace along the southern hill of the site was delineated by comparing day and night time ATM thermal images (Figure 3). Features of subtle elevation differences, such as abandoned terraces and depressions were also identified through lidar images. On the other hand, the possible inland extension of the port, formed by eroded phyllites extended more than 10m below the current surface, was difficult to be detected.

At Rousolakkos, the processing of the lidar imagery is hampered by the presence of vegetation, particularly olive trees in the area immediately around the archaeological site (Figure 4). To overcome the presence of trees and to assess the value of applying airborne lidar to monitor slight variations in topography that could be archaeologically related, an algorithm was applied to effectively remove the vegetation and to produce a bare earth surface model. The outcome of the procedure in the vicinity of the archaeological site at Rousolakkos was inconclusive, due to significant interaction between the lidar and branches resulting in limited return from the ground itself. A further attempt is being made to develop algorithms to study the micro topography under the olive trees.

Beyond the immediate archaeological site both the ATM and CASI imagery indicate linear features extending in a NW-SE direction (Figure 5). These features are not visible in aerial photography or lidar imagery of the area. The linear features are most prominent when the imagery is subjected to directional algorithms. Unfortunately, there is no night time thermal imagery of this area to further interrogate the hypothesis that these may be archaeologically related remains. The area is well beyond existing excavations and may therefore warrant future ground based geophysical research.

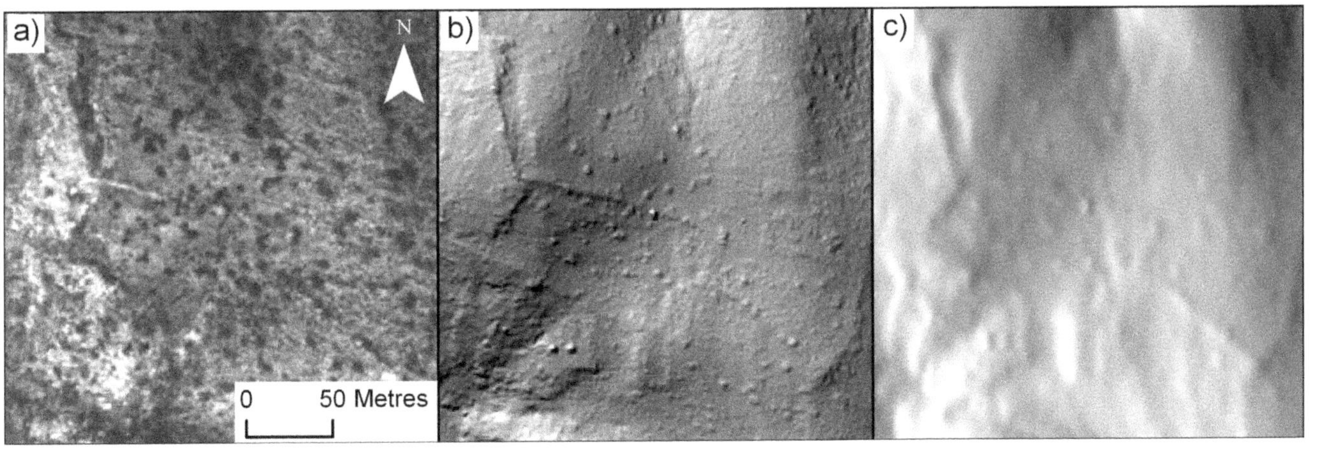

Figure 3 *Imagery focussing on the ancient settlement on the hill to the south of Itanos: a) thermal channel data derived from daytime ATM; b) shaded relief image based on the lidar data; and c) lidar data processed using the a directional edge detection algorithm to highlight the presence of the ancient garrison wall.*

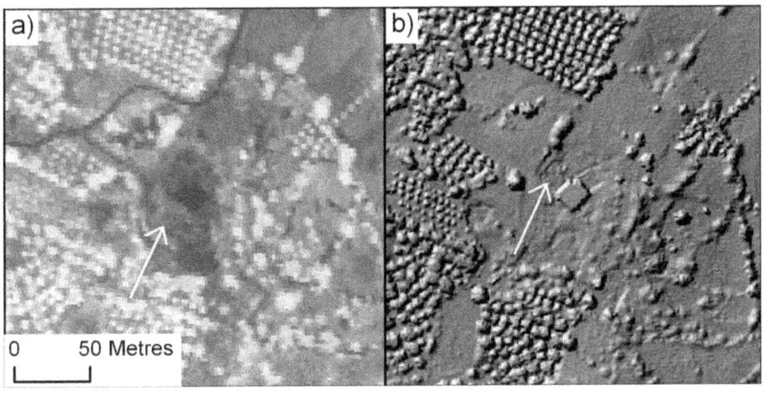

Figure 4 *a) Normalised Difference Vegetation Index (NDVI) image of the Rousollakkos area with the archaeological site placed in the centre of the figure (darker shade) and surrounding vegetation in lighter shades; b) shaded relief image of the same area highlighting excavated areas.*

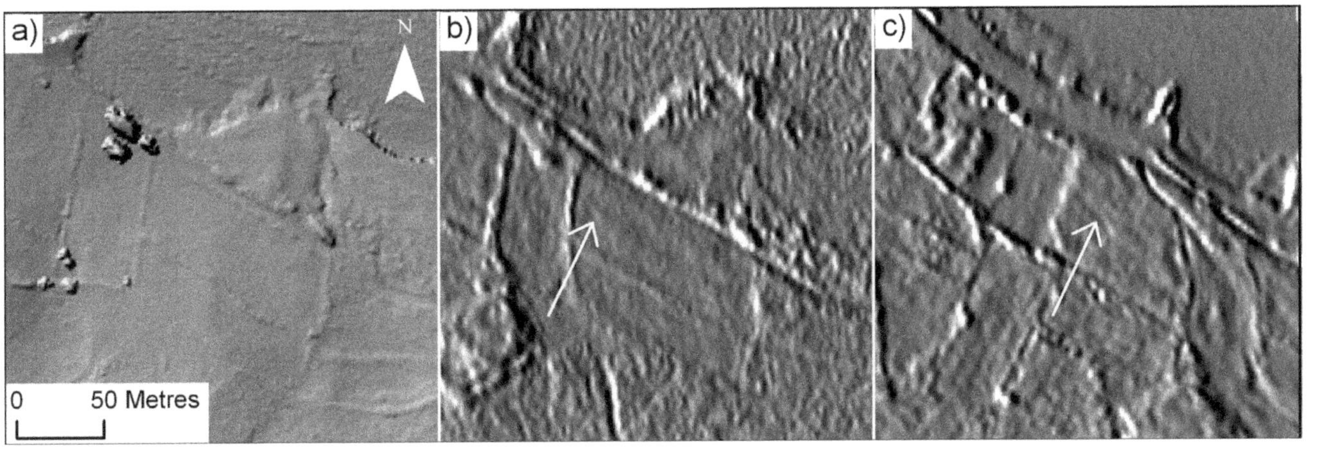

Figure 5 *Area to the SE of the main archaeological site at Rousolakkos. a) Shaded relief image based on the lidar data; b) CASI data of the same area subjected to a directional edge detection algorithm to highlight linear feature running NW-SW and c) ATM edge detection output of an area to the NE of a and b showing another linear feature.*

5 Conclusions

Particular difficulties encountered in the detection of archaeological remains stem from the presence of vegetation and mixed pixels. A key consideration in being able to determine the presence of both surface and shallow subsurface features is that the remote sensing takes place at a spatial scale appropriate to the resolution of the objects. In this particular case study the acquisition resolution was too coarse to detect known subsurface remains. Higher spatial resolution imagery acquired at a spatial scale close to the actual patterns of structures would decrease the number of mixed pixels and assist in the classification. However, the study has indicated zones of archaeological interest that require further investigation. High spatial and spectral resolution imagery of this nature does therefore have merit and can be used as a means of highlighting areas that can then be targeted by more detailed geophysical surveys and/or excavations.

Acknowledgments

The authors grateful acknowledge the support of NERC for the acquisition of the airborne imagery (Award No. MC04/22) and for the loan of the field spectrometery equipment (Loan reference 400.1203). The financial support of the University of Wales Aberystwyth is also appreciated. The authors wish to thank the field teams that have worked towards the collection of ground data, namely: Constantinos Athanassas, Talfan Barnie, Marianna Katifori and Katerina Kouriati.

References

MACGILLIVRAY, J.A., SACKETT, L.H. and DRIESSEN, J. 1984 An Archaeological Survey of the Roussolakkos area at Palaikastro. *BSA.* 79: 129-159.

MASINI, N. and LASAPONARA, R. 2006 Satellite-Based Recognition of Landscape Archaeological Features Related to Ancient Human Transformation. *Journal of Geophysics and Engineering.* 3: 230-235.

MAVRANTZA, O. D. and ARGIALAS, D.P. 2003 "Implementation and Evaluation of Spatial Filtering and Edge Detection Techniques for Lineament Mapping - Case Study: Alevrada, Central Greece" in *Proceedings of SPIE International Conference on Remote Sensing: Remote Sensing for Environmental Monitoring, GIS Applications, and Geology II.* M. Ehlers (ed.). Bellingham, WA, SPIE Press. 4886: 417-428.

ROWLANDS, A.P. and SARRIS, A. *in press* Detection of Exposed and Subsurface Archaeological Remains Using Multisensor Remote Sensing. *Journal of Archaeological Science.*

SANDERS, I. 1982 *Roman Crete: An Archaeological Survey and Gazetteer of Late Hellenistic, Roman and Early Byzantine Crete*, Aris and Phillips, Warminster, England.

SARRIS, A., VAFIDIS, A., MERTIKAS, S., GUY, M., VRONTAKI, E., MANAKOU, M. and KALPAXIS, T. 1998 "Ancient Itanos (Erimoupolis, Lasithi): an Archaeological Site as a Remote Sensing Laboratory" in *Proceedings of the 31st International Symposium of Archaeometry*, Budapest, Hungary, 157-164.

SPYRIDAKIS, S. 1986 *H Archaiki Itanos: Scholia kai Paratiriseis*, Amaltheia, Crete.

VAFIDIS, A., MANAKOU, M., KRITIKAKIS, G., VOGANATSIS, D., SARRIS, A. and KALPAXIS, T. 2003 Mapping the Ancient Port at the Archaeological Site of Itanos (Greece) Using Shallow Seismic Methods. *Journal of Archaeological Prospection.* 10: 163-173.

The potential of LIDAR in assessing elements of cultural heritage hidden under forest canopies or overgrown by vegetation: Possibilities and limits in detecting microrelief structures for archaeological surveys

Benoît Sittler, Sabine Schellberg

Institut für Landespflege Universität D- 79085 FREIBUR
G benoit.sittler@landespflege.uni-freiburg.de; Sabine.schellberg@landespflege.uni-freiburg.de

Introduction

Because conventional remote sensing imagery obtained for the purpose of estimating surface topography is usually constrained by the lack of a third vertical dimension, its use for development of realistic three-dimensional elevation models and morphology assessments was until now difficult.

Recent technology advances such as direct digital imagery acquisition and Light Detection and Ranging (LIDAR)-derived data sets are quickly becoming a promising tool in many disciplines devoted to the study of the earth's surface. Capable of generating digital elevation models (DEMs) more than an order of magnitude more accurate than those currently available, LIDAR data offers the opportunity to assess the shape of the earth's surface in a manner and at resolutions not previously possible especially in situations such as wooded areas.

Application in the field of archaeology however lagged behind and until now, very few examples showing the usefulness and the prospects of such approaches were presented (Bewley et al, 2005). This is especially the case for structures hidden under forests (Sittler, 2004; Devereux et al., 2005) or vegetation canopies.

Ongoing projects aiming the documentation of remains of ancient landscapes in Baden Württemberg (SW Germany) offered a possibility to use laser data collected for altimetric purposes. In the first case, it concerns ridge and furrow hidden under forest canopies, in the second case an abandoned older irrigation system with related structures often no longer visible due to obscuring vegetation. Through observations resulting from these approaches and their comparisons, the purpose here is to provide examples of how LIDAR is performant to improve the assessment and documentation of heritage resources in a way unexpected until now.

The study sites

Ridge and furrow in Rastatt

Ridge and furrow are remnants of past cultivation practices that give the landscape an undulating corrugated appearance, with height differences generally oscillating between 30 and 60 cm. These earthworks were made by ploughing in a clockwise spiral, starting in the middle of the strip, with the plough throwing the soil to the right. They were once a familiar sight across many parts of Europe (Beresford & Saint Joseph, 1966, Ewald, 1969).

Assessing their extent and state of conservation in areas where they have not yet been leveled off by changes in agricultural practices and promoting their conservation is therefore an important issue to landscape and heritage preservation agencies.

Such a site is well preserved under a greater forest (mixed stands of Norway spruce, pines and beeches) near Rastatt in SW Germany, where arable land containing these earthworks was abandoned during the 17. Century and later converted to woodland. A preliminary assessment through field walking (Hauger et al., 2001) had previously identified the site as being one of the largest contiguous area with such remains of the open field system in the whole South-West Germany but many of the features of this site escaped assessment since micro relief structures are hidden under forest vegetation and could not be depicted. Given the fact that these features extend over a wide expanse, remote sensing was regarded as a pertinent approach for mapping them. Because traditional aerial photography is limited by the inability of optical sensors to collect information from beneath tree canopies, the purpose here was to test application prospects of the altimetry laser scanning (Lidar technology) for elevation data capture and for detecting and modeling terrain structure of ridge and furrow.

The ancient irrigation meadows of Central Baden

The other study site includes an ancient irrigation complex along the Dreisam River in the Freiburg Region whose lay out, when seen in plan, displays ladder like patterns.

Parts of the Dreisam floodplain was managed as irrigation water meadow in the mid of the 19th Century. This system of management involved the construction of a complex network of channels and drains which allowed irrigation to the meadows. The system fell into disuse from the mid 20th Century. However, parts of the area are still used as meadows.

A schematic view of how part of this irrigation system was constructed may be taken from the picture in Figure 1.

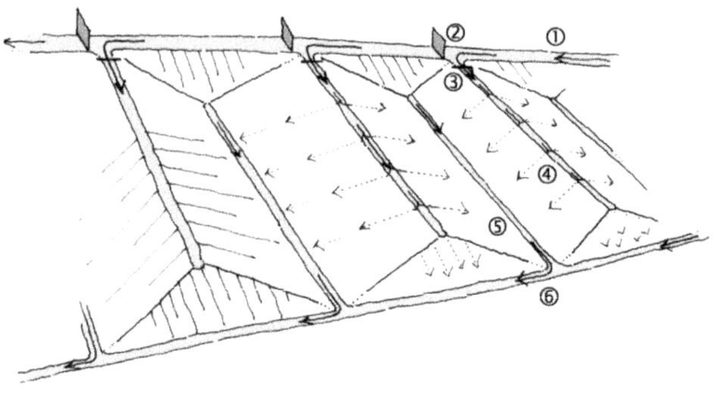

Figure 1 *Simplified model of the overland flow method according Schewior 1941 (Schellberg, 2005)*
① head main, ② sluice, ③ hatch, ④ main, ⑤ drain, ⑥ head drain

Elevation differences generated by this system of channels, ditches and drains were of the order of a few decimetres. Many of these structures were levelled off in the last decades for use as modern arable land, with some ditches and other structures like masonry sluices and hatches getting weathered and overgrown. As a result, crucial pieces of the pattern are no longer recognizable when buried beneath brushes or due to minor elevation differences. To assess how such an ancient landscape worked, use of LIDAR was tested in complement to terrestrial survey methods to reconstruct the overall plan

Use of lidar for 3-d modelling of archeological features

Principle of laser and its application to remote sensing

Lidar, the acronym for Light Distance and Ranging, operates on the principle of reflecting a pulsed laser off a surface and measuring the time of flight to determine the distance traveled by the laser pulse. By knowing the sensor position, the distance and the incidence angle of each measurement, the coordinates of the reflecting object is calculated. The application of this principle to remote sensing approaches (Ackermann 1999) by using aircrafts or satellites as recording platforms now opens new perspectives in the detection and geometric and altimetric assessment of landscape elements.

LiDAR doesn't discriminate, collecting features such as trees, cars and buildings, as well as the ground. The entire data set is known as a "point cloud." Automatic algorithms can remove surface classes such as vegetation, buildings and other structures. Once removed, the processed LiDAR data set is known as a "bald earth" DEM, which is used for orthorectification and contour creation. Limitations arise when the terrain surface is under-sampled. In other words, LiDAR has finite point spacing, and for example it doesn't capture sharp terrain breaks precisely.

Preliminary applications of this new technique were recently considered as promising for digital terrain modeling in wooded areas (Pfeiffer et al., 1999; Hyyppä et al., 2000) while its potential for heritage purposes remained to be demonstrated.

The Laser scanning DEM Project of Baden Württemberg

The Land and Survey bureau of the Baden-Wurttemberg state has commissioned the Topscan Company for the purpose of updating the available DEM for the entire state (Hoss, 1997; Gültlinger et al., 2001). It was aimed at providing comprehensive altimetry data at a resolution of around one meter mesh width in space and less than 50 cm in height. The project started in 2000 and was completed by the end of 2005. Technical details of the flight missions are listed in the table below.

Data processing

The dataset provided by Topscan were first preprocessed by the services of the Land surveying agency. This step included filtering of ground and height points, as well as georeferencing the data in Gauss Krüger coordinate system.

The data made available for the present studies were in ASCII format. By interpolating the data in Erdas Imagine a Digital Surface Model was created from the height points, while the corresponding Digital Terrain Model was obtained from the ground points.

In the case of ridge and furrow, subsets and spatial profiles were established to analyze and assess special characteristics of the prominent features.

Altitude	Scan frequency	Scan angle	Wave length	Strip width	Pulse repetition	Point spacing	Flight velocity
1000 m	25 Hz	+/- 20°	1,55 μm	400 m	25.000 Hz	1,5 m	80 m/s

Results

The detection and assessment of ridge and furrow topography

While the scenery in first pulse (Fig. 2 top) does not reveal the ridge and furrow microrelief, the Digital Terrain Model (Fig. 2 bottom) clearly shows the typical corrugated surface topography. This is especially evident when visualizing in 3-D images with different viewing directions (Fig 3, below). These figures clearly portray the pattern of the earlier medieval landscapes with the strips and furlongs. Additional structures that may be detected at a first glance include also earth mounds seen in the upper right part of figure 1b. In contrast, ridge and furrow structures may no longer be detected in the open landscape with arable land (left side of the scenery) where they have been leveled off.

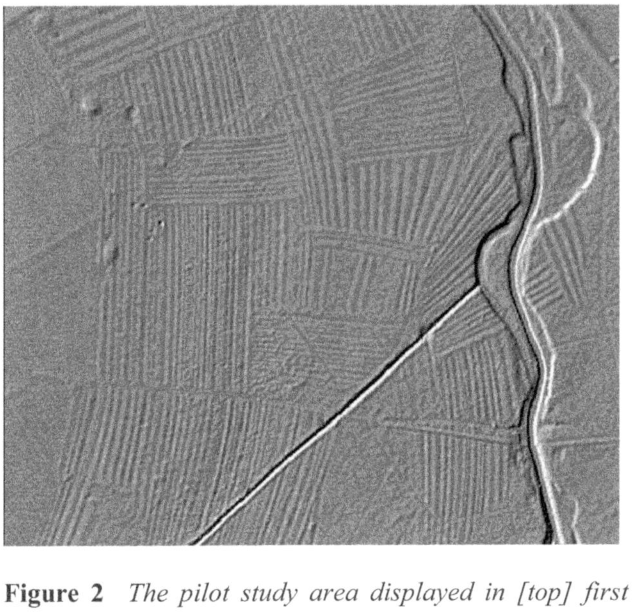

Figure 2 *The pilot study area displayed in [top] first pulse; [bottom] last pulse.*

Besides the visual detection of the patterns (another purpose was to assess the sizes (surface area, length, width) and heights of the ridge and furrow. These parameters were determined using the 3 D analyst extension of ArcView 3.2. It involved manual delineation of single furlongs (Fig. 4).

The 3D analyst extension can measure surface area and volume of surfaces. The surface area is measured taking height into consideration. This parameter is different from the 2D planimetric extent of every model. The 2D planimetric surface is a virtually square area when a surface is viewed from above. Surface area on the other hand is the real surface and gives sufficient information about surface roughness and undulation. The greater the difference between the two parameters the rougher the surface is. The length is determined by measuring the distance across a ridge and the mean width of a furlong calculated by dividing its planimetric area with its corresponding length.

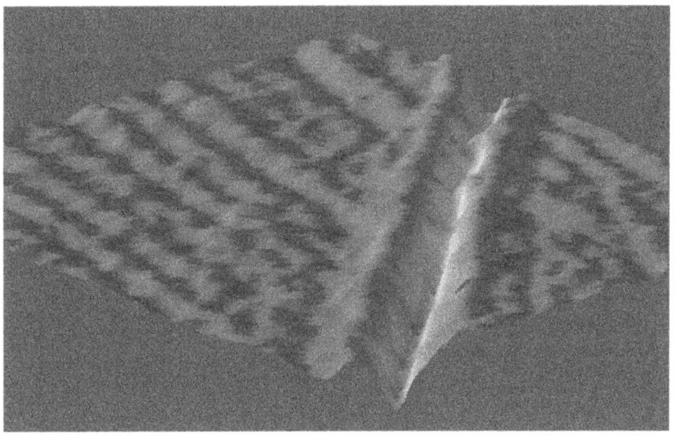

Figure 3 *3-D Visualization of ridge and furrow (Virtual GIS)*

The determination of height proceeded with the establishment of spatial profiles across the ridges. The spatial profiles are later demonstrated in the form of graphs from which height values are easily read (Fig 4). Distance is plotted on the horizontal axis and elevation (height) on the vertical axis.

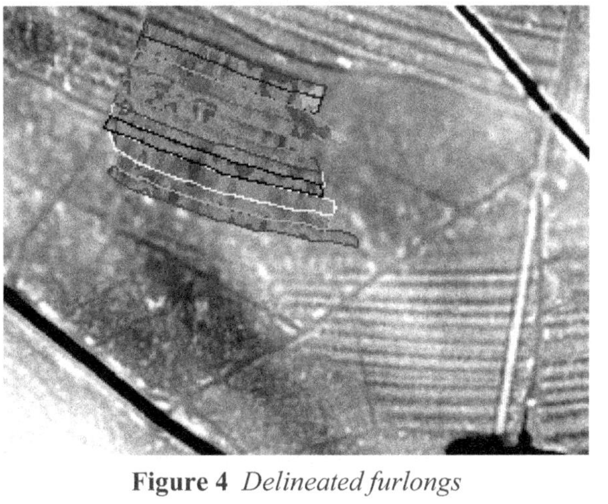

Figure 4 *Delineated furlongs*

The detection and assessment of the irrigation system

The commanding view of the site provided by the Digital Terrain Model clearly illustrates the artificial irrigation structures in the landscape (Fig. 6).

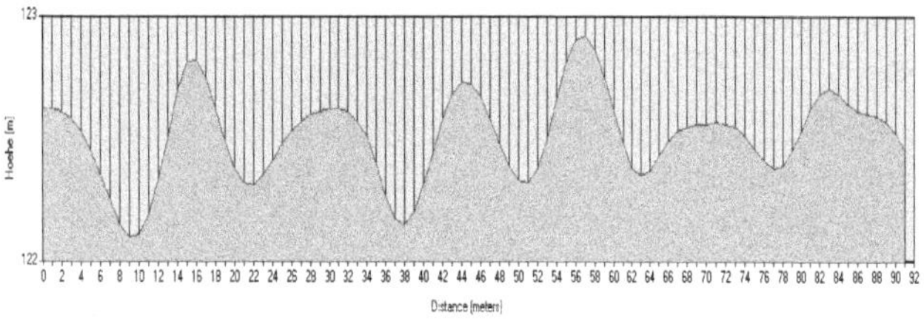

Figure 5 *Spatial profile illustrating height differences (with height ratio exaggerated to highlight the differences between the ridge and the furrow)*

The formal type of irrigation technique – in this case overland flow method - can be comprehended in this digital terrain model. The artificial relief structure has a total difference of a few decimetres. This is being reflected sufficient enough in the laser scanning data to clarify what type of system was in place. As examples cross-cuts were made through the dammed head main and a section of the irrigation system (see Figures 7 - 8).

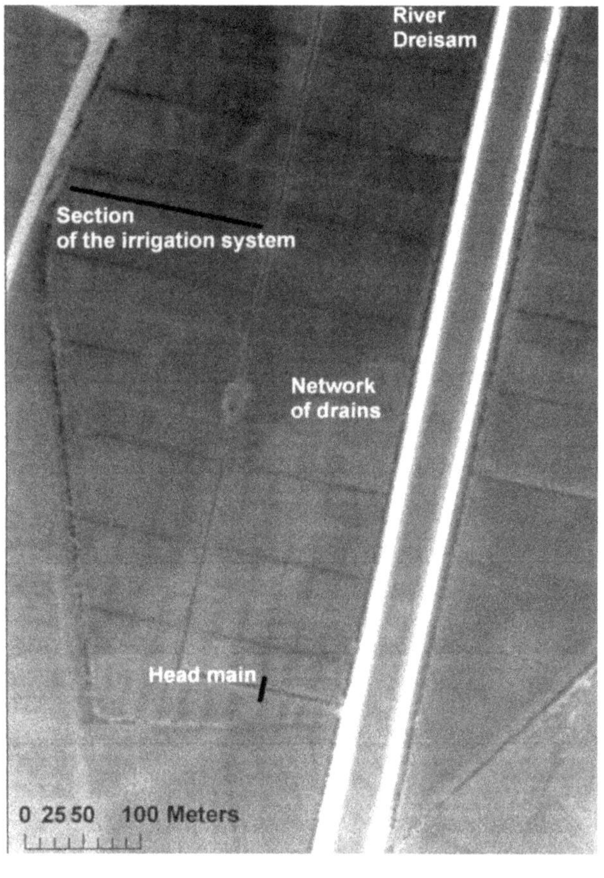

Figure 6 *Digital terrain model showing parts of the ancient irrigation system in Nimberg. The highest identified level in the date is 191,271m the lowest is 182,534m. Black lines map the cross-cuts. Courtesy: Landesvermessungsamt Baden-Württemberg (//www.lv-bw.de///), Az.:2851.9-D/4536/*

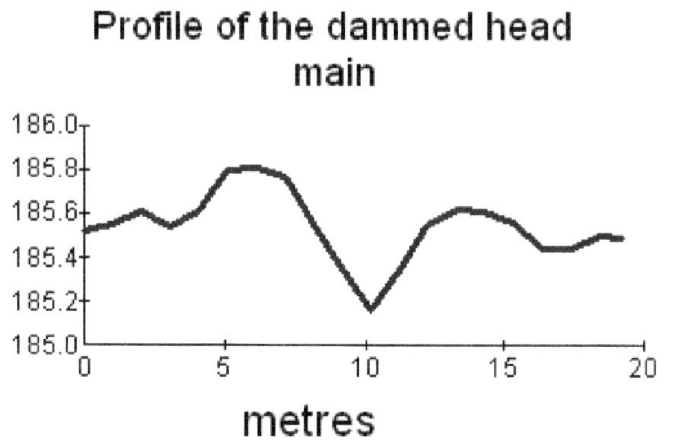

Figure 7 *Cross-cut of head main*

Some structures like profiles from ditches were also assessed as field measurements. They will allow comparisons with laser scanning measurements for the purpose of determining the accuracy of profiles generated by these models but these controls will only be available at a later stage.

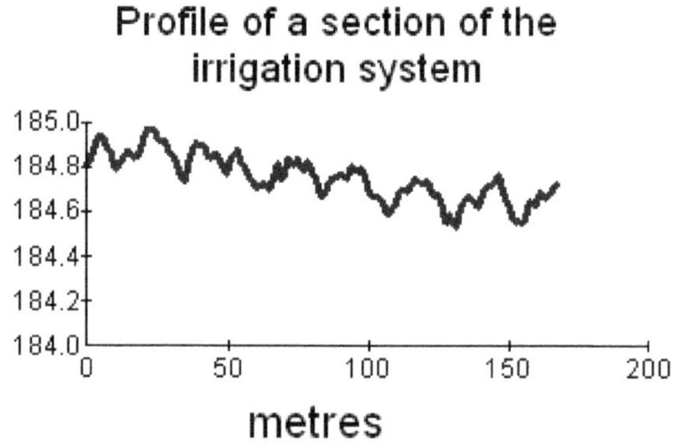

Figure 8 *Cross-cut of a section of the irrigation system.*

Discussion and conclusion

The results presented above confirm that laser measurements do clearly reveal topographic structures like those displayed by these corrugated fields and by this ancient system of irrigation ditches and drains, despite being hidden under obscuring vegetation canopies. The fact that many of the structures are aligned certainly helps to depict and grasp the overall microtopographic patterns. The laser data were acquired in winter during leaf-off conditions to enhance the fraction of the laser pulse energy that will penetrate the canopy and reflect off of the ground surface. Considering that the penetration rate is around 65% for deciduous forests (in winter) against only 30 % for coniferous stands, one can estimate that this rate may be around 50 % for the mixed stands covering the ridge and furrow. As the mean point spacing of these raw airborne lidar data points was around 1,5 meter, possibly that on average only one georeferenced 3-D point was available every two to three meters to generate the final 3-D elevation model. And this proved here to be sufficient for these clear visualizations of the corrugated microtopography, suggesting that other similar shaped structures may become depicted when opting for similar resolution in forested areas. When the morphology of smaller structures of only a few square meters has to be captured, one should then opt for higher resolution for highlighting them (2 to 5 points/m^2). In the more open landscape of the water meadows, many of the structures were less pronounced and narrower, such as the weathered and grown over and partly soil filled channels. Despite partly dense grass vegetation, these elements could still be featured with the same point spacing (1,5 m).

In the case of the ridge and furrow, the commanding view of these structures under forest as provided by this technique recalls also in some respect the aerial oblique photographs of ridge and furrow still visible in unforested open countryside under grassland (Beresford & Saint Joseph, 1979). Compared to the terrestrial mapping (Hauger et. al, 2001) this image unveils many details and structures that were overlooked during these ground surveys. In addition, the position of other structures included in this landscape as revealed by the laser data is quite instructive. As an example, the Sandbach River (right side of the scenery in Fig. 1b) is shown to cut the ancient pattern of ridge and furrow, indicating that it is of anthropic origin. The same applies of course to the very straight ditch (Hardtgraben) crossing the area: In the present case, as written archives suggest that it was built during the 15 century, evidence is provided that ridge and furrow were generated at an earlier date.

Likewise, this technique also enabled new insights into the original lay out and design of ancient irrigation systems by revealing crucial pieces of the pattern that have not previously been recognized. Quite insightful were in this respect also the altimetric profiles reproducing the gradients within the whole complex and that facilitate the understanding of the functioning of the system. In other words, it could be portrayed how water circulated and was diverted from carrier prior getting again out of the system. Conventional remote sensing data could have hardly offered such insights in the control and management of such ancient systems.

As a whole, the resolution of this data and the relative ease of capture compare favorably with existing data sources, allowing large areas of landscape to be captured as three-dimensional surface data facilitating a scientific, analytical approach to the landscape.

In terms of costs, minimum entry prices for smaller projects in Germany range from less than 7000 Euro to upwards 20000 Euro depending upon the data provider. But of course, for projects of only the order of tens of acres, the per acre cost becomes significantly higher due to the fixed sunk costs of mobilizing the lidar sensor. In the case of Baden Württemberg that now disposes on a nearly comprehensive coverage of the land (35000 km^2), data may become available through the surveying agency (LVA) at following conditions: 15 Euro/km^2 for untreated data and 60 Euro/km^2 for end-products usable for DEM generation.

Such conditions provide therefore a very attractive cost-effective alternative for future surveys of remnants of past landscapes. For archeologists in general, it may open historic structures and archeological sites to more visually detailed, accurate and efficient examination. Examples provided here showed how these approaches could save time-consuming fieldwork such as laborious measurements and can therefore be regarded as a very useful complement to classical aerial photographs, especially when forest canopies prevent visual identification of hidden patterns.

To promote such operations, it could be advisable to arrange and share costs for flight missions with other potential users of the LIDAR data like flood protection agencies, urban planners and other institutions interested in updated and fine scaled altimetric data.

Acknowledgments

First and foremost we would like to acknowledge the friendly co-operation given by the Landesvermessungsamt Baden Württemberg (M. Gültlinger) who provided laser data used for these studies. Support in the use of GIS and Erdas Imagine was provided by Rene Siwe from the Remote Sensing Department of the University of Freiburg as well as by Manuel Weis from the Institute of Physical Geography of the University of Freiburg.

Bibliography

ACKERMANN F., 1999, Airborne laser scanning – present status and future expectations. ISPRS *Journal of Photogrammetry and Remote Sensing* 54, 64-67.

BEWLEY R.H, CRUTCHLEY, S.P. & SHELL, C.A., 2005, New light on an ancient landscape: lidar survey in the Stonehenge World Heritage Site. *Antiquity* 79 (305), 636-647.

BERESFORD M., SAINT JOSEPH J., 1979, *Medieval England – an aerial survey.* 2nd edition. Cambridge University Press.

DEVEREAUX, B.J, AMABLE, G.S, CROW, P & CLIFF, A.D., 2005, The potential of airborne lidar for detection of archeological features under woodland canopies. *Antiquity* 79 (305), 637-648.

EWALD K.C., 1969, *Agrarmorphologische Untersuchungen im Sundgau (Oberelsaß) unter besonderer Berücksichtigung der Wölbäcker.*- Tätigkeitsbericht der Naturforschenden Gesellschaft Baselland XXVII, 178 p.

GÜLTLINGER M., SCHLEYER A., SPOHRER M., 2001, Flächendeckendes, hochgenaues DGM von Baden-Württemberg. *Mitteilungen des Vereins für Vermessungswesen,* 48-2. 63-77.

HAUGER K, RIEDINGER R., SITTLER B., 2001. Wölbäcker im Landkreis Rastatt – Auf den Spuren mittelalterlicher Ackerfluren. *Heimatbuch des Landkreises Rastatt*, 163-172.

HOSS H., 1997, Einsatz des Laserscanner-Verfahren beim Aufbau des digitalen Geländehöhenmodells (DGM) von Baden-Württemberg. *DVW Mitteilungen*, 44-1. 5-26.

HYYPPÄ J, PYYSALO U., HYYPPÄ H, SAMBERG A., 2001, *Elevation accuracy of laser scanning-derived digital terrain and target models in forest environment.* 20th EARSel Symposium and Workshops, Dresden, Germany, 14-17 June 2000.

PFEIFFER N., KRAUS K., KÖSTLI A., 1999, Restitution of airborne laser scanner data in wooded areas. *GIS*, 2, 18-21

SCHEWIOR, G. 1941, *Wiesenbau und Bewässerung*, Leipzig

SITTLER B. 2004. Revealing historical landscapes by using airborne laser scanning – A 3-D model of ridge and furrow in forests near Rastatt (Germany). *International Archives of Photogrammetry* Vol. 36 (8/W2) . 258-261.

Archaeological usability of Hyperspectral images: successes and failures of image processing techniques

A. Traviglia

University of Venice, Italy

1 Introduction

In the last decade, particular interest in the use of airborne hyperspectral systems as support in archaeological research has taken place in Italy, where the MIVIS sensor has found large application. The different projects involving its use have tried to demonstrate that MIVIS imagery represents an important complementary source of information for archaeological goals. However, in most of the cases, the limits and the problems generated by using hyperspectral images have never been stressed enough in the research reports, while the spotlight has been focused only on the positive outcomes.

In this paper, moving from a critical review of published reports and through the discussion of the results of a personal project based in NE Italy, in the surroundings of the urban area of Aquileia, an overview will be presented of the main problems encountered when dealing with hyperspectral images together with the presentation of positive aspects and results. Preliminary problems such as resolution, atmospheric corrections, noise reduction along with post-process issues such as redundancy of data and the large number of images to handle will be presented. Focus will be placed also on the fact that standard procedures cannot be largely and automatically applied without consideration of the landscape characteristics, the land coverage and the actual goals of the archaeological research.

With this approach in mind, the paper will expose what one can reasonably expect to obtain by exploiting the spectral content of the MIVIS images and how this can be used to give prominence to the presence of certain ancient buried remains and structures. Starting with the most common technical issues encountered in pre-processing the raw images the paper will go through the enhancement techniques that allow increasing and improving the optic distinction between traces recorded in the scene by generating new images where the useful information is more easily identifiable.

Some years ago in the Introduction to the Proceedings of the IX International School in Archaeology (*"Remote Sensing in Archaeology"*) held at Certosa di Pontignano (Siena, Italy), the editors expressed the hope that archaeologists could fully coordinate RS multidisciplinary projects in a short time, stressing the importance of the creation of an archaeological profession holding the technical capabilities to perform autonomously computer based RS research.

Nowadays, if this process can be considered –partially-accomplished as regards to aerial picture management and even if many steps have been done in the last years toward a "technicalization" of archaeological researchers, still most of the multispectral and hyperspectral RS analysis is performed on the base of the division of roles between RS experts, in charge for the processing activity, and archeologists, in charge for the interpretation of the processed images and detected features. In most the cases the archaeologists have little or no knowledge of the processes applied, of their meaning or of what the derived images really display, which is often made clear by misunderstandings or readiness in the interpretation of detected anomalies. From the other side, in many cases also the RS experts, due to the lack of knowledge in the archaeological needs, are uncertain about what they are looking for and consequently in the processing phase do not always apply those processes that completely fulfill the archaeological needs. This distinction between the research roles strongly affects many research results that could be more significant and incisive if steps were taken in the direction of a reciprocal understanding of the principles and fundamentals of the two different disciplines. Unfortunately, at the moment, the cases of archaeologists who, supported by technical teams or in autonomy, have undertaken the study of the technical principles of multispectral research are very rare. When the archaeological interpretation process is carried out separately from the processing phase, the adopted processes can easily be misinterpreted in their meaning and an unjustified enthusiasm for the multi and hyperspectral analysis can occur. It is common to notice in the literature an easy attribution of archaeological reliability to detected features and anomalies without an accurate report of proof as if the usage of hyperspectral data would determine by itself the certainty of the achieved results. Often, the fact of being able to identify linear features or regular shapes on the landscape is put in direct relation with the presence of archaeological remains, most of the time without giving any explanation of the process of interpretation or explaining the basis for such certainty: the identified features are interpreted as archaeological without any ground truth or without the data (obtained by other sources such as published data, thematic cartography, reports data from field-walking surveys etc) cross processing.

A total confidence in the technology has to be rejected because the role of the visual interpretation in the archeological research of the remotely sensed data is fundamental and it is still based on the classical criteria of the archaeological photo-interpretation: like in the aerial photo-interpretation, extreme caution is due (and opportune) in the explanation of the detected traces and anomalies, which is linked to the continuous comparison with other data sources.

2 Primary issues of MIVIS data: ground resolution, geometrical distortion, atmospheric correction, noise removal

A fact that is never stressed enough in the researches applying MIVIS hyperspectral data is the not always satisfying ground resolution of the sensor, measuring between 3x3 and 4x4 m, dependent on the flight altitude. This level of resolution, already outmoded by other similar products available on the market, is suitable for archeological application only in the case of macro-areas researches, since the features that MIVIS can reasonably detect must be, in the luckiest of the hypothesis, larger than an average area of 27/48 m^2 (so basically at least 3 pixels) or a length of at least 6/8 m (2 pixels). A more reasonable expectation for identifiable feature size averages minimally between 80/100 m^2 and a length of 20 m. As a consequence, the applications of these hyperspectral images within restricted urban areas or small portion of the landscape have to be considered potentially inadequate if not useless: while MIVIS can deal well in the identification of large built structures or settlements, canals or street networks, a large quantity of small archaeological remains can not be detected. For this reason it is important to investigate previously if the resolution at the ground provided by MIVIS images is suitable for the needs of a specific archaeological project on the basis of the expected types of remains and the kinds of archaeological structures that one expects to identify.

When dealing with MIVIS images, a primary technical difficulty that has to be afforded is the rectification and georeferencing of the data. The images in fact are strongly affected from panorama distortions due to scanner geometry and from effects introduced by perturbations in position and attitude of the airborne platform. From one side, a geometric correction that rectifies the images is necessary not only to make more understandable and recognizable the features on the landscape but also to integrate the MIVIS data with other source data in a G.I.S. environment; from the other side, all of the most common procedures followed to rectify hyperspectral images risk to compromise the spectral content of the data set: this is particularly relevant if spectral analyses are going to be applied on the georeferenced data.

These considerations have received little attention in many studies where the geometric deformation issue has been treated as primary, using orthorectified and georeferenced data in the further spectral processes. If not due to the immediate need of positioning data from direct ground survey of spectral signatures simultaneous to the flight recording (for example for SAM classification), the georeferencing process can be postponed and applied after the spectral analysis (Emmolo et al. 2004, p.493; Traviglia 2005a p. 145). Even in the need of referring multi-temporal runs, it can be decided to refer the images to another image used as reference in order to resample only one image (Emmolo et al. 2004, p. 493). It can result useful also to adopt a procedure of dividing each run into an n-number of sub-scenes: this can improve the accuracy in the process of registration, requires the identification of a smaller number of G.C.P. and is suitable for further spectral analysis based on radiometric statistic.

Another aspect that has been taken in consideration during pre-processing of MIVIS data is the atmospheric distortion and the consequent need for a correction. At the current state of the research, final and appropriate algorithms for application to MIVIS images has not been provided yet, so it has not often been possible to perform a specific atmospheric correction to the data used. In addition to that, usually no ground truth information for the time of the shots is available with the images, so that they could be used to implement appropriate corrections. The presence or absence of this pre-processing operation is seldom indicated in the research reports and even when stated little space is given to the process description. The lack of atmospheric disturbance removal can affect certain subsequent processes (e.g. some kind of Vegetation Indices -see Traviglia 2005a, p. 149-) so this should be taken in consideration both when planning the processes to apply and in the discussion of the obtained results that can be invalidated from the data used in the process.

System noise strongly affects MIVIS images as well and its removal is rarely discussed as a pre-processing procedure. The chosen path is often to select a reduced number of non-noise affected bands to use in the further analysis, losing in this way a considerable amount of information and annulling the advantage of having a 102 band image by using only a small part of them.

In order to overcome the shortcomings of the available data sets and improve the accuracy of the obtainable results, a noise removal process has been applied in the Aquileia project obtaining a drastic reduction of the unwanted disturbance due to limitations of the signal digitization and data recording process. In some cases, in fact, noise was not only degrading the true radiometric information content of the images of the case study area but also masking it so that several of them were totally useless. The successful noise reduction process tested consists of two steps: the first operation, the MNF (*Minimum Noise Fraction*), extracts the noise through the inherent dimensionality of image data, segregates it and reduces the computational requirements for subsequent processing (Boardman, Kruse 1994), the second consists

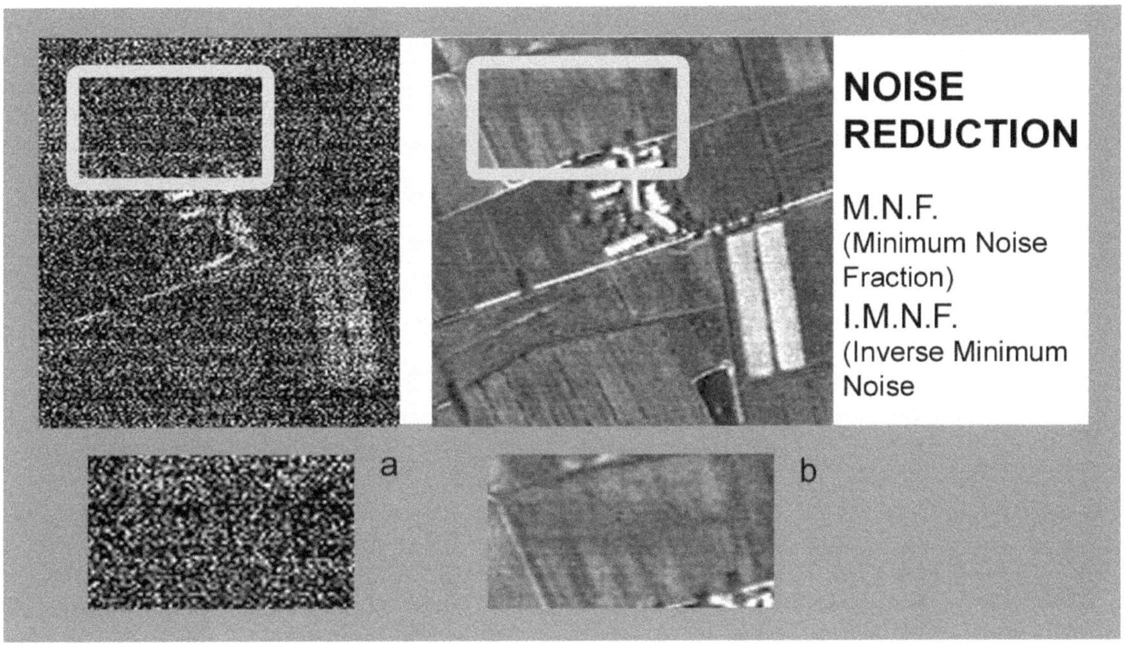

Figure 1 *Comparison of an original MIVIS band (a) to one subjected to MNF/Inverse MNF transformation (b)*

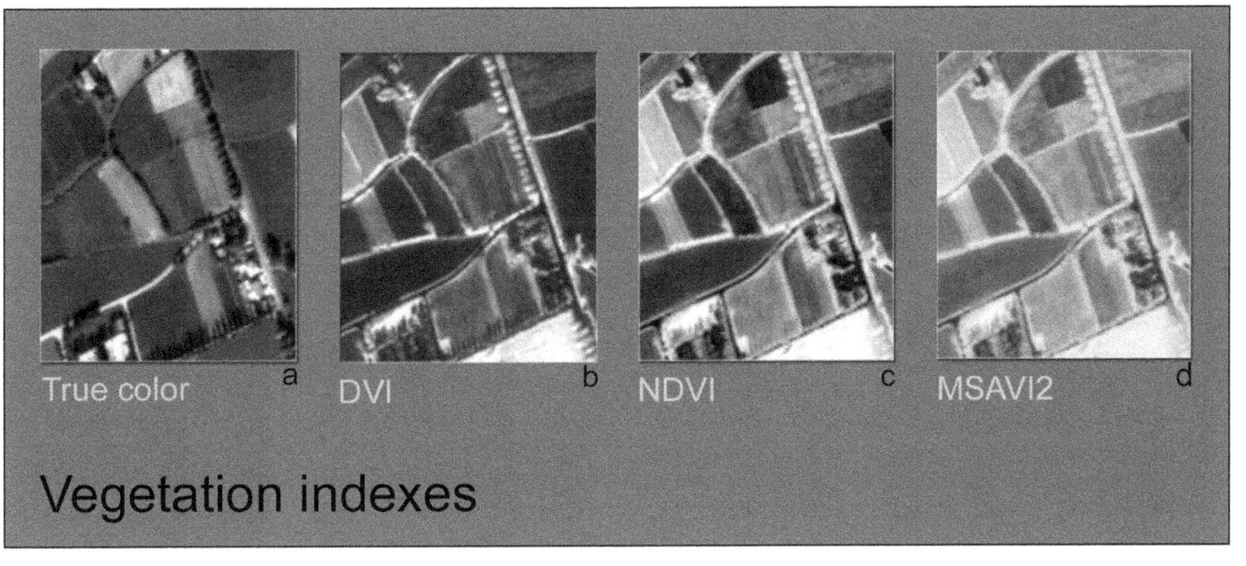

Figure 2 *Comparison among vegetation indexes: a) true color image, b) DVI, c) NDVI, d) MSAVI2*

in the inverse transformation of the previous step (Inverse MNF) that allows for the elimination of the system noise from the original bands by excluding the noise components from the inverse calculation. The images submitted to this process have shown a clear improvement in terms of signal noise and have been used in place of original data for most of the below described analysis processes (Fig. 1).

Still, despite these diminishing factors that necessarily need to be taken in consideration when planning a research using MIVIS shots, the negative aspects of the MIVIS data are overcome by their broad spectral content, which requires some interventions to be fully exploited. Various treatments and processing of the images that have been applied in the many projects involving the use of MIVIS data will be critically discussed and their results compared in order to define the ones that fit better the different research goals.

Vegetation Indices

Vegetation Indices (VI) performed with hyperspectral MIVIS data find wide application in the archaeological research, since studies about the quality of the vegetation, monitoring variations in its vigor, can enable the detection of subterranean archaeological deposits that allow or, in opposition, limit the growth of the vegetation. As well known, heterogeneity of the texture of the subsoil has a strong impact on the growth of the vegetation, creating the manifestation of the so called "crop-marks" over the vegetation canopy. Most of the archaeological researches involving the usage of VIs have used Normalized Vegetation Index (NDVI) more as a default than as the appropriate process based on the specific environmental and morphological characteristics of the target site, applying it in areas where no need for normalization was potentially required. Despite its large use over every type of environmental conditions, other VIs can achieve better results for archaeological purposes, depending on the vegetation coverage type and the density of the canopy: both being primary factors to keep in consideration when selecting the appropriate process to evaluate the biomass of a certain area, since they can strongly affect the results of the process itself. The suggested procedure to follow is to perform and test different vegetation indices after a careful consideration of the study case area characteristics. This means that in studying a vast area containing different vegetation coverage, the VIs must be singularly applied to fields with similar characteristics based on the type of canopy. The results gained can first be compared among themselves to determine the best method for evaluating vegetation health in the target area and secondarily with the original MIVIS images in order to verify the improvement of the visibility of traces they offer (Fig. 2).

This process flow has been tested with good results in Aquileia area, a fertile flat plain mostly devoted to agricultural exploitation, where very different type of ground cover conditions are encountered, ranging from inhomogeneous vegetation growth (even inside the same field) to the heterogeneous density of the canopy. The comparison of the different VIs' results confirmed that the type of surface under examination must previously be taken in consideration and that the same type of vegetation index cannot be applied to all situations. VIs, have been selected from a vast range in order to represent general vegetation indices and indices with correction for soils, one based on simple math (DVI) and two based on ratioing (NDVI and MSAVI2). Further investigation has been executed using orthogonal indices (PVI), but only as a preliminary step in the creation of a soil line (SL) since the limits of the equation mathematically defining a soil line in the PVI theory and the need of empirical data have been largely demonstrated.

The application of the DVI -the subtraction operation involving the use of the Red and NIR bands- appeared to be satisfactory in case of low vegetation cover and in grounds without steep slopes.

In those areas of inhomogeneous crop canopy or steep surfaces, where there is a shadow effect, the NDVI (Normalized Difference Vegetation Index) -the ratio obtainable dividing the difference between IR and red intensities by the total light being reflected- has demonstrated to have very good sensitivity to changes in vegetation cover and compensating for changing illumination conditions, surface slope, aspect and other extraneous factors. However, in those circumstances, it provided better results than any other vegetation index only in case of high vegetation cover, while simultaneously highlighting its sensitivity to canopy background variations in case of low or sparse vegetation. In order to overcame the problems related to interference of the soil background in areas of low plant cover, MSAVI2 (2nd Modified Soil Adjusted Vegetation Index), -a recursion of MSAVI that attempts to reduce soil noise by shifting the place where the isovegetation lines meet (see Qi et al. 1994) - has been successfully applied. However, even if the index has shown encouraging results, it is important to point out that, despite the improvement in terms of accuracy, a difficulty that can happen is an increase in the sensitivity to variations in the atmosphere: this can cause variations in the calculated values of vegetation indices (Qi et al. 1994; Leprieur et al. 1996). Since no atmospheric correction has been performed on MIVIS images, the atmospheric distortion could be the cause of some of the poor results gained in a small number of the runs.

In conclusion, it is important to stress that these results have been reached in a specific landscape context and that different outcomes could be achieved applying similar processes to areas with different morphological characteristics stressing the need for a preliminary analysis of the target area to identify the type of vegetation surface cover.

PCA

The Principal Component Analysis (PCA) has been adopted as an enhancement process in a small number of researches using MIVIS data, not as much as one would expect considering that the goal of the process is to reduce the information previously contained in the original n-band data set into a smaller number of new bands that can be used in place of the original ones, a solution extremely effective when dealing with such an elevated number of bands like in MIVIS. The application of the PCA to archaeological study can be of great utility since it provides supplementary information compared to the original bands, avoids useless loss of time in case of a preliminary surveying of the images and allows a better discrimination of different surfaces that become more distinguishable in visual analysis.

In MIVIS images subjected to PCA, normally the PC 1 and 2 together, in most cases, account for about 98,90% of the total variability in the original bands and usually about 1% is found in 3rd Principal Component: PC1, PC2 and PC3 show virtually all of the variance in the scene (on average 99.6 %) and consequently of the total information. Components from PC4 and higher, usually contain all together only about 0.40 % of the variation in the data. However, some of these higher order components have demonstrated, at least for the Aquileia case study, to contain useful information, recognizable and identifiable only through a visual inspection of the image itself. Technically, high order Principal Components might be considered to be unworthy, having lower signal-to-noise ratio than the low-order Principal Components: here, instead, the presence of significant additional information that was not present in some of the lower orders has been noticed even up to the 30-40th Principal Components, the value varying from run to run. This recognition lead to the necessity for visually check all the produced Principal Component, singularly, rather than relying solely upon the magnitudes of the eigenvalues as an indicator of information content, in order to list the ones to keep for further processes.

The most common procedure of visual inspection in PCA consists of generating a R.G.B. false color composites with the first 3 principal components of a scene (P.C.1, P.C. 2, P.C.3): this creates 6 different combinations. In the case of MIVIS data, applying the same procedure would mean excluding from the composite many of the lower order principal components that turn out to carry interesting information. For this reason, raising to 4 or 6 the number of principal components used in R.G.B. composites generates a total of 24 to 120 different combinations and minimizes the loss of information. Increasing the number of Principal Components combined into color composites was a successful procedure in the Aquileia case, providing interesting results in terms of visibility of specific types of features (such as certain soil irregularity or in terms of water penetration).

However, despite the recognizable advantages obtainable applying PCA, many details that are visible analyzing singularly the original bands cannot always be recognized in the PCs -single components or composites- because they are covered by the overlaying information from other bands. A better result in terms of recognizable features can be reached through the computation of spectral subsets (here referred as SPCA, Selective Principal Components Analysis, that is to say, a PCA computed for groups of bands, detected by a single spectrometer of the sensor or belonging the same spectral region) that can be visually analyzed singularly (see Merola 2005, pp. 311-313 and Traviglia in press) and through their composite (see Traviglia in press). The results obtainable from this technique can in fact be magnified by the combination of the different SPCs instead of relying only on the visual inspection of the first SPCs of each spectrometer.

The selection of the SPC to use in these composites has been based on the Intrinsic Dimensionality (ID) of the data for each SPC, this leads to a quite large number of combination to deal with. The selection of the SPCs to combine has been expedited using a dedicated correlation matrix (Traviglia in press) in order to identify the best usable subset able to ensure no loss of spectral coverage. The use of this type of analysis provided very valuable results in terms of amount of recognizable features: the traces that were better emphasized were lineation of dry or wet soil in contrast with the surrounding soil, alterations in the health of the vegetation and its growth, filled riverbeds, unexplainable alterations in the surface texture.

Soil Line Index

A type of process that is not taken in any account in the archaeological research employing hyperspectral data is one which aims to provide an aid in the identification of anomalies on bare soil, increasing the optical distinction between the wetness or the dryness of a portion of the ground.

The possibilities offered by studying the tonal variations of the bare soil (damp-marks) are well known in archaeology: their existence in fact can be an indication of the presence in the subsoil of archaeological structures or ditches that inhibit or facilitate the absorption of rainwater and of the rising of humidity. The studying of the different degrees of water absorption of a soil becomes therefore particularly useful for the identification of buried archaeological sites. The Soil Line Index (SLI) created for hyperspectral MIVIS data in the course of the Aquileia project (Traviglia 2005b, pp.1970-1973) aims to constitute a support for the identification of traces over non-vegetated soil, emphasizing the wetness or the dryness of portions of the ground. Starting from the well-known concept of the linear relationship between the near-infrared and red reflectance, the SLI produces an image where the driest areas are represented in brighter and lighter colors compared to humid areas. For most of the sample areas, the application of the SLI has shown a clear improvement in the differentiation of the typologies of the soils, accentuating the dry-wet discrimination (Fig.3) and in this way making easier the distinction of lines or zones of different condition from the surrounding ground.

Figure 3 *Result of the application of a Soil Line Index (SLI): SLI (a) compared to an original band (b)*

Classifications

SAM (*Spectral Angle Mapper*) classification is one of the most often adopted process in the treatment of MIVIS data for archaeological goals, the results gained through its application seldom exposed clearly enough in terms of explaining the correlation between the archaeological buried remains and the actual identification of features from RI (rule images) obtained from spectral classes selected in advance (ROI, Region of Interest) and related to modern entities of the landscape.

Due also to ground resolution of the data, the application of classification techniques, both supervised and unsupervised, often deliver invalid and questionable results in archaeological applications. For instance, they are conceptually useless when applied to bare soil areas, being potentially useful only in case of emerging structures (and in this case probably already known, also without RS analysis –unless positioned in recessed areas) or in case of enormous -and extremely rare- concentration of scattered building material or pottery remains. Where it can possibly be applied is in a reduced scale sample area, meaning an area like a single crop or bare soil field, where the spectral signatures of each soil type are known –in case of supervised classification- or strongly marked by archaeological remains -in case of unsupervised classification.
The problem with classification techniques is two part. First is that most of the classifications require some a priori knowledge of the scene: in the case of supervised classification knowledge of the spectral signatures for classes relevant for detection of archaeological features and in the case of unsupervised classification a general notion of the number of distinct classes that are expected to be identifiable in the scene. Most of this information is often unavailable. The second part relates to the reduction of resolution found in the resulting classified image: central to classification is the grouping of pixels into a finite number of classes with no middle grounds. This leads to a flattening of the image into a reduced resolution with significant aliasing, which complicates the visual recognition process used in most of the archeological researches.

For these reasons, in the Aquileia project, classifications, having shown their limits and incapability to provide any original or unique results that cold not be reached through other more credible processes, have been discarded from the routine processes applied to the various runs.

Image number reduction

The amount of processes tested over the hyperspectral shots also clearly generates an issue with the number of produced images. We are dealing here with a 102 bands image; even considering realistically to use 1/10 of the bands for simple TC/FC composites and for the multiple processes, the number of produced images that must be subjected to visual analysis and interpretation is too large. For this reason it is necessary to define a set of procedure to reduce the number of final images to be used, selecting the ones that do not carry redundant information. Consequently, comparison tables, cross process coverage tables, correlation and selection matrixes should be applied in order to reduce the images used in the next step of the research to a small number and be able to insure the complete coverage as regards to the detected traces. This types of solution has been applied in the Aquileia research, both to reduce the usable images in process cycles that could generate multiple images as results or to select the most significant (like in PCA and SPCA) and to reduce in a finale phase the total number of the obtained processed images. For instance, in the case of the SPCA, the spectral subset images were merged into one image and a correlation matrix of the components was calculated to identify a minimal images subset containing all the wanted information (ended up to be four). The consequent series of 24 combinations of the four selected PC images could be reduced by selecting sample areas representative of anomalies (in this example, four sample areas have been used: a sample area of anomalies on dry bare soil, one on wet bare soil, one on vegetation, one on sea water) and attributing to them a quality score from 1 (low visibility quality) to 5 (high visibility quality). The sum of the four grades given for each combination provided the scale index of the quality of the composite

R	G	B	Water	D.b.s.	Veg.	W.b.s.	TOT
1	2	9	3	4	3	4	14
1	2	8	4	3	3	3	13
2	1	9	3	3	2	2	10
2	8	1	3	3	1	2	9
2	9	1	2	2	1	3	8
8	2	1	3	2	2	1	8
1	9	2	1	3	1	2	7
1	9	8	1	3	1	2	7
2	1	8	2	2	1	2	7
8	2	9	3	1	2	1	7
9	1	2	1	3	1	2	7
9	2	8	2	1	3	1	7
1	8	2	2	2	1	1	6
2	8	9	2	2	1	1	6
8	9	2	1	2	2	1	6
9	8	2	2	2	1	1	6
9	2	1	1	1	2	2	6
1	8	9	1	2	1	1	5
2	9	8	1	1	1	2	5
8	1	2	1	2	1	1	5
8	1	9	1	2	1	1	5
8	9	1	1	2	1	1	5
9	1	8	1	2	1	1	5
9	8	1	1	1	1	1	4

Key

Water = sea water
D.b.s. = dry bare soil
Veg. = vegetation
W.b.s. = wet bare soil

Figure 4 *Combinations of SPCs and their quality scale*

Cross Process Coverage Table

Process Name \ Number	0	1	2	3	4	5	6	7	8	9	10	11	12	13	14	15	16	17	18	19	20	21	22	23	24	25	26	27	28	29	30
Ortofoto	253	103	100	96	41	77	54	53	54	50	55	45	38	40	39	37	33	24	36	39	28	31	39	36	22	24	17	16	14	14	12
MIVIS- True Color		132	83	81	16	78	54	59	53	47	43	41	28	42	42	41	37	27	38	39	25	31	34	33	21	24	23	15	14	13	11
MIVIS- TCAP (PCA) Br-NDVI-Wet			124	80	16	67	47	43	46	47	48	42	44	39	38	37	36	31	34	32	31	35	30	28	25	23	23	18	14	14	13
MIVIS- TCAP Br-SLI-Wet				114	18	70	62	55	55	48	52	46	37	46	48	43	39	32	31	40	28	37	36	34	29	23	19	17	12	14	13
MIVIS- PCA (TotRun) 2-4-6					100	10	11	10	13	10	14	8	7	11	8	5	5	6	7	7	6	8	7	7	6	7	5	6	4	3	4
MIVIS- PCA (TotRun) 1-2-5						99	48	51	50	46	42	44	26	37	40	40	38	27	34	38	23	28	31	29	19	23	18	14	17	12	10
MIVIS- PCA S1P2-S1P1-BluP1							68	50	50	33	38	29	22	51	46	36	34	28	30	36	19	34	32	32	21	16	18	16	12	13	12
MIVIS- PCA S1P1-S1P2-S4P1								68	51	32	37	30	17	39	45	39	36	25	33	43	16	29	36	34	20	18	17	15	13	14	13
MIVIS- PCA S1P1-S1P2-BluP1									65	33	38	30	20	42	45	41	36	27	33	42	21	32	35	32	20	18	21	17	14	14	13
MIVIS- PCA R1-G1-B1										60	32	29	17	26	28	25	22	13	39	25	13	22	25	23	16	15	15	13	12	12	10
MIVIS- DVI											59	27	20	31	30	25	25	15	23	32	18	26	32	31	16	16	16	16	13	13	12
MIVIS- NDVI												55	18	28	26	26	24	16	19	26	14	19	22	19	12	19	13	11	17	9	8
MIVIS- MSAVI2													54	21	17	20	18	30	10	15	34	21	12	12	23	9	13	12	4	8	8
MIVIS- SLI														52	38	30	27	23	23	31	18	28	27	28	16	15	16	16	12	12	12
MIVIS- TCAP Wet															51	37	32	26	26	34	16	30	27	27	20	15	16	14	10	13	12
MIVIS- TCAP (PCA) Wet																50	43	30	24	32	22	29	23	21	24	12	19	13	10	12	10
MIVIS- TCAP Green																	48	28	22	29	18	25	21	19	20	14	18	11	11	10	8
MIVIS- TIR b96																		47	11	18	31	22	12	10	30	8	14	10	4	6	6
MIVIS- MIR b57																			45	26	10	20	25	22	12	13	16	12	12	12	10
MIVIS- NIR b18																				45	12	24	32	30	15	16	14	14	12	12	11
MIVIS- Red b13																					43	18	12	10	25	10	14	9	5	5	4
MIVIS- Green b7																						40	21	20	23	12	18	14	10	14	13
MIVIS- Blue b3																							40	35	12	14	10	13	12	13	11
MIVIS- NIR PCA																								37	11	14	10	14	11	13	12
MIVIS- Red PCA																									36	10	13	11	5	6	5
MIVIS- Green PCA																										32	9	9	17	6	5
MIVIS- Blue PCA																											28	11	8	8	8
MIVIS- False Color																												20	5	9	9
MIVIS- TIR PCA																													20	5	4
MIVIS- S4P1-b18-b13																														14	12
MIVIS- TCAP Br-Wet-Gr																															13

Process #	Name of process
0	Ortofoto
1	MIVIS- True Color
2	MIVIS- TCAP (PCA) Br-NDVI-Wet
3	MIVIS- TCAP Br-WWI-Wet
4	MIVIS- PCA (TotRun) 2-4-6
5	MIVIS- PCA (TotRun) 1-2-5
6	MIVIS- PCA S1P2-S1P1-BluP1
7	MIVIS- PCA S1P1-S1P2-S4P1
8	MIVIS- PCA S1P1-S1P2-BluP1
9	MIVIS- PCA R1-G1-B1
10	MIVIS- DVI
11	MIVIS- NDVI
12	MIVIS- MSAVI2
13	MIVIS- SLI
14	MIVIS- TCAP Wet
15	MIVIS- TCAP (PCA) Wet
16	MIVIS- TCAP Green
17	MIVIS- TIR b96
18	MIVIS- MIR b57
19	MIVIS- NIR b18
20	MIVIS- Red b13
21	MIVIS- Green b7
22	MIVIS- Blue b3
23	MIVIS- NIR PCA
24	MIVIS- Red PCA
25	MIVIS- Green PCA
26	MIVIS- Blue PCA
27	MIVIS- False Color
28	MIVIS- TIR PCA
29	MIVIS- S4P1-b18-b13
30	MIVIS- TCAP Br-Wet-Gr

Figure 5 *Cross process coverage table. The i.d. process number is shown in the right table*

and the first 3 or 4 combination of images of the scale were the ones used for archaeological interpretation.

In case of having to manage a large amount of images resulting from different processes another way to compare the results is to look at the coverage between processes, which means comparing the number of traces common to two given processes against the total of one of the processes. This is possible through a table listing the traces found in a process versus the number of those specific traces also found in other processes (Fig.4)

For example, 39 of the 253 traces visible in True Colors are also seen in SPC of Red bands: this gives SPC Red about 15% coverage over True Colors. Reversing it, this means that 39 of the 45 traces found in SPC Red are also seen in True Colors, giving True Colors about 87 % coverage over SPC Red.

GIS

The use of GIS in RS research manages all the archaeological and topographical data necessary to eventually recognize the detected surface anomalies as ancient origin traces, in absence and/or in support of ground truth information. GIS is the obvious choice where the nature and attributes of each feature identified in the remotely sensed images can be stored, investigated and its causes determined by putting it in relationship with the contiguous aspects of the landscape and with all the other data extracted from various sources: the ability to manage this body of information can lead to a correct interpretation of the identified features. An useful interpretation tool that has been adopted in the Aquileia project, where no specific ground truth data was at disposal, was attributing to each of the detected feature, as result of the interpretation process, a value of visibility and archaeological reliability that express an evaluation about the quality of the observed feature. Although unable to substitute the value of field-walking survey, these attributes can at least emphasize those locations on the landscape in which there is a higher probability to find archaeological deposits.

Until now the GIS platforms have been sparsely employed in archaeological RS research or used only to provide and manage archaeological and topographical data matching them through a simple overlay. The increased number of processed images and consequent data amount obtainable in the long term using MIVIS images imposes the usage of a more complex GIS architecture, the new needs pushing in the direction of more advanced structures that allow for more sophisticated storing and visualization techniques. In this way the GIS assumes an indispensable role in the interpretation of remotely sensed data leading to a multi-tools approach which in a multi-scale and diachronic context is the only reliable way to deliver effective contributions in the understanding of settlement patterns.

Conclusions

The introduction of multispectral and hyperspectral data in the archeological research during the last years has created lot of expectations, often not really fulfilled from the results gained. Working with MIVIS images over a long period, evaluating a direct experience and observing results achieved in other similar projects, revealed positive and negative aspects of this sensor. As a general statement it can be asserted that the value of the MIVIS images relies more in their spectral content than in their spatial resolution, which cannot be considered completely adequate for the search of small archaeological remains (e.g. $<35m^2$). Even when identification of small objects happens, they are, most of the times, detected on the base of indications offered from other remotely sensed imagery an/or from the archaeological data, emphasizing the need for an approach of integration of multiple tools and data. However, despite this "physical" limit, the spectral information obtainable from MIVIS provide considerable support in identification of features in the non-visible domain and demonstrated to be a valid instrument for archaeological research especially when integrated in a global information architecture managing various source data.

References

BOARDMAN, J.W. and KRUSE, F.A. 1994. Automated spectral analysis: a geological example using AVIRIS data, northern Grapevine Mountains, Nevada, in *Proceedings of the Tenth Thematic Conference on Geological Remote Sensing, 9-12 May 1994, San Antonio, Texas, Environmental Research Institute of Michigan, Ann Arbor*, pp. 407-418. Ann Arbor, MI: ERIM

CAMPANA S., FORTE M. (eds.) 2001. *Remote Sensing in Archaeology, XI Ciclo di Lezioni sulla Ricerca Applicata in Archeologia (Certosa di Pontignano, Siena, 6-11 dicembre 1999).* Firenze: All'Insegna del Giglio

EMMOLO D., FRANCO V., LO BRUTTO M., ORLANDO P., VILLA B. 2004. Hyperspectral techniques and GIS for archaeological investigation, in *Geo-Imagery Bridging Continents, XXth ISPRS Congress, 12-23 July 2004 Istanbul, Turkey,* Vol. XXXV, pp. 492-497

LEPRIEUR, C., KERR, Y.H., PICHON, J.-M. 1996. Critical assessment of vegetation indices from AVHRR in a semi-arid environment. *International Journal of Remote Sensing.* 17:2549-2564

MEROLA, P. 2005. Tecniche di telerilevamento iperspettrale applicate alla ricerca archeologica. Il caso di Lilybaeum, *Archeologia Aerea.* 1:301-317

QI, J., CHEHBOUNI, A., HUETE, A. R., KERR, Y. H. 1994. Modified Soil Adjusted Vegetation Index (MSAVI). *Remote Sensing of Environment.* 48:119-126.

TRAVIGLIA, A. 2005a. Integration of MIVIS hyperspectral remotely sensed data and geographical information systems to study ancient landscapes: the Aquileia case study. *Agri Centuriati. An International Journal of Landscape Archaeology.* 2:139-170.

TRAVIGLIA, A. 2005b. A semi-empirical index for estimating soil moisture from MIVIS data to identify subsurface archaeological sites, in *9ª Conferenza Nazionale ASITA Federazione delle Associazioni Scientifiche per le Informazioni Territoriali e Ambientali*

15 - 18 novembre 2005, Catania, vol. II, pp. 1969-1974

TRAVIGLIA, A. (in press). MIVIS Hyperspectral Sensor for Detection of Subsoil Archaeological Sites and Interpretation Support GIS: an Italian Case Study, in *Digital Discovery. Exploring New Frontiers in Human Heritage*, CAA2006, Fargo, ND- USA 18-26 April 2006.

From Space to Place: the Aiali project (Tuscany-Italy)

S. Campana,[1] S. Piro,[2] C. Felici,[3] M. Ghisleni[4]

[1] University of Siena, Department of Archaeology and History of Arts, Landscape Archaeology, (LAP&T)
campana@unisi.it
[2] ITABC–CNR, P.O. Box 10–00016 Monterotondo Sc. (Rome), ITALY.
Salvatore.Piro@itabc.cnr.it
[3] University of Siena, Department of Archaeology and History of Arts, Ph.D. Student, (LAP&T)
felicicri@unisi.it
[4] University of Siena at Grosseto, Department of Archaeology and History of Arts, Ph.D. Student, (LAP&T)
ghisleni@lapetlab.it

1 Introduction

The place name Aiali is sited on lowland between the medieval town of Grosseto and the Roman town of Roselle in central Italy. The site discussed in this paper was detected from the air during the Aerial Archaeology Research School organised by the University of Siena in 2001 (Campana et al. 2006). Aerial survey allowed us to recognise an area within which the growth of the wheat varied in such a way as to reveal an articulated group of traces that made up the plan of a complex of structures interpreted as a Roman villa, 4 hectares in extent. In the following years Aiali has become the most important test site for the Laboratory of Landscape Archaeology and Remote Sensing at Grosseto. Since 2001 we have collected, processed and interpreted many different kinds of data: Quickbird-2 satellite imagery, historical and recent vertical coverage (from 1954 to 2001), oblique air photographs in various years, seasons and lighting condition; field-walking survey, and geophysical survey (magnetometry, GPR, EM, ERT).

One intention of the Aiali project is to apply the highest available level and intensity of archaeological prospection methods on a large, complex and stratified site, producing material from the from Etruscan, Roman and Medieval periods. At the same time Aiali forms the starting point for a more wide-ranging approach to the study of the landscape between Grosseto and Roselle. This is an innovative project in Italy but it is clear that we are trying to emulate the strategy applied for more than thirty years by Dominic Powlesland in his Vale of Pickering project in northern England (Tipper 2004; Powlesland 2001; Powlesland 2006).

It is easy to understand, however, that it will never be possible to extend this approach to areas of the scale covered by the University's archaeological mapping projects (Campana et al. 2006). As an example, the archaeological map of the province of Grosseto covers an extent of 4030 Kmsq and that of Tuscany as a whole about 22.990 Kmsq.

As Chris Musson has argued – during one of our long discussions on this topic – the archaeological objective and outcome of the Aiali project has to take account of the critical impact of the kinds of information that are available for recording: to use his own words "in assessing the potential or interpretation of a landscape it is at least as important to know what may *not* be visible as to appreciate what *is* visible".

2 Remote sensing at work

Quickbird-2 Satellite imagery
At the end of spring 2002 we planned the acquisition of 70 Kmsq of Quickbird-2 imagery around Aiali. We ordered the multispectral and the panchromatic data. Unfortunately Digital Globe acquired the data rather later than we had hoped, on 13 June, right at the end on the intended time window.

Even though satellite imagery in this case failed to make a meaningful contribution at Aiali itself, we may consider that – without a multitemporal approach – the input of satellite imagery (and of airborne remote sensing data more generally) will not be effective for intra-site analysis but only for large scale landscape studies. In this case, for instance, on the same image set it is possible to see many other archaeological features, some of them very close to Aiali. One such site lies only 800 m to the south-east, where field-walking survey has revealed the existence of a large medieval mound (Fig.1).

Vertical air photography
As usual we started our work by examining the oldest available aerial photographs, in this case from the national coverage of 1954. Unfortunately no features were visible on this historical flight because the area was at that time used for olive cultivation (Fig.2). Even though the land-use changed to grain cultivation between the 1950s and 1970s we did not find any features on vertical photographs of 1976, 1996 and 2001.

Aerial survey and oblique photography
As noted in the introduction, the site was first detected in early summer 2001 during oblique aerial survey. In the following years we did not have the opportunity to survey the site from the air because of logistical problems until the spring of 2004 (Campana et al. 2006).

Figure 1 *Above, Quickbird-2 imagery, red band. Below, near-infrared band*

In that year, from the end of May to the middle of June, throughout the crop-ripening season, the site was repeatedly monitored from the air to record the aerial visibility of the cropmarks, using flights at intervals of between two and four days to document their development.

This procedure allowed the clear identification of new traces that had not been visible when the site was first discovered in 2001 (Fig.3). Photography was also continued into the later months of the summer. The new evidence included traces of an abandoned river-course and two new structures in the north-western part of the field, adjacent to the main complex of buildings (Fig.4).

The site was also monitored from the air in 2005 but the maize crop proved unhelpful. A similar situation applied in 2006, when the field was not under cultivation. Crop-rotation will return the field to grain in 2007 and this will perhaps present opportunities for the recovery of new information.

S. Campana

Figure 2 *Above, 1954 vertical photograph with the field used for olive cultivation. Centre, 1996 vertical; despite the changed land-use no features are visible. Bottom, 2001 vertical; still nothing is visible but it is interesting to note that, as in the satellite imagery, it is possible to see features relating to the Brancaleta site, 800 m south-east of Aiali.*

Figure 3 *Oblique air photographs taken at different dates in 2004. From top left: 25 June, 30 June, 4 July, 9 July, 13 July, 25 September*

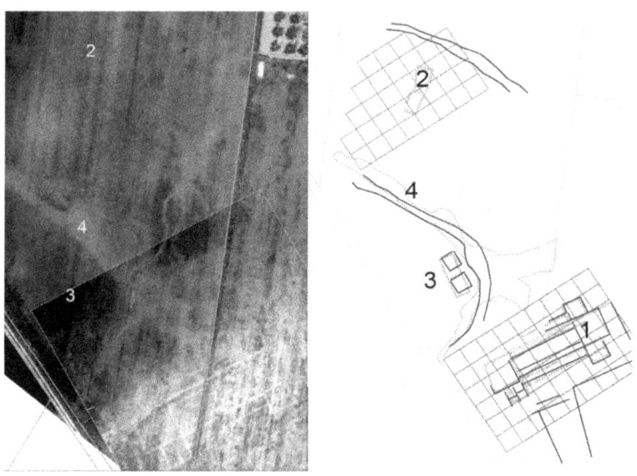

Figure 4 *Interpretation and mapping of the archaeological features*

Field-walking survey and pottery analysis

During the course of 2004 we carried out field-walking survey and surface collection within a 10 m square grid (Fig.5). The ground survey and the study of the collected material has confirmed the archaeological character and interpretation of the site as seen in the air, demonstrating a high level of correspondence between the aerial evidence and concentrations of archaeological finds.

Analysis of the large amount of pottery recovered from the site has permitted us, at the present stage of working-in-progress, to define the chronological range of the villa. The cataloguing of about 1100 pieces of pottery has allowed us to identify a minimum of 1044 different vessels. The artefacts recovered from the grid collection have been catalogued and drawn to characterise useful comparisons so as to define the chronology of each single fragment. Each fabric was analysed under the optical microscope to define products traded over shorter or longer distances through the identification of characteristic mineral inclusions. This analysis confirmed, for instance, the presence at Aiali of medieval pottery from Pisa, containing sand rich in micro-fossils from the Arno. With regard to the imperial phase it was possible to define pottery derived from Africa, in particular amphora body-sherds, all characterised by the presence of the same minerals.

Figure 5 *Grid collection. The spots represent the distribution of medieval pottery within the grids.*

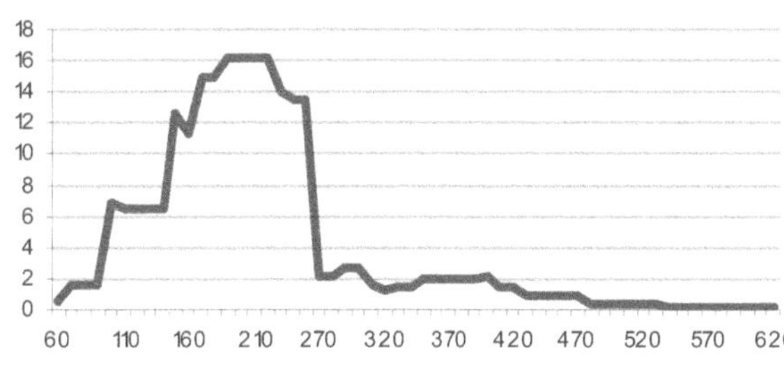

Figure 6 *Distribution of ARSW at Aiali (beginning of second to end of sixth centuries AD).*

The settlement began as an imperial Roman villa in the first century AD. The best marker for the early imperial age is italic red slip ware. The substantial presence of African red slip ware, of types A, C and D, allowed us to estimate that the villa remained in occupation until the sixth century AD, though with reduced amounts of Africa imports from the last quarter of the thirdcentury (Fig. 6).

An interesting feature is the total absence of pottery from North Africa in the fourth and sixth centuries within two adjacent rectangular structures (visible in the oblique air photograph, Fig.4) in the north-western part of the field. The study of the pottery assemblage as a whole allows us to identify a dominant presence of fine red-painted pottery, locally or regionally produced, at the same period in the fourth to sixth centuries (ARSW).

We have identified a minimum of 101 forms of fine red-painted pottery dating from the fourth and sixth centuries, with many different morphologies. Some products imitate forms of ARSW: the bowl Hayes 61, the flanged bowl Hayes 91 and cup Hayes 99; others are bowls, dishes and closed forms of local tradition.

The topographical distribution of this class at Aiali reveals, moreover, a precise coincidence with that of the ARSW. The pottery analysis reveals the abandonment of the villa between the end of the sixth and beginning of the seventh centuries AD, followed by a gap of about three centuries. During the late ninth and tenth century AD there are some vessels which reveal a re-occupation of the site. In particular there are jugs in *acroma depurata* with handles coplanar with the rim, along with cooking pots and red-painted ware from Pisa. This new phase in the site's occupation does not last beyond the twelfth century AD and seems to be concentrated in the area of the villa structures (Fig. 5).

In summary, we can say that the systematic collection within georeferenced grids has been extremely useful in seeking to understand long-term the topographical transformations within the site.

M. Ghisleni

Ground-based remote sensing

The magnetic investigations

In the autumn of 2004 it proved possible to collect 2 hectares of gradiometer data at intervals of 0.50 m along profiles set 1 m apart. The results showed a series of magnetic anomalies which closely match the traces visible on the oblique air photographs (Fig.7). In addition to confirming, very precisely, the evidence seen from the air, the magnetic survey added a series of anomalies that fill in many of the gaps in the main building complex. The central part of the villa consists of a rectangular structure measuring about 70 x 25 m, oriented north-east/south-west, at each end of which are four square rooms 10 x 10 m across. A break in the magnetic data is caused by a disused iron pipe, which obscures the archaeology along its length without reducing the general readability of the structure (Fig.7, no.8). On the evidence of the aerial photographs, which show continuity across the line of the pipe, we can assume that the below-ground archaeological deposits are essentially undisturbed. This interpretation is confirmed by the GPR survey (Fig.8). It is fair to suggest that in the absence of the pipe the gradiometer data would have produced equally positive results. Further magnetic anomalies can be seen in various parts of the field which were previously blank (Fig.7, nos.5-7).

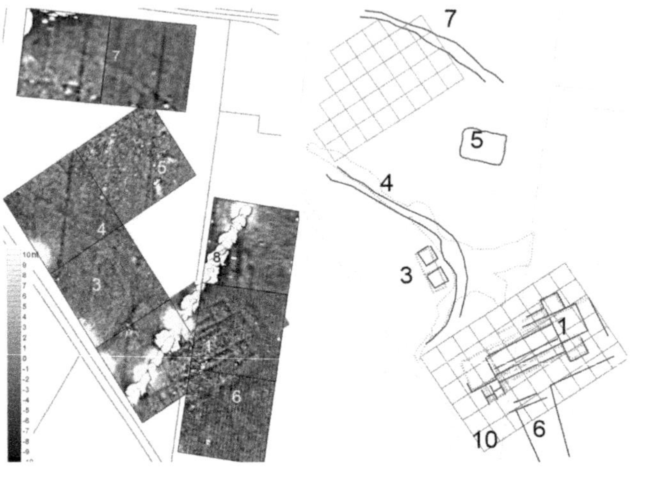

Figure 7 *Magnetic map and interpretation*

Some tens of metres to the north-east and south-east of the main complex a series of linear anomalies (Fig.7, no.6), more or less aligned with the main structure, seem likely to represent an enclosure, perhaps with an entrance-way. To the north there is a weaker anomaly (from -10 to +10 nT/m), approximately rectangular but not aligned on the villa and thus presenting problems of interpretation.

Both the magnetic and aerial surveys produced poorer results in the north-western part of the field, where the ground survey and gridded surface collection yielded considerable amounts of structural material and ceramics, covering a chronological range from the second century BC to the sixth century AD, with a final phase in the tenth to eleventh centuries AD.

C. Felici

The GPR investigation: data acquisition and processing

In 2006 high-resolution GPR surveys (conducted in collaboration with Dean Goodman, Archaeometry Laboratory, Los Angeles, CA, USA) were applied over four areas to test the potential of this technique. For the measurements a GSSI SIR3000, equipped with a 400 *MHz* bistatic antenna with constant offset, was employed. At each site radar profiles were collected alternatively in reversed and unreversed directions across the survey grids. The horizontal spacing between parallel profiles at the site was 0.5 *m*. Radar reflections along the transects were recorded continuously across the ground at 40 *scan* s^{-1}, with a stack = 3; along each profile, markers were spaced every 1 m to provide spatial reference. The gain control was manually adjusted to be more effective. The data were later corrected for a variation in speed to constant 30 *scans* m^{-1} (or 1 scan approximately 0.03 *m*).

All radar reflections within 50 *ns* (two-way travel time) time window were recorded digitally in the field as 16 bit data and 512 samples per radar scan.

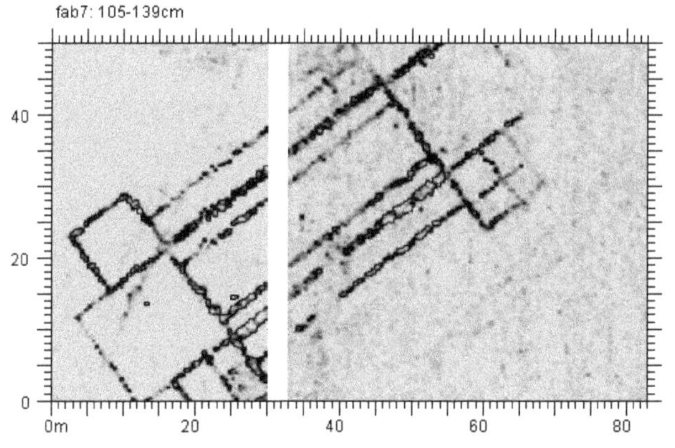

Figure 8 *Time-slice of the Southeast area of the site in the depth range 1.05 – 1.38 m. In this slice are visible many features that is possible to interpret an the walls of the main building of the roman villa still present underground.*

When GPR profiles are collected along closely spaced profiles the data can be processed to display horizontal maps of the recorded radar amplitudes. Referred to as time slice processing, these anomaly maps can be generated at various time/depth windows across the recorded radargram dataset. The time slice maps can provide information regarding the size, shape, location and depth of subsurface archaeological structures buried beneath a site.

Time slice data were created using the spatially averaged square wave amplitudes of the return reflection. These averaged square amplitudes were then gridded using a Kriging routine (Piro et al. 2003, Goodman et al. 2001). The spatially averaged time slice parameter minimizes the effects of line striation parallel to the profile direction.

The larger time window helped to create an understandable visualization of the subsurface. In particular, when structures are not level within the ground or when velocity variations across the site are significant, the larger time window can illuminate continuous features at variable depths (Conyers and Goodman 1997). Other line noises, parallel to the profile collection direction, were removed using a moving filter with customised threshold settings. Filter thresholds were set to signal levels just below the average reflections from buried Roman walls.

In *Area AB* a total number of 162 parallel profiles, stretching S-N, were collected with the instrument configuration indicated above. After the pre-processing, all collected GPR profiles were used to calculate the time-slices related to this area. Fig. 8 shows the time-slice in the depth range 1.05 – 1.38 *m*. In this slice linear radar reflections are clearly visible. These kinds of anomalies are due to the presence of portions of walls still present in the ground.

S. Piro

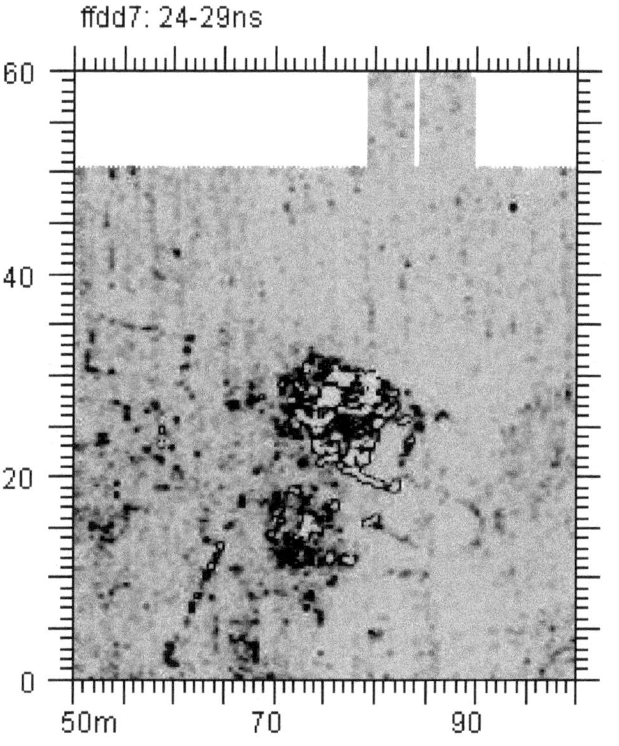

Figure 9 *Time-slice of the Northwest area of the site. The slice shows some features having different dept and different orientation possibly due to different chronological phase of the buildings.*

Conclusion

Bearing in mind the present stage of the research it would be premature to attempt a definitive interpretation. At this stage, however, we are confident enough to outline a few methodological points. Each technique has produced a remarkable increase in both the quantity and quality of the archaeological information available to us. In particular the use of geophysical survey represents an advance of great importance in our researches. In addition to confirming the evidence recovered from the air the geophysical data has without doubt added new and otherwise unseen evidence.

5 Conclusion

The case-study shows that the use of a variety of methods in the search for understanding of the site has made a considerable contribution to the overall result.

Each technique has produced a remarkable increase in both the quantity and quality of the archaeological information available to us. In particular the use of geophysical survey represents an advance of great importance in our researches. In addition to confirming the evidence recovered from the air the geophysical data has without doubt added new and otherwise unseen evidence.

In conclusion, already at the present stage of the research, we believe should be recognized that on large, complex and stratified archaeological site the contribution of an integrated approach is extraordinary.

The next challenge should be possibly to increase the extent of the research the landscape in the valley between the town of Grosseto and Roselle.

Acknowledgments

The authors wish to acknowledge their gratitude and indebtedness to the following:

Riccardo Francovich (University of Siena) for his interest, help and criticism throughout all stages of the project.

Chris Musson, from Wales, for his invaluable comments and for his help with the English version of the text.

Helmut Becker (Bavarian State Department of Historical Monument, Germany) and Iacopo Nicolosi (National Institute of Geophysics and Vulcanology, Italy) for their very useful remarks and for helping us to understand the best configuration for our magnetometer.

Dean Goodman (Geophysical Archaeometry Laboratory, Los Angeles, USA) for the great work done during GPR data acquisition and processing.

Many researchers and students have collaborated, and are still collaborating, in the Siena and Grosseto archaeological mapping projects. Special thanks are due to Willem Vletter (MA student) and to the team of the Laboratory of Landscape Archaeology and Remote Sensing at the University of Siena: Anna Caprasecca, Maria Corsi, Barbara Frezza, Mariaelena Ghisleni, Francesco Pericci and Emanuele Vaccaro.

References

CAMPANA S., FRANCOVICH R., CORSI M., PERICCI F: 2006, Aerial Survey Project in Tuscany: years 2000-2005, in Campana S., Forte M, *From Space to Place, proceedings of the II International Conference Remote Sensing in Archaeology, Rome 4-7 December*, BAR International Series.

CONYERS, L.B. and GOODMAN, D. 1997. Ground-penetrating radar. An introduction for archaeologists. *AltaMira Press. Division of Sage Publications*, Inc.

GOODMAN D., NISHIMURA Y. and PIRO S. 2001. High resolution ground-penetrating radar. In Gaffney et *al* "Forum Novum-Vescovio: studying urbanism in the Tiber valley". *Journal of Roman Archaeology* 14, 5-8, 20.

PIRO S., GOODMAN D., NISHIMURA Y. 2003. The study and characterization of Emperor Traiano's Villa (Altopiani di Arcinazzo, Roma) using high-resolution integrated geophysical surveys. *Archaeological Prospection,* 10(1), 1-25.

POWLESLAND D. 2001, The Heslerton Parish Project :An integrated multi-sensor approach to the archaeological study of Eastern Yorkshire, England', in *Remote Sensing in Archaeology,* Campana & Forte (eds), University of Siena, Firenze 2001, 233-235

POWLESLAND D. 2006, Redefining past landscapes: 30 years of remote sensing in the Vale of Pickering, in Campana S. & Forte M (eds), *From Space to Place, proceedings of the II International Conference Remote Sensing in Archaeology, Rome 4-7 December,* BAR International Series.

TIPPER J. 2004, *The Grubenhaus in Anglo-Saxon England,* Oxford.

Geophysics and the archaeology of an ethnic cleansing: the case of grand-pré national historic site of Canada

Jonathan Fowler

University of Oxford and Saint Mary's University

Introduction

Nova Scotia is one of Canada's smaller provinces, but its geographical position, jutting from the eastern seaboard of North America into the North Atlantic, has given it a large role in the country's history. European whalers and fisherman journeyed to these waters from the 16th century. In earning their livelihood they unwittingly initiated a dynamic process of cultural contact between Europeans and the Aboriginal population of northeastern North America. The trade in furs soon offered an economic basis for sustained contact, and by the turn of the 17th century French commercial operations began to establish a year-round presence on Nova Scotian shores, or, as the French called it, *Acadie*.

The Acadian people, the subject of this paper, were descendents of French colonists introduced to Acadia in the 1630s. Based around the tidal estuaries of the Bay of Fundy, first at Port Royal, and by the late 17th century spreading to all of the major rivers of this system, the Acadians transformed broad tidal marshes into farmland through an ambitious land reclamation program. Wherever the Acadians settled, a network of dams, drains, and sluices soon took shape, as tidal marshes dominated by salt-tolerant grasses gave way to a patchwork quilt of farmers' fields. The uplands immediately adjacent to the reclaimed marshes became dotted with Acadian hamlets, forming a beaded settlement pattern along the river valleys. While their European cousins toiled against land pressure and ever-increasing taxes (Briggs 1998, p. 109), the Acadian population exploded as young families struck out to establish new farms. From perhaps 500 individuals in 1671, and with very little immigration, the Acadian population grew to nearly 15,000 by 1750.

The Acadian experience offers a rich body of evidence through which to examine a variety of subjects of current interest to historians and anthropologists. In addition to their much studied economy, which was a rare adaptation in colonial North America, we find in the Acadians a fascinating example of ethnogenesis; as historian Naomi Griffiths (1973) called it in the subtitle of her ground-breaking study, the 'creation of a people'. Researchers may also be drawn to the relationship that developed between the Acadians and the Aboriginal inhabitants of the land, the Mi'kmaq. Force and coercion have played dominant roles in the history of colonialism, and yet we see in Acadia a rare example of negotiation and peaceful coexistence. The Acadians and the Mi'kmaq were frequent trading partners, and sometimes intermarried (Fowler 1995). What Acadia offers, therefore, is an opportunity to study the creative fusion of cultures that can arise within the framework of intercultural contact.

Rather to its detriment, Acadian historiography has been dominated, ever since its inception with the work of Abbé Raynal (1776), by questions over the events of the years 1755-1763. During this period, which corresponds roughly with an international war known in North America as the French and Indian War, but in Europe as the Seven Years' War, the Acadian people suffered ethnic cleansing at the hands of Britain's colonial government in Nova Scotia. Britain had ruled the colony since seizing it from the French by force of arms in 1710. Since few English or Protestants could be encouraged to settle in Acadia (or Nova Scotia, as the British called it), administering the colony presented a challenge to local British officials, who were chronically under-resourced. True to colonial form, the Acadians reached a compromise with Governor Richard Philipps in 1730 that guaranteed their neutrality. In return, the Acadians swore an oath of allegiance to King George II. However, by 1755, on the eve of an international war, a new acting governor, Col. Charles Lawrence, unilaterally demanded that the Acadians swear an unqualified oath. When their delegates refused, Lawrence moved quickly to remove the Acadians from the colony. Approximately 7,000 Acadians were arrested and deported from Nova Scotia, dispersed in small groups among Britain's American colonies along the eastern seaboard of North America. Thousands of others fled. In all, historians estimate that perhaps one third of the Acadian people died from malnourishment, disease, and ill-treatment (Faragher 2004). This whirlwind left in its path destroyed villages, empty orchards, and stray farm animals wandering a landscape scorched by war.

To the Acadian survivors, this ordeal became known as the *Grand Dérangement*, the Great Upheaval. More clinically, perhaps, the Anglophone community has tended to know it as the Deportation of the Acadians, or the Acadian Expulsion. Whatever the term, it is perhaps ironic that the first writer to effectively give voice to the suffering occasioned by these events was Henry Wadsworth Longfellow, an American romantic poet and author of *Evangeline: A Tale of Acadie*. Published in 1847, this epic poem tells the tale of Evangeline and Gabriel, separated on the eve of their wedding by the *Grand Dérangement*. It follows Evangeline in her fruitless attempts to find her lover as she travels the broad landscape of colonial America, only to be reunited with him many years later in Philadelphia, where she finds him at last on his deathbed. The poem enjoyed immediate success, and was soon translated into French, energizing a national awakening among the Acadian people. The power of this 'identity myth', as some have described it, can still be seen today (LeBlanc 1993).

Figure 1 *The memorial church at Grand-Pré National Historic Site, with statue of Evangeline in the foreground. Photo facing north.*

The irony here is double, for not only is the author of this important work a New Englander – a member of that very community which had carried out the Deportation of the Acadians in the first place – but the main historical evidence Longfellow used to create his portrait of the Deportation derived from the pen of one of its perpetrators. John Winslow, the Massachusetts officer who conducted the Deportation at Grand-Pré, one of the largest Acadian communities, kept a daily record of events through the autumn of 1755. Little could he have known that his words would ultimately come full circle, from detailing his destruction of a community to being wielded as an instrument in its renaissance.

Winslow (1883) relates how he arrived at Grand-Pré with 313 New England volunteer soldiers in August, 1755. Taking command of the parish church, St-Charles-des-Mines, he issued an order to the Acadians, stating that all men and boys over the age of 10 should meet him at their church at 3pm on September 5th to hear a proclamation. Suspecting nothing, over 400 Acadians answered the call. When they arrived, Winslow's men barred the doors behind them and placed them under arrest. Here they would await their forced deportation, which eventually took place between October and December. With the deportation ships under way, Winslow's men put the Acadian villages to the torch, burning nearly 700 buildings in the first week of November alone (1883, p. 185).

As a consequence of Winslow's Journal, and more particularly of Longfellow's poem, the community of Grand-Pré is inscribed in Canada's historical imagination with the date 1755. In the early 20th century, a small group of local citizens purchased the land upon which the old Acadian parish church was said to have stood, in order to protect it for posterity. Gradually, this small parcel of land evolved into Grand-Pré National Historic Site of Canada, the site of our archaeological and geophysical research.

The Archaeology of Grand-Pré National Historic Site

Archaeology was an infant and geophysics had yet to be conceived when Grand-Pré first came to the attention of researchers interested in the physical remains of the past. Consequently, our knowledge of the work they conducted, and our confidence in the conclusions they reached, is limited by the primitive methods employed by these treasure hunters and antiquarians. An article in the *Acadian Recorder*, a Nova Scotia newspaper, attributes the first digging at the site to two young men, John Woodworth and Robert Duncan, who lived near Grand-Pré Station, and who "sallied forth more than forty years ago on an exploring expedition" (8 March 1919, p.1). In a pasture near the train station, "guided by tradition," they uncovered an old well. Continuing their work, they stumbled upon a wooden coffin, which they exhumed. The wood "had become very soft from being in the wet earth so long, and could easily be cut through with a sharp knife." There is little mention of its tenant, and the various 'relics' recovered from the well have since gone missing. At least one other coffin was exhumed in the 1890s from the same general location (Parker 1897, p. 6).

Explorations continued in much the same fashion for a generation, developing in tandem with the growing tourist trade. The well found by Woodworth and Duncan soon became known as 'Evangeline's Well', linking it forever with Longfellow's heroine, and by the late 19th century researchers claimed to have discovered the remains of St-Charles-des-Mines, the Acadian parish church. According to David Otto Parker,

> "West from the well is the site of the chapel from which the exiles embarked to the transports, discovered by its underground foundation. I visited this place a few days since, when the ground had been newly plowed, and now have at my house relics, personally gathered, of old hand-forged nails, fragments of melted glass, charred fragments of wood

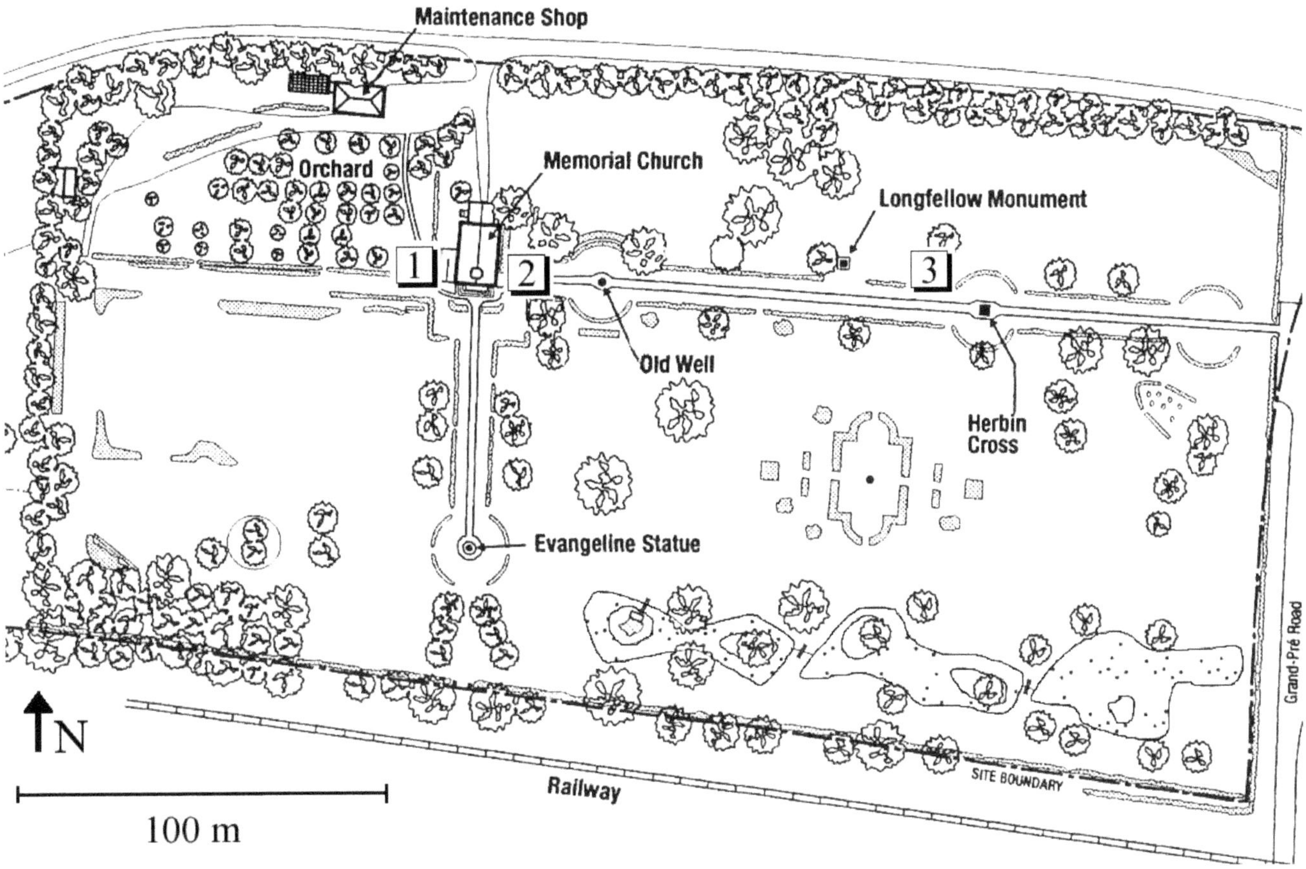

Figure 2 *Plan of a portion of Grand-Pré National Historic Site. [1] and [2] mark the locations of Parker's 'frair's house' and church, respectively (our excavations 8B28 and 8B29). [3] signifies excavation 8B30.*

and remains of bricks and slate from the chimney, which were once a part of this old sanctuary … A few steps west of the chapel, are the remains of a cellar, which without doubt belonged to the house of the friar" (1987, p.5). By 1907, local writer, businessman, and Acadian descendent John Frederic Herbin had purchased the land upon which these remains were located. His intention was to bring an end to destructive treasure hunting and create a suitable memorial for the Acadians of Grand-Pré. The following summer, he conducted excavations to clarify the location of the old church (Dunn 1982). About this time, Herbin erected a stone cross at the site where the Acadian graves had been found. Unfortunately, aside from this gesture, he neither pinpointed his discoveries with a large scale site plan, nor did he record the dimensions of the building he discovered. Therefore, in 1922, when the Acadian community began to construct a memorial church at the site identified by Herbin, oral tradition was literally cemented into historical fact. Until recently, researchers have had no way of confirming whether they had selected the right location.

Fortunately for the archaeologist, one of the antiquarian researchers did take the time to make a scaled plan of Grand-Pré's once visible ruins. The back cover of Parker's *Souvenir of Wolfville and Grand Pre* (1897) contains a line map showing the relationship between the location where the graves had been found, the old well, the church, and the 'friar's house'. A quick check with measuring tapes revealed that Parker's distance from the graves to the well was accurate (Herbin's Cross and Evangeline's Well are still part of the site today). According to Parker, the church and priest's house are located west of the well, at distances of 17.5m and 42.5m respectively. The question then presented itself: how do we test Parker's observations about the church and priest's house in a quick and cost effective fashion?

New Geophysical and Archaeological Evidence

In an effort to ascertain what might lie beneath the manicured surface of Grand-Pré National Historic Site, a geophysical survey was undertaken with the assistance of Mr. Duncan McNeill (retired president of Geonics Limited and an archaeology enthusiast). Parker's two high potential targets? were tested with a Geonics EM38B. This instrument, which has proven very useful in a number of recent archaeological applications in Atlantic Canada, measures both electrical conductivity and magnetic susceptibility simultaneously, permitting the operator to gather a large amount of data in a short period of time. At Grand-Pre simple rectangular survey grids were laid out with reference to the memorial church building, and data was collected in vertical dipole mode at approximately 30cm intervals along lines generally spaced 1m apart. The effective depth of exploration for magnetic susceptibility (which proved to be the most useful response) was approximately 70cm. The EM38B outputs the magnetic susceptibility in parts per million (ppm).

The survey results were most intriguing, and immediately confirmed Parker's observations, if not his interpretations. On the east side of the memorial church, where Parker and others placed the ruins of the Acadian church, a large feature (area approximately 450 sq.m), of variable but consistently high (up to 1000 ppm) magnetic susceptibility was encountered.. Background readings were typically 0 to 100 ppm. On the west side of the memorial church, another substantial anomaly was encountered (measuring approximately 80 sq.m) but less susceptible than the first (150 to 400 ppm). Neither anomaly was sharply enough defined to give evidence of specific architectural features such as foundations, but enough had been revealed to not only justify a program of archaeological excavations, but also to ensure that those excavations could proceed speedily to their intended targets.

The Grand-Pré Archaeological Field School Project, which was the child of this preliminary geophysical survey, commenced fieldwork in May, 2001, and has since developed into an annual event. A collaborative initiative sponsored by Saint Mary's University, Parks Canada, and the Société Promotion Grand-Pré (the Acadian community organization that co-manages the national historic site with Parks Canada), the project's goals are to provide undergraduate field training, bring archaeology to the public, and explore subsurface evidence of the Acadian community at Grand-Pré. The project has made a number of discoveries that challenge traditional interpretations of Grand-Pré's archaeological heritage.Perhaps the most interesting evidence this project has revealed is negative evidence: the absence of clear archaeological traces of St-Charles-des-Mines. In its place, at the site where the antiquarians claimed to have observed its ruins, we see a dry-stone cellar of an Acadian house (8B29), measuring less than 5m square.[1] The lowest level of the cellar fill gives evidence of extensive burning, leaving little doubt that this building – probably a house – met its end by fire. This may speak to the destructive events of 1755, but it equally may not. Textual evidence exerts a powerful gravitational pull in historical archaeology, and bitter experience recommends caution in circumstances like these. On the other hand, the discovery of a coin in the destruction layer, a British halfpenny bearing the date of 1734, ensures that the building came down sometime after that date, and the superimposed fill layers contain artifacts dating to the late 18th century. All the evidence, therefore, is so far consistent with the interpretation of this ruin as an Acadian domestic site, quite possibly destroyed during the Deportation. It would appear that the New England Planters, the English settlers who took up the vacant Acadian lands after 1760, had little use for this ruined structure, and simply back-filled it.

The question as to what was giving the instrument such strong in-phase readings was solved once archaeological excavations proceeded into the cellar. Many of the stones used to construct the cellar wall, and doubtless other parts of the structure's foundation that exist now only as disarticulated fill, are basalt. The EM38B's susceptibility

Figure 3 *Acadian house cellar at 8B29 ([1] in Fig. 1). Top is north. Scale bar is 1m.*

channel responded very positively to this class of mafic rock, making the buildings constructed from this material an easy target for the archaeologist. Ongoing fieldwork is now demonstrating that this phenomenon is also present at other colonial sites, particularly in Nova Scotia's Annapolis Valley, providing researchers a reliable strategy for detecting sites through geophysics.

A Backward Glance at the Sources

Herbin and others were aided in their early explorations by John Winslow's journal, which had been published by the Nova Scotia Historical Society in 1883. After taking charge of the church and church yard in August 1755, the New Englanders quickly set about the task of enclosing the church compound with a wooden palisade, as Winslow put it, "to Prevent a Supprise" (1883, p. 72). Winslow himself occupied the vacant priest's house, and used the church as "a Place of Arms" and as a temporary shelter for his men (1883, p. 71).

[1] Parker's 'friar's house' feature (8B28) was tested in 2001 with inconclusive results. Disarticulated architectural evidence was noted, but the associated ceramics suggested a predominately late 18th century date. As such, work has temporarily halted here in order to focus on the more informative site to the east of the memorial church.

A more recently discovered journal, to which Herbin and others do not appear to have had access, throws much-needed light on the contours of the Acadian settlement. Jeremiah Bancroft, one of Winslow's junior officers, records that the palisade Winslow ordered to be constructed "was Laid out to Enclose the Mass house and ye two Dwelling houses it was [17] rods in Length and nine rods in Bredth" (85m x 45m) (1925, p. 4).[2] The job, as it was initially planned, took approximately 6 days to complete. Bancroft indicates that Winslow subsequently decided to expand his camp to encompass the Acadian cemetery (1925, p. 5). Although this must have enlarged Bancroft's dimensions somewhat, the fact that the additional work appears to have been completed in just one day, even with the reuse of some pickets, suggests that the camp's final footprint would have been of the same order of magnitude as the initial plan.

This information raises difficulties for the traditional view that St-Charles-des-Mines lies underneath the modern memorial church. Not only is the archaeological evidence lacking, but the historical record now also casts its own doubts. Parks Canada archaeologists confirmed the presence of graves near Herbin's Cross in 1982, when the old stone monument underwent extensive renovation. We may have confidence, then, in the notion that this is the site of the Acadian cemetery. The difficulty arises when one then places a footprint of Winslow's camp down around this spot. The height of land, and therefore the driest ground, at the national historic site runs east-west. Assuming that the long axis of Winslow's camp did the same, even at its enlarged extent it seems unlikely to have reached all the way from Herbin's Cross to enclose the memorial church site, let alone Parker's 'friar's house'. This distance would be in excess of 160m (virtually doubling the camp's original length).

Fortunately, following the success of the initial, small scale geophysical survey, McNeill expanded the survey area. Additional susceptibility anomalies were detected just to the west of Herbin's Cross – closer, that is, to the Acadian cemetery and more likely within the bounds of Winslow's palisade. Archaeological testing of these anomalies from 2004-2006 has yielded what appear to be the remains of yet another Acadian building (8B30). Unlike the structures near the memorial church, this structure appears to have been missed by the first generations of researchers at the site. Even more intriguing is the discovery, in and around this building, of more than twenty lead musket balls – the most ever recorded in an Acadian domestic context. Could this, at least, be evidence of the New England military occupation of 1755?

[2] The only surviving version of this source is a typed transcription made in 1925 by Judge William Stevens of Stoneham, Massachusetts. The passage quoted above appears to contain a typographical error in which 117 rods is typed for what I must assume to have been 17 rods in the original manuscript. Every effort to locate the original document has thus far been in vain.

Conclusions

Recent archaeological research at Grand-Pré National Historic Site has amply demonstrated the value of even relatively straightforward geophysical applications in addressing problems that have long vexed researchers. By providing initial survey evidence quickly, cheaply, and in a non-intrusive fashion, the EM38B provided the basis for an ongoing and very fruitful research, education, and public outreach initiative. The fact that geophysics can provide this benefit is well known. Yet geophysical methods have yet to become a regular feature of Canadian archaeology.

Figure 4 *Duncan McNeill conducting a geophysical survey with the EM38B.*

By not only conducting a geophysical survey of Grand-Pré National Historic Site, but also by subjecting the resulting anomalies to excavation over the course of several years, this project has developed a growing understanding of the relationship between the geophysical readings and the material culture they represent. This understanding is now beginning to inform other research projects in Nova Scotia.

As an alternative 'lens' through which we may view past human activity, the geophysical survey at Grand-Pré, quite apart from its utility as a prospecting tool, has demonstrated that it may even offer something by way of an historical perspective. By drawing attention to the significant role of basalt in Acadian domestic architecture, for example, the survey has unexpectedly added a geological dimension to our understanding of Acadian resource procurement strategies.

Archaeological insights have yet to be fully adopted by historians working on the Acadian period. In the short term, archaeology can shed new light on some of the perennial concerns of Acadian historiography, such tracing the New England military occupation at Grand-Pré during that settlement's final days. Yet far greater contributions will be made in the broader study of Acadian cultural landscapes. Our understanding of colonial settlement patterns, economy, and intercultural contact stand to be greatly enriched by this work, which is still very much in its infancy. As research gradually expands to encompass the colonial landscape surrounding Grand- Pré National Historic Site, literally hundreds of as yet unexplored sites will gradually come to light. Just before they do, they will likely come to the attention of an EM38B.

References

BANCROFT, J 'Diary of a New England Soldier, Serving in Nova Scotia in 1755-1756', *Nova Scotia Archives and Records Management (NSARM) MG100* vol. 143, no. 2.

BRIGGS, R 1998 *Early Modern France 1560-1715*, Oxford University Press, Oxford.

DUNN, B 1982, *From Pasture to Posterity: The Development of Grand Pré Park*, Parks Canada, Ottawa.

FARAGHER, JM 2005, *A Great and Noble Scheme: The Tragic Story of the Expulsion of the French Acadians from their American Homeland*, W.W. Norton & Co., New York.

FERGUSON, R 1990, 'The Search for Port La Joye: Archaeology at Ile Saint-Jean's First French Settlement', *The Island Magazine*, no. 27, Spring/Summer, p. 3-8.

FOWLER, J 1995, 'Acadian-Mi'Kmaq Intermarriage in Pre-Expulsion Acadia', Honours Thesis, Saint Mary's University.

GRIFFITHS, N 1973, *The Acadians: Creation of a People*, McGraw Hill, Toronto.

LE BLANC, B 1993, 'Evangeline as Identity Myth', *Canadian Folklore canadien*, vol. 15, no. 2, pp. 139-153.

RAYNAL, GTF 1776 [1770], *A Philosophical and Political History of the Settlements and Trade of the Europeans on the Continent of North America*, vol. 4, trans. J Justamond, William Halhead and John Exshaw, Dublin.

WINSLOW, J 1883 [1755], 'Journal of Colonel John Winslow of the Provincial Troops while Engaged in the Removal of the Acadian French Inhabitants from Grand-Pré', *Collections of the Nova Scotia Historical Society*, vol. 3, pp. 71-196.

Hyperspectral and multispectral perspectives on the prehistoric cultural landscape; the ground-truthed chemical character of prehistoric settlement and infrastructure as identified from space

Ole Grøn,[1] Finn Christensen,[2] Pietro Orlando,[3] Ivar Baarstad,[4] Richard MacPhail[5]

[1] Langelands Museum, DK,
[2] Geokem, DK,
[3] Dipartimento di Rappresentazione, Università di Palermo, IT,
[4]Norsk Elektro Optikk, NO,
[5]Institute of Archaeology, UCL, UK

1 Introduction

A central problem in cultural heritage management is that the use of traditional survey methods (field-walking, test-pits, etc.) for mapping of 'invisible' archaeological sites, with no structural traces preserved above ground, is inefficient and extremely expensive. Since the 1920s aerial photos have been used in archaeological survey with good results in some cases (Crawford 1929). In a number of cases, however, there was no significant

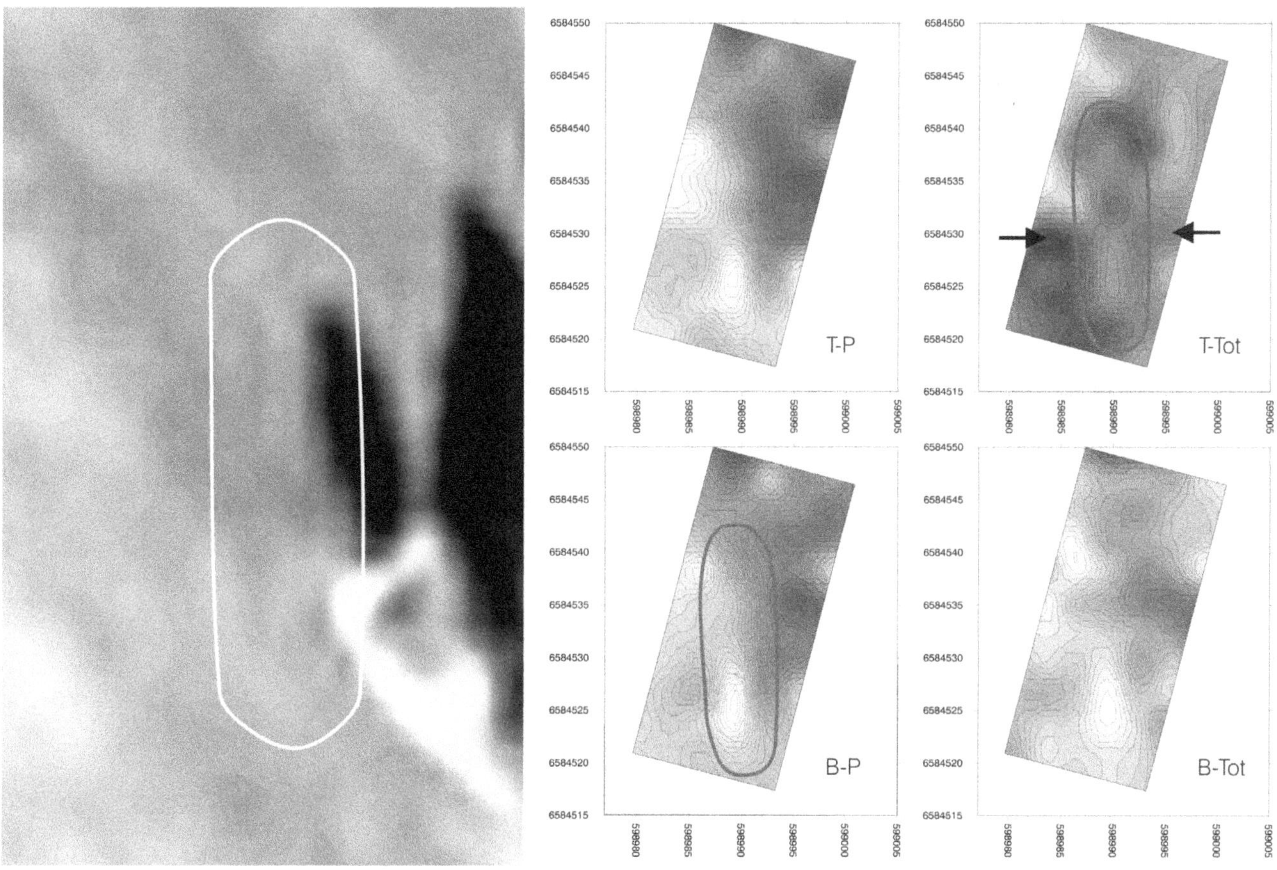

Figure 1 *Feature A, Gipsund Farm, interpreted as a house. Left: seen as a faint anomaly in the Quickbird image from 4.4.2004. Right: chemical data from the feature grid-sampled in 2003; T-P and B-P: phosphate distribution (T-P = plough soil, B-P = sediment, just below the plough soil); T-Tot and B-Tot: distribution of the sum of the five equally weighted chemical components, namely: P, Fe, Cu, Zn, Mn (T-Tot = plough soil, B-Tot = sediment just below the plough soil). The house appears as a significant local minimum with regard to phosphate and with a possible dumping area to the east (T-P, B-P). In T-Tot and B-Tot-, two concentrations may reflect the position of the two entrances common to many prehistoric houses. The supposed eastern dump starts at the suggested eastern entrance (T-Tot). (Quickbird – copyright: Digital Globe and the Satellite Data Archive of the Norwegian Mapping Authority) (Grøn et al. 2005).*

success. A series of factors seem, under certain circumstances, to reduce the possibilities for observation. Equally, these circumstances restrict the optimum period for aerial photography to narrow time-windows that are difficult to predict.

To overcome these problems, development of the use of high-resolution multi-spectral satellite images and hyperspectral images from airborne platforms for cultural heritage monitoring was initiated in 2000 with supported from the Norwegian Space Centre, the Norwegian Directorate for Cultural Heritage, Steigen Municipal and Vest Agder County, Norway. So far these projects have produced promising results indicating that the application of multi/hyper-spectral techniques can serve archaeology considerably better than aerial photos (Grøn & Loska 2002; Grøn et al. 2004a; Grøn et al. 2004b). In this paper we briefly outline our findings and the strategies under development.

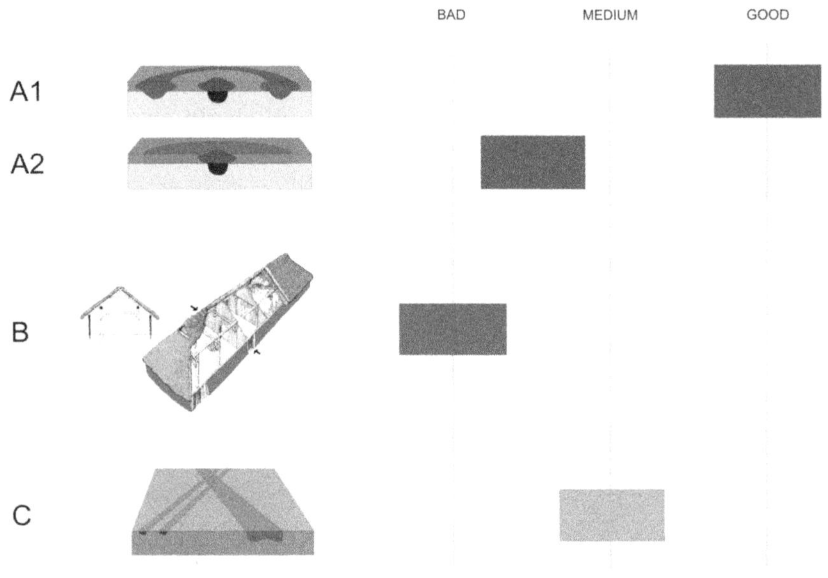

Figure 2 *Generalised representation of the difference in visibility of some cultural heritage categories such as – A1: mounds with a circular ditch around them, A2: mounds with no circular ditch around them, B: houses and C: roads and tracks.*

2 Basic experiences

A number of basic results were obtained with IKONOS and Quickbird images. Many are of an essentially preliminary character and restricted to observations from Norwegian agricultural fields of a limited number of landscape types. In spite of that they appear to have wider implications.

One observed tendency is that the less vegetation there is in an image, the better information it will provide about archaeological features. This indicates that the most important element for archaeological interpretation of multispectral images is the direct ground reflection and not, as we had expected, the reflection from the vegetation. An image from an agricultural area in southern Norway taken during the maximum growth period (June 2003) shows considerably less information than a recording from the same area collected in August 2000; the later also reveals less than one from April 2004. In spite of this seasonal effect some archaeological features seem to be more visible in areas with some vegetation compared to others with no plant cover.

Our chemical ground-truthing shows that archaeological anomalies can be distinguished in multi- and hyperspectral images because of local variations in the chemical/mineral content of the sediment surface. From an archaeological point of view this makes sense. The 'formation' of archaeological sites generally involves moving and manipulating local resources. Hunter-gatherers will accumulate bones from game animals in their settlements that contain phosphate, iron, etc. The construction and use of houses, mounds, etc., often involves manipulation of local materials (clay for walls and floor make-ups that has been imported from off-site, mounds constructed of both turf and (quarried) subsoil materials, midden waste accumulations containing heavy metals and phosphates). The archaeological sites will there-fore tend to appear as relative concentrations of minerals/ chemical components on a geological background that often displays significant small-scale variation in its mineral/chemical components.

Since the method reacts on patterns exposed on the sediment surface one could think that only sites with features directly in contact with the plough layer would be distinguishable. However, many plants have an ability to deal with heavy metals and lift them to the surface. If an archaeological feature is therefore covered by a slow sedi-menttation it should still be possible to find its signature on the surface (e.g. Friedland 1990).

In the analysis of vegetation in multispectral images one can rely on indexes such as for instance the Normalized Difference Vegetation Index (NDVI):

$$NDVI = (NIR - RED)/(NIR + RED)$$

where NIR is the near-infrared band and RED the red band. The experience so far is that, for direct visual categorisation of cultural heritage anomalies, there seems to be no fixed rules for in which bands or in which combination of bands they are best observable. This is probably a consequence of archaeological sites appearing as relative variations on a highly variable geological background. Some indirect indicators of archaeological sites, such as variation in ground-moisture indexes, seem to work well.

There is a marked difference in how easy anomalies representing different types of archaeological features can be distinguished. Mounds with surrounding ditches are generally easy to distinguish, whereas houses for instance can be very difficult (Fig.2).

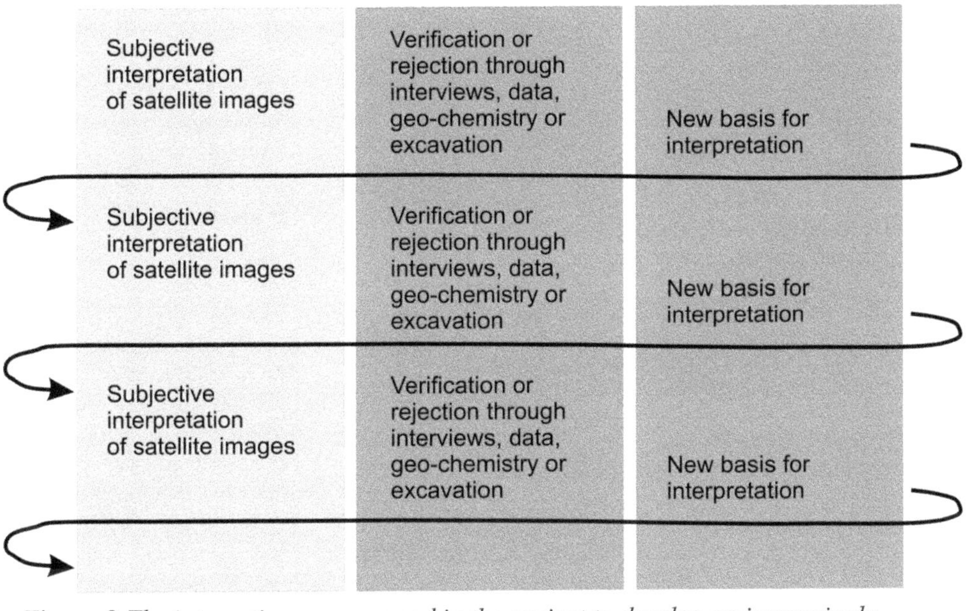

Figure 3 *The interactive process used in the project to develop an increasingly 'objective' categorization of the different types of cultural heritage anomalies, geological anomalies and anomalies reflecting modern activities observed in the satellite images on the basis of phases of subjective interpretation and verification.*

3 Geo-chemical ground truthing

Because variations in the mineral/chemical content of the sediment surface seems to be of crucial importance for the creation of the reflective patterns that make it possible to distinguish archaeological anomalies, it seems logical to develop a ground truthing method based on analysis of the chemical content of the top sediment. Work along these lines have produced promising results indicating that specific 'chemical signatures' may follow the different types of archaeological features.

The focus has been on phosphorus, P (as orthophosphate, PO_4^{3-}), iron, Fe (as Fe^{2+} and Fe^{3+}), copper, Cu (as Cu^{2+}), zinc, Zn (as Zn^{2+}) and manganese, Mn (as Mn^{2+}) because these chemical parameters are cultural indicators that appear in relatively large amounts and generally have a long-time survival in soil. The results obtained in relation to the projects discussed here show that Mn appears to be a more important cultural indicator than expected.

Generally, samples are taken in 'profiles' with a distance of one meter between them or in 2 x 2 m grids. At least in an initial phase it is important to maintain a good resolution to be able to establish 'chemical signatures' of the archaeological features. For each sampling point a sample is taken from the plough soil and one from the sub-soil, directly below it, to gather information on the effect of ploughing on the configurations of the anomalies observable on the top of the sediment.

P is measured by colorimetry. Fe is measured by colorimetry for large concentrations but by atomic absorbtion spectrophoto-metry (AAS) for smaller concentrations. Cu, Zn and Mn are measured by AAS. For both methods the samples must be dis-solved. The characteris-tics of the extract pro-duced depend to a high degree on the chosen extractant, in this case sulphuric acid. A develop-ment of standardization in this area could facilitate comparison and exchange of data between projects.

The choice of extractant is a balance between dis-solving the fraction of the samples that derives from cultural activity and not dissolving the earth-matrix itself. Sulphuric acid with a concentration of 0.1 mol/L has been chosen because it has a number of advantages over other extractants but some problems with heavy dissoluble sulphates, such as for instance $CaSO_4$ (gypsum), that can be generated in calcareous samples. Hydrochloric and citric acid, preferred by some workers, have other disadvantages. EDTA (EthyleneDiamineTetraAcetic acid) is of some interest because it uses a different mechanism with some advantages in the dissolution process.

The samples are dried at 20°C. A sieved fine earth fraction (<1 mm) is separated for analysis, 1.00g per sample. 20.0mL 0.1M sulphuric acid is added and the mixture is stirred for two hours. After some time the essence can be decanted so that the dissolution analyzed contains no particles.

Unlike ground-penetrating radar (GPR) and magneto-meters, that are difficult to apply as ground-truthing methodologies on a general basis because they have problems with some sediment types, the method described here has provided meaningful relative results in all cases where it has been applied. Furthermore it has the advantage that the samples taken from below the plough soil provide an insight into the stratigraphy that in some cases can provide sufficient documentation for the character of the archaeological feature in question.

Figure 4 *Image of beach from Norwegian study area recorded with the hyperspectral scanner HySpex from Norsk Elektro Optikk AS.*

The most important disadvantage is that the analysis is time consuming and expensive. Important for the development of a cost-effective verification strategy will be the design of a faster and cheaper procedure for analysis.

4 Method of analysis – visual inspection or ...

The analysis of the multispectral data are based on visual inspection. This is because the application of semi-automatised pattern recognition to distinction of archaeological anomalies is known to have been unsuccessful in several cases and as yet no convincing evidence has been produced that such a system today can distinguish archaeological and non-archaeological anomalies at a sufficient level of confidence. With regard to small-scale anomalies such as those focused upon in the approach presented here, this renders little space for optimism for an automatised system in the near future.

The central problem met in the other projects was that it was on the one hand impossible to develop allgorithms that were sufficiently precisely defined that they could 'catch' all significant anomalies while on the other hand being flexible enough cope with the variation in shape within the different anomaly categories and background noise. Therefore the automatic distinction between the archaeological and the non-archaeological anomalies did not work as intended. It was necessary to use so much complementary visual inspection that the pattern recognition system did not reduce the manual work-effort to any worthwhile degree.

The strategy that we have developed is to carry out repeated interpretations of the data based on visual inspection of the images followed by verification phases (ground-truthing) (Fig.3). The experience is that such a process can train an observer with a good basis in archaeological fieldwork to obtain an acceptable ability to distinguish between arch-aeological and non-archa-eological anomalies.

5 Discussion

It is difficult to assess the representativity of the data produced by a survey method, but an interesting parameter in such a discussion is that the use of multispectral satellite data in the central study areas seems to begin to produce sensible infra-structural patterns: tracks and roads that apparently connect concentrations of houses and burials.

The data available from the multispectral satellite-based high-resolution scanners today only have four bands and ground-resolutions from 2.6 to 4 m. It is obvious, however, that the technology is developing. Images of the Gipsund Farm archaeological sites have recently been recorded with the airborne hyperspectral scanner HySpex (VNIR-1600 and SWIR-320i) from Norsk Elektro Optikk AS. When operated at 1000m altitude above ground level, this instrument has a spatial pixel size of 20cm by 40cm and 160 spectral bands in the VNIR (400-1000nm) region and a spatial pixel size of 75cm by 75cm and an additional 160 spectral bands in the SWIR (900-1700nm) region.

At this spatial and spectral resolution level it should be possible to refine the analysis considerably. Recent advances in sensor technology also allow the spectral region up to 2500nm and possibly even the MWIR (3000-5000nm) region to be exploited in hyperspectral imagery, which could increase the usefulness of this technology also for archaeological applications, especially in cases where soil and geological parameters are used as indicators.

In general the perspectives for the development of management systems underpinned by multi- and hyperspectral images look promising. A central question is if the main development will go in the direction of satellite based or airborne scanners. The former will have an advantage in that the recording of the images can be more homogeneous, whereas the latter will have an advantageous ability to adapt the data acquisition to the local weather conditions, in addition to increased spatial and spectral resolution.

References

CRAWFORD, O.G.S. 1929 Woodbury. Two Marvellous Air-Photographies. *Antiquity* vol.3, No.12. 452-455

FRIEDLAND, A.J. 1990 The Movement of Metals Through Soils and Ecosystems. *Heavy Metal Tolerance in Plants: Evolutionary Aspects.* (ed.) A.J.Shaw. CRC Press, Boco Raton, Florida.7-19

GRØN, O. & LOSKA, A. 2002 *Development of methods for satellite monitoring of cultural heritage sites.* http://www.riksantikvaren.no/Norsk/Publikasjoner/Utgitt_i_samarbeid_med_andre/filestore/satmon_report_screen2.pdf

GRØN, O., AURDAL, L., CHRISTENSEN, F., TØMMERVIK, H. & LOSKA, A. 2004a *Locating invisible cultural heritage sites in agricultural fields Evaluation of methods for satellite monitoring of cultural heritage sites – results 2003.* http://www.riksantikvaren.no/Norsk/Publikasjoner/Utgitt_i_samarbeid_med_andre/filestore/satellite-Rap.pdf

GRØN, O., AURDAL, L., CHRISTENSEN, F. and LOSKA, A. 2004b Mapping and verifying invisible archaeological sites in agricultural fields by means of multi-spectral satellite images and soil chemistry. *Proceedings and Programme. International Conference on Remote Sensing Archaeology October 18-21, 2004, Beijing, China.* Beijing. 83-90.

GRØN, O., AURDAL, L., CHRISTENSEN, F., LOSKA, A., MACPHAIL, R. 2005 *Locating invisible cultural heritage sites in agricultural fields. Development of methods for satellite monitoring of cultural heritage sites – report 2004.* The Directorate for Cultural Heritage, Oslo.

Roman urban landscapes in Italy: an integrated approach

Sophie Hay,[1] Simon Keay,[2] Martin Millett,[3] Kris Strutt[1]

[1] APSS/University of Southampton
[2] British School of Rome/University of Southampton,
[3] University of Cambridge

Introduction

While there has been much emphasis upon Roman provincial landscapes and rural settlement patterns in recent years, attention is now beginning to turn towards the distinctive methodological challenge offered by Roman town sites. Traditional archaeological approaches have favoured the excavation and study of individual structures, usually public buildings, or the analysis of parts of sites. While much valuable information has been accumulated in this way, it is the contention of this paper that if we are to better understand the character of Roman towns, their dynamics and relationship to surrounding landscapes, then we need to understand them as urban landscapes. Geophysics has made an important contribution to our understanding of Roman towns in recent years (notably Gaffney et al. 2000; Rodríguez Hidalgo et al 1999) not least in Italy, notably the work undertaken by Piro and the Consiglio Nazionale di Ricerca (CNR). However, we believe that the potential of geophysical survey in urban contexts has not yet been fully exploited. In particular, we believe that the integration of a range of geophysical techniques within a broader archaeological approach can significantly enhance our understanding of Roman urban sites. The aim of this paper is to very briefly review some of the results of this work carried out by the British School at Rome in collaboration with the Universities of Southampton and Cambridge since 1997.

Background

The regular application of geophysical and topographical survey techniques to evaluate archaeological sites has been the most recent addition to landscape survey in Italy carried out by The British School at Rome. As a research institution, The British School at Rome has enjoyed a long tradition of promoting integrated landscape survey as a fundamental strategy for academic study. Thomas Ashby, as Director of the School at the beginning of the last century, set the precedent by embarking upon a study of the Roman Campagna; drawing upon topography and extant ancient remains to elucidate the expansion of Rome's territory over time. During the 1950's, John Ward-Perkins conceived the South Etruria Survey, a broad landscape study that incorporated geophysical survey results from Veii (Ward-Perkins 1961, 88-90). Both the methodology and conclusions of the South Etruria Survey (Potter 1979) have formed the basis of many subsequent British and Italian archaeological projects in the area. In particular, the Tiber Valley Project (Patterson 2004) has re-examined the landscape on a much broader scale and initiated new fieldwork using a variety of survey methods. In particular, the Roman Towns in the Middle and Lower Tiber Valley project under the direction of Simon Keay and Martin Millett undertook topographical and geophysical survey on a range of sites (discussed in Keay et al. 2004) including Falerii Novi, Ocriculum, Capena and Portus (Keay et al 2000; Johnson et al. 2004; Keay et al. 2005; Keay et al. 2006; Hay et al. forthcoming 2006). The notable results amassed by this project have demonstrated the success of non-destructive geophysical survey techniques, and underlined the importance of surveying complete urban landscapes and integrating a range of different techniques. Keay, Millett and Hay have moved on from this to undertake geophysical survey on sites to the south of Rome at Fregellae (Ceprano) and Teanum Sidicinum (Teano). Geophysical survey also played an important role in the British School at Rome study of the town of Forum Novum in the Sabina (Gaffney et al. 2004).

Building on these successful applications of geophysical survey techniques to Roman urban sites, a collaborative initiative was established between The British School at Rome and the Archaeological Prospection Services of Southampton (APSS), a unit within the University of Southampton that has successfully undertaken survey on a commercial basis throughout Europe since 2001. This initiative has seen the establishment of a unit of archaeologists and geophysicists based at the British School at Rome in order to undertake surveys primarily in Italy on a commercial basis all the year round jointly with colleagues from APSS. The wide coverage of their work over much of Italy is evident in the diversity of the commissioning bodies; a number of Soprintendenze, Italian and foreign Universities, Provincie, Regione and local Comune. While they undertake survey on a range of sites of different periods, the survey of Roman towns remains a particular speciality.

Our Approach

The methods used on surveys of Roman towns undertaken by the British School at Rome and APSS in conjunction with the Universities of Southampton and Cambridge has been to apply existing techniques rather than to develop new methods of geophysical detection or processing. However, our approach differs from that of many surveys, in recognizing that analysis of Roman urban sites is best served by employing a range of geophysical and archaeological techniques within an integrated archaeological strategy, rather than simply using geophysics in isolation. Thus, for example, while magnetomtery on its own may tell us a lot about the layout of streets and buildings at urban sites, this information alone cannot be properly understood without understanding the topography of the site or the range of ceramics and construction materials lying on the surface. Experience has also taught us that each site offers a different archaeological challenge and that tailor-made fieldwork strategies are required for each site studied. Thus, the magnetic susceptibility of some sites may so poor that magnetomtery results need to be complemented at key points by resistivity or geo-radar. In the case of our own research projects, such as our ongoing work at Teanum Sidicinum, we have designed fieldwork strategies that best answer our research questions. In the case of commercial work, they are designed in close conjunction with the archaeologist or contracting body financing the work to ensure that they meet their respective research or planning questions.

Another key issues concerns large scale survey. While much is being learned about details or aspects of individual sites, it has become clear to us that geophysics – particularly magnetometry – has perhaps most to teach us about large areas, whether it be swathes of urban landscapes, as has been the case at Falerii Novi and Teanum Sidicinum, or about the hinterlands of sites, as in the case of Portus, where we have now surveyed up to 200 Ha and in whose extramural area we have uncovered a complex of canals, a riverside port and an extensive burial area (Keay et al. 2005).

Geophysical Survey Issues

Topographic Survey

The landscape context of sites and the need to create a framework for survey are important prerequisites for geophysical surveys. Since good digital map coverage of 1:2500 or lower is generally difficult to obtain, recourse is usually had to close contour mapping with a total station. Close contour survey of the local topography has consistently proved to be of enormous value when deciphering a site. T The scant remains of the visible past often make us study the landscape closely, leading to an appreciation of how far urban development of many towns is conditioned by the landscape in which they lie. Topographic surveys aim to encompass the recording of both large scale features in the landscape as well as localized earthworks and provide a background onto which other data sets can be draped. While this can be relatively straightforward at sites where land is flat and with few obstacles, uneven topography and dense vegetation can make this a separate project in itself.

The town of Ocriculum (Otricoli) is a good example of a site where this approach has produced important results (Hay et al. Forthcoming 2006). Published plans suggest that this is a relatively flat site, while in reality it lies on a truncated spur overlooking the Tiber that is almost impossible to understand on the ground owing to very dense vegetation. A carefully-crafted network of base stations was set up across the site to enable the survey to be carried out and for the geophysics grid to be established. The survey was undertaken using a total station theodolite and the data processed using LISCAD software. Careful note was taken of present day topography (field boundaries, tracks and buildings) as well as outline plans of existing archaeological structures (such as the amphitheatre and the Grandi Sostruzioni) in sufficient detail to be related to published 1:10,000 and 1: 5000 maps (Fig. 1). Once finished, however, the contour plan was readily converted into a Digital Elevation Model upon which the geophysical results could be draped – and better understood than they would have been in isolation.

Figure 1 *Ocriculum: Close contour survey of central area of the site*

Magnetometry

This is the technique most frequently used on our surveys. This is partly because it is relatively quick, and therefore economical. However, it is also because the presence of volcanic deposits in central and southern Italy means that magnetically susceptible construction material, such as tufa, was often used in ancient buildings, rendering them more readily detectable. The survey was undertaken within a 30m x 30m grid that was established to run at an angle of 30° to potential archaeological features (where known). On the flatter sites, such as Falerii Novi and Fregellae this was a relatively straightforward procedure. The more topographically complex sites, such as Teanum Sidicinum and Ocriculum proved a greater challenge, so that while every effort was made to ensure that grid lines all shared the same orientation with respect to the visible archaeology, they sometimes varied with respect to broader geographical position. In general we have used the FM 36 and 256 produced by Geoscan Research, taking readings at 0.5 m intervals along traverses every 1m, and using an automatic encoding trigger.

In general, the results have been very good, although there was quite a range in the quality of response. In some cases this was probably related to the differing magnetic susceptibility of individual sites that are located in areas with similar volcanic geology. For example, the excellent clarity of the results from Falerii Novi (Fig. 2), where individual rooms and houses can be seen, stands in contrast to those from Ocriculum (Fig. 3), where features are harder to locate even though the presence of material on the surface tells us that ancient structures are clearly present.

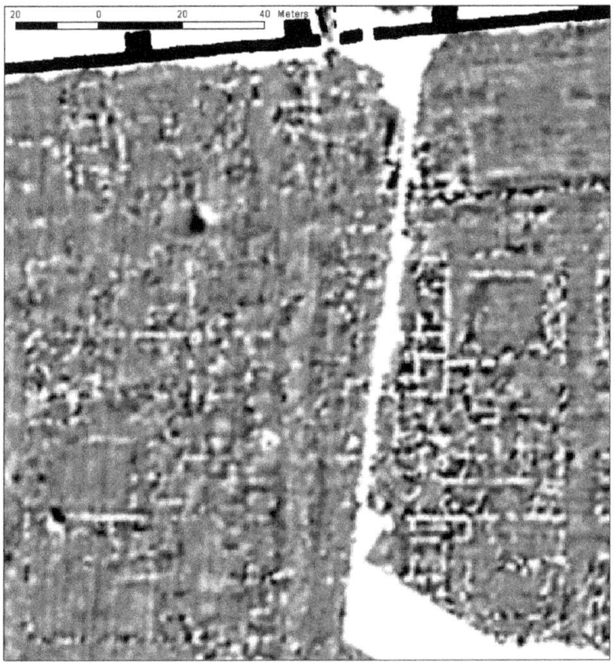

Figure 2 *Falerii Novi: Detail of the magnetometry survey of the area to the north of the forum*

Figure 3 *Ocriculum: Detail of the magnetometry results in the central area of the site*

Figure 4 *Portus: Aerial photograph of the river port*

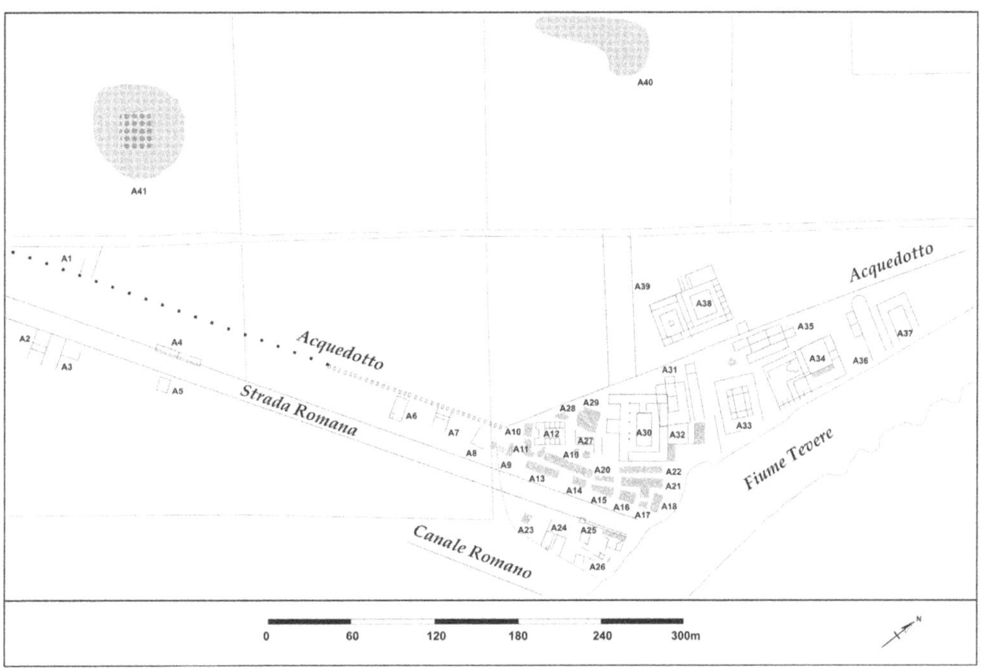

Figure 5 *Portus: Interpretation of the aerial photograph of the river port*

This underlines the desirability of a magnetic susceptibility survey prior to undertaking a survey – although there is often not enough time to do this in a commercial context. Alternatively, a geophysical survey undertaken on gravels at the site of Sentinum (Le Marche) revealed that the local limestone walls were less magnetic than the local limestone gravel geology. An understanding of how archaeological responses compare with those of the geology is thus an important issue to investigate in future work.

However, it is also possible that the lack of clarity in the results from Ocriculum may be related to the presence of large amounts of ceramic material in surface levels, creating a background noise which renders significant magnetic anomalies less visible. Yet again, our understanding of other areas of this site was complicated by the presence of rubble from re-deposited rubble from the 17th and 18th Centuries – a problem also observed on other sites such as parts of Capena and Baccanas. This issue may have also affected the results from Portus – specifically the area of the river port, immediately to the north of the intersection of the Trajanic canal and the Tiber. Here the magetometry results lacked the clarity of areas along the canal and around the Trajanic Hexagon to the west, and proved difficult to interpret. However, aerial photographs of the same area revealed a very clear complex of structures (Fig. 4), whose interpretation contributed greatly to our understanding of this area (Fig. 5). We hope for similar methodological synergies in our ongoing geophysical survey of the town of Potentia (Potenza), which is being conducted in conjunction with a programme of aerial photography and field survey (Vermeulen & Verhoeven 2004).

In other parts of Ocriculum, it was also clear that poor response was due to the fact that major structural features lay beyond the reach of the magnetometer, and could only be traced with Geo-Radar. A major survey by Piro of the CNR revealed the existence of a complex suite of major structures belonging to a building whose orientation followed that of structures detected by magnetometry in other parts of the site.

A different kind of contrast was encountered in the results from surveys of Republican Roman towns such as Fregellae (Fig. 6), where tufa, mud-brick and terracotta were the main construction materials, and towns of early Imperial date such as Falerii Novi and Teanum Sidicinum (Fig. 7) where marble was used and ceramics were considerably more common. At the former, only the outlines of the main public buildings and some details of houses were visible, while at the latter sites, there is evidence for far busier townscapes.

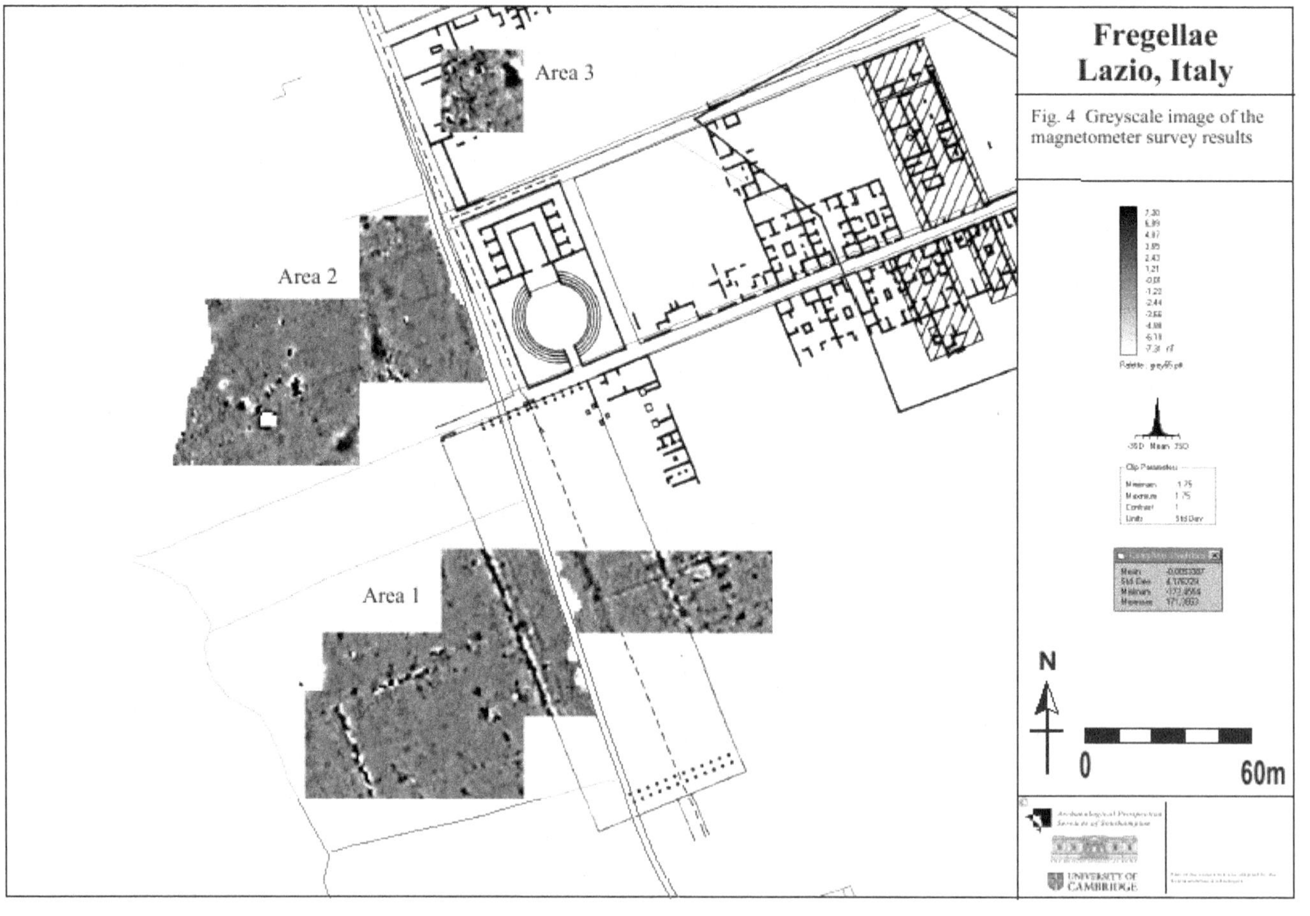

Figure 6 *Fregellae: Magnetomtery survey of the area in the vicinity of the forum (2004)*

Figure 7 *Teanum Sidicinum: Magnetometry survey of the area around the sanctuary of Loreto*

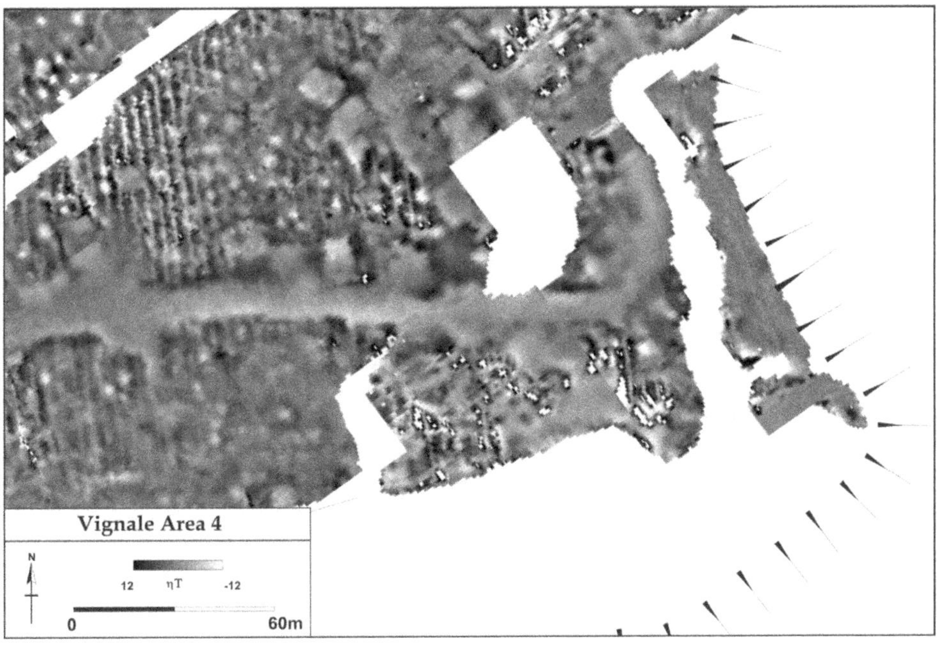

Figure 8 *Vignale: Magnetometry around the cistern and temple at the centre of the site*

Resistivity

While resistivity has many obvious benefits, its main disadvantages are that it is much slower and can only be effectively used at in the Autumn, Winter and Spring. As a consequence we have tended to use it in addition to the magnetometer, as a way of calibrating its results and trying to better understand them at different times of the year. A good example of this can be seen at the Faliscan settlement and sanctuary at Vignale (Carlucci et al. forthcoming 2007). While this was a relatively easy site to survey, it is difficult to interpret, owing to the abundant use of terracotta construction materials for the temple and other structures, the existence of many rock-cut features of ancient and medieval date, and natural cavities in the bedrock. In the area near the cisterns adjacent to the site of the temple, the resistivity (Fig.8) revealed a range of structures not visible in the magnetometry (Fig.9), and which could be worked into the overall interpretative plan of the site.

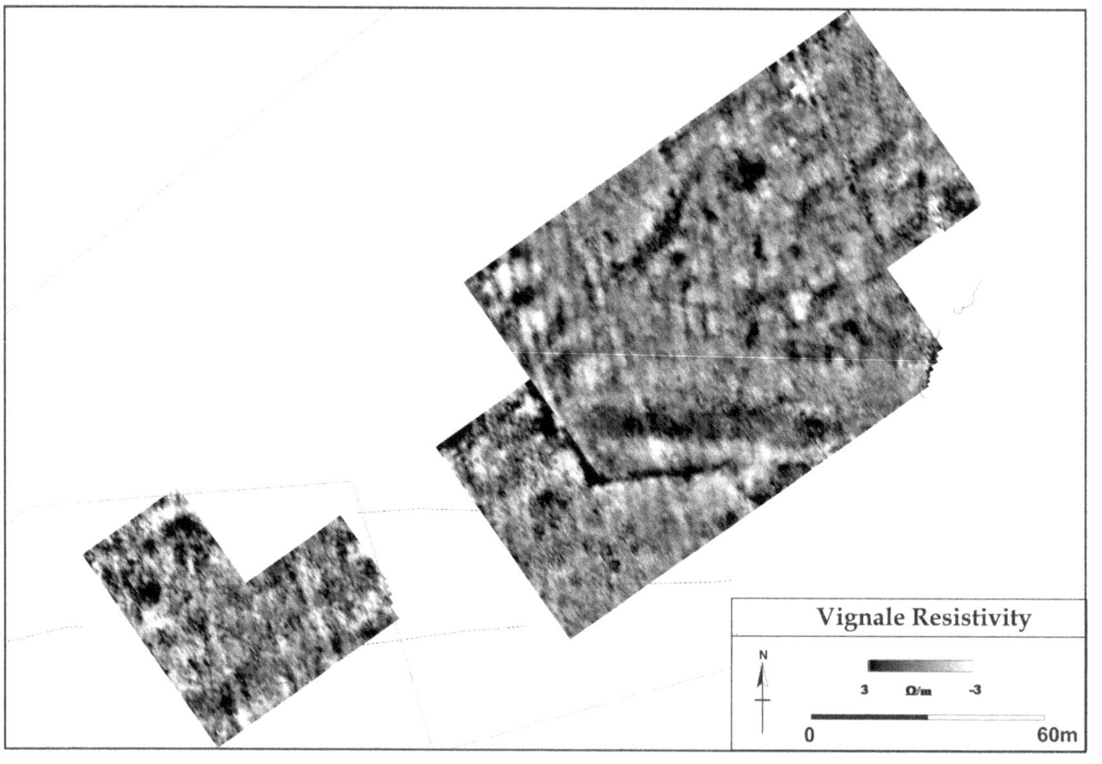

Figure 9 *Vignale: Resistivity survey of the same area around the cistern and temple at the centre of the site*

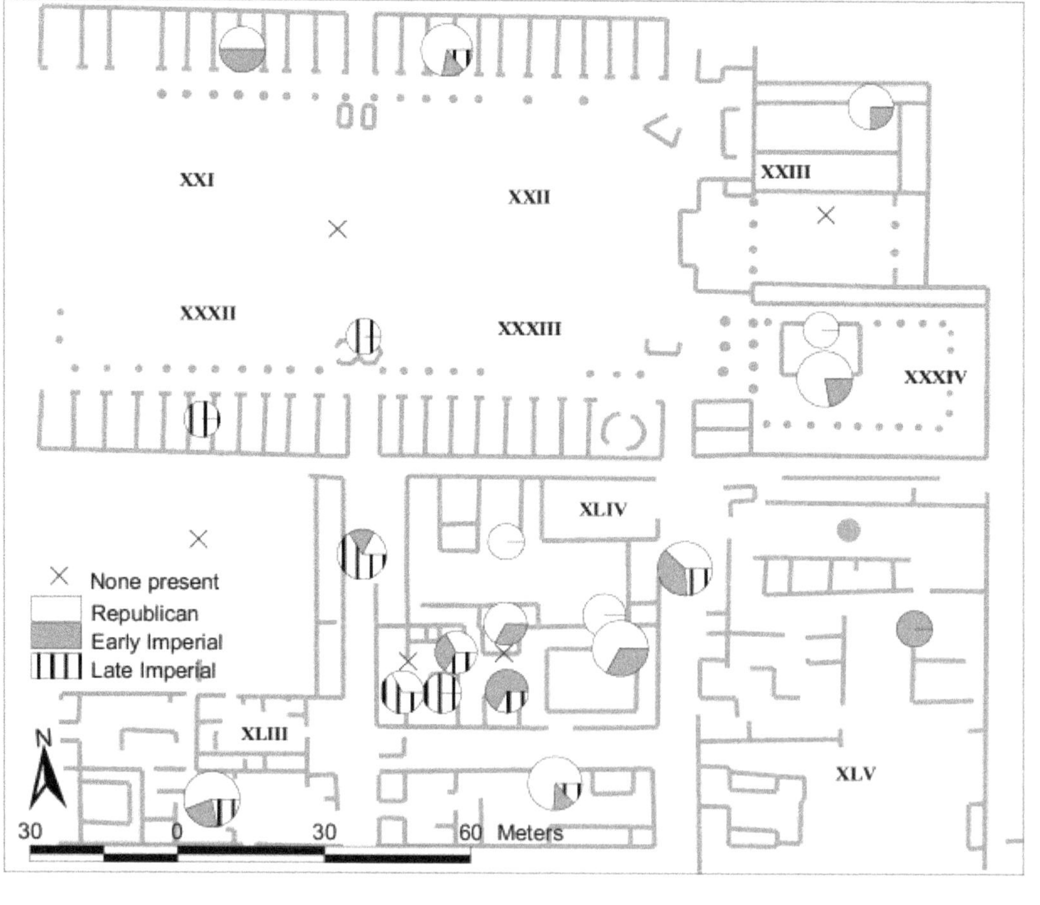

Figure 10 *Falerii Novi: Interpretation of systematic surface collection of surface materials close to the forum*

Surface Materials

While the combined results of topographic survey and geophysics have much to tell us about buried archaeological sites, they tell us relatively little about the chronology (apart from gauging depth with geo-radar and resistivity multi-plexing) and nothing about their function. Many green-field Roman urban sites abound in surface materials that have been generated by the ploughing-up of archaeological strata and which, thus, have a direct relationship with the buried structures detected by geophysical survey. Our approach, pioneered at Italica in southern Spain (Rodríguez Hidalgo et al 1999), and which was developed at Falerii Novi involved mapping the occurrence of different kinds of ceramics and construction material. Material was collected systematically within the framework of the geophysical survey so that the results of both could be integrated. In the case of the latter site, much was learned about the chronology and function of different kinds of building that helped greatly in the overall interpretation of the site (Fig.10). However, the conditions at Falerii were exceptional – matched only by the open area between the Hexagon and the Tiber at Portus. More usually surface visibility is simply not good enough for this kind of work. Even so, some kind of notation of surface materials in the course of a geophysical survey can assist in understanding the chronology and function of structures detected by geophysics – as has been attempted at Capena (Keay et al. 2006).

Conclusion

While we are often asked to undertake the geophysical survey of small 2 to 3 Ha parcels of land, and produce good results *per se*, the greatest potential of geophsyics lies in surveying the large landscapes of Roman towns. The large-scale exploration of urban landscapes affords details of the detailes of the intricacies of town planning and development at an intra-urban level and where possible sets the urban site within its rural context. Whilst the use of any one geophysical survey method may successfully detect and map the buried archaeological remains of a site, the quantity of information retrieved and potency of interpretation is greatly enhanced by the application of more than one technique. This paper has alluded briefly to the success that we have enjoyed in combining magnetometry with resistivity, geo-radar, systematic surface collection, aerial photography and, of course, topographic survey. Each urban site raises a distinct archaeological challenge and it is contingent upon the archaeologist to develop a strategy that involves the appropriate mix of techniques in order to extract the maximum information about the layout and character of buried structures. In doing so, however, there are still many methodological issues that complicate our understanding of what it is that our results actually tell us. Most obviously, it is clear that a good understanding of local geology is crucial if we are not going to confuse natural with anthropically generated anomalies. It is also clearly important to understand the actual values of the signal detected by geophysical surveys. Thus is it possible, for example, to deduce the nature of archaeological remains, such as building materials, from the values of the response given no prior knowledge of the site? To answer these and many other questions it is important to understand site topography, site site-formation processes and the character of the extant (or in process of excavation) archaeological remains, and how these will affect the survey results and our interpretation of them. Achievement of this is contingent upon an ongoing dialogue between geophysicists and archaeologists.

Acknowledgements

We would like to warmly thank the various institutions, funding bodies and individuals that have supported all this research over the past nine years, including the British School at Rome (Prof. A. Wallace-Hadrill), the Universities of Southampton and Cambridge, the Arts and Humanities Research Council, the Society for the Promotion of Roman Studies, the Soprintendenza Archeologica di Etruria Meridionale (Dottssa De Lucia), the Soprintendenza Archeologica per L'Umbria (Dottssa Cencaioli), the Soprintendenza di Beni Archeologici di Ostia (Dottssa Gallina Zevi), the Soprintendenza Archeologica di Napoli e Caserta (Dott. Sirano), Prof. Filippo Coarelli, and Dott.ssa Giovanna Battaglini.

References

CARLUCCI, C., DE LUCIA BROLLI, M., KEAY, S., MILLETT, M. and STRUTT, K. (Forthcoming 2007), An Archaeological Survey of the Faliscan Settlement at Vignale, Falerii Veteres, (Province of Viterbo), *Papers of the British School at Rome LXXIII.*

GAFFNEY, V., GATER, J.A., LINFORD, P., GAFFNEY, V.L. and WHITE, R. (2000), Large-scale Systematic Fluxgate Gradiometry at the Roman City of Wroxeter. *Archaeological Prospection 7*, 81-99.

GAFFNEY, V., PATTERSON, H. and ROBERTS, P. (2004), Forum Novum (Vescovio): A New Study of the Town and Bishopric, in H. Patterson (ed.), *Bridging the Tiber. Approaches to Regional Archaeology in the Middle Tiber Valley. Archaeological Monographs of the British School at Rome 13*, 248-251. London.

HAY, S., KEAY, S. and MILLETT, M. (Forthcoming 2006), Urban Fieldsurvey at Ocriculum (Otricoli, Umbria), in F. Coarelli & H. Patterson (edd.), *Mercator Placidissimus.*

JOHNSON, P., KEAY, S. and MILLETT, M. (2004), Lesser Urban Sites in the Tiber Valley: Baccanae, Forum Cassii and Castellum Amerinum, *Papers of the British School at Rome LXXI,* 69-99.

KEAY, S., MILLETT, M., PAROLI, L. and STRUTT, K. (2005), *Portus. An Archaeological Survey of the Port of Imperial Rome. Archaeological Monographs of the British School at Rome 15*. London.

KEAY, S., MILLETT, M., POPPY, S., ROBINSON, J., TAYLOR, J. and TERRENATO, N. (2004), New Approaches to Roman Urbanism in the Tiber Valley, in H. Patterson (ed.), *Bridging the Tiber. Approaches to Regional Archaeology in the Middle Tiber Valley. Monographs of the British School at Rome 13*, 223-236. London.

KEAY, S. MILLETT, M., POPPY, S., ROBINSON, J., TAYLOR, J. and TERRENATO, N. (2000), Falerii Novi: A New Study of the Walled Area, *Papers of the British School at Rome LXVIII*, 1-93.

KEAY, S., MILLETT, M. and STRUTT, K. (2006), An Archaeological Survey of Capena (La Civitucola, Provincia di Roma), *Papers of the British School at Rome LXXII*, 73-118.

PATTERSON, H. (2004), *Bridging the Tiber. Approaches to Regional Archaeology in the Middle Tiber Valley. Archaeological Monographs of the British School at Rome 13*. London

POTTER, T. W. 1979, *The Changing Landscape of South Etruria*. London.

RODRÍGUEZ HIDALGO, J. M., KEAY, S., JORDAN, D. and CREIGHTON, J. (1999), La Itálica de Adriano. Resultados de las prospecciones arqueológicas de 1991 y 1993, *Archivo Español de Arqueologia 72*, 73-98.

VERMEULEN, F. and VERHOEVEN, G. (2004), The contribution of aerial photography and field survey to the study of urbanization in the Potenza valley (Picenum), *Journal of Roman Archaeology 17*, 57-82.

WARD-PERKINS, J. (1961) Veii: The Historical Topography of the Ancient City. *Papers of the British School at Rome* XXIX.

Detecting Mycenae
Systematic Remote-Sensing Survey in the 'Lower City': Toward the Discovery of the Mycenaean Settlement Outside the Citadel

Christofilis Maggidis,[1] Antonia Stamos[2]

[1] Dickinson College, Athens Archaeological Society
[2] Institute for Aegean Prehistory, Temple University

Introduction

Since the commencement of the systematic investigation of Mycenae by Heinrich Schliemann in 1874, the excavation of the site has focused mainly on the citadel itself, including the fortification walls and gates, the palace, palatial storerooms and workshops, shrines, royal graves (Grave Circle A), the underground fountain, and residence quarters within the wall circuit (Mylonas 1983; Iakovidis 1983), comprising approximately 70% of the total fortified area (30,000 m^2); many of the excavated buildings and contexts have already been fully published, or are currently being prepared for publication (Iakovidis 1996, 1039-1045). Outside the walls (Fig. 2), work has been limited to royal cemeteries (Grave Circle B and three groups of nine tholos tombs) and a fair part of the extant chamber tomb cemeteries stretching farther on the hill slopes around Mycenae (Panagia, Kalkani, Epano and Kato Pegadi, Alepotrypa, Kapsala, and Asprochoma), whereas only a very small segment of the settlement once surrounding the citadel (12-13 completely or partially excavated buildings) has been brought to light (Iakovidis 1996, 1039-1041, 1045-1047). Most of these buildings are located west of the citadel, including the Oil-Merchant complex (West House, House of Shields, House of Sphinxes, and House of the Oil Merchant), the Panagia group (Houses I-III), and the House of Lead; a few more buildings were discovered northeast (House of the Wine Merchant and Cyclopean Terrace Building, Petsas House) and north of the citadel (Plakes House). Only two of these groups, the Panagia group and the Oil-Merchant complex have been fully published (Mylonas-Shear 1987; Tournavitou 1995), while two more buildings are currently being excavated (Petsas House) or await publication (Plakes House). Although some of these structures are clearly domestic units (Panagia group), others appear to have been a combination of residential areas, storerooms, and workshops, which are plausibly interpreted as palatial buildings and annexes connected with palatial economy, bureaucracy and administration (Oil-Merchant complex).

The lack of sufficient evidence for street layouts dividing blocks of structures linked by integrated drainage systems, and the discovery of scattered, isolated building groups of both palatial and residential use, oddly intermingled with tholos tombs and served by separate drain systems, have generated the theory that there was no town around Mycenae in the form of a sizeable urban center with organized closed poleodomic plan; instead, it has been theorized that the citadel of Mycenae was surrounded by several loosely connected small settlements, each with its own cemetery, which must have gradually expanded from original nuclei in an additive open plan and eventually framed the highways and roads leading to Mycenae without obeying to a prior closed town plan (Dickinson 1994, 78-80). The alleged absence of large-scale urbanization has been repeatedly hypothesized for Mycenae and other Mycenaean centers (among others, by Tsountas and Manatt 1897, 32-33, 337; Wace 1949; Mylonas 1966, 80); such views were apparently fueled by the lack of sufficiently extensive excavations of settlements on mainland Greece –most of which range from small towns to villages and farmsteads– and especially of the sites surrounding the main palaces and citadels of Mycenae, Tiryns, Pylos, Thebes, Orchomenos, and Athens (Kilian 1988, 133-136; Hooker 1976, 93-107; Dickinson 1994, 77-88, figs. 4.23-24). This picture, though, is at variance with the textual evidence from the Linear B tablets regarding Mycenaean political geography and administration (e.g. the recorded administrative provinces, districts, and towns of the Pylos state), and socioeconomic organization of the Mycenaean states (Shelmerdine 1997, 557-570; Dickinson 1994, 81-88). In recent years intensive and extensive regional surveys on the mainland have produced important spatial analysis of settlement patterns which further define the diachronic development of settlements and trace synchronic regional patterns of uniformity and variation (Shelmerdine 1997, 550-554); meanwhile, renewed excavation at palatial sites and their surrounding areas (Mycenae, Tiryns, Thebes, Pylos) and systematic extensive excavation of several Mycenaean settlements (Shelmerdine 1997, 541-550, fig. 1), combined with the textual evidence and regional survey data, reveal emerging settlement patterns consistent with varying degrees of localized urbanization.

At Mycenae, an extensive archaeological survey which was jointly conducted by the Athens Archaeological Society (S. Iakovidis) and the British School at Athens (E. French) in the 1990s, located, identified and mapped all visible remains throughout the wider area of Mycenae (32 hectares), including more than 750 sites, structures, buildings, guard towers, beacons, wall remains, tombs, roads, and bridges (Iakovidis and French 2003). The survey retraced the extensive network of roads and highways in the area of Mycenae and established its spatial association with the location of the surrounding cemeteries (Jansen 1994). In 2003 a systematic geophysical prospection and remote-sensing survey was launched at Mycenae under the direction of Prof. C. Maggidis, the general supervision of Prof. S. Iakovidis, Director of Mycenae, and the auspices of the Athens Archaeological Society; the geoprospection survey, which is generously funded by Dickinson College and the Institute for Aegean Prehistory (INSTAP), is conducted by Dr. Antonia Stamos, who leads INSTAP teams of specialists and Dickinson College (D.E.P.A.S.) student teams, in partial collaboration with the University of Patras (Prof. S. Papamarinopoulos). Several ground-based remote-sensing methods (including Ground Penetrating Radar, Electrical Resistivity, Magnetometry, Geotomography, and Geoacoustics) are being combined to map all visible and buried structures in the area around the citadel, to locate buildings, sectors and roads of the settlement of Mycenae outside the walls, and ultimately to create a 3-D digital model of the buried 'Lower City' of Mycenae (Stamos 2006).

Geomorphology and Topography of the Study Area

Figure 1

The citadel of Mycenae, which in its final form comprised an area of 30,000 m^2 surrounded by a 900 m-long circuit wall (Figs. 2-3), was built on a low rocky hill rising 278 m above sea level and approximately 40-45 m above the surrounding plain (Iakovidis 1983, 23). The hill of Mycenae is nestled between two mountains, Profitis Elias to the north and Zara to the south, from which it is separated by two ravines formed by winter torrents, Kokoretsa and Chavos, respectively; it is, therefore, a natural strong-point, protected by deep gorges and steep rocky sides all around, except its western slope which is the only accessible side, and is constantly supplied with fresh water by the Perseia spring which lies 360 m. to the east of the citadel and approximately 13 m. higher than its summit. The hill of Mycenae and the adjacent mountains belong to the western part of the Arachnaion mountain range that divides the Argolid from Corinthia, and rise in the northeastern corner of the Argive plain at the mouth of the only passage connecting the two regions (Tretos gorge or modern Dervenakia) and in the crossroads of the eastward routes to the Hermionid and the Saronic Gulf (Fig. 1). The hill of Mycenae, therefore, combines a strong geopolitical location which controls access points to and from the Argolid, and a commanding view of the Argive plain to the south below. The triangular plain of the Argolid, surrounded and isolated by mountain ranges, stretches approximately 14 km along the coast and 21 km inland. The coastal line of the Argolid Gulf shifted repeatedly in prehistoric times as a result of post-glacial melting and alluvium deposits (Zangger 1993, 83-84; 1994); in the 2nd millennium B.C. the sea was much closer to the site of Tiryns than the present-day coastal line, which must have been the main port of the Argive plain. Finally, a significant fault measuring some 2-4.5 kilometers in length, 1.5 meter in width and a maximum vertical displacement of 3 meters has been located to the east/northeast of the citadel of Mycenae, showing traces of multiple reactivations in the past (Papanastassiou *et al.* 1996, 191-192), which caused intense local seismic activity in the 13th century B.C. and considerable damage at Mycenae, Tiryns and possibly Midea (Iakovidis 1986; Maggidis 2006; forthcoming 2007).

The targeted area (Figs. 2-3) is located immediately south of the citadel on the terraced fields flanking the dry riverbed gorge of Chavos. The terraces, which are retained and delineated by modern low, stone-built walls, were named after their location in regard to the riverbed: South East Bank (SEB), South West Bank (SWB) I, II, III, IV and V (Figs. 6, 17). The study area was selected after careful examination of the local geomorphology (artificially terraced land, thick undisturbed fill, sealed and protected by successive river alluvium deposits) and consideration of several topographical and archaeological variables: the spatial distribution and orientation of visible ruins, especially the alignment of the east façade of the Oil-Merchant complex and the Panagia group which may indicate the presence of a northbound highway leading to the Lion Gate, running almost parallel to and east of the modern road, to which the orientation of both building groups possibly conformed (Fig. 2); the existence of at least two Mycenaean bridges over the Chavos river further south, connecting the two river banks and their respective terraces (Iakovidis and French 2003); the discovery of scant remains of a Cyclopean retaining wall running along the eastern edge of terraces SWB II-IV on the western bank of Chavos; the distinct pattern of successive artificial terraces which was widely used on the hill slopes around Mycenae for palatial and residential or commercial buildings, built on terraces and supported by thick retaining walls on the down slope; the odd lack of any visible architectural remains in the area (Iakovidis and French 2003, Grid Squares D4-E4); and finally, the quantity and types of chance finds during the clearing of the fields, including mainly Mycenaean sherds, stone tools, and figurines, mixed with a few Hellenistic sherds and loom weights, which indicate a strong presence of Mycenaean and Hellenistic domestic contexts.

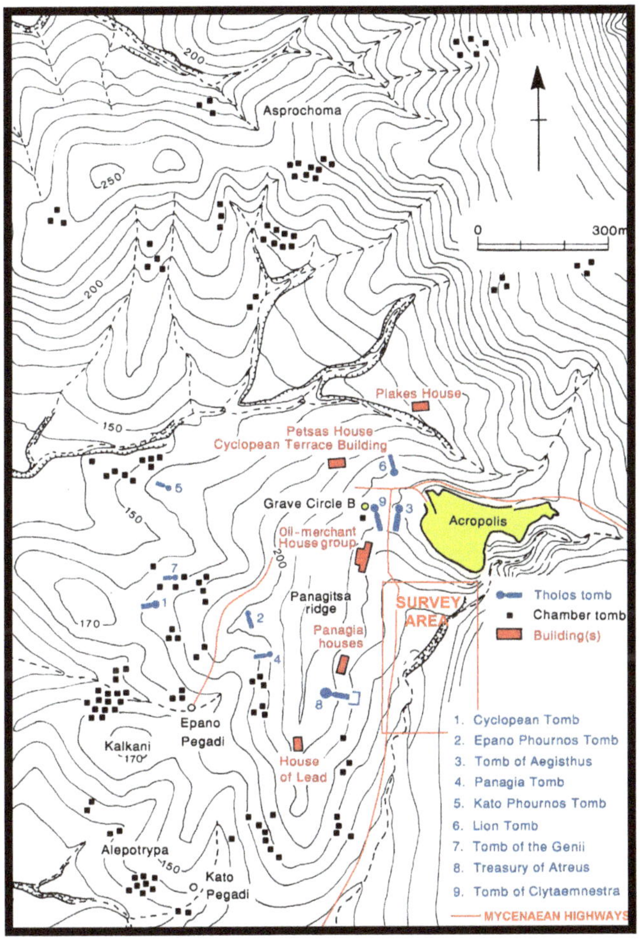

Figure 2 *(based on Dickinson 1994, fig. 4.25; modified and updated)*

In the period of 2003-2005 a total of 32 grids were surveyed with Ground Penetrating Radar and/or electrometer on six separate terraces on both banks, with the South West Bank grids having produced the most interesting results so far (Figs. 7-8). Plotting and mapping of the survey grids was carried out by means of GPS and Total Station; the survey data, the geomorphology of the study area, the archaeological remains and topographical parameters will be eventually integrated in a GIS model.

Aerial Reconnaissance and Site Prediction Model

The software-enhanced analysis of high-resolution aerial photographs of the area around Mycenae, taken in the spring of 1988 and kindly supplied by the Greek Military Geographical Service (G.Y.S.), revealed upon careful inspection distinct "ghost traces" of walls as crop/soil marks on terraces SEB I and SWB I-V of the targeted area (Fig. 5). These aerial photographs of Mycenae were also used in conjunction with a silo-locating military program (ERDAS Imagine 8.4) which compares visible, recognizable features within the photograph to other features that match the set parameters in order to produce the resulting image of a site prediction model. It was, therefore, possible to determine and confirm plausible locations to conduct the geoprospection survey (Fig. 4): the areas highlighted in blue and red indicate sites of possible activity which correspond to similar activity found at known sites such as the Oil-Merchant complex, the Panagia Houses, the Treasury of Atreus, and the citadel itself. The areas in red are the most interesting because they pick up likely features in the very same locations targeted by the archaeological survey and highlighted on the aerial photograph as producing ghost traces (SEB I, SWB I-V): these areas were finally selected as being most likely to produce promising remote-sensing results, and eventually physical remains by excavation.

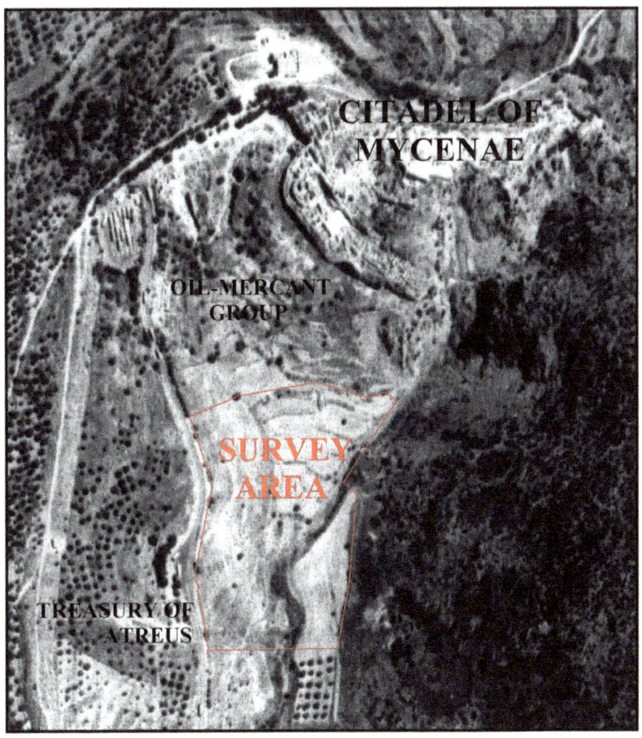

Figure 3

The geoelectric survey was conducted by the Patras University team in 2004 and 2005; data was collected using a digital electrometer (GEOSCAN RM4), special cables and four electrodes in a twin lay-out, which allowed precise measurement of soil electrical resistivity (450-900 Ohms) and detection of subsurface wall ruins to a depth of 1.5 m.

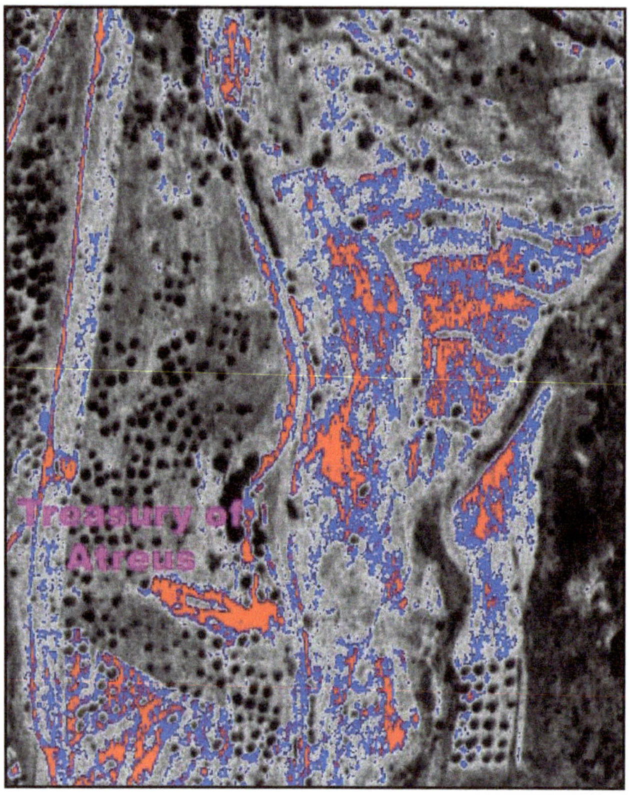

Figure 4

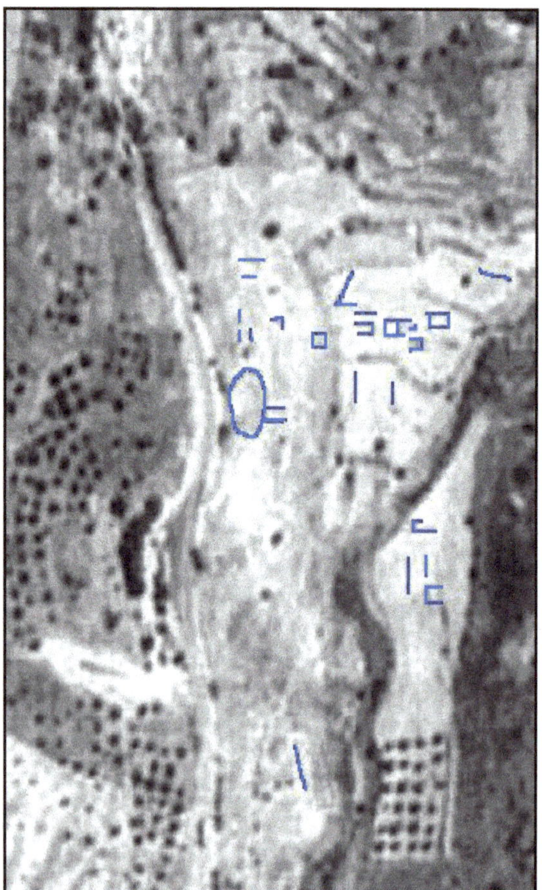

Figure 5

In total, 19 grids were laid out, spatially distributed in terraces SEB I (Grids 1–8), SWB I (Grids 9–10), SWB II (Grids 11–14), SWB III (Grids 15–18), and SWB III (Grid 19).

By overlaying the first fourteen grids, definite geometric features emerged in the northern half of Grids 9-14 (SWB I–II), all of which appear to have the same orientation, being possibly part of the same complex or similar adjacent structures. The most interesting feature in Grids 11–13 (SWB II) is a series of parallel oblong rooms with N–S orientation, reminiscent of palatial storerooms (Figs. 9-11).

Ground Penetrating Radar Survey

The Ground Penetrating Radar survey is conducted by the INSTAP team and the Dickinson student team, headed by Dr. Antonia Stamos; data is collected using a Subsurface Interface Radar or (SIR)-2000 system with a 400MHz antenna and is processed with RADAN XP software (version 6.5) which produces high-quality images and quite accurate depth readings to a depth of approximately 2 m. In the period of 2003-2005 a total of 13 grids were surveyed and processed on five different terraces (SWB I–V), several of which overlap with the grids of the geoelectric survey. The grids were collected in 0.25 m transects in order to produce more detailed results and in two datasets (E–W orientation overlaid by a N–S orientation), thus resulting in an orthogonal cohesive grid which prevents from missing any data while moving in any one direction.

In general, the preliminary results (Stamos 2006, 99-143) reveal wall corners, structures, features, gates and roads at two different depth ranges (0.40–1 m and 1.20–2 m) which consistently pervade most of the surveyed contexts; these depth ranges define possibly two superimposed occupation horizons, including an overlying Hellenistic and an underlying Mycenaean level, accordingly, which are further set apart by the distinct types and size of their structures.

In the northern part of the SWB I grid (Figs. 6-8, 12), various features appear at a depth of –0.58 m which solidify into more regular structures before finally disappearing at the depth of –1 m; particularly interesting is an assemblage of rooms with the same orientation, belonging possibly to a single complex (Fig. 12: Feature A), beneath which emerges a distinctly different, large rectangular structure with thick walls and right-angled corners at a depth of –1.44 m (Feature F).

Three grids were set up in terrace SWB II (Figs. 6-8): in the SWB IIa grid (Fig. 13), a megaron-type building (Feature A) is by far the deepest feature found in any of the grids surveyed to-date (appearing at the depth of –0.20 m and disappearing at –1.5 m), perhaps due to the preserved height of its walls or their deep foundations; to the west lies another rectangular building surrounded by smaller rooms at a depth of –0.66 m (Feature B). In the SWB IIc grid, a series of adjoining rooms with the same orientation emerge at a depth of –0.52 m (Feature A), a very long feature (wall?) running N-S for 10-12 m appears at a depth of –0.68 m (Feature C), and a possible path or road surface, approximately 1 m wide, runs E-W for at least 20 m at a depth of 1.48 m.

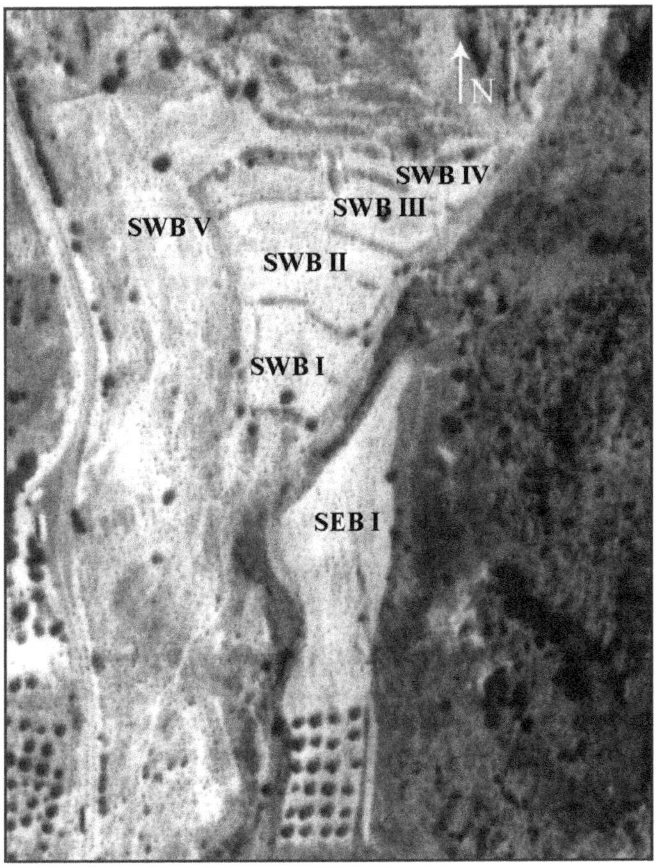

Figure 6

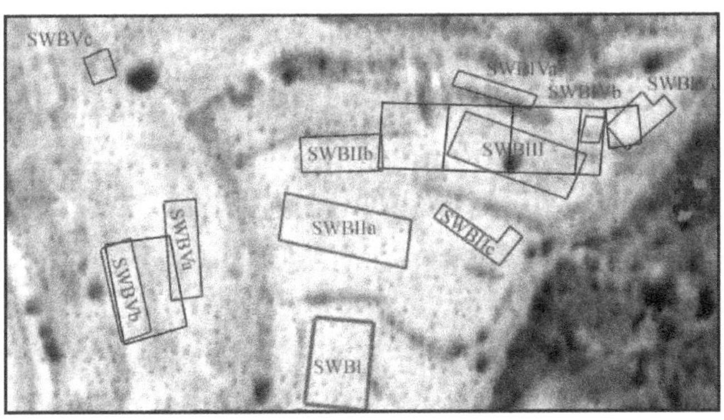

Figure 7

In the SWB III grid (Figs. 6-8), several walls appear in the northern half of the grid at a depth of –0.44 m (Feature A), while in the southern half of the grid emerge a series of adjacent rooms with the same N-S orientation (Feature B).

Figure-8

Three grids were set up on terrace SWB IV (Figs. 6-8): in the SWB IVa grid (Fig. 14), a wall corner (Feature A) emerges at a depth of –0.25 m, acquires a definite shape at a depth of –0.85 m, as does an adjacent set of wall corners (Feature B), and disappears at a depth of –1.30 m. In the SWB IVc grid (Fig. 15), there is an increase in the number of possible features throughout the area of the grid at a depth of –1.04 m, many of which have the same N-S orientation and could be part of a single complex or structure.

The three grids on terrace SWB V (Figs. 6-8) were placed in three separate areas: grids SWB Va-b are parallel to each other and to the modern asphalt road leading up to the citadel, and roughly aligned with the Oil-Merchant complex which lies approx. 50-70 m to the north. The most important single find of the remote-sensing survey so far was identified in the SWB Vb grid (Figs. 16-17): at a depth of –0.68 m emerges a road surface (3 m wide) running N–S (roughly parallel to the modern road), which may well be the northbound Hellenistic or Mycenaean highway leading up to the citadel through the settlement (Feature A). Two Γ-shaped, solid features to the north and on either side of the alleged road may be gates or towers, considering their diagnostic shape and large size (at least 3.5 m wide, preserved to a total height of approximately 1 m). This feature is strikingly similar to a structure located at the extreme edge of the Panagitsa hill and labeled by Steffen in 1884 as a gate: that structure which was cleared again during the Archaeological Survey of Mycenae, consists of two terrace-like buildings set approximately 6 m. apart and framing a possible roadway (French 2002, 68; Stamos 2006, 142). If our interpretation is correct and the evidence eventually bears out our assumptions, then what appears in the radar image on grid SWB Vb is the main gate into the Lower City of Mycenae. The discovery and excavation of this important landmark will certainly shed new light on the elusive topography and road network of the settlement around the citadel of Mycenae.

Geoelectric Survey

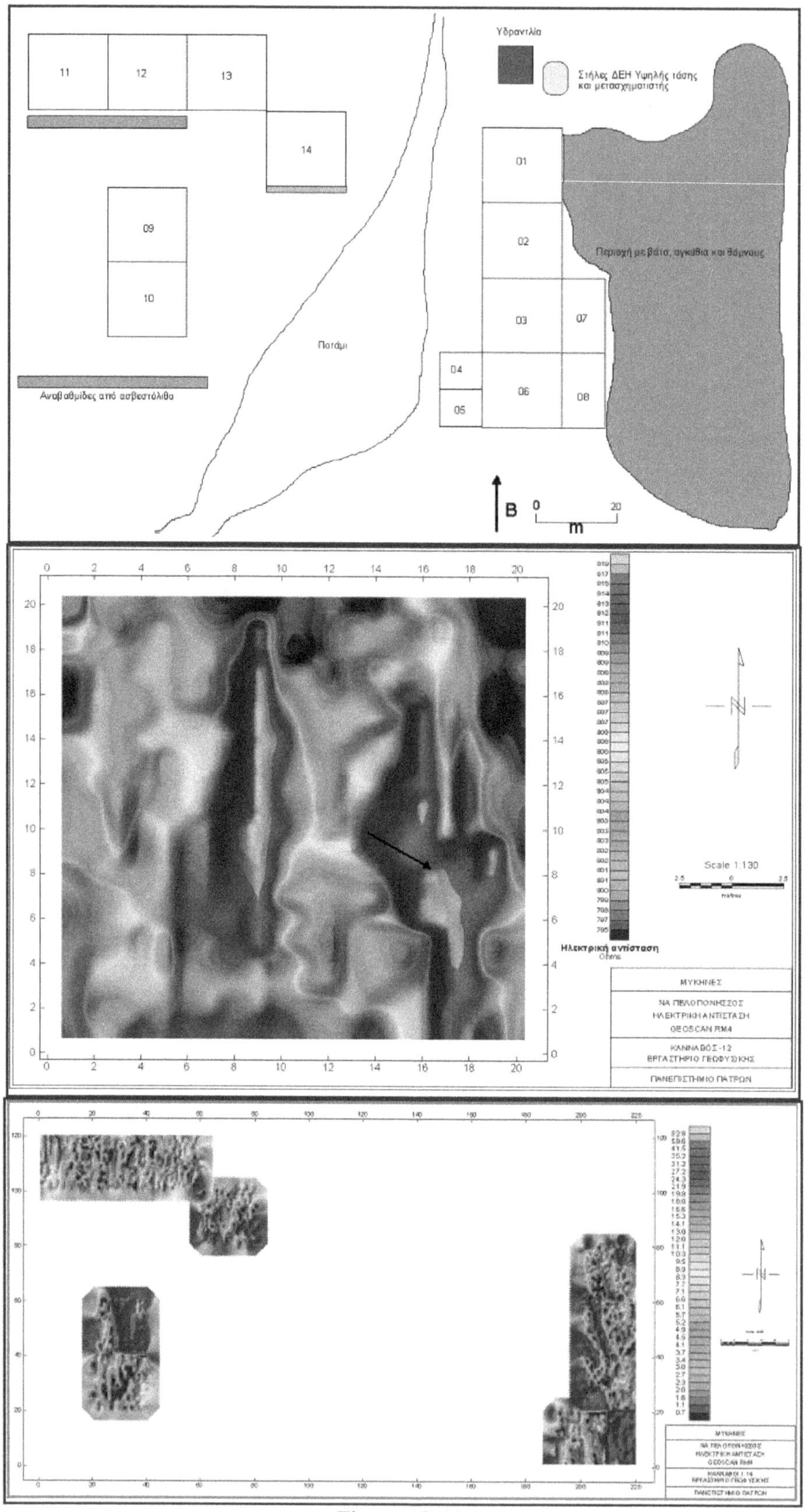

Figures 9-11

Conclusions

Did a 'Lower City' exist at Mycenae? The number, size, type, and spatial distribution of the extant features, structures, and buildings traced thus far by the ground-based remote-sensing survey seem to support the possibility of a sizable urban center surrounding the citadel on the west, south, and north sides: this town may have extended over an area of ca. 30 hectares (with a normal population density of 200 people per hectare, cf. French 2002, 64), consisting mainly of domestic units and commercial buildings, but also including larger palatial structures and annexes (workshops, offices, storerooms) located closer to the main access roads and highways. The potential discovery of the main gate, once associated perhaps with an outer circuit wall, further reinforces the possibility of an organized town plan; to this point also definite town boundary markers that appear to have been in place already by the late 14th century B.C. (French 2002, 64).

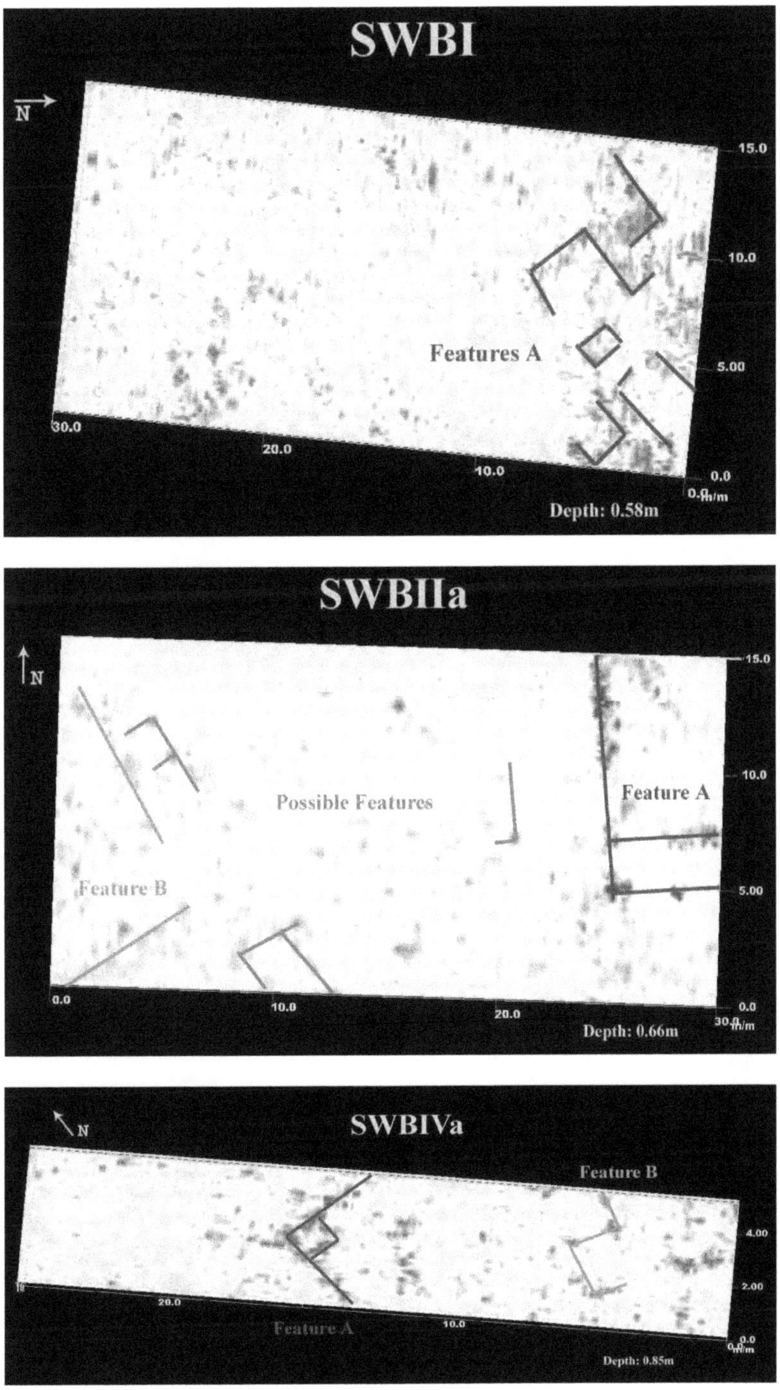

Figures 12-14

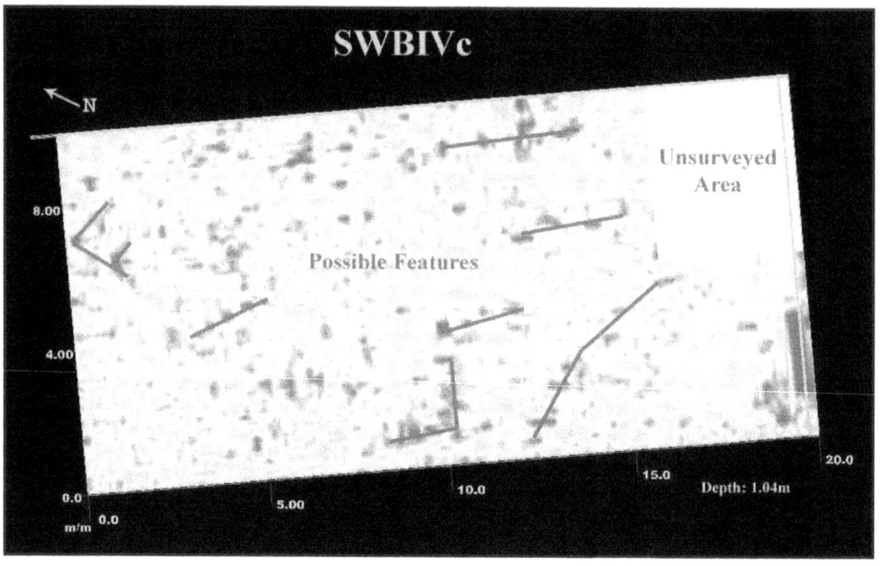

Figure 15

The discovered buildings outside the walls belong to either of two types, the long corridor plan with two levels following the platform line of its terrace, and the square version with rooms arranged around a central court; both types as well as two more types (the megaron type and parallel oblong storerooms) have been identified in the radar images (Stamos 2006, 140) on the south, terraced slope.

Furtermore, it appears that although the palatial annexes outside the walls were destroyed by earthquake and fire in LH IIIB1/2 (ca. 1240 B.C.) and were subsequently abandoned, several residential buildings were repaired and continued to be in use well into the LH IIIB2/LH IIIC period along with a number of chamber tombs (Iakovidis 1986; Maggidis 2006, fig. 4; 2007 forthcoming), a pattern which is consistent with the composition of a town.

The comparative study of the spatial distribution of chamber tomb cemeteries (Wace 1932; Xenaki-Sakellariou 1985), grave circles and tholos tombs may explain the alleged 'odd' intermingling of houses and tombs (Fig. 2): the royal Grave Circle B which was founded at the western end of an extensive MH cemetery stretching from Grave Circle A to the west of the Mycenae hill, was originally lying outside the contemporary MH settlement on the hill; later on, one group of royal tholos tombs (Aegisthus, Lion, and Clytaemnestra Tombs) were gradually built in the area of this earlier cemetery, close to the ancestral royal graves, and occupying the prominent location at the crossroads of the highways leading up to the citadel's main gate, which secured visibility. Meanwhile, the expansion of the town of Mycenae outside the walls eventually engulfed this group of tholos tombs and covered all the adjacent hill slopes; for lack of appropriate space, the very last tholos tomb of another group, the Treasury of Atreus, was built amidst houses (Panagia group) facing the northbound highway leading to the citadel. Other than the latter, though, no other tholos tomb or cluster of chamber tombs breaks the unity of the settlement or infiltrates the central town area; in fact, the peripheral spatial distribution of the cemeteries located farther around the citadel and its surrounding settlement demarcates the borders of the settlement itself, thus revealing a site pattern of three successive and concentric occupation zones (palatial, residential, and cemetery).

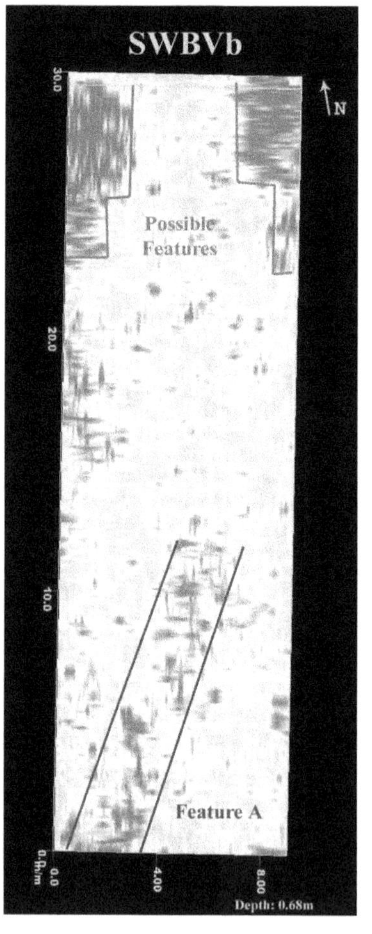

Figure 16

Figure 17

Figure 18

The final results of this on-going geoprospection survey which will expand on several other areas outside the fortification walls, will be published collectively as a second volume to the *Archaeological Atlas of Mycenae*, and will form the basis for the systematic excavation of the settlement of Mycenae.

The real test, though, of the geoprospection survey is the excavation trench: four acres of the surveyed land have already been purchased (Fig. 18), which paves the way for the commencement of the systematic excavation of the Lower City of Mycenae in the summer of 2007.

References

DICKINSON, O.T.P.K. 1994 *The Aegean Bronze Age*. Cambridge University Press.

FRENCH, E.B. 2002 *Mycenae: Agamemnon's Capital. The Site in its Setting*. Charleston. Tempus Publishing Inc.

HOOKER, J.T. 1976 *Mycenaean Greece*. London.
IAKOVIDIS, S. 1983 Late Helladic Citadels on Mainland Greece. Leiden. 23-72.
IAKOVIDIS, S. 1986 Destruction Horizons at Late Bronze Age Mycenae, in: Φίλια Έπη εις Γεώργιον Μυλωνάν διά τα 60 έτη του ανασκαφικού του έργου. Βιβλιοθήκη της Εν Αθήναις Αρχαιολογικής Εταιρείας 103. Athens. Vol. I: 233-260.
IAKOVIDIS, S. 1996 Mycenae in the Light of Recent Discoveries, in: *Atti e Memorie des Secondo Congresso Internazionale di Micenologia (Roma-Napoli, 14-20 ottobre 1991)* (E. De Miro, L. Godart, A. Sacconi, eds.). Rome. Vol. III: 1039-1049.
IAKOVIDIS, S. and E.B. FRENCH 2003 Archaeological Atlas of Mycenae. Library of the Archaeological Society at Athens 229). Athens.
JANSEN, A., 1994 Stations along the Roads in the Area of Mycenae: An Analysis of the Mycenaean Road System and Its Relation to the Mycenaean State. Diss. University of Pennsylvania.
KILIAN, K. 1988 Mycenaeans Up to Date, Trends and Changes in Recent Research, in Problems in Greek Prehistory. Papers Presented at the Centenary Conference of the British School of Archaeology at Athens (Manchester April 1986) (E. French and K. Wardle, eds.). Bristol. 115-152.
MAGGIDIS, C. 2006 Mycenae Abroad. Mycenaean Foreign Policy, the Anatolian Frontier, and the Theory of Overextension: Reconstructing an Integrated Causal Nexus for the Decline and Fall of the Mycenaean World, in: *Moving Across Borders: Foreign Relations, Religion and Cultural Interactions in Ancient Mediterranean. Orientalia Lovaniensia Analecta*. Leuvain, Peeters Publishers.
MAGGIDIS, C. 2007 (forthcoming) Mycenaean Palatial Overextension and the Dynamics of the Systems Collapse of the Mycenaean World, in *Honorary Studies for Spyridon Iakovidis*. Academy of Athens.
MYLONAS, G. 1966 *Mycenae and the Mycenaean Age*. Princeton.
MYLONAS, G. 1983 *Mycenae Rich in Gold*. Athens.
MYLONAS-SHEAR, I. 1987 The Panagia Houses at Mycenae. University Museum Monograph 68. Philadelphia.
PAPANASTASSIOU, D., K. GAKI-PAPANASTASSIOU, and H. MAROUKIAN 1996 Geomorphologic-seismotectonic Observations in Relation to the Catastrophes at Mycenae, in: *Archaeoseismology. Fitch Laboratory Occasional Paper* 7 (S. Stiros and R.E. Jones, eds.). Exeter. 189-194.
SHELMERDINE, C. 1997 Review of Aegean Prehistory VI. *American Journal of Archaeology* 101: 537-585.
STAMOS, A. 2006 Through the Looking Glass: Applications of Ground Penetrating Radar in Archaeology. Diss. Temple University. 95-143.
TOURNAVITOU, I. 1995 The 'Ivory Houses' at Mycenae. Annual of the British School at Athens Suppl. 24. London.
TSOUNTAS, C. and J. MANATT 1897 The Mycenaean Age: A Study of the Monuments and Culture of Pre-Homeric Greece. Boston. Houghton, Mifflin and Co.
WACE, A.J.B. 1932 *Chamber Tombs at Mycenae*. 1932.
WACE, A.J.B. 1949 An Archaeological History and Guide. Princeton.
XENAKI-SAKELLARIOU, A. 1985 Οι Θαλαμωτοί Τάφοι των Μυκηνών, Ανασκαφής Χρ. Τσούντα (1887-1898). Paris.
ZANGGER, E. 1993 *The Geoarchaeology of the Argolid*. Berlin. Gebr. Mann Verlag.
ZANGGER, E. 1994 Landscape Changes around Tiryns during the Bronze Age. *American Journal of Archaeology* 98: 189-212.

Multimethodological approach to study the archaeological park of Maalga Karthago (Tunis) using remote sensing, archaeology and geophysical prospecting methods

S. Piro,[1] M.C. Capanna[2]

1 – ITABC–CNR, P.O. Box 10–00016 Monterotondo Sc. (Rome), ITALY.
Salvatore.Piro@itabc.cnr.it
2 – Dip. Scienze Storiche, Archeologiche e Antropologiche dell'Antichità, Università La Sapienza, Roma.

1 Introduction

It is now widely acknowledged that archaeological sites are almost everywhere increasingly threatened by many natural and anthropic agents. Dealing with essentially non-renewable cultural resources, archaeologists are as a consequence becoming more and more concerned about their husbanding.

The wise management of the cultural heritages calls for non-invasive assessment of unthreatened sites, evaluation of sites in areas to be altered, and identification and recovery of information where sites are to be destroyed. In all cases, any rapid, non-destructive technique for site evaluation, providing archaeologically meaningful knowledges, may become of paramount importance.

Many significant data can be obtained by applying remote sensing, visual and infrared aerial sensing, surface surveying and the more limited methods of pedestrian surveying, surface collecting and shovel testing. With specific symbols this kind of information can be reported on topographic maps. Their combination can provide some close correspondence between the remote sensing indications and the presence of sites frequented by men. The preliminary analysis of materials could also provide a dating, but it is generally not possible to establish extension and depth of any archaeological settlement. Only applying geophysical prospecting methods this further information can be properly achieved.

Nowadays, non-destructive ground surface geophysical prospecting methods are increasingly used for the investigation of archaeological sites, which implies detailed physical and geometrical reconstructions of hidden ambients. The probability of a successful application rapidly increases if a consistent multi-methodological approach is adopted, according to a logic of objective complementarity of information and of global convergence toward a high quality multiparametric imaging of the buried structures.

The fine representation of the static configurations of the explored areas and of the space-time evolutions of the interaction processes between targets and hosting materials, are fundamental elements of primary knowledge in the case of archaeological research. Non-invasive geophysical prospecting methods are to-date the only means available for local reconnaissance and discrimination, prior to any excavation work.

The present work is a portion of the International Project based on the definition of the Cultural and Environmental Park of the Maalga and of the area of Punic Harbours in Karthago (Tunis), which has been developed through the joint contribution of MAE, MBAC, CRAST, CNR, the Appia Antica Park (Italy) and by INP (Tunis). Its aim was to improve, protect and manage the archaeological patrimony of the two selected areas (the Maalga, suburban area of Karthago in Punic, Roman and ancient Ages and the area of the two Harbours of the Punic site) and to better understand the historical context.

The project has foreseen a multimethodological approach based on :

1) The creation of a bibliographical and photographic data base about the Maalga area.
2) 2) The archaeological survey and the creation of a data base of the collected documentation.
3) 3) The interpretation of the images found by the remote sensing.
4) 4) The geophysical prospecting by integrated methods of some among the most significant areas.
5) 5) The integration of all information found in a GIS.

In this work, a combination of Differential Magnetic (Gradiometric) and GPR methods has been employed with the aim of detecting near-surface archaeological structures in the test site of the Maalga Archaeological Park and the Punic Harbours, Karthago (Tunis). The results of all surveys are presented and discussed.

2 The Study Area

The work went through a series of multidisciplinary applications both on the field (archaeological and geophysical surveys and analysis of the materials on the Carthaginian site) and with data processing.

Figure 1

After the archaeological surveys made during 2033, the study has been directed on the Maalga area. This area has been divided in three sectors which are characterised by different remains as walls and wells linked to the Punic-Helenistic period or necropolis and cisterns related to the Roman Imperial era.

Inside an area which is about 3 ha in size, the study of the satellite images (obtained from pan-sharpening of the multispectral bands of the Ikonos 2001 and 2003 with resolution of m 4, with the respective panchromatic band of m 0,80 of resolution) interfaced with the numerical cartography (in the scale 1:25000, 1:5000, 1:2000), had actually already shown, since the beginning of the research, radiometric anomalies which can be interpreted as ancient anthropical "marks", Fig.1.

On the other hand, the study of the existing bibliographical documentation identified in the area the presence of part of the so called Teodosius's wall, of which a very small fragment without outer facing, can be seen at the entrance of the new Mosque. On the basis of this fragment it was very difficult to establish where the real walls of ancient Carthage were situated.

The archaeological recognition of the project (made during November 2003 and June 2004 by Elisa Panero of Centro di Ricerche Archeologiche e Scavi per il Medio-Oriente di Torino, Italy) has enabled to re-locate the cuts of the previous Italian mission and to single a huge concentration of materials on the whole area, especially in the southern part, which, on the basis of

1) the data inferred by the field-walking in the whole Maalga,
2) the location on the map of monumental emergencies of the different periods within the urban perimeter of ancient Carthage,
3) the location of the "historical" carthography,
4) the data taken from the bibliography seems to be pertinent to the urban area of the Roman town at least.

Among the found materials, there are mostly ceramic fragments dating medium-late Imperial age (African Red Slipped Wares, in the most common shapes from the III century A.D., late amphoras such as the Keays XXV and the Late Romans 3 and 7, late lamps)

Besides, it is interesting to see that, along the border north-west /south east of the area, the works for the new road have evidenced a section of the same area 200 mt long, from which it was possible to prove the presence of some isolated wells (unfortunately at present covered with earth).

These wells were perhaps belonging to a watering system of Punic Age present in the whole Maalga, along with some floors and walls belonging to houses that, so far, can be identified as urban areas of the *Colonia Iulia Concordia Karthago*. They show the perfect urban organization of the town in this sector too, on the northern border of the Roman centre, certainly peopled and well structured at least in the first Imperial phases, until the demographic contraction, proved by the teodosian walls, which cut the before mentioned area, took place.

The geophysical prospection by georadar on the south-western border, along the hypotetical axis of the ancient walls demonstrates this hypotesis, since a relevant presence of the walls has been found 60 cm deep (crossing other thinner walls).

All this is coherent with the data provided by previous excavations and with the concentration of late materials found in the survey. It is possible to assume an east-south-east/west-north-west sense of direction orthomorphic to the marks shown by the satellite images.

3 The Geophysical surveys

A range of surface survey techniques, in particular remote sensing techniques (magnetometry and ground penetrating radar) are used to examine the size and layout of the site, combined with intensive field survey to give an indication of the chronology and function of the buried structures.

Ground-based Remote Sensing surveys

From 2003 to 2004, a series of topographic and ground-based remote sensing surveys took place at Archaeological Park of Karthagoo (Maalga) which provide a basic map of the ancient centre. The topographic survey permitted standing monuments and contemporary features to be located within a common grid, while remote sensing surveys allowed the identification of buried structures and features.

Magnetic investigations

Data Acquisition and elaboration - A Geoscan FM 36 Fluxgate Gradiometer was used to carry out the magnetometry survey providing magnetic gradient maps of the site. The magnetometry surveys (undertaken by ITABC-CNR) covered 4 areas of the site. All of the geophysical data were collected following 20×20 *m* grids with 1*m* spacing of transects. Magnetometry survey had a sample rate between two and four data points per metre.

Area_1 (Western of USA war cemetery). This are has been subdivided in 26 squares with dimension 20 *m* × 20 *m*; the measurements of the gradient of Z component of the Total Magnetic Field have been collected along parallel profiles with a sampling interval of 1 *m*.

Area_2 (Eastern of USA war cemetery). This are has been subdivided in 19 squares with dimension 20 *m* × 20 *m*; the measurements of the gradient of Z component of the Total Magnetic Field have been collected along parallel profiles with a sampling interval of 1 *m*.

Area_3 (Eastern of USA war cemetery). This are has been subdivided in 18 squares with dimension 20 *m* × 20 *m*; the measurements of the gradient of Z component of the Total Magnetic Field have been collected along parallel profiles with a sampling interval of 1 *m*.

Area_4 (Western of USA war cemetery). This are has been subdivided in 16 squares with dimension 20 *m* × 20 *m*; the measurements of the gradient of Z component of the Total Magnetic Field have been collected along parallel profiles with a sampling interval of 1 *m*.

The total investigated area is about 3.5 *ha* in size.

Data were processed using Geoscan's Geoplot, GeoWin (a GIS package) and Surfer to produce clear images of the geophysical data for interpretation and presentation.

The analysis of the collected measurements has shown that the 70% of the data present a high level of the noise. This noise is due to the presence, on the surface, of many metallic short objects. These noise sources have masked in some portion of the investigated areas the anomalies due to the presence of archaeological remains. To enhance the signal/noise ratio the following processing has been employed during the elaboration: i) despiking, ii) low pass filter, iii) residual anomalies, iv) representation as 2D contour map of the residual values of the gradient of Z component.

To better locate the investigated area on the known topography of the studied area, a topographyc survey employing the D-GPS (made by Ing. Carlo Alberto Birocco, Torino) has been made.

The results of the magnetic surveys are shown in Figs. 2a, and b.

The GPR investigation

Data Acquisition - In 2004 high-resolution GPR surveys (conducted by ITABC-CNR) were applied over three areas to test the potential of this technique.

For the measurements, a GSSI SIR3000, equipped with a 500 *MHz* bistatic antenna with constant offset were employed. At each site radar profiles were collected alternatively in reversed and unreversed directions across the survey grids. The horizontal spacing between parallel profiles at the site was 0.5 *m*. Radar reflections along the transects were recorded continuously across the ground at 80 $scan\ s^{-1}$, with a stack = 4; along each profile, markers were spaced every 1 m to provide spatial reference. The gain control was manually adjusted to be more effective. The data were later corrected for a variation in speed to constant 30 $scans\ m^{-1}$ (or 1 scan approximately 0.03 *m*). All radar reflections within 65 *ns* (two-way travel time)

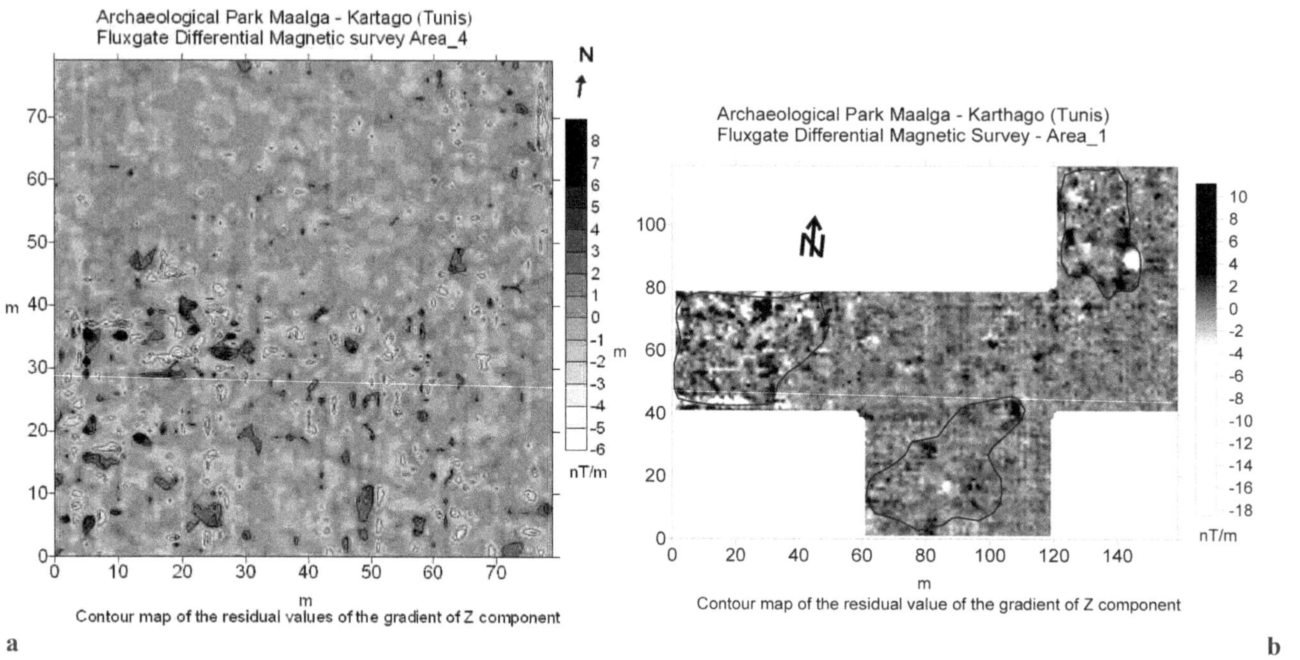

Figure 2

time window were recorded digitally in the field as 8 bit data and 512 samples per radar scan.

When GPR profiles are collected on along closely spaced profiles, the data can be processed to display horizontal maps of the recorded radar amplitudes. Referred to as time slice processing, these anomaly maps can be generated at various time/depth windows across the recorded radargram dataset. The time slice maps can provide information regarding the size, shape, location and depth of subsurface archaeological structures buried beneath a site.

Time slice data were created using the spatially averaged square wave amplitudes of the return reflection. The squared amplitude, which is essentially the energy in the waveform, was averaged horizontally every 0.25*m* along the radargram profile, and in 5 nanoseconds time windows to create a time slice parameter. These averaged square amplitudes were then gridded using a Kriging routine (Piro et *al.*, 2003, Goodman et *al.*, 2002). The motivation for desampling the information i.e. spatially averaging, along the radargram profiles is to help create visually useful time slice maps. In the case when all the data along the radargram is being used, the time slice maps usually show striation patterns which are contiguous with the profile directions. The spatially averaged time slice parameter minimizes the effects of line striation parallel to the profile direction. The rather large vertical averaging window of 8 nanoseconds for this particular site proved to yield the best images of the subsurface archaeology.

Thinner slices were also computed at 2 nanosecond intervals. These images, although they showed better depth resolutions of localized archaeological anomalies, gave less of a clear picture of larger – contiguous archaeological structures. The larger time window helped to create an understandable visualization of the subsurface. In particular, when structures are not level within the ground or when velocity variations across the site are significant, the larger time window can illuminate continuous features at variable depths (Conyers and Goodman, 1997). Other line noises, parallel to the profile collection direction, were removed using a moving filter with customized threshold settings. Filter thresholds were set to signal levels just below the average reflections from buried Roman walls. We have obtained a nominal microwave velocity of about 6 to 7 cm ns^{-1} using an experimental profile carried out in correspondence of a site where the archaeologists know the depth of a wall.

Area_4 (Western of USA war cemetery). In this area a total number of 262 parallel profiles, stretching S-N, have been collected with the instrument configuration indicated above. After the preprocessing, all collected gpr profiles have been used to calculate the time-slices related to this area. Fig. 3 shows the time-slice at the time window (twt-two way time) 10.7 – 13.7 *ns*, corresponding to an estimated depth of 0.5 *m*. In this slice a pseudolinear radar reflections are clearly visible. These kind of anomalies are due to the presence of portion of walls still present in the round.

Area New_Mosque In this area a total number of 128 parallel profiles, stretching S-N, have been collected with the instrument configuration indicated above. After the preprocessing, all collected gpr profiles have been used to calculate the time-slices related to this area. Fig. 4 shows the time-slice at the time window (twt-two way time) 10.7 – 13.7 *ns*, corresponding to an estimated depth of 0.5 *m*. In this slice a pseudolinear radar reflections are clearly visible. These kind of anomalies are due to the presence of portion of walls still present in the round.

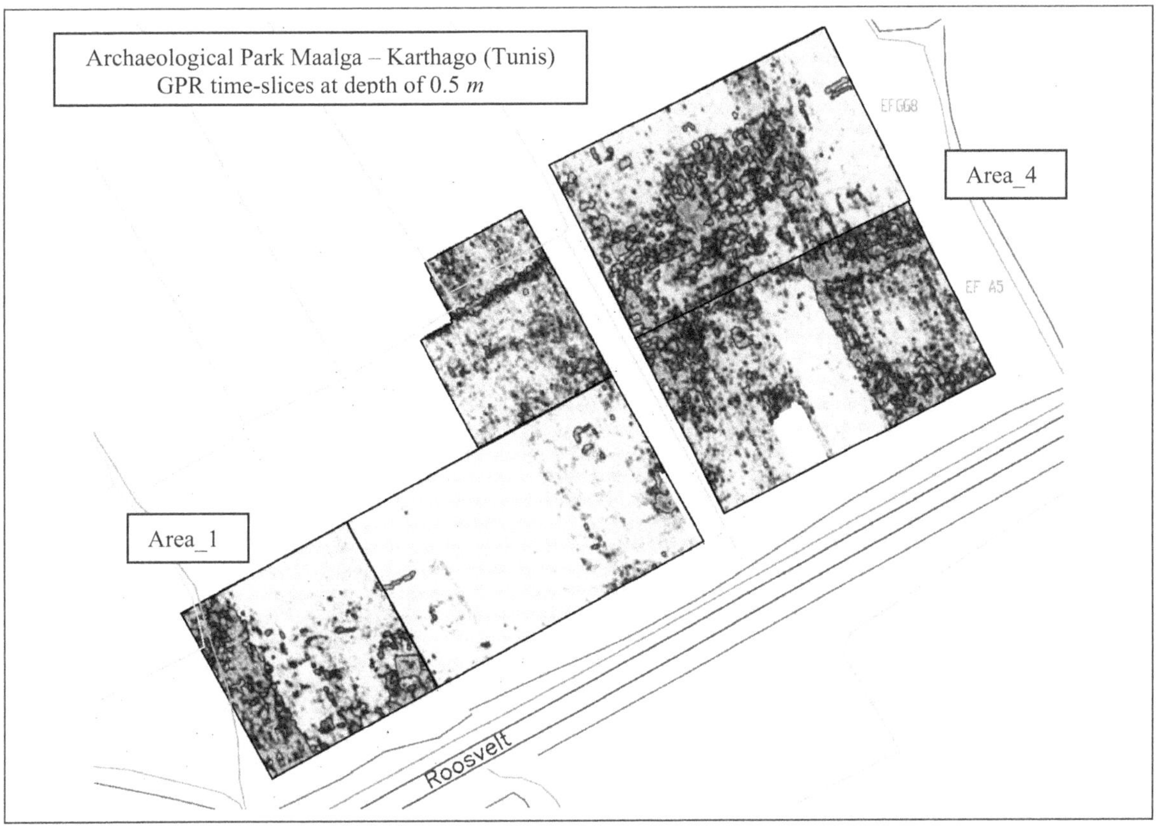

Figure 3

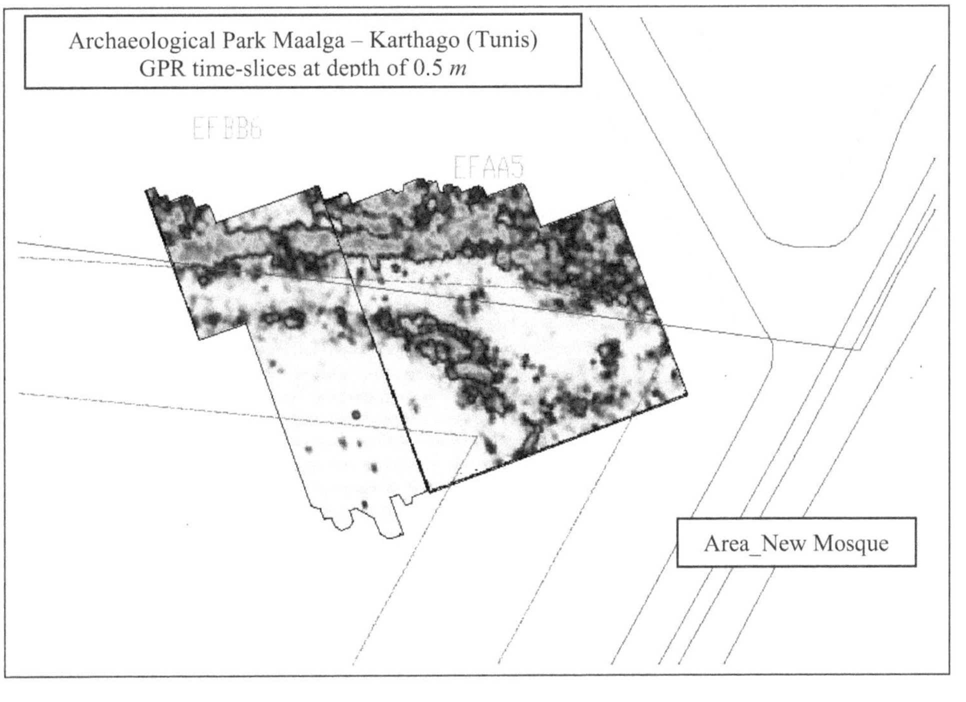

Figure 4

4 Conclusions

The comparison of the results of the two employed methods shows that each single method concurs to give the overall picture of the individuated archaeological structures.

Figs. 3-4 show the positioned and oriented GPR time-slice maps, after the GPS survey, with respect to the known elements of the territory of the Archaeological Park. These figures helped the archaeologists to have a general vision of the individual structures and to make a preliminary correlation between the different located portions of the archaeological structures.

Acknowledgements

The Authors are very grateful to Centro Ricerche e Scavi of Torino and to INP of Tunis for the fruitful collaboration, to Daniele Verrecchia (ITABC-CNR) for the technical support during the surveys.

References

GOODMAN D., NISHIMURA Y. and Piro S. 2001. High resolution ground-penetrating radar. In Gaffney et *al* "Forum Novum-Vescovio: studying urbanism in the Tiber valley". *Journal of Roman Archaeology* 14, 5-8, 20.

GOODMAN D., Piro S., NISHIMURA Y. 2002. GPR time slices images of the Villa of Emperor trajanus, Arcinazzo, Italy (AD 52-117). 9th International Conference on Ground Penetrating Radar. S. K. Koppenjan and J. Lee Editors. *Proceedings* of SPIE Vol. 4758, 268-272.

PIRO S., GOODMAN D., NISHIMURA Y. 2002. The location of Emperor Traino's villa (Altopiani di Arcinazzo-oma) using high-resolution GPR surveys. *Bollettino di Geofisica Teorica ed Applicata*, Vol. 43, n.1-2, 143-155.

PIRO S., GOODMAN D., NISHIMURA Y. 2003. The study and characterization of Emperor Traiano's Villa (Altopiani di Arcinazzo, Roma) using high-resolution integrated geophysical surveys. *Archaeological Prospection,* 10(1), 1-25.

Multi-temporal monitoring of landslides in archaeological mountainous environments using optical imagery: the case of El Tambo, Ecuador

Freddy Yugsi[1], Henri Eisenbeiss[2], Fabio Remondino[2], Wilfried Winkler[1]

[1]Geological Institute, ETH Zurich
E-mail: <freddy.yugsi><wilfried.winkler>@erdw.ethz.ch
Web: http://www.erdw.ethz.ch
[2]Institute of Geodesy and Photogrammetry, Chair of Photogrammetry and Remote Sensing - ETH Zurich
E-mail: <henri><fabio>@geod.baug.ethz.ch
Web: http://www.photogrammetry.ethz.ch

1 Introduction

Ground-based techniques for landslide analysis, as geotechnical characterization, geophysical surveys or field based geodetic monitoring, are often used in preliminary stages of landslide investigations. These methods provide very accurate punctual information, but they do not take in account, neither the anisotropy of the parameters that control the processes nor the asymmetric displacement of the moving masses into the landslide. Besides, the more precise the analysis is required, the more number of tests are needed. Generally, this is not always cost-effective, especially in developing countries where both land use planning and risk policies are under-valued and therefore few funds are assigned for natural hazard investigations.

Remote sensing imagery has proved to be useful for early stages in landslide analysis. Aerial photo-interpretation has been used in landslide detection since several years. Digital photogrammetry has become during last years a powerful tool for landslide detection and monitoring using large scale aerial photographs (1:10 000 or larger) taken with some periodicity. Unfortunately in the case of developing countries archives do not have periodicity and most of the photographs sets have been taken in a small scale (1:40 000 or smaller).

Mountains have always been considered as zones prone to the occurrence of landslides. Its instability could be increased for local geologic conditions like the presence of active tectonic faults. In the case of El Tambo (southern Ecuador), it has been favored by its geologic and climatic characteristics. Hence, large-size landslides have occurred in the area, most of them were active for a long period of time affecting around 25 000 people living in the two principal locations of El Tambo and Cañar.

In our work, field recognition, aerial photo-interpretation and mapping are used to identify the different areas where active landslides are located. A set of three DEMs (Digital Elevation Models) and orthophotos are created from different aerial image sets ranging from 1977 to 1995 to assess the variations of the most active landslides in the area.

2 Geographical Setting and Archaeological Heritage

The villages of El Tambo, Cañar and Ingapirca are situated in southern Ecuador, extending from 2° 28' S to 2° 40' S and 79° W to 78° 50' W around 45 km away of Cuenca, the third important city of Ecuador. The elevation varies from 2700 m.a.s.l., in the vicinity of the Cañar River, to 3900 m.a.s.l. to the north, covering a surface of around 350 km^2 (Fig. 1)

The most important Incan site in Ecuador is at Ingapirca, (Alcina Franch, 1978) towards the east of the area. Here a temple complex incorporating classic Inca stonework was built over earlier Cañari structures. During the Incan empire time its centre was a temple of the sun. The surrounds comprised of a great plaza and of a good number of buildings that must have included official residences, stores, houses and barracks.

3 Geological Setting

The area of El Tambo is settled in the northwestern corner of an inter-mountainous Middle-Late Miocene sedimentary basin (Cuenca basin) limited at both sides for two N-S striking cordilleras forming the Andes: the Cordillera Occidental at the west and the Cordillera Real at the east. The filling activity of the basin was divided into two different periods (Steinmann, 1997): (1) a deltaic to marine/brakish environment (15-9 Ma) with a source of material from the east (Cordillera Real) and (2) a second phase (8-5 Ma) composed by intermontane series whose source was the west (Cordillera Occidental). Between 9.5-8 Ma a major E-W compressive deformation event occurred causing the appearance of large thrust, reverse faults and folds. After the time of activity of the basin, the area was covered by glacier deposits (moraines) and volcanic products, during Pleistocene and Holocene.

Locally, the basin has been divided in three different tectonic units (Lahuathe, 2005) separated by large thrusts (Fig. 2):

The Cocha Huma Unit, comprises sequences of phyllites and cuarcites (Junin Fm.), shales, sandstones, conglomerates and greywakes (Yungilla Fm.) and

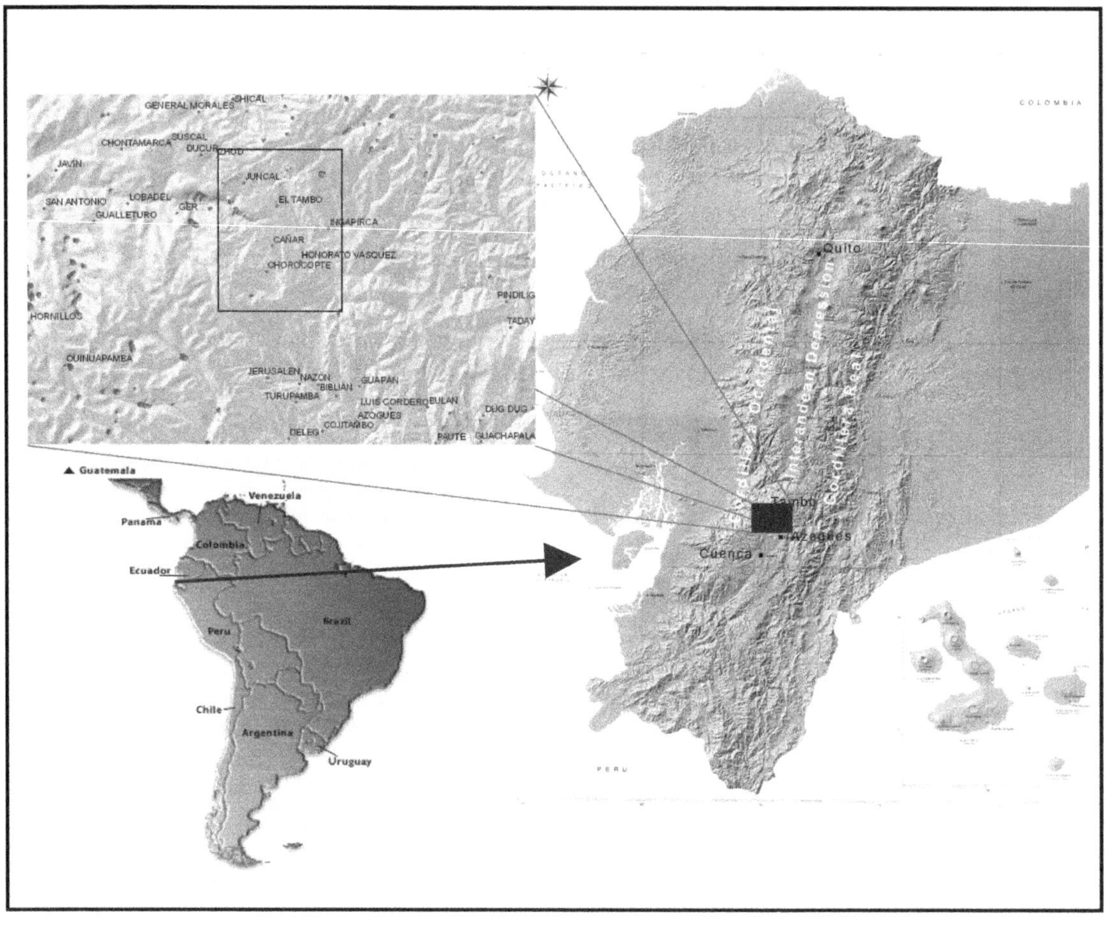

Figure 1 *The area of interest of El Tambo, Ecuador.*

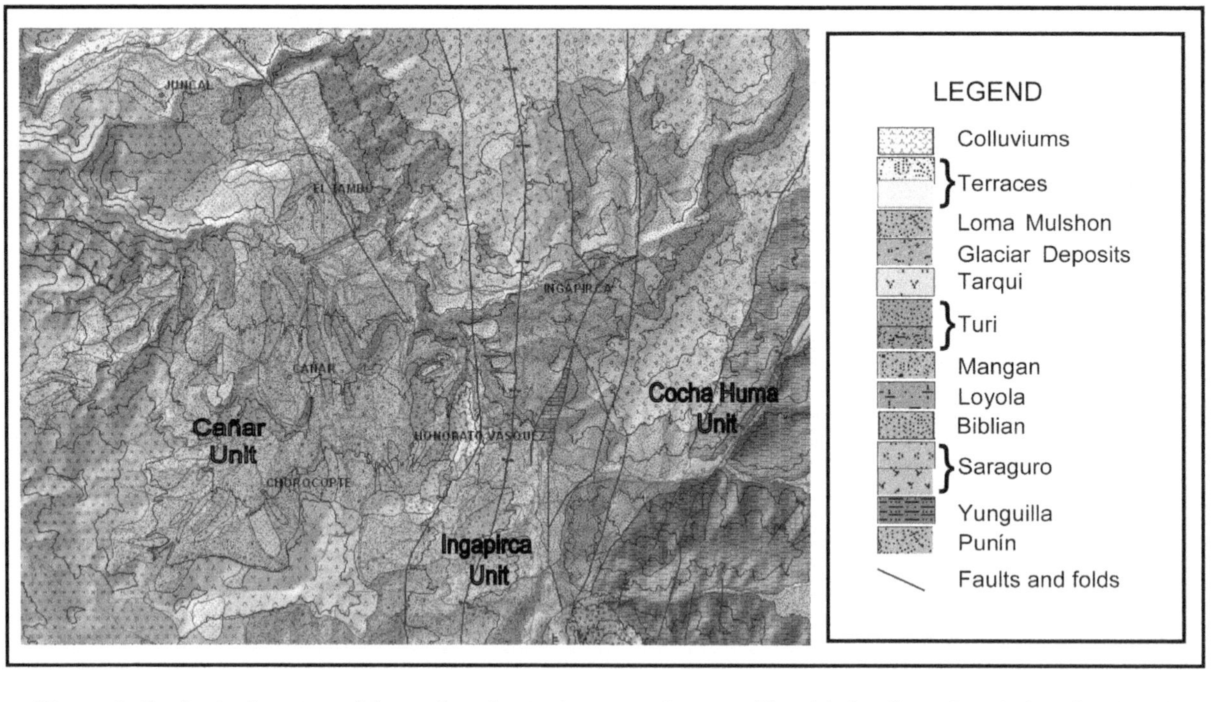

Figure 2 *Geological setting of the archaeological area in danger of landslides (based on Lahuathe, 2005).*

1. andesitic and dacitic breccias (Cerro Tablillas Fm.) ranging since Jurassic to Oligocene.
2. The Ingapirca Unit consists of clays, sands and silts weakly cemented (Biblian Fm.), shales and siltstones (Loyola Fm.), shales, siltstones and sandstones from a metamorphic source (Mangan Fm.) and volcano sediments, andesitic breccias and lavas strongly weathered (Turi Fm.). The age of these materials correspond to Middle-Late Miocene and are covered persistently by quaternary moraines.
3. The Cañar Unit includes riolitic tuffs and breccias hydrotermally affected (Cerro Caucay Fm.), clays, silts and sands poorly consolidated and conglomerates (Turi Fm.). Their age vary from Oligocene to Late Miocene.

Pliocene and Quaternary deposits have been considered as covertures and consist mainly of volcano sediments, glacier deposits, alluvial terraces and colluviums. They are overlaying discordantly the rest of units and are spread along the center west and south of the area.

The Cuenca basin, due to the mechanical characteristics of its constitutive materials, has been affected by several large landslides usually with complex mechanisms. Few local attempts for understanding their distribution and genesis have been made (e.g. Erazo, 1965). Unfortunately most of the efforts have aimed to describe the geometry without taking into account causes and triggering factors.

The PRECUPA Project (Prevention-Ecuador-Cuenca-Paute) carried out between 1994 and 1998 (Basabe & Bonnard, 2002), has been the most complete study for the Cuenca basin. However, the area of el Tambo was not taken in account. Natural hazards were monitored, analysed and mapped into an area of approximately 3700 km^2 including the cities of Cuenca and Azogues and some of the geological units that also exist around el Tambo. Its main activities comprised: geological and geotechnical mapping, landslide inventory mapping, creation of landslides-related hazard maps and pilots studies on vulnerability (Basabe et al, *1998*).

4 DTM Generation and Orthophoto Production

The generation of a precise and reliable Digital Terrain Model (DTM) is a pre-requisite for mapping, landslide documentation and multi-temporal change analysis.

For the area under investigation, 3 sets of aerial photographs, corresponding to 1977, 1989 and 1995 respectively, were used (Table 1). They are part of the National Archive held by the Instituto Geográfico Militar in Ecuador. For the digital processing, the photos were scanned at a resolution of 15 μm. Ground control points were also acquired from the geodetic control archive of the Instituto Geográfico Militar.

Three software packages were employed for photogrammetric image processing: LPS (Leica Photogrammetry Suite) as well as the in-house BUN and SAT-PP (Satellite Imagery Precision Processing) programs. The complete photogrammetric working process goes from project definition, interior orientation computation, measurement of tie and control points, bundle adjustment image triangulation and absolute orientation and finally to DSM generation and orthophoto production (Fig. 3). A similar approach for the documentation of archaeological areas is presented in Eisenbeiss et al (2005).

Bad conditions on the film storage as well as the high relief of the area caused inaccuracy during triangulation phase within LPS Core. Therefore LPS Orima and the in-house program BUN were employed to reduce the triangulation RMSE values by detecting blunders to acceptable levels. With BUN we obtained a RMSE of residuals of 2m in x&y direction and 3m in height.

For the DTM generation, the in-house software SAT-PP was employed. The matching approach is a coarse-to-fine hierarchical solution with a combination of multi image matching algorithms and automatic quality control (Zhang, 2005). After image pre-processing and pyramids generation, three kinds of primitives (feature points, grid points and edges) are detected on the original resolution image and matched between all the available images. To ensure a correct matching in cliff areas, some manually measured breaklines are also imported. Then a regular DTM is derived (Fig. 4a).

Finally, orthophotos for every data set were generated. Cutlines between images as well as color balancing and feathering were corrected with some image processing.

In Fig.4b a 3D visualizations of, the 1989 data set is shown.

Acquisition date	N° Pictures	Scale	Camera	Focal length
1.11.1977	8	1:60000	RC-10 JET	152.68 mm
17 & 25.08.1989	6	1:60000	RC-10	153.03 mm
16.02.1995	10	1:35000	RC-10	153.03 mm

Table 1 *Characteristics of the used aerial images.*

5 Identification and Characterization of Landslides

The analysis of the orthoimages was useful for field recognition and mapping of the area. Information contained in EPN, 1997 and Basabe et al, 1998 was used as a reference. 24 places were identified as affected by landslide processes involving a total area of about 15.5 km^2 (4.4% the whole area), which were clustered in five different groups regarding their geological and geomorphological characteristics, level of activity as well as their geographical position (Fig. 5).

Zone 1, marked with red, is located in the northern flank of the Cañar River. It contains the most active landslides in the area affecting mainly moraines and ancient colluviums poorly consolidated and strongly weathered. Well developed features of water erosion (trenches) can be seen along the area but are stronger concentrated by the river. The river itself has an important role in the instability processes; stream erosion has caused differential acceleration of the in-motion masses and, consequently minor scarps at the toe of the largest landslide (aprox. 4.2 km^2) in zone 1. The scarp of the same landslide has had a retrogressive behavior, forming a stepped surface with an important vertical component of displacement. Indirect indices of movement like roadways and houses show evidence of important displacements.

Zone 2, marked with orange, is settled on the southern flank of the Cañar River, in front of zone 1 and corresponds to glacier deposits (2 different levels of moraines) which are affected by creeping in a large area (5.3 km^2). Based on stratigraphic profiles in vertical excavations carried out by the municipality of Cañar (2005) creeping is a shallow process restricted to glacier deposits, its limit is the contact with the Miocene units (Turi Fm.). Close to the Cañar River, stream erosion causes an increment in the velocity of displacement and also involves underneath deposits.

Green zone, named zone 3, goes along the eastern flank of Quebrada Pucahuaycu, a tributary of Cañar River running in N-S direction trough glacier deposits and conglomerates from Turi Fm. Undercutting is present along the eastern flank by the village of Cañar causing block falls in the steeper parts of the gully. Downwards, close to its mouth, inclination of the slope decreases and gorges gradually disappear. Block falls are replaced by creeping of the glacier materials.

Zone 4 (yellow) is located on unconsolidated glacier deposits and sandstones well cemented with conglomeratic lenses. Morphological features of a relict landslide can be seen to the south. Apparently, no presence of current activity exists. Northwards, creeping (constraint to the vicinity of the Cañar River) on glacier deposits as well as slight water erosion on sandstones are the current phenomena.

Into the zone 5 (light blue), 3 small dormant landslides (less than 0.2 km^2) on well consolidated conglomerates (Mangan Fm.) has been identified. They could have been triggered for undercutting caused by fluvial erosion of the Cañar River. No apparent signals of movement have been observed.

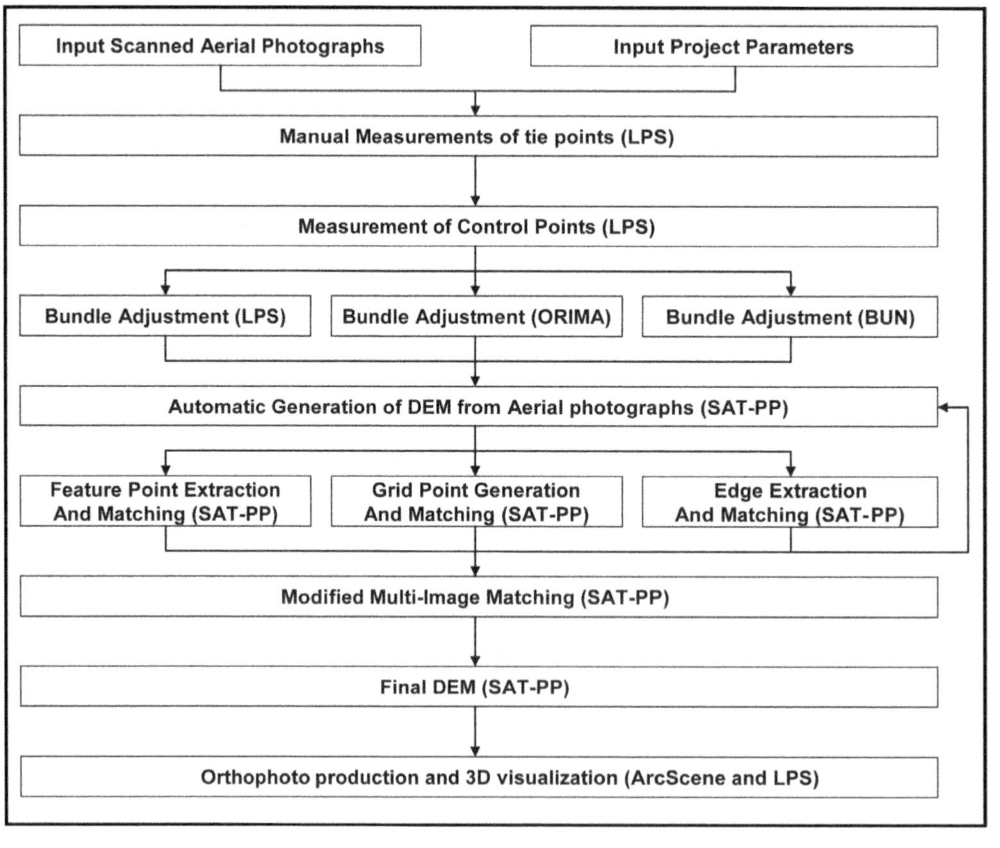

Figure 3 *Photogrammetric pipeline for DTM generation from aerial images.*

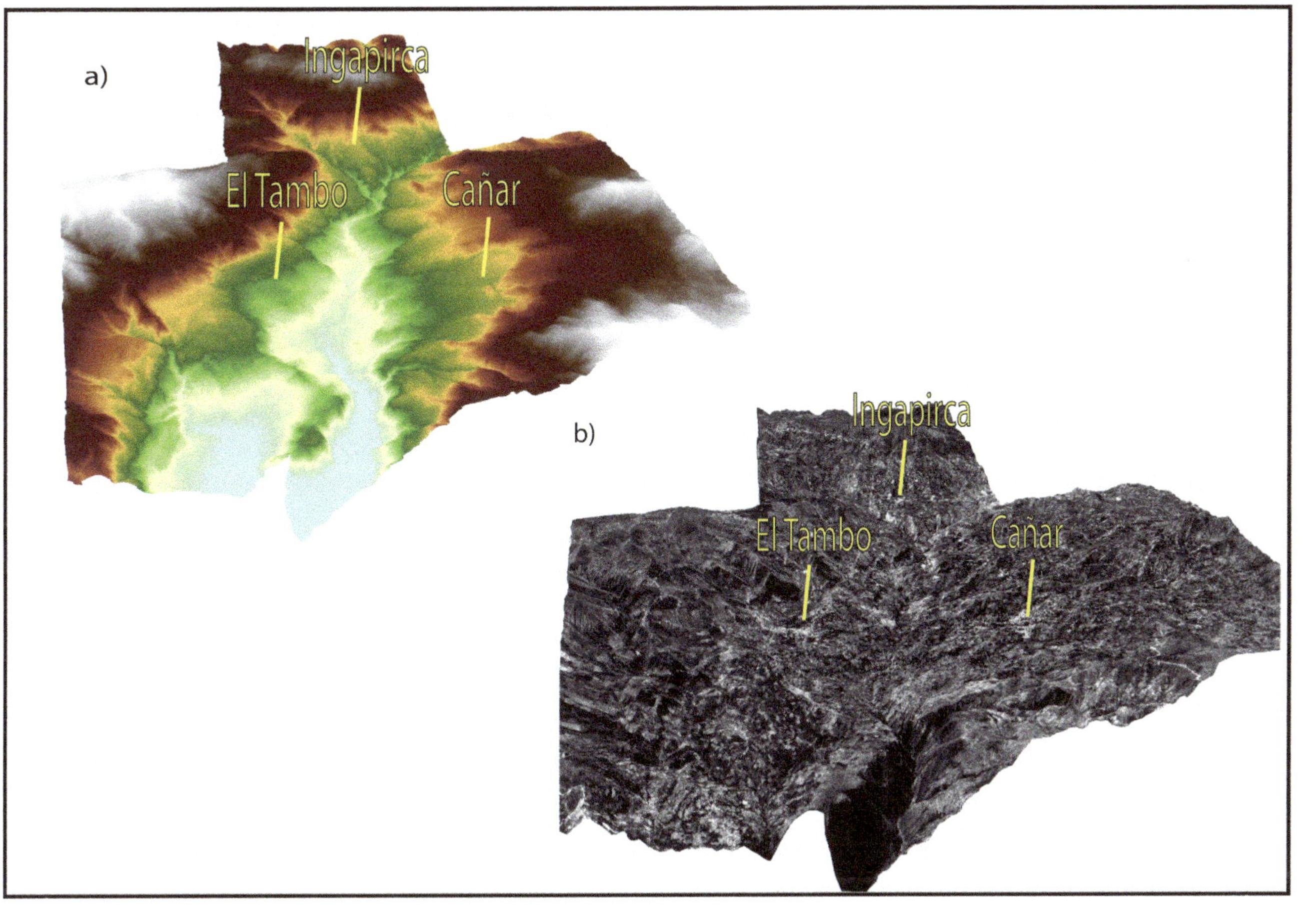

Figure 4 *3D model derived from the aerial image set of 1989; a) DEM; b) Orthophoto*

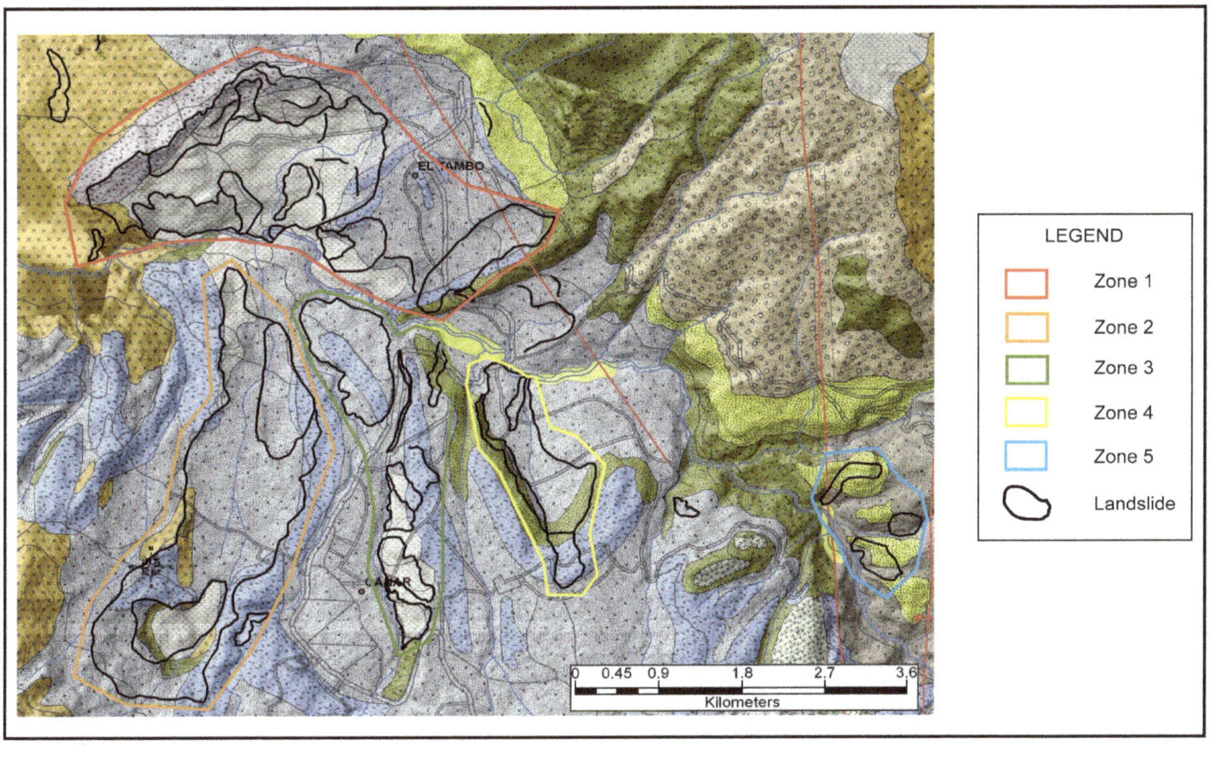

Figure 5: *Classification of the landslide processes using the derived DTM and orthophotos.*

6 Conclusions

In this contribution we have shown how remotely sensing image data can be used for the investigation of archaeological areas and landslide analysis and monitoring. Using 3D visualizations and orthophoto interpretation we have been able to describe most of the slope instability processes that are occurring in the area and to correlate them with other key factors like type and level of erosion or type of material. The multi image matching method used for DSM generation has demonstrated being appropriate to create good quality models based on poor quality images. The entire method would be useful for the same kind of analysis in regions where photographic archives have not been well preserved, which is frequent in most of the developing countries.

From the analysis of results, we can say that water has a strong influence on unstable slopes either as a potential triggering factor or contributing indirectly on unstable areas (by gully erosion or undercutting). Mainly, poorly consolidated materials as glacier deposits or conglomerates have developed instability processes in the area even in gentles hill slopes, just ancient events could be identified in stiffer materials some of them, probably related with different climatic or geomorphological conditions.

In the future, comparison between the different DEMs created during the project will lead to assess rate and direction of displacements in the most active zone.

References

ALCINA, F.J. 1978 Ingapirca: arquitectura y áreas de asentamiento. *Revista Española de Antropología Americana, Departamento de Antropología y Etnología de América,* Madrid, Vol. 8: 127-146.

BASABE, P. & BONNARD, C. 2002 Instability Management in Ecuador - From Policy to Practice, *Proc. Of the Int. Conf. on Instability-Planning and Management, Isle of Wight, United Kingdo*m. Ed. Thomas Telford.

BASABE, P., NEUMANN, A. ALMEIDA, E. GARCIA, E. ONTANEDA, P. 1998 *Informe Final del Proyecto PRECUPA. de Cooperación entre el CSS, DNDC, INECEL, INAMHI, Munic. de Cuenca/ETAPA, U. Cuenca y Consejo de Programación por La Josefina.*

Para la prevención de desastres naturales en la cuenca del rio Paute. Temas: Topografía/geodesia, Geología/geotecnia, Hidro-meteorología, Sismología, Limnología, Capacitación/difusión y apoyo a Defensa Civil., CSS, 1998, pp. 369, Cuenca-Ecuador.

EISENBEISS, H. LAMBERS, K. SAUERBIER, M. ZHANG L. 2005 Photogrammetric documentation of an archaeological site (Palpa, Peru) using an autonomous model helicopter, *Proceedings of the XXth CIPA International Symposium, Torino, Italy In International Archives of Photogrammetry, Remote Sensing and Spatial Information Sciences,* Vol. XXXIV-5/C34: 238-243.

ERAZO, M. 1965 *Estudio de los deslizamientos del suelo en el Austro.* Cuenca-Ecuador.

ESCUELA POLITÉCNICA NACIONAL (EPN) 1997 *Cartografía de Terrenos Inestables. Estudio Regional Zona Cañar-El Tambo. Provincia de Cañar.* EPN-Quito, 41 pp.

LAHUATHE, J.C. 2005 *Levantamiento Geológico y análisis de la deformación tectónica entre las latitudes: 2° 28´S y 2° 37´S. Provincia de Cañar*, Engineering Thesis, EPN-Quito, 124 pp.

MUNICIPALITY OF CAÑAR 2005 *Estudio de mecánica de suelos. Proyecto: Regeneración de plazoleta "El Campanario".* Ciudad de Cañar. 11 pp.

STEINMANN, M. 1997 The *Cuenca basin of southern Ecuador: tectono-sedimentary history and the Tertiary Andean evolution*, PhD Thesis, ETH-Zürich, 176 pp.

ZHANG, L. 2005 *Automatic Digital Surface Model (DSM) generation from linear array images.* Diss. ETH No. 16078, Technische Wissenschaften ETH Zurich, 2005, IGP Mitteilung N. 90

GIS and Web Mapping of S. Leonardo valley and Alesa hinterland

O. Belvedere, M. A. Papa, A. Ceraulo, D.Lauro, A. Burgio

Università di Palermo - Dipartimento di Beni Culturali – Laboratorio di Topografia Antica

1 Introduction

During the last thirty years the Archaeological Institute (today, Dipartimento di Beni Culturali) of the University of Palermo has carried out archaeological survey in the hinterland of Himera and Alesa, ancient towns close to the northern coast of Sicily, whose hinterland is a very suitable area for landscape archaeology. *ArcView* 8.3 and Manifold 7.0 has been employed to carry out thematic maps (visibility, hillshade, geological, morphological and hydrographic layers). The finds have been positioned by a handheld computer (*Recon*) with *ArcPad* software 7.0. An *Access* database (linked to the GIS) allows to query the data of the survey, and to manage a graphic and photographic archive.

The S. Leonardo river is located about 15 km West of Himera, a Greek colony founded on 648 B.C. at the boundary between the Greek and the Punic world (Palermo and Solunto are the most important Punic towns in this area). After the destruction of Himera – ruined by the Carthaginians at the end of V sec. B.C. – a new town, Thermai Himeraiai, was founded close to the mouth of San Leonardo. Thermai was an important town since Hellenistic and Imperial times, through Medieval age and today. The results of Himera archaeological survey have been published on 1988 and 2002, and pre-historic and historic landscapes are showed by thematic maps. About two hundred archaeological sites (***Unità Topografiche***: graves, farms, sanctuary, activity areas etc.) have been identified in the hinterland of Himera and along the Northern Imera valley.[1]

Survey of S. Leonardo valley is another step of Himera project, suitable to investigate economic and political relationship between Punic, Greek and native population in archaic/classical/hellenistic times, and the evolution of landscape from Prehistory to Roman and Medieval ages.[2] Alesa, founded on 403 B.C., is located about 40 km East of Himera, at the foot of Nebrodi mountains, about 1 km from the sea. Around Alesa we have explored part (about 45 kmq) of the Tusa stream valley, between the modern towns of Tusa (SW), Castel di Tusa (N), Motta d'Affermo and Pettineo (E), and, South, towards the Nebrodi, today, and maybe since the past, covered by woods. In the 16th century, among the ruins of the city, the "Table of Alesa" was found, a big marble-slab inscribed in Greek language (*IG* XIV, 352). The "Table" describe in detail the landed property, the distribution and destination of the fields around Alesa, rented by the city to the farmers. Some scholars have tried hypothetical reconstructions of the agrarian landscape, based on topographical and morphological data offered by the "Table"; these assumptions, illustrated by sketches, have been used to restore the distinctive peculiarities of the so-called "mediterranean garden", marked by a specialized arboriculture connected with a woody and pastoral economy.

The archaeological survey has been supported by remote-sensing analysis (the airborne hyperspectral sensor MIVIS, Multispectral Infrared and Visible Imaging Spectrometer), to examine the relationship between physical parameters and the spatial distribution of archaeological finds.[3] About one hundred UT (graves, farms, pipe-line, etc.) have been identified on the ground; the results of the survey are managed by a GIS and showed by historic thematic maps, but the aim of the project is not to compare the archaeological data with the "Table", but to reconstruct the archaeological landscape and add new data to the knowledge of the ancient agrarian landscape.

2 The construction of platform GIS

GIS is a resource available for the "total management" of territorial information that can be gathered from either geographical maps, surveys and actual spatial land representations, plus it associates also all those information that cannot be directly represented on paper. In this document we will not handle GIS by limiting ourselves to explain the *hardware and software,* but we will analyse how the Laboratory of Ancient Topography of the University of Palermo, has developed its *GIS philosophy,* by considering the GIS as a geographical information system planned to acquire, to manage, to analyse and to display alphanumeric and graphic data regarding a territory, thanks to the resources offered by the computers and their applications and by a staff of experts.

The applications of GIS may be employed for the archaeological studies on a regional or sub-regional context or still yet, for careful studies on very limited areas.
The great potentials of managing and preserving coded data, accomplished by the various cross-possibilities and

[1] *Himera* III,1-2; BURGIO 2002.
[2] LAURO 1997 and 2004.
[3] BELVEDERE *et alii* 2004.

virtual simulations, have taken to the use of the GIS in archaeology to attempt some reconstruction of historical reality of the areas analysed and to reproduce in digital format the archaeological research, from the works on the field to the computer processing of the data. So GIS becomes a system-project of research rather than a simple technology of support.

We consider our GIS platforms as a digital ecosystem, an open space, a space within the space, in which the various information coexist, reproduce and evolve in order to generate new contexts.

Therefore, for the two areas in examination, we have tried to reconstruct parallel platforms which, by analysing essential data of the same nature, would result as being similar and conform.

A GIS project, created on the basis of data from an archeological *survey,* is intended as a geo-referenced system of information dedicated to a reasoned filing and to a multi-factors and space-related processing of the information and territorial research.

The data entered in the SIT of Alesa and S. Leonardo Valley, analyse the territorial information in macro-scales and are included in the category of GIS intersites.

The ensuing infomative structure will may be utilized to ask oneself various questions: How is the landscape of the territories analyzed shown? Which anthropic processes have interested the area? Is it possible to reconstruct the ancient settling of these regions? Are all the areas accessible today? Is it necessary to plan interventions of protection or new excavations? Etc...Practically, the work consists in a scientific analysis of the area under study, laying the premises for an assessment of the risks due to the decay and for the future planning of ameliorative interventions.

Therefore, the SIT represent the effective premises for an on-going and efficient monitoring of the territory and a necessary tool today to facilitate the archaeological research.

The first stage of planning a GIS platform, provides for the inputting of various informations inside a single environment, such as: current and historical mapping, aerial photos, satellite images, digital data gathered from files already available or from new filing and database productions. We are showing a table with the respective *data entries* of the two reconstructed GIS platforms, which takes in consideration the knowledge of the data at our disposal necessary for an efficient planning.

Sicilian Region's data have been supplied in Gauss Boaga kilometric coordinates (Datum Rome 1940, oriented on Monte Mario meridian). For the IGM mapping supplied in UTM RD50 coordinates, we have carried out a conversion of the coordinates through a free application (Cartalab 1.0) by inputting the tables in the platform at 25,000 and by following a single system of coordinates.

As you will see, the input data form a *raster* based map, are essential for mapping the Topographic Units surveyed on the field during the archaeological prospecting. During last spring the space information have located directly on the field, thanks to palmtop GPS with the ARCPad software by ESRI.

The two important surveys inside the SIT are produced in two different level of information in SHP format, which can be compared and thus returning interesting suggestions for new methodical approaches for on-the-field research.

The on-site survey files, alias "Archaeological Map", are some unrealistic shapes to which a table is associated and on which the following fields are reported:

OGTS: Identification number of the individual Topographic Unit; PVLC: relevant Township involved in the UT; OGTD: Definition and/or interpretation of the UT; DTF: segment of the chronologic life of the UT; Notes: blank field for generic notes to be inputted in the UT platform.

At this moment, because of being a work in progress, we have believed appropriate to create links through a generic file card for the Units Topographic[4] and with a file card on the materials[5] collected on the field and classified in Access format, which offer images and data of the sites and materials. The survey data have been queried inside the GIS creating queries on the UT chronologies which are produced as stage maps of the territory population settlement.

Other interesting data for the historical reconstruction of the territory have been obtained from the DEM processing, thanks to 3D applications offered by the GIS software. Thus we have obtained maps of the terrain inclines, exposure of the slopes, sun exposure and significant analyses of the inter-visibility among the various sites. Still yet, the works of survey on the field have allowed us to produce a map of visibility on the terrain, which later has been vectored on video and which has turned out to be very useful for in-depth studies of methods and analyses on the physical and anthropic features of the territory[6] and comparing it with the data given by the geologic map.

[4] BELVEDERE-CUCCO 1998 and 1999.
[5] TARDO 2006.
[6] BELVEDERE *et alii* 2005.

Data entry of Alesa						
Type	**Forme**	**Origin**	**Geometrical resolution**	**Optic resolution**	**Scale**	**Year**
CTR	Raster	Regione Sicilia	1mx1m		1:10.000	1994
Ortofoto	Raster	Regione Sicilia	1mx1m		1:10.000	1994
Tavolette	Raster	IGM		300 DPI	1:25.000	1994
DEM Sicilia	Raster	Regione Sicilia	20x20			1994
Carta dei suoli di Sicilia a cura di G. Fierotti	Raster	Regione siciliana; Università degli studi di Palermo.		300 DPI	1:250.000	1988
Volo MIVIS LARA	Raster	CNR		300 DPI	1: 50.000	1994
Lettura del MIVIS a cura di N.Maltese e A.Burgio	Vector (Lineazioni based on NDVI, Thermic Inertness e GTC)	Università degli studi di Palermo			1: 50.000	1997

Figure 1

Data entry of St.Leonardo						
Type	**Forme**	**Origin**	**Geometrical resolution**	**Optic resolution**	**Scale**	**Year**
CTR	Raster	Regione Sicilia	1mx1m		1:10.000	1994
Ortofoto	Raster	Regione Sicilia	1mx1m		1:10.000	1994
Tavolette	Raster	IGM		300 DPI	1:25.000	1994
DEM Sicilia	Raster	Regione Sicilia	10x10			
Carta dei suoli di Sicilia a cura di G. Fierotti	Raster	Regione siciliana; Università degli studi di Palermo.		300 DPI	1:250.000	1988
Carta Geologica dei Monti di Trabia e Termini Imerese a cura di R.Catalano L.Montanari	Raster	Università degli studi di Palermo		300 DPI	1: 50.000	1979

Figure 2

The platform is displayed as a file opened to all information useful for the current research and that can be implemented at any time with new and more detailed data. The SIT manage many geo-morphologic, hydrogeological and use of soils maps, either in *raster* or vectorial format.[7]

An example of this implementation of information, can be supplied by the input in the two platforms of *layers* concerning hydrography and ancient roads practicability. We have chosen to return by vector the hydrography and coastal lines of the territory in order to allow later analyses which place in relation the choices of settlements with the location of water sources.

[7] By en large, we prefer the vectorial data, thanks to which the queries on information appears simple. Obviously the data supplied as images may always be transformed in vectorial format at a later time.

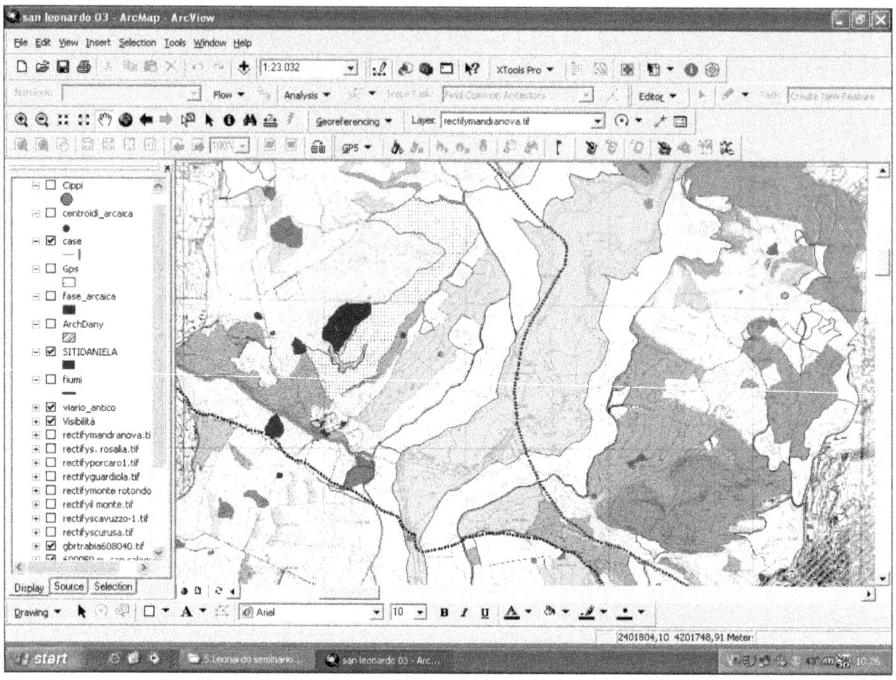

Figure 3 *GIS of S. Leonardo Valley*

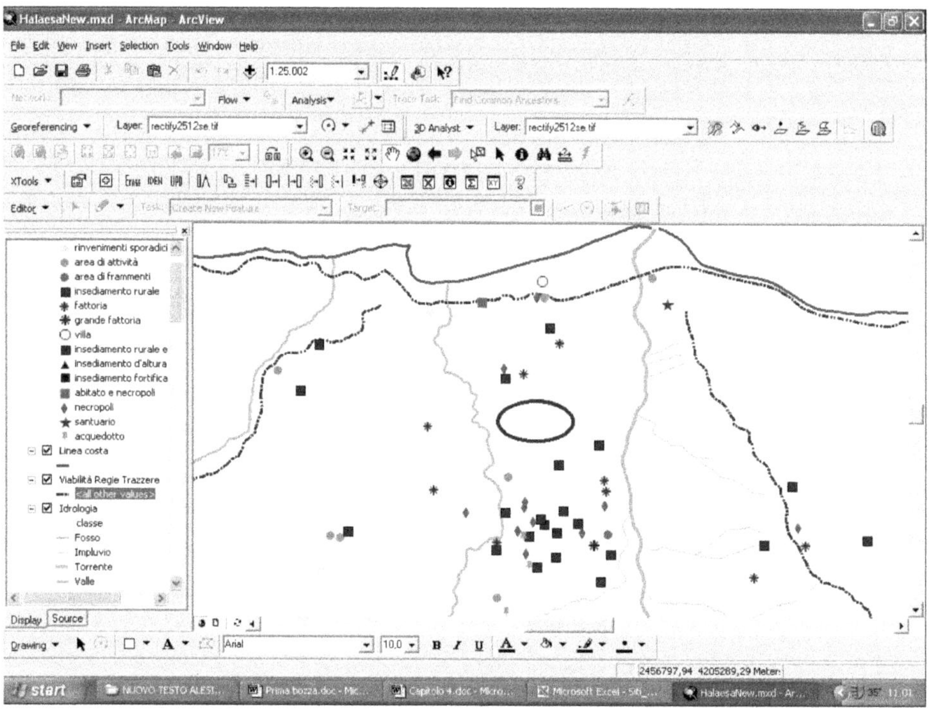

Figure 4 *GIS of Alesa*

3 Softwares

The same working method has been applied in order to reconstruct the hypothetical ancient road system by inputting in the GIS platform the travelling road obtained from the register of the "Regie Trazzere".

At this time, we feel appropriate in supplying information on the software utilized for our project.

Microsoft Access[8] is the most recognized system for managing relational Database or RDBMS (*Relational Database Management System)*. It allows managing the data memorized in Database inside a computer. Microsoft Access is a complete system for managing data, because it places at disposal all the definitions, manipulating and checking functions necessary to

[8] http://www.microsoft.com/

organize large volumes of information. In addition, it has valid tools for research and display of informations.

For the Beta version of the platforms, which the information Laboratory of Ancient Topography of the University of Palermo is developing, we have utilized simple individual file cards aimed at an immediate input of the information.[9]

ArcGIS Desktop by ESRI[10] is the most recognized GIS software developed based on the current information technology standards. ArcGIS Desktop includes Tools: ArcMap, ArcCatalog and ArcToolBox.

The application includes a wide choice of CAD oriented tools and the generating of high quality maps. The utility of GIS platforms allow creation, managing and editing of layers, 3D[11] queries and spatial analyses of the collected data. Moreover, for the work in the field we have utilised an ArcPad[12] software installed in the GPS[13] for the UT surveys and the recording of the data on the geographical maps.

Manifold 7.00[14] is an American GIS software with multiple planning potentials and very cost accessible.

This package offers a pleasant and simple interface and allows the creation of projects based on "maps" which utilize and reprocess vectorial drawings, raster images, 2D and 3D surfaces, maps at several different levels and relational database.

The package, compatible with the ESRI standards, imports and utilizes images, maps and data in more than eighty different formats, the likes of: SHP, MID/MIF, DXF, BMP, GIF, JPG, TFW, TIFF, DBF, MDB, XLS, CSV, etc.

Manifold, thanks to numerous spatial extensions for raster and vector images and thanks to the possibility of creating buffer zones and to several other functions, is a tool that offers excellent studies on territories. The software allows for an elegant editing thanks to its advanced CAD base. In addition, the software is integrated with an *Internet Map Server*. Therefore, for the development of the GIS platform we have worked with two different applications, but perfectly capable of communicating with each other.

It may seem strange that we have chosen to work with two different software of GIS extraction, but the need to utilise two programmes was dictated by the high licensing costs of the ESRI applications. At the Laboratory of Ancient Topography of Palermo we have found in Manifold an excellent and cost effective GIS programme capable of allowing them to work at a fast pace even when other commitments did not permit their constant presence at the Faculty.

The knowledge of the Manifold application has been always appreciated thanks to an efficient clients' service, which has made the learning of the software potentials simpler. The programme has, throughout the years, supplied continuous updates and thus offering an excellent *GIS philosophy* comparable to the ESRI products. The Manifold license has been purchased also by the Laboratory in 2006, interested especially to the easy managing of IMS[15] and to the possibility of utilising gratuitously Demo versions for lecture. The new generation of Web technology has especially developed the GIS applications.

4 The Web Mapping

The Web GIS has thus become an excellent type of web information through HTTP; HTML, ASP and ASP.NET. The publication of a GIS archaeological data on the Web, allows for a constant monitoring of the Cultural Heritage thanks also to the facility of update and use of the data.

The languages processed by the *World Wide Web Consortium* (an organization whose main purposes is the definition of technologies capable of developing the Web potentials) offer a valid starting point for guaranteeing the interoperability of the archaeological information, with languages, which for their intrinsic nature, are open, flexible and easily integrable in the GIS field and territorial analyses.

Exactly in consideration of the utilization of the data, we have felt as proper to test our GIS platforms by utilizing Web Mapping techniques. This has been possible thanks to the presence of the Manifold System of an integrated Internet Map Server. The display of the GIS projects is easily accessible through the web browsing standards, thanks to its pleasant graphic look accompanied by legenda.

The *Map Server*, added in the same Manifold, operates through the automatic update of an ASP file and allowing publication on Internet.

The software interface allows for the creation of web sites starting with projects already developed; this way the information of our platforms are shown on-line and can be consulted and queried by utilizing easily comprehensible standard models by any type of users. The platform data are structured and processed to be placed on-line under the form of maps (prepared according to map-making parameters, vector and raster

[9] Files which we are already organising in organic database and functional to the collection of large quantities of data.. BELVEDERE–CUCCO1999;. TARDO 2006.
[10] http://www.esri.com/
[11] Extensions of software 3D Analyst and Spatial Analyst.
[12] Software GIS and GPS
[13] GPS handheld Tripod Data Systems'Recon Pocket PC.
[14] www.manifold.net/
[15] Integrated in the base version of software.

files, geo-referenced images, list of tables, graphic elements, queries, etc…).

By utilizing the software scripts it is possible to outline the screenshot to display on the web and to consult the maps and the different layers.

The *Web GIS* developed by us, supplies practical tools for analyzing the information published on-line by surfing the databank through report, queries and statistical filters; and by surfing throughout the maps thanks to the zoom, panning, managing levels, querying the subjects, methods of selection, drafting and spatial analyses. It is important to point out that the project of web publishing of the platforms is made possible through Open Source and Free software technologies, thus insuring an easy exchange of information and guaranteeing the quality and integrity of the research. In fact, the application utilized follows the OGC rules and is fitted with WFS and WMS.

Anyway, the software offers the possibility of developing pages available on-line only by administrators, functional to the continuation of the research an management and implementing data to be inputted on the platform (i.e.: the information on findings will be partially published on line and only on those sites believed to be of great importance; even for works of remote-surveying a selection will be done, preferring to place on-line those data easily readable by users not expert in the matter). Currently only some layers of the two platforms have been placed on-line, in order to test the effectiveness of the system, because we believe to be more correct to wait, for the final placement on-line of the two platforms and the scientific publication of the data, according to traditional methods. The GIS platform *Web Mapping* will be offered to the publishing users as a tool for a more immediate comprehension of the territory and its peculiarities and as a filing system of scientific information easily integrable and consultable.

The interoperability that the Web Mapping offers is aimed at a more immediate sharing of data in order to make them available to all agencies which attend to territorial analyses thus preserving the future and past investments.

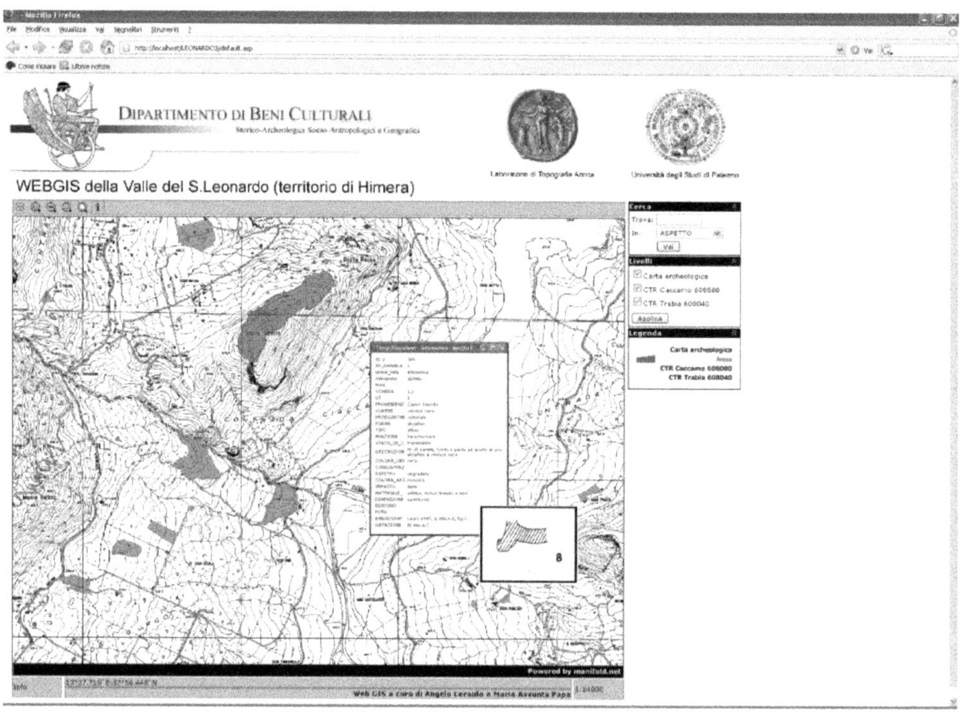

Figure 5 *Web Mapping of S. Leonardo with file card of archaeological find*

Figure 6 *Web Mapping of S. Leonardo with TIN and layer of archaic age*

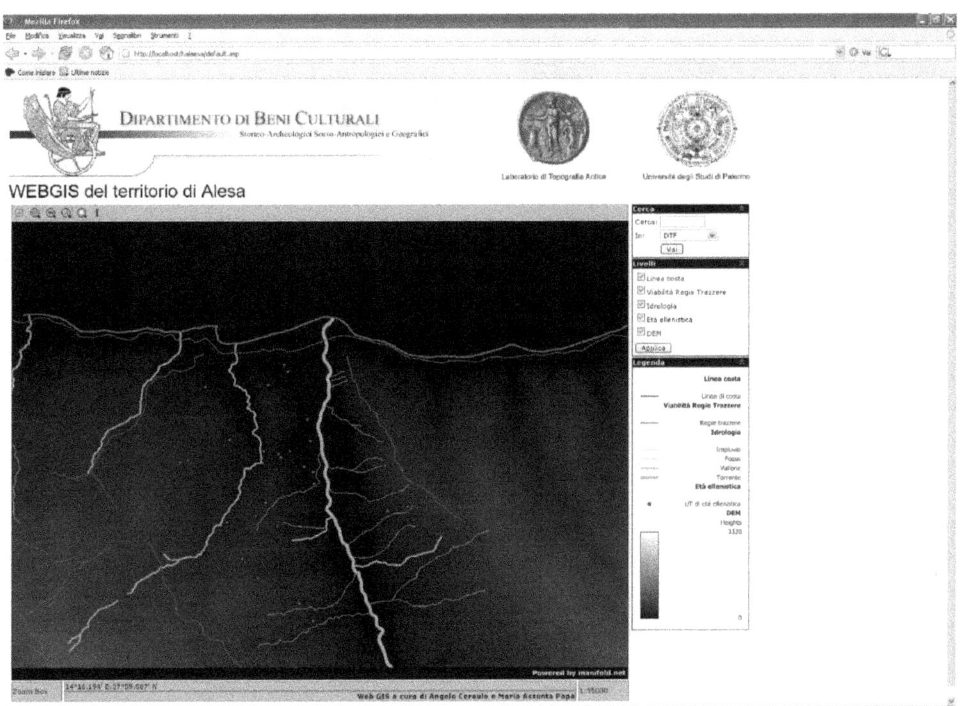

Figure 7 *Web Mapping of Alesa with DEM, layers of hydrography, ancient roads, and UT of hellenistic age*

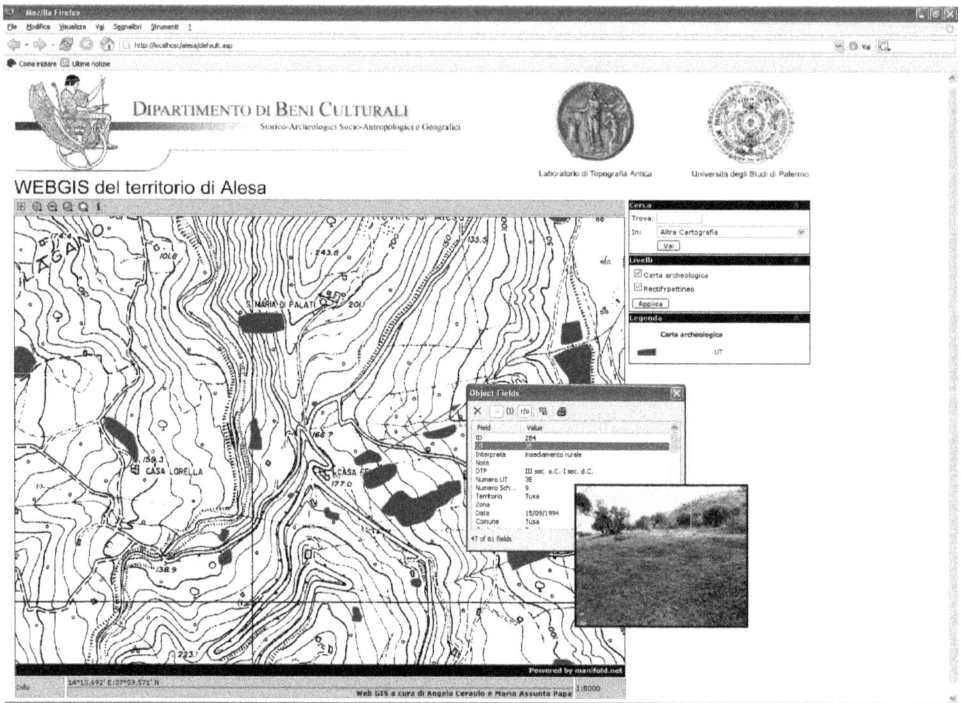

Figure 8 *Web Mapping of Alesa with file card of Units Topographic*

5 References

AL BAYARI 2005: O. Al Baayari, New survey technologies for production of GIS model of the ancient roman jerash city in Jordan, in *CIPA 2005 XX International Symposium,* Torino.

BELVEDERE 1997: O. Belvedere, Prospezione archeologica nel territorio imerese, in H.P.ISLER, O. STEFANI (Edd.), *Wohnbauforschungen in Zentral- und Westsizilien,* Zürich, 91-96.

BELVEDERE-CUCCO 1998: O. Belvedere, R.M. Cucco, B. Gestivo, E. Alessi, Un SIT specifico per l'archeologia e la topografia antica, in Rilevamento, rappresentazione e gestione dei dati territoriali e ambientali (*Atti II Conferenza Nazionale ASITA*) I, Bolzano, 304-308.

BELVEDERE-CUCCO 1999: O. Belvedere, R.M. Cucco, G. Razzino, Progetto ARCHEOSIT: carta archeologica della valle del fiume Torto, in Informazioni Territoriali e Rischi Ambientali (*Atti III Conferenza Nazionale ASITA*) I, Napoli, 291-296.

BELVEDERE et alii 2004: O. Belvedere, A. Burgio, G. Ciraolo, G. La Loggia, A. Maltese, D. Rametta, Telerilevamento di aree archeologiche mediante dati iperspettrali MIVIS, in *Agri Centuriati*, 199-215.

BELVEDERE et alii 2005: O. Belvedere, A. Burgio, R.M.Cucco, D. Lauro, Relazioni tra geomorfologia, processi post-deposizionali e visibilità del suolo nella lettura dei dati di prospezione archeologica, in *Archeologia e Calcolatori* 16, 129-152.

BURGIO 2002: A. *Burgio, Resuttano (IGM 260 III SO)*, Forma Italiae 42, Firenze.

CAMPANA-FORTE 1999: S. Campana, M. Forte, *Remote Sensing in Archaeology (XI Ciclo di lezioni sulla ricerca applicata in archeologia, Certosa di Portignanano),* Firenze.

CAMPANA-FRANCOVICH 2006, S. Campana, R. Francovich, *Laser scanner e tecnologie digitali,* Firenze.

CONOLLY-LAKE 2006: J. Conolly, M. Lake, *Geographical Information System in Archaeology*, Cambridge.

FRONZA 2000: V. Fronza, Il sistema di gestine degli archivi nello scavo di poggio Imperiale a PoggiBonsi, in *Archeologia e Calcolatori* 11, 125-137.

GILLINGS-MATTINGLY-VAN DALEN 1999: M. Gillings, D. Mattingly, J. van Dalen (Edd.), *Geographical Information Systems and Landscape Archaeology* (The Archaeology of Mediterranean Landscapes 3), Oxford.

Himera III.1, V. Alliata, O. Belvedere, A. Cantoni, G. Cusimano, P. Marescalchi, S. Vassallo, *Himera III.1. Prospezione archeologica nel territorio,* Roma 1988.

Himera III.2, O. Belvedere, A. Bertini, G. Boschian, A. Burgio, A. Contino, R.M. Cucco, D. Lauro, *Himera III.1. Prospezione archeologica nel territorio,* Roma 2002.

GOTTARELLI 1997: A. Gottarelli, *Sistemi informativi e reti geografiche in archeologia: GIS.internet*, Firenze.

KORTE 2000: G.B. Korte, *The GIS book*, 2000.

LAURO 1997: D. Lauro, Cozzo Sannita: un insediamento indigeno e punico-ellenistico lungo il corso del fiume San Leonardo, in *Archeologia e Territorio,* Palermo, 349-360.

LAURO 2004: D. Lauro, L'apporto della fotografia aerea alle indagini storico-topografiche lungo la valle del fiume S. Leonardo nel territorio di Himera, in *Archeologia aerea*, 193-214.

LOCK 2003: G.R. Lock, *Using computers in Archaeology,* 2003.

MEHRER-WESCOTT 2005: M.W. Mehrer, K. L. *Wescott, GIS and archaeological site location modeling.*

NARDINI 2000: A. Nardini; La piattaforma GIS dello scavo di Poggio Imperiale a PoggiBonsi, in *Archeologia e Calcolatori* 11, 111-123.

PENG-TSOU 2003: Zhon-Ren Peng, Ming-hsiang Tsou, *Internet GIS.*

TARDO 2006: V.Tardo, Distribuzione delle ceramiche nel territorio imerese. Analisi dei dati nella prospezione archeologica, in S. MENCHELLI, e M. PASQUINUCCI (Edd.) T*erritorio e produzioni ceramiche. Paesaggi, economia e società in età romana* (Atti del Convegno di Pisa), Pisa .139-146.

Multiuser interaction in an archaeological landscape: the Flaminia project

Maurizio Forte,[1] Sofia Pescarin,[1] Eva Pietroni,[1] Claudio Rufa[2]

[1] CNR ITABC – VHLab, [maurizio.forte],[sofia.pescarin], [eva.pietroni]@itabc.cnr.it
[2] SoftLogic, claudio.rufa@fastwebnet.it

1 Introduction. The Flaminia Project

The virtual reconstruction of the archaeological landscape is an holistic process of great complexity, that is made of relations and includes in a virtual ecosystem many kinds of data, according to a multidisciplinary approach. This system of relations, interactions and behaviours assumes cultural, psychological and perceptive relevance.

In this paper we shall present the project dedicated to the Roman "Via Flaminia", an important consular Roman road, built in 220 B.C. by Gaio Flaminio. In that period the road begun in the center of Rome and continued toward north, along the Tiber valley, the Appennini mountains and then it descended toward the Adriatic sea along the Metauro valley until Fanum Fortunae (Fano); from here it continued toward Pesaro and finally reached Ariminum (Rimini) following the coast (Fig 1).

The project, started in 2005 and supported by Arcus Foundation, has the goal of exploiting and diffusing its knowledge and historical context, focusing mainly on five important archaeological areas, in the northern part of Rome: Ponte Milvio (the important bridge where the emperor Constantino defeated Massenzio in October 28th of 312 A.D.); Tor di Quinto and Grottarossa (where the road is partially conserved together with its funerary monuments); Villa of Livia (the superb roman villa of Augustus's wife) and finally Malborghetto (a Roman arch transformed today in a museum) (Messineo, 1991; Messineo, 1993).

The final result of the project is the creation of a Virtual Reality application that will be installed in the National Museum (Museo delle Terme di Diocleziano) in Rome. Here a room will be specifically dedicated to the virtual musealisation of the road "*Via Flaminia*", through a network of four PC and a main, wider, screen, a sort of virtual theater connected with each single PC.

The project[1] will be presented to the public in 2007.

The archaeological observed landscape of this Roman road is very different from other cases such as, for instance, the area around the *Via Appia,* in the southern part of Rome, where it is still possible to find a long, and almost intact, path of the Roman road with its surrounding funerary monuments. From a topographic point of view, the *Flaminia* landscape is characterized by a deep enclosure in the actual urban context, which often has hidden or completely destroyed it. The result is that the Roman path and monuments are not conserved, except for few cases. Its territory is not a closed and continuous entity with a coherent spatial area and delimitation. Thus the problem of its reconstruction is strictly connected with the attempt to establish a linear path in the virtual visit and with the difficulties of imposing arbitrary choices in the case of "empty" or unknown areas, where visible or studied structures can't be clearly located.

For this reason the main problems in the virtual representation are the spatial contextualization of fragmented sites and the cultural and communicative contextualization of "*Via Flaminia*" as a whole entity.
To conciliate these two needs of contextualization, we are implementing two main levels of scale and storytelling:

- a macro scale level, referred to the holistic vision of the archaeological landscape,
- a micro scale level referred to the five relevant and defined sites along the Flaminia road (Malborghetto, Ponte Milvio, Tor di Quinto, Grottarossa, Villa of Livia).

2 The reconstruction of the archaeological landscape

The archaeological landscape has been reconstructed through different techniques and data sources, integrated in a coherent methodology of elaboration and communication. In specific we worked on the reconstruction of

- the actual observable landscape (Fig.2a)
- an interpreted landscape (Fig.2b)
- an hypothetical ancient landscape (Fig.6)

In the first case we used available geospecific data (DTM, geoimages, vector thematic layers) to build a 3d terrain base representing the actual landscape. The digital 3D representation of the terrain was obtained using mainly cartographic data (technical cartography such as Sara Nistri, scale 1:5000) and satellite images (Landsat 30 mt resolution and Ikonos 1m resolution). Cartographic

[1] Still in progress at the time of this publication (Oct.2006).

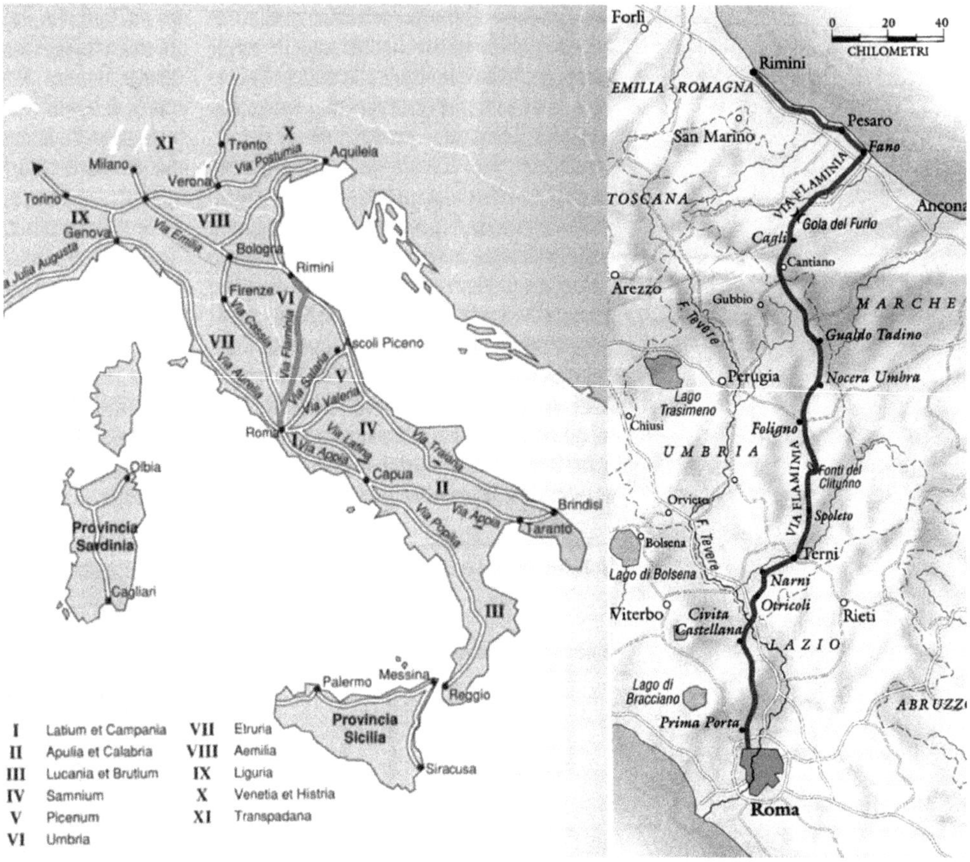

Figure 1 *The Roman* via Flaminia

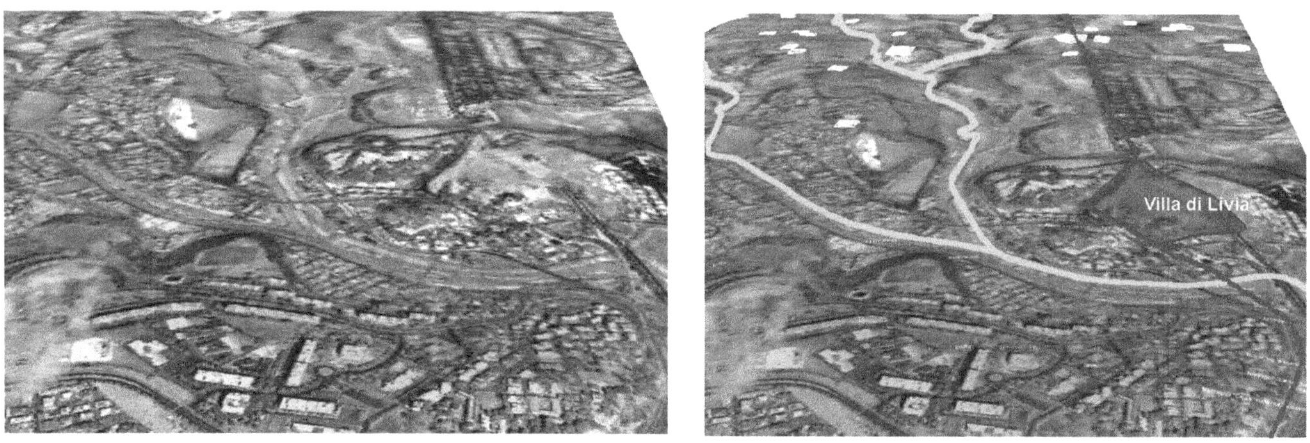

Figure 2 *The observed and interpreted landscape reconstructedIn the second case, thematic layers were overlaid on the 3d terrain in order to help the interpretation process and to allow a transparent transition to the hypothetical reconstruction of the ancient landscape: for this reason vector layers (points, polygons, lines) were taken from archaeological maps and excavation plans and processed to obtain: archaeological areas, Roman villas, Roman aqueducts, hypothetical ancient hydrography, Roman roads, etc. (right).*

maps have been used for elevation data manual digitalization and integrated with a more detailed topographic survey with DGPS of some areas and, in some cases, also with time of flight - laser scanner data used in topographic way (Cyrax 2500).

In many cases the observable landscape is put in relation with interpretative and reconstructive hypothesis of structures and contexts, as they could be in ancient times. This approach is supported by the top down methodology: the critical study of the collapsed areas and decorative fragments, bibliographic references, typological comparisons with similar and contemporary Roman buildings.[2]

[2] For a detailed description of these aspects see L. Vico, A. Moro and V. Vassallo's paper in this volume: *The importance of the relief and sources to interpret and communicate the cultural heritage.*

In the last case, different sources have been analyzed (Latin sources, historical, archaeological, geomorphological data, lithology, natural vegetation, etc.) and used in integration, to try to reconstruct an hypothetical ancient landscape (Traina 1983, Braconi 2004, Funicello Parotto 2001, Volpe Arnoldus Huyzendveld, 2004). Specifically, we defined a series of classes, starting from known data, that were processed in the GIS to obtain new classes that helps in the definition of:

- Cultivated areas
- Non cultivated areas
- Inhabited zones (Villas and other buildings)
- Roman Digital Elevation Model
- Roman hydrography
- Roman roads
- Monumental (funerary) areas
- Roman vegetation

A first test was done on the area of the villa of Livia (Fig.6).

Architectonic surveys were carried on, processed and then integrated in the terrain, choosing the different techniques (scanner laser, photomodelling, photogrammetry, computer vision, total laser station and so on) in accordance with the kind of structures and information we needed.[3]

In order to share available geospecific data with the team and all the partners and co-operate in the construction of the VR application, a first Internet web site has been built, based on a VR webGIS open source platform (www.vhlab.itabc.cnr.it/flaminia). A Mozilla and Explorer plug-in was developed in order to allow a complete interaction with 3d reconstructions, 3d models, vegetation and thematic layers.[4] (Fig.8).

The main aim of the virtual reality application is the creation of a digital virtual ecosystem to integrate geographical, topographical, perceptive, historical interpretative, narrative and symbolic data.

In our approach the spatial three-dimensional reconstruction of an archaeological site is only a first, basic step where it is necessary to organize, link and develop all cultural contents.

Only in these conditions we can communicate the "sense of place" and obtained an impressive cultural result.

The virtual environment should be alive, a space that we can experience, interpret, change, transform and establish relations according to our interests and interpretative approach.

Another important issue for us is "transparency": 3D models, reconstruction of the actual and ancient landscape have to declare the methodology and the sources they come from, so to allow the discussion, the critical awareness of the public and therefore their cultural impact.

3 The use of virtual reality inside museums. A multi-user system to communicate the Ancient "Via Flaminia"

The museum application is based on an innovative methodological approaches in the field of digital virtual heritage:

- the storytelling
- the multi-user domain.

The storytelling is referred to an interactive system based on narrative techniques and metaphor, virtual characters, avatars, speaking objects and sometimes game dynamics and rules . The aim of storytelling is to make the space "alive".

The multi-user domain allows many users to interact at the same time in the virtual environment, recognizing each other and developing shared purposes and joined actions, as it is described below.

In its traditional off-line applications, virtual reality creates quite exclusive relationship between the user and the rules that regulate the exploration. Fruition, interaction, comprehension of the dynamics and behaviours are individual experiences.

Inside a museum or an exhibition, the multiplication of the installations that allows more people to interact at the same time but without relations and contacts among them, is only a partial and not decisive solution.

Museum should encourage a collective and shared communication of cultural contents. For this reason they are generally more interested in communicative formats, that involve a large public at the same time (for instance movies and in very rare cases virtual theaters), but where the conditions of fruition don't stimulate creativity, even if perception can be very impressive.

This project is an attempt to match this kind of needs with the personalization of the cognitive process pushed by virtual reality systems: free exploration of virtual environment in real time, possibility for users to relate themselves with cultural contents in a creative and interactive way, following their preferences, interests and organizing their own "mental map".

Virtual reality systems, virtual storytelling and multi-user domain, combined in an integrated methodological

[3] For a detailed and technical description about this subjects see M. di Ioia, F. Galeazzi and N. dell'Unto in this volume: *The Flaminia project: relief and post processing data techniques*

[4] See for a detailed explanation of the web application Calori et alii, *Sharing interpretation: the challenge of Open Source web approach*, in this volume.

approach, can offer very efficacious and innovative solutions to generate a strong educational impact and involve the public in a collective fruitions of cultural contents.

In the fruition of the Flaminia project inside the museum's room, the focus of the communication system is a large screen positioned in the middle of the wall, in front of the public; it is the attractors of virtual events, addressed to collective fruition. Four users interact at the same time, everyone with one PC, and share the same virtual scene (Malborghetto or Grottarossa or Villa of Livia, and so on). They visualize the virtual environment not only on their monitors, but also on the large screen on which a virtual show is developed in real time according to their actions and behaviours. So they are the creators and the directors of this performance and, at the same time, they are users, observers, together with all the public present in the room (Fig 3).

Figure 3 *3D simulation of the installation for VR system inside the museum's room.(3D model by dell'Unto)*

The multi-user domain will not adopt the traditional channels of communication (such as chats or similar, that are not very suitable to museum contexts), but it will be developed in accordance with the following rules:

- share the same monographic level,
- allow the avatars to meet each other;
- transformation of the virtual environment by discoveries and actions of each user.
- Visualization of avatars actions and cultural contents (objects, characters, structures, storytelling, movies, metadata and so on) discovered in the main screen, that will become an "attractor".

All these elements will create a virtual show for the users and all the people present in the room. The audio contents will be unique and common for players and public, in Dolby surround 5.1.

4 Sharing the same monographic level

The first condition to make the multi-user system effective is the sharing of the same virtual environment. For this reason the users aren't completely free to choose the monographic level to explore, but they will be invited to enter in the scene active at their arrival. The succession of the monographic levels is fixed, settled in advance on the base of the narrative impact, so to stimulate curiosity and interest to continue the virtual exploration until the discovery of all contents in all scenes is accomplished.

Every monographic level has a timeout, according to the complexity of cultural contents, storytelling and metadata that the level contains.

5 Modification of the virtual environment

When a user finds out objects, movies, characters and meta-data, he determines a transformation in the virtual world. On his monitor, and on the large screen, each user visualizes the transformations generated by himself and by the others.

In the "Villa of Livia", for instance, the exploration begins from the actual, observed landscape (Figs. 4, 5, 6). The avatars can enter in the actual villa from four different accesses; they meet characters and objects belonging to the present time and help them to discover and visualize contents connected with the interpretation of the rooms: their structure, their chronological phase, their history and functions and their relations in the context of the ancient villa, some news about the people that lived there in the past.

When the users find out the main cultural contents about that room or group of rooms, the award they receive is the reconstruction of that rooms as they could be in ancient time. We'll consider in particular two more relevant phases: 1st century B.C., referred to Augusto's and his wife Livia's period, and the 1st century A.D. The partial reconstructions will refer to the phases that are more relevant in the actual perception of the villa, in order to allow a clearer interpretation and comprehension of the structures we can see today. But at the end of the exploration, the whole reconstruction of the villa, during Livia's time will be available, to allow a coherent and homogeneous representation of the typology of the roman villa in that chronological phase. We'll distinguish, through graphic solutions and multimedia explanations, the level of reliability of the reconstructions proposed.

Reconstructions offer new interactive contexts and contents; inside them Roman characters will appear while performing their daily activities and they will dialog with the users' avatars.

Also the exploration of Malborghetto begins from the observed archaeological landscape (Fig 7).

Figures 4 and 5 *Nowadays 3D representation of Villa of Livia with avatars exploring it (elaborated by laser scanner data)*

Figure 6 *Hypothetical 3D representation of Villa of Livia's hill during Roman times, processed with Visual Nature software.*

Figure 7 *3D model of Malborghetto as it can be seen today (obtained with Photomodelling techniques).*

Discovering cultural contents, the 3D context will change as it was in medieval time, and furthermore during Roman times, when it was a triumph arc built to celebrate the victory of Constantine over Massenzio.

The final purpose, for users, is to go back to the night before Ponte Milvio's battle, and find out Constantine's military camp in the field of Malborghetto, near Flaminia street, where the arch did not exist yet.

From here, a little later, the events will move to Ponte Milvio, so the battle will be described as if it would happen in real time; it will be represented through virtual storytelling and iconographic resources. We have chosen this event as the most representative of the "identity" of this bridge, from historical point of view, of the collective "sense of place" and the narrative potentiality.

In conclusion the avatars have the following possible behaviours:

- discover characters and interact with them;
- visualize movies, animations, activate audio contents;
- visualize areas through highlighting or colors in order to put in evidence specific themes, relations among structures, interpretative layers, affordances (chronological, typological, functional correspondences for instance);
- transform the environment following an ideal a time line;
- perform actions and 3D behaviours, interact directly with functional objects, manipulate tools to improve the sense of "embodiment" of the users in the 3D space.

Not all contents are narrative. Some of them are visual, as we said just before, put in evidence by highlighting and they are referred to particular relations of meanings (affordances).

In the upper part of the window, in each homogeneous group of rooms, there will be a symbolic Cybermap, with some "branches" or icons corresponding to the various affordances Space, Time, Use, Typology, Similarity, Reliance of reconstruction, Methodology of acquisition and modeling (Fig 8).

The user can visualize the information activating the icons or branches of this symbolic interface.

In this way the informations referred to the archaeological structures and also those relative to the "transparency" of the elaboration processes, will be available and clear for the public, so to allow and improve the critical awareness of users and their capacity of interpretation.

Figure 8 *Flaminia project (CNR-ITABC): Roman Villa of Livia, Rome, where every objects, rooms and so on have relations of contents, such as Space, Time, Use, Type, Similarity. The user can interact with the model also de-constructing it, activating its affordances. Here is a schematic representation, not yet the final symbolic interface of the Cybermap.*

6 Visualization, interaction interfaces and collective fruition of contents

All the actions performed by the avatars and all the cultural contents converge on the main screen, focus of the virtual experience. They are visible through a continuous switch of virtual cameras.
All these events create a sort of movie in real time, destined to both players and public.

Each "player" can follow the development of contents from two perspectives: his single point of view (in his PC he moves with first person camera) and the collective point of view (with third person cameras, dolly cameras, free cameras and so on).

This condition allows less expert users to observe the others players' behaviours and understand the results of their actions. In this way they become able to learn cultural contents and interaction rules and they can defeat the frustration caused by their initial inability.

When users explore the environment without finding out anything, global views and views showing each avatar alternate on the large screen,

When a user discovers a contents, on the large screen, the camera gets near his avatar and near the discovered object (a movie, a character, a reconstruction and so on). In that moment all the local PC will stop and all the users, together with the public, will look, on the large screen, at the event activated by that avatar.

As we said before, the audio comment is common for all the people and it will be spatialized in 3D.

This solution solves the problem of the distribution and management of sound in the room. All other possible

Figure 9 *Navigation on Malborghetto (Costantine Roman arch) in the web site (www.vhlab.itabc.cnr.it)*

solutions (headphones, glass-domes and so on) present some disadvantages from both the economical and spectacular points of view.

When the collective event on the large screen finishes, each user can continue his exploration from the point where he stopped before.

But all the local applications, on each PC, will have loaded the transformations determined by the author of the last discovery. Each user will not be able to discover the contents already discovered by the others but he saw them on the large screen.

In order to make the exploration and the interaction direct, easy and intuitive, no 2D menu are used to access cultural contents or visualization tools. The activation of all the informations inside the 3D space is determined by proximity, contact or crossover of interactive objects.

Navigation and interaction are controlled exclusively by the joystick.

The Museum application is realized with Virtools DEV 4.0 and the Multiuser Server-pack.

7 Conclusions

We think that this new (for museums) methodology can offer many advantages.

First of all it can speed up the time of fruition, because the exploration and the interaction are collective: everything discovered by one user is available and useful for all the public. This is very important also in consideration of the large amount of available information in a quite limited time.

Secondly it enforces a conception of virtual reality that is particularly efficacious inside museum's context.

The innovation consists also in the integration of many media and metaphors of fruition: virtual reality (creativity and personalization of choices and interaction), game-rules (discovery and award dynamics), storytelling and cinema (real time movies performed by all users' interaction on the large screen), enhancement of 3D perception (stereoscopic vision).

Although we know that this field (3d real time environments with multi-user interaction) is at its starting point, we believe that it will be more and more uptaken in the future and it will need more epistemological and psychological researches, involving also neurosciences. In particular we need to study and overcome some problems related to the concentration-attention level, in a communicative and didactic approach, in comparison with the number of behaviours, interactions and windows available. The aim is to create better VR applications where personal analysis capabilities are not lost in the rapid exploration of simple "surfaces", of collection of meaningless objects, of simulations without sources and originals (Turkle, 1995, 2-13).

In the next future we intend to experiment this kind of multi-user VR application also for web application, working on open VR 3d communities in the development of concepts such as sense of place and shared working environments for cultural heritage.

References

ANTINUCCI F., *Comunicare il museo*, LATERZA, ROMA, 2004.

BATESON G. *Steps To An Ecology Of Mind.* San Francisco: Chandler Press, 1972

BATESON G. *Mind And Nature: A Necessary Unit.* New York: Dutton, 1979

BONOMI PONZI L. 1991: *La via Flaminia e l'Umbria, «Viae publicae romanae»*, Roma 1991

BRACONI P., *Territorio e paesaggio agrario nell'alta valle del Tevere*, in The Tiber Valley in Antiquity (New research in the upper and middle river valley), 27-28 febbraio 2004.

FORTE M., PESCARIN S., *Behaviors, Interactions and Affordance in Virtual Archaeology. Advanced ICT Methods Guide*, edited by Lorna Hughes, London 2006, in press.

FORTE M., PESCARIN S., PIETRONI E., DELL'UNTO N.: *The Appia Antica Project. Archaeological Landscapes through Digital Technologies*: Proceedings of the 2nd Italy-United States Workshop, Rome, Italy, November 3-5, 2003 Berkeley, USA, May 2005 edited by Maurizio Forte. BAR S1379, Oxford 2005

FORTE M., PIETRONI E.: *THe Museum's Mind: a cybermap for Cultural Exhibitions,* in Proocedings of VAST 2006, Nicosia, Cipro 30 Oct. - 4 Nov.

FUNICELLO R., PAROTTO M. 2001: *General geological features of the Campagna Romana*, in The World of Elephants - International Congress, Roma 2001.

GIBSON, J.: *The Ecological Approach to Visual Perception*, Lawrence Erlbaum Associates. Hillsdale NJ, 1979.

LUNI M. 1989: *Nuovi documenti sulla Flaminia dall'Appennino alla costa Adriatica*, in *Le strade nelle Marche. Il problema nel tempo. Atti del convegno. Fano, Fabriano, Pesaro, Ancona, 11-14 ottobre 1984*, Ancona 1987, e Urbino 1989.

LUNI M., 1999: *Le fasi di "monumentalizzazione" della Flaminia nella Gola del Furlo*, in Atlante tematico di Topografia Antica I, L'Erma, Roma 1999

MATURANA H. and F. VARELA., *Autopoiesis and Cognition: The Realization of the Living*, Boston Studies in the Philosophy of Science. Edited by Robert S. Cohen and Marx W. Wartofsky, vol. 42, Dordecht (Holland): D. Reidel Publishing Co., 1980.

MESSINEO G.,1991: *La Via Flaminia. Da Porta del Popolo a Malborghetto*, Roma: Quasar.

MESSINEO G., CARBONARA A. 1993: *Via Flaminia,* Roma: Istituto Poligrafico e Zecca dello Stato, Libreria dello Stato.

SCHROEDER R. (1997) 'Networked Worlds: Social Aspects of Multi-User Virtual Reality Technology', *Sociological Research Online,* vol. 2, no. 4. <http://www.socresonline.org.uk/socresonline/2/4/5.html)

TRAINA G. 1983: A*mbiente e paesaggio di Roma antica*, Roma 1983

TURKLE S., *Life on the Screen*. 1995. New York: Simon and Schuster.

VOLPE R., ARNOLDUS HUYZENDVELD A., 2004: *Interpretazione dei dati archeologici nella ricostruzione storica e ambientale del paesaggio suburbano: l'area di Centocelle nel suburbio sudorientale*, in "Roman villas around the urbs. Interaction with landscape and environment", Proceedings of a conference at the Swedish Institute in Rome, September 17-18, 2004

Redefining past landscapes: 30 years of remote sensing in the Vale of Pickering

Dominic Powlesland

The Landscape Research Centre, The Old Bridge Barn, Yedingham , North Yorkshire, YO17 8SL

Introduction

It is now just 30 years since the discovery of an AD 5th to AD 7th century cemetery during aggregate extraction at West Heslerton, North Yorkshire, triggered one of the longest running and most extensive projects in field archaeology in Britain, The Heslerton Parish Project managed by The Landscape Research Centre (LRC). The *ad hoc* discovery of this Early Anglo-Saxon or Anglian cemetery within what, at that time, appeared to be an archaeologically barren landscape, reflected the reactive nature both of the archaeological record and of the archaeological response in Britain during the 1960s and 70's. This cemetery, being progressively destroyed by the machines that converted the past into the building materials of the 20th century was, like so many other 'sites', discovered by accident and documented through a rescue excavation (Powlesland *et al* 1988). This initial excavation undertaken, on a seasonal basis ahead of mineral extraction, between 1978 and 1984 covered nearly 5Ha and included the examination of a range of both burial and domestic features from the Neolithic to Medieval periods. It was clear from the outset that the monumental and domestic evidence did not exist in isolation and that it formed part of a larger landscape which both supported and was changed by human activity from the Mesolithic period at the latest. More importantly the landscape had an important soil history dominated by the movement of blown sands which had in some areas accumulated over time preserving archaeological features in a way rarely found in the British countryside. From the outset it was clear that the apparent lack of archaeological evidence within the landscape setting of the quarry reflected nothing more than a lack of knowledge.

A programme of air photography was begun in 1978 both to document the excavation from the air and identify the context of the excavated areas within the broader landscape. The air-photographic sorties and other remote sensing campaigns that were to follow were designed from the outset to secure information beyond the boundaries of the excavated areas that would enhance our understanding of the excavated features and our ability to interpret the role of the excavated areas within the broader landscape. The research objectives were not constrained by established chronological frameworks nor by any interests in any specific 'period'; that the excavations undertaken in and around West Heslerton have contributed most to our understanding or re-evaluation of the Early to Middle Saxon period is primarily a function of discovery rather than of a deliberate focus, less well known is the considerable contribution made to Late Neolithic and Early Bronze Age studies.

The research agenda developed during the late 1970's and still underpinning the projects developed by the Landscape Research Centre, was driven from an environmentally deterministic viewpoint concerned with the relationship between humanity, the environment, the landscape and its many resources. The ongoing research programme combines many different rescue excavations and independent research projects, the majority funded by English Heritage, within a research framework which relies heavily upon the ability to integrate a wide variety of data from multiple sources at varying scales within a single data management and presentation environment. The distinctive contribution made by the work in and around West Heslerton, situated on the southern side of the Vale of Pickering owes most to the long term support of English Heritage, the scale of excavation and other research undertaken and the exceptional archaeological resource which has been the focus of the work. Incorporating nearly 30Ha of open area excavation, intensive airborne remote sensing programmes covering more than 100 sq km, more than 1000Ha of ground based geophysical survey and more than 400Ha of subsurface deposit modelling undertaken to identify areas with well preserved stratigraphy beneath blown sands; the combined data-set is the most comprehensive of its kind in Britain.

Placing the 'Site' in the Landscape

From 1978 onwards intensive repeat flying, covering an area broadly 10km square, revealed an unanticipated wealth of crop-mark evidence. Air photography or more correctly crop-mark formation is dependent on a large number of variables relating to the types of crops grown, the vagaries of crop rotation, the depth and nature of ploughing and land management being, the nature of the soils and subsoils, the climate throughout the year and possibly over more than one year and, in the case of oblique air photography, the visibility of the crop marks which can be highly dependent on the position and altitude of the plane relative to the crop mark itself. As the air photographic programme continued further very large scale rescue excavations were undertaken to examine a large Anglian settlement associated with the excavated cemetery. This 'site' was not discovered through air photography despite nearly ten years of intensive flying but by deliberate and selective trial trenching. By the early 1990's the returns from the air photographic campaign were diminishing on an annual basis, and it appeared the saturation point had been reached and a large number of 'sites' had been discovered. Continued repeat flying over the same area on a year by year basis was intended to reduce the ad-hoc nature of crop-mark discovery by covering fields under different crops and under different climatic conditions to secure the best chance of seeing crop-marks over the whole of the survey area. Even after 10 years new 'sites' were discovered in areas that had been repeatedly flown many times, it was not initially clear why in some cases these 'sites' remained hidden and then suddenly appeared. In one case, an area that had appeared blank for many years, finally produced magnificent crop marks in 1982, showing extensive areas of settlement enclosures, major landscape boundaries and a trackway linking the wetlands in the centre of the valley to the chalk uplands of the Yorkshire Wolds; the photographs were shown to the landowner and it emerged that the field had been ploughed slightly deeper the year before. Since then the same field has been subjected to further deep ploughing events, the crop marks now appear regularly and it appears that the well defined crop-marks reveal a 'site' that is being actively destroyed by intensive agriculture (Heslerton Parish Project Site 20). The apparently sudden appearance of crop-marks where none had been observed before, despite repeated observation, indicates that crop mark formation had been previously been inhibited either by the presence of a layer of dense blown sand or mineral panning at the base of the plough soil; limited excavation and examination of the plough soil did not indicate the presence of a dense mineral pan and thus we concluded that the ploughing event had cut through the protective cover provided by the dense blown sand layer facilitating root development in the much softer archaeological deposits beneath.

The accumulating body of air photographic data raised key questions about the intellectual frameworks and terminology applied in landscape archaeology, particularly the concept of the 'site'. The data revealed a landscape in which the dot distribution maps, so much loved by archaeologists in the past, and the term 'site' are anachronistic; the palimpsest of features derived from activity from the Neolithic to Post-Medieval periods reflecting funerary, domestic, agricultural and industrial activity extending over hundreds of hectares form the physical and interlinked components of multiple superimposed and evolving landscapes. In the case of the work in Heslerton the term *Site* is still used but it refers simply to a single field defined by the limits when first examined; an often arbitrary definition of space labelled to facilitate the management of any data arising from remote sensing or fieldwork. A key component in the emerging landscape, the 'ladder settlement' so described in 1980 on account of the appearance of the linear arrangement of interlinked rectangular enclosures when viewed from the air, extends for more than 10km. through the LRC research zone, following the edge of the wetlands that once occupied much of the centre of the valley. This type of linear settlement, following a ditched trackway, is also to be found in the Great Wold Valley on the Yorkshire Wolds to the south and also on the northern side of the Vale of Pickering where widely spaced fragments observed from the air are likely to be similarly linked skirting the Vale on its northern side. The term 'ladder settlement' under-represents the complexity of what is clearly a long lived feature of the landscape incorporating multiple overlapping but closely spaced farmsteads dating from the Late Iron Age to the end of the Roman period. Whilst the density of activity along the 'ladder settlement' tended to focus attention on this area the landscape extending to the foot of the Yorkshire Wolds some 2km to the south was covered with other archaeological features from the Neolithic onwards.

Seeing the Invisible

It was clear that in some areas where crop-mark evidence had not been forthcoming it was very likely that the absence of crop marks did not reflect an absence of archaeology. During 1992 an application was made to the Natural Environment Research Council (NERC) to have the area covered by a low level and therefore high resolution multi-spectral (MSS) imaging survey using a Deadelus scanner. It was argued that this approach, which would secure data from the visible, infra-red and thermal wavelengths, may help us to identify new features which were less visible to conventional air photography. A data grant was awarded and the area was flown in June 1992. In addition to gathering multi-spectral data the NERC flight included the acquisition of high resolution vertical colour photography taken using a Wild RC10 camera. The RC10 colour negatives from the 1992 flight have recently been scanned generating digital images with a ground resolution of c.12.5cm.

From an air photographers point of view 1992 was a splendid year for crop mark formation and this is reflected in both the multi-spectral and air photographic returns. An additional benefit of both recording methods by comparison with the oblique photography programme is the generation of contiguous overlapping coverage. The NERC flight added a wealth of new evidence to our landscape plot and revealed a characteristic of crop mark formation which it took more than a decade to fully appreciate; in many cases the NERC flight added detail to areas with known crop marks rather than revealing entirely new feature complexes. Danny Donoghue had demonstrated the potential for archaeological returns using the Deadelus multi-spectral scanner at Morton Fen, Lincolnshire, through a NERC supported survey in 1986 and the 1992 survey provided the opportunity to test this technology over a predominantly cropped rather than soil environment as in the former case. At the simplest level the potential of remote sensing imagery gathered from space or from the air to assist in the identification or mapping of archaeological features is determined by a combination of ground conditions including feature scale and contrast, atmospheric conditions and image resolution. The relationship between resolution and visibility is not at all simple since a small highly radiant feature can often be clearly seen in a relatively low resolution image; the white lines for instance on the principal road through the project area can be seen in the Daedelus data despite the fact that the pixel resolution of the multi-spectral scanner, flown at an altitude of 800 metres, is only 1.5 metres. Although the ground resolution of the NERC multi-spectral data is too low to map small features unless they exhibit a very high contrast it gathers data from the visible, infra-red and thermal parts of the spectrum; the imaging from the infra-red and thermal wavelengths returns additional information not seen in the visible wavelengths either in the multi-spectral or high resolution RC10 images. When using low resolution satellite data such as Landsat or Spot multi-spectral imagery feature extraction through trained or automated classification can be highly successful; crops for instance can easily be identified using multi-spectral data with 30m ground resolution. High resolution multi-spectral imagery reveals subtleties and variations within even a single field that mean that such automated feature extraction is impossible; even classifying the crop present in a single field invariably fails as there is too much variation in the spectral signature within the field. Archaeological feature identification must be undertaken manually, prior to full rectification of the multi-spectral imagery; once the imagery is geo-rectified it needs to be re-examined in detail to ensure that the features identified are neither artefacts of distortion or reduced by the rectification process. Since the 1992 NERC flight changes to the sensor and on board processing have resulted in better quality data which has greatly reduced geometric distortion within the MSS data.

A second NERC flight over the Heslerton research area was undertaken in 2005, designed to compare the returns across two different sand and gravel environments, it generated far better data with minimal geometric distortion, however the data was collected at a time when few visible crop marks were showing. The 2005 programme incorporated two flights, one in the middle of the day and a second during the night designed to secure data that would allow us to test the potential for identifying features through thermal decay. As in the 1992 flight the day flight was coupled with high resolution vertical photography, in this case using a Rollei large format digital camera producing images with a ground resolution of c.15cm per pixel. Once again the Infra-red and thermal wavelengths recorded features that could not be seen in the visible spectrum data, the thermal decay experiment did reveal features but only in a small number of cases. Although in both NERC campaigns the MSS data produced data which enhanced the data set gathered over many years of oblique air photography the limited resolution restricted the returns to large features; where bare earth was exposed rather than crops the thermal data revealed features which were primarily visible on account of enhanced wetness. Quite clearly the potential for returns from multi-spectral and hyper-spectral instruments will improve as new instruments are produced with improved ground resolution; the benefits of 1m resolution satellite imagery gathered using the IKONOS platform are widely appreciated and used in archaeology as for instance at Merv, Turkmenistan, and the returns radically increase when sub-metre data is available as for example with Quickbird imagery.

The most striking discovery of the 2005 NERC campaign was the realisation that features visible in the vertical photography were either not visible or very poorly visible through oblique air photography undertaken a few days before and after the NERC flight. A programme of flying was undertaken during 2006, another year with splendid early positive crop mark returns, to examine the difference between the returns from vertical and oblique photography; in addition to a number of flights combining near vertical and oblique photography, over known crop marks, site visits were made on the ground to investigate the crop marks at ground level. It became very clear from this work that for positive crop marks in particular (where the crop mark in cereal crops ripens more slowly than the crop around it giving a dark mark on a lighter background) the returns from the vertical photography were far greater than from the oblique. An examination of the crop marks on the ground revealed that most of the difference in the crop was present in the lower canopy which whilst visible from the vertical or near vertical angle could not be seen from an oblique viewpoint. Further work and practical demonstration is needed to compare and contrast the results of high resolution vertical photography as opposed to conventional oblique photography taken from a light aircraft; if as we now believe the returns from the vertical photography are indeed greater this has a bearing on future air photographic research strategies. The LRC is currently developing a research programme employing a radio-controlled drone with on-board digital camera and GPS to undertake regular repeat photography over known crop marks as well as for undertaking wider speculative survey. A quick perusal using Google Earth of those areas with high resolution vertical photography such as Puglia in Italy or Norfolk or Cambridgeshire in England reveal large numbers of crop marks and soil marks captured incidentally rather than through an archaeologically defined photographic campaign.

Investigating the surface

Lidar data was also gathered as part of the 2005 NERC campaign covering the northern half of the research area; Lidar data incorporating both elevation and intensity values facilitates the creation of high resolution digital surface and terrain models. Single pass Lidar data such as that collected over Heslerton with a density a little less than one point per metre has provided the necessary data to generate a reasonably high resolution surface model, which is particularly useful for mapping the many relict stream channels in the centre of the valley. The intensity data adds to the 3D model and seems particularly useful for identifying features with enhanced wetness. By modelling the Lidar data in 3D in real-time and manipulating the angle of view, light intensity and direction small topographic features, such as ploughed out barrows and other earthworks can be easily detected. In the almost treeless landscape of the Vale of Pickering the need to map features hidden beneath trees is minimal, multi-pass Lidar surveys with densities as great as one point per quarter metre can, when flown in winter when leaf cover is minimal, allow the mapping of earthworks beneath woodland canopy which may not even be easily accessible on foot.

Enhancing the picture

Although the various airborne campaigns over the Heslerton landscape have revealed hundreds of hectares of fields containing archaeological features the sensitivity of airborne techniques to the vagaries of climate and agricultural process and the continued invisibility of features in fields of permanent pasture or beneath deep blown sands meant that the emerging picture of the landscape still had gaps where logic suggested there should be none. During the early 1980's attempts to map buried features using ground based remote sensing employing magnetic and resistance methods failed to produce convincing results; however, experiments undertaken in association with the ongoing excavations in West Heslerton in the late 1980's using improved instruments produced magnificent results, particularly using a fluxgate gradiometer. During the last six years a series of different projects funded by English Heritage and the Aggregates Levy Sustainability Fund have radically enhanced the picture gained through the various airborne campaigns. More than 1000Ha of contiguous gradiometer survey extending for 9.5km along the margin of the wetland has revealed the most detailed picture of an archaeological landscape in Britain. Unlike airborne survey techniques which are reactive or resistance surveys which are similarly sensitive to ground conditions magnetic surveys by magnetometer or gradiometer are hardly effected by variation in climatic conditions; as such their use can be considered pro-active. If the environment proves suitable for magnetic survey the results will be consistent regardless of climatic or ground conditions, with the exception of subtle distortions arising from recent ploughing. Magnetic surveys which rely on the contrast in the magnetic susceptibility of the buried archaeological features do not reveal everything, in the Heslerton landscape for instance, Neolithic monuments do not show particularly well, the magnetic contrast fades over time such that large features such as round barrows, in which the ditches filled with materials with a low magnetic susceptibility can often be seen clearly as crop marks but do not show as magnetic anomalies. The magnetic data collected at a pixel resolution of .25x1m greatly enhance the picture gained through the airborne surveys; higher resolution .25x.25m surveys over some areas further enhance the picture. High resolution surveys undertaken after removal of the plough-soil as part of the excavation campaign revealed small post-holes and other details that were not even visible on the ground without careful cleaning.

The results of the combined magnetic surveys are staggering, not only do they enhance and add to the picture developed over many airborne campaigns but also they allow us to identify the sheer quantity of evidence that lies beneath the featureless flat fields on the southern margin of the Vale of Pickering. The extent of the magnetic coverage, over two different sand and gravel geologies, and the magnificent response of features, particularly from the Iron Age onwards finally allow us to address the research questions defined in the early 1980's and context of the rescue excavations conducted between 1977 and 1996. The thousands of new features identified through the magnetic surveys and the relationships between them and with the landscape as a whole demonstrate how inappropriate the term 'site' is when we are trying to view landscapes.

Conclusions

The space is too limited here to elaborate on the details of the 30 year campaign of excavation and remote sensing in the Heslerton area. It is abundantly clear that the accumulative effect of many years of multi-sensor remote sensing gives a greater return than can be gained over ad hoc coverage. The results generated both by the very large excavations and even more extensive remote sensing in Heslerton reflect most clearly the benefits of long term funding, even if this was secured across multiple short term projects and the dedication of a small team to the overall research objectives of the LRC. High resolution vertical photography gives better results during the positive crop-mark phase than oblique as it 'sees' deeper into the crop to vegetation differences which, in the case of cereal crops, are mostly at the base of the crop. Even in vertical photography there is a fall off in the crop mark clarity towards the edges of the frame, suggesting that area surveys should be undertaken with a bigger frame overlap than might be required for normal geo-referencing or stereo viewing. MSS imaging can enhance the returns from visible wavelength photography, principally where data is gathered from the infra-red and thermal wavelengths. The real potential of MSS and hyper-spectral imaging will increase as pixel resolutions improve; already we are seeing remarkable results from the QuickBird satellite. When conditions are suitable high resolution geophysical survey, particularly magnetic methods can, if undertaken on a landscape scale, reap tremendous benefits facilitating quantification and analysis of archaeological landscapes with a clarity not possible with other techniques and independent of the environmental factors that determine the occurrence of crop marks. The considerable costs of undertaking very large scale geophysical surveys on foot can easily be justified in the case of Heslerton; new equipment is being developed by a number of institutions that will radically increase the speed of survey and thus the cost per hectare by employing multi-instrument carts of one sort or another pulled by a vehicle.

Simply mapping the evidence is just the first stage in identifying the resource; the challenge now is to develop and apply a suitable research and management strategy that will enable the spectacular discoveries to be understood, protected, interpreted and presented to the wider public. Clearly 30 years is not enough, sadly additional work undertaken to assess the condition of the buried landscapes has shown that action must be taken soon if the exceptionally preserved buried archaeological resource is not to become fundamentally compromised by the effects of modern agriculture.

The excavations undertaken in Heslerton have relied on the dedicated efforts of hundreds of individuals, mostly volunteers; the geophysical surveys have relied almost entirely on the dedicated attention of James Lyall, the LRC geophysics team leader, who deserves particular thanks for his contribution. None of this work would have been possible without the continued support of English Heritage; the airborne surveys could not have been achieved without the help of a number of local fliers and NERC who provided the MSS, Lidar and vertical photographic data sets.whilst Danny Donoghue helped us to get to grips with the MSS data. Last but not least the landowners and farmers of the Vale of Pickering deserve particular thanks; they have given free access to their fields for survey, environmental sampling and excavation, frequently at times that suited us rather than them.

Further Reading

DONOGHUE, D. and SHENNAN, I. "The application of remote sensing to environmental archaeology." *Geoarchaeology* 3.4 (1988): 275-285.

DONOGHUE, D.N.M. "Calibration and Processing of Morton Fen ATM Data." *NERC 1986 airborne campaign workshop Swindon*: Natural Environment Research Council, 1988. 175-186.

DONOGHUE, D.N.M. *"Calibration and Correction of View Angle Effects in Visible / Near Infrared and Thermal Infrared Data."* NERC 1986 airborne campaign workshop Swindon: Natural Environment Research Council, 1988.

HAUGHTON, C.A and POWLESLAND, D.J. 1999. *West Heslerton - The Anglian Cemetery*, Landscape Research Centre Monograph 1,2 vols. Yedingham

LYALL, J. and POWLESLAND, D.J. 1996. 'The application of high resolution fluxgate gradiometry as an aid to excavation planning and strategy formulation'. *Internet Archaeology* 1 (http://intarch.ac.uk/journal/issue1/index.html)

POWLESLAND, D.J. 1988. Approaches to the excavation and interpretation of the Romano-British landscape in the Vale of Pickering, in Price, J. and Wilson, P.R. (eds), *Recent Research in Roman Yorkshire: studies in honour of Mary Kitson Clarke*, 139-151, Oxford

POWLESLAND, D.J. 2001. 'The Heslerton Parish Project :An integrated multi-sensor approach to the archaeological study of Eastern Yorkshire, England' *Remote Sensing in Archaeology* Forte & Campana (eds), University of Siena, Firenze 2001, 233-235

POWLESLAND, D.J. 2003a. 'The Heslerton Parish Project: 20 years of archaeological research in the Vale of Pickering', in *The Archaeology of Yorkshire An assessment at the beginning of the 21st century* T G Manby, S Moorhouse & P Ottaway (eds), 275-292, Yorkshire Archaeological Society, Leeds 2003

POWLESLAND, D.J. 2003b. *25 years research on the sands and gravels of the Vale of Pickering.* The Landscape Research Centre, Yedingham

POWLESLAND, D.J., HAUGHTON, C.A. and HANSON, J.H. 1986. 'Excavations at Heslerton, North Yorkshire 1978-82', *Archaeol. J.* 143, 53-173

POWLESLAND, D.J. LYALL, J. and DONOGHUE, D. 1997. 'Enhancing the record through remote sensing: the application and integration of multi-sensor, non-invasive remote sensing techniques for the enhancement of the Sites and Monuments Record. Heslerton Parish Project, N. Yorkshire, England' *Internet Archaeology* 2 (http://intarch.ac.uk/journal/issue2/pld/index.html)

Towards integration: two prospection methods and some thoughts

Włodzimierz Rączkowski

Institute of Prehistory, Adam Mickiewicz University, Poznań

1 Introduction

The empiricism which is deeply set within archaeologists' traditional thinking plays a major role in decision making when selecting procedures or survey methods. Knowledge of archaeological sites and their location, chronology and function is the most basic and obvious requirement enabling any studies of the past. In the culture-historical approach which advocates an inductionist model of explanation, there is a widespread belief that the more archaeological sites known the fuller our knowledge about the past becomes. However, processual archaeology has developed a number of methods to test scientific hypothesis. Consequently, archaeologists have at their disposal a number of survey methods to discover, record and test more and more archaeological sites (e.g. Renfrew & Bahn 1991). Such an approach leads to the belief that the integration of prospection methods will provide fuller information on a site.

How are these methods actually applied? Archaeologists rate the achieved results in categories of their concordance with the expectations formulated earlier as standard. These expectations usually result from knowledge already held. The use of more than one method brings about the 'threat' of discrepancies in the results achieved. This is a very common phenomenon and a problem which archaeologists have to face. So what then are the accepted solutions? Usually, the results of one method are accepted (the one closest to the expected outcome) and the results of the other method are rejected. In such an approach it is difficult to talk about an integrated survey, regardless of the use of a variety of methods.

In the case of the use of aerial photography in Polish archaeology the situation is a little different (though not completely). The basic problem is the inclusion of aerial photography in archaeology as a competitive method in relation to field-walking survey. The interpretation of the results achieved in categories of concurrence with expectations is only one of the arguments in this 'battle' of methods.

How is the image of archaeological practice from the perspective of the application of both methods in Polish archaeology created? What are the factors of misunderstanding and their consequences for the development of archaeology? Will the integration of the field-prospection methods deepen the divisions or will it add to a more comprehensive analysis of past phenomena? Do the discrepancies in the results 'weaken' the scientific aspect of archaeology? Whilst analysing examples of the application of both methods: field-walking survey and aerial survey, I would also like to consider the usefulness of integrated surveys.

2 Practice of field-walking survey and aerial survey in Poland

For many years field-walking survey was the only fully accepted prospection method in Poland. This is reflected in the Polish Archaeological Record (Archeologiczne Zdjęcie Polski – AZP) program which was begun in 1978 and is still in progress. The AZP program has led to a huge increase in information about archaeological sites, and thousands of sites have been newly discovered. This program has also been of immense value to the heritage protection services. The evaluation of how representative the data recorded is forms one of the major problems which has emerged from the realisation of the AZP program. Even the negative results (those which are barely representative) from some experiments have not, however, brought about any revision of the commonly held opinion that field-walking alone is the best method for site discovery and recording which archaeologists have at their disposal (Rączkowski 2005a).

On the other hand, aerial survey has been regarded with suspicion and has played no part in the gathering of information on a site. The history of the introduction of aerial photographs shows a sequence of waves marking decreasing and increasing interest within archaeology (e.g. Kobyliński 2005; Rączkowski 2005b). Over the last several years efforts have been made to introduce aerial survey as standard in archaeological investigations and are now slowly beginning to bring results.

The very good aerial survey results over the last several years (Nowakowski et al 2005) have been confronted with the results from field-walking so far. This is not surprising because a "newly" introduced method is usually tested at well investigated sites or regions (e.g. application of Lidar in archaeology – Holden et al 2002). So the current process of increasing interest in aerial photography in Polish archaeology is perceived by many as questioning the results of field-walking.

Nevertheless, when we compare field-walking survey and aerial survey results three possible situations are observed:

1) sites recorded during field-walking survey and not recorded on aerial photographs,
2) sites recorded by both methods,
3) sites recorded during aerial survey and not recorded during AZP.

From the very start the problem of how to explain any discrepancies between the two methods arose. When the results concurred all were satisfied, but this additionally underlined the particular cognitive benefits of aerial survey. Due to the need to create a positive image of aerial survey an interpretive scheme was proposed explaining the discrepancies which were found. Within this scheme the first situation indicated sites seriously damaged by ploughing; the second revealed sites only partially damaged and the third one presented sites which were well preserved, had not suffered damage from ploughing (and so lacked shards of pottery on the surface). This interpretive scheme was based on existing opinions about the factors revealing archaeological sites (Mazurowski 1980; Nowakowski et al 1999). It highlights the important advantage aerial survey has in discovering well preserved sites, and therefore of great value to archaeologists. So the proposed scheme is an essential element in the promotion of the potential of aerial survey and, consequently, aerial archaeology.

3 Hermeneutics and cognitive schemes: overcoming simplistic interpretations?

Over the last few years, as a consequence of the persuasive results of aerial reconnaissance, a number of projects have appeared, within which the relation between field-walking survey and aerial survey are arranged differently. Apart from the earlier practice of taking aerial photographs of an area where a field-walking survey has already been made, aerial reconnaissance has been carried out before the AZP survey and/or sites already recorded on aerial photographs have been surveyed again. Also there is a growing number of attempts in which aerial survey is undertaken in the evaluation of archaeological landscapes threatened by new investment projects (highways, pipelines etc.). The results, though they generally reveal the three situations mentioned earlier, pose new questions. An example of which may be concurrence in aspects of the identification/location of a site, but differences in determining its chronology, i.e. where the chronology of the material collected from the surface indicates the presence of a modern site, but the Neolithic features are visible in the photograph. The questions are difficult to answer, as the scheme proposed earlier accentuating how well the site is preserved 'imposes'.

The question arises: why are various relations between the results of both methods always interpreted in a similar way? It seems to me that this problem should be looked at from the perspective of hermeneutics. Hermeneutics, a study of interpretation, points to three components that influence the interpretation process. These are: the hermeneutic circle, pre-understanding and the historical nature of knowledge (e.g. Hodder 1999). I would like to focus on pre-understanding which is a vast concept covering various judgements and directives on ways in which questions and definitions of terms are formulated, the criteria by which significant facts, goals and expected answers are identified, the choice of tools and methods, and on all other aspects of our research. I feel that modern cognitive psychology will help us understand the complexity of knowledge-forming processes – an essential part of pre-understanding.

Cognitive psychology views understanding as a process of integrating newly acquired data with information generated by our minds (Trzebiński 2002). Interpretation of common objects is not controlled, meaning we are not aware of it. Interpretation of complex structures or processes, however, is a more complex process which involves the knowledge we already have and our informed search for additional information. In this process the researcher's knowledge conditions the researcher's focus and thus the level of attention paid to elements of the newly acquired data and how the data is integrated. As a result, an interpretation guided by a researcher's particular framework of knowledge is judged by its author to be "objective". In the case of culture-historical archaeology, a subjectively construed interpretation of different results of two methods offered by "field-walkers" and "aerial archaeologists" is treated by its author as objective knowledge of an aspect of the past.

The interpretation process is neither chaotic nor accidental. The rules applied to interpreting "the" reality are part of an individual's general knowledge and formed during the learning process. Thus they are largely culture-guided rules that make us interpret our environment in similar ways. However, individual experience also shapes our perception and thus some rules of interpretation are characteristic of a particular individual only.

In addition to culture-guided rules, psychologists also recognize the role cognitive schemes play in the interpretation process. A cognitive scheme is a model of a part of reality and it has two inseparable and integrated functions. It is a description or representation of reality and the cognitive scheme is also a set of procedures for processing information about this reality. Of special relevance here is the second function, as the procedures guide our recognition of objects and the way we think about them. When our cognitive scheme is activated with incoming preliminary data, it autonomously, i.e. without us being aware of it, starts guiding our attention, perception and reasoning. Thus activation of cognitive scheme makes our mind:

1) search for pre-defined information,
2) evaluate the incoming information,
3) interpret the information in a particular way (conditioned by one's preliminary hypothesis),
4) supplement the incoming information (with one's own information generated by one's expectations),
5) distort the incoming information to make it match one's expectations.

It follows that one object/process, e.g., prospection results, can be perceived and comprehended in different ways depending on the architecture of one's cognitive scheme guiding data interpretation. The cognitive scheme filters our search and perception making our mind notice pre-selected information, supplement it, as well as combine and categorize the information gathered.

Among cognitive schemes, the narrative scheme is a framework within which we interpret or understand reality. It is our mind which makes reality fit this framework or, in other words, it makes reality fit a story about the world. Like the cognitive scheme, the narrative scheme makes the interpreter expect some information is available and it offers suitable interpretation rules. The narrative scheme also supplies information that 'supplements' the data apparently missing to make the story complete.

Uncertainty tolerance is not unlimited and has threshold values. When the cognitive dissonance between the incoming information and the interpreter's expectations is too large, the narrative scheme applied can be abandoned and another narrative scheme activated instead. In this way the past may be reinterpreted.

The category of narrative scheme (or cognitive scheme) helps explain the unified (within a group of archaeologists) interpretation of mechanisms why the results of two methods differ. The existing interpretive scheme was created based on the knowledge (experience) to-date. The influx of new data, not always in agreement with the current narrative scheme, may lead to changes in it. This will be connected with deeper deliberation on the processes which occur around archaeological sites (formation processes) and the specifics of survey methods. As a result, individualised consideration of the phenomena occurring in the particular conditions of each site may prove necessary.

4 Conclusions: integration as an opening for new interpretations

Two interpretive schemes which have been accepted by archaeologists form a starting point for the explanation of the discrepancies in the results of both methods. However, the latest prospection results where both methods have been used are beginning to indicate the fact that such interpretive schemes are too simplistic. It seems as though the new data has crossed the tolerance limits and the need for the formulation of a new interpretive scheme has arisen. But what does this mean in practice? In my opinion this points to the need for the development of new research into postdepositional processes (see Schifer 1976; 1987). On one hand this should be more in-depth research on the process of how archaeological material appears on the surface, and on the other on the factors affecting the formation of cropmarks. Current knowledge about this last point is based on the research made by A. M. Martin (1971), R. Evans & R. Jones (1977), D. Riley (1979), P. Webster (1978) and D. Wilson (1982) over twenty years ago.

In the process of integrating archaeological site survey methods we should treat them as complementary to each other, rather than competitive. The discrepancies then become an important pretext for further thought on the reasons behind the differences, on the "condition" of the sites, how well they are preserved, the processes of stratification and so on. The integration of methods can and should influence further thought on the essence of formation processes in order to gain a deeper understanding of the features occurring. Different methods allow the disclosure of various types of information about a site. The very foundations of the methods themselves should also be looked at.

Consequently, the integration of methods allows not only for the enrichment of empirical knowledge on a site/sites, but it also encourages critical reflection on the ways of attaining and interpreting data.

References

EVANS, R. & R.J.A. JONES 1977 Crop Marks and Soils at Two Archaeological Sites in Britain. *Journal of Archaeological Science*. 4: 63-76.

HODDER, I. 1999 *The Archaeological Process. An Introduction*. Blackwell, Oxford.

HOLDEN N., P. HORNE, & R. BEWLEY 2002 High-resolution digital airborne mapping and archaeology in *Aerial Archaeology. Developing Future Practice* (R. Bewley & W. Rączkowski, eds.). IOS Press, Amsterdam, 173-180.

KOBYLIŃSKI, Z. 2005 *Archeologia lotnicza w Polsce. Osiem dekad wzlotów i upadków*. Państwowe Muzeum Archeologiczne, Instytut Archeologii i Etnologii PAN, Warszawa.

MARTIN, A.M. 1971 Archaeological Sites – Soils and Climate. *Photogrammetric Engineering and Remote Sensing*. 37: 353-357.

MAZUROWSKI, R. 1980 Podstawy wnioskowania o stopniu zniszczenia obiektów archeologicznych w oparciu o analizę materiału ceramicznego z powierzchni stanowisk. *Przegląd Archeologiczny*. 27: 145-185.

NOWAKOWSKI, J., A. PRINKE, & W. RĄCZKOWSKI 1999 Latać czy nie latać?: zdjęcia lotnicze jako kolejny element standardowej procedury w ochronie stanowisk archeologicznych in *Acta Archaeologica Pomoranica*, T. II: *Konserwatorskie badania archeologiczne w Polsce i w Niemczech – stan prawny, problematyka, osiągnięcia* (M. Dworaczyk, K. Kowalski, A. Porzeziński, S. Słowiński & E. Wilgocki, eds.), Stowarzyszenie Naukowe Archeologów Polskich, Szczecin, 113-152.

NOWAKOWSKI, J., A. PRINKE, & W. RĄCZKOWSKI (eds.) 2005 *Biskupin... i co dalej? Zdjęcia lotnicze w polskiej archeologii*. IP UAM, OODA, MAB, PTP, Poznań.

RĄCZKOWSKI, W. 2005a Tradition in power: vicious circle(s) of aerial survey in Poland in *From the Air. Understanding Aerial Archaeology* (K. Brophy & D. Cowley, eds.). Tempus, Stroud, 151-167.

RĄCZKOWSKI, W. 2005b To Overcome Infirmity. Current Approaches to Aerial Archaeology in Poland in *Aerial Photography and Archaeology 2003. A Century of Information* (J. Bourgeois & M. Meganck eds.). Academia Press, Ghent, 121-135.

RENFREW, C. & P. BAHN 1991 *Archaeology. Theories, Methods and Practice*. Thames and Hudson, London.

RILEY, D.N. 1979 Factors in the Development of Crop Marks. *Aerial Archaeology*. 4: 28-32.

SCHIFFER, M.B. 1976 *Behavioral archaeology*. Academic Press, New York.
SCHIFFER, M.B. 1987 *Formation processes of the archaeological record*. University of New Mexico Press, Albuquerque.
TRZEBIŃSKI, J. (ed.) 2002 *Narracja jako sposób rozumienia świata*. Gdańskie Wydawnictwo Psychologiczne, Gdańsk.
WEBSTER, P. 1978 Cardiff Castle: The Excavation of a Crop Mark Site. *Aerial Archaeology*. 2: 56-57.
WILSON, D.R. 1982 *Air Photo Interpretation for Archaeologists*. Batsford, London.

Tidgrove Warren Farm Archaeological Project: An integrated Approach to the Study of an Archaeological Landscape in Hampshire, UK

Kristian D. Strutt

Archaeological Prospection Services of Southampton, Department of Archaeology, University of Southampton, UK

Introduction

Between November 2003 and 2006, an archaeological project has been under way investigating the pattern of settlement over an area of downland at Tidgrove Warren Farm in Hampshire. The project is an ongoing collaboration between the Department of Archaeology at the University of Southampton, and the Kingsclere Heritage Association, that aims to assess the evidence for past settlement and land use from later prehistory onwards. Current research has focused on a small area of landscape centred on Tidgrove Warren Farm. The farm is located 5 miles (7.5km) to the west of Basingstoke, between the villages of Overton and Kingsclere (Fig. 1). The area is situated on chalk downland (Institute of Geological Sciences 1979), and shows a large amount of evidence for human occupation from the Neolithic onwards. Bronze Age barrows are located on the high ground immediately to the north of Tidgrove Warren Farm. Remnants of Iron Age settlement and land enclosure are demonstrated by traces of extensive Celtic field systems, which divide the downland into terraces and fields. These features are visible as slight earthworks and soil marks, distinguished by the brown loam and white chalk of the local geology, and are particularly prevalent around the Kingsclere area. An advanced network of settlement and trade has been suggested for the area in the Roman period (Cusworth and Dobbs 1999, 31), which included a road system running from the south coast towards Silchester (Fulford and Timby 2000). In the locality of Tidgrove Warren Farm, a continuity of settlement and farming is suggested by the existence of ceramic remains, dating from 1st century BC to 4th century AD. In the medieval period much of the downland was converted either to sheep grazing, or warrening rabbits, or for hunting. This latter use appears to be the case with the land around Tidgrove Warren Farm. Although land to the south at Overton was used by the Bishop of Winchester for sheep grazing (ibid., 44), King Henry II is recorded as having built a hunting lodge and stables at Freemantle, near the line of the Roman Portway road, which was passed down until the 17th century (Cusworth and Dobbs 1999). A number of 17th and 18th century maps also attest to the widespread existence of parkland. Spede's map of 1610 shows ten such parks in the north of Hampshire, and Kitchin's map of 1750 delineates a deer pale, which encircles the area around Tidgrove Warren Farm (see map reproduced in Money 1575), and marks it as *Tremanton Park*, possibly a corruption of *Freemantle*.

Current interest in the archaeology of Tidgrove Warren was initiated by members of the Kingsclere Heritage Association, and the landowner and resident of Tidrove Warren Farm, Mr Raleigh Place. The landowner has found a number of concentrations of ceramic material,

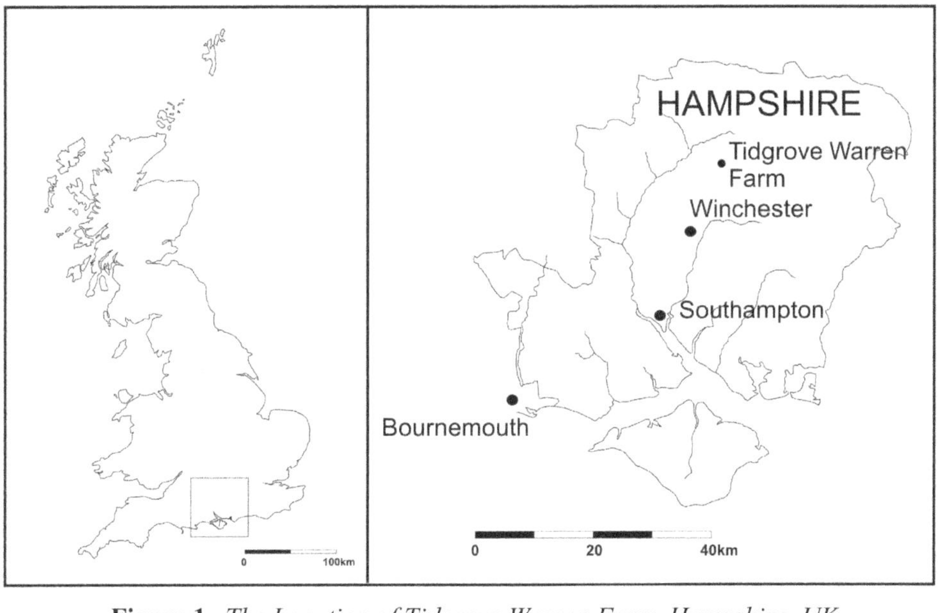

Figure 1 *The Location of Tidgrove Warren Farm, Hampshire, UK*

including 12th-13th century pottery, and 1st century BC to 4th century AD ceramics to the west of the present farm, within the bounds of the Iron Age field system. These concentrations also correspond with soil marks in the harrowed terrain, and slight earthworks over parts of the property. The location of medieval pottery prompted the Kingsclere Heritage Association to research documentary evidence for the area, and the landowner also produced two aerial photographs of the two potential sites. It was on the basis of this evidence that the Kingsclere Heritage Association decided to commission a geophysical survey in 2003 to identify and map the extent of the medieval lodge site, and locate evidence for a Romano-British farmstead on the ridge to the west of the modern farm. This survey successfully located the remains of the settlement (Strutt 2003), and was continued in 2004 and 2005 (Strutt 2004; 2005), with seasons of excavation investigating the sites from 2004 onwards.

The Integrated Survey Methodology

The archaeological survey has involved an integrated strategy of desk-based and field-based methods, drawing together air photographic and SMR data, and the results of geophysical and geochemical surveys, fieldwalking, and excavation. The methodology has been developed to provide high resolution survey data of individual settlements and their immediate surroundings, and to allow the different sites to be placed in their landscape context. Prior to the commencement of the archaeological survey, a basic topographic survey was conducted which included the establishing of a series of control survey stations across the area to ensure that all features and surveys could be geo-referenced to the same arbitrary coordinate system (Bannister et al. 1998; Bowden 1999). The stations were laid out using a Leica total station, with the location of the principal stations being tied into the Ordnance Survey grid using a Leica differential GPS. The grids that were then established for the fieldwork were all geo-referenced to the overall coordinate system. As part of the research, the air photographic archives held by the National Monuments Record offices in Swindon were interrogated. A series of vertical and oblique air photographs were located in the archive, covering the major portion of the study area. The photographs were then scanned and rectified to the base Ordnance Survey map using ArcGIS, and features of archaeological importance were digitised onto the topographic background (Strutt et al. 2004, 7). This has provided an overall plan of the extent and nature of the archaeological features in the landscape, most critically the layout of later prehistoric field systems relating to the Iron Age and Roman-British settlement evidence from the fieldwork that has been conducted in the area.

The fieldwork component of the project utilises a number of non-intrusive survey methods together with excavation. The initial survey of the main settlements in the area was based on a geophysical survey, comprising magnetometry and resistivity (see Clark 1996). Survey was concentrated over the potential area of the Romano-British settlement to the west of the farm, the medieval enclosure to the east, and a possible round barrow situated on the highest point of the farmland to the north (Fig. 2). In all cases the geophysical survey was conducted on 30m by 30m grids laid out across the survey area using a total station. The grids were all orientated at a tangent of around 30° to the perceived alignment of the archaeological features to ensure that no definition of archaeological anomalies would be lost in the survey results. Magnetometer survey was conducted using Geoscan Research FM36 fluxgate gradiometers (see Scollar et al. 1990, 456 for a detailed description of such a system). Readings were taken on traverses spaced at 1m intervals, with readings taken every 0.5m. Measurements were taken at a resolution of 0.1 nT, with the aim of locating and mapping evidence for the ditched enclosures of each settlement, and pit and gully features associated with activity within and outside of the enclosures. Resistivity survey was conducted using Geoscan Research RM15 resistance meters, using the same grids as for the magnetometer surveys, with measurements taken to 0.1 Ω resolution, on 1m traverses at 1m intervals. A twin probe resistance array was used with a 0.5m spacing between the probes (Clark 1996; Gaffney et al. 2002, 2), as the most efficient mode of surveying the different areas.

Figure 2 *Map of the immediate environs of Tidgrove Warren Farm, showing the chalk downland ridges extending southwards towards Overton and the river Test. The Portway Roman road can be seen running to the north. The Bronze Age barrow is also marked, with the Romano-British settlement located on the ridge to the west of the farm, and the medieval enclosure the north-east (©Crown Copyright/database right 2006. An Ordnance Survey/EDINA supplied service)*

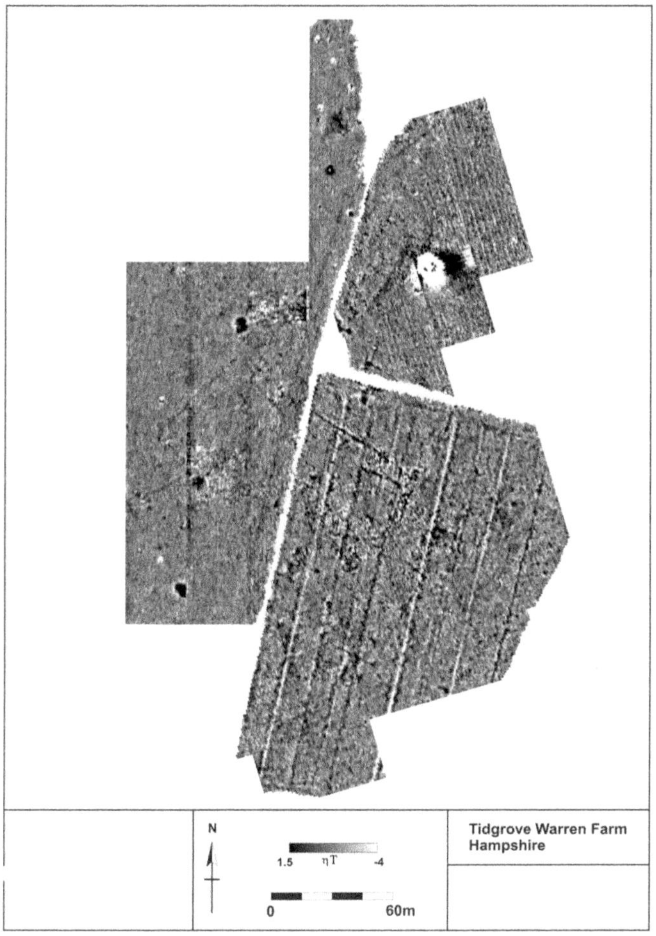

Figure 3 *Greyscale image of the magnetometer survey results of the Romano-British settlement*

Figure 4 *Photograph showing the interior of the Romano-British settlement from the west, under excavation in 2005. Some of the internal storage pits, and a later gully or ditch, are visible in the foreground, with the inner enclosure ditch visible in the background.*

In addition to the geophysical surveys at the site, a geochemical survey has been conducted, together with systematic fieldwalking over the area of the Romano-British settlement. For the geochemistry, soil samples were taken, across the same grid as that utilised for the geophysics, at 10m intervals using a hand auger. An area of approximately 3.5 hectares has been covered in this way. Samples have been processed by Phil Clogg at the University of Durham using an ED-XRF instrument to produce an analysis of the different trace elements in the soil.

An initial phase of systematic fieldwalking has been conducted over the Romano-British site. The work was conducted on the same grid as the geophysical survey, allowing the results of the surface collection to be correlated with the geophysics and geochemical results. Each survey grid of 30m by 30m was walked at 3m intervals by a team of students from the University of Southampton. Finds were bagged by each fieldwalker in a series of three 10m traverses in each grid, allowing individual finds to be placed within a 3m by 10m area within each grid.

Excavation has also been conducted at the Romano-British and medieval sites. The purpose of these investigations is to ascertain the date and nature of different features more comprehensively, and to relate these results to the non-intrusive survey results, allowing a more complete picture of the settlements to be produced. Trenches were targeted over significant features from the geophysical survey results, taking into account the nature of surface material across the area. High resolution geophysical survey has been conducted over each trench prior to full excavation, to enable the identification of features not necessarily visible to the human eye. All features within the trench were then excavated.

The Romano-British Settlement

Results of the integrated survey demonstrate the full plan of the Romano-British settlement on the ridge to the west of the modern farm. The integration of magnetometry and resistivity has revealed the presence of the settlement enclosure (Fig. 3) with a series of pits, features and a second enclosure ditch within the main enclosure. There appear to be at least two phases of feature at the site, consisting of the initial enclosure and settlement, with later rectangular ditches and structures. A series of large maculaic anomalies are also visible to the west of the main settlement, suggesting possible corn-driers or kilns. The geophysical results give an immediate idea of the extent of the settlement, but the data from fieldwalking and excavation provide some more specific evidence in relation to the anomalies in the survey. The fieldwalking defined a concentration of Roman brick and tile across the ridge located over the northern area of the enclosure close to the modern day field boundary. A large quantity of Alice Holt pottery from the surface also indicates that the settlement was continuously occupied into the 3rd and 4th centuries AD (Millett 1979). The excavation, focusing on the enclosure ditch of the settlement together with the inner enclosure and related features (Fig. 4), and the rectilinear ditch located to the south of the main enclosure

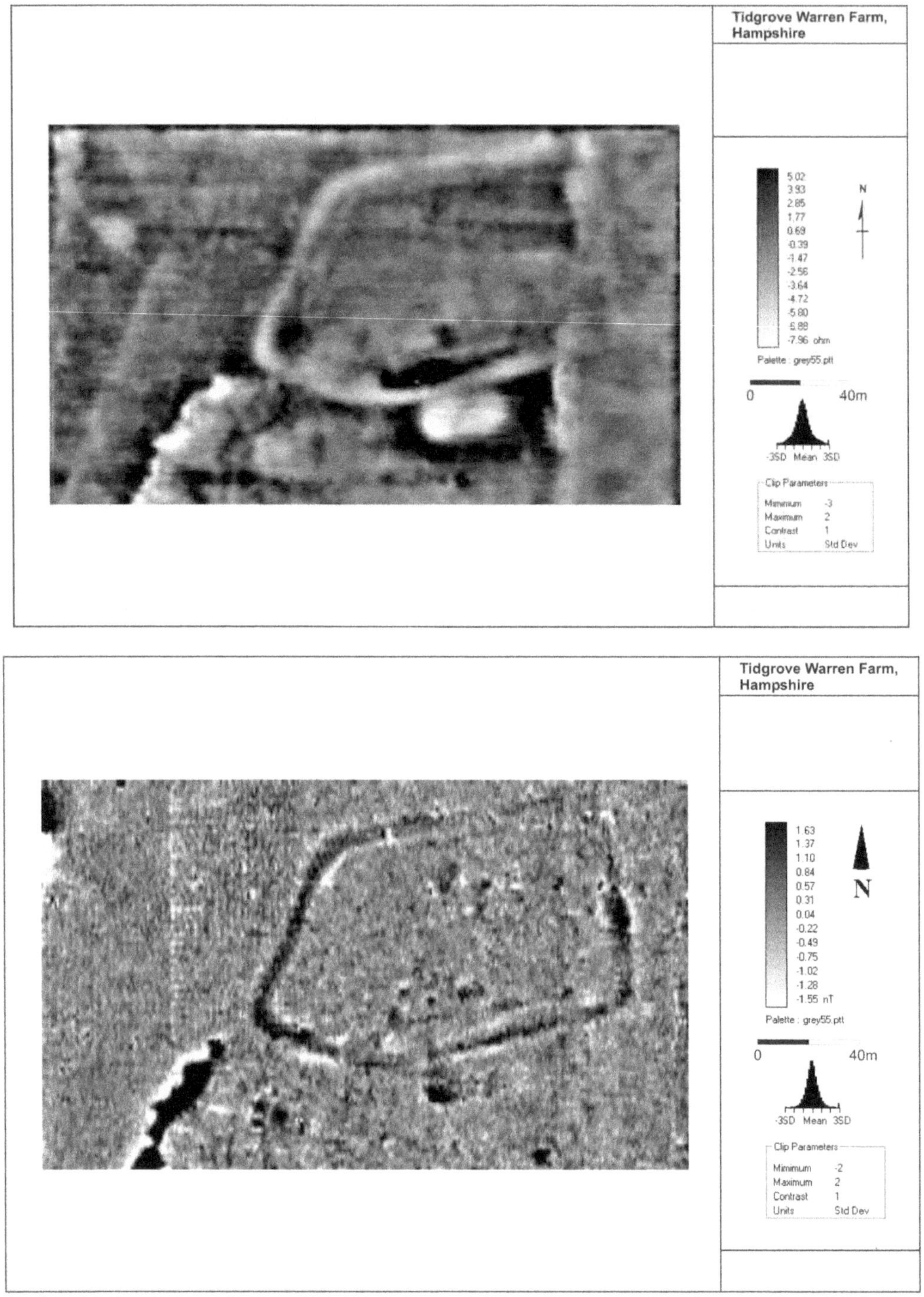

Figure 5 *Greyscale image of the magnetometer and resistivity survey results from the medieval enclosure*

(Strutt et al. 2005), have provided a more clear understanding of the phases of the settlement. The ceramic assemblage from the site reinforces the late Iron Age date of the site, with occupation continuing to the 4th century AD, similar to other rural settlements of the period, such as Viables Farm (Millett and Russell 1984; Gibson 2004). With the air photographic data, the settlement can also be placed within the context of the surrounding field systems. It is situated in one of the open fields straddling the ridge, with a possible route or track running along the ridge northwards to the Portway.

The Medieval Hunting Lodge

The survey at the medieval hunting site was conducted to verify the nature of the enclosure, but faced a different set of conditions to that of the Romano-British settlement in terms of both the extent of the site, and the depth to which the stratified deposits were preserved. Resistivity and magnetometer survey results indicated a substantial defensive enclosure ditch (Fig. 5), encompassing an area of approximately 0.6 hectares, with a series of structures organised within the enclosure. Traces of a possible inner ditch running inside the main enclosure are also visible. In addition, the ditch appeared to cut a large ovate pit to the south. The results from the inside of the enclosure suggested that a significant amount of demolition had occurred, as the plan of each structure was not necessarily clear. The subsequent excavation at this site in 2005 revealed that, instead of the shallow stratigraphy and plough damages structures visible found elsewhere on the downland, the remains of the structures survived to a greater extent, and a massive undercroft or cellar was found to have been cut into the hillslope (Fig. 6). A ground penetrating radar (GPR) survey was conducted early in 2006, and revealed at least one other cellar in the northern part of the enclosure. Because of the limited size of the site, and the recognised boundary ditch around its perimeter, the use of the magnetometry and resistivity with GPR and targeted excavation has proved the most useful approach to investigating the structures and associated features. Results from both the survey and excavation indicate that we have the remains of a hunting lodge similar to the site at Writtle in Essex (Rahtz 1969). However, although the results of the survey suggested that the layout of the site was akin to the ditched enclosure of the site at Writtle, the excavation has proved that the site was probably much more substantial. The depth of the cellar (7.5 ft or 2.5m deep), and the size of the flint and mortar walls (6 ft or 2m high, and 4 ft or 1.2m thick), suggest that the site is similar in scale to initial phases at Clarendon Palace in Dorset (James and Robinson 1988).

Figure 6 *Photograph of the enclosure interior from the west, showing the inter-cutting buildings, and the depth of the cellar of the principal structure along the south edge of the enclosure*

Conclusions

To date the integration of the different datasets mentioned above has provided a significant amount of evidence for the pattern of settlement and development of the landscape in this part of the downland. The combination of air photographic data and the geophysical surveys conducted between 2003 and 2005 places the Romano-British settlement at Tidgrove Warren within a broader pattern of field systems stretching across the landscape. The settlement itself comprises a simple enclosed layout, with structures and inner boundaries associated with a rural settlement similar to those excavated in the region to the east of Basingstoke at Viables Farm (Millett and Russell 1984; Gibson 2004), Cowdery's Down (Millett and James 1984) and Rucstall's Hill (Oliver and Applin 1979). The nature of the medieval site at Tidgrove Warren Farm has relied more heavily on the use of excavation for identifying the exact depth and phases of the structures, although the nature, extent and plan of the site was all provided through the use of integrated geophysical survey methods, and show that a substantial hunting lodge was present at the site.

Acknowledgements

Considerable assistance and advice have been received from a number of sources over the course of the survey and excavation. Primarily thanks are extended to the committee and members of the Kingsclere Heritage Association, whose initial interest and enthusiasm have led to the commencement and continuation of fieldwork at Tidgrove Warren Farm, and whose dedication and hard work led to the funding of the fieldwork. In particular, the work of Janet Bond, Janet Jones, Keith Whiteman, Peter Woodman, Mike Leah and Peter Goff are acknowledged. Funding for the survey and excavation has been granted through the Heritage Lottery Initiative, and My Heritage. The permission and kind assistance of the landowner Mr Raleigh Place has been vital to the smooth running of the project. We are also grateful to Raleigh for the use of his skills and time for the removal of ploughsoil from the trenches, and the final backfilling of the site after completion of the work. The survey and excavation seasons were directed jointly by Dominic Barker, Paul Everill and the author, with Prof. David Hinton and Dr David Wheatley. On site supervision was also provided by Penny Copeland, James Cole, Chris Russel and Lucy Britt. Finally, the completion of the season would have been impossible without the dedication of the field team, including many students from the Department of Archaeology at the University of Southampton, and volunteers from the Kingsclere Heritage Association. Warm thanks are extended to everyone who has taken part. Any errors in this paper remain those of the author.

Bibliography

BANNISTER, A., RAYMOND, S. and BAKER, R. 1998 *Surveying* (6th edition). Harlow: Longman.

BOWDEN, M. 1999 Unravelling the Landscape: an inquisitive approach to archaeology. Stroud: Tempus.

CLARK, A. 1996 *Seeing Beneath the Soil: Prospecting Methods in Archaeology (2nd Edition).* London: Batsford.

CUSWORTH, G. and DOBBS, R. 1999 *Kingsclere Woodlands story : the history of the parish of Ashford Hill with Headley.* Gary Cusworth: Ashford Hill.

FULFORD, M. and TIMBY, J. 2000 *Late Iron Age and Roman Silchester, excavations on the site of the Forum-Basilica 1977, 1980-86*. Britannia Monograph Series No. 15. London: Society for the Promotion of Roman Studies.

GAFFNEY, C., GATER, J. and OVENDON, S. 2002 *The Use of Geophysical Survey Techniques in Archaeological Evaluations*. Institute of Field Archaeologists Paper 6.

INSTITUTE OF GEOLOGICAL SCIENCES 1979 *Geological Map of the United Kingdom: South Sheet*, scale 1:625000. London: Ordnance Survey.

JAMES, T.B. and ROBINSON, A.M. 1988 Clarendon Palace: The History and Archaeology of a Medieval Palace and Hunting Lodge near Salisbury, Wiltshire. London: The Society of Antiquaries of London.

MILLETT, M. 1979 The dating of the Farnham industry (Alice Holt) pottery. *Britannia* **10**, 121-137.

MILLETT, M. and GRAHAM, D. 1986 *Excavations on the Romano-British Small Town at Neatham Hampshire, 1969-1979*. Hampshire Field Club and Archaeological Society Monograph **3**.

MILLETT, M. and JAMES, S. 1983 Excavations at Cowdery's Down, Basingstoke, Hampshire, 1978-81. Archaeological Journal **140**, 151-279.

MILLETT, M. and RUSSELL, D. 1984 An Iron Age and Romano-British Site at Viables Farm, Basingstoke. *Proceedings of the Hampshire Field Club and Archaeological Society* **40**, 49-60.

MONEY, W. 1575 *A perfect booke, of all the landes...within the hundreds of Evenger, Chutlye, Kingsclere, Pastroe and Overton, vewed...1575...by certeine inhabitants...within...County of Southampton...: now first published from the original Ms. in the Wood Library, Whitchurch, in the County of Southampton.*

OLIVER, M. and APPLIN, B. 1979 Excavation of an Iron Age and Romano-British Settlement at Ructstalls Hill, Basingstoke, Hampshire, 1972-5. *Proceedings of the Hampshire Field Club and Archaeological Society* **35**, 41-92.

RAHTZ, P.A. 1969 Excavations at King John's Hunting Lodge, Writtle, Essex, 1955-57. The Society for Medieval Archaeology Monograph Series 3. London: The Society for Medieval Archaeology.

SCOLLAR, I., TABBAGH, A., HESSE, A. and HERZOG, I. 1990 *Archaeological Prospecting and Remote Sensing*. Cambridge: Cambridge University Press.

STRUTT, K. 2003 Tidgrove Warren Farm: Geophysical Survey Report November 2003. Unpublished report (SREP 12/2003).

STRUTT, K. 2004 Tidgrove Warren Farm: Geophysical Survey Report November 2004. Unpublished report (SREP 11/2004).

STRUTT, K 2005 Tidgrove Warren Farm: Geophysical Survey Report September 2005. Unpublished report (SREP 10/2005).

STRUTT, K., BARKER, D and EVERILL, P. 2005 Tidgrove Warren Farm Interim Excavation Report May 2005. Unpublished report (EXCREP 1/2004).

Remote Sensing Archaeology for Ancient Cities Structure of Pingyao and Liangzhu, Yuhang City, Zhejiang Province, China

Li Zhang,[1,2] Jianping Wu[1]

[1]Geography Information Science Laboratory, Ministry of Education, East China Normal University, Shanghai,
[2]China Joint Laboratory of Remote Sensing Archaeology, Chinese Academy of Sciences, Ministry of Education and Cultural Relics Bureau, Shanghai, China

1 Introduction

Liangzhu culture was a Neolithic culture in Taihu area, China. Initially it was discovered in the Liangzhu area, Zhejiang Province, China. Since 1936 a multitude of exquisite artefacts were discovered one after another. Archaeologists locally unearthed many remains. But because the area is very large and the shape of the original earthworks is fragmented, Liangzhu's buildings' layout structure is still a puzzle to be resolved. In 1998 we discovered more sites in same area by remote sensing, some sites were located outside original protectorate (LI ZHANG,2002). In recent years, after processing and interpreting different kinds of remote sensing imagery, we discovered that these sites make a unified structure which has the layout and the shape of the embryo city.

2 Research area and data used

The research area lies in Liangzhu, Zhejiang Province, China. Its scope and boundary of the fundamental materials can be seen in Figure 1.

3 Working method

1) Registration: The registration of aerial photographs and topographic maps was completed in ARCGIS software. Aerial photographs waere corrected on the bases of 1:10000 topographic maps. In order to save space, aerial photographs for drawing were stored with 300-600 DPI resolution (TIF) format according to the complexity of image. Aerial photographs' image resolution for interpretation is 1200 DPI. Few of them are 600 DPI.

2) Interpretation method: Image interpretation was completed by visual interpretation. Stereoscope was the main tool for image interpretation. Besides visual interpretation, computer magnified and produced images in different scale. And then stereoscopy is completed. Through enhancement and fusion we processed some inferior information in images with different temporal, scale and band. On the bases of this, images were comprehensivly interpreted again and drawn on screen using ARCGIS, ARCVIEW software.

3) Restoration: After completing restoration of 1:10000 contour restoration is completed by means of ARCVIEW software.

4 Content of interpretation and investigation

By processing aerial photographs and satellite images and using different temporal, scale and band options, we discovered four important regions of Liangzhu period in the investigated area (Figure 2).

1) Central urban districts

The built up area was near Mojiajiao and was surrounded by a river system. West part was high whereas southeast and north parts were low. At present the simple construction pattern could not be seen on the surface (Figure 3). By image interpretation we discovered that the simple building was divided into three levels. Eastern, southern and northern structures were built on alluvial plain 3 to 4 metres above the sea level, whereas the one on the south end of the western part was built on the hill 9 to10 metres above the sea level. The structures on the west and north ends of the western part were built on a high ground 5 to 6 metres above the sea level. Mojiajiao city is located in the west by northern, artificial flatland, which is based on a natural tableland. Flattop is higher than the east plain by around seven meters. Mojiajiao city is 600 meters long from east to west, 450 meters wide from south to north. City wall surrounds the city. East and south city walls are slightly wider than west and north city walls. City gates are located on all sides (Figure 4).There are 3 artificial rectangle soil platforms in the city of Mojiajiao: Big Mojiajiao-Shan, Small Mojiajiao-Shan and Wugui-Shan. Relative altitude of Mojiajiao-Shan platform base is 3 to 4 meters, 162 meters from east to west, 73 meters from south to north. Building base is divided into eastern and western parts. On the western side the building pattern appears to be of a traditional rectangular courtyard. In the eastern part, the building pattern shows a right angle with broken line. On the north of Big Mojiajiao-Shan platform base there are regularly built steps. Around Big Mojiajiao-Shan platform base there are Protection Rivers intermittently.

Small Mojiajiao-Shan is located tn the right, west of Big Mojiajiao-Shan, 73 meters away. Relative attitude of platform base is 3 meters, 85 meters long from east to west, 25 meters wide from south to north.

Wugui-Shan is located on the right south of Small Mojiajiao-Shan 85 meters away. Relative altitude of

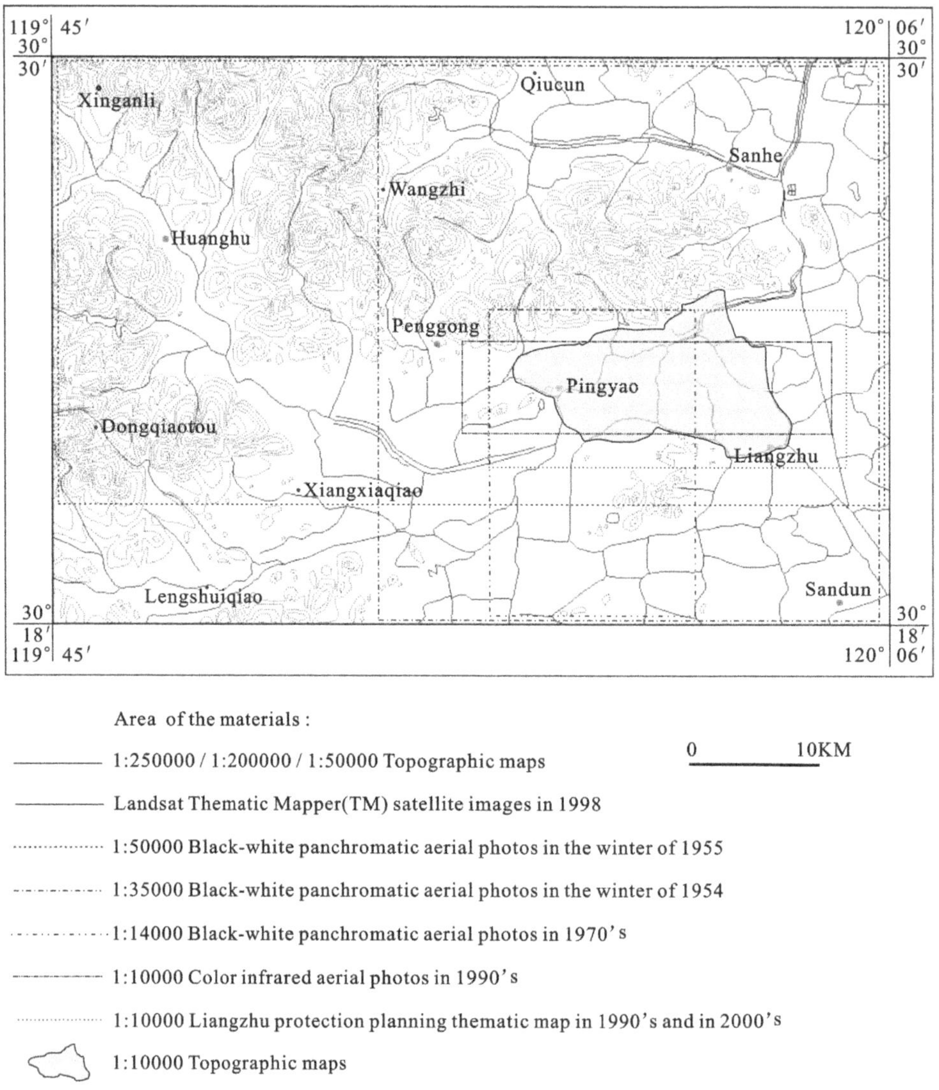

Figure 1 *The investigation area and boundary of the fundamental materials*

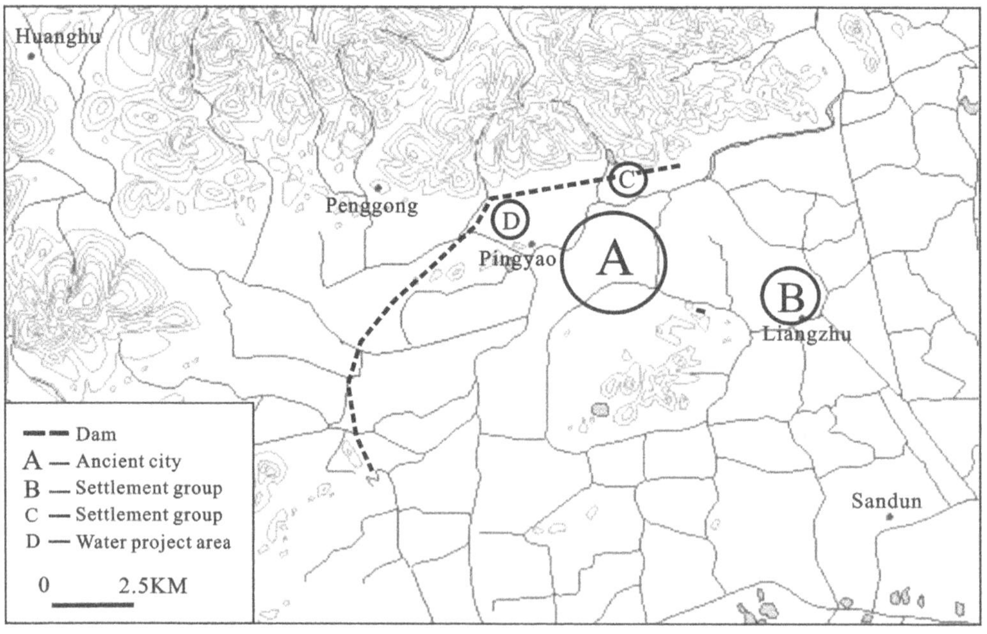

Figure 2 *Four important regions in the investigated area*

Figure 3 *Actual Mojiajiao city*

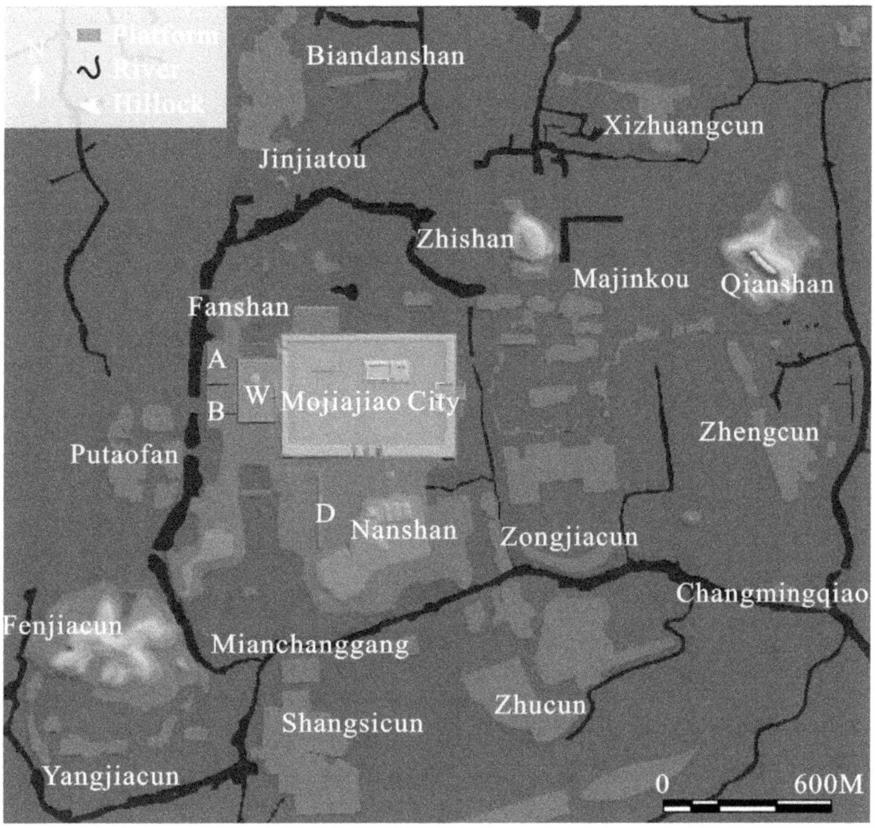

Figure 4 *Layout of the buildings in the centre of Mojiajiao*

platform base is 3 meters, 109 meter long from east to west, 65 meters wide from south to north. There are 3bases from east to west. Around the bases of the platform there are marks of Protection River.

In the centre of Mojiajiao city, a two storey earthen altar is visible. The height of the earthworks is 2 meters. The diameter of lower altar is 25 meters. The diameter of top altar is 12 meters.

There is a long step from south to north up near to west wall of Mojiajiao City. The step may go up to the west gate of Mojiajiao City. In the middle of journey we can enter into Western City. The Western City is a rectangle expanding from south to north. Northern half of the base is slightly lower than the ground surface in Mojiajiao City. Southern half of the base is slightly lower than northern part.

The east and northeast Mojiajiao Cities are built on the lower city, 3 to 4 metres above sea level; many bases of walls and building platform surfaces can be found. The platform is 5 to 6 metres high. The eastern part of the platform is more regular. Because the terrain is lower in the west, these regions have been broken and unendurable.

Near southeast wall of Mojiajiao City there is a samll, narrow river. But the image showed a 20 meters wide beheaded river. The head of river was located at the eastern gate of Mojiajiao City. River water went into main river course in the south Mianchang gang.

Near the west end of northern wall of Mojia, there is a protruding square platform. Platform's height is 7 metres, it is 154 metres long from east to west, and 90 metres wide from south to north. There is no wall.

In the south of Mojiajiao City there is a naturally formed hummock. There are many building contours cut into the block on top. It is obvious on the top of tableland in middle.

In summary, total area of the center city region is around 5 square kilometers. The volume of earthwork of main building base is around 2,400,000 cubic meters (Figure 5).

Figure 5 *Reconstruction of structures of Majiajiao city*

2) Large scale settlement region

This is an area around Xunshan. It encompasses around 2.1 square kilometers. In the middle of Xunshan is a hillock, higher than 30 meters from the surrounding area, in a shape of a pyramid with the four slopes of 30 degrees. The top of Xunshan has been built artificially to make a rectangular platform. There is a barricade on the north of platform, a square mound platform on the middle of top and a spiral pathway going up on the south slope of Xunshan. Half way up the hill are two square platforms. To the south of the Xunshan are three-built pyramid-shaped flat-topped hills. To the south of these hills are dry lands, 5 to 6 metres above the sea level. Ground surface is broken. There are traces of occasional habitation here.

3) Small scale settlement region

Small scale settlement regions are situated mainly along the roots of hills on northeast ot the investigated area. They are located 4 to 5 metres above the sea level. These sites are not very large and are scattered. The largest one is in Yaojiadun. There are 5 platforms with dimensions of 600 x 600 m.

4) Area of irrigation works

Irrigation works are mainly located in the northwestern side of the investigated area. There are many dams in the southeast and north, by the hills. The dams in southeast are uncontinuous and short. The dams in the north are many and long. A dam among them is a continuous dam extending several kilometers, the head of dam is located in Laojiatou, Fengjiatou, Xujiatou, Wengjiatou, the end of dam is at southwestern Chiba. The total length of the continuous dam (including the part formed by natural massif) is about 7.7 kilometers. Out of these, artificial part makes about 3.5 kilometers (Figure 6). The length of whole northern dam from east to west south is circa 14.5km. Man-made better preserved dam (including single dam, double dam and quadruple dam) is 4.2 kilometers long, volume of earthwork is around 1,680,000 cubic metres.

Besides dams, to the north of the investigated area there are a water gathering spots based on natural arc depression. On the bottom of depression there are artificial polders located on different heights. A different volume of water in different polder under different water situation is found. In addition, there are many interception facilities on the bottom of depression to reduce water speed. Among them there is a string of "回" shaped dams. "回" goes bigger from north to south. "回" dam body is 8 metres wide. The clearest dam is in the middle of the string of dams. The dam is square with its sides measuring 40 metres (Photo 1).

Figure 6 *Aerial photo of a typical dam flow*

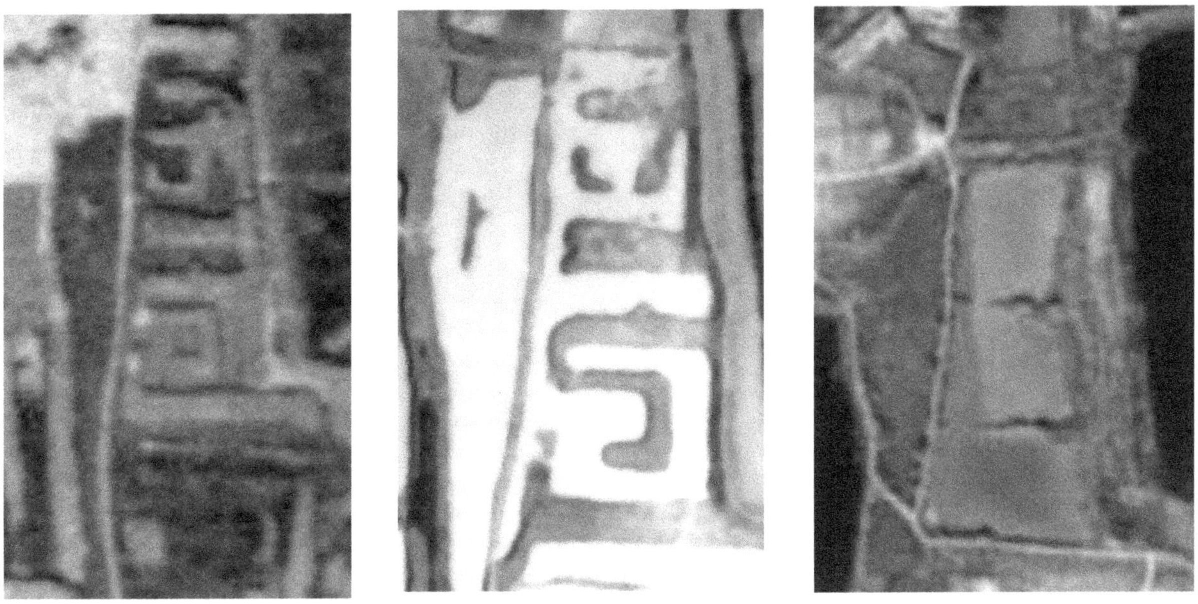

Photo 1 *"回" shape dam from the aerial photos in the fifties, seventies and nineties*

5 Summary

This paper has presented the plans of built structures of Liangzhu Culture. Many details still need to be studied. We have established that the investigated area is the largest one among the embryo cities discovered in China to date. It is an outstanding representative of original cities in the southeast of China.

References

LI ZHANG and SHUREN LIU. 2002 Geoscientific Analysis of the Sites Recorded in the Pingyao and Liangzhu Area of Yuhang City, Zhejiang, by Means of Remote Sensing. *Archaeology* 413: 87-93.

Landscape Archaeology and GIS for the Eco-cultural Heritage Management of the Aksum Region, Ethiopia*

Rossano Ciampalini,[1] Andrea Manzo,[2] Cinzia Perlingieri,[2] Luisa Sernicola[2]

[1]Università degli Studi di Firenze
[2]Università degli Studi di Napoli "l'Orientale"

1 Introduction and Research Design

This paper is based on a research project aimed at reconstructing the diachronic changes in the landscape in the area of Aksum (N Ethiopia, 14° 8' N; 38° 43' E, see Figure 1), the capital city of the Aksumite kingdom, from ca. 400 BC to AD 800 (Conti Rossini 1928; Fattovich 1997 b; Fattovich *et al.* 2000; Phillipson 1997, 1998).

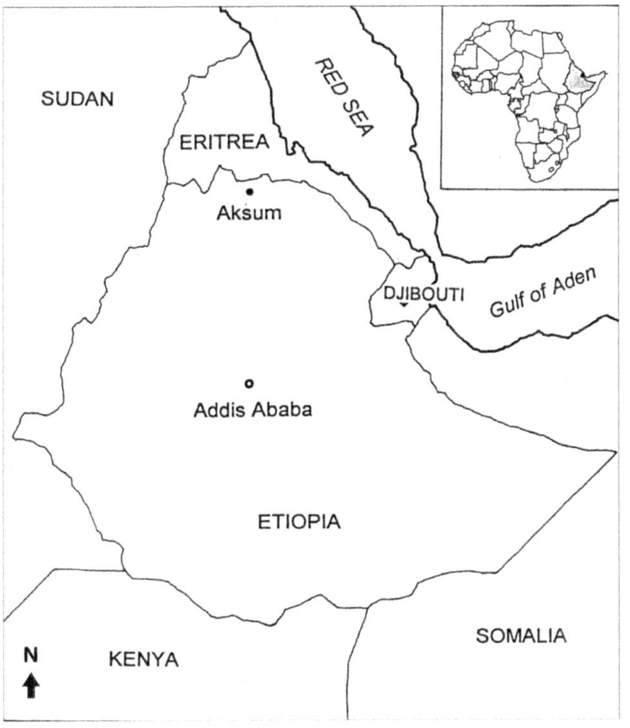

Figure 1

A complementary aim of the project is the understanding of the dynamics of human exploitation, manipulation, and management of this territory also related to environmental stress and to social pressures. As the whole N Ethiopian plateau this area is heavily affected by environmental deterioration and soil erosion (Butzer 1981, 1982; Machado *et al.* 1997; Nyssen *et al.* 2004). Two different sets of data have been integrated in this work. The first set is from excavation, archaeological survey, and geoarchaeological survey on the hill of Bieta Giyorgis and the valley N of it, by the UNO-BU Expedition and by the team of the Soil Science Department of the University of Florence (Bard and Fattovich nd.a, Bard et al. 2003, Ciampalini et al. 2006).[1] These data allowed to generate a very detailed diachronic model of the Man-Environmental relationship at a microregional scale. The second set of data resulted from the survey of the World Bank project "Ethiopian Cultural Heritage Project – Aksum Branch – Site Planning and Conservation Component" covering the whole Aksum region, and was finalized to the site inventory for the heritage management.[2] A predictive model of the archaeological impact on the territory of Aksum was elaborated on the base of the first set of data; subsequently a comparative analysis of the predictive model and the data from the World Bank survey gave interesting and encouraging results.

A systematic collection of archaeological, palaeoagricultural, and palaeoenvironmental data formed a multidisciplinary approach in which the landscape is considered a palimpsest of all past evidences of ecosystemic interaction significant for the reconstruction of many aspects of the Man-Environment relationships (Fattovich 2003). Hence special attention is given to the chronological and functional classification and distribution of the sites, as well to the study of the landscape infrastructures and exploitation marks (ancient fields, roads, terraces, facilities for water management, plough marks, quarries and sources of row materials). The recent pedological geomorphological and palaeoagricultural investigations in the area of Aksum highlighted new and encouraging results concerning the ancient and traditional cultivation techniques. The ancient system of terracing for the conservation/exploitation of the hill slopes was reconstructed, and was as well assessed the impact that modifications in the Man-

---

* Rossano Ciampalini studied the palaeoagricultural evidences; Andrea Manzo contributed with GIS analysis; Cinzia Perlingieri contributed with landscape interpretation; Luisa Sernicola contributed with GIS analysis. All authors contributed to the final remarks.

[1] The project is directed by Prof. Rodolfo Fattovich (University of Naples "l'Orientale"); preliminary reports of the Italo-American Expedition of University of Naples "l'Orientale" and Boston University are published in the Rassegna di Studi Etiopici, and Nyame Akuma since 1993, see also K.A. Bard et al. 2003, the final report of the project, K. A. Bard and Rodolfo Fattovich (ed) nd.a.

[2] The fieldwork was conducted by Prof. Rodolfo Fattovich and Ato Tekle Hagos (ARCCH, Addis Ababa).

Environment relationship had on the whole landscape through times.

2 The many dimensions of landscape

"Landscape is not something in and of itself but the creation or organization of land, a process" (Schwarzer, 2003: 84). In order to outline the diachronic changes in the landscape of the study area, different components of the landscape were distinguished (Ashmore and Knapp 2000; Cambi and Terranato 2001; Fattovich 2003): *Natural landscape*, consisting of all geological, geomorphological, vegetal and animal features that provide information about the spatial distribution of resources; *Economic landscape*, the configuration of artificial features reflecting the different patterns of land use and territory exploitation; *Social landscape*, the configuration of artificial features reflecting the hierarchical organization of the society; *Power landscape*, i.e. the manifestation of the power in the landscape; *Sacral landscape*, the combination of man-made features reflecting the religious and symbolic organization of the territory.

These components have a conceptual value of distinction among the many levels and dimensions at which human dynamics affect the territory, and help in understanding a discernible order of continuities/discontinuities in the landscape changes that, even if not sharply demarcated, can be measured on an empirical ground. As a secondary aim, this paper in fact asks not just what landscape "is" or "means" but what it "does", how it works as a cultural practice; beyond the landscape as a visual background, there is a landscape in the archaeological records that is a source of social identities.

3 Environmental Setting

The study area encompasses the modern town of Aksum, the hills of Bieta Giyorgis and May Qoho, the sloping plains N and S of them, and the hills which surround the area of Aksum and delimit the territory of the ancient capital city. The present agricultural landscape is characterised by scattered households and cultivated fields. Most of the slopes and plains are terraced, and few restricted spots are used for grazing. A massive reforestation affected large areas of the surrounding hills in the last decades. The geology of the region is known by studies at regional scale, Tigrean highlands are partly layers of Tertiary basalts due to volcanic activity forming a distinctive terraced topography with successions of flat surfaces and steep slopes characterized by gully erosion, creeping and sheet wash. Erosive processes have been very active since long time, and mainly associated to rifting and uplifting (Brancaccio *et al.* 1997). A more detailed geological study of Bieta Giyorgis was conducted in the framework of the UNO-BU Expedition (Johnson and Scott Harris nd.a).

The strong lithologic and topographical variation of the Aksum area is reflected in a mosaic of soils. Thin mantels of silt/clay loam reddish soils cover the hilltops; hillsides and piedmonts being more prone to hill wash and soil erosion are left with stony sandy loam. Valley and flat plains are covered by dark grey silt clays (vertisoils) (Koch and Schmidt nd.a). The present soil distribution and the degradated vegetal cover result from ongoing natural erosive and anthropogenetic processes related to climatic changes. The methodological approach of this project highly contributed to the debate concerning the relationships linking these factors.

4 Palaeoagriculture and erosion dynamics[3]

The oldest evidence for cultivation (basically plough marks), has been assumed on the base of the chronology of the terraces associated with archaeological records, thus date back to the Proto or Early Aksumite times (Fattovich 1997, 2000). Agriculture on the footslopes (Butzer 1981) around the syenite plug of Aksum has been performed through a very old technique of terracing (Hurni 1985, Nyssen *et al.* 2004), in order to reduce the slope gradient and to improve the soil drainage. This technique seems to be very effective as the fragile balance between the terrace step and the escarpment is constantly maintained. The terrace system is managed as a whole, like in the Middle East: the upslope terraces are mainly used for grazing and water harvesting, the downslope are mostly exploited for cropping. The terrace sequence is consequently built in the same moment and the upper part of the terracing is not to be considered as an enlargement of agricultural areas under human dynamics.

The traditional plough still used in the region is the *maresha*, an "ard plough" that seems to have been introduced in Ethiopia between 1000 and 400 BC by Semitic tribes (Yemen), or even before by Cushitic speaking peoples from north-eastern Sudan (Gebregziabher et al. 2006). The impact of the plough against the stones is frequent, and the stones are consequently scratched: 1) on the top if rocks are at a lower depth than the tilling depth (8 –16 cm), or 2) on the sides if rocks are protruding (Figure 2). The occurrence of plough marks, thus, is a safe evidence for past cultivation on lands presently cultivated, degraded or strongly dissected.

---

[3] Thanks to Prof. G.A. Ferrari and Prof. P. Billi for the support in "Palaeoagriculture and erosion dynamics".

Figure 2 *Sequence of plough marks on a rock outcrop*

The highest plough marks on the side of the boulder face are actually the oldest. Pedo-geological features (different weathering rinds, patinas, varnishes) are used to reconstruct past ground surfaces. Ancient plough marks may be found at different height levels on big boulders or rock outcrops, thus providing valuable information on the thickness of soil loss. A systematic study of these ancient plough marks suggests that the total soil loss starting from Aksumite times ranges between 1.9 and 4.8 t ha^{-1} y^{-1} (Ciampalini et al. 2006).[4] These mean values are lower than the threshold erosion rate commonly accepted by international soil conservation agencies. The slope of the present terraces results to be lower than the original one, due to aggradation of soil material against the rock bund. In Fig. 3 the inference of old ground surface obtained by plough marks data processing is presented. The uninterrupted line is the present topographic surface; the dashed line connects the upper border of the plough marks belts on boulders of opposite sides of the terrace; the interrupted line is a reconstruction of the old ground surface, considering an average plough depth of 15 cm.

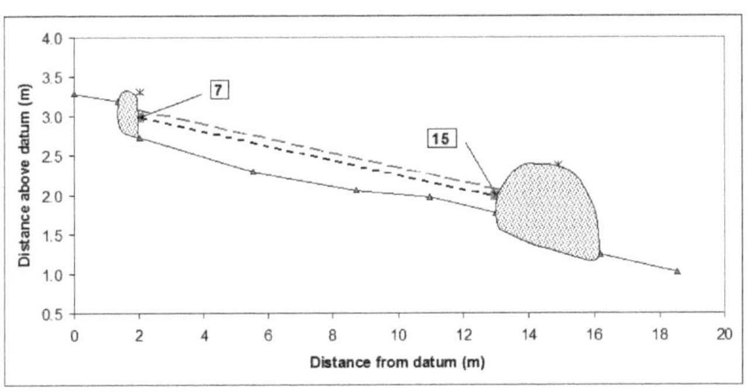

Figure 3 *Scheme of a terrace cross section*

The strike of plough marks on the boulders surface gives information about the ploughing technique. The plough marks orientation is restricted around a marked mode corresponding to the ploughing direction along the contour lines. Archaeological evidences and the study of the plough marks highlighted that the land conservation local practices, were very effective on the long time span. Pedological investigation confirmed this interpretation, since an argillic horizon was found in the terrace soil profiles: this diagnostic horizon is present only in soils developed on very stable geomorphic surfaces of old deposits. The presence of illuvial clay cutans in the Bt horizon demonstrates the stability of the deposits. The lack of skeleton in the soil is an evidence of the "guided aggradation" by farmer activity.

On the contrary, recent changes in the land use from agriculture to grazing, started during the Derg administration (1974-1987), produced an incredibly high soil loss, which strongly accelerated the environmental degradation.

The land degradation is the result of the progressive lack of maintenance of the stone bunds or the abandonment of the agricultural activities. The relationship between land degradation and anthropic pressure in the region of Aksum has to be re-considered under a new perspective. The necessity of high manpower to maintain effective environment conservation structures is a requirement for the preservation of the agricultural capacity (Tiffen 1994).

Preliminary results can be summarized as follows: 1) ancient agricultural exploitation of the upper part of the Bieta Giyorgis etchplain is not to be considered as an expansion of the cultivated valleys, but as the result of an integrated strategy of land use; 2) plough marks are sure evidence for ancient agriculture and help in determine the extension of cultivated areas, as well to evaluate the soil loss; 3) sediment facies and pedological features confirm the validity of the traditional terracing system for the stability and conservation of the land; 4) increasing human pressure is not a cause for degradation, on the contrary it is necessary for a sustainable land use and management; 5) impoverishment of agricultural land does not depend exclusively on degradation for overexploitation or on climatic conditions; 6) variability of soil capability, market fluctuations, wars, pestilences and other plagues or synergy of all, have to be considered as important elements affecting past environment.

5 Quantitative analysis and ancient settlement pattern

One of the first evidences emerged from this work is the occurrence of three main settlement areas: A) the capital city of Aksum, B) the Bieta Giyorgis hill and the plains N of it, and C) the broader surrounding rural area (Figure 4). It is demonstrated by the archaeological evidence that these areas have always been closely connected one to each other, and that the peopling in the capital city of Aksum and in the surrounding regions have always had very tight relationships. In this scenario it must be highlighted that Bieta Giyorgis always had a preminent role directly connected to the development of the capital city, as shown by the occurrence from the 400 B.C.

[4] t=tons, ha=hectares, y=years.

onwards, of strong administrative evidence, monumental buildings, imported goods, and proximity to the main land routes.

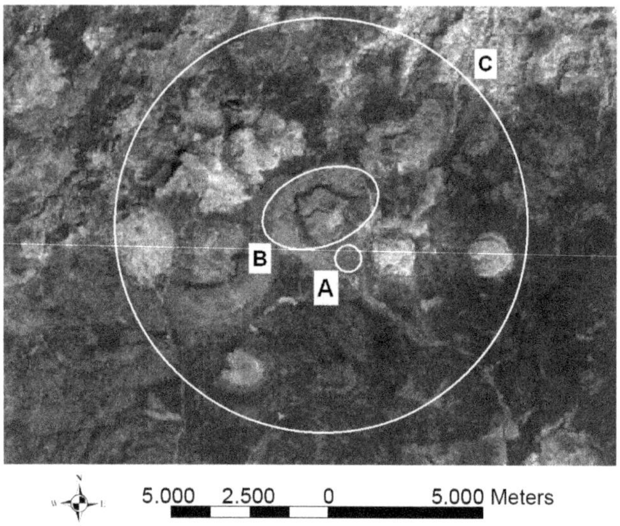

Figure 4

Quantitative analysis allowed the reconstruction of the settlement dynamics occurred in the area, and their relationship with the formation process of the political and economic capital of the Aksumite kingdom. One of the first results was learning that human settlements were not homogeneously distributed: a denser occupation characterises all phases at Aksum and Bieta Giyorgis, comparing to the occupation development in the rural area (as seen in Graph 1).

A constant increase in number and sites average dimensions characterized Bieta Giyorgis and the rural area from Pre-Aksumite (ca. 800 BC) to the end of Early Aksumite times (ca. AD 150) (Graph 1 and Graph 2).
This suggests that a progressive shift occurred from sparse small agglomerates, mainly concentrated around ceremonial sites on top of Bieta Giyorgis and in the SE and SW areas, towards a more intense occupation connected with the emergence at Bieta Giyorgis of elite groups and local polity.

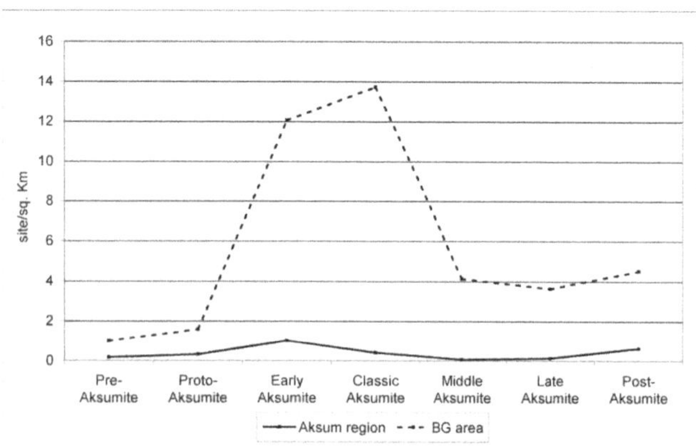

Graph 1

Archaeological evidence for Aksum, though very scanty, seems to suggest that at this time the city was a center with no political and administrative predominance (Figure 5).

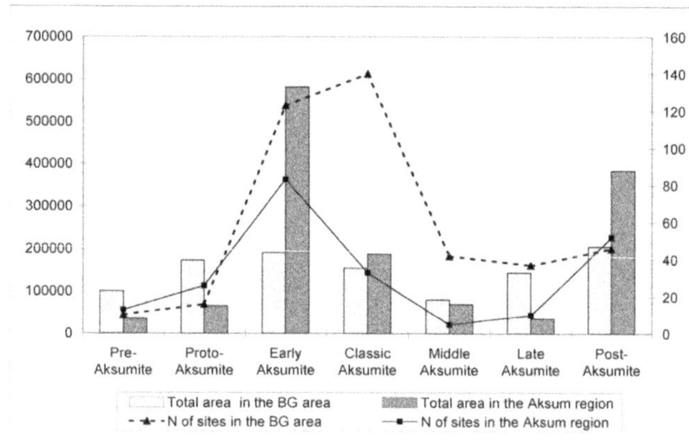

Graph 2

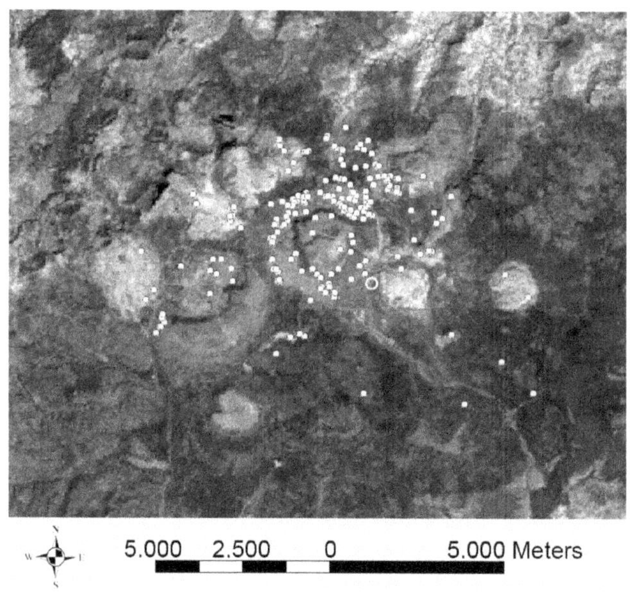

Figure 5

From AD 150, during the Classic Aksumite period, there is a clear differentiation in the occupation of all areas: number and size in the rural area dramatically decrease; at Bieta Giyorgis the total number of sites is continuously increasing but their size decreases (Graph 1 and Graph 2); at Aksum test excavations highlighted that the extension of the city considerably increased (as marked in Figure 6) (Fattovich, Bard, Petrassi & Pisano, 2000, pag. 79, fig. 20-21). The decrease in the number of sites in the rural area and the reduction of the average sites dimensions at Bieta Giyorgis maybe related to the expansion of the urban settlement of Aksum, and may be explained by the urban migration in the capital city rather than by a demographic decrease.

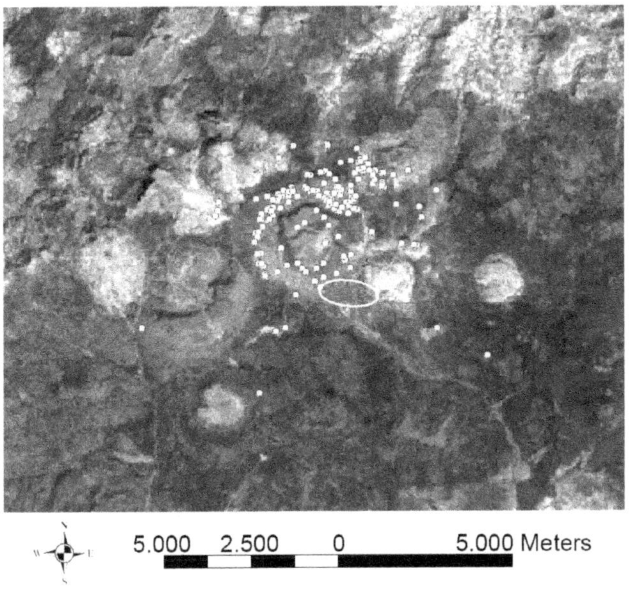

Figure 6

Later (from AD 350 onwards) a similar trend of decrease in the number of sites involved also Bieta Giyorgis. Few sites are recorded both at Bieta Giyorgis and in the rural area, though noteworthy increased in dimensions (Graph 2). This evidence seems to remark a change in the occupation model towards a flourishing of medium to large satellite villages around the expanding capital city (Figure 7). Possibly, this was a consequence of social changes strictly related to a different administrative organization at regional and interregional scale. The beginning of minting at Aksum (already started since AD 270), the erection of monumental buildings and high status funerary structures, as well military expeditions in South Arabia and Nile Valley, witness the rising of a new political scenario based on a centralized power with its political and administrative capital at Aksum.

A completely different trend was remarked from Late Aksumite to Post-Aksumite times: the number and extensions of the sites at Bieta Giyorgis and in the rural area increase despite the contraction of the urban area of Aksum (see Figure 8 and Graph 2) (Fattovich, Bard, Petrassi & Pisano, 2000, pag. 79, fig. 20-21). After its decline as a political and economic capital, Aksum remained an important ceremonial center, reduced in a small area around the cathedral. At the same time there is a reappearing of a considerable number of medium and large sites, possibly scattered households similar to present, in the surrounding areas characterised by the absence of administrative devices and of monumental buildings, and not strikingly proximal to the main land routes.

6 Predictions and verifications: present and future

The first step of this research allowed the elaboration of a predictive model of archaeological sites distribution based on a classification of soils and their fertility, and their relationship with archaeological sites. The major result showed that the archaeological sites are mainly associated with the more fertile and productive soils. This evidence was adopted as a training region to investigate the archaeological impact in non surveyed areas, and produced a map of the archaeological risk.

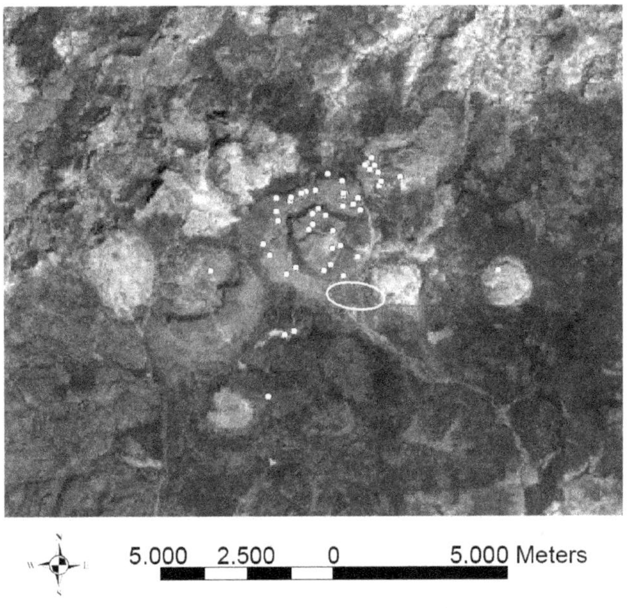

Figure 7

Figure 8

The survey carried out within the World Bank project demonstrated that the predictive model was reliable, as most of predicted area based on the training regions found a precise correspondence in the archaeological areas recorded later. This was an important verification of the validity of the predictive model based on the relationship between soils and sites, suggesting that the model may be extended to other regions of N Ethiopia with the same ecological conditions.

A further step of the project will be conducting further verifications to relate changes in the landscape to the erosional processes. This verification will validate, or not, the correlation between processes of land degradation and de-population of agricultural areas. This may be reflected in the qualitative and quantitative changes of sites in the agricultural areas. Noteworthy, the explanation of the past degradation processes may give parameters for the decision makers and planners of future development (Glantz 1987; Wainwright and Mulligan 2004).

7 Landscape and human geography

Landscape is one of those words that possess many definitions: it can be a scenery of nature, an artistic or architectural design of cities, a patrimony over lands indicative of heritage, the study of the environment or an ecological system (Schwarzer 2003: 85). As landscape expands to vast arrays of forms, through space and time, how does it convey meaning? How cultural practices implant themselves within landscapes and shape the history of places? Natural and property boundaries, roads, commercial networks, patterns of habitation, public meeting places, ceremonial and administrative buildings, and other monuments, all together influence the landscape development over time, but which are the archaeological indicators for the many dimensions of landscape?

Changes in the Aksum landscape as emerged by the analysis above made possible small moves towards the comprehension of some landscape dynamics. Some aspects of human geography as argued by the study of the past are also supported by the strategies observed in the present autonomous household-based communities (similar to the last phase of development of the Aksumite and Post-Aksumite times). In general terms, the decrease in the intensity of the human settlement and the weakening of the social landscape might be related to the changes in the power/ideological landscape and can be archaeologically traced and examined. In turn, the changes in the distribution as well as the re-definition of the monumental/ceremonial buildings are related to the changes in the political and ideological Aksumite identity.

The dispersion and lack of cohesion of the rural agglomerates, differently in respect to the integration of the earlier phases, and the subsequent isolation of households, caused a termination of the integrative process. In periods in which the ideological/administrative presence in the territory is strong (mainly during Classic and Middle Aksumite times), the social landscape is strongly integrated because directly influenced, when not organized, by a centralizing power. The archaeological evidence (based mainly on the distribution of the administrative devices) from Bieta Giyorgis and from Aksum, demonstrates that Sacral and Power landscape are always equated: monumental/ceremonial buildings are at the same time buildings of power and administration. Also the imported materials, important markers for the circulation networks (Manzo 2005), seems to be associated mainly with ceremonial/administrative buildings or sites. The distribution of these buildings on the territory demonstrates the use of extended ritual spaces rhythmically placed among hamlets to ensure social harmony and possibly to deter movements toward ideological/political centralization, as well to assure land maintenance.

8 Cultural Landscape and Heritage Management

The approach adopted here gave the possibility to assess the heritage of the Aksum region not only in terms of single monuments and sites, but in a perspective of eco-cultural impact, according to the up-to-date outlines of UNESCO and ICOMOS (ICOMOS 2005: 42, 46-48, 96-97; UNESCO 1972: art. 1, 2006: 6).

The author's hope is to contribute in shaping the idea of a *cultural landscape* of the Aksum region as "… the total assemblage of visible things that human beings have done to alter the face of the earth …" (Lewis 1993: 115-16). A true and effective way to appreciate, manage and exploit a cultural landscape is considering the reception of landscape as important as its production, and to consider "…the current inhabitants as important in determining the temporal angles of meaning as the original constructors." (Shwarzer 2003: 90). One of the main roles of the archaeological research on the territory is emphasizing the discontinuities of the landscape form through time, and also rise discussions on the importance of inhabitants in making that form. In fact, an important component of the eco-cultural heritage consists of the traditional systems of land management and exploitation, and their development. It includes knowledge of the natural environment, with its cycles and critical moments, which are embedded in the social memory together with possible technological solutions to environmental stresses. Both the traditional systems of land management and exploitation, and the ideological perception of the landscape leave traces that can be detected through a landscape archaeology approach.

The archaeological impact assessment, through the complete site inventory of the region, contributes to the heritage management and planning and decision-making activities. In terms of environmental impact, the degradation and the loss of soil fertility analysed in paragraph 4, allowed us to assess the potential and actual erosion risk. The mapping of the erosion models and locations where rehabilitation is most urgently needed, can play an important role in supporting rehabilitation planning at both the farm and the regional levels. Moreover, the evaluation of the effectiveness of the traditional land management system in response to the environmental and social stresses, will contribute to the planning of a sustainable development for the studied areas (TKWB nd).

A geographic information system outlining the different aspects of the ancient landscape may represent a first step towards the implementation of a sustainable eco-cultural development based on a "presentness of memory" in which the past must not be kept in the past. In this perspective the enhancement of rural development does not collide with people's tradition and sense of identity. In the Tigrean region, where the religious and historical memory is predominant and where the claim for a cultural identity is very strong, past should be revived and lived rather than remembered. This living landscape, where past is kept in the present, is itself a monument that is at risk of invisibility and oblivion.

References

ASHMORE W., AND A. B. KNAPP (eds.) 2000 Archaeologies of Landscape, Oxford.

BARD K.A. (ed.) 1997 The environmental history and human ecology of northern Ethiopia in the late Holocene, Naples.

BARD K.A., M. COLTORTI, M.C. DIBLASI, F. DRAMIS, AND R. FATTOVICH 2000 "The Environmental History of Tigray (Northern Ethiopia) in the Middle and Late Holocene: A Preliminary Outline", African Archaeological Review, 17(2), pp. 65-86.

BARD K.A., M. C. DIBLASI, M. KOCH, L. CRESCENZI, A. C. D'ANDREA, R. FATTOVICH, A. MANZO, C. PERLINGIERI, M. FORTE, M. SCOTT HARRIS, G. H. JOHNSON, S. TILIA, AND B. TRABASSI 2003 The BU/IUO Archaeological Project at Bieta Giyorgis (Aksum), Ethiopia: results, research procedure, and computer applications, in M. Forte and P. R. Williams (eds), The Reconstruction of Archaeological Landscapes through Digital Technologies, Oxford, pp. 1-14

BARD K. A., AND R. FATTOVICH (eds.) nd.a A Cultural landscape through time. Archaeological research at Bieta Giyorgis, Aksum, Ethiopia: 1993-2003, in preparation.

BRANCACCIO L. G. CALDERONI, M. COLTORTI, and F. DRAMIS 1997 Phases of soil erosion during the Holocene in the highlands of western Tigray (northern Ethiopia): a preliminary report, in K.A. Bard (ed.) The environmental history and human ecology of northern Ethiopia in the late Holocene, Naples, pp. 29-44.

BUTZER K.A. 1981 "Rise and fall of Axum, Ethiopia: A Geo-Archaeological Interpretation", American Antiquity 46 (3), pp. 471-495

BUTZER K.A. 1982 Empires, capitals and landscapes of ancient Ethiopia, Archaeology, 35 (5), pp. 30-37

CAMBI F., and N. TERRANATO 2001 Introduzione all'archeologia dei paesaggi.

CIAMPALINI R., P. BILLI, and G. FERRARI 2006. Plough marks as a tool to assess soil erosion: a case study in Axum (Ethiopia)". Poster Session, Highland, Symposium environmental change, geomorphic processes, land degradation and rehabilitation in tropical and subtropical highlands. Mekelle, Ethiopia, 19-25 September 2006.

CONTI ROSSINI C. 1928. *Storia d'Etiopia*, Bergamo.

FATTOVICH R. 1997*a* The Peopling of the Tigray Plateau in Ancient and Mediaval Times (ca. 4000 B.C. – A.D. 1500): Evidences and Syntesis", K.A.Bard (ed.) 1997.

FATTOVICH R. 1997*b* Archaeology and Historical Dynamics: the Case of Bieta Giyorgis (Aksum), Ethiopia", Annali dell'Istituto Universitario Orientale, 57, pp. 46-79.

FATTOVICH R. 2003 The ancient landscape of Aksum (northern Ethiopia), ca. 400 BC-AD 700: some preliminary remarks", in M. Liverani (ed.), Arid Lands in Roman Times, Florence, pp. 123-128.

FATTOVICH R., K.A. BARD, L. PETRASSI, and V. PISANO 2000 The Aksum Archaeological Area: A Preliminary Assessment, Naples.

GEBREGZIABHER, S., A.M. MOUAZEN, H. VAN BRUSSEL, H. RAMON, J. NYSSEN, H. VERPLANCKE, M. BEHAILU, J. JOZEF DECKERSF, and J. DE BAERDEMAEKER 2006 Animal drawn tillage: the Ethiopian ard plough, maresha: A review, Soil and Tillage Research, 89(2), pp. 129-143.

GLANTZ, M.H. (ed.) 1987 *Drought and hunger in Africa: Denying famine a future*. Cambridge.

HURNI H. 1985 Erosion-productivity-conservation systems in Ethiopia", Proc. 4th Int. Conf. On Soil Conservation, Maracay, Venezuela, pp. 654-674.

ICOMOS 2005 The World Heritage List Filling the Gaps- and Action Plan for the Future, Paris.

JOHNSON G. H., AND M. SCOTT HARRIS Geology and Geomorphology",in K.A. Bard, and R. Fattovich (eds.)

KOCH M., AND T. SCHMID Soils, in K.A. Bard, and R. Fattovich (eds.)

LEWIS, P. 1993 Common Landscapes as Historical Documents. In *History from Things: Essays on Material Culture*, Lubar S. and Kingery W. D. (eds), Washington D. C.: 115-116.

MACHADO M., A. PEREZ-GONZALES, and G. BENITO 1997 Erosion Processes and Land Degradation Episodes during the last 3000 yr at the Axum Region (Tigray, Northern Ethiopia), Geografia Fisica e Dinamica Quaternaria, Supplement 3, Vol. 1, 1997, p. 258.

MANZO A 2005 Aksumite Trade and the Red Sea Exchanghe Network : A View from Bieta Giyorgis (Aksum), in J.C.M. Starkey (ed.), People of the Red Sea. Proceedings of the Red Sea Project II, Oxford, pp. 51-66.

NYSSEN J., J. POESSEN, J. MOEYRSONS, J. DECKERS, MITIKU HAILE, and A. LANG 2004 "Human Impact on the Environment in the Ethiopian and Eritrean Highlands – a state of the art", Earth-Science Reviews, 64, 3-4, 273-320.

PHILLIPSON D.W. 1997 The Monuments of Aksum, London.

PHILLIPSON D.W. 1998 Ancient Ethiopia. Aksum, Its Antecedents and Successors, London.

SCHWARZER, M. 2003 The Moving Landscape. In *Monuments and Memory, Made and Unmade*, Nelson

R. S. and Olin M. (eds.), The University of Chicago Press: 83-102.

TIFFEN M., M. MORTIMORE, AND F. GCHUKI 1994 More people less erosion, John Wiley & Sons. Chichester, WS, England.

TKWB Nd. Traditional Knowledge World Bank, http://www.tkwb.org/tkwb/tkwb.pdf

UNESCO 1972 Convention concerning the Protection of the Wordl Cultural and Natural Heritage, Paris.

UNESCO 2006. Revision of Criteria of Properties Inscribed on the Wordl Heritage List according to the Operational Guidelines (2005), Vilnius.

WAINWRIGHT, J., AND MULLIGAN, M. (eds.) 2004. *Environmental Modelling. Finding Simplicity in Complexity*, Chichester.

Reconstructing the Ancient Republics (*Janapadas*) of the Indian sub-continent

Pallavee Gokhale,[1] Shreenand Bapat[2]

[1] D-2 Nirant Colony, Kothari Blocks, Bibwewadi, Pune 411037, Maharashtra, India
[2] "Vasishtha", 15, UCO Bank Colony, Patwardhan Baug, Erandavana, Pune 411 004, Maharashtra, India

Introduction

Landscape dynamics is an integrated part of Archaeology. Archaeological sites become archaic because of either landscape deformations or environmental/climatic changes or invasion and destruction by other human groups. Since the effect of all or any of these factors takes place on the earth surface, forming a 'Site of Archaeological Interest', the landscape interpretation in context of these sites plays vital role for exploration of such sites. Because of this close relationship, Archaeology can be studied better with the tools that are used to study the landscape dynamics. The use of modern geospatial techniques is not only about finding new sites for excavation but also about the evolution of the sites, their spatial importance, their interrelation and their survival. The tools of GIS and remote sensing help to interpret the landscape in many different aspects, providing a solid foundation for the archaeological studies in the area.

Indian subcontinent has been an area of prime interest for the archaeologists all over the world. The sites in this part of the world even date back to few thousand years BC, which show clear evidences of human inhabitation, their movement, and subsequent evolution. In our present study we have made an attempt to analyse a part of this vast subcontinent in order to create the definitive boundaries of the political units that existed in the past. Since the modes of communication were not very advanced, the people used to stay in groups in different areas and their movement and progress was highly dominated by the landscape characteristics. This forms the basis for our present work. Davies (1949: 2) mentions 'The influence of the physiographic environment on the history of a country is nowhere more apparent than in India'.

In the present research paper we have made an effort to delineate the boundaries of republics based on the historical literal references and plotting them on the topography to form the units.

Background and Review of Earlier Work

Except for findings of small, fortified townships and the usual articles of day-to-day usage, one hardly finds any decisive archaeological evidence to identify the Janapadas or trace their borders. Numismatic evidence also proves to be of no help in this regard. The earliest coins in India, the Punch-marked coins (PMCs) possess several symbols being termed as city-, mint-, or guild-marks. They, however, do not reveal any confirmative identity of the issuing authority or agency. Similarly their provenance does not help in bringing out such identity. Numismatists believe that the identifiable coins, especially those bearing legends of the republics, pertain to a later period, second century BC and after (Goyal 1995: 67ff.). Thus, one has to solely depend on the literary data that is available in pieces and full of internal contradictions.

The literary data for the period under consideration comprises the later Vedic texts of Āraṇyakas and Upaniṣads; the Sūtra literature; the texts of grammar: Aṣṭādhyāyī, Vārttikas and Mahābhāṣya; and the Pāli Buddhist texts.

Pāṇini, the celebrated Sanskrit grammarian of the 6th century BC, has presented Sanskrit grammar through 3,993 short aphorisms, in his eight-book work, the Aṣṭādhyāyī. He mentions 'Janapada' both as locality or the inhabitants. He further states (1) Janapadas that have a definite territory, (2) Viṣayas: regions in general, and (3) Gaṇas or Saṁghas: tribes with or without any definite territory, and having a republican government. As auxiliaries to his aphorisms, Pāṇini has given numerous Gaṇas (glossaries) to make his grammar comprehensive and short. Most of these Gaṇas were open for further editing, reducing their authenticity in tracing history (Gode 1952: 203). Kātyāyana, Paṇini's successor of the 4th century BC, has improved upon Pāṇini in his Vārttikas, explanatory and critical aphorisms. Patañjali, the author of the Mahābhāṣya, the first commentary on Pāṇini, has expounded 1,612 aphorisms duing the 2nd century BC. He gives numerous details about Pāṇini's Janapadas. On an overview of these texts one finds that Pāṇini mentions thirty-one Janapadas. Kātyāyana adds four, and Patañjali forty-one. Kātyāyana repeats two of Pāṇini's Janapadas, and Patañjali twenty-seven. Kātyāyana and Patañjali have omitted twenty-one and four Janapadas of Pāṇini respectively. Regarding republics, Pāṇini mentions Uśīnara, Trigarta, Yaudheya, Rājanyaka, and Vṛjji. Kātyāyana adds Kṣudraka and Mālava, while Patañjali puts in Audumbarāyaṇa, Vasāti and Videha. The republics of Yaudheya, Kṣudraka and Mālava shifted from North to South in course of time. Some others turned into monarchical government while some, like Kṣudraka-Mālavas and Vṛjjis have established confederacies. It has to be remembered that Pāṇini was a north-westerner while Patañjali is believed to be an easterner. Kātyāyana's home still remains untraced.

The Buddhist texts, believed to have been compiled during the fourth-third century BC, provide a heap of geographical data described in great details. They bring out a different picture of the period. As mentioned earlier, the Sanskrit grammatical texts have been referred to in original while the others through Cunningham (1871), Dey (1927), Agrawala (1953), Agnihotri (1963), and Law (1943 and 1967). Law has been taken to be the most reliable and confirmatory source in this regard.

Schwartzberg (1992: 166) mentions, 'Though they (the texts) probably contain genuine recollections, opportunities for imaginative interpolation and confusion in the surviving redaction must have been great. Hence, our plotting of names is very tentative and no attempt has been made to depict the spatial limits of the regions presented. However, we are guided in our plotting by literary references to rivers, forests, mountains and other physical features that were set to have been the limits of particular realms'. Schwartzberg, in his work, has created a series of maps. These maps are considered as a most authentic piece of work for plotting the political history of India for the last more than 4000 years. This work is based on a lot of groundwork, and ancient literature support. The plotting of India political, revealed in the plates about Mahājanapadas (8th–6th century BC), Pāṇini's Aṣṭādhyāyī (5th–4th century BC) and Magadhan Hegemony depict the transitional period of these units. The survival and evolution of several groups has been portrayed in this series of maps. We would like to mention that this work has been of immense assistance for depiction of the boundaries of the Janapadas and republics.

Janapadas and Republics

In the Ṛgveda one finds the concept of 'state' and 'kingship'. During this period of pastoral life, two councils of influential citizens, namely Sabhā and Samiti controlled the king. Later on, these councils lost their importance due to the rising central despotic power and organization of small states. These states, named Janapadas, were not larger than a few Tahsils or Districts. The larger ones, conventionally numbering sixteen, were large enough to be called 'Mahā-janapadas'. Most of the Mahājanapadas were monarchical in nature. The ones still ruled by 'tribes' are believed to have been 'republics' or oligarchies. Perhaps franchise was limited and alternatives few. Contentious references in terms of chronology are available about the republics getting converted into monarchies and vice versa. During the first three centuries of the Mahājanapadas (8th-6th century BC) there was no paramount power in India (Law 1951: 1ff.). Small clans governed themselves in republican manner. These were Yaudheyas, Kṣudrakas and Mālavas, Āśvalāyanas in the West, North-West; and Śākyas, Licchavis, Videhas and Mallas in the East.

Slowly, confederacies, namely, Vṛjjis, Kāśī-Kosalas, Kṣudraka-Mālavas, and major states, such as, Magadha, Avanti, Cedi, etc. came up in the form of imperial powers. In the 6th-5th centuries BC the Magadhan hegemony was established in the Gangetic plains, thanks to the efforts of Bimbisāra and Ajātaśatru. This hegemony continued its influence on the region under their successors and the Nandas (Law 1951: 23). While Megasthenes, the illustrious author of the Indika, during the third century BC, reports that many republics had switched over to monarchies, but one finds that the Yaudheyas, Kṣudrakas, Mālavas and others still continuing their tradition in new regions of occupancy. It is notable that the traditional nomenclature of the regions continued in spite of shift of political power. The cause of this co-existence of the Magadhan imperial power and the relatively petty republics was the practice of regular tribute being paid by the latter to the earlier. They also formed frequently occurring matrimonial alliances. The republic of Trigarta seems to have been replaced during the 5th-4th centuries BC by smaller units of Kulūta, Tilakhala, Rājanyaka and Uśīnaras. The advent of the Mauryas during the last quarter of the 3rd century BC, however, shaped the end of most of the republics.

Study Area and Data

The study area of this research is mainly the Gangetic Plain with some extensions in the east-west directions. This area, being the most fertile part of the subcontinent, naturally attracted the groups, tribes to settle and form new societies. These are later referred to as Janapadas and when they agglomerated to larger units, as explained earlier. Later though these groups started invading in each other's territory, until the Magadhan Empire came into existence, these invasions were limited to the Gangetic plain. The Magadhas established themselves as a single dominant force in the subcontinent. The geographical extent of this area is roughly 20° N to 35° N and 65° E to 90° E. Though this area also includes regions outside the Gangetic plain, they played crucial role in the geopolitical history of India.

Since the area is vast, this is a macro level study and is based on the broad descriptions of these political units. The datasets which are created for the study mainly encompass the SRTM (Shuttle RADAR Topographic Mission) 90m pixel data for the entire Indian subcontinent (seamless mosaic of SRTM tiles), administrative boundaries of India upto second lowest administrative unit i.e. Tahasil (1:1,000,000 scale), more than 16000 town locations in India (including state, district sub-divisional headquarters and other towns) and drainage network of India. This vector data is authentic Survey of India data, which is renowned for the positional accuracy. The present administrative unit hierarchy in India is Country > State > District > Tahasil > Village/ Town. Some sample data from ESRI was also used for the administrative boundaries in the neighbouring countries of India. Some of the Janapadas

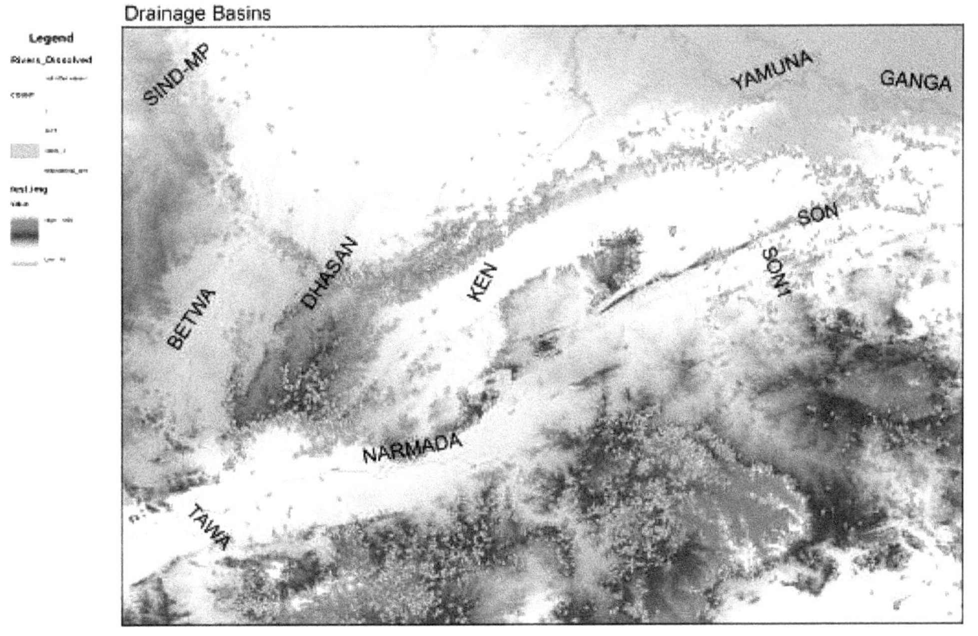

Figure 1

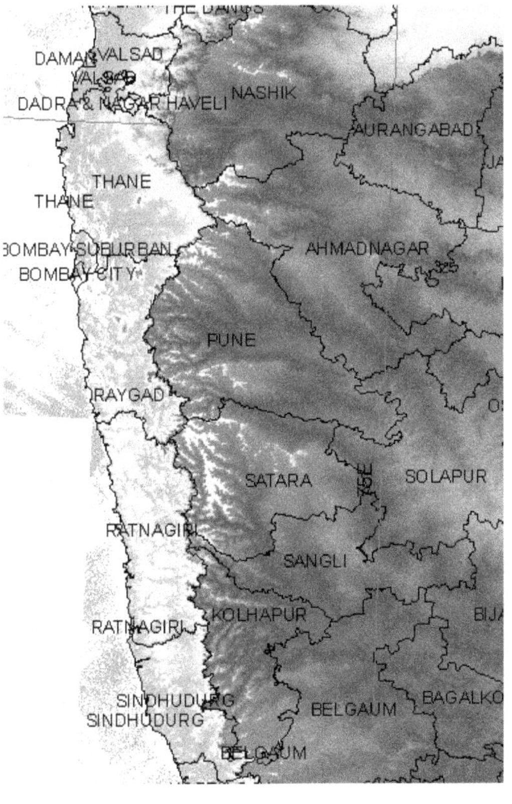

Figure 2

and Mahājanapadas, which existed in the past, have their relics on the present landscape. Their cultural peculiarity or nomenclature of present towns makes it evident that the historical background has percolated through the ages and is visible even today in many aspects of the landscape and culture.

Methodology

The foundation of the present research paper was that the geographical features formed barriers for the human groups and played a vital role in defining boundaries of the Janapadas. They dominated other barriers such as linguistic difference, cultures etc. To create these geographical units, the smaller units were considered which were to be merged later based on different criteria to form the actual political units of Janapadas. By 7th century BC, many of these Janapadas amalgamated to form Mahā-janapadas. The basic geographic unit that was considered in the beginning of the study was a drainage basin. This is because these basins form catchments of water-bodies and they share the same geographic characteristics within themselves. A mosaic covering the most part of the Indian subcontinent was created using SRTM tiles. However, while handling this mosaic in ESRI suite of products, several limitations were faced. These were mainly due to the large data size (>3GB). When ArcGIS9.1 routines were run to create the drainage basins from the SRTM DEM data, a memory allocation problem was encountered and it persisted through out the work. Some work around was explored but due to the time and resource limitations, the solutions were restricted. Grid resampling was one option to reduce the data size but was not preferred. So a smaller portion of the AOI was taken as a subset to run the Basin tool. However, the basins that resulted were not as per our expectations.

This, being a macro-level study, could not be extended to study these micro watersheds in the present scope of work. *Fig.1* shows the basin polygons which were created on the SRTM data.

Since these geographical entities could not be defined, it hampered the basic foundation of the study. These entities formed the basic for forming the larger units. However while comparing the current administrative unit boundaries with the topography, it was revealed that even today, with many advances in modes of communications/ transport/ lifestyle, the topography still plays a vital role in defining these administrative unit boundaries. Of course, their dominance is lesser than what it must have been in the past but still it persists.

Fig.2 shows the Western Ghat overlaid by the district boundaries. It clearly shows the demarcation between Konkan (Western coastal region) and the Deccan Plateau. Noticing this, the Tahasil boundaries, forming second lowest administrative units, were considered as primary entities instead of drainage basin areas (*Fig.3*).

Though this was a significant change in the basic assumption, it was later revealed that even this definition of basic entity gave lot more inputs for the present study. In fact in some cases, it proved to be a better option to the drainage basin concept.

These entities defined above were studied in relation to the drainage network and underlying topography. In the Gangetic plain, since the river courses have been changing drastically, the literal references need to be analysed very carefully as the descriptions deviate from what it exists on the land today. One such example was seen in case of Pāṭaliputra, capital of the Magadhan Empire. It is said that it was situated on the confluence of rivers Śoṇa and Gaṅgā (*Mahābhāṣya* 2.1.16). The present city of Pāṭaliputra (Patna) is situated on Ganga, about 30 km away from the Ganga-Shon confluence. When observed closely in Google Earth, traces of inactive channel meeting Ganga near Patna are clearly visible. However, the use of Google Earth was limited only for observation and was not used for drawing any conclusions.

The Janapadas, as mentioned in the literature review, mainly come from Pāṇini (6th century BC), Kātyāyana (4th century BC) and Patañjali (2nd century BC). For the convenience of the present work, the period of BC 800 to BC 300 was roughly subdivided into two sub-periods on the basis of literary evidence. The list of sixteen Mahājanapadas occurring in various texts varies considerably in content (Law 1951: 1ff). We tried to assimilate these lists to make a more exhaustive and authentic depiction of the Mahājanapadas. Since our interest is mainly in the republics, many of the monarchical Janapadas fall out from the scope of the work. Still, for the spatial consistency and continuity of the boundaries, some of the monarchical Janapadas are also plotted. In some cases, Schwartzberg's maps were georeferred to the present vector data for the ease of plotting. However, since these maps are tentative, the cross referencing of the plotting had to be done with other mentioned sources of literature.

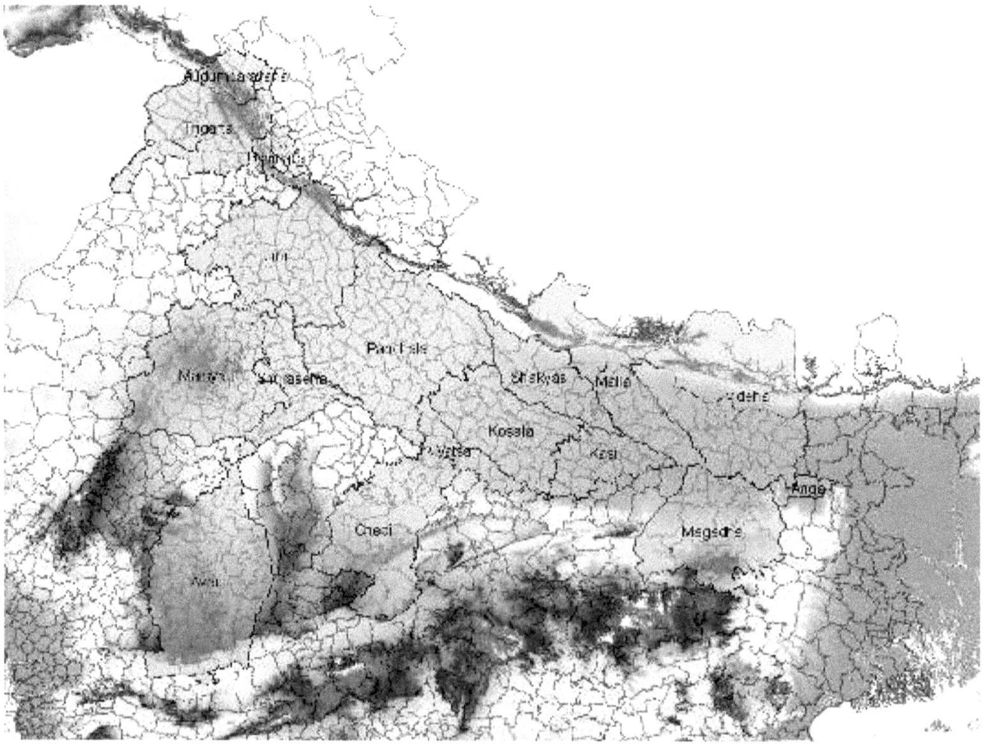

Figure 3

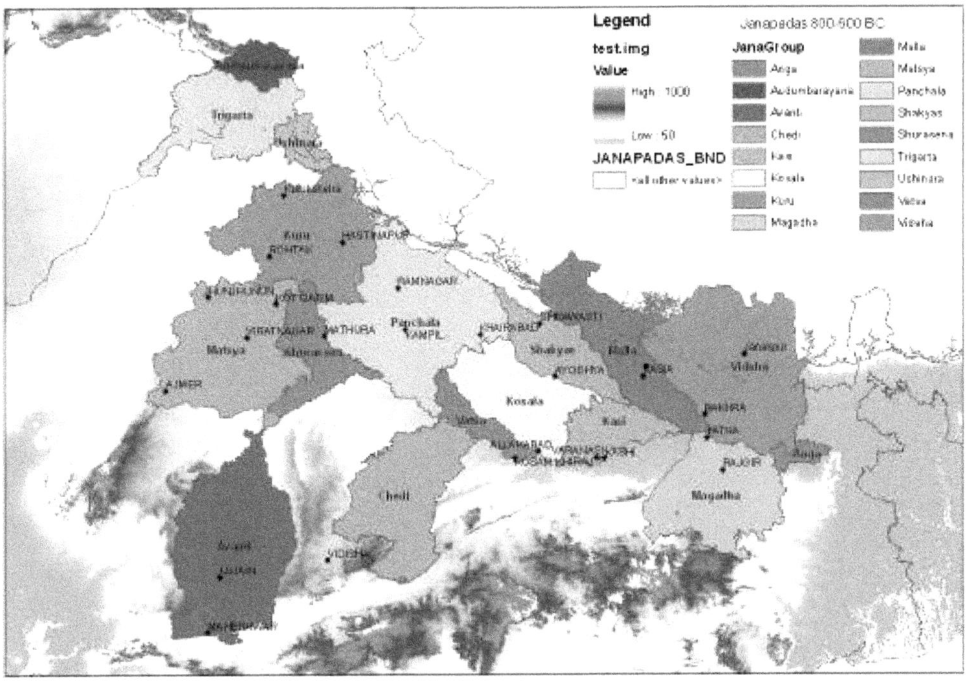

Figure 4

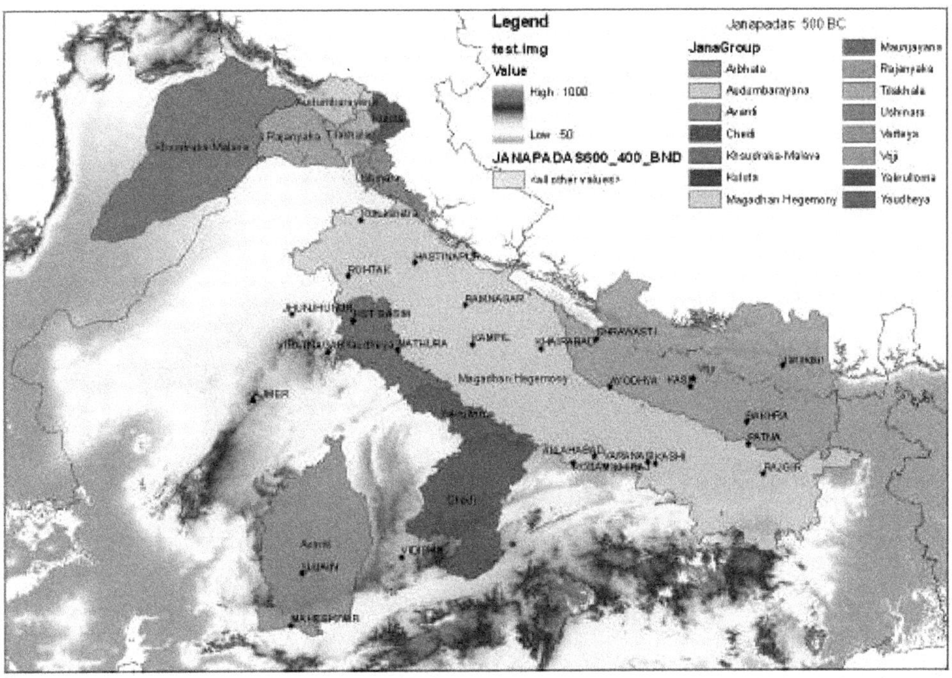

Figure 5

Janapadas identified for the period of 800 to 500 BC are: Aṅga, Audumbarāyaṇa, Avanti, Cedi, Kāśī, Kosala, Kuru, Magadha, Malla, Matsya, Pañcāla, Śākya, Śūrasena, Trigarta, Uśīnara, Vatsa and Videha (*Fig.4*).

Janapadas identified for the period of BC 500 to 300 are: Ārbhata, Audumbarāyaṇa, Avanti, Cedi, Kṣudraka-Mālava, Kulūta, Mauñjāyana, Rājanyaka, Tilakhala, Uśīnara, Vṛjji, Yakṛlloma, Yaudheya, Vārteya and the Magadhan Hegemony (*Fig.5*).

Interpretation and Discussion

Though the above list of Janapadas is quite extensive, we have concentrated on four major republics for the discussion, viz. Trigarta, Vṛjji, Yaudheya and Kṣudraka-Mālava. The exact placing of these Janapadas with respect to the underlying topography has been a challenge.

As already mentioned in the methodology, though river basis were not considered as basic entities and Tahasil polygons were chosen instead, it was revealed that it correlated with the purpose of the present study more closely. If river basin is considered as a unit then the river becomes a part of the unit and the people should be able to commute from either sides of river across the bank. It is possible in cases of small rivers but rather unfeasible for bigger rivers.

In majority of the cases where the literal descriptions of the boundary are there, they mention the river itself as a boundary. So this clearly divides the basin in separate political units. River as a 'barrier' seems to be more prominent than the 'containment' of the basin.

Trigarta: Pāṇini, in his aphorism 4.1.111 and his Gaṇa on 4.1.178 names it as a fighting tribe (Āyudhajīvi-saṁgha). His Gaṇa on 4.2.53 and the epigraphic record shows that the area around Jalandhar was Trigarta (Law 1943: 74). Dey (1927: 205) informs that the town of Tohara or Tihora, near Ludhiana, was the original Trigarta. The Mahābhārata (2.48.13) reports the republic of Trigarta to be situated between Irāvatī (Ravi) and Śutudrī (Sutlej). However, this does not match with the literary meaning of the name of the republic, Tri-garta—situated between three ditches or water bodies. Cunningham rightly infers that Trigarta was situated between the three rivers Irāvatī (Ravi), Śutudrī (Sutlej) and Vipāś (Beas), or in the Kangra valley (*Archaeological Survey Researches*, V: 148, After Law: 1943). An attempt has been made to depict the republic in its right place without causing any harm to its neighbouring contemporaries (*Fig.6*).

The old southern course of Sutlej has been taken into consideration while depicting the Janapada. It seems that during the 4th-3rd century BC, the Buddhist texts attach a great importance to Trigarta. From Patañjali's data, it seems that the republics of Rājanyaka, Audumbarāyaṇa, Tilakhala and Uśīnara came up in this region in spite of the continuation of its cultural nomenclature. (*Fig.7*)

Vṛjji: Eight clans formed the earlier Vṛjji confederacy. It was included in the list of Mahājanapadas. The important ones among them were Videha, Malla and Śākyas. The Brāhmaṇa texts describe Videhas to be situated between the rivers of Kauśikī and Sadānīrā (Kosi and Gandak) (Law 1943: 235). To their east fall the Mallas, between Sadānīrā and Ajiravatī (Rapti), and Śākyas still beyond. The important urban centres in the confederacy were Janakapura, Kuśīnagara, Kapilavastu, Śrāvastī, etc. Liccavi was the most powerful clan of the region, perhaps spread all over the confederacy (*Fig.8*).

Ajātaśatru has been told to have defeated the Vṛjjis. They were further overpowered by the Nandas and totally subdued by Mauryas (Law 1967: 53-54). Shifting of the said rivers in course of time has been considered in depiction.

Yaudheyas: Pāṇini has coined a Gaṇa on their name (4.1.178). As indicated by their name, 'warriors', they form one of the Āyudhajīvi-saṁghas. The Mahābhārata mentions the Yaudheyas as descendents of Yudhiṣṭhira.

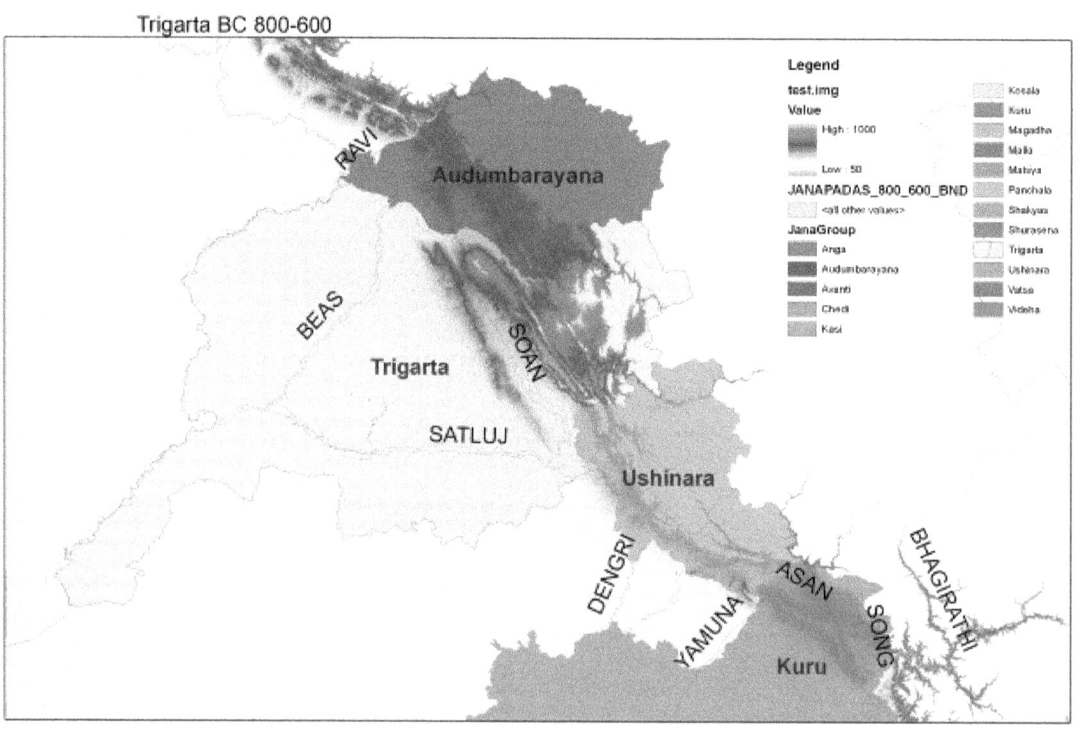

Figure 6

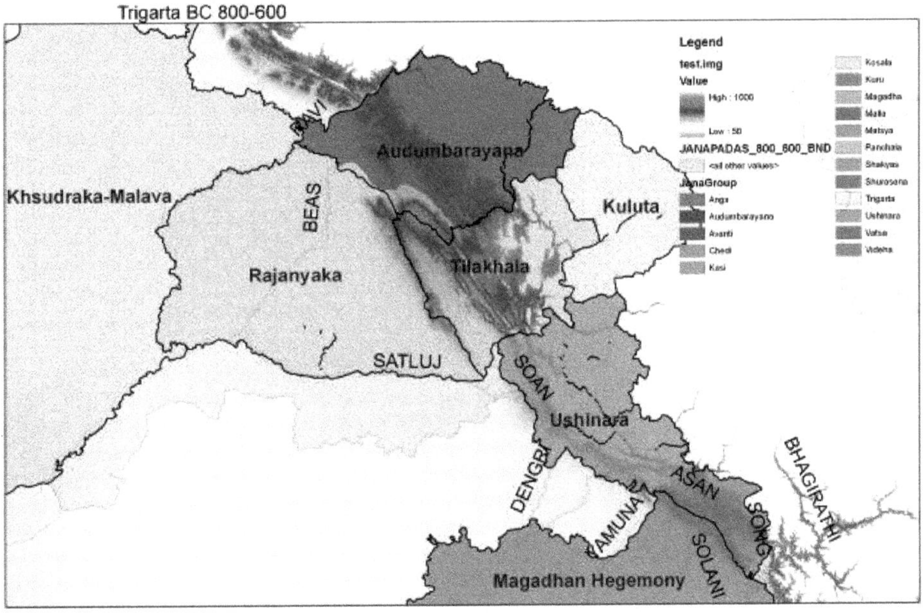

Figure 7

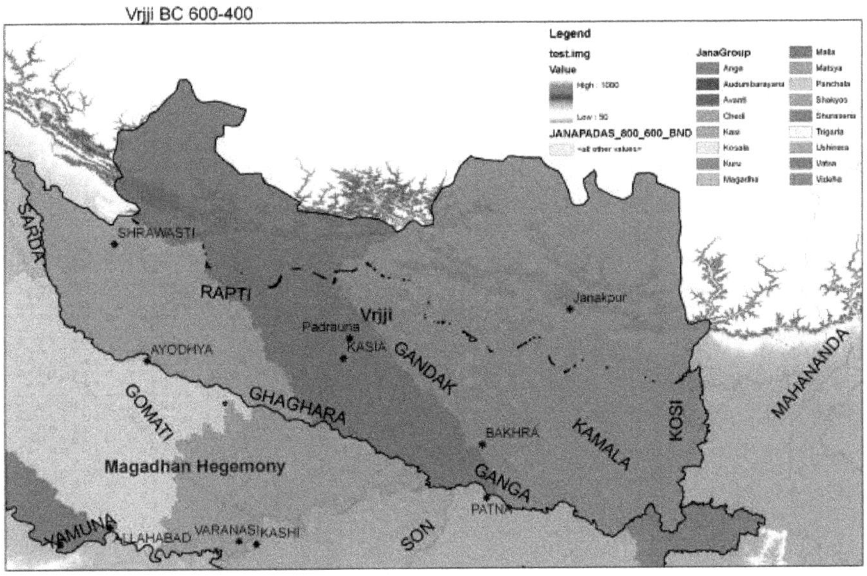

Figure 8

They previously ruled Punjab and adjoining regions. According to some historians, they had their capital at Johiabar, a place near Multan. The coins of the Yaudheyas, however, were found at Bijaygarh near Bharatpur in Rajasthan. Thus it seems that the Yaudheyas shifted from Punjab to Rajasthan sometime between the 5th and the 2nd centuries BC. People named Johiyas, descendents of Yaudheyas, are found in the region of Bharatpur to confirm this shift in their inhabitation.

Kṣudrakas and Mālavas: Kātyāyana and Patañjali compound these peoples for their permanent military confederation (4.2.45.1). They report that the two collectively 'churned' their enemies. The Greek sources locate them between Hydaspes (Vitastā - Jhelum) and Hydraotis (Irāvatī - Ravi). The Mahābharata later locates them with Trigarta. Their inscribed coins are found in the area around Nagara near Kota in Rajasthan (Law 1967: 130). Perhaps they shifted to Punjab and Rajasthan after a stunning defeat at the hands of the Greeks and further to south in view of the absence of any paramount power in that region (Law 1943: 60-66). In the Nashik cave inscription of the 1st century AD, Uṣavadāta locates them in Malwa (Gokhale 1975: 138-139). *Fig.9* depicts their earliest position.

Conclusion

SRTM, that is the source of free topographic data, is sufficient for the macro level study of the present scale. Since the formation of basic units is totally dependent on this data, it has proved to be satisfactory except for the fact that automated basin generation could not be done.

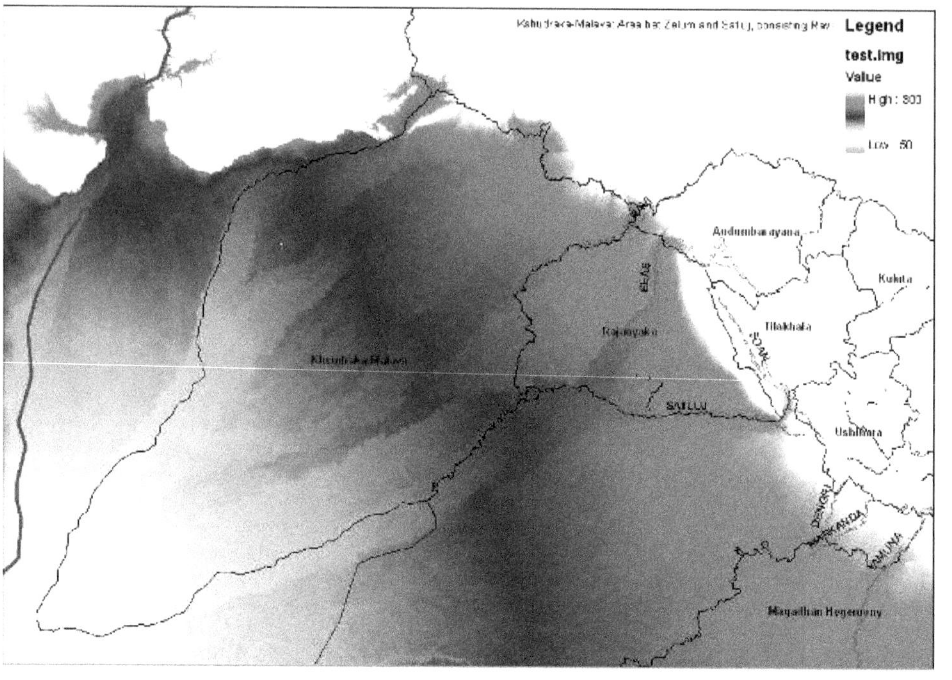

Figure 9

Studying such vast area without this data and without the aid of GIS softwares would have been an unimaginable task.

This wok forms the foundation for defining 'Janapada Era' in the geographical history of Indian subcontinent. Defining the maps for narrower time periods with micro level spatial units (village as administrative unit and basin as geographical unit) will improve the plotting accuracy of these Janapadas.

Acknowledgements

We thank Tilak Maharashtra Vidyapeeth for making library sources available to us and for extending all possible cooperation. We also thank TechniGraphicS (India) for allowing us to use hardware and software for this study.

References

AGNIHOTRI, PRABHU DAYAL. 1963 *Patañjalikālīna Bhārata.* (Hindi). Patna: Bihar Rashtrabhasha Parishad.

AGRAWALA, V. S. (1953) *India as Known to Pāṇini.* Lucknow: University of Lucknow.

CUNNINGHAM, ALEXANDER. 1871 (Reprint 1996) *The Ancient Geography of India.* Delhi: Low Price Publications.

DAVIES, C. COLLIN 1949 (26th impression 1994) *An Historical Atlas of the Indian Peninsula.* Delhi: Oxford University Press.

DEY, NUNDO LAL. 1927 (Reprint 1994) *The Geographical Dictionary of Ancient and Medieval India.* Delhi: Low Price Publications.

ESRI, last accessed on 27/09/06 at URL http://esri.com/.

GODE, P. K. 1952 *Studies in Ancient Indian Cultural History,* Vol. I, Hoshiarpur: Vishwshwaranand Vedic Research Institute.

GOKHALE, SHOBHANA. 1975 *Purābhilekhavidya (Marathi).* Nagpur: Maharashtra Universities Book Production Board.

GOYAL, S. R. 1995 *The Coinage of Ancient India.* Jodhpur: Kusumanjali Prakashan.

LAW, BIMALA CHURN. 1943 (Second Edition 1973) *Tribes in Ancient India.* Pune: Bhandarkar Oriental Research Institute.

LAW, BIMALA CHURN 1951 (Reprint 1990) *Northern India in the sixth century B.C. History and Culture of the Indian People I, The Age of Imperial Unity.* Bombay (Mumbai): Bharatiya Vidya Bhavan, 1-17.

LAW, BIMALA CHURN. 1967 *Historical Geography of Ancient India.* Paris: Société Asiatique de Paris.

SCHWARTZBERG, JOSEPH E. (ed.) 1992 *A Historical Atlas of South Asia.* Second Impression. New York: Oxford University Press.

SRTM, last accessed on 27/09/06 at URL http://srtm.usgs.gov/

Reconstructing an Iron Age and Roman Landscape – new research in the Foulness Valley, East Yorkshire, England

Peter Halkon

Department of History, University of Hull

1 Introduction

Recent analysis of the whole of the catchment of the River Foulness in East Yorkshire, England, (Fig. 1) (Halkon 2006) has enabled the construction of a new model Iron of Age and Roman landscape development within this region, which contains one of the largest Iron Age iron industries yet found in Britain. In an award-winning community-based research project undertaken since 1980, one of the longest-running and most detailed carried out in the North of England, over 1,448 ha have been field walked, large-scale geophysical surveys carried out and 13 seasons of excavation undertaken (Halkon and Millett 1999 and 2003, Halkon 2003, Millett 2006). Within this 20x30 km area, over 1245 crop mark sites from the English Heritage National Mapping Programmes of the Yorkshire Wolds (Stoertz 1997) and Vale of York, and our own aerial survey, have been studied (Halkon 2006). Examination of Sites and Monuments Registers and museum records, together with field survey, has revealed over 600 Iron Age and over 525 Roman sites in this landscape block.

2 Landscape background

Through study of topography, drainage, soils, vegetation, climate, sea-level rise and coastal change, three distinct zones, have been identified within the Foulness Valley (Fig. 2), each with a different trajectory of development:

- Zone 1 - The upper Foulness valley, where feeder streams flow through the Yorkshire Wolds and foothills, with calcareous soils and Brown Earths. Palaeoenvironmental study shows that like the Yorkshire Wolds further to the north (Bush and Flenley 1987; Tweddle 2001) much of this zone was extensively cleared of woodland by the beginning of the periods of study.
- Zone 2 - To the north are the Yorkshire Wolds foothills, with ridges of gravel and sand. The centre of the zone is dominated by sandy soils with some Brown Earths to the north. Many of the streams and rivers are in this zone. Areas of low-lying alluvium fringe some of the rivers, with some larger areas of wetland. The north of this zone seems to have been more open with the amount of woodland increasing towards the lower Foulness Valley, where it provided a resource for an Iron Age Iron industry and Roman pottery industry (Halkon and Millett 1999).
- Zone 3 - the lower Foulness valley containing the Walling Fen, a tidal estuarine inlet in the Iron Age and Roman periods, and lowlands, comprising glacio-lacustrine clays, postglacial silts and alluvium. Although Zone 3 remains wet during most of the study period, there was greater exploitation during a climatic amelioration of the late Iron Age and second century AD, paralleled in the South Yorkshire lowlands and Humberhead Levels (Smith 2002; Fenwick et al 1998; Riley et al 1995) and the Hull valley (Didsbury 1990).

The Walling fen inlet, replaced a forested fresh-water landscape at the beginning of the Iron Age (Halkon and Millett 1999; Long et al 1998) opening up the wider region to the Humber and the North Sea beyond. The tradition of the square barrow (burials covered by a square earth platform enclosed by a ditch) and chariot burial, characteristic of the "Arras Culture", and perhaps iron production itself, (Halkon forthcoming) may have arrived into this region through this route. Like the burial traditions (Stead 1979, 1991), there do appear to be parallels between the iron technology of northern France (Cabboi and Dunikowski 2004) and that of the Foulness Valley. The cache of five swords in decorated sheaths and a bundle of spears, found recently near South Cave, deposited at the time of the Roman conquest of what is now Yorkshire (Evans 2006) may mark the zenith of iron production in this region.

The main focus of this paper is the characterisation and analysis of the crop mark sites of the Foulness Valley, which differ in both density and form, between the three landscape zones partly due to environmental factors (Halkon 2006).

3 Iron Age landscape development

In the Later Bronze Age, a network of long distance linear earthworks was constructed developing through the Iron Age (Mortimer 1905; Fenton-Thomas 2003). These, range in complexity from a single bank and ditch to complex systems of five or six parallel banks/ditches, best exemplified by Huggate Dykes (Mortimer 1905; Halkon 1990). Although more dating evidence is needed, these may correspond with similar features elsewhere in Britain (Bradley 1994). Their distribution contrasts between the three zones, with many more visible in Zone 1 than Zone 2 and none in Zone 3. Many linear earthworks correspond closely with landscape features

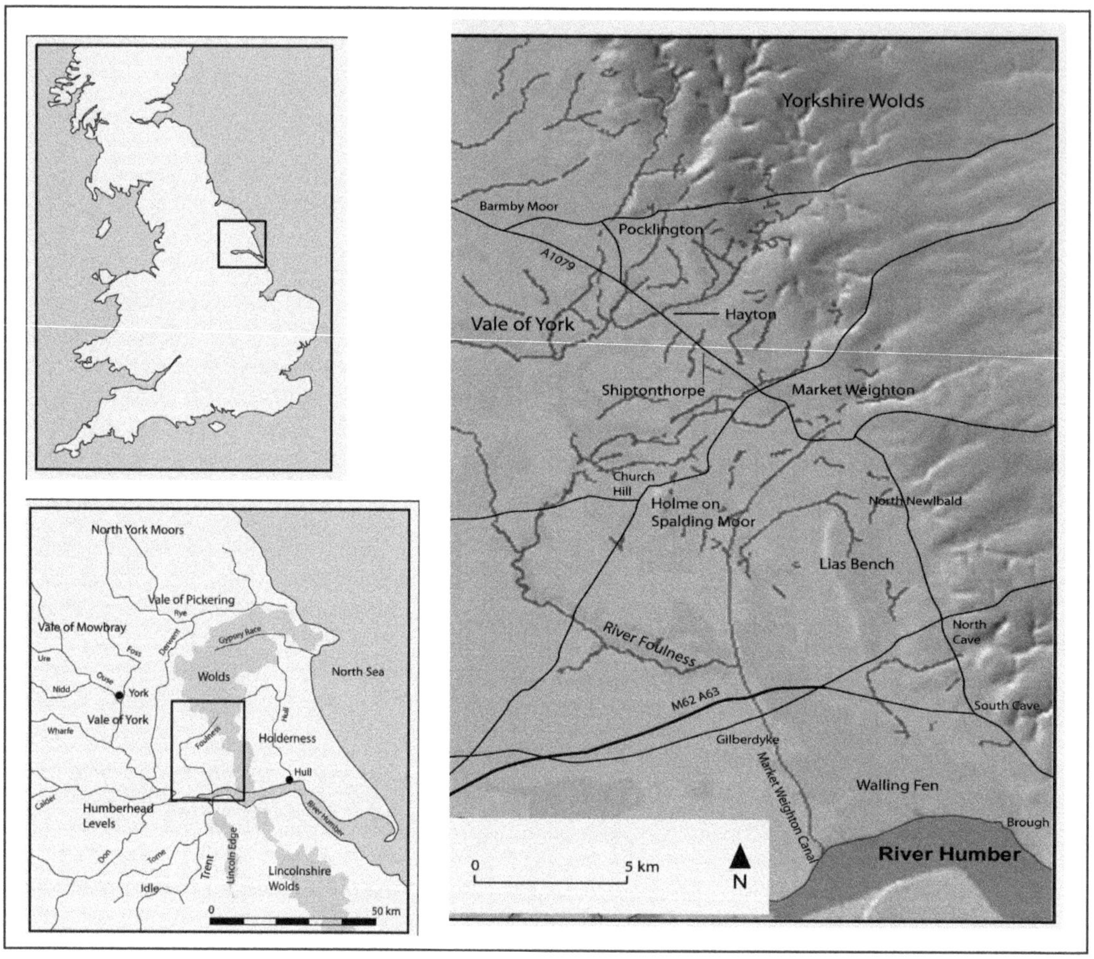

Fig. 1 *Location map*

such as springs and streams. They might have controlled access to water resources, related to routes through the landscape, or have acted as barriers (Halkon 2006). The interrelationship of some systems suggests major landscape realignments.

There is likely to be a link between some linear earthworks and curvilinear enclosures (Manby et al 2003; Powlesland 1988) which may involve the deliberate division of the landscape at a period of climatic instability (Blackford and Chambers 1991; Tweddle 2001; Smith 2002; Whitehouse 2004). The interpretation of hilltop enclosures within the Foulness Valley is difficult as only Grimthorpe (Stead 1968) has been excavated and only one other surveyed geophysically. It has been claimed that unlike other regions where hill-fort use continued into the Roman period, those in East Yorkshire were abandoned in the Middle Iron Age (Millett 1990; Dent 1995). We cannot be certain of this until more research is done. Even if this was the case, it is interesting to note that several enclosures may have continued as significant landscape features. It may not be coincidence that a well-furnished burial with a shield and sword excavated at Grimthorpe Manor (Mortimer 1905), dated typologically to over a century after the *floruit* of the hill-fort, was aligned on its possible southeastern entrance. The burial position of the roughly contemporary weapons cache referred to above may also relate to a hilltop enclosure which overlooks the River Humber. Perhaps we are seeing in the Grimthorpe burial and the weapons cache some form of recognition of the past importance of these places, linking power, memory and landscape. The concentrations of springs close to both enclosures may also be significant.

Several curvilinear enclosures, resembling Staple Howe (Brewster 1963; Stoertz 1997; Halkon and Millett 2000) the Early Iron Age type site, were identified in lowland areas, as well as the Yorkshire Wolds. Some may have been "marsh forts" and others may have related to ritual activity associated with water in a similar way to the site at Sutton Common in South Yorkshire (Parker-Pearson and Sydes 1997; Van De Noort 2004). At Market Weighton Common, three curvilinear enclosures and a curved ditch, which may be part of a fourth enclosure, clustered between a relict watercourse and the present route of the Market Weighton Beck (Ramm1978).

Perhaps the most characteristic Middle Iron Age crop mark features revealed in the East Yorkshire landscape are the square barrows referred to above. The study area contains almost 400; the majority (289), were located in

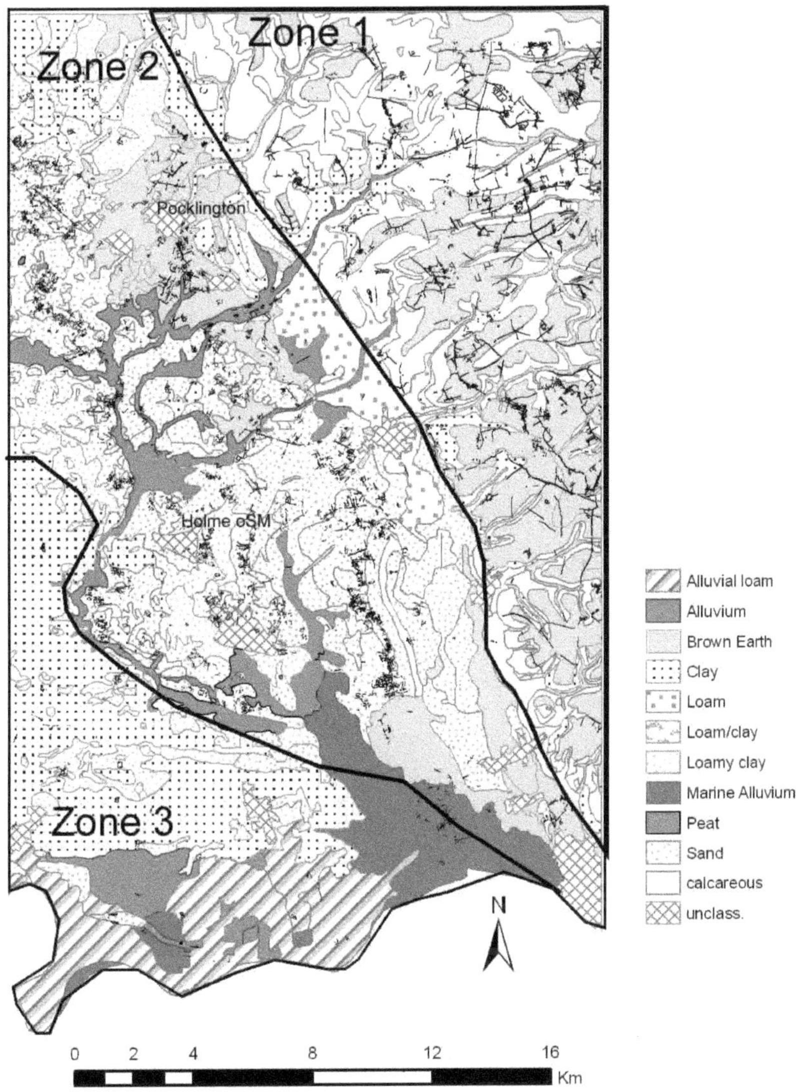

Figure 2 *Crop marks and soils in the Foulness Valley.*
(Crop mark plots courtesy English Heritage. Soil map simplified from King and Bradley 1987).

Zone 1, with 104 in Zone 2 and only 4 in Zone 3. Analysis of square barrow distribution (Halkon 2006) suggests a connection with water supply and routes through the landscape. The positioning of some burials along some linear earthworks may suggest territoriality of some form. The larger square barrow cemeteries at Arras, Warter and North Dalton, (Zone 1) all overlook major valleys, which are likely to have formed route ways through the Wolds. Many of the smaller cemeteries and single barrows are similarly situated. In Zone 2, square barrows seem to be associated with watercourses.

Clear differences could be determined in the distribution of the single rectilinear enclosures of the Middle and later Iron Age and the linear enclosure complexes or "ladder settlements", the strings of enclosures at each side of the droveway resembling the rungs of a ladder (Stoertz 1997) (Fig 3). Single rectilinear enclosures are one of the classic forms of Iron Age settlement type in the north of England (Haselgrove 1981). In the Foulness Valley, 24 of the 31 single rectilinear enclosures plotted are in Zone 2, with only six in Zone 1 and only one in Zone 3. All were situated on either sand or Brown Earths. Some, especially those situated on wider expanses of sandy soil had associated field systems. Where enclosure entrances were visible they seem to relate to watercourses (Fig 3). The biggest concentration of enclosures is in the area between the River Foulness and the Yorkshire Wolds, the ditches necessary for drainage in an area with a high water table. During the course of this project one of these sites at Bursea Grange (Halkon and Millett 1999) was excavated.

Of the 46 linear enclosure complexes recorded in the study area, 60% were situated in Zone 1 in contrast to the distribution of single rectilinear enclosures, with only 20% occurring in this zone. Foulness Valley linear enclosure complexes ranged from 100m to 2.6km in length. As Stoertz (1997, 53) notes, superimposition and inter-cutting makes it difficult to count the number of enclosures which go to make up this settlement type, though the writer counted between 1 and 90, with 86% having less than 20. The enclosures themselves range in

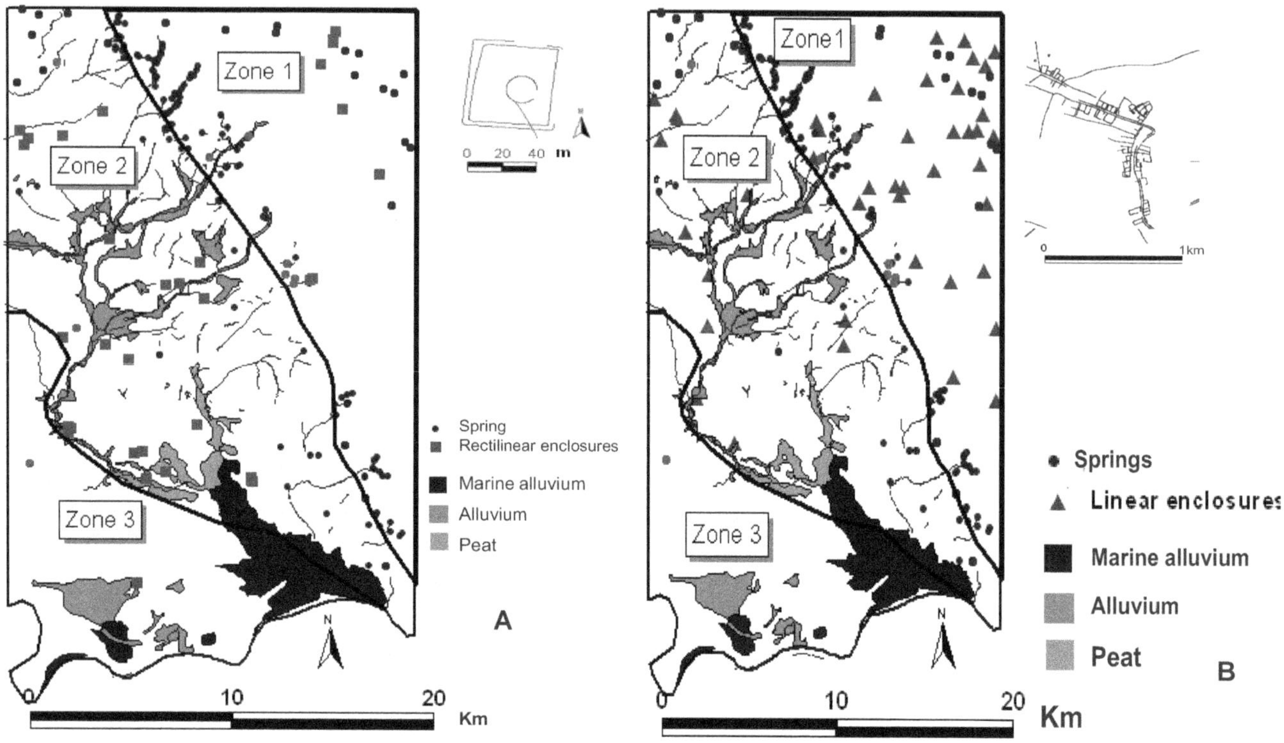

Fig. 3 *Drainage, watercourses and single rectilinear enclosures (A) and linear enclosure complexes (B) The inset in A shows the plot of an example from Bursea (Halkon and Milllett 1999) and in B at Arras (Halkon 2006). (Plots courtesy English Heritage).*

area from 0.08ha to 0.53ha with 83% falling within the range 0.08 to 0.2ha. This is very similar to the area of the single rectilinear enclosures.The droveways at the centre of the linear enclosure complexes may have been used to lead stock to water or have served as a stock management system. The multiplication of enclosures may relate to the agricultural intensification in the later Iron Age or perhaps the apportionment of land parcels amongst families, though it is difficult to be sure about their precise function from aerial photographic evidence alone.

The differences in distribution between Zones 1 and 2 are almost certainly due to environmental factors, the open Wolds landscapes of Zone 1 being more conducive to the development of long enclosure strings. The development of complex enclosure systems, which are often palimpsests of features, also shows settlement expansion in the later Iron Age and Roman periods, though this was not constant. Palaeo-environmental study elsewhere in Yorkshire provides evidence for large scale woodland clearance and an increase in cereal and grass pollen, though woodland and wetland continued in the lower Foulness Valley throughout the Roman period apart from clearance on sandy ridges where most Zone 2 complex enclosure systems were positioned.

3 Romano-British landscape development

In the Roman Foulness Valley, although some crop marks suggest settlement/land-use evolution over a long time-span, there is evidence for major landscape re-alignment, particularly where the Roman roads cut through pre-existing crop mark systems. This is apparent at Hayton (Taylor 2001; Halkon and Millett 2002) where Iron Age, settlement features generally followed the "grain" of the landscape, along the gravel terraces parallel to the Hayton and Burnby Beck. The Roman road crossed the beck, and rather than being orientated on the watercourses, new settlement was attracted to the road itself. In other respects, the strong correlation between site location and landscape constraints seems to have continued. The Roman forts at Hayton (Johnson 1978) and Brough (Wacher 1969) and roads closely relate to soils, terrain and watercourses. In the later Roman period there was a contrast between the lower Foulness Valley, where a pottery industry in a landscape of wetland and woodland continued Iron Age of furnace based production, and the more open, elevated and fertile Wold margins, where "villas" (Taylor 2001) developed. During the fieldwork project "new" villa sites were identified, particularly around Pocklington. The building at Hayton (Halkon et al 2000, Halkon and Millett 2003) was excavated, and geophysics, aerial photography, soil chemistry and detailed field walking have proved effective here. All of the villas so far investigated seem to have developed from Iron Age settlements. Whether this was a gradual evolution, rapid development or merely due to similar environmental constraints at both periods, must await further analysis.

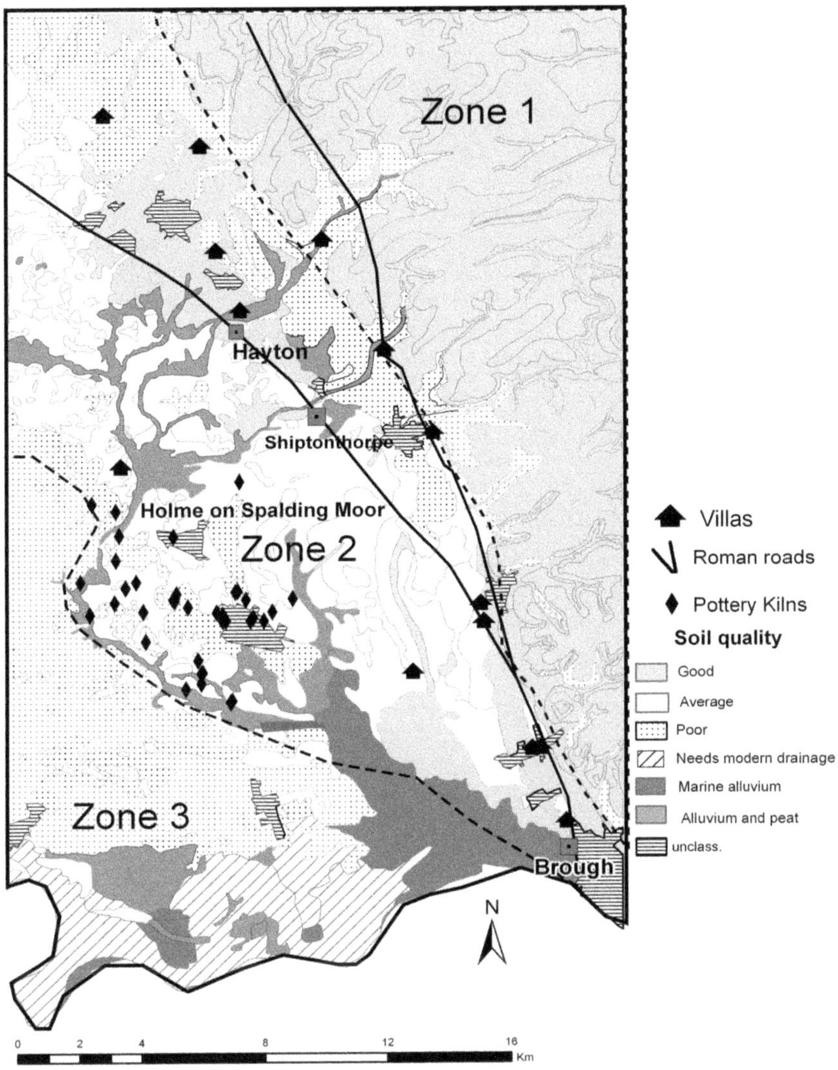

Fig. 4 *The Roman landscape – villas roads and soil quality*

The ability of villas to flourish in the study area, especially at the boundary of Zones 1 and 2, may represent a combination of relatively good agricultural land, convenient watercourses and easy access to the main Roman roads. All but 2 of the 14 Foulness Valley villas recorded are on the best quality land close to the Roman roads (Fig 4). At Shiptonthorpe (Halkon and Millett 2003; Millett 2006) and Hayton (Halkon et al 2000; Halkon and Millett 2003) larger settlements grew up in slightly elevated positions where the Roman road crossed streams. A combination of techniques has proved successful in elucidating the development of these sites and the quantity and variety of surface material recovered demonstrates firmer integration into the Roman system here than at sites in the rural hinterland (Halkon and Millett 1999; Halkon 2006).

Away from the Roman roads, field walking helped to determine the character and chronology of crop mark sites, facilitated by the presence of hard-fired Roman pottery, largely produced in the region (Halkon and Millett 1999).

Intra-site comparison of pottery enabled the identification of settlement shift and highlighted distinct differences between the landscape Zones.

In Zone 1, some linear enclosure complexes were occupied throughout the Roman period. In Zone 2, crop mark superimposition, surface finds and excavation indicated site longevity particularly on the sand ridges (Halkon and Millett 1999). Streams continued to be influential especially in Zone 2. In Zone 3, where moisture retentive soils are not conducive to crop mark visibility, chance finds provided evidence of early Roman activity, possibly relating to lowland exploitation referred to above, resembling that across the North Sea in Sandy Flanders (Vermeulen 1995; 2000) and the Meuse estuary (Brinkkemper 1991). Work elsewhere in the Humber region highlights the importance of rivers for trade (Riley et al 1995; Fenwick et al 1998) and the effect of rising and falling sea levels on occupation. The significance of the Humber waterways has been underestimated, as they provided a route from the North Sea to the legionary fortress and provincial capital of *Eboracum* (York).

As with Iron Age square barrows, a Roman relationship between cult and countryside has also been identified, for example at Millington, in the north of Zone 1. Here what may be rural sanctuary including a circular temple (Halkon et al 2003) resembling Gallic examples (Derks 1998) in its form and location, has been identified. This complex lies at a prominent position at the junction of a number of valleys close to a cluster of springs. The deposition of several large hoards of coins elsewhere in the study area also suggests that rites associated with water seem to have been important throughout the period of study as elsewhere in Roman Britain and beyond (Sauer 2005).

4 Conclusion

Due to the lack of standing remains in this intensively arable landscape, the casual visitor to the Foulness Valley today would be totally unaware of the wealth of archaeological evidence that this region has been shown to contain. The combination of survey techniques employed here has been greatly facilitated by the use of GIS. As well as allowing the rapid and direct comparison of archaeological features against environmental factors. GIS has also provided the platform upon which to promote the better understanding of the area through a website which presents this research (www.ironmasters.hull.ac.uk) (Halkon 2005). In the HIVE (Hull Immersive Visualization Environment) a "pilot" 3D stereoscopic and immersive multi-layered virtual landscape (Pansiot et al 2004) of part of the Foulness Valley has been developed. It is hoped that further funding will allow us to take the exploration of this past world into the future.

References

BLACKFORD, J.J. and CHAMBERS, F.M., 1991 Proxy records of climate from blanket mires: evidence for a Dark Age (1400BP) climatic deterioration in the British Isles. *The Holocene* 1: 63-67

BRADLEY, R. 1984 *The social foundations of British prehistory*. Longman. London.

BREWSTER, T.C.M. 1963 *The excavation of Staple Howe*. East Riding Archaeological Research Trust, Scarborough.

BRINKKEMPER, O. 1991 *Wetland Farming in the Area to the South of the Meuse Estuary during the Iron Age and Roman Periond. An Environmental and Palaeo-economic Reconstruction*. Analecta Praehistorica Leidensia, 24 Leiden.

BUSH M.B. and J.R. FLENLEY 1987. The age of the British chalk grassland. *Nature* 329: 434-436.

CABBOI, L. and DUNIKOWSKI, C Cabboi, L. and Dunikowski, C. 2004 Les systemes de production siderurgiques chez les Celtes au nord de La France. *Pre-Actes,XXVIIIe colloque international AFEAF*, Toulouse.

DENT, J.S. 1995 *Aspects of Iron Age Settlement in East Yorkshire*. Unpublished PhD, Thesis, University of Sheffield

DERKS, T. 1998 *Gods, Temples and Religious Practices: The Transformation of Religious Ideas and Values in Roman Gaul*. Amsterdam Archaeological Studies 2, Amsterdam.

DIDSBURY, P. 1990 Exploitation of the alluvium in the Lower Hull Valley. In S. Ellis and D. Crowther, (eds.) *Humber Perspective*. Hull University Press, Hull: 199-213.

EDIS, J., MACLEOD, D. and BEWLEY, R, 1989 An archaeologist's guide to classification of cropmarks and soilmarks *Antiquity* **63** (238): 112–126.

EVANS, J. 2006 Celtic art revealed: the South Cave weapons hoard. *Current Archaeology* 203: 573.

FENTON-THOMAS, C. 2003 *Late prehistoric and early historic landscapes on the Yorkshire Chalk*. British Archaeological Reports British Series **350**. Oxford.

FENWICK, H., CHAPMAN, H., HEAD, R., AND LILLIE, M. 1998 The archaeological survey of the lower Trent valley and Winterton Beck. In R. Van de Noort and S. Ellis *Wetland Heritage of the Ancholme and Lower Trent Valley*. Humber Wetlands Project, University of Hull, Hull: 143-197.

HALKON, P. 2003 Researching an ancient landscape: the Foulness Valley, East Yorkshire. In T. Manby, S. Moorhouse and P. Ottaway (eds.) *The Archaeology of Yorkshire An assessment at the beginning of the 21st century Yorkshire*. Yorkshire Archaeological Society with English Heritage, CBA, Leeds: 261-275.

HALKON P. 2005 Creating an award winning website for Community Archaeology and Research: "Valley of the First Iron Masters" a case study (http://www.ironmasters.hull.ac.uk). In M. Mudge, N. Ryan, R. Scopigno, *VAST 2005 The 6th International Symposium on Virtual Reality Archaeology and Cultural Heritage*, Eurographics Association Pisa Italy: 44-50.

HALKON P. 2006 *Archaeology and environment in a changing East Yorkshire landscape: The Foulness Valley c. 800 BC to c. AD 400*. Unpublished PhD Thesis, University of Hull.

HALKON, P., CHAPMAN, H., FENWICK H., TAYLOR, J. and WOODHOUSE, H. 2003 The rediscovery of the Roman temple at Millington, East Yorkshire *ARA-The Bulletin of the Association for Roman Archaeology* **5:** 8-9.

HALKON, P. and MILLETT M. 1999 *Rural Settlement and industry: Studies in the Iron Age and Roman Archaeology of lowland East Yorkshire*. Yorkshire Archaeological Report 4. Leeds: Yorkshire Archaeological Society Roman Antiquities Section and East Riding Archaeological Society

HALKON, P and MILLETT, M. 2000 The Foulness Valley - Investigation of an Iron Age landscape in lowland East Yorkshire. In J. Harding and R. Johnson (eds.) *Northern Pasts. Interpretations of the later Prehistory of Northen England and Southern Scotland*. BAR British Series 302, Oxford: 81-93.

HALKON, P. and MILLETT, M. 2003 East Riding: An Iron Age and Roman landscape revealed. *Current Archaeology* 187: 303-309.

HALKON, P., MILLETT M., EASTHAUGH E., TAYLOR, J. FREEMAN, P. 2000 *The landscape archaeology of Hayton*. University of Hull.

HASELGROVE, C. 1981 Indigenous settlement patterns in the Tyne-Tees lowlands. In P. Clack and S. Haselgrove (eds.) *Rural settlement in the Roman north*, Bradford: 9-25.

KING, S.J. and. BRADLEY, R.I 1987 *Soils of the Market Weighton District*. Harpenden: Soil survey of England and Wales.

LONG A.J., INNES J.B., KIRBY J.R., LLOYD J.M., RUTHERFORD M.M., SHENNAN I. and TOOLEY,M.J. 1998 Holocene sea-level change and coastal evolution in the Humber estuary, eastern England: an assessment of rapid coastal change. *The Holocene* 8: 229-247.

MANBY, T., KING, A, and VYNER B. 2003 The Neolithic and Bronze Age: a time of early agriculture. In T. Manby, S. Moorhouse and P. Ottaway (eds.) *The Archaeology of Yorkshire An Assessment at the Beginning of the 21st Century Yorkshire.* Yorkshire Archaeological Society with English Heritage, CBA, Leeds: 35-114.

MILLETT, M. 1990Iron Age and Romano-British settlement in the southern Vale of York and beyond: some problems in perspective. S. Ellis and D. Crowther, (eds.) *Humber Perspectives,* Hull: 347-357.

MILLETT, M. 2006 Shiptonthorpe: archaeological studies of a Roman roadside settlement. Yorkshire Archaeological Report, Leeds.

MORTIMER, J. R. 1905 *Forty years researches in the British and Saxon Burial Mound of East Yorkshire. Including Romano-British discoveries and a description of the ancient entrenchments on a section of the Yorkshire Wolds*. Brown, London.

PANSIOT, J, CHAPMAN P.M, VIANT, W.J. HALKON P. 2004 New Perspectives on Ancient Landscapes: A case Study of the Foulness Valley In K. Cain, Y Chysanthou, F. Niccolucci, N. Silberman *VAST 2004 The 5th International symposium on Virtual Reality, Archaeology and Intelligent Cultural Heritage* (Brussels) The Eurographics Association Aire La Vale Switzerland: 251-260.

PARKER-PEARSON, M. and SYDES R. E. 1997 The Iron Age enclosure and prehistoric landscape of Sutton Common, South Yorkshire. *Proceeding of the Prehistoric Society* 63: 221-59.

POWLESLAND, D. 1988 Staple Howe in its landscape. In T. Manby (ed.) *Archaeology in Eastern Yorkshire: Essays in honour of T.C.M. Brewster,* Department of Archaeology and Prehistory University of Sheffield. Sheffield: 101-108.

RAMM, H.G. 1978 *The Parisi* Duckworth, London.

RILEY, D.N., BUCKLAND, P.C. and WADE, J. 1995 Excavations and aerial reconnaissance at Littleborough on Trent, Nottinghamshire. *Britannia* 25: 253-284.

SAUER, E. 2005 Secrets of a sacred spring. Bourbonne-les-Bains. *Current World Archaeology* 13: 20-27.

SMITH, B. M. 2002 A Palaeoecological study of Raised Mires in the Humberhead Levels. British Archaeological Reports, British Series **336**. Oxford.

STEAD, I.M. 1979 *The Arras Culture*. Yorkshire Philosophical Society. York.

STEAD, I. M. 1991 *Iron Age cemeteries in East Yorkshire*. English Heritage, London

STOERTZ C. 1997 *Ancient Landscapes of the Yorkshire Wolds* RCHME.

TAYLOR, J 1999 The Holme on Spalding Moor Plots. In Halkon, P. and Millett, M. 1999. *Rural Settlement and industry: Studies in the Iron Age and Roman Archaeology of lowland East Yorkshire.* Yorkshire Archaeological Report **4** Yorkshire Archaeological Society. Leeds, 17-42.

TAYLOR, J. 2001 Rural Society in Roman Britain. In S. James and M Millett (eds.) *Britons and Romans: advancing an archaeological agenda*. CBA Research Report **125**, 46-59.

TWEDDLE, J.C., 2001. Regional vegetation history. In Bateman M.D., P.C. Buckland, C.D. Frederick and N.J. Whitehouse (eds) The Quaternary of East Yorkshire and North Lincolnshire Field Guide. Quaternary Research Association. London: 35-46.

VAN DE NOORT, R. 2004 (a) *The Humber Wetlands: the archaeology of a dynamic landscape.* Windgather Press, in association with English Heritage.

VERMEULEN, F., 1995 Sandy Flanders in the Roman period. Towards a regional research strategy. In M. Lodeewijckx (ed.) *Archaeological and historical aspects of western European societies,* Album Amicorum Andre Van Doorselaer, Acta Archaeologica Loveaniesna Monographiae **8.** Leuven: 134-143.

VERMEULEN, F. AND BOURGEOIS, J. 2000 Continuity of prehistoric burial sites in the Roman landscape of Sandy Flanders. In J. Pearce, M. Millett and M. Struck (eds.) *Burial Society and Context in the Roman world.* Oxbow Books, Oxford: 143-162.

WACHER, J. 1969 *Excavations at Brough on Humber, 1958-61.* Society of Antiquaries Research Report 25: London.

WHITEHOUSE N.J. 2004 Mire ontogeny, environmental and climatic change inferred from fossil beetle successions from Hatfield Moors, eastern England. The Holocene 14: 79-93.

Validation Results from the North Carolina Department of Transportation Archaeological Predictive Modelling Project

Scott Madry,[1] S. Seibel[2]

[1] Department Of Anthropology, The University of North Carolina at Chapel Hill
[2] Environmental Services, Inc., Raleigh, NC

1 Introduction

From October 2002 through September 2006, Environmental Services, Inc. and GAI Consultants, Inc. conducted the first stages of a state-wide Archaeological Predictive Model (APM) using a Geographic Information System (GIS) approach. The project was conducted for the North Carolina Department of Transportation (NCDOT) and funded by the Federal Highway Administration (FHwA). The goal of the project is to develop techniques and procedures that will allow for improved planning of new multi-land highways. NCDOT has identified advanced technologies, including GIS, spatial modeling, and Decision Support Systems (DSS) as having the potential for improving the efficiency and effectiveness of their archaeological and cultural resources planning and survey work in support of new highway corridor alternatives (Madry et al. 2003).

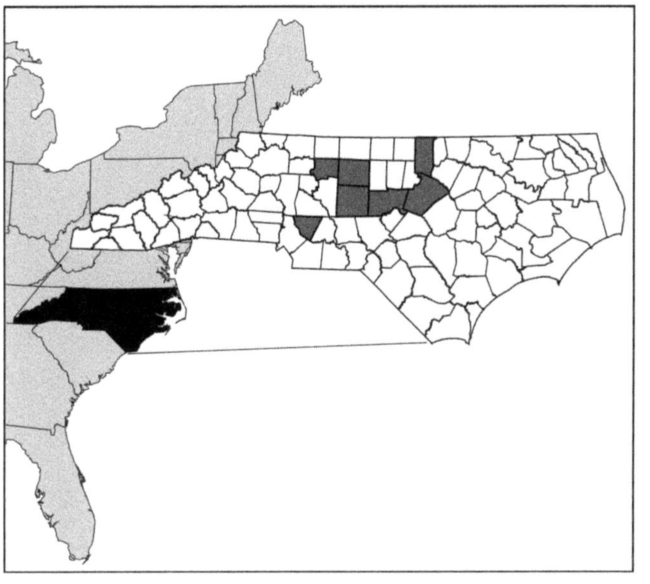

Figure 1

2 The Study Area

The project was conducted in the Piedmont region of North Carolina. The state of North Carolina lies on the eastern seaboard of the United States, south of Virginia, and covers 136,420 square kilometers (Figure 1). The state is composed of 100 counties within three general physiographic provinces: The Atlantic Coastal Plain in the east, the central Piedmont plateau, and Appalachian Mountains in the west; the Piedmont plateau constitutes about 38% of the state. The pilot project includes a project area that encompasses Cabarrus, Granville, Guilford, Randolph, Chatham, Forsyth, and Wake counties within the Piedmont physiographic province, an area containing over 4,000 known prehistoric archaeological sites (Figure 1). Over 360 United States Geologic Service (USGS) 1:24,000 topographic maps cover this region, with 114 covering the project area. The project area was chosen based upon the anticipated locations of NCDOT highway projects and as it covers a wide range of environmental variability. This provides a sufficient data sample for the initial model development.

3 Summary of Project Work

Task 1

The project was divided into two tasks. The main goal of Task 1 was to collect the necessary data covering the project area required for the development of the predictive model for Task 2 and for serving on the DSS. A total of 114 archaeological site maps (1:24,000 scale) from the North Carolina Office of State Archaeology (OSA) covering the project area were scanned, georeferenced, and had archaeological features and surveyed areas extracted into the GIS. Four hundred twenty-one historic North Carolina maps of different dates, scales, and extents were also scanned; 44 were georeferenced, and 11 had cultural features extracted into the GIS.

A new MS Access archaeological site database with a variety of new forms, queries, and report capabilities was designed for OSA, giving OSA a true relational database system. This new MS Access database became the new database for all future OSA activities. A new version of the OSA archaeological site form was also created in consultation with OSA that will be more compatible with the new MS Access database. A total of 327 environmental GIS data layers covering the project area and the immediate surrounding area were acquired, reviewed, and integrated with the new archaeological GIS data created for this project.

Task 2

Task 2 included the included creation of the GIS database, the manipulation and analysis of the archaeological and environmental data, statistical

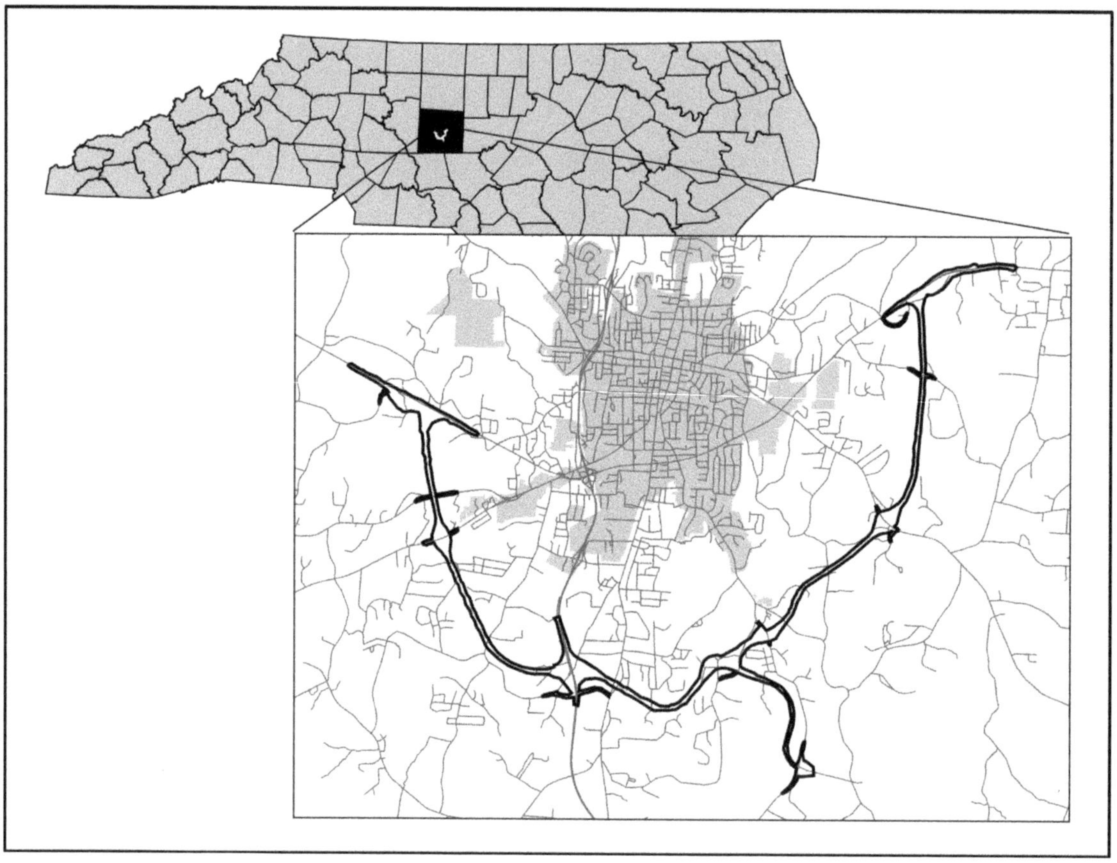

Figure 2

analysis, generation and testing of archaeological predictive models, further refinement of the OSA archaeological site database, and development of a DSS for sharing and analysis of the data (Madry et al. 2005). The final products delivered include a robust archaeological predictive model developed using logistic regression, containing high, medium, and low probability zones, a non-queryable nine-category model for easy visual analysis, an ArcIMS-based DSS to support future NCDOT and OSA work, delivery of all data, software, training, and manuals and the final report.

4 Results of Task 2 Modeling

A total of nine archaeological predictive models were created and eight models validated within the boundaries of the project area. Only two of the models covered the entire project area; the other models were created for evaluation of model scalar issues. The two models with complete coverage were an 11-variable model and a 10-variable model. The 11-variable model included the elevation variable; whereas elevation was excluded from the 10-variable model because of high variability within the variable and the difficulty of normalization of the data.

The analysis determined that the model using 10 environmental variables was the most appropriate final model. All models were validated by comparison with four control data sets, including prehistoric site data not included in the construction of the model. The validation datasets included the 90% Archaeological Site Sample used to create the model, a 10% Withheld Sample not used in the analysis, the UNC Diagnostic Lithic Database, and a full-coverage intensive archaeological field survey of the Asheboro Bypass, a new highway alignment project by NCDOT.

When comparing the results of the validation for the seven county models, the models showed similar results except when comparing them with the UNC Database, which is a relational database of approximately 48,000 projectile points from North Carolina representing the Paleo-Indian, Archaic, and Woodland periods maintained by the Research Laboratories of Archaeology (RLA) of the University of North Carolina at Chapel Hill (UNC-CH), and modified by Theresa McReynolds for her Ph.D. research (McReynolds 2003). The final 10-variable model captured 71% of the diagnostic lithic points in the high category, which covers 20% of the project area (Madry and Seibel 2006).

Model Validation using Asheboro Bypass Data

Model validation and field verification testing are sometimes excluded from predictive modeling methodology, even though these are a vital aspect of the process. One major criticism of this type of inductive archaeological predictive modeling is that many such models are never subjected to quantitative validation. The research design for the NCDOT APM specifically included multiple model validation and testing activities in order to understand the relative predictive power of the different models, the ability to extrapolate the model out to a larger area, and to determine the extent of the utility of the final products for use by NCDOT.

Specifically, independent test samples were identified for the sole purpose of model testing and validation, as discussed above: The 90% Sample used to create the models, the 10% Sample, the independent UNC Database, and field collected data from the Asheboro Bypass survey which accurately follows the actual NCDOT work for which this work was developed. The first offered a large sample size and therefore a good indication of performance, but since the models are fit to those data they would likely be biased. The second and third test samples, each somewhat different in character, were to provide less biased views of how these models might perform in practice. The fourth provided a real-world test of the models.

Of the 3,183 shovel tests excavated at 30-meter intervals and mapped using differential GPS across the 486-hectare Asheboro Bypass survey corridor (Figure 2), 285 were positive for prehistoric artifacts. A total of 79 archaeological sites containing prehistoric cultural material covering an area of 28.04 hectares within the project corridor were recorded as a result of the survey. The distribution of archaeological sites in the Asheboro Bypass corridor was statistically compared with the 10-category breakouts of four of the scaled models: The 10-variable model, the Randolph County model, the Asheboro Roving model, and the Asheboro Quad model. The results of this analysis are shown in Table 1.

Table 1 *Numerical comparison of Asheboro Bypass results to four APM models*

Prob. Cat.	Final	Randolph Co.	Asheboro Quad	Asheboro Roving
1	0.00	0.00	0.00	0.00
2	0.00	0.00	0.00	0.00
3	0.37	0.06	0.00	0.31
4	0.92	0.55	1.17	0.11
5	1.01	0.82	0.87	0.87
6	1.20	1.15	0.75	1.50
7	1.00	1.02	1.32	1.23
8	0.64	0.76	1.07	0.87
9	0.00	0.49	0.84	0.49
10	0.00	0.00	1.99	0.00

These analyses show that all four models performed better than random, and all followed a generally similar pattern, with Categories 4-8 containing the majority of the prehistoric archaeological sites. It also shows a clear trend with the most local model, the Asheboro Quad model being the most powerful, followed by the Asheboro Roving, the Randolph County model, and the 10-variable model. There is some concern by the team that this specific area does not represent an unbiased sample. Many of the archaeological sites in the Asheboro Quad in particular, and Randolph County in general, are lithic extraction sites (quarries), which have a different environmental pattern from the overall site sample.

While the predictive power of the final APM model has been demonstrated, it does not serve any practical value unless it can be utilized by NCDOT for the quantitative assessment of highway alternative corridors and for the scoping of archaeological survey projects. Two practical applications of the APM model for NCDOT were developed, both using the Asheboro Bypass as a test case.

Alternative Corridor Comparison

To allow NCDOT to quantitatively rank highway alternative corridors for the likelihood of impacts to prehistoric archaeological sites, an analytical method was developed. This compared NCDOT's final chosen corridor for the Asheboro Bypass (which was chosen before the modeling work was completed) and the seven other final alternative corridors.

A comparison was conducted of the eight alternatives where the total amount of high/medium/low area was measured to compare the total potential for archaeological resources in the different alternatives. A simple index of the medium value of all cells was computed,

$$\text{Index} = \sum (\% \text{ area in category X}) * (\text{category X})$$

where X is 1 through 3 for the three-category breakout or 1 through 9 for the nine-category model.

This analysis (Table 2) shows that the final, preferred corridor chosen for construction (Alternate # 29) actually had the second to highest percentage of high probability areas of the eight alternatives. In comparison, the lowest ranking corridor, Alternate # 1, contained a significantly lower overall site probability, with very minor alterations in the corridor location.

Table 2 *Archaeological site sensitivity ranking of the eight final alternate routes*

Alt Route	Index	Ranking	Notes
1	3.596065	1	Least high probability areas
13	3.706089	2	
2	3.73172	3	
4	3.750767	4	
14	3.836214	5	
10	3.837305	6	
29	3.886563	7	Actual preferred corridor
33	3.995184	8	Most high probability areas

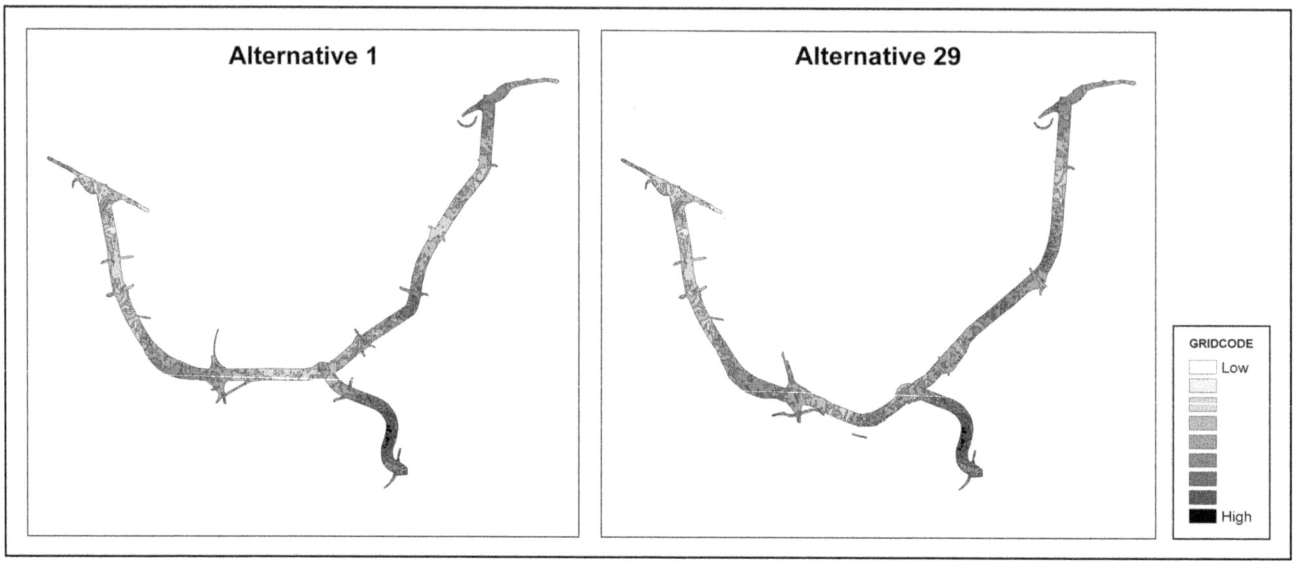

Figure 3

Figure 3 allows for a visual comparison of this. At the top is Alternate # 29, the preferred corridor surveyed a part of this project, showing the three-category predictive modeling zones (red=high, yellow=medium, green =low). At the bottom is Alternate # 1, with the lowest total overall predictive model sensitivity. The variation is clearly visible. Although it is not possible to quantitatively compare the actual site distributions without field survey of Alternative # 1, this comparison demonstrates the potential that access to the predictive model and GIS data can have in picking the alternative with the lowest archaeological site potential.

Scoping of Archaeological Surveys

While the final APM model will be very useful at the front end of the preferred corridor selection process by allowing the quantitative ranking of alternative corridors by probability zones, it will also allow for the estimation of positive 30-meter shovel tests and areal site coverage, both for preferred corridor selection and the scoping of the survey of the preferred corridor. Until now, the way archaeologists in North Carolina quantitatively estimated the number of archaeological sites that could be encountered in a project area was to add up the hectarage of previously surveyed areas and divide by the number of archaeological sites recorded in those areas. While this produces a site per hectare estimate that can be used to scope new projects, it is filled with biases, the most significant of which are environmental and topographic variability.

The existing Asheboro Bypass survey can be used as an example of the old method of project scoping. Novick (1997) compiled data from previously conducted archaeological surveys in Randolph County. While the data included historic sites, the average archaeological site density for those surveys was 0.014 sites per hectare. The bias inherent in this method of calculation is shown when only the surveys of the Randleman Reservoir are examined. These surveys encountered a site density of 0.058 sites per hectare, 408% greater than for the county as a whole. When scoping the survey of the Asheboro Bypass survey corridor, the county average was used to estimate that 42 prehistoric and historic archaeological sites would be encountered. However, the actual survey encountered 82 prehistoric and historic archaeological sites, a density of 0.028 sites per hectare, nearly 94% greater than the county average. This could have happened for two reasons. One, that the comprehensive survey strategy has a greater potential to encounter prehistoric archaeological sites than a targeted survey strategy or, two, that the Asheboro Bypass corridor passes through an area that is not representative of prehistoric archaeological site distribution in Randolph County. All told, it is probably a result of both.

This exercise shows that NCDOT needs a much better way to estimate the amount of fieldwork that may be necessary to survey any given corridor. Towards this goal, another means of analysis was developed specifically for NCDOT. This was a recategorization of the archaeological sites into number of sites per hectare in each of the three probability zones. This was thought to be specifically useful for NCDOT in terms of planning, costing, and survey purposes.

The model uses an algorithm to determine the probability of a 30-x-30 meter area to contain prehistoric cultural material. When previously recorded archaeological sites were queried for environmental variables, the areal coverage of each individual site was based on the boundaries lying within a 30-x-30 meter pixel, and thus the environmental variables were captured on this same 30-meter grid. Following this methodology, the number of 30-x-30 meter pixels per hectare predicted to contain prehistoric cultural material in each of the probability areas (Low/Medium/High) was calculated for the 90% Sample and the UNC Database.

For the 90% Sample:
- Low – 0.12 pixels/hectare (1 pixel/8.33 hectares)
- Medium – 0.15 pixels/hectare (1 pixel/6.67 hectares)
- High – 6.45 pixels/hectare (1 pixel/0.16 hectares)

For the UNC Database:
- Low – 0.20 pixels/hectare (1 pixel/5.00 hectares)
- Medium – 0.72 pixels/hectare (1 pixel/1.38 hectares)
- High – 4.49 pixels/hectare (1 pixel/0.22 hectares)

The 30-meter shovel testing grid for the Asheboro Bypass survey was purposefully designed to match with the scale of model analysis; presence/absence of prehistoric cultural material was based on positive and negative 30-meter grid shovel tests. Using this positive pixel per hectare estimate, the number of positive 30-meter grid shovel tests recorded as part of the Asheboro Bypass survey was predicted:

- 90% Sample – 265.5 positive shovel tests predicted
- UNC Sample – 398.6 positive shovel tests predicted
- Asheboro Bypass – 285 positive shovel tests recorded

This number can then be used to predict the areal coverage of all prehistoric archaeological sites that may be recorded in a corridor. To calculate this number, one multiplies the acreage of a 30-x-30 meter area (0.089 hectares) by the predicted number of positive shovel tests. Table 3, below, shows the predicted coverage of all prehistoric archaeological sites within the project corridor using the 90% Sample, the UNC Sample, and the number of positive 30-meter shovel tests recorded during the Asheboro Bypass survey. It also shows the actual areal coverage of the recorded prehistoric archaeological sites within the project corridor.

Although the new and the old methods produce different estimates for different analytical categories (Sites per Hectare versus Number of Positive Shovel Tests and Areal Coverage of Sites), this exercise still shows that the final APM model gives a much better predictive power than the old method, described above (Table 4). Using the old method, it was predicted that the survey of the Asheboro Bypass would record 42 sites, but 82 sites were recorded, an error of nearly 49 percent. Using the new method, it was predicted that the survey would encounter 265.5 positive prehistoric 30-meter shovel tests covering 58.91 acres, while the survey encountered 285 positive prehistoric 30-meter shovel tests covering 69.29 acres, errors of 7 percent and 15 percent, respectively.

Table 3 *Predicted areal coverage of prehistoric archaeological sites*

Coverage	Area
90% Sample Predicted Coverage	58.91 acres
UNC Database Predicted Coverage	88.47 acres
Asheboro Bypass Predicted Coverage	63.27 acres
Asheboro Bypass Encountered Coverage	69.29 acres

Table 4 *Differences in predictive power*

Old Method		
Prediction	**Actual**	**Error**
42 sites	82 sites	49%
New Method		
Prediction	**Actual**	**Error**
265.5 positive STs	285 positive STs	7 %
58.91 acres of coverage	69.29 acres of coverage	14%

5 Conclusions

As has been demonstrated in the Final Report for Task 2 of the APM (Madry and Seibel 2006, Madry et al. 2003, 2005), the final predictive model developed for NCDOT is robust. The model performed well overall against the validation datasets, and performed extremely well against the UNC diagnostic lithic site database which was not used to create the model. The model contained 71% of the known diagnostic lithic sites in the high probability category, which covered only 20% of the project area. A total of 22.8% of the known diagnostic lithics were located in the medium probability zone, which contained 40% of the project area, and 6% were in the low probability zone, which also contained 40% of the area. Additionally the model and the other data sets will be served through the ArcIMS Decision Support System designed for this project. This will be of great utility to NCDOT, as well as the OSA and the rest of the archaeological community. This utility should become more and more evident as the system is utilized on a day-to-day basis.

References

MADRY, S., S. GOULD, B. RESNICK, and M.T. WILKERSON 2003 "A GIS-Based Archaeological Predictive Model for the North Carolina Department of Transportation." in *Landschaftsharchaologie und geographishe informationssystems- The archaeology of landscape and geographic information systems: predictive maps, settlement dynamics and space and territory in prehistory*. Forschungen zur Archaologie im Land Brandenburg 8, Von 15, Bis 19, pp. 161-170. Brandenburgisches Landesmuseum für Ur- und Frühgeschichte. Potsdam, Germany.

MADRY, S., and S. SEIBEL 2006 *North Carolina Archaeological Predictive Modeling Project: Results of Task 2 – Cabarrus, Chatham, Forsyth, Granville, Guilford, Randolph, and Wake counties.* Environmental Services, Inc., Report of Investigations No. 1000. Raleigh, North Carolina.

MADRY, S., S. SEIBEL, B. RESNICK, M. COLE, and S. GOULD 2005 "Development of a State-wide Archaeological Predictive Model for the North Carolina Department of Transportation and Computerized Archaeological Database for the NC Office of State Archaeology." in *GIS and Archaeological Site Location Modeling.* Mark. W. Mehrer and K. L. Wescott (eds). Taylor and Francis Books, Boca Raton, Florida.

McREYNOLDS, T. 2003 *Patterns in the Distribution of Archaic and Woodland Projectile Points in North Carolina*. Paper Presented at the 60th Annual Meeting of the Southeastern Archaeological Conference, Charlotte, North Carolina.

NOVICK, L. 1997 *Archaeological and Historical Background Report: Improvements to NC 311 from High Point East Belt to US 220 (R-2606) Study Area, Randolph County, North Carolina.* North Carolina Department of Transportation, Raleigh, North Carolina.

A (GIS)-based predictive mapping to locate prehistoric site locations in the Gamasb River Basin, Central Zagros, Iran

K. A. Niknami,[1] M. R Saeedi Harsini[2]

[1]Department of Archaeology, University of Tehran, Iran
[2]Department of Archaeology, University of Tarbiat Modarres, Tehran, Iran

1 Introduction

The analysis of archaeological site–environment relationships has always been a central issue in archaeology. The importance of environmental variables to explain site distribution was recognized early on (Carr 1985; Kohler and Parker 1986; Kvamme 1989; Kvamme 1992). Environmental factors have been much used to explain the main archaeological site patterns around the world The quantification of such site–environment relationships represents the core of predictive archaeological modeling in archaeology. These models are generally based on various hypotheses as to how environmental factors control the distribution of ancient communities. Besides its prime importance as a research tool in archaeology, predictive modeling recently gained importance as a tool to assess the impact of accelerated land use and other environmental change on the distribution of archaeological sites to set up site conservation priorities (e.g. see references in Wescott and Brandon 2000). A variety of statistical models is currently in use to simulate either the spatial distribution of sites (e.g. Kvamme 1992; Warren 1990a; 1990b; Warren and Asch 2000). Such static, comparative, models are opposed to more mechanistic models of archaeological site formation processes, since only very few sites have been studied in detail in terms of their dynamic responses to environmental change. Static distribution modeling often remains the only approach for studying the possible consequences of a changing environment on site distribution. The development of predictive models is coherent with (Dalla Bona 1994:16-17) view of correlations between prehistoric archaeological site locations and environmental variables which reflect spatial organization across the landscape during the prehistoric period. The use of and theoretical limitations of static models compared with dynamic approaches have been described in several papers (e.g. Allen et al. 1990; Wescott and Brandon 2000; Mehrer

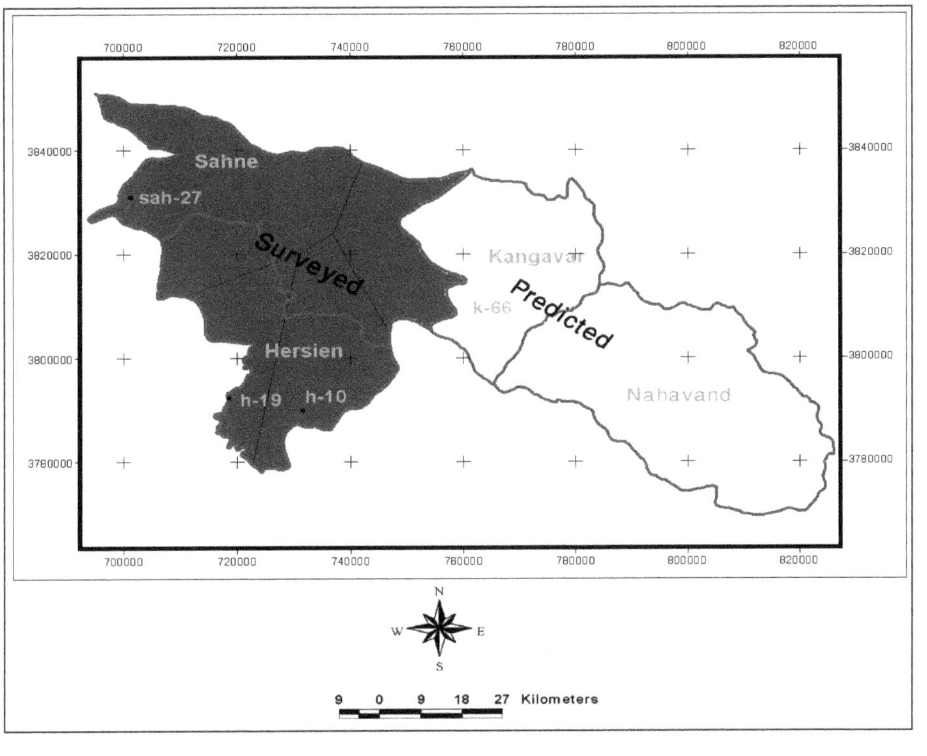

Figure 1 *Location of the study area in the Central Zagros, Western Iran, showing the distribution of archaeologically surveyed land areas*

and Wescott 2006) provide a review of some currently used statistical techniques in the spatial site modeling. Particular aspects of model development (e.g. verification, evaluation and qualification) have been covered in more specific papers, with very special attention given in recent years to evaluation and its usefulness for testing spatial models (e.g. Hodder and Orton 1976). Since then, new statistical techniques for testing predictive models have emerged (see e.g. Schwarz and Mount 2006). The aim of this paper is to review the various steps of predictive modeling, from the conceptual model formulation to prediction and application. We discuss the importance of differentiating between model formulation, and model application. Additionally, we provide an overview of specific analytical, statistical methods currently in use.

The GIS database covers an area of about 2500 km^2, covering the majority of the Gamasb River Basin and its immediate environs (Fig. 1). The current basic raster and vector layers of the GIS data base include: elevation (derived from the various relevant for example geological, hydrological and topographic 1: 50.000 maps) as well as archaeological sites and field survey transects (from project surveys and other sources). Additional derived data layers showing different distance categories, or buffer zones, from: roads, streams, faults, archaeological sites, and ancient roads were then generated from the data above. Additional data have recently been added that were derived from the 1: 50,000 maps, including reclassifications and distance measurements from sites, ancient roads, and hydrology. In all there are currently over 100 point, vector, and raster data layers in the database. These data are used to conduct a variety of analyses of the prehistoric site locations and the development of predictive models of archaeological sites of Calcolithic periods.

2 Concepts

Over the past two decades, there has been a resurgence of interest among cultural resource managers in constructing predictive models for prehistoric site locations. This renewed interest is the result of both rapid developments in GIS technology that make the modeling process more efficient (Kvamme 1992; Allen et al. 1990; Dalla Bona 1994; Marschner 1996) and public pressure to conduct cultural resource activities in a cost-effective manner.

The model is based on the assumption that the most important factors controlling prehistoric settlements and activity locations were physical and biotic attributes of the landscape (Dalla Bona 1994: 17). This is in contrast to more technologically complex societies, where social, ideological, and political forces can take precedence over environmental factors in influencing settlement location. As a result of this assumption, the model and many other contemporary modeling efforts rely on a series of biophysical variables (e.g. slope, elevation, soils, proximity to various water sources, vegetation) to construct models. The reliance on these types of variables has led to criticism that the resulting models are environmentally deterministic (Kohler 1988: 9). Although archeologists who construct predictive models recognize the importance of cultural factors in the location of settlements, they contend that the temporal control needed to establish contemporaneity between sites is usually lacking. Consequently it is difficult, if not impossible, in most situations to include these variables in the modeling process (Brandt et al. 1992:269: Kvamme 1997: 1-2).

Other basic assumptions of the project are that (**1**) environmental attributes of the prehistoric period are still identifiable, at least in two dimensions, in current data sources; and (**2**) correlations between prehistoric archaeological site locations and environmental variables reflect spatial organization across the landscape during the prehistoric period (Dalla Bona 1994: 16-17). The model, through its geomorphological landscape component, incorporates the third and fourth dimensions of time and buried landscapes/surfaces for areas with higher probabilities for buried cultural resources. This was done for Gamasb River valleys and a limited number of small upland areas. This is a unique aspect of the model that has not been previously attempted by other modeling efforts on this scale.

4 Methodology

The model is designed to be an empiric correlative (inductive) rather than a deductive predictive model Dalla Bona 1994: 5). This is the most popular approach in predictive modeling. In empiric correlative models, no presuppositions are made other than: **1**-in non-complex societies the most important economic transactions for most people were with the environment; **2**-humans tend to minimize the time or effort expended in their economic transactions with the environment; and, **3**-by implication, human activities and settlements tend to be located close to environmental resources.

The results of empiric correlative models are developed inductively by exploring associations between specific environmental variables and archaeological site locations. In these models, the dependent variables are archaeological sites and the independent variables are biophysical characteristics of locations, such as slope, soil type, elevation, plant community, and distance to water. Statistical analysis is used to identify relationships between the environmental and archaeological attributes, generating combinations of environmental variables that correlate with the presence or absence of sites.

The other approach to predictive modeling relies more heavily on deductive reasoning. Sometimes referred to as explanatory or systemic modeling (Weatley and Gillings 2002) deductive models rely on hypothesized relationships between various biophysical and/or cultural factors to predict site locations. The ultimate goal is to explain the distribution of settlements across the landscape in terms of various social, political,

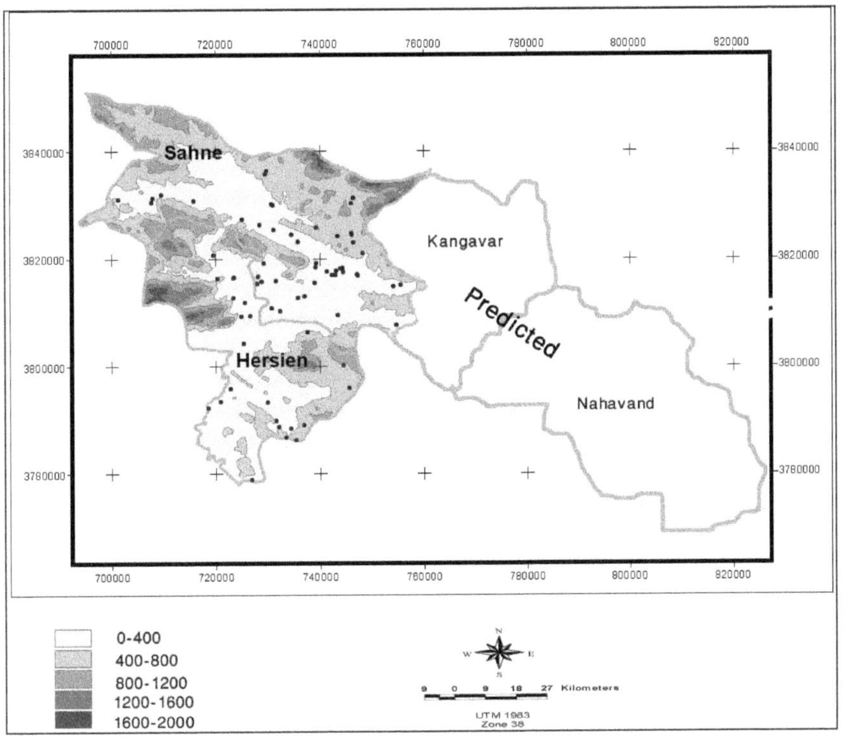

Figure 2 *GIS coverage of the study area showing topographic features within 0m.-2000m. ranges and distribution of the known archaeological Calcolithic sites*

ideological, and physical factors that are assumed to be significant (Wandsnider, and Dooley 2004). The location of sites and their contents are placed within a larger organizational system where past and anticipated uses of the land intertwine with how intensively resources are exploited and how they are distributed through time and space. The models range from informal intuitive ones developed by most archaeologists through experience, to somewhat more formal ones (Cassell et al. 1997, Dalla Bona and Larcombe 1996). This approach is sometimes taken when there is insufficient data to build inductive-correlative models like this model.

To accommodate the technical requirements of the project, the model used Geographic Information Systems as the generative framework. GIS incorporates the essential elements of computer cartography and relational database management into one system. It efficiently handles very large databases, maintains links between maps and tabular data, and allows for the analysis of spatial relations. This includes analytical and modeling functions that are not practical or possible with other methods. The study area is about 2500 square km in extent and has many recorded archaeological sites. Each environmental database had to contain more than thousands data values to cover the entire state at the required resolution. Without GIS, the storage and manipulation of the required high resolution geographic databases would have been unmanageable. Archaeological site information was obtained from archaeological survey. These data were carefully filtered for quality control. Archaeological survey, a probabilistic survey conducted in 2004 contributed data gathered through a variety of survey procedures, including stratified random sampling, sampling by landscape type, and professional intuition (Niknami 2004). These surveys resulted in large samples of site and non-site locations. Site location modeling is a useful GIS product (Fig. 2). It allows us to model where archaeological sites of a given period may be located, based on the known site locations and various environmental and cultural data in the GIS. Statistical analyses were run on the archaeological sites of different periods with most layers in the GIS to look for pattern. Various predictive models were developed using the site data generated from field survey. This model was then generalized to include a much larger area surrounding the transects. New layers in the GIS containing these locations were produced and new maps showing the areas with the highest probability of site locations were created. These areas of higher probability of archaeological sites have a high correlation with areas that are threatened by current mining activities in the area.

The project was divided into three phases. Phase **1** involved basic data accumulation and model development. Environmental and archaeological attributes for study area, with site location data gathered through the surveys, were incorporated into the GIS in conjunction with new information gained from the model. Prototype GIS models were derived, using logistic regression for the Phase 1 study area.

Phase **2** consisted of incorporating the remaining parts, which had site location data collected by non-random procedures, into the GIS. Phase 2 field surveys in study

Table 1 *Unadjusted effects of categorical predictor variables obtained from Logistic Regressions*

Categorical predictor	Range	Likelihood Ratio test	Odd-Ratio	95% Confidence Interval for Odd-Ratio
Distance to permanent water	0-50 m. 0-100 m.	X^2 (2) = 19.2 P<0.001	6.23 3.16	4.23-7-80 2.16-4.32
Elevation	+1200m. 1600-1800m. 1200-1600m. 800-1200m. 400-800m. 0-400m.	X^2 (6) = 78.42 P< 0.001	1.23 1.02 1.45 1.56 4.08 6.98	0.876-3.43 0.732-3.05 0.56-2.83 0.67-2.98 2.21-6.11 4.62-9.56
Slope	0-30° +30°	X^2 (2) = 127.8 P< 0.001	4.405 0.89.3	1.72-11.23 0.332-2.67
Relief		X^2 (1) = 7.78 p> 0087	0.003	0.0001-0.098
Distance to ancient roads	0-100m. +100m.	X^2 (2) = 6.07 p> 0.0138	1.02 0.541	0.324-4.54 0.332-2.01

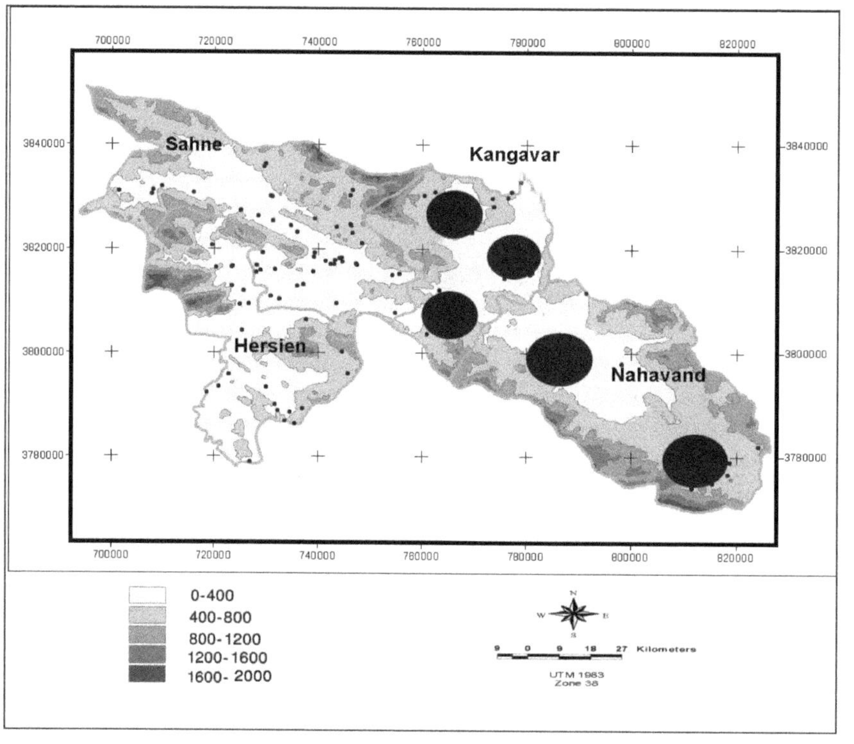

Figure 3 *Map of predicted site presence probability in the Gamasb River Basin area*

area were initiated to generate additional archaeological data, this time from truly probabilistic surveys. Modeling procedures were refined. Preliminary models were generated for the entire area. For some areas, additional models were built using data which are not yet available to test the utility of these data sets. Finally, environmental data were developed and evaluated for their ability to contribute to the project. Phase **3** of the project involved refining the modeling procedures further. This included adopting the Ecological Classification System subsections as the regionalization scheme and incorporating more sites into the database used to build the models. Modeling was extended to using information about locations surveyed to build models of survey bias. Additional procedures for evaluating model performance were adopted. As a final product of this phase, site probability and survey probability models were combined into composite survey implementation models.

5 Results

Predictive models applied with the aid of GIS can provide accurate probability estimate of Calcolithic site location in the surveyed area. This review of prehistoric site variability with respect to some of the environmental variables has revealed a number of patterns in the GIS integrated data relating to the occurrences of Calcolithic site locations. For instance, sites show a significant trend toward a moderately laying elevation as well as a close proximity to water resources (Table 1). These trends generally fit the expected pattern based on our observation of site distributions and settlement pattern of the region. Logistic regression and univariate statistical analysis indicate that the site and non-site locations in the study area tend to occur in different environmental setting. In the Gamasb River predictive model 5 independent variables. An assessment of logistic analysis indicated that 3 variables had significant predictive power, from which powerful variable was elevation at 0-400m providing greater separation between sites and non-sites. (Fig. 3)

6 Conclussion

The primary product of the model is a series of high resolution, predictive models for archaeological site and survey potential across the study area. A number of secondary products were generated, most importantly a single repository of environmental data in GIS format. One other essential product of the model is the generation of a series of geomorphological maps of study area. These maps define the location, age, and integrity of archaeolandscapes when practical. The geomorphology maps do not predict site locations, but rather predict for suitable landscapes to contain in situ archaeological deposits. Site probability models are presented as low, medium, and high probability areas for site presence. Survey probability models are represented as low, medium, and high potential that a landscape is similar to the places with particular environmental characteristics have been surveyed. Portions of the landscape where prehistoric sites are expected not to be located (under water, in mines, on steep slopes) were excluded from these probability classes. Survey implementation models also identify areas where surveys have been inadequate and, consequently, the probability of finding sites is unknown.

Besides providing a scientific, cost effective planning tool, the model can be used to generate hypotheses of the causal relationships between environmental variables and prehistoric site locations. These relationships can be used to generate theories about spatial settlement patterns across the landscape.

7 Acknowledgement

This project was supported by a grant from University of Tehran. The author gratefully acknowledges the useful comments on every stage of this research by Dr. H. Laleh the head of the Department of Archaeology of the University of Tehran.

8 References

ALLEN, K.M.S., GREEN S.W. and ZUBROW, E.B.W (eds.) 1990. *Interpreting Space: GIS and Archaeology*. London: Taylor and Francis.

BRANDT, R., B.J. GROENEWOUDT and K.L. KVAMME. 1992. An Experiment in Archaeological Site Location: Modeling in the Netherlands Using GIS Techniques. *World Archaeology* 24:268-282.

CARR, C. 1985. Introductory Remarks on Regional Analysis. In: C. Carr (ed.), *For Concordance in Archaeological Analysis: Bridging Data Structure Quantitative Technique and Theory*. Kansas City: Westport Publisher. 114-127.

CASSELL, M.S., H.D. MOOERS, C.A. DOBBS, T. MADIGAN, M. COVILL, J. BERRY, and D.A. BIRK. 1997. *An Archaeological Sensitivity Model of Prehistoric and Contact Period Settlement at Camp Ripley, Morrison County, Minnesota*. Reports of Investigation No. 397. Institute for Minnesota Archaeology, Minneapolis.

DALLA BONA, L. 1994. *Methodological Considerations. Cultural Heritage Resource Predictive Modeling Project*. Vol. 4. Centre for Archaeological Resource Prediction, Lakehead University. Thunder Bay Ontario.

DALLA BONA, L. and L. LARCOMBE. 1996. Modeling Prehistoric Land Use in Northern Ontario. In: H.D.G. Maschner, (ed.), *New Methods, Old Problems: Geographic Information Systems in Modern Archaeological Research*. Center for Archaeological Investigations Occasional Paper No. 23. Southern Illinois University, Carbondale. 252-271

HODDER, I and ORTON, C. 1976. *Spatial Analysis in Archaeology*. Cambridge, Cambridge University Press.

Kohler, T.A. 1988. Predictive Locational Modeling: History and Current Practice. In: W. J. Judge and L, Sebastian, (eds.), *Quantifying the Present and Predicting the Past: Theory, Method, and Application of Archaeological Predictive Modeling*. U.S. Government Printing Office Washington, D.C. 1119-59.

KOHLER, T.A. and PARKER, S.C. 1986. Predictive Models for Archaeological Resource Location. In: M.B. Schiffer (ed.), *Advances in Archaeological Method and Theory*, Vol. 9, New York: Academic Press, 397-452.

KVAMME, K.L. 1989. Geographic Information System in Regional Archaeological Research and Data Management. In: M.B. Schiffer (ed.), *Archaeological Method and Theory*,Vol.1, Tucson: University of Arizona Press, 139-203.

KVAMME, K.L. 1992. A Predictive Site Location Model on the High Plains: an Example with a Independent Test. *Plains Anthropologist* 37: 19-40.

KVAMME, K.L. 1997. Bringing the Camps Together: GIS and ED. *Archaeological Computing Newsletter* 47:1-5.

MASCHNER, H. D. G. (ed.), 1996 *New Methods, Old Problems: Geographic Information Systems in Modern Archaeological Research*. Occasional Paper No. 23, Center for Archaeological Investigations. Southern Illinois University at Carbondale.

MEHRER, M.W. and WESCOTT, K.L. (eds.), *GIS and Archaeological Site Location Modeling*. London: Taylor and Francis

NIKNAMI, KAMAL A. 2004 Application of Remote Sensing and Geographic Information System (GIS) for the Study of Prehistoric Archaeological Site Locations: Case Study from Garrangu River Basin, Northwestern, Iran: a Preliminary Report. *Proceedings of the International Conference on Remote Sensing Archaeology* Beijing China, 208-215.

SCHWARZ, K.R. and MOUNT, J. 2006. Integrating Spatial Statistics into Archaeological Data Modeling. In: Mehrer, M.W. and Wescott, K.L. (eds.), *GIS and Archaeological Site Location Modeling*. London: Taylor and Francis, 167-189.

WANDSNIDER, L. and MATHEW, A.D. 2004. Landscape Approaches to Regional Archaeological Variation, Paper prepared for the Electronic Symposium Survey Methodologies in Global Archaeological Contexts at the 2004. *Annual Meeting of the Society for American Archaeology. University of Nebraska-Lincoln,* 1-30.

WARREN, R.E. 1990a. Predictive Modeling in Archaeology: a Primer. In: K.M.S. Allen, Green S.W. and Zubrow, E.B.W (eds.) 1990. *Interpreting Space: GIS and Archaeology*. London: Taylor and Francis.90-111.

WARREN, R.E. 1990b. Predictive Modeling of Archaeological Site Location: a Case Study in the Midwest. In: K.M.S. Allen, Green S.W. and Zubrow, E.B.W (eds.) 1990. *Interpreting Space: GIS and Archaeology*. London: Taylor and Francis.201-215.

WARREN, R.E. and ASCH, D.L. 2000. A Predictive Model of Archaeological Site Location in the Eastern Prairie Peninsula. In: K.L. Westcott and R.J. Brandon, (eds.), *Practical Application of GIS for Archaeologists: a Predictive Modeling Kit.* London: Taylor and Francis. 5-32.

WEATLEY, D, and GILLINGS, M. 2002. *Spatial Technology and Archaeology: The Archaeology Applications of GIS*. London: Taylor and Francis.

WESCOTT, K.L. and BRANDON, R.J. 2000. *Practical Application of GIS for Archaeologists. A Predictive Modeling Toolkit*. London: Taylor and Francis.

Paleorelief detection and modelling a case of study in eastern Languedoc (France)

Krištof Oštir,[1] Laure Nuninger[2]

[1] Scientific Research Centre of the Slovenian Academy of Sciences and Arts (Slovenia)
[2] Chrono-Ecology, UMR 6565 CNRS/University of Franche-Comté (France)

1 Introduction

The main objective of the paper is paleo digital elevation model (paleoDEM) detection and modelling, a multidisciplinary approach involving geodesy, archaeology, and paleoenvironmental studies. The digital elevation model (DEM) is an important data layer to understand settlement pattern and socioenvironmental contexts and evolution that archaeologists are using together with GIS analysis. It can, for example help to analyze viewsheds, to define travel time between places, to simulate ancient paths, etc. Considering a period of several millennia, some topographical areas can be considered without fundamental change. Within littoral zones or fluvial areas, however, the situation is different. Sedimentation and erosion change the relief permanently, due to alluvial and sea level dynamic. A nowadays-flat area could be hilly in prehistoric times and the application of present DEM for the analysis is therefore unsuitable, since it would distort the expected result completely.

Our case of study area is situated in Languedoc (France), on the littoral plain of Mauguio between the cities of Montpellier to the west and Nîmes to the north (Fig. 1). This area covers around 250 km^2. From an archaeological point of view, the zone is very well explored with almost 350 sites inhabited from Bronze Age to Middle Ages, showing also land division evidences obtained from aerial photo interpretation. In addition, in 1990s several geoarchaeological surveys were done in the main deltas of Lez and Vidourle rivers. Unfortunately the geoarchaeological coring, performed to understand paleochannel profiles in the Vidourle delta, is concentrated in three zones only. Considering a regional scale, only three average points could be used in further analyses. The geoarchaeological approach in itself brings indices about paleorelief but since the measurements are represented as points they are not adequate to produce a proper historical DEM. Additionally the approach is slow and relatively expensive. We have therefore used remote sensing, optical image processing and radar interferometry, in combination with spatial analysis to detect the paleorelief.

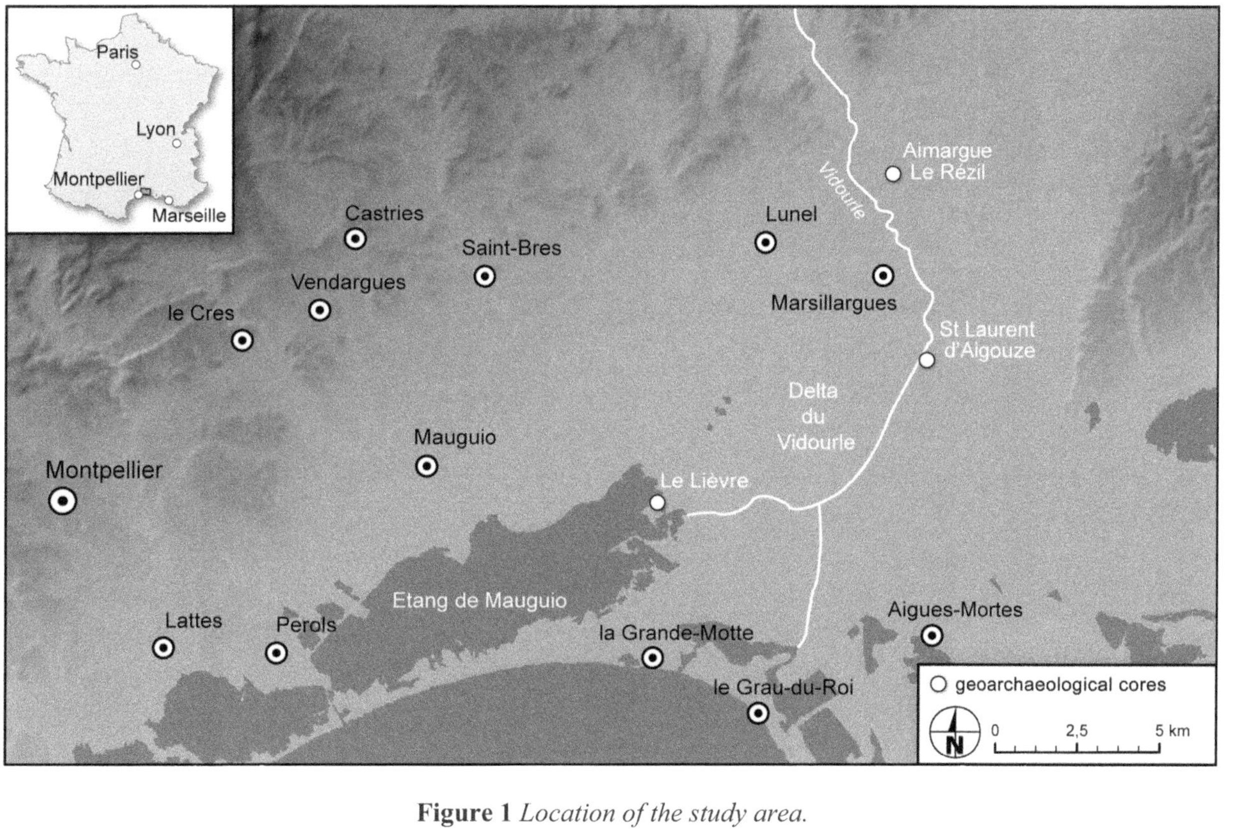

Figure 1 *Location of the study area.*

The first step was to create an accurate present DEM with a resolution suitable for analyses. Afterwards, the DEM, as well as SPOT and Landsat satellite images, was processed with several filtering techniques to detect "paleofeatures" – evidences of paleorelief. Features were digitalised and their spatial distribution was statistically tested according to archaeological data distribution. At last, we compared and tried to qualify the paleofeatures with the results obtained by photo-interpretation, classification and simulation.

2 Paleorelief interpretation

Existing technology does not allow a direct detection of paleorelief. Nevertheless, satellite imagery and current digital elevation models can be used to detect the "paleo-features" and to relate them to past relief.

The digital elevation model of the study area has been made as a combination (weighted sum) of all available sources (Podobnikar et al. 2000). Firstly, a DEM has been produced with radar interferometry (InSAR, Oštir et al. 2004) from two ERS radar images, acquired on 1996-04-04 and 1996-04-05. The images were acquired in tandem mode by ERS-1 and ERS-2 with a time difference of only one day. The coherence of the image pair is very high in the whole study area, with exception of water bodies (sea, rivers, channels, etc.). Interferometric processing produced a digital elevation model with a resolution of 25 m. The vertical accuracy of the relief is approximately 8 m and the positional accuracy is almost 5 m (one fifth of pixel size). The model has been only moderately filtered and prepared for combination with other DEMs.

Additionally the following DEMs were used: IGN DEM (with the resolution of 50 m, provided by IGN), Aster DEM (30 m, made from Aster stereo satellite images), SRTM DEM (90 m, produced with radar interferometry from the Shuttle Radar Topography Mission data). These models have different resolution and different coverage of the area: from ~60% for Aster to almost 100% for InSAR and SRTM (both are produced with interferometry and contain some areas with missing data) and complete for IGN. Standard deviations of all DEMs for the study area are less than 10 m (Table 1); Aster and InSAR are systematically higher because they include vegetation cover.

The final (combined) model is a weighted sum of partial models (Podobnikar et al. 2000). It takes the rough shape of IGN DEM and adds details of others, particularly InSAR. Its resolution is 25 m and it is very close to the control points (-0.2 m, standard deviation 3.7 m, Table 1, Fig. 1). The model is rather homogenous and could be used for further analysis and paleorelief interpretation.

Optical satellite images were an additional important interpretation source. A panchromatic image acquired by SPOT 4 in 1999-04-08 and a Landsat 7 ETM+ image from 2001-08-13 were used. Optical images (Landsat and SPOT) were enhanced prior interpretation. We have been trying to observe paleofeatures indirectly through detection of edges, particularly related to humidity and vegetation anomalies. Landsat with its seven bands gives much better spectral information than SPOT while SPOT has a better spatial resolution (SPOT 4 pan with 10 m resolution has been used only).

Simple and advanced edge detection has been performed, including Sobel and high-pass filtering supplemented with edge thresholding. Special attention has been given to filtering in the direction of the Sun to expose small variations in the direction of incidence radiation. The position of the Sun has been computed from the image acquisition time and the filter has been oriented in the selected direction.

To increase the spatial and preserve spectral resolution of Landsat resolution merge of panchromatic and selected multispectral channels has been performed (Švab and Oštir 2006). A similar procedure has been repeated with panchromatic SPOT and multispectral Landsat data. It has to be mentioned the huge difference between 15 m resolution of panchromatic Landsat and 10 m of SPOT. The latter is much sharper, what is probably the influence of different sensor types, simplifying the observation of anomalies.

Table 1 *Statistical comparison of available DEMs and the final combined DEM*

	Control points	IGN DEM 50	Aster DEM	SRTM DEM	InSAR DEM	Combined DEM
Number	110	110	67	108	105	110
Percentage	100%	100%	61%	98%	95%	100%
Average	81.4	80.9	82.4	81.6	90.0	81.2
Min Diff		-13	-13	-25	-15	-15
Max Diff		8	24	11	29	8
Average Diff		-0.5	9.1	-1.2	4.7	-0.2
StdDev Diff		3.7	8.4	5.0	7.8	3.7

Figure 2 *Enhanced SPOT (a) and Landsat (b) satellite images were used together with DEM (c) to detect paleofeatures.*

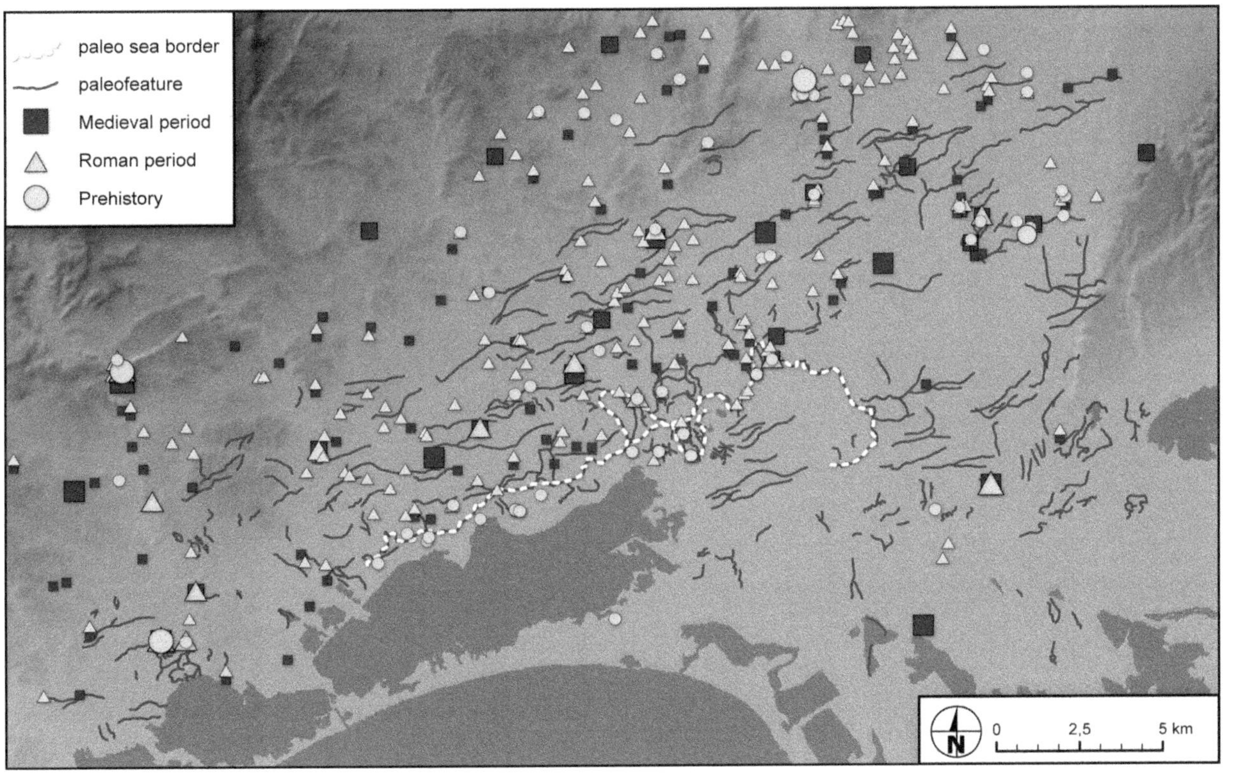

Figure 3 *Digitized and cleaned paleofetures are shown in relation to archaeological sites.*

The pre-processed DEM and satellite imagery enabled paleofeature delineation. Manual on screen digitalization has been performed and all possible features have been included. The following data has been mainly used to digitalize potential features (Fig. 2):

- panchromatic and edge filtered SPOT,
- multispectral and pan-sharpened Landsat, edge filtered panchromatic Landsat, and
- shaded and edge filtered combined DEM.

All possible anomalies have been considered and the features have been over-digitalized and later cleaned (Fig. 3). In the iterative procedure different images were used to digitize and recognize features. Maps in the scale 1:25.000 were used to verify all the vectors by eliminating existent relief characteristics (roads, buildings, etc.) and water courses (rivers, channels, etc.). SPOT and shaded combined DEM were most useful in feature delineation and all other layers were used only to update or check the results. Much better results are expected with a digital elevation model produced from lidar imaging (Challis 2006) that is scheduled for December 2006.

3 Thematical identification

To test if the detected paleofeatures can be linked to real paleorelief, an analysis of archaeological sites in the area has been performed (Fig. 3). The distribution of sites in relation to the features (buffer zones, site proximity analysis) has been observed for Prehistory, Roman period and medieval period. The sites were compared to random point distribution. For all known sites the distance from digitalized paleofeatures has been determined. Only sites that are less than certain distance (2.5 km) from the features have been considered. It has been statistically proven that the average distance of sites for all periods, especially for Prehistory, is smaller than for random points (Fig. 4a). For Prehistory the average distance is only approximately 450 m, while for random points it is more than 800 m, almost twice as large. One can also observe that the majority of the sites lie less then 500 m from the detected features (Fig. 4b).

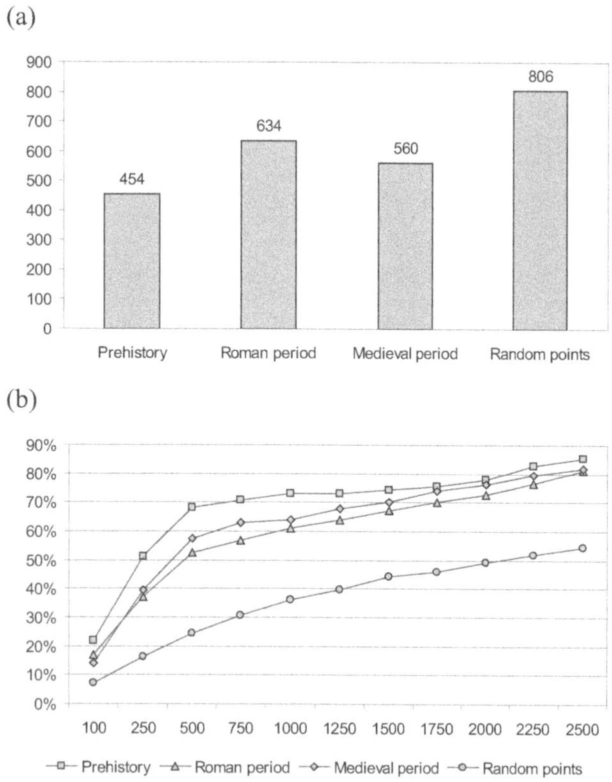

Figure 4 *Average distance to of archaeological sites to paleofeatures for different periods (a) and percentage of the sites within certain distance (b) compared with random points.*

Despite of the connection between evidence of paleofeatures and archaeological sites, we do not know their exact nature. Three different approaches were performed to define types of paleofeatures. First, the digitized layer was overlaid with geology, pedology, cadastre with contour lines at 1 m interval, and historical aerial photography (from allied missions in 1944). By observation of small areas chosen as test sites, several paleofeatures were identified as paleochannels, edges of natural paleodepressions, ancient roads, terraces, dried valleys etc. Nevertheless, it was noticed that some of the features are still related to the artefacts observed on the IGN DEM (stages caused by contour line interpolation) and some features remain unresolved.

The second approach was related to paleocoastline or sea border (Fig. 3). To ascertain if any of the features can be interpreted as coastline, we used simulations taking into account geoarchaeological information. A small number of cores with an acceptable spatial distribution permitted the generation of a very basic model with an approximated subsiding of the alluvial floor (Nuninger and Oštir 2005). The result shows a rather good general connection and by adding archaeological sites distribution, we were able to notice a relation between chronology and local variation. In fact, reconstitution of paleo-shore during antiquity suggests that the sea-level was approximately 1 m above the present level. However, the archaeological site locations suggest less elevated paleo-shore which is better connected to detected paleofeatures. Even if the scale of layers is different, we have to consider the local variation of sea level, due to hydro sedimentary dynamics. Thus, features detected (Fig. 5) show two steps of shore evolution from Bronze Age to Antiquity. In this case, a shore seems to have moved back until Antiquity and Bronze Age settlements were probably under water during that time.

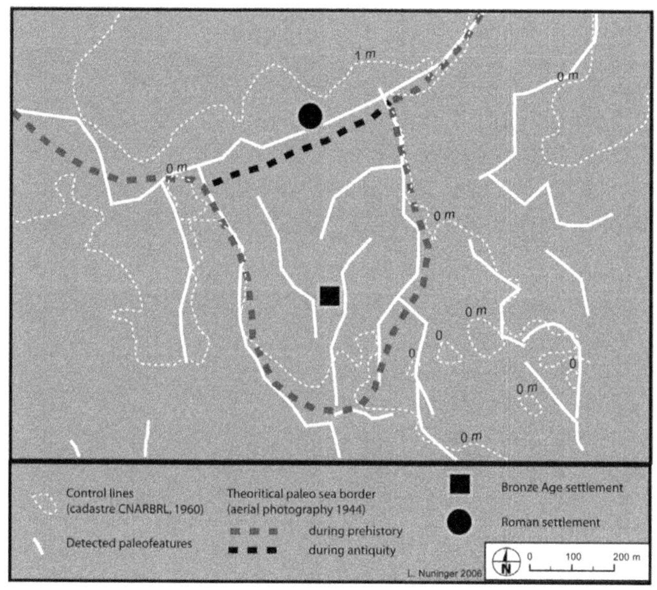

Figure 5 *Evolution of paleo sea border from Bronze Age to Antiquity*

At last, in order to automatically classify the detected evidence of paleorelief, a multivariate analysis (factor analysis with classification[1]) was processed (Coulon 2006). First, the detected features were rasterised and filtered to delete each cells corresponding to DEM artificial stages and thus to avoid noise due to IGN DEM artefacts. Remaining cells were grouped as new entities (1020) according to neighbourhood and their thematic homogeneity (mostly soils). Then, each entity was described by several criteria:

- Archaeological density was calculated as kernel density taking into account duration of

[1] The factor analysis and classification used are an « Analyse Factorielle des Correspondances - AFC » with a « Classification Ascendante Hiérarchique – CAH »

occupation (settlements). E.g. 1 settlement occupied during period 100 to 199 AD and period 200 to 299 AD counts as 2 occupations for density calculation.

- Compactness is a criteria linked to morphological shape of checked palaeorelief. Entities with strong compactness can reflect geomorphological types as depression, micro-reliefs (small hills). Entities with low compactness trend to show linear shapes as terrace, embankment, dried valley or road for example.
- Aspect is based on slope calculation. This criterion is important to detect entities linked to the general morphology of the littoral plain having a dip north-east / south-west.
- Soil types were grouped in 4 classes according to their age and potential accumulation. Recent alluvial soils, from flood plain of major streams, were isolated since they count almost no detected features. In addition, few features connected to this type of soil are probably due to artefact of DEM or imagery noise, recent sedimentation being very dip in such area.

The factor analysis results grouped the detected paleofeatures in 7 classes for 1020 entities. We will focus to the most significant only. In spite of containing dip 6% of all entities, class 1 is very interesting. It groups features with a strong archaeological density, a very strong compactness, a principal aspect to the east and a concentration on the oldest soils (61% "fersiallitiques"). Features from this class can be interpreted as micro-knoll with a high probability. Class 7 and 8 (14% and 6%) can be interpreted as class 1 with lower probability. On the contrary within class 6 (12%), one can observe a weak archaeological density (78% from null to mean value), a weak compactness (88%), principal aspect to north or north-east (67%) and most recent soils (70%). Features from this class are probably in connection with alluvial terrace or dried valley. At last, the class 5 (10%) is represented mostly by no archaeological occupation (82%), a very strong compactness (70%) and most recent soil (85%), especially soils with hydromorphological characteristics (41%). This class should bring together small depressions and paleomeanders.

The multivariate analysis is a first stage to sort automatically the detected paleofeatures. Nevertheless, such assumptions should be verified by control on the original data (IGN map, cadastre, aerial photograph) and in the field. Then, if the classification can be validated, criteria should be integrated directly within the process of detection.

4 Conclusions

With the case study we have proven that remote sensing can help in the detection of paleorelief. While it is not possible to detect past environment directly, several data layers – in our case most successfully the enhanced panchromatic SPOT 4 and the combined DEM – can be used to observe paleofeatures and to relate them to relief. The described approach will not replace precise work of archaeological and paleoenvironmental investigation, but can simplify the interpretation.

The results obtained present a first step and should be improved with further processing and additional data. The project will continue with advanced processing of high resolution satellite imagery and aerial photography. Furthermore, lidar scanning is scheduled to provide better elevation data in the flat areas.

Acknowledgments

The project was supported by CNRS within the framework of a four months associate researcher position for Krištof Oštir (Laboratory ThéMA, UMR 6049, Besançon, France). We would like to thank Jean-François Berger, Cécile Jung and M. Coulon for their collaboration and access to their data. Ziga Kokalj helped with figures and text corrections.

References

CHALLIS, K. 2006 Airborne laser altimetry in alluviated landscapes, Archaeological Prospection, 13(2): 103-127.

COULON, M., 2006 Contribution à la recherche des paléoreliefs dans la plaine littorale de Mauguio : analyse spatiale et statistique des données pour l'élaboration d'une typologie des indices de paléoreliefs. Master pro, training stage report « Géomatique et conduite de projet en développement », University of Avignon, unpublished.

NUNINGER, L., OŠTIR, K. 2005 Contribution à la modélisation des paléo-reliefs de la plaine littorale de l'étang de Mauguio (Languedoc, France): premières approches par télédétection. In Temps et espaces de l'Homme en société. Analyses et modèles spatiaux en archéologie. Actes des XXVe Rencontres internationales d'Archéologie et d'Histoire d'Antibes (J.-F. Berger, F. Bertoncello, F. Braemer, G. Davtian, M. Gazenbeek eds.), Antibes: APDCA: 123-134.

OŠTIR, K., Z. STANČIČ, T. PODOBNIKAR, T. VELJANOVSKI 2004 Producing digital elevation models with radar interferometry, in Making the connection to the past CAA 99 (K. Fennema and H. Kamermans, eds.), Leiden, Leiden University: 97-102.

PODOBNIKAR, T., Z. STANČIČ Z. and K. OŠTIR K. 2000 Data integration for the DTM production, in International Cooperation and Technology Transfer (M. Kosmatin Fras, L. Mussio and F. Crosilla), Ljubljana, Institute of Geodesy, Cartography and Photogrammetry (International archives of photogrametry and remote sensing), vol. 32 (6W8/1): 134-139.

ŠVAB, A. and K. OŠTIR 2006 High-resolution image fusion: methods to preserve spectral and spatial resolution, Photogrammetric Engineering and Remote Sensing, 72 (5): 565-572.

Exploring the archaeological landscape through a local perspective: spaces and places in the prehistory of the Florentine plain

Giovanna Pizzaiolo,[1] Lucia Sarti[2]

[1]University of Siena, Department of Archaeology and History of Arts
giopiz@yahoo.it
[2]University of Siena, Department of Archaeology and History of Arts
sarti@unisi.it

Introduction

The on-going recovery excavations[1] in the territory of Sesto Fiorentino (Florence, Italy) reveal that the plain area has been extensively settled ever since prehistoric times. Our research project is oriented to read the complex palimpsest of this archaeological landscape within a GIS environment. Different sources of information have been integrated into the GIS; in particular aerial photographs and ancient cartography have been used to reconstruct the different phases of landscape transformations in this ancient wetland area. Through the study of infrared aerial photographs, we strive to illustrate their contribution in investigating the relationship between environmental and archaeological perspectives. We focus on the research potential offered by IR aerial photographs and their accuracy in providing detailed information on ancient wetland areas. The exploration of prehistoric places in Sesto Fiorentino area is deeply related to the analysis of prehistoric landscape at a micro and macro scale. We believe that detailed information derived by intrasite analysis have to be integrated in an archaeological landscape investigation throughout the adoption of a local perspective.

1 The study area

Sesto Fiorentino is a town located in an alluvial plain on the northwestern side of Florence (Tuscany, Italy) at the foothills of Mount Morello. The area is part of the large system of Firenze-Prato-Pistoia plain originated on a lacustrine basin developed during Villafranchian time (Fig. 1). Such a basin, around 10 km wide and 30 km in length, developed on a NW-SE elongated tectonic depression since Late Pliocene in a fluvial lacustrine setting, receiving large amount of sediments from the eastern and western rocky margins. The Arno River is the main tributary of the basin and during Middle Pleistocene started to drain the area, leading the development of a wide alluvial plain in the central areas, and well developed alluvial fan along the margins. Due to this environmental settings the Sesto Fiorentino plain during Holocene prehistory appeared as a wetland area characterised by frequent alluvial esondations (Fig. 2). Geomorphological studies (Condera and Ercoli 1973, Capecchi et alii 1975) suggested that the inner basin of the Florentine plain was characterised during the prehistoric period by creeks, marshes and bogs which dominated the landscape.

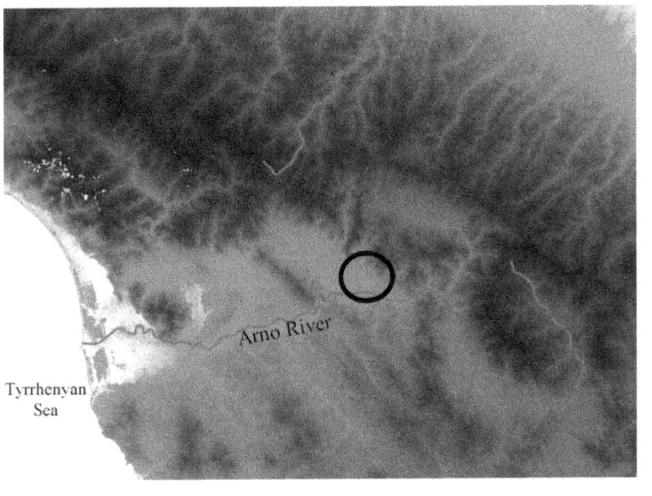

Figure 1 *The study area (black circle)*

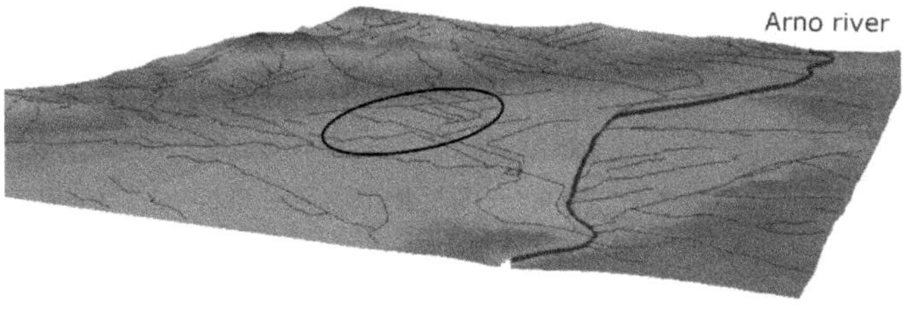

Figure 2 *3D View of the Florentine Plain (in black oval the study area).*

The lacustrine sedimentation has been covered by recent alluvial deposition laid down by the Arno River and other tributaries. The old town of Sesto Fiorentino is located on the major of the three alluvial fans characterising the foothills of Mount Morello and derived by the palaeostream action. Nowadays the watercourses are mostly regularised and transformed into channels especially when they flow in the plain area. Therefore the present day landscape is characterised by the evidence of

[1] This work has been carried out in full collaboration by the authors. In particular paragraphs 1 and 2 have to be attributed to Lucia Sarti and paragraph 3,4,5,6,7, have to be attributed to Giovanna Pizziolo.

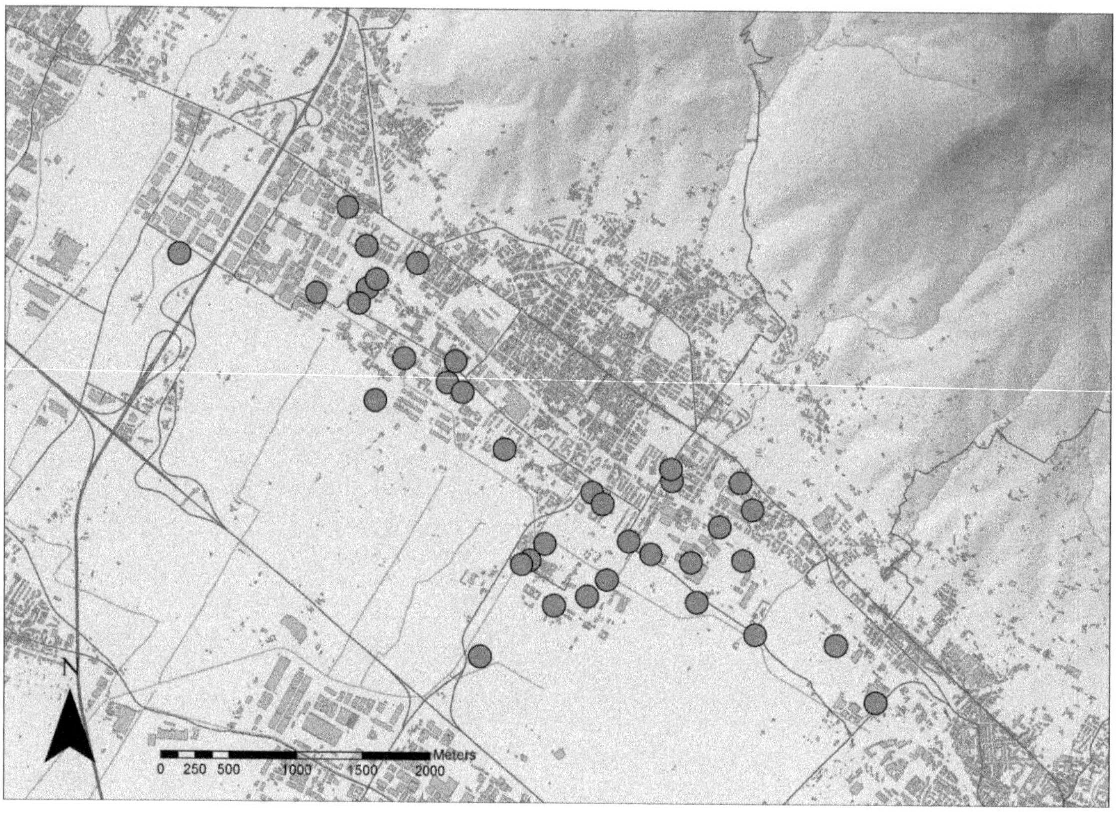

Figure 3 *Distribution of prehistoric sites in Sesto Fiorentino Plain*

different reclamation activities undertaken several times in the past to drainage a territory with a naturally wetland predisposition. In the last decades the area has been interested by an urban expansion stretching in the inner part of the plain which led to a progressive transformation of the land-use settings.

2 Prehistoric investigation in Sesto Fiorentino Area

Since 1982 the archaeological heritage policy adopted by the Municipality of Sesto Fiorentino in collaboration with Soprintendenza per i Beni Archeologici della Toscana includes systematic preliminary archaeological test pits which were executed in all the area interested by building or infrastructure constructions. This punctual monitoring condition associated with survey activities allowed archaeologists of the Universities of Siena and Florence to carry out several rescue excavations in the Sesto Fiorentino territory. More than 2500 test pits have been dug revealing archaeological evidence approximately for the 10% of them. These small trenches are disseminated irregularly in the study area. In the majority of cases the positive archaeological test pits have been investigated throughout systematic excavations bringing to light more than 30 prehistoric evidence of settling or production activities.

The general overview of Sesto Fiorentino prehistoric site distribution (Fig. 3) shows the high density of archaeological evidence (Martini, Sarti 1993; Martini et alii 2000). From the end of the V to the II millennium BC prehistoric sites expanded towards south-east and slightly into the inner part of the plain. The Neolithic evidences are concentrated on four settlements characterised by structures probably in use for a short period. The Copper Age culture is attested in six sites, three of them in continuity with the previous Neolithic ones. Until the first phase of Copper Age the settlement structures were built directly on the soil without shaping important pre-structures on the ground. In the second half of the III millennium the settlement units became more elaborate especially in regards to the depth of the depression over which the structures were installed; moreover the presence of little channels surrounding them becomes significant. A first substantial change in the history of the peopling of the plain becomes manifest at the end of the third millennium when the phenomenon of the Bell Beaker culture was widely revealed on Sesto plain. In particular the increase in population growth is testified by abundant archaeological evidences. Furthermore during the Evolved Bell Baker phase it is attested the presence of some large settlements. Some of these sites offer a consistent chronological continuity showing the repeated occupation of the same structures till the first phase of the Bronze Age. However the diffused presence of multiperiod sites, even if they may show some chronological breaks in their cultural sequence, testifies a transversal general interest for some specific locations. The observation of settlement evidence related to prehistoric period in the area highlights a significant difference in settlement strategies emerged in the Evolved Bell Baker sites. At the end of the third millennium people started to exploit palaeoriver beds using them as drainage structures. In some case in multiphase settlements the use of the palaeoriver is attested till the complete filling in of the bed (Sarti 1997). In particular the top of the draining layer constituted by gravel deposits have been exploited and represents the palaeosurface on which the inhabitants lived. Some artificial enlargements of the dry river bed banks have been recognized. The postholes are generally dug into the natural very compact gravel. The hypothetic covered areas generally show an irregular elliptical perimeter, with an extent of about 10 to 16 sq.m.

The Bell Beaker inhabitants of Sesto plain had a specific interest for the drainage facilities provided by the palaeoriver.

The Bell Beaker culture appears in Sesto as an original and autonomous core which shows an uninterrupted development of economic strategies, organization of settlements and artefact production until the Early Bronze Age. Despite differences in lithic and ceramic productions the settlement strategies appear the same (Sarti and Martini 1998). The main economic activities were related to agriculture and breeding; the palaeobothanical analysis available by now indicates a gradual increasing – from Neolithic to Bronze Age – of deforestation and cleared areas in the Sesto plain. During Middle Bronze Age, people changed their settlement strategy building up simple structures on not-prearranged floors (Sarti and Martini 1993). The distribution of these settlements (from the second half of the II mill.) shows the expansion of population which settled also in the eastern side of the studied area and at the same time testifies the occupation of the lowest part of Sesto plain (i.e. Bronze Age site of Dogaia located at 37m. on the sea level).

Evidence of the occupation of the area is well testified also during Iron Age and the first transformation of the territorial settings may be testified by Etruscan drainage activities. However the strong imprint to the Florentine plain is due to Roman centuriation (De Marinis and Salvini 2000), organization imposed upon the Sesto Fiorentino area during the first century B.C. (Bacci and Giachetti 1995), which deeply transformed and marked the landscape of the plain (De Silva and Pizziolo 2004).

3 Exploring "places" in prehistoric landscape: aims and methodology

The sense of place (Tilley 1994) in prehistoric context of Sesto Fiorentino is quite difficult to be investigated and defined. Undoubtedly the study area has testified a wide and persistent occupation: the analysis of archaeological data shows that the area was constantly inhabited and with particular intensity during the third and second millennia when Bell Beaker people settled recurrently in the plain; nevertheless it is not clear how we can interpret these data from a "place" perspective. Certainly the attraction that the Sesto area exerted since the fifth millennium onward is largely attested by archaeological evidence, but in order to understand their sense of place we have to increase our knowledge about the prehistoric landscape and consequently we need to fully contextualize our data.

Thus the attempt to explore the prehistoric landscape moves from the reconstruction of micro and macro morphologies which constituted the prehistoric living context.

In our research perspective the interpretation of the archaeological contexts and of the ancient spatial appropriation of territories, is strongly related to the analysis of landscape transformation which can be delineated by a detailed framework of the geographical and environmental settings in which the peopling process occurred.

In the case of the Sesto Fiorentino area a continuous dynamic process of erosion and deposition phenomena , typical of alluvial plains, has frequently modified the morphology and the inhabited landscape.

During different chronological phases there were people who decided to exploit, settle or simply cross the plain being attracted by some of the characteristics of this region, often adopting a different relationship with the landscape, according more or less to the modification of the morphology.

Which are the attractive elements of the plain? Considering the different nature of the prehistoric evidence (Martini et alii 2000) we can infer that a variety of space perception has taken place.

On large scale of investigation we can highlight that the inner basin of Florence probably has been crossed by a network of communication pathways through Apennine region and by waterways to coastal area following the River Arno or its tributaries: in this perspective the plain may constitute a crossing space connecting different regional subset; moreover the natural resources and in particular the presence of raw material like native copper and gabbro crops (Agostini et alii, in press) may have played an important role also for the neighbourhood regions. Furthermore the wetland characteristics might have represented an important resource in the subsistence strategy of prehistoric people.

But moving from a regional scale to a local perspective it might be more challenging to analyse the inhabited landscape of Sesto Plain from the point of view of a settling space and dwelling approach (Ingold 2000). A Local perspective includes a different scale of investigation and a more comprehensive examination in terms of settlement characteristics.

In order to explore and recognize the prehistoric evidence in a perspective which includes the concepts of place and landscape we need to use all the different sources that can provide information on settlement and territorial settings shifting dynamically from a micro to a macro scale and vice versa. It appears obviously necessary to built up such kind of research within a GIS system.

4 Sources for landscape archaeology analysis in Sesto Fiorentino

Different kind of sources are nowadays available for our study area: in particular we are focusing our attention on stratigraphic information (derived by archaeological excavations and a large number of preliminary test pits), aerial photographs (historical, present day color and infrared photos) and historical cartography. The research is oriented to explore the various circumstances of landscape transformation by taking advantage of the potential offered by these sources analysed within GIS and other analytical tools (i.e. image elaborations provided by Erdas Imagine 8.5)

The analysis of cartographical sources (De Silva and Pizziolo 2003,2004,2005) is a fundamental step in the individuation of the different historical phases that the area has passed through confirming either continuity or changes in the territorial settings of Sesto Fiorentino. The cartographical sources used for this project are extensive and take into account actual topographical and technical maps (from a scale of 1:100.000 to a scale of 1:2.000) as well as historical cartography dating back from the end of nineteenth century to the sixteenth century. The contribution of ancient maps is particularly significant from several points of view. Once historical documents have been interpreted and verified, they can provide a considerable amount of meaningful information regarding the previous landscapes and the way in which they were perceived by ancient cartographers (De Silva and Pizziolo, 2005).

Among the ancient maps examined for the project the "Catasto Leopoldino" (1817-1835) represented a significant support for reconstructing landscape transformation. The cadastral maps have been geo-referenced (De Silva and Pizziolo 2003, De Silva and Pizziolo 2004), and information on hydrography, road networks, settlements, buildings, place names, and field extensions has been inputted into different vector layers.
Other essential sources are historical and present-day aerial photographs. The photos of 1954 have been used to analyse land-use and to individuate continuity with previous settings. Historical aerial photos also supply the necessary background to elaborate a well-informed geomorphological interpretation of the plain area.

The individuation of alluvial fans and paleo-rivers is clearly readable on those photographs portraying the Sesto territory as it appeared before the radical transformation typified by urban expansion. The geomorphological studies have been undertaken by means of photo-interpretation by Conedera and Ercoli (1973) using those black and white photographs. Thus the input into the GIS of the 1954 aerial photos has proved invaluable for the individuation and validation of data derived by geomorphological studies. The project is based on the integration of different sources with the aim of correlating the multiple elements, which may have played a role in the constitution of the archaeological landscape. The comparison with present day settings has been undertaken using the ortophoto dating to 1997 and the colour aerial photographs dating to 2004 at an approximate scale 1:6000 made available by the Comune di Sesto Fiorentino. Among aerial photographs examined the most interesting ones are the false colour infrared sensitive aerial photographs available at a scale 1:6000. The photographs have made accessible to the Dipartimento di Archeologia e Storia delle Arti of the University of Siena by CAVET and FIATENGINEERING, two companies in the process of developing a new high-speed rail transit system in the territory of Sesto Fiorentino. These photos were taken by CAVET as a study aid in the planning of the railway track, which is to traverse the southern part of the recently expanded urban part of Sesto Fiorentino. The photos have been shot using the KODAK AEROCHROME III Infrared Film 1443, which is a medium resolution, fine grain, infrared-sensitive and false-colour reversal film. FIAT ENGENEERING provided us with both the transparency and print copies of the photographs.

On false-colour films (in this case obtained through KODAK AEROCHROME Infrared Films) colours are reproduced falsely. In the transparency photographs the resulting colours after exposure and processing are as follows: infrared radiation appears as Red, Green reproduces as Blue, Red reproduces as Green, and Blue in the original subject has not been recorded because of the yellow filter (minus blue) always used over the camera lens, and is therefore rendered as Black. Numerous other colours will be formed, depending on the proportions of green, red, and infrared reflected or transmitted by the original subject (Eastman Kodak Company 2003). Due to the characteristics of the Kodak film, we scanned in a RGB mode. The RGB acquisition allowed us to separate the specifics related to reflected and transmitted colours of the Sesto Fiorentino territory on three levels. The idea is to exploit at the best the potentiality of information provided by these infrared photos elaborating them through analytical tools and then comparing the results within the GIS system. In fact when comparing and contrasting the different sensibility characteristics of the film in relation to the Red, Green, or Blue bands, and finally when detecting the infrared component in the photo, it is possible analyse the humidity, wetness, and saturation characteristics of our study area (Pizziolo, in press).

5 Progress in local perspective, phase 1: the contribution of infrared aerial photographs

In order to perform the investigation in a local perspective it is important to define in detail the geomorphology and in particular the evolution of wetland conditions which, in the case of Sesto, strongly characterized the environmental settings.

The evolution of the lacustrine basin of Florence has been previously studied in larger picture incorporating Florence, Prato, and Pistoia. The absence of detailed data for the subset of Sesto Fiorentino precludes an exhaustive and meticulous palaeoenvironmental reconstruction of our study area. The first step of analysis has been oriented to data acquisition: palaeo riverbeds, alluvial fans, and colluvium previously identified by means of photo-interpretation (Conedera and Ercoli 1973) have been mapped into the GIS and transformed into vector layers. In order to devise a more comprehensive assessment of the Landscape Archaeology of the Sesto Plain, it is essential to delineate the palaeohydrology aspect. What sources can we access in order to investigate this wetland environment?

The historical aerial photos and cadastral maps provide interesting information on field patterns, which can generally be used to individuate hydrological anomalies (De Silva, Pizziolo 2003; Pizziolo, Sarti 2005; Pizziolo, in press). Moreover they contain, in themselves, accurate details about water networks that we can use to compare with other particulars on the region, having either a direct or indirect bearing on the archaeological data (see for example Bondesan and Meneghel 2004). The opportunity is also there to compare these historic layers with the present day situation.

As a further step our research examines the essential contribution of infrared aerial photo analysis and in particular its accuracy in providing information concerning ancient wetland areas. We have performed several elaborations but we consider that the most effective one has been the calculation of the NDVI (Normalized Difference Vegetation Index). In fact in our case, as we are dealing with photos and not with satellite images, the index is obtained by determining the ratio between the Red image and Green Image obtained by RGB scanning of each photograph (Pizziolo, in press). Our study of infrared aerial photos and of the particulars revealed by our application of the NDVI permitted to detect the presence of vegetation intensity, which probably correspond to the palaeo river zones. Furthermore the existence of several interesting anomalies emerged: small and very sinuous features with an extremely high concentration of vegetation, which show a different morphology when compared with palaorivers identified by Condera and Ercoli (1973) (Fig. 4). We are actually testing if these features conform to the exact nature of a palaeoriver.

Some other anomalies may be interpreted by comparing them with features previously identified: they appear for example an extension of those palaeorivers (Fig. 5).

However the infrared photo-interpretation permitted us to detected small elements, morphologically different from the main palaeoriver features, not detectable through the previous means of investigation.

It is always important to integrate the new features highlighted by infrared anomalies within the GIS; in fact it is throughout a comparison with other data in a unique system that we can augment the meaning of our results and get further in the reconstruction of landscape at small

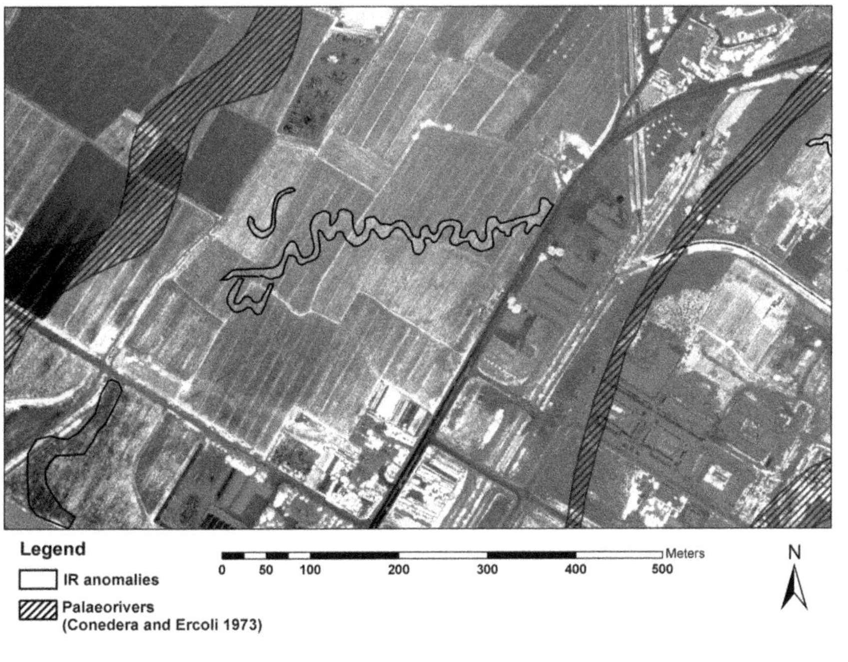

Figure 4 *The NDVI photo-interpretation (a detail)*

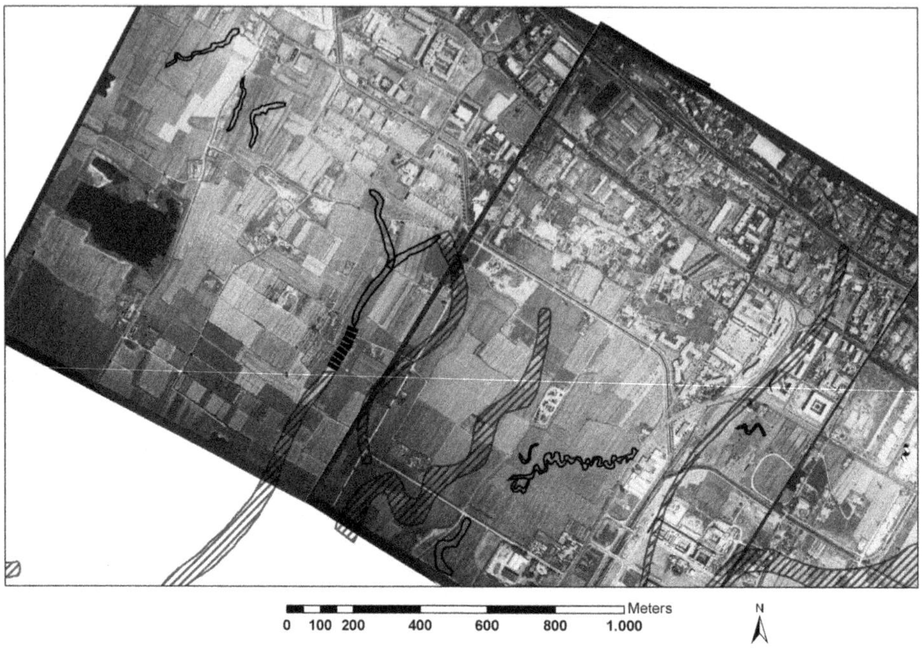

Figure 5 *The IR aerial photos with NDVI anomalies and geomorphological features*

and large scale. New encouraging results are coming from stratigraphical interpretation of test pits performed in the same area.

6 Progress in local perspective, phase 2: the value of a dynamic change of scale

According to this research direction a peculiar attention has been dedicated to data derived by archaeological excavations. Challenging results are coming from the analysis of paleosurfaces and stratigraphies once input into the system. In particular the reconstruction of settlements which occupied ancient palaeoriverbeds (Sarti 1997; Sarti and Martini 2000, 2001) provides information regarding the morphology of these features which match with data derived by infrared photo-interpretation. Small streams sometimes very sinuous and characterised by a similar bed profile have been detected and dated before the end of the third millennium (Fig. 6). Thus throughout an integration of data, ranging in scale from 1:50 to 1:50.000, it has been possible to get a more detailed framework of palaeohydrology and to characterise in a local perspective the landscape in which prehistoric people lived. It is important to explore this scenario at different zoom levels taking into consideration the role of local elements as part of the perception of a wide territory. As a result we know that the landscape was structured with main geomorphological features, as rivers that flow in a NE-SW direction but at a local scale we also highlighted that the landscape was moreover characterised by small water courses or palaeo streams which might flew in different directions and that provide variability at a local morphology level: from a general geomorphologic point of view they are not significant but they may have played a considerable role in the perception of the Sesto plain when prehistoric settling occurred. A dynamic shift of scale of investigation (site, near-site, territory and vice versa) allows us to better explore the different elements that characterise the landscape of the past defining both the general and local perspective: by mixing these views we may get closer to the characteristics of environmental and morphological settings of the prehistoric phases and thus to explore the relationship between man, environment and its dwelling and settling approach. All this may increase our information about the idea of place belonging to prehistoric inhabitants of Sesto Plain.

7 Exploring prehistoric places in Sesto Fiorentino: further steps

In our research perspective of Sesto Fiorentino context the sense of place is based on the environmental characteristic of the territory as perceived by local community and it is obviously fully imbued with cultural and social factors. As we know the same space can be lived as different places by different human communities. In the prehistoric context of Sesto Fiorentino significant examples are represented by the zones which revealed multiperiod occupations. These zones testify continuity of settlement and they show a transversal interest on that areas. This interest has been renewed by different population since Neolithic to Bronze Age. The same area has been chosen for settling activities in different periods, however every paleosurface is stratigraphically separated and the local morphology might have been lightly changed in every living phase, nevertheless the general geographical settings remains the same. As a case study we present the area of Podere delle Gore. In this context the archaeological excavations has revealed a ditch – which is actually under study – dug during the early phase of Copper Age. At a distance of m 30 the

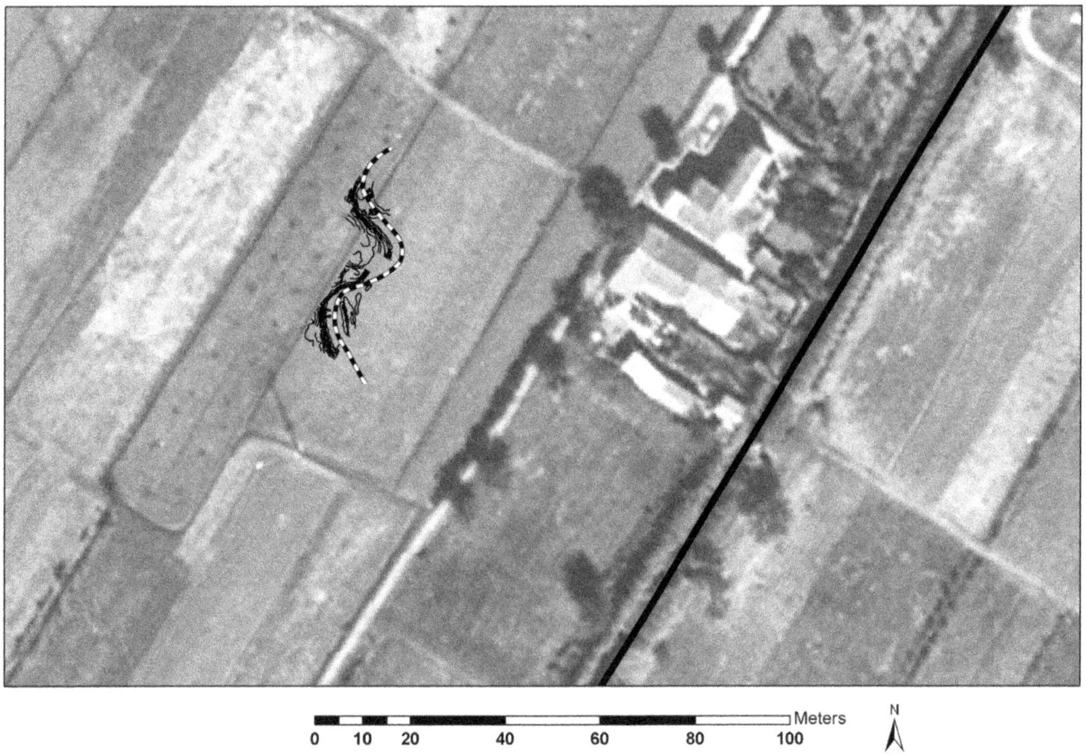

Figure 6 *Contour lines of an Epi Bell Beaker inhabited palaeoriverbed compared to 1954 aerial photo. (in continuous black line the present day hydrology; in interrupted black and white line the small palaeoriver)*

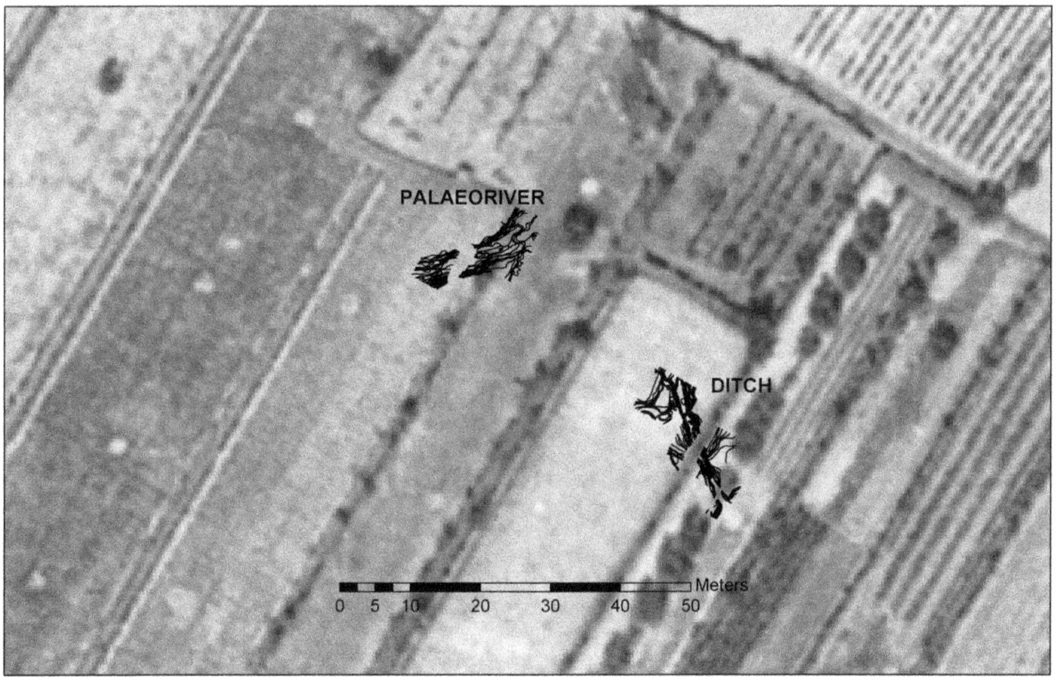

Figure 7 *Podere delle Gore. Contour lines of the Bell Beaker inhabited palaeoriverbed (left side) and of the Chalcolithic artificial ditch (right side).*

excavations revealed a settlement related to Bell Beaker culture. In this last case Bell Beaker people perceived the local characteristics in an opposite way in respect of the previous occupation of the area: instead of digging a ditch they chose to exploit the abandoned bed of a palaeoriver and settled inside this natural morphology (Fig. 7). This case study shows clearly two different approaches in perceiving and living the same space related to Copper Age cultures: these archaeological evidence express two different sense of place related to the same environment.

We actually do not provide an interpretation of these data but we underline the importance of keeping a local perspective in the analysis of prehistoric places. In order to appreciate this kind of differences and explore their spatial meaning it is important to anchor archaeological data to morphological and environmental settings analysing landscape at a detailed scale.

8 Conclusion

The reconstruction of archaeological landscape of Sesto Fiorentino takes advantage of an integrated approach. The comparison of different type of sources and their elaborations provide further meaning to our data and helps in defining prehistoric landscape settings in details. The research is in progress but in order to analyse archaeological and geographical data in terms of prehistoric landscape and explore the sense of place of prehistoric communities we have to manage archaeological record in a local perspective considering detailed morphologies in a complementary assessment with general landscape framework. We believe that challenging results can be obtained when intra site and inter site perspective are merging in a close vision of the archaeological landscape.

Acknowledgement

We sincerely thank CAVET, in particular the Eng. P. P. Marcheselli, for providing access to the infrared aerial photographs; we are also grateful to Comune di Sesto Fiorentino, in particular the Arch. G. Beni, for providing access to archive data and in particular to colour aerial photographs.

References

AGOSTINI L., BRIANI F., GHINASSI M., MORANDUZZO B., PALLECCHI P., in press, "Bell Beaker lithic, pottery and metal in the Florentine area: an archaeometric characterisation, in *Atti del Convegno Bell Beaker days in Italy: Bell Beaker in every day life,* Firenze, Siena, Villanuova sul Clisi, Gavardo, 12-15 Maggio 2006

AZZARI, M., DE SILVA, M. AND PIZZIOLO G. 2002. Cartografia del passato e GIS per l'analisi della trasformazione del paesaggio – Cartography of the past and GIS for the analysis of landscape transformations, *Geostorie*, Aprile-Agosto 10: 30-31 and CD Rom.

BACCI, M., GIACHETTI, M., 1995. "Insediamenti romani nella piana fiorentina e il loro rapporto con la centuriazione di Florentia", *L'Universo* 76: 546-561.

BONDESAN A., MENEGHEL M. 2004, *Geomorfologia della Provincia di Venezia*, Esedra, Padova.

CAPECCHI, F., GUAZZANE, G. AND PRANZINI, G., 1975. Il bacino lacustre di Firenze-Prato-Pistoia, *Boll. Soc. Geol.It.* 94:637-660.

CONDERA C., ERCOLI A. 1973. Elementi geomorfologici della piana di Firenze dedotti da fotointerpretazione, *L'Universo* 54 (2):255-262.

DE MARINIS, G. SALVINI M. 2000, "Dal VII sec. A.C. alla romanizzazione", in Martini F., Poggesi G., Sarti L. (eds), *Lunga memoria della piana. L'area fiorentina dalla preistoria alla romanizzazione*, Centro Stampa 2P, Firenze :97-98.

DE SILVA, M. PIZZIOLO, G., 2003. Cartografia catastale lorenese per la ricostruzione del paesaggio storico: problematiche e stimoli relativi all'uso di una fonte complessa all'interno di un sistema informativo geografico, il caso di Sesto Fiorentino, in Azzari, M. (ed), *III Workshop Beni Ambientali e Culturali e GIS, GIS e Internet,* Firenze 2002, CD Rom, Firenze University Press, ISBN 88-8453-117-9.

DE SILVA, M. PIZZIOLO, G., 2004. GIS analysis of historical cadastral maps as a contribution in landscape archaeology, in Magistrat der stadt Wien-Referat Kulturelles Erbe- Stadtarchaeologie Wien (eds), *Enter the past. The E-way into the Four Dimensions of Cultural Heritage, CAA 2003 Computer Applications and Quantitative Methods in Archaeology, Proceedings of the 31th Conference, Vienna, April 2003*, BAR Int. Series 1227, Oxford 2004: 294-298.

DE SILVA M., PIZZIOLO G. 2005, ""Signs", place, continuity and changes: chronological and typological integration of sources for landscape archaeology investigation in Tuscany", in Forte M. (ed), T*he reconstruction of Archaeological Landscapes through Digital Technologies", Proceedings of the 2nd Italy-United States Workshop, Rome, Italy, November 3-5, 2003. Berkeley, USA, May, 2005*, BAR International series, 1379:193-203.

EASTMAN KODAK COMPANY 2003. Aerial Data AS-77, http://www.kodak.com (visited 10/10/2003).

GHEDINI, F., BONDESAN, A. AND BUSANA, M.S., 2002, *La tenuta di Ca'Tron*, Verona, Cierre ed.

INGOLD T. 2000, *The perception of the environment: essays on livelihood, dwelling and skill*, London, Routledge.

MARTINI, F., SARTI, L., 1993. *Costruire la memoria,archeologia preistorica a Sesto Fiorentino*, Firenze, Garlatti e Razzai.

MARTINI, F., POGGESI, G. AND SARTI, L., 2000. *Lunga memoria della piana*, Centro 2P, Firenze.

PIZZIOLO G., SARTI L. 2005,"Landscape archaeology in Sesto Fiorentino: a GIS analysis for investigating settlement strategies in wetland area", in Berger J.F., Bertoncello F., Braemer F., Davtian G., Gazenbeek M., *"Temps et espaces de l'homme en société, analyses et modèles spatiaux en archéologie"* XXV Rencontres Internationales D'archeologie Et D'histoire D'antibe, Editions APDCA, Antibes: 443-452.

PIZZIOLO G., in press, "Landscape Archaeology at Sesto Fiorentino: the contribution of aerial photographs to the study of archaeological contexts within an integrated approach.", *Proceedings of CAA 2003 Computer Applications and Quantitative Methods in Archaeology, CAA 2004, Beyond the artifact, digital interpretation of the past*, Prato April 2004.

SARTI, L. 1997 (ed), Querciola. *Insediamento campaniforme a Sesto Fiorentino.* Ed. Garlatti e Razzai, Firenze.

SARTI, L. MARTINI, F. (eds) 2000, *Insediamenti e artigianati dell'età del Bronzo in area fiorentina. Le ricerche archeologiche nei cantieri CONSIAG (1994-1996),* Millenni, Firenze.

SARTI, L. MARTINI, F. 2001, "Strategie insediative del Campaniforme nell'Italia centrale tirrenica", in Nicolis F. (ed.) 2001, *Bell Beakers today. Pottery, people, culture, symbols in prehistoric Europe. Proceedings of the International Colloquium, Riva del Garda (Trento, Itlay) 11-16 may 1998,* 2001: 187-198.

TILLEY C. 1994, *The phenomenology of landscape: places paths and monuments*, Oxford.

A simulation of the medieval environment and its change around medieval castles – special case in Finland

Kari Uotila,[1] Isto Huvila,[2] Anna-Maria Vilkuna,[3] Terttu Lempiäinen,[4] Elisabeth Grönlund,[5] Pentti Zetterberg[6]

[1] Archaeology, University of Turku. Finland
[2] Åbo Akademi
[3] History, University of Jyväskylä. Finland
[4] Centre for Biodiversity, Herbarium, University of Turku
[5] University of Joensuu, Karelian Institute, Dept. of Ecology
[6] University of Joensuu, Karelian Institute. Laboratory of Dendrochronology

Introduction

At the end of the European medieval period in the 13th - 16th centuries Finland constituted the northeast periphery of the catholic church and hanseatic communities and Europe as a whole. Finland was at the time a frontier area of extreme building conditions (eg.stonechurches and castles), where the harsh climate stipulated strict terms for the use of lime mortar, which cannot be used in sub-zero temperatures.

According to the traditional research the history of building in stone in Finland is dated to a rather long period ranging from the late 13th century to the beginning of the 16th, with each large building project (castles and large churches like Dom of Turku) lasting for several generations. A newer approach, which has developed during the last fifteen years, dates practically the entire Finnish stone building mass to the period between 1400 - 1520, with the bulk of the buildings dated to the period between 1450 - 1500. This research model would practically set the main medieval settlement areas in Finland as one great building project for a period of less than a fifty years.(e.g. Drake 2001, Hiekkanen 1994, Uotila 2000, 2001 and 2004)

One of the key scientific questions in the research project (Seats of Power in Medieval and Early Modern Finland) under presentation is, whether this kind of a short term but - under the prevailing circumstances - large scale stone building activity could be regarded as a change and strain factor on the environment and communities of the time, and whether the simulation and 3D- modelling of the research material of various principles could be used in the research of such a historical process.

The modelling is based on a notion of continuous process instead of the typical approach of preparing the data first and creating the model as a final outcome of the research. The transparency of modelling process advocated especially by Forte (2000, 2004) expands to grasp not only stepping back in the reasoning process, but also integrating the reasoning process to the modelling itself. The general tendency of seeing a model as a representation of a final hypothesis is rather remarkable, because the approach is clearly different from the typical way of archaeological reasoning (Gardin 1981). In the present approach, a model is not a representation of a theory, but it is a theory about the past state of affairs, which develops throughout the research process. The notion compares with a typical archaeological research process during which a scholar iterates a cycle of building a hypothesis, consulting evidence and literature. A comparable process has been made explicit in the grounded theory approach (Glaser & Strauss 1967) used in a qualitative social science research and therefore the present paradigm may be denoted as *grounded modelling*.

Besides emphasising the embeddedness of the modeling and archaeological reasoning processes, the proposed approach underlines also the horizontally layered character of the information, which is available for modeling. In exemplary case of Kuusisto castle, the information from the site and its surroundings were conceptualised in three scalar layers. The information about the activities concerning the different layers is very different by its scope, detail and context. The layered approach keeps the different levels of information distinguished and emphasises the significance careful consideration while making inferences of their convergencies. We argue that in these three spatial layers or circles, it is necessary to use different approaches and levels of detail in the modeling work, because the scientific and scholarly information, which is available in the different layers is not directly comparable. Therefore it has to be modeled and simulated in different manners.

Medieval Stone Castle and its environment

The scientific Seats of Power in Medieval and Early Modern Finland is a multidisciplinary research project combining the latest achievements in the humanities, the natural sciences, the information sciences and teleinformatics. The subjects of research will be Kuusisto Castle (Kustö) in Varsinais-Suomi (Finland Proper) and Häme Castle (Tavastehus) located in the Heart of Häme. The project combines research results of history, archaeology and natural sciences to produce simulated

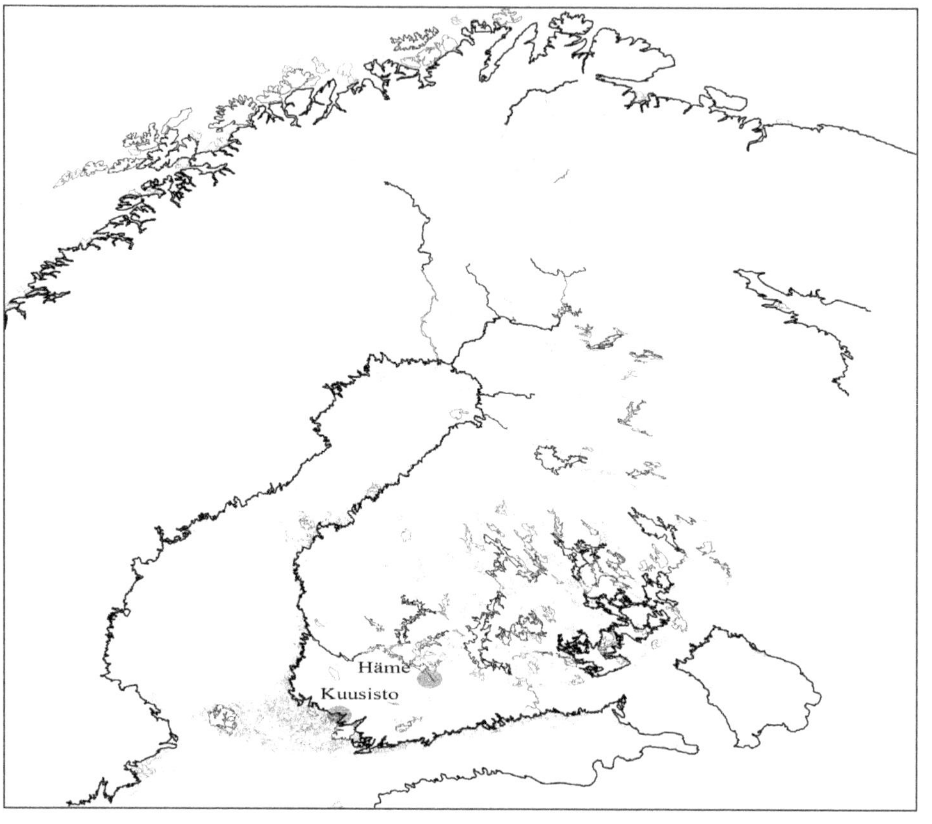

Figure 1 *Kuusisto (sw Kustö) and Häme (sw Tavastehus) castles in southern Finland. (Map Kari Uotila)*

Figure 2 *The ruins of Kuusisto Castle from SW (photograph) with an indicative 3d-model representing the vastle in the year 1500 (photo and model by K. Uotila). The Baltic sea was probably some 2 m higher in the medieval period. There was a wooden palisade around the castle in deep water. The modern waterfront is following the line of that palisade.*

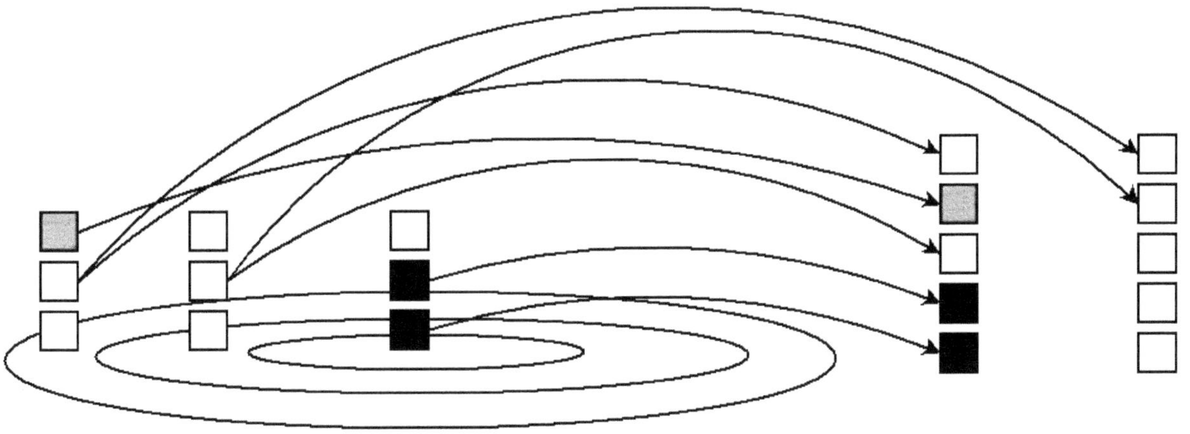

Figure 3 *Besides emphasising the embeddedness of the modeling and archaeological reasoning processes, the proposed approach underlines also the horizontally layered character of the information, which is available for modeling. In exemplary case of Kuusisto castle, the information from the site and its surroundings were conceptualised in three scalar layers. The information about the activities concerning the different layers is very different by its scope, detail and context. (Drawing Isto Huvila)*

3d-models of the Finnish Middle Ages (1200-1600). Research results from different science faculties are to be simulated within the framework of a historical context. The multivariate utilization of the different source types will provide an excellent basis for the study and understanding of both long-term and short-term phenomena and processes.

It can be stated that the objective is to examine changes in the environment of the medieval stone castles in the different stages of the castles' construction and occupation. In medieval (1200-1600) history research in Finland, the construction dates of significant structures (churches and castles) have been vigorously debated in recent years. (Drake 2001, Hiekkanen 1994) In this project an attempt will be made to see the castles' construction, in a sense, by way of a mirror or silhouette: how the construction and habitation are seen in the environment and have affected it and how to model that silhouette in 4d.

One can assume that, during the initial castle-construction stage, the emphasis will be on the procurement of good-quality construction material and different large scale pre-industrial production processes, such as the firing of bricks and lime and mortal. One can presume the reflection of these in changes in surrounding nature. When the castle construction has reached a certain level of activity the changes effected by the sphere of the castle on the environment, that is to say how, under medieval conditions, an exceptionally large group of people has affected the environment by their lives. For example, heating a medieval stone castle during a Finnish winterseason (at least 6 month/year) was certainly a considerable undertaking and required a really large amount of wood annually.

One research issue will be to supplement the deficiencies of medieval history sources by examining the construction and everyday acquisitions of the best known 16th and 17th century castles from the standpoint of sources (in this case, Häme Castle) and verify them with the research data obtained from the environment. These results can then be simulated by GIS-analyses for the medieval situation, which has been, in regard to the castles, a significant period of construction and occupation (Pukkila & Uotila 2005; Vilkuna 2003).

The intent of the research group is to examine simultaneously comparison data from the development during the medieval and modern periods (1000-1700) of some southwestern-Finnish agricultural areas (for example, Laitila and Eura-Köyliö) as well. With the aid of the model subjects and the wider research data, it will be possible to separate the "background noise" attending the growth of general agriculture, industry, and population from the particular situations attaching to the castles. (Grönlund 1996, Pukkila & Uotila 2005)

Historical background

Kuusisto Castle was, during the Middle Ages, the castle of Turku's Catholic bishop. The earliest documentation of Kuusisto is from the year 1295. The castle was in possession of the Church continuously to the beginning of the 1520s, when it was transferred in connection with the Reformation from the Church to the Crown. Kuusisto became an agricultural estate under Turku Castle, and agriculture directed toward the castle's needs was practiced thereat. (Uotila 2000)

Häme Castle was established as a base for the Swedish royal power in the midst of a prosperous Häme settlement at the turn of the 13th and 14th centuries. The castle retained its position as administrative center of Häme Province until the 1630s, when it was transferred to

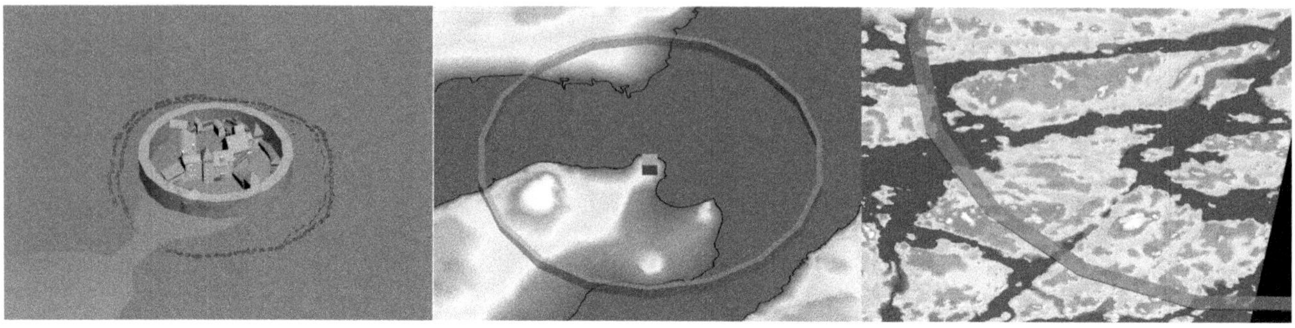

Figure 4 *The three different scalar layers of modeling a medieval castle based on the different spheres of use of the castle in the Middle Ages. In the inner layer is the castle itself (in Kuusisto case the circle is some 100 m). The middle layer comprises the main economical and military surroundings (in Kuusisto about 500-1000 m). The outer layer encompasses farms, hunting areas, forests, and mortar and brick production sites (in Kuusisto, 5-15 km). (Models Kari Uotila)*

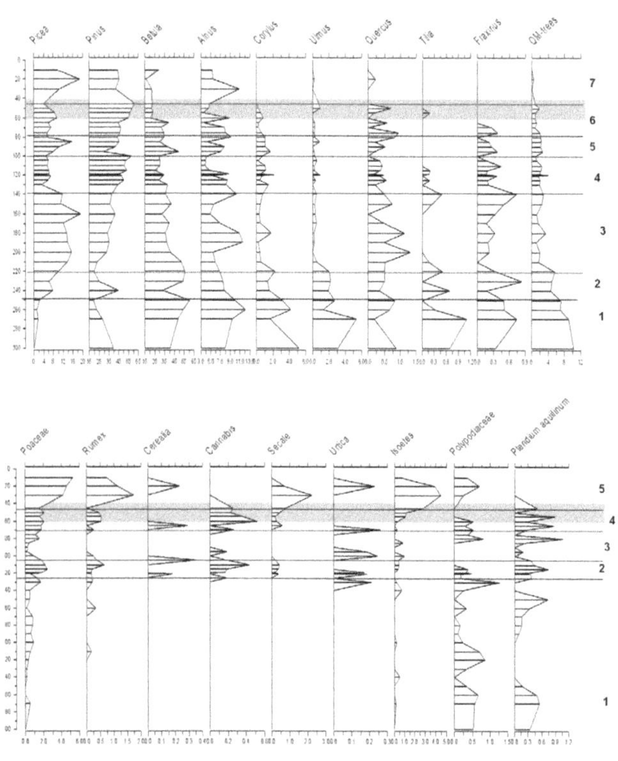

Figure 5 *First pollen-analysis studie from Häme castle. In grey colour is marked 1200-1400 AD – the time of Häme Castle. (Analysis Elisabeth Grönlund and editig Kari Uotila)*

provincial management under governors and the national authority moved from the castle to the newly-established town of Hämeenlinna. (Vilkuna 2003)

Earlier source and research data, which are available from Häme and Kuusisto Castles, provide a good point of departure for study of the habitat. The multiple ways in which the settlements, castle construction and everyday activities affected the environment will be illustrated by the aid of the research results. In addition to the castles' role in representing the state with their physical being, they were also significant economic centers where the affairs of several dozen persons were managed on a daily basis and large-scale agriculture was practiced. (Uotila 2000, 2004 and Vilkuna 2003)

Medieval written source data are sparse all the way to the beginning of the 16th century in Finland. The significance of historical source data becomes more marked from the 1520s, from which time written sources prepared for the needs of both public administration and private households are available. On the part of the economy of individual manors, the 16th-century sources provide a possibility to follow the development of agriculture, livestock production, food economy and construction activities. The documents prepared for Crown taxes reveal the regional tendency of the agriculture of the peasantry. Historical map material is available from the beginning of the mid-seventeenth century, and will be utilized wherever possible. (Pukkila & Uotila 2005, Uotila 2002a, 2002b, Vilkuna 2003)

Objectives and methods

The habitat of the medieval person will always be modeled at Kuusisto and Häme Castles by way of his immediate environment to the broader environmental whole. Thus the models will move from the medieval houses and yards to the nearby fields and meadows, from which they will progress to the broader agricultural totality. With the aid of the research results of the humanities and natural and social sciences, it will be possible to model and simulate different variables and their effect on the environment and on people's everyday choices. Thus we will be able to uncover, for example, rapid and local climatic changes and their effects on the people's livelihood and ways of life (e.g. Uotila 2001, Zetterberg et al 1995).

In order to obtain the greatest possible benefit from their environment, the castle builders started an active and systematic crop cultivation activity in their new area of settlement, where the people had brought along plant and animal species from their original home areas. These new, imported species, in turn, became naturalized in the area. The basic elements of the landscape were fields, meadows, slash-and-burn fields, forests, mills,

Figure 6 *Laitila village in Southern Finland from 100-1500 AD. Information in models is based GIS-analyses, old maps and historical sources and digital elevation model and the level of Baltic see and rivers amd lakes in ancient time. (Models Kari Uotila)*

neighboring villages, highways, trails and water routes. With the aid of research information, it will be possible to bring out the historical dimensions preserved in the landscape, if often perceived with difficulty.

The only locality that has been rather well studied so far by natural scientific methods is the city of Turku, from which completed scientific data may thus be obtained for an environmental model. The macrofossil analysis soil samples from the cellar of Häme Castle, taken by Doc. Terttu Lempiäinen, have already shown encouraging results. For example, several dozen seeds of henbane, commonly used as an analgesic in the Middle Ages, were found in the analysis. There is no previous knowledge of henbane utilization at the Häme Castle.(e.g. Lempiäinen 2002)

Macrofossil analysis has proven especially useful in conjunction with pollen analysis. Pollen-analysis studies aim to provide as detailed as possible information about vegetation changes, especially land-use related changes in forest cover, cattle grazing and crop cultivation, during the study period. The first task in the present project will be to review the already existing and available pollen data sets in the study areas. Preliminary lake study sites for paleoecological reconstructions in Varsinais-Suomi (Finland Proper) are Lake Kaukjärvi (Kalanti) and Lake Pitkäjärvi (Laitila), where data and dating information are available (Grönlund 1996). Former Lake Kuusistonjärvi (Kaarina) and Lake Kakskerranjärvi (Turku) will be analyzed during the present project. In the Häme area, three or four suitable lake sites will be selected at a 1-10 km radius from Häme Castle. The intention is to find lakes with annually deposited sediment layers that provide reliable datings for the vegetation and land-use signs, which will be identified in the pollen analysis.

The University of Joensuu's dendrochronology laboratory will make tree-ring information for the Turku and Hämeenlinna areas available to the project. The regionally representative annual tree-ring calendar for the region of Turku and Hämeenlinna covers the entire medieval 500-year period. The annual tree rings include new, and annually precise, information for the period for which written sources are very scarce.

Pine-tree annual-ring data starts in the Turku region in the year 1012 and in the Hämeenlinna region in the year 1094, so they cover the research period, 1100 to 1600. The data collected from both research areas are, however, used very seldom in other than ordinary dating. Thus, there is great research potential in the data. To this can be added, as entirely new material, graph data being prepared for oak. The dated wood material also indirectly reveals variations in human activity according to the date of felling: for example, in the 1350s and 1360s a lot of trees were felled, while again, during the 1340s, for example, hardly any trees were felled.

During the period of the project, additional tree-ring information will be gathered at Häme Castle, some of it by means of the funding obtained. In preliminary studies, it has already been confirmed that there are nearly one hundred objects at Häme Castle suitable for dating. The main scientific benefits of dendrochronological studies in this project are: 1) precise datings of wooden material will give an exact absolute time-scale for human activities and events in the natural environments, 2) precisely dated annual rings give exact annual data of variations in climatic growing conditions (mainly temperature of the growing season) in the study areas Varsinais-Suomi and the Heart of Häme, 3) the fluctuations in the number of dated trees gives knowledge of the changes in the amounts of trees used by man.

The information of written sources on agriculture's annual yields gives indirect indication of possible variations in climate. According to the yield information contained in Häme Castle's account books, for example, an exceptionally good rye harvest was gathered at the castle in the 1550s. On the Crown farms located in Varsinais-Suomi and Uusimaa, moderate harvests were also gathered during these years, but they were, however, clearly smaller than at Häme Castle. An analysis of annual pine-tree growth rings can clarify whether the conditions for growth were exceptionally good at Häme Castle during the years in question. Additionally, good or bad crop periods may also in some cases be detectable by pollen-analysis methods. (Vilkuna 2003)

Summary

The central aims of the project are to use and develop 3d-modeling in different scientific questions (e.g. Uotila 2006), the use of 3d-modeling in simulations, the utilization of research information and create solutions to questions in connection to the application and control of new types of research material. The 3d-model and simulation material will typically differ significantly from earlier research material in the natural and social sciences (the raw data, research reports and published material) in relation to material type, amount and method of application. Control of the material in the research situation, its archiving for future use and, especially, its effective and sufficiently detailed representation require new methods and processes. One objective is a relatively extensive so-called domain analysis, a need mapping, as well as, especially, development of a suitable method for constructing information and development of a complete information-control process on the basis of the data and its use during the project period.

References

DRAKE, K. *Der Meister des Hochhores zu Turku. Backsteinarchitektur in Mitteleuropa.* Studier zur Backsteinarchitektur 3. Berlin 2001.

FORTE, M. About Virtual Archaeology: Disorders, Cognitive Interactions and Virtuality. In Barceló, J.A.; Forte, M. & Sanders, D.H. (ed.) *Virtual Reality in Archaeology. Proceedings of the Computer Applications and Quantitative Methods in Archaeology (CAA) 1998, Barcelona.* Oxford: BAR Publishing, 2000, 247-259.

Forte, M. Realtà virtuale, beni culturali e cibernetica: un approccio ecosistemico. *Archeologia e calcolatori,* 2004, 15, 423-448.

GARDIN, J. *Archaeological constructs : an aspect of theoretical archaeology.* Cambridge 1980.

GLASER, B.G. & STRAUSS, A.L. *The discovery of grounded theory: Strategies for qualitative research.* Hawthorne 1967.

GRÖNLUND, E.. Palaeoecology of land use and settlement history of Laitila Kalanti area, Finland Proper, SW Finland. *Arkaeologiske Rapporter fra Esbjerg Museum* 1:177-188.1996.

HIEKKANEN, M. The Stone Churches of the Medieval Diocese of Turku. *SMYA* 101. 1994.

LEMPIÄINEN, T. Plant macrofossils from graves and churches. In: Nordic Archaeobotany – NAG 2000 in Umeå (Ed. Karin Wiklund) – *Archaeology and Environment* 15: 144-161, Umeå 2002.

PUKKILA, J & UOTILA, K. From Ancient Monument to Virtual Model – GIS as a content production tool in 3D visualization. Mäntylä. Sari (ed.) *Rituals and Relations. Studies on the history and material culture of the Baltic Finns.* The Finnish Academy of Science and Letters 336. 2005.

UOTILA. K. Die Bischofsburg Kuusisto und ihre nächste Umgebung im Mittelalter. (English summary) *Archaeologia Medii Aevii Finlandiae V, Castella Maris Baltici* 3-4. Gdansk 2001. pp 185-192.

UOTILA. K. *The collapse of defence in Finnish castles around 1500. Chateau Gaillard XIX (Graz 1998).* Caen 2000. s. 297-303.

UOTILA et al 2002a. Changes in Natural and Human Landscape from 500 BC to 1500 AD - Modeling large landscape areas in prehistoric and medieval South-West Finland. Uotila; Petteri Alho; Jouko Pukkila, Carita Tulkki. Niccolucci, F. and Hermon, S. (eds.) *Multimedia Communication for Cultural Heritage. Proceedings of the workshop held in Prato, 1 October 2001.* Budapest: Archaeolingua, 2002. s. 119-126.

UOTILA et al 2002b. Uotila, Kari, Alho, Petteri, Pukkila, Jouko & Tulkki, Carita 2003: Modeling Natural and Human Landscape in Prehistoric and Medieval Southwest Finland from 500 BC to 1500 AD – Computer Based Visualization. Doerr, Martin & Sarris, Apostolos (eds.), *The Digital Heritage of Archaeology, CAA2002.* BAR, International Series: 191-194.

UOTILA, K. *The changing roles of outer baileys in*

Finnish castles.Chateau Gaillard XXI (Maynooth 2002). Caen 2004.

UOTILA. K. 2006. *Vyborg Castle and Town Wall : 3D Modelling in Castle Research. Kari Uotila. Chateau Gaillard XXII.* Grenoble - Voiron 2004.Caen 2006.

VILKUNA, A-M. *Financial Management at Häme Castle in the Mid-Sixteenth Century (1539 – about 1570). At Home within Stone Walls: Life in the Late Medieval Häme Castle.* Archaeogia Medii Aevi Finlandiae VIII. Saarijärvi 2003.

ZETTERBERG, P., ERONEN, M. and BRIFFA, K.R. Subfossil forest limit pines from northern Finland: evidence on climatic variability, growth variations and prehistoric human activities betveen 165 B.C. and A.D. 1400. In Ohta, S., Fuji, T., Okada, N. Hughes, M.K. and Eckstein, D. (eds.) *Tree Rings, from the past to the future, Proceedings of the International Workshop on Asian and Pacific Dendrochronology.* Forestry and Forest Products Research Institute Scientific Meeting Report, 1:134-144. 1995.

The Kargaly Project: modelling Bronze Age landscapes in the steppe

Juan Vicent,[1] Santiago Ormeño,[2] MªIsabel Martinez-Navarrete,[1] Julian Delgado[2]

[1] Laboratory of Remote Sensing Archaeology (LabTel), Institute of History, CSIC, Madrid
[2] Research group of Remote Sensing and Photogrammetry, EUITTO, Universidad Politécnica, Madrid

1 The Kargaly project and landscape archaeology

This paper presents the main lines of an experience in the incorporation of techniques of Remote Sensing (RS) with the aim of modelling ancient landscapes and solving the problems involved. The specific model proposed here was designed to cope with the research problems resulting from the context of the Kargaly project.

Kargaly is a mining field located in the *oblast* of Orenburg, in the Southeastern limit of the Russian Federation (Fig.1). During the Bronze Age (BA) its extensive copper deposits (about 500 km^2) were object of an intensive mining and metallurgical activity. The mining activity is recorded from the Yamnaya cultural phase (4th milBC), reaching its phase of maximal intensity during the Late Bronze Age (LBA, Srubnaya cultural phase, 2nd milBC). At *ca.* 1400 BC mining and its related settlements were abandoned rather abruptly. The region was peopled by nomads until its incorporation to the Russian empire in the first half of the 18th century AD. Since then, the renewed mining operations lasted until the early-20th century. During the Soviet period, Kargaly became an agricultural land.

During the LBA, thanks to its huge copper production and its strategic position in the limits of the Volga and Ural basins, the Kargaly complex plays an outstanding role in the process of early development of metallurgy in steppes. Since 1995, a Russian-Spanish research team under E.N. Chernykh's direction runs archaeological, archeo-metallurgical and paleo-environmental investigations in Kargaly (Chernykh 2002, 2004).

The historical interpretation of the Kargaly complex raises some relevant questions as regards the landscape, the environment and their changes through time. We emphasize two of them: the availability of forest resources and subsistence practices in the BA.

Kargaly is at the moment a typically steppe landscape. The forests cover less than 2% of the territory. The scale of mining and metallurgical activities, particularly during the Srubnaya phase, suggests an intensive demand of fuel. To understand the balance between these two activities and the kind of copper production exports (metal, copper ore, or lumps of slag-like material) requires determining whether the limited amount of woodlands in the region today can be extrapolated to the

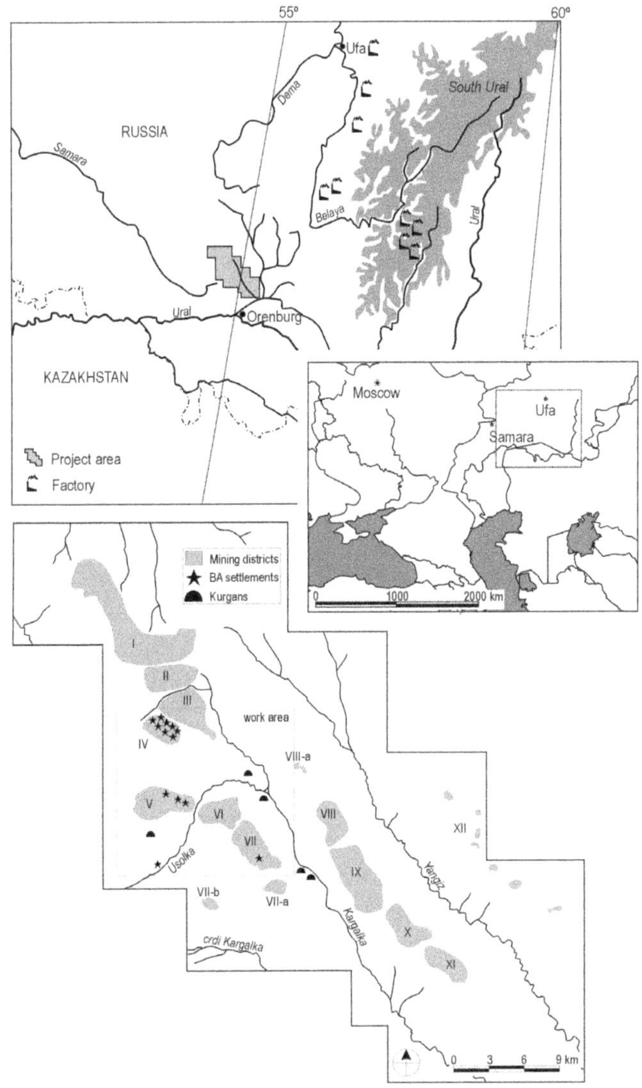

Figure 1 *Location of Kargaly region and project work area.*

LBA, or on the contrary, if it is the result of a dramatic change in the distribution of forests.

The existence of permanent settlements during the LBA poses the problem of the subsistence practices of their inhabitants. The available archaeobotanical evidence suggests the absence of agriculture during that period from eastern Don to the Transurals (Lebedeva 2005); thus the interpretation of food supply for those permanent settlers demands alternative hypotheses. The data recovered in the systematic excavations at the metallurgical settlement of Gorny (LBA, Kargaly project) suggest that subsistence was based on the exchange of ore and/or -if it were the case- metal products. The complete absence of botanical and archaeological evidences of any agricultural practices is complemented with the massive accumulation of domestic *bovidae* remains. Their number, consumption patterns and provenance suggest that the exchange of ore by animals constituted the main food base of Gorny population (Antipina and Morales 2005). However, it is essential to determine if this peculiar subsistence pattern fits in a transregional, regional or local scheme of exchange and labour division. Achieving that aim requires evaluating if there are or not any indicators of farming activities in the paleoenvironmental record at Kargaly, and, eventually, also the type of scale relevant for their interpretation.

In this context, the recent identification of cereal-type pollen grains in BA levels of a palynological sequence coming from a natural deposit in the hill of Gorny itself is a far reaching find (López et al. n.d.; Chernykh and Martinez Navarrete 2005). Their classification either as cultivated cereals or wild *Poaceae* with large pollen grains depends on the development of palynological researches that require non-conventional procedures. Whatever the result, however, their interpretation as evidence of agriculture on a local, regional or transregional scale depends on their recurrent association with other palynological indicators of agriculture. That means that the sole presence of cereal pollen grains do not necessarily evidence a local practice of agriculture. Further indicators should accompany them.

Archaeological palynology is the main source of data on vegetational distribution and, by extension, on past land use on a regional scale. The Kargaly project has made an effort to the effect. We developed a total of 6 paleopalynological sequences from the BA to present time supported by radiocarbon dates in the study area (3 in archaeological deposits and 3 in natural deposits) (López et al. 2003, López-Sáez 2002, Lopez- Sáez et al. 2002). Nevertheless, the use of the information coming from these sequences requires the adoption of a nonconventional approach in archaeological palynology to cope with the problems we aforementioned.

The observable change in palynological sequences used to be interpreted in terms of general environmental change disregarding the variability originated by changes in the local morphology of ancient landscapes. In order to use palynological data here we must be able to identify the determinants operating on a local and regional scale in the formation of the palynological record. Such determinants allow explaining the variability in the qualitative and quantitative composition of palynological spectra in terms of vegetational communities distribution and land uses.

This is indeed the aim of the postulated analysis of the present day landscape: to calibrate the local, regional and global components of spectra variability, analising the formation processes of the palynological record in a controlled environment. In this way, it will be possible to a) recognize how the palynological record represents the local distribution of vegetational zones and land uses in a controlled context, b) determine to which extent the qualitative and quantitative variability of the palynological spectra over the time reflects local, regional or global changes, and c) identify the meaning of these changes in terms of the concrete morphology of the landscapes.

2 Palynology, remote sensing and landscape modelling

The methodology developed for the program of landscape archaeology embedded in the Kargaly project proposes the application of the "Modern Analogue Technique" (MAT) approach (Gavin et al. 2003) using RS for modelling the factors of the landscape. Thus, it creates an experimental framework to test interpretative hypotheses on the ancient landscape.

The first step is to develop a model as detailed as possible of the present day distribution of determining factors for the emission, dispersion and settling out of pollen, mainly, the spatial variability of vegetation, land uses, and topographic factors. The combined use of RS techniques, DTM processing and analysis and GIS technology offer the possibility of an efficient and precise tackling of these problems.

The designed experimental device consists of (1) a subactual pollen rain sampling and (2) a cartographic modelling of present-day distribution of vegetational zones in the study area. Both elements are combined for their analysis by a statistical model aimed at testing hypothesis on the spatial correlation between the qualitative and quantitative composition of the palynological spectra and the distribution of land cover classes.

(1) Subactual pollen rain sampling (Fig.2): two adjoining transects were defined in areas with and without mining activity to establish the effect of mining on the landscape. These transects were defined around the sampling points where the paleopalynological sequences were obtained. Each of those transects was divided in sampling units of 250m of radius. A 10% of these units was randomly selected in each transect. In the selected units we took samples in the upper 10cm of the topsoil. There was no difference in pollen preparation between those samples and the paleopalynological ones to compare results. A total of 55 randomly distributed palynological spectra was obtained in the studied area. Moreover, we took 24 samples from outside the sampling model, in locations selected according to botanical criteria. This sample in itself suitably represents the space variability of present-day pollen rain. Field work took place in July 1997 and August 1998; processing and identification of samples was carried out by P. López García and J. A. López Sáez (Laboratory of Archaeobotany, Institute of History, CSIC).

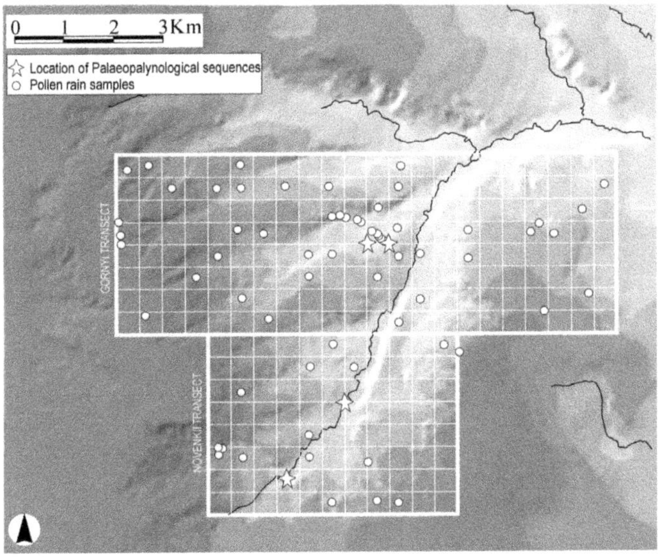

Figure 2 *Palinological sampling model.*

(2) Modelling present-day land cover classes: maps of present-day vegetation distribution were obtained by the joint processing of three satellite images. In a first approach we analysed a single Landsat TM image from September 1994. The preliminary results were presented in the First International Conference on Remote Sensing Archaeology (October 2004, Beijing).

Nevertheless, crop rotation in the study area created difficulties for separating the natural steppe vegetation from the fallow fields temporarily reverted to grasses and forbs. At the same time, the space resolution of TM sensor was not sufficient for the scale of the study area. With the purpose of overcoming these ambiguities and limitations, a diachronic-multisensor approach was recently introduced, consisting of the incorporation of two ASTER TERRA images of different months and years (June 2002 and July 2004). In this way we covered the complete plant life cycle (spring-summer-autumn) and a lasting enough interannual lapse to reflect the effects derived from crop rotation as well. At the same time, the spatial resolution of ASTER (15m) duplicates the precision in the delimitation of land cover classes boundaries, especially relevant for the analysis of cereal pollen grains dispersion.

In order to simplify the integration of ASTER and TM data we selected 5 bands from the ASTER VNIR sensor in visible (blue), near and mid-infrared spectral regions and we analysed them by three parallel processes.

In the first place, we calculated the Normalized Difference Vegetation Index (NDVI) for each single image. The RGB composition of the three NDVI bands (Fig. 3) allows recognizing the changes in vegetational zones distribution during the period of analysis. It turns

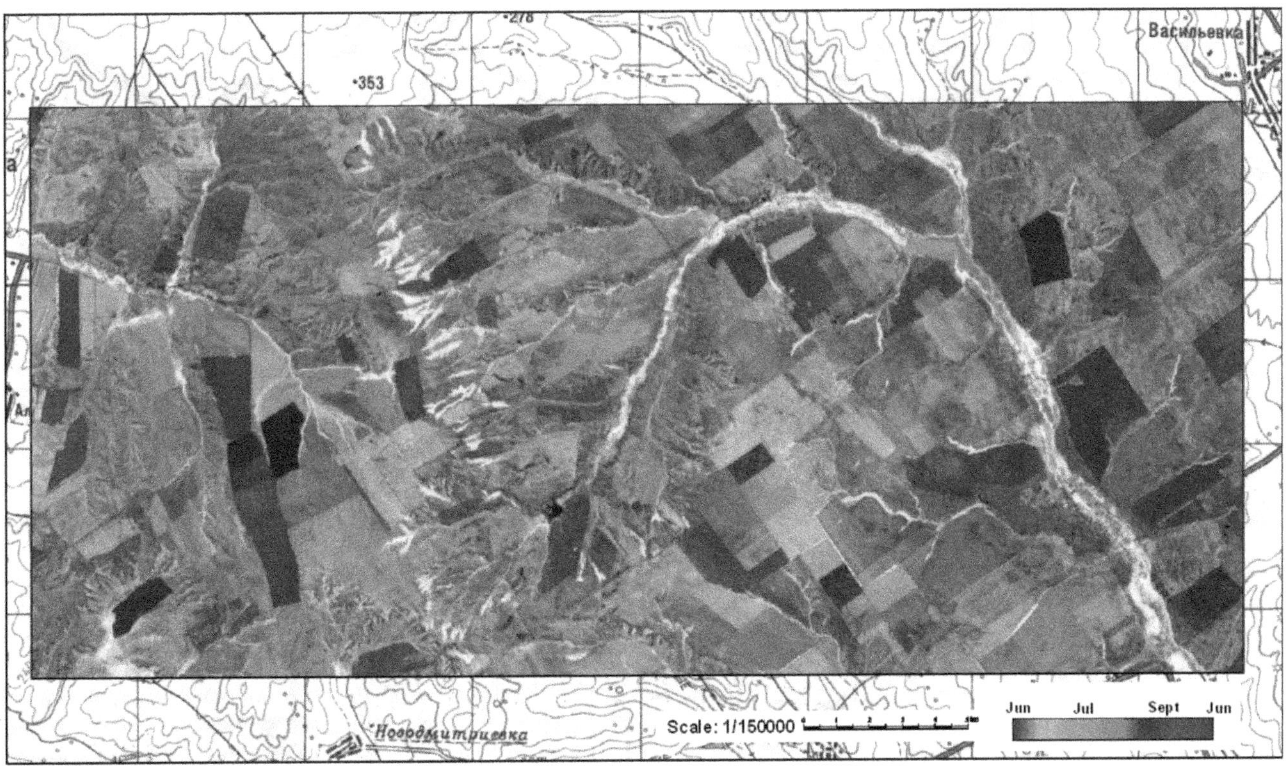

Figure 3 *RGB composition of calculated NDVI (ASTER, June 2002; ASTER, July, 2004; Landsat TM, September, 1994). Reference frame: Russian topographical cartography, esc. 1:200000; Zhdanovka sheet 14-40-31/ N-40-XXXI; Coordinate system: Pulkovo, 1942; GUGK SSSR, ed. 1989.*

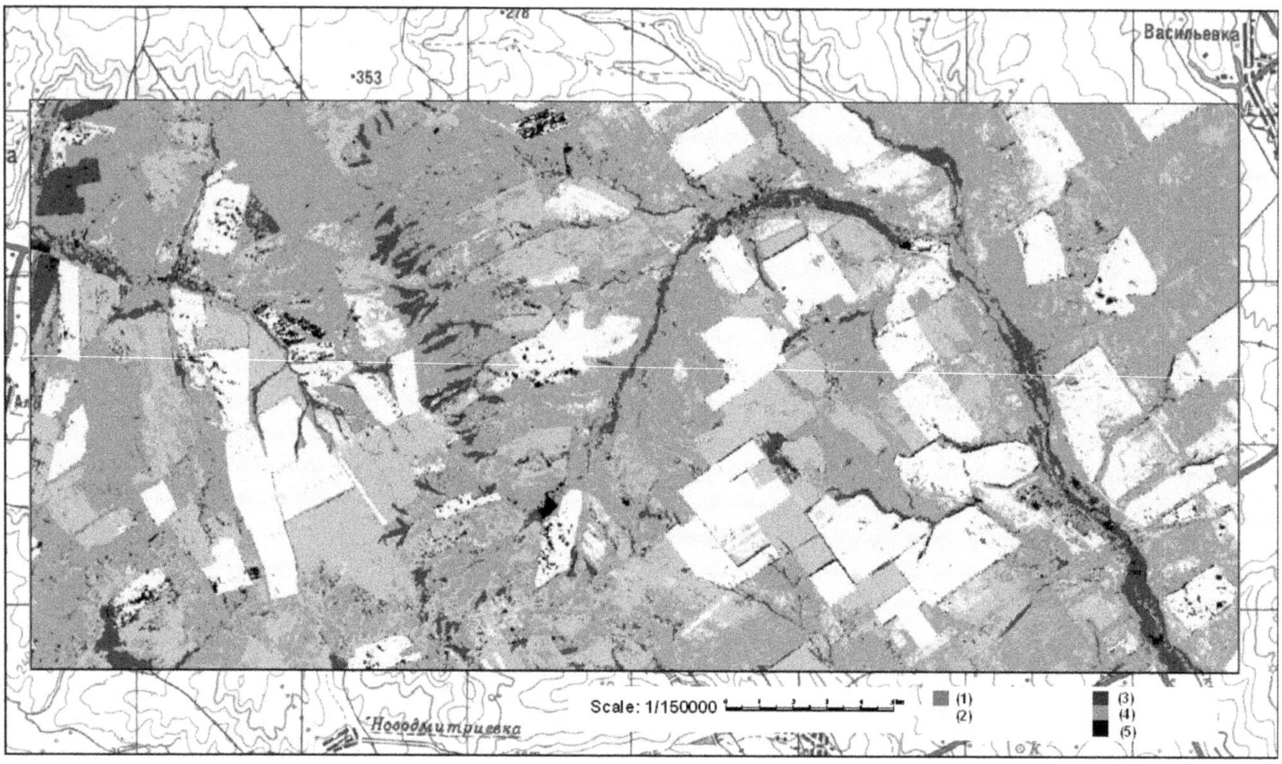

Figure 4 *Information classes. 1. Natural steppe. 2. Active crop fields. 3. Forest. 4. Fallows. 5. Not classified. Same reference frame as Fig. 3.*

out specially useful to determine and to delimit the highest expansion of cultivated fields, excluding the variations due to crop rotation and the progressive abandonment of agriculture in the area as a result of the socioeconomic changes in the region during the last decade.

Some parallel direct classifications (Duda and Hart 1973) of the selected bands of each one TM and ASTER images were elaborated to settle the specific conditions of vegetation in each moment of the series. These classifications used an unsupervised method to determine the variability in the spectral response of land covers. The resulting spectral classes were interpreted in terms of land cover (vegetation types) on the basis of the *a priori* knowledge on "ground truth" gathered during the field work: inventory of the flora, photographical reports and direct descriptions of the landscape.

Finally, the comparative analysis of single spectral thematic documents supervised after the NDVI diachronic series allows (allows siempre va seguido de – ing o de un sustantivo) the elaboration of a final synthetic classification in land-use classes (forest, crops, natural vegetation of steppe, etc.) (Fig. 4).

The documents based on interest land cover classes resulting from this process are the starting point for the elaboration of specific statistical models to test the different hypotheses on the meaning of present-day pollen rain in relation to the problems that the interpretation of the paleopalynological sequences poses.

By means of GIS techniques, we can elaborate a statistical model that facilitates the analysis of covariance between the spatial distribution of the types of vegetation and land uses, and the quantitative and qualitative variability of the palynological spectra composition. In this way we define a controlled comparative framework to understanding the variability of this composition in the case of paleopalynological sequences.

3 Applying the model

The modelling already discussed can be applied to diverse purposes. We suggest three of them: the evaluation of the taphonomical meaning of palynological indicators, the testing of hypothesis on the distribution of past land uses, and the definition of predictive models on the morphology of past landscapes.

An example of the first type of application is the analysis of the taphonomical meaning of arboreal pollen proportions in Kargaly samples based on the new data from the natural deposit in Gorny. We produced a detailed map of present distribution of forest coverage in the study area. A GIS generated "distance model" allows the analysis of the correlation between the spatial distribution of forest areas and its palynological representation. Using the model we may fix the critical levels of pollen representation to discriminate local against regional background variance sources. The analysis of the spatial distribution of the two main local species with enough calorific power for metallurgical production (*Betula* and *Quercus*) exhibits significant differences suggesting the interpretation of the long term changes seen in the paleopalynological sequences of locational patterns. Thus, for example, whereas the levels of representation of *Betula* observed in the samples of the BA of Gorny are within the limits of their present distribution (Fig. 5), those of *Quercus*, a residual species at present time, show levels corresponding to a strong local implantation in the surroundings of the settlement (Fig. 6). This suggests a relevant change in the availability of fuel resources that the previous palynological data did not show (Díaz-del-Río et al. 2005).

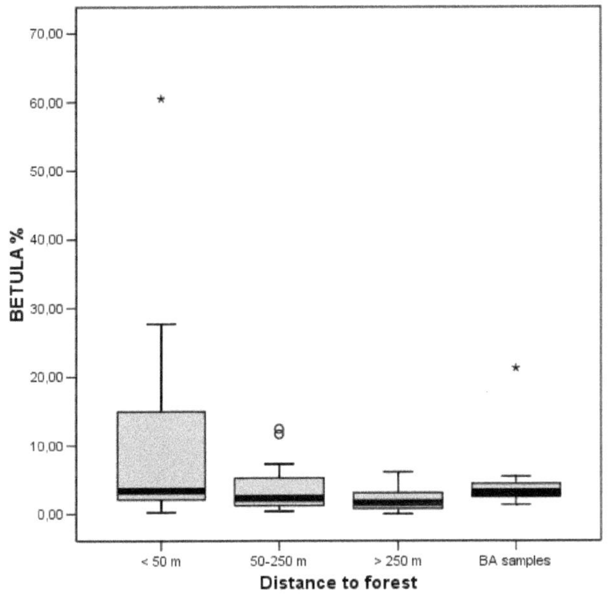

Figure 5 *Distribution of* Betula.

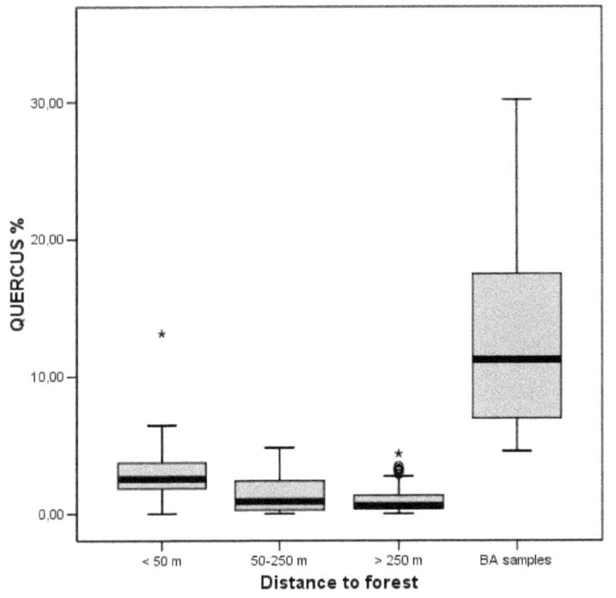

Figure 6 *Distribution of* Quercus.

The research on the subsistence practices during the BA requires a more complex approach, involving the location patterns and covariance of many palynological indicators. We approach this problem as a process to test hypothesis analysing the present distribution of these indicators on the basis of the cartography of land uses. The analysis of correlations between crop-fields location and patterns of cereal pollen dispersion demonstrates a strong space restriction: the probability of cereal pollen occurrence becomes almost insignificant 50m away from the crop field limits. The "palynological signature" of the local agriculture can, however, be followed investigating the spatial variability in the representation of other taxa associated to agriculture (weeds, etc.) to set the critical levels of representation that denote the local presence of cultivated fields.

Among the 137 taxa identified in Kargaly, in only 11 cases the analysis of variance (ANOVA) shows a statistically significant correlation between its representation in samples and the proximity of crop fields. 7 of those taxa are absolutely absent in BA samples from the Gorny natural deposit. Among the rest, the analysis of variance demonstrates significant differences between BA and subactual samples. The distribution of *Boraginaceae* and *Polygonum aviculare t.* in BA samples is below the critical levels of the class distance of the amounts of pollen of potentially cultivated Poaceae observed in those samples, while *Rumex acetosella t.* and *Plantago lanceolata t.* fall inside those limits. Since the last two taxa are in fact generic indicators of anthropic activity, those results do not support the hypothesis of the location of crop fields in the proximities of the settlement during the BA.

Consequently, the contextual data coming from the palynological record, interpreted with reference to their "present analogue", discard that the cereal-type pollen grains in samples of the BA in Gorny witness a local practice of agriculture. This obviously demands alternative hypotheses to explain their presence, such as long distance transport by winds in convective storms documented in other contexts.

Finally, one of the most suggestive lines for the interconnected operation of satellite images and palynological data is the elaboration of predictive models on the specific conditions of the vegetation in the past correlating the composition of the palynological spectra and the local values of the NDVI (D'Antoni and Spanner 1993). These authors demonstrated that, from an extensive enough number of palynological spectra dispersed through a territory, it is possible to find linear combinations of taxa that explain the NDVI local variations in the sampling points. The resulting regression equations might, in theory, predict the local values of the index in the past from the palynological data. These predictions, in combination with present data, might specify the local variations in the morphology of past landscapes in terms of specific vegetation conditions.

One of the objectives of the landscape archaeology programme in Kargaly is the application of D'Antoni and Spanner's approach to the regional context, to cope with the problem of the formation process of the steppe landscape. We presented the preliminary results, based on experiments on Landsat TM image, in the Beijing Conference in 2004. We demonstrated that the pollen samples of the BA obtained during the project define index values corresponding to the areas of steppe landscape altered by mining works, frequent in the regional surroundings. The recent incorporation of the ASTER images opens new possibilities in this framework, which will be the object of future communications.

Beyond the concrete results that we present as examples in this paper, we emphasize the potentiality of a combined use of RS techniques and the archaeological palynology as a methodology for modelling past landscapes.

References

ANTIPINA, E.E. & MORALES, A. 2005 'Cowboys' of Easteuropean steppe in the Late Bronze Age. *OPUS.* 4:45-49

CHERNYKH, E.N. (ed.) 2002 *Kargaly* I. Moskva: Yazyki slavyanskoi kyl´turi

CHERNYKH, E.N. (ed.) 2004. *Kargaly* III. Moskva: Yazyki slavyanskoi kyl´turi

CHERNYKH, E.N. & MARTINEZ-NAVARRETE, MªI. 2005 "Raspredelenie radiouglerodnyj dat v kulturnom sloe i za ego predelami (poselenie Gornyi, Kargaly)" in *Arjeologiia i estestvennonauchnye metody.* E.H. Chernyj & V.I. Zavyalov (eds.). Moskva: Yazyki slavianskoi kultury

D'ANTONI, H. & SPANNER, M.A. 1993 Remote sensing and modern pollen dispersal in Southern Patagonia and Tierra del Fuego (Argentina): Models for Palaeoecology. *Grana* (32):29-39

DÍAZ-DEL-RÍO, P.; LÓPEZ GARCÍA, P.; LÓPEZ SÁEZ, J.A.; MARTINEZ NAVARRETE, M.I.; RODRÍGUEZ ALCALDE, A.L.; ROVIRA, S.; VICENT GARCÍA, J.M. & ZAVALA, I. DE 2005 "Understanding the productive economy during the Bronze Age through archaeometallurgical and palaeoenvironmental research at Kargaly (Southern Urals, Orenburg, Russia)" in *Beyond the Steppe and the Sown: Proceedings of the 2002 University of Chicago Conference on Eurasian Archaeology.* D.L. Peterson, L.M. Popova & A.T. Smith (eds.). Leiden: Brill

DUDA, R.O. & HART, P.E. 1973 *Pattern classification and scene analysis.* New York: John Wiley and sons

GAVIN, D.G.; OSWALD, W.W.; WAHL, E.R.; WILLIAMS, J.E. 2003. A statistical approach to evaluating distance metrics and analog assigments for pollen records. *Quaternary Research* 60:356-367

LEBEDEVA, E. Yu. 2005 Archaeobotany and study of the bronze age agriculture in Eastern Europe. *OPUS.* 4:64-68

LÓPEZ, P. et al. N.d. Problems of diagnostics of ancient agriculture by methods of archaeobotany: The LBA in Kargaly (Orenburg, Russia). Oral presentation *ESF Programme: Early Agricultural Remnants and Technical Heritage. The dynamics of non-industrial agriculture. 8000 years of resilience and innovation.* East Kilbride, Scotland 11-13 March 2005

LÓPEZ, P.; LÓPEZ-SÁEZ, J.A.; CHERNYKH, E.N.; TARASOV, P. 2003. Late Holocene vegetation history and human activity shown by pollen analysis of Novienki peat bog (Kargaly region, Orenburg Oblast, Russia). *Vegetation History and Archaeobotany* 12, 1:75-82

LÓPEZ-SÁEZ, J.A. 2002. 'Glossarii: sovremennaya flora Kargalov', in *Kargaly* II (E.N. Chernykh, ed.). Moskva: Yazyki slavyanskoi kyl´turi

LÓPEZ-SÁEZ, J.A.; LÓPEZ-GARCÍA, P. & MARTINEZ-NAVARRETE, Mª.I. 2002. 'Glava 10. Palinologicheskie issledovaniya na kholme Gornogo', in *Kargaly* II (E.N. Chernykh, ed.). Moskva: Yazyki slavyanskoi kyl´turi

The Fourth Dimension of Places: Landscape as an Environmental and Cultural Dynamic Process in the Maremma Regional Park

Michele De Silva

Università degli Studi di Siena

1 Introduction

The landscape should be considered the result of a historical sedimentation process where different natural events and human activities have all left traces in forming an interwoven and complex whole. In order to understand, preserve and appreciate a landscape it is not sufficient to study places as they are, or as they were, in all three dimensions, but it is necessary to diachronically explore, in the fourth dimension of time, the natural and human processes which constitute its identity from a historical and environmental point of view.

Through the analysis of aerial photographs and historical sources - in particular medium to large scale historical cartographies - correlated with survey data, it is possible to reconstruct the territorial settings of the past and subsequently identify the dynamics and transformation processes of the landscape. The diachronic approach to the Landscape Archaeology allows us to recognize the role of natural events and human intervention in the transformation processes that constitute the matrix of the present territorial setting.

Whereas for ancient periods only archaeological evidence from surveys, excavations or aerial photographs and little historical documentation is available, for more recent periods old maps – especially historical cadastres – can provide detailed information on territorial settings.

The case study presented refers to the Alberese Farm area[1] which constitutes a representative portion of the Maremma Regional Park in Southern Tuscany (Fig. 1). Although today the Maremma Regional Park may appear an example of natural wilderness as opposed to other manmade areas of Tuscany, we should not undervalue the role of human activity in shaping this landscape.

2 Aims and goals

The research presented here is part of a project that is the result of a collaboration between the Department of Archaeology and History of Art and the LIAAM (Laboratory of Computer Science Applied to Medieval Archaeology) of the University of Siena, the State Archive of Grosseto, and the MEDCORE (European project for Mediterranean Coast-River Ecosystems).[2]

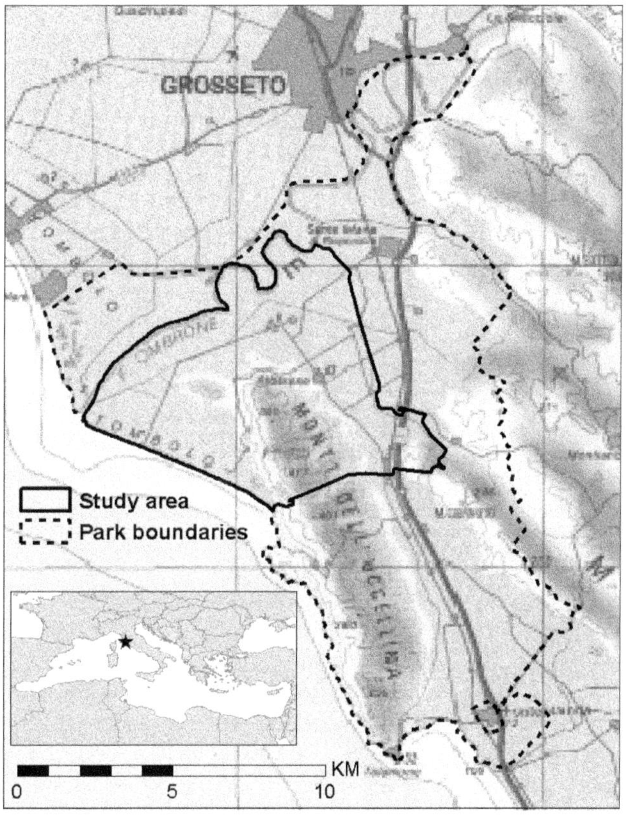

Figure 1 *Location of Maremma Regional Park and Alberese Farm study area (base map from the 1:250,000 physical map edited by the Regione Toscana).*

The main objectives of the research project are, on the one hand, to set up a methodology for the study of landscapes of the past using different kinds of sources integrated in a GIS environment with special reference to historical cartography and aerial photographs (recent and historical), on the other, to be able to apply this methodology for a greater in depth understanding of landscape transformation processes and the historical evolution of settlement patterns in the Grosseto coastal belt.

---

[1] Study area has an extension of about 5500 hectares, ranging from 42°37'42" to 42°42'49" of latitude North and from 11°00'26" to 11°09'12" of latitude East (WGS84).

[2] The research is partially founded by the Mediterranean Coast-River Ecosystems (MEDCORE) EU Project (ICA3-2002-10003, 5° FP, INCO-MED Programme).

A few of the results and methodological aspects of the research have already been presented on other occasions (De Silva 2006; De Silva and Pizziolo 2004, 2005, 2006). In the case study presented here we focus on the analysis of the territorial setting of the Alberese Farm. A comparison of the land use at the beginning of the 19th century with the current land use not only indicates the more recent environmental transformations but enables us to more fully comprehend the interaction between nature and human activities in the shaping of the current day landscape.

3 Study area: the Alberese Farm in the Maremma Regional Park

The Tuscan regional law of 5th June 1975 n. 65, relating to the 'Institution of the natural park of the Maremma', reports in its first article:

"The natural park of the Maremma has been established. The aim of the park is the tutelage of the natural, environmental and historic characteristics of the Maremma in relation to the social use of these values, as well as the promotion of scientific research and naturalistic didactics."

From the very outset, the park explicitly and essentially takes on the form (despite the timid and dutiful allusion to the historical values) of a natural park. This is fully justified by the presence of a scarcely urbanised landscape, articulated on a naturalistic level, where the traits, colours and smells of a wild habitat predominate, in close affinity with the common present day image of a coastal Maremma region populated by cowherds, marshes, pinewoods, Mediterranean vegetation and the sea. The image must have certainly been very different in the past when the 'bitter Maremma' was populated by transhumant shepherds, lumberjacks and seasonal labourers who were forced by necessity to live in a mosquito-and-malaria ridden habitat.

The tormented historic vicissitudes, in addition to the presence of a 'forceful' nature, have allowed this marginal strip of Tuscany to reach us today almost uncontaminated. But this must not lead to a common misconception that a natural landscape, in the sense of a landscape where the natural elements prevail over human ones, is the fruit of a natural process in which man has taken no part. In these places, as is the case virtually throughout the rest of Italy and in most densely populated countries, even the most 'natural' landscape is the result of a slow process whereby the elements of nature and the work of man over thousands of years are deposited through historical events that unfold over time. Thanks only to a close reading of the signs left by man throughout the territory, from the most obvious to the most deeply hidden ones, coupled with an analysis of historic documents, is it possible to understand the processes that have led to the formation of the present day landscape. Just as the numerous canals in the plains of the Park bear witness to the incessant work of man in draining of the marshes and swamps which today appear as meadows or cultivated fields, signs of charcoal burning and terracing which can be still today perceived in the woods of the Monti dell'Uccellina document past activities that affected these areas at different times.

4 Methodology

The study of the area of the Farm of Alberese initially concentrated on the acquisition and analysis of historical data, and in particular data deriving from surveys and old or modern aerial photographs and maps. The integration of data in a single system, namely a GIS, allows us to reconstruct various phases of the past. The subsequent comparative analyses between present and previous settings constitute the basis for an improved understanding of the landscape making process (De Silva M. and Pizziolo G. 2004, 2006).

In order to appreciate transformations that have occurred over the last two centuries in the land use of the study area, we have adopted the *Catasto Leopoldino* as our data source for the beginning of the 19th century and the Land Use Map edited by the Regione Toscana in 1987 for the recent period.

The *Catasto Leopoldino*, a geometric parcel cadastre created between 1817 and 1835 for the whole mainland territory of the Grand Duchy of Tuscany, represents a fundamental source thanks to its geometric accuracy and its richness in detailed information. The Cadastre is composed of maps and inventories linked together by parcel numbers. The maps (those referring to the study area are in the scale of 1:5000) represent buildings, road networks, hydrographic elements, place names and cadastral parcels whereas the inventories provide information regarding ownership, land use, surface area and other information relating to the parcels. Every Commune has its own index map (the *Quadro d'Insieme*) which can be subdivided into cadastral sections. Each section may be represented by one or more cadastral maps (the *Fogli*).

The general research project has involved digital acquisition using a large format scanner of the index map and 80 detailed cadastral maps referring to the former administrative territory of the Commune of Grosseto.[3] These have subsequently been geometrically rectified, accurately georeferenced and inserted into the GIS. Maps referring to the same section have been furthermore blocked out and mosaiced (De Silva 2006).

Moreover, single features relating to the hydrography, road networks, buildings, place names and parcels of the Alberese area have been acquired from the Cadastre in a vector format. Each parcel feature has been associated

[3] The *Catasto Leopoldino* of the Grosseto territory is located in the State Archive of Grosseto.

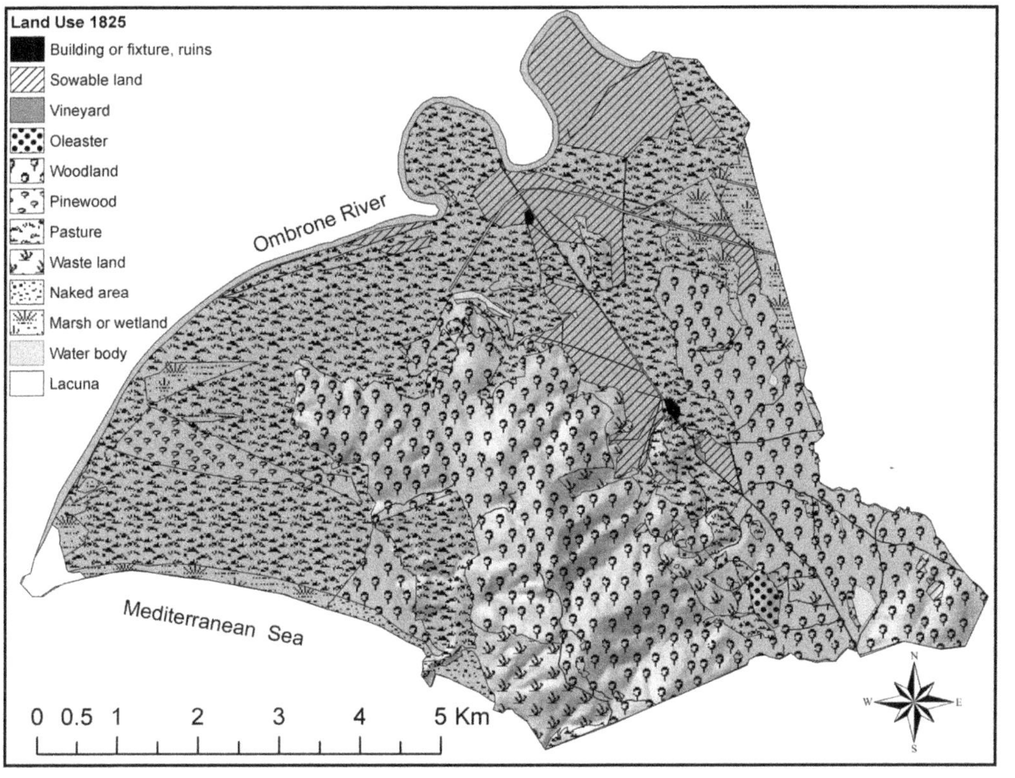

Figure 2 *Simplified land use map and chart of the Alberese area of 1825 deriving from the* Catasto Leopoldino.

with its descriptive data obtained from the inventories to create a land use map.

The analysis of this set of data, which constitutes a detailed outline of the Farm of Alberese at the beginning of the 19th century, allows us, on the one hand, to make comparisons with the general coeval setting of the Grand Duchy, highlighting the distinctive characters of this area, on the other hand, it enables us to diachronically compare the data with the present setting on a topographic level of investigation so as to recognize continuity and changes in the Alberese territory that constitute the historical matrix of this landscape.

5 Landscape transformation processes through the analysis of land use changes

In order to simplify our analytical and comparative investigation, the original 55 land use classes reported by the *Catasto Lorenese* for the Alberese area have been reduced to 11 classes through a generalisation process grouping together similar classes and omitting parcels referring to the road network.

The simplified land use map obtained from this procedure provides a detailed and easily comprehensible picture of the territorial setting at the beginning of the 19th century (Fig. 2).

At first glance we can clearly observe that settlements are rare whereas woodlands and pasture occupy a large portion of the area (38% each) as opposed to sowable lands (8%). Furthermore, wetlands are still present in the area (5%) despite the fact that several reclamation works had already been undertaken at that time. Pinewoods are present only in a restricted area (2%) and olive groves seem to be altogether absent. A few small areas classified as 'oleasters' may represent relicts of former degraded olive groves.

The land use in recent times has been ascertained from the Land Use Map based on aerial photographs of 1978 and edited by the Regione Toscana in 1987. These paper maps have been scanned and georeferenced and land use parcels have been acquired in vector format and inserted into the GIS. Also in this case we have generalised data by reducing the 23 original classes to 11 new classes.

The map obtained (Fig. 3) shows a very different situation to that dating to the beginning of the 19th century. We may observe that sowable lands extend throughout the northern and eastern areas (46%) whereas marshes and wetlands have virtually disappeared and pinewoods cover the flat coastal belt (11%). Woodlands are reduced to 26% of the territory, pastures are limited to small areas (3%) near woodlands, and olive groves reach a significant coverage (7%). New rural settlements and small parcels of vineyard and olive groves are scattered throughout the inland plain.

Visual comparison between the representations of the historical and recent land uses aid us in understanding more fully the qualitative changes that have occurred in the landscape in relation to the landform. The flat plain

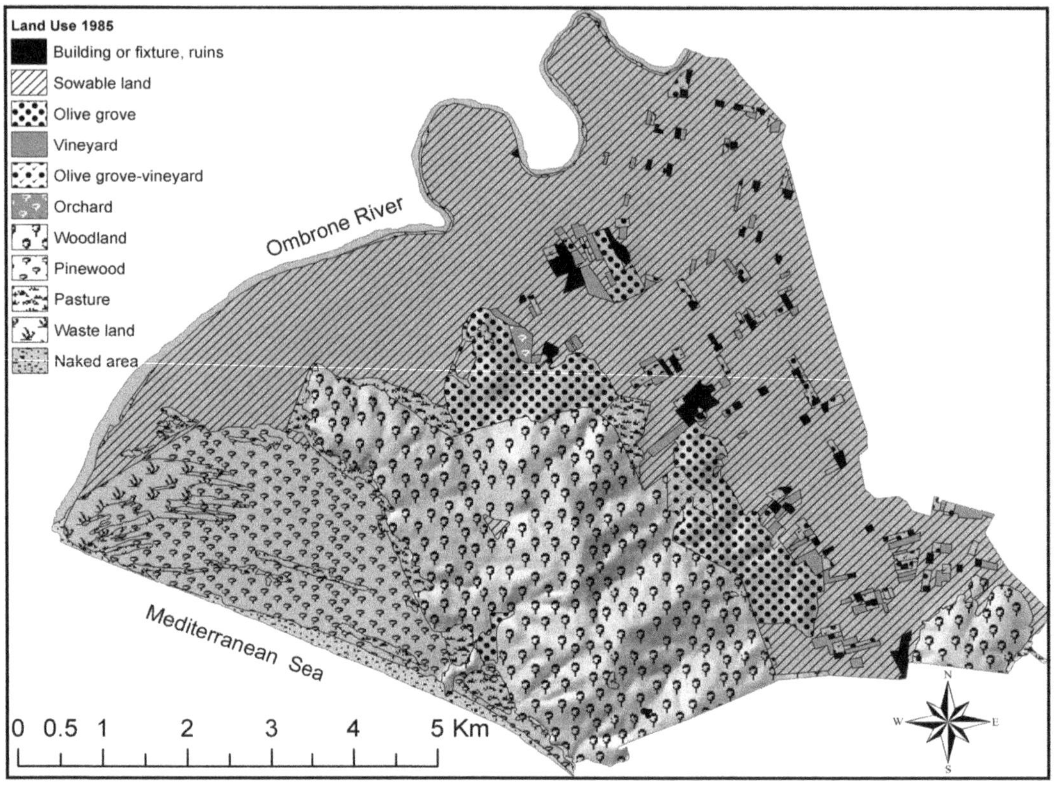

Figure 3 *Simplified land use map and chart of the Alberese area of 1985 deriving from the 1:25,000 Land Use Map edited by the Regione Toscana.*

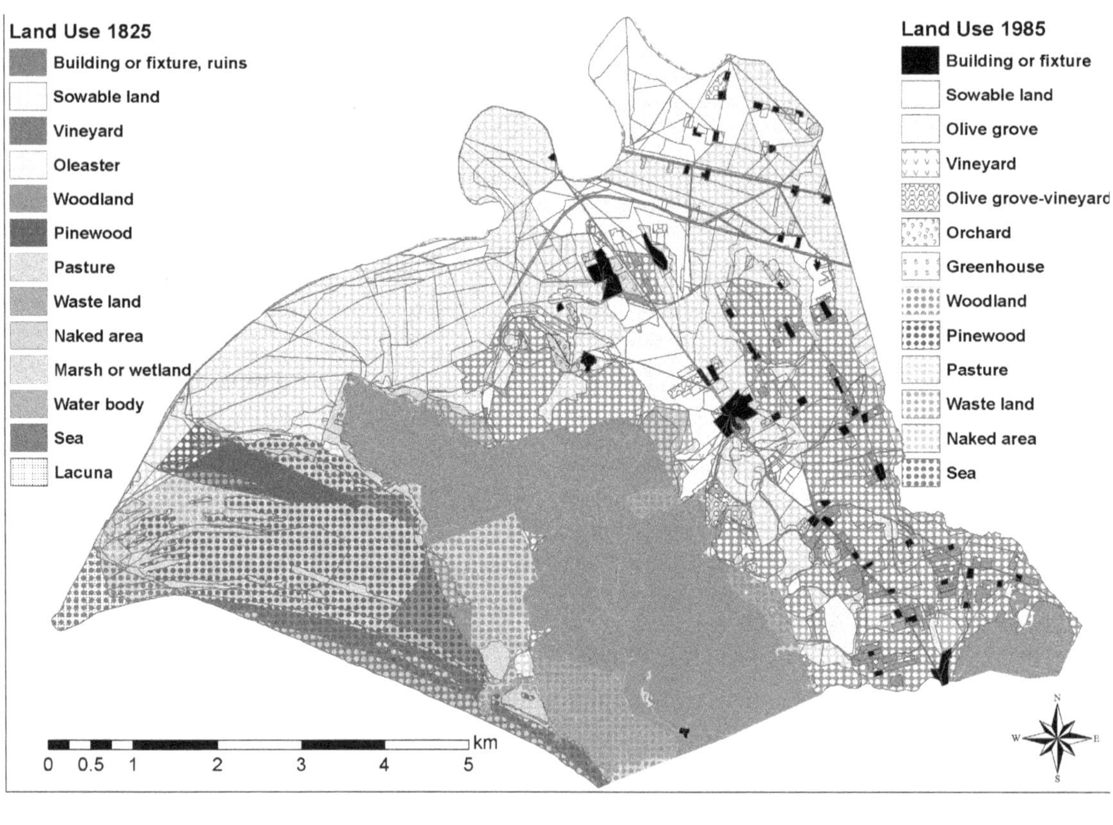

Figure 4 *Synthesis map of changes in land use of the Alberese area between 1825 and 1985.*

areas have been heavily transformed whereas the hilly zones have maintained greater continuity over time. A good example of this is the total disappearance of woodlands in the inland plain.

The map of synthesis obtained through overlaying the historical and recent land uses by means of a cross-tabulation analysis in the GIS, demonstrates in greater detail the qualitative and quantitative transformation that has occurred throughout the area (Fig. 4). The solid coloured polygons on the map generally represent areas of land use continuity, whereas the dotted polygons in two different colours highlight changes in land use between 1825 and 1985. As we can see, the main areas of continuity in land use are represented by the hilly woodlands, certain sowable lands and a small area of pinewood. Sowable lands and pinewoods have replaced pastures and woodlands in the plains, and olive groves have covered the inland lower hills. In addition, the map clearly illustrates the dramatic changes of the coastline. The coastline has, indeed, regressed by approximately 1 kilometre near to the river mouth, whereas it has advanced up to 400 meters in other areas (De Silva and Pizziolo 2004, 2006; Ciampi 2004).

Conclusions

The Alberese Farm at the beginning of the 19th century is characterized by the prevalence of woodlands and pastures, bearing witness to the existence of an economy based mainly on breeding and woodland exploitation. The role of grain farming seems to be marginal considering the limited extent of sowable land.

These distinctive characters, constituting the historical matrix of the landscape, emerge more clearly if we compare the territorial setting of the study area in 1825 with the general setting of the whole mainland of the Grand Duchy of Tuscany in the same period (Fig. 5). In the latter the total extent of cultivated areas are proportionally greater (33% versus 8%) denoting a more significant role of agricultural activities[4].

Analysis of the evolution of the land use in the study area over the last two centuries highlights that changes have mainly occurred on the plains. From an ecological point of view, the most relevant changes concern the disappearance of the plain woodlands and wetlands, the substitution of pastures by sowable land and the expansion of pinewoods.

We would, furthermore, like to stress the role that human activity has played in shaping the landscape. Economic and social choices such as reclamation works and farming activities, or in other words historical events, are undoubtedly the main causes of landscape transformations.

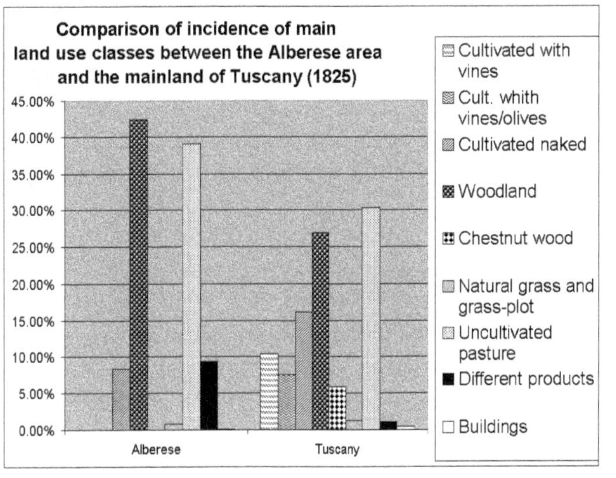

Figure 5 *Comparison of incidence of main land use classes between the Alberese area and the mainland of Tuscany dating to c.1825.*

In due consideration, study of the current landscape is insufficient for competent territorial management and must be coupled with a diachronic investigation, in other words, an analysis of the historical dynamic processes of transformation that constitute the fourth dimension of landscape. Only in this way can more convincing and tenable choices be made relating to the planning of activities and interventions that are compatible with the preservation and appreciation of our natural and cultural heritage.

Acknowledgements

We are grateful to the Director of the State Archive of Grosseto, Dr. Maddalena Corti, for her kind and precious collaboration. Our deepest thanks to Svitlana Hluvko for her intervention in linguistic support in the writing this paper.

References

AZZARI M., DE SILVA M. and PIZZIOLO G., 2002. «Cartografie del passato e GIS per l'analisi delle trasformazioni del paesaggio – Cartography of the past and GIS for the analysis of landscape transformations», in M. Azzari (ed.), *Beni culturali e ambientali e Geographic Information Systems*, Firenze University Press, Firenze, CD-ROM.

BIAGIOLI G., 1975. *L'agricoltura e la popolazione toscana all'inizio dell'Ottocento – Un'indagine sul catasto particellare*, Pacini, Pisa.

Catasto della Toscana, 1819. *Istruzioni e regolamenti approvati dall'I. e R. governo sul catasto della Toscana*, Tipografia Piatti, Firenze.

CIAMPI G., 2004. «Il delta dell'Ombrone: indizi sui fattori della sua dinamica desunti dalla cartografia», *Bollettino della Società Geografica Italiana*, 4, pp.

[4] Data referred to the whole mainland of Tuscany was obtained from the "Relazione finale al Granduca della Deputazione sopra il Catasto" (manuscript) located in the Archivio di Stato di Firenze, *Segreteria di Gabinetto – Appendice*, N. 244.

991-996.
CIUFFOLETTI Z. and GUERRINI G. (eds.), 1989, *Il parco della Maremma – Storia e natura*, Marsilio, Venezia.
DE SILVA M., 2006. «Landscape of the Past: the Maremma Regional Park and the Grosseto Coastal Belt. Methodology and Technical Procedures», in *Beyond the Artifact – CAA 2004 – Computer Applications and Quantitative Methods in Archaeology – Proceedings of the 32nd Conference, Prato, Italy, April 2004*, BAR International Series, BAR Publishing, Oxford, (in press).
DE SILVA M. and PIZZIOLO G., 2003. «Cartografia catastale lorenese per la ricostruzione del paesaggio storico: problematiche e stimoli relativi all'uso di una fonte complessa all'interno di un Sistema Informativo Geografico, il caso di Sesto Fiorentino - Lorenese Cadastral Cartography and GIS for the reconstruction of historical landscape: problems and issues in using a complex historical source, the Sesto Fiorentino case study», in M. Azzari (ed.), *III Workshop Beni Ambientali e Culturali e GIS – GIS e Internet, Firenze 2002*, Firenze University Press, Firenze, CD-ROM.
DE SILVA M. and PIZZIOLO G., 2004. «GIS Analysis of Historical Cadastral Maps as a Contribution in Landscape Archaeology», in Magistrat der Stadt Wien – Referat Kulturelles Erbe – Stadtarchäologie Wien (ed.), *[Enter the Past] – The E-way into the Four Dimensions of Cultural Heritage CAA 2003 – Computer Applications and Quantitative Methods in Archaeology - Proceedings of the 31st Conference, Vienna, Austria, April 2003*, BAR International Series 1227, BAR Publishing, Oxford, pp.294-298 and download.
DE SILVA M. and PIZZIOLO G., 2005. «"Signs", place, continuity and changes: investigating the "Landscape Perception" through the integration of chronological and typological sources in the Tuscan plains», in M. Forte (ed.), *The Reconstruction of Archaeological Landscapes through Digital Technologies – Proceedings of the 2nd Italy-United States Workshop, Rome, Italy, November 3-5, 2003 – Berkeley, USA, May 2005*, BAR International Series 1379, BAR Publishing, Oxford, pp. 193-203.
DE SILVA M. and PIZZIOLO G., 2006. «Processi storici di trasformazione del paesaggio toscano: alcuni esempi di integrazione cronologica e metodologica delle fonti in ambiente GIS», in Azzari M., Faretto G. (eds), *Atti del IV Workshop Beni Ambientali e Culturali e GIS, Firenze 2003*, Florence University Press, Cd-Rom.

High resolution DTM for the geomorphological and geoarchaeological analysis of the city of Padua (Italy)

F. Ferrarese,[1] P. Mozzi,[1] F.Veronese,[2] F. Cervo[1]

[1]Dipartimento di Geografia "G. Morandini", Università degli Studi di Padova (Italy)
[2]Dipartimento di Scienze dell'Antichità, Università degli Studi di Padova (Italy)

Introduction

Digital Terrain Models (DTMs) are usually used in the analysis of mountainous or hilly areas. Efforts of building DTMs of alluvial plains started about 20 years ago in Italy, following the publication of large scale topographic maps by the Regional Cartographic Departments, the so-called Regional Technical Map (C.T.R. *Carta Tecnica Regionale*) at scale 1:5000. These maps have a much larger number of elevation spots than the 1:25,000 topographic maps of the Italian Army (IGM *Istituto Geografico Militare*). The interpolation of these points through GRID or TIN methods allows the reconstruction of the plain microtopography with a good accuracy. Some control by the operator is needed in the choice of the elevation spots, depending on the aim of the investigation. For the analysis of natural landforms, only the points representative of the alluvial plain surface are usually selected, while those on dikes, artificial levees, embankments or other artifacts are discarded. The use of DTMs has proved to be a worthy tool for geomorphological investigation in the Po Plain (Tellini 2001). If the goal is oriented to applicative tasks, such as the modelling of floods or the dispersion of pollutants in surface runoff, the DTM must be as similar as possible to the actual surface topography, comprehensive of all the artificial positive (elevated) or negative (depressed) breaklines. In this latter case, all elevation points will be used and particular effort will be devoted to the definition of the 3D geometry of the artificial landforms (Bondesan et al. 2003).

These two different approaches to the building and use of DTMs have brought about the attitude of considering urban areas in alluvial plains as places where the modern buildings and infrastructures have completely deleted the evidence of the fluvial landforms on which they sit, thus losing interest for the geomorphological research. This, of course, may happen, but it can in no way be considered as a rule. Even more important is the evidence that some cities with a long settlement history, such as Padua or Treviso in the Venetian Plain, have a centre which is several metres higher than the outskirts. The cores of these cities consist of large mounds which derive from the accumulation of archaeological layers during several millennia, locally intercalated with alluvial deposits.

In this study we focus on the city of Padua and surroundings, where the Municipality provides a 1:1000 topographic map in vector format, which has allowed the building of a DTM with higher resolution than those based on the Veneto Region C.T.R. (Fig. 1).

The geomorphological setting

Padua lies in the fine grained alluvial plain of the Brenta river, a major Alpine river which drains the western portion of the Dolomites. The Brenta alluvial plain underwent several geomorphic changes during the last 20,000 years, in response to changing climatic conditions (Mozzi 2005). As a consequence, in Padua and surroundings the fluvial sediments deposited during the Last Glacial Maximum (LGM) are buried at depths of several metres under the Holocene sequence, while some kilometres to the north and west the LGM deposits are outcropping (Castiglioni et al. 1987; Iliceto et al. 2001; ARPAV-REGIONE DEL VENETO 2005).

The Padua plain is crossed also by another river, the Bacchiglione. This river has a limited mountain watershed in the Venetian Prealps and is partly fed by springs located in the plain, at the transition between the gravelly piedmont plain and the clay-silty-sandy lower plain. As a consequence of the different magnitude of the two rivers, the sedimentary activity of the Brenta has generally overwhelmed that of the Bacchiglione. While the Brenta deposits constitute the majority of the plain, the sedimentary and geomorphic activity of the Bacchiglione is apparently restricted to its present meander belt. Upstream of Padua this river has built a shallow ridge while downstream it is slightly incised in the plain. Since the second half of the 19th Century the Bacchiglione has been crossing the city centre following two large meanders. In the second half of the 19th Century the main river course has been artificially confined out of the city in order to prevent flood events. In the 1950s, long tracts of the river in the city center were partly infilled and covered by roads.

The ancient Padua and its river

The city of Padua arose along the banks of a major watercourse; more precisely, it grew up between opposing eastward and westward bends of the river called, in antiquity, the *Meduacus*.

Figure 1 *a) The DTM of the Padua Municipality obtained from ca. 150,000 elevation spots (in black not surveyed areas). b) Geomorphological sketch of the study area, derived from DTM and aerial photos.*

So we can see that for centuries the life of the city and the life of the river have been interwoven (Gasparotto 1951; Bosio 1981a, Rosada 1993; Balista 2004; Balista, Ruta Serafini 2004; Balista, Rinaldi 2005).

Only relatively recently has the city lost its identity as "citta d'acqua", "water city" (*Padova città d'acque* 1989; Bonarrigo 1992). It is not by chance that in ancient texts Padua is often associated with water (Plin., *NH*, III, 121; Strab. V, 1, 7) to the extent that their authors did not hesitate to consider the city as land encircled by water, defining it as an *insula* (Strab. V, 1, 5), an island city, therefore almost entirely (Rosada 1993, 65-66) delimited by the waters of the *Meduacus*, which, meandering on its winding course, finally surrounded the urban settlement and divided it in two, passing through its centre.

The *Meduacus* was that *flumen oppidi medium* about which the historian Tito Livio left his testimony: a major river with sufficient flow as to necessitate the construction of stone bridges more than forty metres long, with three or five arches (Galliazzo 1971), creating a river port comparable in dimensions to the *portus Tiberinus* of Rome, and with infrastructure and port buildings on both banks (Tosi 2002, 99). Indeed, during celebrations mock naval battles were even enacted on its waters (Liv. X, 2, 14-15). However, in the end such ambitious developments required the construction of huge levees and structures along the banks to contain the frequent floodwaters (Balista, Ruta Serafini 1993; Tosi 2002, 91-94), and drainage ditches to improve the low lying areas which were too damp and subject to frequent flooding (Tosi 2002, 94). The waters of the river have conditioned and moulded the growth of the settlement by both demarcating its northern and western boundaries, and by acting as a fulcrum around which the city developed (this development can be identified in the descending parts of the bend). All these features constitute an urban hydrographic situation unusual in the ancient world (Tosi 2002, 90-91).

However this configuration, recognisable during the Roman era from textual evidence and amply confirmed by archaeological data, existed well before the historical era when the settlement that was to become in the second century BC the Roman *Patavium* was simply a collection of small nuclei developing in higher topographic position. The most ancient traces of protohistoric habitation, datable between the end of the 9th and the beginning of the 8th Centuries BC (Gamba et al. 2005; Ruta Serafini, Vigoni 2006, 85-92), are in fact still in evidence inside this system of meanders, from which also come the few traces of the preceding Bronze Age (Gamba et al. 2005, 31; Ruta Serafini, Vigoni 2006, 86-88). This growth of one city on top of another has never allowed us to form a complete and detailed picture, given the complexity of the archaeological palimpsest, with stratification that goes down to seven metres, plus the frequent incompatibilities between archaeological excavations and the normal activities of city life.

The articulation of the protohistorical settlement, dispersed on the highest points, known today thanks to the stratigraphic excavations of the last thirty years and the location of the small areas of artisan activity generally discovered along the river bank (Gamba et al 2005, 23-31; Gamba, Gambacurta, Sainati 2005), confirms that the river was already at that time the spine of the settlement, determining its shape and structure - the numerous drainage ditches, and the position of the fords - a structure which remained the same in succeeding ages (Tosi 2002, 95).

If therefore the river had always been the fundamental element for Padua in antiquity, and had allowed the city's development around this precious water resource, it has served also as a defence. Moreover, it has determined the border, physical and symbolic, of the settlement, a border beyond which the world of the living gave way to the world of the dead. Whether in the protohistoric (Michelini, Ruta Serafini 2005) or Roman epochs (Ruta Serafini 2002, 70), the necropoli are in fact located outside the area delimited by the river; and from these necropoli those routes fanned out which connected

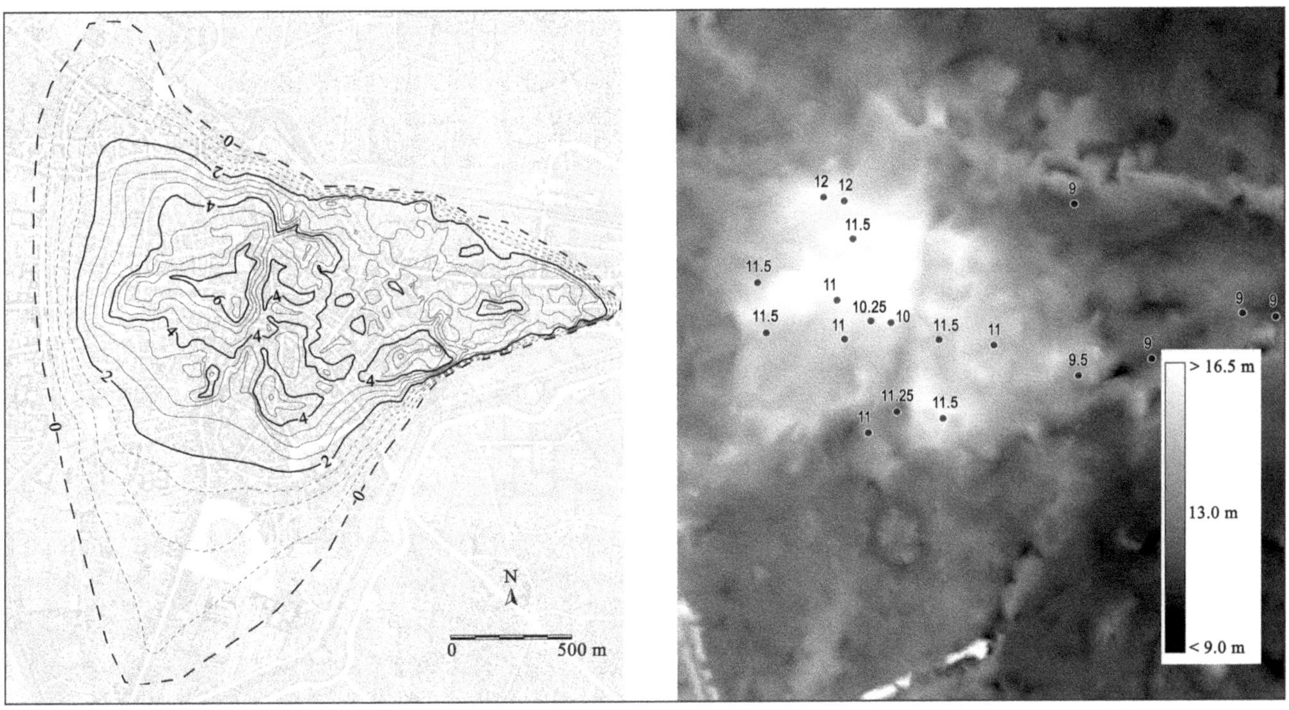

Figure 2 *Left: depth contour lines (interval of 0.5 m) of the Iron Age surface, derived from elevation spots of archaeologic stratigraphies (dots in the right frame). Dashed contour lines are in areas with no stratigraphic control. Right: DTM of the city with elevation spots of the Iron Age surface (values are metres a.s.l.). It is evident the shape of the city mound, with the western side higher than the eastern one, divided by the river. This latter was partly artificially infilled in 1950s.*

ancient Padua, first by track and then by road, with the surrounding territory (Bosio 1981b, 239).

Methods

The DTM has been elaborated from data of the 1996 Municipality Technical Map (CTC *Carta Tecnica Comunale*), scale 1:1000, using a very large number of elevation spots (over 240,000 points for 92 km^2, with a density of 2642 spots/ km^2) having 35 cm of nominal precision. These points are available in digital format and are importable in most common GIS softwares. To interpolate a terrain surface from spots we use Idrisi15™ TIN (Triangular Irregular Network) application. Two models were realized, the first with all 243,000 points, and a second one with about 150,000 points. This latter DTM has the advantage of less noise elevation derived from buildings and represents better the ground surface. No particular filters were applied to DTMs: only a mask of "null data" was draped over the DTMs, that have a 2 m grid resolution. Data were georeferenced in national kilometric system (Gauss-Boaga). Historical maps of Padua were imported in the GIS in digital format, in particular maps that have a good representation of the urban area during the last centuries. The aim of this operation is to acquire geo-referenced information concerning the spatial evolution of the city, possibly related to the artificial elevation of the ground surface. The so-called "Valle map of Padua", a detailed map of the 1784, at 1:2144 scale, has been very useful, being the first geometrically correct map of the city centre. Important information was also provided by the so-called Morello's map at scale 1:10,000 of 1880 and the historical 1:25,000 maps of the IGM, starting from 1910. Geo-referenced aerial photos were also imported in the GIS and were used for the integrated analysis of the DTM. Palaeohydrographic and archaeological traces are most evident in photos of 1956 and 2004, due to the optimal ground conditions.

The Iron Age level can be regarded with a good approximation as equivalent to the alluvial plain surface before the building up of the mound, even though the real extension of the Bronze Age settlement is not yet known. Twenty four archaeological excavations have reached the Iron Age level in the mound underground, at elevations ranging from 9 to 12 meters above sea level (Balista, Rinaldi 2005). Using these data as elevation spots it was possible to build the DTM of the Iron Age surface in part of the city centre. The subtraction of the Iron Age surface DTM to the present day DTM allows the calculation of the volume of the mound (Tab. 1).
We consider that the mound pinches out inside the 16th century city walls. In order to be sure to comprise the whole mound in this calculation and, on the other hand, not to include the city wall embankments, the "0 m" thickness of the archaeological deposits was supposed to be at the feet of the walls (Fig. 2). The river channel was also masked in order not to bias the operation. The thickness of the archaeological deposits is represented through 0.5 m interval contour isopachs. Dashed contours 0-2 m correspond to the sectors where no stratigraphic control was available.

area (m^2)	volume (m^3)	max depth (m)	average depth (m)
3,201,976	7,053,436	6.8	2.2

Table 1 *Morphometric parameters of the Padua mound. The calculation is performed in the area shown in Fig. 2, leaving out the river channel.*

Discussion

The DTM shows very neatly the mound in the city centre (Fig. 1). The most elevated areas, which reach at the maximum 17.5 m a.s.l., are located inside the two meanders. In the West meander, there are two ridges which are constantly higher than 17 m; one is oriented WSW-ENE, the other is almost perpendicular and is directed towards NNE. The higher area in the East meander is adjacent to the river bank and have a semi-lunated shape which mimics the river bend.

Draping historical maps on the DTM and elaborations, as shown in Fig. 2 with Valle Map, it can be seen that the area of the mound corresponds to the city area until the beginning of the 20th century. In fact, Padua has kept on living strictly within the two river bends for more than two millennia, without filling in the space comprised between the boundary of the ancient city and the 16th century walls. The areas where the archaeological deposits posterior to the Iron Age are thicker generally correspond to the maximum elevations in the DTM (Fig. 2). Nevertheless, the place of maximum thickness, located in the West meander, lies where the early Iron Age surface (that is, the "pre-mound" surface) presents a depression. This hollow was infillled before the mound started to rise above the surrounding plain.

Outside the city centre, the network of fields, ditches and canals shows up very well in the DTM, as much as the embankments of modern roads, railway tracks and bridges. To the purposes of this investigation, the most outstanding landforms recognizable in the DTM are the traces left by the abandoned river beds. These are partly detectable in aerial photographs (Baggio et al. 1992; Castiglioni et al. 1987) but their geometry and overall extension are much more evident in the DTM (Fig. 3).

The meandering trace located NW of the city centre has cut a 2 m high scarp in what is probably an LGM sector of the plain. Radiocarbon datings (Castiglioni et al. 1987) indicate that this river channel was already abandoned by the beginning of the 4th Millennium B.C. A linear channel departs from one of these meanders; due to its morphology it was interpreted as an artificial canal.

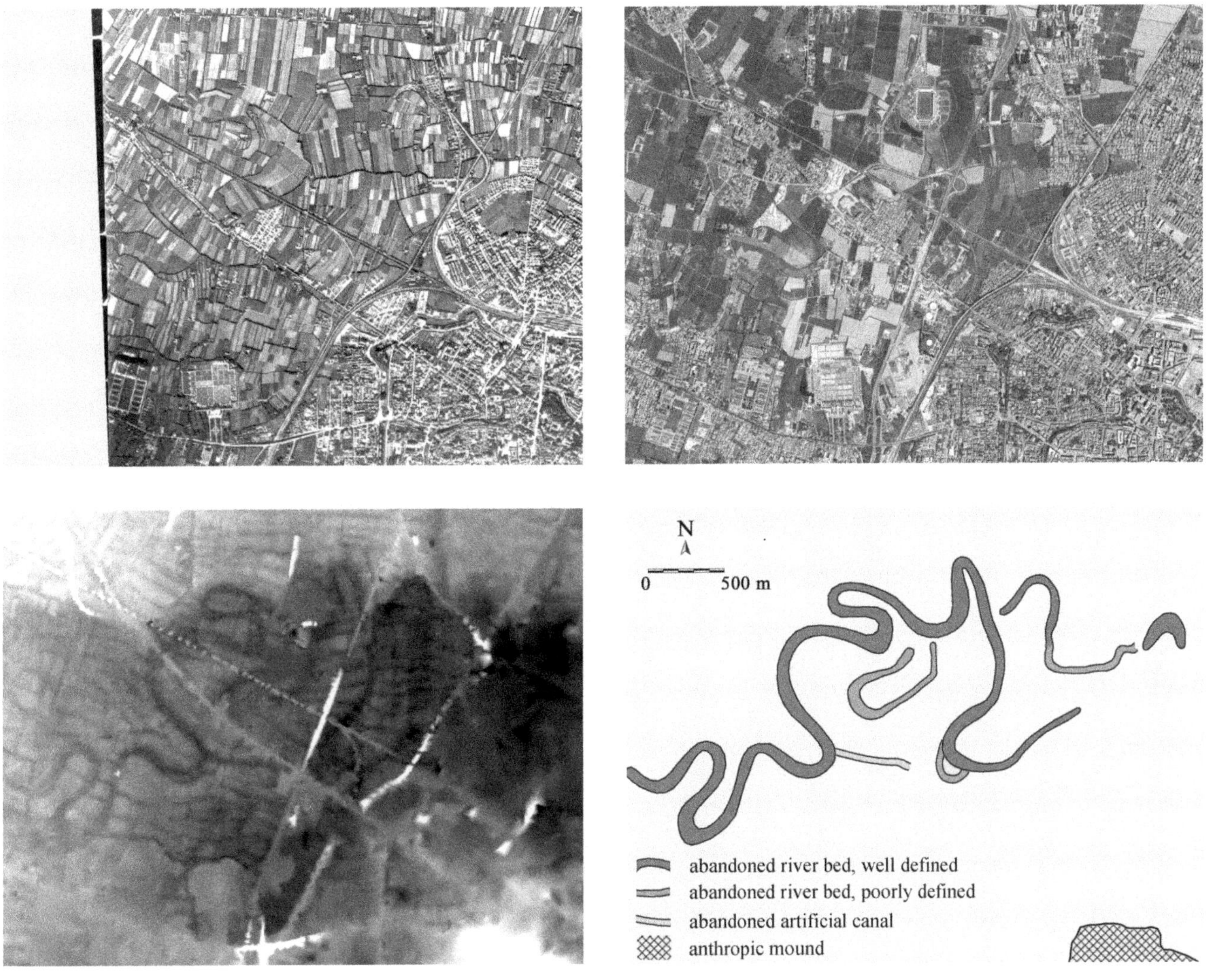

Figure 3 *Aerial photo of 1956 (above, left), aerial photo of 2004 (above, right), DTM (below, left) of an area North West of Padua. From these data derive the geomorphological interpretation (below, right). The main abandoned river bed is defined in a continuos shape in DTM, while in aerial photos only limited tracts are evident.*

The abandoned river beds South of the city follow a NW- SE drainage direction of unknown age. The channel widths and the meander radii suggest that all the palaeohydrographic traces pertain to the Brenta River. Part of the shallow fluvial ridge built by the Brenta river in present position can also be recognized in the DTM.

Conclusions

The use of a high resolution DTM has provided new insights in the geoarchaeology and geomorphology of Padua. The interpretation of the DTM has led to the recognition and mapping of the major fluvial landforms around the urban area. It has also allowed the detection and morphometric evaluation of the large anthropic mound in the city centre. This mound corresponds to the area of ancient settlement, comprised within the two meanders presently followed by the Bacchiglione River. Historical cartography shows that the city expanded out of this limited sector just during the 20th century. The calculation of the volume of archaeological deposits inside the city walls and the mapping of their thickness have been performed through to the subtraction of the present day DTM to the reconstructed DTM of the early Iron Age.

Acknowledgments

We wish to thank Fiorenza Colombo of the Cartographic Office of the Municipality of Padua, who provided the elevation data. Giovanni De Luca of the ARPAV (Environmental Protection Agency of the Veneto Region) provided fundamental help in the pre-treatment of these same data. Jenny and Ronald Guariento provided the revision of the translation to English.

References

ARPAV-Regione del Veneto (2005). - *Carta dei suoli del Veneto alla scala 1:250.000.* ARPAV - Agenzia per la

Prevenzione e Protezione Ambientale del Veneto, Padova.
BAGGIO P., G.B. SIGALOTTI, C. ZAMBONI 1992 Analisi territoriale di aree perturbane: il nord-ovest di Padova, in *Padova nord-ovest, archeologia e territorio*, 19-33.
BALISTA, C. 2004, Il contesto geomorfologico e paleoidrografico, in RUTA SERAFINI A., S. TUZZATO, La necropoli patavina di via Umberto I. *Quaderni di archeologia del Veneto*, 20: 91-102.
BALISTA, C., L. RINALDI 2005 I percorsi pre-protostorici del fiume Brenta a Padova, in *La città invisibile. Padova preromana, trent'anni di scavi e ricerche* (M. De Min, M. Gamba, G. Gambacurta, A. Ruta Serafini eds.), 11-21.
BALISTA, C., A. RUTA SERAFINI 1993 Saggio stratigrafico presso il muro romano di Largo Europa a Padova. *Quaderni di archeologia del Veneto*, 9: 95-111.
BALISTA, C., A. RUTA SERAFINI 2004 Primi elementi di urbanistica arcaica a Padova. *Hesperìa*, 18: 291-310 (*I Greci in Adriatico*, 2).
BONARRIGO, M. 1992 *Padova. La città, le acque*, Padova.
BONDESAN A. ET AL. 2003 Modelli d'elevazione del terreno per la definizione di carte del rischio idraulico: un esempio nella bassa pianura padovana. *Geologia dell'Ambiente, numero speciale Atti del Convegno "La geologia ambientale strategie per il nuovo millennio"*, 113-116.
BOSIO, L. 1981a Padova e il suo territorio in età preromana, in *Padova antica. Da comunità paleoveneta a città romano-cristiana*, Padova-Trieste, 3-18.
BOSIO, L. 1981b Padova in età romana. Organizzazione urbanistica e territorio, in *Padova antica. Da comunità paleoveneta a città romano-cristiana*, Padova-Trieste, 231-246.
CASTIGLIONI G.B., A. GIRARDI, G. RODOLFI 1987 Le tracce degli antichi percorsi del Brenta per Montà e Arcella nei pressi di Padova: studio geomorfologico. *Mem. Sc. Geol.*, XXXIX pp. 129-149.
GALLIAZZO, V. 1971 *I ponti di Padova romana*, Padova.
GAMBA, M., G. GAMBACURTA, A. RUTA SERAFINI, C. BALISTA 2005 Topografia e urbanistica, in *La città invisibile. Padova preromana, trent'anni di scavi e ricerche* (M. De Min, M. Gamba, G. Gambacurta, A. Ruta Serafini eds.), 23-31.
GAMBA, M., G. GAMBACURTA, C. SAINATI 2005 L'abitato, in *La città invisibile. Padova preromana, trent'anni di scavi e ricerche* (M. De Min, M. Gamba, G. Gambacurta, A. Ruta Serafini eds.), 65-75.
GASPAROTTO, C. 1951, *Padova romana*, Roma.
ILICETO V., F. MELONI, P. MOZZI, F. RIZZETTO 2001 Il sottosuolo della Cappella degli Scrovegni a Padova. *Geologia Tecnica e Ambientale*, 9 (4).
MICHELINI, P., A. RUTA SERAFINI 2005 Le necropoli, in *La città invisibile. Padova preromana, trent'anni di scavi e ricerche* (M. De Min, M. Gamba, G. Gambacurta, A. Ruta Serafini eds.), 131-143.
MOZZI P. 2005 Alluvial plain formation during the Late Quaternary between the southern Alpine margin and the Lagoon of Venice (Northern Italy). *Suppl. Geogr. Fis. Dinam. Quat.*, 7: 219-229.
Padova città d'acque 1989, guida alla mostra, Padova.
ROSADA, G. 1993 Patavium: note di archeologia del paesaggio e di topografia urbana. *Journal of Ancient Topography*, 3: 63-76.
RUTA SERAFINI, A., 2002 L'archeologia urbana: nuovi dati, in *Padova romana* (H. Hiller, G. Zampieri eds.) 57-73.
RUTA SERAFINI A., A. VIGONI 2006 (eds.) Lo scavo archeologico nel cortile della Casa del Clero, in *Casa del Clero, Padova. Recupero di un luogo nel centro storico di Padova*, Padova, 85-111.
TELLINI C. 2001 Altimetry, in *Illustrative notes of the geomorphological map of the Po Plain.*, (G.B. Castiglioni, G.B. Pellegrini eds.), *Suppl. Geogr. Fis. Dinam. Quat.* 4: 21-32.
TOSI, G. 2002 Aspetti urbanistici ed architettonici di Padova antica alla luce delle fonti storiche e di vecchi e nuovi rinvenimenti. *Antenor*, 3: 87-127.

"Valle d'Agredo": a paleoenviromental and geoarcheological reconstruction based on remote sensing analysis

Bruno Marcolongo,[1] Andrea Ninfo,[2] Matteo Simone[3]

[1] National Research Council of Italy, Ist. of Applied Geology, Padova
[2] Department of Geography, University of Padova
[3] National Research Council of Italy, Ist. of Applied Geology, Padova

Introduction

"Agredo's Valley", awarded by Hugh of Provence, king of Italy, to Adalberto, bishop of the Treviso's diocese on Christmas 926, is historically remembered as a wide area including the territories of the present communes of Castelfranco, Camposampiero, Noale and Mirano.

Nevertheless its boundaries are not better defined, due to the lack of clear physiographic limits in a relatively homogeneous plain. Moreover heavy anthropic pressure, characterized by a high settlement density and spread extractive industry, interferes with the water resource's quality and contributes to veil many landscape features.

Therefore, the present environmental situation does not help the attempt of reconstructing and defining the extension of the ancient "Agredo's Valley", which without doubt has represented for long time a solid reality, before physiographic and then cultural one, fixed through generations either in the common popular feeling or in the learned literary tradition.

The Study Area

"Agredo's Valley" entirely belongs to the "morphological unit" of the ancient alluvial fan of Brenta river, the so called "Bassano's relict cone" after AA.VV. (see A. Bondesan *et alii*, 2002; P. Mozzi, 1998 and 2003), which stretches from the outlet of Valsugana to the South-East till the Venetian lagoon (see Fig. 1)

This wide fan, mainly built during upper Pleistocene (main phase between 25 and 18 ka BP, followed by a minor phase around 14 ka BP), clearly shows an average slope of about 5‰ in the upper portion of the plain till the line Cittadella-Castelfranco, which after lowers down to 1‰ and even less near the lagoon inner border.

The middle and lower plain is marked by a series of shallow fluvial levees separated by narrow depressions, representing past flow directions of the prehistoric Brenta. These natural levees are rising less than 2 m over the surrounding flat terrain with a variable width from several hundreds of meters up to 1-2 km. Their longitudinal development can reaches even 15-20 km.

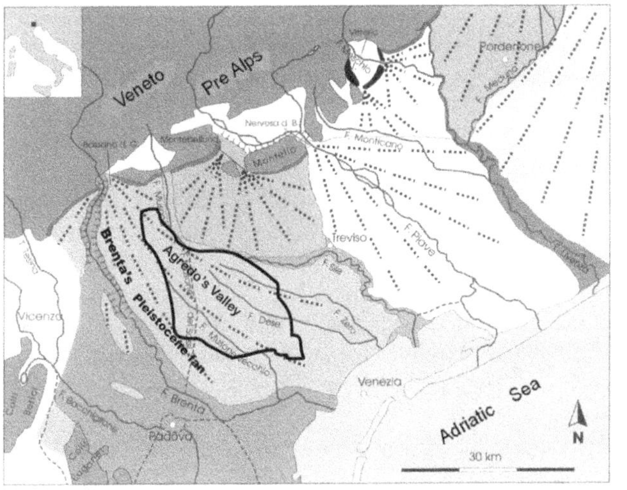

Figure 1

The sandy levee, on top of which are today lying the settlements of Piombino Dese, Trebaseleghe and Scorzè, appears with its 20 km of length the most continuous and extended of the whole studied area. As it will be seen later on, the microrelief analysis on bibliographic data (see P. Mozzi, 1998; A. Bondesan, 2003) exactly represents one of the starting points to understand and define the physiographic identity of the historical "Agredo's Valley".

Beside that, nature of the soils and their typological distribution are in this approach important elements for improving the reconstruction of the extension of "Agredo's Valley" itself. Here a detailed pedological maps at 1:25.000 scale has been utilized, covering the territories of Piombino Dese and Trebaseleghe and edited by the "Venetian Agricultural Development Organization" (P. Mozzi et alii, 1996). In this thematic maps a tight correlation is very well shown between bands of soils with sandy texture and the highest natural levees.

Contextually to collection and compared analysis of bibliography, elaboration of multispectral and panchromatic images taken from different platform (aircraft and satellite) has been done. Their geomorphological and geoarchaeological interpretation, superimposed to the digital cartography produced by Region Veneto, has provided further information on the area, like soil humidity pattern and pseudo circular features in the agrarian parcels mosaic, referable as anthropic "mound" ("motte"), so enriching the "Agredo's Valley" description.

Microrelief Analysis

Microrelief analysis is fundamental for the characterization of: fluvial ridges, depressed areas and escarpments; often these morphologies can be detected only with this method for the low gradient that distinguishes them from the rest of the plain.

To produce microrelief map with a suitable algorithm of triangulation (TIN "triangular irregular network") the quoted points given by CTR's maps (regional technical cartography at the scale 1:5.000) have been considered and contour lines of 1 m equidistance have been generated. Figure 2 synthesizes the pattern of the major natural levees, expression of the Brenta's palaeo courses during tardo-glacial period marking peculiarly the "Agredo's Valley" territory. These morphological positive features determine the limits of the narrow elongated depressions, today occupied by the main rivers of "risorgiva" (originating from plain springs) of the zone, as Dese, Marzenego and Musone Vecchio.
The most developed and continuous branch is that passing through the communes of Piombino Dese, Trebaseleghe and Scorzè. This relief feature represents the most outstanding element of a flat landscape and has to be identified with the longitudinal axis of the historical "Agredo's Valley", physiographic entity whose record has lasted for long time deeply routed in the memory of the local inhabitants. It appears clearly evident the localization of the Trebaseleghe settlement, just on top of that sandy levee, encircled by lower alluvial terrains of finer silty texture.

Satellite Images Analysis

The first satellite image utilized has been a cosmic spectro-zonal photo Sojuz KFA 1000, taken on June 1990 from a Russian platform placed in a lower orbit (270 km of altitude) with a good geometric resolution of 5x5 m^2. On it one can easily identify a large patch of terrains with darker tonality, showing an elongated shape NW-SE (see Fig. 3). It is very significant to note that all the four villages named in the Ugo of Provenza's donation to bishop Adalberto fall in this portion of plain area. Most likely the dark tone is expression of a soil higher humidity, in connection with preferred directions of groundwater flow still active at present, which follow the prehistoric branches of Brenta river.

Successively, Aster multispectral images taken on December 2001 and February 2002 (see Fig. 4) have been analysed, having a geometrical resolution of 15x15 m^2 in the visible/near infrared range (VNIR), kindly furnished for the present research by "Ministero delle Infrastrutture e dei Trasporti – Magistrato alle Acque di Venezia/Concessionario Consorzio Venezia Nuova". Their interpretation has confirmed the presence of a central band more humid and homogeneous as soil texture along the direction Piombino Dese – Trebaseleghe – Noale – Mirano, corresponding to the longitudinal axis of the historical "Agredo's Valley". Moreover the elaboration of Aster data permitted to go into details of the soil use of the area, thus contributing to give a more precise delimitation of the valley itself.

Last but not least, various Ikonos images, taken on July 2002 with a very high geometrical resolution (4x4 m^2 in the multispectral mode and 1x1 m^2 in the panchromatic one), have been suitably elaborated and interpreted especially for the commune of Trebaseleghe, which is occupying a barycentre position in the area of interest (see Fig. 5). These data describe much better the agrarian plots mosaic, which is the best witness of the existence in that area of an ancient roman cadastre (Altino's "centuriazione", cfr. P. Furlanetto, 1998; J. Bonetto, 2003) today surviving in the network of channels, roads and limits of properties.

All the remote sensing data have been linked to a precise geometrical base, constituted by the technical regional cartography (CTR) produced by Region Veneto at the scale 1:5.000 in the nineties.

It is interesting to add that in the territory of Trebaseleghe three anomalies of the dense and regular mosaic of agrarian parcels have been recognized on these detailed Ikonos images. The shape of these feature is sub circular or elliptic one and they are localized on the sandy natural levees already well identified. Their space distribution forms a kind of alignment cutting transversally the "Agredo's Valley" axis, being the first near Fossalta, the second near the centre of Trebaseleghe and the third one in a locality called S. Ambrogio. Traces of these landscape anomalies recalling the morphology of mounds ("motte") are reported by an old cartographic document of the beginning of the nineteen century ("Topographisch-geometrische Kriegs karte von dem Herzogthurns Venedig", Topographic-geometric Military Map of the Venice's duchy, General Anton von Zach, 1978-1804). In particular, the structure of Trebaseleghe still possesses on this map some morphological prominence, even if it appears partially eroded by the agrarian labours.

Interpretation of cosmic photo Sojuz crossed with that of aerial photos and Ikonos data allowed the identification of the roman agrarian network which imprinted the settlement model of the entire area (see Figs 6 and 7). Already in the middle of the seventies researches on this topic had been done utilising Landsat images (*Marcolongo and Mascellani, 1978; Marcolongo, Mascellani and Matteotti, 1978*), which underlined the presence of such pattern. Nevertheless the today improved geometrical and radiometric resolution of a greater number of remote sensing images has sustained a very punctual reconstruction of the whole reticule of

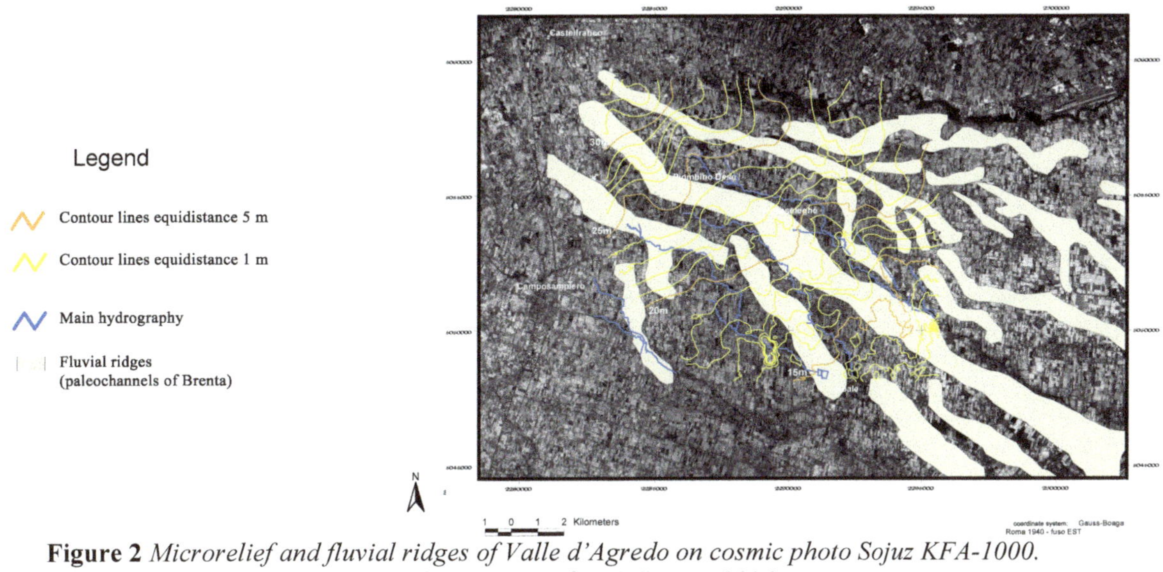

Figure 2 *Microrelief and fluvial ridges of Valle d'Agredo on cosmic photo Sojuz KFA-1000.*
B. Marcolongo, A. Ninfo, M. Simone 2006

Figure 3 *Alluvial plain between Brenta and Sile recorded by Sojuz KFA-1000.*
B. Marcolongo, A. Ninfo, M. Simone 2006

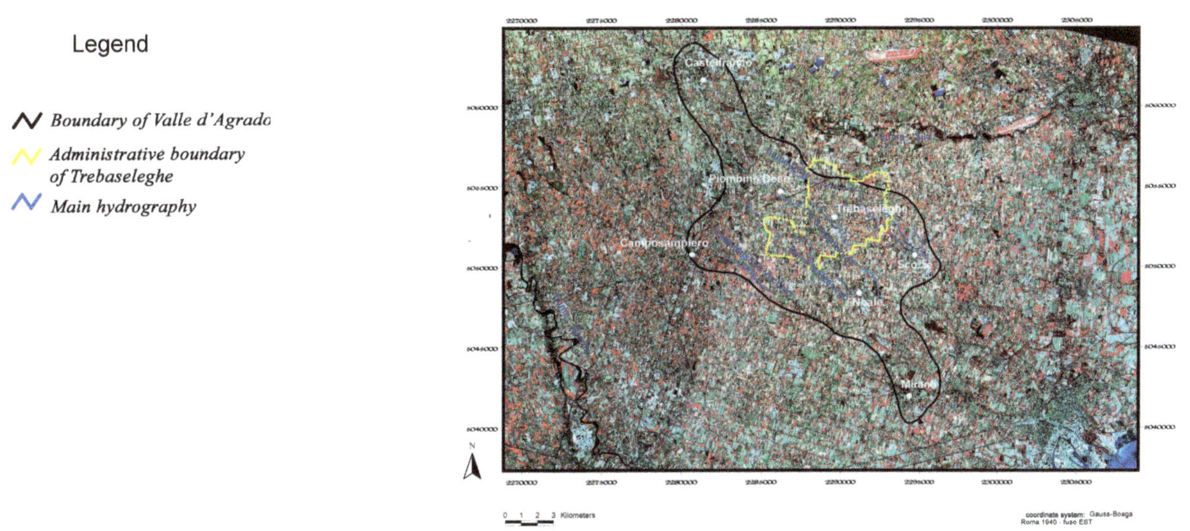

Figure 4 *General view of Valle d'Agredo on Aster Fcc 321vnir image elaboration.*
B. Marcolongo, A. Ninfo, M. Simone 2006

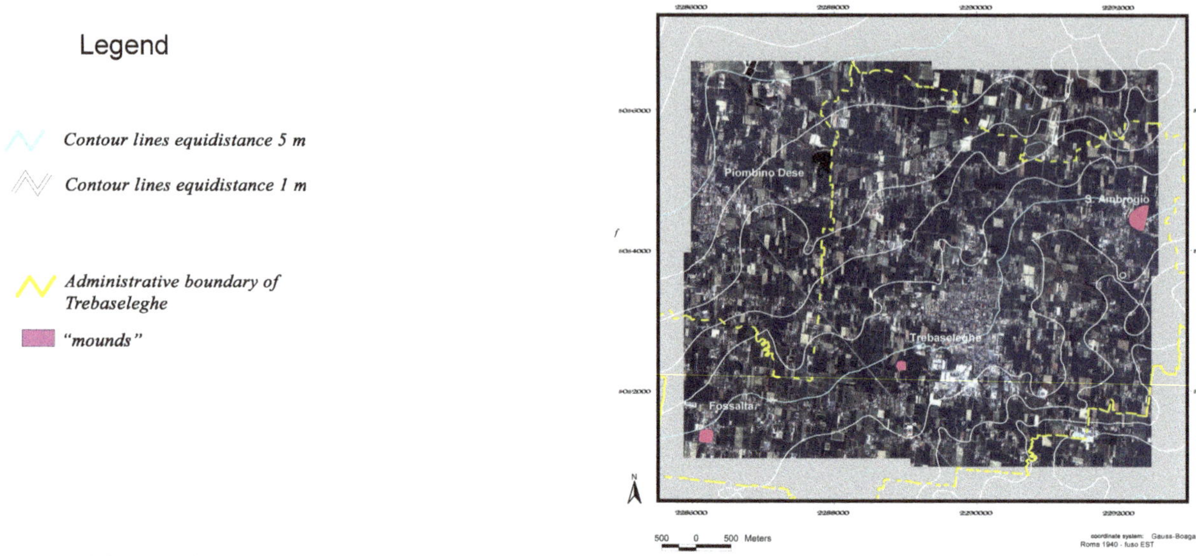

Figure 5 *Microrelief and anthropic mounds in Trebaseleghe's territory on an Ikonos Fcc321 images mosaic. B. Marcolongo, A. Ninfo, M. Simone 2006*

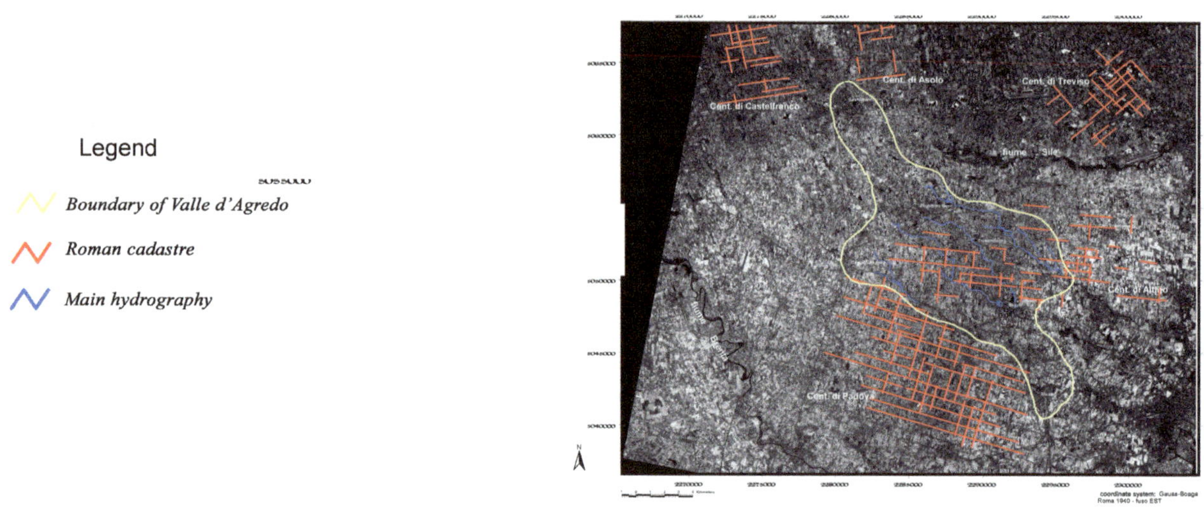

Figure 6 *Elements of Roman cadastre in Valle d'Agredo on Sojuz KFA-1000 B. Marcolongo, A. Ninfo, M. Simone 2006*

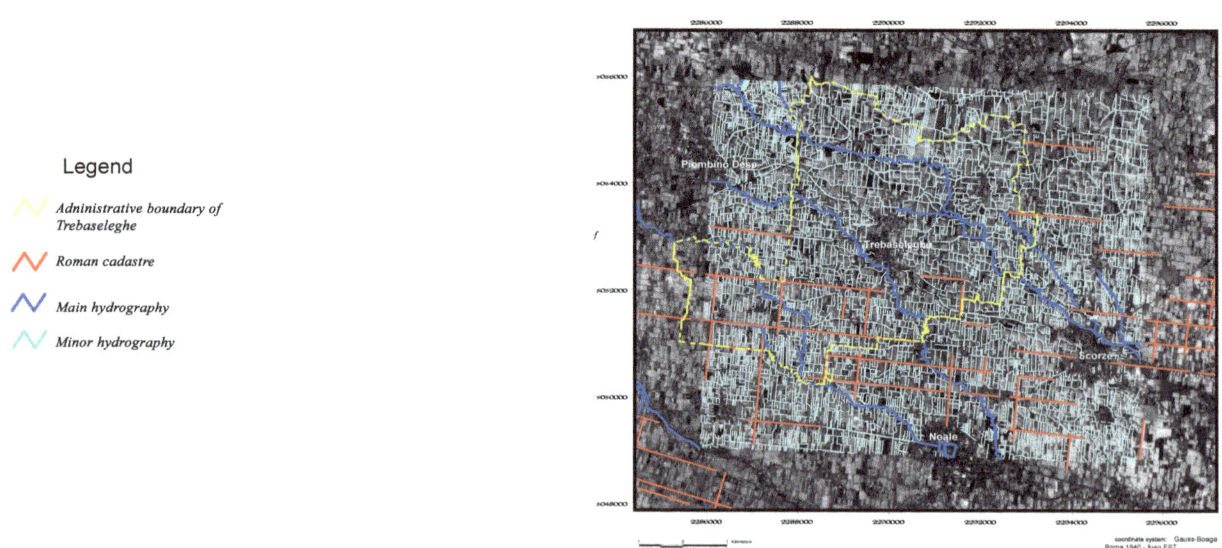

Figure 7 *Detaled elements of Roman cadaster in Valle d'Agredo coinciding with minor hydrography network, on Sojuz KFA-1000. B. Marcolongo, A. Ninfo, M. Simone 2006*

linear elements (canals, roads and property limits) inside the systems of Asolo-Castelfranco-Cittadella, Paese-Treviso, Altino and Padova-Campodarsego (*Marcolongo and Mascellani, 1978; Marcolongo, Mascellani and Matteotti, 1978; A. Costi, L. Lazzaro, B. Marcolongo, J. Visentin, 1992; P. Furlanetto, 1998; J. Bonetto, 2003*). "Agredo's Valley" appears to be interested by the roman subdivision of Altino, which extends as far North as to Sile river and to Musone Vecchio to the West.

Figure 6 proposes an extract of the minor hydrographic network of Trebaseleghe, which shows how evident is still today the mark left on the agrarian landscape organization by the old subdivision of the land through perpendicular and regular limits. To a careful lecture it emerges that the roman cadastre is relatively more rarefied and disgregated in correspondence of the Piombino Dese-Trebaseleghe-Scorzè levee, probably because this band attracted more than the other ones the residential and industrial settlements linked by a road network, only partially superimposed to the older system.

Conclusions

The historical "Agredo's Valley", initial objective of the present research, is emerging more and more as a true territorial entity which attracted settlements since prehistorical times, leaving then a strong cultural identity to the local population. Reconstruction of fluvial paleo-forms and other landscape features and comprehension of their dynamic evolution, through an integrated remote sensing approach (bibliography, various images interpretation, field survey), made it possible. Sophisticated tools like remote sensing data with high geometric and spectral resolution, elaborated and interpreted in accordance with a consistent material coming from conventional studies, have given demonstration and validity to an old common popular feeling: the existence in the past of a well defined territorial unit like "Agredo's Valley".

The effects of the Brenta's ancient branch continued to be influent in the plain environment long after the abandonment of the palaeo-watercourse itself, through its control on the groundwater flow direction, on the discharge of minor rivers fed by the "risorgive" belt and on the distribution of soils of different texture.

The "Agredo's Valley" axis, superimposed upon the tardo-glacial Brenta riverbed, has renewed a peculiar physiographic direction connecting the peri-lagoon fringe to the Valsugana entrance. This activated in historical periods an important trade and culture communication way between the Adriatic sea area and the continental alpine area.

It has to be underlined also the probable role of control on traffic and communication exercised by the three mounds ("motte") and in particular by that central one of Trebaseleghe.

Along and around this direction Venetian lagoon-Valsugana mouth (SE-NW) a more and more complex settlement organization developed. It spontaneously followed the traces of the past better than cancel them, thus giving new vitality to an old landscape organization in a right equilibrium between human settlements and natural resources.

References

BONETTO J. (2003) – *Il medio e basso corso del Brenta in età romana*. In "Il Brenta" a cura di A. Bondesan *et alii*, Cierre Ed., Sommacampagna (VR)

BONDESAN A. (2003) – *Natura antica e idrografia moderna del basso corso*. In "Il Brenta" a cura di A. Bondesan *et alii*, Cierre Ed., Sommacampagna (VR)

COSTI A., LAZZARO L., MARCOLONGO B., VISENTIN J. (1992) – *La centuriazione romana fra Sile e Piave nel suo contesto fisiografico*. CNR, Padova

FURLANETTO P. (1998) – *Fluvius Silis ex montibus tarvisanis*. In "Il Sile" a cura di A. Bondesan *et alii*, Cierre Ed., Sommacampagna (VR)

MARCOLONGO B., MASCELLANI M. (1978) - *Immagini da satellite e loro elaborazioni applicate alla individuazione del reticolato romano nella pianura veneta*. "Archeologia Veneta", anno I, Soc. Archeologica Veneta, Padova

MARCOLONGO B., MASCELLANI M., MATTEOTTI E. (1978) - *Significato storico-ambientale di antiche strutture topografiche sepolte nella pianura veneta*. "Archeologia Veneta", anno I, Soc. Archeologica Veneta, Padova

MOZZI P. (1998) – *Nascita e trasformazione della pianura del Sile*. In "Il Sile" a cura di A. Bondesan *et alii*, Cierre Ed., Sommacampagna (VR)

MOZZI P. (2003) – *L'alta e media pianura del Brenta*. In "Il Brenta" a cura di A. Bondesan *et alii*, Cierre Ed., Sommacampagna (VR)

MOZZI P., ORTOLANI R., RAGAZZI F., VINCI I. (1996) – *I suoli di Piombino Dese e Trebaseleghe*. ESAV, Giunta Regionale del Veneto, Padova

Using Vegetation Indices to study archaeological areas

Pasquale Merola, Alessia Allegrini, Daniela Guglietta, Simone Sampieri[1]

[1]CNR Atmospheric Pollution Institute

1 Introduction

In recent years there has been a growing interest for the use of remote sensing in archaeology.

Since 1994, in Italy, the CNR-LARA (Airborne Laboratory for Environmental Research) has carried out research activities involving both remote sensing and archaeology, by using Daedalus AA5000 MIVIS (Multispectral Infrared and Visible Imaging Spectrometer) airborne hyperspectral scanner to recognize the soil discontinuities connected to buried anthropic elements. The MIVIS spectrometer is a passive remote sensing instrument which collects the earth surface radiation from an airborne platform. The detector separates incoming radiation into four optical ports covering the Visible (0.43-0.83 μm), the Near-InfraRed (1.15-1.55 μm), the Mid-InfraRed (2.0-2.5 μm), and the Thermal-InfraRed (8.2-12.7 μm) region, a total of 102 spectral bands. It operates with an Instantaneous Field Of View (IFOV) of 2 mrad (Bianchi *et al.* 1994, Bianchi *et al.* 1996). The MIVIS data, that can be enhanced, rectified and reclassified using several algorithms and specific software, permit the identification of buried archaeological structures because any buried ruins, either of human or natural origin, affects, over time, characteristics of the soil surface and creats anomalies. These anomalies are due to different factors, such as physical and chemical soil features and the state of vegetation. The above factors are strictly connected and are responsible of surface spectral responses.

In particular the spectral response of areas under vegetation presents a complex mixture of vegetation, soil brightness, environmental effects, shadow, soil colour and moisture. The scientists have developed vegetation indices for qualitatively and quantitatively evaluating vegetative cover using spectral measurements.

MIVIS remote sensing permits us to simultaneously analyse a wide range of different wavelengths: these studies have shown that the use of the red and near-infrared channels of the MIVIS sensor are particularly useful for the study of cover vegetation.

Spectral Vegetation Indices are an important tool in observing spatial and temporal variations of vegetation, its biophysical properties and photosynthetic activities, through which it is possible to analyse the effects of buried ruins on the state of vegetation cover.

In normal conditions, at the same time, chlorophyll is degraded and produced in continuation, while the action of various factors, such as the presence of anthropic structures in the first subsoil layers, could alter the

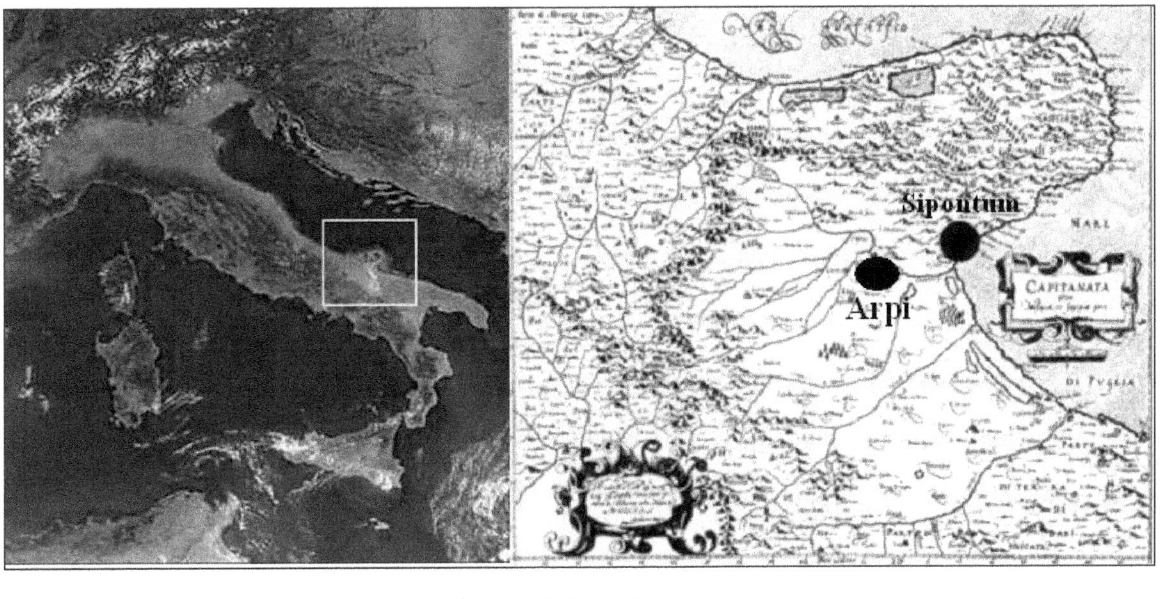

Figure 1 *The study areas.*

normal process of vegetation growth (vegetation stress). All buried structures induce anomalies in the soil moisture, creating zones of differentiated humidity, where the vegetation can grow more luxuriant: in the image these variations appear in different colours to the surrounding areas (vegetation traces) (Piccarreta and Ceraudo 2000).

Situations of vegetation stress can be found through the use of Vegetation Indices, which are quantitive measurements indicating the vigor of vegetation (Campbell 1987).

In this work, we will illustrate the results obtained studying the archaeological area of Arpi and *Sipontum* (Puglia Region, South of Italy) using the main Vegetation Indices.

2 The Study Areas: historical and topographic context

MIVIS data acquired are very useful for the study of several archaeological sites: Selinunte (Cavalli and Pignatti 2001), Villa Adriana, *Lilybeaum* (Marsala) (Merola 2004 and 2005), Arpi (Cavalli et al. 2005) and *Sipontum* (Fig. 1).

Arpi

Arpi is situated 8 km North-East of Foggia. The place was inhabited from the Neolithic age (fourth millenium B.C.); the ancient sources attribute the foundation of the city to Diomede.

In 320 B.C., during the second sannitic war, Arpi stipulated an alliance with Rome in order to besiege the city of Lucera. The good relationship with Rome was conserved strongly until the explosion of the Punic war, when the Romans were defeated in the Trasimeno battle (summer of 217 B.C.). In 213 B.C. Rome decided to reorganize the territory, depriving Arpi of its coastal line where the colony of *Sipontum* was founded.

From the eleventh century the territory of Arpi was divided between the Siponto and Troia dioceses, while the other part, minor, was incorporated in the Lucera diocese (Mazzei 1995).

The major part of Arpi was defended by majestic walls (Fig. 4), described in its entirety by Bradford (Bradford 1957), which enclosed an area of about 1000 hectares. The walls were running approximately 13 km in length, with the traces of probable defense towers in near-by fields corresponding to the layout of the gates.

Wide areas were destined to crop cultivation for the supply of entire population, also during the long periods of war; other areas were designated to the pastures or were simply left uncultivated. Inside the walls there were also the necropolies, which would imply that, in case of danger, the dead were being protected as well as the living (Guaitoli 2003, Piccarreta and Ceraudo 2000).

During the Daunian period, the houses consisted of huts, with the floor in battered soil and the structure formed by a series of wood poles, covered with branches and straw, and made waterproof with clay. From the fourth century B.C. it is possible to recognize a reorganization of the city, with the smaller area used. In this area, structures were built of bricks and stones, probably organized in a more regular road network, inspired by the criteria of the urban planning, changed from the Greek world, with an orthogonal formulation of the city.

The necropolis were located inside (Montarozzi e Arpinova) and outside (San Nicola d'Arpi) to the city perimeter (Marin 1970).

Sipontum

The ancient *Sipontum* (the Roman colony, certainly the second foundation) is situated near S. Maria Maggiore church's of Siponto, at about 3 km south of Manfredonia (Foggia, Southern Italy).

The *Sipontum* Roman city was built on a small rocky rise in front of a Manfredonia gulf. The archaeological research and the ancient sources show that in the S. Maria Maggiore area of Siponto only Roman-medieval *Sipontum* center was located whereas the Daunian city was situated in another area, on the old coast, in "Masseria Cupola-Beccarini" locality.

The "Masseria Cupola-Beccarini" settlement lived until the beginning of the third century B.C., when in 194 B.C., Rome deduced the first Roman colony of *Sipontum.* In 185 B.C. Rome built the new *Sipontum* colony in the area of Saint Maria Maggiore. *Sipontum* was a Daunian trading emporium, extremely important since pre-Roman period, when it was a maritime port of an important and powerful city like Arpi. The harbour had an intense activity during imperial and medieval age (Danti 1999).

The earthquake of 1223 destroyed it nearly completely and induced Manfredi in 1265 to construct a new city, Manfredonia, situated about 2 km to the north of the ancient city (Mazzei 1999, Mazzei 2003).

Although the knowledge of the maritime colony is extremely limited, it is possible to suggest that it was very similar to the traditional Roman colony. The walls, 3000 metres long, had square towers of trapezoid shape: the blocks are regular and of local stone; the walls are two to three metres wide. A gate was located in front of the sea. The *decumanus maximus* can be recognized in the line of ex-main road, the so called "ex-statale 89". It crosses the city in the E-W direction, while the *cardo maximus* could be identified in the line that went throughout the city from the gate in the N-S direction.

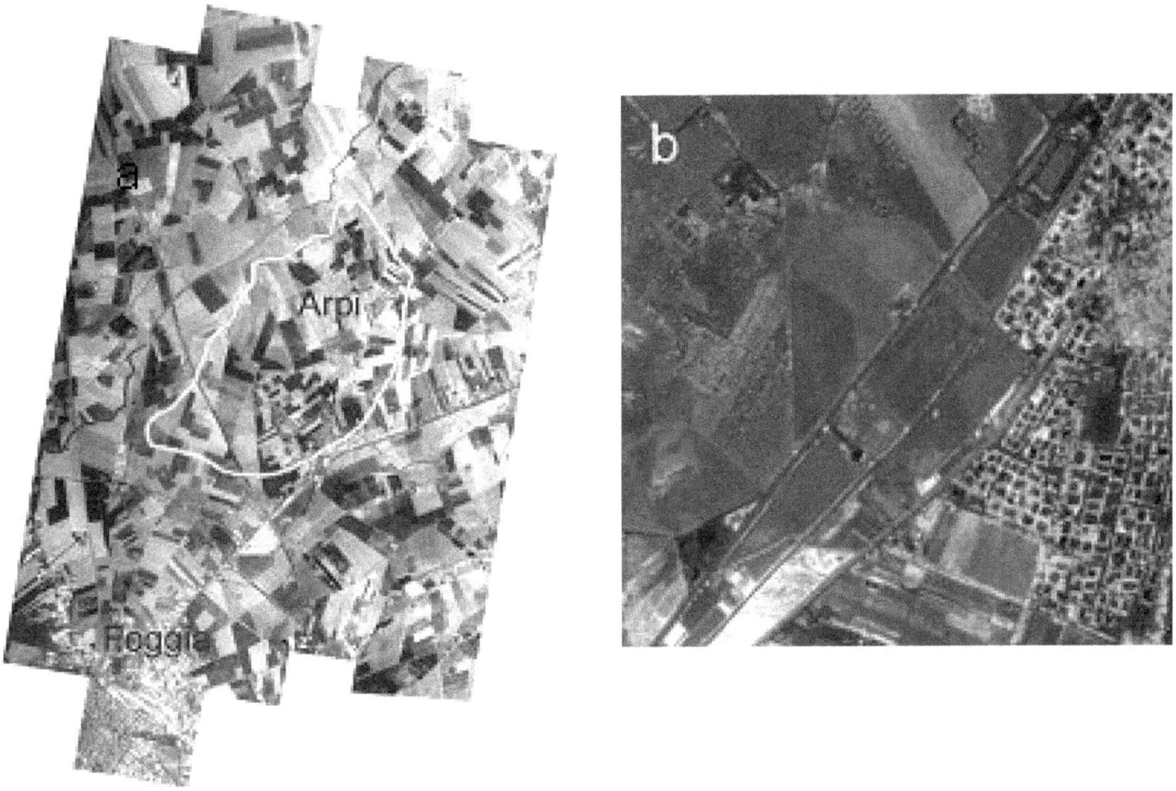

Figure 2 *MIVIS images: a) Arpi; b) Sipontum.*

In the Augustan age important buildings were built: the amphitheatre, two thermae, whereas the forum has not been located although it supposed in the S. Maria Maggiore area.

The amphitheatre was constructed inside of the urban area, along the N-E side of the city. It was built at distance of five metres from the walls: questions their function as the defences at the time of building of the amphitheatre. Today the wall in *opus reticolatum* with semicircular form is visible; in modern age it was used as the base for Masseria Garzia. One thermae are located in northern urban area, the other in southern area.

The necropolis is located in the central area of the ancient southern side, and more tombs were found in the are of paleocristian basilica (Mazzei 1999, Mazzei 2003, Schmiedt 1973)

3 MIVIS Data

MIVIS spectral data were collected during two flight campaign carried out on:

- Arpi: 27th June 2002 at 2:05 pm (local time) from an altitude of 1500m a.s.l., along five flightlines with an orientation NNE-SSW (Fig. 2, a), and obtaining a pixel ground resolution of 3m;
- Sipontum: 20th August 2002, at 10:06 a.m. (local time), from an altitude of 2000 m. a.s.l., thus achieving a ground resolution of 4 m/pixels (Fig. 2, b).

The acquired data were radiometrically calibrated and geocoded (Avanzi 1996, Bassani 2002)

- MIVIS data were normalized as regards the spectral signature of a region with well known reflectance (for example a square in concrete), by *Flat Field Calibration method*, in order to obtain an apparent reflectance;
- MIVIS images were geocoded by PARGE software for the orthorettification of the images. We obtained a good geometric accuracy, with errors less than 3-4 pixel (9-12m).

In this work we analysed different vegetation indices, calculated on MIVIS hyperspectral images, to characterize and recognize the buried archeological structures. In particular, our attention was focused on the calculation of ratio indices (Huete 1984, Huete 1989).

4 Vegetation Indices

Vegetation Indices are quantitative measurements indicating the vigor of vegetation (Campbell 1987).

The different vegetative covers can be distinguished according to their unique spectral behaviour in relation to overall ground elements: visible radiation in the Red (630-690 nm) is absorbed by chlorophyll while radiation in the Near-Infrared (760-900 nm) is strongly reflected by leaf cellular structures. When examining the general reflectance curve of vegetation, the deviation observed the Red and Near-Infrared constitutes a variable sensitive to the presence of green vegetation. The spectral response of vegetation in the Red is strongly correlated with chlorophyll concentration, while the spectral response in the Near-Infrared is controlled by the leaf area index and green vegetation density (Major *et al.*). The combination of these two spectral domains permits to differentiate vegetation from soils and to determine photosynthetically active biomass through vegetative cover density (Huete 1984, Huete 1989).

Huete (Huete 1984, Huete 1989). has classified vegetation indices into two categories:

- Ratio indices;
- Orthogonal vegetation (Bannari *et al.*, 1995).

5 Data Processing and Analysis

To highlight the surface anomalies identified in the MIVIS images, due to variations of texture, humidity and vegetation brought about by the presence of buried structures, several Vegetation Indices were used. The bands 20 (0,82 μm) and 12 (0,66 μm) have been used for all indices analysed:

- Ratio Vegetation Index (RVI);
- Vegetation Index Number (VIN);
- Normalized Difference Vegetation Index (NDVI).

Ratio Vegetation Index and Vegetation Index Number

The first two indices were developed in the form of ratio:

- Ratio Vegetation Index (RVI) = RED/NIR
- Vegetation Index Number (VIN) = NIR/RED

where NIR is the mean reflectance in the Near-Infrared channel, Red is the mean reflectance in the red channel. These indices enhance the contrast between the soil and the vegetation (Bannari *et al.*, 1995). The relationship between the reflectances of the two bands permits to eliminate disturbances from factors, affecting in the same manner the radiances of each band (Holben and Justice 1981)

NDVI: Normalized Difference vegetation Index

Remotely sensed vegetation indices are generally a very effective tool for environmental resource monitoring and for quantifing different ecological events.

The Normalized Difference Vegetation Index (NDVI) has become the primary tool for accurate description of vegetation changes and interpretation of the impacts of environmental phenomena and it can be calculated from the data of Channel 20 (0,82 μm) and 12 (0,66 μm) of the MIVIS.

Extensive research has shown that the NDVI can be used for accurate description of continental land cover due to its strong relationship with the amount of absorbed photosynthetically active radiation, total biomass and fractional vegetation cover.

The NDVI is especially well suited for monitoring broad-scale phenological characteristics of terrestrial ecosystems because normalization partially compensates for changing illumination conditions and surface terrain effects.

The NDVI is calculated from the reflectance of red and near-infrared radiation:

$$NDVI= (NIR - RED)/(NIR + RED)$$

where NIR and RED are respectively the spectral radiances in the Near-Infrared and Red bands.

The computed NDVI ranges are included from -1.0 to +1.0: increasing positive values of this index indicate increasing green vegetation and negative values indicate unvegetated areas.

As evidence of its general relation to vegetational biomass variations, this index is closely related to the amount of green leaf biomass and leaf area index (Gong Dao-Yi and Shi Pei-Jun 2003).

6 Interpretation

The photointerpretation of MIVIS images was supported by a methodical bibliographic research (excavating information and previous photointerpretations) (Guaitoli 2003, Schmiedt 1973) and by survey carried out in different periods in the archaeological area. In the MIVIS images, numerous archaeological structures derived from the calculation of vegetation indices have been characterized. In particular, in the vegetated areas the recognition of the archaeological traces is the result of the spectral difference between the surfaces above the archaeological structures and those in the surrounding area; this result is due essentially to the different structure of the absorbed and reflected peak of vegetation in the Visible and Near-Infrared wavelengths (Fig. 3).

Arpi

Today the archaeological area of Arpi is not an urban area and it is destined for an extensive agricultural activity, where the cereal cultivation represents the main source. The actual landscape conserves only little signs of the ancient city: besides the "toponymy" (San Nicola d'Arpi, Arpetta, Masseria Arpi, Arpinova), the only mark of the ancient city is represented by the wall that delimited the settlement.

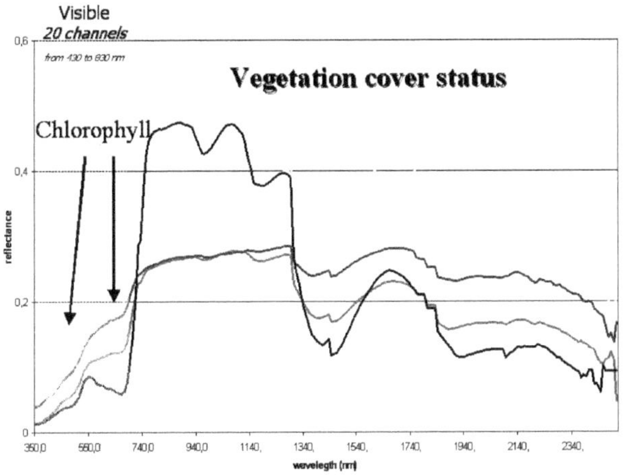

Figure 3 *Examples of vegetation spectral signatures.*

The perimeter of the city is visible as a whitish cord/sting/line placed side by side to a darker one, attributed to the moat. Also, the north side of the settlement was protected by walls, which follow a non linear course, while along the south side, the walls form a semi-circumference that is combined in its two extreme points with the north side (Fig. 4).

The continuity of the walls is interrupted by the presence of various urban gates, still identifiable today. Gate represents one of the elements more feeble in a defensive structure: it requires garrisons, bastions or defence towers.

One structure has been identified along the north side of the walls: it is characterized by one clear trace, placed side by side a dark one, of triangular shape.

This structure can be considered to be a tower or a bastion of defence or a structure of custom and control of the men and the animals of passage towards the water river basin.

On the MIVIS images, many large communication lines were located: more than 15 linear traces are very evident, much darker than the surrounding soil, and bordered by bright traces in both sides that, from the surrounding territory, come into the urban area.

These lines are characterized by their particular width (from 20 to 40-50 meters) and they can be interpreted as *tratturi*, ie the principal artery of the so called "transumanza phenomena" (movement of large herds). These great communication lines meet in the city area, mainly towards the centre of the settlement, where some researchers (Guaitoli 2003, Marin 1970) locate the acropolis and, in the intersection point with the walls, they endure a narrowing, that reorganizes the amplitude of the traces; beyond the walls the trace of the lines is increased (Fig. 4).

These traces are particularly visible on the MIVIS images analysed, and allow us to assume that they have been used before the construction of the walls (sixth century B.C.), since the walls seem to have been erected over them.

From the analysis of the MIVIS images, we also found traces of buildings and anomalies which would imply a particularly complex and dense internal road network. The twisting course of the ways, singled out in more points of the archaeological area, recall the Livio's sentence (XXIV, 46-47) that defined the ways of the Daunian city *tenebris angustisque viis est.*

In more zones of the settlement a series of circular traces of darker tonality to the surrounding soil have been characterized: these circular traces could be connected to the presence of a Neolithic entrenched village (Marin 1970, Mazzei 1995).

There are different traces found outside the city perimeter pertinent to villas, ancient roads or other structures of anthropic origin. Between these an anomaly is marked to

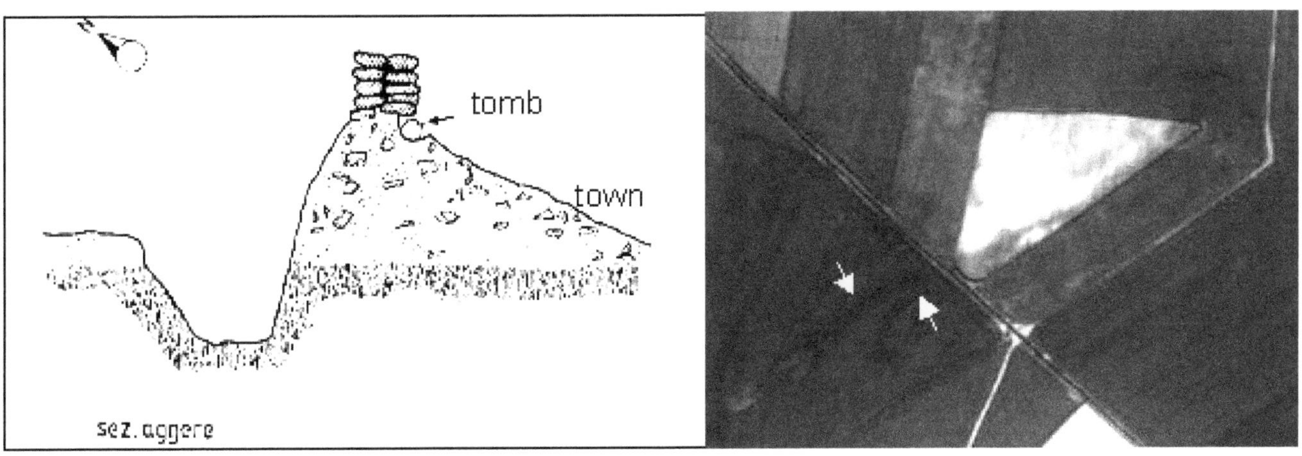

Figure 4 *Arp,i section of walls and NDVI image of the walls.*

the South of the walls: it is a clear trace of rectangular shape, characterized inside by a series of bright features, that represent an element of disturbance in the contest of our study area and could be the survival of one important inhabited structure. Only a punctual and direct control on the study area can confirm the formulated hypothesis.

Sipontum

To the south of the archaeological area the modern Siponto city is risen. The archaeological area was barely affected by the buildings: the romanesque church, risen near of the paleochristian Basilica; the line-keeper's house, situated on the margins of a road called "ex-statale 89"; the "Masseria Garzia" constructed on the remains of the amphitheatre near to the N-W angle of the village and one crumbling structure in the centre of the city area.

Two profound scars cut the city from the N-W to S-E, dividing the ancient system city in three great fields: the ex-statale 89 and the line of railway connection Foggia-Manfredonia constructed towards the end of the XIX century.

A part of the area is destined to intensive agricultural activities (cereals), and another was left uncultivated.

The examination of MIVIS images has allowed us to correct the traditional trapezoidal shape of the city: the reconstructive hypothesis of the S-E side of the walls can be only accepted for the feature to north, while in the south side the walls would have to follow the axial railroad direction, beyond the line that marks the course of the same relief. The walls in that point, centre of a remarkable depression, could not have covered one effective defensive action: in fact it was more logical to construct fortifications on the height than on the feet of it. The N-W side of the city can be seen in a dark trace that slides beyond Masseria Garzia, including the amphitheatre inside of the city area. Such thesis finds confirmation in the recent excavation tests carried out along this side where the rests of the town-walls have been recovered.

Also the hypothesis of reconstruction of the line of walls on the north side of the city would be modified: if in past the researchers claimed that the walls were closed on themselves in order to form a right angle, today by MIVIS Images the reconstruction of Danti (Danti 1999) can be accepted, which suggests that on the northeast and the N-W sides part of walls were oblique: a trace of dark colour can be identified on the MIVIS images (Fig. 5).

During the survey near the amphitheatre some great stone blocks have been recovered which presumably belonged to the town-walls. At last the limit of the city is visible, on the south-western side, delimited by a clear line trace.

Once the problem of the existence and the course of the town-walls was cleared, it is necessary to face the problem of the inner partition of the city. An accurate study of the traces, found inside the city's perimeter, allows us to assume the viability of the city before verifying it in the field by archaeological exploration: in the point where sampling tests have recovered the rests of a probable gate, a clear linear trace that follows a perpendicular direction respect the ex-statale 89 is visible. It can be identified as *cardo maximum*.

The forum has not been localized, while the traces of the two *thermae* are clear (at south of urban area).

The ancient coast line has been identified by MIVIS images. It is visible in different colours and tonality of the soil, and the images explain the relation of *Sipontum* centre with sea: in fact, the ancient port was located in the modern agricultural area.

Figure 5 *Sipontum, walls identified in the NDVI image.*

7 Conclusions

MIVIS hyperspectral data processing enabled us to acquire information useful for the recognition of the ancient topography of Arpi and *Sipontum,* located on the basis of anomalies due to the variations of textures, humidity and vegetation of the earth's surface caused by the presence of buried human activities.

The anomalies, correlated to the archaeological structure, have been characterized on the bases of variations in the state of the vegetational cover and only indirectly from the soil moisture.

The anomalies found in areas covered by vegetation, studying images derived from vegetation indices, have been extracted making to stand out the minimal spectral variations of the various states of the vegetation.

The obtained results confirm and sometimes change archaeological hypotheses based on traditional investigation methods. They must obviously be ascertained directly *in situ*, in order to formulate a precise strategy for future excavation campaigns.

References

AVANZI G., R. BIANCHI, R.M. CAVALLI, L. FIUMI, C.M. MARINO AND S. PIGNATTI 1996 Use of MIVIS navigation data for precise aircraft positioning and attitude estimation. *SPIE 2960*, pp. 184-192.

BANNARI A., D. MORIN AND F. BONN 1995 A review of Vegetation Indices. *Remote Sensing* 1995, Vol 13, pp.95-120.

BASSANI C.,. R.M. CAVALLI, A. PALOMBO AND S. PIGNATTI 2002 La calibrazione dello scanner MIVIS (Multispectral Imaging Visibile and Infrared Spectrometer) nelle campagne di misura realizzate dal CNR nel 2001-2002. *Atti VI Conferenza Nazionale ASITA "Geomatica per l'ambiente, il territorio e il patrimonio culturale"*, vol. I, p. 337.

BIANCHI R., R.M. CAVALLI, L. FIUMI, C.M. MARINO AND S. PIGNATTI 1996 Airborne imaging spectrometry: a new approach to environmental problems. *Proceeding of the XVIII ISPRS* Congress. Vienna 9-19 July 1996, vol. I, pp. 128-132.

BIANCHI R., C.M. MARINO AND S. PIGNATTI, 1994 Airborne Hyperspectral remote sensing in Italy. *Europto*. Rome 27-29 September 1994, *SPIE* V 2318, pp. 29-37.

BRADFORD S. 1957 The ancient city of Arpi in Apulia. *Antiquity, XXXI*, pp.167-169.

CAMPBELL J.B. 1987 Introduction to Remote Sensing. *The Guilford Pres*. New York.

CAVALLI R.M. AND S. PIGNATTI 2001 Il telerilevamento iperspettrale da aereo per lo studio dei Beni Archeologici: applicazione dei dati iperspettrali MIVIS. *Remote Sensing in Archeology, XI Ciclo di lezioni sulla ricerca applicata in archeologia*, Certosa di Pontignano (Siena). 6-11 dicembre 1999, pp. 221-232.

CAVALLI R.M., P. MEROLA, S. PIGNATTI AND M. POSCOLIERI 2005 Telerilevamento iperspettrale MIVIS per lo studio delle testimonianze antropiche nell'area archeologica di Arpi (FG). *Rivista Italiana di Telerilevamento* n. 33/34, Firenze, pp. 109-117.

DANTI A. 1999 L'area portuale. *Siponto antica*. Foggia.

GONG DAO-YI and SHI PEI-JUN 2003 Northern hemispheric NDVI variations associated with large-scale climate indices in spring. *International Journal of Remote Sensing* 24, pp. 2559-2566.

GUAITOLI M. 2003 Arpi. *Lo sguardo di Icaro: le collezioni dall'Aerofototeca Nazionale per la conoscenza del territorio (catalogo della mostra)*, Roma.

HOLBEN B.N. and C.O. JUSTICE 1981 An examination of spectral band ratioing to reduce the topographic effect on remotely sensed data. *International Journal of Remote Sensing,* 2. pp. 115-133.

HUETE A.R. 1984 Soil spectral effects on vegetation discrimination. *Ph. D. Thesis*. Department of Soils, Water and Engineering; University of Arizona, USA.

HUETE A.R. 1989 Soil influences in remotely sensed vegetation-canopy spectra. *Theory and Applications of Optical Remote Sensing*. Wiley, Washington, USA, pp. 107-141.

LIVIO XXXIV, 45: <*Sipontum item in agrum qui Arpinorum fuerat coloniam civium Romanorum alii triunviri D. iunius Brutus, M. B. Tamphilus, M. Helvius deduxerunt*>.

MAJOR D.J., F. BARET AND G. GUYOT 1984 A ratio vegetation index adjustied for soil brightness. *International Journal of Remote Sensing 11 (5)*, pp. 727-740.

MARIN M.D. 1970 Topografia storica della Daunia antica. *Civiltà della Daunia*, pp. 64-68.

MAZZEI M. 1995 Arpi. L'ipogeo della Medusa e necropoli. Bari.

MAZZEI M. 1999 Siponto antica. Foggia.

MAZZEI M. 2003 Siponto. *Lo sguardo di Icaro: le collezioni dall'Aerofototeca Nazionale per la conoscenza del territorio (catalogo della mostra)*. Roma.

MEROLA P. 2004 Tecniche di telerilevamento iperspettrale applicate alla ricerca archeologica. Il caso di Lilybaeum (Marsala). *Archeologia Aerea. Studi di Aerotopografia Archeologica*. pp. 301-318, Roma.

MEROLA P., S. BAJOCCO AND A. ALLEGRINI 2005 Hyperspectral MIVIS data to investigate the Lilybaeum (Marsala), Archaeological Park. *Proceedings of SPIE (International Society for Optical Engineering): Remote Sensing for Environmental Monitoring, GIS, Applications and Geology V*, Vol. 5983, 59830W.

PICCARRETA F. and G. CERAUDO 2000 Manuale di aerofotografia archeologica. Metodologia, tecniche e applicazioni. Italia.

SCHMIEDT G. 1973 Contributo della fotografia aerea alla ricostruzione dell'antica laguna compresa fra Siponto e Salaria. *Archeologia Storica Pugliese XXVI*.

Palaeohydrography and ancient settlements in the Adige river plain, between Rovigo and Adria (Italy)

Silvia Piovan, [1] Raffaele Peretto, [2] Paolo Mozzi[1]

[1] Department of Geography – University of Padova,
[2] Museum of Grandi Fiumi – Rovigo

1 Introduction

This research aims to study the relationship between man and the environment in the Adige and Po fluvial systems and their influences on the natural and artificial landscapes during the centuries. Case studies represent two good examples in which these relations produced particular morphologies and anthropic settlements of the past that can firstly be studied with remote sensing. Field survey, stratigraphical study and archaeological survey have been driven by photo aerial analysis and interpretation.

The Villadose case investigates a situation in which the fluvial landforms influenced the choice of an ancient settlement. From the archaeological viewpoint the area is of great interest due to the presence of important Bronze and Iron Age sites and the existence of an exceptionally well-preserved Roman centuriation, which extends 250 km^2 from the city of Rovigo almost to the Lagoon of Venice (Figure 1).

The Pettorazza case is an example in which man modifies the natural fluvial system once it has become a danger with its avulsions and crevasses. In order to clarify the stratigraphy of the alluvial ridge of the Adige river which crosses the area between Rovigo and Adria, a transect of boreholes has been made in the Pettorazza site.

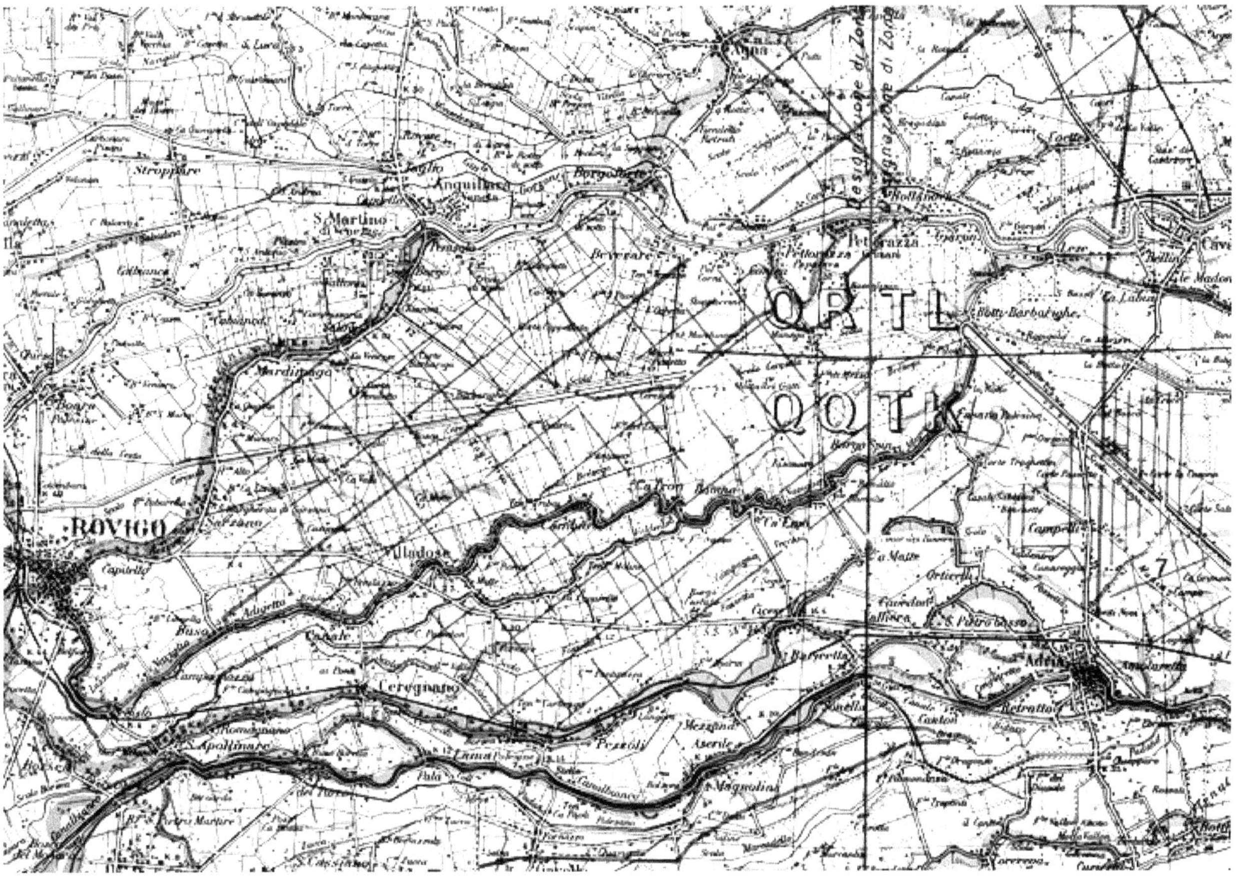

Figure 1 *Palaeohydrography and ancient anthropic structures in the east aerea of Rovigo (Peretto, 1986)*

2 Geological and geomorphological context

The Adige river is the second longest Italian river (410 km) and the third by catchment area (12.200 km²). It enters the Venetian plain near the city of Verona, after following a long, north-south alpine valley cut in carbonatic and porfiric rocks. During the Last Glacial Maximum the Adige valley was occupied by a glacier down to the valley mouth. In the so-called "high plain", near the Alpine piedmont, the river is of the braided type and the deposits are manly gravels and sands. Downstream, in the "low plain", it is meandering and the sediments are finer. The sandy-silty-clay Adige alluvial plain interfingers with the adjacent sedimentary systems of the Po, Tartaro and Brenta rivers. During the 16th century the Venetian Republic started a land reclamation program with embankments and artificial cut-offs. Recurrent floods, avulsions and crevasses which characterized the Adige have greatly diminished in the last centuries due to river management (Turri, 1992; Peretto, 1992).

The study area is delimited respectively to the north and to the south by the Adige and Po rivers, between the cities of Rovigo and Adria. This stretch of low plain consists of Late Holocene fine sediments. Some portions belong to the Adige system while others to the Po one.

3 Archaeological context

Late Bronze Age (12th-9th century BC) archaeological sites have been discovered in Sarzano and Saline, both situated along the northern branch of the "Po di Adria" alluvial ridge. Also Late Bronze Age-Early Iron Age sites of Frattesina, near Fratta Polesine and of Villamarzana are important in the Adige-Po plain. These latter are locatated along the "Po di Adria" alluvial ridge at the west of Rovigo (AA.VV., 1986; Bietti Sestieri, 1990; Arenoso Callipo, Bellintani, 1996), out of the study area.

Adria was the most important site of the 6th-5th century BC: it was a relevant commercial and social etrusco-padano centre. Le Balone di Rovigo, Borsea Romanina di S. Cassiano, Cantone di Crespino, Gavello, S. Basilio, Ca' Zen di Taglio di Po, Vallone di Loreo sites date back to the same period (Peretto 1994).

In the second part of the 2nd century BC the area surrounding the city of Adria was struck by Roman expansion that embraced all of the north-eastern Venetian Po plain (Masiero, 1999). As usual, romanization processes started with the building of roads (Postumia, Popilia, Annia). Roman politics encouraged friendly indigenous population or Roman citizens (for example ex-legionnaires) to settle in this area. An efficient road web allowed communications between biggest cities and small villages so that relationships increased. Besides, a heavy land restructuring program was planned in order to fully exploit the potential of the rural landscape.

A web of *limites*, perpendicular to each other, was imposed to create the *centuriae*, from which the singles plot for colonials were taken out. Adria centuriation extends 250 km² from the city of Rovigo almost until the Lagoon of Venice (Peretto, 1986). Each *centuria* had a side of 27 *actus* equivalent to about 965 m (mean value): this measure is not usual: in other centuriation areas, each *centuria* had a side of 20 actus (about 710 m). This is probably due to the particular geomorphological and palaeoenvironmental conditions; on the other hand, Roman rural subdivision morphology was very differentiated (Peretto, 1986).

The most important road of the centuriation area between Rovigo and Adria is the *decumanus maximus* called "via di Villadose" (Peretto, 1986; Maragno, 1993), revealed for the first time by Rodolfo Peretto through the interpretation of an aerial photo (1967). The track is visible by vertical and oblique aerial photos and sometimes in the field in particular condition of soil humidity and sunlight, because of the presence of two lateral ditches (Figure 2).

Figure 2 Decumanus maximus *near Villadose. Low altitude aerial photo by Raffaele Peretto.*

The road has been measured by aerial photo and by stratigraphical analysis: the roadway is 19-26 meters wide or 30-34 meters if we consider the ditches (Peretto &

Zerbinati, 1984). Systematic archaeological survey of all the area, performed by Gruppo Archeologico di Villadose (since 1988) revealed the presence of a Roman Age villa very close to "via di Villadose" in Ca' Motte. Archaeological studies performed by Prof. Facchini's team (University of Verona) in the area are still in progress (Facchini & Pisano, 2003).

The *decumanus maximus* was, for Bottazzi & Calzolari (1990), a part of the track which joined via Emilia Altinate with via Popilia and via Annia. Via di Villadose crosses via Annia about 1 km north from the Adige river, south-est of the town of Agna.

Populating of the centuriation area continued until the Roman Empire was divided into Western and Eastern parts, during the 5th century AD. In that period the Empire came to decline. The absence of maintenance of rivers, channels and lands and a period commonly believed to be characterized by "climatic deterioration" (Early Middle Age) (Castiglioni, 2001), caused a progressive swamping of the area.

Since 16th century AD noble Venetian landowner reclaimed the improvement of the territory. The government of Venetian Republic started a reclamation program in the area, in order to improve the landscape conditions. In 18th century many meander artificial cut-offs and embankments of Adige river were made, in order to decrease the frequency of disastrous breaches (Peretto, 1992). One of the most important cut-offs of the meander is the Pettorazza Grimani case, carried out in 1783.

4 Methodologies

The investigation takes advantage of several methods: microrelief analysis, remote sensing and boreholes down to the depth of 4-6 m with the Edelman hand auger. Microrelief analysis is one of the main tools for a geomorphological study in a flat land.

A digital elevation model of the investigated area has been developed (Figure 4), based on contour lines with the spacing of 0.5 m. For the construction of the contour lines, a manual interpolation of spot heights of the "Carta Tecnica Regionale del Veneto" at scale 1:5000 has been adopted. This method allows to discard the points that are related to present-day artificial structures and to reconstruct the natural swale-and-ridge systems.

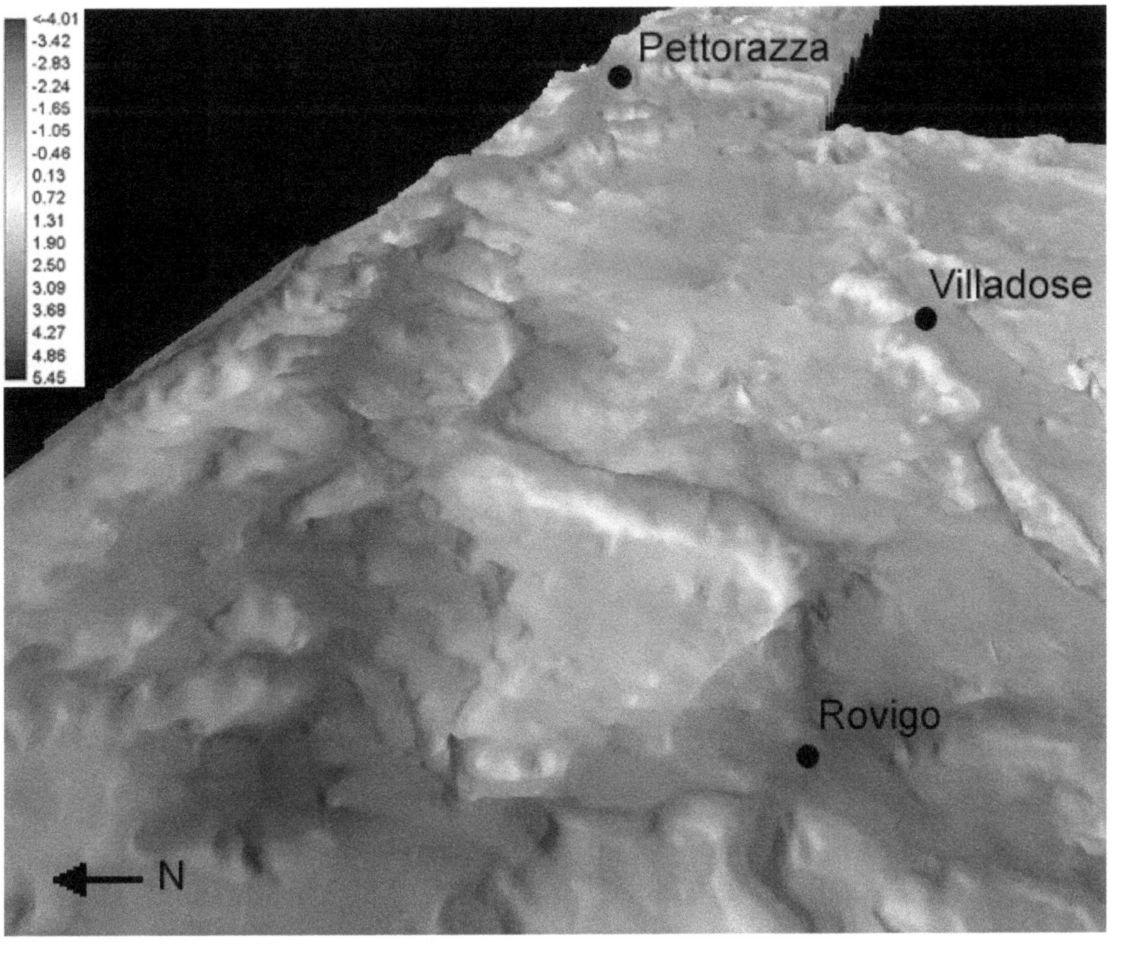

Figure 3 *Digital Elevation Model of the studied area.*

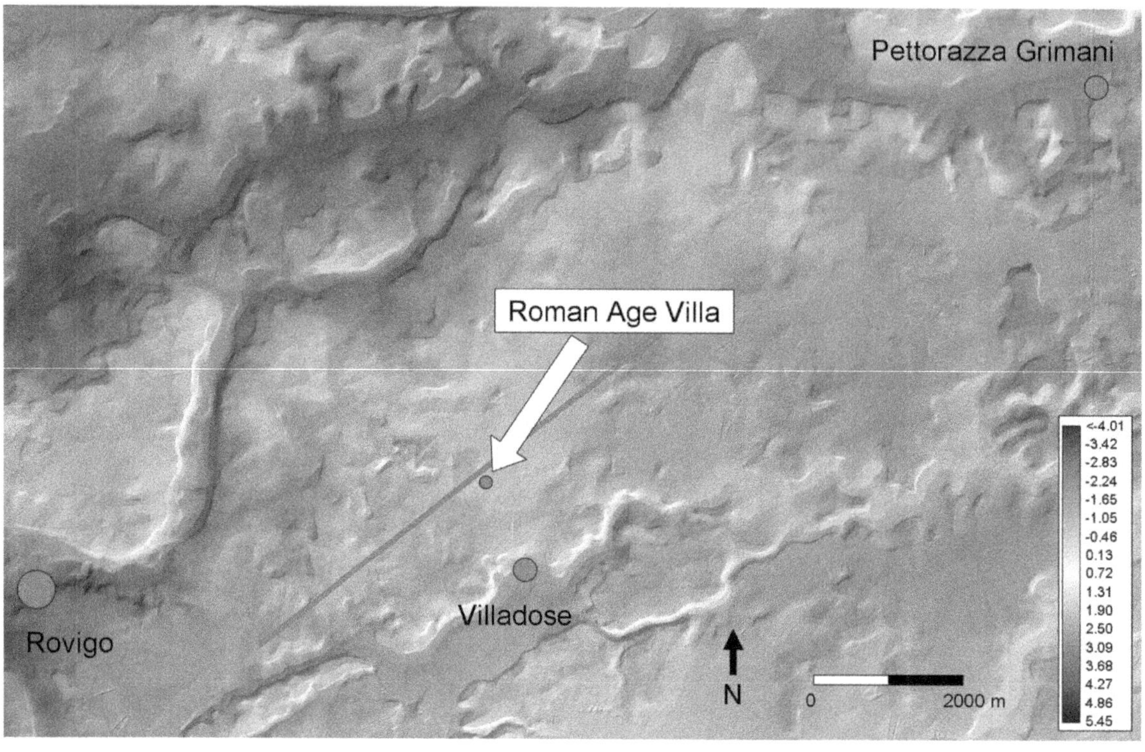

Figure 4 *Digital Elevation Model of the area between Rovigo and Adria. Decumanus maximus is represented by the straight line near Roman Age Villa.*

Landsat TM images (Figure 5) have been used for the analysis of major landforms while photo-aerial interpretation has been essential for finer detection of natural morphologies and anthropic structures.

Both oblique and vertical panchromatic aerial photographs were used to map the palaeochannels and crevasse splays, and to analyse their interaction with ancient roads, ditches and field patterns.

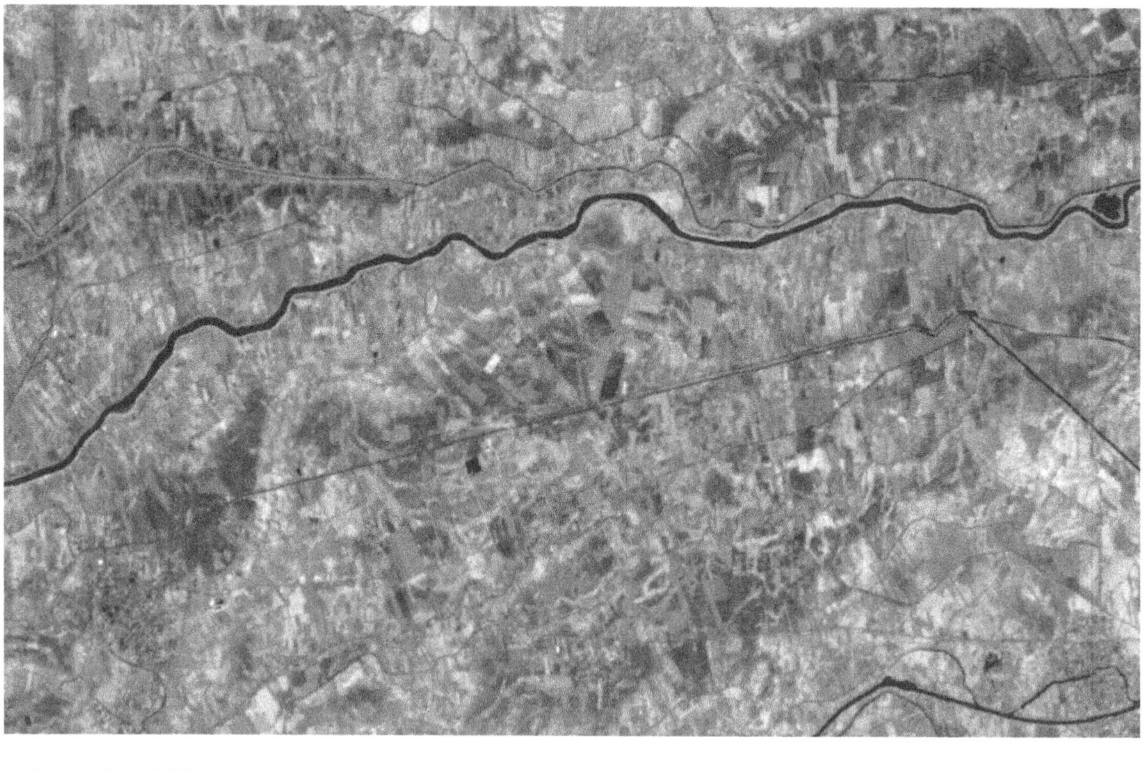

Figure 5 *Landsat 5 TM image of the area between Rovigo and Adria, along Adige river. At left, the trace of alluvial ridge of the "Po di Adria" palaeochannel.*

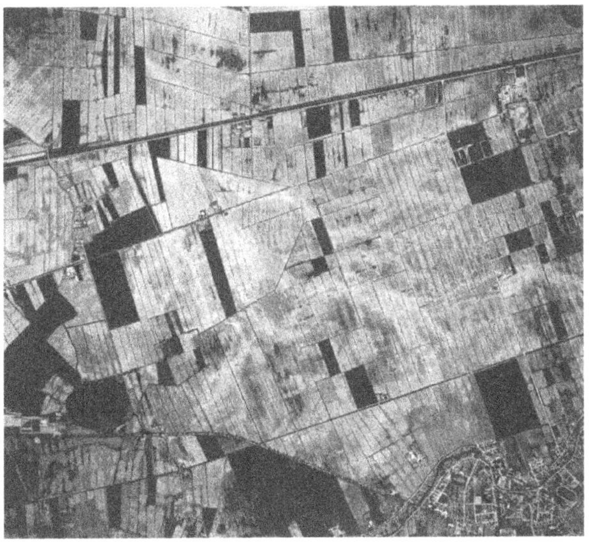

Figure 6 *Aerial photo of crevasse-splay in Villadose.*

Figure 7 *Aerial photo of Pettorazza palaeomeander.*

In the Villadose case, aerial photos (Figure 6) and microrelief indicated the convenience of drilling a set of boreholes across a crevasse splay placed north of the *decumanus maximus*, nowadays called "via di Villadose" (Figure 8). The cross section shows the interactions between the crevasse splay and the Roman road; it also clarifies the environmental conditions before local centuriation took place.

Pettorazza Grimani meander site was chosen because it is a good example of an alluvial ridge that is not influenced by anthropic dikes (the meander cut-off dates back to 1783). Photo aerial interpretation (Figure 7) suggested the direction of the palaeomeander cross-section: the transect includes the whole alluvial ridge and it is as much as possible perpendicular to the palaeochannel (Figure 9).

Radiocarbon dating of the backswamp peats sampled in Pettorazza, currently under way, will help to date the major avulsion which, during the Middle Ages, led the terminal stretch of the Adige river to follow the present direction.

5 Discussion

Villadose

The cross-section across the crevasse-splay shows three main sand bodies, interpreted as crevasse channels deposit (Figure 10). They incised a more ancient alluvial plain, characterized by the presence of a peat layer ca. 40 cm thick, at the average depth of 4 meters. The presence of these crevasse deposits is in accord with the photoaerial interpretation and the terrain morphology, characterized by a relief about 1.5 m higher than the surrounding plain.

The Roman villa is located very close to the *decumanus maximus*, near the top of the crevasse-splay. V02 core has been drilled nearby. The archaeological site sits over the sand deposit that corresponds to a crevasse channel.

Pettorazza Grimani

The cross section of the meander shows a complex sedimentary history characterized by different depositional environments. Periods of aggradation due to the upbuilding of fluvial ridges and crevasse splays alternated with periods when overbank clays and peats accumulated with lower depositional rates.

The stratigraphic section (Figure 11) has revealed one main central sand body at least 6.5 m deep that can be interpreted as the main channel deposit.

After the Adige river changed its path from north to a more southern area, both parts of the south-western Venetian Plain, the river incised a more ancient alluvial plain, characterized by the presence of two peat layers around 3 and 4 m depth levels.

Field survey, driven by historical cartography and aerial photo interpretation, allows to reconstruct the palaeogeography of the meander and to collect geomorphological data on the Adige palaeochannel in Pettorazza. The anthropic landmarks, recognized during the field survey, allow us to establish where the palaeochannel flowed with better precision in comparison to the historical maps.

Following is the list of landmarks (see the numbers in Figure 9).

1. Outer natural levee relief, about 3 m higher than surrounding plain.

2. "Border stone" (16th century) which signified the border between the provinces of Padova and Venice (Antonio Litamè, interview).

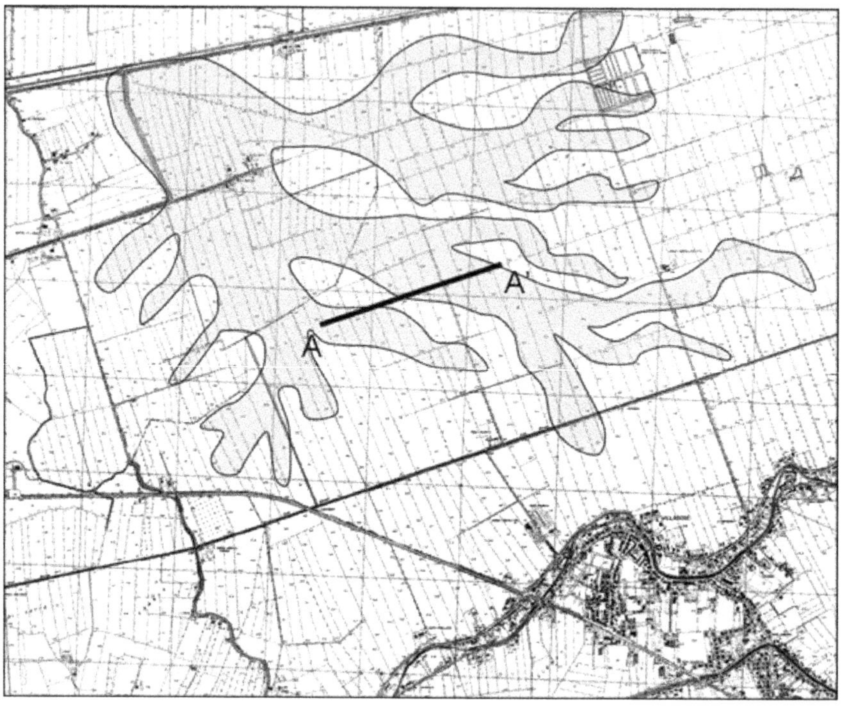

Figure 8 *Schematic photoaerial interpretation of Villadose crevasse-splay area. AA' indicates the cross section*

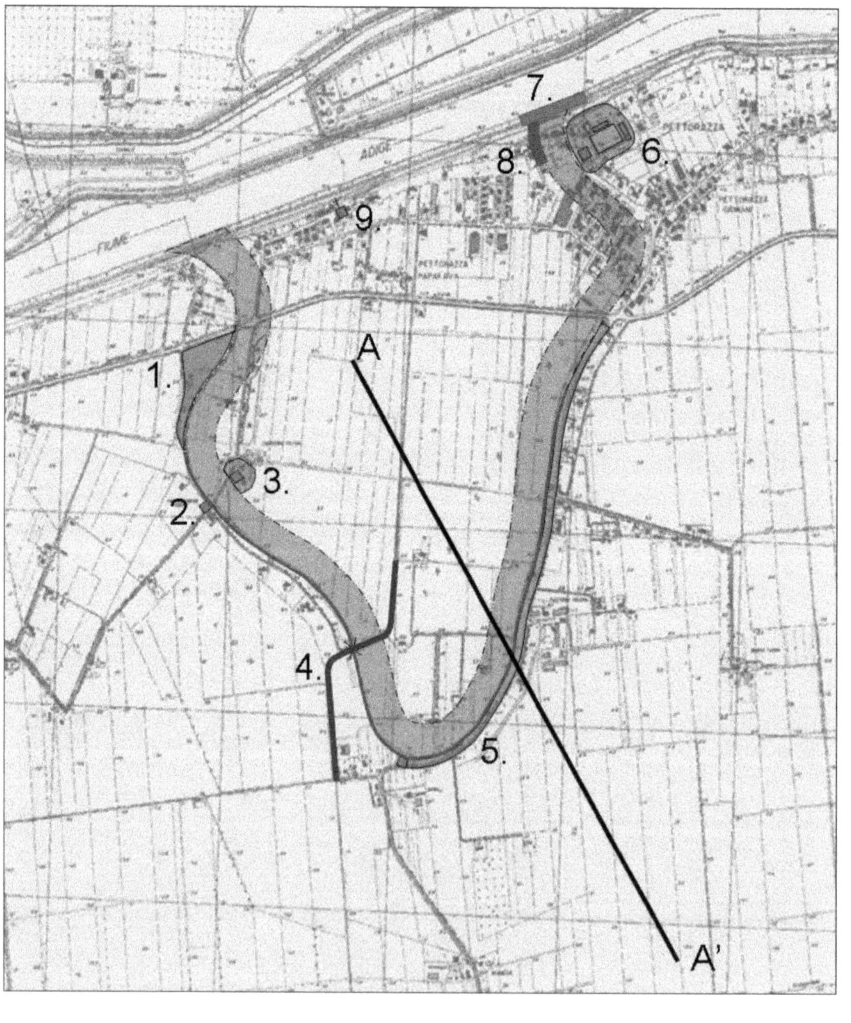

Figure 9 *Palaeogeographical map of Pettorazza palaeomeander. AA' indicates the cross-section.*

3. Living house at the present day that was an hydro-depot and boats' dock since 15th century: in 1782 it was turned into a horseshed during the cut-off works. Beneath the present ground floor there are three brick arcs on which the house lies (Antonio Litamè, interview). Old iron rings hung on the arcs column have been also found. It was built about 3 m above the alluvial plain.

4. Transversal ditch about 3 m wide and 90 m long from the bridge to the inner bend. It was also present in the 1783 cut-off map.

5. Outer levee relief, about 3 m higher than surrounding plain.

6. Grimani Villa (16th century - Venetian Republic). It was built on the right bank very close to the water. There is historical evidence of a pier in front of the Villa (Michele Fugalli, interview; web site http://www.pettorazza.it).

7. Anthropic embankment 100 m long, about 3 m high.

8. Remains of ancient ditch now partially buried due of urbanization. In current land records it is called "Scolo pubblico" that means "Public drainage". It begins from the artificial embankment of point 7.

9. Before the cut-off, Pettorazza was divided in two villages: Pettorazza Grimani (Venetian province) and Pettorazza Papafava (Padova province). Now there is a memorial tablet of the cut-off event (1782-83) on the bell-tower of Papafava. The origin of the church dates back to 1691 AD (web site http://www.pettorazza.it).

Geomorphological data about Adige palaeochannel in Pettorazza:

- Width: 90-100 m.
- Circumference lenght: 3,700 m.
- Width of the alluvial ridge: nearly 500 m.
- Evidence of crevasse splay out the outer levee.

The radiocarbon dating of the backswamp peats, gives a conventional age of 920+/-60 years BP, corresponding to 1010-1260 AD calibrated with a confidence of 95%. This dating helps to date the major avulsion which, during the Middle Ages, led the terminal stretch of the Adige river to follow the present direction.

6 Conclusions and improvements

In the Villadose case, the results suggest that the position of the ancient settlement is not casual. In fact, the settlement is located in a place characterized by high geomorphological condition and good drainage. The cross section indicates also that the *decumanus maximus* lies over the crevasse splay deposit.

In Pettorazza Grimani case, the artificial cut-off of the meander modified the palaeogeography of the area in order to improve the hydrographical condition. However, landmarks survived and have been recorded by field survey.

An improvement in the definition of the position and the geometry of the palaeochannel was carried out with better precision than the historical maps.

The cross-section of the palaeomeander allowed to clarify the sedimentological and stratigraphical characteristics of the alluvial ridge of the Adige river and its relation with the plain.

In the future, other selected peat layer samples will be radiodated in order to date the crevasse-splay event in Villadose and to define with better precision the avulsion that caused the led of the Adige river in the southern part of the plain. Further boreholes are being planned, in order to improve the resolution of the stratigraphy. Petrographical analysis of selected sand samples are under way in order to attribute the crevasse to a specific palaeo-river is in course.

Acknowledgements

- Prof. G. M. Facchini, Department of "Discipline storiche, artistiche e geografiche" – University of Verona.
- Mr. F. Fermon, photogrammetry laboratory IRPI – CNR Padova.
- Dr. F. Ferrarese, Department of Geography – University of Padova.
- Mr. M. Fugalli, Social Service – Pettorazza Grimani.
- Mr. A. Litamè and Mrs. L. Borella for interview about memories of Pettorazza.
- Dr. E. Maragno, Gruppo Archeologico di Villadose.
- Dr. A. Ninfo, Department of Geography – University of Padova.
- Prof. C. Stefani, Department of Geology – University of Padova.

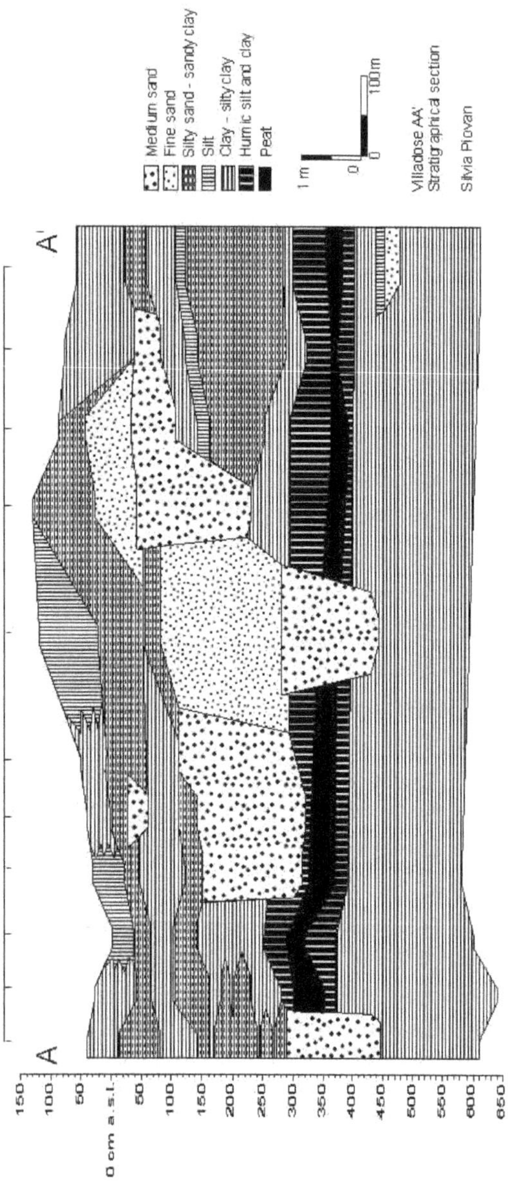

Figure 10 *Stratigraphic cross-section of Villadose crevasse-splay.*

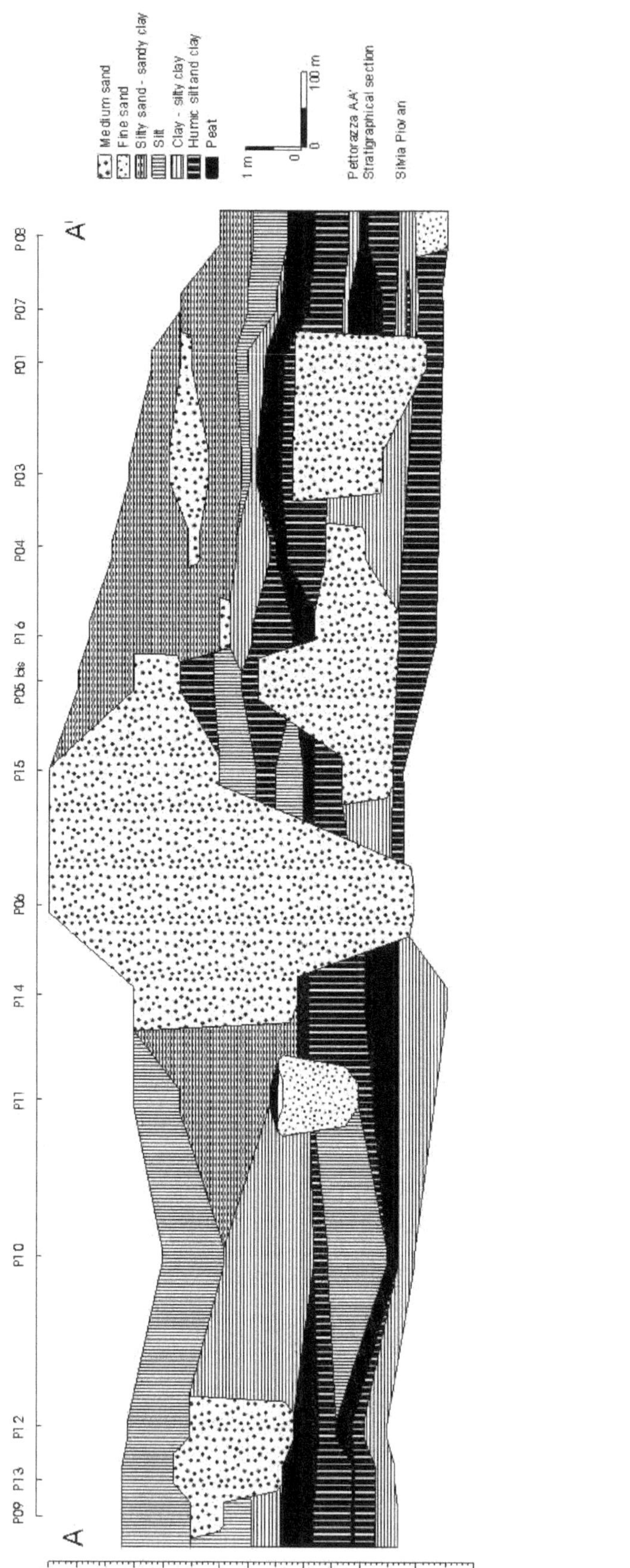

Figure 11 *Stratigraphic cross-section of Pettorazza Grimani palaeomeander.*

References

AA.VV. 1986 – *Preistoria e protostoria nel Polesine*. PADUSA, 549 pp.

ARENOSO CALLIPO C.M.S., BELLINTANI P. 1996, *Dati archeologici e paleoambientali del territorio di Frattesina di Fratta Polesine (RO) tra la tarda età del Bronzo e la prima età del Ferro*, PADUSA, XXX.

BIETTI SESTIERI A.M. 1990, *La campagna di scavo 1989 nell'abitato protostorico di Frattesina di Fratta Polesine*. Quaderni di Archeologia del Veneto, VI.

BOTTAZZI G. & CALZOLARI M. 1990, *Vicus Varianus* (Vigarano) e la strada romana dal Modenese ad Este. In: *"Quaderni della Bassa Modenese"*, IV, 1, pp. 11-24.

CASTIGLIONI G. B. 1991, Le risposte del sistema fluviale alle variazioni ambientali. In: Castiglioni G. B. & Pellegrini G. B. (edited by) "Note illustrative della carta geomorfologia della Pianura Padana". *Supplementi di Geografia Fisica e Dinamica Quaternaria, suppl.* IV-2001, pp. 178-182.

FACCHINI G. M., PISANO A. 2003, Scavi archeologici dell'Univerisità degli Studi di Verona a Villadose (RO): osservazione sui materiali rinvenuti nell'US 107. *PADUSA, Bollettino del Centro Polesano di studi archeologici ed etnografici Rovigo, Anno XXXIX Nuova Serie, Istituti Editoriali e Poligrafici Internazionali*, Pisa-Roma, pp. 143-144.

MARAGNO E. 1993, La mostra archeologica e la centuriazione. In: Maragno E. (edited by) *"La centuriazione dell'agro di Adria"*, Linea AGS Edizioni, Stanghella, pp. 10-45.

MARCOLONGO B. 1987, Ricostruzione paleoidrografica attraverso interpretazione di immagini telerilevate. In: Marcolongo B. (edited by) *"Paleoidrografia tardoquaternaria della pianura Veneta sud-occidentale e il suo significato in una ricostruzione paleoambientale"*, Grafiche Erredici, Padova, pp. 6-8.

MASIERO E. 1999, La strada *in levada* nell'agro nord-occidentale di Adria. In: *"Journal of Ancient Topography"*, IX, Roma, pp. 107-120.

MIALL A.D. 1996, *The geology of fluvial deposits: sedimentary facies, basin analysis, and petroleum geology*. Springer-Verlag, Berlin, 582 pp.

PERETTO R. 1986, Ambiente e strutture antropiche nell'antico Polesine. In: *"L'antico Polesine, testimonianze archeologiche e paleoambientali, Catalogo delle esposizioni di Adria e Rovigo, Febbraio-Novembre 1986"*, Antoniana, Padova, pp. 21-100.

PERETTO R. 1992, La bassa pianura. In: Turri E. & Ruffo S. (edited by) *"L'Adige il fiume, gli uomini, la storia"*, Cierre Edizioni, Verona, pp. 71-85.

PERETTO R. 1994, La scoperta del paesaggio. Il territorio tra Protostoria e Romanità. In: *"Balone. Insediamento etrusco presso un ramo del Po"*, Padova.

PERETTO R. & ZERBINATI E. 1984, Aspetti del popolamento in età romana tra Bassa Padovana e Polesine. Gli interventi dell'uomo sul territorio. In: Lucani Alberti *et al.* (essay) *"Territorio e popolamento in Bassa Padovana"*, Museo Civico Etnografico, Stanghella.

TURRI S. 1992, Lo spazio atesino. In: Turri E. & Ruffo S. (edited by) *"L'Adige il fiume, gli uomini, la storia"*, Cierre Edizioni, Verona, pp. 3-19.

Surface modelling of complex archaelogical structures by digital close-range photogrammetry

Gabriele Bitelli,[1] Valentina Alena Girelli,[1] Fabio Remondino,[2] Luca Vittuari [1]

[1]DISTART Dept. – University of Bologna, Italy – (gabriele.bitelli, valentina.girelli, luca.vittuari)@mail.ing.unibo.it
[2]Institute of Geodesy and Photogrammetry – ETH Zurich, Switzerland – fabio@geod.baug.ethz.ch

1 Introduction

Surveying and representation of archaeological sites and objects represents today a very interesting contest where the potentialities of the new digital technologies of Geomatics can be fruitfully expressed. The tackled problems span from the data acquisition and integration processes to the automatic or semi-automatic data handling, or to new methods for data representation and exploration. One of the most interesting questions in this sense is related to the 3D modelling of surfaces and complex archaeological structures, where different problems can of course emerge with respect to modelling of terrain or of simple artificial objects.

In archaeology, the use of 3D models for documentation and visualization purposes is generally applied in few case studies for different reasons: (i) the high "cost" of 3D; (ii) the difficulties of non-experts in achieving easily good 3D models; (iii) the consideration that 3D is mainly an additional "aesthetic" factor; (iv) the difficulty to integrate 3D worlds with other classical 2D data.

Currently we can distinguish three main approaches for the recording, documentation and visualization of cultural heritages sites and objects: (1) image-based methods (Remondino and El-Hakim, 2006) (2) range-based methods (Blais, 2004; Böhler, 2005) (3) a combination of image- and range-based methods (Beraldin et al. 2002). The requirements specified for many applications, including digital archiving or mapping, involve high geometric accuracy, photo-realism of the results and the modeling of the complete details, as well as some automation, low cost, portability and flexibility of the technique. Therefore, selecting the most appropriate 3D modeling technique to satisfy all requirements for a given application is not always an easy task.

The paper describes an experience carried out in the framework of an archaeological activity, where various geomatic techniques were applied and integrated, tying together disciplines such as topographical surveying, geophysics, photogrammetry, remote sensing, etc. During the work carried out at the archaeological site of Tilmen Höyük, in Turkey, the approach was to firstly define and

Figure 1 *Geographic location (on colour composite imagery and on SRTM surface model) and a picture of the site.*

maintain a common shared reference system among the different surveying activities and in that realize the data acquisition phase. During the 2005 campaign some experiences were realized in close-range photogrammetry using non-metric cameras, with subsequent data modelling and representation. Aim of the paper is to show some results and discuss problems resulting from the field campaign and data processing.

2 The archaeological site of Tilmen Höyük

The Archaeological Mission of the University of Bologna in Turkey, directed by N. Marchetti, started in 2003 at the ancient town of Tilmen Höyük (Marchetti, 2004). The area of investigation (Fig. 1) is an ancient settlement located 10 km East of Islahiye town within Gaziantep province in the South Eastern Turkey and dating back to 3400 BC. The palace complex with a temple, inner and outer strong defence walls surrounding the city, bring Tilmen Höyük to an important position from an archeological point of view not only in the region, but in the whole country.

In order to produce a complete and accurate metric survey of the site, a multi-disciplinary approach was choosen, also with the aim of supporting all the research activities in the area (archaeological, geophysical, agrarian, etc…). The surveying activity involved different methodologies simultaneously applied.

The first problem adressed was the absolute georeferencing of the site, performed by GPS. Long period GPS observations were conducted using dual frequency Trimble geodetic receivers on a reference point, named ST01, signalized on the site hill. This allowed the connection of this vertex to the IGS permanent stations of Mersin and Ankara, in order to insert every other local reference point or survey into International Terrestrial Reference Frame ITRF2000 (Fig. 2). In particular, the vertexes coordinates of the polygonal were determined by fast-static GPS method and used by the archaeologists for the survey of the structures with total station and for further geophysical investigation. A kinematic survey was besides conducted to complete the description of the site surface morphology, mainly accomplished by the archaeologists by total station surveying (a local geoidal undulation value was necessary for combining the two kinds of heights). The resulting DTM is of very high detail, describing with good accuracy the natural morphology of the site.

The results from this activity were also used to perform a orthorectification of high resolution satellite imagery. A panchromatic QuickBird image (DigitalGlobe Inc.) was orthorectified by adopting associated Rational Polinomial Coefficients (RPC), with ground control points and using a local Digital Terrain Model. The last was realized for the archaeological site using the data acquired during the Mission (total station and GPS) and for the outside territory using the SRTM surface model. A DTM accomplished by ASTER imagery, by means of the stereoscopic capabilities of the VNIR bands, was also considered. The projection chosen is UTM (37) - datum WGS84. A little portion of the digital orthophoto of the ancient town is shown in Fig. 3, where is evidenced the location of the terrestrial photogrammetric survey described in the next paragraphs.

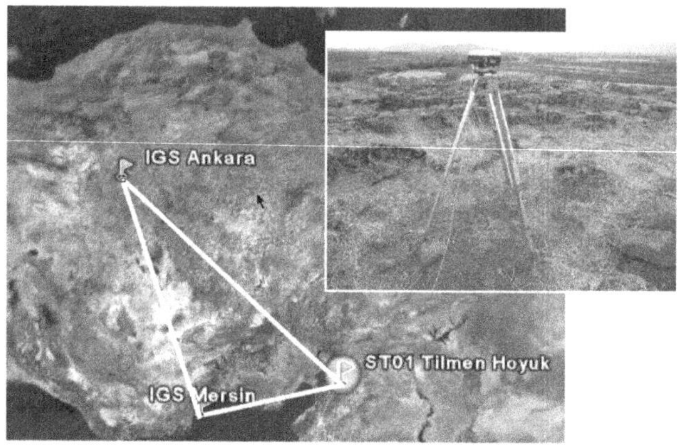

Figure 2 *The reference point and its connection to the near GPS permanent stations.*

In order to experiment the potentiality of image-based methods for the metric documentation and the 3D reconstruction of interesting objects, a close-range photogrammetric survey of some structures was also conducted during the 2005 mission.

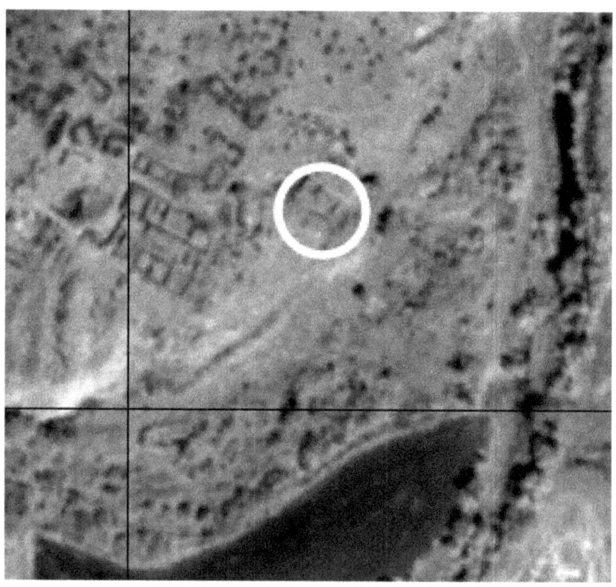

Figure 3 *Panchromatic orthophoto from QuickBird imagery with location of the structures subject to the close-range photogrammetric processing.*

The images were acquired by means of some amateur digital cameras: Canon EOS350D (8 Mpixel) and Nikon Coolpix 5400 (5 Mpixel). The choice of using this type of cameras is often justified in archaeological contexts by the cheapness, easiness and manageability in their use, but leaves open the important problem of camera

calibration, that is directly related to the realization of an accurate metric object restitution and the problem of detailed object reconstruction.
The next paragraphs report about the photogrammetric data processing and its problems.

3 Digital close-range photogrammetric modeling

The generation of textured 3D model of objects has nowadays become a very important research field not only for industry or robotic, but also for Cultural Heritage applications, for the completeness of metric and descriptive informations that a product of this type can offer to studiouses.

3D modeling of a scene should be meant as the complete process that starts with the data acquisition and ends with a virtual model in three dimensions visible interactively on a computer. The interest in 3D modeling is motivated by a wide spectrum of applications, such as animation, navigation of autonomous vehicles, object recognition, surveillance, visualization and documentation (Remondino & Zhang, 2006).

A 3D virtual model of an object, textured with a digital photo, could be considered an evolution of the orthophoto; if the generation of the model is performed by means of photogrammetry, the data process passes through the same steps and encountered the same problems (camera calibration and image orientation) at least until the production of the object DSM.

In the last years different solutions for image-based 3D modeling have been developed. Most of the current reliable and precise approaches are based on semi-automated procedures. In fact, even if the introduction of automated algorithms is a key goal in the photogrammetric and vision communities, fully automated surface reconstruction methods seem not able to achieve a high level of metrical accuracy, a key factor in evaluating this kind of products together with the capability of reconstruct all the object details. The user interaction, in term of point measurements, image triangulation or editing, is therefore so far the most important factor to achieve a precise and reliable 3D model.

Photogrammetry is nowadays much more frequently used to generate computer 3D models, for its generally easy and cheap image data acquisition procedure as well as for the availability of a wide array of new technologies to support the generation and analysis of such models.

On the other hand, data processing is still a difficult task, in particular if the images are uncalibrated or are acquired under a non-conventional geometrical configuration, or the object is very complex, as often in case of archaeological ruins or buildings. One of the critical phases is certainly the camera calibration process (Remondino & Fraser, 2006) and the detailed surface measurement, explained in the next paragraphs.

3.1 Consumer digital cameras calibration

In order to obtain inner orientation parameters of the employed consumer digital cameras, a self-calibration approach was adopted by the bundle adjustment implemented in PhotoModeler Pro v5.0 (EOS Systems Inc.). Working with unstable non-metric cameras, it should be necessary to perform the self-calibration on the field as a part of the photogrammetric project, contemporary to data acquisition. Unfortunately this is not always possible, especially when the photogrammetric survey has to be performed in a short time, like archaeological work conditions often impose and as the typical network for camera calibration is very different from the conventional image network for scene reconstruction. Furthermore, for precise calibration procedures, targets or well signalized points are necessary as natural points are not well measurable in the images.

The used software can apply a bundle adjustment analytical model for calibration by using several images of a plane test-field, supplied with the software. In this case the process is fully automatic, requiring nothing more than images recorded in a suitable multi-station geometry, an initial guess of the focal length and image-identifiable coded targets which form the object point array. The real important point is a favourable network geometry: (1) photos must be taken with the same focusing condition of the application case; (2) in order to eliminate the high correlation present among some of the inner orientation parameters and increase the precision of the adjustment, a convergent scheme of acquisition including orthogonal rolled images is recommended (Fig. 4); (3), to compensate the planarity of the test-field, the images should be acquired at different distances from the object.

Once the camera(s) is calibrated, the interior orientation parameters can be used to retrieve the exterior orientation parameters of the data set and derive accurate surface models by applying dense image matching algorithms.

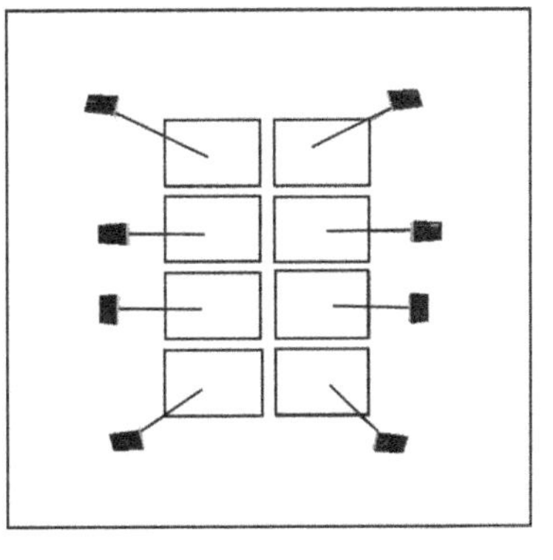

Figure 4 *Example of acquisition scheme of test-field for camera calibration process.*

3.2 Data processing and final products

The methodology in terrestrial photogrammetry changed significantly in the last years thanks to the improvement of matching algorithms and the introduction of simplified digital tools that permit, also without stereoscopic skills, a 3D object model reconstruction.

Different solutions can be experimented using digital photogrammetric systems, in terms of data acquisition (semi-metric and digital cameras), data processing (monoscopic and stereoscopic plotting, with and without the use of Control Points, in the last case requiring at least the measure of one distance on the object), automated surface model reconstruction, generation of different representation products (e.g. orthophotos, VRML models ...).

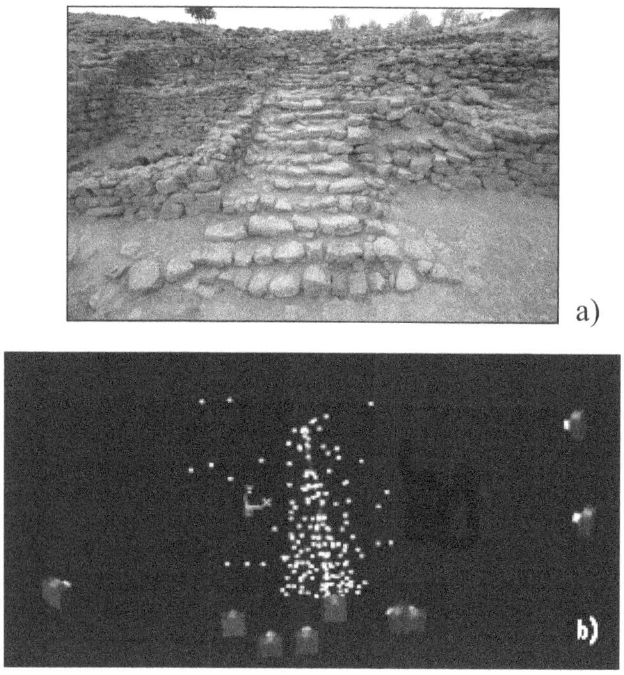

Figure 5 *The big stair of Tilmen Höyük (a) was acquired by different cameras and positions and distances. The recovered camera poses together with the object coordinates of some measured tie points (b)*

But, even if new technologies and software offer different solutions and final products nowadays a commercial package specific for terrestrial image able to derive automatically precise and reliable 3D models doesn't exist. Some of the most known softwares for monoscopic close-range photogrammetry, e.g. iWitness (Photometrix) or PhotoModeler (EOS Systems), perform the restitution of objects based on manually measured sparse points, permitting to mix in the same project images acquired from different cameras and creating a metric environment where is possible make measurements, sections, vector drawing and 3D modeling of regular geometric shapes. But these systems still require an important manual intervention by the operator and cannot perform dense matching on irregular shapes. Other photogrammetric stations, as for example Socet Set (Bae Systems) and Leica Photogrammetric Suite (Leica), perform the automatic extraction of very dense Digital Surface Models but a post-editing phase is in general necessary and they find difficult to work with highly convergent terrestrial images.

One of the products that the new digital photogrammetric systems could permit to obtain in a simple and fast way, is the orthophoto. The use of orthophotos in Archaeology and in general for Cultural Heritage applications is so becoming quite common, due to their property of combining metrical characteristics together with an high level of photographic detail, useful to evaluate colours, material, decay, textures, etc.

Fig. 6 shows the orthophoto of a wall in the Tilmen Höyük site. It was produced using, Socet Set v.5.3 orthophoto production tool, with a bilinear transformation as resampling method and a dense DSM (5 mm medium post-spacing) as base. The DSM was automatically extracted by the software, but as already said, a manual editing was necessary, especially at the wall bottom and top, when the surface morphology changes suddenly. The Ground Sample Distance of the orthophoto is 10 cm.

Figure 6 *Orthophoto of a site wall.*

3.3 A typical case study

In this section we report the 3D modeling results of a little wall in Tilmen Höyük site (ca 6.4 square meters), chosen as example of object of archaeological interest (Fig. 7).

Figure 7 *The chosen object for 3D modeling application.*

A detailed 3D model is required to derive different kinds of representation and documentation products, to perform stability and structural analysis or simply for visualization of the archaeological site.

An in-house surface measurement program developed to match (convergent) close-range images and based on multi-photo geometrically constrained least squares matching was used (Remondino & Zhang, 2006). The multi-image matching approach was originally developed for the processing of the very high-resolution images and afterwards modified to process other image data, such as the traditional aerial photos or close-range images.

The program, starting from the known interior and exterior orientation parameters, performs firstly an image pre-processing and generates image pyramids. The images are enhanced combining an adaptive smoothing filter and the Wallis filter, in order to reduce the effects of the radiometric problems, such as strong bright and dark regions and optimizing the images for subsequent feature extraction and image matching. Then the DSM is created by means of a Multiple Primitive Multi-Image (MPM) matching.. All available images are matched simultaneously, without having to match all individual stereo-pairs and then merge the results. Primitives like feature points, grid points and edges are extracted and matched using together area-based, feature-based and relational based matching procedures. Moreover, at each pyramid level, a TIN is reconstructed from the matched features and used to constraint the search in the next pyramid level. This type of approach permits to generate a very accurate surface model that can be also textured with a digital photo of the object and is characterized by a high level of detail representation.

Figure 8 shows the obtained 3D model of the wall, which contains approximately 200.000 points. Some simple examples of the geometric study of the object are also presented.

In order to evaluate the potentialities of commercial software in recovering detailed DSM from convergent terrestrial images, a test was also conducted using one of the most advanced photogrammetric workstations.

The result is shown in Figure 9. Compared to Fig. 8, it's evident that the process was not completely successful. The reasons could be multiple.

As said before, commercial software especially destined for 3D modeling by terrestrial images are virtually not existing today, so in many cases they fail when working with a geometrical conformation far from the classical photogrammetric block acquired by aerial or satellite platform. For instance, in close-range photogrammetry convergent images with different scale are very common.

Furthermore commercial software often use only stereo-pair surface measurement, while, exploiting the multi-image concept, highly redundant matching results are obtained and this allows automatic blunder detection. Finally, it is important to underline two concepts about the area-based matching algorithms that some commercial software usually adopt:

- the image patches are assumed to correspond to planar object surface patches and this assumption is not valid along edge or conrners, therefore the features are smoothed out;
- smaller image patches could theoretically avoid or reduce the smoothing effect, but they may be not suitable for the correct matching, containing not enough image signal content.

Figure 10 shows the comparison, related to a single stone of the wall, between the two recovered DSMs. The green colour is when the points have the same Z value (depth coordinate), obviously considered the error associates at this survey, estimated in about 3mm; yellow colour is when the Z value of DSM by commercial software is lower, in an order of about 1 cm; it's evident that the central part of the stone was completely smoothed. These discrepancies emerge by a study at a local level, while considering the overall dataset the mean of these difference values is in the order of 2 mm.

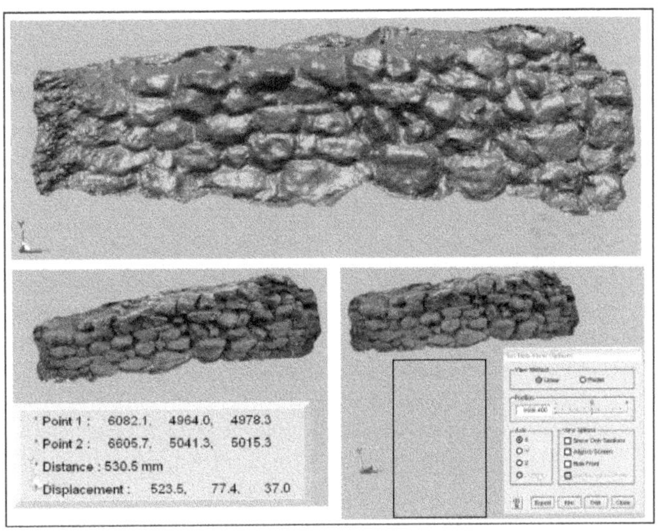

Figure 8 *The 3D model of the wall and simple applications: distance measurements and sections.*

Figure 9 *3D model of the same wall by commercial software.*

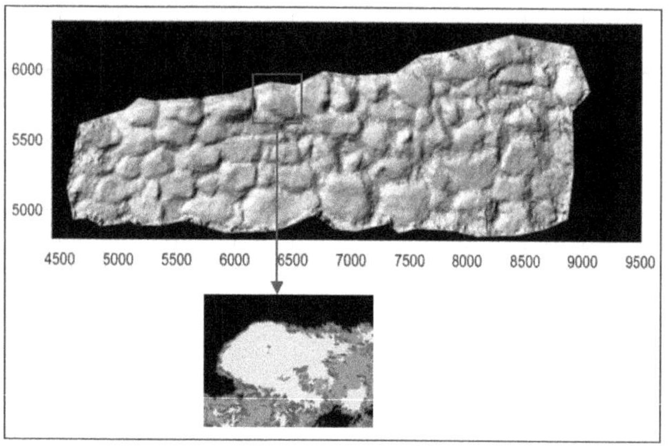

Figure10 *Particular of the comparison between the two DSMs.*

Conclusions

In this contribution we briefly reported on different geomatic techniques employed in the Tilmen Höyük site, with particular emphasis on the problems encountered using terrestrial photogrammetry. Between the available surveying methods, images provide all the information useful to obtain 3D geometry and textured results. Image-based modeling techniques can really increase quality and quantity of 3D information in archaeology, with low cost results and an integrated approach. The choice on a specific software or method depends on the final aim, on the complexity of the object or monument to document and the type of representation/communication we want to obtain.

Questions related to 3D photogrammetric modelling by non-metric cameras and the use of commercial and scientific software for surface matching were addressed in the paper. Results show that one of the crucial points is the generation of accurate 3D models and that there is the need of a larger development of packages specifically devised for this purpose. Results are not to be considered only as representation products, but they can be of valuable interest also for specific analysis of the structures and for engineering applications.

Acknowledgements

The authors would like to thank the members of the Archaeological Mission of the University of Bologna in Turkey, and in particular the Director Nicolò Marchetti and the archaeologists involved in the topographical survey (M. Zanfini et al.).

References

BERALDIN, J.-A., PICARD, M., EL-HAKIM, S., GODIN, G., LATOUCHE, C., VALZANO, V. and BANDIERA, A., 2002 Exploring a Byzantine crypt through a high-resolution texture mapped 3D model: combining range data and photogrammetry. *Proceedings of ISPRS/CIPA Int. Workshop Scanning for Cultural Heritage Recording*. Corfu, Greece, pp. 65-72.

BITELLI, G., M. A. TINI & L. VITTUARI, 2000 Close-range photogrammetry, Virtual reality and their integration in archaeology *Int. Arch. Of Photogrammetry and Remote Sensing,* B5, 872-879, Amsterdam.

BITELLI, G. 2002 Moderne tecniche e strumentazioni per il rilievo dei Beni Culturali *Atti 6° Conferenza Nazionale ASITA,* Perugia, Volume 1, IX-XXIV.

BITELLI, G., V. A. GIRELLI, M. A. TINI & L. VITTUARI, 2005 Integration of geomatic techniques for quick and rigorous surveying of cultural heritage *Proceedings of CIPA International Symposium*, 124-129, ISSN 1682-1750, Torino.

BLAIS, F. 2004 Review of 20 years of range sensor development. *Journal of Electronic Imaging*. 13(1): 231-240.

BOEHLER, W. 2005 Comparison of 3D laser scanning and other 3D measurement techniques. *Int. Workshop on 'Recording, Modeling and Visualization of Cultural Heritage'* - E.Baltsavias, A.Gruen, L.Van Gool, M.Pateraki (Eds), Taylor & Francis / Balkema, ISBN 0 415 39208 X, pp. 89-99, May 22-27, Ascona, Switzerland.

MARCHETTI, N. 2004 La cittadella regale di Tilmen Höyük. Palazzi, templi e fortezze del II millennio a. C. in un'antica capitale dell'Anatolia sud-orientale (Turchia) *"Scoprire. Scavi del Dipartimento di Archeologia"* (M. T. Guaitoli, N. Marchetti, D. Scagliarini eds.). Studi e Scavi nuova serie, 3, ed. Antequem, Bologna.

REMONDINO, F. and L. ZHANG, 2006 Surface reconstruction algorithms for detailed close-range object modeling *International Archives of Photogrammetry, Remote Sensisng and Spatial Information Sciences*, Vol.36(B3), Bonn, Germany.

REMONDINO, F. and C. FRASER, 2006 Digital camera calibration methods: consideration and comparisons *International Archives of Photogrammetry, Remote Sensisng and Spatial Information Sciences*, Vol.36(B5), Dresden, Germany.

REMONDINO, F. and S. EL-HAKIM, 2006 Image-based 3D modeling: a review. *The Photogrammetric Record*, 21(115).

Architectural lectures trough three-dimensional point cloud model: Villa Adriana in Tivoli

Sergio Di Tondo

University of Florence, Doctorande, Sergio Di Tondo c/o Giorgio Verdiani, Facoltà di Architettura, Piazza Ghiberti, 27 50100 Firenze – Italy, Office phone +39 055 20007237, Fax +39 055 20007236, Cell. phone +39 347 4748644
E-mail sergioditondo@inwind.it

Introduction

This research is still under development; its main purpose is indepth study of the architectonical and compositive rules that guided the design of a monumental complex such as the Villa of the Emperor Hadrian at Tivoli. Our intention was the surveying of whole of the villa, erected during the rule of Hadrian (117 a.C.-138 a.C.). The surveying instruments are those typical of the architectonic discipline: the survey, the analysis of the measurements, the analysis of the building structures and their relative chronological collocation, the study of the compositive and distributive structures of the entire aggregate of the villa and the single buildings, in order to understand the chronological development of Hadrian's complex and a methodology finalized to the understanding of the compositive structures that govern the "dispositio".[1] The latest surveying technologies used in the cultural heritage (laser-scanner, time of flight, TOF) allowed us to quickly obtain metric data and to generate three-dimensional digital model (points cloud) related to each monument. The job carried out in laboratory has been concentrated not only on the survey result, but also on the methodologies for the treatment of the data in relation to our main objectives. During September 2004 and 2005 the *Dipartimento di Progettazione dell'Architettura di Firenze,*[2] in accordance with the Premio Piranesi, and Premio Piranesi DARC,[3] had the possibility to lead some digital survey relative to monumental complexes inside the Villa.[4] This work has produced substantial data for the use not only by personal researches, but also to help the students during the Premio Piranesi seminar in the work on their hypothesis of plan and for graduate thesis[5] derived from that experience.

Research method and the state of art

The research has been divided during this first session of work in different surveying plans. The objective of the first one was the entire complex of the Villa and environmental evidence along the axis of the Tiburtina way; it was necessary to circumscribe the problem inside of the temporal arc in which Emperor Hadrian supervised the construction of the Villa, in order to this reason a map is still under writing course, in this map is possible to distinguish: (1) the present buildings situated in the area before the arrival of Hadrian in Tivoli (118); 2) the buildings erected by Hadrian's successors who have used the Villa (Antonine dynasty, after 138 a.C.); (3) the chronological development of the villa. This job is based on the considerations expressed by G. Lugli[6] and later, in a more exhausting way, by H. Bloch.[7] The purpose is to give precise explanations of differnet phases of the Villa and its modifications between Republican Age[8] and 138 d.C. The second level of research, more specific,

---

[1] In this case we referred to the concept of *Dispositio* expressed by Vitruvio:"*Giusta Disposizione delle cose*" Vitruvio *De Architectura* Libro III Capo I, L'architettura di Vitruvio nella versione di Carlo Amati (1829-1830) a cura di Gabriele Morolli, Firenze, Alinea 1988.

[2] *Dipartimento di Progettazione dell'Architettura di Firenze*; director prof. Marco Bini, the person coordinating the research. dott. Giorgio Verdiani: survey squad: Carlo Battini, Michele Cornieti, Sergio Di Tondo, Filippo Fantini, Mauro Giannini, Francesco Tioli.

[3] The survey was conducted during the *Premio Piranesi*, seminar that face museology theme in the archeological area of *Villa Adriana*. Director prof Luca Basso Peressut, coordinator prof Pier Federico Caliari.

[4] During September 2004 was made the survey of the *Grandi Terme* monumental complex and some parts of *Grande Vestibolo*; during September 2005 was made the survey of the *Palestra* and parts of *Cento Camerelle*.

[5] "*Il Grande Vestibolo, Progetto Museografico di Ricostruzione*" graduate thesis of Davide Giovagnoli e Sebastiano Longaretti, supervisor prof. Pier Federico Caliari, co-supervisor Giorgio Verdiani, *Politecnico di Milano, Facoltà di Architettura e Società, Corso di Laurea in Architettura*, a.a. 2004-2005; "*Le Grandi Terme di Villa Adriana: rilievo e indagine sulla genesi geometrica*" graduate thesis of Raffaele Mencucci supervisor prof. Marco Bini, co-supervisores dott. Giorgio Verdiani, dott. Cecilia Luschi, *Università degli studi di Firenze, facoltà di Architettura*, a.a. 2005-2006.

[6] G. Lugli "Studi Topografici intorno alle ville suburbane, Villa Adriana, Una villa di età repubblicana inclusa nelle costruzioni imperiali"in "Bollettino Commissione Archeologica comunale di Roma" 1927 54-55. pp 139-204; "Studi Topografici intorno alle ville suburbane, Villa Adriana, Le fasi della villa da Adriano al Tardo Impero" in "Bollettino Commissione Archeologica comunale di Roma" 1932. pp 111-176.

[7] H. Bloch "*I Bolli Laterizi e la storia edilizia romana; la villa di Adriano a Tivoli*" in "Bollettino *Commissione Archeologica comunale di Roma 65*" 1937 pp 113-181.

[8] G.Lugli "*Non pretendendo in questo studio esaurire l'argomento che è troppo vasto: per far questo bisognerebbe eseguire un nuovo rilievo generale della villa, a strati, e in piccola scala, distinguendo per ogni edificio le varie epoche: infatti quasi in ogni luogo, escon fuori muri di età anteriore o posteriore ad Adriano, e si notano pentimenti di costruzione, adattamenti e restauri che ci additano una storia quasi ininterrotta di secoli.*" op.cit.1932 p. 111.

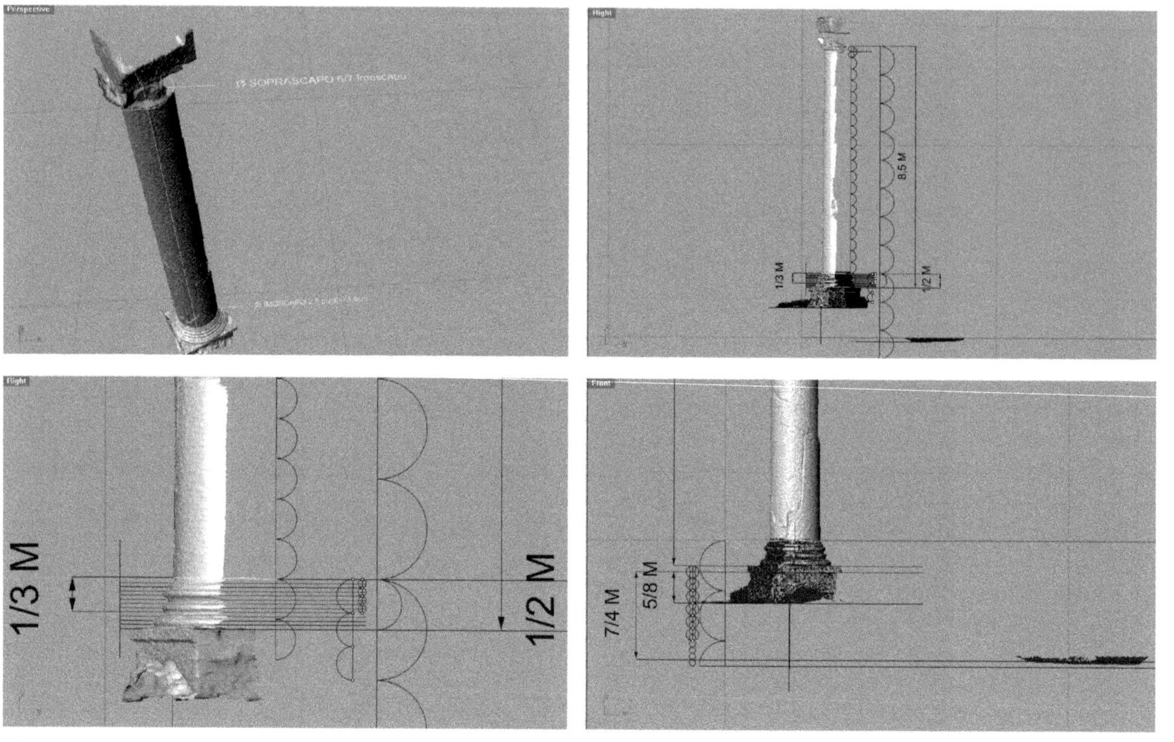

Figure 1 *Metrological considerations using the three-dimensional polygon surface model of the column.*

concerns the study of the measurements of the *Grandi Terme* complex; finally, the last one deals with the treatment of the data acquired by the digital survey. To date it is possible to show only some results of these studies; the one related to the measure of the *Grandi Terme* and the other related to the methodology employed for the treatment of the data acquired during the survey. However it is important to consider these first experiences as necessary steps pertaining to a more complex research towards the understanding of a complex archeological site as Villa Adriana is.

Metrological surveying applied to the monumental complex of the *Grandi Terme* in Hadrian's Villa

A lot of architects and students were interested in the *Grandi Terme*, not only because this monument maintained a good state of conservation during the centuries and the capacity to bring still emotional potential, but mainly for its strategic position inside the Hadrian's residence. The monumental complex is a fundamental junction between the Grande Vestibolo, Pretorio, Piccole Terme and is the obvious alignment that order the Canopo Serapeo. The Grandi Terme has been for a long time subject of studies, mainly in order to the date of its construction.[9] Today the dating proposed by A.C.G. Smith[10] and Salza Prina Ricotti[11] seems most credible, which places the G.T. in the first phase of the construction of the Villa (from first half of the 118 a.C. to the second half of the 125 a.C.)[12] in accordance with the chronological development proposed by H. Bloch. Thus it would have been erected in the same time as the *Piccole Terme*, the *Edificio con tre Esedre*, the *Pecile*, without masonry structure of *Cento Camerelle* and *Palazzo d'Inverno*.[13] If we leave the dating *querelle* begind, it is now interesting to face the analysis of the measurements for better understanding of the design of the Villa and to verify the compositive basic module used by the Hadrian's architects. The research took off from the data acquired from the digital survey executed in September 2004 by the *Dipartimento di Progettazione dell'Architettura di Firenze*. It is necessary to specify that the raw data does not allow an easy and immediate reading of the metric characteristics of the subject, consequently it was suitable to produce two-dimensional

9 H. Bloch "*E' evidente che questo grandioso monumento non può essere anteriore alle piccole terme. Rimangono solo due possibilità; o le Grandi Terme sono contemporanee o sono successive. Il Winnefeld le ritiene senza dare spiegazioni successive, il Lugli le attribuisce ad una stessa fase (la terza), mentre il Wirth colloca le Grandi Terme per lo stile della volta a stucco senz'altro sotto Antonino Pio. In complesso tra i 76 bolli provenienti dalle terme se ne trovano solo 7 datati, nessun timbro è stato rinvenuto nella cortina dei muri che sono in alcuni punti discretamente conservati*" op. cit. 1937 p. 158

10 A.C.G. Smith "The date of the Grandi Terme of Hadrian's Villa at Tivoli" in Papers of the British School at Rome 1978 46 pp. 73-93.

11 Salza Prina Ricotti "*Nascita e sviluppo di Villa Adriana*" *Rendiconto della Pontificia Accademia di Archeologia vol. LXV Anno Accademico* 1992-1993.

12 H. Bloch, op. cit. 1937.

13 Zaccaria Mari,Anna Maria Reggiani, Roberto Righi "*Il grande Vestibolo di Villa Adriana*" in "*Villa Adriana: Paesaggio antico e ambiente moderno, elementi di novità e ricerche in corso*" Electa, Milano 2002.

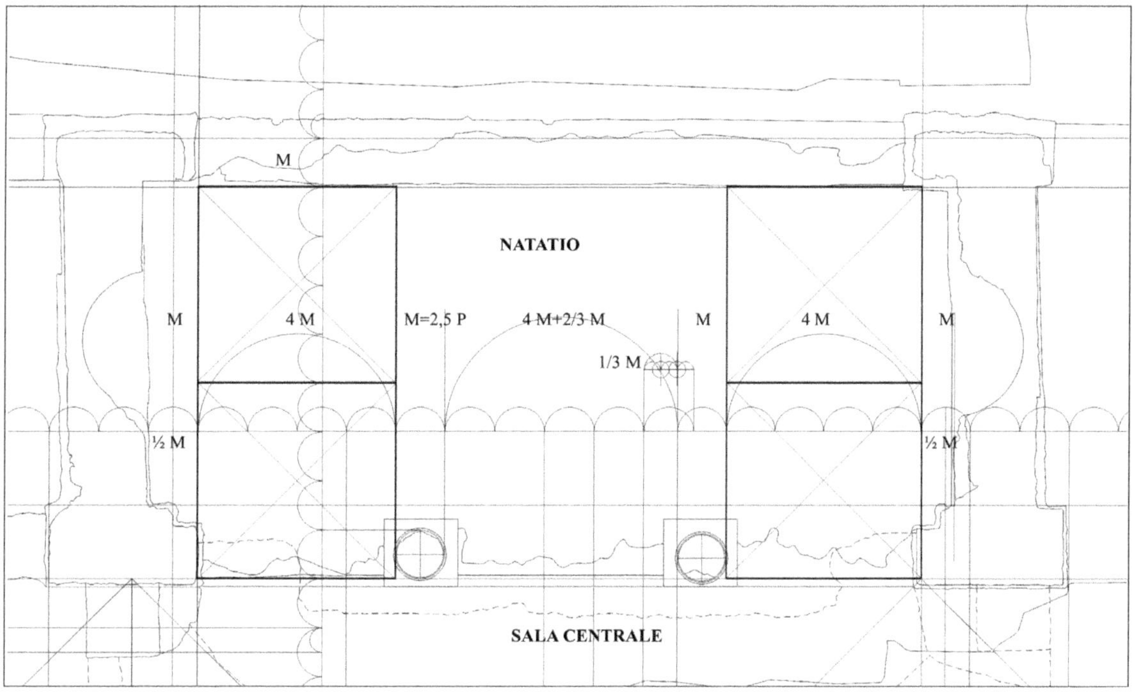

Figure 2 *The module on which the area of* natatio *was based.*

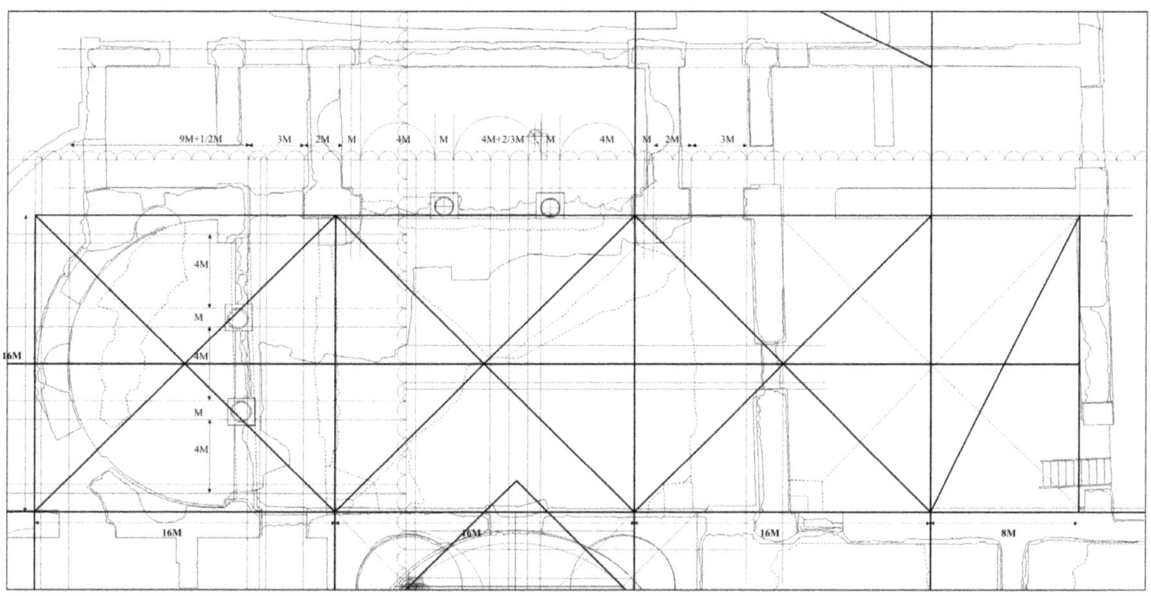

Figure 3 *The sequence of base module used in the arrangement of the main hall of the* Grandi Terme. *The hall width is equal to one macro-module=16 M (M=2,5 R.F.), and the length is equal to 3+1/2 macro-module (16 M).*

drawings from 3D digital survey data. The experience has been carried ahead using not only two-dimensional drawings but also three-dimensional models obtained converting in polygonal surface portions from the point cloud model. The metrological investigations need the consulted data to be the more near possible to the surveyed subject, in our case more near to the three-dimensional point cloud model.[14] So the conversion of point cloud model into polygonal surface model, which is more efficient in swift management and a number of

[14] For widening an inquiry: G.Verdiani "*Il battistero di Pisa: rilievo e rappresentazione digitale tra ricerca e innovazione" Università degli studi di Firenze, Dipartimento di Progettazione dell'architettura, Dottorato di ricerca in Rilievo e rappresentazione della città e dell'Ambiente* coordinator prof. Emma Mandelli.

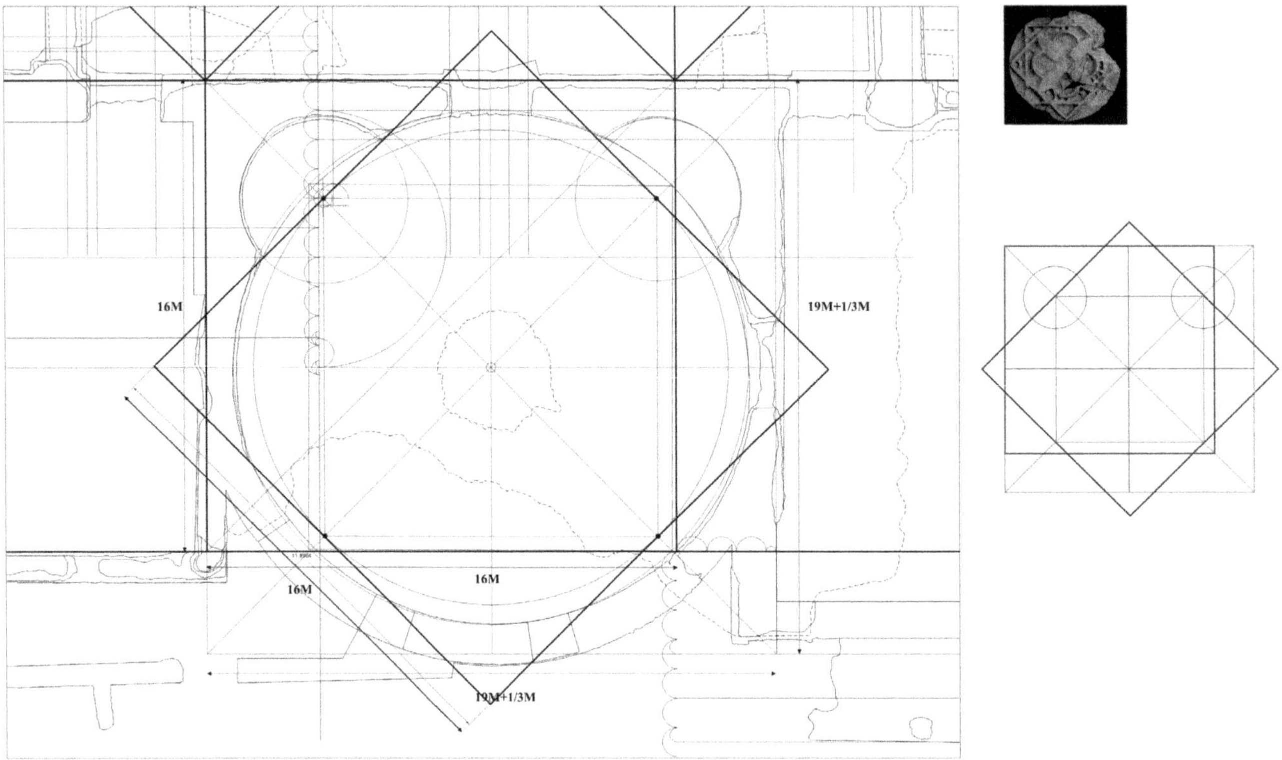

Figure 4 *The pattern of the geometrical rules used in the* sudatio *area. The image of a decorative element discovered during the excavation of* Edificio con tre Esedre, *in which the geometrical rule of the rotated square is shown.*

application, will have to be executed with extreme accuracy to avoid alterations of the metric data. It is necessary during this phase to exclude all the operations such as the merging between surfaces and the reducing of the number of polygons in the surface. Clearly this method is only acceptable for small portions of point cloud model. First of all it was suitable to define the unit of measure to trace the base module that articulate the space of the *Grandi Terme*. W. L. MacDonald and J. Pinto[15] suggested the Roman foot as the unit of measure used for the construction of a large part of the Villa, a unit corresponding to 29,4cm.[16] The measure expressed from the two authors turned out singular because usually, during the imperial age, the Roman Foot (the Pantheon unit of measure) corresponds to 29,56cm. However, metrological analysis based on our survey seems to confirm the thesis expressed by the two authors. The search of the base module has been faced taking in consideration one of the columns of the *natatio* inside the large main hall of *Grandi Terme*, as Vitruvio suggests in *De Architettura*. The use of the digital polygonal surface model of the column has simplified the operations of location of the diameter of the *imoscapo;*[17] diameter *imoscapo* corresponds to 73,5 = 2,5cm Roman feet (with roman foot = 29,4 cm). The verification in the position of the *soprascapo*[18] of the column has confirmed the measure of the base module,[19] diameter soprascapo 63cm

[15] W. L. MacDonald, J. Pinto "*Villa Adriana; La costruzione del mito da Adriano a Louis Kahn*" Electa, Milano 1997.

[16] W. L. MacDonald, J. Pinto " *Spesso si è fatto ricorso al reticolo con moduli di 5 piedi (1,47 m per lato), benché sia possibile riscontrare delle leggere divergenze dal canone, dovute alle tecniche di progettazione, costruzione e rifinitura adottate dai romani*" op. cit. 1997 p.58.

[17] Vitruvio, De Architectura, Libro III, Capo II,"Di queste parti poi, siano di tetrastilo, di esastilo, e di ottastilo, se ne prende una e questa sarà il Modulo, e di un modulo debb'essere la grossezza di una colonna". [...] "L'altezza di queste colonne sarà di 8 Moduli e mezzo; e così da questa distribuzione si avrà la giusta misura degli intercolumni, e dell'altezza delle colonne."[...]"L'altezza della colonna poi nel tempio eustilo si divide, come nel diastilo, in otto parti e mezzo, con una di queste parti poi si determini l'imo scapo, e così si avrà partitamente la regola per ciascuna specie di intercolumni." L'architettura di Vitruvio nella versione di Carlo Amati (1829-1830) a cura di Gabriele Morolli, Firenze, Alinea 1988.

[18] Vitruvio, *De Architectura*, Libro III, Capo II, "*Quanto poi al restringimento delle colonne nel sommo scapo, sembra esso doversi fare colla seguente regola: cioè, se la colonna sarà da piedi quindici in sotto, si divida la grossezza inferiore in parti sei, e se ne diano cinque alla parte superiore: se la colonna sarà fra i quindici piedi e i venti, l'imo scapo si divida in sei parti e mezzo, e si faccia di cinque parti e mezzo la grossezza superiore della colonna. Così in quelle da venti a trenta piedi si divida l'imo scapo in parti sette e se ne diano sei al restringimento superiore.*[...]" L'architettura di Vitruvio nella versione di Carlo Amati (1829-1830) a cura di Gabriele Morolli, Firenze, Alinea 1988.

[19] Column height is equal to8,5 M, in the 8,5 M (considering the base and a portion of capital -½ Roman feet-) without the plinth (1/8 M), if the plinth is considered a part of the basement, it rise

= 6/7 of 73,5. Once base module M (= 73,5 that is 2,5 Roman Feet) was defined it has been possible to verify the congruence between the module and sub-module and the other parts of the column directly on the three-dimensional digital model of the column. To verify completely the base module and to understand the design that regulated the Grandi Terme monumental complex, there was the opportunity, as Vitruvio[20] suggest in the *De Architectura* (*iconografia*), to extend the base module to the whole layout of the thermal building. This operation, executed on the two-dimensional drawings[21] obtained from the digital survey, has shown one effective reliability of the adopted measure. The drawings and images, included with this paper, explain clearly the experience that we have conducted. However, I would like to point out the possible geometric origin of the sudation hall in which macro-module is equal to 16 M (with M = 2,5 Roman feet) and is used in an unusual way. Every type of metrological evaluation usually has been faced considering the free space between the walls because spatial distribution was designed thinking about the needs for free, usable, areas. The composition of the planar layout is ordered by a macro-module, but the great circular bathtub of the sudatio, considering the masonry thickness, is inscribed into a square (side=19 M+1/3M), which is constructed along the diagonal of one of the macro-modules side equal to 16 M. Tracing the diagonals of square (side=19 M+1/3) it is possible to find the center of the bathtub of the *sudatio;* if we rotate on his center of 45° angle one of the macro-modules (side=16 M), it is possible to trace, along the diagonals of the square (side=19M+1/3), the centers of the circumferences that describe the two large *esedre* inside the *sudatio* hall. These considerations related to geometric features are extremely precious not only for the verification of the measurement of the base module but also for understanding the principles that govern the composition of Hadrian's Villa. The results obtained, even if not fundamental for the understanding of the general design layout of the Villa, show a possible methodological approach which faces the compositive topic, using the architect's own instruments: the measure and the functional, structural and intellectual rules (intended as a group of instances) that organize the composition of the layout.

from the central hall floor as 3/4 M and from the *natatio* foor as 15/8 M.

[20] Vitruvio, *De Architectura*, Libro I, Capo II, "Così *la Simmetria è un accordo uniforme fra i membri della medesima opera, ed una corrispondenza di ciascuno de'medesimi, presi separatamente a tutta la figura intiera, secondo la proporzione che gli compete: siccome nel corpo umano vi è Simmetria tra il braccio, il piede, il palmo, il dito, e tutte le altre parti, così addiviene in ogni opera perfetta.*" L'architettura di Vitruvio nella versione di Carlo Amati (1829-1830) a cura di Gabriele Morolli, Firenze, Alinea 1988.

[21] The two-dimensional drawings are elaborated by Francesca Granci during the course of Survey of the Architecture prof. Stefano Bertocci. In this case the three-dimensional point cloud model was elaborated using Cyclone to produce the main section line, and in a second time it was used a high definition bitmap to complete the drawing.

Bibliography

AA.VV 2000 *Adriano: architettura e progetto* Electa, Milano.

AA.VV. A cura di Anna Maria Reggiani 2002 *Villa Adriana: Paesaggio antico e ambiente moderno, elementi di novità e ricerche in corso* Electa Milano.

BETTINI, S. 1992 *Lo spazio architettonico da Roma a Bisanzio* Dedalo Bari.

BASSO, P., F. GHEDINI 2003 *Subterraneae Domus: ambienti residenziali e di servizio nell'edilizia privata romana* Cierre Edizioni, Caselle di Sommacampagna (VR).

BLOCH, H. 1937 "I Bolli Laterizi e la storia edilizia romana; la villa di Adriano a Tivoli" *Bollettino Commissione Archeologica comunale di Roma*, 65.

ECOLE NATIONALE SUPÉRIEURE DES BEAUX-ARTS *Italia Antiqua, Envois degli architetti francesi 1811- 1950, Italia e area del Mediterraneo* servizio delle pubblicazioni dell'Ensba Parigi 2002.

FALSITTA, M. 2000 *Villa Adriana, una questione di composizione architettonica* Skira Editore: Milano.

GROS, P. 2001 *L'architettura Romana: dagli inizi del III secolo a. C. alla fine dell'alto impero*, Longanesi Milano; titolo originale "L'architecture du début du III siécle av. J.-C à la fin du haut-empire".

LUGLI, G. 1927 Studi topografici intorno alle antiche ville suburbane: Villa Adriana, una villa di età repubblicana inclusa nelle costruzioni imperiali *Bollettino Commissione Archeologica comunale di Roma,* 54-55.

LUGLI, G. 1932 Studi topografici intorno alle antiche ville suburbane: Villa Adriana, le fasi della villa da Adriano al tardo impero, *Bollettino Commissione Archeologica comunale di Roma.*

MacDONALD, W. L., J. A. PINTO 1997 V*illa Adriana. La costruzione del mito da Adriano a Louis Khan* electa Milano.

ORTOLANI, G. 1998 *Il Padiglione di Afrodite Cnidia a Villa Adriana, progetto e significato* Dedalo Roma, 1998.

Active and Passive 3D survey merging. The case study of the water channel system in Al Habis castle, Jordan.

P. Drap,[1] R. Franchi,[2] R. Gabrielli,[3] D. Peloso,[3] A. Angelini[3]

[1]UMR 694 MAP – CNRS- Marseilles School of Architecture, France
[2]Istituto di Geodinamica e Sedimentologia & Centro Studi Archeologici "CESAR", University of Urbino, Italy
[3]Institute for Technologies Applied to Cultural Heritage – CNR, Rome, Italy

1 Introduction

The Petra monument site is located on the left rim of the Rift Valley in central-southern Jordan (Raikes, 1985). The entire valley, which is tectonic in origin, rests on Late Cambrian quartzarenite rock formations of continental origin. This sandstone is composed principally of quartz (in percentages ranging as high as 95%) and of kaolin, hematite, goethite and – to a lesser extent – calcite cement rock. These different types of cement rocks often coexist arranged in patches which are generally made up of a single mineral. The physical characteristics of the arenites are extremely variable depending on the grain size of the rocks and on the type and quantity of the cements (Franchi et al., 2004).

The only part of the valley's architectonic patrimony that is still visible consists in a series of tombs and temples dating back to the Nabataean period, chiselled out of and sculpted in the rock walls. Past research conducted by the members of this team has allowed us to identify the following main causes of the weathering processes that have led to deterioration of these monuments:

- thermal expansion and shrinkage of rocks caused by the great differences in day and night temperatures contribute to the loosening of quartz grains and lead to rock disruption. This process has been shown to be active at depths of up to approximately 20 cm;
- the effect of gravity is one of the principal factors contributing to removing single grains of quartz or more or less coarse-grained aggregates from the rock surface;
- heavy rainfall (rainstorms are not rare here and tend to be violent) and above all surface runoff contribute significantly to removing weathered materials from the surface;
- the presence of salts (halite, silvite, polyhalite, etc.) along fractures and joint systems is another important contributing factor in other forms of intense rock alteration.

Both fieldwork and laboratory analyses performed in the past in a climate chamber have underlined that the products usually used in conventional restoration work are not very effective, due to the 'extreme' climate and the chemico-physical characteristics of the rocks. Moreover, the vast dimensions of the surfaces requiring restoration would make these procedures extremely expensive and anti-economical. Indeed, past inspections carried out in the field have highlighted the key role played by rainwater run-off flowing down the facades of the monuments in causing their advanced deterioration.

This risk had been hypothesized by the Nabataean builders of these monuments, who succeeded in mitigating its effects by setting up an efficient network of drains and rainwater collection cisterns upslope from the monumental structures. Moreover, these complex hydraulic engineering projects also ensured an adequate water supply for the local population. Today, this drainage system is no longer operational, as it has been blocked by both the accumulation of debris and collapses caused by landslides and earthquakes.

A typical example of this situation can be seen at the Palace Tomb site (Fig.1), the top right-hand side of which was built using blocks of rock placed into a deep cut in the rock wall where a deep cleft was converted into a rainwater reservoir. With the passing of the centuries, this structure was slowly filled with debris, to such an extent that today the runoff, which is no longer collected, drives large amounts of sediment against the monument's facade.

Figure 1

This phenomenon has already caused a partial collapse of the man-made part of the monument, and rebuilding efforts carried out just a few years ago have been thwarted by the absolute lack of a suitable drainage system.

One of the main objectives of the architectural and patrimonial survey is to provide a precise documentation of the *status quo* of the surveyed objects (monuments, buildings, archaeological objects and sites) in order to preserve and protect them, to study and restore the monuments and to present them to the people. Complex objects, not planar or with ornaments and decoration, require the high-density and the high-resolution of the spatial data.

The laser scanning techniques and close range photogrammetry can offer two complementary sets of instruments and technologies able to answer to the specific requirements of architectural and archaeological survey.

2. Surveying Methods

Several methods which will be used for studies on the entire Petra valley have already been tested in the ambit of a research project funded by the Italian Ministry of Research and University and the Foreign Ministry, entitled, "*Petra medioevale. Archaeologia dell'insediamento crociato in Transgiordania*". Size 2004 the University of Urbino team worked at the Al Habis Castle site, chosen because it not only has characteristics which make it a significant model system upon which to test the method chosen but also due to its relatively small size. The Al Habis hill (Fig.2), which dominates the central part of the valley, is known above all for the numerous Nabataean tombs dug into its flanks and for the castle which occupies its upper slope.

Figure 2

Exploration of the entire archaeological site allowed us to identify the main functional areas of the castle by means of an initial classification of its topographical elements (battlements, observation deck, walls and passageways, different walls, towers, accessways, residential areas, the lower court, and the monumental cistern) and and provided evidence leading us to hypothesize that building took place on this site in different eras, one of which was certainly the Byzantine period. The castle, whose perfect position makes it a sort of natural acropolis, was used intensively in even more ancient times, as seen by the numerous examples of stonework and by the decorative ceramic-work dating back to the Nabataean period which can be seen throughout the site. The complex water drainage and collection system probably dates back to this period, and was still in use in the Byzantine era.

A complex network of channels of varying depths and lengths runs throughout the entire site, from the rock platform of the so-called "lower court" to the summit;a long series of channels carved into the rock and which run along the outer perimeter of the summit of the hill are particularly interesting in that they follow the natural slopes of the hill and end in cisterns of various sizes (Fig 3).

Figure 3

2.1 Topography

As mentioned above, the most important water collection channels run around the Al Habis site along a curve positioned at a level coinciding with that of the third plateau. These channels were chiselled into the rocky slopes, and are still visible today even from the foot of the hill. It was therefore possible to quickly survey their course with the total station equipment and map their position on the 3-D model of the castle (Fig.4).

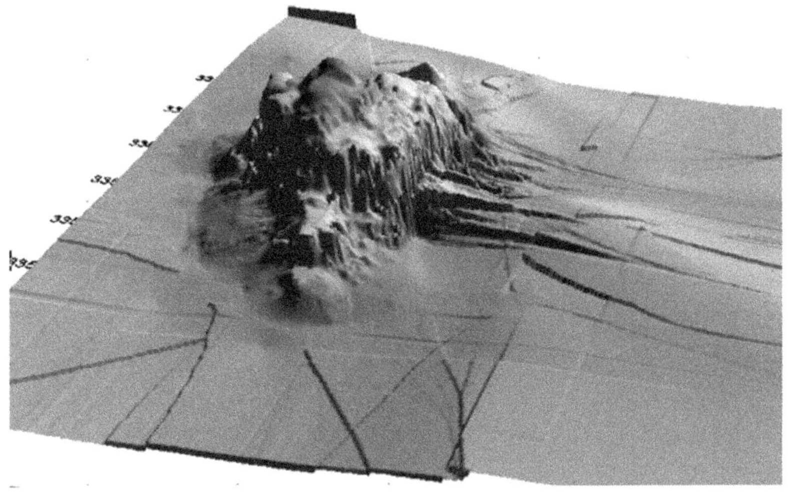

Figure 4

Given the inaccessible nature of the site, which makes it impossible to use Global Positioning Systems (GPS) to rapidly acquire a cinematic view of the points, the measurements, for the 3-D model reconstruction of the castle, were made using the Total Station Trimble 5600.

Using a closed polygonal network with ten stations positioned around the Al Habis castle, the different slopes of the entire hill were measured. This automatic equipment was used with a laser pointer, acquiring points at distances every five meters horizontally and two meters vertically. In this way it was possible to record the trend along the rock surface with a certain regularity and in a relatively short period of time.

The data obtained were downloaded and processed using the Toposoft topography software, yielding a cluster of points (shown as a TIN - Triangular Irregular Network to facilitate interpretation) that provide a detailed reproduction of the different slopes of the hill. Further processing of these data with specific 3-D software[1] yielded a high-definition 3-D model of the castle made possible thanks to the large number of points acquired.

As mentioned above, the most important water collection channels run around the Al Habis site along a curve positioned at a level coinciding with that of the third plateau. These channels were chiselled into the rocky slopes, and are still visible today even from the foot of the hill. It was thus possible to quickly survey their course with the total station equipment and map their position on the 3-D model of the castle.

This automatic equipment was used with a laser pointer, acquiring points at distances every five meters horizontally and two meters vertically. In this way it was possible to record the trend along the rock surface with certain regularity and in a relatively short period of time.

2.2 Photogrammetry

The site is not easily accessible, due to the terrain morphology and to the collapse of some structures of the castle. It was not always possible to use traditional and time-consuming procedures and to apply traditional topographical instrumentations.

This is one of the main reasons to use digital photogrammetry. The 3D reconstructions of the geometric shape of each single section of the channels were obtained by photogrammetry. The photographs were taken with a Nikon D100 digital camera, without a tripod and for the most inaccessible place we used an extensible telescopic arm with remote shooting system.

Commercial software (PhotoModeler 5.2) was used for *bundle adjustment*, while the ARPENTEUR and ROMA software are used for automatic or semiautomatic measurements of the surface of the object.(Drap, Grussenmayer, Gaillard, 2001).

The ARPENTEUR (Architectural Photogrammetry Network Tool for Education and Research) is a set of software tools developed by the MAP research Group, a French National Research Council (CNRS) laboratory. ARPENTEUR is developed in Java using the library JAI (Java Advanced Imaging) and X3D/VRML for 3D visualization.

The software is based upon a process guided directly by the available knowledge regarding a specific class of objects. In this way the experts can use their knowledge to facilitate the measuring process, thanks also to a series of tools provided by the software. The ARPENTEUR system can be used by professional architects and archaeologists with only minimal assistance from a specialist in photogrammetry.

2.3 Automatic generation of 3D measures

Whereas in aerial photogrammetry automatic generation of DSM (Digital Surface Model) is becoming mastered, it remains a research topic in close range photogrammetry because of the greater complexity of the scenes. A relevant result can be obtained provided that a pre-existing network of 3D-measures is to be densified. With the aim of measuring new points automatically, two multi-image approaches have been explored: area-based and feature-based matching.

2.3.1. Roma, 3-D Automatic Measurement Principles

ROMA, **R**epresentation of **O**riented **M**odel for **A**rpenteur, is the first tool built on the I-MAGE method (standing for Image processing and Measure Assisted by GEometrical primitive) developed in the framework of the ARPENTEUR Project. (Drap, et all, 2003) Roma allows automatic measurement using a set of oriented photographs and a mesh visible on these photographs.

We use four steps in this Semi-automated Primitive Measurement Method, considering that a mesh has been measured and computed from a set of 3-D points visible on at least two images (Fig.5)

- ✓ For each triangle of the mesh we scan triangle and get point Π. Each point Π is projected as p1 on to the photograph 1;
- ✓ Π is projected as p2 onto the second image;
- ✓ Point p2 is used as an approximate position to initiate the area based correlation process with p1;
- ✓ Point p3 is the result of the correlation; p1 and its homologous p3 are used for the computation of the 3-D co-ordinates of Π1.

This first implemented algorithm generates automatically regularly 3D-points through area-based matching. Initially a set of points was measured by the user.

This collection of points is triangulated and the regular scanning of the 3D-surface of each triangle provides theoretical 3D-points which are projected on a reference image. The semi-automated Primitive Measurement process called I-MAGE supplies measured 3D-points thanks to automatic correlation with other images. (Drap, Grussenmeyer, Curtinot, Seinturier; Gaillard 2004).

As a geometric-construction-based method, it gives a regular grid but the correlation process tends to fail when it works on low-textured image zones.

2.3.2. Other Simple algorithms

On the contrary, the second algorithm uses the featured parts of images. Firstly, interest points are extracted by Harris' improved algorithm (Harris, Stephens, 1988) considered as the most efficient according to (Schmid, Mohr, Bauckhage, Bauckhage 2000). Then the homologous points of the reference image interest points are searched among other images. But the correlation process is limited by two geometric conditions. On the one hand, out of continuity constraint, when a point belongs to the projection of a triangle from the initial network, this homologous point must be in the projection of the same triangle in the other image. Moreover, each homologous point is checked by exchanging reference and search matrix.

To improve the number of matched points, an iterative method that applies this principle has been implemented.

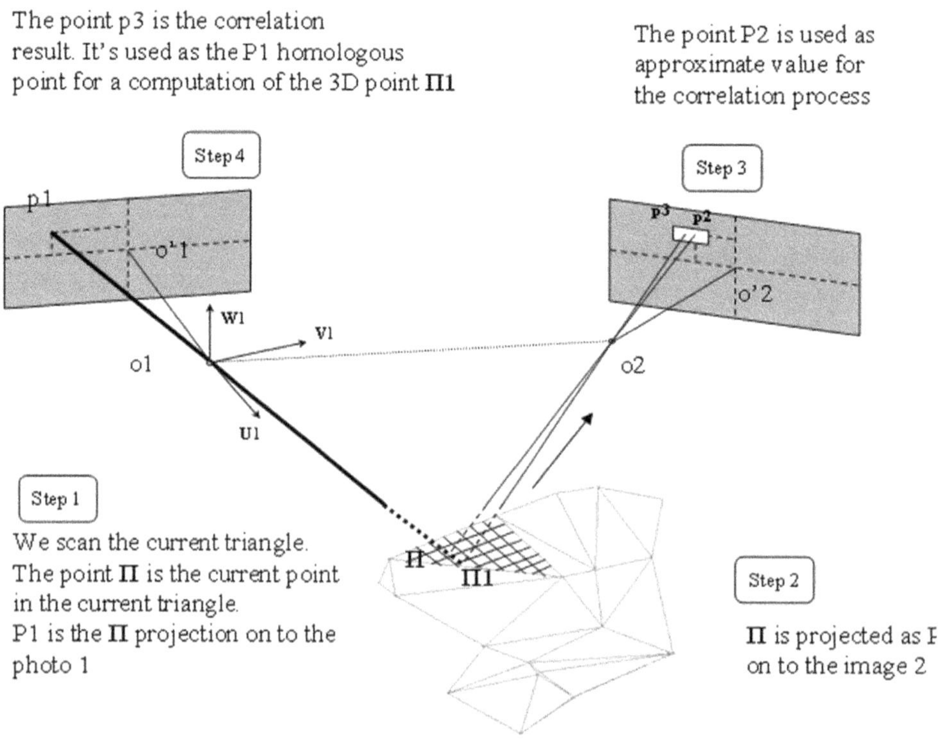

Figure 5

Indeed, corner detector can classify the points according to their degree of interest. Usually only the best points are conserved. It would not be judicious to take a larger part of points because it would raise the number of wrong matching. However we observed that if a point is repeated, its degree of interest has approximately the same rank in the other image. Then the algorithm can match feature points progressively by rank.

2.3.3. Combined algorithms

Even though the feature-based matching is more robust than the area-based matching, it depends entirely on the repeatability of the detector. Another idea is to combine the two principles to avoid this drawback. For each triangle of the network of measures chosen by the user, a reference image is determined and the feature points are calculated in this area.

Then supposing that those points lie on the 3D-triangle, their homologous points are found by correlation in the other images through the I-MAGE process.

If this method provides many successful matching, the repartition of the points can be inhomogeneous on some scenes if interest points are clustered.

The solution used here is a mixed method using this algorithm iteratively.

3. Conclusion and Future work

The result of this operation is an extremely detailed and measurable 3-D model of the channels that can be used both for classification and study purposes as well as for virtual tests and simulations on the waters flow (Fig.6).

The creation of this detailed model allowed us to extract information on sections along the course of a channel and revealed the degree of deterioration of the side walls of the water channel.

The work presented in this paper will be the first step and the background of a GIS (Geographic Information System) on the AL Habis castle where the topographic and photogrammetric survey will be used as a support to manage all the archaeological data produced by archaeologist on this castle. This interdisciplinary approach gives a new interest to this work and will produce a very innovative Information System from archaeological study to patrimonial restoration.

Future work will involve immersive visualization and interaction of the 3D models in collaboration with the team of Dr Paul Chapman from the University of Hull, UK. Through the use of state of the art stereo display devices provided by the HIVE (Hull Immersive Visualization Environment), we aim to gain both greater insight into the data and also to provide the general public with a previously unattainable immersive view of the archaeological site.

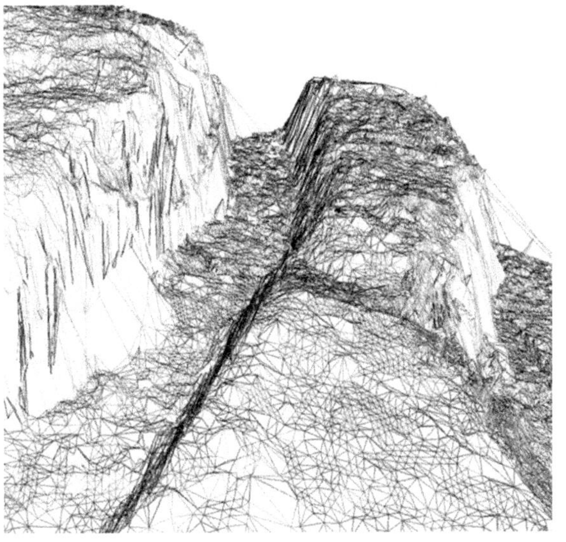

Figure 6

References

DRAP P., GRUSSENMEYER P., CURTINOT P.Y., SEINTURIER J., GAILLARD G., 2004, *Presentation of the web based ARPENTEUR tools: towards a Photogrammetry based Heritage Information System*, in XXth ISPRS Congress, ed. IAPRS, Vol.XXXV, ISSN 1682-1750, Istanbul, Turkey, pp.123-128.

DRAP P., SGRENZAROLI M., CANCIANI M., CANNATA G, SEINTURIER J., 2003. *"Laser Scanning and close range photogrammetry: Towards a single measuring tool dedicated to architecture and archaeology"*. CIPA XIXth international symposium, antalya, Turkey, October 2003.

DRAP P., GRUSSENMEYER P., GAILLARD G., 2001, *Simple photogrammetric methods with arpenteur. 3-d plotting and orthoimage generation : the I-MAGE process*. CIPA 2001 International Symposium, Potsdam University (Germany) September 18 - 21, 2001. ISPRS International Archives of Photogrammetry, Remote Sensing and Spatial Information Sciences, Volume XXXIV – 5/C7, ISSN 1682-1750, p. 47, 54.,

FRANCHI, R., SAVELLI, D., AND MORETTI E., 2004, *Petra and Beida (Jordan): two adjacent archaeological sites deeply integrated in an impressive geomorphologic landscape (abs)*: 32nd International Geological Congress, Florence – Italy, August 20-28 2004, Scientific Sessions, v. 1, pp. 138

FRANCHI, R., 2002, *A study of the natural environment and of the problems of conservation of the Historical-artistic heritage in the area of Petra*, in Jedrkiewicz, S., ed., Civilizations of the Past, Dialogue of the Present: Italian Research Mission in Jordan: Amman, Ambasciata d'Italia, pp.67-92.

FRANCHI, R., PALLECCHI, P., 1997, *The sandstone of Petra, Petrography and problems in conservation*, in Pancella, R., ed., Preservation and restoration of cultural heritage: Losanna, Pancella, pp. 679-689

HARRIS C., STEPHENS M., 1988. *A combined edge and corner detector*. Proceedings of the 4th Alvey Conference, Manchester, 1988, pp. 147-151.

RAIKES, T., 1985, The Character of Wadi Araba, in Hadidi, A., ed., Studies in the History of Jordan: Avon, Bath Press, v. 2, pp. 95-101.

SCHMID C., MOHR R., BAUCKHAGE C., 2000. *Evaluation of Interest Point Detectors*. International Journal of Computer Vision, 37(2), 2000, pp. 151-172.

ZAYADINE, F., 1985, *Caravan Routes Between Egypt and Nabatea and the Voyage of Sultan Baibars to Petra in 1276*, in Hadidi, A., ed., Studies in the History of Jordan: Avon, Bath Press, v. 2, pp. 159-174.

3D Visualization of archaeological place of Corzano

Paola Puma, Carlo Battini, Lorenzo Bianchini, Francesca Concas, Michele Cornieti, Francesco Tioli

Facoltà di Architettura di Firenze, Dipartimento di Progettazione dell'Architettura, 50122 Firenze
tel 055/200071 fax 055/20007236, paola.puma@unifi.it

Extents, purposes and context of the research

The report we are to present is the result of a former research, which was commissioned in order to constitute the documental basis for the material and signification recovery of Corzano – a fortified settlement on mountains between Toscana and Emilia-Romagna. This research was conducted through the study and the documentation of surviving fortifications in a place, which, during the centuries, has represented an important presence in the life of the community and of the surrounding landscape.

Therefore, it is the result of a complex work of reconstructing an organic documentary picture, which has been obtained through the integration of multiple informative repertoires: from material documentation on the site's existing buildings (created during three different surveying campaigns on the surviving fortifications) to territorial and land documentation of its context, to the individualization of realistic evaluating potentiality for material and meaning restoration of the site of Corzano.

The ranging and the proliferation of fortified settlement and garrisons along the main roads interests the communities on the Tosco-Romagna mountains especially between the twelfth and the fourteenth centuries. The seigniorial powers with military origins and the same monastic orders rise and take roots (until the Florentine conquest, which is completed at the beginning of the fifteenth century) in peripheral territories of the imperial and ecclesiastical jurisdictional seats, faraway from big urban centres, but with an important strategic role for the viability connecting the Po valley to central Italy.

Here, the exercise of power is executed controlling the ancient *romee* roads, which pilgrims and armies from the north ride to reach the central Italy through the passes of the *Alpes appenninae*: in fact, along these roads are settled many *castra* and *rocchae*, as well documented by the contemporary sources and the current remains of material structures.

At the end of 2005, the Municipality of Bagno di Romagna, in the province of Forlì-Cesena, has commissioned a research to the Department of *Progettazione dell'Architettura* of the University of Florence[1] for the creation of documents and the formulation of hypothesis on the recovery of constructional works in Corzano, an outstanding example of a garrison with a military eminent character that controlled important trans-Apennine roads conducting to the Val di Bagno.

Articulation of the research

The documentation and the survey of material and immaterial aspects that connote cultural heritage represents more and more the tool to try keeping knowledge control at least of rapid land transformations (where it is impossible to act on every single case), and, more, the necessary operative assumption for all of the situations where it is possible to concretely intervene.

In this strategy, the result of surveying operations assumes the value of representing model of the investigated object, a model for which the thematic elaboration of basic information represent a fundamental critical aspect to manage the building.

In fact, the higher is the technological level of surveying procedures (by means of the most sophisticated digital tools and methodologies) the greater results the exigency of critically rule the sense attribution of acquired information and the deep understanding of formal, functional, building and spatial meaning of the surveyed/revealed object.

For this reason, during the survey planning of the site of Corzano, the surveying project preceding the field operations has carefully considered the exigency of thematic and critical elaboration of data, looking further the objective of precision and data exactitude, to concentrate on determining the complex formal, social and cultural organization of the settlement and its constituting elements to try and go along the steps of its creation again.

---

[1] The scientific chairman for this Agreement is Doc. Paola Puma, who coordinates the working group including: Michele Cornieti, Carlo Battini, Lorenzo Bianchini, Diego Cacciamani, Elia Carli, Francesca Concas, Sergio Di Tondo, Silvia Mantovani, Francesco Tioli, Giorgio Verdiani.

Il Santuario della Madonna di Corzano

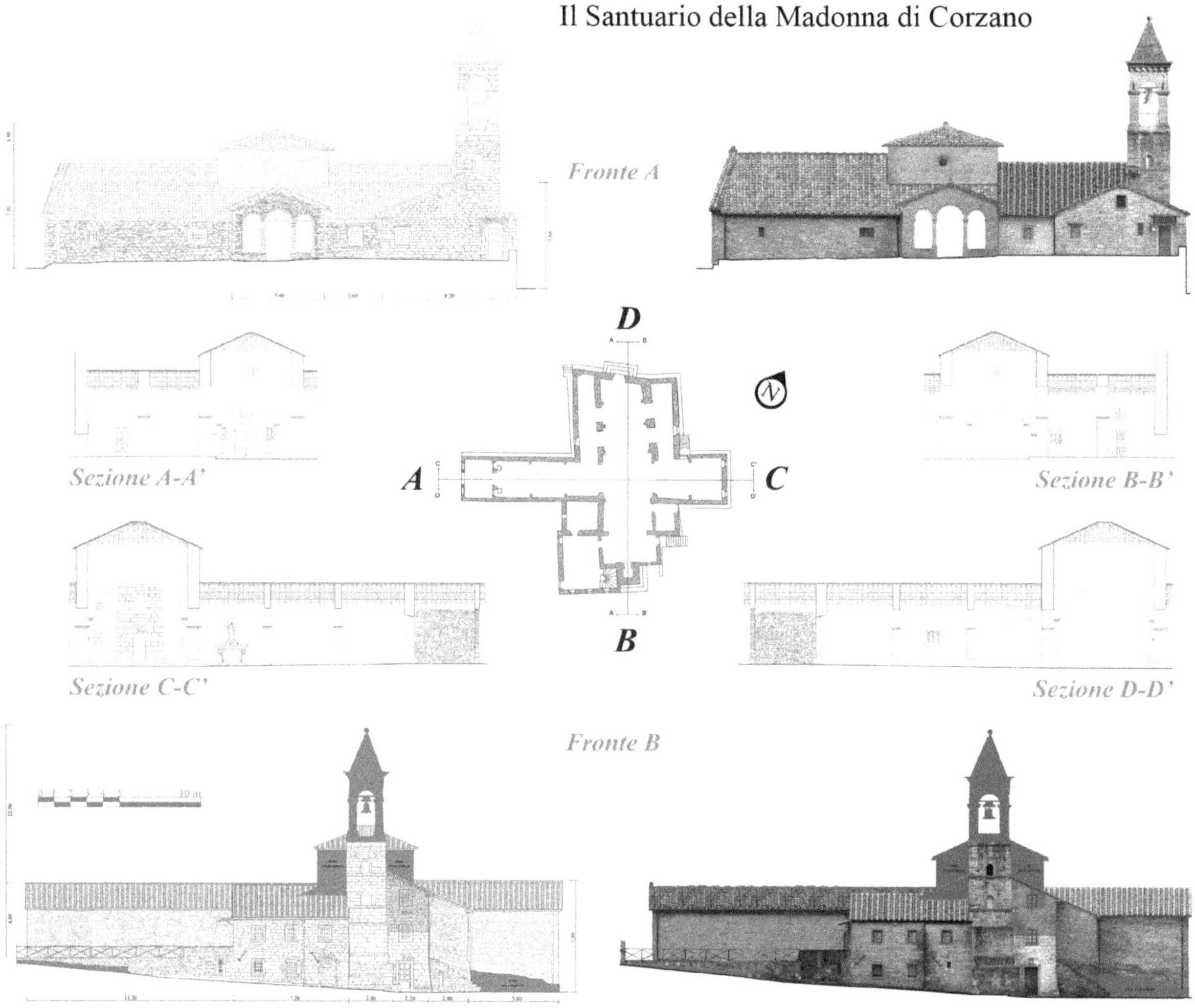

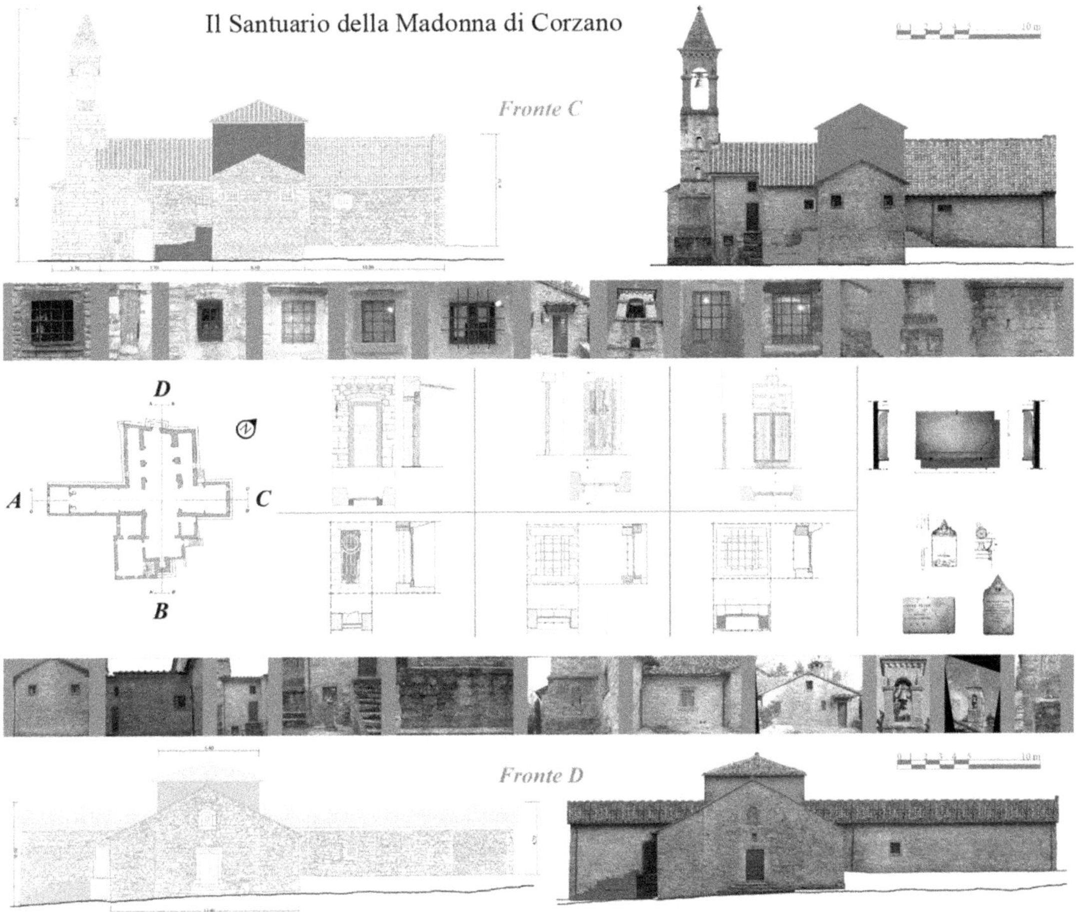

In such sense, we have tried to effect a *good* survey, a scientific survey which could give reasons to metrical, formal, spatial and material characteristics, but which could also allow the reconstruction of the place's historical transformations, the reflection of its chronological steps, the verification of formal problems, the underlining of time successions, the collecting of its essence by recording anomalies or static reasons and, finally, its real comprehension.

The surviving constructional works in this place, the ruins of the Fortress (which is datable to the twelfth century) and the Sanctuary (which is documented since the thirteenth century), have been surveyed at the beginning of 2006 during three thematically articulated campaigns, with following base restitutions and thematic elaborations aiming to the critical evaluation of a range of aspects in connection with the possible individuation of remains, which can conduct to the typical settling fortified typologies of that territorial context, the specific individual settling and territorial control modalities, morphology, building techniques, the possible original configuration of the castle. The plan of work has been drawn considering different steps their own methodology, tools and proceedings: but these steps are characterised by strong integration, aiming to the production of descriptive elaborations of the elements complexity, though they are intentionally simplified and friendly designed for not specialists final users.

The researching project of the work group had different objectives and steps:

1 data acquirement: creation of a data base, documentation and integrated survey through the realization of an updated and complete morphological, dimensional survey[2] of surviving buildings and their characters: the rests of the fortress, the Sanctuary and their close context;
2 thematic restitutions of morphologic, metrical and material data through the following production of basic graphical restitutions and thematic elaborations for the critical evaluation of the building[3] ;
3 Critical elaborations: structural, building, geometric and preservative consistence of the constructional works aiming to chose useful elements for the possible individualization of signs that could help us to verify if the few current surviving structures allow to give hypothesis on formal analogies with other typical fortified places in that territorial context, on the specific settling and land control modalities and on the possible original configuration of the castle;[4]

---

[2] Specialist deepening on surveying methodologies is by Francesco Tioli and Giorgio Verdiani.

[3] Specialist deepening on thematic elaboration is by Carlo Battini.

[4] Specialist deepening on critical evaluations is by Francesca Concas, Michele Cornieti, Sergio Di Tondo.

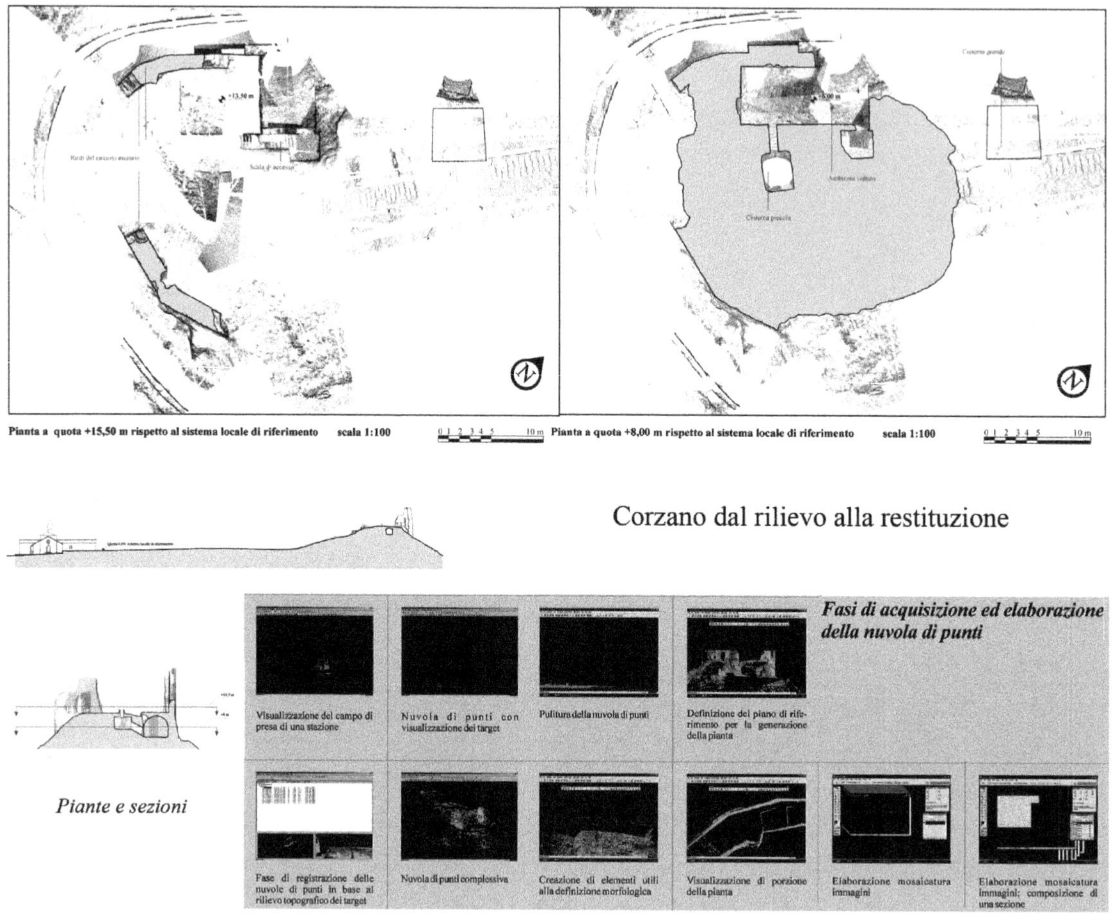

4 Planning and restoration proposals consisting on the verification of the possibility to interpret an evaluating project on Corzano not as a sequence of additional building interventions but as the evaluation of resources already present in such a rich territory, and, if ever, limiting interventions on the site to the few necessary maintaining and recovering works.[5]

Data collection

During the first organizing step, the structure of the data base has been created: at first the historical documentation had to be collected in this data base, in order to be systematically ordered, configured and referred to for the following creation of links among historical, surveyed, reinstated and elaborated data.

Integrated survey

The planning of the integrated survey campaign has been necessary to the collection of data from:

- Digital survey of the fortress, by means of Leica HSD 2500 laser scan with ToF (time of flight) technology;
- Photogrammetric survey of the Sanctuary;
- Topographic survey, by means of a total station Leica 706 no-prism, for the necessary references;
- Direct survey, supporting the photogrammetric survey.

Thematic restitutions and critical elaborations

In this case, the particular morphological and environmental conditions of the place, with the intrinsic characteristic of the constructional work and its grade of preservation, asked for and indispensable integration among the different tools, surveying and data interpretation procedures. In fact, every methodology, from the most traditional to the most technically evolved one, has been chosen because it was the most suitable to investigate specific aspects. The surveying campaign and the following interpretation of the Corzano Castle's structures have put in evidence the necessity of a systematic confrontation during the operations between the data that were collected by means of the laser scan technology and those collected with the traditional investigating tools for architecture and archaeology, such as the analysis of geometrical configuration and stratigraphy of the fronts.

The undifferentiated and massive collection of the metric data through the laser scan technology on one hand gives a complete, precise and fast survey as unimaginable

[5] Specialist deepening on project proposals is by Anna Lambertini, Silvia Mantovani, Paola Puma, Gianfranco Corzani.

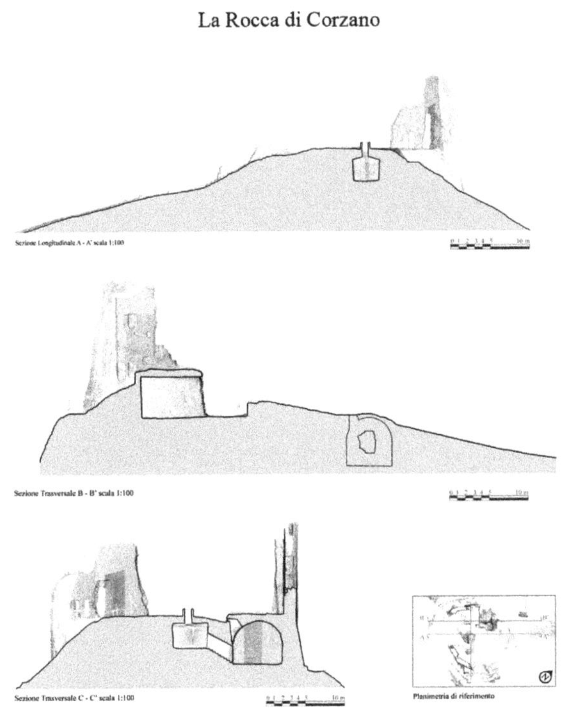

before, on the other hand shows the lack in the surveying action of the fundamental intellectual operation: the critical discrimination of the constructional work.

During this necessarily autopsic step, the surveyor is immediately bound to read and to formulate interpretations on the investigated object, while with the laser scan technology the operator risks to remain at a superficial understanding of the object: the point cloud model that has been generated through the laser scan technology digital survey, in fact, does not supply an exhaustive picture of the possible information of the constructional work, but its efficiency appears as it is supported by other investigative tools, so that it can clarify complex situations through the three dimensional view.

The same accurate and massive characteristics of its contained data put the point cloud model in an intermediate position between the investigated object and its representation through conventional systems and elaborations, which are still indispensable to conduct critical analysis.

Once that the geometrical data are acquired, it is necessary to effect a "virtual survey" through a selection of sensible points, in order to describe the geometry of the object; this operation can be equalized to discernment, although the point cloud itself is a discontinue object.

In any case, as every following elaboration of data and digital geometry causes a loss or an alteration of the original information, such a model is the most precise and congruent under a metric and morphologic point of view, although perceptively distant from the investigated object and difficultly manageable.

If, at the level of usual representation, the data gained from digital survey has to be treated through more or less laborious applications and processes, the rough point cloud is a valid tool during the stratigraphical analysis, as it allows, for example, verifying at a three dimensional level interfaces and relations among the different actions, or eventual alterations of parameters' co planarity and variations in the alignment of constructive elements, as index of meaningful stratigraphical events. More, it is immediately possible to extend the verification to interpreting hypothesis on the whole building, in order to evaluate its coherence.

In the investigation on the Castle of Corzano, the interpretation of the complete three-dimensional cloud of points model has put in evidence discontinuities and anomalies in the plans and in the walls alignment, which could not be perceived with traditional tools, due to the characteristics of the place and the inaccessibility of some structures. Their congruence has been verified through the information of the traditional reading of wall disposition.

This has helped in the individuation of late or restoration interventions, which sometimes have modified the perimeter of the original plan in the fortified settlement. In those contiguous walls portions, which have been realised in different times, the different superficial degradation of the material do not seem to affect the view of the cloud of points with false colours, due to the material reflectivity (a local sandstone). Instead, the multi resolution and NURBS digital geometry three dimensional model, which has been elaborated starting from a cloud point with the integration of data collected from a following detailed topographical survey, is useful to deepen the investigations the composing and spatial matrixes of the fortress, through the individuation of geometrical and proportional relations. Particularly, the possibility of extrapolating aimed cross sections in different zones of the whole area and spatially extending the hypothesis about surviving original walls has permitted to desume information on the planning origin of the settlement and to give hypothesis on the building protocols based on empirical measurements used by the makers of the fortress. Finally, such a model, shaped on the selection of the surviving original structures, has given support and reference to some reconstructing hypothesis that, even if showing evident limits, have been formulated on the base of typological and morphological confrontations on similar buildings in the same geographical area and as a consequence of the analysis of

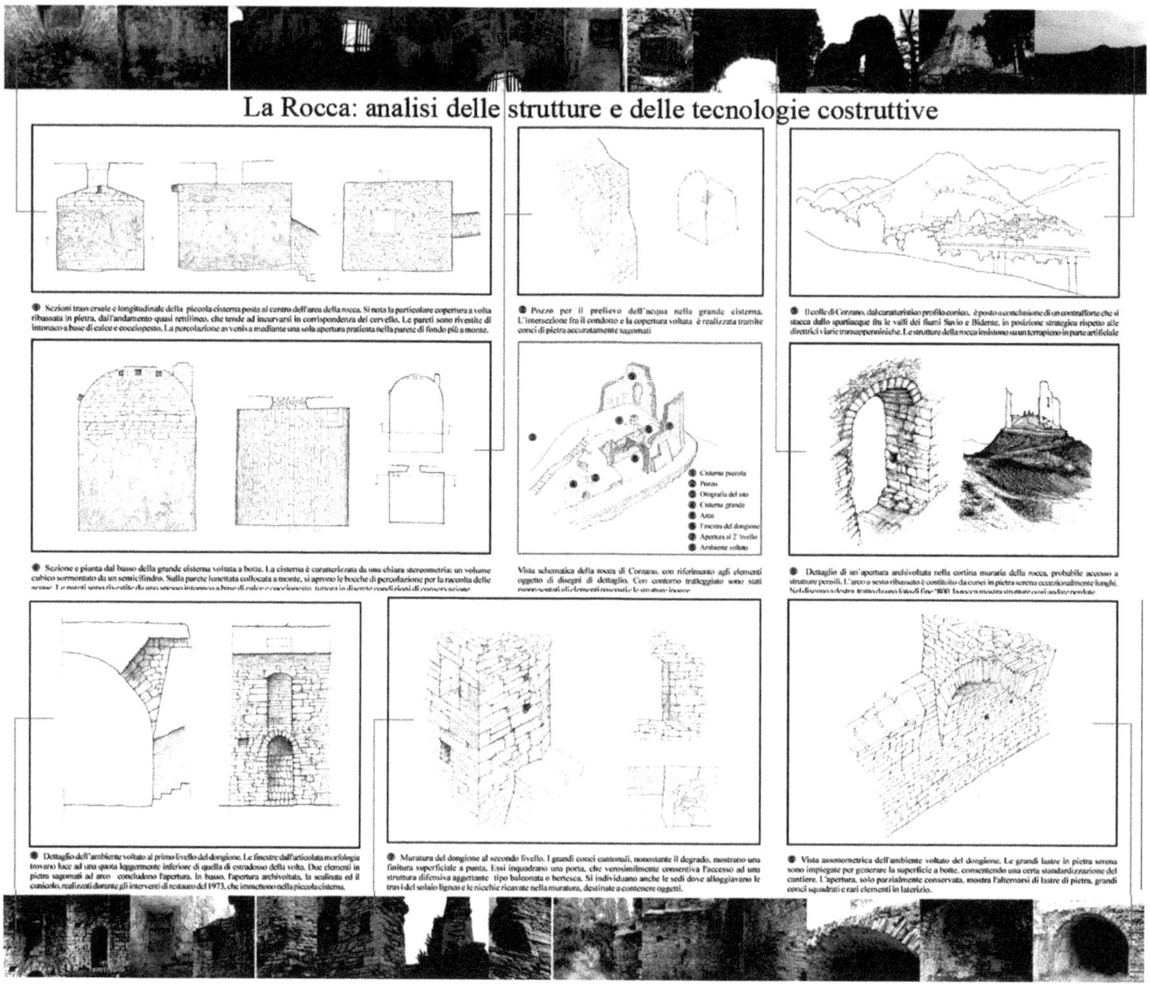

indirect, written and iconographical documentary sources.

From the drawing to the 3d model: three-dimensional view as analysing, verifying and communication tool

The collected data that have been collected during the steps 1 and 2 have been elaborated and divided into different database, which have been related under finalized rules.

The first use of the collected information has allowed the production of a wide range of graphical elaborations, constituting the necessary documental base for the understanding and the interpretation of data. More, they represented synthetic documents of the analysis that have been conducted in a direct contact with the place, such as the stratigraphic interpretation of walls, the geometric and measure relations among the parts, interpretative keys which have contributed to clarify the historical and constructive facts of the structures that have determined its current configuration.

The second use of the collected data has involved the regulation of tools allowing the analysis of the examined object by means of interactive systems that is manageable by final user too. These tools have been defined in a plan that foresees the creation of three-dimensional digital models with different scale of detail: textured VRML model, average detail definition model, high detail definition model.

An important goal of the working group has constantly been the will of using an interpreting and representing system of surveyed data where the three dimensional visualization is not a simple representation of both the current state and the hypothesis of reconstruction, but a tool for investigation, analysis and scientific control.

Under this profile the study of the place has been conducted creating an easy manageable general model, which is indispensable for the morphologic and geometric analysis of the constructional work. Later, the detail definition has been improved to investigate the building, stratigraphic and structural characteristics: this has allowed the formulation of a former hypothesis on the genesis and the development of the building, which has converged on the final model. The final model has been set on an average scale of representation, in order to better suit for a fast visualization within a web browser.

The third use of the surveys has been the production of a virtual model of Corzano place, as a synthesis of the study and allowing the proposal of the castle's original configuration.

The purpose of the differently scaled three dimensional representations rises from the necessity of producing elaborations that can be used not only on dedicated systems with suitable software and hardware, but also within a wider project that indicates the web as source of information exchanges and cultural growth.

In fact, the planned research included the creation of informative web pages.

The elaborated three-dimensional models have been used as a tool for immediate visual communication, with links to other pages with more definition and information. This simple and intuitive tree-structure allows both the simple visitor and the expert to concentrate his attention also on detailed portions of the constructional work, without loosing the starting general conuration.

The plan foresees the possibility of interaction with data banks for admitted people, who can insert annotations and results of researches on analogous themes.

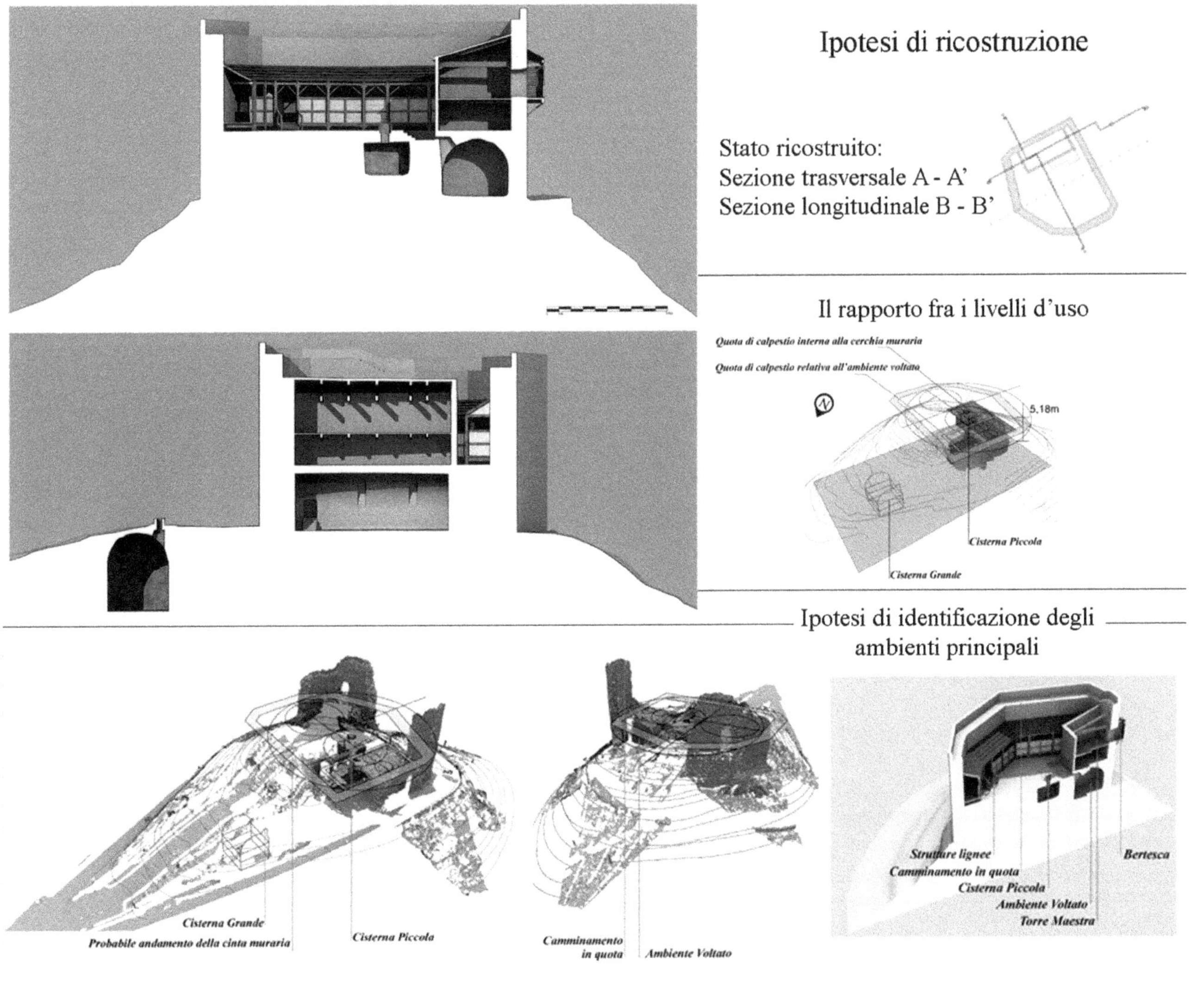

Laser scanner, quick stereo-photogrammetric system, 3D modelling: new tools for the analysis and the documentation of cultural archaeological heritage

Paolo Salonia,[1] Serena Scolastico,[1] Valentina Bellucci[1]

[1]CNR - Istituto per le Tecnologie Applicate ai Beni Culturali
Rome Research Area - Via Salaria, km 29.300, 00016 Monterotondo St. (Rome) ITALY
tel. +390690672384, fax +390690672373, e-mail: paolo.salonia@itabc.cnr.it

1 Introduction

The critical process of ancient monuments analysis and conservation has been facilitated by the introduction of Computer Science in the field of documentation and survey of architectural heritage.

Indeed, computer-based techniques have strongly modified the acquisition phase as well as the next operations such as computation and management of all the information necessary to usefully describe the features and peculiarities of buildings and environments.

New survey technologies allow to acquire and elaborate a great deal of geometric data; processing data software guarantee the control and the utilization of the same geometric data in 3D visualization environments both raster and vectorial.

The set of all these innovative tools lets administrative bodies, entrusted to preservation, safeguard and valorisation of cultural heritage, to avail themselves of great levels of documentation and then of knowledge.

What above said becomes more important in the case of archaeological artefacts reconstruction, where the difficulties of survey are diversified and the complexity of representation, communication and transfer of acquired information, directed or not to a specialist user, is of particular value to the goals of *sharing* knowledge.

This paper focuses on multimodal approaches, through the presentation of experimental results deriving from the application of innovative instrumental survey systems, i.e quick stereo-photogrammetric systems and laser scanner, to the documentation of the ancient monuments of the Appia Antica archaeological park, in the framework of a joint research project carried out by the Institute for Technology Applied to Cultural Heritage of the National Research Council (CNR-ITABC) and the Municipal Archaeological Superintendence of Rome.

The Public Administration has required a reliable scientific documentation finalised to the realisation of an Informative System and 3D explorable and navigable environments of the Appia Antica archaeological heritage. The necessity of this scientifically reliable documentation, leads to the experimentation of the integration of different survey technologies.

2 Quick stereo-photogrammetric system application

One of the survey technologies used is a quick non-conventional stereo-photogrammetric system applied for the complete stereoscopic coverage of different ancient archaeological artefacts. It is *CyclopII* produced by *Menci Software*. It consists of a digital mono-camera system which has the same advantages of a bi-camera based system and allows stereoscopic acquisition that can be used directly for restitution.

This system guarantees the same accuracy of geometric data acquisition as a traditional stereoscopic method, which is fundamental in the state of conservation process of diagnosis.

It consists on a digital calibrated photographic camera (in this case a Nikon D-100 digital photocamera, with a 24 mm lens was used) applied to a sleigh bound so that it can flow on a rectilinear guide realized in a special steel and mounted on an aluminium bar.

The sleigh flows freely on the guide and can be set in specific holes which represent different settled position in the guide. The system so constituted can be positioned on a common topographical tripod.

The system allows to use every type of camera, provided with calibration parameters, depending on the characteristics of the object to relief and on the precision to obtain. Thanks to the use of a single photocamera, the system has numerous advantages as: cheapness, accuracy (a single calibration), constructive homogeneity (same response to colour and same defectology in general).

The distance between the two shots, is variable and it can be optimized considering the shot distance, the artefact dimension and the objective of the camera.

For the realization of the stereoscopic pair it will be enough to determine the optimal distance from the object and between the two shots, to move the sleigh on the first position taking the first photo and finally to move the

Figure 1 *Laterizio I: raster 3D stereoscopic model (anaglyphic image)*

sleigh on the other position taking the second photo. In this way a stereoscopic pair can be realized, with all the characteristics that make it optimal for three-dimensional restitution (Fig. 1).

Although the system does not require topographical support, the latter was nevertheless provided in order to verify the reliability of the test system, the accuracy of the results obtained and to set the stereoscopic models properly in real space. The topographic survey of coordinates of significant points, natural points or special targets, was carried out with a total laser station. In this way it has been possible to obtain data of control for the following phase of 3D stereoscopic models creation in order to minimize errors in overlapping stereoscopic models. Moreover the use of control points in a dedicated software, *Stereo View of Menci software*, allows to link together the stereoscopic couples in such a way that they can be explored successively passing from one model to another without interruption (Menci, 2000). The final result is composed of metric *raster* stereoscopic strips, explorable in a stereoscopic environment, where it is possible to perform three-dimensional measures and to survey conservation status. In this way the vectorial phase, which often represents a subjective abstraction from the truth, can be avoided: the *raster* strips provide a corpus of information deriving from the geometrically controlled stereoscopic pairs, which are qualitatively and quantitatively richer than a traditional survey (Menci, 1999).

For the quick creation of single stereoscopic model it is possible to use an other software, *Cyclop II*; this software actually displays all the characteristics and functions of a digital photogrammetric renderer, but does not require any preparatory phases of photogram orientation. The software allows the pair of images of each stereomodel to be loaded immediately; the only data required, in addition to the couple of images chosen, are the stereoscopic pair base used (distance between two shots), the distance from the object and the selection of the calibration parameters of the lens used.

Besides the treatment of digital images for the stereoscopic model creation the software allows measure and three-dimensional vectorial restitution.

Using a computer provided with dedicated hardware characteristics (3D video card, glasses for stereoscopic vision, or Z-screen) it is possible to pass directly from shooting to measure and sketch.

In this way geometrically controlled 3D models of the artefacts surveyed has been realized in order to be successively inserted in an archaeological landscape reconstruction (Fig. 2).

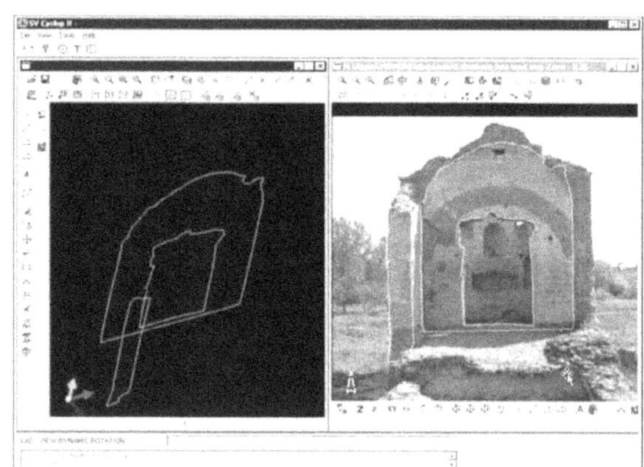

Figure 2 *Colombario: example of three-dimensional vectorial restitution*

3 Laser scanner technology application

Laser scanner technology has been applied for the documentation of ancient artefacts that have a complex and difficult to survey geometry. Indeed this technology allows to obtain points clouds that give an objective and discrete representation of the artefacts and that can easily and quickly describe complex geometry surfaces.

A punctual three-dimensional survey, as that it is possible to acquire by the laser scanner, allowed to take the x-y-z spatial coordinates of a lot of points of the studied surfaces, whose reciprocal distance has been fixed in relation to the needed survey precision and detail representation.

The processing of this great number of points and the subsequent definition of the surveyed surfaces permitted to describe, analyze, and "communicate" the studied monuments through digital models, from whom it's possible to obtain measurements, projections, sections that can be utilized for historical-architectural analyses or for other specific studies of the artefacts themselves.

For the survey activities a Leica HDS2500 laser scanner (former Cyrax 2500) has been used. It is a *time of flight* laser scanner, able to scan objects up to 100 metres with a single point accuracy of 6 millimetres. Indeed the acquisition distance can be changed from 1.5 to 100 metres, although maximum recommended range is 50 metres (Johansson, 2002). For his characteristics it is widely applied in the field of architectural survey: it can be used not only for single monuments documentation, but also for urban environments acquisition.

The scanner functionality is controlled with the help of a laptop computer, connected to the system, and through the use of Cyclone software, of Cyra Technologies. Since the laser scanner field of view has a limited angle of 40 degrees and the laser pulse reflect off the object with a specific angle-shot, the hidden surfaces won't be digital acquired. Indeed, in the acquisition phase, several scans have to be taken in order to obtain the documentation of the whole object, changing many times the scanner position. Cyclone software allows the acquisition, the registration and the alignment of single points clouds, captured from the different scanner position, and necessary for the complete coverage of the different part of a monument.

Through the use of this laser scanner the complex geometry of ancient artefacts, which remains are represented only by the internal construction in *opus caementitium*, has been realized.

The use of this advanced technique has guaranteed optimized working times and thus costs optimization and, above all, more rigour and precision of the geometric data acquired than that one that would be possible to obtain by the use of traditional topographic survey systems. The acquired points clouds were in fact imported in an AutoCAD environment, where, by the use of a special plug-in, furnished with the software Cyclone, it was possible to obtain a great number of horizontal and vertical sections of the artefacts, necessary for geometry object understanding.

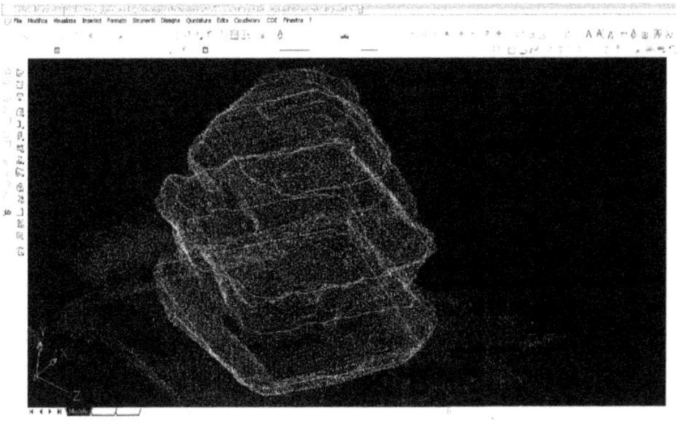

Figure 3 *Creation of horizontal and vertical sections from acquired points clouds of the Appia Antica park monuments*

For example, in the case of the so called "mausoleum with a pyramid shape" survey (Fig. 3), sectioning the points cloud acquired with a great number of horizontal planes, positioned to a fixed distance, it was possible to show the variability of plan morphology and to bound up this variability to the monument state of conservation, in such a way as to suggest to public administration priorities of preserving interventions.

From the acquired data it is finally possible to realize three-dimensional virtual wireframe models and then to make render of them, applying materials information. An experimentation of 3D modelling was carried out in collaboration with the Virtual Heritage Lab (ITABC-CNR), co-ordinated by Maurizio Forte, regarding the study case of the Ninphaeum of Egeria situated in the Caffarella Park (Forte et al., 2003). Other 3D modelling experimentations are still in progress.

In order to obtain photorealistic model, i.e. an as realistic as possible affinity between real and virtual materials, we proceed to texture the models, obtained by typical operations of points clouds modelling (points clouds cleaning, noise filtering, triangulation and construction of the mesh), with the photos obtained from the stereoscopic acquisition: we derive the monoscopic orthorectified images, necessary to realise the geometric models texture, by processing the whole stereoscopic coverage of the artefacts, provided with the topographical support.

The final result is a very precise overlapping between geometric and photographic documentation of the monuments surfaces (Fig. 4): in this way a complete description of the whole geometric, dimensional, chromatic and material information of the architectural building is possible.

Figure 4 *Ninphaeum of Egeria: orthorectified images texturized on the 3D geometric model*

4 A study case: San Nicola church on the Appia Antica

Both the described technologies, photogrammetry and laser scanner, have been applied for the documentation of the Appia Antica San Nicola church, which survey is still in progress. This building is part of the monumental architectural whole formed by the mausoleum of Cecilia Metella and the *castrum Caetani*, situated at the third mile of the Appia Antica (Fig. 5).

Figure 5 *San Nicola church in the Appia Antica*

The church, which dates back to the beginning of XIV century, that is to the foundation of the medieval *Caetani* fortress, can be considered a late example of Cistercian ecclesiastical architecture and was probably destined to be utilized from all the *castrum* villager (Paris, 2000). It is included in the remains of the fortress wall system and it is an unique aisle church with a semicircular apse at the end and a bell gable in the front.

The inside of the church is characterized by narrow ogive arch windows, between which peperino wall brackets recur: they are the support of the ancient ribs of the "diaframma" arches that held up the wooden beams of the saddle roof, nowadays completely lost.

The survey activities have been planned in such a way as to realize, side by side, the stereoscopic coverage and the laser scanner acquisition of the monument. Having the same condition of data acquisition was considered important in order to correctly integrate this two technologies and to obtain a scientific comparison of the survey results. Particular attention was devoted to the topographic survey of significant points coordinates (carried out with a total laser station), necessary either to perform the chaining of the different stereoscopic models, or to realize the alignment and global registration of the single points clouds or, finally, to produce the orthorectified images for geometric models texture. Where possible the same topographic points were acquired to perform all the next processing phases, in order to optimize the quality and accuracy of results

Figure 6 *San Nicola church: global point cloud of the surveyed part of the monument*

integration and to guarantee a perfect overlap between geometric and photographic documentation.

Actually only the surveys of the inside and of the back-front of the church were completed (Fig. 6). As regards the laser scanner acquisition a general resolution of 1 cm was adopted; however some detailed scans of the decorative elements, as for example of the peperino wall brackets with the decorative motif of goblet with leaves, were realized, adopting a point to point spacing of 6 mm (Fig. 7). The alignment and registration of the points clouds, obtained in Cyclone software through the topographic data use, produced a global points cloud with an error threshold of 5 mm.

The survey of the semicircular apse external façade was particularly complex, since the limited free area in front of, where the equipment could be positioned. "Limit conditions" of the systems use have been experimented.

As regards the stereo-photogrammetric acquisition a large number of shots were taken, proceeding with a coverage of vertical strips as well as horizontal ones; afterwards some shots were realized with an axis inclined towards the bottom and/or the top, but with a rotation respect to the vertical of not more than 15°, in order to avoid excessive deformation.

Figure 7 *San Nicola church: detailed scans of the decorative elements*

As far as scanner laser concerned, a limit acquisition distance (about 2,5 metres), with a great angle-shot inclination, was tested; moreover, near the maximum of the apse semi circumference, the overlap between the two subsequent scans had to be smaller than the 30%, making the choice of the two scan common reference point, for carrying out the alignment with, very limited. However a good alignment result was obtained, with few not acquired hidden surfaces (Fig. 8).

The processing in AutoCAD environment of the global points cloud till now created is starting, in order to obtain 2D polylines, formed by hundreds of vertex, that represent different level plans and vertical profiles of the artefact (Fig. 9).

The aim is to obtain geometric information about the generating curves of the few portions of ancient "diaframma" arch ribs still well conserved. The integration of these geometrical analyses results with all data derived from historical-critical investigations or typological-morphological studies, could help in the definition of reconstructive hypothesis of the ancient saddle roof, not longer preserved, and of the internal spatial configuration of the church.

Figure 8 *San Nicola church: global point cloud of the semicircular apse external façade.*

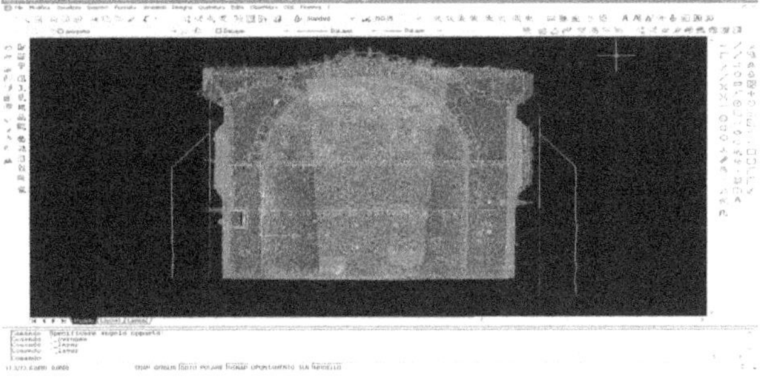

Figure 9 *San Nicola church: example of the global point cloud vertical section.*

These kind of geometric elaborations could be very important too for inquiring into the presence of possible monument static damages and for verifying the verticality of the outside main walls, not any more stiffened and connected together trough the church covering structure. The complete stereoscopic coverage of the church also represents an instrument of great help to evaluate the conservation state of such a monument by now reduced to a ruin state, for which a constant *monitoring* and *programmed maintenance* are essential.

The stereoscopic survey can be considered as a rigorous representation of the artefact state of conservation in a specific time, that can be analyzed by expert users, tanks to all the qualitative, morphological and colour information specific of the raster document, with the purpose to manage and to program correct conservative interventions.

The research project has been a valid occasion to experiment the integration of traditional survey techniques with innovative systems, developed in the last years.

Different ancient artefacts have been chosen for being surveyed both by photogrammetry and laser scanner in order to compare the results of these measurement techniques, to define which method would be the best for specific kind of monuments and at last to experiment the possibility of combination of these two technologies. This combination is certainly important to be pursued for reaching a dept artefacts knowledge, since each technique has a specific field of application and allows the acquisition of different kind of geometric and descriptive data.

On the one hand 3D laser scanner technology allows to automatically obtain very detailed discrete geometric model of architectural buildings, independent from subjective user choices. On the other hand the stereo-photogrammetric survey produces 3D raster models, metrically correct, in which morphological details, architectural materials and colorimetric definitions are easily singled out.

The Appia Antica Park particular experience, refers to another research project which ITABC-CNR is actually carrying out in collaboration with the Superintendence for Cultural and Environmental Heritage of the Valle d'Aosta, in order to perform the 3D survey and documentation of the roman triumphal arch of Augusto in Aosta. It is a question of applying the same methodologies for acquiring a scientific base of geometric data for which it will be possible to reconstruct, in this case, an archaeological landscape in a urban context, in such a way to compare the performances of the same techniques applied to different fields (architecture, archaeology, etc.) and at different scale (town-planning, architectural, etc.).

References

FORTE M. et al. 2003, "Integrating technologies: the Appia Antica project", Proceedings of the *Italy-United States Workshop: The reconstruction of archaeological landscapes through digital technologies*, Rome

JOHANSSON M. 2002, "Explorations into the Behaviour of Three Different High-Resolution Ground-Based Laser Scanners in the Built Environment", Proceedings of *CIPA WG 6. International Workshop on Scanning and Cultural Heritage Recording*, Corfu

MENCI L., CECCARONI F., SALONIA P. 1999, "The stereoscopic exploration of 3D-models as instrument of knowledge, documentation and measurement for mural painting", Proceedings of *ICCROM Graphic Documentation Systems in Conservation of Mural Paintings GraDoc*, Rome

MENCI L. 2000, "StereoSpace: an idea for photogrammetric data collection", Proceedings of *XIX ISPRS Congress*, Amsterdam

PARIS R. (edited by) 2003, "Via Appia. Il mausoleo di Cecilia Metella e il Castrum Caetani", publisher Electa, Milan

Multi-Resolution Image-Based Visualization of Archaeological Landscapes in Palpa (Peru)

Martin Sauerbier,[1] Gerhard Schrotter,[1] Henri Eisenbeiss,[1] Karsten Lambers[2]

[1]ETH Zurich, Institute of Geodesy and Photogrammetry, CH-8093 Zurich, Switzerland
Email: [martin.sauerbier], [gerhard.schrotter], [henri.eisenbeiss]@geod.baug.ethz.ch
[2]German Archaeological Institute, DAI-KAAK, Endenicher Str. 41, D-53115 Bonn, Germany
Email: lambers.kaak@gmx.de

1 Introduction

In recent years the increasing availability of digital maps, elevation models, and aerial as well as satellite images has provided archaeologists with a new, highly attractive and powerful tool for the study of archaeological landscapes. When combined and visualized in an appropriate way, these kinds of geocoded data allow enriched impressions of the region under study. Visualization plays an important role in two main fields: (1) archaeological research and (2) presentation and dissemination of results.

Virtual 3D models allow a given landscape to be appreciated in much more detail than paper maps due to the scalability and three-dimensionality of the data layers, and the possibility to integrate texture and to assign attributes to elements of the model. When visualized in an interactive mode and in real-time, moving through such a 3D model offers different perspectives on the landscape, while zooming in and out enables studies at any desired scale. A major constraint in this regard is that available digital data usually reflect the current state of the study area, which may strongly differ from the state of the time period under study. This limitation, however, may itself become an important aspect of archaeological research if the study of environmental, and possibly man-made change is integrated into the research design. Virtual 3D models are especially helpful when complementing rather than replacing archaeological fieldwork as they allow certain details of the study region to be reviewed during analysis back in the laboratory. They may furthermore serve as a starting point for a GIS-based analysis of spatial patterns and relations and are thus a primary tool for archaeological research.

At the same time, their visual attractiveness makes virtual 3D models a valuable means of disseminating the results of archaeological research to interested colleagues, the wider public and, last but not least, funding agencies and sponsors. A visualization of the study area combined with archaeological data, such as the location of sites, allows the audience a much more palpable impression of the research undertaken. While the aforementioned limitations of visualizations in an archaeological context should always be made transparent, virtual 3D models nevertheless constitute a powerful tool for publication and dissemination as well.

From a technical point of view, limiting factors when visualizing archaeological landscapes are often scale or resolution and texture. If several data layers from different sources are combined, their scale or resolution may not match up, resulting in unsatisfying visualizations. While high resolution photo texture allows full advantage to be taken of the capability of zooming into a 3D model, it increases computing times and storage requirements considerably. Many virtual 3D models therefore represent a compromise between user requirements and available data.

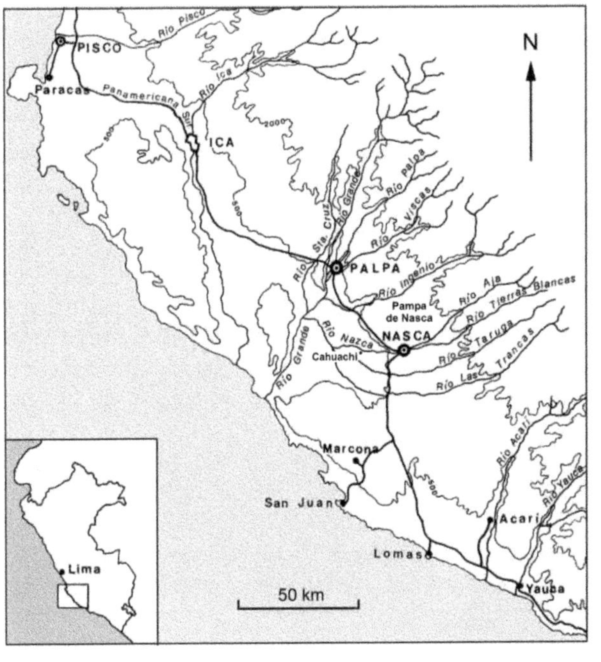

Figure 1 *Location of Palpa on the south coast of Peru.*

In this paper we present a new method for multi-resolution image-based visualization that aims to overcome these problems. The data used to demonstrate this new method have been acquired in the framework of the Nasca-Palpa Archaeological Project on the south coast of Peru (Fig. 1). Since the first field season in 1997, this research project has been a joint venture between archaeologists from the German Archaeological Institue (KAAK Bonn) and the Andean Institute of Archaeological Studies (INDEA Lima) and geomatic engineers from the Swiss Federal Institute of Technology (ETH Zurich). While the primary goal of this project is to investigate the prehispanic cultural history of the Palpa region in the northern part of the Nasca basin (Isla, Reindel 2006; Lambers 2006), at the same time it has offered us many opportunities to develop, test, and adapt new technologies and methods for archaeological research (Reindel, Gruen 2006).

This interdisciplinary approach has allowed us over the past decade to produce different virtual 3D models of subareas of our study region for well defined research purposes (Sauerbier, Lambers 2003; Eisenbeiss et al. 2005; Lambers, Sauerbier 2006). The scale and scope of these models range from a supra-regional overview model, to subareas of our study area, and finally to single sites. While we mostly used data acquired by ourselves to generate these models, we also integrated data from other sources where necessary. Eventually we decided to integrate the different 3D models into a single one. This was achieved by multi-resolution image-based visualization, the details of which are described in the following sections.

2 The Data Basis

For this first attempt towards a combined visualization of 3D models from the Palpa region, three datasets were chosen. Based on images at a scale of 1:4,000 acquired in 2004 from an autonomous mini helicopter, a so-called UAV (unmanned aerial vehicle), which carried a digital CMOS camera Canon D10 with a resolution of 6 megapixels, photogrammetric processing resulted in a high resolution digital terrain model (DTM) as well as a high resolution orthomosaic of the Late Intermediate Period settlement of Pinchango Alto (AD 1000 to 1450) with a footprint of 0.03m (Eisenbeiss et al. 2005). This dataset covers an area of approximately 200 x 300 m^2. It is hereafter referred to as the "UAV model".

Prior to this, in 1998 a series of B/W aerial images had been acquired and manually processed on an analytical plotter WILD S9, resulting in a DTM with a mesh size of 2m and an orthomosaic with a footprint of 0.28m (Sauerbier, Lambers 2003). This photogrammetric data set, hereafter called the "aerial photogrammetric model", covers an area of 89 km^2, representing the core area of our study region around the modern town of Palpa.

Furthermore, in 2003 a digital surface model (DSM) with a mesh size of 30m was generated from an ASTER scene for the purpose of conducting regional visibility studies beyond the limits of the core area (Lambers, Sauerbier 2006). The DSM that covers an area of about 810 km^2, hereafter the "ASTER model", was generated in PCI Geomatica 8.2 using the bands 3N (nadir) and 3B (backward), which allow for stereo measurements. While the DSM was generated in an automated mode by image matching, the resulting grid was then edited manually and semi-automatically using Geomagic Studio 6. Additionally, from the ASTER bands 1 (0.52 - 0.60μm), 2 (0.63 - 0.69μm), and 3 (0.78 - 0.86μm) of the VNIR spectral range of the ASTER sensor, a composite orthoimage with a footprint of 15m was generated. Due to the fragmentary coverage of the visible spectrum by the ASTER sensor, the three bands were combined in order to achieve a natural color composition in RGB color space according to the following formulas:

$$R = b_2$$
$$G = 0.75\ b_1 + 0.25\ b_3$$
$$B = 0.75\ b_1 - 0.25\ b_3$$

Afterwards, the color balance was adjusted manually in Adobe Photoshop.

The UAV model represents a subset of the area covered by the aerial photogrammetric model, which in turn is a subarea of the ASTER model. The three terrain datasets were co-registered even though they had already been available in a common UTM coordinate system (zone 18S, horizontal and vertical datum: WGS 84) in order to avoid visible steps along the border regions. In a first step, the UAV model was registered to the aerial photogrammetric model using the iterative closest point algorithm implemented in Geomagic Studio 6.0. In a second step, the ASTER model was then registered to the aerial photogrammetric model using least squares 3D surface matching (Gruen, Akca 2005).

Prior to the visualization described here, the UAV model and the aerial photogrammetric model had been visualized in different ways, e.g. in real time and as virtual flyovers and high resolution still views (Sauerbier, Lambers 2003). Various commercial software tools had been used for this purpose, such as ERDAS Imagine (Leica Geosystems) for still views and virtual flyovers, Skyline Terra Explorer (Skylinesoft Inc.) for real time navigation and virtual flyovers, and Scene Viewer using the OpenInventor format on a SGI platform for real time navigation in stereo mode. Nevertheless, all of these software products had reached certain limits with respect to size or resolution of the data sets already when dealing with the individual 3D models. Therefore, a combined visualization of all three models was not feasible at the original resolution of each data set. Another shortcoming of most commercial visualization softwares concerning the output format of image sequences is their restriction to certain standard formats, such as NTSC or PAL. Our new approach of multi-resolution image-based visualization strives to overcome these shortcomings.

3 The Workflow

One goal during the development of the visualization approach was to ensure the capability of each module to be modified and enhanced on purpose. Therefore, it was decided to create our software in a modular platform-independent way based on command line philosophy. As operating system for development we chose Redhat's Open Source Fedora Project. The modules were written in standard C++ code (Norman 2001a, b) using the

OpenGL library-API (OpenGL 2006) to handle the 3D processing and visualization tasks. Scripts were wrapped around the modules to lead and widely automate the workflow. In addition to the well known on-screen XWindow-OpenGL mode (Kilgard 1996) we used the GPU functions of this library in combination with NVIDIA drivers (NVIDIA 2006a, b, OpenVIDIA 2006). These hardware based functions can be easily switched on and off to ensure the claimed platform and hardware independence. This design allowed us to stack the modules together in a flexible way, distribute tasks even over different operating systems, combine the modules with native image and video processing programs provided by the open source community, and handle error exceptions at the operating system level. The modules can be divided into three main tasks:

1. Design of the flight path
2. Processing of a single frame
3. Combination of multi image frames to a movie

The following three subsections deal with these tasks. The entire workflow is illustrated in Fig. 5.

3.1 Design of the flight path

A graphical user interface (Polack 2002) is applied to define viewpoints of interest and to save every single hotspot in a compact 3x4 projection matrix. These hotspots serve as nodes (key-frames) for the flight path. In order to achieve a smooth movement throughout the whole virtual flyover, a 3D Bézier curve is interpolated between the points of interest, and for every 3D sampling point an orientation vector is spherically interpolated.

3.2 Processing of a single frame

The above described projection matrix defines the field of view and allows the reduction of the huge amount of data. In our case the whole data set consists of around 3.6 million points from the UAV model, 22.2 million points from the aerial photogrammetric model, and 0.9 million points from the ASTER model, resulting in about 27 million object points in total.

In order to determine the image content of each single frame of a flightpath, the terrain data has to be projected from object space into the image coordinate system of the virtual camera at each position. This projection can be implemented in a single loop framework which reads data in and writes it out, and is therefore independent of the workstation's memory. To further increase the speed of this task, the terrain data is organized in bounding boxes.

The 2½D TIN is created by means of a Delaunay-triangulation (Fig. 2) using the qdelaunay function of the QHULL package (Barber et al. 1996).

At this processing step we do not yet consider the available image data. As described before, we use three different kinds of data sets. We have thus to decide for each viewpoint which orthoimage to use for terrain texturing. The orthoimage derived from the UAV data has a size of 10692 x 10149 pixels with a depth of 8 bit per color RGB, while the orthoimage derived from the photogrammetric flight consists of 42006x52154 pixels at 8 bit grayscale, and the ASTER orthoimage has a size of 7040x7430 pixels with 8 bit per color RGB depth. In a first step, we produce a so-called Facet-ID image (Fua, Leclerc 1994) to choose the most suitable of the existing image data sets (Fig. 5).

Figure 2 *Goraud-shaded DTM after view-dependent cropping of data and triangulation.*

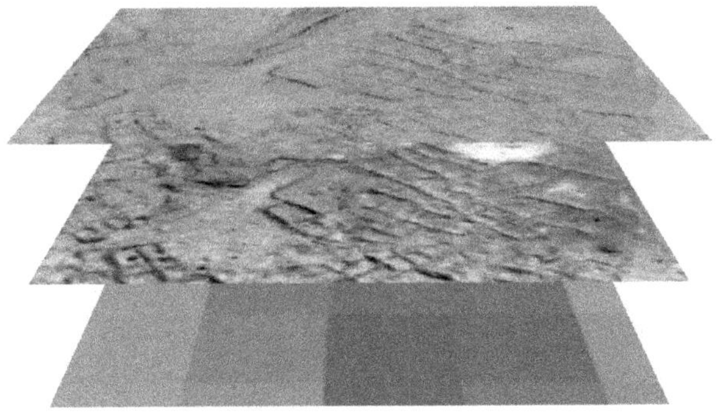

Figure 3 *Fully automated choice of orthoimages of different resolution from aerial photogrammetric, UAV and ASTER models.*

For each frame of the defined flight path, the Facet-ID is generated by encoding the index of each facet as a unique color depending on the distance and viewing angle of the virtual camera, and projecting the surface into the image plane using a standard Z-buffer algorithm.

We use RGB images to store the Facet-ID image. A RGB image with 8 bit per channel can store 2^{24} unique colors. We can thus handle up to 16777216 facets for every point of view. We use these ray-casted (synthetic) images to

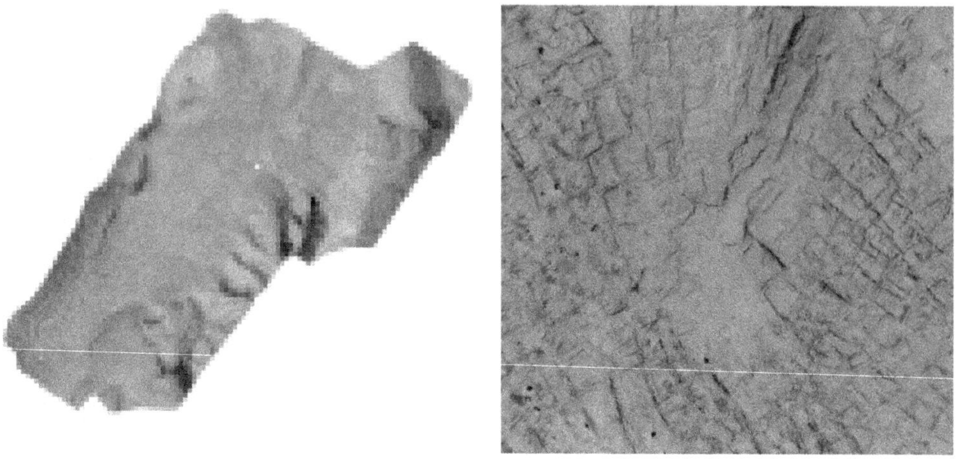

Figure 4 *Using the ImageMagick tool "convert" to obtain the pixel area of interest*

determine very quickly which surface points are occluded in a given view and which orthoimage should be used for texturing. The histogram of the facet area in the corresponding image plane is used to choose the source orthoimage (Fig. 3).

It should be noted here that the Facet-ID image in addition offers much more useful information, which we do not take advantage of in this workflow (Fua, Leclerc 1994). The known georeference and pixel size in object space of the orthoimages allow us to introduce a translation vector, which can be used to cut out the pixel area of interest from the chosen orthoimage. For this purpose, we apply the convert function with crop flag from the Open Source package ImageMagick to implement this key function in an efficient way (Fig. 4).

The cropped part of the orthoimage along with the points of field of view and the TIN of facets provide the required information to start a GPU-based off-screen planar orthographic texture mapping (Fig. 6). The texture coordinates obtained that way define the connection between the cropped image and the 3D model. The field of view is rendered using OpenGL and hardware acceleration.

3.3 Combination of the multi-image frames into a movie

For a flight path of around three minutes we need an amount of 4320 frames, assuming a standard frame rate of 24 frames per second for a smooth transition between the frames. Every single frame has to be integrated into a video file. For this task we apply mencoder, which is included in the mplayer package (MPLAYER 2006). As a video file format we chose mspeg4v2 (pre-standard MPEG-4 by Microsoft, version 2), which is well suited for our purpose due to the relatively high compression rate, which yields a small file size. Regardless of the high compression rate the visual quality is well-preserved, and furthermore the format is supported by most commercial multimedia players such as RealPlayer or Windows Media Player. The optimal bit rate and further settings of the mencoder tool, which need to be taken into account for a high quality flyover, can be found in Valle (2006).

The flyover generated on the basis of the data described above allows us to fly into the area of archaeological interest while the resolution of both geometry and texture increases. Nevertheless, a smooth transition along the border regions of the different data set was not implemented in our case because the image data changes from RGB to grayscale and back to RGB, which is why smoothing would not be useful.

4 Conclusions and Outlook

The above described series of software modules and the workflow presented here are very useful tools for a high quality and efficient 3D visualization of large terrain data sets and corresponding texture by means of image sequences. A graphical user interface for viewpoint determination increases user-friendliness by allowing for an intuitive flightpath generation. Further steps in the workflow are largely automated by means of implemented scripts, such that the input of various parameters is sufficient for the production of a movie file. Furthermore, rendering of high resolution still views and image sequences with arbitrary image format size can be performed independently of the size of the data sets.

The modules described above are scalable to any kind of data set, as long as any single view does not exceed the memory capacity of a single workstation.

This network idea allows for future enhancements, such that the scripts may run in a server network, which can be extended to a web application framework to provide client computers access to the software via internet. The user defines the area of interest in his browser, which shows the orthoimage as an overview. The developed scripts are then executed on the server side and provide the client with a textured 3D VRML data set of the area of interest. In the future we will provide this additional service and provide the implemented modules to the open source community.

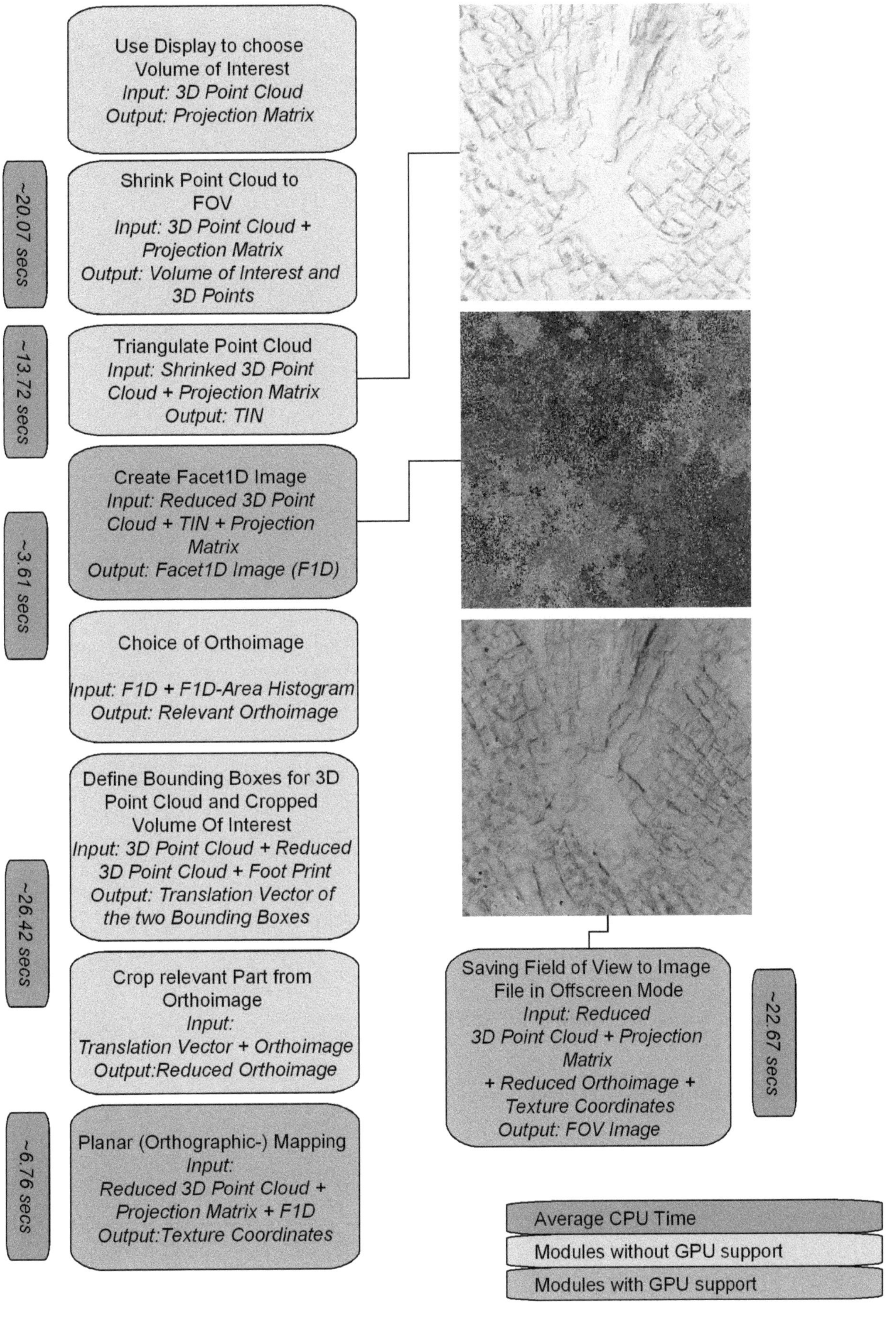
Use Display to choose Volume of Interest
Input: 3D Point Cloud
Output: Projection Matrix
~20.07 secs
Shrink Point Cloud to FOV
Input: 3D Point Cloud + Projection Matrix
Output: Volume of Interest and 3D Points
~13.72 secs
Triangulate Point Cloud
Input: Shrinked 3D Point Cloud + Projection Matrix
Output: TIN
~3.61 secs
Create Facet1D Image
Input: Reduced 3D Point Cloud + TIN + Projection Matrix
Output: Facet1D Image (F1D)
Choice of Orthoimage
Input: F1D + F1D-Area Histogram
Output: Relevant Orthoimage
~26.42 secs
Define Bounding Boxes for 3D Point Cloud and Cropped Volume Of Interest
Input: 3D Point Cloud + Reduced 3D Point Cloud + Foot Print
Output: Translation Vector of the two Bounding Boxes
Crop relevant Part from Orthoimage
Input:
Translation Vector + Orthoimage
Output:Reduced Orthoimage
~6.76 secs
Planar (Orthographic-) Mapping
Input:
Reduced 3D Point Cloud + Projection Matrix + F1D
Output:Texture Coordinates
Saving Field of View to Image File in Offscreen Mode
Input: Reduced 3D Point Cloud + Projection Matrix + Reduced Orthoimage + Texture Coordinates
Output: FOV Image
~22.67 secs
Average CPU Time
Modules without GPU support
Modules with GPU support

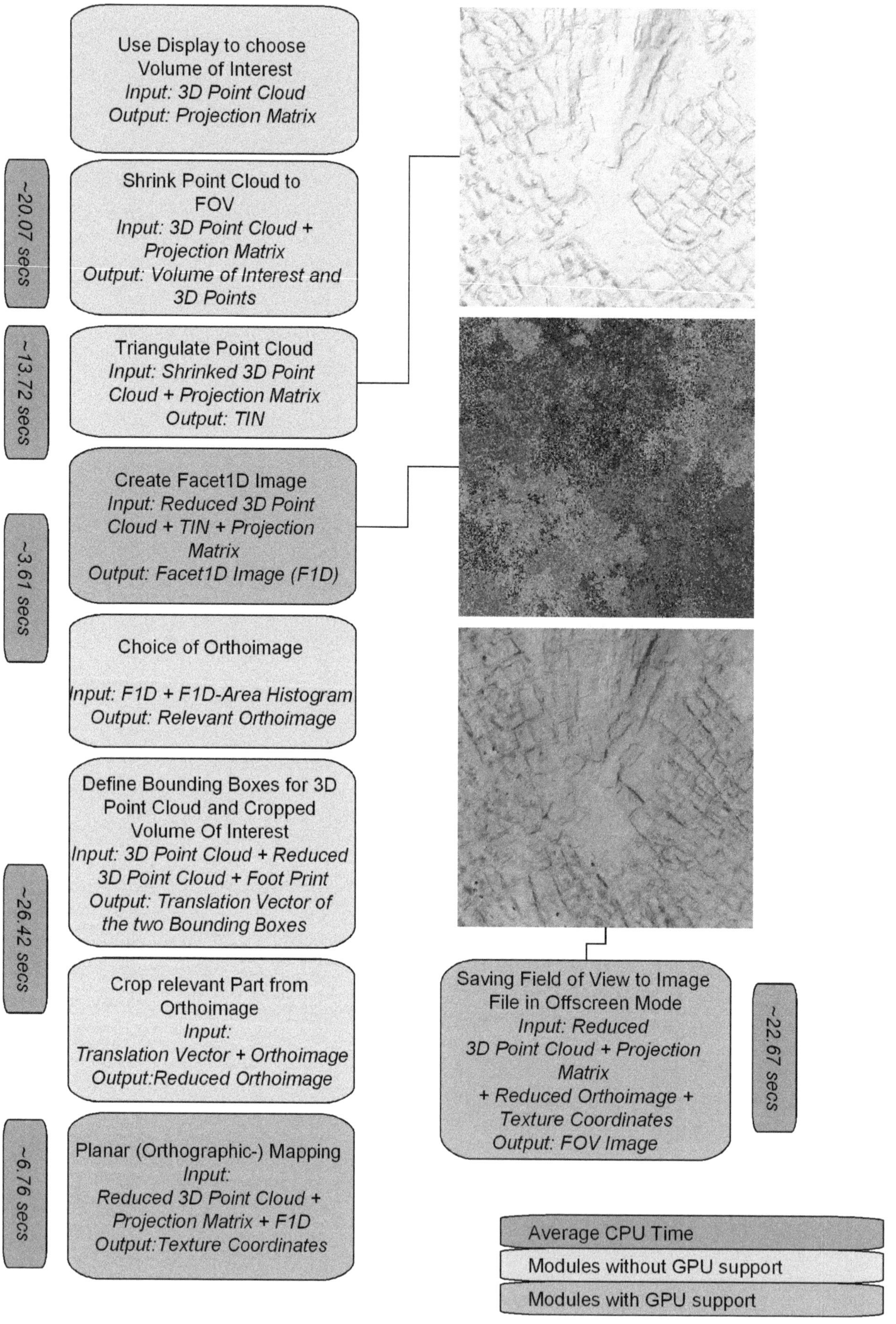

Figure 5 *Workflow for the processing of a single frame with required computation times*

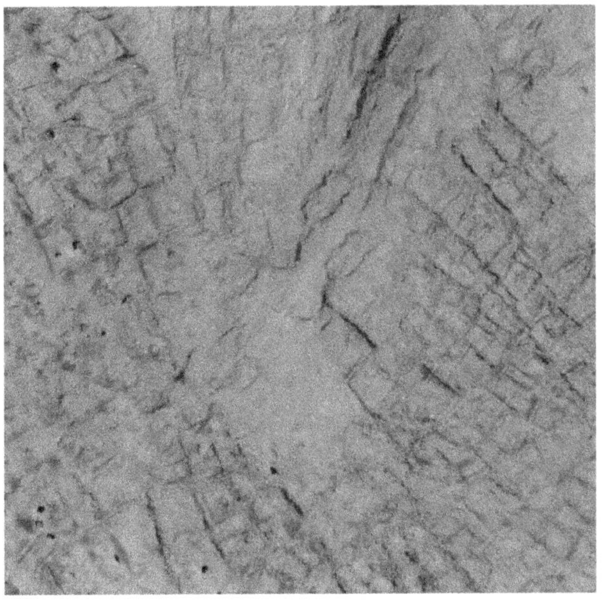

Figure 6 *3D model textured depending on view in off-screen mode*

With regard to our ongoing archaeological research in Palpa, the combined visualization of different 3D models that had originally been developed from different data sets, for different purposes, and at different times, provides us with a versatile tool for research and presentation as described in the introduction. Each of the three available 3D models can now be visualized with the other two providing additional information regarding area covered (scope) and detail (resolution). At the same time, additional 3D models that may be generated in the future can be smoothly integrated into the visualization. All this can be achieved without actually generating a new, additional model from the existing ones, which would considerable increase the required disk space for data storage. The visualization approach described here is thus a valuable tool for our work in Palpa and for archaeological research at a regional scale in general.

References

BARBER, C.B., DOPKIN, D.P., AND HUHDANPAA, H.T., 1996. The Quickhull algorithm for convex hulls. *ACM Trans. on Mathematical Software,* 22(4), pp. 469-483.

EISENBEISS, H., LAMBERS, K., SAUERBIER, M., AND ZHANG, L., 2005. Photogrammetric documentation of an archaeological site (Palpa, Peru) using an autonomous model helicopter. *International Archives of Photogrammetry, Remote Sensing and Spatial Information Sciences,* XXXIV-5/C34, pp. 238-243.

FUA, P. AND LECLERC, Y.G., 1994. Using 3-dimensional meshes to combine image-based and geometry-based constraints. *Proceedings of the European Conference on Computer Vision,* Stockholm, pp. 281-291.

GRUEN, A. AND AKCA, D., 2005. Least squares 3D surface and curve matching. *ISPRS Journal of Photogrammetry and Remote Sensing,* 59(3), pp. 151-174.

ISLA, J. AND REINDEL, M., 2006. Burial patterns and sociopolitical organization in Nasca 5 society. In W.I. Isbell, H. Silverman eds., *Andean Archaeology III: North and South*, pp. 374-400. New York: Springer.

KILGARD, M.J., 1996. *OpenGL Programming for the X Window System.* Boston: Addison-Wesley Professional.

LAMBERS, K., 2006. *The Geoglyphs of Palpa, Peru: Documentation, Analysis, and Interpretation.* Aichwald: Linden Soft.

LAMBERS, K. AND SAUERBIER, M., 2006. GIS-based visibility studies of the Nasca geoglyphs at Palpa, Peru. In E. Baltsavias, A. Gruen, L. Van Gool, M. Pateraki, eds., *Recording, Modeling and Visualization of Cultural Heritage,* pp. 249-262. Rotterdam: Balkema.

MPLAYER, 2006. Media Player/Mencoder, GPL license, Available at: http://www.mplayer.org [accessed 22 September 2006].

NORMAN, L., 2001a. *Linux 3D Graphics Programming.* Plano (TX): Wordware Publishing.

NORMAN, L., 2001b. *Advanced Linux 3D Graphics Programming*. Plano (TX): Wordware Publishing.

NVIDIA, 2006a. *Programmable Graphics Processor Technology*. Available at: http://www.nvidia.com [accessed 22 September 2006].

NVIDIA, 2006b. *NVIDIA GPU Programming Guide.* Available at: http://developer.nvidia.com/object/gpu_programming_guide.html [accessed 22 September 2006].

OPENGL, 2006. OpenGL environment for developing portable, interactive 2D and 3D graphics applications, Available at: http://www.opengl.org [accessed 22 September 2006].

OPENVIDIA, 2006. Parallel GPU Computer Vision. Available at: http://openvidia.sourceforge.net/ [accessed 22 September 2006].

POLACK, T., 2002. *Focus On 3D Terrain Programming.* Boston: Thomson Course Technology PTR.

THE GEOMETRY CENTER, Minneapolis MN, 2003. Qhull code . Available at: http://www.qhull.org [accessed 22 September 2006].

REINDEL, M. AND GRUEN, A., 2006. The Nasca-Palpa Project: a cooperative approach of archaeology, archaeomtery and photogrammetry. In E. Baltsavias, A. Gruen, L. Van Gool, M. Pateraki, eds., *Recording, Modeling and Visualization of Cultural Heritage,* pp. 21-32. Rotterdam: Balkema.

SAUERBIER, M. AND LAMBERS, K., 2003. A 3D model of the Nasca Lines at Palpa (Peru). *International Archives of Photogrammetry, Remote Sensing and Spatial Information Sciences,* XXXIV-5/W10 (CD-ROM).

VALLE, M., 2006. Mencoder Documentation, Swiss National Supercomputing Center, Available at: http://www.cscs.ch/~mvalle/mencoder/mencoder.html [accessed 22 September 2006].

A Multilevel Banded Intelligent Scissors Method for Fast Segmentation in Large Virtual Terrains

M. Schneider and R. Klein

University of Bonn
Institute of Computer Science II,
Römerstr. 164, 53117 Bonn, Germany
e-mail: {ms,rk}@cs.uni-bonn.de

1 Introduction

A fundamental task in remote sensing archaeology is the recovering and identification of places. The result of this process is usually a map of the landscape serving as a valuable resource for various kinds of further archeological investigations. Traditionally, the creation of an archaeological map is based on field work supplemented by the interpretation of aerial photography and the corresponding elevation model. By combining these data in a 3d visualisation system the landscape can be represented in its natural three-dimensional form simulating the perception in the field. Moreover, by navigating in the 3d environment, objects of interest can be inspected from arbitrary views including perspectives hardly possible in reality. In addition, data from various additional sources like survey results, GPS measurements, laser scanner data, etc. can be used to further augment the landscape. However, up to now, three-dimensional visualisation systems are restricted to a simple data exploration without the possibility to interact with the virtual environment or to extract information from it. Instead the actual segmentation is performed on the aerial photography, elevation data or their derivatives restricting perception to a simple two-dimensional bird's eye view. Even worse, the segmentation in common GIS usually has to be performed completely manually, i.e. the user has to define each object individually with pixel accuracy. This pixel-exact work makes object extraction a particularly time-consuming task for users that is often perceived as frustrating.

We present a semi-automatic segmentation tool integrated into a 3d terrain visualisation system. The combination of a realistic visualisation of the landscape with a computer-assisted segmentation tool facilitates the visual recovery and identification of objects and simplifies their extraction. The semi-automatic nature of the segmentation method allows the user to specify an object faster and with less user interactions than manual marking without reducing accuracy. Thus, the high level cognitive task of object identification is left to the user whereas the segmentation algorithm performs the low level task of capturing the fine details of the object boundary. The object boundaries obtained with our method are completely georeferenced and can be saved in the common shapefile format and are therefore importable by other GIS for further processing.

2 Previous Work

Our idea of combining a semi-automatic segmentation tool with a 3d terrain visualisation engine is based on work in the field of terrain rendering and semi-automatic image segmentation. Multiresolution algorithms for fast rendering of large terrain data sets with viewpoint adaptive resolution have been an active area of research in the field of computer graphics and GIS for many years. Since giving a complete overview is beyond the scope of this paper, we refer to recent surveys (Lindstrom 2002, Pajarola 2002). Semi-automatic segmentation methods can basically be divided into region-based and boundary-based approaches. Since we present a boundary-based technique, we will briefly review previous methods belonging to this category.

Active contour models (Kass 1988), also called snakes, require a user to input an initial curve that approximates the target boundary. After initialisation the points of the curve are iteratively adjusted by minimising an energy functional. The energy functional is a combination of internal forces (such as boundary curvature) and external forces (like image gradient magnitude). Since snakes follow a pattern of user-initialisation followed by an automatic energy minimisation the user does not know in advance what the resulting boundary will look like. If the resulting boundary is not satisfactory, the entire process has to be repeated or the boundary must be postprocessed manually.

Another well-known group of boundary-based techniques are those based on Intelligent Scissors (Falcão 1998, 2000, Kang 2002, Mortensen 1995, 1998, 1999, Wong 2000). Intelligent Scissors (Mortensen 1995, 1998) is a highly interactive tool based on a global graph search. Later, Falcão et al. presented a slightly different version called Live Wire (Falcão 1998). By planting an initial seed point, a path map is constructed to provide the minimum-cost path from the seed to every pixel in the image. Moving the cursor near the boundary of an object causes the current path to be extended according to the path map forming a boundary segment. Whenever the path deviates from the true object boundary the user can insert an additional seed point thereby fixing the old boundary segment and creating a new one starting from the newly created seed point. Whenever a new seed point is selected the path map has to be recomputed with regard to the new seed replacing the previous map. While Intelligent Scissors provides highly interactive visual feedback once the path map is computed, it is time-consuming to recompute the path map, especially for large images.

Based on Live Wire a method called Live Lane (Falcão 1998) was presented that aims at increasing the efficiency of the boundary construction. Live Lane restricts the search domain to construct the path map only within a local window centered at the current seed point. As the cursor moves in this window, the corresponding boundary segment is interactively displayed according to the path map. When the cursor crosses the window, the boundary segment from the seed point to the crossing point is automatically frozen. The crossing point becomes the new seed and a new path map is constructed within the window centered at the new seed, i.e. a new seed point is added whenever the cursor leaves the current window. Hence, Live Lane requires more seed points than Live Wire, especially when the window size is small. Moreover, the automatically inserted seed points may not lie exactly on the target boundary which degrades the accuracy and repeatability of Live Lane.

Enhanced Lane (Kang 2002) adopts the idea of the local search from the Live Lane, i.e. finding a boundary segment within a small window. However, in Enhanced Lane the window is moved together with the cursor, incrementally extending and updating the path from the current seed point to every pixel in each successive window, forming a cumulative path map. Like Live Wire, Enhanced Lane selects a new seed only when the current path digresses from the desired boundary and does not need additional seed points to be inserted like Live Lane. Assuming that the window sequence completely contains the target boundary in the right order, it can be proven that Enhanced Lane always produces the same result as Live Wire.

3 Overview

In our system the user can navigate freely within the virtual landscape and explore objects of interests from arbitrary views. By providing simple mouse gestures directly on the 3d terrain surface, objects can be extracted easily from the landscape. Therefore, hidden to the user, the input hints are projected to the underlying aerial photograph serving as input to the segmentation algorithm. The actual object boundary is then extracted by the semi-automatic image segmentation algorithm working on the aerial photograph. The resulting boundary is afterwards projected back onto the terrain surface providing immediate feedback to the user giving him the impression to work completely in 3d.

The Visualisation Engine

A vital requirement for our method is to ensure interactive response. To this end, an efficient high quality, real-time visualisation of the landscape as well as interactive feedback by the segmentation algorithm is indispensable even for very large data sets comprising several gigabytes of raw data. We use the terrain rendering system presented by Wahl et al. (Wahl 2004) which is based on a quadtree data structure (see Figure 1). The system has proven to be able to visualise very large data sets efficiently and with high quality, e.g. data sets with a resolution up to a few centimetres for the aerial photography together with elevation models of about 1m covering areas of hundreds of square kilometers have already been visualised with real-time frame rates. Due to the good scalability of the system the real-time visualisation of upcoming data sets of even higher resolution as a result of improved acquisition methods can be expected to run at real-time frame rates as well.

As it will be used later, we briefly review the quadtree data structure as it is used by the visualisation engine to represent the aerial photography as well as the elevation model. The root of the quadtree holds the entire domain of the data set in a single tile. Up to four children partition their parents' domain into equally sized quarters where each quarter has the same size as the parent tile, i.e. the four children represent the same domain but with twice the resolution. The quadtree is created by initially partitioning the air photo into equally sized square tiles of a given size representing the base level of the tree. The remaining levels of the quadtree are created by recursively downsampling the original air photo by a factor of two and merging 2x2 tiles from the lower level into one tile of the same size at the higher level. The segmentation tool operates on the same quadtree data structure used for rendering. This is advantageous in several respects: image data for rendering and segmentation has to be held in memory only once, the quadtree data structure allows fast access to spatial subparts of the terrain and different levels of detail of the data set are already available.

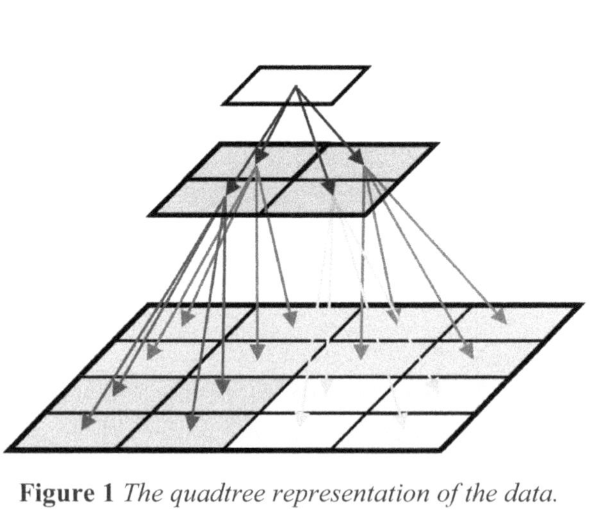

Figure 1 *The quadtree representation of the data.*

Review of Intelligent Scissors

Our segmentation algorithm is based on the basic idea of Intelligent Scissors to formulate the boundary detection problem as an optimal path search in a graph. The objective is to find the optimal path from a seed node to a destination node where pixels in the image represent nodes with directed and weighted edges connecting its eight adjacent neighbours. An optimal path is defined by the minimum cost path, i.e. a path with the smallest sum of edge costs. Since a shortest path in the graph should correspond to an object boundary in the image, pixels with strong edge features in the image should lead to low local costs in the graph and vice-versa. Hence local costs are created as a weighted sum of the following edge features: Laplacian zero-crossing f_z, gradient magnitude f_g and gradient direction f_d. Letting $L(p,q)$ represent the local cost on the directed edge from node p to a neighbouring node q, the local cost function is:

$$L(p,q) = w_z \cdot f_z(q) + w_g \cdot f_g(q) + w_d \cdot f_d(p,q),$$

Where the w's are empirically chosen weights of the corresponding feature function. The laplacian zero-crossing is a binary edge feature used for edge localisation. Since the laplacian zero-crossing creates a binary feature only it does not distinguish between strong, high gradient edges and weak, low gradient edges. However, gradient magnitude provides a direct correlation between edge strength and local cost. The gradient direction or orientation adds a smoothness constraint to the boundary by associating a relatively high cost for sharp changes in boundary direction.

4 Description of the segmentation

Despite the fact that the Intelligent Scissors technique provides a powerful tool for image segmentation its speed and memory consumption constrain its feasibility when large data sets need to be processed. Since in the case of terrain data we usually have very large data sets the direct application of the original Intelligent Scissors approach is not possible.

In order to ensure interactive response to user input there are several issues concerning speed and memory requirements we have to address: the sheer size of the air photo often exceeding main memory capacity, the fast calculation of the image edge features and the efficient computation of the shortest path. To this end, we show how the quadtree representation of the image data that is also used by the visualisation engine for rendering can be utilised to drastically improve performance with respect to memory consumption and speed. Our segmentation algorithm ensures interactive response even on very large data sets by exploiting the quadtree data structure in two ways: Localising the search domain within a quadtree level and employing a multilevel banded heuristic to exploit the hierarchical structure. The multiscale nature of our approach cannot only be considered as a heuristic to speed up the segmentation process but can also be used to further simplify the segmentation of objects taking into account their scale. In addition, the computation of the edge features is accelerated by utilising the GPU for the respective convolutions.

Localising the boundary search

Our approach of localising the search domain adapts and enhances the idea of Enhanced Lane to work on tiled images as given by a quadtree (see Figure 2). Loading the air photo completely into memory and computing the edge costs on startup, as in Enhanced Lane, is extremely costly in terms of time and memory requirements or even not possible at all. Precomputing the edge features and storing them to disk would help to minimise computation time at runtime but requires large amounts of disk space even more than the data set itself.

In contrast to Enhanced Lane we extend the idea of incrementally extending the search domain to the loading of the necessary parts of the image, to the computation of the edge features and to the construction of the corresponding parts of the graph. Instead of the entire air photo only the image tiles covering the current search domain are hold in memory. When the mouse is moved and the search domain needs to be incremented the necessary tiles are loaded, their edge features are computed and the graph is extended. The tiles needed to increment the search domain are selected by considering a virtual circle centered at the current cursor position. The circle is moved along together with the mouse cursor and all tiles at least partly within this circle are added to the search domain. The insertion of a new seed point causes all previously loaded tiles, computed edge costs and the graph to be deleted in order to minimise memory requirements.

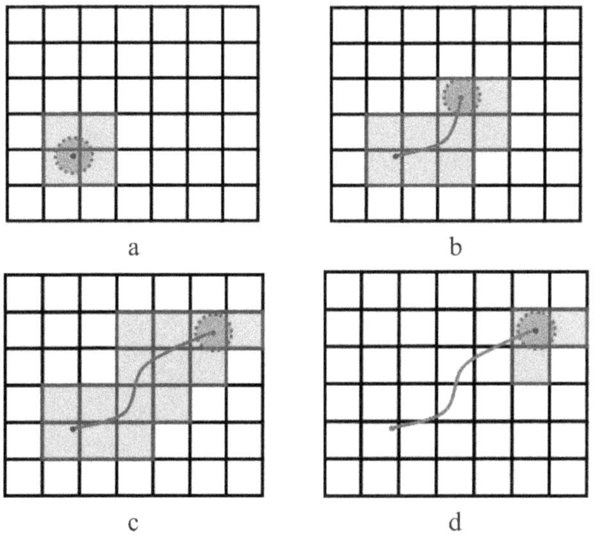

Figure 2 *Localising the search domain using the given quadtree tiling. In a) an initial seed point is planted causing tiles being activated (grey). When the mouse is moved additional tiles are loaded b) and c). Planting a new seed causes previously activated tiles to be deactivated.*

The computation of the edge features is accelerated on modern graphics hardware by performing the convolution with the appropriate filter kernels on the GPU using vertex and fragment shader. The gain in speed that is achieved depends on the number of convolutions performed and the support of the applied filter kernels. Generally, the more convolutions are performed and the larger the support of the kernels, the bigger is the advance in speed since the fixed costs of uploading the image tile to the GPU are getting relatively smaller.

In contrast to Enhanced Lane we do not use a dynamic version of the Dijkstra algorithm that computes the shortest path from the seed to all nodes. Instead we use the D* algorithm (Stentz 1994, 1995), a dynamic version of the A* algorithm, that only computes the shortest path from the seed to the current node. There are two main advantages of this algorithm over a dynamic Dijkstra algorithm. With D* only the shortest path to the nodes we really need are computed. In addition, D* usually does not need to recompute large parts of the path map as it is often necessary with a dynamic Dijkstra, but in most cases only a much smaller subset of it.

Placing seed points directly on an object's boundary is often difficult and results in spikes if the seed is not localised exactly. To facilitate seed point placement image snapping (Gleicher 1995) is supported which automatically moves the mouse pointer to nearby image features within a small, user-defined neighbourhood.

Multilevel banded approach

Our multilevel banded Intelligent Scissors approach is similar to the approach used to accelerate graph cuts (Lombaert 2005) that was inspired by the well-known narrow band algorithm in level set methods (Adalsteinsson 1995) as well as the multilevel graph partition method (Karypis 1998). We exploit the hierarchical structure of the quadtree by starting segmentation not on the finest quadtree level, i.e. the original air photo, but on a coarser level. The segmentation on the starting level is performed with the approach presented in the previous section. The result is propagated to the next higher resolution level where the segmentation is performed only within a narrow band surrounding the projected result from the coarser level (see Figure 3). This procedure is iterated until the highest resolution level is reached. Since the algorithm is only run on the subgraph that comprises the narrow band the additional computation at the fine resolution is drastically reduced compared to running it on the full graph. Additionally, since only the subgraph has to be constructed, memory consumption is also reduced.

The size of the band can be adjusted by an optional dilation parameter d that is important in practice. If d is large, the computational benefits are decreased and the wider band may also introduce additional outliers far away from the desired object boundary. On the other hand, if d is small, the algorithm may not be able to capture the full details of object boundaries with complex shapes or high curvature. In our implementation we use $d=1$ pixel which seems to be a reasonable compromise.

The multiscale heuristic proposed here is an approximation that does not guarantee exactly the same results compared to performing the algorithm on the finest resolution level. However, this observation can not only be interpreted as an approximation error but also as a view at the landscape at a different scale. Considering multiple resolutions can help to reduce uncertainty in segmentation because at lower resolution the boundary might be better defined while higher resolutions are needed to obtain accurate boundaries. Specifying low frequency boundaries located in the direct vicinity of high frequency boundaries is often difficult since the segmentation algorithm has to choose from many possible alternative boundaries nearly equally likely. Even worse, high frequency boundaries often have stronger responses to the applied edge detectors resulting in even lower costs and therefore shorter path in the graph. In practice this leads to snapping of the free path segment to unwanted boundaries and therefore requires many seed points to be placed in order to obtain the desired boundary. However, on a coarser resolution level where the high frequencies are removed or at least attenuated the object boundary of the large scale object prevails and can be detected more easily at the expense of resolution. The propagation to finer resolution levels can then be interpreted as a boundary estimation procedure that gradually refines the object boundary.

5 Results and Discussion

In order to prove the general-purpose applicability of our segmentation tool we tested it with several objects from different categories and of various sizes and complexity.

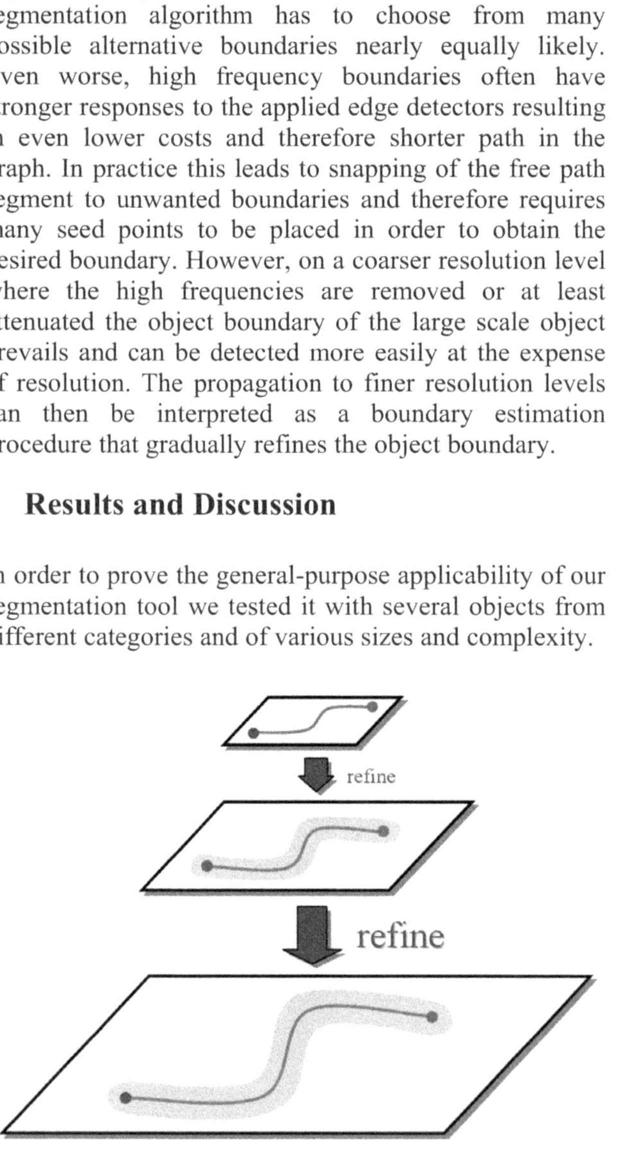

Figure 3 *The banded Intelligent Scissor heuristic. The path computation is started on a user-defined level of the quadtree. The resulting path is successively projected down to the finer resolution levels restricting the search domain to a narrow band surrounding the projected result.*

Two large HRSC data sets served as basis for the visualisation as well as the segmentation. The first data set contains the Turtmanntal valley in Switzerland covering an area of about 300km^2 with a resolution of 1m for the aerial photography and elevation model each. The second data set comprises the Wettersteingebirge including the Zugspitze. It also covers an area of approximately 300km^2 but with a resolution of 0.25m for the aerial photography and 2m for the elevation model. As a result of our incremental and hierarchical algorithm, segmentation can be performed at interactive speed nearly independent of the size of the underlying data set. Figure 4 depicts some of the test objects we used to asses the efficiency, robustness and generality of our method. In Figure 4 a) and b) the segmentation of a forest boundary is shown, starting at different levels of the quadtree. In a) the segmentation is performed solely on the finest resolution level resulting in several deviances from the desired boundary caused by fine scale features. Many seed points need to be placed in order to obtain the desired result. However, in b) the segmentation is started from three levels above the finest resolution leading to the correct boundary since the disturbing small scale features are not present at higher levels of the quadtree.

In general boundary identification on aerial photography is influenced to a great extent by the lighting conditions during the acquisition process. Therefore segmentation results are biased, especially by shadows, causing the segmentation algorithm to follow shadow boundaries instead of true object boundaries. One approach to tackle this problem is to consider the elevation data in the boundary estimation procedure in addition to the aerial photography. Unfortunately, high resolution elevation data is required for this which is often not available.

6 Conclusions

In this paper we presented an interactive image segmentation technique inside a terrain visualisation engine. Our method lifts the usual restriction of current systems that allow either a data inspection within a 3d environment or a segmentation of the data performed on the base images separately. It is important to note that this was realised only through the ability to provide interactive response to user input even when processing very large data sets. To this end, we utilised an efficient visualisation engine and adapted and enhanced already existing segmentation techniques. In conclusion, when compared to tedious manual boundary tracing on images, semi-automatic segmentation in a 3d environment offers increased insight into the structure of the landscape and the objects it contains and provides quicker and more accurate segmentation results at the same time.

References

ADALSTEINSSON, D. and SETHIAN J. A. 1995 A Fast Level Set Method for Propagating Interfaces, *Journal of Computational Physics,* 118:269-277

FALCÃO, A. X., UDUPA J. K., SAMARASEKERA S., SHARMA S., HIRSCH B. E. and LOTUFO R. 1998 User-Steered Image Segmentation Paradigms: Live Wire and Live Lane *Graphical Models and Image Processing,* 60:233-260

FALCÃO, A. X., UDUPA J. K. and MIYAZAWA F. K. 2000 An Ultra-Fast User-Steered Image Segmentation Paradigm: Live Wire on the Fly *IEEE Transactions on Medical Imaging,* 19(1):55-62

FALCÃO, A. X. and UDUPA J. K. 2000 A 3D Generalization of the User-Steered Live-Wire Image Segmentation *Medical Image Analysis,* 4:389-402

GLEICHER M. 1995 Image Snapping *SIGGRAPH 95 Proceedings,* 183-190

KANG H. W. and SHIN S. Y. 2002 Enhanced Lane: Interactive Image Segmentation by Incremental Path Map Construction *Graphical Models,* 64(5):282-303

KARYPIS G. and KUMAR V. 1998 Multilevel K-Way Partitioning Scheme for Irregular Graphs *Journal of Parallel and Distributed Computing,* 48:96-129

KASS M., WITKIN A. and TERZOPOULOS D. 1988 United Snakes *Proceedings of IEEE International Conference on Computer Vision,* 933-940

LINDSTROM P. and PASCUCCI V. 2002 Terrain simplification simplified: A General Framework for View-Dependent Out-Of-Core Visualization, *IEEE Transaction on Visualization and Computer Graphics,* 8(3):239--254

LOMBAERT H., SUN Y., GRADY L. and XU C. 2005 A Multilevel Banded Graph Cuts Method for Fast Image Segmentation *Proceedings of Tenth ICCV,* 1:259-265

MORTENSEN, E. N. and BARRETT W. A. 1995 Intelligent Scissors for Image Composition *SIGGRAPH 95 Proceedings*, 191-198

MORTENSEN, E. N. and BARRETT W. A. 1998 Interactive Segmentation with Intelligent Scissors *Graphical Models and Image Processing*, 60:349-384

MORTENSEN, E. N. and BARRETT W. A. 1999 Toboggan-Based Intelligent Scissors with a Four Parameter Edge Model *Proceedings of IEEE Computer Vision and Pattern Recognition,* 2:452-458

PAJAROLA R. 2002 Overview of quadtree based terrain triangulation and visualization, *Technical Report UCI-ICS TR-02-01*, University of California Irvine

STENTZ, A. 1994 Optimal and Efficient Path Planning for Partially-Known Environments. *Proceedings of the IEEE International Conference on Robotics and Automation,* 3310-3317

STENTZ, A. 1995 The Focussed D* Algorithm for Real-Time Replanning. *Proceedings of the International Joint Conference on Artificial Intelligence,* 1652-1659

WAHL R., MASSING M., DEGENER P., GUTHE M. and KLEIN R. 2004 Scalable Compression of Textured Terrain Data *Journal of WSCG,* 12(3):521-528

WONG K. C., P. -A. HENG and WONG T. T. 2000 Accelerating 'Intelligent Scissors' Using Slimmed Graphs *Journal of Graphic Tools,* 5(2):1-13

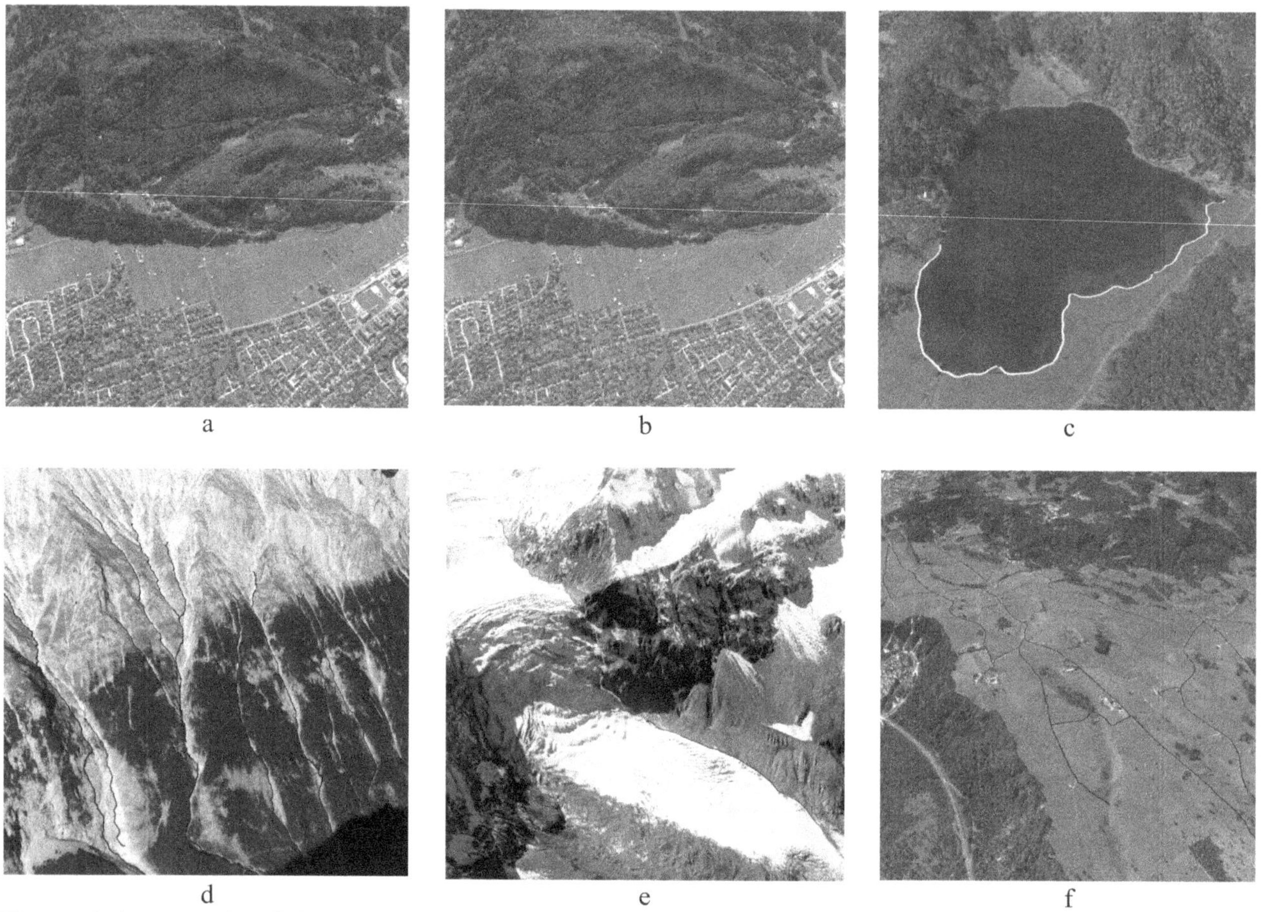

Figure 4 *An example of the results obtained when starting segmentation on different levels of the quadtree is illustrated in a) and b). In a) only the finest resolution level is used whereas in b) segmentation is started at an eight times lower resolution and is afterwards refined. Figure c)-f) show the segmentation of a lake near Krün, debris flow tracks in the Reintal close to the Zugspitze, the Turtmann glacier in Turtmanntal valley and roads near Mittenwald.*

Watermarking of 3D digital models for IPR protection

F. Uccheddu, V. Cappellini

Department of Electronics and Telecommunications, University of Florence, Italy

1 Introduction

In the last decade, watermarking has been one of the most active research topics, attracting the interest of researchers with different backgrounds, such as signal processing, communication and information theory, cryptography, and computational vision (Cox et al. 2001, Barni et al. 2004). However a great deal of this research effort has focused on digital watermarking of audio, images and video data. The result is that watermarking technology for this kind of media has now reached a good maturity. On the contrary, watermarking of 3D objects is far from this level even if 3D models are diffused in several applications such as virtual prototyping, Cultural Heritage, and entertainment industry (film and video-games). One of the reasons for this gap is that it is difficult to extend common processing algorithms used in signal processing to 3D data. This is the case, for example of basic tools such as filtering, frequency analysis, and compression.

Even if 3D objects can be represented in several different ways (e.g. NURBS, voxels, implicit surface, polygonal meshes) most of the existing 3D watermarking algorithms work on polygonal meshes since this representation is the lowest common denominator of the other ones (i.e. it is easy to convert the other representations to meshes). For example the watermark may be inserted by altering mesh attributes such as vertex coordinates or vertex connectivity. In this paper we follow the same approach, i.e. we assume mesh representation for the 3D object to watermark.

A strategy which is successfully adopted by many watermarking algorithms designed to deal with still images and video sequences, consists in first describing the host signal by means of a multiresolution framework and then inserting the watermark at a resolution level ensuring a satisfactory trade-off between perceptibility of the watermark and robustness against attacks.

The extension of this multiresolution approach to the 3D case, however, is not straightforward. The main reason for this difficulty, as well as for the difficulties encountered when trying to extend 2D processing tools to the 3D mesh case, is that the essentially 2-manifold structure of 3D surfaces has to be taken into account. In particular it is not possible to define equi-spaced sampling patterns on general 2-manifolds thus making the extension of Fourier and other multiresolution analysis and synthesis tools very difficult. As a matter of fact, in order to extend common signal processing algorithms to geometry data we need to use sampling patterns which are as regular as possible.

Subdivision surfaces (Schroder et al. 99) have recently attracted the attention of computer graphics researchers since by providing a semi-regular sampling of surfaces, they are likely to become a fundamental block of many multiresolution algorithms for mesh processing (Schroder 2002). This is the case, for example, of 3D wavelet decomposition (Lounsbery et al. 97). In the following, a polygonal mesh obtained by regularly subdividing an irregular coarse one will be referred to as a semi-regular mesh. In this paper, we present a novel multiresolution mesh watermarking algorithm particularly designed to work with semi-regular meshes with subdivision connectivity.

The proposed algorithm embeds the watermark by modifying the wavelet coefficients of 3D models obtained by decomposing the host mesh by means of the algorithm proposed by Lounsbery et al. (Lounsbery et al. 97). Particular attention is paid to ensure that the embedding algorithm preserves the visual integrity of the models. The watermark is recovered by means of a correlation detector designed according to statistical detection theory.

A multiresolution framework has been used for watermarking of polygonal meshes by other 3D watermarking systems (Kanai et al. 98, Praun et al. 99, Yin et al. 2001, Ohbuchi et al. 2002). The main distinguishing feature of the watermarking algorithm proposed in this paper with respect to these methods regards watermark detection. In fact all of these algorithms require the original non-marked model to detect the watermark, i.e. they are non-blind techniques. On the contrary our technique is blind thus resulting in a much more flexible system easily adaptable to practical applications.

This paper is organized as follows. In section 2 some issues on 3D watermarking are given. Section 3 gives an overall idea about the perceptual aspects of a watermarked 3D model. Section 4 describes the proposed watermarking algorithm, with regard to the embedding phase, and watermark detection. Experimental results are presented in section 5. Finally some conclusions are drawn in section 6.

2 3D Watermarking technology

In digital watermarking, a digital code, or watermark, is embedded into a digital media, so that a given piece of information is indissolubly tied to it. This information can be used to prove ownership, identify a misappropriating person, trace the model dissemination through the network, and so on. The main goal of 3D watermarking is the production of a stego-model, i.e. a 3D model containing some hidden data, mainly in order to protect the intellectual property rights.

A watermarking algorithm is called *blind*, if it does not require the original data to recover the embedded information (on the contrary, the algorithm is called *non-blind*); *readable*, if the algorithm is capable to read the watermark without knowing it in advance, *detectable* if it is able to establish if a certain watermark is present or not in the watermarked data.

Watermarking technology can be used to embed public or private information. A *private watermark* may contain information to prove ownership of the digital media in dispute. *Public watermark* usually replaces header or other user-defined information. This last type of watermark requires high capacity, i.e. the watermarking system has to be capable to embed a lot of bits in the digital data. On the contrary, the first type of watermark has strong robustness requirements but less constrains about capacity; in fact malicious users, the *attackers*, could want to do a misappropriate use of the digital media eliminating the watermark by modifying it or a part of it. So, a private watermark has to resist to these modifications, that are called *attacks*.

An easy way to develop a 3D watermarking technique is to extend an image watermarking algorithm to 3D, since image watermarking is an already mature research field. However, the nature of the data itself make complicated this extension. In fact, an image is a bi-dimensional, regularly sampled collection of values, while a mesh is a collection of 3D space points (not regularly sampled) with intrinsic curvatures and a particular topology. So there is more than one degree of freedom of difference between mesh and images: it is not a simple 2D to 3D extension!

2.1 Three-dimensional Models and Representations

One of the characteristic of 3D models that 3D watermarking technology has to take in account is the fact that a model can be represented in different way. For example, by a collection of parametric curves (e.g. NURBS), by a set of implicit surfaces, or by polygonal meshes. In the following we assume *triangular mesh* representation, i.e. a mesh composed only by triangles. This is a common assumption because meshes can be seen as the lowest common denominator of other representations because it is very simple to convert other representation to mesh.

2.2 Embedding Features

The features used to embed the information depend on the media to watermark. For example, in image watermarking the information can be hidden by changing the value of a subset of image pixels, or of the coefficients of some mathematical transformation (e.g. the Fourier Transform). A mesh can provide more features to manipulate such as vertex positions, connections between vertices (topology) and other surface properties like texture or vertex colours. This characteristic apparently allows a lot of possibilities for watermark embedding, but we have to take into account that most of the model information is encoded by vertices. For this reason we are interested in the features related to the geometry of the model, i.e. *geometric* and *topological features*. The main *geometric feature* of a mesh are its vertices. One possible way to embed the watermark is to modify the position or the normals of vertices. Both this entities are altered by perturbing the coordinates of the mesh vertices. *Topological features* are related to the connectivity of the mesh vertices. Usually, a set of connected vertices is selected by using some proper geometric feature. Then, the connections of this set are altered to encode one or more bits. In the following, when we are talking about generic 3D watermarking algorithms, we assume that these algorithms embed the information in the model by perturbing mesh vertices.

2.3 Attacks on a 3D model

One of the main problems in 3D watermarking is that a lot of complex attacks can be carried out on a mesh. Additionally, due to the more complex nature of the data itself, an attack carried out on 3D data is more complicated to prevent than one on image or video data. In fact, as previously noticed, a mesh is not a collection of regularly sampled values, such as audio, images or video, but a collection of unorganized points in 3D space with intrinsic curvatures and a particular topology defined by the connections between vertices. It is difficult to imagine all possible attacks on a polygonal mesh, some of these are:

- *Translation/Rotation/Uniform Scaling.* These geometric transformations are very used in computer graphics to position a 3D model inside a scene.
- *Noise.* For noise attack we intend the random perturbation of mesh vertices.
- *Re-triangulation.* This attack concerns the changes between the connections of the mesh vertices.
- *Mesh smoothing.* A smoothing of the surface represented by a polygonal mesh can be obtained by mesh filtering. One of the most popular mesh filter is Taubin filter(Taubin 2005) that acts as a low-pass filter on the mesh attenuating the roughness of the surface that it represented.
- *Polygonal simplification.* This operation is often used to transmit a low-level version of the

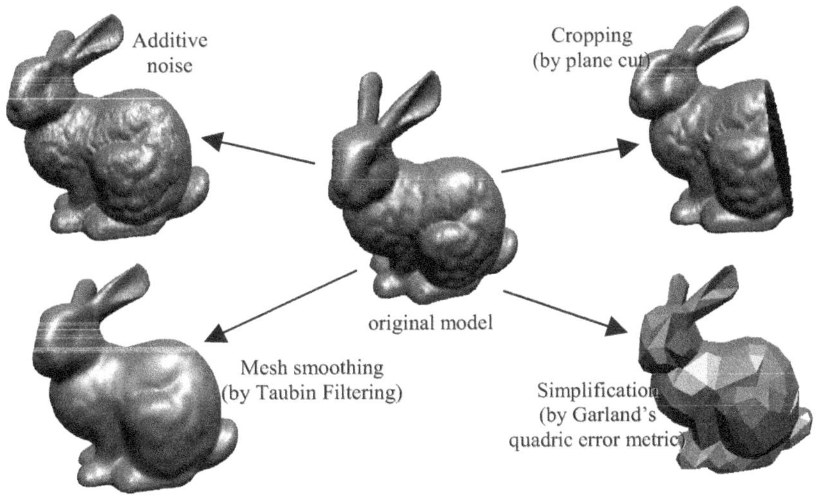

Figure 1 *Examples of 3D watermarking attacks*

model or to optimize a model eliminating most of the non-salient faces.

- *Cropping.* An attacker can discard the part of the model that he does not need. For example the hand of a statue.
- *Remeshing.* This operation can be described as a geometric resampling of the shape of the mesh followed by a re-definition of the connections between vertices.

3 Perceptual aspects

An important issue related to 3D watermarking concerns the perception of the watermark by the user. This crucial point has been deeply studied in image watermarking where a lot of work to make the watermark invisible to human eye has been done (Podilchuk et al. 98, Wolfgang et al. 99,Bartolini et al, 98). For many reasons this aspect radically changes when we treat 3D watermarking. For example, in interactive applications, the user can see the 3D model from every point of view he likes, so, if the user has an original version of the model a visual comparison with the watermarked one will point out the geometric deformation introduced by the watermarking algorithm. On the contrary, if the user does not know anything of the original model, even high deformations could not be noticeable. For example, consider a human head. If the watermarking algorithm strongly deforms facial features, for an observer it is impossible to say if the head he is seeing is watermarked or not.

Nevertheless, in some applications (e.g. Cultural Heritage preservation) the watermarked model and the original one must be very close each other from a geometrical point of view, despite the perceptibility of the watermark. One way to give an objective measure of the degradation introduced on the model by a watermarking algorithm it is to use some metric (e.g. Hausdorff distance) to quantify the geometric difference between the surface of the original model and the watermarked one. In this work we use a method to measure the difference between a model and its watermarked version from a subjective point of view, in order to evaluate the perceptibility of the watermark. To achieve this goal we have realized a software to test if a user is able to distinguish between two version of the same model; the original and the watermarked one. During our studies to develop 3D watermarking algorithm we have investigated a multiresolution-based approach in order to improve both robustness of the watermarking algorithm and visual quality of the watermarked model. The framework of our algorithm is sketched in Figure 2. Our approach provides a *class of 3D watermarking algorithms*, in fact the information in the coarse version can be embedded by different watermarking algorithms. Additionally, several kind of mesh multiresolution analysis tools(Lee et al. 98, Eck et al. 95,Lounsbery et al.97) can be used.

The solution we have investigated consist on applying the watermarking algorithm on a multiresolution version of the input model and re-obtain the watermarked model at the original resolution by re-adding the wavelet coefficients extracted during the multiresolution analysis process.

4 The proposed system

The 3D watermarking system presented implements an algorithm that embeds a numeric code into a semi-regular mesh with subdivision connectivity using a multiresolution framework. An important feature of the algorithm is the detector blindness. To obtain this result, and yet preserve robustness against geometrical transformations, it is necessary that the watermark embedding and detection phases work on a normalized model, i.e. on a model re-oriented and scaled as the original one. This normalization allows to ignore

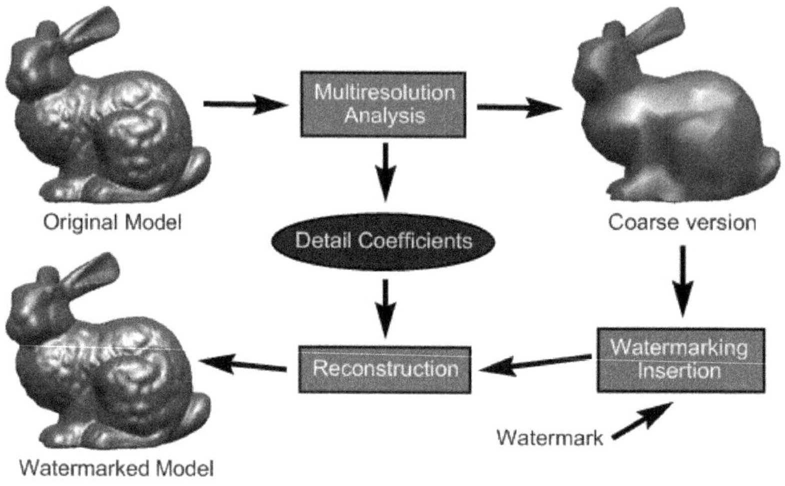

Figure 2 *A multiresolution framework for 3D watermarking algorithms.*

translation, rotation and uniform scaling modifications. Mesh normalization works in two stages. In the first one, model orientation is normalized by means of Principal Component Analysis (PCA). Then the model is fitted to a bounding box consisting of a cube of dimensions 1.0*x*1.0*x*1.0 centred in the barycentre of the model. Fitting is obtained by applying a translation and a uniform scaling. It goes without saying that model normalization at the detector must be accomplished without the use of the original model, not to compromise the blindness of the overall watermarking system.

Embedding rule

The watermarking embedding algorithm works according to three parameters:

i) a secret key K;
ii) the resolution level l that hosts the watermark, and
iii) a coefficient γ determining the strength of the watermark.

An overall picture of the watermark insertion process is shown in Figure 3.

In particular by means of a wavelet analysis we obtain a coarse version of the input model with a set of wavelet coefficients representing the level details of the model. Depending on the word we want to encode, a numeric key is generated and by means of a map the code has embedded, by adding a value to all vertices of the coarse model. Subsequently detail coefficients have added to the coarse watermarked model to obtain the high resolution

$$\rho = \frac{1}{n} \sum_{w_i \in W_l} \langle w_i, \gamma f_i \rangle$$

watermarked model.

Readers interested in the details of the embedding algorithm are referred to (Uccheddu et al. 2004).

Watermark Detection

The detection procedure works as follows: the user specifies the numeric key K and the level of resolution l where he wants to verify the presence of K and the detector provides a positive or negative answer. To do so the watermarking map W_{MAP} is generated by starting from K. V_l and W_l are obtained by wavelet analysis, then the correlation between the watermarking signal and the wavelet coefficients is calculated as follows:

where $f_i = f(w_i, W_{MAP})$ and n is the cardinality of W_l. If ρ is greater than a certain threshold T_ρ the watermark is

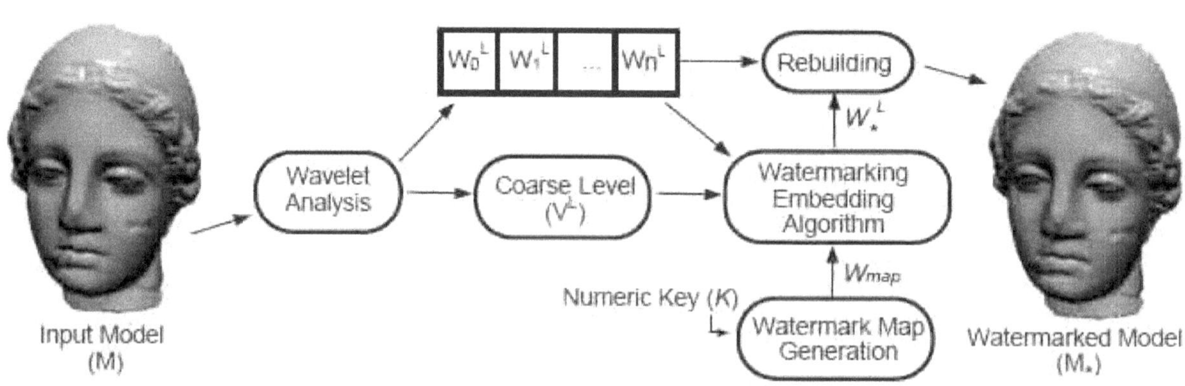

Figure 3 *Scheme of the watermark embedding system*

present, otherwise the model is declared non-marked. The value of T_ρ is obtained by means of statistical considerations. It is important to observer that correlation-based detection, is proven to be optimal by Neyman Pearson theory (Barni, Bartolini 2004) for supposed Gaussian and linearly independent features.

Choice of T_ρ

The most popular approach to set T_ρ consists in resorting to statistical detection theory. Specifically the Neyman-Pearson criterion is usually adopted, which consists in maximizing the missed detection probability for a given probability of falsely revealing the watermark in a non-marked host model. We use the same approach here. To do so, we need to compute the probability of falsely detecting the watermark presence, i.e.:

$$Pf = P\{\rho > T_\rho | H_0\}$$

where H_0 indicates the hypothesis that the examined model is not marked.

Similarly we define the probability of missing the watermark as:

$$Pm = P\{\rho < T_\rho | H_1\}$$

where H_1 indicates the hypothesis that the examined wavelet coefficients w_i are marked with the watermark whose presence is under verification. In both cases, the error probabilities are obtained by fixing the host model and averaging over different watermark codes K. By fixing a certain false alarm probability P_f we obtain a value for T_ρ.

5 Experimental results

In this section we report a selection of the results that we obtained while testing the validity of our watermarking algorithm. Specifically two aspects were considered: watermark invisibility and robustness.

Watermark invisibility

In order to achieve high visual quality of the watermarked model we have carefully considered the problem of watermark perceptibility. Since it is very difficult to evaluate in a objective way the perceptibility of the geometric distortions introduced by the watermarking process, we evaluated such distortions by visual inspection; an human user compares the original and the watermarked model using a software expressly designed for this purpose and find the maximum watermark strength (γ) that results in a non-visible watermark.

We used this analysis to determine a value of γ that ensured invisibility in all the tested models. All the experimental results we provide in this section guarantee the imperceptibility of the watermark, i.e. a high visual quality of the watermarked model.

bunny		venus	
l	γ_{max}	l	γ_{max}
3	0.0005	3	0.0003
2	0.0008	2	0.0005
1	0.0030	1	0.0009

Table 1 γ_{max} *increases at lower resolution levels*

In table 1 it is shown how the maximum perceptually tolerable value of γ (indicated with γ_{max}) changes while varying the model and the resolution level l. In particular γ_{max} increases at lower resolution levels.

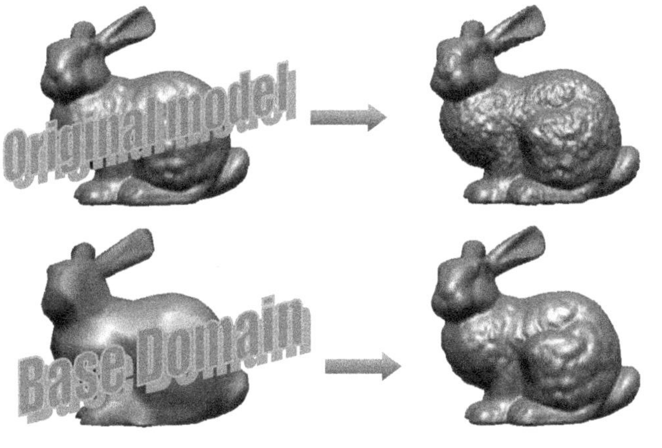

Figure 4 *Watermark visual effects*

By embedding the watermark directly on the original full resolution model the deformation introduced by the watermark algorithm are very distinguishable on the resulting model. On the contrary, the embedding in a low resolution version (the base domain) of the original model preserves the original high quality of the model as we can observe in Figure 4.

Watermark robustness

One of the main problems of 3D watermarking is the wide variety of different attacks possible on a polygonal mesh. This is one of the main reasons why many 3D watermarking algorithms use the original model in the extraction phase. Since our technique is specific for semi-regular meshes with subdivision connectivity, we do not take into account those attacks that alter this properties of the mesh such as re-triangulation, simplifications or re-meshing. Instead we tested the robustness of the algorithm against geometric transformations such as translation, rotation, uniform scaling, additive noise, filtering, cropping and a combination of the above.

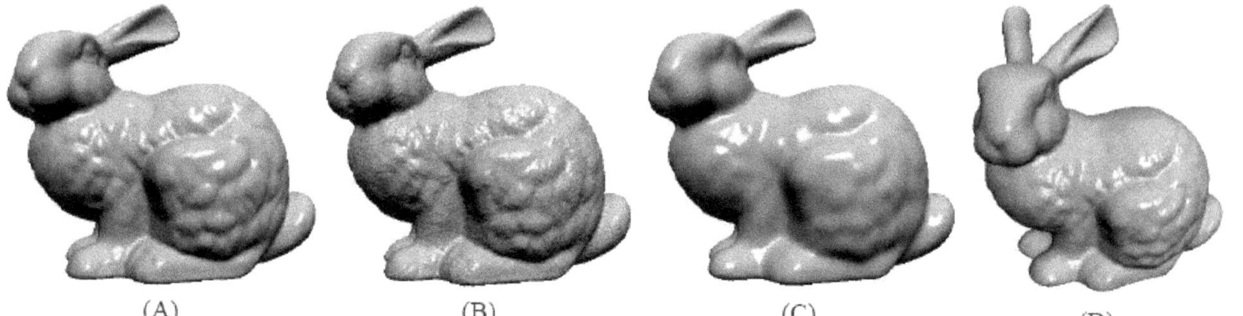

(A) (B) (C) (D)

Figure 5 *Robustness against Additive Noise, Smoothing and Combined attacks. (A) Original bunny model. (B) The model attacked with the maximum amount of noise for which the watermark is detected. (C) The model after 15 applications of smoothing filter. (D) Robustness against combined attacks; the model has been smoothed by Taubin filter, attacked with noise and then rotated by 22° around x axis and 11° around y axis. In all cases the detector is able to recover the watermark.*

6 Conclusion

In this paper we presented a blind watermarking algorithm for 3D models. In order to cast the watermarking problem in a multiresolution framework, the algorithm is expressly designed to work with semi-regular meshes, thus making 3D wavelet analysis feasible. A particular mapping strategy is proposed to take into account the non-regular sampling of he 3D mesh. Correlation-based and geometric normalization allow the blind detection of the watermark and a good robustness against several attacks.

Several directions for future work remain open. First of all, we are planning to apply the watermark to subparts of the mesh (Kats et al 2003). By applying geometric normalization to each subpart, robustness against combined cropping and geometric manipulations should be achieved. We are also going to evaluate the robustness of the watermark when the semi-regular mesh is converted to an irregular one, edited, and brought back to a semi-regular format by remeshing. Finally, to further diminish watermark visibility, the possibility of modulating the watermark strength according to perceptual considerations will be investigated.

References

COX I. J., MILLER M. L., BLOOM J. A.: *Digital Watermarking*. Morgan Kaufmann, 2001.

BARNI M., BARTOLINI F.: *Watermarking Systems Engineering: Enabling Digital Assets Security and other Applications*. Marcel Dekker,2004.

SCHRÖDER P., ZONIN D.: Course notes: Subdivision for modeling and animation. In *Proc.SIGGRAPH* '99 (1999).

SCHRÖDER P.: Subdivision as a fundamental building block of digital geometry processing algorithms. *Journal of Computational and Applied Mathematics* 149, 1 (Dec. 2002), 207–219.

LOUNSBERY M., DEROSE T. D., WARREN J.: Multiresolution analysis for surfaces of arbitrary topological type. *ACM Trans. Graph*. 16, 1 (1997), 34–73.

KANAI S., DATE H., KISHINAMI T.: Digital watermarking for 3d polygons using multiresolution wavelet decomposition. In *Sixth IFIP WG* 5.2 GEO-6 (1998).

PRAUN E., HOPPE H., FINKELSTEIN A.: Robust mesh watermarking. *SIGGRAPH* '99 (1999), 49–56.

YIN K., PAN Z., SHI J., ZHANG D.: Robust mesh watermarking based on multiresolution processing. *Computer and Graphics* 25 (2001), 409–420.

OHBUCHI R., AKIO M., SHIGEO T.: A frequency-domain approach to watermarking 3d shapes. *EUROGRAPHICS* 2002 21, 3 (2002).

TAUBIN G.: A Signal Processing Approach to Fair Surface Design, *ACM SIGGRAPH 95 Conference Proceedings* (Aug. 1995), pag. 351-358, 1995.

PODILCHUK C. I, ZENG W: Image-adaptive watermarking using visual models, *IEEE Journal Sel. Areas Commun.*, vol. 16, no. 4, pp. 525-539, May 1998.

WOLFGANG R. B., PODILCHUK C. I. and DELP E. J.: Perceptual watermark for digital images and video, *Proc. IEEE*, vol. 87, no.7, pp. 1108-1126, July 1999.

BARTOLINI F., BARNI M. PIVA A. and CAPPELLINI V.: Mask building for perceptually hiding frequency embedded watermarks. *In Proc. ICIP98 Int. Conf. Image processing,* Chicago, IL, Oct. 1998, vol. I, pp. 450-454.

UCCHEDDU F., CORSINI M., BARNI M.: W*avelet-Based blind Watermarking of 3D Models, Multimedia and Security, Workshop 2004,* Magdeburg, Germany

LEE A.W.F.,SWELDENS W, SCHRODER P., COWSAR L. and DOBKIN D., MAPS: Multiresolution Adaptive Parameterization of Surfaces. In *Proceedings of SIGGRAPH* 1998, pp. 95-104.

ECK M., DEROSE T., DUCHAMP T., HOPPE H., LOUNSBERY M. and SWELDENS W. Multiresolution Analysis of Arbitrary Meshes, *Computer Graphics (SIGGRAPH 96 Proceedings),* pages. 325-334, 1995.

BARNI M, BARTOLINI F: *Watermarking System Engineering: Enabling Digital Assets Security and other Applications,* Marcel Dekker, 2004

KATZ S., TAL A.: Hierarchical mesh decomposition using fuzzy clustering and cuts. *ACM Trans. Graph.* 22, 3 (2003), 954–961.

Recording and modeling of cultural heritage objects with coded structured light projection systems

Devrim Akca, Fabio Remondino, David Novák, Thomas Hanusch, Gerhard Schrotter, Armin Gruen

Institute of Geodesy and Photogrammetry, ETH Zurich, CH-8093 Zurich, Switzerland
(akca, fabio, hanuscht, schrotter, agruen)@geod.baug.ethz.ch, dnovak@student.ethz.ch
http://www.photogrammetry.ethz.ch

1 Introduction

Active sensors (Blais 2004), based on coherent (laser) and non-coherent light, are nowadays used for many kinds of 3D reconstruction tasks and recently very much for the recording and 3D documentation of cultural heritage objects. They have become a very common source of documentation data, in particular for non-expert users, as they easily provide range data of surfaces in high resolution and with high accuracy. Compared to passive image-based approaches (Remondino and El-Hakim, 2006), active sensors provide directly and quickly 3D information of the surveyed object in form of range data (point clouds). Active sensors are suitable for different scales and objects. While the recording devices are still relatively expensive, important progress has been made in recent years towards an efficient processing and analysis of range data.

Structured light systems consist of one (or more) camera(s) and an active light source, which illuminates the object with a known pattern of light sequence. Based on the triangulation principle, the 3D object coordinates are generally recovered in ca. 2-3 seconds with a potential accuracy of 50 microns or even better.

This paper reports about two case studies where a coded structured light system (optoTOP-HE™ and optoTOP-SE™, Breuckmann GmbH) is used for the precise 3D digitization and documentation of Cultural Heritage objects. It includes all essential steps of the 3D object modeling pipeline from data acquisition to 3D visualization. The first study is the 3D modeling of a part of a marble Herakles statue, named "Weary Herakles" (Fig. 1a), which is on display in the Antalya Museum (Turkey), digitized with an optoTOP-HE system. The second study is about the 3D modeling of a Khmer head sculpture (Fig. 1b), which is in the collection of Rietberg Museum, Zurich (Switzerland), digitized using an optoTOP-SE sensor.

The next chapter introduces the scanner with emphasis on the working principle and technical specifications. The following third and fourth chapters explain the data acquisition and modeling workflow of the projects. The fifth chapter addresses the capabilities and the limitations of the used hardware and software.

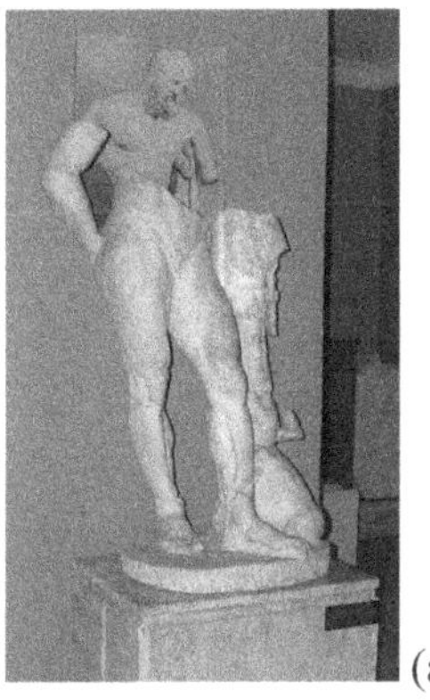
(a)

(b)

Figure 1 *(a) Weary Herakles statue (ca 1 m height) in the Antalya Museum, (b) the Khmer Head (ca 30 cm height) in the Rietberg Museum of Zurich.*

A comparison of the used reverse engineering software (Geomagic Studio™ and PolyWorks™) is also reported. Another comparison performed on larger datasets is presented in (Boehm and Pateraki, 2006).

2 Data Acquisition System

2.1 Coded Structured Light System

The key feature of a structured light system is the replacement of one of the cameras with an active light source, which illuminates the object with a known pattern. This solves the correspondence problem in a direct way and many variants of the active light principle exist (Beraldin et al 2000, Beraldin 2004, Blais 2004).

The coded structured light technique, also called topometric technique, is based on a unique codification of each light token projected onto the object. When a token is detected in the image, the correspondence is directly solved by the de-codification technique. It requires a complex light projection system and many codification methods have been developed (Batlle et al 1998, Salvi et al 2004, Dipanda and Woo 2005).

The time-multiplexing, also called temporal codification, with a combined Gray code and phase shifting is the mostly employed de-codification technique. The used optoTOP-HE and -SE sensors apply the same technique.

A Gray code is a binary numeral system where two successive values differ in only one digit, i.e. 000, 001, 010, 011, … in natural (plain) binary codes, and 000, 001, 011, 010, … in Gray binary codes. It was invented and patented by Frank Gray (Gray 1953) in Bell Labs. For the case of coded structured light (or fringe projection) systems it is superior to the natural binary codification, since it resolves the ambiguity better at the edges of consecutive patterns (Fig. 2b and 2c).

A sequence of Gray coded binary fringe patterns is projected onto the object (Fig. 2a). This divides the object into a number of 2^n sections, where n is the number of pattern sequences, e.g. 128 sections for n=7. Thus each pixel is associated with a codeword, which is the sequence of 0s and 1s obtained from the n patterns. The codeword establishes the correspondences relating the image pixels to the projector stripe numbers. The object space point coordinates are calculated using the spatial intersection provided that system calibration is known. All pixels belonging to the same stripe in the highest frequency pattern share the same codeword. This limits the resolution to half the size of the finest pattern.

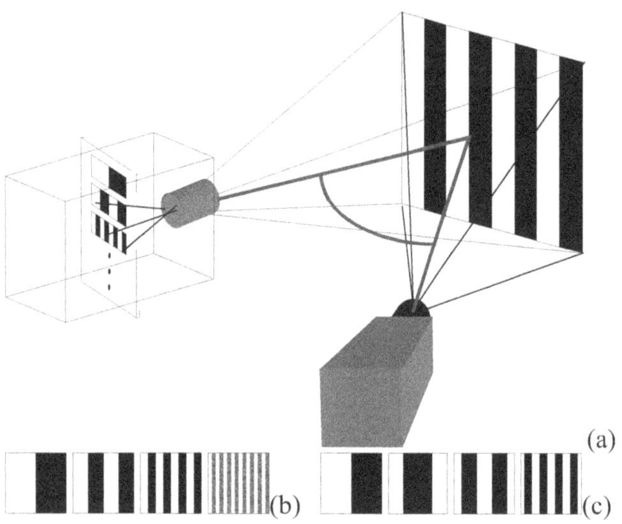

Figure 2 *(a) Setup of a fringe projection system with the natural binary codification (courtesy of Dr. B. Breuckmann), (b) natural binary code, (c) Gray binary code.*

Generally a periodical pattern is projected several times by shifting it in one direction in order to increase the resolution of the system. For each camera pixel the corresponding projector stripe number with sub-stripe accuracy is yielded by a phase shift method (Burke et al 2002).

2.2 Breuckmann optoTOP-HE and optoTOP-SE

The optoTOP-HE system (Fig. 3), as a high definition topometrical 3D-scanner, allows the 3-dimensional digitization of objects with high resolution and accuracy. The optoTOP-HE system uses special projection patterns with a combined Gray code and phase shift technique, which guarantees an unambiguous determination of the recorded 3D data with highest accuracy (Breuckmann 2003). The sensor of the optoTOP-HE system can be scaled for a wide range of Field of Views (FOV), by changing the baseline distance and/or lenses, typically between a few centimeters up to some meters. Thus the specifications of the sensor can be adapted to the special demands of a given measuring task.

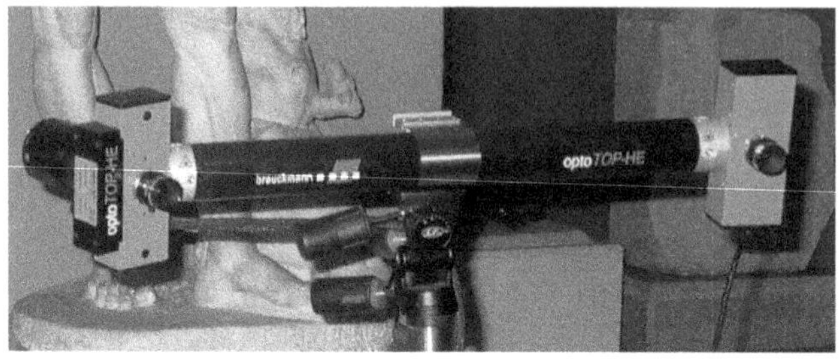

Figure 3 *The optoTOP-HE sensor.*

The optoTOP-SE (Special Edition) series are the identical systems with the major difference that the sensors have only three different FOV with a fixed 300 mm base length. More details are given in Table 1.

Table 1 *Technical specifications of optoTOP-HE and –SE sensors that were used in both projects*

	optoTOP -HE	optoTOP -SE
Field of View (mm)	480x360	400x315
Depth of View (mm)	320	260
Acquisition time (sec)	<1	
Weight (kg)	2-3	
Digitization (points)	1280x1024[1]	1280x1024
Base length (mm)	600	300
Triangulation angle (deg)	30^0	
Projector	128 order sinus patterns	
Lamp	100W halogen	
Lateral resolution (μm)	~360	~340
Feature accuracy (relative[2])	1/15000	1/10000
Feature accuracy (μm)	~45	~50

([1]) Current optoTOP-HE has a 1380x1040 dimension.
([2]) According to image diagonal

3 Weary Herakles project[1]

The Weary Herakles is a marble statue of the Greek demigod which dates back to the 2nd century AD (Fig. 1a). It is a copy of an original bronze statue of Herakles sculptured about 330-320 BC by the Greek master Lysippos of Sikyon. Many artisans devoted their skills to replicating this original around that period. This particular example was probably carved in the Hadrianic or Antonine (Roman) period. The version is identified as the "Herakles Farnese" type on the basis of its similarity to a more complete copy in the Naples National Archaeological Museum (Italy).

[1] http://www.photogrammetry.ethz.ch/research/herakles/

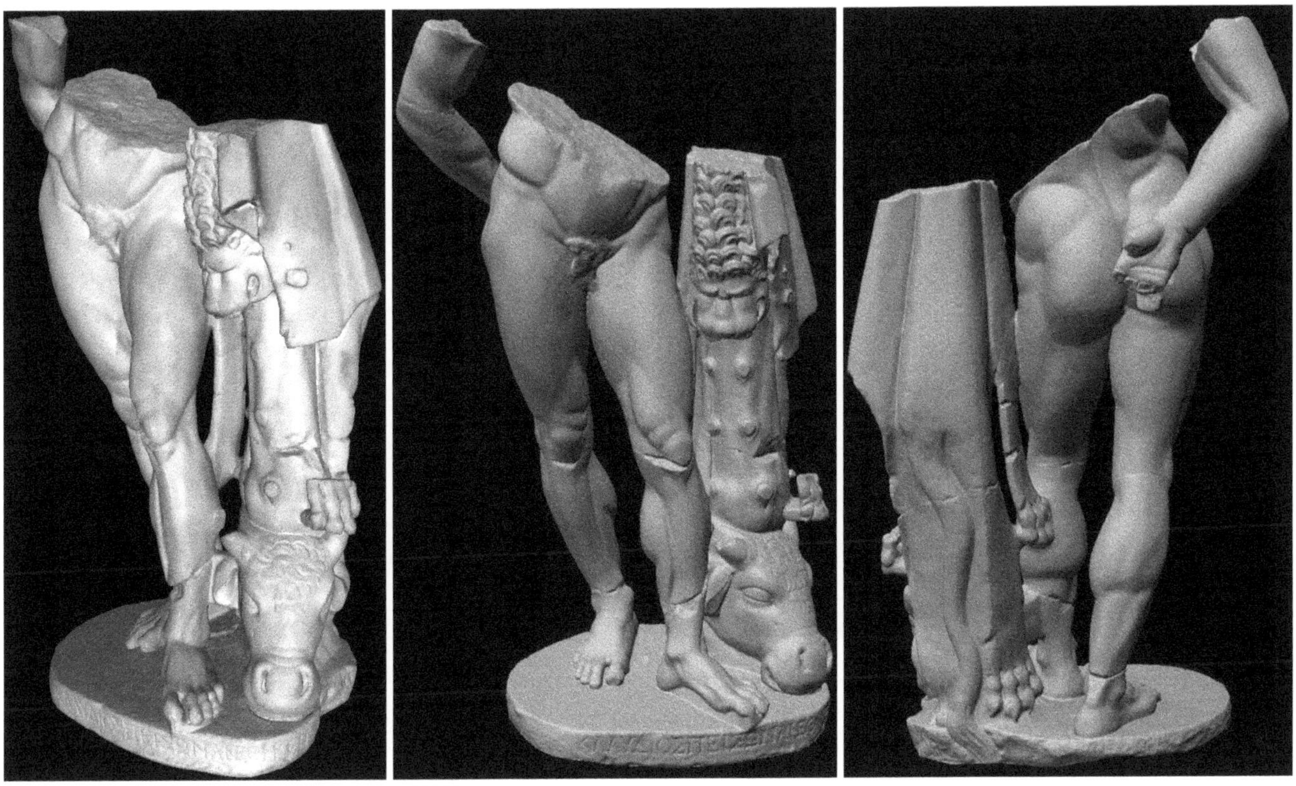

Figure 4 *Frontal view of the textured model (left), frontal view (central) and back view (right) of the grey shaded model.*

The statue was broken in two parts. We do not know when and by whom it was done. The upper half was first seen in the USA in the early 1980s and nowadays it can be found at the Boston Museum of Fine Arts. The lower part was found by Prof. Jale İnan (İnan 1981, 1992) at an excavation site in Perge (Antalya, Turkey) in 1980. It is now shown in the Antalya Museum, along with a photograph of the top half (Fig. 1a). According to the Turkish law, Turkish antiques are state properties since Ottoman times 1906. The Turkish government has requested the upper part of the statue so that the two fragments could be joined together. The Boston Museum has refused to consider the Turkish petition. Nevertheless, in 1992, the two fragments were placed together and they matched perfectly. The Boston Museum says that the statue may have been broken in ancient times and that the upper torso may have been taken from Turkey before the Turkish law established the state ownership of archaeological finds (Rose and Acar 1995, Brodie et al 2000, Brodie 2003, Gizzarelli 2006).

Since both parts are unfortunately geographically separated, our aim was to record and model both the lower and the upper parts and bring them together in a 3D computer model, to be able to see the complete statue, appreciate and analyze it. With the help of the Turkish authorities and the Antalya Museum we were able to complete our work on the lower part, but access to the Boston Museum was denied.

The digitization of the lower part of the statue was done on 19-20 September 2005 in the Antalya Museum with a Breuckmann optoTOP-HE coded structured light system (http://www.breuckmann.com). The system was kindly provided by the Turkish reseller InfoTRON Co. (http://www.infotron.com.tr), Istanbul.

3.1 Scanning in the Antalya Museum

The scanning campaign was completed in one and a half days of work. The statue is around 1.1 meters in height. The whole object was covered with 56 scans of the first day work. The remaining 11 scans of the second day were performed to fill the data holes and occlusion areas. Totally 83.75M points were acquired in 67 scan files. The average point spacing is 0.5 millimeter.

The optoTOP-HE 3D digitization system is able to acquire one point cloud in nearly less than one second. However, orienting the scanner and planning the scan overlay needs careful preparation, especially for this kind of object with many concave and hidden parts. Due to the sensitivity of the sensor to ambient light, special attention was paid to environment lighting conditions. Two ceiling halogen lamps looking at the statue were turned off.

3.2 Point Cloud Registration

The registration of the acquired scans was done using an in-house developed method, called Least Squares 3D Surface Matching (LS3D) (Gruen and Akca 2005). The mathematical model is a generalization of the Least Squares image matching method, in particular the method given by Gruen 1985. It provides mechanisms for internal quality control and the capability of matching multi-resolution and multi-quality data sets.

The pairwise LS3D matchings are run on every overlapping pairs (totally 234) and the average precision of the registration was 81 microns. In the successive global registration step, performed with a block adjustment by independent models, orientation procedure well known in photogrammetry, gave 47 micron a posteriori sigma naught value.

3.3 Point Cloud Editing

After the registration, all scan files were merged as one XYZ file containing ca 36M points, discarding the scanner detected blunders. The file was imported to Geomagic Studio™ 6 (Raindrop Geomagic Inc., release 2004_05_11_B) for further editing procedures. The data set was cropped to include only the area of interest (AOI), i.e. deleting the background wall or other non relevant parts, concluding with 33.9 M points. A low level noise reduction was applied using the "Reduce Noise" function.

3.4 Surface Triangulation and Editing

As a first attempt, the surface mesh generation was performed with the original data resolution. The operation could not be performed, since the memory requested by the software exceeded the physical memory limit of 2 GB of the computer. Therefore, the number of points was reduced to 9 million by applying the "Curvature Sampling" function of Geomagic Studio™. This operation eliminates points in flat regions but preserves points in high-curvature regions to maintain detail. Afterwards, the surface triangulation was done by setting the number of target triangles to 5 million. Because of the complexity of the statue and occlusions, some inner concave parts could not be seen by the scanner. This resulted in several data holes on the wrapped surface. They were interactively filled with the "Fill Holes" option of the software. The final model contains 5.2 M triangles. In the final 3D model we were able to achieve a high level of realism, which can make a one-to-one scale production of the statue possible, if required (Fig. 4).

Worth to be mentioned is that the main portion of the editing effort was dedicated to the holes filling. It is a tedious work which takes the longest time among all the steps of the project.

3.5 Texture Mapping and Visualization

For a photo-realistic visualization of the 3D model, images acquired with a 4M pixel CCD Leica Digilux 1 camera were used for the texture mapping phase. The Weaver module of the VCLab's 3D Scanning Tool (ISTI-CNR, Pisa, Italy) was used. The VCLab's Tool comprises different modules for the 3D modeling from range data (Callieri et al 2003), including registration, editing and texturing.

The visualization of the final shaded model was done with the IMView module of PolyWorks™ (InnovMetric Software Inc., version 9.0.2) as it gives a better shading than Geomagic Studio™. The textured model was instead visualized with the viewer of the VCLab's Tool (Fig. 4).

4 Khmer head project

The earliest examples of Buddhist art on the mainland of Southeast Asia date from the 4th and 5th centuries and emerged under the influence of Indian and Sri Lankan art. During the 6th century the Khmer people established themselves in the fertile tropical plains of Cambodia, and as the dominating power in Southeast Asia in the 12th and 13th centuries. They built the stunning group of temples at Angkor. The Khmer rulers supported both Hinduism, displayed most magnificently at Angkor Wat, and Buddhism, whose most important monument is the Bayon temple, still admirable in the central area at Angkor Thom, Cambodia.

In our work, we modeled a bodhisattva head from the late 12th or early 13th century, carved in the Bayon style. It represents Lokeshvara or Avalokiteshvara, the "Lord of compassion who looks down (on the suffering of the world)," an emanation of the Buddha Amitabha as demonstrated by the seated Buddha on his hair ornament. Its serene expression and transcendent smile convey better than any words the sublime essence of the Buddhist teachings (Museum Rietberg 2006).

4.1 Data acquisition in Museum Rietberg

The head is made of sandstone and 28 centimeters in height. The data acquisition was done in Museum Rietberg on 4 May 2006. A Breuckmann OptoTOP-SE fringe projection system was used for this purpose. The scanning and imaging took four hours. The head was covered with 18 point clouds, totally 23.6 million points.

Figure 5 *Scanning in Rietberg Museum.*

4.2 Point Cloud Registration

The point cloud registration was done again with the LS3D surface matching method. 52 pairwise LS3D matchings for all overlaps gave an average sigma naught value of 60 microns. The global registration with the block adjustment by independent models solution concluded with 28 microns sigma naught value.

4.3 Surface Triangulation and Editing

The surface modeling was done using two commercial packages, namely Geomagic Studio™ and Polyworks™. The aim was to compare the capabilities of both software. The registered point clouds were imported in the proper formats. Accordingly, the registration steps were skipped in both software. Both software packages have different processing pipelines (Table 2).

Table 2 *Modeling workflows in Geomagic Studio and PolyWorks.*

Geomagic Studio™	PolyWorks™
Importing the point clouds	Importing the point clouds
Point cloud merging (??)	Surface triangulation
Defining the AOI	Surface merging
Noise reduction	Defining the AOI
Down sampling	Surface editing
Surface triangulation	
Surface editing	

Geomagic Studio™ offers fully automatic data import functionality provided that data is given in one of the appropriate point cloud formats. Totally 18 point clouds were imported, merged into one, which gave a very dense (denser than 50 microns inter-point distance at some locations) point cloud. After discarding the no data or scanner signed erroneous points and points belonging to background and other non relevant objects, 3.2 million points remained. The noise reduction ensures that points coming from different views in different quality will finally have the similar signal-to-noise ratio. In our case a slight (low level) noise reduction was applied. After this step, the model contains highly redundant points coming from the multiple views. The "Curvature Sampling" function with a 60% reduction rate reduced the number of points to 1.9 millions. Intentionally, a restricted reduction rate was selected, so that small details can be preserved.
The surface triangulation in Geomagic Studio is a fully 3D and automated approach, with limits the user interaction. Hence, the resulting mesh will have some topological errors and holes. On the other hand, it can preserve the high frequency details of the object geometry successfully by considering all points in one processing sweep. In general, surface triangulation quality is highly related to the point density and homogeneity.

PolyWorks™ has a significantly different workflow. Each step is represented as a module inside the package. Data import is not automatically performed. Each point cloud is individually imported, subsequently converted to the surface form by applying a 2.5D triangulation, similar to the terrain modeling case. Therefore, the user should interactively rotate the point cloud to a position where the viewing angle is close to the one at the acquisition instant. This substantially reduces the topological errors. On the other side, such a stepwise surface generation strategy does not utilize all the available information properly. For example, there might be some object parts with thin point distributions in individual views, whereas the combination of all views together provides a good solution. In the next step, separate surfaces were merged as one manifold using the IMMerge module. This part is highly automated, and additionally offers a noise reduction option. During the process, triangulation is also optimized especially at the overlapping regions by associating dense triangles to high curvature areas and sparse at flat areas. Finally the IMEdit module offers many surface editing functions, e.g. cropping the AOI, filling the data holes, correcting the wrong triangles, boundary cleaning, etc. However, it is less flexible and user friendly than Geomagic Studio™.

The resulting models from both software packages meet the project requirements. PolyWork model (0.6 million triangles) has substantially less number of triangles than Geomagi model (3.9 million triangles), thus having a better and optimized triangulation algorithm. However, the model from Geomagic Studio preserves the small details and structures slightly better than the model of PolyWorks (Fig. 6). Nevertheless, the shader of Polyworks represents better the geometry of the modeled object. More details can be found in Novák 2006.

Figure 6 *Shaded view of the model generated with Geomagic Studio (top-left) and PolyWorks (top-right). Closer view on the left side of the lip: the different meshes produced by Geomagic Studio (bottom-left) and PolyWorks (bottom-right).*

4.4 Texture Mapping

Some digital images of the Khmer head were acquired with a photographer-type professional illumination system consisting of two diffuse lights on a tripod (Fig. 7). The system was used to reduce the radiometric differences between the images and shadow effects at the complex parts or object silhouettes. Images were taken with a Sony DSC-W30 6 megapixel digital camera. The 3D model generated with PolyWorks (at the original resolution) was used to produce a photo-realistic 3D result.

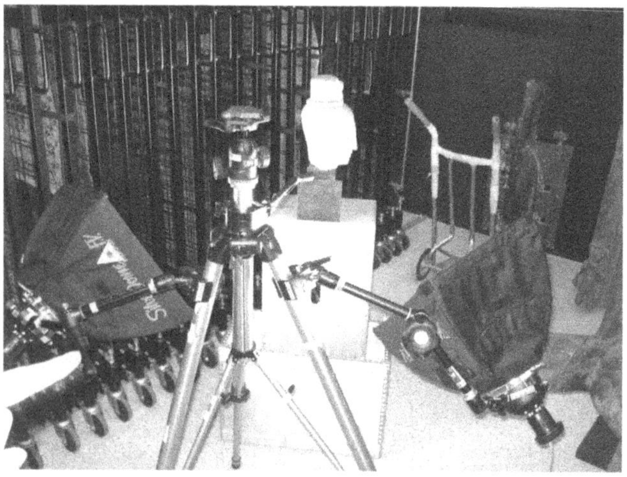

Figure 7. *Illumination system used for* the texture mapping.

Figure 8 *Texture mapped model of the Khmer head.*

Internal and external orientations of the images were computed using a photogrammetric bundle adjustment. The object space coordinate system was defined as the coordinate system of the 3D model. The common points were interactively identified both in digital images and in the intensity images of the scanner. Afterwards we used this information, in addition to the geometric model and the images, to conduct a visibility analysis for every camera position. The algorithm calculates the visible parts for every acquisition position, resolution independent and fully triangle-based in object space. Partly occluded triangles are subdivided and re-meshed into fully visible or fully occluded triangles. The resultant list of visible triangles is used to calculate the texture coordinates for every vertex of the mesh. The underlying algorithm uses a "best view" algorithm to evaluate the optimal texture for every triangle. The mesh consists of approximately 295 000 vertices and 585 000 facets. With the open source software "Blender", we found an adequate software package to handle this huge number of triangles to produce high quality renderings and movies. However, for online and real-time visualizations, this full-resolution model cannot be used. For this application area, we reduce the number of triangles, without degradation of the visual impression. The high quality textured model (Fig. 8) is useful for presentations on high performance computers (concerning the physical memory, CPU and graphic card) and basically to preserve the object for e.g. reconstruction purposes.

5 Capabilities of the used Hardware and Software

The optoTOP-HE and -SE sensors as a coded light projection system meet the project requirements satisfactorily. They have some distinctive advantages over the triangulation-based laser systems (Table 3).

Table 3 *Triangulation based systems: Laser versus coded light.*

	Laser	Coded light
Weight and price	Identical	Identical
Speed		Faster
Sensitivity to ambient light	Less	
Speckle noise		Less
Penetration into object surface		No
Imaging for texture mapping		Yes
Depth of view	Larger	
Eye safety		Better

The use of incoherent light reduces speckle noise and provides better surface smoothness (Blais 2004). Furthermore it does not penetrate into the object surface, unlike laser light whose penetration property is well known, e.g. for marble (Godin et al 2001). All these reasons make the system a suitable choice for Cultural Heritage applications. Unfortunately most of the coded light systems are very sensitive to ambient light, requiring almost darkness during the acquisition. Nevertheless, new developments and digital projectors allow 3D data acquisition also under normal light conditions.

Although surface digitization is a very easy and straightforward task, the surface triangulation and editing, which is the key step of the whole modeling chain, is still cumbersome and needs heavy semi-automatic or manual work. The management of large data sets is another aspect. Geomagic Studio™ crashed several times while filling the holes interactively, whereas PolyWorks™ did not. Geomagic Studio™ gives better details in surface geometry with the cost of large number of triangles. Table 4 gives a comparison of both software packages.

Table 4 *PolyWorks versus Geomagic Studio.*

	PolyWorks™	Geomagic™
Data import	Manual	Automatic
Surface generation		
Type	2.5D	3D
Optimality	Better	
Detail preservation		Better
Topological correctness	Better	
Automatisation		Better
Editing capabilities		Better
Performance	Better	
Visualization	Better	
User friendliness		Better
Stability	Better	

6 Conclusions

Nowadays active sensors are often used for many kinds of 3D object reconstruction tasks, one important area of which is 3D documentation of cultural heritage objects. This study presents the results of 3D modeling of two cultural heritage objects, where a close-range coded structured light system was used for digitization. In fact active sensing with coded structured light systems is a mature technology and allows high resolution documentation of cultural heritage objects.

The used instruments have acquired high quality point cloud data of the statues. The results of the processing (accuracy of about 50 micron and better) are in good agreement with the system specifications and project requirements. The heaviest user interaction is needed in the editing steps, e.g. for filling the data holes. We have used two commercial software packages in order to carry out the modeling. Each software package has its own particular advantages and functions. A unique package, which fulfills all requirements with sophisticated and automatic editing capabilities, is not yet available. In our opinion, the use of both packages can give the optimal modeling results. Texture mapping is another issue, which is not fully supported by either software.

Acknowledgement

The helps of Prof. Dr. Orhan Altan (Istanbul Technical University) and Mr. Lorenz Homberger (Museum Rietberg) are greatly appreciated. We thank the personnel of the Antalya Museum and Museum Rietberg for their kind help and permission. We also thank the Visual Computing Lab of ISTI-CNR for providing their Weaver software for the texture mapping.

References

BATLLE, J., MOUADDIB, E., and SALVI, J. 1998 Recent progress in coded structured light as a technique to solve correspondence problem: A survey. *Pattern Recognition*. 31(7): 963-982.

BERALDIN, J.-A., BLAIS, F., COURNOYER, L., GODIN, G., and RIOUX, M. 2000 Active 3D sensing. *Modelli E Metodi per lo studio e la conservazione dell'architettura storica*. University: Scola Normale Superiore, Pisa, 10: 22-46.

BERALDIN, J.-A. 2004 Integration of laser scanning and close-range photogrammetry – The last decade and beyond. *International Archives of the Photogrammetry, Remote Sensing and Spatial Information Sciences*, 35(B7): 972-983.

BLAIS, F. 2004 Review of 20 years of range sensor development. *Journal of Electronic Imaging*. 13(1): 231-240.

BOEHM, J. and PATERAKI, M., 2006. From point samples to surfaces - on meshing and alternatives. *IAPRS*, Vol.36(5)

BREUCKMANN, B. 2003 State of the art of topometric 3D-metrology. *Optical 3-D Measurement Techniques VI*, Zurich, Sept. 22-25, Vol. II, pp. 152-158.

BRODIE, N., DOOLE, J., and WATSON, P. 2000 Stealing history: The illicit trade in cultural material. *The McDonald Institute of Archaeological Research*, Cambridge, pp. 31-32.

BRODIE, N. 2003 Stolen history: looting and illicit trade. *Museum International*. 55(3-4): 10-22.

BURKE, J., BOTHE, T., OSTEN, W., and HESS, C. 2002 Reverse engineering by fringe projection. *Interferometry XI: Applications*, SPIE vol. 4778, pp. 312-324.

CALLIERI, M., CIGNONI, P., GANOVELLI, F., MONTANI, C., PINGI, P., and SCOPIGNO, R. 2003 VCLab's Tools for 3D range data processing. *VAST'03*, Brighton, pp. 13-22.

DIPANDA, A., and WOO, S. 2005 Efficient correspondence problem-solving in 3-D shape reconstruction using a structured light system. *Optical Engineering*. 44(9): 1-14.

GIZZARELLI, C. 2006 'Weary Herakles': Looters vs. Archaeologists, www.bellaonline.com/articles/art28239.asp (accessed September 2006).

GODIN, G., RIOUX, M., BERALDIN, J.-A., LEVOY, M., and COURNOYER, L. 2001 An assessment of laser range measurement of marble surfaces. *Optical 3-*

D Measurement Techniques V, Vienna, Oct. 1-4, pp. 49-56.

GRAY, F. 1953 Pulse code communication. *US Patent no. 2,632,058*. March 17, 1953.

GRUEN, A. 1985 Adaptive least squares correlation: a powerful image matching technique. *South Afr. J. of Photogrammetry, Remote Sensing and Cartography*. 14(3): 175-187.

GRUEN, A., and AKCA, D. 2005 Least squares 3D surface and curve matching. *ISPRS Journal of Photogrammetry and Remote Sensing*. 59(3): 151-174.

İNAN, J. 1981 Perge kazısı 1980 çalışmaları. *III KST, DSİ Basım ve Foto Film İşletme Müdürlüğü Matbaası*, Ankara, sayfa 43-48 (in Turkish).

İNAN, J. 1992 Heraklesstatue vom Typus des "Herakles Fernese" aus Perge. *Festschrift für Max Wegner Zum 90. Geburtstag*, Bonn, pp. 223-232.

ROSE, M., and ACAR, O. 1995 Turkey's war on the illicit antiquities trade. *Archaeology*. 48(2): 45-56.

MUSEUM RIETBERG 2006 Exhibition Brochure.

NOVÁK, D. 2006 3D Object modeling by use of a close range active sensor. *B.Sc. Thesis in Photogrammetry and Remote Sensing*, Summer Semester 2006, Institute of Geodesy and Photogrammetry, ETH Zurich.

REMONDINO, F. and EL-HAKIM, S., 2006 Image-based modelling: a review. *The Photogrammetric Record*, 21(115).

SALVI, J., PAGES, J., and BATLLE, J. 2004 Pattern codification strategies in structured light systems. *Pattern Recognition*. 37(4): 827-849.

WebGIS and tourist personalized itineraries for exploitation of calabrian cultural and archaeological heritage

P. A. Bertacchini,[1] A. Dell'Accio,[2] S. Giambò,[3] G. Naccarato,[2] P. Pantano[2]

[1] Department of Linguistic, University of Calabria
[2] Department of Mathematics, University of Calabria
[3] Department of Mathematics, University of Messina

1 Introduction

Calabrian History and Culture have very ancient roots, so that Southern Italy could be considered the first territory to weave a net of relationships with the Mediterranean populations. The region of Calabria offers a wide variety of its cultural heritage, witness of its many rulers (Greek-Hellenistic, Byzantine, Norman, and so on). In fact, these hegemonies marked in a determining way the Calabrian culture, enriching the land of architectural and artistic works. Moreover, some important archaeological finds can be admired in several museums and in the archaeological sites of Sibari, Locri, Roccelletta di Borgia and Capo Colonna.

In order to promote the protection, the preservation and the exploitation of these cultural resources, many projects were started, such as the "Virtual Museum System of Magna Graecia" project, financed by the Calabria Regional Operational Programme (ROP) 2000 - 2006. Its objective was the offering of original procedures for the fruition of Calabrian cultural heritage, in particular of Magna Graecia period, using the most advanced technologies. In fact, the system foresees the access and the visit of a remote site: the Web becomes a way to spread knowledge and to promote historical places and a specific culture, with obvious development of tourist and economical aspects of the region. A central role is given to the multimedia technologies: modules of virtual visits, three-dimensional reconstructions of archaeological finds, documentary films are the frame of the traditional information, enriching its own semantic content.

A WebGIS module, a system of navigation through a dynamical interactive map, was specifically designed by open source systems, for an on line promotion of the cultural heritage (a monument as well as a church, a find or an archaeological site), that was not separated from the territory determining the localization. The variations of scale allow to pass from a panoramic sight connecting the relationships among the objects present in the territory to a detailed perspective, in which the specific object is important. Moreover, the user has the possibility to access to different levels of the informative content in an immediate manner.

The use of the above-described tools and technologies allowed the building of real world models and the creation of bi-dimensional scenarios, but they also allow the simulation of three-dimensionality: the aim was to offer to the user the opportunity to test virtually a possible future tourist visit.

In this work we present the development activity of a tool that can create tourist-personalized itineraries for the user of the "Virtual Museum System of Magna Graecia". It is a relatively complex optimization problem involving several parameters which are not well defined in some cases; the solution involves the utilization of AI (Artificial Intelligence) and the support of tools like the GIS (Geographical Information System), because of the geographical data.

2 Tourist itineraries

Itinerary is the central part of any tourist visit and it is an important strategic element for the exploitation of the territory. Its planning includes defined objective elements, for example the distance between two places or the limit of speed on a road, as well as subjective elements, for example the importance of a monument for a specific tourist or the necessary time to visit it. In this case, the definition of these elements is more articulated and complex. The possibility for a computer to develop the typical reasoning of the human mind is an application of the Artificial Intelligence.

We can suppose that a tourist interested in the archaeological sites of Calabria wants to visit the most important ones in a specific number of days. The words "most important" create a problem of interpretation: in fact, the concept of importance is not only based on recognizable objective pointers, but it is linked to the sensibility of the users, that could have different opinions about the design of the itinerary. In general terms, we assign to every cultural place, a possible visit in the itinerary, a degree of importance or "value" that we call "v". From a mathematical point of view v is a function of different variables $x1, \ldots, xp$, that is:

$v = f(x_1, \ldots, x_p)$ with $x \in X$

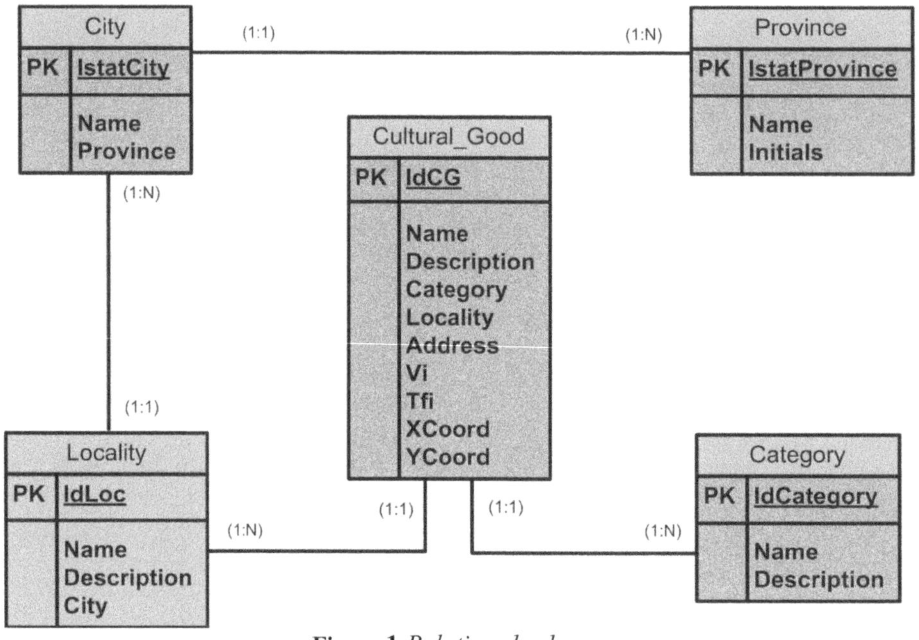

Figure 1 *Relational scheme*

where X is the set of pointers that describe the features of a cultural good. Particularly, v depends from the age of construction, the state of preservation, the degree of accessibility of the site, the artistic quality, and so on.

Besides v, other considerable parameters for the definition of the problem are the optimal duration of the fruition, tf, that could be determined for example from the visit time obtained by a sample of users, and the cost c when the fruition requires a ticket.

The information needed to fill the indicators, result of a consistent research effort, has been stored in a database, whose optimal management is insured by the use of a DBMS (Data Base Management System) that guarantees reliability and security. In particular, these stored data concern the immovable goods that belong to the calabrian cultural heritage, including architectural monuments, archaeological sites, monumental centers, as well as buildings where the movable cultural goods are exposed and/or preserved, such as museums, libraries, archives. Every cultural immovable good has a descriptive format that defines its own characteristics and location.

3 Formulation of Mathematical Model

A tourist itinerary consists of a set of steps, coinciding with the places of interest, and of the possible connections among them. Therefore, it is convenient to represent it through a graph-based formalization (see Fig. 3). A graph G is defined as an ordered pair, $G = (V,E)$, where V is a set of vertices (the steps) and E is a set of links between pairs of vertices, called edges (communication channels).

Initially, the parameter tij, that represents the time needed to cover a link $(i,j) \in E$, has been assigned to each edge of G. The time tij depends on a certain number of factors, including distance, road type, speed limits, mean of transport. The value vi and the fruition time tfi of the cultural good has been assigned to each vertex. Another important parameter for travel planning is the user time T, that is not necessarily boundless.

Our purpose is to find the itinerary that maximizes the travel performance in terms of importance of the visited goods with respect to the time limits or to any other constraint defined earlier (i.e. cost of the trip, existence of forced layovers, etc.), taking into account a certain number of preferences expressed by the user, such as category of the cultural good to be visited or geographical zone of interest.

Once the relevant parameters have been chosen, we need to formulate a model able to describe the problem. If one forgets about the connections between steps, the problem can easily be solved using the known Knapsack model.

This model allows to find the set of objects that maximize the profit without overcoming the maximum capacity of the knapsack, for a given collection of objects characterized by a value vi and a volume ci, and for a given knapsack of fixed capacity C.
The Knapsack problem belongs to the class of NP-hard optimization problems whose solution can be found through an heuristic approach. Including connections between steps complicates the analysis, moving the problem to the application field of Network Optimization. Indeed, in formulating the constraints on the time span that user can employ, one has to take into account the time to cross the connections between different steps.

So, it is possible to propose a first formulation of the itinerary personalized problem in the following way:

$$max\ z_1(x) = \sum_{i=1}^{N} v_i x_i$$

s. a.

$$\sum_{i=1}^{N} tf_i x_i + \sum_{(i,j)\in E} t_{ij} y_{ij} \leq T$$
$$x_i \in \{0,1\} \quad \forall i = 1,\dots,N$$
$$y_{ij} \in \{0,1\} \quad \forall (i,j) \in E$$

For the application of the method described above, the following hypoteses have to be taken into account:

(1) $t_{fi} \leq T \qquad \forall i = 1 \dots N$

(2) $\sum_{i=1}^{N} t_{fi} > T$

Condition (1) says the fruition time of each cultural good, that is a step of the tour, must be smaller than the total time *T* (capacity of the knapsack), needed by the user to complete the tour. Otherwise, the selected object would be automatically excluded, that is the cultural good alone could not be considered in the knapsack.

Condition (2) says the sum of the weighs *tfi* of all objects must be bigger than the capacity of the knapsack, otherwise, neglecting for now the time needed to cover the link (i,j), the application of the knapsack algorithm to a problem of this kind would be useless.

This formulation can be, obviously, generalized by the addition of further constraints or objectives, in such a way to guarantee a high level of personalization from the user.

4 Some possible approaches

Once the mathematical model has been defined, one can develop appropriate resolution algorithms for the particular problem under study. From an introductive survey of the state of the art, it is easily seen that the problem can be faced in different ways.

A traditional approach consists in using the optimization techniques. In this case, the solution for a given optimization problem *P* is based on research, use and modification of the approximate algorithm *A*, present in literature. For a given instance *x* of *P*, this algorithm gives an approximate solution *A(x)*, in a time that is a polynomial function of $|x|$.

Another approach foresees the use of Fuzzy Logic.

Fuzzy Logic is basically a multi-valued logic: this can be considered as an extension of the conventional Boolean Logic that allows to handle the concept of "uncertainty" of a proposition, rather than the absolute values and categories in Boolean logic.

In other terms this allows more human-like interpretation and reasoning in machines by resolving intermediate categories between notations such as yes/not, interesting/uninteresting, and so on, used in Boolean logic.

Using the Fuzzy Linear Programming techniques allows to take in account the not well defined ("fuzzy") nature of many involved parameters. The advantage of this method is that it does not force the decision maker to define the problem in precise terms, as it would be needed for a not fuzzy approach.

An alternative approach of which it is also interesting to appraise the applicability consists in using the Disjunctive Logic Programming (DLP).

DLP is an advanced formalism for knowledge representation and reasoning, which is very expressive in a precise mathematical sense: it allows to express every property of finite structures that is decidable in the complexity class NP^{NP}.

Thus, under widely believed assumptions, DLP is strictly more expressive than normal (disjunction-free) logic programming, whose expressiveness is limited to properties decidable in NP. Importantly, apart from enlarging the class of applications which can be encoded in the language, disjunction often allows for representing problems of lower complexity in a simpler and more natural fashion.

In other terms DLP is also able to simply and naturally represent way of thinking, incomplete knowledge and, in general, "difficult" problems. It is currently applied in different fields, such as the expert systems, the artificial intelligence and the evolved systems for database management (Leone et al. 2006).

5 Conclusions

The inclusion of a tool for tourist personalized planning in the Virtual Museum System of Magna Graecia

constitutes an innovation in the field of exploitation of Calabrian cultural and archaeological heritage. In this context, WebGIS will constitute the natural interface through which the consumer can express his/her preferences, also in terms of geographical chooses, regarding the itinerary and by which this can be returned and visualized.

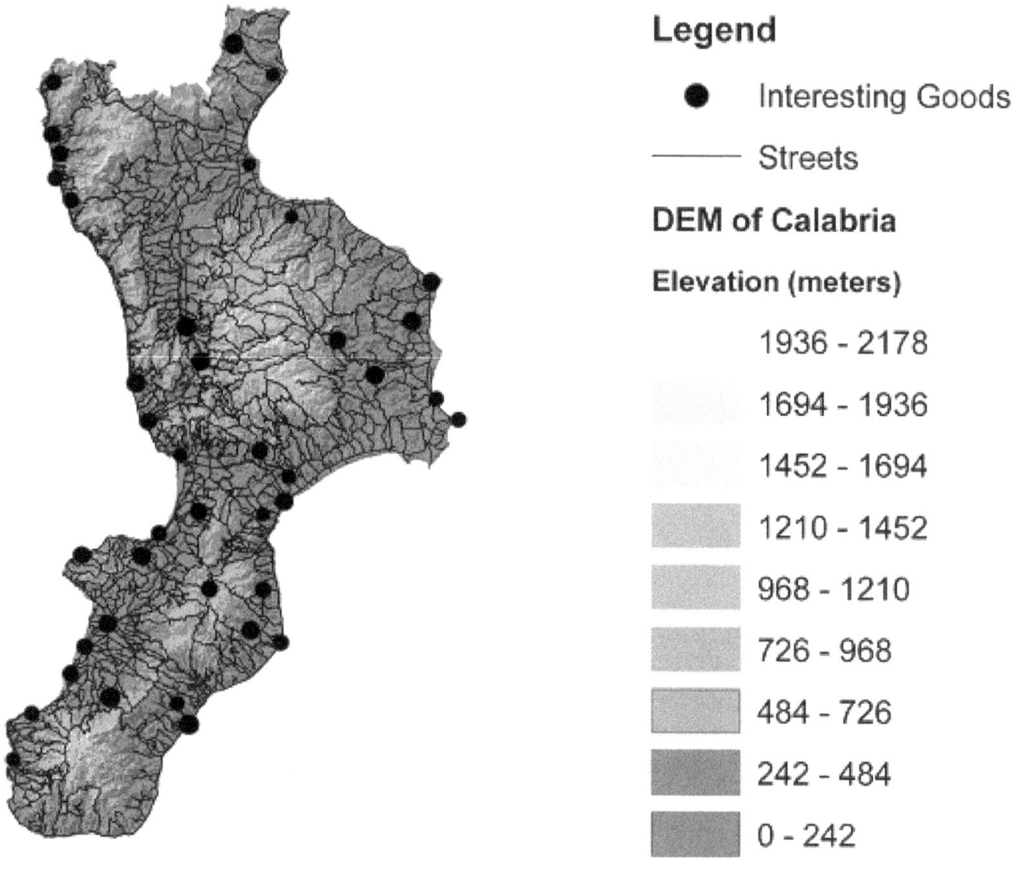

Figure 2 *Streets network of calabrian territory and localization of interesting goods*

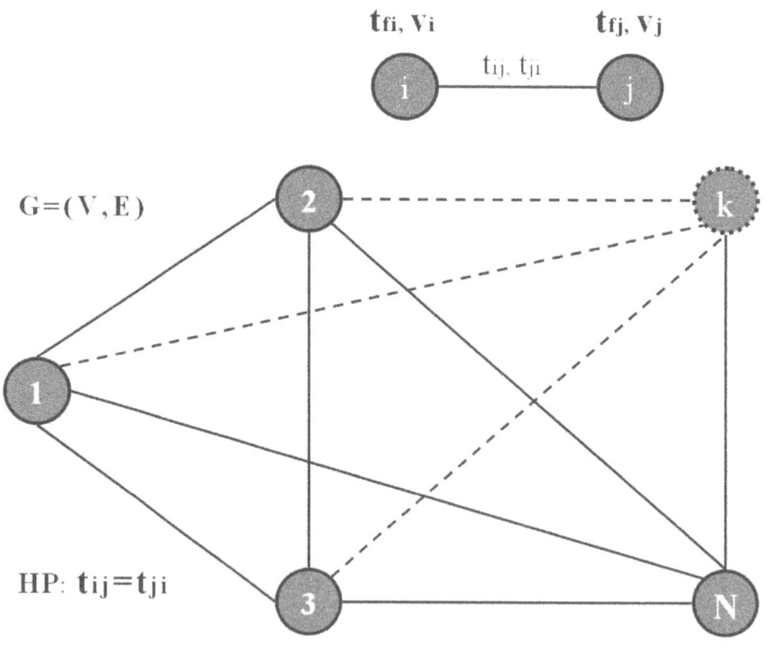

Figure 3 *Representation of interesting goods and connection among them through a graph-based formalization*

References

BARTH P. et al. 1998 "Modelling Discrete Optimization Problems in Constraint Logic Programming" in *Annals of Operations Research*, Springer, 81:467-496

BALDACCI, R., et al. 2002 *Fondamenti di Ricerca Operativa*, Pitagora Editrice Bologna, Italy

CATALDO A. et al. 2003 "Le tecnologie Web-GIS nella definizione di itinerari turistici personalizzati" in *Atti della 7a Conferenza Nazionale ASITA. L'informazione territoriale e la dimensione tempo.* Verona, Italy

FASOLO, B. et al. 2005 "Il processo decisionale on-line: rassegna di studi empirici e confronto tra siti internet per l'aiuto alle decisioni negli Stati Uniti e in Italia" in *Giornale italiano di Psicologia a. XXXII, n. 4: 693-710*

FISCHETTI, M. 1995 *Lezioni di Ricerca Operativa*, Edizioni Libreria Progetto Padova, Italy

GOTTLOB G. 1994 "Complexity and Expressive Power of Disjunctive Logic Programming" in *Proceedings of the International Logic Programming Symposium (ILPS'94),* M. Bruynooghe ed.*:23-42*, Ithaca NY, MIT Press.

LEONE, N. et al. 2006 "The DLV System for Knowledge Representation and Reasoning" in *ACM Transactions on Computational Logic,* ACM Press, New York, USA. 7(3): 499-562

MARTELLO, S. et al. 1990 *Knapsack Problems: Algorithms and Computer Implementations*, John Wiley & Sons, Chichester.

MARTELLO, S. et al. 1987 "Algorithms for knapsack problems". S. Martello, G. Laporte, M. Minoux, C. Ribeiro (eds.). *Surveys in Combinatorial Optimization*, Annals of Discrete Mathematics, 31, North-Holland, Amsterdam, 213–258

MARTELLO, S. et al. 1999 "Dynamic programming and strong bounds for the 0-1 knapsack problem" in *Management Science* 45:414–424

MINKER, J. 1994 "Overview of Disjunctive Logic Programming". in *Annals of Mathematics and Artificial Intelligence*, Springer Netherlands 12:1-24

NEMHAUSER, G.L. et al. 1998 *Integer and Combinatorial Optimization*, Wiley

ZADEH, L.A. 1965 "Fuzzy Sets" in *Information and Control,* 8: 338-353

ZIMMERMANN, H.(2001) Fuzzy Set Theory and its Applications, ISBN 0-7923-7435-5

The Volumnis' Hypogeum in Perugia, Italy
Application of 3D survey and modelling in archaeological sites for the analysis of deviances and deformations

Daniel Blersch,[1] Marcello Balzani,[1] Gennaro Tampone[2]

[1]University of Ferrara, Department of Architecture, Centre for Development of Integrated Automatic Procedures for Restoration of Monuments D.I.A.P.Re.M., Ferrara, Italy [2]University of Florence, Department of "Restauro e Conservazione dei Beni Architettonici" D.I.R.E.S., Florence, Italy

1 Hystorical Notes and Introduction

The Hypogeum of the Volumnis, titled family native from Perugia, has been built almost around the III and the II century B.C., coinciding with the Hellenistic period of Etruscan civilization.The tomb was discovered in 1840, still in original conditions and preserving incineration urns, fact that allowed Scholars to reconstruct a possible genealogic tree. The Hypogeum of the Volumnis represent a quite rare example, together with Cerveteri Tombs (Rome), Puttu Codinu (Sassari) and others, of sepulchre, completely excavated in the soft sand stone and simulating an anthropomorphic environment. The planimetry, in fact, is composed in accordance with the typical features of a civil house: it consists of a large atrium introducing to a final tablinum, two alae and three couples of lateral cellae.

This kind of arrangement of the plan was made to ensure the defunted a comfortable afterlife, similar to the temporal one.The atrium is covered by a two-slope ceiling, realised to resemble the intrados of a wooden hut roof; the entire carpentry is simulated in its components as ridge beams, rafters, joists and boards, carved in the rock in true scale. The studies on the Hypogeum started along with the discovery of the sepulchre and, until 1930's, consisted in several sketches and summary drawings.

The first survey is owed to Armin von Gerkan, German historian of Greek and Roman architecture. Anyway, the measurement work revealed critical because of subterranean conditions and plan's characteristics: consequently von Gerkan's observations and interpretations were consciously limited.

2 3D Laser Scanner Survey Procedure

The 3D survey has been realised with a Leica Geosystem HDS 3000, setting a 1x1 cm grid as point cloud density. This operation required two operators for fourteen hours of work, acquiring 10,7 millions of points in 23 scans by 13 positions. The environmental humidity rate was of 97% in subterranean rooms; the survey has been served out without interfering with the affluence of tourists. The survey of the sepulchre was anchored to a topographic one, consisting in 19 targets and realised with a Leica TCR 1101. In phase assembly of the total model, the overlap of all the partial registrations caused a non homog-enous density of the point cloud, which required the appliance of redundancy filter with minimum radius of 0,5 cm, operation carried out with Leica Cyclone.5.1 software. The model reduced so its amount of points to 7,7 millions. The final point cloud was, then, cleaned of mobile installations such as furniture, alarms, urns.

Figure 1 *View from the atrium towards the tablinum. Photograph and point cloud carried out from 3D survey*

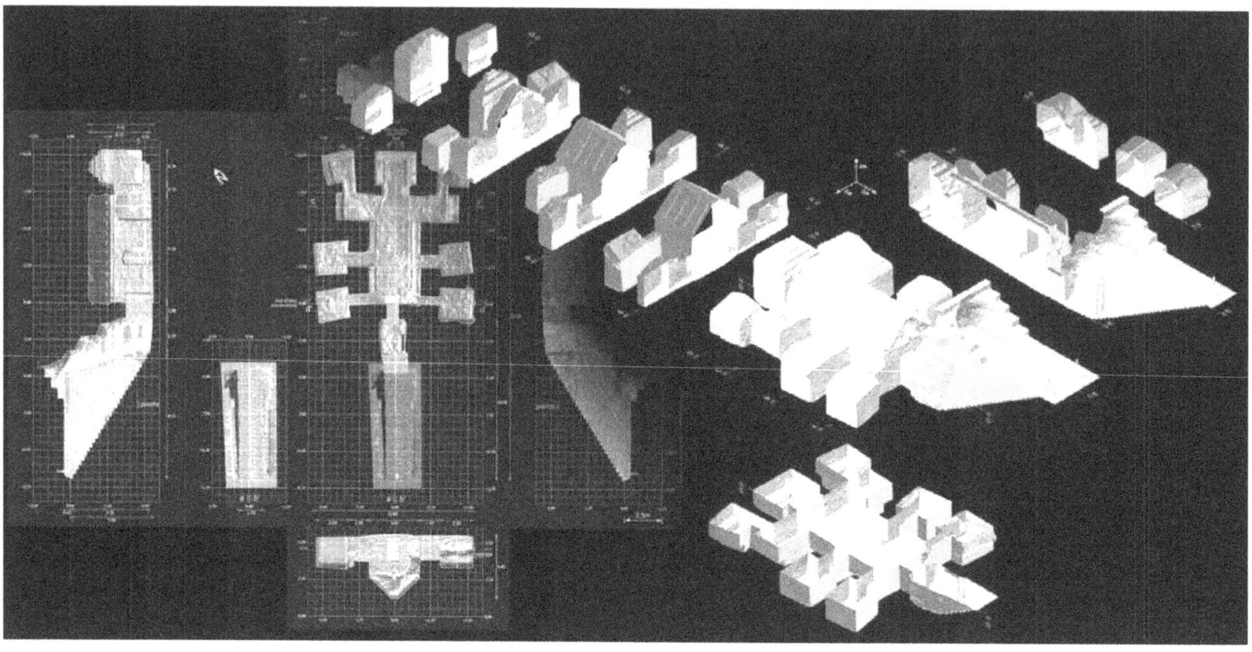

Figure 2 *The organic triangle meshed model. Plan set and exploded axonometric view*

3 The Organic Model Mesh

The firs step to evaluation of distortions and deviations of the structure was to create a mashed model, following the 3D modelling procedure with the Technology RapidForm 2004, running on a standard PC platform. That meant the triangulation of the point cloud, previously divided in 11 parts, due to the logical morphology of the monument, cutting through the section of the short corridors. This division was made necessary to avoid overloading computer memory during the mesh calculation, which nevertheless remained critical. During the meshing operations it has been possible to integrate scan shadow-areas introducing homogeneous point grids adopting Rhinoceros software. After that, the meshed model portions were gathered, decimating to 60% flat areas; than the sewing procedure required an overall decimation to 80%, in order to flattern the joints and obtain a handy model of 2,9 millions triangle surfaces, corresponding to a 48% reduction of the original.

4 Summarizing 3D Surface Model of the Main Ceiling

In order to put in evidence and quantify distortions and distinguish between decay of a 2000 years old architecture and deviation tendencies, a summarizing model was built on the main ceiling of the organic triangle meshed model with straight surfaces. These were created by intelligent choice of anchor points, considering each carpentry member and its logical deformed configuration under load condition, in order to intermediate the surface on its restraints (pondered average) and not on intermediate and eventually flexed parts (arithmetic average). To ensure the correct orientation of the straight surfaces on single conditioned anchor points, an average position of three points per short edge was chosen. This method consented to obtain an homogeneous global overview, resulting flexing carpentry curving below the ideal (straight) carpentry members.

5 Application of Industrial "Deviation Check" to the Archaeological Site

The overlay of global meshed and summarizing models allowed to put in evidence distortions and deviances in the structure. The "deviation check" procedure, usually adopted in industrial production to control the effective quality between CAD model and final product, was in this circumstance experimented on architectural scale to visualize global deviation tendencies in the elements constituting the ceiling. Overlaying the models of the complete two-slopes ceiling the presence of flexion under load condition appears clearly.The same steps were applied to each category of carpentry, which rafters, ridge beam, joists, boards and closures. Evident global tendencies of flexion could be noticed in the main categories. The rafters (visible dimensions: approximately 230x41x8 cm) show an average flexion of 2,1 cm and a maximum of 3,2 cm with the peak in the centre line area. The section line on the selected section appears close to a circular segment similar to the lay under condition of vertical distributed load with hinge re-straints at top and bottom.The ridge beam (visible dimensions: approximately 745x33x7 cm; 671 cm length between corbels) shows a flexion peak of 3,1 cm, the centre line section shows a deformation analogous to a wooden beam with hinge or maybe joint restraints subjected to distributed load. Nevertheless the curve isn't symmetric, in fact the peak is located at 105 cm from the centre of the beam: this lets suppose a further decentred

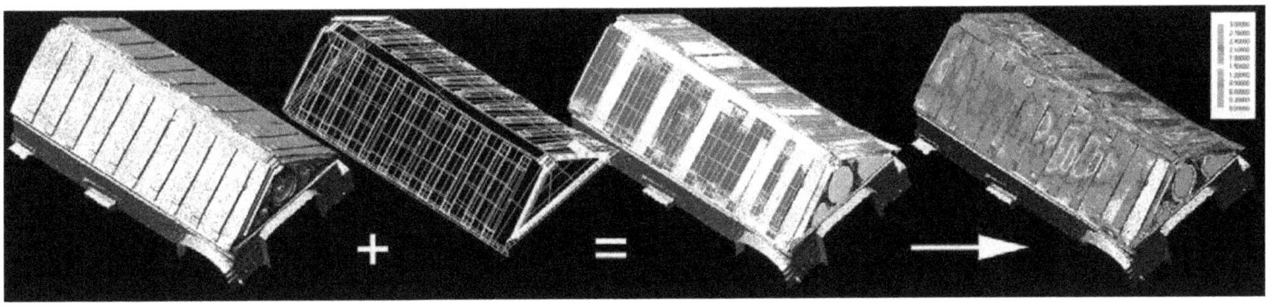

Figure 3 *Inspection procedure. Overlap of the two models and application of "deviation check"*

punctual load.This could be consisted in a hanging oil lamp of the atrium, found at the moment of the discovery, as attested by Giovanni Battista Vermiglioni in 1840s through his own description about the condition of the original interior set of the tomb. Vermiglioni clarified that the terracotta lamp, alike another one situated between the atrium and the tablinum, was found crashed on the ground, but, hanging above it, there still was the suspension stick fixed to the ceiling. Anyway, it is discussable if a six flames oil lamp, traditionally made in bronze in Etruscan houses, could cause a flexion in the beam. It is not possible to notice a clear tendency of deformation in joists and closures.In Armin von Gerkan 1930's survey are already reported some observations about a not perfectly vertical disposition of the rafters,due to execution error or a partial collapse of the soft sand stone surface, happened already during the construction works, in the hypothesis proposed by the historian himself. If this was correct, than we should suppose to notice in the laser scanner survey a gradual error reduction until the last couple of rafters, which should lay on the vertical plane. Nevertheless the overlaying of the two models showed that the inclination, departing from the inner tympanum, starts at 3,2° (omitting the first couple, with 1,3°), gradually increases to 4,7° and ends at the opposite tympanum reducing 3,4°.The detailed investigation on the cross section of the ridge beam showed that it changes, on 745 cm length, of about 1,8 cm, starting at the outer tympanum with 33,6 cm and finishing at the opposite one with 31,8 cm. It runs against the logical expectation, considering that the smaller cross section poses on its corbel at 1,4 cm lower than the major one, causing an inclination of the ridge beam. This particular can be explained considering the symmetrical raising of the floor from the entrance in direction of the tablinum, amounting to 30 cm on its overall length: extending the centre line of the lower surface of the ridge beam and the corresponding on the floor plane it is possible to individuate a focal point, determined at the intersection of the two layer, at 127 m in the mountain, measured from the entrance. In the mean time, the extended centre line on the side faces of the ridge beam converge in a theoretical focal point 125 m far in the mountain.

6 Studies on the Covering System

The Hypogeum of the Volumnis present a two-slope roof on the atrium and a square corbelled coffered ceiling on the tablinum and the right ala.

The two-slope ceiling of the atrium

The slopes have an unusual emphasized inclination of 47° and are supported by lateral beams at the bottom (725 x 23 left / 27 right x 12 cm, 4 cm overhanging from the walls) and at the top by a ridge beam (725 x 33 x 7 cm) which itself is laid on double corbels.The slopes have an extension of about 730 x 230 cm and can be divided in four categories of carpentry members (with average visible dimensions): 11 raf-ters (230 x 43 x 8 cm), 2 joists next to ridge and bot-tom beams (visible in 10 spots between the rafters, 25 x 25 x 4 cm), boards (visible in 10 spots between the rafters, 166 x 25 cm) and 20 small closures at the ridge and the bottom beams. The pediments have rafters that directly rest on the ridge beam and have a bottom beam (inner pediment: laid above the bottom beams of the slope; outer pediment: no level change) and are 25 cm high in section, with 4 cm overhang from the walls. The experimentation of an ideal reconstruction of the two-slope roof carpentry based on the true scale summarizing model, together with observations of Armin von Gerkan, conduced to the following thesis: the main supporting structure is given by a non-joint disposition of bottom beams fixed with the larger face of the cross-section on the surrounding walls, and the triangular shaped ridge beam supported by double corbels. The corbels are anchored directly in the wall structure and the rafters of the pediments work as spacers to disable the horizontal movement of the ridge beam. The previously identified rafters and boards could both consist of equal boards (280 x 43 x 8 cm) staggered in two layers. The joists result as inserted boards between the first layer of rafters in order to keep them in position during the execution of the work. The small closures are shaped boards and work as spacers between ridge beam and joists.

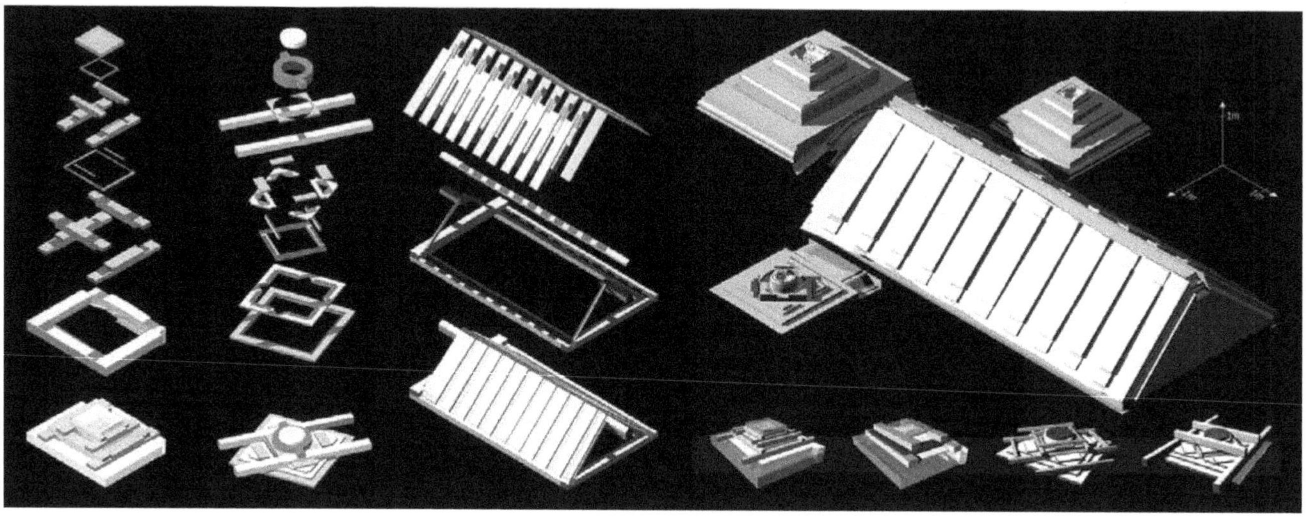

Figure 4 *Solutions for ideal carpentries' reconstructions. Summarizing model of the main ceilings*

The squared corbelled coffered ceilings of the tablinum and the right ala

The ceilings built on rectangular plan are arranged with two overhanging orders of each two parallel couples of joint beams (4 cm level change per couple) and a top closure with a framed sculptured head. The transitions to the orders are closed with shaped frames. During the investigation on the coffered ceilings it was clear that the data available today, based on a simulation in rock-cut sandstone of the inner surface of a wooden structure, was not sufficient to determine a final solution of the original carpentry set. Therefore it was reasonable to work out a number of solutions which differ not only by arrangement of elements, but as well by system and material. Some of the proposals for the original carpentry set could be discarded due to the direct dimension control handling with a true scale model, but are still considerable as functional carpentry systems. Finally, three different carpentry systems for the squared corbelled coffered ceilings have been worked out:

-visible squared partially joint beams, wooden frames and closure stone plate;
-roughly squared eventually joint beams, panelling, wooden frames and closure stone plate;
-visible squared beams and partial panelling, interposed stone plates and closure stone plate.

The diagonally corbelled coffered ceiling of the left ala

The ceiling built on an approximate square plan is arranged with two overhanging orders without level change between the parallel couples, a 45° rotated diagonal framework which remains in the same level of the second order and a joint centre torus with a framed sculptured head. All resulting triangles are sheet covered and framed. The diagonally corbelled coffered ceiling is to be considered a very similar system with the difference that the supporting structure can be done by partially visible squared beams with panelling or roughly squared beams with complete panelling. The last one lets open several variations of the supporting structure.

7 Statistical Analisys of Repeated Elements in the Two-Slope Roof of the Atrium

The last step in the analysis of the two-slope ceiling consisted in the statistical analysis based on the Gaussian distribution, thanks to the repetition of several elements in the two-slope roof, which the rafters. The non linear rotation around a perpendicular direction to the axis of the ridge beam induce a different interaxis between the couple of rafters, at their bottom level, while this keep a substantially constant value at the top; that shows how the ceiling has been carried out starting from the ridge beam, prosecuting to the bottom ones. Considering that a real wooden roof is executed arranging the rafters along the ridge beam at a determined distance, it appeared correct to effectuate the statistical analysis by means of three logical measures, which the interaxis of the rafters measured in the summarizing model above the centre line of the ridge beam, the left distance between rafters and the width of the rafters themselves. The interaxis results of 679 mm with a standard de-viation of ± 14 mm calculated with 19 imputs; the distance between rafters is 253 mm ± 18 mm with 20 imputs; the width of the rafters is 428 mm with 20 imputs.The measurement system chosen to check the values were Roman foot (296 mm), proposed by Matteini-Chiari, the Etruscan foot (324 mm), proposed by Hallier, and finally the nameless unit noticed in several tombs in the area of Chiusi (200- 210 mm), reasoned by the big influence of this city on Perugia.No good results of compatibility that allow to prefer a unit more than the others for rafter distance and width have been found. A good compatibility was found for the rafter width 428 ± 14 mm with the units from Chiusi 400- 420 mm.

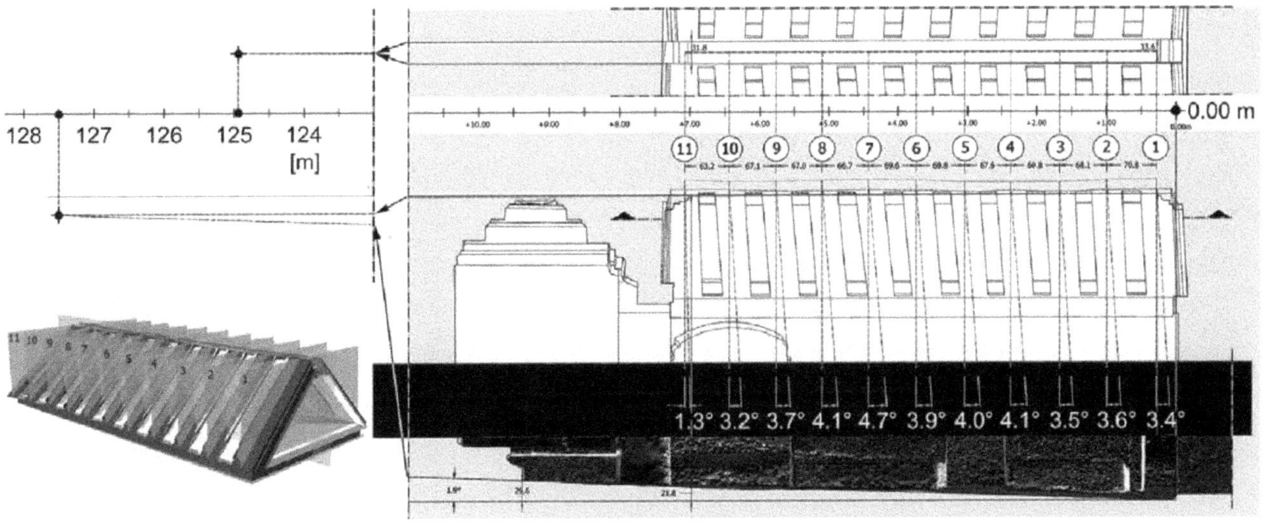

Figure 5 *Two-slope ceiling. Inclinations of rafters, focal points and determination of average planes*

8 Conclusions

The combined appliance of 3D laser scanner survey and triangular meshed model allowed to update the 2D canonical documentation elaborated by Armin von Gerkan in 1930's, showing the variation of regularity between atrium, alae and tablinum. The followed steps have been chosen according to the principle of simplicity and easy execution in the elaboration and handling of databases and models. The "deviation check", methodology proper to industrial activity but applied in archaeology through laser scanner survey and triangular meshed model, permitted to appreciate global deformation tendencies for ancient architecture with overall phenomenon of decay. This procedure has been fundamental to define the representative section planes. No univocal results have been found to exclude many hypothesis about different questions, but the informations here acquired represent a starting point to discuss them. It's not possible to exclude to attribute the flexion of the elements to a systematic error due to execution techniques of Etruscan workers, but it's still compatible with analogue funeral architectures of even older communities of Central Italy and surrounding, such as Cerveteri Tombs and Domus de Janas in Sardinia. Even the truth of the enhance perspective effect, given by the ridge beam and floor inclination, is not verifiable: to verify this we should notice the behaviour of bottom beams, but this is not possible. In fact in some photograph of the early 1900's, it is possible to see that the right bottom beam is missing because of probable decay; by means of this we can guess that the actual one is a modern reconstruction and we have no information about the disposition of the original one. The same problem exist for several other elements. The rotation of the rafters gives a slim, further optical effect of depth of the atrium, but it is not possible to totally exclude an accidental error. Anyway the thesis of optical effect in the Hypogeum of the Volumnis family is supported from investigations on contemporaneous architecture as the Parthenon in Athens, realised 1996 by the Greek archaeologist Manolis Korres, who revealed optical corrections handled by light inclin ation of the outer columns with different focal points vertically above the temple. Similar correction haven't been observed in Etruscan architecture yet and this thesis should be discussed by archaeologists. Since optical corrections in Etruscan architecture hadn't been observed, yet, this thesis should be further discussed by archaeologists. At least, we can state that in the case of the Volumnis' hypogeum it could be possible, from a chronological point of view.

References

AA.VV. 2003. *RapidFormTM 2004 User Guide & Tutorial*. Seoul: INUS Technology Inc.

BLERSCH D., 2006. *The simulated timber structure of the Volumnis hypogeum in Perugina, Italy*; *paper for Structuran Analisys of Historical Construction 2006 Conference in New Delhi.*

CENCIAIOLI, L. 2002. *Ipogeo dei Volumni e Necropoli del Palazzone*. Perugia: TSM.

CONESTABILE, G. 1855. *Dei monumenti di Perugia etrusca e romana, della letteratura e bibliografia perugina. Nuove pubblicazioni precedute da un discorso intorno alla vita, agli studi ed alle opere di Giambattista Vermiglioli: Il sepolcro dei Volumi* 2. Perugia: Bartelli.

DI GUGLIELMO, E. 1998. *Soffitti intagliati e architettura domestica in Etruria*, Florence: Olschki.

FRÖHLICH, C. METTENLEITER, M. 2004. *Terrestrial laser scanning. New perspectives in 3D surveying.* Wangen: Z+F.

HODGE, A.T. 1960. *The woodworks of Greek roofs.* Cambridge: CUP.

KORRES, M., PANETSOS, G.A. AND SEKI, T. 1996. Parthenon. In M. Korres (ed.), *The Parthenon. Architecture and conservation.* Athens/Osaka: CCAM Publisher.

OLESON, J.P. 1978. Technical aspects of Etruscan rock-cut tomb architecture. *Römische Mitteilungen* 85(1). Mainz: Zabern.

OLESON, J.P. 1982. *The sources of innovation in later Etruscan tomb design (ca. 350 - 100 B.C.)*. Rome: Bretschneider.

TAMPONE, G. 1996. *Il restauro delle strutture di legno*. Milan:Hoepli.

TAMPONE, G. 2002. *Le strutture di legno. Cultura, conservazione,restauro*. Milan: De Lettera.

VON GERKAN, A. MESSERSCHMIDT, F. 1942. Das Grab der Volumnier bei Perugia. *Römische Mitteilungen* 57, p. 122-235. Munich: Bruckmann.

Digital reconstruction of archaeological objects using hybrid sensing techniques – the example Porta Nigra at Trier

F. Boochs, A. Hoffmann, U. Huxhagen, D. Welter

Mainz, Institute for Spatial Information and Surveying Technology

1 Introduction

Several buildings and monuments, especially historic sites, have a certain history which has to be investigated. In the past, the most common proceeding has been analogue measurements assisted by simple tools and documentation media. Today, the work of art historians is still characterised by those methods.

An important benefit to analyse a monument is implemented by hand-based measuring methods which provide documentation for different levels of detail (geometry and text). Those surveys allow giving a variety of answers to the complex questions surrounding an ancient monument.

In the environment of monument conservation hand-based measurements are kindly regarded as standard. The procedure of those surveys makes use of the observers contact with the object and are usually leading into coloured analogue contour plans or digital raster information due to the applied simple measuring instruments.

Figure 1 *The Porta Nigra in Trier, partly mapped with a digital surface model*

In the field of surveying, the application of digital workflows is well progressed. In view of surveying engineers, data acquisition and its processing as well as the presentation of multimedia content is a well established practice – however, their use for the preservation of historical monuments is not well developed.

In this context modern technologies like Tachiometry, 3D laser scanning and digital photogrammetry have been tested at the Prota Nigra with regard to the requirements of historians, in order to show their potential for these aims. The presentation of the results is part of this paper which shows pros and cons of digitally acquired and handled geometric data for the use of monument preservation.

2 History

The Porta Nigra was built as the northern city gate of the city Augusta Treverorum (today's Trier) at the end of the 2nd century AD. At the time it had the name Porta Martis. Later it became the name Porta Nigra (black gate) because of the dark colouring of the sandstone caused by the weather. The durability of the construction method is of particular interest, because the building resists weather and enemies since more than 1700 years without the use of mortar. It's the only holdover of the Roman city wall.

After the retreat of the Romans, the Porta Nigra was unused for 600 years. Then, in the middleage (11the century), it was rebuilt to a so-called double church. In the upper floor was an abbey church, underneath a church for the people. About 1150 the archbishop extended the building with an apse at the Eastern tower. Nowadays, this polygonal choir is of art historical interest and ranks among the main works of the Romanesque architecture of the 12th century.

After the conquest of Trier by the French at 1794, the revolution troops took the roof of the church off and carried off the whole interior decoration. Because of the subsequent disrepair of the fane, the constructions of the middleage has been largely removed, except for the apse in view of static reasons. The basis of the building has been laid open to the level of the Romans time and all parts of the building have been roofed. Since 1986, the Porta Nigra is World Cultural Heritage of the UNESCO.

3 Tacheometry

Although often used for the complete surveying of objects the most important task of tacheometric measurements is the establishment of a stable and precise geometric frame. This allows to build up a base for all subsequent measuring techniques and was provided by a geodetic network derived from over determined measurements from inside and outside the monument. To these data a least squares adjustment was applied, verifying a sufficient accuracy for all observations (3D point accuracy of about 1mm).

As a first survey in the project tacheometry was used to record ground plots of the Porta Nigra, which obviously is one of the most easiest methods of geometric data acquisition.

The discussion of the results with art historians showed, however, that a data capture in the opinion of a surveyor, trying to find an idealized and generalized ground plot, is not acceptable for the use of a detailed building research analysis. It turned out, that in contrast to usual hand measuring methods that explicitly go into detail, the recorded results showed too large deviations to the buildings true shape.

As a result, this experience pointed out that the work of an art historian always demands for a contact with the object to investigate important and characteristic sections of the monument. For this reason, a collaboration of surveyors and art historians seems to the best choice to provide valuable and accurate measurement results.

The use of tacheometry in the context of a detailed building analysis helps at most for the geometrical registration of trivial building structures, at best in 2 dimensions. This could be done with the help of a specialized interpretation by experts.

Tacheometrical instruments have a big potential of geometrical accuracy. On the other hand, it has to be considered, that only a limited number of discrete points can be recorded. With increasing the number of measured points, the tacheometry gets economically unattractive compared to other techniques like 3D laser scanning and photogrammetry.

A great profit of tacheometry is the determination of reference points for the applied 3D laser scanning and photogrammetry. The targets can be acquired with the help of reflectorless distance measurements – a state of the art implemented in modern tacheometers.

4 Laser scanning

Laser Scanning Survey

Laserscanning is an increasingly popular measuring technology, allowing to simply generate the full geometry of even complex objects. We used a *Cyrax 2500 from Leica Geosystems* (*Cyra*), a so called camera-view-scanner with a 40°x40° field of view. The angle sensing is realized with two rotative mirrors which are situated in the optical path of the laser, which divert the rays in horizontal and vertical direction. Here, the reflectorless distance measurement is implemented with the time-of-flight principle. The scanning speed is about 1000 points per second.

According to manufacture's data the accuracy of a coordinate measurement is about 6mm@50m whereas the dependence to the characteristics of the surface must be noticed. Influencing factors can be color and roughness of the material as well as the convergence angle of the measurement beam and the geometry of the surface.

The effect of the material color is small and is situated within the measurement accuracy. Of greater influence is the reflectance characteristic of the surface, resulting in failures in case of low reflected energy. The roughness of the surface affects the measurements because the distance sometimes will be interpreted not correctly. Significant influence to the reliability and accuracy of the distance measurement has the convergence of the measurement beam and the geometry of the surface (Kern 2003, pp. 40-46).

With the 3D laser scanner applied, almost the whole basement of the Porta Nigra could be acquired with a resolution of about 0.5 cm. Additionally a large section of the upper parts of the monument was recorded with a resolution of 1cm and also some parts of the apse. Altogether 51 viewpoints were needed to acquire in total about 132 million points. This corresponds to a space requirement of 5.1 GB on the data medium.

Significant problems occurred with regard to the dark, in some places nearly black surface of the monument. From a distance of 40-50m to the monument, the scanner acquired only few points, as the reflecting signal was too poor.

Another problem in this special case was the public-transit-traffic. During the acquisition we got to know, that the data which were matched from the eastern position of the Porta Nigra have been very noisy, because of the traffic amount in the adjacent street. The orientation of these point clouds resulted in a four times higher standard deviation as in those without these influences. One reason for that effect is the impact of vibrations on the instrument due to public service vehicle traffic.

To transform the point clouds in a reference coordinate system, reference points (*HDS*-targets, spherical targets) were fixed on the object and were coordinated with a reflectorless measuring tacheometer.

Analysis

At first, all point clouds had to be freed from erroneous points which were measured on passing tourists or past the object. Afterwards a global registration was accomplished with the help of the reference points. These manipulations have been accomplished with the *Leica* software *Cyclone*.

The next modelling steps have been realized with the software *Geomagic* from *Raindrop* within smaller clouds of about 10 million points, in order to keep the workload for the hardware in acceptable limits during the following calculation.

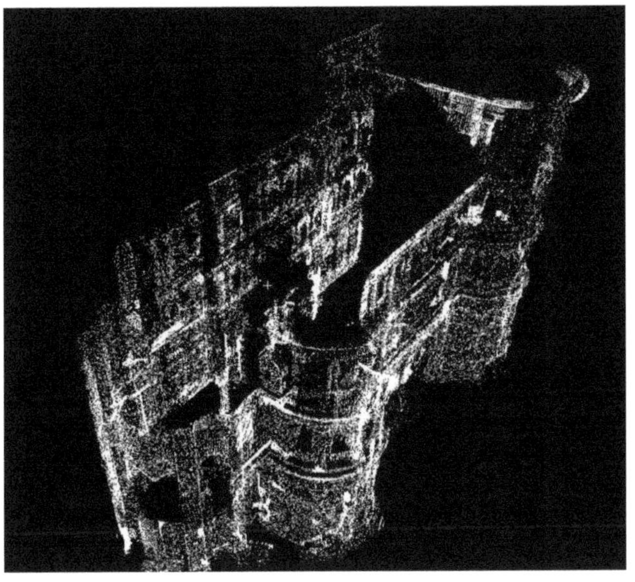

Figure 2 *3D point clouds after orientation*

The data had to be cleaned from meshes, which have been placed on the monument as protection against doves. This needed a high manual effort whereas it is not guaranteed that the data are clean of defects. A better solution would be to get off the dove meshes previously to get correct data.

After the cleaning of these obstacles, all outliers have been analysed and deleted with an automatic algorithm. This filter searches all points exceeding a certain distance to the point clouds, assuming a larger distance for wrong points.

Another kind of degradation is caused by the reflectorless distance measurements being used from 3D laser scanner. This is a random type noise, which is superimposed to the true distance measurement. It describes the deviation of the point from its true position. The deviation depends on the reflection characteristics and the geometry of the measured surface, as well as the angle of incidence of the measurement beam. The spot of the laser is not infinitive small, which means it hits the surface in a certain area. Especially at edges measurements cause a high noise, because only a part of the spot hits the edge. *Raindrop Geomagic* includes a filter to clean up the noisy data.

However, it has to be noticed that a loss of details results from a noise reduction. It is always a compromise between a smoothed and easy to model surface and the visibility of details. Unfortunately some important details like exact edges of the sandstone cuboids have been flattened, which downgrades the usage of the data for detailed building research. In this concrete aspect it is of importance how a scanner works. There are large differences between the technical characteristics of existing instruments, what might lead to strong variations in the quality of the data resulting in correspondingly strong variations in the potential for further usage.

After noise reduction, the point clouds have got orientated among themselves by means of a best-fit-algorithm. Afterwards all point clouds were merged to a single one. Additionally the quantity of points had to be reduced, as the density of points in the overlapping area was much too high. Because of a homogeneous density of points, one gets a homogeneous point cloud used to create a triangulated surface.

Despite detailed editing, several holes resist in the surface. A time intensive manual post processing of the triangulation network is therefore needed. Only smaller holes on less curved surfaces can be filled automatically. As the surface is really irregular it is quite impossible to assume a correct course of the surface by an algorithm.

Such big holes emerged from shadows in view direction on one hand and on the other hand mostly due to the sparse density on the dark surface.

Finally a complete 3D model has been received representing the closed surface of the object. However, this surface model does not fulfil all requirements of art historians. They need a very detailed description as especially for areas of importance like edges and joints. But here the geometric accuracy and the level of detail necessary cannot be guaranteed. Because of the irregular surface, there is no sense to construct edges.

Nevertheless there are several advantages for the work of art historians like the possibility to quickly produce of a rough surface model allowing to give an object's overview, as base for other measurement techniques as well as for the analysis of small scaled relations and proportions.

5 Structured light projection

Structured light projection is another scanning technique allowing to generate very precise and detailed 3D models. In order to show the potential of such a technology a high resolution recording of the memorial plaque of St. Hieronymus insides the monument has been performed. The size of the plate is 2.84m x 1.63m x 0.05m. As projector the *GOM ATOS II* was used, a digitising system that works with white structured light. The light pattern is projected on the object's surface while two digital cameras record the reflected light. For good measuring conditions (bright object's surface, few ambient light) 3D coordinates can be calculated for every pixel.

The scan volume can be changed by modification of the systems configuration. For this large object we used the maximum scan volume of 800mm x 640mm x 640mm, giving a point density of about 0.7mm for one measurement.

To record an object that is many times larger than a single scan volume it is necessary to define precise reference points for a global orientation of the individual scans. That was done by a photogrammetric triangulation. The photographs were taken with a calibrated *Fuji FinePix S1* digital camera (17mm lens), the subsequent evaluation was done with the photogrammetric software *ImetricS*, which detects the coded targets automatically and solves the bundle adjustment.

Figure 3 *High resolution 3D surface model of the memorial plaque of St. Hieronymus*

With the data of the reference points, the sensor can be moved as necessary to record all surface parts. The final result is a high-resolution 3D surface model which can serve for different aspects. One potential use could be the production of an accurate replication by means of molding tools Furthermore it is very well adapted, as opposed to the terrestrial laser scanning measurements, for the research of the historians due to the high accuracy. Drawback is the usability in just small areas and the high amount of data.

6 Photogrammetry

Photogrammetry plays a major role in terms of preservation of ancient monuments and architecture. On one hand the visual information in the images is of high value for an interpretation and on the other hand the geometric description is guaranteed. Furthermore images are easy to archive. The interpretation on a computer screen is possible by using effective computer systems with comprehensive functionality but they require mostly the knowledge of surveying technique. A fact, which needs to be reviewed during planning of a measuring campaign and which leads to a close collaboration between experts from various faculties.

Single image evaluation

In the range of architectural photogrammetry the method of single image evaluation is often used. The most widely-used method, which furthermore can be applied from non-specialised user, is the 2D photo rectification. The pixels of the images are reshaped to an rectified representation by means of a perspective transformation of points measured at the object. The result is a true to scale digital image with exact geometrical information. But the described method is only useful, when the recorded object is approximately flat. This is the case for house fronts which are the most likely application of this method.

Another simple usage is possible if the object contains more regular shapes like cylinders, for example. Then it is possible to compute a wrapped-off image of the measured picture with help of a resampling process. To use this method, the knowledge of the regular shape of the object as the interior and exterior orientation of the image are strongly needed to re-establish the relation between the position of the picture and the reference surface.

But also in general cases accurate results are possible when detailed information to the surface is available. Such a surface model can be derived from a 3D laser scan of the surface, for example. Then, the triangulated network represents the reference surface for a differential rectification. A product of this computation-consuming evaluation method are digital ortho-images which might be produced with high precision and valuable information, as . high resolution photographs can be rectified very accurately.

With methods from the virtual reality it is possible to map rectified images onto a point cloud or on a reference surface. With this, very impressive observation possibilities can be given.

Rectifications are used during the building survey as mapping basis or as a preliminary product for modelling work. They include a high density of information in terms of a pictographic representation, which should be used with certain reserves as they are often employed by non-specialized persons and are often underestimated in terms of accuracy.

Stereo image evaluation

For non planar objects a proper geometrical description needs a 3D data acquisition., what is supported by stereo photogrammetry. By means of two images and their known relative and absolute orientation it is possible to reconstruct object points by two intersecting rays. With his perception the observer gets a three-dimensional impression of the images and 3D coordinates can be determined like with analytic plotters. The result is a graphical representation of the object, reduced to significant contents in form of the representation of contour lines.

If the interactive evaluation is performed by surveyors, this might be critical due to the lack of specialized knowledge concerning the application. Therefore results are more valuable if the specialists for monument research do the interpretation themselves. This needs a close interdisciplinary collaboration between surveyors and historians in order to provide correctly oriented stereo images, because these technical steps might be too demanding for non surveyors.

Multi image evaluation

High accuracy and feasibility is provided by a multi image evaluation. In opposition to the stereoscopic interpretation a defined geometric adjustment and not an exact number of images is needed. With the help of a triangulation all coordinates and orientations are calculated simultaneously. This well knwon photogrammetric task produces accurate object points which might be used for further steps and gives most reliable orientation values also being useful for following tasks like a 3D evaluation for example.

Interpretation of the Porta Nigra

The result of photogrammetric interpretations is mostly a CAD product. The user gets two dimensional plans and simple surface models being useful to show the results. Clear structured surfaces can be divided into their basic geometries with the help of which distribution of edges and contours is easy to understand.

The situation at the Porta Nigra is more complicated in this point of view. Continuous geometries are rare and nearly not existing for the view of a surveyor. The choice of significant points to document the characteristic geometry of the building must be done with the help of an expert.

A gapless digital 3D reconstruction model of the Prota Nigra can not be realised be any of the applied measuring techniques in an economic and scientific way. The combination of every technique's peak power is the key to a solid documentation of the monument, whereas high resolution photogrammetry can serve as base for further investigations. With the help of an high resolution image bundle it will be possible to find more sections of interest among guidance of experts of other profession. Those investigations could lead into a stereoscopic analysis at small-sized areas or to an evaluation of collected measurements at the building's surface.

Discussions with user from the preservation of monuments showed a necessary resolution of the images of at least 1mm/pix at the object to get a useful profit from the data. This only can be achieved by high resolution metric cameras. To show the potential we used a ICam28 from IMETRIC, providing high quality digital images (cf. Fig. 4).

Figure 4 *Icam28 in action at the Porta Nigra, the resolution corresponds to 1mm/pixel at the object*

In addition, a combination of photogrammetry with an information system would give the possibility for a combined use of geometric data and content documentation allowing to simultaneously use geometric and semantic information.

7 Conclusions

All applied measurement techniques in this project have their respective strengths and their limitations as well. There is no overarching procedure for the digital reconstruction of archaeological objects. In fact a bundle of combinations of the several measurement techniques is needed to get a monument documentation of high quality. In the following, the strengths of the several procedures with regard to the particular tasks by means of the obtained cognitions of the Porta Nigra project are illustrated.

To establish a rough model of the whole object, as overview and base for following activities, photogrammetry and laser scanning can both be used. For further tasks and for the accurate evaluation of contours photos are needed anyhow. For that reason a photogrammetric evaluation for the creation of a grid model is to prefer for economical purposes, because the evaluation of the laser scanning data emerged as very time-consuming at complex buildings, like Porta Nigra.

The advantage of 3D laser scanning is the detailed modelling for the complex, irregular surface for small scales. In addition more accurate analyses of proportional relations of several parts of the building can be done. Even the comparison of floor situations can be done as well. Height relations of arches and cupolas are good to interpret, at least in fact of the possibility of observation in different perspectives. For those analyses a partly acquisition of the Porta Nigra is recommendable.

Figure 4 *3D model of vertical intersection between apse and roman construction useful for coherence and proportion evaluation*

If the requirements demand for a more accurate modelling of the surface's structure at large scaled space, such as ornaments or original vestiges of processing, the use of the high-resolution structured light projection is needed.

For additional analyses, concerning the purpose of building research, the results of laser scanning are not useful. The required details can not be detected, a disadvantage in contrast to the other measurement techniques applied. In addition, the geometrical accuracy does not achieve the historian's resolution requirements of the monument of 5mm at critical areas like edges and joints.

Advantage of photogrammetry and structured light projection is the comparatively short working time at the object. The analysis can take place in the office. Especially photogrammetry provides the opportunity to control image data capture and the subsequent photo triangulation by a geodesist, while the analysis and the consequent interpretation might be done by a briefed historian.

In spite of all modern measuring techniques it should never be forgotten, that the contact to the user (historian, conservator etc) never can be left off. A lot of relevant details concerning the purpose of building research can only be detected in contact to these specialists. These details are often acquired with a hand-based measurement or plotted into a 2D contour plan.

The different methods of measuring techniques for building documentation have to be seen from different angles and according to different requirements. From the high resolutive modelling of small building details up to the extensive documentation of geometrical and attributive data by hand measuring methods or the complete recording of the building it is important to proof the purposes and constraints of the acquisition at the beginning of any activity in order to get a result of high quality.

There is a strong correlation between content wise completeness and geometrical accuracy. The best geometrical accuracy is not sufficient when the data extracted doesn't match the needs of the specialist as well as a qualified specialised evaluation doesn't help without the adequate geometrical relation. The outcome of this is the imperative commitment to a interdisciplinary and methods combining strategy of monument surveying, within which the building survey is done in closely teamwork between the professional groups, which are interested in the result, and the specialists of the applied methods (Weferling 2002, pp. 149-152).

References

GOM Optical Measuring Techniques, 2005, *User Manual.*

HOFFMANN A., WELTER D., 2006, *3D-Aufnahme der Porta Nigra mit unterschiedlichen Messverfahren unter denkmalpflegerischen Gesichtspunkten,* Diploma thesis at Universitiy of Applied Sciences Mainz.

KERN F., 2003, *Automatisierte Modellierung von Bauwerksgeometrien aus 3D-Laserscanner-Daten,* Dissertation at Technical University at Brunswick, Geodätische Schriftenreihe der Technischen Universität Braunschweig, Nr.19.

MARBS A., 2002, *Experiences with Laser Scanning at i3mainz,* CIPA, Heritage Documentation - International Workshop on Scanning for Cultural Heritage Recording, pp. 110-114, Corfu, Greece.

WEFERLING U., 2002, *Bauaufnahme als Modellierungsaufgabe*, Dissertation at Brandenburg Technical University at Cottbus, Retrieved 2002 from Deutche Geodätische Kommission at Verlag der Bayerischen Akademie der Wissenschaften, München.

Incallajta, Performance Center of the Inkas: A Digital Reconstruction and Virtual Reality Analysis

Lawrence S. Coben

University of Pennsylvania, and Paul John Boulifard

Imperial control may be exercised and manifested in a number of ways. Strategies of such control and expansion include overt military occupation as well as more subtle forms of ideological dominion such as the construction and manipulation of architectural space and the sponsorship and performance of rituals, spectacles and ceremonies. While proficient armies are critical to the expansion and maintenance of an empire, archaeologists have frequently given insufficient analytical weight and effort to the roles of performance and ritual theater in constituting, extending, and manipulating political power in premodern complex societies of performance (Inomata and Coben 2006). In contrast, Coben (2006) has argued, utilizing the Inka as a case study, that empires and states constructed and placed ritual architecture and performance spaces as a principal strategy to achieve and exercise control, particularly in contested frontier areas.

In this paper, we explore some of the methodological issues surrounding the collection of data in the field for and the creation of digital reconstructions and virtual reality walkthroughs of archaeological sites. we then demonstrate the potential roles of such reconstructions and walkthroughs in the archaeological analyses of such architecture and spaces. we employ data from my work at the monumental Inka site of Incallajta in central Bolivia as a case study.

Incallajta and Other Cuzcos-Sacred Performance Centers

The site of Incallajta (Figure 1) is located near the modern and Inka town of Pocona, about 120 kilometers east of Cochabamba. The site is believed to have been constructed in the late fifteenth century during the reign of the ruler Topa Inka, and rebuilt by the ruler Huayna Capac in the early 16th century. (Cobo 1990, Sarmiento de Gamboa 1947, Cabello de Balboa 1951, Betanzos 1996) The site is located 2900-3200 meters above sea level, on an elevation above the Rio Machamarca, between two *quebradas,* or streams (Coben 2006, Gonzalez and Cravotto 1977). This river arises a short distance to the west of the site and passes to the south of the site east through the rich agricultural valleys of Pocona. The Machamarca Valley is thus an access route into this area, one of several such valleys in the region. This region is near the eastern frontier of the Inka empire (D'Altroy 2002).

An enormous single roomed, multidoored roofed structure or *kallanka*, 78 meters long by 26 meters wide, dominates the central area of the site (Coben 2006, Nordenskiold 1957). The highest of the remaining standing walls of this structure reaches over 12 meters high, and was certainly higher when the complex was utilized (Figure 2). The walls of the *kallanka* are niched, and the southern wall contains 12 entrances. The *kallanka* opens on a large, two-level plaza, with the higher level to the north abutting the *kallanka*. A wall separates the two sections of the plaza. Adjacent to the *kallanka's* central door is an *ushnu*, a small ritual platform with a large stone upon it.

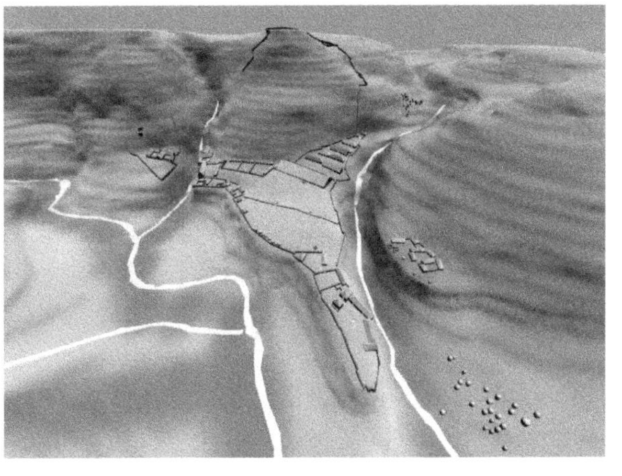

Figure 1 *Digital overview of Incallajta*

Figure 2 *The* kallanka, *with a portion of the upper plaza at right*

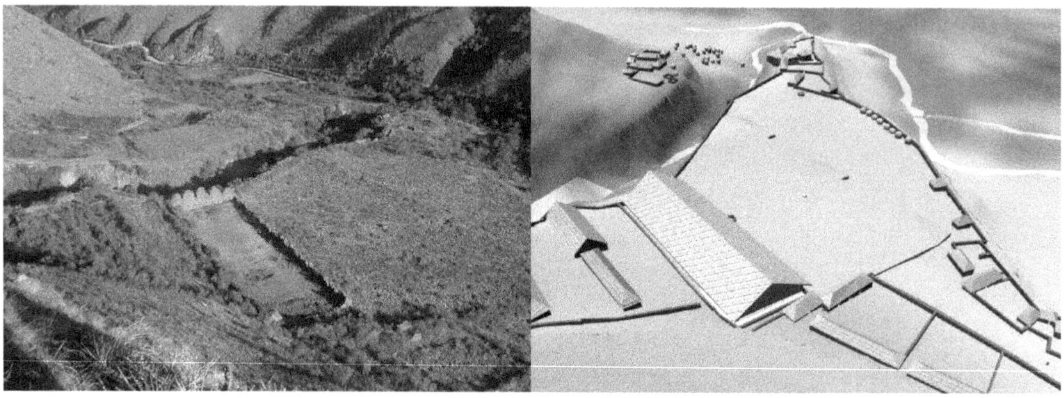

Figure 3 *The site of Incallajta and a digital reconstruction, from the Northwest*

Figure 4 *Incallajta from the North*

A zigzag wall is located several hundred meters above the *kallanka* on a hill overlooking the site. This single wall, evocative of Cuzco's Sachsahuaman in both form and location within the site, is 4-5 meters high with a single baffled entrance that restricts access to the hill's summit and the site. Immediately inside the zigzag wall is a second wall with two doorways. Numerous other large rectangular structures are spread throughout the site, as are smaller structures, plazas and stairways (Coben 2006, Querejazu 1998).

Until recently, little was known about Incallajta, the result of both a dearth of references to the site in the chronicles and the absence of any systematic archaeological investigation at the site or in the region. The chroniclers Pedro Sarmiento de Gamboa (1947), Miguel Cabello de Balboa (1951) and Bernabe Cobo (1990) in similar language mention a "*fortaleza*" near the Inka administrative center of Pocona. Most commentators have assumed this *fortaleza* to be Incallajta (Querejazu 1998). Erland Nordenskiold (1957) mapped and briefly conducted excavations at the site in 1913. Prior analyses of the monumental Inka site of Incallajta, Bolivia have suggested that it was an administrative, political, ceremonial, or military center (Querejazu 1998). Most recently, the author (Coben 2003, 2006) has identified the site as a Peircean replica (Peirce 1931-5) of Cuzco, and as such the locus for the performance of the most sacred Inka state ceremonies and rituals that had been restricted to Cuzco itself. These Cuzcos (Guaman Poma 1980[1603], Cieza 1959[1552], Garcilaso 1987[1609], and the spectacles and rituals within them, were an important strategy in Inka imperial expansion, as the empire grew in a fifty year period from a single valley in Peru to encompass Peru and significant portions of Bolivia, Ecuador, Chile and Argentina.

According to the historical chronicles, the Inka constructed their capital as a physical representation of their worldview (Rowe 1944, 1946, Zuidema 1990). Cuzco was a sacral theater, the site of an elaborate series of massive structured spectacles performed according to a ritual calendar (Betanzos 1996 [1551]: Book 1, Ch. XV, Cobo1990: Ch. 25-30, D'Altroy 2002:152-153). Special ceremonies and rituals were also performed in Cuzco in times of war or crisis (Cobo 1990[1653]:112, Ch.31). The idols of newly conquered peoples were frequently transported to Cuzco, and local elites were required to worship their idols there in temples built for such purpose (Cobo 1990[1653]). The Inka later built a limited number of "new or other Cuzcos" in specified regions of their empire, including one in a region now defined approximately by modern day Bolivia (Guaman Poma 1980[1603], Cieza 1959[1552], Garcilaso 1987[1609]). Employing primarily an analytical framework drawing upon theories of semiotics and pragmatics which have been successfully utilized in the study of ritual in many parts of the world, (Parmentier 1994; Preucel and Bauer 2001), Coben has argued that new or other Cuzcos contain buildings, landscapes and other features arranged in a particular pattern, and that the site of Incallajta is one such Cuzco, serving as a theatrical space and locus of performance as discussed above. Coben (2006) suggested that new or other Cuzcos were strategically placed only in areas of war and rebellion where the utilization of ritual performance to maintain, reinforce, and manipulate Inka ideology and identity was a critical element of imperial strategy as the polity expanded from a single valley in highland Peru to encompass most of the Andes.

Integral to the analysis of the performance and theatrical qualities of Incallajta is a digital reconstruction of the site, utilizing the latest in computer technology. In this paper, we will briefly describe the methods and processes utilized for such reconstruction. Through a series of digitally reconstructed "animated walk-through" simulations, we analyze how access to the site's core was orchestrated through narrow and controlled entry points and winding passages. Similarly, we explore the architecture of the kallanka, the largest single room structure in the Inka empire, and demonstrate how its construction, physical scale, and relationship to adjacent areas, particularly large plazas, would have been shaped both a performer's and a viewer's the experience of the space and the performances within it.

Digital Reconstructions and Virtual Reality Simulations

Coben's field crew shot over 2,600 points with standard survey equipment. These points were then inputted into AutoCAD Architectural Desktop software to create an accurate two-dimensional digital map of the site's structures, as well as a topographic map. This software then interpolated the building and topographic data to convert the two-dimensional representation into a three dimensional one. Ruined walls were then rebuilt, and roofs and other building features consistent with Inka architecture were added to produce a series of reconstructed representations of the site (Figures 3 and 4).

After completion of this reconstruction, we prepared a series of animations to explore the performatic qualities of the site. Routes were traversed virtually from various points of the site to reach the *kallanka, ushnu* and central plaza area. These animations revealed that access to such area from both the northern and western points of the site was highly convoluted and restricted, and likely highly secured and managed (Figures 5 and 6). These highly constricted interior series of passageways open dramatically into a broad and expansive area for gathering and performance. On all routes, the *kallanka*'s enormity and grandeur are slowly revealed as an actor moves from the site's perimeter into the central plaza zone. By comparison, most Spanish colonial plazas of the post-Inka period have more open and straightforward approaches.

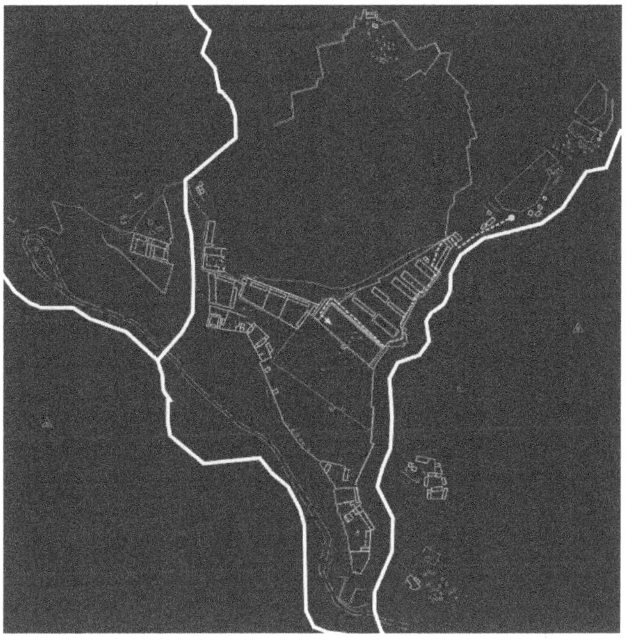

Figure 5 *Passage to the plaza from the north*

A second form of animation involves the recreation within digital Incallajta of ritual movements documented at other locations.

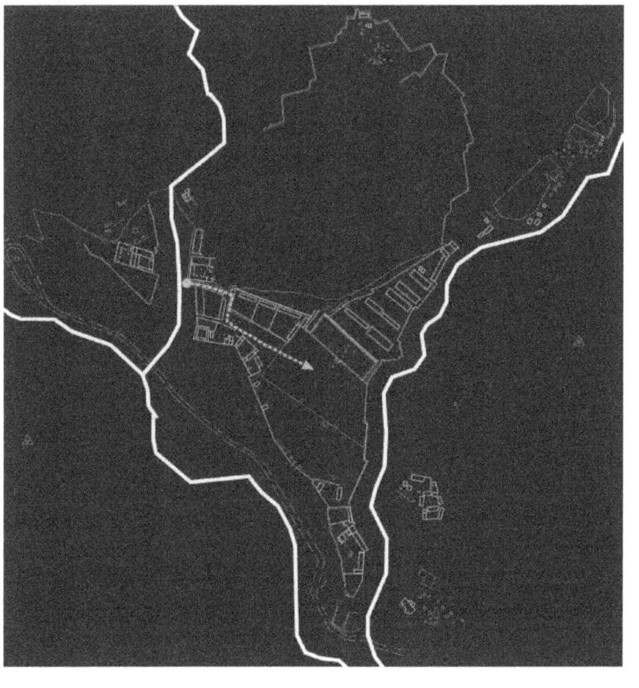

Figure 6 *Passage to the plaza from the west*

Unfortunately, most descriptions of Inka ritual do not include movement with sufficient specificity for such an analysis. However, Garcilaso (1609) documents an Inka ceremony through a series of lanes or turns at the Temple of Viracocha, the Inka creator deity, at Raqchi (Figure 7).

Figure 7 *The temple of Raqchi, interior wall with doorways*

As many scholars have suggested that the *kallanka* is also a Temple of Viracocha, we simulated such ceremony therein. At Raqchi, the building has a central interior wall with several doors that create the lanes, while in the *kallanka*, where no such wall exists, we hypothesize that the lanes are created by the rows of columns which support the roof (Figure 8). While exact simulation is not possible due to the location of the respective buildings' entrances, a high degree of correlation is possible, further supporting the identification of the *kallanka* as a Temple of Viracocha.

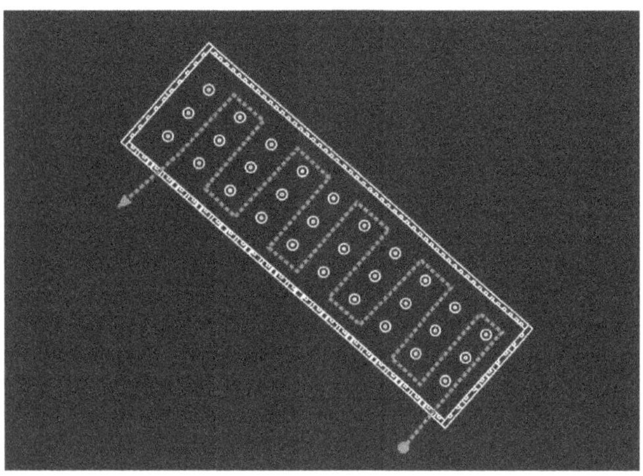

Figure 8 *Ritual Path through the "lanes" of the kallanka*

Lastly, we have combined panoramic simulations and individual building reconstructions to test hypotheses regarding Inka construction techniques and building uses. These simulations are created through the linkage and overlapping of a series of rendered images that simulate the view and experience of a particular environment from a pre-selected point. We utilized this technique both within the *kallanka* to examine the quantity and volume of trees required to support the roof, as well as to examine the lighting at various points of the day (Figure 9).

Figure 9 *The passageway blocked by the roof and made impassable by its runoff*

This technique also enabled us to identify a canal immediately behind the *kallanka* and a large pool of water at its western end in an area previously identified as a passageway. Reconstruction and simulation revealed that this pathway would have been impassable due to the presence of the *kallanka's* roof as well as the enormous water that would have run off such roof into the pathway during the rainy season (Figures 10 and 11).

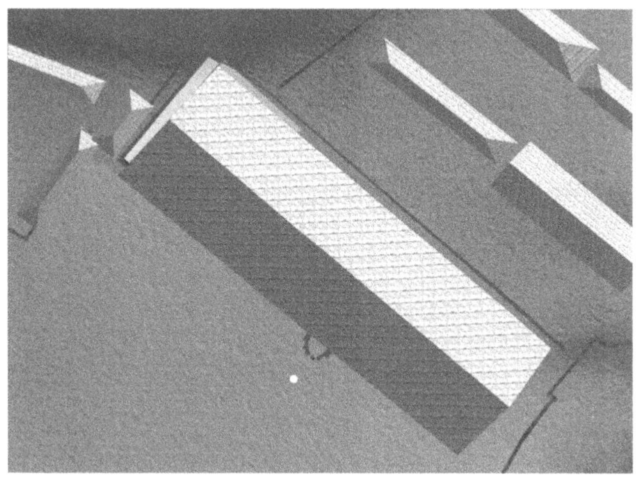

Figure 10 *The pool adjacent to the* kallanka

Conclusion

This paper demonstrates some of the potential uses of digital reconstruction and virtual reality simulations in examining performance and ritual archaeologically. Utilizing only off the shelf software and a standard laptop, such simulations allow for the fuller consideration of pathways, access, lighting and other performatic qualities, as well as for building techniques and viewsheds.

With respect to the Inka empire, our digital reconstructions suggest that Inka performance, and access thereto, was highly controlled and regimented. The power of such spectacles was reinforced through the use of convoluted access routes that opened into large plazas surrounded by large, grand structures. At Incallajta, such structure was likely a Temple of Viracocha, the creator diety and one of the most important in the Inka pantheon, which temple was a locus of important ceremonies and required enormous amounts of material and labor to construct.

References

BETANZOS, JUAN DE 1996 [1551] *Narrative of the Incas.* Austin: University of Texas Press.

CABELLO BALBOA, MIGUEL 1951 [1586] *Miscelánea Antárctica.* Lima: Instituto Etnológico de la Universidad Mayor de San Marcos.

CIEZA DE LEON, PEDRO DE 1959 [1552] *The Incas of Pedro Cieza de Leon.* Norman: University of Oklahoma Press.

COBEN, LAWRENCE S. 2003 Incallajta: El Otro Cuzco en las Charcas. Paper given at the *31st Annual Midwest Conference Andean on Andean and Amazonian Archaeology and Ethnohistory*, Chicago.

COBEN, LAWRENCE S. 2006 Other Cuzcos: Replicated Theaters of Inka Power. In *Archaeology of Performance: Theater, Power and Community.*T. Inomata and L. Coben eds Walnut Creek, CA: AltaMira.

COBEN, LAWRENCE S. AND CHARLES STANISH. 2006 Archaeological Reconnaissance in the Carabaya Region, Peru. In *Advances in the Archaeology of the Titicaca Basin.* Charles Stanish, Amanda Cohen, and Mark Aldenderfer, eds. Los Angeles: Cotsen Institute of Archaeology Press

COBO, BERNABE 1990 [1653] *Inca Religion and Customs.* Austin: University of Texas Press.

COBO, BERNABE 1979 [1653] *History of the Inca Empire.* Austin: University of Texas Press.

D'ALTROY, TERENCE N.1992 *Provincial Power in the Inka Empire.* Washington and London: Smithsonian Institution Press.

GARCILASO DE LA VEGA, EL INCA 1987 [1609] *Royal Commentaries of the Incas and General History of Peru.* Austin: University of Texas Press.

GUAMAN POMA DE AYALA, FELIPE 1980 [1603] El *Primer Nueva Corónica y Buen Gobierno.* Critical edition by John V. Murra and Rolena Adorno; translation and textual analysis of the Quechua by Jorge L. Urioste. Mexico City: Siglo XXI.

INOMATA, TAKESHI, AND LAWRENCE COBEN 2006 An Introduction to the Archaeological Theater. In *Archaeology of Performance: Theater, Power and Community.*T. Inomata and L. Coben eds Walnut Creek, CA: AltaMira.

INOMATA, TAKESHI, AND LAWRENCE COBEN 1986 *The Sources of Social Power.* Cambridge: Cambridge University Press.

NORDENSKIOLD, ERLAND 1957 [1915] Incallajta, Ciudad Fortificada Fundada por El Inca Tupac Yupanqui. *Khana* (*Revista Municipal de Arte y Letras*) 25/26: 6-22.

PARMENTIER, RICHARD, *Signs in Society: Studies in Semiotic Anthropology.* Bloomington: Indiana University Press.

PARMENTIER, RICHARD 1997 The Pragmatic Semiotics of Culture. *Semiotica* 116:1-115

PEIRCE, CHARLES SANDERS 1931-5, *Collected papers of Charles Sanders Peirce.* Charles Hartshorne and Paul Weiss eds. Cambridge: Harvard University Press

PIZARRO, PEDRO 1978 [1571] *Relacion del descubrimeiento y conquista del Peru.* Lima: Pontificia Universidad Catolica del Peru.

POLO DE ONDEGARDO, JUAN 1990 [1571] *Relacion de los fundamentos acerca del notable dano que resulta de no guardar a los indios sus fueros.* Madrid: Historia 16

PREUCEL, ROBERT W. AND ALEXANDER A. BAUER 2001, Archaeological Pragmatics. *Norwegian Archaeological Review* 34:85-96

QUEREJAZU LEWIS, ROY *Incallajta y la Conquista Incaica del Collasuyu.* Cochabamba: Editorial Los Amigos del Libro.

ROWE, JOHN H. 1944 An Introduction to the Archaeology of Cuzco. *Papers of the Peabody Museum of American Archaeology and Ethnology* 27:2

Inca Culture at the Time of the Spanish Conquest. In T*he Andean Civilizations.* Julian H. Steward ed., pp. 183-330. Handbook of South American Indians, Vol.2, Bulletin 143. Washington, D.C.: Smithsonian

SARMIENTO DE GAMBOA, PEDRO 1947 [1572] *Historia de los Incas.* Buenos Aires: Emece Editores

ZUIDEMA, R. TOM 1990 *Inca Civilization in Cuzco.* Austin: University of Texas Press.

Reflection Transformation Imaging on Larger Objects: an Alternative Method for Virtual Representations

Massimiliano Corsini, Matteo Dellepiane, Marco Callieri, Roberto Scopigno

Visual Computing Lab, ISTI – Area della ricerca CNR
Via G. Moruzzi 1, 56124, Pisa, Italy
Email: {dellepiane, corsini, callieri, scopigno} @isti.cnr.it

1 Introduction

The term *Reflection Transformation Imaging* has been originally used by Tom Malzbender of the HP laboratories to describe an image-based method to acquire the reflectance properties of objects' surfaces. This technique, as other image-based relighting techniques, creates a "special" image of the object which encodes a reflectance function per-pixel. These images are called Polynomial Texture Maps (PTMs) (Malzbender et al. 2001). PTMs are generated starting from a set of photos of the object of interest taken under controlled illumination. Thanks to the reflectance functions encoded in the PTMs, it is possible to generate new images of the object under different lighting conditions. Due to the ease and effectiveness of this method, PTMs have been used in several applications in the field of Cultural Heritage in recent years. One of the main motivations to use PTMs to represent artefacts, is that this kind of visualization allows the user to change interactively the effects of the illumination on the object's surface, and hence the exploration of the object is performed in a new and interesting way. In particular, the use of contrast enhancement mechanisms related to the PTM viewing technology (specular enhancement, diffuse gain) have proven to be very useful not only in terms of documentation, but also in terms of analysis of the surface details which are hard to reveal with classical viewing systems. Another reason is that PTMs may be an interesting alternative to the other well-known 3D acquisition techniques, like 3D scanning, which are "expensive" in terms of equipment, acquisition time and post-processing. Moreover, visualization of detailed 3D models is problematic in an online environment, while PTMs offer a good representation, making data more accessible to the public. Finally, PTMs seem to be a better solution for the visualization of certain objects, like bas-reliefs, where the information provided by re-illumination is more important than the one provided by geometry. Most of the existing work concerning PTM acquisition is in regard to small size objects. In this paper we present a new low-cost system for producing high quality PTMs of medium to large objects, from 60-70 cm wide by 50-60 cm high up to 2m wide by 1m high. During the description of our system, we will also outline some differences with respect to the classic 3D scanning pipeline. We will also present an analysis of the quality of the PTMs acquired by our system and a novel software to browse high resolution PTMs.

Following an overview of the application of PTMs in section 2, we analyse our method for PTM acquisition in all its aspects in Section 3. In Section 4 we propose an analysis of the quality of PTMs obtained in comparison with 3D scanning, and a study of the quality degradation relative to the number of shots used to calculate the reflectance function. This analysis is not limited to our system and it is a good starting point to optimise PTM acquisition techniques. Some examples of Cultural Heritage objects acquired with our system are presented in Section 5. Conclusions and future work are outlined in Section 6.

2 PTMs and Cultural Heritage

The appearance of a surface can change significantly when lighting conditions change. Typically, global illumination rendering methods are used to render such appearances (Debevec et al. 2004; Goesele 2004). This kind of method requires both the geometry and the properties of the material of the object, and hence a lot of effort in the acquisition and modelling phase. Another way to reproduce global illumination rendering effects like sub-surface scattering, inter-reflections, shadowing and refraction, that does not need the geometry of the model and an accurate reflectance estimation of the surface object, are the so-called image-based relighting (IBRL) methods (Choudhury and Chandran 2006). These kinds of methods start from a set of images used to compute a new image with different lighting conditions. Polynomial Texture Maps are one of these methods. As stated in the introduction, PTMs are created starting from a set of images used to compute a new image with different lighting conditions. For a static object and a fixed camera, per-pixel reflectance functions can easily be estimated. More specifically, the reflectance function is approximated by a biquadratic polynomial in the following way:

$$
\begin{aligned}
L(u,v,l_u,l_v) &= a_0(u,v)\, l_u + a_1(u,v)\, l_v + \\
&+ a_2(u,v)\, l_u l_v + a_3(u,v)\, l_u + \\
&+ a_4(u,v)\, l_v + a_5(u,v) \qquad (1)
\end{aligned}
$$

where (l_u, l_v) is the direction of the incident light and (u, v) are the pixel coordinates. Hence, each pixel of a PTM is composed of the RGB values and the six coefficients of the reflectance function. In order to estimate the coefficients $(a_0, \ldots, a_5)$ the light positions have to be known.

Several Cultural Heritage projects have been using PTMs for the inspection of artefacts. One of the first works in this field was the representation of cuneiform epigraphy (Malzbender et al. 2000). The PTM viewer of HP Laboratories (Malzbender et al. 2001) was used to inspect clay cuneiform tablets under different (optimal) light conditions. Reflection transformation tools were used also in Palaeonotology, to provide noticeable improvement in imaging of low colour contrast, high relief fossils (Hammer et al. 2002). The application of PTM technology on ancient stone tools revealed fine details of conchoidal knapping fractures, use scarring and stone grain (Mudge). A joint work done by the National Gallery and the Tate Gallery of London showed that PTMs under specular enhancement provided additional information about the surface textures of oil paintings (Padfield et al. 2005). Cuneiform tablets were analysed using both 2D (PTM) and 3D (structured light scanner) information. The PTMs were texture mapped on the model, and a special 3D viewer was created (Mudge 2004). Recently, the application of PTMs and scanning techniques on a large numismatic collection permitted the creation of a "virtual exhibition" (Mudge et al. 2005). Moreover, the use of specular enhancement and diffuse gain produced an improvement in data discernment.

3 Acquisition and Visualization

Standard PTM acquisition devices work by positioning the object of interest inside a light dome of fixed size. In this way the photos under controlled illumination can be easily acquired in a completely automatic fashion. However, PTM acquisition for objects of medium to large size has several specific issues and the overall pipeline has to be re-designed. Since the size of our objects is too big to create a fixed dome, we use a "virtual" light dome, as explained in the next section. In particular, the acquisition process is subdivided into three steps. First, we consider the physical acquisition planning. The size of the objects, and the fact that in most cases they cannot be moved from their place, led us to the necessity of planning carefully the acquisition phase in order to make it faster. After the acquisition planning, the effective acquisition is performed using a simple system as described below. The third step consists of the post-processing of the acquired data, in order to improve the estimation of the per-pixel reflectance function. These three steps are described in detail below.

3.1 Acquisition planning

Selecting the correct lighting point is an important step in the PTM acquisition of large objects; given the size and position (in the majority of cases, on a wall) of an object, in general we do not have the possibility of using a physical dome to illuminate the object. Instead, we have to manually place the light in different positions, forming a "virtual" illumination dome. The size of this illumination dome and its light distribution will depend on the size of the target object and on the number of light directions we want to use to sample the reflectance function of the object. To simplify the light placements we developed a software tool, called the *PTM Planner*. With this tool it is possible to define the properties of the lighting dome, to visually check its correctness and to automatically generate the coordinates for the light placements. The tool usage is quite simple; the scene setup is generated as the user inputs the size of the object to be acquired, its height from the ground and the distance of the camera. Objects in the scene are scaled according to user specifications; the camera is pointed towards the centre of the object. The next step is the definition of the acquisition pattern. The array of light can be generated by choosing the light distance and two angles (vertical and horizontal step). The tool can automatically exclude the light positions that are near to the "wall" (there can be problems in positioning the light source in such a position) and that are aligned with the camera axis (light will be shadowed by the camera or will occlude the camera).

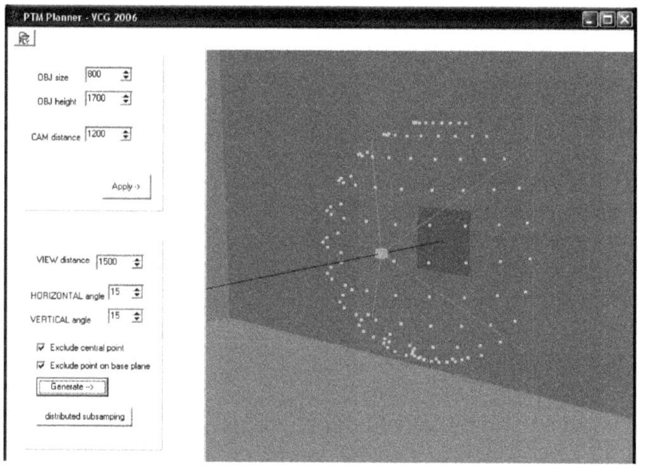

Figure 1 *PTM Planner tool*

The points are generated using a parallel-meridian grid as shown in Figure 1. This method does not guarantee a uniform distribution over the sphere, but, as we will show when describing the acquisition procedure (see section 3.2), having a series of light positions at the same height results in a much faster acquisition. The user can also manually *turn off* (by clicking on the 3D view) the light positions that will probably be impossible to use due to occlusions. Finally, given a complete dome, the program can perform a light pruning following the *distributed scheme* described in Section 4. This scheme, by generating a more uniform distribution, greatly reduces the number of required light positions without influencing excessively the PTM quality. When the light setup has been completed, the PTM Planner tool can save a written description of all the points where the light should be positioned. This information is also saved in the format required by the transformation tool that creates the PTM starting from the input photos and light positions. Even though this tool is quite simple software, it greatly helped us in speeding up the acquisition process both during planning, by giving visual feedback and instant parameter editing, and during acquisition, by providing step-by-step instructions on light placement.

(a)

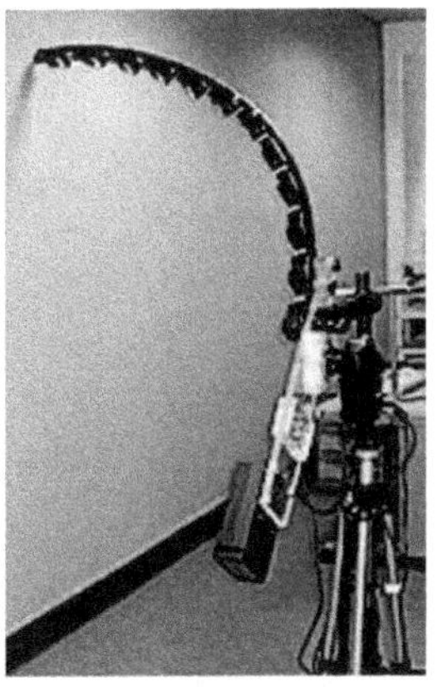

(b)

Figure 2 *(a) PTM acquisition dome; (b) acquisition arc (Photos: HP Labs)*

Figure 3 *Our acquisition setup*

3.2 Acquisition
Several experimental devices have been created to acquire PTMs. Two of them are shown in Figure 2.

The object in Figure 2a is suitable for sampling small objects (nearly 15 cm). It is a 90 cm diameter black plastic hemisphere, with fifty evenly distributed strobe lights mounted so that they illuminate the hemispheric dome's interior. The digital camera is positioned at the top of the hemisphere and photographs the PTM subject through a view port cut in the dome. In Figure 2b a device designed for larger objects is shown. A 90 degree arc 1.50 m in diameter is mounted with 12 strobe lights facing towards the centre of the arc. One end of the arc is connected to a circular bearing race in the shape of a doughnut. This allows the arc to spin in a 360 degree circle around the bearing race.

These two experimental devices work very well but they are not suitable for our target. Various reasons support this statement:

- The diameter of the "hemisphere" formed by all the light positions depends on the size of the object, since for each photo the light must completely cover the target. For the object shown in the results section, the minimum diameter for the virtual dome was 3 m.
- To completely illuminate a large sized object we need powerful lights.

- In most cases, the target object cannot be moved from its place, so we have to deal with the fact that it is not always possible to exploit all the light positions, due for example to the height from the ground.

We have designed the system following this reasoning. Our solution is shown in Figure 3. Since it is not feasible to use a big number of lights, we decided to use only one, and to change its position for every photo of the set. The time needed to position the light is minimized thanks to the acquisition planning and by using some references placed on the floor. We make the acquisition faster using a printed scheme of the angle directions (it helps in placing the references on the floor very quickly), and a plumb line attached to the light in order to facilitate the positioning. Our acquisition equipment is composed of an 8Mpixel Canon Digital Camera, a 1000W halogen floodlight, a tripod and a boom stand. The fact that we use only one light explains also the parallel-meridian placement of lights: with this configuration we need to set the height and direction of the light only once for each level of height. The acquisition can be summarized in this way:

- Take the measurements of the object, find the centre of it and its height from the ground.
- Using this data, generate the "virtual dome" and choose the position of all the lights.
- Position the digital camera on the tripod. Measure aperture and shutter speed under the illumination of the central light. Keep these values fixed for all the photos, in order to have a constant exposure.
- With the help of the output of PTM planner, mark the reference marks related to each light.
- For each level of height, set the height and the direction of the light, then put it on each reference mark related to the level, and take the photo.

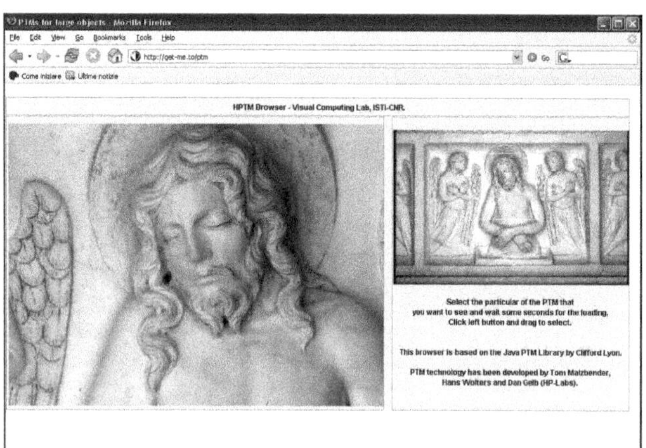

Figure 4 *A screenshot of the HPTM Browser in action*

Following this approach we are able to acquire several PTMs of an object in a relatively short time (see Section 5). Other big advantages of this equipment are that it is quite cheap (nearly 1000 Euros in total) and easily transportable.

3.3 Data processing

In order to calculate an accurate reflection function, a critical factor is that the digital camera must no move from one photo to the other. In fact, from our experience even a misalignment of a few pixels (corresponding to a millimetric misalignment) can produce visible aliasing. In our experimental acquisition set it was almost impossible to avoid small movements of the camera. This led to the necessity of aligning the set of photos before building the PTM. To deal with this problem we align the photos automatically before the reflectance function estimation using a freeware tool for panoramic images. This is the only data processing we nee, in fact there is no need of any image for image calibration, since all the photos have the same exposure. This step is very different with respect to the usual post-processing required by 3D scanning data. We can state that one of the main advantages of PTM technology compared to 3D scanning, is that PTMs require trivial post-processing. In fact in 3D scanning the most time-consuming phase is the post-processing phase, when the range images acquired have to be aligned and merged together in order to build the final 3D model. Additionally, after the alignment and merging phase, the complex task of colour mapping using the photos also has to be performed. After this post-processing the PTM is generated with the *PTMfitter* tool developed by the HP Laboratories and publicly available.

Figure 5 *Comparison between the normal maps of the 3D scanning and the PTM: full model and particular*

3.4 Visualization

The remote realization of the virtual representation of the object obtained is an important aspect, especially in Cultural Heritage applications. In fact, it is desirable that the virtual representation can be inspected by a large number of users through the Internet. For this purpose we developed a Java Applet capable of efficiently browsing PTMs with huge resolutions. We call this browser *HPTM Browser*. The PTM is first broken down into a set of sub-ptms by another tool that works in conjunction with the browser. The browser efficiently retrieves these sub-ptms

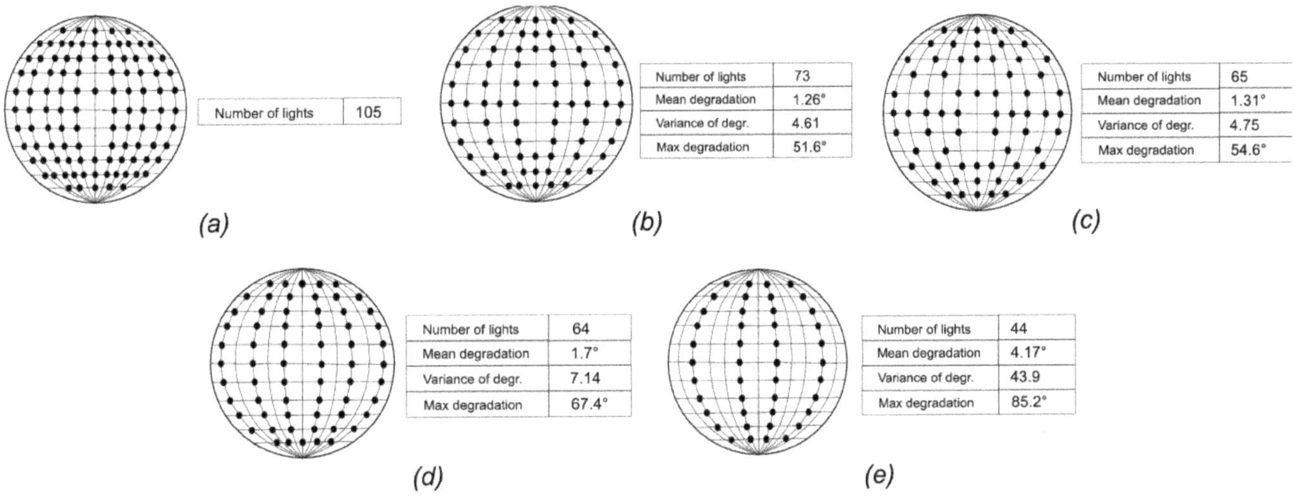

Figure 6 *Quality degradation: (a) Best quality PTM (b-e) Differences in dihedral angle of normals. The Sphere shows the lights placement.*

in a multi-resolution fashion, in order to visualize quickly the huge PTM in the case of zoom operation. Figure 4 shows a screenshot of the HPTM browser. This software is now part of the *ptmviewer* Java open source project regarding PTM visualization originally created and developed by Clifford Lyon (http://ptmviewer.dev.java.net).

4 Quality assessment

In this section we discuss and analyse some issues regarding the quality evaluation of the acquired PTM. As a case study we consider a 70 by 80 cm section of the 14th century tomb of Archbishop Giovanni Masotti. We performed a very accurate PTM acquisition using a large number of lights (105 light positions, 11 angles and 11 height levels) and we acquired the same object also with a triangulation Scanner (Minolta 910i). We used the 3D scanned model as a reference for our quality evaluation. In fact, for larger objects 3D scanning is a very reliable technique, in terms of accuracy (Bernardini and Rushmeier 2002). The output 3D model is composed of nearly 2.4 million faces, the accuracy is about 1/3 of a millimetre. Our first comparison concerns the quality of the normals calculated from the PTM data. To do this, we aligned the 3D scan model to the PTM (Franken et al. 2005) and we calculated the normals of both the model and the PTM. In Figure 5 a comparison of the normal maps is shown. The variation of the normals in the PTM is smoother than in the corresponding 3D scan, but their values are coherent. This test demonstrates that, even thought PTM provides an approximation of the objects' geometry, estimated by analysing the per-pixel reflectance functions, the obtained data is reliable.

It also demonstrates that our setup does not introduce significant errors in the PTM acquisition. The second analysis is related to the degradation of the PTM quality with respect to the number and position of lights. For this purpose, we compute four PTMs starting from subsets of the original lights. Then we compare between the normal maps of the "best" PTM (the one with 105 lights) and the "subsampled" ones. This comparison was made calculating the difference in dihedral angle between the normals of each pixel. In Figure 6 we show the analysis of the difference between the best PTM and four possible subsets. In terms of number of lights, we can observe that we can considerably reduce the number of lights without having an excessive degradation of quality. In particular, concerning the lights placement, we notice (Figure 6c and 6d) that a more "distributed" position of the lights brings lower mean degradation and peak error. Considering these facts, we can conclude that a set of 60-70 properly distributed photos can be used to produce high-quality PTMs.

5 Experimental Results

Several objects have been acquired with the developed system. In this section we present three examples; all these PTMs are available for real-time exploration at http://get-me.to/ptm. The first example is one face of a small (30 x 30 x 20 cm) medieval capital from the Museum of S. Matteo in Pisa. With the help of a professional photographer, we created a set of 36 high resolution (5440 x 4080) photos. In this case, we did not use the equipment described in section 3.2, but a 20 MPixel Monorail View Camera and a professional flash.

We produced a very detailed horizontal PTM of the capital: some snapshots are shown in Figure 7a. The acquisition time for this object was nearly one hour. The second example has already been shown in section 5: it is a part (70 x 80 cm) of the 14th century tomb of the Archbishop Giovanni Scherlatti, by Nino Pisano (Museum of the Opera Primaziale in Pisa). We performed a very detailed acquisition (105 light positions, 3496 x 2280), which lasted about 3.5 hours. In this case we used the acquisition system described in Section 3. In Figure 7b we show a snapshot of the acquired PTM.

(a) (b) (c)

Figure 7 *Acquired PTMs. The images are generated from different lighting conditions. (a) The Museum of San Matteo Capitel. (b) Placed Christ from the Museum of Opera. (c) Roman sarcophagus from Csmposanto Monumentale.*

The third object is a 2nd century AD Roman Sarcophagus in the Camposanto Monumetale of Pisa, representing the Phaedra and Hippolitus myth. We chose this particular example since the Sarcophagus is situated outdoors. In this way we tested if the proposed system could produce good results even when the ambient light is considerably high with respect to the light equipment we used. In this acquisition we also used the considerations made in the quality assessment in order to perform the acquisition with a lower number of lights (an consequently a shorter acquisition time). The acquisition time was about 2.5 hours. The size of the portion is 90 x 60 cm (66 photos, resolution 3496 x 2280, see Fig. 7c). In this time we acquired also two horizontal PTMs (10 photos each) of the two halves of the Sarcophagus. The obtained PTMs are more than satisfying, considering the non-ideal condition of lighting.

6 Conclusions and Future Work

In this article we presented a new, low-cost system to acquire Polynomial Texture Maps of large objects. Due to the impossibility of utilizing the usual fixed acquisition dome, we have had to rethink the normal PTM acquisition pipeline. The comparison and the differences with respect to the 3D scanning acquisition pipeline are also analysed and described. In order to preserve the high detail and the interactivity of exploration, we developed new software to progressively browse PTMs. Finally, we presented novel studies about the quality assessment of polynomial texture maps. These studies could be very useful to design a more efficient acquisition scheme without compromising quality in the final result. The examples produced with our system gave satisfying results, showing that PTMs can be an alternative method for documenting and communicating cultural heritage information for large objects. In particular, PTMs could be a technology particularly suited to analysing and representing bas-reliefs and paintings.

Currently we are working to estimate the light direction starting from the photo set in order to eliminate the need of manual positioning and making the system more automatic and accurate. Another desirable feature is that at the moment the possibility of removing the ambient lighting contribution is missing, which lowers the quality of the results, especially with outdoor subjects.

Acknowledgements

We would like to thank the museums of San Matteo and Opera Primaziale in Pisa for their support. This work was partially funded by the European Research Network on Excellence in Processing Open Cultural Heritage (EPOCH) and the Cultural Heritage Department of the Italian National Research Council.

References

F. BERNARDINI and H. E. RUSHMEIER, "The 3D models acquisition pipeline." *Comput. Graph. Forum*, vol. 21, no. 2 pp. 149–172, 2002

B. CHOUDHURY and S. CHANDRAN, "A survey of image-based relighting techniques," in *Proceedings of the 2nd International Conference on Computer Graphics Theory and Applications*, March 2006.

P. DEBEVEC, E. REINHARD, G. WARD, and S PATTANAIK (eds), *High Dynamic Range Imaging, (ACM SIGGRAPH Course Notes No. 13*) ACM Press, 2004.

T. FRANKEN, M. DELLEPIANE, F. GANOVELLI, P. CIGNONI, C. MONTANI, and R. SCOPIGNO, "

Minimizing user intervention in registering 2D images to 3D models," *The Visual Computer*, Vol. 21, no8-10, pp. 619–628, Sep. 2005, special issues for Pacific Graphics 2005. [online] Available: http://vcg.isti.cnr.it/Publications/2005/ CDFGMS05

M. GOESELE, *"New acquisition techniques for real objects and light sources in computer graphics*," PhD dissertation, Universitat Saarbrucken, 2004. [online] Available: http://scidok.sulb.uni-saarland.de/volltexte/2004/338

O. HAMMER, S. BENGTSON, T. MALZBENDER and D. GELB, "Imageing fossils using reflectance transformation and interactive manipulation of virtual light sources", in *Paleontologia Electronica*, no. 1 2002.

T. MALZBENDER, D. GELB and H. WOLDERS, "Polynomial texture maps", in *SIGGRAPH '01: Proceedings of the 28th annual conference on Computer graphics and interactive techniques*. New York, NY, USA: ACM Press, 2001, pp. 519–528

T. MALZBENDER, D. GELB, H. WOLDERS and B. ZUCKERMANN, "*Enhancement of shape perception by surface reflectance transformation*", Hewlett-Packard Laboratories, Palo Alto, California, Tech. Rep. HPL-2000-38R1, 2000.

M. MUDGE, "*Siggraph 2004 conference presentations, web graphics/special sessions/panels, cultural heritage and computer graphics panel"*, 2004.

M. MUDGE, J.-P. VOUTAZ, C. SCHROER and M. LUM, "Reflection transformation imaging and virtual representations of coins from the hospice of the Grand St. Bernard", in *Proceedings of 6th International Symposium on Virtual Reality, Archaeology and Cultural Heritage (VAST2005)*, M. Mudge, N. Ryan and R. Scopigno (eds), Eurographics Association, November 2005, pp. 29–39.

J. PADFIELD, D. SAUNDERS and T. MALZBENDER, "Polynomial texture mapping: a new tool for examining the surface of paintings", *ICOM Committee for Conservation,* pp, 504–510, 2005.

3-D Visualization of Ancient Processional Landscapes in Costa Rica

Payson Sheets,[1] Tom Sever[2]

[1] Anthropology, University of Colorado, Boulder CO USA
[1] NASA Huntsville, AL, USA

1 Introduction

For three decades one of us (PS) has been collecting cases of ancient Middle American societies affected by explosive volcanic eruptions. Sheets began with cases in southern Mesoamerica because he was doing his doctoral research in El Salvador. The sample of cases expanded to cover much of Mesoamerica because there were so many examples of eruptions affecting ancient societies in Central Mexico and the Gulf Coast area. He found that complex societies were often quite vulnerable to the unanticipated great stresses. Because complex society emerged early in Mesoamerica, compared to the surrounding culture areas, the sample was weak in egalitarian societies. Sheets therefore turned to Costa Rica to try to correct that imbalance, by adding cases involving less complex societies. Egalitarian societies predominated in the Arenal area of Costa Rica. We have yet to find clear evidence of that egalitarian/ranking boundary having been crossed at any time in the Arenal area.

We began the "Proyecto Prehistorico Arenal" in the 1980s with the objective of exploring the reactions of egalitarian societies to the sudden massive stresses from the ten explosive eruptions of Arenal volcano during the past 4000 years. We were surprised to find these "simple" but clearly very sophisticated societies were exceptionally resilient to the disruptions caused by the big eruptions. We have expended considerable efforts to try to understand why and how these egalitarian societies could have developed such resiliency. Based on our efforts to date, we believe the key factors were (1) low regional population densities and availability of uncontested refuge areas in times of emergency, (2) alliances from exogamy and from multiple community feasting, (3) reliance on wild foods rather than a domesticated dietary staple, and (4) the household or sometimes the village was the decision-making unit rather than having centralized authority. Sedentary villages have been dated to as early as 3000 BC, with sophisticated ceramics and at least some maize agriculture. Secondary burials were placed with the households. A major cultural change occurred around 500 BC when villagers began burying their dead in cemeteries that were located at a distance from the village. Ethnographic accounts by traditional Cuna and Bribri Indians of lower Central America describe the spirits of the deceased needing separation from the noise of villages. That may provide a reason for that separation. And the spirits of the dead are powerful, and are better kept at a distance from the functioning community. The archaeological evidence is clear that people walked single-file in their processions to and from the cemeteries, but the reasons for that are unknown. The routes they followed were strikingly straight, as we found from mapping the ancient paths, and traditional ethnographic accounts may shed light on the reasons for that. Bribri accounts mention stretching a string to help guide the spirit from the community to the grave. An unanticipated consequence of generations of people following the same path is entrenchment, often to 2 to 4 meters below the surrounding ground surface. Using entrenched entrances into special places must have become a cultural standard. And such entrenched paths certainly focus one's attention on the objective, as the surrounding countryside disappears on both sides of the path. When one enters the cemetery it opens up to view, very much like emerging from a tunnel. All Arenal area societies apparently were egalitarian from the earliest (Clovis horizon, about 11,500 years ago) to the Spanish conquest about 500 years ago.

More complex societies did emerge to the east of the Arenal area, by about AD 500, as exemplified by the Cutris chiefdom. What could be called the "mentality of monumentality" develops when egalitarian principles no longer govern society, and the chiefly elite seeks ways to physically express their centralized authority. The chiefs at Cutris apparently chose the entrenched straight entryways that became so important in the Arenal area as something to magnify immensely in their search for monumentality. Chiefs mobilized large workforces to construct and maintain deep and wide entryways that connected satellite communities some 6-10 km distant with their site center. And we can only imagine the processions of people using the huge sunken entryways into the site center. The entryways are almost 50 meters wide, and many meters deep, as they disgorge their participants into the site center. Ring roads allow processional participants to connect with the radial entryways without being seen from the site center. The radial principal entryways and the ring roads are almost exclusively of earthen construction. As this is a tropical wet rainforest environment, with some 3000 mm of annual mean precipitation, it is unknown how erosion and water management were achieved.

The emergence of complex society at Cutris was followed by other chiefdoms emerging, including Fortuna, Las Mercedes, and Guayabo de Turrialba. These largely date from AD 1000 to 1500. The favored construction material of the entryways changes to stone, perhaps because of possible erosional problems of the earlier earthen construction. This later monumentality was achieved by slightly different means, but the net effect was quite similar to Cutris.

The paths were discovered in remote sensing imagery provided by Tom Sever of NASA. Particularly effective are color infrared aerial photography and IKONOS satellite imagery for discovery of the ancient paths. Other imagery that contributed include LIDAR, Radar, color and black-and-white aerial photography, and the Thermal Infrared Multispectral Scanner (TIMS). We have developed a field and lab methodology for confirming, and disconfirming, anomalies in the remote sensing imagery as ancient footpaths. The detailed tephrochronology of the ten large explosive eruptions of Arenal volcano assists in our dating the initiation and the cessation of use of a particular path or path segment.

One problem we have had is adequately conveying the nature of our discoveries to our colleagues and to the general public. In books, articles, and other printed media we use maps, drawings and photographs of stratigraphic sections, airphotos, and satellite imagery. However all these illustrative media are static. The 3-d visualization of the terrain, with paths and excavated features included, is the most dynamic and successful means of education that we have found to date. That is the focus of this paper. Unfortunately, there is no way to include an interactive, dynamic 3-d reconstruction in a static printed paper, at least not with any technology presently in existence.

2 Previous Research and Current Issues

The Arenal Research Project has been operating in northwestern Costa Rica since the early 1980s, with Sheets directing the archaeology (summarized in Sheets 1994, 2003) and Sever directing the remote sensing (summarized in McKee et al. 1994, and Sever et al. 2003). Funding from the National Science Foundation, National Geographic, and the University of Colorado has supported field and laboratory research. NASA has provided abundant remotely sensed imagery from aircraft and satellite platforms, in both analog and digital formats (McKee et al. 1994, Sever et al. 2003).

We believe this case of the small footpaths with the simple egalitarian Arenal societies being later "writ large" into monumental sunken entryways in chiefdoms is consistent with the practice theory espoused by Pierre Bourdieu (1977). He argues that repeated daily practices can result in large scale social structures. Thus the understanding of small-scale activities is necessary to understand the origins and functioning of large-scale social operations.

During the many season of field work we have discovered dozens of village and cemetery sites, along with some campsites, lithic sources, and other special-purpose loci. We have not detected any footpaths from our earliest time spans, from PaleoIndian through Archaic and the early sedentary (Tronadora phase) occupations. That constitutes 79% of the time that we have documented people living in the area, from approximately 10,000 to 500 BC. It is apparent that the cultural prescription that certain kinds of travel needed to be along the same path did not exist during these occupations. Our assumption is that task-oriented travel predominated, wherein people perceiving a need to obtain a food or other resource, visit kin, or conduct a ritual, generally would travel least-cost routes on an individual task basis, therefore sufficiently randomizing foot travel across their countryside so that entrenched paths would not develop. It is indeed fortunate that every footprint of people moving across the landscape does not preserve, or the life support capacity of a totally trampled world would be miniscule. Tronadora phase (c. 2000-500 BC) villages were small, evidently less than 100 people, characterized by round houses with (presumably) thatch roofs. Ceramics were highly sophisticated, and were accompanied by basic and efficient chipped and ground stone tool assemblages. The manos and metates probably were used for grinding maize and other seeds, and both macrofossils and pollen indicate minor gardening of domesticates was done. However, the bulk of the diet probably derived from wild species of trees, bushes, vines, and other plants, along with hunting and fishing. The deceased were placed in secondary burials, in small rectangular pits just outside the drip lines of house roofs, and often accompanied by ceramic vessels, including nicely decorated drums. It is important to emphasize the association of the deceased with a particular household in the Tronadora phase, because this changes so dramatically in the next phase. We have closely examined all film and digital remote sensing imagery in and around these early villages, as well as all earlier sites, and walked the territory around each early site, and found no linear anomalies that could be ancient footpaths. In terms of social organization, all material indicators consistently point toward egalitarian societies throughout the thousands of years from PaleoIndian through Tronadora times.

The population density during all Precolumbian phases of occupation of the Arenal area remained very low by Mesoamerican, Andean, and even by overall Costa Rican standards. It probably never exceeded a few people per square kilometer. However, comparatively within our area, the Arenal phase (500 BC to AD 600) had the largest settlements and the highest regional population density of any phase. Instead of burying the dead adjacent to each household, villages created separate cemeteries, in some cases only a few hundred meters away, but usually a kilometer to many kilometers distant. In terms of social organization, societies during all phases of occupation prior to the Arenal phase were egalitarian, based on the uniformity of housing, artifacts, and grave goods in the Tronadora phase as well as evident uniformity in the two earlier phases. Housing and household artifacts do not show any differentiation during the Arenal phase, supporting the interpretation that egalitarian societies continued. However, some evidence from Arenal phase cemeteries indicates that egalitarian societies might have been pushing the egalitarian boundary.

In the Bolivar site, as an example of an Arenal phase cemetery, the burials at the very top of the hill had more formal rock coverings and more feasting evidence and a lot more broken pots and metates than did the burials just downslope (Hoopes and Chenault 1994). If these differences are not reflective of variation in gender, age-grade, or other similar factors, they could be indicative of the beginnings of social inequality. The lacks of any skeletal preservation in this, and in all other Arenal phase cemeteries, is due to high soil acidity and mean precipitation of approximately 3000 mm per year. Unfortunately, that means gender or age studies of the deceased cannot be performed. The habitation area was approximately 150 meters down slope to the northwest of the cemetery. No path was found connecting them, probably because the plowing for agriculture that has been done on this peninsula in recent decades destroyed it.

Although we detected no remains of a path at Bolivar, the cemetery itself is instructive as to burial procedures and post-interment rituals (Hoopes and Chenault 1994). The ridge-top cemetery received considerable pre-and post-interment attention. The primary burials were placed in pits dug about a meter below the ground surface, and occasionally accompanied with grave goods (stone axes). The pits were filled in with dirt and then outlined with elongated stones on the surface, and then rounded river stones were harshly smashed onto the entire surface, creating a low mound of rock. That was followed by extensive feasting and smashing of hundreds of complete pottery vessels and dozens of decorated metates and other artifacts in post-funerary rituals throughout the cemetery. Clearly this was the most prestigious area of the cemetery. Only ten meters down the gentle slope was a different kind of cemetery, but one that appeared similar at first glance as it was also covered with broken pottery and a few broken ground stone artifacts. However, the project ceramicist and the lithic analyst could get virtually no rejoins, and artifacts were smaller and had more rounded break edges than those atop the ridge. We concluded that people had scavenged broken artifacts from a midden, and scattered them over the burials, rather than being complete artifacts broken in place, and thus appears to be a "poor person's" imitation of the more elaborate cemetery with conspicuous consumption nearby. The closest midden was adjacent to the small village 150 meters away, and that presumably was the source of the "shortcut" grave-covering artifacts.

In other areas paths that date to the Arenal phase have been found. The shortest known path is only 250 meters long, and its positioning helped us understand how important entrenched paths were to ancient people in this area. It is between a cemetery (G-184) and a village (G-180) that are 1.1 km apart (McKee, Sever, and Sheets 1994:144-146). The easiest transit between the village and cemetery would be an almost straight line along the flat floodplain of the Rio Piedra. However, they routed their path up the hill, over the top, and down the other side, compellingly showing the deep cultural need for an entrenched path. By choosing the route up and over the hill they increased the distance of travel slightly, and the effort significantly. Traveling from the cemetery, one would be in the entrenched path up and down the hill, but as one reached the bottom of the hill and the entrenched path disappears, the village would open up to view. At Trench 37 the path had eroded down to about 1.5 m below the surrounding ground surface, while other parts of the trench on both sides of the hill had eroded down to an estimated 2 m deep below the surrounding ground surfaces.

3 Ethnographic Considerations

In our Western culture we view a footpath as providing the most practical and economic (least-cost) route for people to walk from one place to another. However, ethnographic research has indicated that native peoples can imbue paths with powerful meaning, and we should not underestimate such meaning in the past. Why travel such a precisely prescribed route? James Snead's study of ancestral Pueblo trails of northern New Mexico (2002) suggests an answer. He found that much more than practical and economic factors were extant while people established and used ancient paths in the Bandelier area. The generations of use had endowed the paths with meaning. In a paradigm-changing insight into how differently Westerners and Native Americans can view a trail, Snead (2002:756) provides a quote from Waterman about the Yurok of California: "Trails are sentient, and must be traveled with urbanity. If you step out of a trail and in again, and fail to preserve decorum, the trail becomes resentful." This special sense of a trail, that channeled transit within a learned social memory, is what evidently developed in ancient Costa Rica. Generations of Arenal people were constructing meaning by single file path processions following precisely the same route. The spiritual aspects of the Arenal area paths probably were more important than their practical aspects, and the unforeseen consequences of sustained use developed into a treasured cultural value of entrenched entry into special places.

An initial necessary condition to form entrenched Arenal paths must have been the emergence of the cultural/ritual prescription of single-file travel along precisely the same path, year after year, from village to cemetery and back. That apparently began about 500 BC with the separation of cemetery from village. Ethnographic accounts of the Bribri (Skinner 1920, Bozzoli de Wille 1975) provide possible clues as to why people would have done these processions. The Bribri believe the soul-spirit, after separation from the deceased, needs guidance between village and cemetery. The Cuna in Panama believe the soul-spirit appreciates the body being buried at a distance from the noise and busyness of the village (Dillon 1984). Whatever the reason, the actual first effect of sustained path use was linear compression, and where inclinations were greater than 5°, water from strong rainstorms began eroding and entrenching the path. The actual walked path surface is consistently only about ½ meter wide, and that can only be formed by single-file use. Because the tephra layers and the juvenile soils on them are so unconsolidated in the Arenal area, their angle of repose under these conditions is a slope on either side of the path of approximately 30° from horizontal. Thus, as the path surface erodes downward, it takes a broadened "V" shape of surrounding surface down with it. Centuries of use often result in the path entrenching to two, three, or up to seven meters below the landscape. Thus the prescription on path use in sloping areas resulted in the inadvertent entrenchment, which I believe became a cultural standard of the proper way to traverse between special places. To date, not a single instance of a constructed feature of any path itself has been found in the Arenal area. That contrasts dramatically with the constructed roads-entryways of the later chiefdoms. Only a brief mention of the later chiefdoms is warranted here.

4 Later Chiefdoms and Constructed Monumentality

Recent research by Vazquez et al (2003) discovered the emergence of the Cutris chiefdom between AD 500 and 1000, to the east of Arenal. The chiefs created monumentality by organizing large work gangs to dig four 6-10 km long deep roadways to enter the site center. We suggest the value of entrenched entryways was developed by egalitarian societies centuries earlier in the Arenal area and probably many other areas. Later chiefdoms continued making monumental entryways, such as at Guayabo de Turrialba (Fonseca 1981, Murillo 2002). They were less entrenched, and surfaced with stone, perhaps because of the difficulties in water management of entrenched earthen entryways. When the mentality of monumentality developed, it is not surprising that chiefs would turn to something that had deeply embedded itself into the egalitarian value system: entrenched entryways. People had for dozens of generations "bought into" the significance of those entryways. And the use of them, seeing important places open up in front of the person arriving, must have carried intense emotional reactions.

5 3-D Visualization

Three-dimensional interactive visualizations have been constructed of the Arenal landscape, including the sunken pathways linking cemeteries with villages. The paths also link cemeteries with the sources of stone used in their construction. And the paths link cemeteries with springs, as people needed large amounts of water to support themselves in construction efforts and particularly in feasting and ritual. The 3-D visualization allows one to fly along or over a pathway and experience the topography in ways that can not be done by any other technology.

The Arenal 3-D visualization was part of a broader program including Mesoamerica and Central America. To create the terrain file for Mesoamerica and Central America the research team selected 15-meter resolution imagery from Landsat 7 that covered the entire region from southern Mexico to northern Colombia, plus selected IKONOS scenes, at one-meter resolution, for special areas of interest. The imagery was draped over 90 meter resolution elevation data from the Shuttle Radar Topography Mission (SRTM).

In order to better understand the context, function and relationship of ancient footpaths in the Arenal Region of northern Costa Rica and their evolution to the natural terrain and connectivity to ancient villages and sites, a three dimensional model of the terrain with overlaying features was developed. The resultant visualization allows the investigator to "fly over" the footpaths at various altitudes, directions, and tilt angles and zoom in upon features of interest upon the landscape. The aerial vantage point leads to the realization that everything observable in the ancient landscape has meaning and interconnection, factors that sometimes can not be appreciated from the ground. This gives the archeologist the ability to understand the complex evolution of human factors for settlement patterns, subsistence strategies, use of resources, and the social, political, religious, and economic relationships that connect the study of the past with the present.

The terrain model was produced from several data sources. High resolution, pan-sharpened IKONOS data were processed using ERDAS software to produce both natural and color infrared images at one-meter resolution. These high resolution IKONOS scenes and 30-meter resolution LANDSAT imagery were draped over a Digital Elevation Model (DEM) derived from digitized contour lines of the 1:50,000 scale topographic maps of Costa Rica using TerraBuilder - a 3D terrain construction application from Skyline Software Corporation.

TerraBuilder provides a three-dimensional spherical Earth with latitude/longitude coordinates as a base for holding imagery, elevation and vector data types. Because the IKONOS data were formatted using a UTM coordinate system, Terra Builder had to convert the files into latitude and longitude using built in conversion procedures. (Terra Builder is quite versatile in it's acceptance of a wide range of data formats including GeoTiff, .img files output by ERDAS Imagine, and MrSID compressed files to name a few.) The multi-resolution terrain file produced by TerraBuilder combines the imagery and elevation data with built-in imagery of the entire globe.

Using the resultant multi-resolution terrain file another Skyline application – Terra Explorer Pro (TEP) - was used to interact with the file, add and remove vector data, and add or remove other assets such as labels, images, and icons. For instance, the location of the ancient footpaths, determined from previous analysis of NASA airborne scanner data and color infrared photography, and ground verified using Global Positioning System (GPS) measurements, were input as vectors onto the terrain file. In addition, the location of ancient sites and cemeteries were also incorporated as vector layers. Terra Explorer Pro was also used to create and record animated "flights" over the terrain and support analyses of the terrain through a series of tools that measure distance, produce contour maps, and locate best path to name a few. Due to the flexibility of TEP other GIS data can be added to the mosaic and any of the existing data layers can be toggled on or off.

6 Summary and Conclusions

The Arenal Research Project, operating intermittently in northwestern Costa Rica since the 1980s, has documented human occupation for the past 10,000 years. During most of that time, people walking across the landscape were presumably task-oriented, and thus sufficiently randomized to leave no traces that we have been able to detect today. However, from about 500 BC to about AD 1200, people separated their cemeteries from their villages. The processions that regularly linked the two were by people traveling single-file in as straight of lines as possible. The unanticipated consequence, path entrenchment, apparently became a highly valued cultural standard. That was "writ large" when complex societies developed, and chiefs mobilized massive construction efforts to make monumental sunken entryways.

Until three-dimensional interactive visualizations were developed, we had only paltry, static ways of sharing the landscape, the locations of villages and cemeteries, and the paths that linked them. Now with the visualization one can fly over the processional landscape, following a path to its endpoint. One can sense the feasting done at the cemeteries and see how people got to them. One can see from where people obtained construction stone, and see the spring from which they obtained water for cooking and drinking at the cemetery. The landscape comes alive in a way not matched by any other graphic displays at our disposal.

The three-dimensional landscape of the Arenal area was constructed by draping high-resolution IKONOS satellite imagery over a Ditigal Elevation Model (DEM). The DEM was constructed from the 1:50,000 scale topographic maps from the Instituto Geografico in San Jose. Specific locations of paths, cemeteries, and village sites were verified with the Global Positioning System (GPS). Terra Explorer Pro software allows us to fly over the terrain and sense movement from village to cemetery and back, along the ancient footpaths.

References

BOURDIEU, PIERRE 1977. *Outline of a Theory of Practice*. Cambridge: Cambridge University Press.

BOZZOLI DE WILLE, MARIA 1975. *Birth and Death in the Belief System of the Bribri Indians of Costa Rica.* Unpublished Ph.D. dissertation, Anthroplogy, University of Georgia, Athens.

DILLON, BRIAN 1984. Island Building and Villages of the Dead: Living Archaeology in the Comarca de San Blas, Panama. *Journal of New World Archaeology* 6:2:49-65.

FONSECA, OSCAR 1981. Guayabo de Turrialba and its Significance In Abel-Vidor, Suzanne, Ed. *Between Continents, Between Seas, Precolumbian Art of Costa Rica*. New York: Abrams. Pp. 104-111.

HOOPES, JOHN, AND MARK CHENAULT 1994. Excavations at Sitio Bolivar: A Late Formative Village in the Arenal Basin In *Archaeology, Volcanism, and Remote Sensing in the Arenal Region, Costa Rica,* edited by P. Sheets and B. McKee, pp. 87-105. Austin: University of Texas Press.

MCKEE, BRIAN, TOM SEVER, AND PAYSON SHEETS 1994. Prehistoric Footpaths in Costa Rica: Remote Sensing and Field Verification. In *Archaeology, Volcanism, and Remote Sensing in the Arenal Region, Costa Rica,* edited by P. Sheets and B. McKee, pp. 142-157. Austin: University of Texas Press.

MURILLO HERRERA, MAURICIO 2002. *Análisis crítico de las investigaciones en el sitio Guayabo (UCR-43), de Turrialba y las repercusiones sociales con relación al manejo de sus recursos culturales.* Documento inédito. Trabajo Final de Graduación para optar por el grado de Licenciado en Antropología con énfasis en Arquelogía. Escuela de Antropología y Sociología. San Jose: Universidad de Costa Rica.

SEVER, TOM, PAYSON SHEETS, AND DANIEL IRWIN 2003 Satellite and Airborne Remote Sensing Analysis for the Detection of Ancient Footpatys in Costa Rica. Vinculos_28:9-18.

SHEETS, PAYSON 1994. Summary and Conclusions. In *Archaeology, Volcanism, and Remote Sensing in the Arenal Region, Costa Rica,* edited by P. Sheets and B. McKee, pp. 312-325. Austin: University of Texas Press.

SHEETS, PAYSON (Ed.) 2003 VINCULOS, Revista de Antropologia del Museo Nacional de Costa Rica. Special issue on Arenal Research Program. Volumen 28, # 1-2.

SKINNER, ALANSON 1920. Notes on the Bribri of Costa Rica. *Indian Notes and Monographs*, V. VI, # 3. New York: Museum of the American Indian, Heye Foundation.

SNEAD, JAMES 2002. Ancestral Pueblo Trails and the Cultural Landscape of the Pajarito Plateau, New Mexico. *Antiquity* 76: 756-765.

VAZQUEZ, RICARDO, JUAN VICENTE GUERRERO, Y JULIO CESAR SANCHEZ 2003. Cutris: Descripcion, cronologia, y afiliacion de un Centro

Arquitectonico con Caminos Monumentales en la Llanura de San Carlos, Costa Rica. *VINCULOS* 28:1-2, pp. 149-174.

The importance of the relief and the sources to interpret and communicate the Cultural Heritage

Valentina Vassallo,[1] A. Moro,[1] L. Vico[1]

[1]CNR-ITABC, Istituto per le Tecnologie Applicate ai Beni Culturali, VHLab, Montelibretti, Roma, Italy

1 Introduction

The cultural heritage must communicate with its addressees; therefore, it is impossible not to take into consideration its expressive and emotional ability, or to forget that scientific knowledge is at the base of a reliable and transparent communication.

For this purpose, in this paper we analyse the importance of the relief and the study of the sources finalized to the knowledge, interpretation and communication of the cultural heritage.

Our project, carried out at the Virtual Heritage Lab of the Itabc at the CNR (commissioned from ARCUS), looks at the "reconstruction of the archaeological landscape of the ancient Flaminia with virtual reality systems".

It is a digital ecosystem, oriented towards a real time desk top Direct X 9.0 application, that individualizes two levels of perception. The first one represents a holistic vision of the road from Rome to Rimini , based on historical maps, information on the archaeological excavation, technical cartography and aerial and satellite photos able to support the realization of a web gis. The second one, which dips into a micro-space vision, focuses the attention on five sities, with the aim to represent and contextualize the local entities within the "ancient landscape" entity. The areas of monographic interest are Grottarossa, Ponte Milvio, Malborghetto-Tor di Quinto and Villa of Livia.

2 Study area

2.1 The ancient via Flaminia

It starts from Rome, ends at Rimini, and takes its name from the constructor C. Flaminio, consul in 223 b.C. and censor in 220-219 b.C. The road was built for the conquest of the Valle Padana and of the Gallia Cisalpina and acquires a greater importance through the monumental restoration made by August. Inside of the Flaminia road we have localized five areas of interest.

2.2 Villa of Livia

The great archaeological complex of Villa of Livia, situated at Prima Porta, at the IX mile of the Flaminia, it is traditionally identified with the imperial villa of Livia Drusilla, wife of Ottaviano Augusto. The villa was built between 30 and 25 b.C., on a previous structure of the republic age and it was occupied at least until Severian age.

2.3 Malborghetto

It rises along the ancient Flaminia, close to the XIII mile of the road. It's a great triumphal arch, dated to the first half of the IV century, which becomes, after various transformations, a museum.

2.4 Grottarossa

The aerea is at the VI mile of the ancient via Flaminia and is constituted by a necropolis (dated between first century B.C. to the second century A.D.). It reaches an extraordinary monumentality also thanks to the nearby large residential area (the fish farm, in fact, is surely related to a villa of *otium*).

2.5 Tor di Quinto

The necropolis, at the V mile of the road, is mostly constituted by *columbaria* and *mausoleii*. It is datable between late republican age and second century and it is characterized by military depositions of the *coorts pretoriae*, with association of *stelae* and *ollae cinerariae*. Actually the site is not accessible by the public because is in a military area.

2.6 Ponte Milvio

It rises at III mile of the Flaminia road. It is the departure point of some main streets that go to the north of Italy and it is a strategic junction to control the access to Rome. We have not precise details about the date of building of the bridge; but undoubtedly, its first phase is of the same period of the Flaminia road (220 B.C.).

3 Aims

The present study is proposed, only after having obtained accessible data, in order to attain knowledge of the cultural heritage and to spread it through new technologies.

Only through the collection of high level data it is possible to achieve a scientific instrument able to understand and communicate the archaeology where human being is the maker and the user of the history.

Figure 1 *Plan of Villa of Livia with historical phases*

4 Methodologies of the 3D reconstruction

A firm knowledge of the cultural heritage is at the base of all the successive processes of interpretation and communication.

To design a reliable and verifiable plan of reconstruction in all its parts it is necessary to begin from a first phase gathering of high level data, both by the collection of texts, instances, archives etc. relative to actual places and on study cases morphologically and temporarily comparable ("top down") and through survey of the vestiges ("bottom up").

4.1 Top down

The first category of data has put into evidence the problem of the interpretation of the ancient architecture that, connected with the limitations of the literary sources, make the architecture become the main way to understand the roman architecture, placing the problem of how the public interpret it. The history of the buildings is not a collection of irremovable facts, but is a continuous process, a set of various inclinations and interpretations which transform the object, as always, everyone transforms the past adapting it to its own experiences.

The study of the sources (a), the observation of its geometry and materials, the static verifications and the arithmetical calculations (b), the study of typological parallels (c), become the fundamental instrument in order to understand a building.

(a) The sources

Through the study of texts, *itineraria*, ancient iconography etc., kept in archives and libraries, it's possible to collect the first kind of information.

The only historical surviving treatise of the roman age is *De architectura Libri Decem* of Vitruvio. Unfortunately, it is not sufficient to reconstruct the roman architecture because there are many gaps due to the Greek architectural model. In fact, the careful observation of architecture and design remains the only solid foundation.

Figure 2 *Engraving of L. Prosseda, 1828. View of the Villa of Livia' hill. From Messineo G. 1991.*

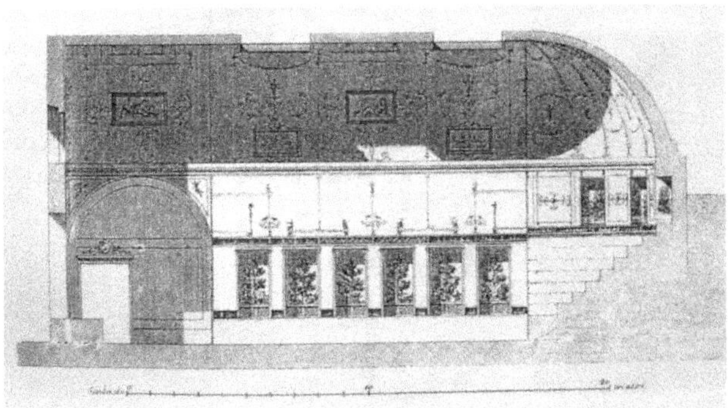

Figure 3 *"Horti Maecenatis." Auditorium". Hypothetical vertical section by C. Vespignani (Bcom 1874, tav. 11).*

Figure 4 *Particular of the parietal decoration from Auditorium Maecenatis (left) and from villa of Livia (right).*

(b) Evaluation of the state of art

The analysis of the state of the places allows us to draw conclusions about possible logical 3D reconstructions, tied to the constructive rules and ours cultural background. In classical architecture, the relationship of interdependence between plant and prospect has one independent logical. They serve in order to organize the space, and often in the archaeological diggings, we find only partial horizontal section and nothing of the prospects. The original architecture morphology can be understood through the typological and material analysis. The structural verification, through the calculation of graphical static, represents a fundamental instrument in order to ratify the reconstructive hypotheses. Roman architecture is based on geometric elementary shapes like the regular circle, polygons and rectangles. The Arithmetic determines the shapes and the dimensions of the space and the architectonic elements with measurable proportions, like the proportion 1:1, 2:1 (both geometric and Arithmetic), 2:3, 5:4, √2, √3 etc. This approach is valid not only for the structure but also for the decorations. The study of materials supplies another important aspect. The mechanical abilities of the materials condition architectonic aspects like height of walls, systems of cover, etc. The design is a process, a dynamics interaction between concept and contingency, generic and specific that evolves progressively in a variety of decisions that come assimilated "into the whole".

(c) Lack of data

The sources and the material of study often turn out incomplete in order to assume a reliable and verifiable architectonic reconstruction. Therefore, the problem is resolved through the historical and architectonic analysis of comparisons and typological parallelisms (composite, constructive, structural, distribution and repetition's elements).

In the same way, parallelisms for the reconstruction of the decoration are fundamental.

4.2 Use of transparencies

Since the acquisition phase, we make of the choices that determine the result. A reconstructive process must be visible and transparent.

The type of transparency information will be different depending on the aims and type of 3D visualisation method and outcome being used is according to the type and level of knowledge and understanding.

Transparency information requirements may therefore differ at different phases within a project. Transparency information may change as levels and complexity of understanding of particular 3D visualisation methods rise, and will vary from different groups.

In the project of virtual reconstruction of the Via Flaminia we establish three levels of reliability about the reconstructive hypotheses that will be complete with three different colours (high, medium and low reliability). Optionally audio and video informative schedules are foreseen on the reconstructive hypothesis with information about reconstruction's reliability, type, aim, similarity, space, time, and affordability in reconstruction.

5 Bottom up

The first category of information about the cognition of the actual state of the sites is based on the possibility of using advanced technologies for the data acquisition that is intimately connected to the typology and the particularities of the entities to examine.

In the case of Malborghetto, for the simplicity of the geometry, the technique of photo-modelling was chosen, and, together with the relief of strategic points and the total station and the Gps, has concurred to refer the site geographically and reconstruct the landscape.

Figure 5 *Malborghetto - technique of photo-modelling*

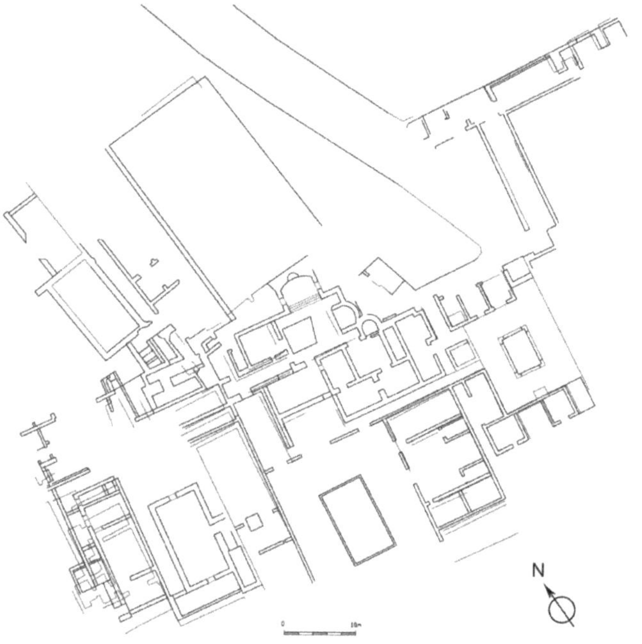

Figure 6: *Villa of Livia comparison of S.A.R. plan with our laser plan*

Actually, Malborghetto is a museum that collects materials coming from Flaminia road and in particular from the necropolis of Tor di Quinto. Our project realizes a virtual museum of Malborghetto through the 3d reconstruction of the interior of the museum and of the guide-objects (through computer vision and laser scanner) and is possible to replace them virtually in their original sites.

Particularly interesting is the case of Tor di Quinto because it is at present not accessible to the public.

The importance of starting from objective and accessible data, in the case of the Villa of Livia has turned out even more necessary, due to the complexity of the architectonic system and the insufficient and partial credibility of the plan in public archives and of scarce information caused by incomplete excavations.

In fact, the great archaeological complex of Villa of Livia has made use of advanced technology in the laser scanner to obtain 3d reliefs with a resolution of 6mm.

Figure 7 *Villa of Livia - point cloud registred with Cyclon software by Cirax Tecnology*

This technology represents an enormous potential of publishable, up-to-date, available information and, above all, georeferenceble, fundamental for an archaeological contextualized knowledge of the territory and in a position to transmit a strong evocative and communicative component.

The acquisition of 3D data with the laser scanner generates a high-resolution model with questionable and exact data, with generated and interpolated faces. The same technique is used to detect both the architectonic compound of the Villa of Livia and the hill on which it exists.

The acquired and elaborate models (Rapidform by Inus Technology) are further optimized within other three and bidimensional graphic softwares (3D Studio Max, Photoshop CS2), used for texturing, decimation and the improvement of the graphical quality of the rendering (techniques of multi res, normal maps, svolgi UVW, render to textures).

Moreover, this complicated model gives the possibility to refer to 2D reliefs that enable an accurate and deeper analysis.

In fact, by the cloud of points method, through the cloudwork tool, it is possible to elaborate horizontal and vertical sections in CAD format that constitute an analyzable documentation from the architectonic point of view as a base of the typological and structural study of the building.

Figure 8 *Villa of Livia - 3d mesh realized with Rapidform software by Inus Technology*

In the case of Grottarossa, because of its great extension area, we utilize the aerial photographic technique with aerostatic balloon (lab. Directed by F. Gabrielli, Itabc/Cnr).

On the obtained plan, georefered with Gps points, it is possible to place the various architecture elements that are reconstructed with different techniques (laser scanner for the fish farm, 3d modelling for the *mausoleii* etc.)

Figure 9 *Grottarossa - aerial photographic technique with aerostatic balloon*

Regardin Ponte Milvio, for our project pourpose, the symbolic and emotional evaluation takes on particular meaning because the space represents and communicates the place of important historical events. So, for the reconstruction of the building we can base on the pre-existent vertical and horizontal sections to realize a low polygon resolution model (CAD, 3D Max).

Figure 10 *Villa of Livia - rooms textured and optimized with 3dS Max*

Figure 11 *Villa of Livia - rooms textured and optimized with 3dS Max*

6 Conclusions

The via Flaminia project is still in progress but the work done up to now already allows to draw methodological conclusions for the creation of a virtual reality system based on a unique model of high accuracy and detailed data, that is an example of integration of spatial data, different technologies and methodologies.

Once the virtual reconstruction is finished, all information generated becomes part of the documentation regarding the archaeological site. The output model will be a synthetic, accessible and suggestive document that contributes to progress in analysing the building and, beyond, in art history.

The method can also generates a low polygon resolution version meshes (oriented towards real time desktop Direct X) without modifying the perception of it.

The cultural wealth at our disposal and the application of advanced manageable technologies give the possibility to communicate and share the common heritage rendering it completely available to everyone in its historical, artistic and scientific valences.

Through the 3D reconstruction of the architectonic heritage aimed at real time fruition, we approach an effective and sustainable interpretation that involves the public and transmits the importance of the conservation and communication of the cultural heritage. They decode data and information that often belong to the scientific ambit that are not accessible or comprehensible to the majority of the unspecialized public.

Acknowledgements

The authors thank Archaeological Superintendence of Rome and for the scientific collaboration Polytechnic University of Catalonia (Spain), ETH of Zurich (Switzerland) , IMT of Lucca (Italy), Miralab of Geneva (Switzerland) and the lab directed by R. Gabrielli of Itabc-C.N.R. of Rome (Italy).

References

ADAM, J.P. 1988, *L'arte di costruire presso i Romani,* Milano, Longanesi.

DE FRANCESCHINI, M. 2005 *Ville dell'Agro Romano,* Roma: L'Erma di Bretschneider.

DE VOS, M. 1993 Horti Maecenatis. 'Auditorium' , in *Lexicon Topographicum Urbis Romae*, Roma: Ed. Quasar.

FRIGGERI, R. 1988 Tor di Quinto, in *BullComArch*, XCII

GALLIAZZO, V. 1994 *I ponti romani*, Treviso: Canova.

GAZZOLA, P. 1963 *Ponti romani*, Firenze: Olschki

GIULIANI, C.F. 1997 *L' edilizia nell'antichita*, Roma: NIS.

GROS, P. 1987 *Architettura e società nell'Italia romana*, Roma: Curcio

MESSINEO, G, CALCI, C. 1984 La *Villa di Livia a Prima Porta,* Roma: SAR.

MESSINEO G., SCOPPOLA F. 1988 Scoperte nei nuovi insediamenti militari della via Cassia e della via Flaminia, *QUAEI, ArchLaz* IX,

MESSINEO, G., CALCI, C. 1989 *Malborghetto*, Roma: Ed. De Luca.

MESSINEO, G.1991 La *Via Flaminia. Da Porta del Popolo a Malborghetto,* Roma: Quasar.

MESSINEO, G., CARBONARA, A. 1993 *Via Flaminia*, Roma: Istituto Poligrafico e Zecca dello Stato, Libreria dello Stato.

MESSINEO, G. 1997 Grottarossa o Saxa Rubra, in *Forma Urbis* 12.

MESSINEO, G. 1998 *Malborghetto (il monumento e l'antiquarium),* Roma: SAR.

MESSINEO, G. 2001 *Ad Gallinas Albas: Villa di Livia*, Roma: L'Erma di Bretschneider

PERKINS J.B.W. 1974 *Architettura romana*, Milano: Electa Stampa.

QUILICI L., 1990 *Le strade: Viabilità tra Roma e Lazio*, Roma: Quasar.

VITRUVIO, *De architectura libri decem.*

VESPIGNANI, C., VISCONTI, L. 1874 *Antica sala da recitazioni ovvero Auditorio scoperto tra le ruine degli Orti Mecenaziani sull'Esquilino*, BCom.

WILSON JONES M. 2003, *Principles of Roman architecture.* London: Yale University Press.

The Flaminian Way in Umbria: an integrated survey project for the study and conservation of the historical, architectural and archaeological features

Stefano Bertocci, Sandro Parrinello

University of Florence Architectural Planning Department Faculty of Architecture Viale A. Gramsci 42, Firenze

The Umbrian territory crossed by the ancient Roman road, the Flaminian Way, which goes through the municipalities of Massamartana, Acquasparta and Sangemini, presents monuments and art works connected with the old road network. During the centuries, the Roman road track, which was marked in 223 a.C., has coagulate an amount of settlements and monuments, like the Massamartana catacombs, rural churches, monasteries and castles. All these products prove the remarkable anthropological activities of the last two millenniums and they embody therefore a great interest for the archaeology, but also for the history, the architecture and the environment.

The conservation and the accessibility of all the sites lay down problems not only concerning merely the management of the survey, study and conservation compaign, but also, if the aim is to create a museum network, those related on one side with the improvement of a correct use of the sites, on the other with the realisation of an executive program and planning of the works in order to guarantee the best conservation of the heritage itself. These considerations must be taken especially regarding the open-air archaeological fields, which are constantly exposed to stress caused by tourism and by the necessity of keeping pace with the economical and social development of the whole region.

The aim of this project of research is to evaluate the area that actually cover the administrative territories of Sangemini, Acquasparta, and Massamartana and to analyze the disposition of such structures connected with the old road Flaminia network in this territory; the disposition of the remains can now appear casual at first, but otherwise it is configured as the natural composition of human activities that accumulate and stratify in this place.

Museums, with a functional specificity to divulge their informative contents, are containers of collective memories and icons, often considered as the instruments through which a community identifies with its territory, tangible traces of relations between human beings as well as between them and the territory in which they live.

The complexity of preparation and museum building implies cognitive and research activities, based on some studies which focus on the heritage itself represented inside the museums by multiple elements that are ordered through research activities, generally linked with a specific direction of social "marketing".

The current strategies of preparation and museum building are oriented towards a re-examination of the past and addressed to a non-specialized public in order to answer the growing demand for less ideologized preparation and museum building.

The "territory museum", intended as socio-cultural space, serves to articulate a common conceptual data as the different themes and data around the territory result as an open museum where the objects and the themes of the course are presented within the specific social context and surrounded by their original physical environment.

There is an intense bond between the disposition and, therefore, the arrangement of the elements in a place, with the consequent world of relations established between those two, and the dynamics that determine the activities and the cultural capacities of a certain museum system.

The aim of a territory's museum, as well as museum didactic, is to highlight and facilitate the construction of such connections within the cultural patrimony.

The project of a territory's museum is a process that implies a series of events that allow an element, or a general context, to be recognized as museum worthy; this operation often coincides with a re-qualification process.
A preventive conscience is emerging towards goods and towards the patrimony ad the territory that must be safeguarded; this conscience resorts, through social organizations and institutions, in order to safeguard a specific set of goods or an environment which is exposed to devaluation or de-signification.

Therefore, structures or systems of relations are being projected in a way to be able to transform a place into a museum, focal points of cultural aggregation which sometimes transform the qualities of an environment or a territory in order to have a museum.

The object to be safeguarded is even moved to another dimension, spatial or virtual; some of its aspects are captured and transformed to allow its availability and "net-surfing".

In order to make the information accessible, the image or the real status of the elements to be safeguarded can be

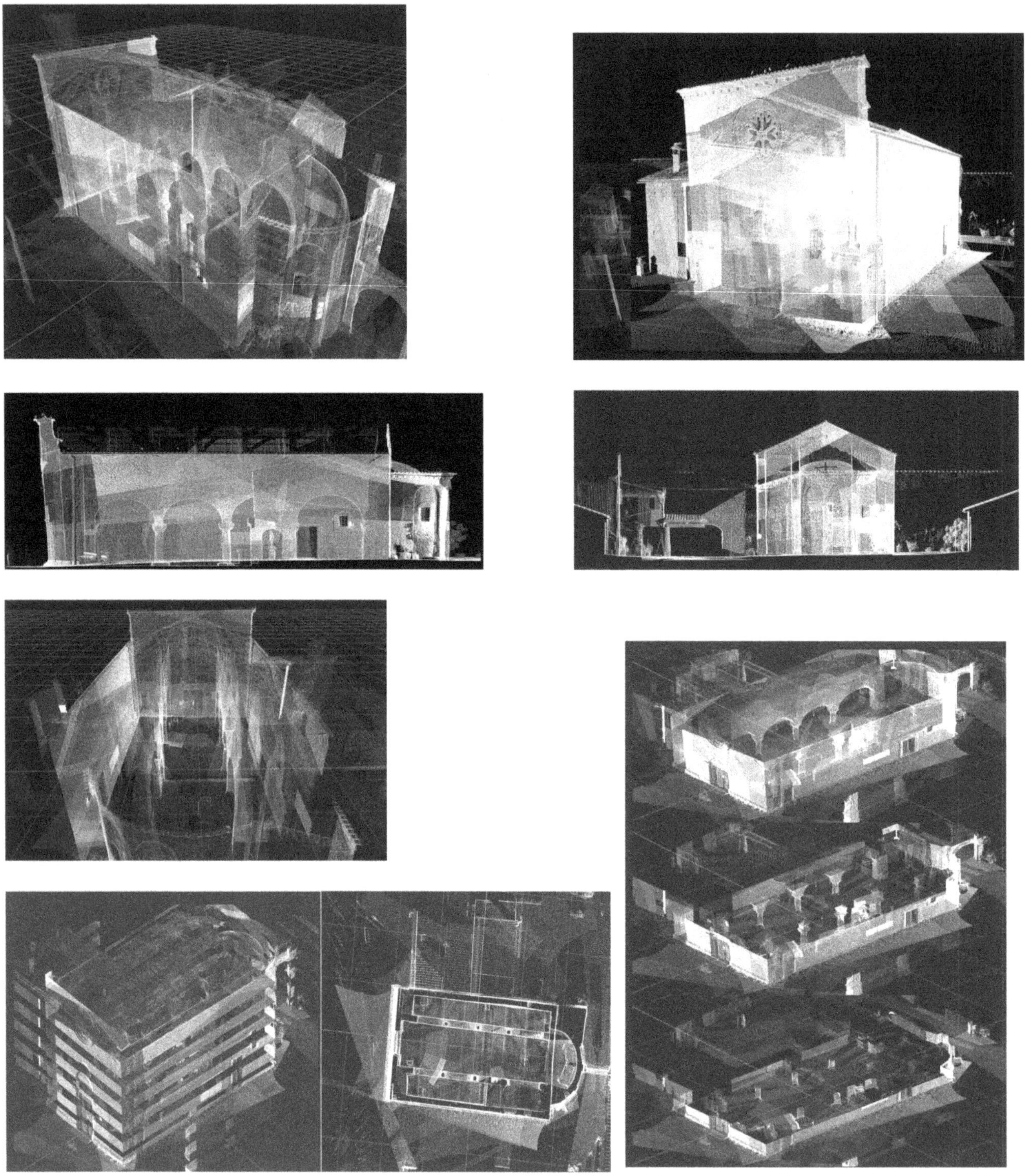

Figures 1-8 *View of 3D Laser scanning survey of Santa Maria in Pantano at Massamartana*

manipulated; this is all probably part of the same process to conserve collective memory, as well as research and "social target or marketing".

The first steps of this process are the documentation and the survey for conservation and interpretation, using new technologies for interpretation, for territory's museums management and for qualifying the architectural space and the territory of the remains.

The two methodologies of the remot sensing and the photo-interpretation with the integration of site survey and documents are used during the research in order to gain the definition and the structure of the connective principles of the revealing material and the digital data bases, in order to get an aimed use for the project purposes.

The air photo-interpretation, the study and the creation of virtual models with the support of the laser scanning are

Figures 9-14 *View of 3D Laser scanning survey of S. Faustino at Massamartana.*

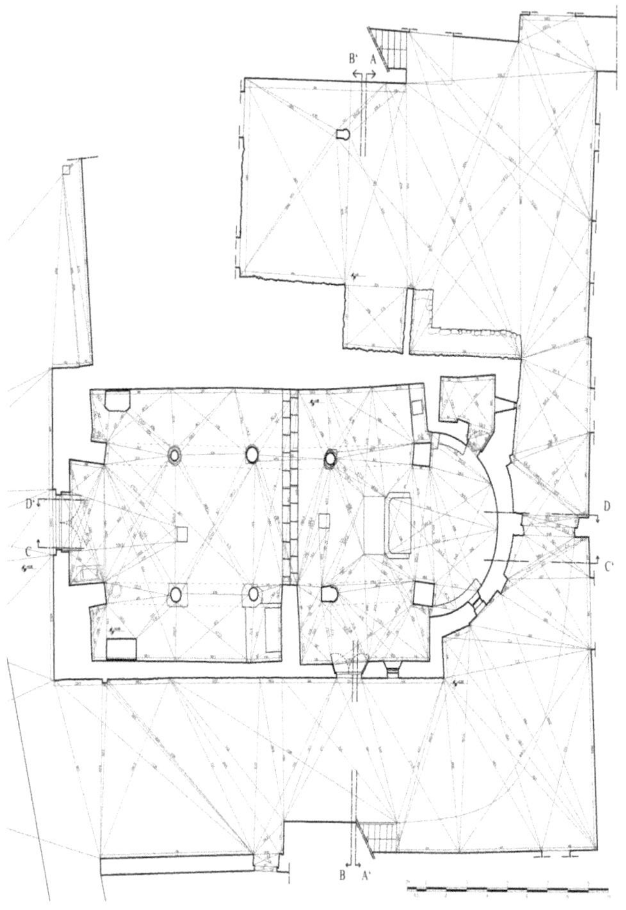

Figure 15 *Plans of the Santa Maria in Pantano's church at Massamartana*

determined steps for the generation of a connecting system, which supports the drawing records, the cognitive research and the conservation of the heritage.

One of the principal aims of this project is the consitution of a GIS (Geographic Information System) that with its characteristic of flexible use, also on different levels, becomes a basic support for the development of an open-air museum network which tries not only to preserve the image and the survey data of the single finds and the monuments related to the Flaminian Way, but also to build a procedure for the management, the accessibility, the protection and the conservation of the heritage itself.

References

BERTOCCI S., S. PARRINELLO (edited by), *From the survey to the project: the identity of the towns. The contribution of new technology in remote data management*, Firenze, Edifir, 2006

BERTOCCI, S., G.PANCANI and P. PUMA (a cura di) *Ville e parchi storici. Strategie per la conoscenza*, Firenze, Edifir, 2006.

BERTOCCI S., S. PARRINELLO, A. NICCOLI and R. TIBERI, *The management of green areas in the urban environment, p.p.45-47* in: Atti del convegno *9th European Forum on Urban Forestry, May 21 – 27,* F. Ferrini, F. Salbitano, G. Sanesi, (edited by), 2006, Florence – Italy

La via Flaminia nel territorio di Acquasparta e Massamartana. Comune di Massamartana, s.d.

Figure 16 *Landscape view from Google Earth*

Figure 17 *Landscape view from Google Earth*

BERTOCCI, S., L. BIANCHINI and S. PARRINELLO *La scedatura del centro storico di Montepulciano*, Piano Strutturale, Comune di Montepulciano, Montepulciano (SI),Aprile 2004, p.p.7-12.

BERTOCCI, S., L. BIANCHINI and S. PARRINELLO, *La documentación y la valorización del centro histórico de Montepulciano. Nuevas tecnologías al servicio de la planificación urbana*, Noveno Seminario Internacional Forum UNESCO, Buenos Aires, 11-15 Ottobre 2004, pag. 162.

Firenze Architettura. Castelli medievali a Petra e nel vicino oriente tra rilievo e archeologia, numero speciale "eventi" del periodico del Dipartimento di Progettazione dell'Architettura, Università degli Studi di Firenze, 2004.

BERTOCCI, S. *Indagine sulle murature di una torre del lato orientale della cinta di Iasos di Caria*, è in corso di stampa a cura dell'Accademia delle Scienze di Ferrara.

BERTOCCI, S. *Analisi archeologica ed architettonica di edifici campione a Gravina in Puglia* (con la collaborazione di E. Donato), in AA. VV. *Il Parco della pietra e dell'acqua*, Gravina in Puglia, 2000, pp. 165 – 184;

BERTOCCI, S. *Luce colore e geometrie nell'architettura rupestre a Petra*, in Colore Luce e Materia in Architettura, a cura di E. Mandelli, Alinea editrice, Firenze, 2000, pp. 23 – 28; S.

BERTOCCI, S. *La torre di Galatrona*, Comune di Bucine, Università degli Studi di Firenze, Dipartimento di Progettazione dell'Architettura, S. Giovanni Valdarno, 2001.

BERTOCCI, S. *Proporzioni e geometria nell'architettura rupestre a Petra*, in *Matematica e Architettura - Metodi analitici, metodi geometrici e rappresentazione in Architettura*, Alinea editrice, Firenze, 2001, pp. 47 – 54.

BINI, M. and S. BERTOCCI, *Castelli di Pietre- Aspetti formali e materiali dei castelli crociat nell'area di Petrai in Transgiordania*, Polistampa, Firenze, 2004

BINI, M., S. BERTOCCI, L. BIANCHINI, A. MRAKIC and S. PARRINELLO, *Information Technolgy with the menagement of Archaeological Heritage: Petra (Jordan), Iasos (Turkey), Park of 100 Roman Farms at Lucca (Italy) and the Archaeological park of Rocca San Silvestro (Italy)*, in Atti del Convegno "International Conference on Remote Sensing Archaeology", Joint Laboratory of Remote Sensing Archaeology in affiliation to the Chinese Academy of Sciences, Ministry of Education and State Bureau of Cultural Relicts, Beijing (Pechino), Cina, 18 –21 Ottobre 2004, p. 310.

PARRINELLO, S. and R. TIBERI, *New surveyng technologies for phytosanitary assessments and environmental forest monitorino*, in R. La fortezza, G. Sanesi (edited by), *Patterns and Processes in Forest Landscapes. Consequences of Human management, international conference IUFRO,* p.p. 487-492, Bari, Accademia Italiana di Scienze Forestali, 2006

PARRINELLO, S. (con presentazione di M.Bini e testi di S.Bertocci), *Il centro urbano di San Piero a Sieve*, Firenze, Edifir, 2006

Sharing interpretation: the challenge of Open Source web approach

Luigi Calori,[1] Carlo Camporesi,[2] Augusto Palombini,[2] Sofia Pescarin[2]

[1] CINECA Visit Lab, Bologna, Italy , l.calori@cineca.it
[2] CNR ITABC, VHLab, Rome, Italy, [carlo.camporesi].[augusto.palombini],[sofia.pescarin]@itabc.cnr.it

1 Introduction

The impact of digital technologies is going to cause a drastic change both in the working process and in theoretical and epistemological settings of archaeological science and cultural heritage management, forcing everybody to rethink the traditional ways of information validation and diffusion.

Today, digital technologies let a large amount of information to be available and transmittable in short time. Almost everybody can potentially create and diffuse data to the whole world population, or at least to that who can reach the web, setting up relevant problems in terms of scientific validation. Science, in this sense, seems to have to change its tools: from a narrow scientific community who controlled the access to publication, to an *open* approach in which everybody can be an actor, and information reliability relies on the *transparency* of its sources and of the development process (e.g. Camporesi et al. 2006, Feller and Fitzgerald 2002; Forte, 2005).

According to this approach, an *open process* can not only be focussed on the adoption of Open Source software, but it implies also the application of an *open approach* in the use of interpretation and validation structure.

A famous example that explains the aspects of *Open Source* is the one described by Eric Raymond in his famous book "the cathedral and the bazaar", where he describes *Commercial* and *Open* software development philosophies as *Cathedral* and *Bazaar* models (Raymond 1999). In the *Cathedral* model, software projects are developed by small groups of experts, while in the *Bazaar* model everybody can look at the people working, participate, give suggestions and criticism.

It is interesting to think that this metaphor could be also applied to a general scientific interpretation process, distinguishing a traditional *Cathedral* way from a different *Bazaar* process, open to public or at least to a wider evaluation by the transparency of its creation process.

In this latter approach, the open dimension of information and the use of open source software are two elements of the same process, that recall each other.

2 VR WebGIS Project of CNR

In accordance with these considerations, we started a new research direction in our lab, the Virtual Heritage Lab (VHLab) of the National Council of Researches, Institute of Technologies Applied to Cultural Heritage (CNR ITABC), testing the the potentiality of such tools. We were aimed to face some problems related to the broad diffusion of archaeological scientific data through the web, as well as to consider the state-of-the art of open source software in our field and its suitability for a large community of non-programmers, as that represented by archaeologists, historians, art historians, etc.

We tried to use open source software in our projects, trying to analyse the possibility to convert or extend commercial tools already part of our digital pipeline into open ones. Moreover we tried to develop new open tools, building a bridge with IT experts and IT institutions, such as the Visualisation Lab of Cineca Supercomputing Center, signing a research agreement. Finally we tried to apply to our projects and to our applications (based both on open and commercial tools) an open and transparent approach, focused on metadata publications, on interaction and on information communication.

Our effort was focussed on the realization of a VR WebGIS based on the 3D open engine OpenSceneGraph (www.openscenegraph.org), on a modified version of the Virtual Terrain library (www.vterrain.org) and on a visualization Plugin, the OSGvisWeb completely developed by the team, for Mozilla Firefox and for MS Internet Explorer, for 3d web data interactive fruition.

The software development was planned by CNR (mainly 2d webGIS and plug-in for Mozilla) and CINECA (terrain generation and plug-in for Explorer).

The main feature of this tool is the opportunity to create and browse in real time virtual reality environments and terrains. In order to achieve an open approach also in term of transparency policy, we have also developed specific features for data exchange and verification in the VR webGIS environment.

These tools and features were partially developed and applied for the first time in the *Appia Antica* project (Forte et al., 2005). Others were better defined and newly

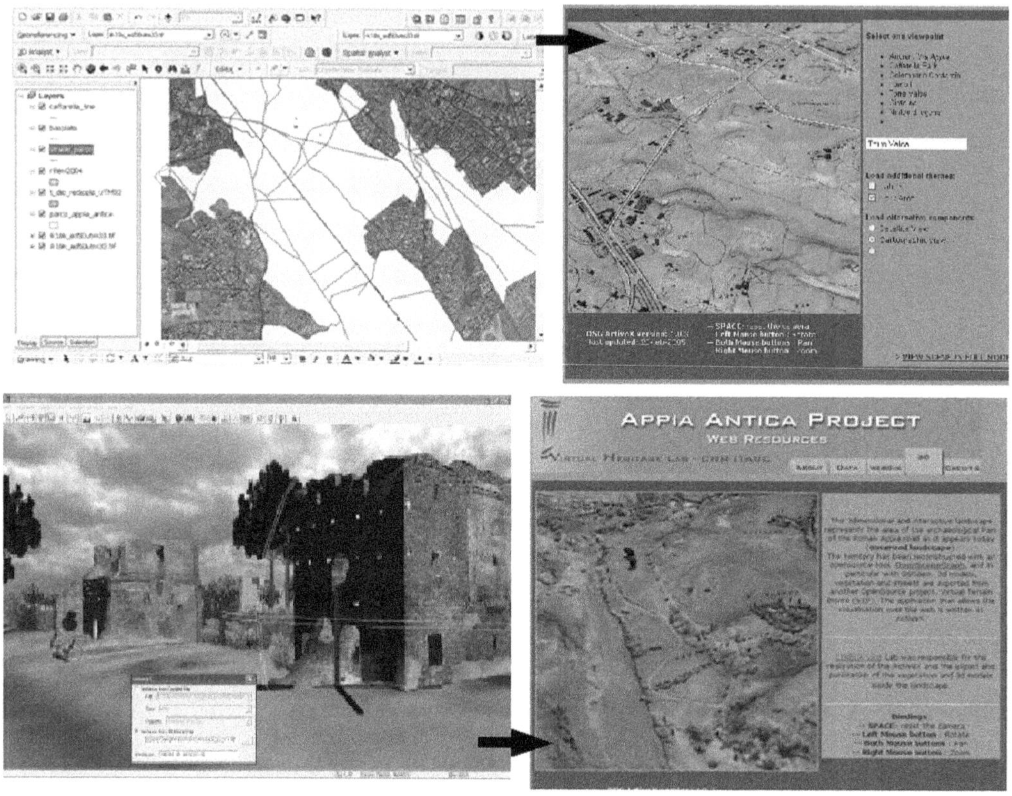

Figure 1 *The Appia Antica project, on the left: From GIS project to web 3d navigation and interaction; 3d dynamic off-line environment for landscape reconstruction and 3d data published over the web.*

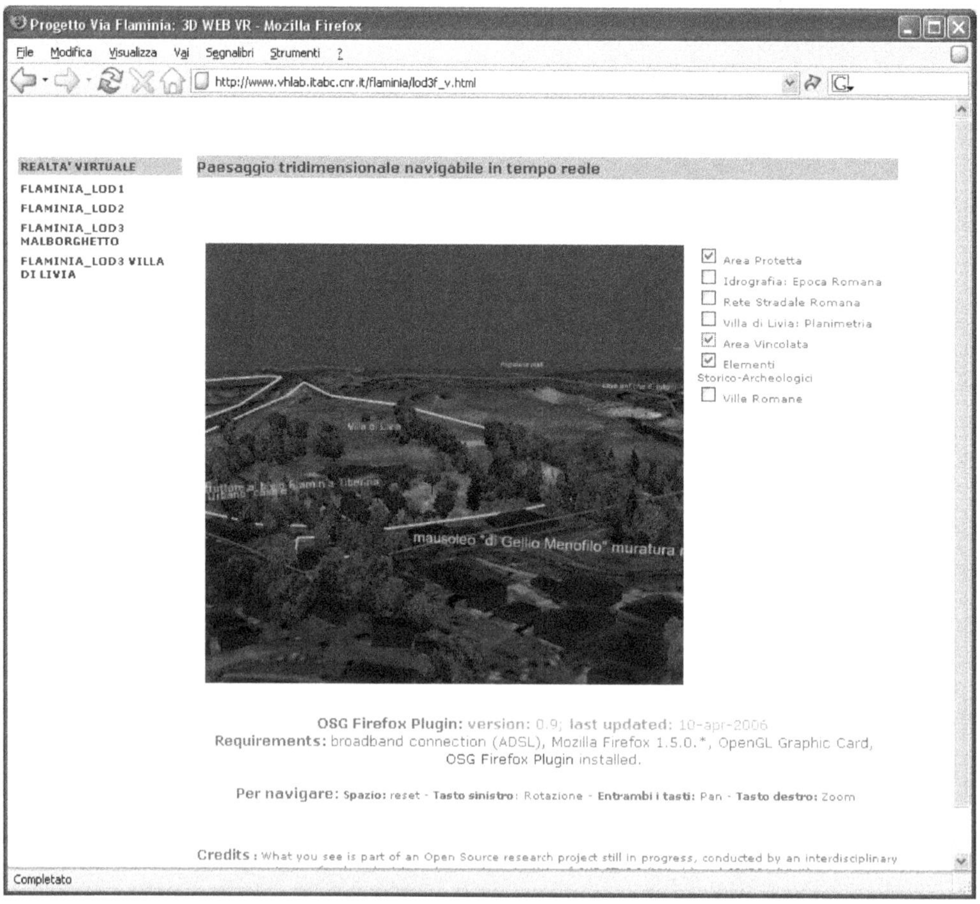

Figure 2 *Screenshot of the OSGVisWeb interface (on Mozilla Firefox) and the Flaminia Project*

developed in the case of the *Flaminia* project, in which the open approach become a sort of guideline during its development.

In the first case study, the Appia Antica Project (Fig. 1), the primary goal was the realization of a spatial 2d and 3d archive for Roman Superintendency of a wide archaeological park, in the southern part of Rome.[1] Data were collected directly on the field in topographic (DGPS, PDA, Laser Total Station) and architectonic survey (Total Station, Laser Scanner, digital camera, etc.). Terrain information were acquired and processed in a GIS and published in a webGIS; we used the open source MapServer for this purpose, working on the interface with the tool Chamaleon. 3D terrain and models where then published in a separate VR webGIS section, together with vector information, thematic layers, vegetation attributes and alternative switchable 3d terrain models.

In the second case study, the Flaminia Project, we worked on a website with the aim of creating, since the very beginning, a real working tool for the team and all other partners (Fig. 2). In the web project we tested different solutions for interpretation and communication. Data transparency was also considered a fundamental goal, achieved through the publication of metadata connected to switchable models.

3 Open source and virtual archaeology

The use of Open Source software is certainly the first goal in our approach. As a matter of fact, the empiric premise of any validation process is based on the reproducibility of the data production chain. Then, whoever presents a set of data through the web should give a sufficient information both on original sources and on the software used, to allow everybody to check the products. Only dealing with open source software we can really define such a process as *open* to everybody's checks.

This step is still conditioned by a general problem. Why *Open Source*, despite of its well-known advantages in terms of cost and availability, is still far from a general diffusion?

Without a deep analysis of such a complex and studied topic (e.g. Bonaccorsia and Rossia 2005, Economides and Evangelos 2005), we can see that in most of people vision, *Open Source* suitability problems are related to some specific items, such as: lack of information on products and features, difficulties in installation and use, lack of user-friendly interface and – above all – the need of a gradual shift from their traditional commercial software to the open packages, so that the path to a diffused use seems to pass through:

1- The adoption of an open source working chain
2- A complete portability Windows/Linux of the process and a search for applications available on both the operating systems.
3- The selection of user friendly open source software.
4- A particular effort to move technical problems from the user side to the server side, through the creation of pre-compiled software packages.

An overview of the available applications for the production of GIS layers, DTMs, and 3D models reveals that many of these problems are today overcome. Figure 3 shows that the first three steps of the list above are substantially overcome.

	Commercial sw (Win)	Open software (Linux)	Open software (Win)
Raster editing	Adobe Photoshop	 Gimp	Gimp
Vector editing	AutoCad	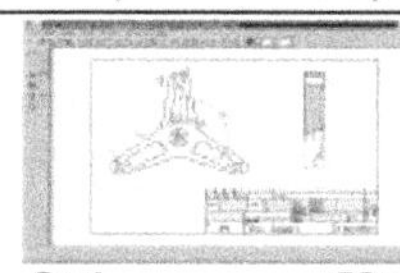 Qcad	QCad
GIS	ArcView	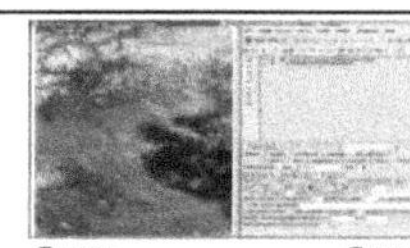 Grass	Grass
3D Modeling (point processing)	RapidForm	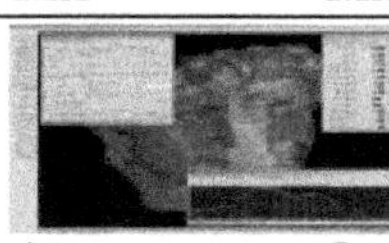 Scanalyze	Scanalyze
3D Modeling (texturing)	3D Studio Max	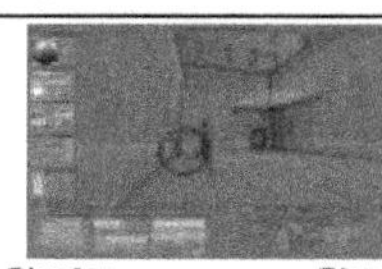 Blender	Blender
Terrain generator	TerraVista, Visual Nature	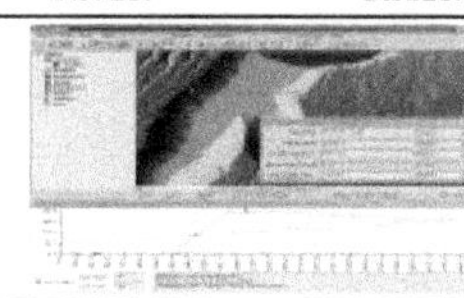 VTP, OSG VisWeb	VTP, OSGVisWeb
Web publication	MS IIS	Apache	Apache

Figure 3 *Table of open and commercial software available on different operating systems.*

As shown in Fig. 3, the whole working chain of a VR WebGIS creation process is currently possible on an open source working-chain.

The open source applications working on Linux and Windows are exactly the same, so that the portability of

[1] The project was developed by CNR ITABC, under the direction of M. Forte and P. Salonia for Roman Superintendence, with the cooperation of technical sponsors and the research contrinbution of CINECA (www.cineca.it). See Forte, Pescarin, Pietroni, 2005, pp.

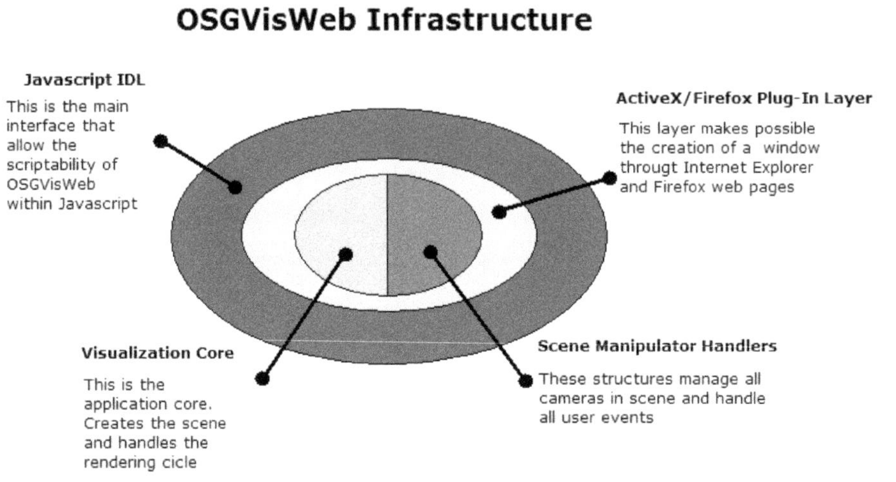

Figure 4 *A simple scheme of OSGVisWeb architecture.*

the VR WebGIS creation process for different platforms is total.

The figure also contains snapshots of the open source applications, showing that such a software works almost totally by user-friendly graphical interfaces.

The further step was the specific aim adopted during the developing of the Flaminia VR WebGIS project.

In order to completely remove technical problems from the user experience, the creation of pre-compiled software packages is needed. A similar experience has already been done in a Linux environment, which has a complete archaeological working suite: the ArcTeam group developed in the last year a Linux release devoted to archaeologists (ArcheOS "Akhenaton" www.arc-team.com) (Bezzi et al., in press). It is a Debian-based Linux distribution containing all the software needed for the 3DWebGis working chain.

The situation is currently more complex on Windows platforms, but the realization of a complete suite containing self-extracting software binaries, tutorials and links, to let the end-user to avoid all installation problems, is one of our next goals.

The current step has been devoted to the realization of a VR WebGIS totally based on open source software, absolutely easy and friendly for the end user, shown in Figs. 1 and 2.

4 OSGVisWeb: Visualization infrastructure for Mozilla and Explorer

OSGVisWeb is a platform created to view and manipulate 3D terrain models through web pages. It's completely written in C/C++ language and based on the middleware OpenSceneGraph (a real-time OpenGL rendering library).

OSGVisWeb is easily adaptable to HTML pages (published through MS Internet Explorer and Mozilla Firefox browsers), and easy to control by JavaScript instructions. It doesn't need dedicated services on the Server side.

4.1 Architecture

OSGVisWeb's architecture is composed by 4 principal elements (Fig. 4):

- Scene Manipulator Handlers (Ext Manipulator): it's a scene manipulator which manages the user events derived from the allocated visualization window. Moreover, it performs modifications on ViewCore's states. ExtManipulator contains, at the moment, two manipulators which can manage in different ways user movements and interactions: LandManipulator allows a macroscopic and fast interaction with selected terrains; Walk Manipulator allows an immersive interaction simulating a walk-through on terrain (still in a testing phase).
- Visualization Core (ViewCore): The main feature of such a structure is managing the 3D rendering cycle and a correct creation and modification of the scene tree. Moreover, it manages, at a high level, the user events deriving from the selected Scene Manipulator as well as the web loading of all the models which will compose the scene. To do that, the extension module OSG NET is used. To avoid the problem of the freeze effect on the scene in the previous version (during the loading), a new method of thread-loading has been implemented.
- ActiveX / Firefox Plug-In Layer (OSGActiveX / npOSG4 Moz): such a middle-range conceptual layer (Fig. 4) allows the creation and visualization of 3D rendering windows inside the web pages published by Mozilla Firefox (version 1.5 and newer) and MS Internet Explorer (version 6.0 and newer). It addition, it can be defined as a control flux-manipulator

between the 3D render engine and the web pages render engine. Physically, two structures are present to act on these topics: the first is npOSG4Moz creates the windows instances, through WIN32 requests (MS Windows 32/64 bit architectures) and X11 (Linux 32/64 bit architectures), to connect the 3D visualization engine to the Mozilla HTML renderer (Gecko). The second is OSGActiveX; like npOSG4Moz it integrates the 3D visualization engine with the MS Internet Explorer HTML publication engine by MS ActiveX Component technology.

- Javascript IDL: a Javascript Interface Definition Language which publishes a common method-interface between the component/plug-in to the web pages, to have a bilateral control flux. Such an abstraction layer allows to send and receive events through a series of control methods. Javascript IDL is completely transparent and allows a web pages development in an independent way if pages are published through Firefox or MS Internet Explorer.

4.2 Features

OSGVisWeb is a viewing platform useful and adaptable to different terrains and 3D models. Among the new features and functions there are:

- Visualization of web paged terrains and complex 3D models.
- On-demand terrain loading and switching capabilities during the rendering process of already rendered scenes.
- Adding and switching of 2D and 3D vector layers in the scene.
- Adding and switching labels on terrain.
- Adding of external 3D models and positioning on terrain by scripting functions.
- Visible scene visualization by a little 2D map
- JavaScript "FlyTo" flying function to defined coordinates position.
- Main menu of interaction with the scene, with "rotation", "pan", "zoom", "FlyTO" (arriving by cursor pointing) and "picking" options and content visualization.
- Fast content visualization through a pop up window inside the scene (using the "picking" function)
- External detailed content web page loading (using the "picking" function), for active object on the scene.
- Javascript function running option by event of Fire type (Only for OSGActiveX) or by HTTP GET requests to the browser.
- The availability of data tables for each GIS layer, containing all the realization references as well as (downloadable) software and methods used.

4.3 Future development

OSGVisWeb is continuously in evolution, and there are studies on possible extensions in many directions. Herby we report some possible next areas of development.

- Studies on ways of maximizing data transfer and implementation of a methodology of catching on disk.
- Implementation of advanced 3D objects on the scene as avatars, and interaction with them.
- Multimedia files viewing.
- Implementation of a methodology of automatic software updating through the web.
- Linking objects to external databases and publication of informations inside the 3D area.
- Generation of more 3D layers by informations obtained from spatial databases.
- Implementation of ways to improve realistic effects on the scene.

5. Concluding remarks

The VR WebGIS Project is an interactive platform aimed at web fruition and based on open source libraries and tools, provided with interface and behaviours. The methodological approach in the use of the platform is oriented towards the public sharing of data, originally available as raw data to a restrict group, with metadata attached (complete references on sources, on software and methods used, etc), for a complete transparency of scientific process, in order to keep reference of the different steps which have led to the final processed information.

Furthermore visibility and availability of data and information are two central concepts, connected to each other, in the open approach. The diffusion of a truly scientific approach to archaeological information is based on the repeatability of processes and on the data and knowledge exchange. On the other side, software availability can represent a good opportunity to enlarge open source community of non programmers, to use and modify specific tools, allowing a faster and further development.

6 Acknowledgments

This paper synthesizes a work developed in the context of a highly interdisciplinary team of CNR ITABC and CINECA Supercomputing Center. For CNR ITABC, the Virtual Heritage Lab team is directed by Maurizio Forte. In the Appia and Flaminia projects worked also Nicolò Dell'Unto, Fabrizio Galeazzi and Valentina Vassallo (archaeologists), Marco Di Ioia, Alessia Moro e Lola Vico (architects), Eva Pietroni (cultural heritage media expert), Claudio Rufa (musician and computer scientist) and Bartolomeo Trabassi (audio/video operator).

References

ECONOMIDES, N. and E. KATSAMAKAS, 2005. *Two-sided competition of proprietary vs. open source technology platforms and the implications for the software industry.* Electronic document, MIT 2005. http://opensource.mit.edu/papers/economideskatsamakas2.pdf

BONACCORSIA A. and C. ROSSIA, 2005. *Why Open Source software can succeed.* Electronic document, MIT 2005. http://opensource.mit.edu/papers/rp-onaccorsirossi.pdf

CAMPORESI, C., PALOMBINI, A. and S. PESCARIN, in press, "Revolution OS" in archeologia. Esempi di interfacce web per l'archeologia del paesaggio in *Open Source, Free Software e Open Format nei processi di ricerca archeologici. Atti del I Workshop (Grosseto, 8 maggio 2006)* R. Bagnara and G. Macchi Janica (eds.)

BEZZI, A., BEZZI, L., FRANCISCI, D. and R. GIETL, in press, ArcheOS Ahenaton versione 1.1.0 in *Open Source, Free Software e Open Format nei processi di ricerca archeologici. Atti del I Workshop (Grosseto, 8 maggio 2006)* R. Bagnara and G. Macchi Janica (eds.)

FELLER J. and B. FITZGERALD, 2002 *Understanding Open Source Software development*, Addison-Wesley, London

FORTE M. 2005 A Digital "Cyber" Protocol for the Reconstruction of the Archaeological Landscape: Virtual Reality and Mindscapes in *Recording, Modeling and Visualization of Cultural Heritage* E. Baltsavias, A.Gruen, L.Van Gool, M.Pateraki (eds.) Taylor & Francis / Balkema, pp. 339-351.

FORTE M., PESCARIN S., PIETRONI E., DELL'UNTO N. 2005. The Appia Antica Project. In *Archaeological Landscapes through Digital Technologies*: Proceedings of the 2nd Italy-United States Workshop, Rome, Italy, November 3-5, 2003 Berkeley, USA, May 2005 M. Forte (ed.) BAR S 1379 2005.

PESCARIN S., 2006 Open Source in archeologia nuove prospettive per la ricerca, *Archeologia e Calcolatori*, 17, All'insegna del Giglio, 2006

PESCARIN S., CALORI L., CAMPORESI C., FORTE M. 2005, Interactive landscapes reconstruction: a Web 2D and 3D opensource solution, in M. Mudge, N. Ryan, R. Scopigno (eds.), *The 6th International Symposium on Virtual Reality, Archaeology and Cultural Heritage, VAST'05 Short Presentations*, STAR, Pisa : 33-38

RAYMOND E.S. 1999, The Cathedral and the Bazaar: Musings on Linux and Open Source by an Accidental Revolutionary, Paperback Ed. Rev. O'Reilly, Sebastapol, CA.

Project links:

CNR Virtual Heritage Lab.: http://www.vhlab.itabc.cnr.it
Appia Project: *http://*www.vhlab.itabc.cnr.it/appia
Flaminia Project:
*http://*www.vhlab.itabc.cnr.it/flaminia
OpenSceneGraph website:
http://www.openscenegraph.org
Virtual Terrain Project website: *http://www.vterrain.org*

Operative action for the conservation of the Archaeological Complex of Chan Chan, Perù

F. Colosi,[1] G. Fangi,[2] R. Gabrielli,[1] R. Orazi,[3] D. Peloso[1]

[1] CNR- Istituto per le Tecnologie Applicate ai Beni Culturali (ITABC), V. Salaria Km 29,300 – 00016 – Monterotondo Sc. (Roma)
[2] Università Politecnica delle Marche – DARDUS
[3] Director of the Italian Mission in Perù

1 Introduction

The town of Chan Chan represents the most important witness of the Chimù culture, one of the many civilizations that arose in the valleys and along the northern coast of Perù since the beginning of our Era.

Though it is inscribed in the UNESCO "List of Word Heritage" and "List of in danger Word Heritage", the town is suffering a dramatic process of material and urban degrade due to the building technique, the extension of the settlement and the luck of funds.

The "Plan Maestro de Conservacion y Manejo del Complejo Arqueologico Chan Chan" (Master Plan for the Conservation and Management of the Chan Chan Archaeological Complex), was prepared by the Peruvian institutionally competent body, the INC, and approved by a Supreme Decree of the President of the Republic of Peru In the year 2000. It thus represents a unique legal tool for conservation, enhancement and management of the site. The Plan Maestro actually provides for a series of programmes, sub-programmes and projects, mostly focused on archaeological investigation, documentation and conservation of earthen architectonic structures and the associated decorated surfaces.

Following the directions of the Plan Maestro, the MIPE (Missione Italiana in Perù) is operating in Chan Chan since 2001, carrying on a wide action of documentation, conservation and exploitation.

The most important purpose of MIPE's work is represented by the restoration of Palacio Rivero and by the protection of the site from the aggressive and uncontrolled growth of the near town of Trujillo.

The project is also pointed to the creation of a Documentation Center, a Museum and a wide Archaeological Park in order to affect the tourist sector and, consequently, the socio – economic structure of the region.

The analysis and documentation phases regarding the site of Chan Chan have been carried on by using modern techniques and research procedures for surveying and cataloguing as remote sensing, GPS, laser scanner and GIS.

2 The Archaeological area

The archaeological complex of Chan Chan is located in the northern part of Peru not far from the town of Trujillo and about 550 Km. north of Lima. The town, that represents the largest mud brick pre-Columbian settlement covers an area of 14 km^2 and is placed on a sedimentary terrace a few hundred meters from the Pacific Ocean between 16 and 40 m. above the sea level. Chan Chan, the political and economic capital of the Chimu culture, was built around the half of the IX century and was particularly flourishing from the XIII to the XV century when it became part of the great Inca empire (1470).

Regarding the site of Chan Chan we can divide the large area in two main parts: the central one and its surroundings. The former, almost rectangular in shape, contains three architectural typologies: the palaces or *Ciudadelas*, the *Huacas* or stepped pyramids and intermediate architecture. The latter represents agricultural land where structures and buildings, often interrelated, are widespread such as granaries, warehouses, channels, fencing walls, cemeteries and orchards (Fig. 1).

The most important architectural typology of the urban area is represented by nine *Ciudadelas* that are characterized by some recurrent architectural features. In almost all the *Ciudadelas* there are three main sectors the first of which is reserved for public ceremonies, the second and third sectors are reserved for the king and his court. In each sector there is a great quantity of small yards, warehouses, *Audiencias* (yards for ceremonial purposes with a small U shaped structures) and lodgings. In most of the palaces the second sector contains the *plataforma de entierro* that is the pyramidal – shaped structure where the king would be buried at his death together with his servants, concubines and possessions that had accompanied his earthly life.

3 Operative project

The strategic goals of the project are related to the architectural and territorial scale and to the training sector.

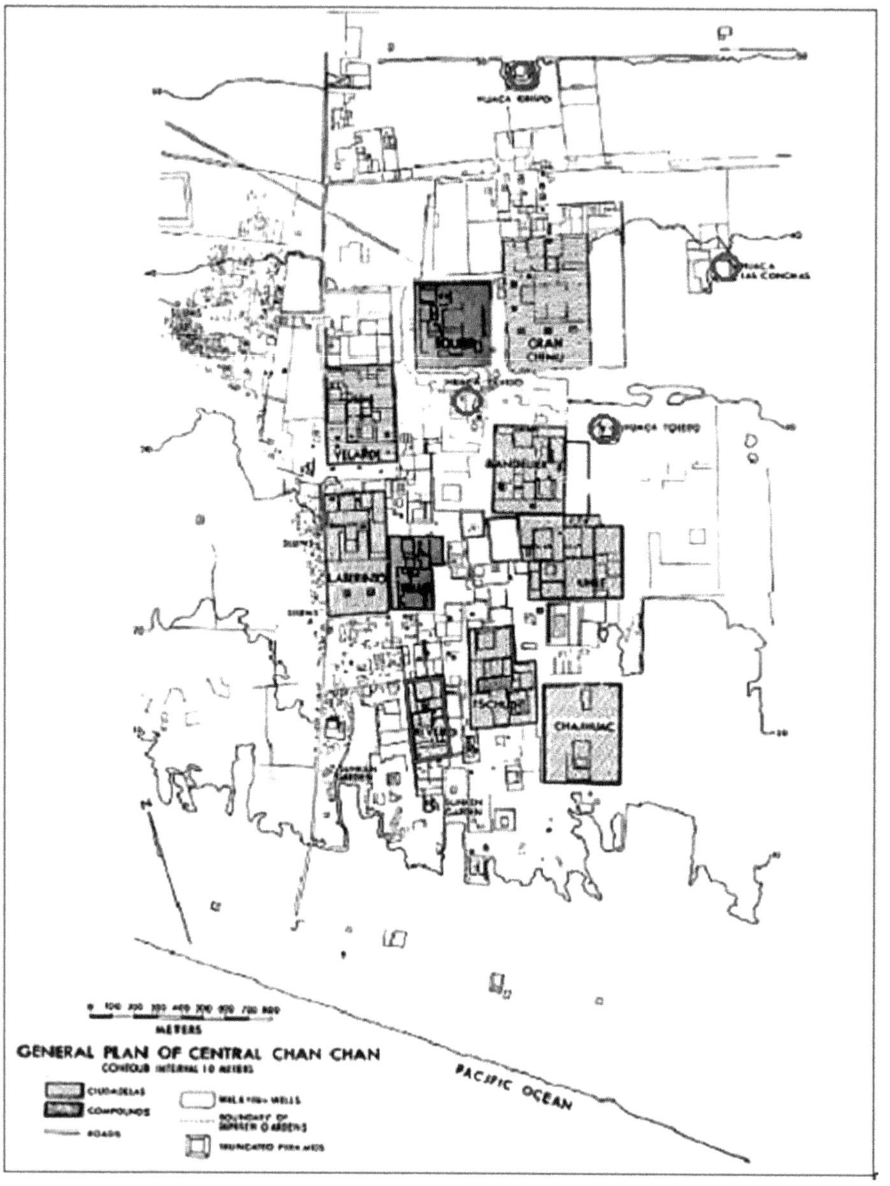

Figure 1 *Map of the site*

Planning of the Archaeological Park of Chan Chan
The main purpose is to stop the process of urban decay by creating an archaeological park and by pointing out, on the territory, its different elements:

Preparing the restoration project of Palacio Rivero
The project is focused on the development of an action model to be tested on one of the palaces of Chan Chan and subsequently extended to the entire archaeological complex. The building selected for the test is the *Palacio Rivero* which, although one of the smallest in Chan Chan, has an area of 70,000 m^2. Besides Rivero is near Palacio Tschudi, the only one open to tourists and under restoration, so that it can be envisaged a larger public area.

Creation of a Centre for the documentation and managing of the Archaeological Park of Chan Chan
By mean of a specific GIS the centre will be operating in the cataloguing of all the existing documentation on a digital support and on the reconstruction of virtual visits which become everyday more important in consideration of the progressive lost of the material evidence.

4 Analysis of the environmental and architectonic decay by using the Quickbird image

The extensive settlement of Chan Chan cannot be analysed and studied without using great scale images. The MIPE is in possession of some recent (2003) high resolution images from the Quickbird satellite.

These images can be extremely useful for two different aims: the first regarding a quantitative measurement of the deterioration suffered by the structures in these last 35 years and the second regarding a detailed analysis of the territory of Chan Chan.

The Quickbird satellite is operative since 2002. It is able to acquire both in multi – spectral and in panchromatic mode with geometric resolution about 0,6 m. and about 2,4 m. in the XS mode.

The image used for the present study has been acquired in the XS mode in 10-05-2003 as standard level (corrections of the system, radiometric corrections, geometric corrections, organization in a specific cartographic system of reference: Datum WGS84, UTM projection, 17 Fuse, South Hemisphere). The satellite imagery is immediately available, the quality is so high that it constitutes a tool extremely useful, since the very strong resolution allows us to work at different scales of investigation (Fig. 2).

The satellite image shows us two different factors of territorial decay due to direct anthropic activity (new paths crossing architectural structures, spontaneous vegetation, uncontrolled agriculture) and to the aggressive approaching to the archaeological area by the near town of Trujillo and by the village of Huanchaco and its airport (Fig. 2).

The process of degradation of the structures of *Palacio Rivero* is due to different causes as illegal excavations, marine salts and climatic periodical actions as the natural

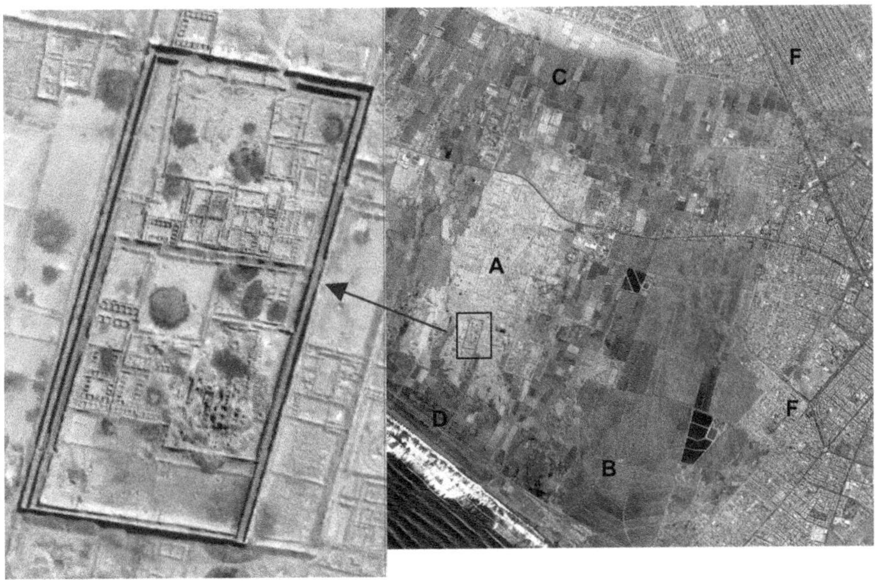

Figure 2 *Quickbird panchromatic image. On the left a zoom of Palacio Rivero. The letters show the different classified zones*

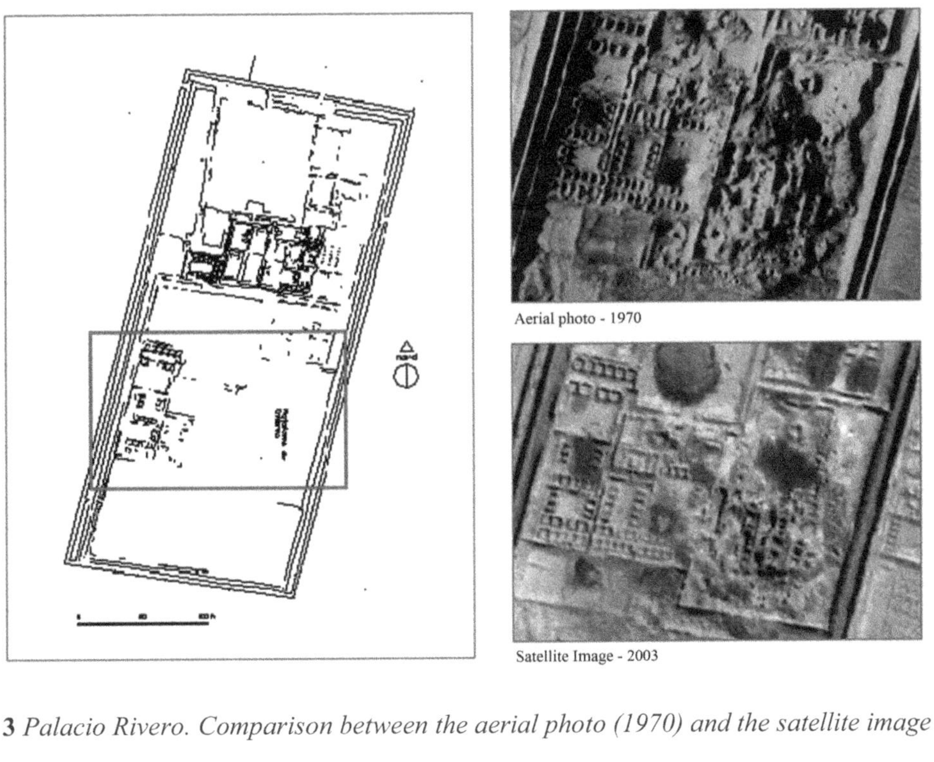

Figure 3 *Palacio Rivero. Comparison between the aerial photo (1970) and the satellite image (2003)*

phenomenon of *Niño*. The comparison between the aerial zenital picture of Palacio Rivero, which was taken in the seventies by the University of Harvard (MOSELEY 1975; 1982), and the Quickbird satellite image is particularly helpful for the evaluation of this process: many structures of the second sector of the palace still visible in the Harvard picture, are almost completely disappeared in the satellite image (Fig. 3)

5 Remote sensing for the study of the territory and the Archaeological Park of Chan Chan

The visual analysis of the syntheses RGB 3 –2 – 1 (true colours) and 4 – 3- 2 (false colours), facilitated by the elevated geometric resolution of the images allowed to find some zones with diversified features shown in Fig. 2: a) the archaeological site of Chan Chan, b) a band of intensive cultivation characterized by bills of elevated extension (South – East), c) a band of mixed cultivations characterized by bills of redoubt dimension (North), d) a costal band of vegetation characterized by a less evident parcels (South), e)parcels without vegetation and modified by antropic phenomena (West and North), f) areas densely urbanized (West and North – West) (FANGI et AL 2005). The classification, carried out in ENVI software, clearly separates the central archaeological site from the area intensively cultivated to South – East and from those less intensive to North. The costal area is characterized by the presence of damp zones, the so called *humedales* where countrymen cultivate the *totoral* can, spontaneous and perennial grassy formation from which small boats, used by craftsmen (*caballitos de totora*), and small handicraft objects were obtained. Moreover, the classification highlighted the perimeter of the buildings and some stains of vegetation more developed that colonized the archaeological area.

With the aim to improve the quality of the information a fusion between the real colour and the panchromatic (architectural and archaeological structures) and between the infrared and the panchromatic (soil and vegetation) images has been performed. The new images keep all the information of the real colour and the infrared shots but achieve the high resolution of the panchromatic one.

The new images, beside the visible archaeological structures, show also architectural alignments which are not visible on the ground, providing therefore an important tool for the identification of buried structures.

6 Archaeological Park of Chan Chan: Creation of a Geographic Information System

As mentioned above, the project, proposed as a supporting activity to the implementation of the *Plan Maestro*, is strongly tilted in favour of the use of IT for the documentation and cataloguing of the building heritage. It is a fact that the subsequent management of an archaeological complex of these proportions cannot be approached without an adequate electronic (and no longer paper based) documentation system which will allow real-time switching among the various levels of use (research, conservation, promotion, exploitation, administrative management) and the location of the above activities in space (town plan, sector map, building plan, elevations, construction and decorative details, etc.).

The final objective of such a system is the gradual creation of a multidisciplinary data archive that will allow the various research and conservation activities to be updated and to schedule and implement the improvement activities, as well as to regulate the complex economic activities related to the management of the archaeological complex (tourist flow, multimedia products, constraints and integration with the modern city).

7 Restoration of Palacio Rivero: geometric and photogrammetric documentation

The main objective of the present project is namely to use the case study of *Palacio Rivero* as an example of Chan Chan architecture to create an operating model allowing the execution and relative application of IT to all the phases of documentation, research, action and the interrelations among them.

Topographic survey of emerging structures
The first technology used for recording geometrical data was based on the use of an electronic total station. This system was used to measure the buildings and the position of the long external walls, of the corridors and of the internal rooms. Since the decay of the upper parts of the structures had settled along the base of the walls it was often impossible to identify the original architectural alignments. Therefore the total station was used to record the coordinates of those walls whose connection with the ground was clearly visible (Fig. 4).

Survey of the morphological surface of the terrain
The great deterioration of the walls, of which we have just written, has created an almost lunar landscape characterized by the slight sinking and elevation of the ground from which erect walls emerge. In order to perform this kind of survey we have used a differential GPS (Leika 530) with a single frequency in a cinematic mode , that is with one of the antennas on the shoulders of an operator who, walking around the area to be surveyed, has been able to register the coordinates of one point every two seconds. The great quantity of data which were gathered in this way has allowed us to draw the shape of the ground by means of contour lines and so to enrich our information about the structures and the rooms (Fig. 4).

The photogrammetric survey
The aerial photogrammetric survey has been carried out flying on the palace by mean of a small Cessna 172 with a precalibrated digital camera Canon Eos D1s, 50 mm focal length at the relative altitude of about 200 m, getting a photo scale 1/4000. The strip is formed by nine images from which a first 3D reconstruction of the Palace has been obtained (Fig. 5).

Due to the high level of decay, a more accurate survey and restitution were dedicated to the *Plataforma de Entierro.* The plotting of the platform has been performed in a digital stereoplotter RFD (Restitutore Fotogrammetrico Digitale) by Geopro, Ancona. In fig. 4 the plotting with the contours lines is visible. The contour interval set at 25 cm. From the digital models a DTM and a true – orthophoto were produced (FANGI et al. 2005)

The laser scanner survey
The employment of innovative technologies like three dimensional surveying, for the documentation of archaeological sites and monuments, has produced new

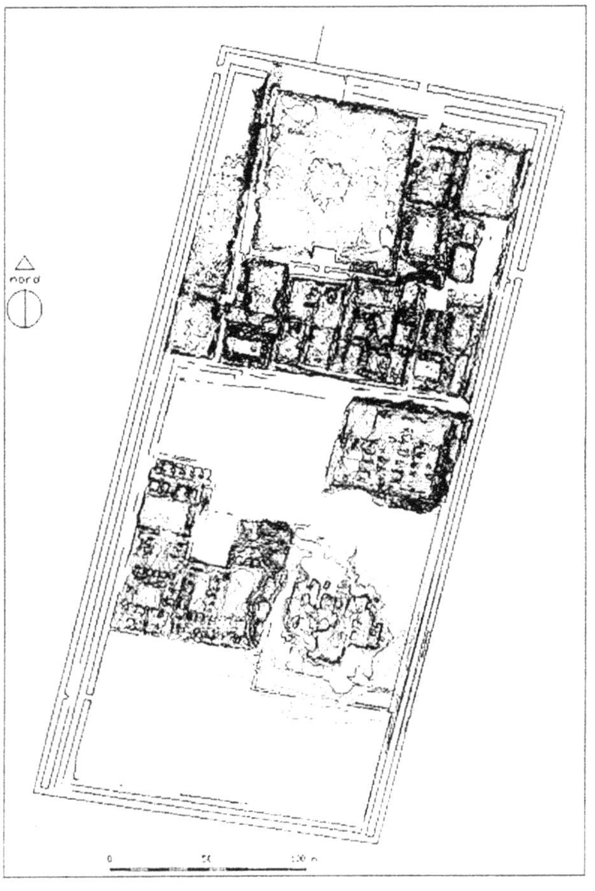

Figure 4 *Palacio Rivero. Survey of the structures by mean of different techniques. The Plataforma de Entierro is visible on the south-east side*

occupational fields. The use of 3D scanning technology, high-resolution and three dimensional computer tomography is becoming every day more important in the documentation of cultural heritage.

A direct advantage of laser scanning is that it can be employed in archaeological areas like Chan Chan and other sites where, due to the complexity of the structures and the size of the settlement, a 3D recording is highly helpful. The surveying can be carried out over a range of several hundred metres and it is therefore ideal for documenting poorly accessible sites or architectural structures which, as in the case of Chan Chan, have been particularly damaged by the degradation so to assume a shapeless appearance.

A part of *Palacio Rivero* regarding the wide area of warehouses and *Audiencias* was recorded using the above mentioned procedure in a very short time compared with the time needed for the post processing of data.

The Laser Scanner survey was performed with Callidus CP3200 (Trimble) laser sensor. It works on the time-of-flight principle (TOF) which is the duration of the time needed by the monochrome beam to hit the object and be reflected. By scanning at a speed of several thousand measurements per second, a dense mesh of the object is built up. Several million points can be registered during a procedure lasting only a few minutes. TOF laser scanners usually have an accuracy ranging from a few millimetres to about one centimetre.

The scanning of the area in a part of the first sector of *Palacio Rivero* was performed in seven working days, using 33 different view points. In some of these the scanner was mounted on a tripod while for the other it was just placed on the floor.

The scanning survey has been articulated in the following steps:

- Planning the distribution of viewpoints
- Laser data acquisition
- Scan merging and orientation in an absolute reference system
- Meshing to produce a triangulated 3D model
- Editing operations meshes.

Figure 5 *Palacio Rivero. 3D reconstruction (elaboraton by C.A. Bozzi)*

- Texturing the photograph images on the geometric model

In most cases it is not possible to have a complete 3D representation with data acquired from a single viewpoint therefore, in order to solve the problem deriving from the difficulty of watching the object, it is necessary to acquire data from different viewpoints. The images are required to have a sufficient overlapping region between two scans in order to allow their integration.

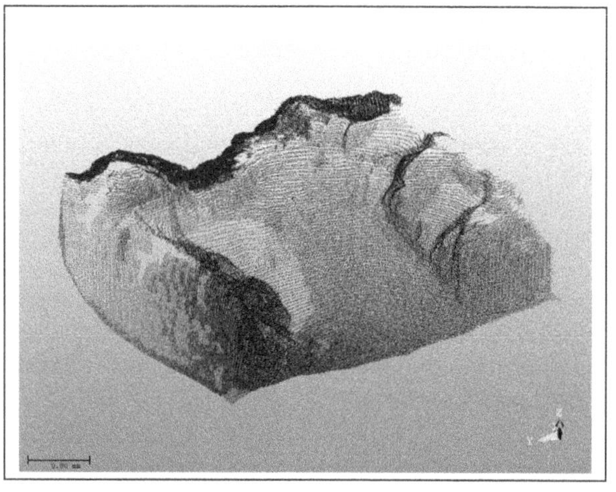

Figure 6 *A detail of the U-shaped structure of an* Audiencia *rappresented with points cloud*

To carry out this procedure in a successful way, the software employed (3D-Exctractor) must be capable of containing hundred of millions of scanned points. The individual scans are registered and transferred into a common coordinate system. The points clouds, which can be converted in DXF and in DWG format, create different models up to the generation of meshes (Fig. 6).

Following this procedure we have been able to create a 3D model by using the Real Works software (provided by Menci) which administrates the vectors data, by employing algorithms in order to display and edit point clouds. This software allows the creation of solid models from which it is possible to extract profiles, sections, elevations and all the other drawings needed for the planning of the restoration project of *Palacio Rivero* (Fig. 7).

The target of the project is to create virtual environments that offer the possibility to expand our own abilities in creating and experiencing new planning procedures and in creating a kind of virtual tourism extremely useful for what is regarding sites that have been closed to the public or that are not known as some may want. The possibility of creating models similar to reality allows us to create virtual restoration actions and to envisage specific applications to the field of cultural heritage.

Conclusions

The new documentation techniques for the analysis and the managing of the territorial data are particularly useful for the study of complex and large sites like the one of the Chimu capital of Chan Chan.

The high resolution Quickbird images allow us to carry on an in-depth analysis of the huge territory, to draw the map of the buildings of the ancient city, to perform a three-dimensional model of the terrain and to identify buried structures.

In absence of an upgraded cartography, the georeferenced image, operated by a GIS, forms a base on which we can place the constituent data of a territory (archaeological findings, roads, modern buildings, vegetation etc.) and on which we can simulate planning actions in order to realize the Archaeological Park.

With reference to a more detailed scale and considering the characteristics of earth architecture and the bad conservation state of the structures, it couldn't have been possible to realize a geometrical survey of the Rivero Palace without the help of modern and integrated technologies that have allowed us to realize the first three dimensional restitution of a Chan Chan palace (up to now the only available documentation was represented by aerial photogrammetric survey on paper support realized by the Harvard University).

The possibility to realize three-dimensional copies of buildings is particularly important in the case of earth reconstructions represent a unique and precious architecture which, as we have seen through the comparison of aerial images, is suffering a continuous and irreversible process of decay. The virtual documentation of buildings bound to die and save their memory for future generations.

Figure 7 *3D rappresentation of a small Audiencia*

Acknowledgements

We warmly thank the Euroimage S.p.A. for the supply of the Quickbird image and the Geosystem Group S.p.A. (Trimble) for its availability regarding the laser scanner Callidus.

The MIPE also tanks all the specialists and the students who have been and are working to the project: Cinzia Bacigalupo, Carlo Alberto Bozzi, Enzo Francaviglia, Alessandro Grandoni, Eva Malinverni, Mario Mascellani, Alessandro Mazzaferro, Gaetano Pappalardo

References

ANDREWS, A., 1974, The U-Shaped Structures at Chan Chan, Perú,. *Journal of Field Archaeology*, 1, nn. 3-4, 241-264.

COLOSI, F. and ORAZI, R. 2003. Integrated Technologies for the Study, Documentation and Exploitation of the Archaeological Area of Chan Chan, Peru. In the *Proceedings of the International Conference on the Study and the Conservation of Earthern Architecture - Terra 2003*, Yazd, Iran, 29 Nov - 2 Dic. 2003.

COLOSI, F. E ORAZI, R. 2004. New technologies for the conservation of the Archaeological Park of Chan Chan. In the *Proceedings LEHM 2004, Trade fair and Conference on building with earth. Lipsia*, Germania, 27-31 Ottobre 2004.

FANGI G., MALINVERNI E.S., SCHIAVONI A., 2005 Integrated surveying techniques for the Archaeological Park of Chan Chan in Perù, *XX International Simposium CIPA,* 26 sept – 1 oct, Torino

KOLATA, A., 1990, The Urban Concept Of Chan Chan, in Moseley M. - Cordy-Collins A. (a cura di) The Northern Dynasties: Kingship and Statecraft in Chimor, Dumbartón Oaks research Library and Collection, Washington , 107-145.

MOSELEY, M., Chan Chan: Andean alternative of the preindustrial city, *Science,* 187, 219 -225

MOSELEY, M., and KENT D., (a cura di), 1982, Chan Chan, Andean Desert City. University of New Mexico Press, Albuquerque.

ORAZI, R., 2001. Action model for research, documentation and conservation of the archaeological complex of Chan Chan in *Proceedings of the 3rd International Conference Science and Technology for the Safeguard of the Cultural Heritage in the Mediterranean Basin*, Alcalà de Hinares, Spagna, 9-14, Luglio 2001.

Use of remote sensing and GIS in the management and conservation of heritage properties at Agra

D. Dayalan

Archaeological Survey of India, Agra, INDIA

Remote sensing and GIS are complementary technologies. The remote sensing systems are used to observe the earth's surface through the sensors mounted on the satellite and aircraft and to collect information of an object, area or phenomenon. The geographic information systems (GIS) can be used to interpret remote sensing data more accurately by integrating them with data derived from various other sources and can analyze large volumes of geographically referenced data that would be overwhelming to process manually. The satellite pictures can also help to update efficiently certain data layers in the GIS.

The field of remote sensing took shape during the 1960s as an outgrowth of aerial photography. William Pecora, Director of the US Geological Survey (USGS), took the lead in pushing for a spacecraft that could cover the entire world with false colour images. The first of them is Landsat 1 by the National Aeronautics and Space Administration (NASA) which reached the orbit in 1972. The French government launched the SPOT program, the Satellite Pour I'Observation de la Terre in 1978 and in 1982 established the firm of SPOT image to market its photos. Lockheed was the first company which obtained the license from the USA authorities for building commercial remote sensing satellite system and launched its IKONOS satellite in 1994.The other system followed this are: Orbview of Orbital Sciences, Quickbird for the firm of DigitalGlobe, etc. The photos taken by Quickbird which flew in 2001 show the detail as small as two feet i.e. 0.62 metre in size. The entire surface of the earth, including the most remote and inaccessible places can be observed by satellites. The new development in the technology like, microwave remote sensing, Global Positioning System (GPS), etc are also being utilized for various studies. As the satellites images are obtained in digital form it can be processed directly in the computer. Moreover the satellite data are more reliable, repetitive, non-invasive, rapid and cost-effective information. All information is exactly localized and gathered in one tool. The direct extraction of topographic and thematic maps for terrain use is also possible with this technology. The geographic information system (GIS) is a combination of hardware, software and procedures for storing, managing, processing and reproducing spatial data with data coupled to them. The data can be obtained from ground surveys, GPS measurements and from remote sensing data.

Notwithstanding the remote sensing and GIS technologies are used in the fields such as town and country planning, watershed management, agriculture, forestry, conservation, mineral resource development and prevention and mitigation of natural disasters, the necessary for the utilization of remote sensing, GIS and related geo-information technologies for the conservation and maintenance of the cultural and natural properties has been felt only in the recent years. In fact the GIS and remote Sensing are the excellent tools to support the monitoring process of cultural and natural heritage sites

The activities wherein the remote sensing technology can be utilized for the management and conservation of cultural and natural properties are:

1. Creation of data-base and digital maps of monuments and their environments, ancient settlements, etc.
2. Evaluation and management of the monuments and sites
3. Study of chronological changes in the nearby area of the monument/site.
4. Measuring and monitoring of encroachment or any other changes in the ecosystem in the surrounding of the properties in order to take early corrective actions.
5. Preparing base maps for drawing master plan for integrated development in and around the heritage properties
6. Survey and monitoring of natural properties
7. Development of Natural Disaster Management System and monitoring of river pollution.
8. Survey and monitoring of natural properties
9. Development of Natural Disaster Management System and monitoring of river pollution

Study Area

Agra (27^0 11' N; 78^0 1'E) in the Uttar Pradesh State, India is the prime destination of tourism in India, located approximately 200 Kilometres south of New Delhi. It is not only an important urban centre in the medieval period but also blessed with three world heritage monuments including the world famous Taj Mahal.

The miracle in marble known as Taj Mahal (1631-1648) constructed by the Mughal Emperior Shah Jahan (1628-1658) for his favorite wife Anjuman Banu Begam popularly known as Mumtaz Mahal is the culmination of the sepulchral architecture of Shah Jahan and indeed of

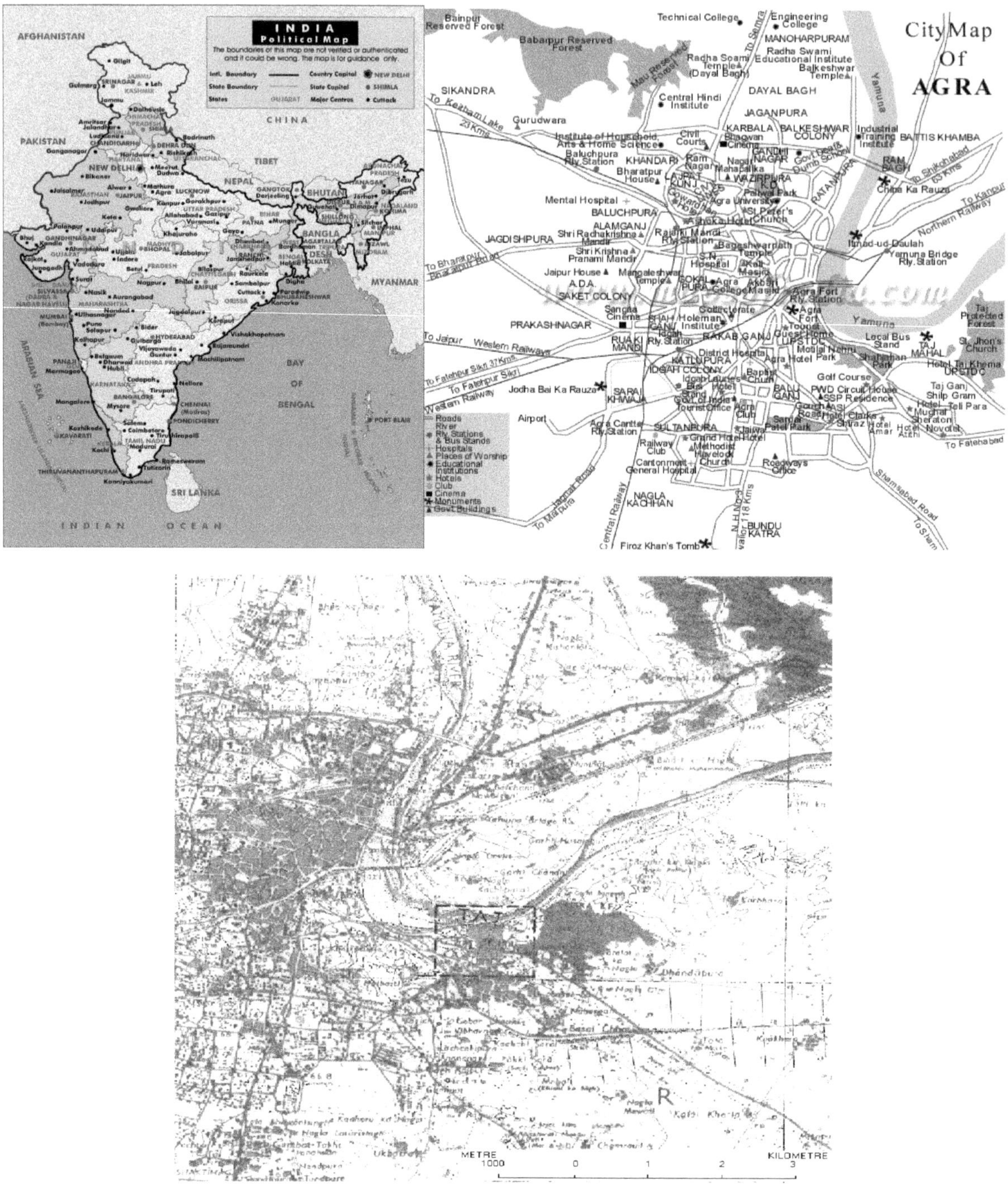

Figure 1 *Study area*

the Mughals. Notwithstanding the contemporary sources refer to the mausoleum as "Illumined Tomb" (*rauza-i-munavvara*), it derived its name from the title of Mumtaz Mahal, distorted by popular etymology to Taj Mahal. The Taj Mahal was inscribed on the World Heritage List in 1983. Its inscription as a cultural property of universal significance was justified in the following words: *"The Taj Mahal is the greatest architectural achievement of the whole range of Indo-Islamic architecture. Apart from its tremendous romantic appeal, it is a masterpiece of architectural style, in conception, treatment and execution, all alike".*

Spread to an area of 22.44 hectare, Taj Mahal complex is planned in the form of a rectangle, aligned north-south and measuring 579.12 metre by 304.8 metre. The central area is divided in to square garden leaving an oblong space at either end of the rectangular. In the south this is taken up by the gateway whereas on the north consists of a raised platform containing the main marble mausoleum and the auxiliary buildings like Mosque and Mehman Khana, all overlooking the river below. Taj Mahal has to be visualized along with the river bank, river Yamuna and adjoining monuments, gardens and landscape as they are the integral part of this monument.

The remote sensing and GIS techniques are applied for creating data-base, maps and various other purposes.

These techniques are also used for the management of the buffer zone, preparation of the integrated scheme in and around the monument, conservation activities, etc. The satellites images are also useful to visualize the overall urban settlement pattern during the medieval period.

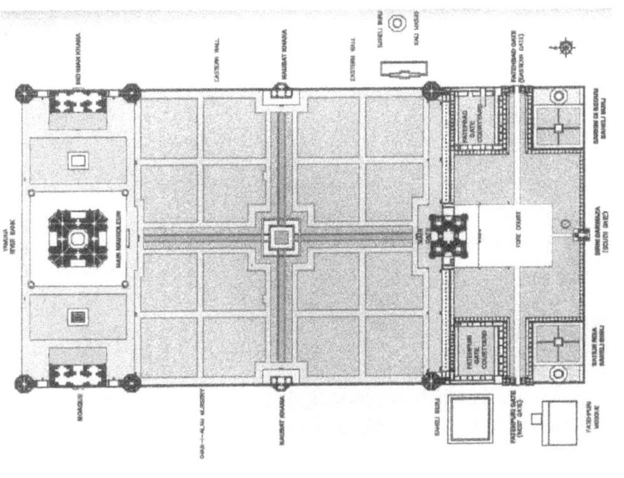

Figure 2 *Plan of Taj Mahal Complex*

Fatehpur Sikri, another world heritage monument near Agra is the planned city of the Mughal Emperior Akbar (1556-1605) constructed on the bank of a large natural lake. The ancient city wall, which in its 9 kms run is pierced by nine gates, holds a large area with undisturbed archaeological remains in the form of palaces, mosques, *sarais, hammams*, water works and many other secular, religious and defense edifices. The department is the process of applying the remote sensing and GIS techniques for mapping the various structures, monitoring the monumental and the periphery areas and for the conservation activities.

Data used

1. High resolution satellite images such as the Taj Mahal and its surrounding areas including Agra Fort, another World Heritage Monument and ancient gardens and edifices and Taj Mahal overlooking the southwest bank of the River Yamuna collected by Space Imaging's IKONOS satellite.
2. The maps prepared by means of Topographical survey of the whole area by using the Total station. All the final survey drawing have been realized in AutoCAD or compatible industry standard digital vector files, *i.e* ESRI shape.
3. One inch Topography map by Survey of India
4. Architectural Maps prepared by the Archaeological Survey of India

System Requirements for GIS

Hardware

- RAM-256 MB (512 MB recommended)
- Preferably P (4) CPU 3 GHz
- Hard drive free space required 4 GB

Figure 3 *Satellite view of Agra City and its surounding areas*

Software

- Arc GIS 9.0 or higher version.
- i_view32.exe (freeware exe already present in

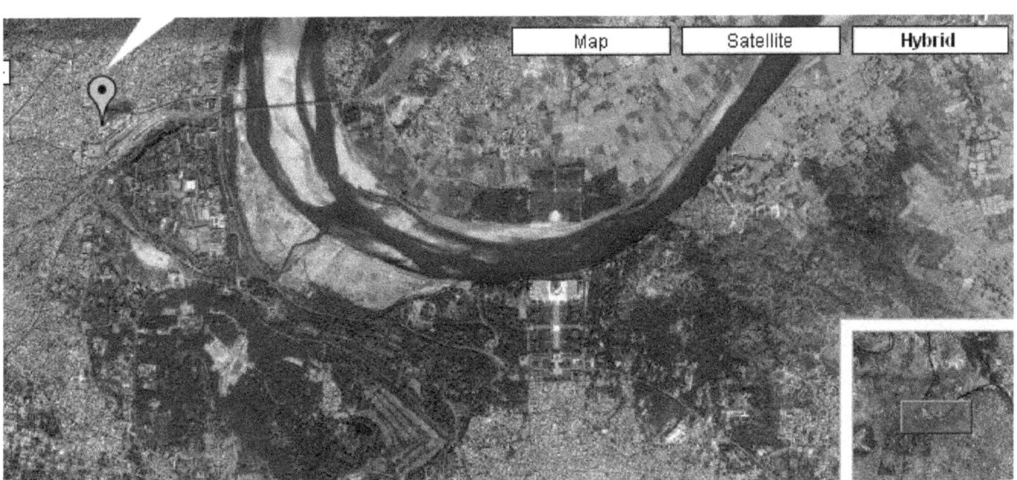

Figure 4 *Taj Mahal and its surrounding areas including Agra Fort and ancient gardens and edifices*

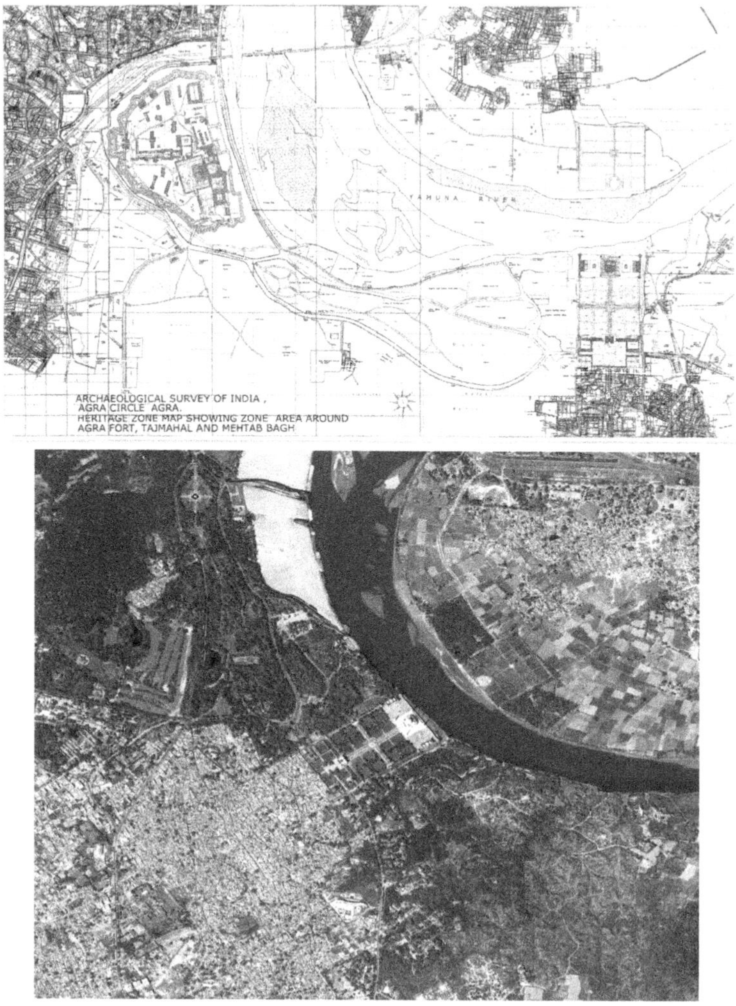

Figure 5 *One-metre resolution Satellite image of Taj Mahal overlooking the southwest bank of the River Yamuna*

Images folder)

An Integrated digital database for Taj Mahal on designated GIS platform is created after site survey for proper monitoring of this World Heritage Monument.The dataset is having three levels of details.

Location:

Overall area of interest has been divided into the following manner.

1. Fore Court
2. Char Bagh (NE, NW, SE, SW and Central)
3. Naubat Khana
4. Museum
5. Khan I Alam Nursery
6. Mehman Khana
7. Mausoleum
8. Mosque
9. Backyard

Land use Classification:

The entire site has been classified in the following 22 land use classes:

1. Ancient Water System
2. Backyard
3. Drainage System
4. Fence
5. Flower Beds
6. Fountain System
7. Lawns
8. Platforms
9. Steps
10. Walls
11. Cement Floor
12. Bench Mark
13. Built Structure
14. Facilities
15. Furniture and Fixtures
16. Foot Path
17. Garden
18. Pillar Post
19. Shrubs
20. Drainage
21. Water Features
22. Stone Lining
23. Inlay Works

Figure 6 *Aerial View of Taj Mahal Complex*

Figure 8 *Aerial view of Fatehpur Sikhri monuments*

Figure 7 *Aerial view of Taj Mahal*

The above classification is represented at two different levels as follows:

1 Land use classification at Macro level:

At this level the layers represent details at a Macro level. For instance, in the Footpath layer the complete footpath from edge to edge is represented as a single polygon without differentiating between the internal sub-divisions of different design patterns within the footpath (Fig. 9).

2. Land use classification at Micro Level;
This level of information provides further detailed differentiation of the Macro representation by clearly identifying various internal patterns within the macro classification (Fig. 10).

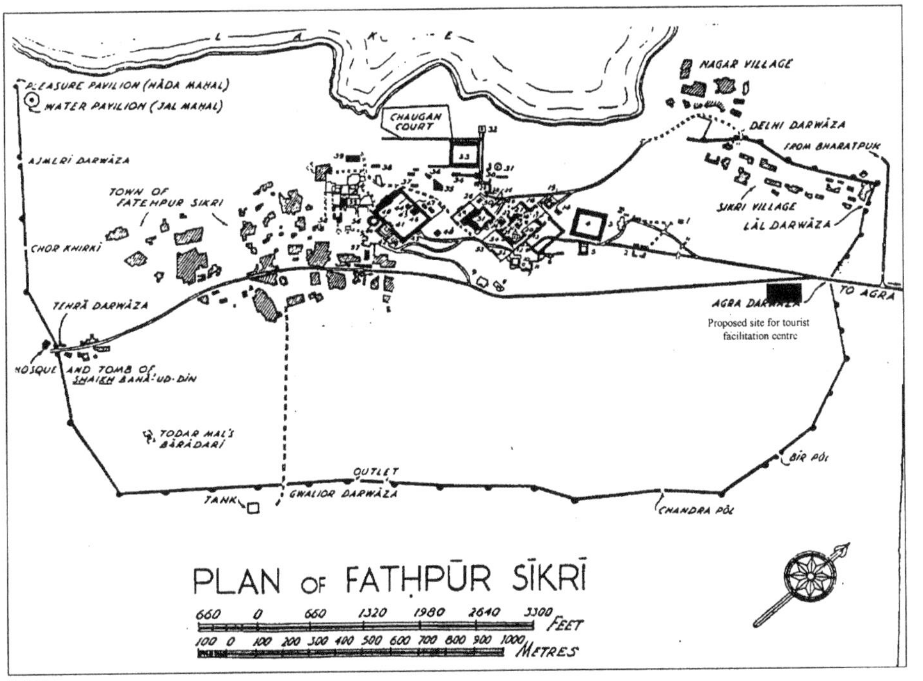

Figure 8

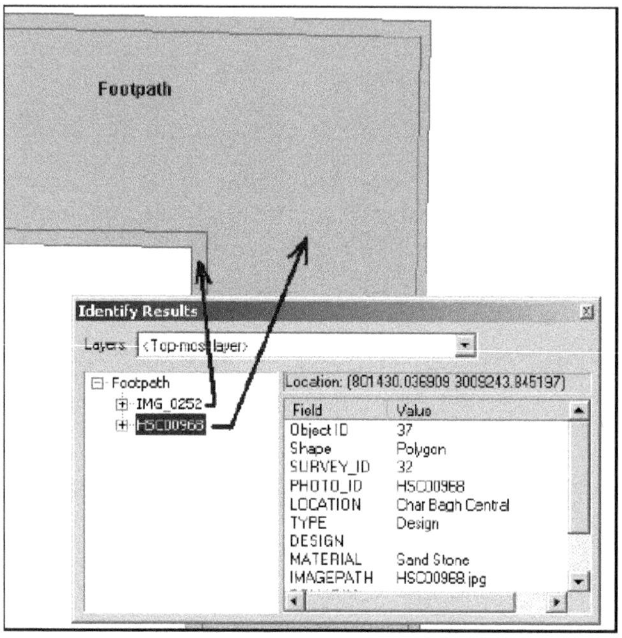

Figure 9

Schema

Following is the Schema of the GIS dataset for Taj Mahal. It shows the layer classifications along with the database field structures depicted with the corresponding colour of the boxes (Fig. 11).

Operational Guidelines of the GIS Programme

1. File structure

\Tajmahal\Images*.jpg (Contains photographs in jpg format)

\Tajmahal\Images\ I_view32.exe (Irfan viewer application)

\Tajmahal\TajMahal_24022006.mdb (ArcGIS Personal Geo-database)

\Tajmahal\TajMahal_24022006.mxd (ArcGIS project file)

User Manual Ver 1.0

2. Open **<drivename>:\Tajmahal** folder and right click on TajMahal_24022006.mdb fil

and then click Properties. Click to clear the Read only check box (if present), and then click Ok. 3. Following are the steps for setting I_view32.exe as a default image viewer.

1) Start ***Windows Explorer*** and open **<drivename>:\Tajmahal\Images** folder and right click on any jpg file (in right hand side window in the explorer) and click on **'Open with'** and then click on **'Choose Program'** and then click on **'Other'** button and now select the application **I_view32.exe** present in the **<drivename>:\Tajmahal\Images** folder as shown in the picture below.

4. Open **<drivename>:\Tajmahal** folder and double click on the Tajmahal_2402006.mxd file to open your map project (Or open *ArcMap* and direct it to open an existing .mxd file which is located in **<drivename>:\Tajmahal** folder in the local machine) User Manual Ver 1.0

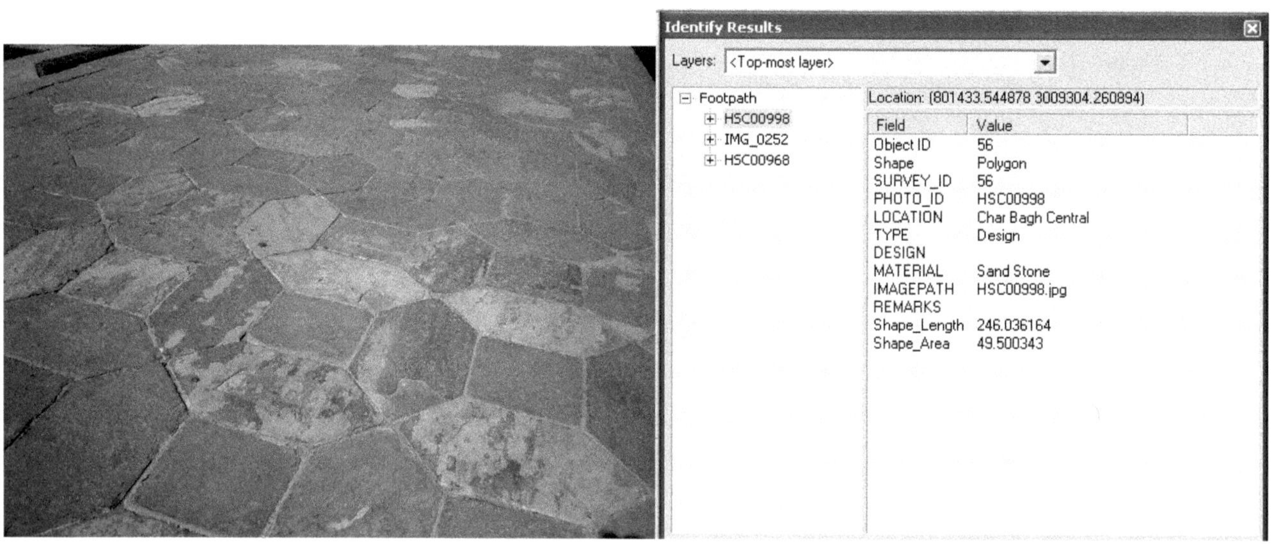

Figure 10

5. On File menu, click Map Properties, to set the values for 'Hyperlink base' to **drivename>:\Tajmahal\Images** for e.g. if the data has been copied to d: drive then write 'D:\Tajmahal\Images' (It defines the location of the folder in which the Taj Mahal snapshots are lying) and press **"OK"** button as shown in the picture below

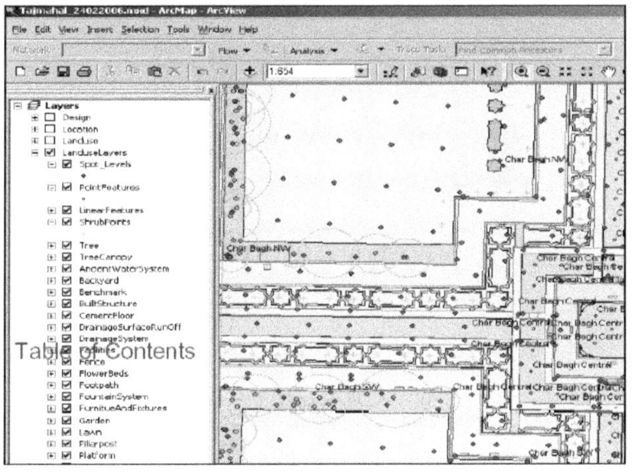

6. Select the layers in the Table of Contents and check the layers on for display in the Arc Map.
7. To view the attributes information of the features, first select "i" button and then click on the features as shown in the picture below.

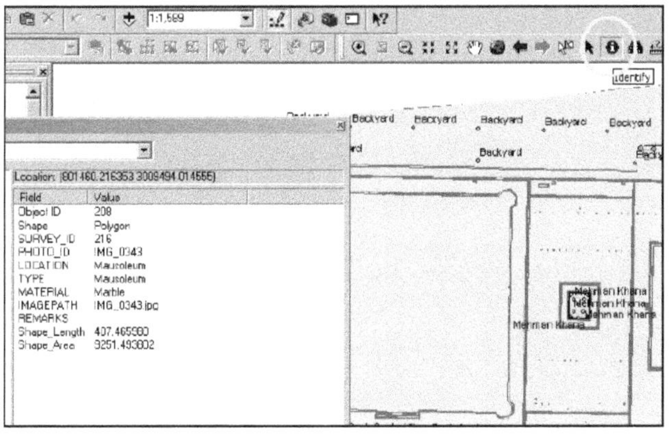

8. To view the image of the feature, first select lightning button and then click on the features as shown in the picture below.

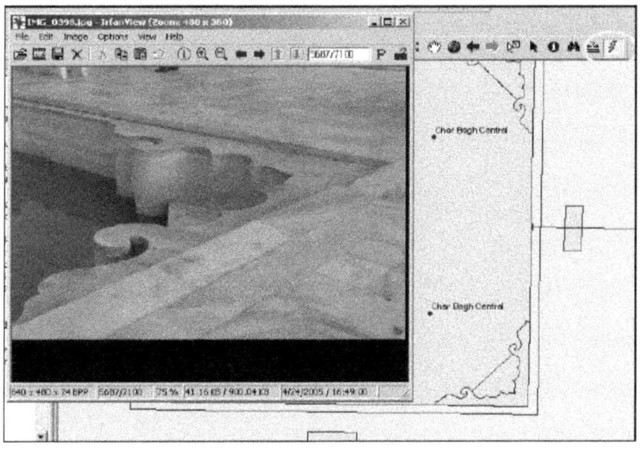

9. View showing different features like Design, Location, Land use and Land Use Layer

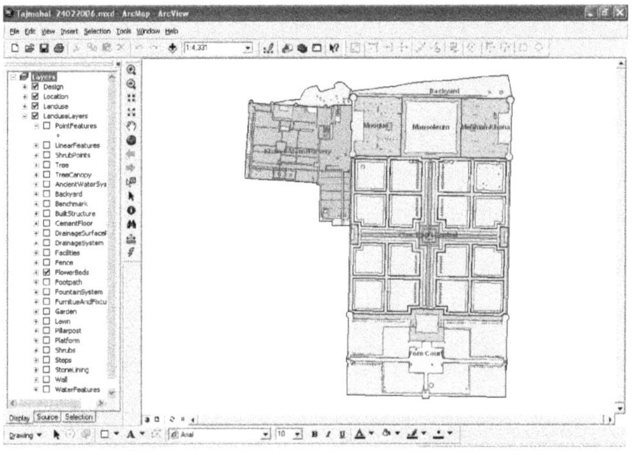

10. View showing Location and different Land Use Layer such as Shrub features, Tree, Ancient Water system, Build structure, Drainage surface, Drainage system, Fence, Flower beds, Foot path, Fountain system, Garden and Lawn.

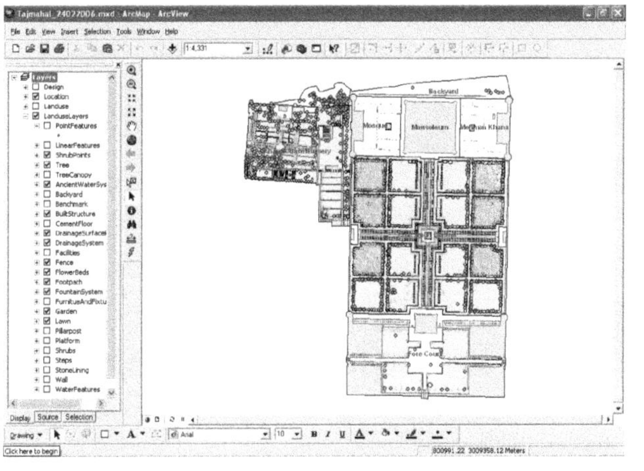

11. Flower beds layer along with attribute and image

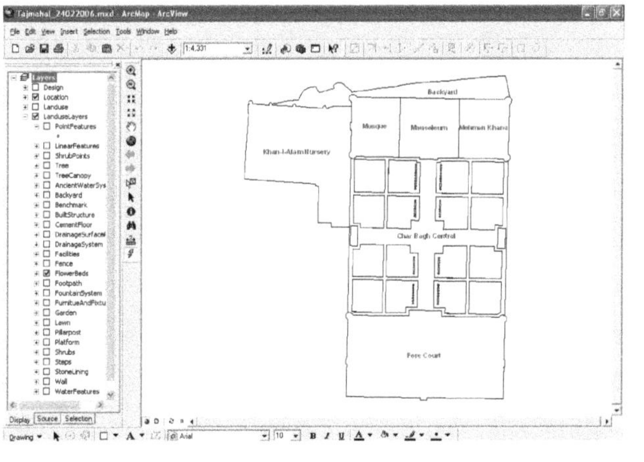

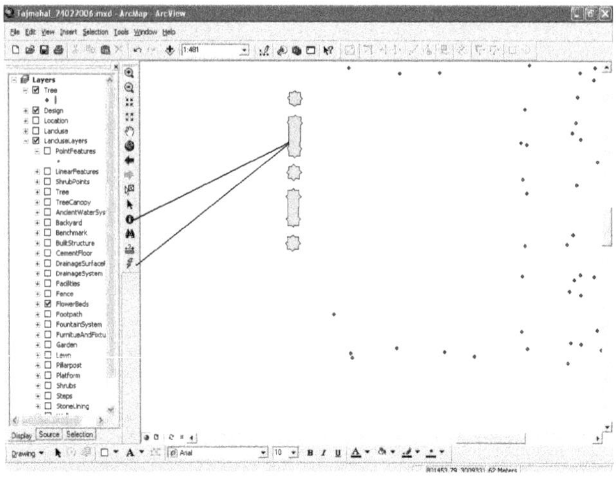

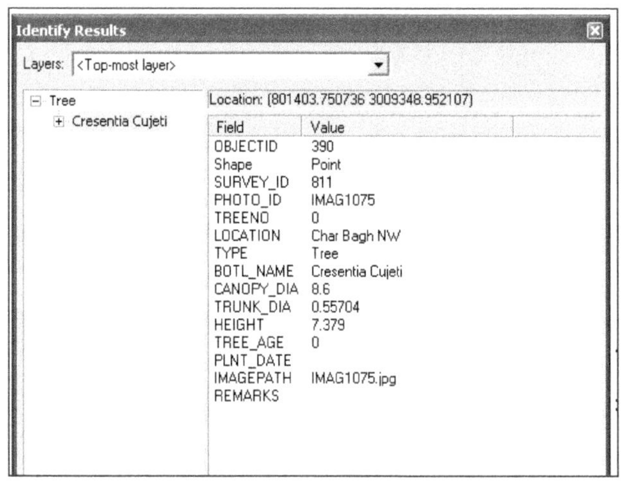

Identify Results

Layers: <Top-most layer>

Tree

Cresentia Cujeti

Location: (801403.750736 3009348.952107)

Field	Value
OBJECTID	390
Shape	Point
SURVEY_ID	811
PHOTO_ID	IMAG1075
TREENO	0
LOCATION	Char Bagh NW
TYPE	Tree
BOTL_NAME	Cresentia Cujeti
CANOPY_DIA	8.6
TRUNK_DIA	0.55704
HEIGHT	7.379
TREE_AGE	0
PLNT_DATE	
IMAGEPATH	IMAG1075.jpg
REMARKS	

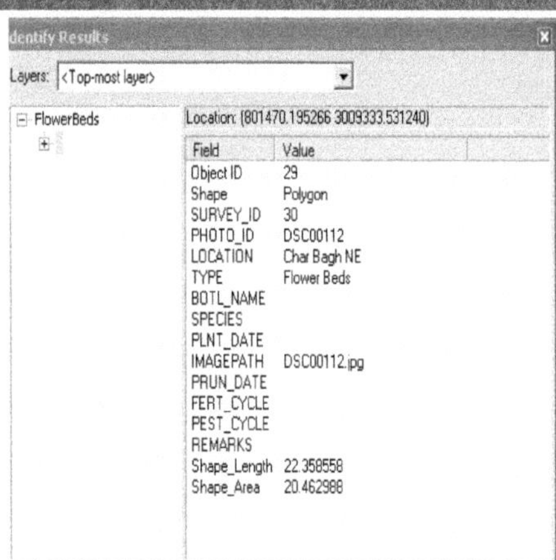

Identify Results

Layers: <Top-most layer>

FlowerBeds

Location: (801470.195266 3009333.531240)

Field	Value
Object ID	29
Shape	Polygon
SURVEY_ID	30
PHOTO_ID	DSC00112
LOCATION	Char Bagh NE
TYPE	Flower Beds
BOTL_NAME	
SPECIES	
PLNT_DATE	
IMAGEPATH	DSC00112.jpg
PRUN_DATE	
FERT_CYCLE	
PEST_CYCLE	
REMARKS	
Shape_Length	22.358558
Shape_Area	20.462988

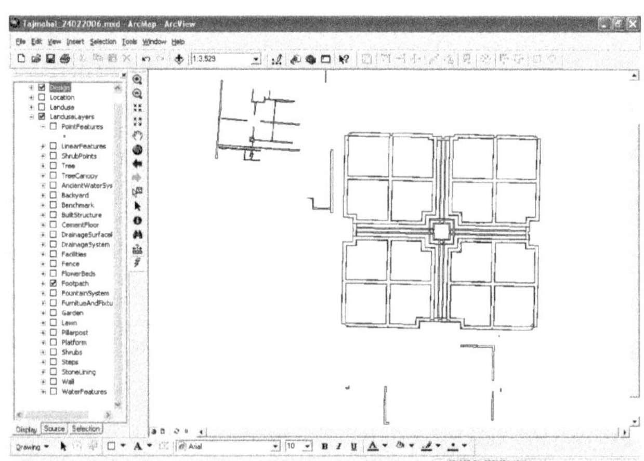

12. Tree layer along with attribute and image

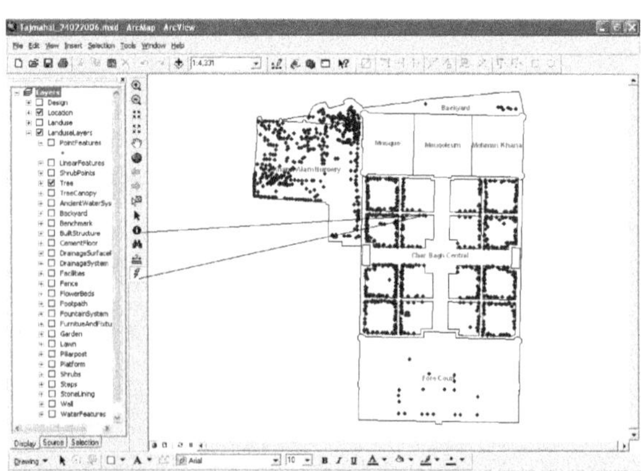

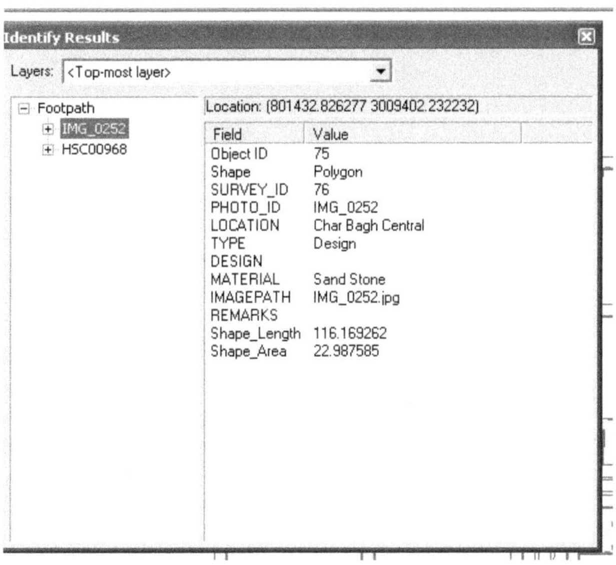

The remote sensing and the GIS are the effective tools for the management of the heritage properties particularly for mapping, monitoring and analyzing. GIS is the integrated data-base management system that is increasing its use in many form of management. It gives the possibility of storing, analyzing and displaying large amount of data at the same time. Moreover this rapid growing technology is more user-friendly due to its advanced software and also provides more geographical information.

Benefit of Remote sensing and GIS

The comparison of the Satellite images of recent years and the Aerial photographs taken few decades before provide excellent idea on the periodical land-cover changes in the surrounding area of the monument. At present Taj Mahal is surrounded by shops, green belt, emporium, offices, hotels etc. on the east; restaurant, shopping complex, parks, emporium etc. on the west; thickly populated Taj Ganj area, shopping complex, etc. on the south and Yamuna river, Mehtab Bagh, green belt area on the north. The aerial photograph taken few decades before shows that the greenery on the north and east side of Taj Mahal is not so dense in comparison to the situation today. But the building activities on the south side is increased alarmingly in the recent years than what was the position earlier as depicted in the old aerial photography. Government of India in consultation with UNESCO proposed to formulate the Taj Mahal, the riverbank, river Yamuna and adjoining monuments and landscape as heritage zone to safeguard the landscape and the environment from any unregulated and inappropriate development. Archaeological Survey of India has done detail survey of the modern buildings within prohibited and regulated areas of Taj Mahal and prepared a data base by using the GIS tool. Each and every minute detail of the modern structures and other features in the surrounding area is stored in the GIS for analyzing and monitoring the activities in the area. The construction, mining, vehicle movement and pollution in this area are regulated by various govt. rules, gazette notifications and Supreme court directions from time to time. Remote sensing and GIS are very much useful in implementing the government and court orders.

2. The inlay of floral arabesques fashioned from precious/semi-precious stones in white marble and the inlaid black marble inscriptions that frame the central vaulted portal or *iwan,* are some of the prime attraction in Taj Mahal. The inlay work is seen on the cenotaphs, the borders of the marble screen, the dado panels, the spandrels of the arches, bottom of the dome and other places. In order to achieve the desired tonal variations the inlay workers used as many as forty-eight pieces of precious and semi-precious stones to compose a single flower. There are nearly 45 precious/semi-precious stones are used for inlay work. They are: 1 .*Ajuba/ Ujuba* (It is mottled with green and yellow concretions in a brown ground-mass); 2. Alabaster; 3 Agate; 4. Black marble; 5.Bloodstone; 6.Carnelian of Arabia Felix; 7.Chrysolite; 8.Conch-shell; 9.Coral; 10.Cornelian; 11.Crystal; 12.Diamond; 13. Emerald; 14.Garnet; 15.*Ghar;* 16.*Ghori;* 17.*Godar;* 18.Golden stone; 19.Green Marble; 20.*Gulabi;* 21.*Jade;* 22.Jasper; 23.*Khara;* 24.Lapis Lazuli; 25.Load stone; 26.Magnet Stone; 27.Melachite; 28.Mother-of-pearl; 29.*Nakhud;* 30.Onyx; 31.Onyx Marble; 32.*Pai-Zahar;*33.*Pankhni;* 34.Pearls; 35.Pink Marble; 36.*Pitunia;* 37.Redstone; 38.*Reg;* 39.Ruby; 40.Sapphire; 41.Slate (Calcareous shale); 42.*Sumaq* ; 43.Turquoise; 44.Topaz; 45.Yellow Marble. The data base comprising the details of location, number of pieces, name of precious/semi-precious stones, size and colour of stone and other details has been created with help of GIS programme for the purpose of conservation, documentation and checking of the inlay works.

3. Few years before there were big news that the minarets in the Taj Mahal is tilting and a portion of the monument is shrinking. The first scientific study regarding the foundation stability and verticality of the minarets was carried by the Survey of India in 1941-42 and fixed 104 bench marks in Taj Mahal complex for periodical cross checking. It was observed in the first scientific survey that all the minarets inclining outwardly perhaps they were designed originally like that and the detail of minaret wise inclination is as follows: South-west-21.6 cm; South-East-11.4 cm; North-East- 4.8 cm and North-West- 3.5 cm It was decided to check the bench marks ones in 3 or 4 years to ascertain the verticality of minarets and stability of structure. All the 104 bench marks fixed in the Taj Mahal complex are surveyed and documented. On this basis, a separate data layer has been created in the GIS programme which provides all details like location, size, year wise reading of the levels, present status and other details of the bench marks in nutshell. Now with the help of GIS it is scientifically proved with relevant data that all the minarets are stable as there is no change in the inclination of the minarets.

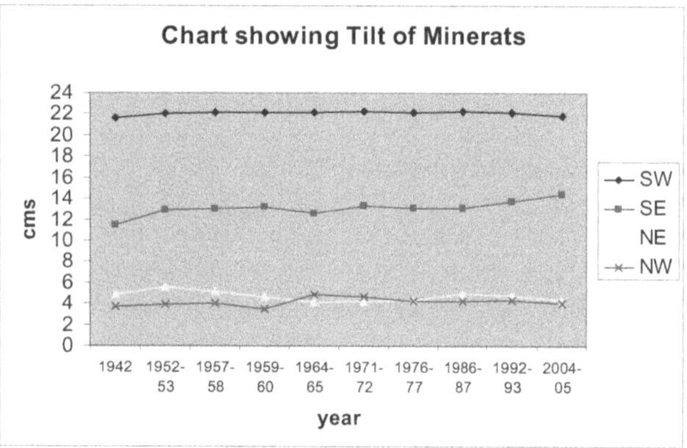

4. The details of the flora within the Taj Mahal complex including the Khan-i- Alam, the ancient water channel system associated with Taj Mahal has been documented and following information are stored in the GIS for the purpose of monitoring and analyzing. The details of the trees and shrubs available in the GIS are: 1.Object, 2.Shape, 3.Survey ID, 4.Photo ID, 5.Tree No., 6.Location, 7.Type, 8.Botanical name, 9. Canopy dia., 10.Trunk dia., 11.Height, 12.Tree age, 13.Planted date, 14.Image path, 15.Remark

Tourist inflow at Taj Mahal (1997-2005)

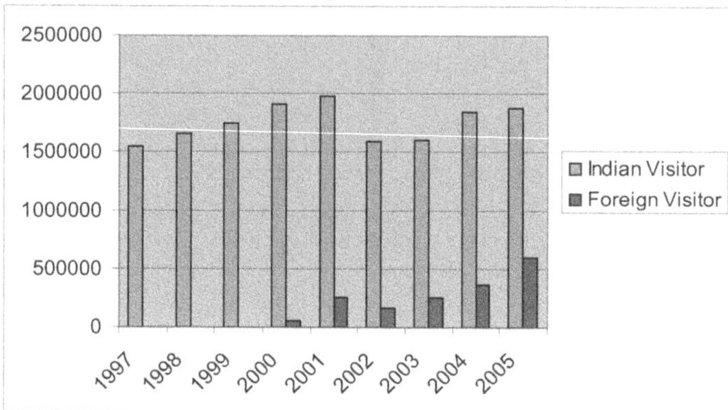

Note: The data does not include children below 15 years, VVIPs and VIP,Licensed Photographers and Guides and personnel on duty.

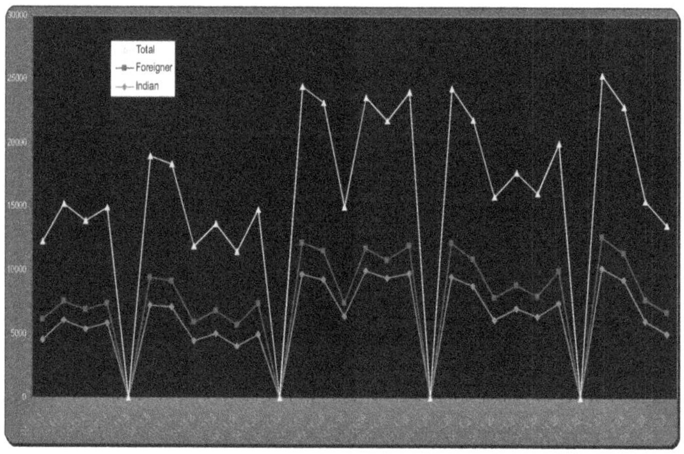

Space and Time: Virtual Reality and the Art of Experience

Judith van der Elst, Jack Ox

The University of New Mexico

1 Introduction

In anthropology and archaeology we often study cultures or communities that have/had a different notion of space and time than what is accepted in a scientific – EuroAmerican- framework. Yet, we impose our Western frame of reference where space is Cartesian and time is linear, to understand all human behavior (Laituri 2002, Wylie 1995). This paper addresses emerging technologies and their role in cultural studies in two major ways. First by discussing the possibility of integrating other knowledge systems, and second by advocating collaboration between artists and scientists and how this can result in better visualization techniques and visual metaphors for tangible as well as intangible components in digital heritage reconstruction. Finally, it is argued that in order to create visual metaphors for virtual environments we need to include both high and low-tech methods and actively seek participation with (research) communities that work from other epistemological viewpoints. A cutting edge art project –*Gridjam* - by Jack Ox, David Britton, and collaborators provides an innovative model, or stepping stone for new avenues in anthropological/ archaeological research. *Gridjam* is a real-time, geographically distributed, networked multimedia event. It is an experimental project that brings together visual artists, composers, musicians, and (computer) scientists, while using the new high-speed international LambdaRail network (http://www.jackox.net, Leigh et.al 2006). In the course of the paper, the approach will be described and then used to explain how such methods can help enrich our current digital representations of cultural experience.

Several researchers from diverse backgrounds have recently addressed the issue of forgotten, ignored, warped, or eliminated histories (Schmidt and Patterson 1995, Smith 2005). The dominant research paradigm, one that employs the scientific method and derived from Western thought, has been carefully constructed and maintained. Smith (2005:48) writes, "When confronted by the alternative conceptions of other societies, Western reality became reified as representing something 'better' – the last stage on the way to civilization -, reflecting 'higher orders' of thinking, and being less prone to the dogma, witchcraft and immediacy of people and societies, which were so 'primitive'." Ideas stemming from Western thought seek out dualisms such as the Mind vs. Body and Space vs. Time. Each is individually assessed and as such mind is divided from body and space is often viewed as being static or divorced from time. According to Smith what makes ideas 'real' is the system of knowledge, formations of culture and the power relations in which these ideas are located not universalized conceptions. This leads current research strategies typically to often give only part of the picture and this partial picture results in part from the historical circumstances in which the dominant research paradigm was developed.

The practice of archaeology has for a long time been focused on material remains, i.e. material objects and this emphasis often translates into a positivist practice even though our theories may, and often do, account for other kinds of ideas (Hodder 1992). Expanding the toolkit for cultural exploration to include visualization and virtual technologies has been suggested and implemented by a number of researchers (Baltsavias et.al. ed.2006, Forte 1997). What is still problematic is that our visual language and technology are driven by western ideas, representing a specific kind of reality, despite the fact that visualization and virtual technologies "can provide the ultimate 'what if' scenarios by allowing the user to create images and then manipulate them in various ways" (Williamson 1999/2001). Virtual environments are representations of physical reality; 'virtual reality' is then reality in the sense that it provides an abstraction of the physical world as perceived by an artist or a (group of) scientist(s). In essence, virtual reality as it is often presented suffers from similar biases as traditional science, in that it fails to use its full potential for exploration of *different realities.* However, there is more than one (hi)story to tell and whether we ever 'get it right' is of less importance than opening our minds to different ways of interpretation of human experience or creating alternative histories.

2 The practice of archaeology

One of the tenets of North American archaeology is that archaeology is anthropology, reflecting a sensibility and understanding of the importance of thinking about a wide range of societal and cultural expressions. Unfortunately, this understanding is often not integrated in our methodological practice. In reality these cultural expressions may be acknowledged, but they seldom guide innovative methodologies that arise out of understanding and perceptions of local communities and investigators (Basso 1996, Schmidt and Patterson 1995). The lack of technological expertise is one of the reasons western researchers ignore ideas that are proposed by local research communities. The fact that we are able to integrate low-tech (such as hand-drawn images) data into virtual environments should help overcome this hurdle and incorporate local voices into research.

In recent years, landscape oriented applications have become more and more common in archaeology and the introduction of Geographic Information Systems (GIS) has transformed the way archaeologists analyze past human behavior. Even though the integration of geospatial technologies into archaeological practice has changed the way we can do research, such technological advancement comes with shortcomings that continue to remind us of the criticisms associated with positivist practice. Geographic Information Systems (GIS) and other geospatial approaches tend to be static and environmental deterministic as it is still difficult to incorporate cultural/social variables, which tend to be qualitative, into predominantly quantitative systems (Gillings and Goodrick, Wheatley and Gillings 2000). Processes that take place often do not have a physical manifestation, and as such cultural experience, learning, and perception are all intangible elements of the social landscape. Archaeologists need to understand the intangible as well as tangible cultural landscape to investigate past human behavior; however, they have limited means to do so. The advent of geospatial technologies has provided a visual, and at times more dynamic framework to carry out cultural studies; however, we have yet to come up with visual metaphors for analyzing and understanding human experience. Current directions in digital heritage and archaeological research seek the integration and development of visualization and virtual environments to explore both the intangible and tangible within cultural landscapes. Our task then becomes to "find new visual metaphors for representing information" that has no obvious physical representation (Gershon et.al. as quoted in Forte 2003).

Gillings and Goodrick advocate for the development of a theory of GIS and to expand our analytical toolbox to include virtual technologies to better understand human experience in the past. Exploring the use and experience of space and time by different groups has been the subject of investigation within the discipline of geography for the last decade, bringing critical theory into the practice of geography (Harvey et.al. 2005, Kwan and Lee 2004). Critical GIS, as it is termed, crosses many disciplinary boundaries and it is concerned with the power embedded in the use and production of technology. It is focused on integrating different perspectives and transient data; a common practice is participatory GIS in which communities are actively involved in planning and research process (Craig et.al. 2002, Laituri 2002). This practice has lead to rethinking of how to combine different epistemologies within archaeological practice. We suggest to expand the idea of Gershon et.al (in Forte 2003) to incorporate elements from critical theory that state that the physical world is as much a construction of the mind as the world constructed from Virtual Reality cues. It is also important to account for the idea that technological developments are not value free, but are developed within a specific knowledge system. Many technological developments are seldom initiated with archaeology or any other social sciences in mind. This in itself brings up problematic issues; however, even more troubling is that researchers often adopt these technologies uncritically. The fact that our discipline is unique in that we seek to understand past human behavior and experience derived from limited –material- data requires a different visual language than those created for other research fields. Reconstructions (or interpretations) that we create within an archaeological framework or often far removed from our perceived reality. This means that if the data that are used to create a landscape, reconstruction of a building, movement path, are limited, interpretation or interpolation/ extrapolation is involved. That is, our minds fill in the gaps based on our (personal) prior knowledge or experience. Visual language or metaphors can help us make clear the level of interpretation that was involved in creation. For instance, hand-drawn images can provide a level of depth that exceeds that of the slick atmosphere of 3D software programs (Conroy et.al.2003).

3 Art and Science collaboration

The purported usefulness of collaborations between artists and scientists has prompted many questions and discussions. However, the authors agree in large part with other researchers who claim that science can learn from art and vice versa (Arends and Thackara 2003, Casti and Karlqvist ed. 2003). Both artists and scientists are driven by curiosity, an urge to find out; sharing their methods and practices can lead to new insights into human experience. For instance, the benefit of drawing, a method belonging to the arts, has been successfully used to better understand and solve complex problems in science (Hjort 2003). Similarly, collaboration with artists has led to better design solutions in participatory urban planning projects (Al-Kodmany 2002). In this study the artist translated the ideas verbally expressed by community members into visual sketches. The visual language proved to be universally understood, which greatly improved the design process and the final outcome. Furthermore, as is stated on the UNESCO website on digital art, "the history of art further suggests that communication is not simply just a sending of sensual cues as many engineers would like to imagine. An artist is practiced in how subtleties and nuances within a given medium may achieve a richer sense of presence than one created by someone who lacks this specific type of knowledge (http://digitalarts.lcc.gatech.edu/unesco/vr/home.html)." These ideas suggest that as archaeologists, venture into visualization and virtual technologies to help us better understand past human behavior it is essential that other types of knowledge be integrated in our practice.

For some 30 years, Jack Ox's artistic research has been focused on the "translation" of musical language, composition, orchestration, notation and sound-structure

into original visual representations—using models spanning the Romantic symphonies to Schwitters post-Dada Sound-Poems, to pure Digitally generated structures; these "paintings" and drawings always recast her sound-sources in extremely rigorous but nonetheless fanciful visual creations. *GridJam* will make use of artist Jack Ox and programmer David Britton's Virtual Color Organ™, visualizing Alvin Curran's electronic music, which will be performed by musicians located in distant locales but connected via next generation networking technologies (Ox and Britton 2000). The Virtual Color Organ™ (VCO) is a 3D immersive environment in which music is visually realized in colored and image-textured shapes as it is heard. The visualization remains as a 3D graphical sculpture after the performance. The colors, images, shapes and even the motions and placement of the visualized musical shapes are governed by artist-defined metaphoric relationships, created by hand as aesthetic and symbolic qualities rather than algorithmically. The VCO visually illustrates the information contained in the music's score, the composer's instructions to the musicians, and the musicians contributions to the score as they improvise in reaction to each other's performances and to the immersive visual experience. Illustrative of metaphoric synesthesia and intermedia (the inclusion of structural elements from two or more different media into one medium), the VCO displays the emergent properties within the meaning of music, both as information and as art.
The aesthetic vs. algorithmic quality is exemplified in the extremely high-resolution detailed hand-drawn imagery of desert landscape formations that comprise the current "visual organ stop" to be used for *GridJam*. The images are designed to provide suggestive metaphors for the representation of timbre-based contemporary music. In this case, composer Alvin Curran's library of nearly two hundred found sounds provides the starting point for the electronic composition. For *GridJam*, each of these sound files is provided with a hand-made 3D object whose shape is based on the amplitude and pitch waveforms of the sound file's audio analysis.

The original building blocks of GridJam come from a list of almost 200 collected sounds belonging to Curran. The sound files include John Cage reciting short phrases, Maria Callas singing a high note, animal sounds, objects like coins being tossed, and various musical sounds etc. Ox put them through a Max/msp program creating graphs of melody and dynamics. She sorted them into 8 groups based on these simple visualizations and made 3D models reflecting melody from the front and dynamics on the top of the object. There are eight groups because there are eight landscape texture maps. Each object in a group has as the base of the texture map the same hand drawn landscape picture. The colors are based on the timbre of each specific sound sample. These colors can be rather complicated gradients, often in layers. If the sound sample has a sequence of vowel sounds then the colors in the gradient will show these based on Ox's color system, which defines how and where vowels are made in the vocal track. Other timbre colors come from the extensive list of timbre created by Ox.The 3D objects made by Ox from Curran's collected sound files will be able to appear in small bits of the whole, based on the duration of the played sound. If the Disklavier key remains depressed for longer than the actual sound file, the object will begin to appear for a second time, moving along the time line. The time line follows a path from the center of the virtual desert out, until it curves up and back towards the center. Before reaching the center it will curve up again and move outward once more, repeating this pattern for as many times as necessary in order to accommodate the playing. Because of this path, time will move both horizontally and vertically. At the end of the performance the visualization will remain and can be more thoroughly inspected as a place with moving objects in a black and white desert environment. A planned recorded version of the entire multimedia experience enables playback in digital environments, such as the ever more ubiquitous digital planetarium theaters.

Figure 1 *View from the desert color organ stop from the Virtual Color Organ ™. Modeled from original drawings made by Ox from real places in the deserts of Arizona and California. The models are by Richard Rodriguez and the texture maps are by Ox.*

Figure 2 *3 Quarter view of coins being tossed; a sound from Alvin Curran's found sound library.*

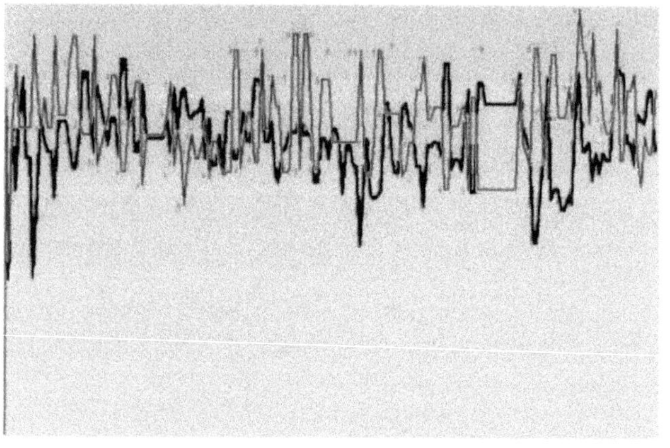

Figure 3 *Graph made from file in Max/msp; greyline delineates pitch, the form of the model follows this line and the black line shows changes in dynamics/velocity. The top of the model is scaled in and out according to this line.*

Figure 4 *3 models of Alvin Curran's collected sounds in the desert organ stop, made by Ox. They show pitch, dynamics, and timbre in the colored and materialed texture maps.*

GridJam will employ 4 musical performers at 4-6 networked nodes. Each node will be a system of one or more computers providing the following services:

- Real-time immersive display of the VCO world,
- Receiving and processing video avatar streams from other nodes and displaying them in the VCO world,
- Digitizing and processing video camera data in real time into the local node's video avatar, displaying this avatar image locally within the VCO world and transmitting it over the network,
- Receiving and processing of MIDI data in and out of the network,
- Generating very high quality audio based on MIDI input data, including applying real-time transformations to the basic "found sound" library. The real-time MIDI performance data will indicate which sounds to play, which transforms to apply and the degree or amount of each transform to apply.
- Processing local viewpoint control data, to change the viewpoint arbitrarily in response to a mouse, wand, joystick or other controller,
- Building video streams from virtual camera viewpoints within the VCO world and transmitting them for network broadcast, including receiving and processing instructions to move or reposition the virtual camera viewpoints.
- Networked encapsulation of uncompressed as well as multi-channel audio compression schemes (Ac3, DTS, MPEG3, etc.) and their applicability to this project will be studied.

These nodes are attached to the Lambda optical networking for lowest possible latency in transmissions among each other. The data being transmitted consists of low bandwidth MIDI control signals, the digitized audio stream and the high bandwidth video avatar streams, as well as the high bandwidth captured video imagery from the observational viewpoints (Leigh et.al. 2006).

4 Change in practice: virtual environments and making of alternative histories

The project by Ox/Britton shows an innovative way to represent transient data (such as sound and time) The visual vocabularies for sounds and the landscapes that were created by Ox serves as model for how we can approach and represent different realities in ancient environments. Cultural experience exceeds the memory, perception, and imagination that we currently represent in archaeological applications based on physical data, such as remotely sensed data. By integrating high and low-tech methods into virtual environments such as drawing, we can incorporate methods from the Arts that is not widely used in the sciences but, as was shown here, have some advantages for anthropology, as it can represent intangible and qualitative elements of cultural experience, provided that we create relevant visual metaphors. Within the 3D world of Ox and Britton, time and space have taken on new meaning, as the geographical distance is disintegrated. Within virtual environments, space and time is different from the physical world we live in, especially if communicated over large networks.

Following Forte (2003), there is a significant difference in the maps we create as archaeologists, with respect to those that represent the current archaeological landscape and those that represent the ancient landscape. Therefore the same methods cannot be used to represent these different landscapes. The archaeological landscape, which shows us where cultural remains are present in the current landscape, can be created and visualized using traditional remote sensing technologies. The ancient landscape however is a *memorization* for lack of a better term, it is a collage of different kinds of data and information created by a single or a group of researchers of an interpreted past experience. Visual language and metaphors can help us distinguish different kinds of information – is it information that is based on a wide

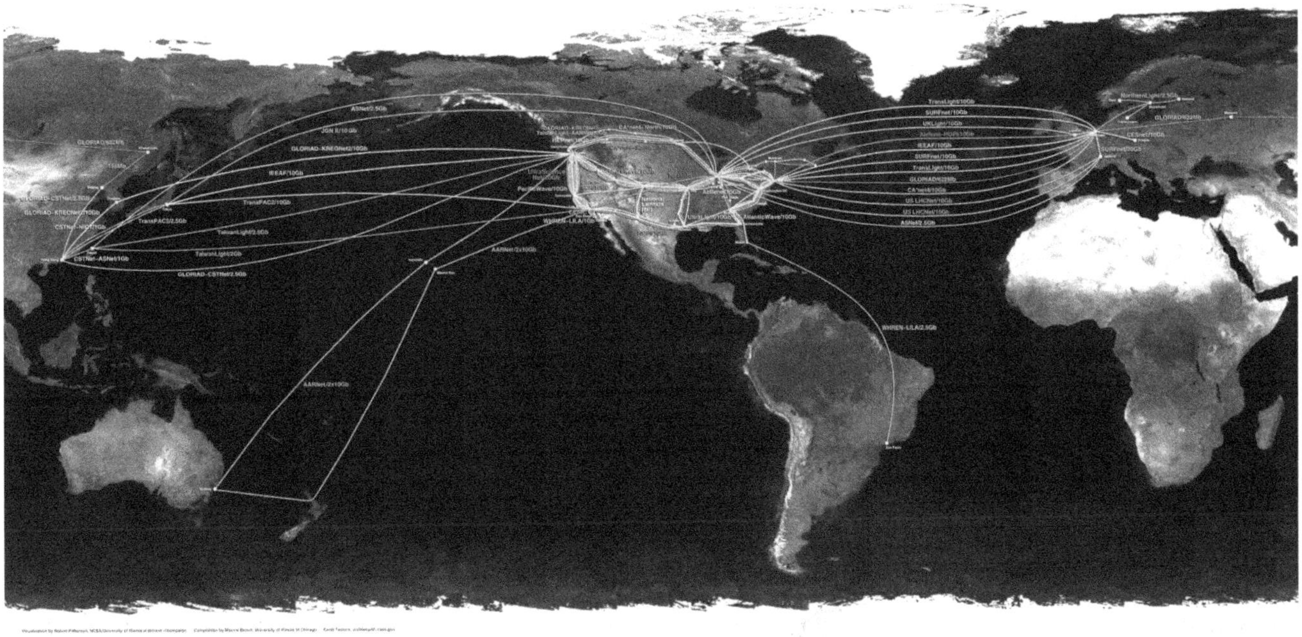

Figure 5 *Global Lamba Integrated Facility*

consensus or does it represent an interpretation by a small group of investigators. Careful consultation with local communities can inform us on 'how they see their world', which will subsequently aid in creating multidimensional cultural sensitive virtual environments.

References

AL-KODMANY, K. 2002 GIS and the artist: shaping the image of a neighbourhood through participatory environmental design, in: *Community Participation and Geographic Information Systems*, Taylor and Francis, London and New York.

ARENDS B. and D. THACKARA Eds. 2003 *Experiment: Conversations in Art and Science*, The Wellcome Trust, London

BALTSAVIAS E., A. GRUEN, VAN GOOL L., and M. PATERAKI, Eds..2006, *Recording Modeling and Visualization of Cultural Heritage,* Taylor and Francis/Balkema, London.

CASTI J. and A.KARLQVIST Eds. 2003 *Art and Complexity*, Elsevier, Amsterdam.

CONROY C., J. PEDLEY, L. CAMFIELD, and N. FOREMAN 2003 Navigating Memories, in: *Experiment: Conversations in Art and Science* (Arends B. and D. Thackara Eds.), The Wellcome Trust, London

CRAIG W.J., T.M. HARRIS, and D. WEINER Eds. 2002 *Community Participation and Geographic Information Systems,* Taylor and Francis, London and New York

FORTE M. and A. SILIOTTI Eds. 1997 *Virtual Archaeology, Re-creating Ancient Worlds,* Harry N, Abrams, Inc., Publishers, New York

FORTE M. 2003 Mindscape: ecological thinking, cyber-anthropology and virtual archaeological landscapes, in: *The Reconstruction of Archaeological Landscapes through Digital Technologies,* (M. Forte and P.R. Williams, eds.) Oxford: BAR S1151

GILLINGS M. and G. GOODRICK, Sensous and Reflexive GIS Exploring Visualisation and VRML, http://intarch.ac.uk/journal/issue1/gillings/toc.html (accessed Oct 1, 2006)

HARVEY F, M. KWAN, and M. PAVLOVSKAYA 2005 Introduction: Critical GIS, In: *Cartographica,* Special Issue: Critical GIS, Volume 40, Number 4, Winter 2005: 1-4.

HODDER I. 1992 *Theory and Practice in Archaeology*, Routledge, London

HJORT B. 2003 Drawing, Knowledge, and Intuitive Thinking: Drawing as a Way to understand and Solve Complex Problems, in: *Art and Complexity* (Casti J. and A.Karlqvist Eds), Elsevier, Amsterdam

KWAN M. and J. LEE 2004 Geovisualization of Human Activity Patterns Using 3D GIS: A Time Geographic Approach. In: *Socially Integrated Social Science.* (Janelle D. and M. Goodchild eds.), Oxford University Press.

LAITURI M. 2002 Ensuring access to GIS for marginal societies, in: *Community Participation and Geographic Information Systems* (Craig W.J., T.M. Harris , and D. Weiner Eds.), Taylor and Francis, London and New York

LEIGH J. et.al., COX D. et.al., TALANDIS J. et.al., COFFIN T., DAVIS B, WIELINGA P & B. STOLK, KOO G.B., KIM J. et.al., CORRIE B. & T. ZIMMERMAN, BOULANGER P. & M. GARCIA 2006 The Global Lambda Visualization Facility: An International Ultra-High Definition Wide-Area Visualization Collaboratory. In: *Future Generation Computer Systems,* Volume 22, issue 8: 964-971, Oct. 2006

OX J. and D. BRITTON 2000 The 21st Century Virtual Reality Color Organ, in: IEEE, July- September 2000

SCHMIDT P.R. and T.C. PATTERSON Eds. 1995 *Making Alternative Histories, The Practice of Archaeology and History in Non-Western Settings*, School of American Research Press, Santa Fe, New Mexico

SMITH L.T. 2005 *Decolonizing Methodologies: Research and Indigenous Peoples*, Zed Books Ltd., London & New York, University of Otago Press, Dunedin

WHEATLEY D. and M. GILLINGS 2000 *Spatial Technology and Archaeology: The Archaeological Applications of GIS,* Taylor and Francis, New York

WILLIAMSON R. A. 1999/2001 The Opportunities and Challenges of Preservation Technologies. In: *Archives and Museum Informatics* 13: 211-225. Kluwer Academic Publishers, the Netherlands.

WYLIE A. 1995 Alternative Histories: Epistemic Disunity and Political Integrity. In: *Making Alternative Histories, The Practice of Archaeology and History in Non-Western Settings,* (Schmidt P.R. and T.C. Patterson Eds.), School of American Research Press, Santa Fe, New Mexico

Applications of remote sensing archaeology technologies in China

Guo Huadong[1], Wang Changlin[1], Nie Yueping[1], Fan Xiangtao[1], Yang Lin[2]

[1] Institute of Remote Sensing Applications
Chinese Academy of Sciences
P.O. Box 9718, Beijing, 100101, China
Tel: 86-10-68597231, Fax: 86-10-64879740
Emails: [guohd, wcl, Nieyp]@irsa.irsa.ac.cn, Fangxt@163bj.com
[2] China National Museum
East Side of Tian An Men Square, 100006 Beijing, China
Email: lbyanglin@263.net

1 Introduction

China is a country with historical records stretching over 5000 years. Cultural heritage is found throughout the country, and Chinese culture and technology spread to many countries of the world along the silk road, a route being blazed in 2 BC from the business routes of land and sea. Therefore, to acquire the information on human cultural heritage is of great importance for the study of ancient China and the complete human history. Archaeology, born in the beginning of the 19th century in the western countries, came to China in early 20th century. Up to now, several branches of archaeology have been established, such as environmental archaeology, remote sensing archaeology, underwater archaeology, archaeometry etc. These have exhibited a vigorous development of archaeology, and increased depth and width of archaeology, in particular the combination of human sciences and natural sciences in theory, methodology and technology. Spatial information acquisition and processing technology represented by earth observation technology, geographic information system, global positioning system, and virtual reality is an important means for understanding the laws of spatial and temporal distribution of cultural heritage, for the reconstruction of the history of ancient civilization, for building of the cultural heritage information management system, and for displaying spatial information of cultural heritage using multi-scale and full dimensional digital earth technology.

With the rapid development of remote sensing, particularly the high resolution optical remote sensing satellites and high resolution multi-mode synthetic aperture radar satellites, the importance of applying airborne and spaceborne earth observing technology to natural cultural heritage conservation and archaeological research has caught a great amount of attention worldwide. Application of aerial photos to archaeology in China is around 70 years behind that in western countries. Satellite images were applied to archaeological interpretation in late 1970s, but they had little use due to very coarse resolution. With the advance of remote sensing sensors, both airborne and spaceborne platforms, the importance of space technology for archaeology has been realized, and organizations dedicated to archaeological research have been established. In 1993, the Open Research Laboratory of Remote Sensing in Urban and Environmental Archaeology was set up in East China Normal University. In 1996, Center of Remote Sensing and Aerial Archaeology was established in National Museum of Chinese History. In November 2001, in order to enhance the strength of space technology application in archaeology, the Joint Laboratory of Remote Sensing Archaeology (JLRSA), affiliated to the Chinese Academy of Sciences, Ministry of Education, State Bureau of Cultural Relics, was founded. The JLRSA has three departments: Research Department at the Institute of Remote Sensing Applications, Chinese Academy of Sciences; Information Department at the National Museum of China; and Training Department at the East China Normal University. At present, there are 10 provincial centers of the JLRSA. In April 2004, the JLRSA became a subordinate organization of the National Center of Remote Sensing in China.

The aims of establishing the JLRSA are, by implementing complementary advantages and fully exploiting the strong disciplinary strengths in each organization,

a) to promote the integrated use of natural sciences and humanistic sciences,
b) to undertake national key archaeological project and achieve high-level scientific results with remote sensing technology,
c) to foster experts on remote sensing archaeology at different levels,
d) to become a base for international academic exchange, and,
e) to be a center of remote sensing archaeology in China.

Since the establishment of JLRSA, some activities on remote sensing archaeology have been conducted. In December 2002, the first National Remote Sensing Archaeology Symposium was held in Beijing. In November 2003, China's top science forum named "Xiangshan Science Meeting" was held with the theme of "Understanding Historic Cultural Heritage with Space Technology", and participants, archaeologists and remote sensing experts, discussed the roles of remote sensing archaeology and problems encountered in applying remote sensing to archaeological research. The International Conference on Remote Sensing Archaeology was successfully convened from October 18 to 21, 2004, on which the participants have shared experiences of their research work, and interests on international cooperation in the field of remote sensing archaeology were discussed.

2 Role of Remote Sensing in Archaeology

Compared with traditional field archaeology, remote sensing archaeology can acquire much more information unavailable or difficult to be obtained in many aspects. As remote sensing provides images for the study area at different scales and resolutions, applications of remote sensing in archaeology range from environmental and landscape archaeology to detailed excavation site mapping. Especially, as a complementary source of airborne remote sensing, the availability of high resolution satellite remotely sensed data such as Quickbird, IKONOS, or in the future Radarsat-2 has facilitated the use of remote sensing data for archaeological research (e.g. Fowler, 2001; Evans, et al. 2003; Blom, 2001). The wide spatial and spectral coverage of remote sensing makes it a powerful means in multivariate environmental archaeological site inventory, geophysical prospecting, and cultural heritage management (Sheets, 2004; Becker et al., 2004; Grøn et al., 2004; An 2004). The penetration capability of synthetic aperture radar is particularly useful in detecting archaeological features underneath a homogenous sand cover or in palaeo-environmental study (MaCauley et al., 1982). The non-destructive advantage of remote sensing, for example ground penetrating radar, to archaeological objects makes the technique useful for detecting and studying the structure and distribution of ancient remains without exposing them, thus, it saves manpower, time, and money (Williams et al., 2004). Close-range photogrammetry enables us to capture information of historical cultural heritage in digital ways, and reconstruct the archaeological matters with computer technology (Sechidis et al., 2004). Virtual reality technology allows the computer reconstruction of palaeoenvironment and ancient scenes with its strong 3-D spatial representation, man-computer interface ability (Forte and Siliotti, 1997; Balzani et al., 2004; Miyatsuka et al., 2002). Remote sensing plays a key role in providing spatial information in building archaeological database for virtual reality. The products from both close range photogrammetry and virtual reality can be put on the internet, therefore, archaeological information and views of the archaeological scenes can be acquired from far-distance.

3 Achievements of Remote Sensing Archaeology in China

Nowadays, China has 30 cultural heritage sites inscribed by the UNESCO, and more historical sites are going to submit applications. Remote sensing images, in particular the high resolution satellite imagery or aerial imagery, are necessary documents for submitting the application proposal. In this sense, remote sensing images provide an important role in monitoring the conservation status of the world cultural heritages. They also serve as a necessary information source for the decision-makers in government for planning and construction of tourism projects for the purpose of advancing economic development.

Remote sensing has been exploited for archaeological research in China, especially in recent years. The first national conference on remote sensing archaeology, held in December 2002 in Beijing, exhibited many achievements in this field. As early as in the 1960s, aerial photos were used to analyze the distribution of ancient tombs and other remains in building San Men Xia reservoir of the Yellow River. In the 1970s, aerial photogrammetric mapping was made for "Shouchu City" of the Warring States Period (403-221 BC) in Anhui province. The application of remote sensing archaeology in the lower reach of the Yangtze River area has revealed 28,087 sites suspected of containing historic remains of the Wuyue culture from Western Zhou (1100-771 BC) period to the "Chun-Qiu" period (722-481 BC). Archaeologists investigated the sites as determined from remotely sensed data and confirmed that an accuracy of 95% was achieved (Liu, 2003). After establishment of the Center of Remote Sensing and Aerial Archaeology in National Museum of Chinese History, a number of aerial photo surveying activities were conducted for many archaeological sites in China, such as Yanshi Er Li Tou and Shixianggou Shang City, Luoyang Dongdu remains in Henan Province. Through cooperation between Shandong Institute of Cultural Relics and the Institute for Pre-and Protohistory of the Ruhr-University Bochum Germany, detailed study was made for ancient cities and tombs in Linzi county. This project published China's first aerial atlas of archaeology (Li and Pingel, 2000), which constitutes an archaeological information system. Such a system integrated satellite and airborne remote sensing images with excavation reports, investigation reports, archaeological maps and terrain maps. The positioning accuracy of the distribution of cultural relics has been greatly improved through the use of this system, and archeologists found that spatial information index and analysis were more convenient for their research. In 2000, the Institute of Remote Sensing Applications, Chinese Academy of Sciences took visible and color infrared images for a tomb area of Han Dynasty in Lao Shan, Beijing, and found 4 abnormally suspectable sites for further verification. Aerial photo surveys for the status of the Great Wall in Ningxia and Beijing were investigated by the Aerial Geophysics Survey and Remote Sensing Center for Land and Resources, Ministry of Land Resources (Li, et al., 1997; Zeng and Gu, 1998), so as to make a detailed investigation for the Great Wall conservation and protection. The joint work made by the National Museum of Chinese History and Inner Mongolia Institute of Cultural Relics produced a book on the atlas of aerial archaeology for the southwest part of Inner Mongolia (Yang Lin and Ta La, 2002). The aerial photogrammetric surveying work has been extended to other parts of Inner Mongolia and Shaanxi provinces in recent years, and achieved satisfactory results. In 2002, the National Museum of Chinese History purchased from the United States over 10,000 aerial photos taken by

Japanese during the Second World War. Digitization, documentation, and analysis of these photos are in progress. These materials will provide valuable information for the study of history and archaeology, as China's rapid economic construction has caused the surface great changes and damaged or buried plenty of historic cultural remains.

Close range photogrammetry has played an important role in preservation and reconstruction of ancient remains. Digital photos, a form of remote sensing, of Mogaoku frescos in Dunhuang area have laid a preliminary foundation for the reconstruction of sculpture and frescos. With laser and holographic imaging, the plundered and damaged cultural relics of the silk-road could be vividly recaptured and displayed by using virtual reality techniques (www.xinhuanet.com, 2004). Remote sensing provides an innovative information technology for repairing and protecting cultural relics, and recording the excavation scenes and reconstructing the environmental setting of ancient sites.

Applications of satellite images for archaeological research have also been practiced since 1970s. The Archaeology Institute of Chinese Social Sciences utilized the processed Landsat TM images and identified construction sites and tombs of Yin Xu ruins in An Yang city of Henan province (Yang et al., 2001; Liu, 1999). The Great Wall segments in a desert region of north central China, at the boundary of Ningxia and Shaanxi provinces, were revealed by SIR-C/X-SAR images. Three Walls were identified, two of them were built in the Ming Dynasty, and one in the Sui Dynasty. The Great Wall of the Sui dynasty was built in the year 585 with rammed earth. It is now discontinuously elongated, about 4m in width and 1 to 3m in height. The Great Wall of the Ming dynasty was built in 1513 with rammed earth. The remnant wall is about 6 to 8m in height and 6 to 8m in basal width. As the radar beam was perpendicular to the Great Wall, resulting in a corner reflector effect, which enables SAR to detect two generations of walls effectively. In the area where sand covers the Wall segments, the penetration ability of long wavelength SAR reveals the Wall's feature and displays them on the SAR images. In terms of wavelength, L band is obviously better than C band for revealing the Wall's feature. In terms of polarization, HH polarization is better than HV polarization. On the L-HH and C-HH images, the Great Wall of both Ming and Sui dynasties is shown, but it is hard to see if both of them existed simultaneously on both images. Trees lining a road parallel to the Wall show up as a bright feature because the L-HV and C band channels are sensitive to complex vegetation structure. The Great Canal of Sui and Tang Dynasties has also been studied with Landsat TM images and Radarsat images, and good results have been achieved (Wang, 2004). Recently, with the launch of Beijing No. 1 Satellite (also known as DMC satellite launched by China), which has 4m panchromatic resolution data and 32m multispectral data (B1, Green, 0.52 ~ 0.62μm; B2, Red, 0.63 ~ 0.69μm; B3, NIR, 0.76~0.90μm), the capability of extracting the Great Wall information was compared with Spot 5 image. The results show that although DMC data has lower resolution than SPOT 5, it basically meets the needs of acquiring the Great Wall information, which is valuable for the ongoing of the Great Wall inventory program aiming to have a detailed data for the Great Wall, in which space information is an important source for the Great Wall database including GIS, RS, GPS and ground surveying data.

Aerial photo and high resolution Quickbird image were used to study the ancient Hun Capital "Tongwancheng Town" for its environmental change (Yin, 2004; Deng and Xia, 2001). The Tongwancheng Town is the world's only ruins of ancient Huns, ancient Chinese nomadic tribe which fought across northern China, central Asia and Europe. The town was built between 413 and 418, and was devastated by Song Dynasty troops. It is located in Jinbian county in the Northern Shaanxi province and in the southern edge of Maowusu desert. The Tongwancheng Town comprises three parts: the palace section and the inner and the outer sections. The palace section is where the imperial palace was located, the inner section consists of government offices and the dwelling region for officials and royal relatives, while the outer section was the residential area for the common people. Built following to the terrain, the northwestern part of the town is higher and the southeastern part is lower, which effectively impeded cold winter wind blown from the north and fully took advantage of the river resources to the north of the town. It can be seen from the Quickbird image and landcover classification map that 76% of Inner section is occupied by moving sands and the entire area is dominated by desert, whereas the flat area is used as farmland. As the town site is under the threat of desertification, the State Council designated Tongwancheng town as a cultural relic under top state protection in 1996, and the Shaanxi provincial government selected the Tongwancheng capital site as a candidate for the world cultural and natural heritage list. A recent research of the plant remains at Tongwancheng site indicates that in the past 1600 years, the Maowusu desert has advanced to the south about 200 km, and average desertification rate is about 125m per year. Cypress, arbor and shrub were grown in this area (Science Time, 2004). This further confirms a famous Chinese archaeologist view that the area once was a beautiful place with rich water, grass and trees (Hou, 1973).

Integrating geophysical prospecting, aerial remote sensing, and satellite remote sensing has been utilized in several research projects. In 2002, the National High-Tech Program of China sponsored a project that used integrated technology for detecting and identifying the Qinshihuang mausoleum, which was made for the first emperor of Qin Dynasty (259-210BC) who unified the countries in the Warrior States period and became the first emperor of China. The objective of the project is to use remote sensing and geophysical methods to conduct non-destructive prospecting, and to detect relic distribution of Qinshihuang mausoleum, and to provide scientific evidence on the presence or absence of the underground palace in the Qinshihuang mausoleum (Yu and He, 2003). The research teams completed hyperspectral imaging of 60 km^2 in the mausoleum area, and collected panchromatic, color infrared and thermal images of early years, and conducted field geophysical surveying in 6 categories (23 sub-categories). Satisfactory results have been obtained and the presence of underground palace was confirmed (Tan et al., 2004; Tan et al., 2006, Zhou et al., 2004, Wan et al. 2004). A new research project with integrated technology for Jinsha remains in Sichuan Province is on agenda, supported by China's Ministry of Science and Technology. Based on literature study of the history of Western Zhou dynasty (Zhu 2003), the area of undiscovered mausoleum of that period (1100-771 BC) in Shaanxi province has been delineated, and the integrated technology will be used in the research.

Environmental evolution and causes of the disappearing of ancient cities (e.g., Loulan City in Lop Nur) were studied with remotely sensed data. The Lop Nur River, which supplied water for Loulan, a busy commercial city on the ancient Silk Road, has dried up and civilization moved from there to elsewhere in China. The kingdom of Loulan was ruled by the government of the Han Dynasty. Troops of the Han Dynasty were stationed in Loulan. Researchers at the Aerial Geophysics Survey and Remote Sensing Center for Land and Resources, Ministry of Land Resources of China, used Landsat TM images to study the causes of the vanishing of Loulan city, and they found there were two lakes in the upper reach of Kong Que River (Peacock river), and formation of the lakes is probably due to landslides which occurred in ancient times (He and Sun, 2001). The landslides blocked the river supply to Loulan city, which resulted in the perishing of the ancient city.

Spatial information acquired with space technology is vitally important for reconstructing ancient landscape and palaeo-environment. For geological and environmental research in an arid area, a unique advantage of radar remote sensing is that radar waves can penetrate a certain layer of dry sand (a few centimeters to meters thick) to reach the buried bedrock. The penetration capability allows us to reveal the subsurface geological structure and old drainage paths (McCauley et al., 1986; McCauley et al., 1982, Blom et al., 1984). Based on the analysis of SIR-A, SIR-B, SIR-C, Radarsat Scan SAR, Landsat MSS and Landsat TM images acquired on different dates and the investigations made in several field trips in Alashan Plateau of Inner Mongolia, a number of old river valley and lake basins buried by wind-blown sand were recognized. Two parallel old drainage systems in the north and middle of the study area are delineated. The study suggests that the moving sand belts mainly follow the old drainage courses. This study also establishes a preliminary drainage evolution model for an area of about 300 000 km^2 since the Tertiary, and finds that the Alashan Plateau was once an area with many rivers and lakes with a warm and humid climate. The relief reversion caused by neotectonic movement since "Qinghai-Tibet movement" is also analyzed (Guo et al., 2000).

Virtual reality technology is now being applied in several research projects. Many organizations, culture and relic departments and institutions have also paid great attention to the application of virtual reality. One of noticeable achievement is "Digital Forbidden City", which utilizes high performance graphic computer workstations, and produces 3-D scene in reality on a large screen (www.dmp.com.cn). With this technology, the staff at the Palace can monitor the status of exhibition, the rare articles and buildings in the Forbidden City can be shown vividly, and the audiences can access each place of the Palace to see every scene and article. With virtual reality technique, the Shanhaiguan Great Wall has been constructed with a 3-D model on the basis of Quickbird image, topographic terrain data at 1:50,000 scale, and on-site photos (Fan et al., 2004). The reconstruction of the Great Wall for Mutianyu section in Beijing was also completed. The virtual reality technology can be used in many aspects in archaeology, such as reconstruction of the lost cities (e.g. Loulan city), and to rebuild the ancient environment and civilization.

4 Problems Concerned with Remote Sensing Archaeology

Although there are many successful stories with remote sensing archaeology at home and abroad, many questions related to this subject still exist. Many people, especially those working in the social sciences, still have doubts about the ability of remote sensing in archaeological research. There is certainly incorrect information with ancient remains extracted from remotely sensed data. We don't know much about the interacting mechanism between ancient remains and remotely sensed data, and how to correctly understand the ancient landscapes from remotely sensed information. We believe that with fusion of social sciences and natural sciences, a better understanding to the applications of remote sensing in archaeology on the theories, methods and techniques will be made. We not only should conduct remote sensing archaeology research at small scale (micro-scale), but also at macro-scale, including Silk Road, Great Wall, and Great Canal. For micro-scale remote sensing archaeology, emphasis should be placed on the important mausoleum, such as the finding of the mausoleum of Western Zhou dynasty. Through research on remote sensing archaeology, the status of this branch in remote sensing could be improved in China, and more contribution could be made to China's 5000 years civilization.

References

AN JIN JIE, 2004, Spatial Querying Method of Remote Sensing Archaeology Data, *In Proceedings of International Conference on Remote Sensing Archaeology, 18-21 October, Beijing,* pp. 113-120.

BLOM, R. G., R. J. CRIPPEN AND C. ELACHI, 1984, Detection of Subsurface Features in Seasat Radar Images of Means Valley, Mojave Desert California, *Geology,* Vol. 12, pp. 346-349.

CHEN SHUPENG, GUO HUADONG, 2003, Remote Sensing and Digital Earth, *Science Progress in China.* Science Press.

DAMIAN EVANS, ROLAND FLETCHER, 2003, Remote Sensing at Angkor: Mapping Settlement Structure Using Imaging Radar, *Aerial Survey and GIS* . In: The Fifth World Archaeological Congress. Washington D.C.

DENG HUI, XIA ZHENGKAI, 2001, From the rise and decline of Tongwancheng to see the impact of man's activity on the area *with* vulnerable ecological environment, *Journal of China Historical Geography,* 16(2): 104-113.

FAN XIANGTAO, HOU TAO, LI HONGGA, WANG CHANGLIN, NIE YUEPING, 2004, Study on the Reconstruction of Ancient Great Wall of Shanhaiguan Area Using Virtual Reality Technology, *In Proceedings of International Conference on Remote Sensing Archaeology, 18-21 October, Beijing.*

FORTE M. AND SILIOTTI, A. (EDS.), 1997, *Virtual Archaeology: Re-Creating Ancient Worlds, New York,* 294pp.

GUO HUADONG, LIU HAO, WANG XINYUAN, SHAO YUN, SUN YAN, 2000, Subsurface old drainage detection and paleoenvironment analysis using spaceborne radar images in Alxa Plateau, *Science in China,* 43(4), 439-448.

BECKER, H, M. CIMINALE, D. GALLO, 2004, Integrated Technologies for Archaeological Investigation: the Celone Valley Project, *In Proceedings of International Conference on Remote Sensing Archaeology, 18-21 October, Beijing.*

HE YUHUA, SUN YONGJUN, 2001, Exploring the disappearing mystery of the Loulan Ruins with satellite data, *Land Resources Remote Sensing,* vol. 2, 64-67.

Hou Renzhi, 1973, From the ancient ruins of ancient city tobeside Hongliuhe river to see the Maowusu desert migration change of Maowusu desert, *Cultural Relics,* 1: 35-41.

SECHIDIS L., D. KAIMARIS, S. SYLAIOU, O. GEORGOULA, P. PATIAS, 2004, From 3D Model to Stereoscopic Video: Example of an Ancient Theatre, *In Proceedings of International Conference on Remote Sensing Archaeology, 18-21, October, Beijing.*

LI CHUANGRONG, VOLKER PINGEL, 2000, *The Archeological Aerial Photo_Atlas of Linzi, China, The Topographical Publishing House of Shangdong Province,* 424P.

LI FENG, GU WEI, CHAO CANXIA, 1997, Aerial remote sensing investigation for Great Wall in Ningxia, *Land Resources Remote Sensing,* 3:7-17.

LIU JIANGUO, 1999, Remote Sensing Archaeology Research for Yin Xu in Anyang, *Archaeology,* vol. 7.

LIU SHUREN, 2003, From the Understanding of Historical Cultural Heritage "The Great Wall" and Others to view the science content of space archaeology, *Proceedings of Xiangshan Science Meeting, No. 216.*

FOWLER M. J. F., 2001, A High-resolution Satellite Image of Archaeological Features to the South of Stonehenge. *International Journal of Remote Sensing,* 22□7□1167□1171.

MACAULEY, J. F., G. G. SCHABER, C. S. BREED, M. J. GROLIER, C. V. HAYNES, B. ISSAWI, C. ELACHI, R. G. BLOM, 1982, Subsurface Valleys and geoarchaeology of the eastern Sahara Revealed by Shuttle Radar, *Science,* Vol. 218, pp. 1004-1020.

MARCELLO BALZANI, NICOLA SANTOPUOLI, ALESSANDRO GRIECO, NICOLA ZALTRON, 2004, Laser Scanner 3D Survey in Archaeological Field: the Forum of Pompeii, *In Proceedings of International Conference on Remote Sensing Archaeology, 18-21, October Beijing.*

MCCAULEY J., C.S. BREED, G.G. SCHABER, ET AL., 1986, Paleodrainages of the eastern Sahara-the radar rivers revisited (SIR-A/B) implications for a Mid-Tertiary Trans-African Drainage system. *IEEE Trans. Geoscience and Remote Sensing* □ GE-24(4):624-648

MIYATSUKA, Y., K. UNO, K. SAKAMOTO, 2002, Archaeological Visualization for Yangtze River Civilization in China. *ISPRS Commission V, WG V/6.*

NIE Yue-ping, ZHANG Xian-feng, Dai Jingjing, YANG LIN, WANG CHANG-LIN, 2006, The application Potentials of "Beijing No.1"satellite data in investigation of The Great Wall, *In the Proc. Second Conference on Remote Sensing Archaeology.* This volume.

OLE GRØN, LARS AURDAL, FINN CHRISTENSEN, ANKE LOSKA, 2004, Mapping and Verifying Invisible Archaeological Sites in Agricultural Fields by Means of Multi-Spectral Satellite Images and Soil Chemistry, *In Proceedings of International Conference on Remote Sensing Archaeology, 18-21 October, Beijing.*

PATRICK RYAN WILLIAMS, NICOLE COUTURE DEBORAH BLOM, BENJAMIN VINING ,CHRISTOPHER DAYTON, JOHN JANUSEK, 2004, Ground-Based Remote Sensing and Early State Development in the South-Central Andes, *In Proceedings of International Conference on Remote Sensing Archaeology, 18-21 October, Beijing.*

PAYSON SHEETS, 2004, The Origins of Monumentality in Ancient Costa Rica, Revealed by Satellite and Aircraft Remote Sensing", *in Proceedings of International Conference on Remote Sensing Archaeology, 18-21 October, Beijing.*

RONALD G. BLOM, ROBERT E. CRIPPEN, AT EL., 2000, Remote sensing, shuttle radar topographic mapper data, and ancient Frankincense trade routes. *IGASS 2000.*

SCIENCE TIME, 2004, The Chinese Academy of Sciences Quantitatively Study the Desertification Rate with Plant Remains, Dec. 12.

TAN KELONG, WAN YUQING, DUAN QINGBO, ZHOU XIAOHU, SONG DEWEN, NIU XINLONG, 2004, The Application of Remote Sensing Technology in Archaeological Study of Qinshihuang's Mausoleum, *In Proceedings of International Conference on Remote Sensing Archaeology, 18-21 October, Beijing*

TAN, K.L., Y. WAN, X. ZHOU, D. SONG, Q. DUAN, 2006, The application of remote sensing technology in the archaeological study of the Mausoleum of Emperor Qinshihuang, *International Journal of Remote Sensing,* Volume 27, Number 16 / 20, 3347 – 3363.

WANG XIN-YUAN , ZHOU YING-QIU, LU YINGCHENG, GAO CHAO 2004, On Remote Sensing Archaeology Analysis for Transformation of Traffic Function of Sui dynasty Tongji Canal, *In Proceedings of International Conference on Remote Sensing Archaeology, 18-21 October, Beijing*

YANG LIN, TA LA, 2002, *Report of Aerial Photographic Archaeology in Southeastern Inner Mongolia,* Science Press, 235p.

YANG RUIXIA, GUO YANGSHAN, ZAN ZHIMING, 2001, Application of Remote Sensing Technology to Archaeology in Henan Province, *Land Resources Remote Sensing,* vol. 2, 19-24.

YING NING, 2004, Environmental Changes and Archaeological Study Using Multi-source Remote Sensing datasets of Typical area in West China, *M.Sc Thesis, Institute of Remote Sensing Applications, Chinese Academy of Sciences,* 63p.

YU WENXIAN, HE KAITAO, 2003, New Earth Observing Technology and Information Extraction to Ancient Remains, *Proceedings of Xiangshan Science Meeting, No. 216.*

ZENG CHAOMING, GU WEI, 1998, Aerial Remote Sensing Investigation for the Great Wall in Beijing, *Journal of East China Normal University, Special Issue of Archaeology Remote Sensing,* pp. 131-136.

ZHU FENGHAN, 2003, Exploration and Study of the Spatial and Temporal Distribution of Historical Cultural Heritage, *Proceedings of Xiangshan Science Meeting, No. 216.*

Teaching and using remote sensing in Argentine Archaeology: evaluating the University of Buenos Aires curriculum and the graduation theses of the last decade

Débora M. Kligmann

CONICET (Consejo Nacional de Investigaciones Científicas y Técnicas)
UBA (Universidad de Buenos Aires, Argentina)

Introduction

Remote sensing in archaeology can be addressed from three different angles:

1) Creation of technologies (developers).
2) Transmission of knowledge (teachers).
3) Use of technologies (researchers).

In this paper, I will deal with remote sensing education and research because they are still underdeveloped in argentine archaeology. These two areas are closely related because we need to acquire knowledge before we can use it.

Advantages and disadvantages of remote sensing techniques as applied to archaeology

The principal **advantage** of remotely sensed imagery in archaeology is that it provides general information pertaining to very large areas, revealing widespread spatial distributions and patterns of vegetation and landforms (Hester *et al.* 1997: 178).

Aerial photographs and satellite images can be very helpful when, as a result of scale, pedestrian surveys are ineffective at discerning subtle patterns produced by archaeological remains and where visibility is low because of heavy vegetation (Hester *et al.* 1997: 63).

Archaeologists can obtain valuable data about where sites might be found, in addition to getting images of the sites themselves (Staeck 2002: 94). Aerial photographs and satellite images can be used for prospecting, measuring, mapping and monitoring changes over time. Information can be collected for both surface and buried sites.

Images are also helpful to locate outcrops of raw materials (both minerals and rocks) available in the study area and potentially used in the past to manufacture artifacts (e.g. ceramic vessels and projectile points).

Geologic analysis of satellite images can offer useful information about areas of erosion and deposition, thus allowing predictive models on archaeological site location to be generated.

Among the **limitations**, both the spatial and temporal scales have to be mentioned.

As far as the *spatial scale* is concerned, remote sensing instruments often cannot detect anything as small as individual artifacts, unless they are deposited in rather dense clusters. Besides, archaeological sites are usually too small to be seen in "low resolution" images.

Regarding the *temporal scale*, the challenge is how to use present information to study areas that were occupied several hundred or even several thousand years ago. We cannot assume that the landscape is stable because the environment certainly does change through time. These changes can be triggered by both natural and anthropic processes.

To sum up, remote sensing techniques were not developed for archaeological purposes and thus have some limitations. If used carefully, however, they can be very useful. The key issue is to know what kind of archaeological questions these techniques can answer. Although an interpretation of the archaeological sites located is sometimes possible, excavation is the only real way to test the truth of what remote sensing instruments are detecting.

When working in archaeology, we should take into account:

- the characteristics of the archaeological remains (e.g. artifact raw material, artifact density, structure size, height, shape and structure location),
- the characteristics of the landscape where these remains are located (e.g. ground cover and vegetation cover) and
- the characteristics of the remote sensing techniques used (e.g. spatial and spectral resolution).

Remote sensing in Argentina: conae and the earth observation satellites

Argentina is one of the few countries in the world that has a National Space Agency. CONAE (National Commission on Space Activities), created in 1991, is in charge of the national satellites (entirely designed and built in the country) as well as of a ground station, located in the province of Córdoba. In addition to data from one national satellite (SAC-C), data from 12 foreign orbiting satellites (including Landsat 5 and 7, Radarsat, SPOT, the NOAA series, EROS, Orbview 2, Terra, Aqua and ERS) are downloaded, transmitted and stored (see CONAE's web page for details).

CONAE's satellite missions are devoted to fill vacant areas of space information at the local and regional levels, thus fulfilling the needs of argentine users.

The SAC-C (Satellite of Scientific Applications), launched in 2000, is fully operational and two more satellites called SAOCOM 1A and 1B will be launched in the near future.

SAC-C (with sensors in the visible and infrared range of the electromagnetic spectrum) is the first argentine Earth observation satellite and is part of the AM Constellation of Earth observation together with Landsat 7, EO-1 and Terra US satellites.

SAOCOM 1A and 1B (with sensors in the active microwave range) integrate the SIASGE Constellation (Italian-Argentine Satellite System for Emergency Management) together with four satellites of the COSMO-SkyMed Italian Constellation. The aim of this cooperation program is to help in emergency situations by preventing, detecting and monitoring natural and anthropogenic disasters such as fires, floods, hurricanes, earthquakes, volcanic eruptions, landslides, draughts, oil spills and agricultural plagues. The information generated can be used by archaeologists, agronomists, biologists, geologists and other scientists interested in the environment.

It is important to mention that CONAE cooperates with argentine researchers by providing training, technical support and images free of charge for scientific projects.

Argentina and UNESCO's "open initiative"

In 2001, UNESCO and ESA (European Space Agency) launched the "Open Initiative on the Use of Space Technologies to Monitor Natural and Cultural Heritage of UNESCO Sites". The "Open Initiative" is a framework of cooperation to assist countries to improve the observation, monitoring and management of natural and cultural sites as well as of their surroundings, through space technologies (see UNESCO's web page for details).

Argentina adhered to the World Heritage Convention in 1978. Of the eight argentine properties inscribed on the list, four are cultural: Cueva de las Manos (Patagonia), Jesuit Block and Estancias of Córdoba (province of Córdoba), Jesuit Missions of the Guaranis (Northeast Argentina, property shared with Brazil) and Quebrada de Humahuaca (northwest Argentina). Some more cultural properties (such as the Qhapaq Ñan or Main Andean Road) are currently being considered to be included in this list.

One of the main goals of the "Open Initiative" is that of capacity building. This is carried out through the implementation of workshops and the development of joint research projects with local people in order to provide training.

Looking back into the past: the history or remote sensing in argentine archaeology

The first paper dealing with the use of airborne technologies as an aid in archaeological research in Argentina was published 50 years ago (González 1956). This author used two planes provided by the argentine Air Force to examine different areas located in the provinces of Catamarca and Tucumán, northwest Argentina. The aerial photographs obtained became a helpful tool to locate and map archaeological sites.

It was not until almost 30 years later that the next paper about remote sensing and archaeology was published in the country (Albeck and Scattolin 1984). Here, aerial photographs were used to map the drainage pattern as well as the location of both modern settlements and archaeological sites in the province of Catamarca, northwest Argentina. The goal of this study was to characterize prehispanic agricultural structures and understand their relationship with available watercourses.

A few years later, the same authors (Albeck and Scattolin 1991) used aerial photographs to establish the boundaries and measure the extension of agricultural fields in the province of Jujuy, northwest Argentina.

In this same decade, two archaeology graduation theses were defended at the University of Buenos Aires (Sosa 1994; Nastri 1995) and two more papers were published (Sosa 1996-98, 1999). In all of these, aerial photographs were used for discovering and mapping sites in the provinces of Catamarca and Tucumán, northwest Argentina. This produced an increase in the chronological and functional variability of the archaeological sites known at the time as well as in their distribution. Thus, remote sensing proved to be extremely useful because some of the areas of interest were quite difficult for pedestrian prospection.

In only one paper, satellite images were used in Argentina for archaeological purposes (González de Bonaveri 1989). Here, a visual interpretation of two LANDSAT images was carried out in order to study the geomorphology and delineate areas where archaeological sites were likely to be found. A pedestrian prospection was later performed in these areas to locate sites in the Pampean Region.

In summary, even though some argentine archaeologists use aerial photographs or satellite images, very rarely they publish papers or carry out research projects entirely devoted to remote sensing. Except for one case where satellite images were used in the Pampean Region, all the remaining papers reviewed in this section deal with the use of aerial photographs in northwest Argentina. This can be explained by two reasons: 1) the sites in question have stone structures, which are easily visible in aerial photographs and 2) the areas where the sites are located are not heavily vegetated, a key factor for the success of photointerpretation.

Remote sensing in Argentine archaeology: education and research

1) Teaching remote sensing in argentine archaeology: an analysis of the University of Buenos Aires (UBA) curriculum

Degree: M.A. in Anthropology (specialty in Archaeology)

In order to become an archaeologist, students enrolled at the College of Arts and Sciences of the UBA need to take 6 introductory classes (common to many different careers), 11 general anthropology classes, 11 archaeology classes and one optional class. Finally, they have to produce and defend a graduation thesis. The whole process takes a minimum of 5 years.

The detailed curriculum is presented below:

1. Introductory classes

- Politics.
- Economy.
- Philosophy and Logic.
- Semeiology.
- Sociology.
- Anthropology.

2. Anthropology classes

2.1. Classes shared by both specialties (sociocultural anthropology and archaeology)

- Systematic anthropology I (social and political organization).
- Systematic anthropology II (economic organization).
- Systematic anthropology III (symbolic systems).
- Sociocultural systems of America I (hunters, gatherers and early farmers).
- General social history.
- History of anthropological theory.
- Linguistics and semiotic.
- Folklore.
- Statistics.
- Bioanthropology and paleoanthropology.
- Introduction to prehistory.

2.2. Archaeology classes

- Old World prehistory.
- American and Argentine prehistory I (hunters-gatherers).
- American and argentine prehistory II (farmers).
- Argentine archaeology.
- Geology.
- Methods of archaeological research.
- Introduction to archaeological materials.
- Contemporary archaeological theory.
- Models and methods of analysis in prehistoric economy.
- An optional archaeology seminar (the offer of courses taught changes every year).
- Research design seminar.

3. An optional class that can be chosen from:

- Classes of the sociocultural specialty.
- Classes taught by other careers of the College of Arts and Sciences.
- Classes of other Colleges of the UBA.

4. Modern Languages

- Three levels (introductory, intermediate and advanced) of Italian, French or Portuguese.
- Three levels (introductory, intermediate and advanced) of either English or German.

5. Graduation thesis

An analysis of the curriculum reveals that just a few hours are devoted in only two of these classes to aerial photographs: Geology and Methods of archaeological research. It is obviously not enough because students need to understand not only why remote sensing is useful for archaeology but also how to interpret aerial photographs and satellite images.

2) Using remote sensing in argentine archaeology: an analysis of the graduation theses defended at the UBA in the last decade

I chose the archaeology graduation theses produced by the students of the College of Arts and Sciences of the UBA as a "thermometer" to evaluate what is going on in the remote sensing arena in the country. These theses were produced as a result of the participation of the students in many different research projects and answer questions posed by these projects. Thus, the scientific production mentioned somehow reflects a general tendency in the argentine archaeological community.

A total of 104 archaeology graduation theses was defended at the UBA in the last decade. As it can be seen in figure 1, there is a tendency towards an increase in the number of students graduating every year.

The theses were classified according to the region from where the archaeological material remains analyzed come (see Figure 2) and to their research topic (see Figure 3).

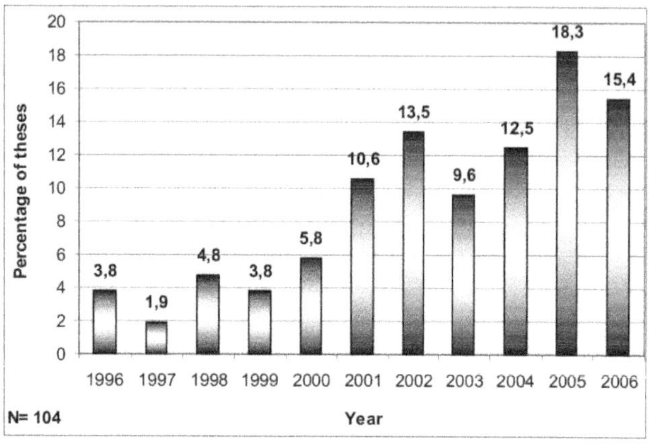

Figure 1 *Percentage of archaeology graduation theses defended at the UBA in the last decade*

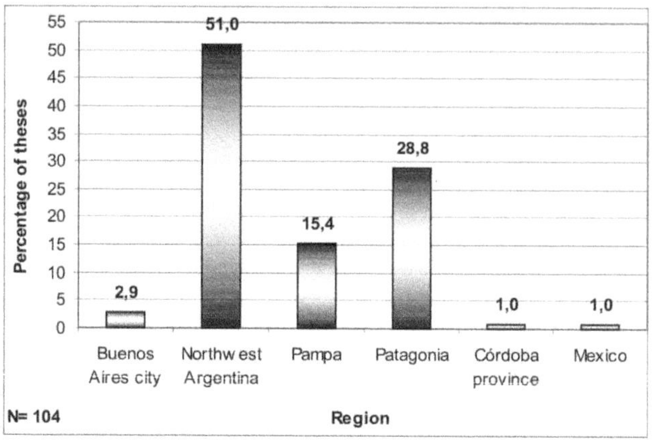

Figure 2 *Percentage of archaeology graduation theses classified by region*

As Figure 2 shows, not all the regions of the country are equally "productive" as far as archaeological research is concerned. Half of the theses are devoted to northwest Argentina, followed by theses dealing with other regions such as Pampa and Patagonia.

As it can be seen in Figure 3, a wide variety of topics are covered. However, in the period considered for this paper (1996-2006), only one thesis deals with sensors carried on airborne platforms (Villegas 2006). Unfortunately, none of them deals with spaceborne sensors.

The goal of Villegas' thesis was to study the late prehispanic settlement pattern and space use in the middle Calchaquí Valley, Province of Salta, northwest Argentina. The archaeological sites located were preliminarily classified according to their size, functionality (e.g. houses, agricultural fields, roads), chronology, distribution, vegetation cover and relationship with geomorphological units. The results of this first step were used to guide the pedestrian prospection in the field.

In some of the theses considered there are a few aerial photographs or satellite images of the study area. However, they are only used as "maps" to show the location of different sites. Very rarely, information about the photographs or the images is provided (e.g. date, type of sensor, spatial, temporal, spectral and radiometric resolution). Thus, the potential of these techniques is not fully taken advantage of. For example, instead of choosing overlapping photographs to create a stereoscopic effect, so that sites and landscapes are seen in three dimensions, only one photograph is used. When it comes to satellite images, no digital analysis is performed. This means that the images are not enhanced

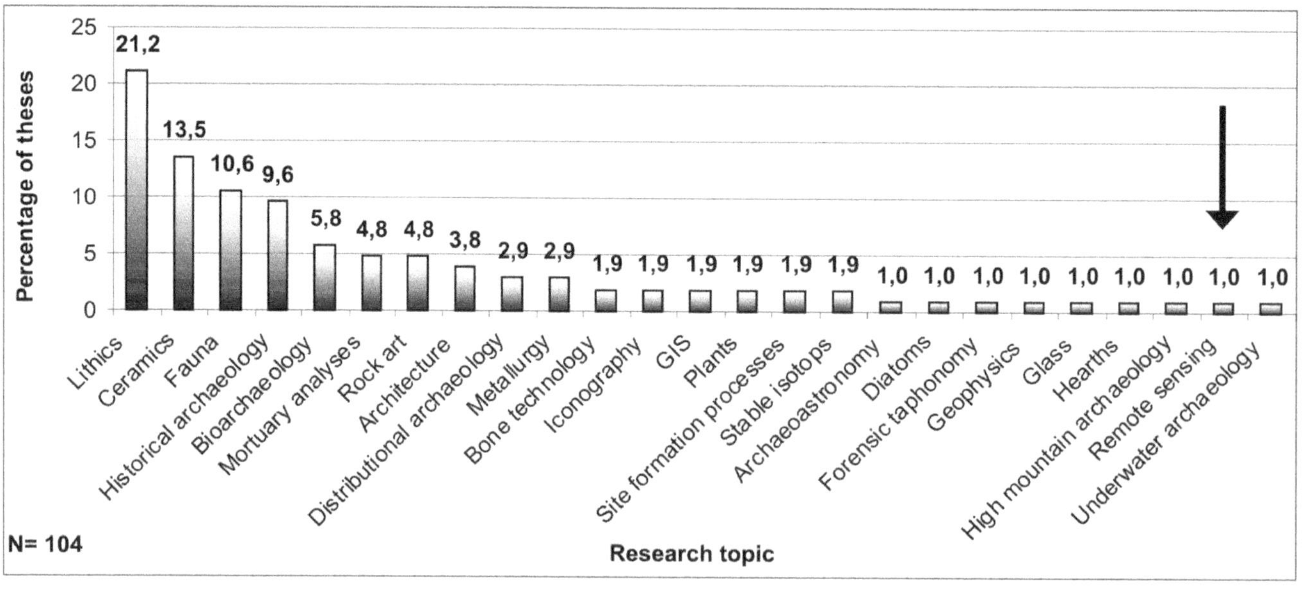

Figure 3 *Percentage of archaeology graduation theses classified by research topic*

and that bands are not combined in such a way as to highlight the features of interest.

Conclusions

The huge technological development of the last decade in the area of remote sensing has not been followed by an intensification in the archaeological applications of these technologies. This is happening worldwide and is also happening in Argentina.

A detailed analysis of the UBA curriculum shows that remote sensing is not being taught. When looking at the archaeology graduation theses defended at the same university an interesting issue emerges: very few of them deal with remote sensing. A similar situation can be noticed while reviewing the papers presented at national scientific meetings or published in peer-reviewed journals in Argentina. I believe this is a consequence of the lack of proper education on the topic.

In summary, we have a National Space Agency that is willing to cooperate with argentine researchers, we have free access to some satellite images, we have many sites to protect and we have lots of problems that can be solved through remote sensing technologies. What we do not have is the knowledge.

The question that follows is why argentine archaeologists are not getting the proper training so that we can take full advantage of what we do have. At the University of Buenos Aires, the absence of remote sensing classes can be explained by the following issues:

- lack of trained archaeologists in remote sensing technologies and
- lack of adequate infrastructure, including hardware, software and scanners.

This probably happens in all other universities of the country as well.

Even when students can take remote sensing classes outside the archaeology department (in other departments of the same university or even in other universities), these classes are not applied to archaeology. This means that although students can learn theoretical and methodological issues, they still need to read and discuss specific applications in order to understand how archaeology can benefit from remote sensing.

Also, they have to learn that remote sensing products have different characteristics. This will allow them to choose the product that best suits their problem. Let me illustrate this point with a few examples. A certain image can be useful for somebody doing site-specific archaeology but not for another one doing regional archaeology. Two archaeologists, one interested in the vegetation cover and the second one interested in mineral resources, will surely use different band combinations. It is not the same to work with high mobility hunter-gatherers than with complex sedentary societies. In this case, redundancy in space use will certainly be different. This, in turn, will affect the number of sites in a given area as well as their distribution.

This can be partially solved by teaching a remote sensing course for archaeologists at the undergraduate or graduate level. Thus, future argentine archaeologists can be trained to use remote sensing technologies both for their own research as well as for conservation of the National Heritage.

The challenge of choosing a research topic that very few other archaeologists are dealing with is great. People feel lonely because they have no colleagues to discuss with and they have to find their own way in a vacant area. However, and because of this same reason, they have to be really creative and can come up with something unique and original that will show the way to those who follow. I think it is worth the effort.

Acknowledgements

To Beatriz Diaz Pereyra, head of the Office of Student Affairs of the College of Arts and Sciences (UBA), who very kindly gave me access to the official records dealing with the archaeology graduation theses.

To all the employees of the library of the College of Arts and Sciences (UBA) who found graduation theses that were "hidden" in shelves long ago forgotten, thus making my job much easier.

To the Agencia Nacional de Promoción Científica y Tecnológica and the Universidad de Buenos Aires for their financial help.

References

ALBECK, M.E. AND M.C. SCATTOLIN 1984 Análisis preliminar de los asentamientos prehispánicos de Laguna Blanca (Catamarca), mediante el uso de la fotografía aérea. *Revista del Museo de La Plata* N.S. VIII(61): 279-302.

ALBECK, M.E. AND M.C. SCATTOLIN 1991 Cálculo fotogramétrico de superficies de cultivo en Coctaca y Rodero, Quebrada de Humahuaca. *Avances en Arqueología* 1: 43-58 (Instituto Interdisciplinario de Tilcara, FFyL, UBA).

CONAE's web page http://www.conae.gov.ar

GONZALEZ, R.A. 1956 La fotografía y el reconocimiento aéreo en las investigaciones arqueológicas del N.O.A. *Anales de Arqueología y Etnología* XII: 41-62.

GONZÁLEZ DE BONAVERI, M.I. 1989 El uso de la imagen satelitaria como técnica de investigación arqueológica. *Primeras Jornadas de Becarios de la UBA*, pp. 137-139. Buenos Aires.

HESTER, T., H.J. SHAFFER AND K.L. FEDDER 1997*Field Methods in Archaeology*. Mayfield Publishing Company, Mountain View.

NASTRI, J. 1995 *Distribución de instalaciones prehispánicas en el sudoeste del Valle de Santa María (NOA)*. Tesis de Licenciatura en Ciencias Antropológicas, Facultad de Filosofía y Letras, Universidad de Buenos Aires.

SOSA, J.A.1994 *Teleprospección arqueológica en Amaicha del Valle y sus alrededores*. Tesis de Licenciatura en Ciencias Antropológicas, Facultad de Filosofía y Letras, Universidad de Buenos Aires.

SOSA, J.A. 1996-98 Arqueología de Amaicha del Valle (Tucumán): prospección aerofotográfica. *Palimpsesto* 5: 183-188.

SOSA, J.A. 1999 Teleprospección arqueológica en Amaicha del Valle (Departamento de Tafí del Valle, Tucumán). *Actas del XII Congreso Nacional de Arqueología Argentina*, Tomo III, pp. 358-365. La Plata.

STAECK, J.P. 2002 *Back to the Earth. An Introduction to Archaeology*. Mayfield Publishing Company, Mountain View.

UNESCO's web page http://whc.unesco.org

VILLEGAS, M.P. 2006 *Teledetección arqueológica en el Valle Calchaquí medio (Salta, Argentina)*. Tesis de Licenciatura en Ciencias Antropológicas, Facultad de Filosofía y Letras, Universidad de Buenos Aires.

Integrated Technologies for the Reconstruction of the Ancient Landscape of Poseidonia-Paestum

A. Pontrandolfo,[1] A. Santoriello,[1] L. Sirangelo,[1] F.U. Scelza,[1] M. Marinaro,[2] L. Pugliese,[2] S. Scarpetta,[2] A. Esposito[3]

[1] Università di Salerno – Dip. di Fisica – Via S.Allende 1, Baronissi (SA) silvia@sa.infn.it
[2] Università di Salerno – Dip. di Beni Culturali – Via Ponte Don Melillo, Fisciano (SA) asantori@unisa.it
[3] Istituto Internazionale per gli Alti Studi Scientifici "E.R.Caianiello" Via G.Pellegrino 19, Vietri sul mare (SA) iiass.luca@tiscali.it

Introduction

This work has been developed within the Innova Project (Regional Centre for the Development and Transfer of the Innovation applied to Cultural and Environmental Goods, www.innova.campania.it) whose leading subject is the University of Salerno, which has stipulated an agreement for the purpose with the "Soprintendenza Archeologica delle Province di Salerno, Avellino e Benevento". Its main goal was the definition of mentioned patterns relevant to the ancient landscape of Poseidonia-Paestum and of its territory, and to the forms and ways of employing of the human groups.

The present work also provides monitoring features for the tutelage and valorisation of the cultural resources. The attention was particularly focused on environmental analyses carried out through the use of integrated GIS test. Among the several features the so-structured system uses multi-spectral satellite images on which some algorithms have been applied in order to confirm cognizance and to point out the environments which humanity faced in the course of time. This work aims to illustrate some techniques of analyses applied to Ikonos II imagery.

The Study Area

The interested area is the territory of the city of Poseidonia-Paestum and especially the southern part of the alluvial plain of Salerno gulf. This area is bordered northerly by the Sele river, easterly by hilly reliefs – such as Calpazio, Sottano, Soprano – and southerly by Agropoli hills and the Capodifiume river – which laps the city on its southern part – and the Solofrone and Testene rivers. The urban centre of Poseidonia, which arose around 600 B.C. corresponded to the calcareous platform slightly lifted above the sea level and situated in the middle of the southern plain (Sele left bank side). The boundaries of this area are defined by the necropolis, which, from their foundation until Roman colonization (273 B.C), surrounded the city both in the North (Arcioni-Gaudio and Andriuolo) and in the South (Santa venera-Spinazzo on SE and Licinella on SW) side.

Environmental Factors analysis: a Neural Net classification of an Ikonos image

One of the relevant objective of the Project, is the study of environmental factors evolution and how this evolution has influenced settlements in the area (Landscape Archaeology). In this respect, processing of satellite supports are conducted to boost the identification over the Poseidonia-Paestum territory of anomalies , that are good candidates for the successive archaeological interpretation.

As stated before, the main objective of this paper is the discussion of some results obtained in the satellite environmental analysis part of the project. While the value of satellite imagery to archaeology is increasingly apparent, most current applications involve its use for environmental reconstruction.

Landsat Thematic Mapper imagery has been used to aid the identification of environmental factors (Pugliese et alii, 2005), particularly useful in landscape archaeology approaches (Rothaus, De Morett, 2003). Because of the relatively low spatial resolution of Landsat multispectral data (e.g. 30 m for Landsat multispectral data), these are of limited applicability for the analysis in sites where human activity has radically transformed (and in most cases fragmented) the aspect of natural landscape. However the continuing improvement in the resolving power of the last generation of satellites is changing the possible uses of satellite imagery so that in the right circumstances the information drawn from, for example, Ikonos-2 imagery is beginning to stand comparison with that of vertical air photography. In the present research, in order to take advantage of both regional information driven from Landsat support and territorial details contained in Ikonos and air photograph and having to bridge the two different scales, we acquired also a Spot 5 panchromatic scene.

One of the useful approaches to the analysis of satellite multispectral images involves the grouping of image data into a finite number of discrete clusters or classes that identify the distribution, over the land, of environmental factors such as soils, vegetation types, waters, and artificial surfaces (Mather, 1999). However classification of intricate, high-dimensional spectral signatures has turned out to be far from trivial (both in the environment analysis and in the archaeological field). Discrimination among many surface cover classes, discovery of spatially small, interesting spectral species proved to be an insurmountable challenge to many traditional clustering and classification methods. By customary measures (for example by means of linear transformation such as the Principal Component Analysis, PCA) the spectral dimensionality of multispectral images can be reduced. Yet dimensionality reduction to such low numbers has not been often successful in terms of preservation of important class distinctions. A faithful topological mapping is necessary to preserve the information contained in the multispectral image.

The spectral bands, many of which are highly correlated, may lie on a low-dimensional but non-linear manifold, which is a scenario that push research into advanced and novel approaches for remote sensing image analysis and, in particular, neural networks. That's the reason why, having to face an environmental analysis on a territory, the Paestum Area described above, which appears strongly fragmented nowadays, the neural classifier issue, and in particular the Self Organzing Map (SOM) one seemed to be the right attempt.

The Kohonen's Self-Organizing Map (SOM) is an unsupervised neural network algorithm that has turned out to be an efficient tool, in various applications, for data exploration tasks (Kohonen, 1997). Recently it has been used also for the analyses of remote sensing multispectral images with quite successful results (Villman et alii 2003). It carries out a nonlinear mapping of input data onto a two-dimensional grid of neurons or nodes (see the schematic architecture of a SOM depicted in Figure 1). The mapping preserves the most important topological and metric relationships of the data.

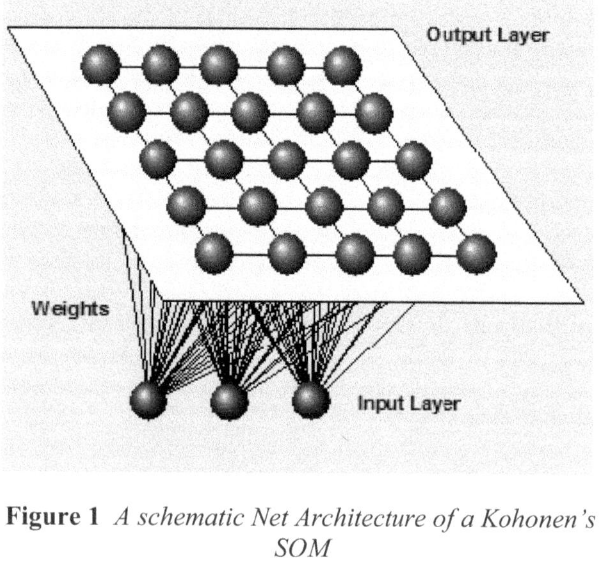

Figure 1 *A schematic Net Architecture of a Kohonen's SOM*

In this paper we employ the Self Organizing Map (SOM) as a strategy for an unsupervised analysis of IKONOS MultiSpectral (MS) images over the archaeological area. The segmentation goal was to identify both environmental factors (soils, vegetation types, water...) and manmade structures.

In order to ease the successive photo-interpretation of the image segmentation, the Sammon Algorithm was employed to project the SOM prototypes into a three dimensional colour space, thus defining a Look Up Table (LUT) that preserves in its colour composition the topological information exploited by the SOM strategy.[1]

A IKONOS 2 Multispectral (MS) image from January 2004 licensed from Planetek Italia ®(www.planetek.com) was processed for this study using IDRISI Andes ®(version 15, released April 2006) geo-analytic and image processing system (http://www.clarklabs.org), which includes among its New Features, a SOM module conceived to perform unsupervised classification tasks. A successive analysis- interpretation step was performed by importing the segmentation results in a PCI Geomatica ©(version 10.1) geo-processing environment (http://www.pcigeomatics.com). For this study we employed the four multispectral bands of the Ikonos 2 image. The product was at GEOretification level, i.e. was rectified to a Geographic projection System; in this case to UTM 33N projection with WGS84 Ellipsoid and Datum. The cloud coverage was zero percent. Data are quantized at 11 bits. The size of an image, for each band, is 899 lines x 851353 pixels.

The Area Of Interest (AOI) is centered on the Paestum Temples site located in the South of Italy. The AOI is identified in the UTM (Universal Transverse Mercator) reference system, zone 33 N, with the following coordinates: Upper Left Corner(m): (498363.2890349; 4477310.5027174); Lower Right Corner(m): (502150.9004424; 4473714.8718062).

Comparing the results of the SOM Classification (see Figures 2a, 2b) with the prior knowledge of the territory we found that the thesis that the final distribution of colors over the segmented scene is such that similar hues identify objects that are homogeneous in a physical sense (for example different types of vegetation, or different types of rocks) is fairly confirmed.

For instance, a group of only three similar gray tone (classes 15 12 and 16 in Figure 2b, which are shades of the same color in the colored version) isolate all artificial surfaces (temples, buildings, greenhouses, streets, and rails). In the same manner, the sea (lower left corner of figure 2a) is represented with three very similar shades of dark gray (classes 3,4 and 8 in Figure 2b), and these different shades are probably due to different sea levels.

[1] The coloring procedure described herein has been adapted in order to accomplish an editorial constraint. In particular, the SOM classes have been projected in the gray-tone subspace. Clearly all the potentialities of the coloring procedure is not entirely exploited when limited to gray tones.

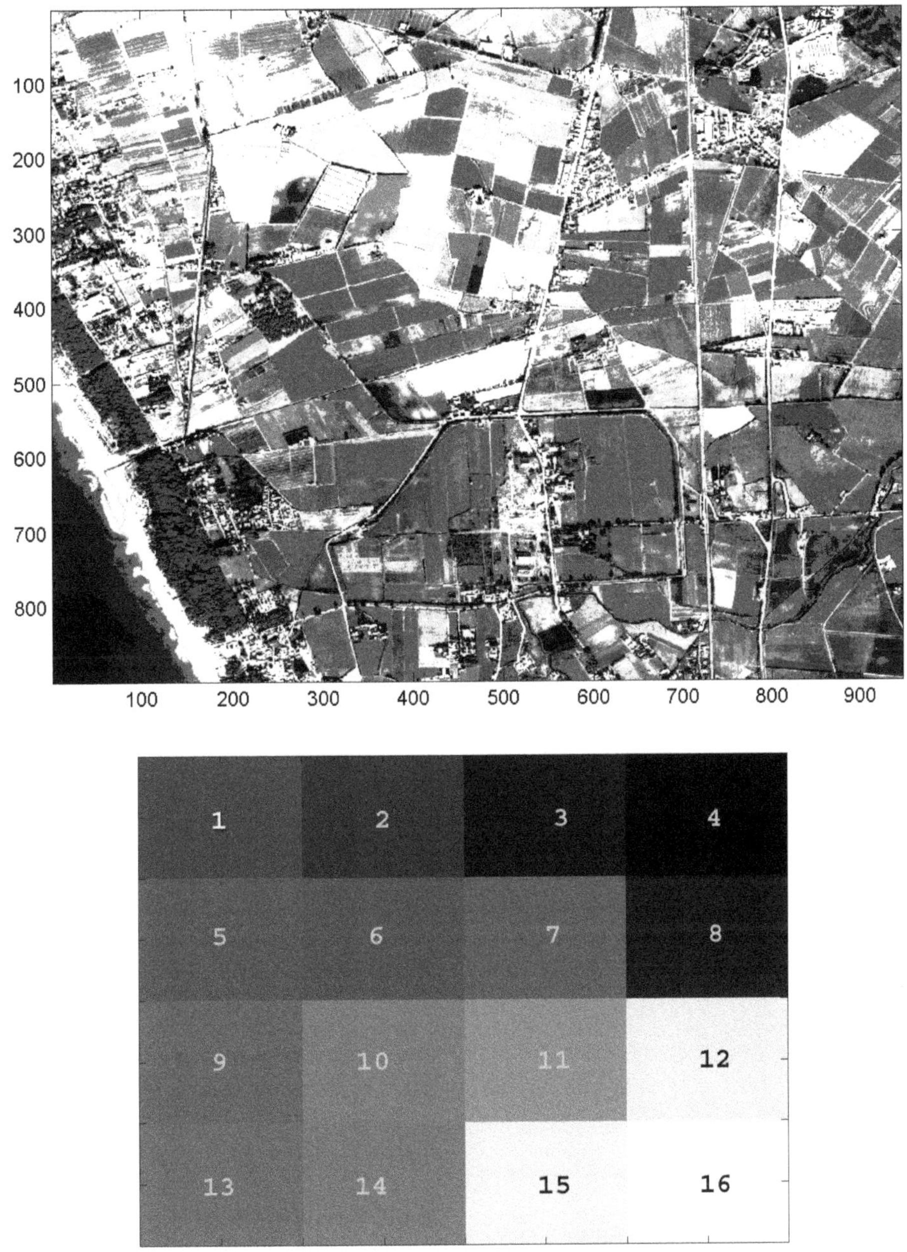

Figure 2 *a) The output of the SOM classification - a 16 cluster segmentation of the multispectral image*
b) The Look Up Table is organized in a square grid in order to visualize similarities among clusters revealed by the SOM net.

Clearly in the gray-tone figures, segmentation is less clear than in the colored ones; as it is well known, the human eye can distinguish more colors than gray-levels; certainly this led to a reduction in the possibility of a successive photo-interpretation of the classified product.

Some few mixed classes still remain. We believe that with the insertion of additional information such as textural information it will be possible to better separate these mixed classes. Indeed the insertion of textural information is one of the future step of our project, also in view of the ability of the SOM network to extract information from high-dimensional dataset.

With the same aim of refining the outcomes of the Ikonos 2 segmentation step, the macro classes obtained have been compared with the other satellite analysis products realized from Landsat and Spot supports and also with the results of photo-interpretation conducted on air photos at different scales and taken in different dates. This refining task also takes advantage from the combination of the above analysis results with information extracted from literature and archives.

The Information System

All the information derived from the excavations of this site - which have been carried out from the beginning of

'900 onwards - led to the creation of an Information System. This system was able to elaborate – on a regional and territorial scale – all data coming from several sources and their relevant analyses' methods and techniques which led to the creation of synthetic products, such as thematic geo-referential maps, both territorial and detailed, with both recognized and interpreted elements. The numerous and different used tools were managed within the GIS (Geographical Information System), by using a unique cartographic projection system (Gauss-Boaga, East Area) and by realizing an informative topographical basis through several multi-temporal and multi-scale tools and the use of different informatics treatments.

A GPS-Glonass High precision positioning system (at subcentimetric location accuracy) has been employed in ground-truth and survey campaigns. The whole area of the Paestum Plane has been framed in a topographic grid; this one is built over a well distributed set of fiducial ties that belong to the IGMI95 topographic satellite net.

A list of data sources is traced hereafter, departing from widely detailed (great scale) data to small scale data:

- plans of excavations from the Archive of the Archaeological Supervising Body;
- vectorial reproductions of the ancient inhabited Poseidonia-Paestum area, resulted from the excavations, plotting and restoration from past century up to nowadays;
- cadastral maps of the historical records, on a 1:2000 scale;
- detailed photogrametries on 1:5000 scale provided by Capaccio municipality;
- Italy geological maps on 1:100.000 scale, whose information were used as a departure point for defining the geo-environmental analysed territory;
- air photographs from past and present flights. We especially focused on photograms and films relevant to the flights made in 2002 and 1944;
- Landsat satellite imagery with a soil resolution of 30X30m, Spot with soil resolution 5X5m, Ikonos 2 with soil resolution 1-4m

Archaeological surveys have been conducted and they are still in progress, to confirm the outcomes of the laboratory analysis cited above.

Conclusions

The Classification algorithm cited above in combination with the other tools of the Information System has proven his effectiveness in widening the knowledge useful for the reconstruction of the ancient landscape of Poseidonia-Paestum. In particular, the classified image is a synoptic view of the territory, abridging past and present archaeological knowledge bases.

As an illustration of the usefulness of the complete Geographic Information System (GIS) and in particular of the Classification Product, an interpretation of the relationship between the distribution of the so called "Urban Necropolis" and the position of the Ancient Town is given. The analyses of the funerary areas is indeed fundamental in understanding the organization of the entire Greek Polis.

Till now, there was a lack of detailed plans of the archaeological findings and in most cases the position itself of entire Necropolis was uncertain. Sometimes the location of a funerary area was reduced to a merely cadastral registration data. Sometimes the graphic documentation was largely insufficient to make any kind of plan metric reconstruction.

Passing from symbolic maps without any spatial rigorous information to a specifically realized territorial GIS product has allowed us to make new hypotheses about the aforementioned relationship between the Necropolis and the Ancient Town and also among the funerary areas themselves.

As for the classification product outcomes, besides the environmental general analysis conducted above, a preliminary hypothesis to be verified is made about the presence of an extended funerary area next to the northern part of the town-walls, at present undiscovered. To strengthen this hypothesis there is the anomalous presence of patterns in the processed satellite scene, probably due to huge movement of terrain.

Finally, the Geographic Information Management System represents the best solution to accommodate data about new findings; moreover, all the data are easily shared and distributed to all the interested players in the field of tutelage, monitoring and planning of Cultural Resources.

Acknowledgements

We are thankful to Professor G.Avagliano belonging to the Directorate of The National Archeological Museum of Paestum

References

AA. VV. (1992), *Poseidonia-Paestum, Atti XXVII Convegno di studi sulla Magna Grecia, (Taranto – Paestum 1987),* Napoli 1992.

CIPRIANI M.-AVAGLIANO G. (1987) Gli insediamenti antichi nel territorio di Poseidonia-Paestum. In *Paestum*. Taranto: Istituto per la Storia e l'Archeologia della Magna Grecia (Collana di monografie a cura dell'Istituto per la Storia e l'Archeologia della Magna Grecia; I), p. 17-54.

DELEZIR J.-GUY M. (1992), Les conditions géographiques du site e du terroir de Paestum étudiées d'aprés des images de satellites (Landsat TM et Spot). In *Poseidonia – Paestum*, Atti del XXVII Convegno di studi sulla Magna Grecia, (Taranto – Paestum 1987), Napoli 1992, p. 463-470.

GRECO E. (1979), Ricerche sulla *chora* poseidoniate: il paesaggio agrario dalla fondazione della città alla fine del sec.IV a.C., in 'DialArch', n.s. 1, 2, 1979, pp.7-26.

GRECO E. (1992), La città e il territorio: problemi di storia topografica, In *Poseidonia – Paestum*, Atti del XXVII Convegno di studi sulla Magna Grecia, (Taranto – Paestum 1987), Napoli 1992, p.471-499.

GRECO E. - LONGO F. (2000) (a cura di), P*aestum. Scavi, studi, ricerche. Bilancio di un decennio (1988-1998),* Paestum 2000.

GRECO E. –STAZIO A. –VALLET G. (1987), *Città e territorio nelle colonie greche d'occidente,* I. Paestum, Napoli 1987.

KOHONEN, T. (1997). *Self-Organizing Maps*. Series in Information Sciences, Vol. 30. Springer, Heidelberg. Second ed. 1997.

MATHER, PAUL M. (1999) *Computer Processing of Remotely Sensed Images. An Introduction* – Second Edition – John Wiley Press

PONTRANDOLFO A. (1992) - Le necropoli dalla città Greca alla colonia latina. In *Poseidonia – Paestum, Atti del XXVII Convegno di studi sulla Magna Grecia, (Taranto – Paestum 1987),* Napoli 1992, p. 225-267.

PONTRANDOLFO A. –ROUVERET A. (1992), *Le tombe dipinte di Paestum*, Modena 1992.

PUGLIESE, L., SCARPETTA, S., ESPOSITO, A., MARINARO, M. (2005). An Application of Neural and Probabilistic Unsupervised Methods to Environmental Factor Analysis of Multi-spectral Images. *Lecture Notes in Computer Science* 3617, pp. 1190 – 1197.

ROTHAUS, R. M., DE MORETT, A. A. (2003): *Landsat TM Imagery in Landscape Archaeology: Detection and Modeling,* p. 149 – 180. ©2003 Edizioni all'Insegna del Giglio

VILLMANN T., MERENYI E., HAMMER B., (2003). Neural maps in remote sensing image analysis, *Neural Networks* 16 389–403

The application Potentials of "Beijing No.1"satellite data in investigation of The Great Wall

Nie Yue-ping,[1] Zhang Xian-feng,[2] Dai Jingjing,[1] Yang Lin,[3] Wang Chang-lin[1]

[1]Institute of Remote Sensing Applications Chinese Academy of Sciences , Beijing 100101
[2]Peking University, Beijing 100087
[3]The National Museum of China, Beijing 100006

1 Introduction

Historical cultural heritage is the information carrier of human history. Remote Sensing Satellite can observe the earth from the space, and contains abundant historic traces and objectively reflects their spatial distribution patterns and characters. The technology of remote sensing, geographic information system (GIS) and global positioning system (GPS) developed in the 20th century enables us to investigate ancient sites over ground and underground from a new angle, and become more and more popular with archaeological scientists. And it forms a new inter-discipline named: Spatial-Archaeology. The application of Spatial-Archaeology mainly includes these aspects: (1) detect ancient sites using aerial or satellite images through the abnormity information; (2) investigate underwater ancient sites, ancient ports, ancient watercourse, and their sedimentary environment using remote sensing data; (3) analyze ancient environment from remote sensing data, to rebuild ancient environment and study social, cultural and natural situation in that era; (4) make maps and build spatial archaeological information system, to provide service for management of archaeological recourses; (5) rebuild ancient sites and ancient environment using GIS and virtual reality techniques.

The Great Wall is the precious cultural heritage which represents the historical evolution of China, and has great impact in the world. From the aspect of culture, it is not only an important military barrier that was developed through the warfare and compromise between agricultural nationality mid of China and north nomads, but also a facility that protects the trade communication. It developed through Spring and Autumn Period, Warring states time, the Qin and Han Dynasty, Northern Wei Dynasty, Sui Dynasty, and finally Ming Dynasty. And it extends from the south foothill of Hu Mountain in Liaoning provice Dandong city in the east to Jiayu Pass in Gansu Provice in the west. It is an emergent task to effectively develop this precious cultural heritage without destroy.

Our government has prepared a project named "The Great Wall project" which indicates that the National Cultural Relic Bureau will take charge of the Great Wall investigation of its distribution and reserved condition. Because of the great spatial scale, traditional filed archaeology can hardly finish such a huge investigation job. What's more, some sites are not available because its complex landform condition. So this project emphasizes the application of spatial techniques such as remote sensing, GIS and GPS. And now it is planned to make some experiments and trainings in Shanhai Pass and Jiayu Pass.

"Beijing No.1" small satellite was launched by China in 2005, and it has three multi-spectral bands whose spatial resolution is 32 meters, and one panchromatic band whose spatial resolution is 4 meters. Compared with other satellites, "Beijing No.1" small satellite has wider imaging range (600 kilometers), and has higher temporal resolution (about 2-3 days). What's more, its

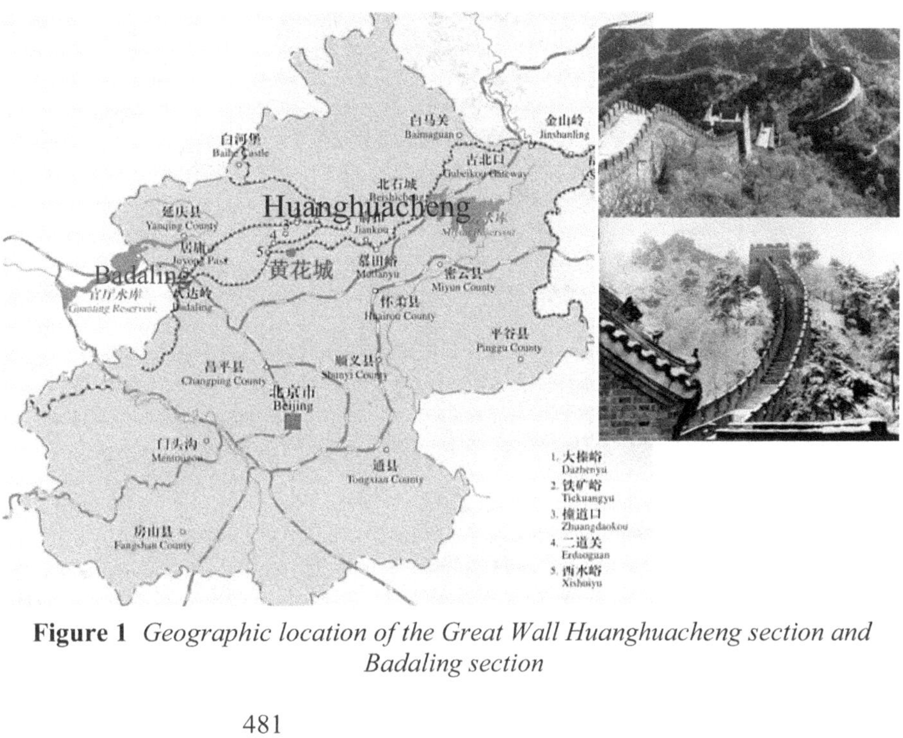

Figure 1 *Geographic location of the Great Wall Huanghuacheng section and Badaling section*

panchromatic bands can provide high spatial resolution data. All of these characteristics are very useful to archaeological studies.

This paper chose the Great Wall Huanghuacheng section and Badaling section in Beijing area as the example, mainly discussed the application potential of "Beijing No.1" satellite data on the Great Wall study, especially on interpretation and auto-extraction of the Great Wall.

2 Extraction of the Great Wall

2.1 The characteristic of the Great Wall

The Great Wall which was conserved well was mostly built in Ming-Dynasty, and it was mainly made of soil or brick. Moreover, it was usually composed of relatively narrow rampart and relatively wide war-flame platform. It was told that the Great Wall in Beijing area was built in Ming-Dynasty by the general Xu Da. And it mainly distributed in north, west and east of Beijing mountain areas.

The Great Wall is several meters high, and six meters wide. So it presents as linear characteristic on remote sensing images whose spatial resolution is less than ten meters, but when the spatial resolution is over ten meters, it is hardly to detect the Great Wall. With regard to spectral respond, since the Great Wall is made of man-made substances and its density is lager than the environment, it presents high echo information on microwave remote sensing data, and presents high spectral value on visible light remote sensing data.

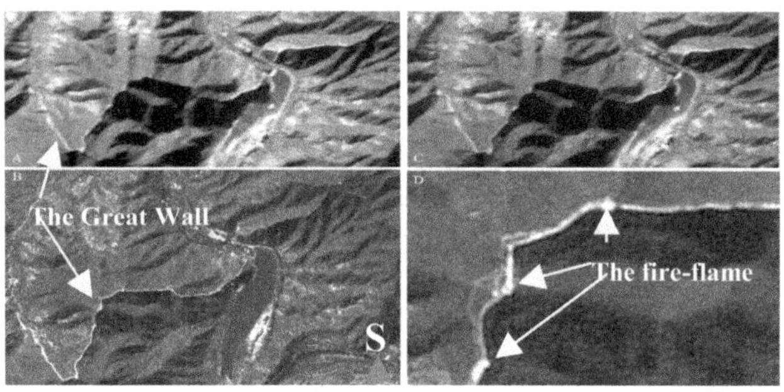

Figure 2 *Fusion image of multi-spectral and panchromatic bands of "Beijing No.1" vs. "SPOT" panchromatic image in Huanghuacheng section (A: "Beijing No.1" panchromatic band; B: "SPOT" panchromatic band; C: Fusion image of "Beijing No.1" multi-spectral and panchromatic bands; D: SPOT panchromatic band image after magnified 4 times)*

The Great Wall Badaling section was mostly repaired in contemporary era and was relatively reserved in good condition. At the same time, it is easy to identify on the remote sensing images. However, Huanghuacheng section was repaired in about 2004 only for a small part, and mostly collapsed, so different parts may present different characteristic on the images.

2.2 Interpretation keys and results

Spectral radiation value:
For "Beijing No.1" satellite data, on the panchromatic image (4 meters of spatial resolution), the Great Wall in Beijing area presents high light, and easy to be distinguished from other substances (Fig.2 and Fig.3). While on the three multi-spectral images (32 meters of spatial resolution), it is difficult to identify the Great Wall because of low spatial resolution.

Geometry characteristic:
Since the Great Wall is composed of relatively narrow rampart and relatively wild war-flame platform, it usually presents high lightness on high spatial resolution images compared with the environment, and the war-flame platform can even be identified, such as on the SPOT and QuickBird images (Fig.2-D). On the panchromatic band image of "Beijing No.1" satellite, the Great Wall can be identified as linear characteristic. But compared with SPOT image, the fire-flame platform can be identified in the Badaling section, but not clear in the Huahuacheng section (Fig.2 and Fig.3).

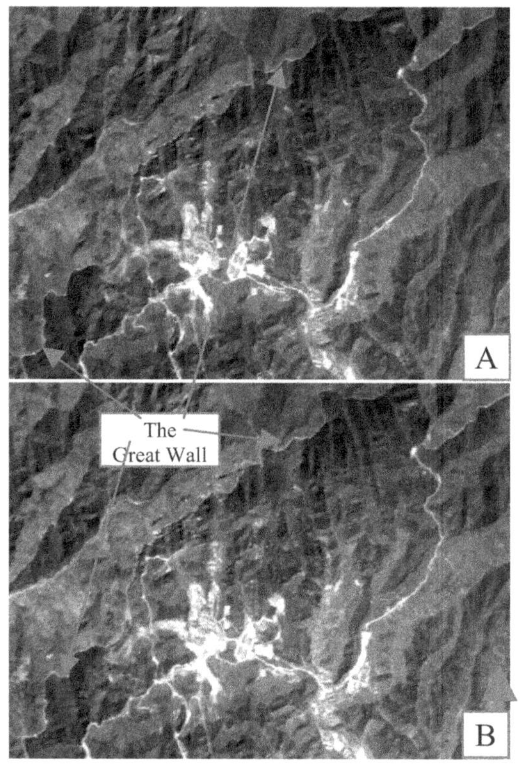

Figure 3 *"Beijing No.1" satellite image of Badaling section (A: Panchromatic band; B: Fusion image of panchromatic band and multi-spectral bands)*

Location and relationship with the environment:
Because the Great Wall was primarily built as the military barrier, the Beijing section was mostly built along the ridge of the mountain, and slope grade is greater on the north slope than the south slope. When span the channel, there are some high defend buildings, such as the famous pass Juyong Pass.

Figure 4 *Overlay of the interpretation result (the white line) and the DEM image*

There is usually vegetation, bare rock or sand surround the Great Wall. On the remote sensing image, when it is north upward and east rightward, the Great Wall present in the channel, and when rotate 180 degree, the Great Wall present in the ridge of the mountain.

According to the interpretation key above, we firstly interpret for the Huanghuacheng section, and then overlay the result on the DEM image (Fig.4). Fig.5 presents overlay of the fusion image and the DEM. From the three-dimension presentation, we can see the distribution of the Great Wall.

Figure 5 *Overlay of fusion image (panchromatic band and multi-spectral bands) and DEM image*

2.3 Auto-extraction experiment of the Great Wall

According to the analysis above, the extraction of the Great Wall is similar to linear objects extraction. The methods of linear objects extraction include: various filter methods (such as edge enhancement), grade direction analysis (GDPA) (Wang and Zhang 2000), template matching (Park and Kim, 2001), SNAKE Model (Gruen and Li, 1997; Kland, 1998), and GIS method, etc. Linear objects extraction plays an important role in road extraction and geological linear structure extraction, but now there are even not a very powerful method for linear road auto-extraction. In a word, high spatial resolution and spectral resolution data is useful for the Great Wall study.

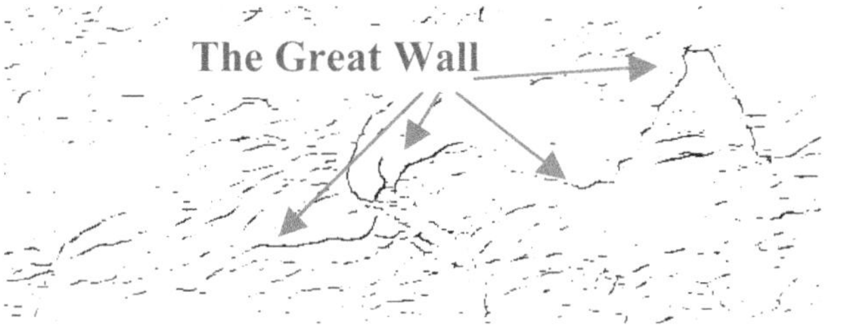

Figure 6 *The extraction experiment using spectral curvature method near Huanghuacheng section*

If we compare the spectral radiation value on the panchromatic image with the altitude value on DEM image, we can get the curvature of the spectral radiation value as well. This variable can reflect the change speed of the spectral radiation value. After put the threshold, we get the spectral curvature image (Fig.6). On the image, the Great Wall presents high light because high spectral curvature value, but it is difficult to distinguish it from other objects like road, dry river valley etc. Some other methods and data need to be studied to eliminate non-Great Wall information. For example, based on different physiognomy position and topography characteristic , the non-great Wall information can be wiped off.

3 The compration of "Beijing No.1" small satellite data with "SPOT 5" satellite data

3.1 Data characteristic

From Table 1, it can be seen that "SPOT 5" data excels "Beijing No.1"satallite data in spatial resolution and Signal-to-Noise Ratio (SNR). For remote sensing archaeology, spatial resolution is usually the key index. For example, the width of the Great Wall is usually six meters, it presents 2 to 3 pixels on "SPOT" panchromatic image whose spatial resolution is 2.5 meters, but presents 1 to 2 pixels on "Beijing No.1" panchromatic image whose spatial resolution is 4 meters. Although it may be interpreted less information on the "Beijing No.1" panchromatic image than on the "SPOT" panchromatic image, the "Beijing No.1" panchromatic image can basically fulfill the needs of the Great Wall investigation, and has some other special advantages.

Figure 7 *"Beijing No.1" panchromatic band vs. "SPOT" panchromatic band of*

3.2 Image intuitionistic compration

On Fig.7, it is more distinguishable on "SPOT" panchromatic image than "Beijing No.1" image for A section and C section. However, it can be identified well on both "SPOT" panchromatic image and "Beijing No.1" image for B section. For Badaling section on Fig 8, we found that 92.4% of the Great Wall can be interpreted from "Beijing No.1"satellite image. It indicates that although spatial resolution of "Beijing No.1" satellite is 1.6 times less than "SPOT", it can also fulfill the need of the Great Wall investigation.

3.3 Temporal resolution

The temporal resolution of "SPOT" is 26 days, while the temporal resolution of "Beijing No.1" is 2-3 days. So "Beijing No.1" is more advantageous for emergent conditions like disaster investigation. In the mean time, "Beijing No.1" was launched by China, and the data is far cheaper than "SPOT" for us, what's more, we can adjust the take-over range and gesture of the satellite according to different special application needs, which is very useful to remote sensing archaeology.

3.4 Auto-extraction result compration

The spectral curvature method is also applied to "SPOT" panchromatic image, see Fig.9. And form the compration of Fig .9 and Fig.6, it is clear that it presents more coherent linear structure on Fig 9, because of higher spatial resolution of "SPOT" (2.5 meters). However, another problem takes place that more linear structures are extracted because of higher spatial resolution such as mountain alley and the bank of reservoir. There are no such problem on "Beijing No.1" satellite image.

Table 1 *"Beijing No.1" satellite data vs. "SPOT 5" satellite data*

Remote Sensor	Bands (SNR)	Spatial Resolution (m)	Wavelength Range (μm)	Orbit Width(km)
DMC("Beijing No.1")	Panchromatic band: 140 B1 :Green 140 B2 :Red 140 B3 : Near Infrared 140	4 m 32 m 32 m 32 m	Panchromatic 0.52 – 0.62 μm 0.63 – 0.69 μm 0.76 – 0.90 μm	24 600
SPOT 5	Panchromatic band: 170 B1 : Green 240 B2 : Red 240 B3 : Near Infrared 240 B4 : Middle Infrared 230	2.5 m or 5 m 10 m 10 m 10 m 20 m	0.48 - 0.71 μm 0.50 - 0.59 μm 0.61 - 0.68 μm 0.78 - 0.89 μm 1.58 - 1.75 μm	60

4 Field investigation and rectification

The purpose of field investigation and rectification is as follows:

(1) understand the geometry feature characteristic, and combine with the sensor feature of "Beijing No1", to erect the interpretation key of the Great Wall;

(2) validate the result of interpretation, and estimate the precision; (3) survey the doubtful point, and find out the reasons and answer. So we investigate in Huanghuacheng section on March 25th 2006.

Through field investigation, the Great Wall in Beijing area was mostly built and repaired in Ming-Dynasty, and mainly made of brick, some was made of stone (Fig.10). The Huanghuacheng section was repaired from 2004 to 2005, so it is more recognizable on "Beijing No.1" satellite image. While before the reparation, vegetation coverage was very clear, and it is more difficult to recognize the Great Wall on images (Fig.11). In fig.2, we found that it is more recognizable for the section in the west of Huanghuacheng reservoir on "Beijing No.1"

Figure 8 *"Beijing No.1" panchromatic band (A) vs. "SPOT5" panchromatic band (B) of Badaling section (both are not after geometric rectification)*

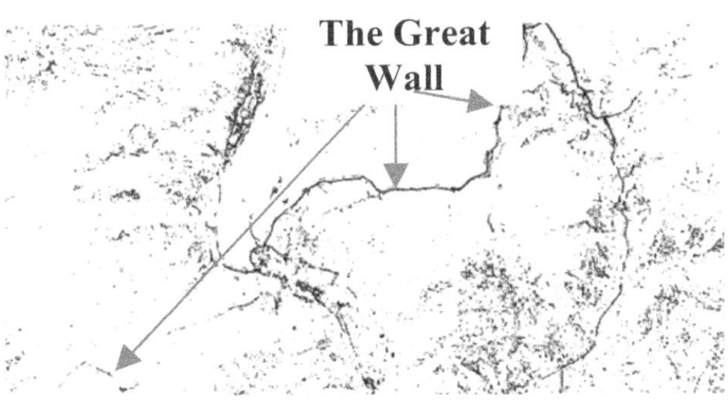

Figure 9 *Spectral curvature method on "SPOT" panchromatic image near Huanghuacheng section*

Figure 10 *The Great Wall made of brick (right) and stone (left) near Huanghuacheng section*

image than "SPOT" image, it is because that "SPOT" image was received in October, 2004, while "Beijing No.1" image was received in February, 2005 (Fig.12).

5 Conclusions and discussions

5.1 Conclusions

"Beijing No.1"satellite is one of the DMC series moonlets that our country launched independently. Its spatial resolution is 32m (multi-spectral band) and 4m (panchromatic band), and can be used in many fields such as resources and environment survey, disaster monitoring, archaeology etc. Our research focuses on the application potential of the "Beijing No.1" satellite data in the Great Wall extraction in northern Huairou area of Beijing. And we can draw some conclusions:

Figure 11 *The Great Wall near Huanghuacheng section before reparation (Photo took in April,2004)*

(1) The panchromatic band of "Beijing No.1" with 4m resolution can recognize the brick Great Wall of northern Beijing, which can be used for Great Wall investigation. After data fusion, the multi-spectral data can recognize the environment of Great Wall and improve the precision of interpretation. The Great Wall of Huanghuacheng section isn't protected very well, but most of it can be recognized easily. The data of "Beijing No.1" has many advantages such as data capture and temporal resolution, so it can provide very good data source and will play an important role in archaeology in the future.

(2) Compared with "SPOT" data, the panchromatic band of "SPOT" can recognize the Great Wall of Beijing well, even the rampart and war-flame platform. This thanks to the scale coupling of high spatial resolution of sensor and the width of the Great Wall. The rampart and war-flame platform can be recognized in the images captured by "Beijing No.1", which demonstrates that "Beijing No.1" data can rival "SPOT5" data in information extraction.

(3) The auto-extraction of Great Wall based on the panchromatic band of "Beijing No.1" need more research. Though radiation curvature method used in the research can extract the Great Wall information whose lightness is higher than its environment, it can be mixed with other high light objects such as road. In order to extract the Great Wall, we need more information.

(4) Considering the fusion experiment, because the resolution of multi-spectrum is eight times less than panchromatic band. If we use Principal Component Analysis(PCA), Gram-Schmidt Transform, HIS transform, Wavelet transform or Brovey transform, the image after fusion will have great changes and distortion, so more new arithmetic need to be discussed. Among these, PCA and Gram-Schmidt transform can achieve better results than others. Moreover, the fused image of "Beijing No.1" don't fit to quality and half-quality analysis such as ratio.

5.2 Questions and discussions

(1) Observe and attain the shadow images of Great Wall away from nadir;
We cannot recognize representative Badaling Great Wall completely from the first panchromatic band data received (February 12th 2005) because of satellite pose. The reverse question of this question is that the Great Wall will have visible shadow if we observe away from nadir. The shadow can be used to measure the altitude and capture information of Great Wall.
(2) Whether panchromatic band of "Beijing No.1" can recognize the brick Great Wall that is made of soil or not protected well need more exemplification research and demonstration. It has great guiding significance for the Great Wall investigation.

Figure 12 *The Great Wall near Huanghuacheng section after reparation*

(3) We should pay more attention to how to make use of other non-remote sensing information or applicable arithmetic to study automatic and semi-automatic extraction of the Great Wall information.
(4) No matter "Beijing No.1" data or "SPOT5" data can not discern the details of Great Wall. Though we can't capture its geometry feature, we can capture its linear feature to find out the geographic location and distribution. If the conservation condition of Great Wall should be investigated, aerial remote sensing of higher resolution even field investigation assisted by GPS is needed. Hence, in the task of Great Wall investigation, we should apply multiple remote sensing means associated with the non- remote sensing methods. One effective way is that using "Beijing No.1"data for macroscopical investigation, then use aerial remote sensing and field investigation assisted by GPS for detail investigation.

"Embanked roads" of the Southern Valli Grandi Veronesi Between airphoto interpretation and ground evaluation of buried stratigraphy: a positive feedback

A. Betto,[1] G. De Angeli,[1] F. Sartor[1]

[1] C.I.S.A.S. – Centro Italiano Studi di Archeologia di Superficie

1 Introduction and topics

This note[1] reports the results of a study carried out in the southern area of the Valli Grandi Veronesi as a part of a wider project, called 'AMPBV – Alto Medio Polesine Basso Veronese', co-ordinated by C. Balista, A. De Guio and A. Vanzetti. This area is characterised by the presence of some Bronze Age (second millennium B.C.) *terramara*-type settlements, the most important of which (Castello del Tartaro, Fondo Paviani, Fabbrica dei Soci) have been widely investigated and published (Balista, De Guio 1997, also for previous bibliography). Thanks to its geo-morphological history – a widespread marshland, reclaimed only in the 19th century - it is an extraordinary example of palimpsestical and multiphase fossil landscape, particularly rich of well preserved buried natural and anthropic marks (Balista et alii 1998, Ferri 1992, Tozzi, Harari 1990).

Specifically, this landscape is marked by the tracks of two large embanked roads:

- a northern road, called 'Northern Embanked Road' – N.E.R. (known in literature as 'Strada del Cavariolo'), investigated by core borings and pedo-alluvial trenches, realised near the site of Castello del Tartaro (BALISTA 1996), and dated up to the Romanisation period.
- a southern road, named 'Southern Embanked Road' – S.E.R., uncertainly dated, but probably related to the bronze-age anthropic landscape because of its relative chronology relations.

Our research concentrated on the latter structure, in order to fix a strict dating and to make a proposal for a formative model inside the ancient natural and anthropic landscape, in which it is perfectly inserted.

2 Airphoto Interpretation

The choice of the target-points to investigate has been pointed out by the analysis of a covering of some aerial photographs (CGR 1990, GAI 1955-56). The result was that, in two different areas, the feature of a paleochannel (the so called Ponte Moro palaeochannel – Balista in print), NW/SE oriented (Fig. 2), showed the same relationship between the road of interest southwards and the Cavariolo road northwards. The airphoto interpretation, therefore, focused on these two main points. It has been possible to elaborate a first stratigraphic diagram model (De Guio 1992, 1996, 2001) explaining the relationship between each identified unit (Fig. 1).

Particularly, units referred to the Southern Embanked Road and its lateral ditches, coming from West, are interrupted near the intersection with the Ponte Moro paleochannel, and then continue eastwards (Fig. 3): this fact let us deduce that the paleochannel was actually cutting the road.

On the reverse, the airphoto interpretation of the intersection between the Northern Embanked Road and the paleochannel showed an opposite relationship (Fig. 4): the fluvial units interrupted near the road, determining the posteriority of the road as regards to the Ponte Moro paleochannel.

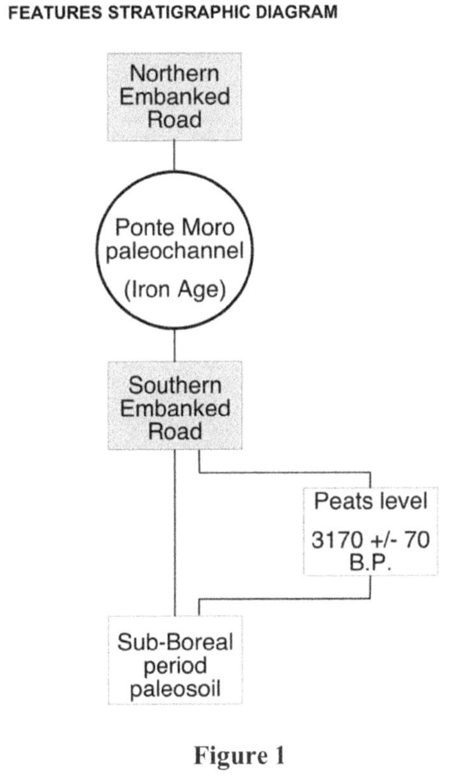

Figure 1

[1] This work has to be equally referred to the authors. We wish to thank prof A. De Guio, dott. A. Vanzetti and dott C. Balista for giving the chance to develop this research; a special thank to C. Balista also for his continuous discussion and precious suggestions.

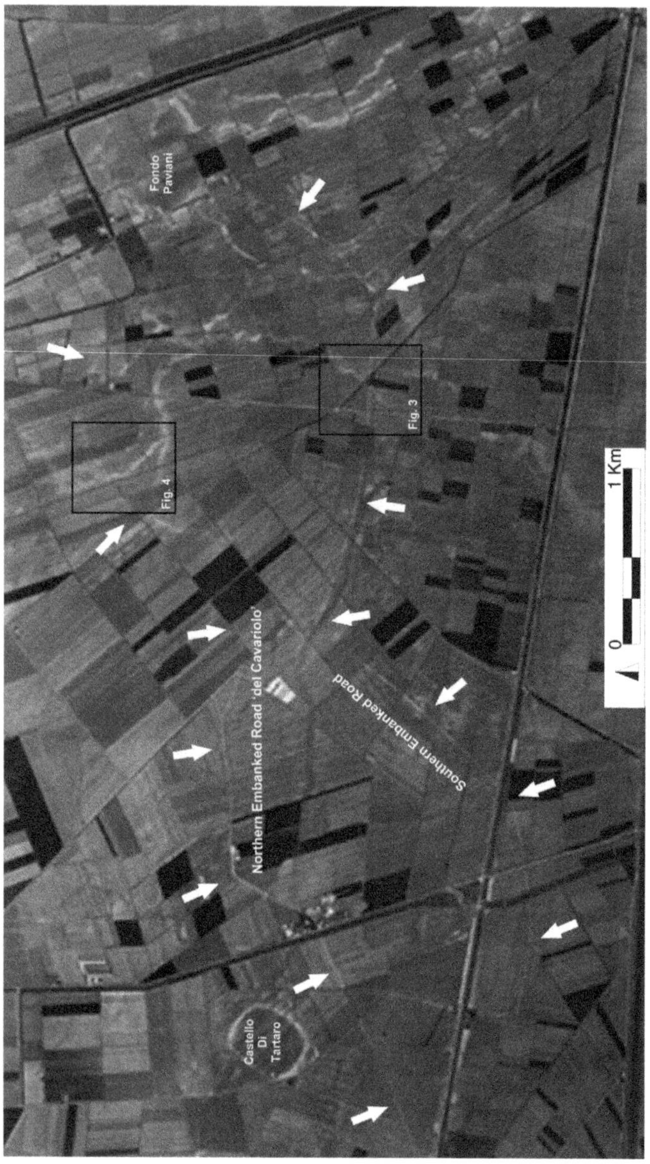

Figure 2

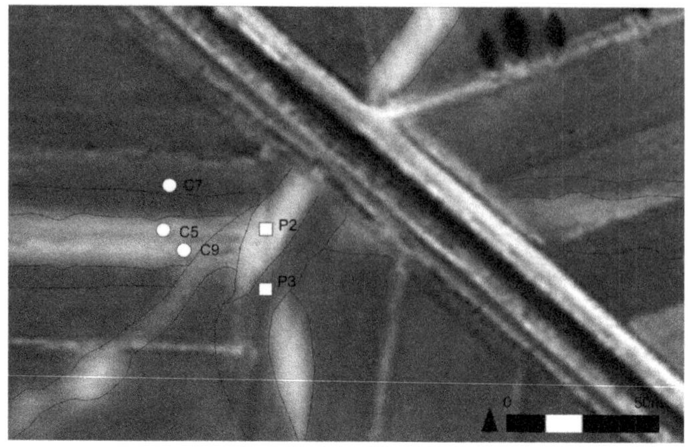

Figure 3

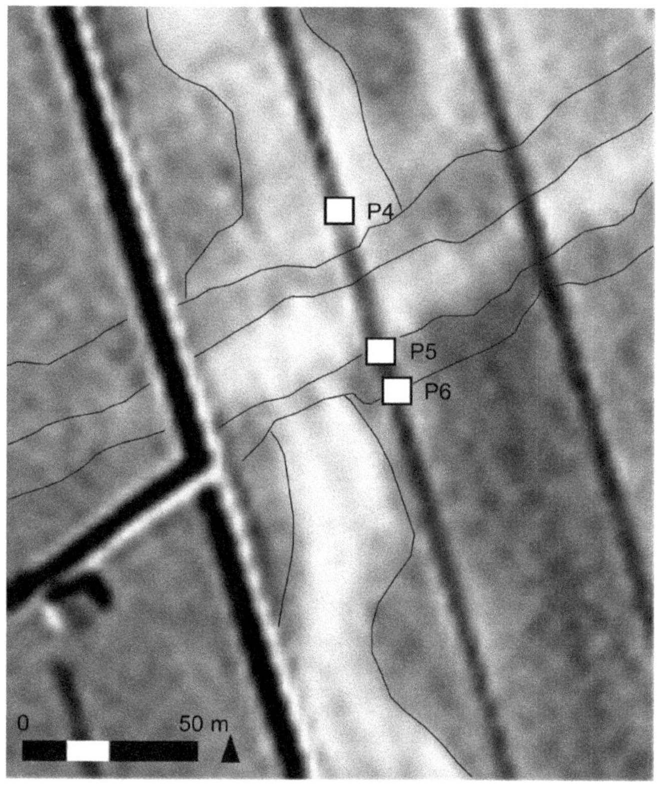

Figure 4

3 Field investigation

These hypotheses have therefore been checked on the field by means of the documentation of stratigraphic sections, and of manual core borings in the most sensitive points.

In Ponte Moro area, the field analysis has been carried out by twelve core borings and three sections[2] exposed on the vertical surface of already existing drains through the paleochannel trace. The latter brought to light only flooding layers that can refer to an abandoned river corresponding with the feature observed in the aerial photographs.

The core borings, lined up in order to carry out a North-South cross section trough the road (Fig. 3), could let us gather some data about the sedimentologic characteristics of the road embanking and the stratigraphic relationships between the latter and the below and lateral units.

The data synthesis (graphically illustrated in Figs. 5, 7) proved a posteriority relationship of the paleochannel to the road by cutting/erosion, and a posteriority of the latter to a sub-Boreal period (third millennium B.C.) paleosoil (Sorbini et Alii 1984).

A similar method has been adopted northwards to check the hypothesis about the relationships between the above-mentioned paleochannel and the Northern Embanked Road: on the reverse, differently from the previous case,

[2] Out of all the carried out sections and core borings here are reported only the most representative to understand the buried stratigraphy.

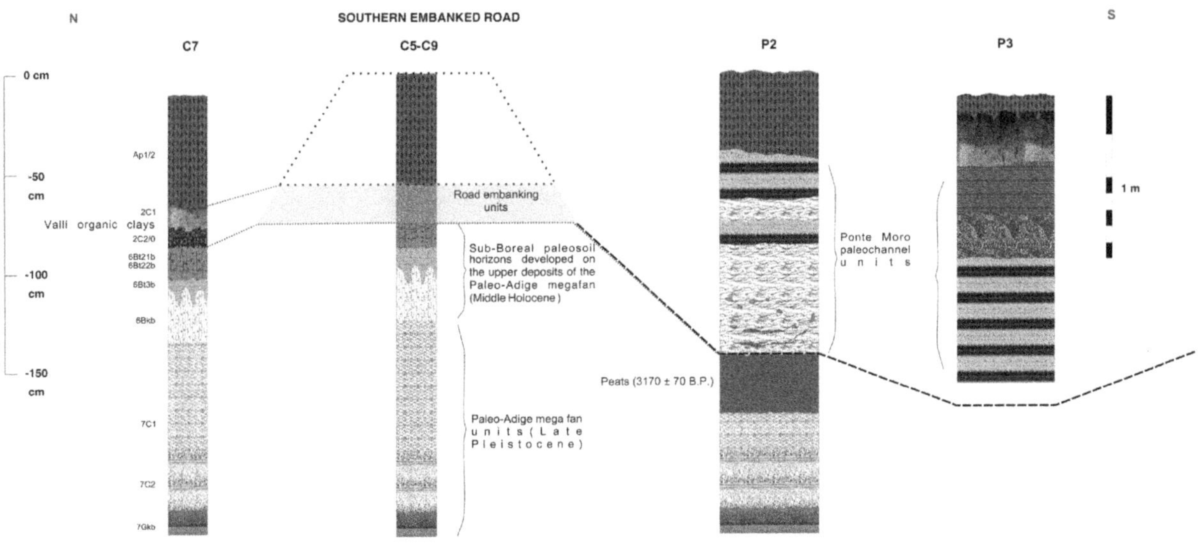

Figure 5

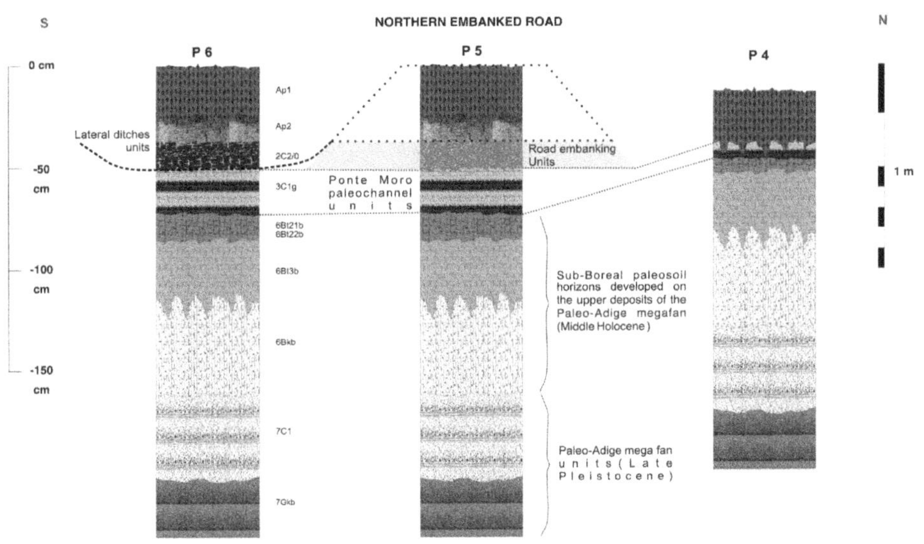

Figure 6

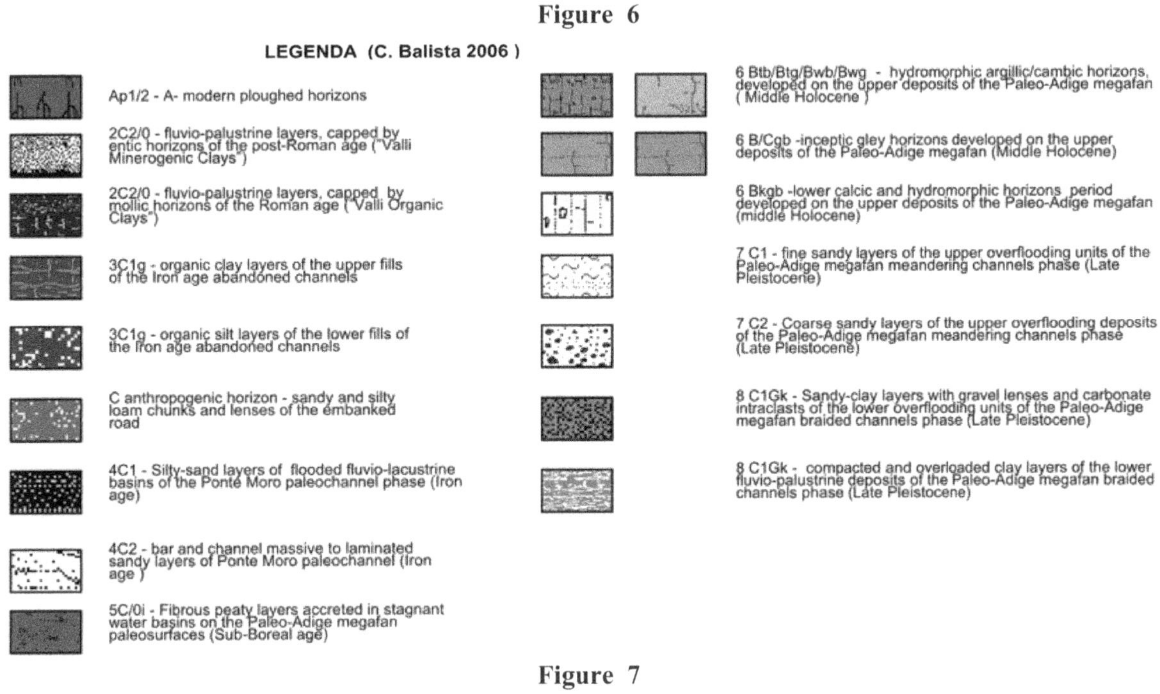

Figure 7

the four profiles that have been realised show the recentness of the road embanking as regards to the fluvial layers by means of a physic covering relationship (Figs. 6, 7). This latter analysis completes and confirms the frame of the relationships among the features shaped in the aerial photographs interpretation phase.

4 Conclusions

The gathered data, so far exposed, derive from an interaction work (with a positive feedback phenomenon) between the aerial photographs analysis and the subsequent ground evaluation of the hypothesis by means of a detailed analysis of the buried stratigraphy.

The remote observation put the merk of the Southern Embanked Road in direct stratigraphic relationship with the Northern Embanked Road thanks to the ground feature of the Ponte Moro[3] paleochannel: this latter is covered by the northern track and its ditches, and on its turn it covers the southern track.

The field stratigraphic study and the synthesis of the so gathered data allowed us giving also a vertical dimension (both physically and chronologically) to a landscape that in the remote sensing analysis looks palimpsestic and bidimensional; in particular, the ground investigation allowed us to verify the stratigraphic relationship between the different elements analysed in the aerial photographs interpretation phase (Figs. 3, 4). The result has been setting a relative chronological sequence which, in two distinct points of the area puts the features of two embanked roads in relationship with that of an only paleochannel (Ponte Moro paleochannel). This sequence has one of the two roads features, that is the Northern Embanked Road, as a known chronological term, already referred to Romanisation period by C. Balista (Balista 1996).

Once submitted to ground truth, the hypothesis of a priority of the southern road building to early Iron Age flooding events suggested the chance of its collocation in the Bronze Age anthropic landscape (the most ancient infra-structured anthropic landscape known so far).

A further confirmation of the suggested model derives from a peat horizon set below the silty and sandy layers that form the flooding cycle of Ponte Moro paleochannel (Figs. 5, 7), cut by this latter, and stratigraphically set in phase with the sub-Boreal paleosoil covered by the road. A sample of peat, picked out from this horizon and dated by means of ^{14}C, reported a date to 3170± 70 B.P. (1452 B.C. central calibrated date[4]). This datum represents an important *post-quem* term for the object under examination, and forms a useful support for the so far suggested model that fixes chronologically the road handmade in the range of the local *terramara*-type sites, in the moment of their maximum development (Middle-Recent Bronze Age).

[3] A recent work by C. Balista even more strongly relates this latter to the early iron age (Balista, in print).

[4] Stuiver, Reimer 1993.

References

BALISTA 1996, Geoarcheologia delle formazioni superficiali: linee guida e casi studio dal progetto AMPBV. I risultati di una ricerca volta al definitivo inquadramento stratigrafico delle strade su argine delle Valli Grandi Veronesi, in *La Ricerca Archeologica di superficie in area padana*, a cura di E. Maragno, Stanghella, pp. 319-349.

BALISTA, DE GUIO 1997, Ambiente ed insediamenti dell'età del bronzo nelle Valli Grandi Veronesi, in *Le terramare la più antica civiltà padana,* catalogo della mostra, a cura di M. Bernabò Brea, A. Cardarelli, M. Cremaschi, Modena 1997, pp. 137-165.

BALISTA ET ALII 1998, Bronze-Age "Fossil Landscapes" in the Po Plain, Northern Italy, a cura di Hansel B., *in Man and Environment in European Bronze Age*, Oetker-Voges Verlag, Kiel, pp. 493- 499.

DE GUIO A. 1992, Progetto Harris, in *Tipologia di insediamento e distribuzione antropica nell'area veneto – istriana dalla protostorica all'alto medioevo*, atti del seminario di studio, Asolo, 3-5 novembre 1989, Monfalcone, Edizioni Della Laguna, pp. 120-121.

DE GUIO A. 1996, Archeologia della Complessità e Pattern Recognition di Superficie, in *La Ricerca Archeologica di Superficie in Area Padana*, Ags Edizioni, pp. 275-317.

DE GUIO A. (ed.) 2001, "Superfici di rischio" e C.I.S.A.S. se lo conosci non lo eviti, in *Rischio archeologico*, atti del convegno, Ferrara 24-25 marzo 2000, pp. 265-306.

FERRI 1992, Paleogeografia di un settore delle Valli Grandi Veronesi tra il Naviglio Bussè e il Castagnaro a nord del F. Tartaro:nuovi contributi e metodologie dell'indagine aerofotografica, in "*Tipologia di insediamento e distribuzione antropica nell'area veneto-istriana dalla protostoria all'alto medioevo*", Atti del Seminario di studio, Asolo, 3-5 novembre 1989, Edizioni della Laguna, Monfalcone/GO, pp. 111-115.

SORBINI L., ACCORSI C.A., BADINI MAZZANTI M., FORLANI L., GANDINI F., MENEGHEL M., RIGONI A., SOMMARUGA M. 1984, Geologia e Geomorfologia di una porzione della pianura a sud-est di Verona, *Memorie del Museo Civico di Storia Naturale di Verona (II^a serie), sez. di Scienze della Terra N°2*, pp. 1-9, Verona.

STUIVER M., REIMER P.J. 1993, Extended ^{14}C data base and revised calib 3.0 ^{14}C age calibration program, in *Radiocarbon*, 35, pp.215-230.

TOZZI P., HARARI M. 1990, *Tempi di un territorio. Atlante aerofotografico della Valli Grandi Veronesi*, (edizione a cura della Provincia di Verona), Comp. Generale Riprese Aeree Ed., Parma.

Remote Sensing and Ground-Truthing of a Medieval Mound (Tuscany - Italy)

S. Campana,[1] R. Francovich,[2] L. Marasco[3]

[1] University of Siena, Department of Archaeology and History of Arts, Landscape Archaeology, (LAP&T)
campana@unisi.it
[2] University of Siena, Department of Archaeology and History of Arts, Medieval Archaeology
francovich@unisi.it
[3] University of Siena, Department of Archaeology and History of Arts, PhD Student, (LAP&T)
marasco@unisi.it

1 Introduction

The site which we propose to discuss in this paper was 'discovered' from the air in the summer of 2005 during the University of Siena's XIVth International Summer School in Archaeology. It lies close to the castle of Scarlino, facing the sea on the west coast in the province of Grosseto.

The landscape of this area has been studied by the University of Siena's Department of Archaeology since 1979, when Riccardo Francovich started archaeological excavations at the castle of Scarlino (Francovich 1985). Shortly after the start of the excavations, work also began on a programme of field-walking survey that has helped to make this one of the best-studied landscapes in the province of Grosseto (Cucini 1985).

The site had already been identified at an earlier stage through the air photo interpretation of vertical historical coverage (Cosci 2005). Unfortunately the absence of a detailed map of the features and the unavailability at the time of these earlier studies of GPS navigation made the identification of the evidence in the field extremely difficult. Field-walking survey and ground-truthing variously suggested a positioning either one side or other of the apparent air photo location, without ever revealing its true position or scale (Chapman 2001; Campana 2005). Although analysis of the aerial photographs allowed the site to be interpreted as a triple enclosure, the lack of its correspondence on the ground with any artefactual evidence prevented us from obtaining any understanding of its chronology or of its potential contribution to the development of new settlement patterns in the area. In this context we should perhaps

Figure 1 *Overview of the landscape*

note the difference in familiarity with this kind of site in Italy as compared with other parts of Continental Europe or the United Kingdom. In Germany or England, for instance, archaeologists would find it quite easy to interpret this kind of feature without the need for ground-truthing, but in Tuscany this kind of evidence has no immediate parallels or comparisons.

In this paper we will show how the quality and the quantity of the information was supplements by more recent studies. Using this new evidence we will then discuss the new perspective which this opens up for archaeological research in the surrounding landscape.

2 The local landscape and historical background

The site, whose placename Castellina indicates its possible relationship with a castle, is set at the centre of a large area of lowland fronting onto the Gulf of Follonica. The area may have been characterised in the past by a wide brackish lake. Attempts at reconstruction of the ancient lake boundary would set our site close to the ancient coastline (Bartolini et al 1977).

The shoreline extended approximately from our site to the base of the hill on which stands the castle of Scarlino. The navigability of this salty lake is still documented in the late Middle Ages but was impeded in the following centuries by a progressive dessication of the area, creating an un-navigable swamp. The archaeological study of the site is still in its early stages but there appears at first glance to be a clear relationship with the surrounding landscape context and with the development of economic activity in the area. Past researches have concentrated on interpretation of the medieval landscape as being focused, from the early Middle Ages, around the hilltop village of Scarlino. Here the archaeological excavations, above settlement evidence from the Bronze Age to the days of imperial Rome, revealed firstly the remains of an open village, later converted into a *curtis* in the ninth century AD and finally transformed into a castle settlement from the end of the tenth century onwards (Francovich 1985).

More recently a new study of the archaeological data from the Scarlino excavation has furnished interesting comparisons for the pottery recovered from the site (Marasco forthcoming). Furthermore, the archaeological evidence has been supplemented by a rich archive of early medieval documentary sources relating to the Bishopric of Lucca. The integration of the various information sources has allowed us to recognise that, after the crisis of the Late Roman period, the rural landscape of this area was characterised between the seventh and eighth centuries by the presence of villages. The development of the settlement pattern at the beginning of the early Middle Ages followed two main paths: the concentration of population centres and the beginnings of hilltop villages.

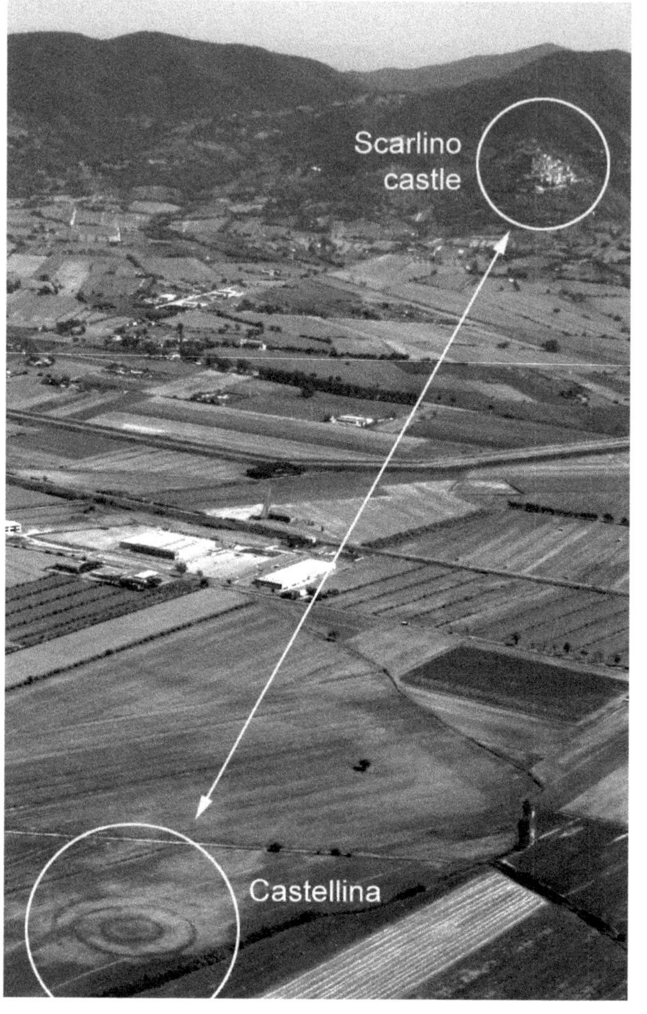

Figure 2 *Oblique air photograph of the cropmark features and their relationship to the central place of Scarlino*

The archaeological excavations at the castle of Scarlino – and at many other excavations – allow us to describe some aspects of these villages during the seven to eighth and ninth centuries AD. They consist mainly of wooden huts, organised into a central nucleus and sometimes defended by palisades or other forms of enclosure. During the eighth century the documentary sources attest to the growth of a new form of resource-management and exploitation, the *curtis* system. The archaeological investigations show the overlap between these new system and the preceding settlement pattern. In the same century the archaeological sources show the transformation of the villages and it is now possible to recognize evidence of spatial organisation related to the exercise of management and power.

However, many hilltop villages did not succeed in taking on the role of centres of human aggregation, perhaps because of the weakness of the local aristocracy or because of competition with other developing centres. The site that we discuss in this paper is one of these latter cases – a failure site. The site evidence represents an early Middle Age survival – not necessarily a direct one –

of the lowland population pattern, probably related to specific natural or man-made features – swamp-land or road systems etc.

Lorenzo Marasco

3 Air photo survey to study the site

The new episode of research in this area started with the XIVth International School in Archaeology, organised in collaboration with the Culture 2000 project *European landscapes: past, present and future* (www.e-landscapes.com). Within this project, in the spring of 2005, the University of Siena, in collaboration with English Heritage and the international Aerial Archaeology Research Group (AARG), conducted an Aerial Archaeology Research School at Grosseto

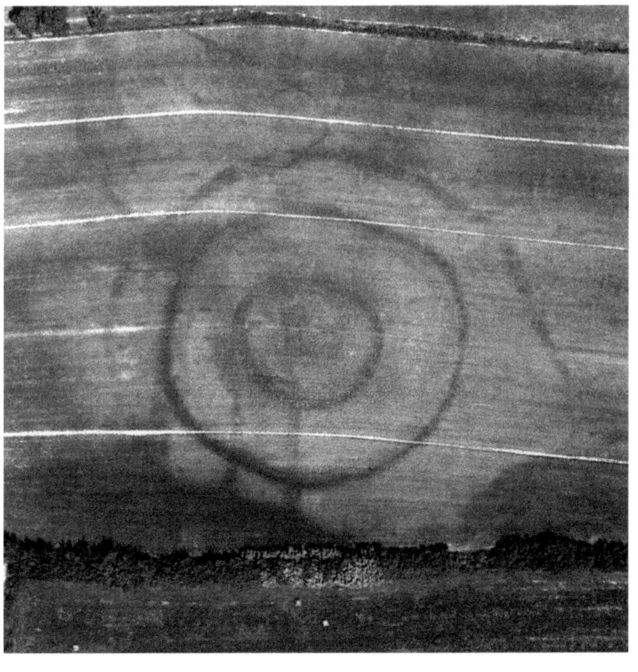

Figure 3 *Detail of the features, from an oblique air photograph taken in the summer of 2005.*

During the School, through repeated flights in light aircraft, oblique air-photo records were made of a distinctive group of archaeological cropmarks, illustrated in Fig.3 above. The features consisted of a triple concentric circular enclosure with a fourth mark in the centre and a fifth on the left-hand side of the second enclosure. In addition, there were traces of a number of parallel linear cropmarks, overlapping the enclosures. Mapping of the new air-photo evidence against the historical and technical national cartography (IGM 1:25.000 and CTR 1:10.000) revealed the that the closest placename (less than 500 m away) was Podere Castellina, now attached to an old agricultural building but possibly preserving the memory of a former medieval castle somewhere in the area. This interpretation could be in keeping with the reading of the air photo data. The circular enclosures – encompassing an area of almost one hectare – could be interpreted as concentric ditches and/or palisades. The marks inside and outside the smallest enclosure could be related to buildings. The linear features – visible also on the IGM 1:25.000 historical maps – are probably related to an ancient road system.

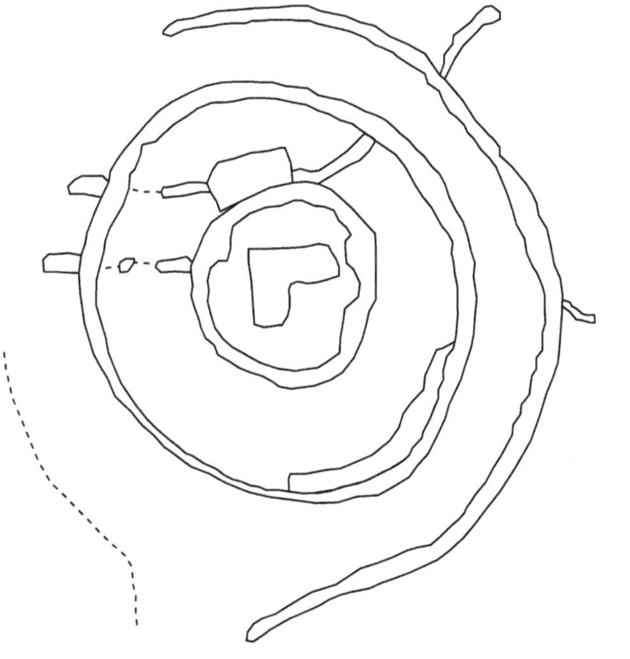

Figure 4 *GIS based interpretation (polygon) and hypothesis (line)*

The mapping process permitted a better understanding of the site and overcame the problem pointed out at the beginning of this article – the absence of evidence on the ground. Earlier examination of vertical air photographs from 1954 had been successful in detecting the site but the absence then of appropriate mapping procedures, along with the use of a base-map at the relatively small scale of 1:25.000, had shifted the site-positioning 150 metres to the north-east of the true location revealed by the more recent studies.

4 Field-walking and DGPD survey

Ground-examination in 2005 brought an immediate understanding of the high archaeological and historical value of the site. The survey was carried out in the month of September, immediately after ploughing and in good lighting conditions, creating excellent archaeological visibility and allowing the recovery of a large quantity of artefacts. The first survey results showed that some of the pottery could be dated to the early Middle Ages. But bearing in mind the poor understanding of the material culture of this period we felt the need for caution and therefore organised a more intensive grid-collection within 10 m by 10 m squares (Fig.5).

The investigation allowed us to confirm the first evaluation regarding the chronological period and helped in the general interpretation of the site. The pottery studies permitted identification of the first phase of the site as starting from the eighth century AD, with an apparent cessation of occupation in the eleventh century. Furthermore, ground survey was very helpful in revealing the site's distinctive morphology. On the ground it is quite easy to recognise, in correspondence with the area of the features seen on the air photographs, a slightly higher area, roughly circular, rising up to a metre above the surrounding terrain.

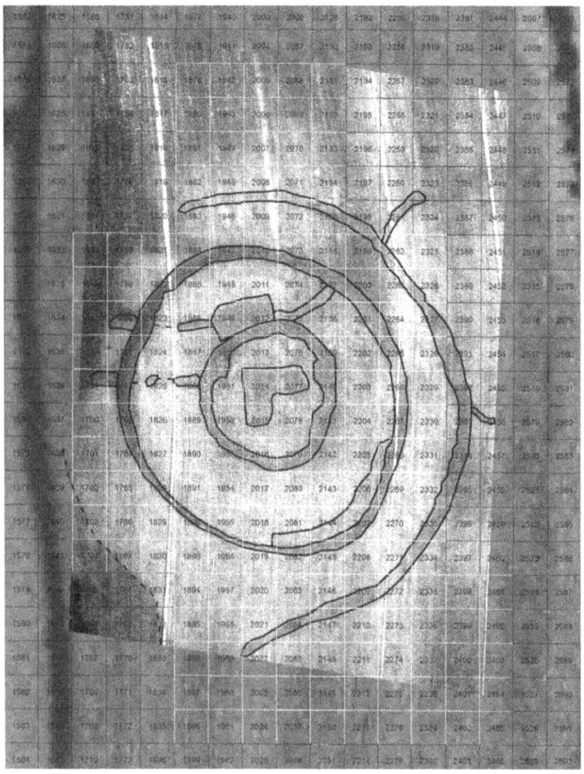

Figure 5 *The grid layout for field-walking (within 10 m squares)*

Following this we decided to make a full survey of the area using a DGPS system to collect precise information on the local morphology. The data was collected using a Leica 500 device as the base station and a Leica 1200 as the rover device, with sampling rate of 1 second. The post-processing of the data was carried out by the Geophysics Laboratory of the University of Bologna. The result is a very detailed cloud-point representation, interpolated using the circular krieging algorithm and then imported into the GIS to facilitate data integration and interpretation. The resulting DEM (Fig.7) shows clearly the presence of a slightly raised area cut by a number of deep plough-furrows.

5 Magnetic survey

The magnetic survey was carried out immediately after the field-walking and artefact-collection from the ploughed surface. These conditions are not inherently favourable to this technique because they introduce noise into the data, making the post-processing significantly more complicated (Fig.6). The measurements have a resolution of 1 m between the profiles and between 30 and 50 cm along the traverses. The surveyed area was 2 hectares in extent and corresponded to the main area of the features visible on the oblique aerial photographs and to the general extent of the artefact scatter.

Interpretation of the magnetic plot allowed us to recognise a number of features, with a degree of detail roughly similar to that of the oblique air photograph. The magnetic map shows three concentric enclosures and slightly below and to the left of the centre of the inner enclosure in the centre a smaller circular one that can probably be interpreted as a building.

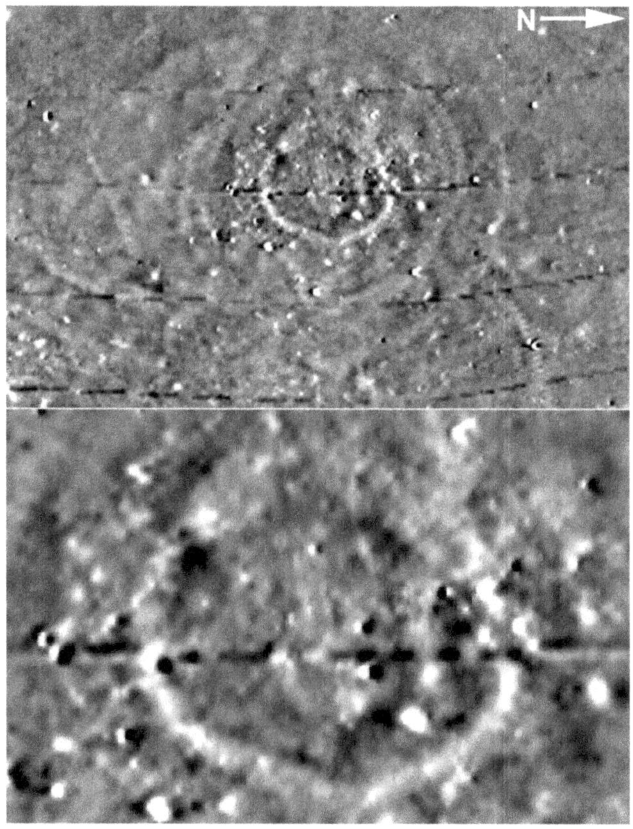

Figure 6 *Above, map of 2 hectares of magnetic data, clearly showing three enclosures and many other features. Below, detail of the central part of the magnetic map, showing the higher strength of the signal of the smaller enclosure and the large number of dipoles in this area.*

Other potential features are represented by small dipoles, higher in density inside and immediately outside the smallest enclosure (Fig.6). These latter features are difficult to interpret, mainly because of the low resolution of the survey. Bearing in mind the particular nature of the site – hypothetically that of a medieval castle mound – this type of evidence could find parallels in other case studies, perhaps relating to the existence of post holes in the underlying deposits (Gaffney, Gater 2003; Tipper

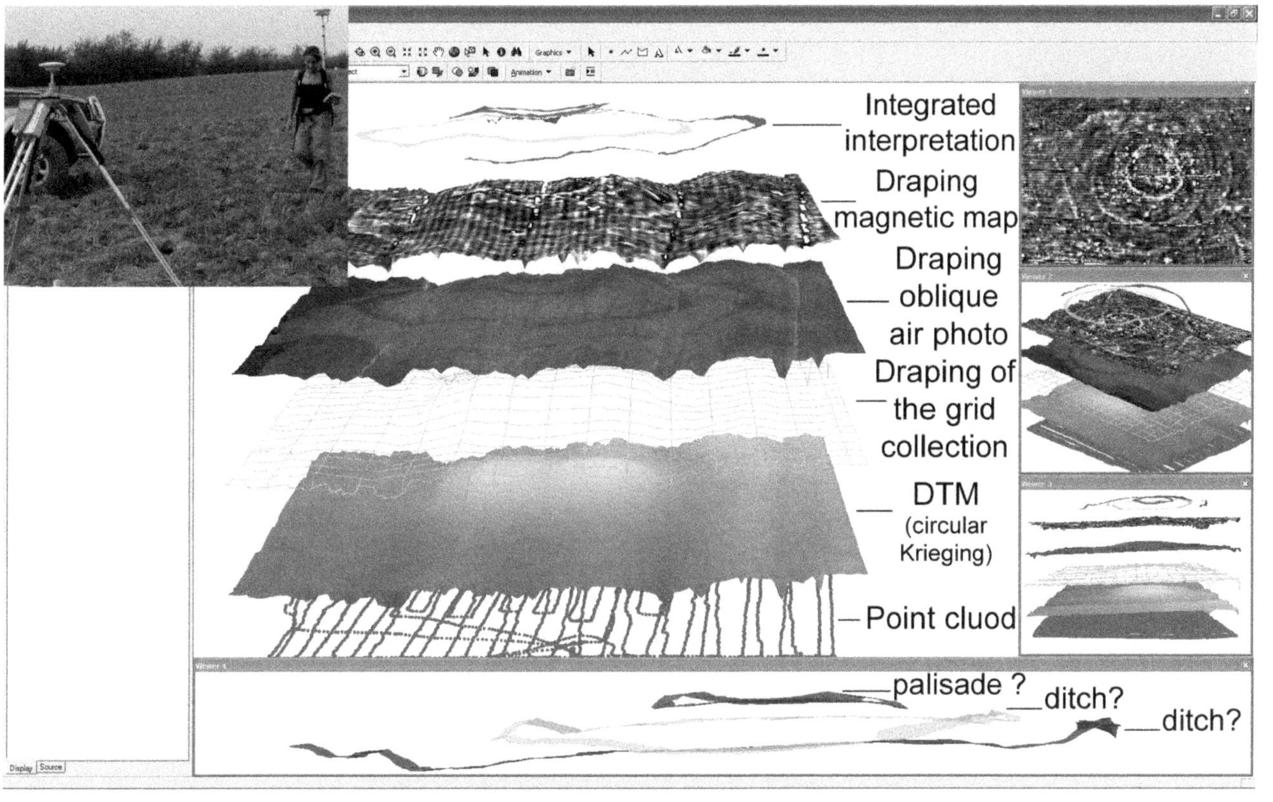

Figure 7 *Integrated analysis of the site*

2004). To obtain a better understanding of the site it would be useful, in the near future, to repeat the magnetic survey, increasing the resolution between profiles to between 50 to 20 cm.

Stefano Campana

6 Data integration and conclusion

On the evidence so far collected we feel confident about our general interpretation. We have probably located the first known lowland castle-mound in Tuscany.

The chronology suggested by the finds – from the eighth to ninth centuries through to the eleventh century – is also consistent with the data drawn from the air photographs, magnetic mapping and DGPS survey (Fig.7). Taken together, the various kinds of evidence point to features that it is possible to interpret as a lowland timber castle characterised by at least three concentric defensive features, presumably ditches or palisades.

As is well-known from other parts of Europe, a mound or *motte* of this kind is a typical form of medieval fortification, consisting of a raised earthen mound, resembling a small hill, usually artificial or at least artificially enhanced and topped with a wooden or stone structure. The earth for the mound would be taken from the ditch around the *motte* itself and perhaps from other ditches around the castle as a whole. The outer surface of the mound might in some instances have been covered with clay or strengthened with wooden supports (Hingham, Barker 2004). This kind of castle, however, can exhibit a wide range of variability.

In our case the aerial photographs show evidence that, even without other information, we can interpret with some confidence as ditches. The magnetic map is superficially similar in appearance to the aerial photograph, showing three circular traces. But we should be carefully in drawing conclusions from this similarity – aerial reconnaissance relies on phenomena that are quite different from those involved in the interpretation of magnetic data. Archaeologists are mainly used to interpreting features through their shape and size. We should bear in mind, however, that there is a significant difference between the meaning of the two soil measurements, the first related to the growth and colour of the overlying crop, the second to the *magnetic* properties of the underlying deposits. While acknowledging this difference there is a remarkable degree of similarity between the air photo evidence and the magnetic map. The magnetic measurements for the smaller enclosure show a very high and consistent signal, probably indicating the presence of a higher concentration of organic material, perhaps related to the former presence of a timber palisade or to a concentration of organic or other rubbish thrown out from the central enclosure.

The air-photo evidence in the central part of the mound, possibly indicating a masonry building, is also reflected in the magnetic map and even more so in the results of the field survey – artefact collection in the central area showed a high density of stones and mortar.

The available evidence allows us to interpret the site as an early medieval fortified settlement – perhaps in relationship to a *curtis* – starting from at least the ninth century AD. Its main peculiarity is its fortress typology and its lowland siting. Such a location should probably be linked with lake of Scarlino and with the ancient road system of the Aurelia. The settlement does not appear to have lasted beyond the eleventh century, to judge by the pottery and the absence of building materials that indicate any kind of Romanesque stylistic tradition.

In conclusion, we should emphasise that without an integrated approach it would never have been possible to present such a rich and articulated interpretation of the site. The isolated application of each prospection method on its own would have revealed only *some* aspects of the site, limiting our overall understanding of the evidence. For instance the field-walking survey would have shown a clear pottery concentration no more than 20 by 20 m across. That evidence, on its own, could have been interpreted as the remains of a very small settlement – a completely different picture from the one-hectare timber castle revealed by the several techniques in partnership.

Stefano Campana, Lorenzo Marasco

Acknowledgments

The authors are indebted to Dr. Salvatore Piro (ITABC – CNR) and Dr. Iacopo Nicolosi (National Institute of Geophysics and Vulcanology) for their invaluable comments and criticisms on geophysical survey techniques and data processing. Special thanks are also due to Chris Musson for help with the translation into English and for his valuable comments on the practice of archaeological research, both in the air and on the ground. Many researchers and students have collaborated, and are still collaborating, in the Siena and Grosseto archaeological mapping projects. Special thanks are due to the team of the Landscapes Archaeology and Remote Sensing Laboratories of the University of Siena at Grosseto: C.Felici, A.Caprasecca, M.Corsi, B.Frezza, M.Ghisleni, F.Pericci and E.Vaccaro.

References

BARTOLINI, E. PRANZINI, E. LUPIA PALMIERI, C. CAPUTO 1977 C., *Studi di geomorfologia costiera: IV L'erosione del litorale di Follonica*, «BS Geol.I», 96.pp.87-116.

CAMPANA S., 2005, Tecnologie GPS e Personal Data Assistant applicate all'archeologia dei paesaggi, "Archeologia e Calcolatori", 16, , Firenze.

COSCI 2005, Ricerche aeree in Toscana, in MUSSON C., PALMER R., CAMPANA S.

CHAPMAN H. 2001, *Understanding and Using Archaeological Topographic Surveys – The "Error Conspirancy"*, in *Computing Archaeology for understanding the past*, Proceedings of CAA 2000, BAR International Series 931, Oxford, pp.19-23.

CUCINI C., 1985, *Topografia del territorio delle valli del Pecora e dell'Alma*, in FRANCOVICH R. 1985, pp. 147-335.

FRANCOVICH R. (ed.) 1985a, *Scarlino. I. Storia e territorio*, "Ricerche di Archeologia Altomedievale e Medievale", 9/10, Firenze.

FRANCOVICH R., HODGES R. 2003 *Villa to village. The transformation of the Roman Countryside in Italy, c.400-1000*, Duckworth, London.

GAFFNEY C., GATER J. 2003, Revealing the buried past. Geophysics for Archaeologists, Tempus

HINGHAM R., BARKER P. 2004, Timber castle, Exeter.

MARASCO L. forthcoming, *Scarlino. Nascita e sviluppo di un castello medievale.* L'indagine archeologica, Firenze.

MUSSON C., PALMER R., CAMPANA S. 2005 *In volo nel passato. Ricognizioni aeree e aerofotografia obliqua,* Firenze, pp.328.

TIPPER J. 2004, *The Grubenhaus in Anglo-Saxon England,* Oxford

Aerial Survey Project in Tuscany: years 2000-2005

Stefano Campana,[1] Riccardo Francovich,[2] Francesco Pericci,[3] Maria Corsi[3]

[1] University of Siena, Dept. of Archaeology and History of Arts, Landscape Archaeology, (LAP&T)
campana@unisi.it
[2] University of Siena, Dept. of Archaeology and History of Arts, Medieval Archaeology
francovich@unisi.it
[3] University of Siena, Dept. of Archaeology and History of Arts, (LAP&T)
<corsi>;<pericci>@lapetlab.it

1 Introduction

The idea of developing an aerial survey project in the Department of Archaeology and History of Arts at the University of Siena was first conceived in 1991. On the occasion of the IVth International Summer School in Archaeology, on the subject of Landscape Archaeology, I invited Chris Musson and Dr. Rowan Whimster, two British aerial archaeologists, to take part in the School. Unfortunately the Italian domestic law on aerial photography current at that time did not allow us to develop this idea and we had to wait almost another ten years before we were able to put this method of research into practice.

Nowadays aerial survey represents and involves an important section of the university's Laboratory of Landscape Archaeology and Remote Sensing at Grosseto. This method is part of a wider and integrated strategy which we have been developing over the years to achieve a better understanding of the landscapes of Tuscany.

The aims of the Aerial Survey Project in Tuscany are mainly concerned with:

- Documentation though oblique aerial photography of all of the region's monuments and archaeological features, from prehistory to the Middle Ages and on to the present day.
- The research programme forms part of the Archaeological Mapping Project of the province of Grosseto, Siena and Livorno.
- Landscape monitoring for heritage conservation.

Aerial survey is currently part of a research design aimed at the progressive introduction of remote earth observation systems, along with the enhancement of surface collection techniques through the application of new instrumentation and methods of data collection and documentation, for both the archaeological and the environmental records. So far, we have put in train the following approaches:

- *High-resolution satellite imagery: Ikonos and Quickbird*. The introduction of satellite imagery was aimed in the first place at providing a total, continuous and objective view of chosen study areas at a particular moment of the year, as planned by the archaeologist. A second feature is the capacity to provide multispectral data to monitor plant health and to detect water-stress in vegetation where it cannot be seen by the naked eye (Campana 2002).
- *Historical air photo coverage*. Vertical aerial photographs, spanning a wide temporal range, represent an irreplaceable source for the analysis of the Tuscan landscape (Guaitoli 2003). Anyone interpreting the photographs through a digital photogrammetric workstation will see a 3D replica model of the landscape as it was in 1938, 1954, 1976 or 1994. In addition to their historical content, vertical photographs are of course an important source for the conduct of 'aerial reconnaissance' (Campana 2005*a*).
- *Enhancement of field-walking survey techniques*. The technique of field-walking is aimed at the systematic investigation of sample areas and the verification of remotely sensed evidence. The application of GPS and PDA technology greatly improves the quality and effectiveness of the research (Campana, Francovich 2005).
- *Large-scale magnetic survey*. In 2003 we tested a system of data acquisition that allowed us to cover one hectare per day at a resolution of 60 cm along traverses set 1 m apart. So far we have only acquired data for 20 hectares, covering a total of 22 archaeological sites (Campana et al 2005).
- *Field Data integration: PDA technology*. The merging of PDA and GPS technologies goes far beyond the level of increased fieldwork efficiency, enabling data integration directly in the field and at its best allowing the systematic and economical application of strategies and methodologies shown to be effective in the past (Campana 2005*b*).

Although we are not yet in a position to present definitive results, we have started to use additional methods and instruments that we are confident will play a significant future role in offsetting the present gaps in our strategy for the study of the ancient landscapes of Tuscany. These include:

- *digital photogrammetry*.
- *lidar* (in partnership with the Airborne Research and Survey Facility of the Natural Environment Research Council in the United Kingdom).
- *GPR survey* (in particular the testing of large scale systems such as GSSI TerraVision)

- *Geo-electrical survey.*
- *Electromagnetic survey.*

2 Archaeological visibility from the air

It is widely acknowledged that the 'aerial visibility' of archaeological features is not constant throughout the year. There are favourable time-windows, often difficult to forecast with any certainty and of a duration that can vary from year to year.

In a large region such as Tuscany the various factors that influence archaeological visibility interact with one another to produce areas that are characterised by high levels or low levels of visibility, with a wide variety of gradations in between. The factors that determine archaeological visibility are many, but three take precedence over the rest: historical factors such as land-use and agricultural practices; soil-type; and seasonal climatic conditions.

The first problem that we encounter in central-southern Tuscany is historical in origin. Both in Italy and elsewhere the countryside of this part of Tuscany is known for its lack of industry or other modern development, its fields of waving corn, its olive groves and vineyards, its winding white roads and groups of farm buildings. This is an image of a rural landscape, its natural character virtually 'fossilised' to the present day. But in practice this characterisation is entirely false, since it ignores the dramatic and sometimes violent changes that have transformed the original character of the area. A striking example can be seen in the Val d'Orcia, a district that covers an area of about 500 Km^2 to the south-east of Siena, today characterised by an expanse of gently rolling hill-country dominated by cereal cultivation and (in the southern parts) by vineyards. This kind of monoculture, however, is the result of the great and often radical transformations, starting in the Fascist era and continuing through the post-war years and up to the present day (Felici 2004). The Val d'Orcia, in the past, was an area of summer drought and winter floods, bleak and dominated by clay (Mangiavacchi 2004). Its morphology and heavy clay soils (too hard for widespread cultivation by traditional pre-mechanised methods) restricted agricultural activity and productivity. Similar patterns can be seen in the areas of Maremma and Chianti, now radically changed by land-improvement and extensive monoculture.

In relation to land-use one must also bear in mind that Tuscany, of all the regions of Italy, has the highest percentage of its surface covered in woodland, about 50% in all. Woodland constitutes a serious impediment in terms of aerial visibility.

The areas given over to arable crops for the most part occupy heavy clay soils, occasionally interspersed with sand and subjected in the past to heavy land-improvement projects. These types of soils are generally unfavourable to most types of remote sensing because of their poor drainage.

The areas with a higher level of archaeological visibility are principally those of the alluvial plains and river valleys, in particular those of the Arno, Serchio, Cecina, Ombrone, Chiana, Orcia and Albegna etc. In some of these areas there are further problems because of the depth of the alluvial soil and recent human intervention, such as land-drainage schemes and present-day developments in industry and housing (Agnoletto 2002). These factors lead us to say that Tuscany is characterised by relatively low levels of archaeological visibility from the air. Our experience over the past few years shows us that the landscapes characterised by the best level of aerial visibility are concentrated in southern Tuscany, in the Province of Grosseto and more particularly in the following areas:

- The Lower Ombrone Valley
- The Albegna Valley
- The Oro Valley
- The Chiarore Valley
- The Lower Pecora Valley

At the moment we have not found areas which display stratified landscape systems like those that can be seen on the Tavoliere in Puglia or in northern Lazio (Bradford 1974; Guaitoli 2003; Musson, Campana 2005). Despite the limitations that this background is bound to impose, it has been demonstrated elsewhere that aerial survey can also be successfully applied in less favourable contexts (Grosman 2000; Musson 1994). In order to obtain good results in these conditions it is necessary to design a purpose-made strategy and not simply to apply those that have proved effective in more favourable contexts. It is necessary to develop an approach which responds to the physical and cultural peculiarities of the landscape under study.

3 Aerial survey programme: years 2000-2005

The first realistic opportunity to develop a project based on aerial survey presented itself during and after the University of Siena's XIth International School in Archaeology in 1999. Anticipating the imminent modification of the Italian law regarding air photography and filming activities over the Italian homeland, we discussed with two British aerial archaeologists, Chris Musson and Robert Bewley, the possibility of organising a training course in Siena. Following preparatory survey in year 2000 in cooperation with Dr. Otto Braasch from Germany and Mrs. Darja Grossman from Slovenia, the research and training school eventually took place in the early summer of 2001, with the help of Culture 2000 and other funding (MUSSON *et al* 2005).

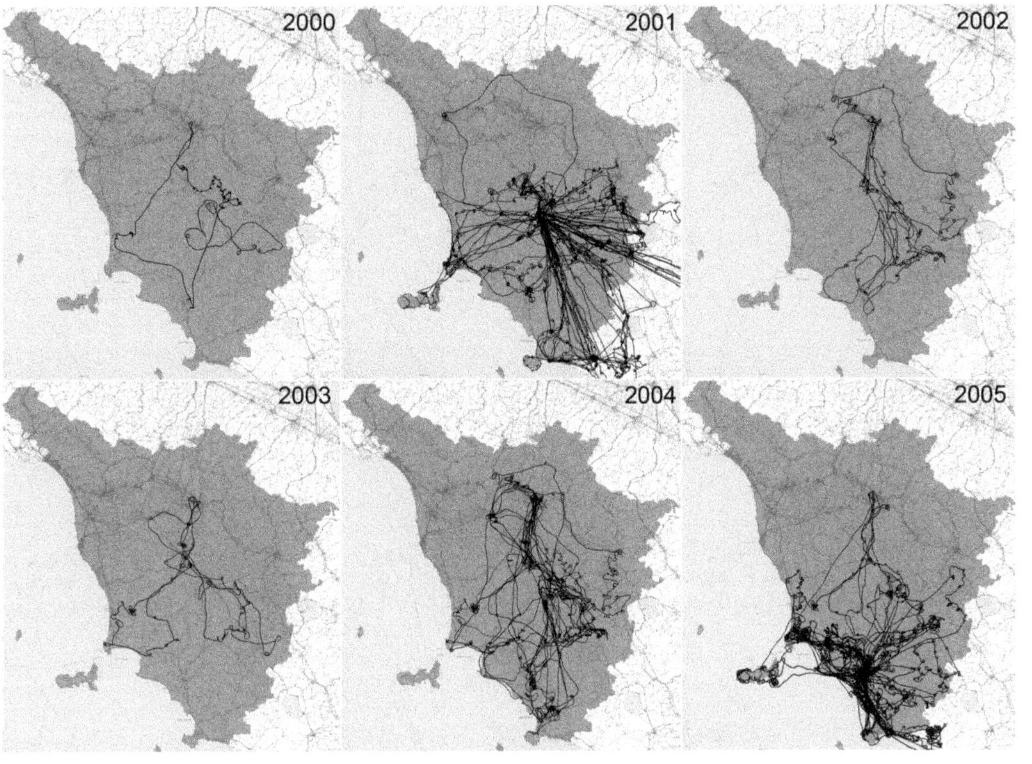

Figure 1 *GPS track of aerial survey from 2000 to 2005*

During the first two years of aerial work we flew a total of 132 hours, surveying a large part of Southern Tuscany (Fig.1).

In 2002 and 2003 aerial activity continued at an average of about 30 hours per year. In total, during the first four years of survey, we were able to collect approximately 9500 oblique air photographs, covering about a thousand archaeological sites.

The most heavily represented site types are known monuments, in particular medieval castles (Fig.2) and historic city centres (Fig.3). In part this is due to the academic interests of the Department of Medieval Archaeology at the University of Siena, in part (as we will explain below) a by-product of reliance on British and German colleagues in the early development of the survey programme.

Figure 2 *Examples of recorded monuments*

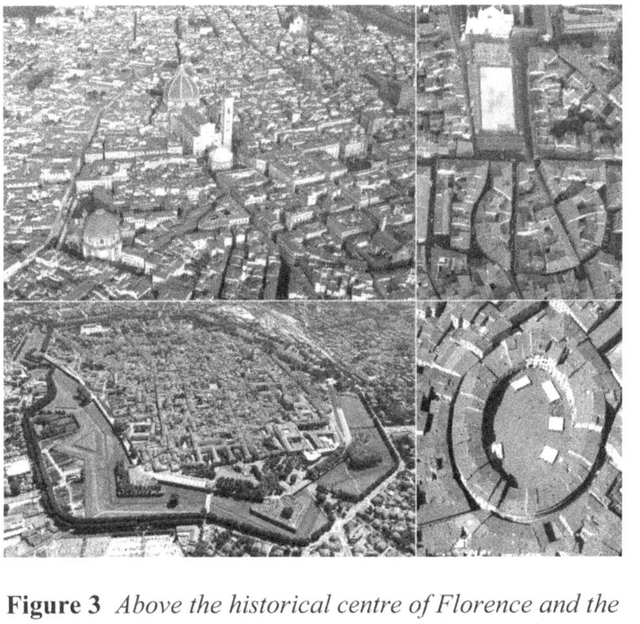

Figure 3 *Above the historical centre of Florence and the particular of the amphiteather. Below the city of Lucca and the detail if the amphiteather.*

It is a well-known principle that the success of any programme of aerial photography is based on careful observation of the 'visibility' of archaeological features. Flights need to be planned so as to coincide with times of year or day when the archaeological features will appear to best effect.

If the aircraft or the pilot is not available at these times (as was inevitably the case with our colleagues from Germany and the UK) the effectiveness of the aerial survey will be correspondingly reduced.

Despite the absence in many cases of ideal conditions (logistical or in terms of visibility) it has nevertheless been possible to recognise a significant number of previously unknown archaeological sites through the traditional instruments of air photography and air-photo interpretation in response to variations in vegetation-growth, soil humidity, the physical characteristics of the land and micro-morphological variations.

The presence of cropmark or soilmark evidence, especially in poor conditions, is closely connected to local differences in soil moisture and the local character of the landscape (land-use, post-depositional processes etc). Similar factors determine the visibility of archaeological features in vertical air photography (Alvisi 1989).

The archaeological features which we have discovered or documented from air cover a wide chronological range, from prehistory to the modern age, and encompass a wide range of activities – cemeteries, settlements, Roman villas, mounds, castles, farms, walls and boundaries of various kinds, road-systems and field-systems etc (Fig.4).

From the beginning of 2004 we made significant improvements in our aerial research work, principally through establishing a collaboration with the Aero Club of Florence, thereby placing at our disposal locally-based aircraft and pilots. We were now in a position to make our own choice about when to fly.

Freedom to plan the timing of flights has not only made it possible to monitor the growth and ripening of the grain, and therefore to operate at times when the visibility of buried archaeological remains is at its highest, but also to fly in every part of the year so as to explore the particularities of the Tuscan landscape from the air.

During 2004 we flew a total of 51 hours, at fairly regular intervals between January and August, with a peak of activity around the end May and beginning of June. We collected 5000 oblique photographs and documented about 500 archaeological and paleoenvironmental sites.

Aerial survey has proved to be highly effective for the location of new features and for increasing information about previously-known sites.

Figure 4 *Sample of the features recorded in the Tuscan countryside. Above from left to right: Etruscan necropolis, enclosure, two examples of Roman villas, road systems, ancient river bed, field system.*

In total, over six years of aerial work, we have flown for 375 hours and acquired a collection of approximately 30.000 oblique air photographs.

6 First thoughts on the contribution of aerial survey in the study of archaeological landscapes in Tuscany

The results of aerial survey projects are often measured through data quantification, in particular through the number of new features photographed or the extra evidence recovered for known sites. In some cases, as on the Tavoliere in Puglia or in the northern part of Lazio, this type of evaluation can be effective (albeit strongly reductive) in giving an idea of the extraordinary contribution that aerial survey can offer to archaeological research.

We believe, however, that to make an assessment of the first six years of our activity in Tuscany on such criteria would lead to a modest but extremely misleading result. In explanation we would point out that the objective of the aerial survey project in Tuscany is not simply the recovery of cropmark and soilmark evidence but also the photographic documentation of ancient and medieval monuments together with their rural or urban landscape contexts.

Figure 5 *Left, archaeological features of a Roman villa in the Orcia Valley. Right, archaeological features of a small church in front of the Romanesque building Pieve di Corsano.*

Moreover the aerial work is part of a wider survey strategy for the study of past landscapes in which aerial reconnaissance plays a central role, but one which is complementary to – and complemented by – other survey techniques. The implementation of our strategy for landscape analysis through the discovery of new evidence or the improvement of information on already-known sites must adapt to the particular context in which it is applied (a principle that remains valid for all of the methods which we apply).

For instance, we can say with certainty that the Province of Siena represents the least favourable area in Tuscany for the aerial recognition of features defined by cropmarks and soilmarks.

In the past six years we have documented in the countryside of Siena approximately 450 archaeological sites, including a near-total coverage of medieval castles. But in the whole of this area we have so far identified only two new sites through cropmark evidence (Fig.5).

The role of exploratory aerial survey in such a context is clearly marginal, although the two new sites that have been discovered are extremely interesting and were not revealed by other methods of research.

The opposite end of the range, in terms of aerial visibility, is represented by the landscape around Grosseto, already mentioned above. In this area the role of our air-survey work is more balanced between the documentation of known monuments and the identification of new features. In the past two years we have turned our attention to this area, obtaining encouraging results in particular for Etrusco-Roman and medieval features. Roman sites are mainly to be seen in the Albegna, Oro and Ombrone Valleys and consist principally of settlements (Roman villas and small farms) and ancient road systems (Fig.6).

Figure 6 *Sample of the features recorded in the Grosseto countryside. Above from left to right: Roman building, roman farm and different kind of Roman villas.*

To the west of the ancient city of Roselle we have identified an area which to date is the most favourable in Tuscany. Here there are visible traces of several buildings, of differing sizes, as well as field-systems and ancient roads. We believe that in the near future this area could provide a very interesting study area, with many positive factors in its favour: good visibility from the air, excellent conservation of archaeological features and a high density of archaeological remains. It seems likely that in the next few years aerial survey will play a primary role in the reconstruction of stratified archaeological landscapes in this area (Fig.7).

Figure 7 *Landscape of the Roselle Valley showing the most meaningful features*

Figure 8 *Triple medieval enclosure placed in the plain between Follonica and Scarlino (GR). Features of an hillfort in the Amiata countryside (GR).*

In the Maremma Plains, also, aerial survey has enabled us to obtain valuable results for the understanding of the area's medieval landscapes, extending beyond the simple documentation of known monuments in their landscape contexts. The last few years here have seen significant results, modest in number but extremely interesting and typologically meaningful.

In Fig.8, to the left, there are visible three concentric enclosures that it is possible to interpret as ditches or palisades (Campana et al 2006). Field survey allowed us to interpret the site as a medieval castle mound or motte. This kind of settlement is present in southern and northern Italy but is virtually unknown in Tuscany. The importance of the discovery lies not so much in the novelty of its detection but in its implications for the possible existence of similar sites in areas with the same basic geographical characteristics. In the past year we have indeed found three other mound of this kind on the coastal plain of Grosseto and Piombino.

In concluding this short paper we would like to emphasise that although an area like Tuscany, with its acknowledged limitations in aerial visibility, will never rival more favoured landscapes like those of the Tavoliere or northern Lazio, our researches have become progressively more fruitful over time. We feel confident that our increasing experience in understanding the local territorial characteristics, along with the year-on-year incorporation of new sites and features into the department's GIS, will undoubtedly increase the role of aerial survey in our exploration, documentation and understanding of the landscapes of Tuscany.

Acknowledgments

The authors are indebted to many researchers for their invaluable comments and criticisms on aerial survey. We would particularly like to thank Bob Bewley, Prof. Giuseppe Ceraudo, Michael Doneus, Chris Musson, Prof. Marcello Guaitoli, Darja Grosman, Rog Palmer and Prof. Fabio Piccarreta.

Special thanks are also due to Chris Musson for his help with the English translation of the text and for his valuable comments on the practice of archaeological research. Lastly, we owe a special debt of gratitude to the pilots and aerial archaeologists Dr. Otto Braasch and Klaus Leidorf, and to the Aero Club of Florence, its pilots and its President, Pietro Banci.

References

ALVISI G. 1989. *La fotografia aerea nell'indagine archeologica*, Rome, pp. 28.

AGNOLETTO M. 2002. *Il paesaggio agro-forestale toscano. Strumenti per l'analisi, la gestione e la conservazione,* ARSIA, Firenze.

BRADFORD J. 1974. *Ancient Landscapes*, London.

CAMPANA S. 2002. Ikonos-2 multispectral satellite imagery in the study of archaeological landscapes: an integrated multi-sensor approach in combination with "traditional" methods. In M. Doerr, A. Sarris (eds.), *The Digital Heritage of Archaeology, Proceedings of CAA02 Computer Applications and Quantitative Methods in Archaeology*, Athens: pp. 219-225.

CAMPANA S. 2005*a*. Looking to the future: una strategia per l'archeologia dei paesaggi toscani. In Musson, Palmer & Campana, 2005, pp. 221-249.

CAMPANA S. 2005*b*. Mobile Computing in Archaeological Prospection. *Proceedings of the 6th International Conference on Archaeological Prospection, Rome, 14-17 September 2005*, Rome.

CAMPANA S., FRANCOVICH R. 2005. Linking remote sensing and infra-site analysis to the reconstruction of rural settlement and landscape patterns. In. Forte (eds) *The Reconstruction of Archaeological Landscapes through Digital Technologies, Proceedings of the 2nd Italy-United States workshop, Rome 3-5 November 2003 Rome,* BAR International Series No. 1379.

CAMPANA S., FRANCOVICH R., FELICI C. 2005. Large scale magnetic survey and archaeological mapping: the Grosetto and Siena projects. In S. Piro (eds.), *Proceedings of the 6th International Conference on Archaeological Prospection, Rome,* 14-17 September 2005, Rome.

FELICI C., 2004. *Carta archeologica della provincia di Siena. Pienza,* Volume VI, Nuova Immagine, Siena.

CAMPANA S., FRANCOVICH R., MARASCO L. 2006. Remote Sensing and Ground-truthing of a Medieval Mound (Tuscany - Italy), in S. Campana, M. Forte (eds), *From Space to Place, Proceedings of the IInd International Conference on Remote Sensing in Archaeology, Rome 4-7 December 2006*, BAR International Series.

GROSMAN D. 2000. Two examples of using combined prospecting techniques, in M. Pasquinucci, F. Trément, (eds), *Non-destructive techniques applied to landscape archaeology*, Oxford, pp. 245-255.

GUAITOLI M. 2003. *Lo sguardo di Icaro. Le collezioni dell'Aerofototeca Nazionale per la conoscenza del territorio,* Rome.

MANGIAVACCHI M. 2004. *Archivio del Consorzio di Bonifica della Val d'Orcia. Immagini fotografiche per la lettura del territorio,* Aska Edizioni, Florence.

MUSSON C. 1994. *Wales from the Air – Patterns of Past and Present,* RCAHMW, Aberystwyth.

MUSSON C., CAMPANA S. 2005. Fotografare i segni del passato, in Musson, Palmer & Campana 2005, pp. 172-229.

MUSSON C., PALMER R., CAMPANA S. 2005. In *Volo nel Passato: aerofotografia e cartografia archeolo-gica*, Insegna del Giglio, Florence

A proposal for the digital storage and sharing of remotely sensed archaeological data

Stefano Campana,[1] Barbara Frezza[2]

[1] University of Siena, Department of Archaeology and History of Arts, Landscape Archaeology, (LAP&T)
campana@unisi.it
[2] University of Siena, Department of Archaeology and History of Arts, (LAP&T)
frezza@lapetlab.it

Introduction

The Department of Medieval Archaeology at the University of Siena has been actively engaged in programmes of landscape archaeology and remote sensing for over twenty-five years. This research activity has produced a large collection of data and in particular a vast number of photographs with related information. For instance twenty years of aerial reconnaissance has resulted in a total of about six thousand vertical air photograph, while in the last six years our own aerial survey programme has produced an archive of about 30000 oblique photographs. It is obvious that the management and communication of this quantity of data can only be handled through the use of digital databases. We started our own experiments with multimedia databases in 1999, using Canto Cumulus software, but this product soon showed many problems of inflexibility and difficulty in the sharing of data.

Digital technology, of course, in addition to offering powerful tools for the cataloguing, storage, and analysis of archaeological data, has become a common and speedy platform for the communication of information. Some years after our first experiments we had the opportunity to move to a database that could share information through the Internet. A first attempt was made in 2004 and 2005 to build a web-based database (Donati 2005). The result was excellent for interrogating and managing the data but unfortunately it was not possible to use the system for the everyday storage and use of the data. (See http://www.paesaggimedievali.it/atlante/index.html). This prototype system required us first to catalogue the data in Canto Cumulus and then to export the whole database each time we wished to update the web-based system. In a new development, under design more recently, the revised system has been required to fulfil the following tasks:

- Real time update.
- Quick and easy communication of data within the laboratory as well as with the general public, private individuals and other institutions.
- First steps towards a web-based GIS that could manage the data integration of graphical, geographical and alphanumeric data

It is the development of this system which is discussed in the remainder of this article.

Stefano Campana

Database management systems

Today the Internet represents one of the more effective instruments for data exchange and it will undoubtedly remain one of the most effective channels for the communication of information in the future. Through agreed communication standards its architecture guarantees compatibility of interchange between the different sites that make up the net, facilitating the sharing of experience, services and information sources despite the physical separation of the users.

Our appreciation of the enormous potential of the Internet has led us to put in train a project which aims at the creation of an Internet-enabled information system based on a Data Base Management System (DBMS). What we envisage is a distributed system that allows for the storage, up-dating and recovery of information from a database of aerial photographs of Tuscany, managed through an interactive graphical interface.

A DBMS is in essence an instrument that treats the data as the shared resource of an organisation and of its various users, thereby meeting the ever more obvious need for the distribution and sharing of information and knowledge. The essential properties of a DBMS are:

- The recording, organisation and management of the data according to criteria that answer directly to the needs of the context in which system has been developed.
- The easy communication and sharing of information – the users should be able to access the database by interrogating it in a variety of ways.
- Reliability, in the sense of the capacity to conserve the content of the DBMS through rescue functions and restoration procedures.
- Privacy, guaranteed through the use of authorisation protocols and control procedures.

In addition, the system must be efficient, in that it is able to operate within the resources of space and time available to the users. It must also be effective in helping to make the user's work more productive.

A DBMS is in essence a layer of software interposed between the user and the information stored within the system. This intermediate layer prevents the user and the applications from interacting directly with the data, which can therefore be stored, physically, in the most effective way possible without preventing the user from having access to it. This permits a high degree of independence between the applications and the physical storage of the data. These characteristics ensure that any application that is developed can draw on a data source that is secure, reliable and economical in scale. Such properties are ideal for applications that use the Internet in their infrastructure and which therefore have an obvious need for security and economy of scale.

The decision to place the DBMS on the net also greatly widens the system's capacity for increased effectiveness in the communication and sharing of information. The net is capable of making available to the public, virtually instantaneously, a wide variety of data, thereby allowing them to consult the stored information and to take part in critical appraisal and re-interpretation of its content. In the archaeological context, for instance, this could offers to conservation agencies a simple and immediate means of checking information about sites and other features with which they might be concerned. Internet access also makes information available to users in all parts of the country, and to people who are not necessarily specialists in the field of archaeology. It also permits the database to be enhanced or up-dated in real time from mobile locations or from a variety of different work stations.

The research design

The development now in hand envisages the use of an MSQL database connected to the net through a series of PHP (PreProcessor Hypertext) routines. Access to the database is achieved through the use of the codified language SQL (Structured Query Language). This requires every query to be expressed in terms of commands, or phrases, written in conformity with strictly codified rules. These in turn trigger the execution of operations for the up-dating or interrogation of the data. The DBMS acts at three levels:

- Storage and management of the data.
- Communication and exchange of the information
- Making the information available to non-specialists within the community.

The structure of the DBMS will use a simple bi-dimensional database, of the kind often described as 'flat-file' in that it is composed of a single table. The table in our case is composed of a field for the individual photo and thirteen related fields – year of photography, type of film or sensor, region, province, district, community, place-name, type of evidence (cropmark, soilmark etc), documentary references, interpretation, reliability, site type, chronology – plus an identification reference for each record.

The fields selected for inclusion in the DBMS are those which we feel will most readily supplement the user's understanding of the individual images. It was necessary, therefore, to make judgements about which of many available kinds of information would best help the user to form a more complete picture and to understand the ways in which the information is organised and managed.

In planning the database an aspect which has been given particular attention has been standardisation of the language used, especially that relating to the fields which involve one form or another of interpretation. Ordinary spoken or written language has a vast capacity for expressing shades of meaning, depending on the context, the tone of communication or the subject under discussion. The computer is, by its nature, incapable of coping with such subtleties and it becomes necessary to formulate strict standards for the language used in the compilation and interrogation of the DBMS. This is done by establishing a series of fixed terms for use in fields where there might be several different ways of expressing the ideas or factual information involved.

The operations involved in the compilation and interrogation of information within the table, and more generally the proper operation of the database as a whole, can be assured only if these basic rules of planning are respected. This formal clarity of expression has a critical impact on accessibility and on the correct use of the database, both for enquirers and for those responsible for up-dating its content. The best solution lies in the creation of restricted lists of terms in the entry and search screens of the DBMS, chosen as required from pull-down menus. In creating these lists of terms it has been necessary to consider all the possible variations that might present themselves in up-dating or interrogation of the archive, taking account of clearly defined criteria and of the degree of detail that one intends to add to or extract from the system.

The descriptive fields which require the use of fixed terms are those that deal with type of evidence, interpretation, reliability, site type and chronology. The routines for operations involving the addition or extraction of information from the record force users to employ only the pre-determined formulae, allowing no scope for variations or additions. Only the administrator of the database is allowed, in particular circumstances, to add to the lists, modify the definitions or insert new terms.

All the graphical documentation in our own system, covering approximately 35000 between vertical and oblique air photographs of Tuscany, is stored in a folder outside the data bank. In the photo field there is recorded only the path through which the image can be recovered from this folder. In this way the size of the DBMS itself is kept more economical. Recovery of the images from their original positions, in response to each enquiry, also increases the speed with which the system can obtain and deliver the images to the user.

Another element that directly influences the effectiveness of any DBMS with a user interface is the need for it to have a clear and linear architecture, so that the user can benefit to the full from the potentiality of the data. The interface which we have been developing meets this requirement and offers the user various ways of securing access to the data. The interface has been designed so that users can include in their interrogation almost all of the fields present in the table, either by keeping to the lists of fixed terms or by digitising the text directly when these are not pre-determined. The only fields excluded from enquiries are those concerned with the reliability of the information and the presence or otherwise of bibliographical references, plus of course the photo field itself. Once users have made codified entries in the fields which they wish to use in their enquiry, they simply click on the search button to start the interrogation process. The structure of the query routine provides for the interrogation on several fields simultaneously. Setting the system up in this way increases the speed with which users can carry out enquiries on specific problems.

The same interface screen is used for searching and for up-dating the DBMS.

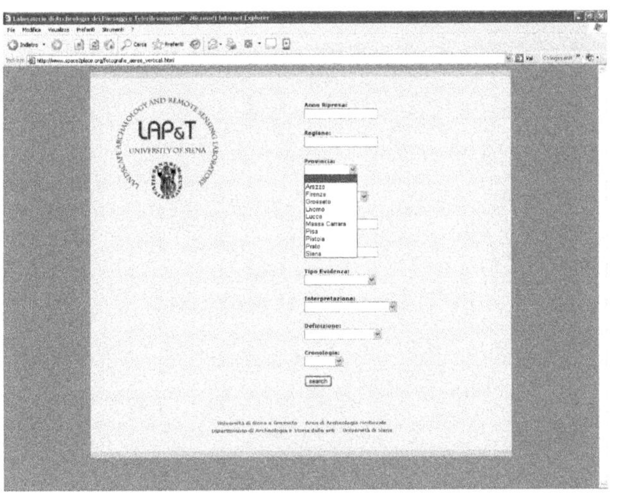

Entry of new data takes place directly on the net, in real time, from any computer connected to the Internet. The added text information is transferred automatically to the DBMS, creating new records, while the photographs, through an up-loading routine, are transferred to the image folder. Access to the entry page for the creation of new records is controlled by a system of authorisation and protection, using passwords. Control protocols are also under development to guarantee secrecy and to prevent manipulation of the data bank's contents by unauthorised users. Operations on the database itself can only be carried out by operators who hold authorisation for the particular operations or kinds of information concerned.

The development of the DBMS has also involved work on design of the pages through which the results are presented on the user's screen in response to enquiries. Once a successful search has been made, the results are presented on-screen. It is possible, of course, that the records extracted from the database will be too numerous for presentation on a single web page. To ensure clarity of presentation in such cases it is necessary to 'format' the results through the SQL code.

'Formatting' means the subdivision of the query results into smaller parts, pages in this case. Once transmitted to the user these can be presented on-screen through a navigation system which allows the user to move easily from one page of results to the next.

Failure to use this kind of formatting would be risky because it could:

- compromise readability of the information (a long page, rich in data, has less chance of being read to the end);
- lengthen the time taken to load the page;
- increase the quantity of data to be exchanged between server and user for each request;
- take up an excessive amount of server time, especially if more than one query is being made about the same record at any one time.

For these reasons the division into pages achieves an improvement in overall performance of the operation. It is also worth remembering that the query results often follow extended journeys on the net and it is therefore best to limit their number. It was therefore decided to subdivide the data extracted from the database into groups of five pages, and to attach, at the foot of the last page, a link to the previous or next page so as to allow the user to navigate through the results on-screen. In effect the page is always the same but the contents change according to user's choice of page number for scanning, the relevant number being passed on through a QueryString.

The presence of a photo field containing exclusively graphical data has made it necessary to introduce thumbnails, or miniature previews of the images concerned. A thumbnail is simply a copy of a larger image, modified to reduce the size of the file. Through a simple link to the larger image the user can see the image in its original form.

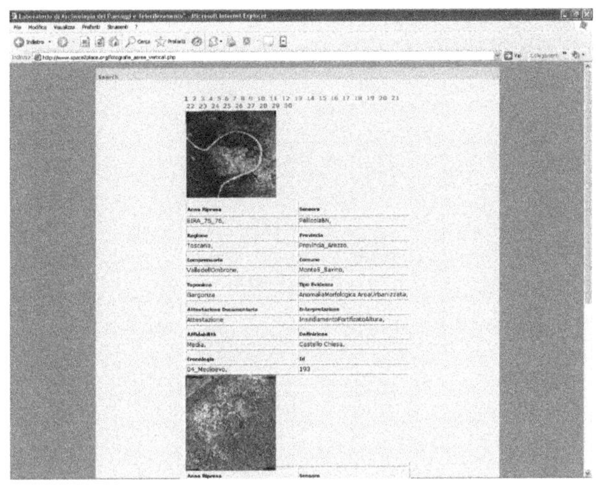

The 5000 or so air photographs in our database are all in JPEG format, with an average a size of about 2 MB, too big to be downloaded and presented on screen relatively quickly. We have therefore made provision for the creation of thumbnails of 200 x 200 pixels and a file-size of about 90 GB, using SQL code to create them as and when needed, without taking up any permanent storage space. In response to each query the database simply collects the image from its original position and creates an instantaneous thumbnail that can be seen by the user on the results screen.

Conclusion

The DBMS described in this short article has proved to be an extremely effective instrument for the transmission and exchange of all types of archaeological data in alphanumeric form. It is not our intention to present this as a finished product, but rather as a prototype to be tested in the future through a suitable number of case-studies. It could represent, however, a first basic model for the organisation of easily exportable information, readily adaptable to all types of data. The system should be seen as in a state of continuous evolution, in that it must necessarily adapt to new products coming onto the software and hardware markets, and to the opportunities presented by future advances in our researches.

Barbara Frezza

Acknowledgements

The authors wish to acknowledge their gratitude to the following:
Giancarlo Macchi (University of Siena) for help and criticism at the beginning of the project.
Enrico Donati for the development of the first prototype web-based database system.
Chris Musson, from Wales, for his invaluable comments and for his help with the English version of the text.

References

ATZENI P., CERI S., PARABOSCHI S.,TORLONE R. 2002, *Basi di dati: modelli e linguaggi di interrogazione*, Milano.

ATZENI P. [et al.] 2003, *Basi di dati: architetture e linee di evoluzione,* Milano.

BUYENS J. 2002, *Sviluppo di database per il Web: passo per passo,* Segrate (Mondatori informatica).

CAMPANA S., FRANCOVICH R. 2003, Landscape archaeology in Tuscany: Cultural resource management, remotely sensed techniques, GIS based data integration and interpretation. In M. Forte, G. Williams (eds.), T*he Reconstruction of Archaeological Landscapes through Digital Technologies, Proceedings of the I*[st] *Italy-United States Workshop, Boston, Massachusetts, 1-3 November 2001*, BAR International Series 1151, pp. 15-28.

D'ANDRIA F. 1997, *Metodologie di catalogazione dei beni archeologici, Lecce-Bari* (Beni Archeologici - Conoscenza e Tecnologie, Quaderno 1.1)

DOERR M., SARRIS A., *CAA 2002 The Digital Heritage of Archaeology, Computer Application and Quantitative Methods in Archaeology. Proceedings of the 30*[th] *Conference, Heraklion, Crete, April 2002.*

DONATI, E. 2005, *Un database on-line per le fotografie aeree oblique dell'Università di Siena. Prototipo di contenitore dinamico.* Undergraduate thesis in the Department of Medieval Archaeology at the University of Siena. Unpublished.

GOTTARELLI A. 1997, *Sistemi informativi e reti geografiche in archeologia: Gis-Intenet*, Quaderni del Dipartimento di Archeologia e Storia delle Arti, Sezione archeologica – Università di Siena, VII Ciclo di Lezioni sulla Ricerca applicata in Archeologia, Certosa di Pontignano (SI) 11-17 dicembre 1995.

Mobile computing in archaeological prospection: an update

Stefano Campana,[1] Matteo Sordini[2]

[1] University of Siena a Grosseto, Department of Archaeology and History of Arts, Landscape Archaeology, (LAP&T)
campana@unisi.it
[2] University of Siena at Grosseto, Department of Archaeology and History of Arts, (LAP&T)
sordini@lapetlab.it

1 Introduction

We have emphasized in the past that we have been conscious of an increasing schism between methods and technologies in the laboratory as compared with work in the field. While the availability of advanced technologies has been rapidly growing, activities in the field have continued to make use of instruments and methodologies developed in the 1970s. The problems arising from this situation are mainly inherent in the collection of data that lack the accuracy required by GIS, or which work to a different kind of rational logic. There is also the problem of the unavailability in the field of the large amount of data stored in the desktop GIS.

The laboratory of Landscape Archaeology and Remote Sensing at the University of Siena started working on mobile computing in 1999 (Campana, Francovich 2003; Campana 2005). In the early years we mainly concentrated on studying the application to landscape archaeology of various research methods, and in testing different configurations of mobile systems (PDA, Tablet PC). From 2005 we have concentrated our efforts on improving the mobile software to make it better-related to our specific needs.

Stefano Campana

1 Field-waking survey

In the past two years our digital achives have been moved from a FileMaker database to Microsoft Access. We were forced into this choice mainly because the FileMaker software, which we started using in 1995, has never achieved compatibility ArcGIS software. This has meant that our laboratory has never been able to use a genuine and sophisticated GIS.

Our new database 'Archaeological Map' has been structured intentionally in a simple way, in an effort to eliminate data redundancy and to make use of the Microsoft Access software and its compatibility with ESRI GIS software.

The DATABASE is composed altogether of six tables:

Environmental conditions. This section is directed at the storage of data on morphological characteristics and land-use. This gives us the background to each topographic unit in the archaeological evidence.

Topographic unit. In this table we store all the descriptive attributes of the topographic units recorded in the landscape – mainly artefact scatters but also off-site material and individual monuments.

Chronologlcal unit. This section describes the main characteristics of every historical phase present in the topographic unit. The table records attributes that have necessarily been the subject of interpretation. This section is required in order to draw a distinction between the different functions that the site may have served at different periods. It records for each chronological phase the relative function, and viceversa. A typical example in the Tuscan landscape might be a Roman villa. This kind of settlement appears in the field as a widespread artefact scatter, usually indicating a main period of building construction between the first century BC and the beginning of the second century AD. But in many instances Roman villas overlie earlier settlements, and also show significant transformations as rural settlements in the course of their long existence, which commonly finishes around the end of the fifth century AD (Francovich, Hodges 2003).

Finds: The database for archaeological finds has remained substantially unchanged. The most meaningful innovation relates to the addition of a field concerned with grid collection. If the artefacts are collected by this method it will be sufficient to specify the reference number of the grid in order to relate this to the finds recovered from it.

Pottery paste. This database has been completely revised and is now based on a new method of analysis of the ceramic body under the optical microscope.

The changes introduced in our laboratory instrumentation have a direct influence on the devices that we use in the fieldwork. The mobile GIS has been updated to Arc Pad version 7, the ACCESS database of which is perfectly compatible with Windows Mobile Pocket PC OS. The database is automatically converted into Pocket Access format through Activesync software.

The main limitation of Pocket Access is the unavailability of an editing tool. This means that you can seek information on your mobile but cannot update it.

To overcame this problem we have created a low-level tool for the PDA which allows us to edit the database. We are now, at last, able to share the same data bank in the laboratory and in the field, and to update both databases through the synchronization tool Activesync every time we connect the mobile device to the Desktop PC or the server.

The current phase of our work is aimed at progressively reducing the *hiatus* between some of the functions in the desktop information system and those available in the mobile version. With the help of ESRI Italy we have been able to start working with a new software, Arc Pad Application Builder (version 7). This software allows us not only to implement visualization masks and data entry but also to build a real mobile GIS.

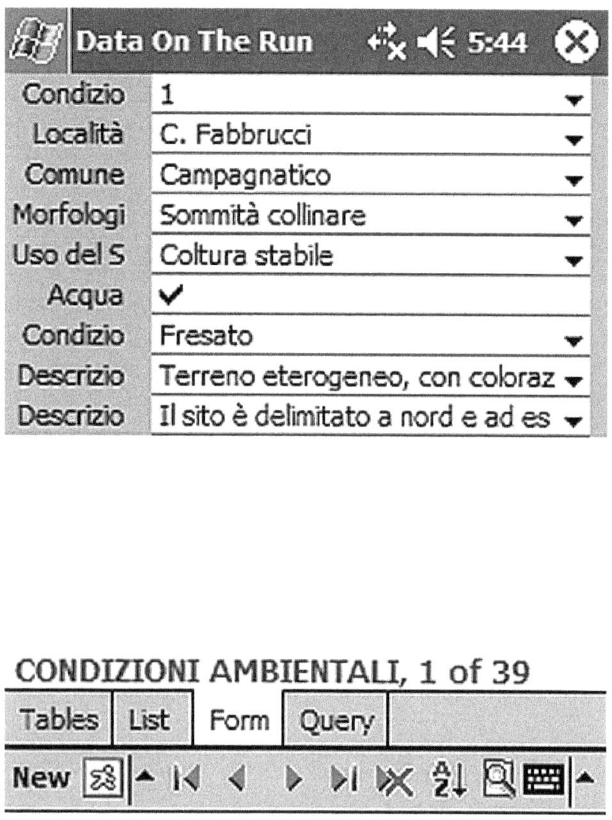

Figure 1 *Interface of the mobile database 'use of soil'*

This latest development will also facilitate the link between the geographical interface and the database or more detailed, the relationship between the shapefiles stored in the mobile GIS and those recorded in the mobile database (Tripcevich 2004).

This will further narrow the gap between the desktop and mobile systems. The user will also be able to access information which is progressively more similar to the amount available in the laboratory. We should emphasize here the compatibility between the ESRI and Microsoft software, which can be elaborated through use of a similar programming language – ESRI VBscript is in fact based on Microsoft Visual Basic.

Stefano Campana, Matteo Sordini

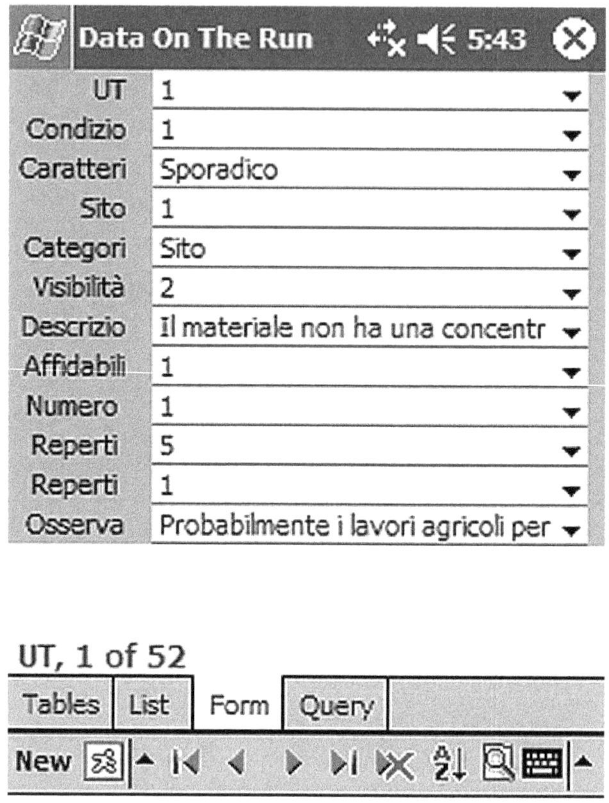

Figure 2 *Interface of the mobile database 'topographic unit'*

2 Aerial survey

We began experimenting with mobile systems for aerial use in 2004. Initially we employed a mobile PDA originally developed for field use but we quickly turned to an integrated system based on the Tablet PC.

This pattern of devices, involving an HP 1100 Tablet PC linked to a Socket Bluetooth GPS, has been designed to undertake both navigation and data management tasks (Campana 2005). While this is still 'work in progress' three significant aspects have emerged during the first stage of use in the air. The capacity to take a PC into the air, in addition to satisfying the primary needs of survey and navigation, also provides an instrument that can be used in a variety of other ways. For example, aerial archaeologists are making ever greater use of digital cameras and the Tablet can be used to download data if the camera's memory becomes full. Also, in cases where the quality of the recorded image is of prime importance (for instance in the recording of historical monuments or excavations in progress), the results can be downloaded and examined on-screen in real time. It is also possible to install software such as Jeppesen Flitestar, giving access to aeronautical maps and assisting route planning. A second consideration is that of writing. Unlike the PDA, the Tablet has a system for the recognition of writing, which means that it is possible to make entries in lists and tables with Tablet's own pen, just as one would with an ordinary biro. In addition to facilitating work in an environment which (we must not forget) is subject to constant movement and disturbance, this allows us to record both 'standardized' and 'free-form' information. It is practically impossible to do this with a traditional digital keypad, even less so in the air than on the ground. A final consideration, before describing the system, lies into the tasks that need to be performed when using this equipment. Aerial work is very difficult and intensive. The archaeologist has to follow the route, read and interpret the landscape below, keep cameras and other equipment under control so as to arrive at the next target with the right lens already in place, make sure that there are enough exposures still available, check the lighting-conditions and write detailed flight reports and site records – and these are just *some* of the many tasks that need to be done.

Our aim is to aid significantly the work in the air, simplifying navigation as well as the recording of (and access to) information in ways which will allow the archaeologist to give more attention to the landscape below.

During 2005 and 2006 we have been attempting to solve a number of problems which we encountered in our earlier work. As noted above, our main objective has been to develop instrumentation and software that will make the work of the archaeologist in the air easier, quicker and more efficient.

After each flight the features (or 'topographical units') recognised in the air and documented through oblique photography must be entered and mapped in the GIS, so that they can be interpreted and integrated with all of the other data available.

In order to make these operations easier it is necessary to link each series of photographs, or 'shots', to its related geographical coordinates. Only if we are able to perform this link systematically will it become possible to access the information and to process the data in the GIS.

The systems available on the market did not take account of the conditions encountered in the cockpit of a light aircraft: cramped, noisy and constantly affected by vibration. To help the airborne archaeologist to record the geographical coordinates of each archaeological site or landscape feature we decided to write a purpose-made software using Microsoft Visual Basic. The resulting tool has been named *Aerialsurveytool* and it has been made compatible with the ESRI ArcGIS software.

During aerial survey this tool allows us to link every photograph or set of exposures, identified by the number shown on the camera's display, with the current GPS position. The tool's architecture has been designed so as to be used with Tablet's own pen. Its operation is closely related to the GIS/GPS system, from which it receives the North/East coordinates. The resulting data are stored in a Microsoft Access database.

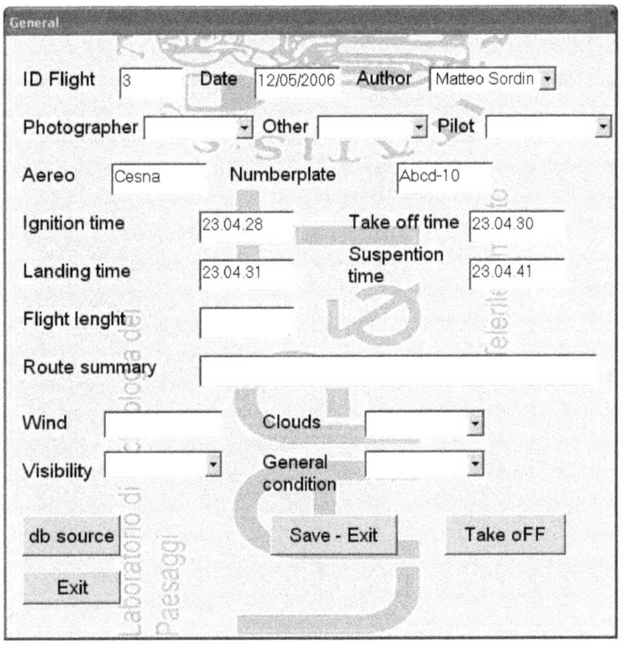

Figure 3 *Starting interface of* Aerialsurveytool

The tool uses three interfaces and the data are entered by pushing buttons or by selecting for each field pre-programmed terms from the relevant menus, greatly simplifying the updating of the database. Our aim, at every stage, is to make each task simpler by making the greatest possible use of automated actions or entries.

The first set of entries records specific information about the flight, such as atmospheric conditions, participants (pilot, photographer, assistant etc), flight time, flight path (track), and so on. Once this information has been entered the user presses the button 'Take-off' and there appears the ESRI ArcGIS screen with the bar 'Position'. This is another interface of the *Aerialsurveytool*. It is worth noting that this bar does not prevent use of the normal navigation aids of the GPS. Furthermore, ArcGIS can also be used at the same time.

The bar always appears at the top of the screen and includes two buttons, named 'GPS position' and 'Schedule'. Using the Tablet's own pen and clicking the button 'GPS position' the software automatically acquires the geographical coordinates of the aircraft's current position and stores the data in a table. Pressing Schedule allows the user to return to the starting interface to make corrections or to add final information at the end of the flight.

Once the geographical coordinates have been recorded the software asks the user to link the GPS position to the reference numbers of the related photographs (shots) through an interface called 'image'.

We are also considering to possible use of the scrollbar to avoid the need to write the numbers. The selected numbers, for the first and last shots recorded at each site or location, will through this process be linked to the GPS record. By selecting pre-programmed terms from the menu it is also possible to record a simple site interpretation, such as enclosure, ditch, building, mound etc.

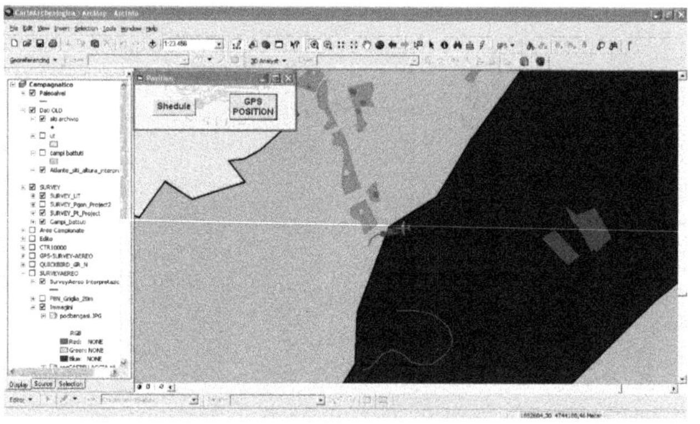

Figure 4 *Interface* Aerialsurveytool, *for entering the GPS coordinates overlaid on the ArcGIS software showing the real time position of the aircraft*

Figure 5 *Detail of the interface in* Aerialsurveytool *used to enter GPS coordinates*

The tool automatically resolves a common problem of image numbering through a simple mathematical function. The display on most digital cameras (in our case a Canon 20D) shows the number of photographs *still available on the memory card*, in decreasing order. Unfortunately after downloading the images to a PC the numbering becomes inverted, so that they appear in increasing rather than decreasing order. This destroys the correspondence between the photo (shot) reference numbers taken in the air and the related GPS position. *Aerialsurveytool*, however, automatically converts the numbering from decreasing to increasing order, thereby saving the user from having to carry out calculations in flight or to waste time during the already lengthy post-flight processes in the laboratory.

Finally, we have addressed the question of changing the memory card without creating problems in the linking of shots to GPS coordinates. All that the user needs to do is to remember to download the images to the PC using a *different* folder for each memory card.

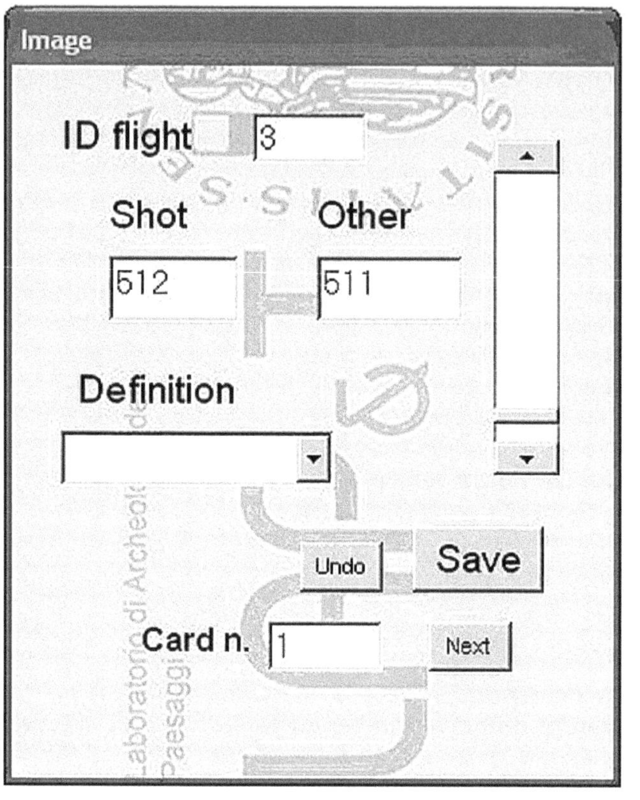

Figure 6 *Detail of the interface* Aerialsurveytool *to enter photo (shot) numbers through the scrollbar, along with a preliminary interpretation of the site photographed*

Using this system, all of the information recorded during flight is stored in the database and is it possible to create, export or print a record in the form of a flight report that shows the the GPS track of the flight and the GPS position from which the key photograph for each site was taken.

In conclusion we would like to emphasise that this short paper does not present the final result of our experiments, but simply the present state of work-in-progress that will be updated year on year as our experience grows.

Matteo Sordini

Acknowledgments

The authors are indebted to ESRI Italy and Pitro Coffaro for their invaluable help and for providing the Arc Pad and Arc Pad Application Builder software.

Special thanks are also due to Chris Musson for helping with the English version of the text and for his valuable comments on the practice of aerial research.

Special thanks are also due to the team of the Laboratory of Landscape Archaeology and Remote Sensing of the University of Siena at Grosseto: C.Felici, E.Vaccaro, F.Pericci, M.Ghisleni, L. Marasco, B.Frezza, A.Caprasecca and M.Corsi.

References

BALTSAVIAS E., GRUEN A., VAN GOOL L., PATERAKI M. 2006, *Recording, Modeling and Visualisation of Cultural Heritage,* London, Taylor & Francis Group plc.

CAMPANA S. 2005, Mobile Computing in Archaeological Prospection. In *Proceedings of the 6th International Conference on Archaeological Prospection, (Rome 14-17 September 2005),* Institute of Technologies Applied to Cultural Heritage, Rome, pp. 427-431.

CAMPANA S., FRANCOVICH R. 2003, Landscape Archaeology in Tuscany: Cultural resource management, remotely sensed techniques, GIS based data integration and interpretation. In M.Forte, G. Williams (eds.), *The Reconstruction of Archaeological Landscapes through Digital Technologies, Proceedings of the 1st Italy-United States Workshop, Boston, Massachusetts, 1-3 November 2001*, BAR International Series 1151, pp. 15-28.

CAMPANA S. FORTE M. (eds.) 2001, *Remote sensing in archaeology*, Firenze.

CAMPANA. S., FRANCOVICH R. 2006, *Paesaggi Archeologici e tecnologie digitali 1: Laser scanner e GPS,* Città di Castello, All'insegna del Giglio.

CAMPANA S. 2001, *Carta archeologica della Provincia di Siena - Murlo,* Siena.

FELICI C. 2004, *Carta archeologica della Provincia di Siena – Pienza,* Siena.

FORTE M. 2002, *I Sistemi Informativi Geografici in Archeologia*, Rome.

FORTE M. (ed.) 2005, *The reconstruction of Archaeological landscapes through digital technologies,* Oxford, BAR International Series 1379.

FRANCOVICH R., HODGES R. 2003 *Villa to village. The transformation of the Roman Countryside in Italy, c.400-1000*, Duckworth, London.

MUSSON C., PALMER R., CAMPANA S. 2005, *In volo nel passato. Aerofotografia e cartografia archeologica*, Insegna del Giglio, Florence.

TRIPCEVICH, N. 2004, *Flexibility by design: how mobile GIS metts the needs of archeological survey*, Cartography and Geographic Information Science, VOL.31 (3), pp.137-151.

Geo-archaeological site location modelling in the territory of a classical town - Case Study Sagalassos (southwest Turkey)

Véronique De Laet,[1] E. Paulissen,[1] H. Vanhaverbeke,[2] M. Waelkens[2]

[1] Physical and Regional Geography Research Group (KULeuven, Belgium),
[2] Eastern Mediterranean Archaeology (KULeuven, Belgium)

1 Introduction

The objective of site location modelling is not only to identify the main factors that dedicated a settlement location, but also to make predictions for settlement locations from a sample of known locations to a much broader area. Predictive site location models exploit the contrasts between environmental and cultural characteristics of places where sites do and do not occur (Warren 1990). According to these authors, the key to the success of site location modelling is the fact that archaeological sites tend to occur in favourable environmental settings (Warren and Asch 2000). Most archaeological location models rest on two fundamental assumptions: first, settlement choices and utilization within the study area were influenced by local environmental characteristics and second, environmental conditions that influenced these choices are portrayed in the actual environment and consequently in modern maps. An underlying assumption is that environmental changes were uniform throughout the entire study area and affected settlement patterns similarly (Custer 1989).

Since 1993 more than 250 archaeological sites are identified in the territory of Sagalassos, dating from Epipalaeolithic to Ottoman times (BC ~18000–AD 1923) (Vanhaverbeke and Waelkens 2003).

The general objectives of the settlement research in the vicinity of Sagalassos are to (1) identify the environmental factors playing a major role in the location of archaeological settlements, (2) to describe the evolution of factors that dedicated these settlement locations and (3) to create a settlement location model for each relevant archaeological time period using stepwise rare event logistic regression analysis (King and Zeng 2001[a-b]), in order to find locations likely to contain archaeological sites.

This paper attempts to illustrate the method that was used to create different site location models. Besides information collected during various field campaigns also topographical and digital environmental data are used in a GIS environment.

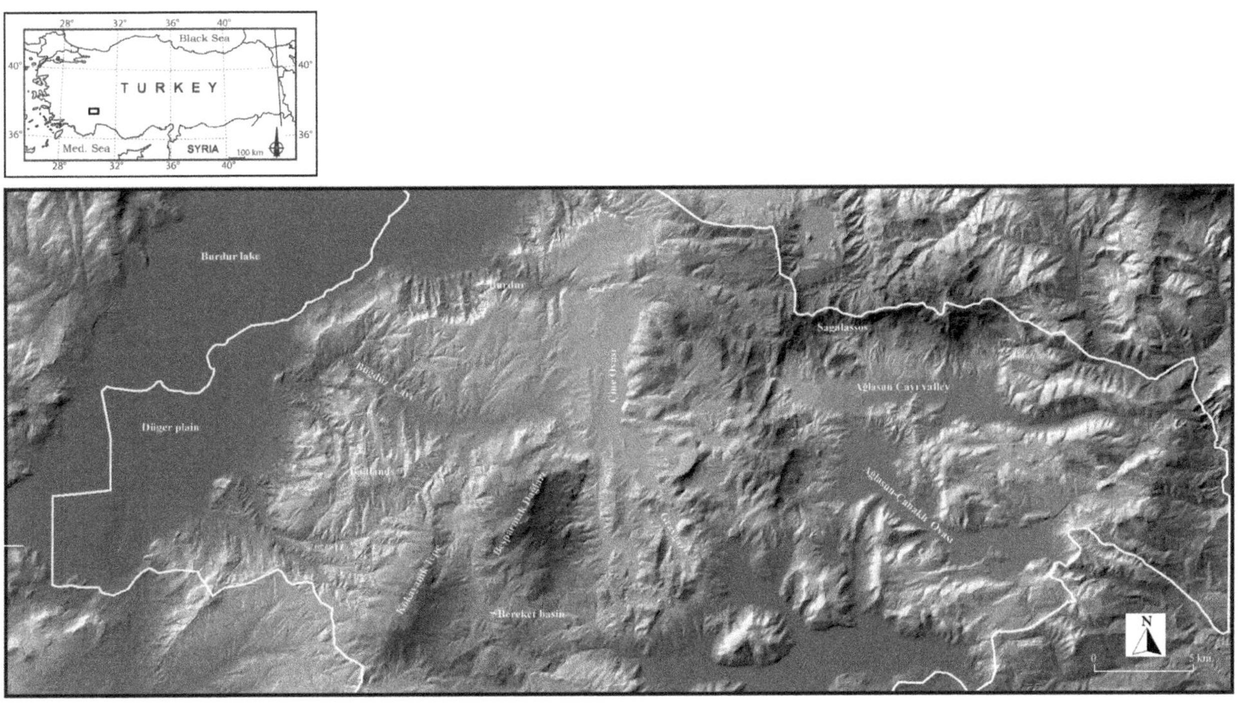

Figure 1 *Situation of the study area*

2 The Study Area

The study area, situated in a region in ancient times known as Pisidia (Bracke 1993), covers the territory (*Chora*) of the antique city of Sagalassos (1176 km²) and is located in the Western Taurus Mountains, roughly 100 km to the north of Antalya in southwest Turkey (Fig. 1). The boundaries of the territory are established based on inscriptions, milestones, literary evidence and topographic observations (Waelkens et al. 1997, 2000). As a result the territory of Sagalassos is largely determined by natural borders and by the existence of urban settlements whose origins may go back to (pre)-Hellenistic times (Vanhaverbeke and Waelkens 2003).

3 Data

3.1 Archaeological data

For this research a chronological subdivision, based on pottery, architectural characteristics and cultural or climatic changes is used. Even though a modeling approach is applied to all archaeological time units presented below, the emphasis of this paper is laid on Middle Bronze Age till Early Byzantine sites. Prehistoric sites are not included because this time unit is relatively long and encompasses various types of settlements. Also Post-Early Byzantine sites are excluded as present-day villages should be added to this time unit in the future. One can argue that methodologically also Middle Bronze Age till Early Iron Age sites should be eliminated, since only a limited number of sites are detected during the archaeological survey. Though, sites belonging to this period are very specific and consequently the analysis of a limited number of sites provides reasonable results.

1. Prehistoric sites (mainly BC 9000-BC 2000, 35 sites)
2. Middle Bronze Age till Early Iron Age sites (MBA-EIA, BC 2000-BC 330, 11 sites)
3. Hellenistic sites (BC 330–BC 25, 24 sites)
4. Early Roman till Middle Roman sites (ER-MR, BC 25-AD 450, 61 sites)
5. Late Roman till Early Byzantine sites (LR-EB, AD 450-AD 650, 90 sites)
6. Post-Early Byzantine sites (Post-EB, AD 650-AD 1900, 20 sites)

All site positions are located by GPS measurements without differential correction or indicated on topographic maps. Archaeological sites located within modern villages are not taken into account, since these archaeological remnants are not necessarily located in situ or related to clustered inhabitation. Finally locations of archaeological material or architectural remains without any evidence of reliance to an archaeological site are eliminated from the list.

An important consequence of such a coarse chronological framework is that site distribution maps based on survey data are to some extent "palimpsests" of sites belonging to the same period that were not all in use simultaneously (Cherry 1983).

A representative site area is defined for each of the archaeological sites surveyed, depending on the period of occupation and is based on survey information, literature or defined in deliberation with the archaeologists. In case no specific information for site size is available, we have defined an area of 2 ha (Bintliff 1997; Diacopoulos 2004; Fall 1998). As a result 113 MBA-EIA site pixels are taken into account, 2053 Hellenistic positions, 925 ER-MR site pixels and 1470 LR-EB locations (Table 1). The application of site area implies that the risk of a certain degree of autocorrelation is real, but inevitable in a holistic rare event logistic regression model for archaeological applications.

Time Unit	Number of sites	Site area pixels	Calibration / Validation	Non-site area pixels	Calibration / Validation	τ	y
MBA-EIA	11	113	91	452	362	0.0002	0.25
			22		90		
Hellensitic	24	2053	1643	6159	4928	0.0044	0.33
			410		1231		
ER-MR	61	925	740	4625	3700	0.0020	0.2
			185		925		
LR-EB	90	1470	1176	7350	5880	0.0031	0.2
			294		1470		
Total study area (pixels):				**470940**			

Table 1 *Archaeological data*

3.2 Environmental data

According to Altschul and Jones (1990) 70% of archaeological site locations are defined by 3 variables: *SLOPE GRADIENT, ALTITUDE (*m a.s.l.) and *ASPECT*. Other literature mentions also distance to water as an important location factor for Neolithic settlements (Warren and Asch 2000). A location factor we fully agree with, also for other archaeological periods. Hence, those 4 factors are included in our analysis. The variable distance to water is subdivided into two sub-factors, *DISTANCE TO MAIN RIVERS* and *DISTANCE TO AFFLUENT RIVERS*. Because no information is available about the distribution of natural springs, the assumption is made that natural springs are located on the *CONTACT BETWEEN PERMEABLE (OPHIOLITIC MÉLANGE) AND LESS-PERMEABLE (LIMESTONE) LITHOLOGIES*, something which is confirmed by field work in the region. Additionally *DISTANCE TO NATURAL CORRIDORS*, *SOLAR INSULATION*, *GEOMORPHOLOGY*, *GEOLOGY* (Şenel 1997[a-c]) and *VISIBILITY* are taken into account. The GIS-derived variables are based on a 50 m resolution contours-derived DEM. Regarding all variables including distance, friction is taken into account (Marble and Machovina 1997; Nuninger 2003; Van Leusen 1999) and the DISTANCE algorithm in Idrisi® is used. Table 2 provides an overview of all variables used in the statistical analyses with indication of their subdivision, interpretation, type, information source and an indication of the category used as reference in the logistic regression analysis.

Variable	Interpretation (reference =bold)	Units	Information Source
ELEVATION	Height a.s.l.	m	TIN interpolation isolines ⇒ raster generation ⇒ mean filter applied
SLOPE	Slope gradient of soil surface	%	TIN ⇒ Raster ⇒ SURFACE algorithm Idrisi© (steepest slope)
ASPECT	**NW-NE**; NE-SE; SE-SW; SW-NW, flat (<5%)	C	TIN ⇒ Raster ⇒ Surface algorithm Idrisi©
D CONTACT OPHIO-LIMESTONE	Distance to contact ophiolitic mélange - limestone	m	Contact delineated on the geological map ⇒ Cost algorithm Idrisi© using friction
D MAIN RIVER SYSTEM	Distance to main rivers	m	Runoff algorithm Idrisi© >25 km² ⇒ Cost algorithm Idrisi© using friction
D AFFLUENT RIVERS	Distance to affluent rivers	m	Runoff algorithm Idrisi© (7.5-25 km²) ⇒ Cost algorithm Idrisi© using friction
D NATURAL CORRIDORS	Distance to natural corridors	m	Roads defined by archaeological survey ⇒ Cost algorithm Idrisi© using friction
VISIBILITY	Number of pixels that a certain pixel can see	N	Tin ⇒ Raster ⇒ viewshed algorithm Idrisi©
GEOMORPHOLOGY	Plain, Footslope, Midslope connected to footslope, **Midslope connected to ridge**, Ridge / plateau	C	TIN ⇒ Raster ⇒ Edge filter ⇒ reclass ⇒ crosstable with slope map (Dalrymple et al. 1968)
SUN INSOLATION	**Negative**, Positive, Flat (gradient <5%)	m	Shade map summer conditions, Shade map winter conditions (shadow, normal sun, extensive sun) ⇒ Crostable (flat areas gradient <5%)
GEOLOGY	16 categories : reference=**marls**	C	Geological map of Şenel (1997[a-c])

Table 2 *Environmental data*

Because actual *LAND COVER* may probably not be related to ancient vegetation distribution, this variable is not incorporated into the logistic regression analysis (Kaniewski et al. submitted). Neither are social factors included because of the lack of accurate information.

4 Stepwise rare event logistic regression

Stepwise rare event logistic regression makes it possible to get a holistic view on the factors influencing the location of archaeological sites. An individual predictor variable may be significant, but in comparison to others, may have a minor influence on the location of archaeological sites. Since all predictor variables are integrated into one holistic model, we are able to produce probability distribution maps of locations likely to contain archaeological sites belonging to different archaeological time units (Figs. 2a-d).

With relation to stepwise logistic regression, variables are selected either for inclusion or exclusion from the model in a sequential way, based only on statistical criteria (significance level of 0.15 to enter the model, 0.05 to stay in the model) (Hosmer and Lemeshow 1989). Ordinary logistic regression (OLR) describes the relationship between a dichotomous response (dependent) variable (Y, i.e. the presence or absence of archaeological sites) and a set of categorical or continuous explanatory (independent) variables (X_1, X2, …, X_n), which do not need to fit a normal distribution (Allison 1999; Hosmer and Lemeshow 1989). The logistic response function can be written as follow (Allison 1999):

$$P_{i,Y_i=1} = \frac{1}{1+e^{(-\alpha-\beta_1 X_{1i}-\beta_2 X_{2i}-\ldots-\beta_n X_{ni})}} \qquad \text{Eq. (1)}$$

Where $P_{i,Y=1}$ is the probability of occurrence of an archaeological site for pixel i, α the intercept and β_k the regression coefficient for the explanatory variable X_{ki} estimated by maximum likelihood.

The occurrence of archaeological sites can be seen as a "rare event" with dozens to thousands of times more 0s (no archaeological site locations) than 1s (archaeological site locations) and as a result logistic regression techniques adopted from the social science should be used (King and Zeng 2001[a-b]). If in this case OLR is used, the probability for the occurrence of archaeological sites is sharply underestimated. In order to overcome the problems related to rare event data, King and Zeng (2001[a-b]) proposed to incorporate three correction techniques into the OLR methodology. These corrections include stratified sampling (1-5 times more non-site pixels than site pixels) (Table 1), the calculation of an adjusted intercept (Eq. 2) and adjusted probability value (Eq. 3 and 4).

$$\alpha' = \alpha - \ln\left(\frac{(1-\tau)}{\tau} * \frac{\hat{y}}{(1-\hat{y})}\right) \qquad \text{Eq. (2)}$$

Where τ is the actual fraction of 1s in the dataset (**Table 1**), ŷ the observed fraction of 1s in the sample (**Table 1**), α the estimated intercept and α' the corrected intercept

$$P_{i,Y_i=1} = P_i' + C_i \qquad \text{Eq. (3)}$$

$$C_i = (0.5 - P_i') * P_i' * (1 - P_i') X V(\beta) X' \qquad \text{Eq. (4)}$$

Where X is a 1× (n+1) vector of values for each independent variable, X' is the transposed vector of X, V(β) the variance-covariance matrix, P_i' the estimated probability value and P_i the corrected probability value.

Since a model does not have any scientific significance without validation, the entire sample of archaeological site and non-site locations was split into a calibration and validation dataset with 80% calibration and 20% validation data (Table 1) (Addiscott 1995; Gong 1986). Validation of the modelling results is performed using the area under the Receiver Operating Characteristic curve (ROC-curve) and Bias Adjusted Kappa (PABAK) statistics (Byrt 1993; Hoehler 2000).

A ROC-curve is a graphic representation of the predictive accuracy of the fitted model and plots the proportion of 1's correctly predicted by the model (TP/TP+FN; sensitivity) against 1 minus the proportion of 0s correctly predicted by the model (TN/FP+FN; specificity) using different cut-off values. In this case TP are the True Positive, FP the False Positive, FN the False Negative and TN the True Negative observations (Brown & Davis 2006) (Fig. 3). For a perfect model the ROC-curve runs vertically from (0, 0) to (0, 1) and further horizontally to (1, 1). A model not performing better than chance (Brown & Davis 2006) follows the 1-1 line (dotted line in Fig. 3). The closer a curve gets to the ideal situation, the better the model performance will be.

PABAK statistics are applied when the modelled and real classification differ in proportion of positive results, and are calculated using Eq. 5 and 6.

$$P_{obs} = \frac{TP + TN}{N} \qquad \text{Eq. (5)}$$

Where P_{obs} is the observed agreement

$$PABAK = 2P_{obs} - 1 \qquad \text{Eq. (6)}$$

In the ideal case of a perfect mach between the model and reality, the PABAK value will be 1. If on the other hand a model is not performing better than chance, the PABAK value will be 0. In reality PABAK values range between 0 and 1 with:

- <20%: slight agreement
- 20-40%: fair agreement
- 40-60%: moderate agreement
- 60-80%: substantial agreement
- 80-100%: nearly perfect agreement.

In the worst case, if the agreement is worse than chance, the PABAK value will be negative.

5 Results

Table 3 presents the equations used to calculate the corresponding probability maps presented in Fig. 2. The symbols X1 to Xn in this table refer to the specific categories (variables) that are indicated in Table 4. Time units indicated in bold in Tables 3 and 4 are periods for which no model convergence could be reached (Allison 1999). Nevertheless, the results for these time units correspond well with the qualitative interpretations of Vanhaverbeke and Waelkens (2003).

Time unit	Probability formula (Eq. 1)
MBA-EIA	**-8.6250-1.3858*X24+1.4943*X26-0.00006*X30+0.000508*X36**
Hellenistic	-10.1673+2.2767*X1+1.6287*X2+3.3047*X7+4.3966*X10+1.653*X12+4.3114*X14+1.9853*X15-0.2928*X19-0.3989*X21-1.5044*X22+0.3463*X24+1.0439*X25-0.00003*X30-0.00005*X31-0.00451*X32+0.00334*X33+0.00117*X34+0.034*X35+0.00084*X36
ER-MR	-6.7017-0.8451*X1-0.8806*X3-0.6453*X7+0.9121*X10+0.7131*X14-0.755*X15+0.5062*X18+1.2178*X22+1.4302*X24+1.4853*X25+1.0441*X26-0.00001*X30-0.00001*X31-0.00176*X32+0.00145*X33-0.011*X35+0.000158*X36
LR-EB	-5.9417+0.3066*X1+0.4768*X7+1.4701*X10+0.5931*X12+1.4454*X14+1.1265*X15+1.1458*X8-0.6424*X22+0.2349*X25-0.00000792*X30-0.00002*X31-0.00049*X32+0.00129*X33-0.0094*X35

Table 3 *Probability formula*

Table 4 shows the odds ratios for each continuous variable or category of a specific categorical variable. Odds ratios are used to adequately interpret the stepwise rare event logistic regression results. An odds ratio of 50 (-50) for a categorical variable indicates that it is 50 times more (less) likely to find sites on that specific category of a categorical variable relative to the reference category. An odds ratio of 1.20 (0.20) for a continuous variables indicates that is 20% (80%) more (less) likely to find archaeological sites with each unit increase of that specific continuous variable.

Variable	Category	Archaeological time unit							
		MBA-EIA		Hell.		ER-MR		LR-EB	
		Odds	sign	Odds	sign	Odds	sign	Odds	sign
Geology									
X1	Quaternary alluv.			9.744	+	0.430	-	1.359	+
X2	Pleist conglomerate			5.097	+				
X3	Pleistocene mixture					0.415	-		
X4	Pliocene slope debris								
X5	Pliocene volcanics								
X6	**Marls**								
X7	Flysh			27.240	+	0.525	-	1.611	+
X8	Antalya nappes limestone							3.145	+
X9	Angular unconformity limestone								
X10	Bey dagl. Sandstone + Tort. Conglom.			81.171	+	2.489	+	4.300	+
X11	Bey dagl. Limestone miocene								
X12	Bey dagl. Limestone cretaceous			5.222	+	2.040	+	1.810	+
X13	Bey dagl. Olistostrome								
X14	Lycian nappes limestone			74.543	+			4.244	+
X15	Lycian nappes ophiolites			7.282	+	0.470	+	3.085	+
X16	Lycian nappes olistostrome								
Geomorphology									
X17	Valley								
X18	footslope					1.659	+		
X19	Midslope connected to footslope			0.746	-				
X20	**Midlsope connected to ridge**								
X21	Ridge			0.671	-				
Aspect									
X22	Flat			0.222	-	3.380	+	0.526	-
X23	**NW-NE**								
X24	NE-SE	0.250	-	1.414	+	4.179	+		
X25	SE-SW			2.840	+	4.416	+	1.265	+
X26	SW-NW	4.456	+			2.841	+		
Solar insulation									
X27	Positive								
X28	**Negative**								
X29	Flat								
Continuous variables									
X30	D ophio-limestone (200 m units)	0.9881	-	0.994	-	0.998	-	0.9984	-
X31	D natural corridors (200 m units)			0.99	-	0.998	-	0.996	-
X32	D main river system (200m units)			0.4058	-	0.7033	-	0.9066	-
X33	D affluent rivers (200 m units)			1.9503	+	1.3364	+	1.2943	+
X34	Altitude (50 m units)			1.0602	+				
X35	Slope (10%) units			1.4049	+	0.8958	-	0.9103	-
X36	Visibility (100 pixel units = 5 km)	1.052	+	1.0876	+	1.0159	+		

Table 4 *Odds ratios*

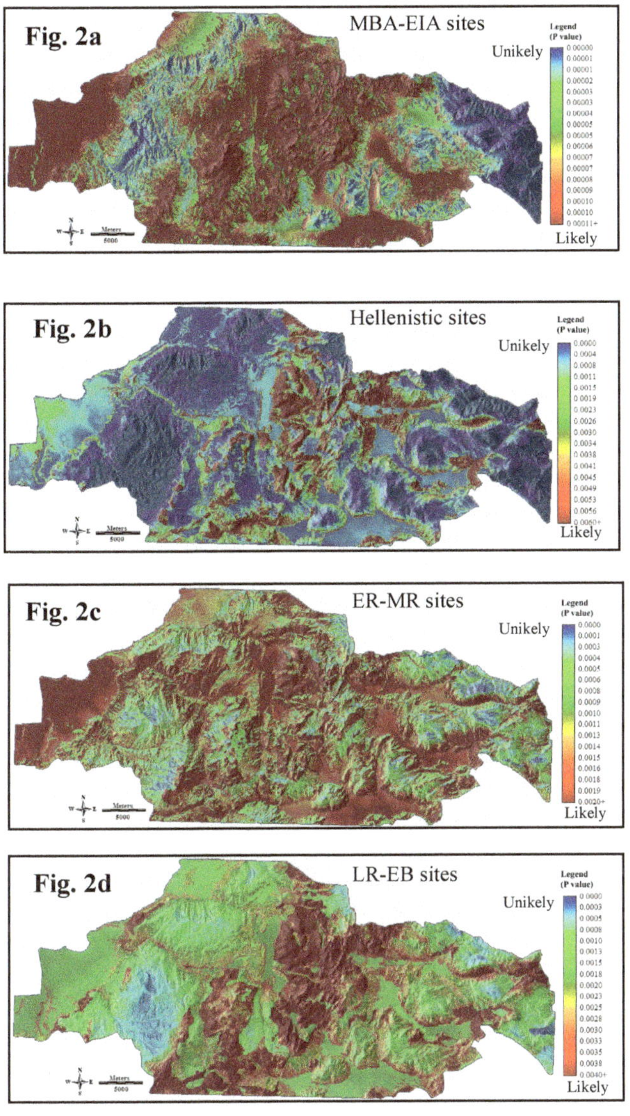

Figure 2a-f *Site location probability maps*

Field data show that colluvation and to a lesser extent alluvation are important in the region and therefore a number of sites could not be discovered during field survey. In the future, both colluvation and alluvation should be integrated in our analyses.

Validation of the stepwise rare event logistic regression probability maps (Fig. 2a-d) shows that for most time units the prediction model is doing moderate to substantial (Fig. 3, Table 5).

6 Discussion and conclusion

Rare event logistic regression modelling is able to quantify and amplify the initial archaeological settlement research of Vanhavebeke and Waelkens (2003).

Table 6 presents an overview of the factors possibly influencing the location of archaeological sites according to different archaeological time units, using the modelling strategy outlined before.

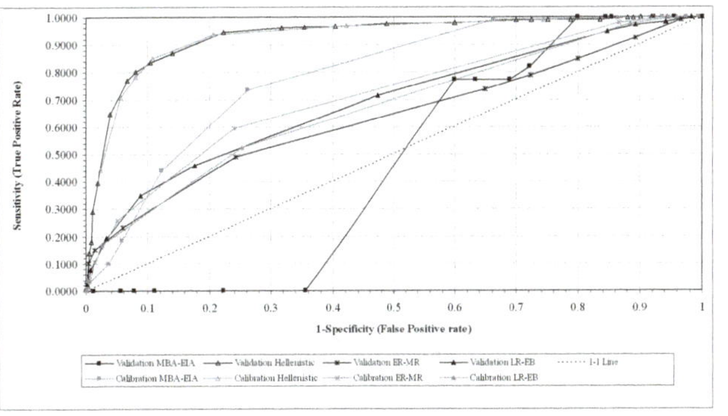

Figure 3 *ROC curves*

Time unit	τ	Optimal Cut-off	% Correctly Classified	PABAK (%)
MBA-EIA	0.0002	0.0001	47	-0.05 = no agreement
Hell.	0.0044	0.006	89	77 = substantial
ER-MR	0.002	0.002	71	43 = moderate
LR-EB	0.0031	0.004	77	53 = moderate

Table 5 *PABAK values*

Time unit	Results
MBA-EIA	Univar. Non-"significant" behaviour of distance to spring, multivar. Slightly closer, "long" distance to nat. corridors, far from affluent/main rivers, strategic flat areas, high visibility, ridge/plateau positions, SW-NW positions
Hellenistic	outside flat areas, Tortonian conglomerates most important, springs closer than in Roman times, Nat. corridors easy to reach (strategic), main rivers short distance, far from affluent rivers, steep slopes, high visibility , SE-SW positions
ER-MR	Footslopes, No extreme high values for substrate, Springs slightly farther than in Hellensitic/PostEB times, Nat. corridors easy to reach (trade), main rivers short distance, closer to affluent rivers than before (irrigation), low slope gradient, low visibility,NE-SW positions
LR-EB	Only slight differences with ER-MR times: away from plain, short distance to main rivers is less pronounced and slightly shorter to affluent rivers than in ER-MR, slightly increased importance of substrate (conglomerate, limestone) + more divers

Table 6 *Summary rare event logistic regression results*

The application of rare event logistic regression analysis for the territory of Sagalassos shows that during the MBA-EIA period extreme defensive settlement strategies are practised with sites on flat hilltops and a high visibility. The preference for defensive locations continues during Hellenistic times, although less pronounced. During this time unit natural water sources and "roads" were near by. Subsequently during ER-MR times increased settlement diversity is noticed and during this period settlement strategies reflect the great significance of crop cultivation in relatively flat areas as well as trade. Finally, during LR-EB times a slight decline of the region is noticed. People again preferred a settlement location away from the plain and affluent rivers were close by.

Even though rare event logistic regression analysis provides accurate results, currently, it does not answer the question: "What is the most important location factor for a specific time unit", because no partial correlations are supplied. As a result, presently we are not able to make a distinction between primary and secondary abiotic location factors.

Logistic regression analysis is the only method that provides a holistic view on settlement strategies and therefore also other methods cannot solve the former question. The intention is to do some research on the subject in the near future. In the future also social and cultural factors should be included in the model, because the location of archaeological sites is not purely based on characteristics and constraints of the physical environment.

References

ADDISCOTT, T., J. SMITH & N.BRADBURY 1995 Critical Evaluation of Models and Their Parameters. *Journal of Environmental Quality*. 24(5): 803-807.

ALLISON, P.D. 1999 *Logistic Regression Using the SAS System: Theory and Application*. SAS Institute Inc., Cary North Carolina.

ALTSCHUL, J.H. and B.A. JONES 1990 *Settlement trends in the Middle San Pedro Valley: A Cultural Resources Sample Survey of the Fort Huachuca Military Reservation*. Tuscon, AZ: Statistical Research, Technical Series 19.

BINTLIFF, J. 1997 Further considerations on the population of Ancient Boeotia. *Recent Developments in the History and Archaeology of central Greece* (J. Binliff, ed.). Oxford: BAR 666: 231-252.

BRACKE, H. 1993 Pisidia in Hellenistic Times (334-25BC) *in Sagalassos I, First general report on the survey (1986-1989) and excavations (1990-1991)* (M. Waelkens, ed.). Acta Archaeologica Lovaniensia Monographiae 5, Leuven University Press, Leuven: 15-37.

BROWN, C.D. and H.T. DAVIS 2006 Receiver operating characteristics curves and related decision measures: A tutorial. *Chemometrics and Intelligent Laboratory Systems*. 80: 24-38.

BYRT, T., J. BISHOP & J.B. CARLIN 1993 Bias, Prevalence and Kappa. *Journal of Clinical Epidemiology*. 46(5): 423-429.

CHERRY, J.F. 1983 Frogs around the pond: Perspectives on current archaeological survey projects in the Mediterranean Region in *Archaeological Survey in the Mediterranean Area* (D.R. Keller & D.W. Rupp, eds.). Oxford: BAR 155: 375-416.

CUSTER, J. 1989 *Prehistoric Cultures of the Delmarva Peninsula: An Archaeological Study*. University of Delaware Press, Newark.

DIACOPOULOS, L. 2004 Investigating Social Complexity through Regional Survey: 'Second Generation' Analysis of Bronce Age Data from the Canadian Palaipaphos Survey Project, Southwestern Cyprus. *Journal of Mediterranean Archaeology*. 17(1): 59-85.

EASTMAN, J.R. 2004 *Idrisi Kilimanjaro®*. Clark University, Worcester MA

FALL, P.L., L. LINES & S.E. FALCONER 1998 Seeds of civilization: Bronze Age Rural Economy and Ecology in the Southern Levant. *Annals of the Association of American Geographers*. 88(1): 107-125.

GONG, G. 1986 Cross-validation, the jackknife, and the bootstrap: excess error estimation in forward logistic regression. *Journal of the American Statistical Association*. 81: 108-113.

HOEHLER, F.K. 2000 Bias and prevalence effects on kappa viewed in terms of sensitivity and specificity. *Journal of Clinical Epidemiology*. 53(5): 499-503.

HOSMER, D.W. and S. LEMESHOW 1989 *Applied Logistic Regression*. John Wiley & Sons, New York.

KANIEWSKI D., V. DE LAET, E. PAULISSEN, M. WAELKENS Submitted Long-term effects of human impact on mountainous ecosystems, western Taurus Mountains, Turkey. *Biogeography*.

KING, G. and L. ZENG 2001[a] Explaining rare Events in International relations. *International Organization*. 55: 693-715.

KING, G. and L. ZENG 2001[b] Logistic Regression in rare Event Data. *Political Analysis*. 9: 137-163.

MARBLE, D.F. and B. MACHOVINA 1997 A GIS-based approach to estimating the human effort involved in movement over natural terrain. *Iskos 11 - Proceedings of the VII Nordic Conference on the Application of Scientific Methods in Archaeology*. Savonlinna, Finland.

NUNINGER, L. 2003 Peuplement et territoires protohistoriques en Languedoc oriental. *Ecole Thématique Systèmes d'Iformation à Référence Spatial et Archéologie*. Tours.

ŞENEL, M. 1997[a] *1:250 000 scale geological maps of Turkey - Isparta N°:4*. General Directorate of Mineral Research and Exploitation, Ankara.

ŞENEL, M. 1997[b] *1:100 000 scale geological maps of Isparta - J10 Quadrangle*. General Directorate of Mineral Research and Exploitation, Ankara.

ŞENEL, M. 1997[c] *1:100 000 scale geological maps of Isparta - J11 Quadrangle*. General Directorate of Mineral Research and Exploitation, Ankara.

VANHAVERBEKE, H. and M. WAELKENS 2003 *The Chora of Sagalassos (Pisiadia, southwest Turkey). The evolution of the settlement pattern from prehistory until recent times*. Brepols Publishers, Turnhout.

VAN LEUSEN, M. 1999 Viewshed and Cost Surface Analysis using GIS (Cartographic Modelling in a Cell-Based GIS II) in *New Techniques for Old Times* (J.A. Barceló, I. Briz & A. Vila, eds.), Proceedings of the 26th Conference on Computer Applications and Quantitative Methods in Archaeology. BAR Int. Series S757. Oxford, BAR Publishing.

WAELKENS, M., E. PAULISSEN, H. VAN HAVERBEKE, I. ÖZTÜRK, B. DE CUPERE, H.A. EKINCI, P. VERMEERSCH, J. POBLOME & R. DEGEEST 1997 The 1994 and 1995 Surveys on the territory of Sagalassos in *Sagalassos IV, Report on the survey and excavation campaigns of 1994 and 1995* (M. Waelkens & J. Poblome, eds.). Acta Archaeologica Lovaniensia Monographiae 9, Leuven University Press, Leuven: 11-103.

WAELKENS, M., H. VANHAVERBEKE, E. PAULISSEN, J. POBLOME, J. REYNIERS, W. VIANE, J. DECKERS, B. DE CUPERE, W. VAN NEER, H.A. EKINCI & M.O. ERBAY 2000 The 1996 and 1997 survey season at Sagalassos in *Sagalassos V. Report on the survey and excavation campaigns of 1996 and 1997* (M. Waelkens & L. Loots, eds.). Acta Archaeologica Lovaniensia Monographiae 11A, Leuven University Press, Leuven: 17-216.

WARREN, R.E. 1990 Predictive modeling in archaeology: a primer in *Interpreting Space: GIS and Archaeology* (K.M.S. Allen, S.W. Green & E.B.W. Zubrow, eds.). Taylor & Francis, London: 90-111.

WARREN, R.E. and D.L. ASCH 2000 A Predictive Model of Archaeological Site Location in the Eastern Prairie Peninsula in *Practical applications of GIS for archaeologists. A predicitve modeling kit* (K.L. Westcott & R.J. Brandon eds.). Taylor & Francis, London: 5-25.

Via Flaminia project: relief and post processing data techniques

Nicolò Dell'Unto,[1] Fabrizio Galeazzi,[2] Marco Di Ioia[2]

[1] IMT Insititute of Advanced Studies, Lucca, Italy
[2] CNR-ITABC, Istituto per le Tecnologie Applicate ai Beni Culturali, VHLab, Montelibretti, Roma, Italy

1 Introduction

In autumn 2005, CNR-ITABC signed a research agreement with ARCUS (Art & Culture Society), in collaboration with the Archaeological Superintendence of Rome, for the "reconstruction of the archaeological landscape of the ancient road Flaminia with virtual reality systems". The aim is to build a digital ecosystem that individualizes two levels of perception: the first one represents a holistic vision of the road from Rome to Rimini, based on historical maps, information on the archaeological excavation, technical cartography and aerial and satellite photos able to support the realization of a web gis; the second one represents and contextualizes the local entities within the "ancient landscape", going into a micro-space vision that focuses the attention on four sites: Villa di Livia, Malborghetto-Tor di Quinto, Ponte Milvio, Grottarossa.

This paper describes the methodologies and the problems faced during the acquisition and the elaboration of the elements which compose the virtual scene of the Villa di Livia case. These processes were studied to produce a virtual reality application.

The instruments used to acquire and generate the data are: laser scanner, photogrammetry, computer vision, Total Station and differential GPS. All the data acquired with these tools have been elaborated to exist in the same system.

The different nature of all the archaeological sites studied leads to the problematic relationship between optimization and details of every single structure. These two elements are strongly connected together: the greater is the detail of an object (in number of polygons), the more difficult it will be to use it in a virtual reality environment.

Due to the complexity of the archaeological structures, we have developed different strategies of data acquisition, i.e. the integration of different techniques and the use of many softwares.

2 Villa di Livia case

The Villa of Livia is the most important and complex part of our project. The archaeological area is situated at the ninth mile of the Via Flaminia, traditionally identified with the villa of Livia Drusilla, wife of Ottaviano Augusto. The villa was built between 30 and 25 B.C. using the rests of a previous republican construction, and was occupied until the Severian Age, when important restauration works were made, especially in the thermal area. Four construction phases are identified, the last one in the third century A.D. The Villa was discovered in 1863. Unfortunately, the villa was bombed and damaged during the last world war.

The archaeological site of Villa di Livia is one of the most difficult areas faced in the project: all the archaeological site is covered by a roof which is supported by many 20x20 cm square pillars, positioned in the whole excavation area (Fig.1). This leads, first, to a very low illumination of the structures, thus creating problems in order to obtain good pictures, necessary to generate textures; second, it limits the use of the laser scanner, so that the only way to get the total topography of the site is to put together a series of different micro-reliefs of the single rooms.

Figure 1 *Roof that covers Villa di Livia*

We used a laser scanner (Cyrax 2005) in order to acquire the entire 3D relief of the structures. This technology allows a very detailed digital acquisition of three-dimensional objects as point clouds. The model obtained is characterized by high geometric resolution that is fundamental for the correct interpretation of the structure. Through the successive phases of elaboration of the point cloud, we obtain a mesh characterized by a millimetric precision.

The 3d meshes of the single rooms have been mapped using a particular method that allows to keep the textures of the pictures taken.

Figure 2 (a, b) *Texture applied to the mesh in RapidForm, detail of the puzzle-effect*

We adapted the images to the surfaces of the objects using the 3dsMax "unwrap modifier", which is a special tool of mapping that allows to directly apply the image on the mesh. With this technique it is possible to link the plain of projection to the position of the digital camera.

A different method had been used previously: the cloud of points was acquired through a software (rapid form) that allowed to map the surfaces using a specific algorithm with the dedicated aim of generating textures. This fast method produced images composed by a series of triangles, impossible to correct with photographic modelling programs (Fig 2a).

The possibility to manipulate the textures permits to adjust brightness and colours of every single image of the total scene, conferring its realism and great likelihood.

Moreover, the algorithm of rapid form introduced a monochrome background to the unused zones, thus causing an annoying puzzle effect to the meshes (Fig 2b).

Once mapped, the single rooms has been imported in the same file to be rotated and translated in the correct position, using like spatial reference the plant, obtained with the Total Station (Fig 3).

In order to import the complete model inside the virtual reality application and to obtain a fluid movement of the camera inside the system, it is necessary to keep the mesh polygons' number very low.

We optimized the meshes with the 3dsMax modifier "multires", which - despite the modifier "optimize", previously used - is made by a very sophisticated algorithm, that allows to have a great control over the reduction of the polygons' number.

The MultiRes modifier reduces the memory, needed to render models, by decreasing the number of vertices and polygons. This tool is useful not only within 3dsMax but even for game, virtual reality applications and Web content creators who export models to use them outside the program. Moreover this method keeps the original UVWmap coordinates unmodified.

Models realized with techniques of photomodelling have been also integrated inside the system.

These techniques are used in order to create simplified 3D metric models with photorealistic aspect.

Photomodeling techniques are used to calculate measurements and build 3d models through digital pictures. With a digital camera we take photos all around the object, in order to acquire images from every point of view. The acquired pictures are imported into a software in which it is possible to correct the perspective distortion of the camera's lens through specific functions of calibration. Moreover, it is possible to recognize and fix common control points directly on the images. Furthermore, the points are projected in a virtual space through the calculation of the optical lines coming from each picture.

Models created through these techniques are textured, scaled and spatially correct; in addition it is possible to export them in every desired format (Fig 4).

This method (if applied to regular geometries from a morphological point of view) is certainly the most suitable, because it produces very light models endowed with textures.

Figure 3 *Villa di Livia, room 51 realized with laser scanner*

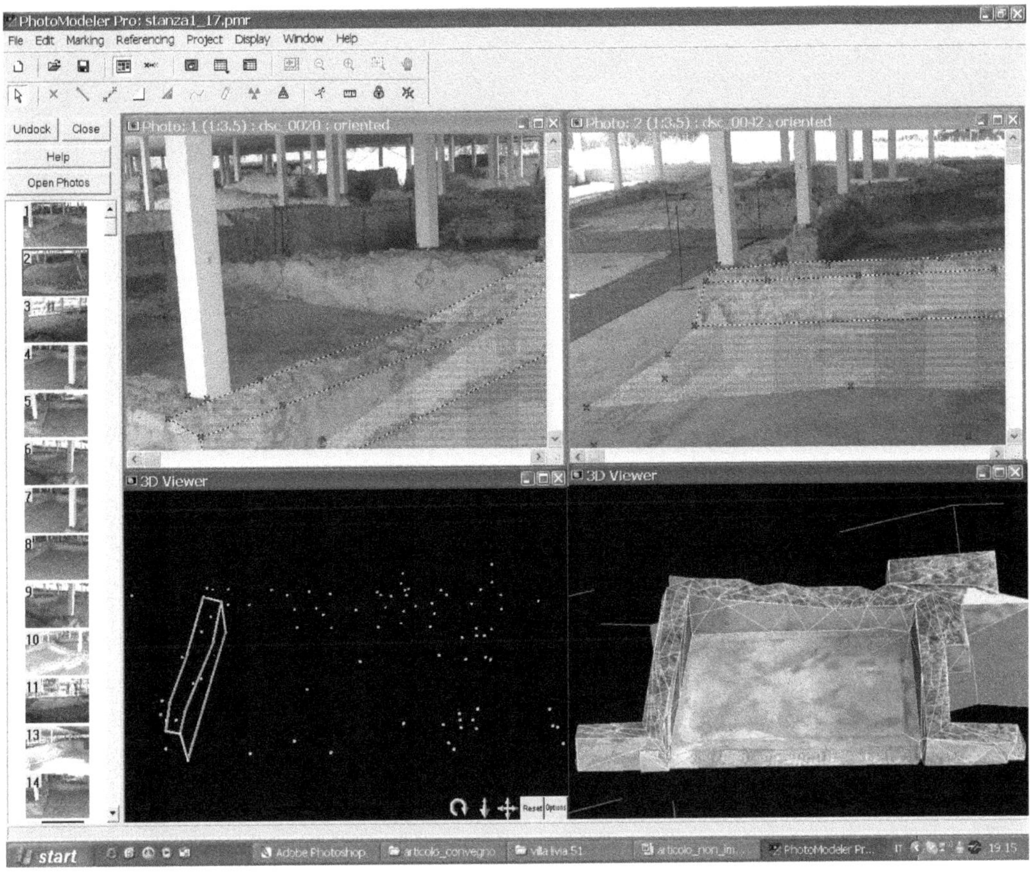

Figure 4 *Villa di Livia, room 51 realized with photomodelling techniques*

This technique has many advantages; first of all, the speed of data acquisition and model processing. It is necessary to have enough pictures to completely document the object otherwise the software will not be able to calculate a 3d metric model. The level of details depends on the number of control points taken on the images; in any case, models obtained with this kind of processing are optimized, in terms of number of polygons, and ready to be implemented in real time applications.

In the same application there will be even a complete reconstruction of the whole Villa as it was in the first century B.C. and in the first century A.D. (Vico, Moro, Vassallo 2006).

3 Techniques Comparison

We tried to apply both techniques (photomodelling and laser scanner) to relief the same room of the villa and compare the different models obtained (Figs. 3-4) .

The aim of this test was to value the best relief technique for this particular case.

The high irregularity of the structures and the closed spaces between the rooms forced us to choose the laser scanner technology .

Although the model realized with the photomodelling techniques is composed by 234 polygons, its resolution is not enough high to represent properly the area. In order to have more detailed models it is necessary to improve the number of control points, causing an exponential growth of the working time.

The archaeological site is composed by irregular structures: this forced us to choose the laser scanner technology, acquire and document all the elements with a very high resolution, obtaining a model closer to the reality.

Through the use of the laser scanner, we acquired the same structure realized with the photomodeling technique, we obtained a very detailed model composed by 48.832 polygons (Fig. 5).

The final project foresees the building of a virtual reality system where it is possible to explore in real time all the relevant archaeological sites along the Flaminia road, near Rome.

Figure 5 *This image shows the different level of detail using two different techniques (laser scanner, photomodelling)*

4 Construction of the avatars

The realization of the avatars has been achieved by using a different approach. These characters are used for communicating the cultural contents inside the system and represent the users that explore the virtual world.

Most part of the avatars has been realized starting from statuary models of historical characters.

For the realization, we have used specific techniques of modelling and of computer vision, which permit to acquire a 3D model through the elaboration of pictures taken all around the object itself.

In this specific case, we compare the result obtained from the elaboration of two different 3D models: the emperor August and his wife Livia.

For the realizations of these chracters we use different techniques and compare the results in term of time and visualization.

August was realized through the computer-vision web-service, developed by Katholieke Universiteit Leuven (VISICS) in Belgium (fig 6). This tool permits to obtain 3D models through the elaboration of digital images and is very effective in the presence of bas-reliefs. Anyway, it

Figure 6 *Statue of Augustus realized with computer vision technique*

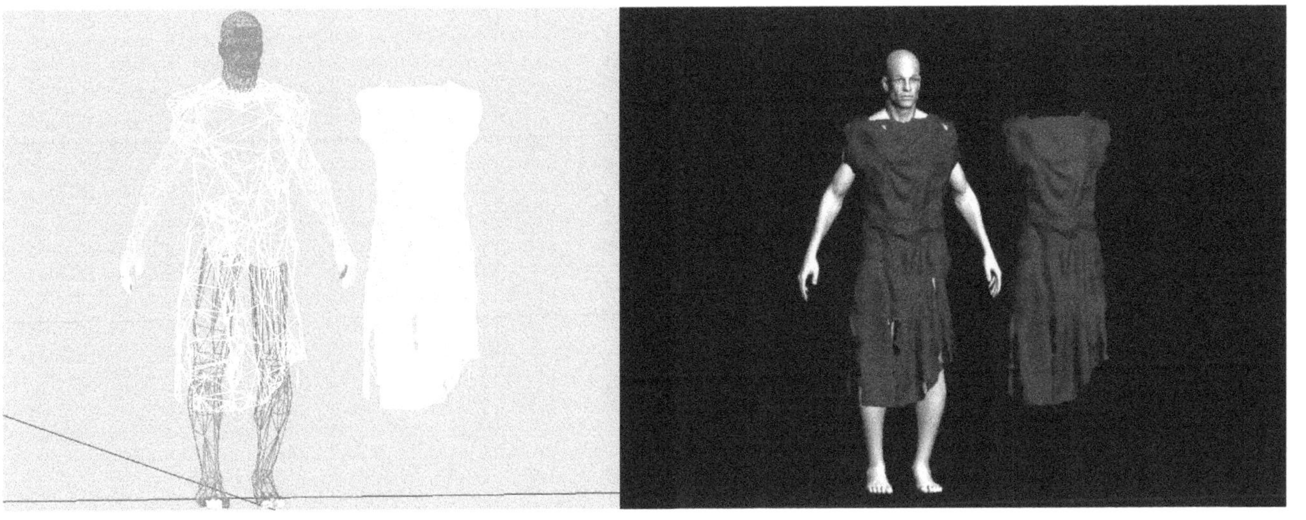

Fig 7 *The slave's tunica acquired through laser scanner and visualized with normal maps*

has underlined some strong limits in the elaboration of round objects, so it has been necessary to study a methodology to acquire single parts of the same object, recomposed later with 3dsMax.

The final model showed very high quality particulars, as the clothes (drapery). The disadvantage resides in the elaboration time, which is too long. Moreover, due to the high number of polygons, the model had been strongly optimized, thus loosing details in many parts.

To construct the 3D model of Livia we used traditional techniques of modelling, starting from the statue kept in the national roman museum of Palazzo Massimo in Rome (Fig. 8). Although we encountered difficulties in modelling the typical roman clothes. The traditional modelling technique is faster and preferible to model the body parts, because allows more control in order to manage vertexes position and count.

At the end of this test we decided to create the body of the characters with the modelling technique, while the clothes with computer vision techniques.

For the slave character we created clothes using the laser scanner. We placed a real *tunica* on a mock-up and we acquired it with the laser scanner. We obtained a high poly mesh model of the *tunica*, and we optimized the model through the "multires modifier" afterwards.

A high resolution model was necessary to represent properly the drapery of the toga, so we used the "normal maps", interpolating the mesh information of the high resolution model with the optimized model (Fig. 7).

Normal bump mapping is a way of adding high-resolution details to low-polygon objects. It is especially useful for real-time visualization devices such as game engines, and it can also be used in rendered scenes and animations.

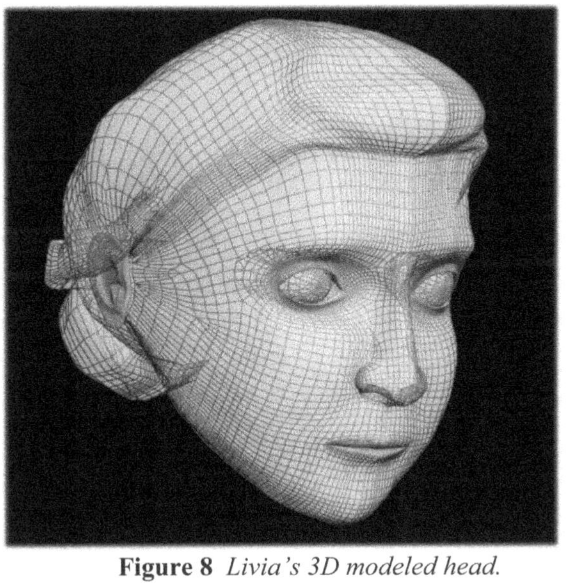

Figure 8 *Livia's 3D modeled head.*

Unlike the greyscale maps used for regular bump mapping, a normal map is a RGB color map; the red channel encodes the left-right axis of normal orientation, the green channel encodes the up-down axis of normal orientation, and the blue channel encodes vertical depth.

5 Conclusions

The complexity of this site allowed to integrate many different techniques of relief and modelling in order to experiment new methodologies for the production of 3D models for virtual reality systems.

Moreover, by comparing new techniques, it was possible to find new strategies of interaction among different softwares and obtain the best compromise between visualization and number of polygons.

We found a methodology based on the comparison of the results obtained. Moreover, the integration of different figures (archaeologists, architects, computer scientists, etc.) in the project was the key to build a system full of different contents and virtual behaviours

Moreover, the complexity of the case of study allows to find new techniques and create a strategy in order to develop a multy-user system (Forte, Pescarin, Pietroni, Rufa 2006).

References

BUCCELLATI F., DELL'UNTO N., FORTE M.: *The Mozan/Urkesh Archaeological Project: an Integrated Approachof Spatial Technologies* 2st Italy-United States Workshop, Rome, Italy, November 3-4, 2003, BAR International Series, Oxford

FORTE M.: About virtual archaeology, in "Virtual Reality in archaeology", Barcelo J., Forte M., Sanders D., 2000 (eds), Oxford, BAR Publishing (BAR International Series S 843), pp. 247-263

FORTE M., 2003: Mindscape: ecological thinking, cyber anthropology and virtual archaeological landscapes, in "The Reconstruction of Archaeological Landscapes through Digital Technologies", ed. M.Forte, P. Ryan Williams (eds.), 2003, Oxford, BAR Publishing (BAR International Series S 1151), pp.95-109

FORTE M., PIETRONI E., RUFA C., BIZZARRO A., TILIA A., TILIA S., *DVR-Archaeology Project: Reconstruction of Archaeological Contexts in Desktop –Opengl Environments,* 46-48. EVA 2001, Electronic Imaging and Visual Arts, Florence, 26-30 March 2001

FORTE M., GOMEZ L., PESCARIN S., PIETRONI E., VICO L.: *Integrating Technologies: The Appia Antica Project*, 2st Italy-United States Workshop, Rome, Italy, November 3-4, 2003, BAR International Series, Oxford.

FORTE M., PESCARIN S., PIETRONI E., RUFA C.: *Multiuser interaction in an archaeological landscape: the Flaminia project,* in proceeding *From Space to Place*, Rome, Italy, December 4-7, 2006.

GOODRICK G., GILLINGS M.: *Constructs, simulations and hyperreal worlds: the role of Virtual Reality (VR) in archaeological research*, Lock, G. and Brown, K. (eds.), On the Theory and Practice of Archaeological Computing, Oxford, Oxbow Books 2000.

MESSINEO G., CALCI C.: *La villa di Livia a Prima Porta*, Roma, De Luca, 1984.

MESSIINEO G.et alii.: *Ad Gallinas Albas: villa di Livia*, Roma, l'Erma di Bretschneider, 2001.

MESSINEO G.: *La via Flaminia (da Porta del Popolo a Malborghetto)*, Roma, Istituto poligrafico e Zecca dello Stato, 1993.

MESSINEO G., SCOPPOLA F.: *Scoperte nei nuovi insediamenti militari della via Cassia e della via Flaminia*, QUAEI, ArchLaz IX, 1988

PIETRONI E., PESCARIN S., FORTE M., DELL'UNTO N.: *Integrating The narrative approach of virtual heritage real-time applications*, 2st Italy-United States Workshop, Rome, Italy, November 3-4, 2003, BAR International Series, Oxford.

QUILICI L.: *Le strade: Viabilità tra Roma e Lazio*, Roma Quasar, 1990.

VICO L., MORO A., VASSALLO V.: *The importance of the relief and sources to interpret and communicate the cultural heritage,*in proceeding *From Space to Place*, Rome, Italy, December 4-7, 2006.

The Ferento Project.
The Role of GIS and Databases for the data integration and analysis

E. De Minicis,[1] R. Gabrielli,[2] D. Peloso[2]

[1]Dipartimento del Mondo Antico, University of Tuscia (Viterbo)
[2]Institute for Technologies Applied to Cultural Heritage (CNR), Rome, Italy

1 Introduction

During the last archaeological excavations carried out in the city of Ferento (Fig.1), within of the "Ferento Project", an intra-situated GIS has been developed. It is an example of how it could be possible to proceed with innovative methods at any single part of an archaeological dig, by using survey techniques, to ensure that the control of archaeological information is always greater through the acquisition and elaboration of the documentation of digging in real time. It is necessary to optimize survey and documentation; moreover, with the slender financial resources, the best possible relationship between investments and results must be obtained. In order to achieve these aims, a simple system consisting of photographic survey associated to a computer with image elaboration programs with stratigraphic data has been set up. In this way it was possible to create a 3D topographic informative system of the excavation.

2 The "Ferento Project"

The city of Ferento (Etruscan, then Roman) is the heir of the near Etruscan centre of Colle S.Francesco-Acquarossa abandoned in 500 B.C.

Raised to *municipium* after the social war (first century B.C), the town extended to the upland of Pianicara (30 h). During the first imperial age, in time of Augustus, when the theatre and the baths were built, the city had a flourishing development.

A rapid succession of architectural and urbanistic interventions confirmed richness and development of the city during the following centuries.

In 487 A.D. the city was made a diocese, until 649 A.D when the bishop's seat was transferred in Bomarzo. During the Greco-Gothic war, the area in the western part of upland was fortified, encompassing the theatre with the houses nearby. With the Longobard invasion, Ferento had a strong period of economic and demographic crisis. Taken by gastaldato of Tuscania, it became a stronghold, near the Longobard-Byzantine border. In the ninth century it was named as *civitas,* among the Tuscian cities given to Carlo Magno. We do not have news about the political-administrative attitude of Ferento between eleventh and twelfth centuries, but we surely know about its expansion which represented a rebirth of the city.

A new tier of outer walls was built on the eastern part of pianoro, on which ruins of three churches (Santa Maria, Santa Croce and S. Bonifacio) were placed. Ferento's economic development (thanks to utilization of agricultural and mineral resources) brought on the war with Viterbo in 1170 that ended in 1172 with Viterbo's victory. Although completely destroyed, the city was slowly abandoned. From thirteenth century we do not have any further news about the site. In the following centuries it was frequented for the extraction of iron and vitriol, an activity remembered in Statutes of Viterbo of 1251-52 (Romagnoli 2001, Maetzke 2001).

3 The documentation exporting system

The requirement to control the archaeological remains from above has always been one of the objectives that have seen a lot of attempts realized at the time of the development of photography and in particular of photogrammetry. Other similar systems have been introduced in the past and the main problem has been the realization of fast systems for easy use in connection with the quality of the goals and in particular with the surface covered with the operation of the photomosaic (Buchsenschutz 1989, Gottarelli 1987, Gottarelli 1992). This paper does not represent an absolute innovation in archaeological field, but it would like to demonstrate the employment facility of the informative systems applied to archaeological digging without using expensive and excessive equipment and without resorting to complicated elaborations in post excavation analysis.

The system of photographic survey by using a telescopic rod, as we have done, allowed good exporting shots aiming for the punctual survey. It is important, therefore, that images have a very good definition and that they are perpendicular to the land as much as possible. Further corrections can be made in a second time through procedures of georeference, orthorectification and mosaic of the single photograms. Moreover, it is to possible treat the images, when they are transferred to the computer by modifying levels, brightness, contrast, and dimensions. The images are usually modified with a high resolution saved in an uncompressed format (*.TIF) for being straightened with the appropriate software.

The ordinary procedure to georeference photo consists of

Figure 1

a homogeneous disposition of a series of sights on the land at the distance of approximately two meters, trying to distribute the sights geometrically in every single photogram: for example, four to the tops and one in the centre.

This phase needs a great attention; it is necessary to make a difference if the surface is flat, without particular altimetry differences and without objects or parts that are over the medium quota, the only thing to do is trying to cover the exacavtion area with the camera; the operation of the photographic survey is more complex when there is a high presence of obstacles or of structures that can cover the surface angle of the frame. If there are high structures more of a photo has to be carried out in order to avoid the parts of the surface that remain hidden.

The sights dutifully numbered are taken for a second time topographically, by a Total Station. The acquisition of every type of relief starting from certain points placed in the excavation area, allows attributing geographic coordinates to every single point and to georeference every single photography by means of an ortho-azimutal algorithm. This process is necessary because the digital images are deformed by the camera lens and by the not perfect verticality of the exposure point and so they must be corrected geometrically through the use of a specific software.

In this case Photometric software by Topcon has been used, a system for digital exporting of photogrammetric restitution for the 2D restitution of plain objects.

This program transforms the prospective of the plain object images in a parallel projection that coincides with a straightening in a correct scale.
Beside the geometric correction the single images, the program allows to obtain photomosaic made on the overlap of a multitude of images that in this way become part of unique georeference photography. This new photo is exportable in the most common graphic format with all the relative geometric information (Fig.2).

Figure 2

In addition to the necessary point to straighten the images, the acquisition of altimetric point is basic, so as to allow a punctual reconstruction of the excavated area.

In fact it is necessary that all the altimetric differences particularly the more considerable difference, are represented by a right number of points, so that the next elaboration of the digital model of the soil (DTM) is as similar as possible to the real one [Colosi et al 2003].

Mosaics are realized and used like real plain rightly positioned and set in the space and fated to include in the informative system.

Before this however, it is basic that photomosaics are printed not only as paper documentation for archives and didactic aims, but also that on these copies a report of the limits of stratigraphic units is directly inserted by hand and at the same time that all the elements for which a better identification is necessary such as the kind of material that need to be differentiated from the predominant element and all descriptive notes of archaeological information visible on the area are reported (Cattani 2004).

4 Informative system of the excavated data

The operative phase of management and control of the archaeological documentation has been based on the use of geographic informative system (GIS) with software ESRI ArcView9. The methodological goals in the realization of an intra-site GIS were to achieve structure with specific purposes, that had scientific and correct contents organized and found with innovative technologies, but with a simple access and a very easy use for students as well.

When every georeferenced photomosaic is imported, the first operation that is performed is the vectorialization of architectonic structures and of ruins, and the delimitation of every stratigraphic unit, directly in *.shp format.

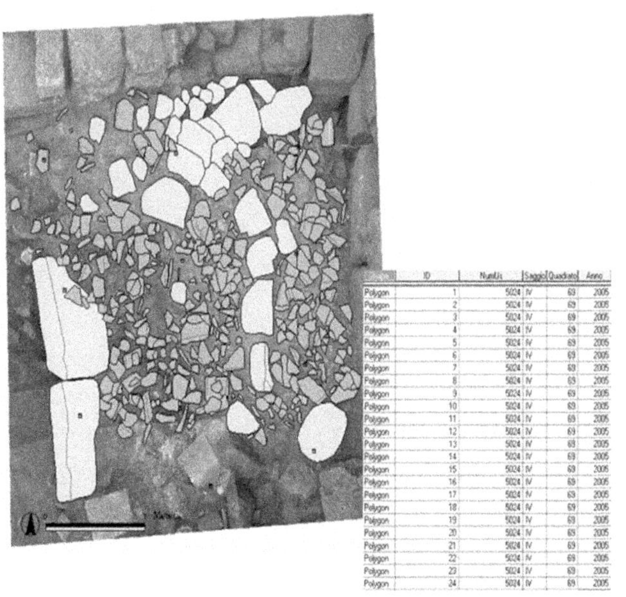

Figure 3

All the information that is possible to get out of plans (limits, altimetric informations, details related to materials recovered inside ruins or in walls in evidence) are subdivided in classes and transferred to a unique layer and expressed as numbered points located in the exact position in which every one was collected (Fig.3).

The position of every artifact found during the dig is transferred on the relative US and identified with a punctual simbology. In this way it is possible to access the window describing artifact classes list and all the detailed information on every object.

In arc view, in fact, every vectorial layer has an attribute table that contains as many records as there are features represented.

It is possible to select an object on the map and visualize the correspondent line in the attribute table as well; but at the same time it is possible to start from a record of the table to obtain the evidence of the right vectorial feature.

In any case descriptive data can be visualized as a data sheet structured in lines and columns or for single records inside the appropriate window.

All other descriptive information is conserved inside tables of a relational DataBase realized in Access and these tables are connected to attribute tables of the ArcView layer through punctual functions.

A data bank in Access 2000 was planned, having a complex structure able to collate more archives and to put together different information. A lot of tables have been created inside the db to manage information about US, USM, about interments and artifacts of every US.

The customer server for data-entry and the management of information is realized with specific masks; the US and USM forms are based on the model of the ICCD. For every US it is possible to insert and visualize many records, as many as needed to define components and stratigraphic relationships with every US punctually (Fig. 4).

Other information masked reference, instead, to curtain menu with a closed vocabulary.

It is possible to pass from US forms to the filling in and consultation of all found artifacts in the layer.

So the experience allows the establishment of the correct and effective way of data management and interpretation based on a cross control among matrix, features and overlays.

5 Laser scanning applied to archaeological heritage

The recording of the structural medieval of the archeological site Ferento's was used to test the laser technology. The Laser Scanner Callidus CP 3200 (Trimble) was used to make scans (Fig. 5).

It took about 2 hours to acquire, in 5 scans, an aproximate surface of $30m^2$. The measurements have been collected with a sampling of 0.125°, for the horizontal resolution and a sampling of 0.25° for the vertical resolution. Similar exercise method took several days using the traditional methods, and the level of precision and details observed was definitely lower.

The technique is well suited to the fields of archaeology and heritage in which complex, amorphous objects must be recorded which would otherwise be extremely difficult to survey. Although large and complex areas can be recorded in a very short time using this procedure, it should be noted that the post-processing of the resulting data takes much longer than the actual surveying (Peloso 2005).

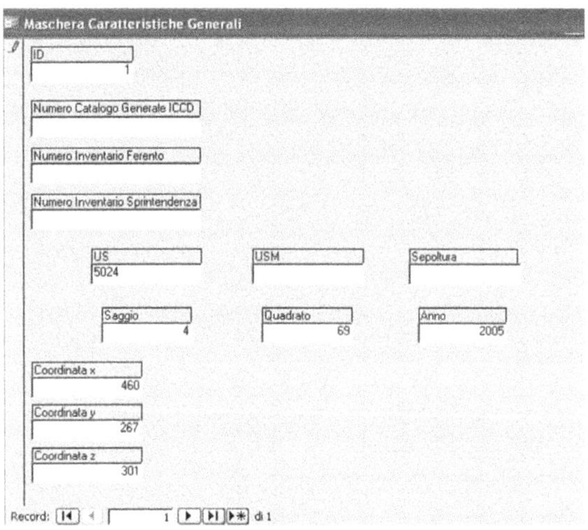

Figure 4

Figure 5

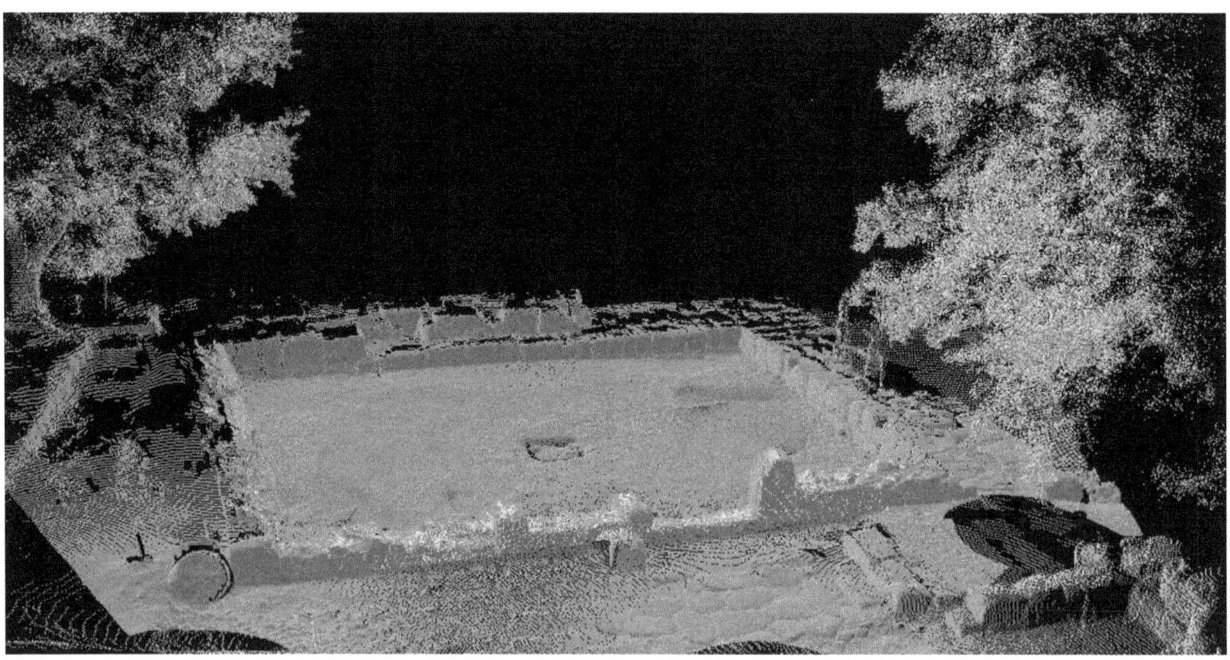

Figure 6

The clearest advantages of 3D laser scanning are the reduction of the time spent in the field and the reduction of surveying costs. Once the task of data acquisition has been concluded, all further measurements are carried out on the model (Fig. 6).

Using further, systematic scanning, it is also possible to document the development of complicated, stratigraphic features in all their various stages of excavation.

5 Conclusions

These methodologies are the bases of a multidisciplinary and very important system of analysis and interpretation of archaeological layers.

Each one of these processes gives the possibility of a quicker procedure in every phase of the dig, but they are expensive. It is right to specify that the laser scanner has been used only as a comparison system with the 3D restitutions got out of photogrammetric images.

These experiments have been really valid in the didactic excavations because the students learn how this technique is valid and how it is important to choose better methodologies for each working goal.

References

BUCHSENSCHUTZ O., 1989. *Méthodes de recherche. Experimentation sur le site du Mont Beuvray*, in Le courier du CNRS, 73, 30

CATTANI M., FIORINI A., RONDELLI B., 2004. *Computer Applications for a Reconstruction of Archaeological Stratigraphy as a Predictive model in urban and territorial context*, in Computer and Quantitative Methods in Archaeology, CAA Wien, pril 4-9 2003, Wien 2004, pp. 299-303 & extended version in CD rom

COLOSI F., GABRIELLI R., MAURIELLO P., PELOSO D., 2003. *Cerveteri: Topografia della Vigna ParrocchialeII. Metodologie integrate per lo studio di un'area archeologica* in Archeologia e Calcolatori, 14, 2003, pp.177-197

GOTTARELLI A., 1987. *Tecniche di documentazione dello scavo archeologico : introduzione alla videometria digitalizzata*, in F. D'Andria (ed.) Informatica e Archeologia classica. Atti del Convegno (Lecce 1986), Galatina 1987, pp. 63-90

GOTTARELLI A., 1992. *La video-documentazione elettronica dello scavo archeologico (V.M.D.)*. Studi ed esperienze per il progetto di una periferica dedicata, in Archeologia e Calcolatori, 3 1992, pp. 77-99

GÜLL P., SCAIA F. (a cura di), 2002. Ferento Civitas Splendidissima, Catalogo della Mostra, Dipartimento di Scienze del mondo antico. Università degli Studi della Tuscia, Ed. Quatrini, Viterbo

HARRIS E.C. 1979, *Principles of Archeological Stratigraphy*, Academic Press, London.

MAETZKE G. ET ALII, 2001. *Ferento (Viterbo), indagini archeologichenell'area urbana (1994-2000),* "Archeologia Medievale" XXVIII, pp.295-322

PELOSO D., 2005. *Tecniche Laser Scanner per il rilievo dei Beni Culturali* (Peloso D.), Archeologia e Calcolatori, 16, 2005, Edizioni all'Insegna del Giglio, pp.199-224

ROMAGNOLI G., 2001. *Ferento (Viterbo). Storia degli scavi e delle ricerche archeologiche*, in Daidalos. Studi e ricerche del Dipartimento di Scienze del mondo antico. Università degli Studi della Tuscia, 3, 2001, pp.273-300

Transparency, interaction, communication and open source in Virtual Archaeology

M. Forte,[1] S. Pescarin,[1] E. Pietroni[1]

[1]CNR ITABC, Virtual Heritage Lab, Rome, Italy, [maurizio.forte],[sofia.pescarin],[eva.pietroni]@itabc.cnr.it

1 Knowledge and communication through transparent VR applications

In the field of cultural heritage there are a few virtual reality applications oriented to scientific communities, interested to the dynamic exchange and sharing of data, methodologies, cultural contents. In the most of cases, VR projects are dedicated to a wide public fruition; even for this reason, their main goal is to communicate, reconstruct and interact, using the virtual environment, at least in the best cases, as final result of a long process of study and interpretation integrating bottom-up and top-down methodologies [FP05]. Very rarely they encourage a real discussion about methodological approaches, interpretative processes followed in the reconstruction of present and ancient cultural contexts. They show only the final result, selecting few types of data, omitting which kind of sources were used, which data and methodology, considering in this way communication as totally separated from knowledge.

VR applications realized with this approach can produce and show just models and environments closed and not transparent. Users have no possibility to understand how those representations have been obtained. Therefore they can not decode information they are receiving, nor analyze it or give any evaluation with critical awareness. This situation is quite peculiar in the virtual heritage domain; other fields such as medicine, industrial prototyping, military simulations can produce today virtual reality applications in a more transparent way.

In fact, it is important to allow scientific community and experts in virtual heritage to participate more actively in the process of creating and re-creating VR applications for communication, didactics and research.

Transparency is becoming one of the key-concepts in the discussion about virtual reality and cultural heritage, especially inside the scientific community, even if it is difficult to find VR applications and case studies following this issue.

An initial discussion, preceded by a first initiative called CVRO(http://www.cs.kent.ac.uk/people/staff/nsr/cvro/workshop.html) [FNRB04], was started in February 2006 during an international meeting that was held in London and organized by King's College 3D Visualisation Laboratory. At the end of the meeting, where the topic was discussed from different points of view by worldwide researchers, a first document was produced, the so called "London Charter". It is still an in progress document, presented also in Eva2006 in Florence, and it will published together with the proceedings of the conference in *"Behaviors, Interactions and Affordance in Virtual Archaeology"* [FP06]. It was an important opportunity where a representative part of the scientific community who works in the field of virtual heritage, expressed the need to identify some guidelines for the future, starting from a common consideration. Until now, in fact, we have produced and we are still producing thousands different digital models for many purposes. Most of these models are forgotten or not available outside very restrict groups, with no benefit for the worldwide community. There is obviously a problem of access of something that should be common patrimony. This problem is then strictly connected with communication issues: without communication there is no access to information. But there is also another question that should be considered, that has a strong impact on the future of scientific research: how can a digital model can be defined as "scientific"? How can we give any opinion on it and how can be used to help interpretation? And finally how can we introduce it fully in the scientific process, that uses, at present, just printed publications?

Some suggestions will be discussed here, oriented toward two possible solutions that could be integrated: transparency intended as an interactive process in VR and as an open source process that can help data sharing.

2 Transparency in 3d real time environments

How digital technologies and virtual reality can support organization, visualization, integration of complex cultural contents? Which are the best metaphors to interact with virtual environments, with 3D models and all connected data? How can we exchange data? How can a digital environment become an ecosystem [Gib79]? How can we make "transparent" a 3D representation?

In our approach, making a model transparent means to communicate and to share data source and methodological processes so to develop the educational value and the cultural impact of the VR project.

In the virtual reconstruction of a cultural context, a central issue should be the connection between 3D models and the informative system, made of levels of detail, metadata, tools for visualizing, measuring, connecting objects to databases or to multimedia resources; each of this interactions can be thought as a behavior.

We can imagine the communication process like a network of relations where at the top, visible, we find the object/artifact (the "attractor" of cultural content) and in the successive deeper levels all the relations connected with its context (psychological, social, philosophical, historical, symbolic, and so on). In order to explain all these connections we have to conceive and organize a VR application as an integrated communication system.

3D research outcome is not only visual but it is the result of a multimodal and multisensorial dynamic interaction. A model, indeed, represents a complexity of *relations*, named "affordance". The word "affordance" was originally created by the perceptual psychologist J. J. Gibson to refer to actionable properties between the world and an actor [Gib79]. According to Gibson, affordances are environmental *relationships* to refer to the actionable properties between the world and an actor; so it is a property of an object, or a feature that indicates how to interface with that object or feature. In a virtual environment we define affordances all the possible relations between users, avatars and digital ecosystems; hence an affordance draws the possible interactions/behaviours linked with an object or an information.

Unfortunately, in particular in the scientific domain, VR applications are often limited to a simple 3D representation with no behaviors or relations: nothing is told about life, perception, human activities, ecosystem. This approach makes these kinds of applications a useful exercise of graphic and aesthetic reconstruction, but not a useful support for cultural transmission, because we cannot imagine, understand, create relations of meanings. Therefore we could consider transparency as connected with a double evidence:

1. the evidence of contextualizing objects, structures, artifacts, landscapes, through the connection between 3D models and cultural contents;
2. the evidence of data sources, methodological approaches, interpretative processes.

Through interaction it is possible to achieve these goals, especially if we use an open approach. Interaction in fact could be more effective if open to contributions, to data or information sharing, to skills and experience exchange. A process, in order to be defined scientific, needs to be validated and results should be checkable, transparent and documented. If we use close formats (that could also easily die) or we don't give the access, at least to the scientific community, to the code of programs developed in our field, there's a risk for the future of Research itself in the future [FP06].

The Virtual Heritage Lab of CNR - ITABC (www.vhlab.itabc.cnr.it) has been orienting its research activities in the creation of VR applications dedicated to cultural heritage, in particular to the reconstruction of archaeological landscapes. The via Appia Antica and via Flaminia projects, the virtual musealisation of the Scrovegni Chapel are VR applications dedicated to the public use, conceived as integrated informative systems.
In most cases our work begin in the field, documenting, through topographical survey, the territory and the structures of the site, using integrated technologies (DGPS, total laser station, laser scanner, photogrammetry, remote sensing, digital photographic acquisition, and so on). Acquired data are then processed in order to obtain 2D and 3D representations at many levels of detail, holistic and monographic (*bottom-up process*). At the same time we have to interpret structures and contexts, so we employ comparative examples and typologies, bibliographic resources, literary descriptions and so on, all those references that can help us to understand and propose a virtual reconstruction of the ancient context (*top-down process*) [FPP*05].

This methodological approach allows to collect many kinds of data and offers, at the same time, a good opportunity to respect the issues related to transparency in the communication choices. We would like to propose some possible solutions we adopted in different VR projects in which transparency was carried out in concrete, presenting also some examples referred to the projects mentioned before:

1) Comparison between virtual (3D model) and real (photos, movies accessible from the 3D environment).

The possibility to switch or activate different modes adds an inter-active opportunity to perceive and interpret, through differences [Bat72] [Bat79], increasing the knowledge. An example could be to add a movie connected to the 3d monument showing it as it appears nowadays (Figure 1);

Figure 1: *The Appia Antica narrative museum (VR application, CNR ITABC, 2005): 3D model and movie of a cistern in the Caffarella valley*

2) Connection between models and metadata, storytelling, characters, multimedia contents, database.

In order to make the virtual environment a digital ecosystem and a territory to "experience" it's also possible to connect virtual object with characters for further explanations with a storytelling approach.

3) Creation of tools and metaphors of visualization.

Adding the possibility to switch to different visualizations of the same object, for instance seeing it as point cloud, monochrome mesh, textured mesh or wireframe can increase the value of the interpretation. Even measuring tools or lighting tools could help to give correct proportion to virtual environment.

4) Loading of vector layers on 3D model

Considering the three-dimensional object as a sort of 3d GIS, where spatial component is kept, overlaying thematic layers, can put in evidence "cultures" and thematic contents, as shown in Figure 2.

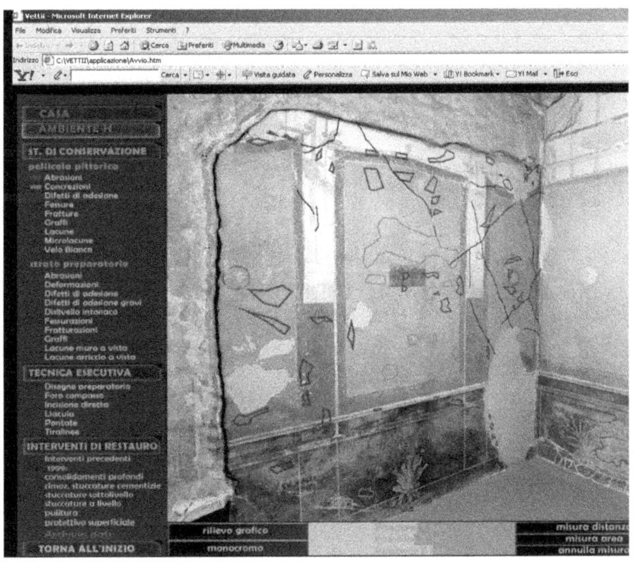

Figure 2: *Domus of Vettii in Pompei, VR application, room H elaborated by laser scanner data: vectorial layers describing conservative problems, visualised in overlay on 3d model (ICR in collaboration with Treerre, Softlogic, CNR-ITABC, 2001-2005).*

5) Possibility to change materials, loading realistic or symbolic textures on 3D models according to the kind of information we want to suggest (for instance on the territory satellite/aerial photos or cartography) (Figure 3).

6) Connection of 3d objects with web pages with explanation on sources used (Figure 4)

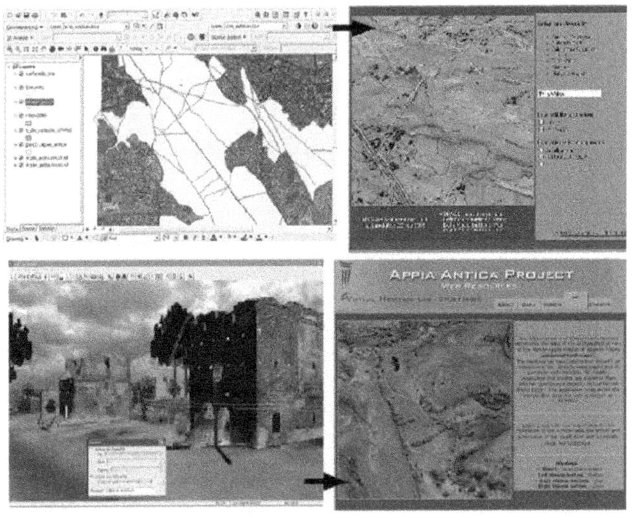

Figure 3. *The Appia Antica project, on the left: From GIS project to web 3d navigation and interaction; 3d dynamic off-line environment for landscape reconstruction and 3d data published over the web.*

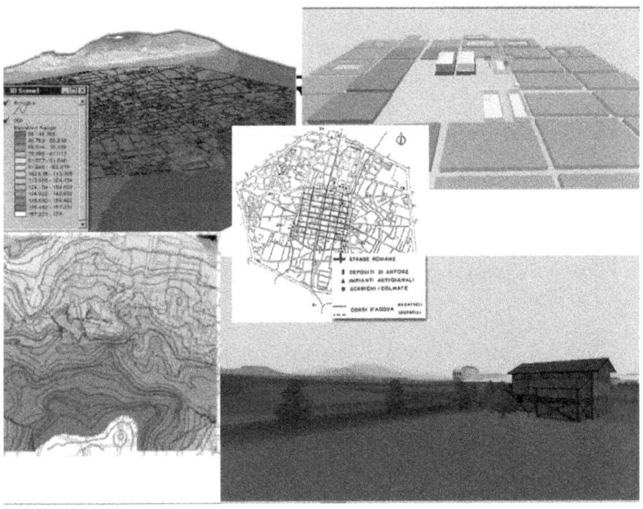

Figure 4. *The Bononia project with sources and post-processed data created (DEM, realistic models, didactic models).*

7) Explanation of data sources and methodological processes for the elaboration of 3D models (through movies, photos or audio files explaining acquisition techniques, showing the main phases of post-processing, literary and iconographic references, comparative cases used in the interpretative process) (Figure 5);

8) Possibility to add and update information, thanks to an open architecture of the informative system;

9) Comparison between observed, interpreted, reconstructed context;

10) discussion about problematic reconstructions, showing different possible versions or doubtful

elements (through colored transparent areas, wireframe, monochrome, icons representing the level of probability).

11) Possibility to deconstruct and reconstruct models.

12) Possibility to change the map of relations among contents.

Figure 5 *Ninpheum of Egeria in the Caffarella valley, 3D model and movie explaining the methodology of acquisition by laser scanner, total laser station, photogrammetry and the main phase of elaboration. In the image on the right a movie shows the reconstruction of ancient structure (Lola Vico, Lucia Gomez)*

Figure 6. *Flaminia project (CNR-ITABC): Roman Villa of Livia, Rome, where every objects, rooms and so on have relations of contents, such as Space, Time, Use, Type, Similarity. The user can interact with the model also de-constructing it, activating its affordances*

We have tested these examples in different VR applications, both off line and on line, using commercial or open source software. One of this project, the *Appia Park Narrative Museum* was also shown in an international exhibition on virtual archaeology that was held in Rome in 2005 (*Building Virtual Rome:* Rome, Trajan Markets, 14 September-14 November 2005). One of the most significant problems pointed out by this exhibition is that users, while interacting with virtual environments, appreciate particularly an external support to completely understand and also enjoy cultural contents, metaphors, linguistic choices, even if applications have an attractive power. Beyond storytelling, beautiful scenarios and visualizations there are a lot of contents that remain not explicit to which people are very interested; this contents are relative to historical information (what is it) and also to the methodology followed to construct the virtual environment (how it was done).

Therefore even in this public event, transparency has came out as a central issue and need not only for the scientific community but also for the future of communication of cultural heritage through digital techniques.

3 Conclusions

In the next future (FIRB Project: I*ntegrated technologies of robotic and virtual reality in archaeology, 2005-2008*), we are going to realize a multi-user domain based on virtual reality application for the web, oriented to scientific and archaeological communities, were each user will be able to personalize, change, edit the virtual environment, according to his own interpretation of the cultural context.

Transparency can produce a sort of feeling of "affection" toward virtual reality, due to the possibility to understand, to "enter" in the system with a good level of awareness and consequently to approach cultural contents in a more personal and creative way.

Similar impressions and power of attraction can be experimented for instance when an art critic explains us a painting, a director explains the meaning and the stylistic choices of a movie or an artist reveals how special effects have been realized in the backstage. Our impression is to penetrate something of special, reserved, and to become able to decode the deep reasons of those communicative, expressive, formal languages. The capacity of learning and communicating information depends on the affordances we are able to identity and to transmit. The key aim of transparency is to create multiple dynamic processes of interaction where many and different users can explore and share virtual environments so that to have an open space of cultural and multimodal interpretation. The new affordances will produce new ways of learning and of virtual familiarization, validating the scientific content of models and contexts.

References

[Bat72] BATESON G. Steps To An Ecology Of Mind. San Francisco: Chandler Press, 1972

[Bat79] BATESON G. Mind And Nature: A Necessary Unit. New York: Dutton, 1979

[FNRB04] FRISCHER B., NICCOLUCCI F., RYAN N. AND BARCELÓ J., From cvr to cvro: The past, present, and future of cultural virtual reality. In *Virtual Archaeology between Scientific Research and Territorial Marketing, proceedings of the VAST*

EuroConference, Arezzo, Italy, Niccolucci F. editor, November 2000.

[FP05] FORTE M., PESCARIN S.: The virtual reconstruction of the archaeological landscape. *XXIV Rencontres internationales d'Archéologie et d'Histoire d'Antibes: temps et espaces de l'homme en société, analyses et modèles spatiaux en archéologie*, Ed. Jean-François Berger, Frédérique Bertoncello, Frank Braemer, Gourguen Davtian and Michiel Gazenbeek, 2005

[FP06] FORTE M., PESCARIN S., Behaviors, Interactions and Affordance in Virtual Archaeology. *Advanced ICT Methods Guide*, edited by Lorna Hughes, London 2006, in press.

[FPP*05] FORTE M., PESCARIN S., PIETRONI E., DELL'UNTO N.: The Appia Antica Project. *Archaeological Landscapes through Digital Technologies*: Proceedings of the 2nd Italy-United States Workshop, Rome, Italy, November 3-5, 2003 Berkeley, USA, May 2005 edited by Maurizio Forte. BAR S1379 2005. ISBN 184171819X.

[Gib79] GIBSON, J.: T*he Ecological Approach to Visual Perception,* Lawrence Erlbaum Associates. Hillsdale NJ, 1979.

[MV80] MATURANA H. and F. VARELA., *Autopoiesis and Cognition: The Realization of the Living,* Boston Studies in the Philosophy of Science. Edited by Robert S. Cohen and Marx W. Wartofsky, vol. 42, Dordecht (Holland): D. Reidel Publishing Co., 1980.

Aerial archaeology in Daunia (Northern Puglia, Italy). New research and developments

Roberto Goffredo[1]

[1] Università di Foggia

More than four decades have passed since the premature death of J. Bradford (Piccarreta, Ceraudo 2000, pp. 80-81; Musson, Palmer, Campana 2005, pp. 23-24; Franchin Radcliffe 2006), the British military officer-archaeologist, famous for the interpretation of WWII aerial photographs of the Tavoliere (Bradford 1949; Id. 1950; Id. 1957a; Id. 1957b; Bradford, Williams Hunt 1946): his pioneer investigation method was based on the principle that ancient landscapes were to be looked in their totality, as a whole.

His methodological approach is still today a reference for any modern research projects which now use sophisticated investigation methods and technologies.

Looking at how Landscape Archaeology has progressed over the years, it appears clear a fast and explosive growth of interest for the potentialities of aerial archaeology as investigative tool. A proof of this awakening of interest is the recent flow of publications in Italian on the subject of aerial archaeology (Piccarreta, Ceraudo 2000; Musson, Palmer, Campana 2005), the activity carried on by the Aerial Research Group in Italy (Musson 2004), and the organisation of Summer Schools entirely dedicated to Aerial Archaeology first at Siena, and then at Foggia and Grosseto. Even those who always had a sceptic approach towards the potential of oblique photography, have finally begun to accept the benefit of an integration between oblique and vertical ones: one need only mention the catalogue (by M.Guaitoli) of the 2003 exhibition "*Lo Sguardo di Icaro. Le collezioni dell'Aerofototeca Nazionale per la conoscenza del territorio*" and the appearance in 2004 of the first issue of the journal '*Archeologia Aerea. Studi di aerotopografia archeologica*' by G. Ceraudo e F. Piccarreta.

The concept of landscape as a palimpsest of survived traces of the ancient human activities pushes modern researchers to look at the stratified landscapes of a territorial context as a whole and, consequently, to follow the way of *total archaeology* as multi-disciplinary integration of different competences (Mannoni 1997, p. 21-25; Manacorda 2004, p. 136-139).

In light of this principle, the current research project 'Landscapes of North Puglia' carried out by the Department of Archaeology of the University of Foggia and directed by Giuliano Volpe, tries to have a methodological approach which is flexible and can be modified to suit best the area under investigation. If this can be of more efficiency, there is integration between traditional methods and innovative and experimental investigative techniques. This flexible approach gives the opportunity to examine remotely the landscape and potentially fill the gap (which, according to G. Barker, is 'typical of Italian landscape archaeology') between the study of individual settlements and the study of the territory and its transformations (Barker 1986; Campana, Francovich 2005).

For its geology, geomorphology and its arable nature, the Tavoliere di Puglia represents an ideal location to test, on both small and large scale sites, the potential of aerial survey as an investigative tool (Musson, Palmer, Campana 2005, pp. 206-207).

In 2003, following a brief but prolific aerial survey in spring 2002 (in collaboration with Chris Musson and Otto Braash of the Aerial Archaeology Research Group), was held the first Summer School by the University of Foggia. The school aimed to introduce students to both theoretical and practical applications of aerial survey and oblique photography, with a European wide panel of tutors and pilots from the local Foggia Aeroclub.

During this intensive 8 day-long course 80 flights were carried out, with an astonishing flying time totalling 76 hours that produced 4195 photographs of 987 archaeological sites; Further flights were then carried out in 2004, 2005 and 2006 as part of the Project *Culture 2000 "European Landscapes: Past, Present and Future"*. Without doubt, the most important achievement of the school was the demonstration of the potential of exploratory aerial survey for the discovery of new archaeological sites. Furthermore, at the same time was also carried out systematic monitoring of the archaeological heritage of Northern Puglia, published or still unknown, currently threatened by extensive and poorly planned urban developments, environmental changes, agricultural and industrial activities as well as illegal excavations.

Following the photo reading, interpretation and siting, the most informative oblique photographs were rectified and then imported into a GIS system on a 1:10,000 scale base map available for the whole province of Foggia; the drawing was created in *ArcMap* on themed layers selected according to the archaeology mapped. The whole process produced a series of archaeological maps used in later stages for studying the historical landuse of the area and planning suitable conservation strategies for the heritage sites identified.

At present, field survey remains the technique for which the majority of resources and time is spent: this is the crucial instrument that provides a constant flood of reliable and verified data. Our method is based on an integration of extensive surveys on large scale and intensive surveys on targeted areas, combining systematic analysis over broad areas and tests on sites of specific interest (for instance those identified on air photographs). The complexity of this approach and the huge amount of data produced made the creation of a GIS system for cataloguing, management and access of the information acquired essential.

It is undoubtedly difficult to identify and illustrate all the results gained from aerial reconnaissance activities from 2002 onward: not all the archaeological evidence has been yet verified on the ground and there are still sites whose interpretation is in doubt and that wait to be clarified with ground inspection. Nevertheless, despite this status of 'work in progress', at this stage it is possible to illustrate some significant results and demonstrate how important is the contribution of aerial survey for the study of the archaeological landscapes in Daunia.

In spite of the multi-period approach to the reading of landscape – from Prehistory to the late Medieval – it's opportune to focus this paper on the data which have enable us to increase our understanding, in particular, of the Roman period and Late Antiquity, especially regarding field systems and different site types.

Important contributions (especially on topics of Roman centuriation and road systems in Daunia) were already obtained with the interpretation by G. Schmiedt (Schmiedt 1974; Id. 1985, p. 260-304; Id. 1989), R. Compatangelo Soussignan (Compatangelo 1994a; Ead. 1994b; Compatangelo Soussignan 1999a; Ead. 1999b), G. Alvisi (Alvisi 1970) of vertical photographs taken by IGM from the second half of the last century onward.

For areas not altered by modern land division and agricultural activities the new results are extraordinary. An example is the accurate reconstruction of the *limitatio* of 20 *actus* module visible between *Aecae- Luceria-Arpi* (Troia, Lucera, Foggia), much wider than initially thought by G.Schmiedt, realised by A.V.Romano as a result of flights carried out between 2003 and 2005. And also the example of the *ager Canusinus* (Canosa di Puglia): air photo evidences illustrates that the Antoninian centuriation, mentioned in the second edition of *Liber Coloniarum* and identified along the river Ofanto by R.Compatangelo, was in reality much broader than initially thought. The features represent the remains of an extensive late *divisio et adsignatio* that involved the entire surroundings of Canosa.

The most important benefit from low-altitude aerial reconnaissance and its integration with field survey results is however represented by the precious contribution to the understanding of the planimetric organisation and typological definition of rural settlements.

In absence of sufficient excavation-data, it has already been proved that field survey on its own does not offer enough information to show the complexity of plans and structures of sites; conversely, aerial photography sometimes allows us to easy identify the plan of features buried underneath the ground.

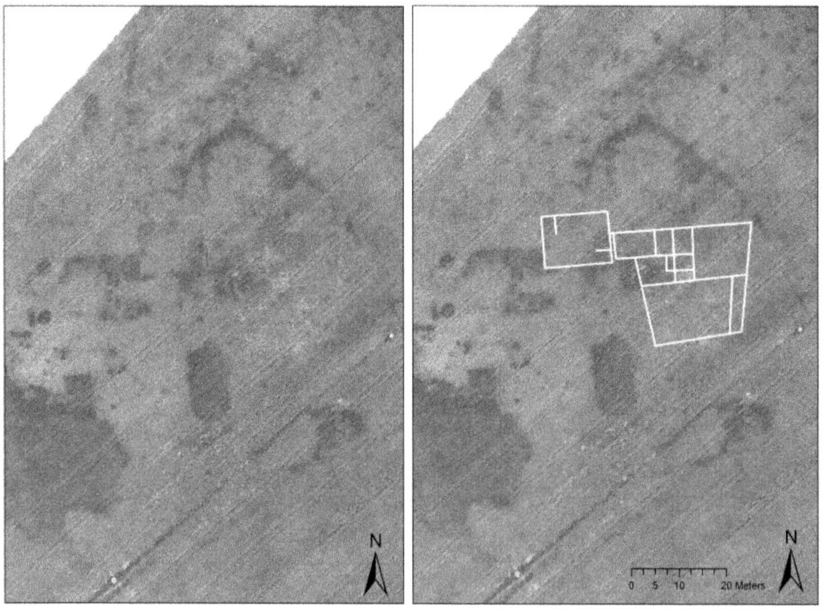

Figure 1

Figure 2

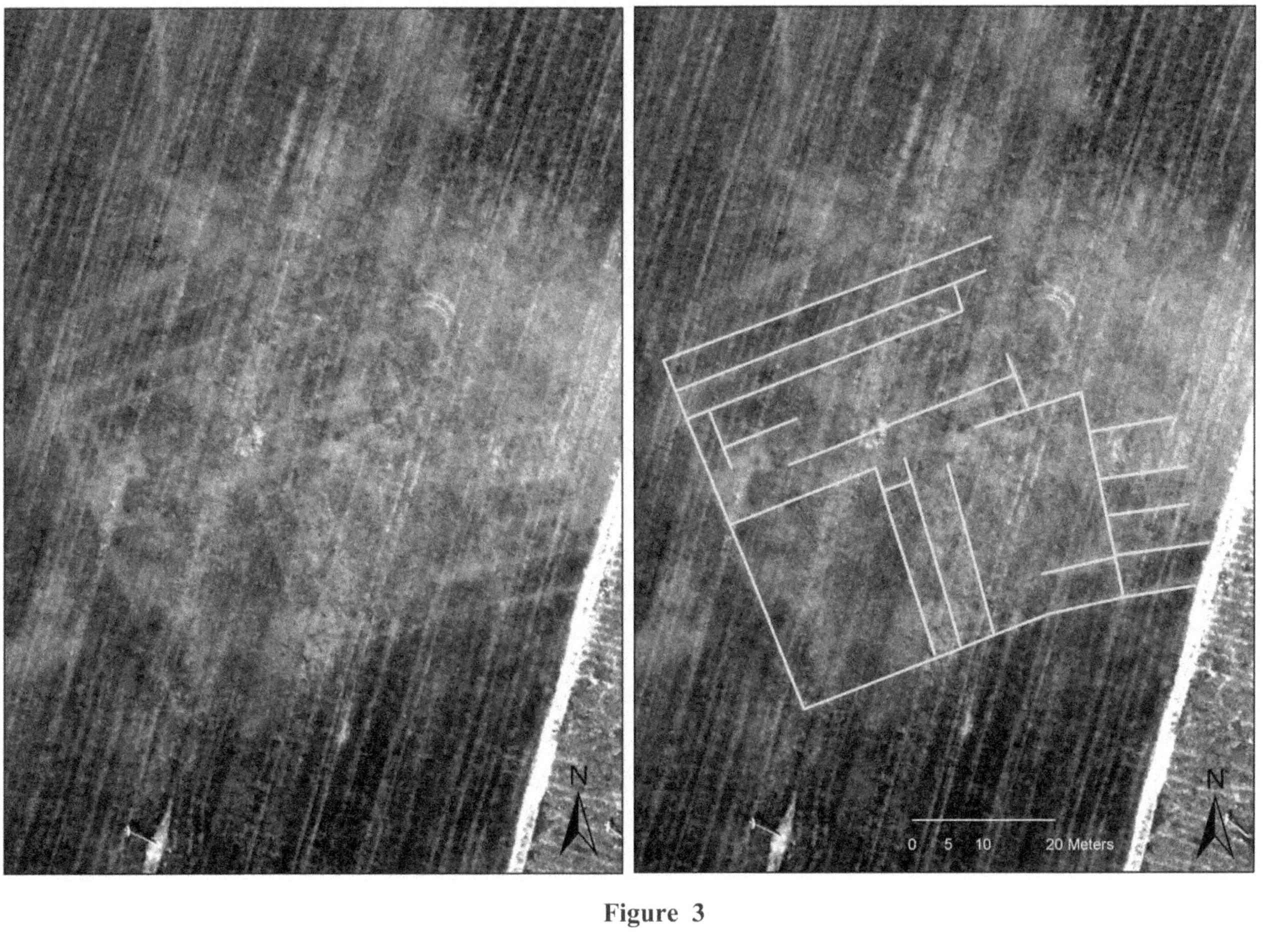

Figure 3

It is the case of small-size farmsteads rectangular in form, identified at San Marco (Fig. 1), Torretta, Montagna Spaccata (Fig. 2), Ripalta near Cerignola (Foggia), and Mezzana Grande near Ascoli Satriano (Foggia).

More complex plans and much bigger sizes characterize on the contrary sites which can be interpretated as *villae*. This site type is shown on air photographs as a compacted self-contained feature with elaborate internal divisions, commonly between 4.5 and 6ha in size: in this way appears the *villa* 'Troia North' photographed by J.Bradford in the *ager Aecanus* (Romano, Volpe 2005, p. 248), and the ones at Masseria Vocale (Fig. 3) and Posta Carrera near Ordona (Foggia) (Fig. 4).

The above examples do not illustrate in full the complexity and variety of what is visible from the air in north Puglia: this region offers a wide-ranging assortment

Figure 4

of sites of different date and function, scoping between settlements, field systems, funerary and religious sites; there are also traces whose origin is still obscure and for which archaeologists have yet to formulate reasonable interpretations.

The study of the medieval landscape of Capitanata has also benefited from aerial photography, both vertical and oblique. As a result of a fruitful collaboration between the Ecole Francaise de Rome and the University of Bari, a research project aimed to investigate deserted medieval villages started at the beginning of the 1980s (Martin, Noyé 1988, pp. 501-526; Iid. 1991, p. 53-58) but unfortunately any interest for the 11th and 14th century medieval villages of Tavoliere was soon gone.

Motte and bailey sites are generally well known in the Tavoliere, often surviving on ancient maps as place-names or, even more frequently, because their earthworks appear as distinctive features that can be easily recognisable from the air.

Emblematic is the case of the *casale* in locality Masseria Petrullo (San Marco in Lamis, Foggia) (Fig. 5), for which it is possible to assume a frequentation dated between the 13th and 14th centuries on the base of the surface evidence. The site appears on air photographs as a double ditched feature with associated trackways forming a road network that links the site to nearby settlements.

In terms of present-day research, there are 22 medieval *casalia* whose location in the Tavoliere is currently known, others are mentioned by documentary evidence but have not been yet identified.

This last line of research finds now a link with the excavation, from 2005, of the *casale* of San Lorenzo in Carmignano (Foggia). This abandoned settlement in the surroundings of Foggia is a clear example of what kind of threat modern landuse in the Tavoliere represents to archaeology: the landscape, almost completely preserved until the 1950s, has been the victim of unremitting alterations that, in most cases, have compromised forever any traces of past human activities. Not surprisingly, in these conditions it is difficult to identify the archaeological landuse from the Norman period onward.

Most terraces, banks, ditches, motte and bailey sites have all been levelled to give space to arable land. These features, so important to understand in full the historical landuse, have now totally disappeared from the countryside.

Landscape Archaeology is no longer a study of 'sites', individual settlements and their phasing, road networks and field systems but it is *tout court* an archaeology of *the* landscape for *the* landscape. Landscape Archaeology is, quoting B. Jones (Jones 1985), a 'total archaeology' and must be a comprehensive study of the evidence from

Figure 5

the past. This study of the landscape and its phasing is also a major tool in monitoring, helping to formulate effective planning for heritage protection, valuing the historical heritage, and most of all, promoting an organic cohesion of modern society needs and conservation of the past.

The standard methods employed in landscape research – from air-photo interpretation to geophysical and field surveys, from cataloguing to database and GIS management – all of these can offer a precious contribution so that signs and traces not erased by centuries, which are the archive of the history and identity of a country, can be preservate.

References

ALVISI, G. 1970, *La viabilità romana della Daunia,* Bari.

ALVISI, G. 1979, *Problemi di topografia tardo antica nella zona di Siponto. La rete viaria,* in *Puglia paleocristiana e altomedievale III,* Bari, pp. 1-45.

BARKER, G. 1986, *L'archeologia del paesaggio italiano: nuovi orientamenti e recenti esperienze,* AMediev, 13, pp. 7-30.

BRADFORD, J. 1949, *Buried landscapes in southern Italy,* Antiquity, 23, 89, 1949, pp. 58-72.

BRADFORD, J. 1950, *The Apulia expedition: an interim report,* Antiquity, 24, 93, 1950, pp. 84-95.

BRADFORD, J. 1957a, *Ancient landscapes. Studies in Field Archaeology,* London 1957.

BRADFORD, J. 1957b – J. Bradford, *The ancient city of Arpi in Apulia,* Antiquity, 31, 1957, pp. 167-169.

BRADFORD, J., WILLIAMS HUNT, R. 1946, *Siticulosa Apulia,* Antiquity, 20, 77, 1946, pp. 191-200.

CAMPANA, S., FRANCOVICH, R. 2003, *Landscape archaeology in Tuscany: cultural resource management, remotely sensed techniques, GIS based data integration and interpretation*, in M. Forte, P.R. Williams (eds), *The reconstruction of archaeological landscapes through digital technologies*, Proceedings of the 1st Italy-United States Workshop, Boston, Massachusetts, USA (November 1-3 2001), Oxford 2003, pp. 15-28.

CAMPANA, S., FRANCOVICH, R. 2005, *Linking remote sensing and infra site analysis to the reconstruction of rural settlement and landscape patterns*, in M. Forte (eds), *The Reconstruction of Archaeological Landscapes through Digital Technologies*, Roma 3-5 novembre 2003, Bar Series, Oxford.

COMPATANGELO, R. 1994a, *Canosa e la Puglia settentrionale: produzione agricola e catasti rurali,* in Actes du Colloque International *Le revitaillement en blé de Rome et des centres urbains des début de le republique jusqu'au Haut Empire (Naples 1991),* Naples-Rome, pp. 167-176.

COMPATANGELO, R. 1994b, *Recherches sur l'occupation du sol et les cadastrations antiques du territoire de Canosa,* DialHistAnc, 20, 1, 1994, pp. 199-243.

COMPATANGELO SOUSSIGNAN, R. 1999a, *Sur les routes d'Hannibal. Paysages de Campanie et d'Apulie,* Paris.

COMPATANGELO SOUSSIGNAN, R. 1999b, *Centuriazione senza coloni? Il caso di Canosa nel quadro della regio Apulia et Calabria*, in A. Gravina (a cura di), Atti del XVII Convegno sulla Preistoria, Protostoria e Storia della Daunia (San Severo 1998), San Severo, pp. 167-183.

FRANCHIN RADCLIFFE, F. 2006, *Paesaggi sepolti in Daunia. John Bradford e la ricerca archeologia dal cielo (1945-1957)*, Foggia.

JONES, G.D.B. 1980, *Il Tavoliere romano. L'agricoltura romana attraverso l'aerofotografia e lo scavo,* ArchCl, 32, 1980, p. 85-100.

JONES, G.D.B. 1985, *Introduction*, in S. Macready, H. Thompson (eds), *Archaeological Field Survey in Britain and Abroad*, Society of Antiquaries Occasional Papers 6, London.

MANACORDA, D. 2004, *Prima lezione di archeologia*, Bari.

MANNONI, T. 1997, *Archeologia globale e archeologia postmedievale*, Archeologia Postmedievale, I, 1997, pp. 21-25.

MARTIN, J.M., NOYE, GH. 1988, *Habitats et systémes fortifiés en Capitanate. Première confrontation des données textuelles et archélogiques,* in Gh. Noyè (ed), *Structures de l'habitat et occupation du sol dans les pays méditerranéens. Les méthodes et l'apport de l'archéologie extensive*, Roma-Madrid, pp. 501-526.

MARTIN, J.M., NOYE, GH. 1991, *La Capitanata nella storia del Mezzogiorno medievale*, Bari.

MUSSON, C. 2004, *The Aerial Archaeology Research Group (AARG) and Aerial Survey Training Schools in Hungary, Poland and Italy, 1996-2003*, in *Archeologia Aerea I*, pp. 81-90.

MUSSON, C., PALMER, R., CAMPANA, S. 2005, *In volo nel Passato. Aerofotografia e cartografia archeologica*, Firenze.

PICCARRETA, F., CERAUDO, G. 2000, *Manuale di Aerofotografia Archeologica. Metodologia, tecniche e applicazioni*, Bari 2000.

ROMANO, V., VOLPE, G. 2005, *Paesaggi e insediamenti rurali nel comprensorio del Celone fra Tardoantico e Altomedioevo*, in G. Volpe, M. Turchiano (eds), Atti del I Seminario sul Tardoantico e l'Altomedioevo in Italia meridionale, *Paesaggi e insediamenti rurali in Italia meridionale fra Tardoantico e Altomedioevo* (Foggia, 12-14 febbraio 2004), Bari, pp. 241-264.

SCHMIEDT, G. 1968, *Le fortificazioni altomedievali in Italia viste dall'aereo,* in Atti della XV settimana di studi del CISAM, Spoleto 1967, pp. 860-927.

SCHMIEDT, G. 1974, *Atlante aerofotografico delle sedi umane in Italia: parte III, la centuriazione*, Firenze.

SCHMIEDT, G. 1975, *Contributo della fotografia aerea alla conoscenza delle strutture fortificate altomedievali,* in *Metodologia nella ricerca delle strutture fortificate nell'Alto Medioevo,* V Tavola Rotonda Nazionale (Udine-Cividale-Trieste 26-27-28-29 ottobre 1967), pp. 35-66.

SCHMIEDT, G. 1985, *Le centuriazioni di Lucera e Aecae,* L'Universo, 65, 2, 1985, pp. 260-304.

SCHMIEDT, G. 1989, *Atlante aerofotografico delle sedi umane in Italia: parte III, la centuriazione romana,* Firenze.

3D Blog – A New Way of Supporting Communication about Cultural Heritage

Rieko Kadobayashi

National Institute of Information and Communications Technology
rieko@nict.go.jp

Introduction

The amount of 3D digital recording, modeling, and visualizations of cultural heritage has been rapidly increasing. These 3D models and visualizations are utilized for a wide variety of purposes such as research, preservation, and presentation. Although there are many case studies and applications, most of them provide only limited interactivity and basic communicative functions.

When people visit a museum, they can usually share the experience with other people such as their family, friends, or schoolmates. They can watch how other people interact with exhibits. This observational or social learning is thought to be very important (Falk et al 2000). The sharing of knowledge, interest, impressions, and experience can deepen each person's understanding and appreciation of the exhibits. In addition, there are a lot of hands-on exhibits in real museums that enable visitors to interact, which facilitates their understanding.

However, existing 3D applications support only limited sharing and interactivity. This is especially true of applications for the general public. Experts can interact with 3D models using proprietary or expensive commercial software, but the general public are allowed only to view them passively, e.g. to see still images or animation in most cases. The experts may convey their interpretation of cultural heritage via 3D visualization, but the general public does not have such means. The predominant communication in the cultural heritage sector is among experts or from experts to the general public as in museum exhibitions.

As a result, 3D models and visualization have been made available basically to experts. However, the general public or local community should be included since they are also stakeholders in cultural heritage, and the public or community's utilization of 3D models and visualization is very important.

Of course, there are 3D applications targeting the general public. There have been many studies on Networked Virtual Environment (NVE) where multiple users can share a 3D virtual space and collaborate on many kinds of tasks (Barbieri 2000). Among the promising NVE applications are virtual 3D museums that provide on-line chat functions so that users can communicate with each other (Paolini et al 1999). They can also get help to deeply understand content posted by those with more experience. The main problem in virtual 3D museums, from the viewpoint of communication and collaboration, is that users need to be in a virtual museum at the same time. Practically, this may be an overly restrictive requirement, and other means should be provided (Barbieri and Paoloni 2000).

To overcome these problems, we propose a new way of using 3D models of cultural heritage items. We have been developing "3D Blog", which allows a user to interact with a 3D model, annotate any spot on the model, and publish the annotations in a blog so that the user can easily communicate with other people and share interests, ideas, questions, and so on about 3D models of interest to the user (Kadobayashi 2005).

One of the main features of a 3D Blog is that each annotation is saved as a blog entry with 3D information on the user's viewpoint, points of interest, and direction of gaze. Readers of a 3D Blog can reconstruct the exact same 3D scene as the blogger saw when reading the blog entry and, therefore, the readers may understand it better. Another feature is that the 3D Blog system detects similar entries based on distance of viewpoint, point of interest, and similarity in direction of gaze and sends trackback links automatically to enable users to easily find people who may have similar interests and to promote emergence of communities of people who share interests.

The 3D Blog System

Features

3D Blog enables users to interact with a 3D model, annotate any spot on the 3D model, and publish the annotations as a blog. It does not require any expensive proprietary software to make annotations or read blog entries. It only requires a 3D model viewer that runs as a browser plug-in for interacting with 3D models. The current 3D Blog system uses Cortona VRML Client (Cortona).

The main feature of 3D Blog is that it uses the 3D viewpoint information of the 3D scene that the user (blogger) sees when adding an annotation. The 3D viewpoint information is automatically added to the annotation by the system and stored in the database. It is then used to reconstruct the same 3D scene when a blog reader reads entries, which is thought to be a great help for readers to understand the content of the entries.

This system is based on the assumption that impressions

may change if the user's point of view changes, even if the user looks at the same thing. This may enable users to convey their feelings more accurately if they can show the object as it looked when they saw it.

Another feature is that similar entries are detected based on the similarities of the 3D viewpoint information and linked automatically using the TrackBack mechanism. This automatic trackback helps users find other users with similar interests and create communities, and facilitates communication among users.

How it works

We have created a prototype system using lily (lily) as a blogging tool and Cortona VRML Client as a viewer for the 3D models. Lily is an implementation in Ruby of blosxom (blosxom) written in perl. Users need to download and install the plugin, i.e., Cortona VRML Client, to their web browser (currently Internet Explorer only) before starting 3D Blogging.

Writing 3D Blog Entries

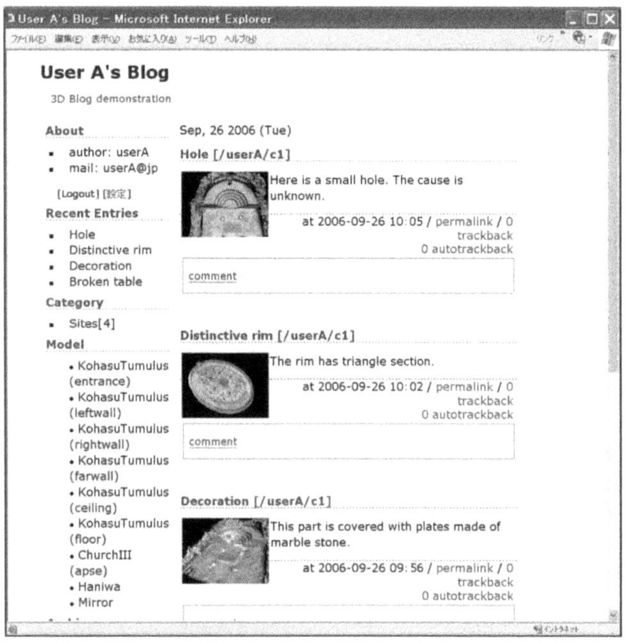

Figure 1 *Top page of a 3D Blog. Annotations to 3D models are listed as blog entries in reverse chronological order.*

When a 3D Blog user logs in, the top page of the user's blogsite will open and show recent entries in reverse chronological order, as in ordinary blogsites (Figure 1). By selecting a 3D model from the list on the left-hand side of the window and clicking on the model's name, the 3D model viewer will appear at the top of the list of entries so that the user can browse and interact with the model in either the walk-through, fly-through, or study mode in order to add notes about the model (Figure 2).

Every time the user clicks on an interesting spot that he/she wants to annotate, a small ball, i.e. a marker, appears that covers the spot in the 3D model viewer. The user types an appropriate title and a comment in the dialogue box below the 3D model viewer and then clicks the 'post' button.

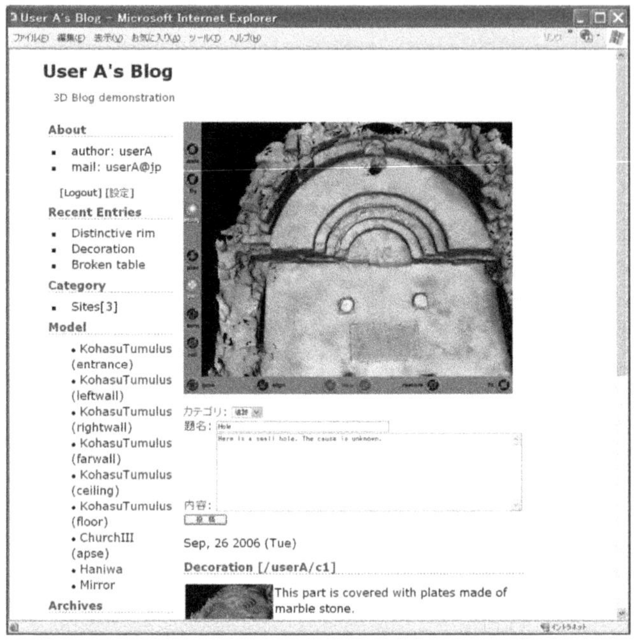

Figure 2 *Typical page layout for interacting with and annotating 3D models. The 3D model viewer is displayed at the top of the list of entries.*

Along with information such as the position of the annotated area, the viewpoint of the virtual camera, the date and time, and the snapshot of the 3D model with the marker, the annotation (i.e., the title and comments) is converted into a blog entry and stored in the blog database. The annotation is then added to the top of the entries of the user's Blog site as the latest entry.

The user can annotate a 3D model as often as desired. All annotations are automatically added to the blog site as separate entries. The user can change a 3D model by clicking on a model name and continuing to blog on a different model.

Reading 3D Blog

At first glance, the top page of a 3D blogsite looks similar to ordinary blog top pages, but there is a big difference. Every entry in the 3D Blog includes information about the 3D viewpoint, i.e., the viewpoint from which the 3D model was annotated and that of the spot to which the annotation was added.

The snapshot included in each entry is an anchor to the 3D scene and, once it is clicked by a reader, the 3D model viewer pops up to render the 3D scene using the 3D viewpoint information while at the same time the reader can read the entry (Figure 3). This is a great help to readers in understanding the content of entries, as they can visually confirm it. Note that the 3D model viewer opened by clicking the snapshot allows readers to interact with a 3D model but not to annotate it.

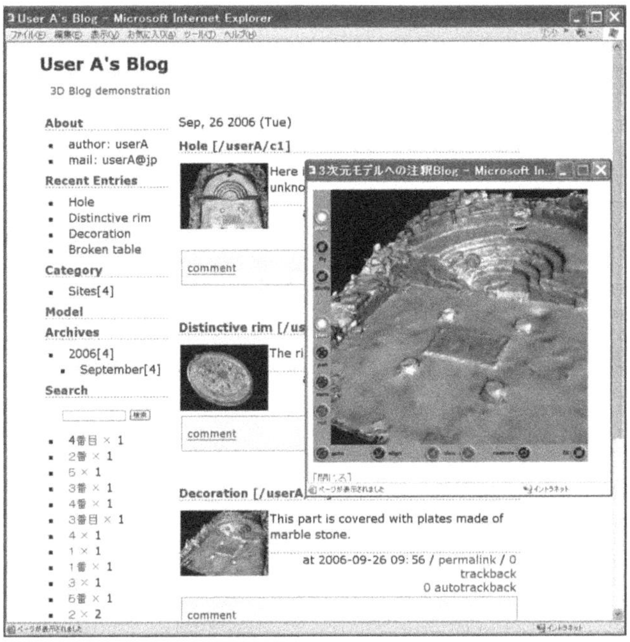

Figure 3 *The 3D model viewer pops up when the snapshot of an entry is clicked. The user can read the entry and confirm how the 3D model looks.*

Readers can add comments to any entry by clicking the 'comment' button at the bottom of each entry to open a dialogue box. Other basic functions such as viewing old entries and searching entries are also available.

Automatic TrackBack

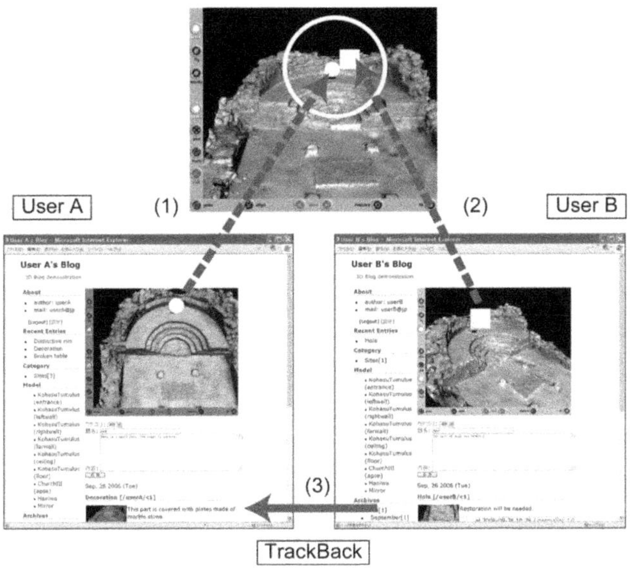

Figure 4 *Automatic TrackBack is done by the system when similar entries are found. Users may easily find other users with similar interests.*

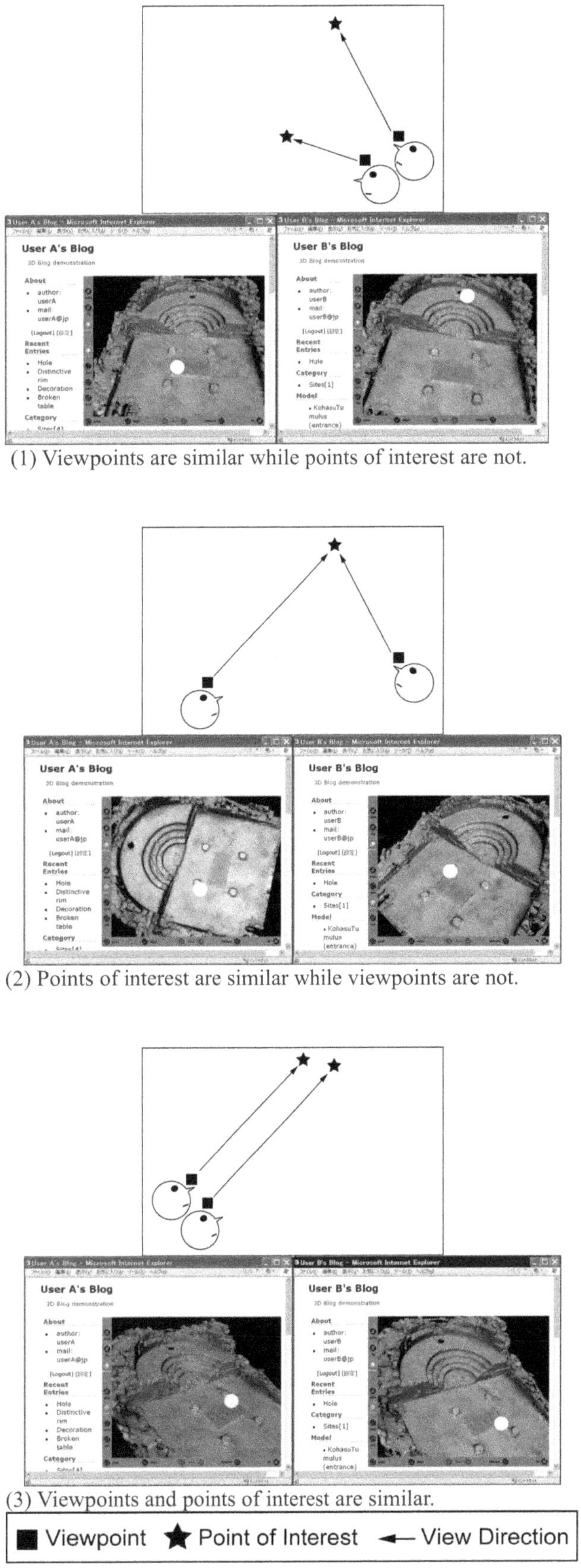

Figure 5 *Similarities between two entries are determined based on the distance of viewpoints (top), the distance of annotated spots (middle), and the cosine similarity of view direction (bottom).*

Usually the TrackBack link is used by a blogger who writes an entry referring to another blogger's entry in order to explicitly notify the other blogger of the reference. In the 3D Blog system, the system also automatically sends TrackBack to entries in the database if it finds similar entries as new entries are written. Similarities are determined based on the distance of the viewpoint of each entry, the distance of the annotated spot of each entry, and the cosine similarity of the view direction, i.e., the line of vision from the viewpoint to the annotated spot.

Assume that user A has written an entry about a 3D model and the entry is stored in the database. After user B writes an entry about the same 3D model (see (2) in Figure 4), the system checks for similar entries in the database. In this case, the system finds that an entry written by user A is similar to the entry that user B has just posted because the distance between the annotated spots of both entries is very short and is less than the threshold that user A specified. It then sends TrackBack to user A's entry so that user A will notice user B's entry (see (3) in Figure 4).

This automatic TrackBack mechanism helps 3D Blog users find similar entries and others who may have similar interests or perspectives. Users can set the threshold according to their preference. For example, some people may want to find people who may have similar interests and set the threshold for the distance between annotated spots low. In contrast, others may want to find people who have a similar (physical) perspective and set that distance threshold low. Note that currently, the system uses only physical (spatial) viewpoints. Figure 5 shows the difference among the similarity measures.

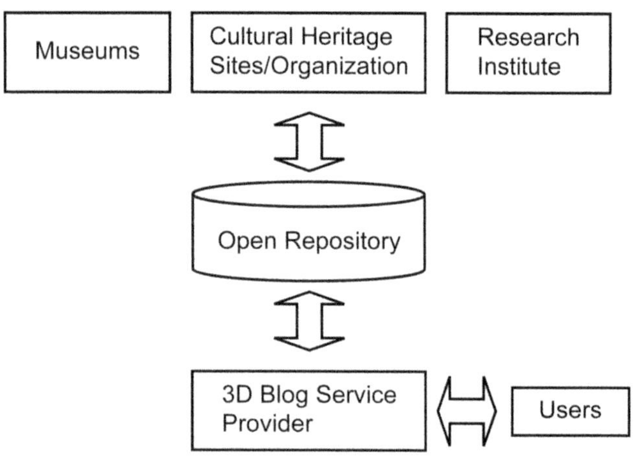

Figure 6 *Possible 3D Blog Service Model in actual use. Museums and other institutions in the cultural heritage sector deposit 3D models in the open repository. People access the 3D models by subscribing to 3D Blog services.*

Service Model

There will be multiple implementations of the 3D Blog system for supporting communication in a wide variety of communities depending on things such as the purpose, scale, and policies of the stakeholders. Figure 6 shows one of the possible service models. There are 3D model providers such as museums, cultural heritage sites and organizations, and research institutes that create 3D models of their cultural heritage. They deposit the 3D models in the open repository, which is maintained by a neutral institution. 3D Blog service providers provide the general public with 3D Blog service of the cultural heritage by allowing their members access to the 3D models stored in the open repository. A 3D Blog service provider may provide service only to those who live in a particular area in order to support communication in that community.

Discussion

To make the 3D Blog system fully functional, there are many topics that need further research and development.

First of all, the number of 3D models of cultural heritage items open to the general public will greatly affect the success of 3D Blogs. Access to many attractive 3D models will help people get accustomed to 3D blogging, open up a way to create a community, and change the power balance among stakeholders. However, an open repository of 3D models of cultural heritage is far from being realized.

Also, development of an excellent user interface for 3D models, especially for annotating and comparing 3D models, is crucial. It is difficult, intuitively and otherwise, to select an arbitrary part of a 3D model, but is essential when adding annotation. In addition, easy-to-use interfaces that compare 3D models in a single viewer should be provided since comparing objects is fundamental to research and discussion.

Finally, the annotations and comments made by the public, researchers, and others involved in the cultural heritage sector are themselves valuable properties as well as the cultural heritage and its digital data. We should consider how to store, manage, search, and use these annotations and comments, as they could supplement cultural heritage and help us understand cultural heritage.

These topics are closely related to standard format and digitization procedure for 3D models, which are already being discussed in forums such as the EPOCH project. When designing the standards, we should consider "technical seeds" or "new applications" as well as "needs" or "known workflow" so that the standards contribute enormously to wider stakeholder/user for sharing open cultural heritage.

Conclusion

We discussed the 3D Blog system, which allows users to exchange ideas and feelings about cultural heritage by annotating 3D models of cultural heritage and publishing them as blogs. With this tool, the general public can actively join the discussion on the particular cultural heritage of their local community. For example, when discussing preservation of a historical area or an archaeological site, the 3D blog system could be used as a tool to facilitate consensus-building efforts among the general public as well as among cultural heritage experts. It may also encourage people who live far from the area to join the discussion. In short, the 3D blog system can appeal to a wider audience than ever and involve them in preserving valuable cultural heritage.

Although we need further investigation on additional functions of the system to support user communication and knowledge sharing via 3D blogging, we believe 3D blogging has a great potential and can be applied to many domains.

References

BARBIERI, T. 2000 Networked Virtual Environments for the Web: The WebTalk-I and WebTalk-II Architectures. In *Proceedings of International Conference on Multimedia & Expo*. IEEE

BARBIERI, T., and PAOLINI, P. 2000 Cooperative Visits for Museum WWW Sites a Year Later: Evaluating the Effect. In *Proceedings of Museums and the Web 2000*, Archives & Museum Informatics 173-178

blosxom: http://www.blosxom.com/

Cortona VRML client: http://www.parallelgraphics.com/products/cortona/

EPOCH: http://www.epoch-net.org/index.php

FALK et al 2000, FALK, John H. and DIERKING, Lynn D.,*Learning from Museums*. AltaMira Press.

KADOBAYASHI, R. 2005 Automatic 3D Blogging to Support the Collaborative Experience of 3D Digital Archives. In *Proceedings of the 8th International Conference on Asian Digital Libraries (ICADL 2005)*, LNCS 3815 109-118

lily: http://www.mikihoshi.com/lily/

PAOLINI et al 1999, PAOLINI, P., BARBIERI, T., LOIUDICE, P., ALONZO, F., GAIA, G., ZANTI, M., Visiting a Museum together: How to share a visit to a virtual world. In P*roceedings of Museums and the Web 1999,* Archives & Museum Informatics 27-35

Application of Remote Sensing in the Detection of Maya Archaeological Sites in South-Eastern Campeche, Mexico

Krištof Oštir, Žiga Kokalj, Ivan Šprajc[1]

[1] Scientific Research Centre of the Slovenian Academy of Sciences and Arts (Slovenia)

1 Introduction

Some of the most extensive gaps on the archaeological map of the Maya area were found, until recently, in the central parts of the Yucatan peninsula, particularly in the vast southern portion of the Mexican federal state of Campeche. Until the early 1980s, when intensive research in the Calakmul area began (Folan et al. 1995), Ruppert and Denison's monumental work (1943) was practically the only source of information on some archaeological sites in this region. However, these were only a few of the largest and best preserved (Ruppert and Denison 1943). Information on some additional sites reported afterwards is so scanty that they cannot be identified even in the field. Indeed, only two decades ago Adams (1981) affirmed that sites in the area, except for El Palmar, "were all found by Ruppert and his Carnegie Institution surveys of the 1930s".

In an attempt to improve the situation, archaeological reconnaissance works were carried out in south-eastern Campeche in 1996, 1998, 2001, 2002, 2004 and 2005. The surveys, on which preliminary reports and synthesis have been published (Šprajc 2002-2004, Šprajc in press), resulted in a relatively satisfactory coverage of the area between the Río Bec region to the north (around the modern town of Xpujil) and the border with Guatemala to the south, and between the border with Belize and Quintana Roo to the east and the Calakmul Biosphere Reserve to the west. Apart from this sparsely populated territory, where the land is allotted mostly to ejidos (rural communities enjoying usufruct rights to the land held in common), our surveys included the southern part of the Calakmul Biosphere to the west, which is now without permanent population. The recent human activity in the area of ejidos, particularly the slash-and-burn cultivation, which continues to be the most common agricultural technique, has reduced considerable extents of tropical forest and has also contributed to the destruction of archaeological vestiges. On the other hand, and more importantly, the archaeological sites are threatened by systematic and professional looting, which normally results in irretrievable loss of important archaeological data and irreparably damaged architectural structures. It can thus be affirmed that the surveys carried out so far helped rescue a considerable amount of significant archaeological information.

Apart from rediscovering most of the previously reported archaeological sites in the area, we recorded more than 60 formerly unknown sites. Most of them are settlements with architectural remains dating to the Late Preclassic and Classic periods (ca. 300 B.C. – A.D. 900). Their number, different size and characteristics, and the presence of major urban centres with monumental architecture and sculpted monuments reveal that both the population density and the complexity of social and political organization in the territory surveyed were similar to those observed in the surrounding, better-known areas. While this is true particularly for the Classic period (A.D. 250 – 900), a number of earlier complexes of monumental architecture were also found, indicating that south-eastern Campeche was part of the area that witnessed important Late Preclassic developments, most clearly attested in the adjacent northern Guatemalan Petén. Major centres are regularly located on the edges of large bajos (low wetlands or seasonally flooded swamps), conforming to the pattern observed elsewhere in the Maya Lowlands.

The results of the surveys represent a basis not only for planning future archaeological research focused on particular sites and areas, but also for analyses that are expected to shed light on correlations between settlement distribution, environmental peculiarities and sociopolitical factors, on the validity of the proposed models of territorial organization, as well as on other questions concerning cultural development in the central Maya Lowlands.

2 Remote sensing of the Maya region

Remote sensing surveys provide a rapid means of data collection that can achieve complete coverage of large areas, with far lower costs than those associated with excavations or field surveying. Traditional invasive methods generally damage or destroy the resources they were designed to investigate. Remote sensing methods, on the other hand, are non-destructive and leave the resource intact. Archaeological remote sensing allows large regions to be rapidly investigated for archaeological features, at relatively low cost; it can detect features unseen on the surface, map them accurately, and offer interpretations based on their form, distribution, and context (Kvamme 2005).

The application of remote sensing in Maya studies has changed substantially since 1929, when Charles Lindbergh made the first photographs of Mayan remains in the Yucatan peninsula. In the late 1970s archaeologists were employing Landsat images to help determine the location of Maya ruins in the dense jungles of the Yucatan Peninsula (El Baz 1997). Even though the resolution of Landsat images is low for archaeological prospection, their use has provided some interesting results (Adams et al. 1981). Great potential has been shown by newer radar systems, with their ability to penetrate desert sands and vegetation and to unveil ancient topography. Several systems, both airborne and space borne, have provided evidence of archaeological sites in the central Maya area (Holcomb 1998, Sever and Irwin 2003). Saturno et al. (2006) have combined a broad dataset of mid- and high-resolution optical and radar data, i.e. Landsat TM and ETM+, Ikonos, STAR-3i radar imagery, to identify ancient causeways, sites, water storage areas, reservoirs, and human-altered hydrologic drainage patterns. The exact form and size of bajos, bajo islands, and multiple courses of the Holmul River and its tributaries have been defined by the use of satellite and airborne images. In the near future, a great deal of research will be performed with high-resolution airborne systems, e.g. lidar.

The primary function of archaeological remote sensing is to find contrasts between archaeological features and the natural background. When archaeological deposits or features possess physical properties different from those of the surrounding matrix, a distinction can be noticed. There have been several methods employed for identifying the locations of archaeological sites in the Maya area from remote sensing data. Sever and Irwin (2003), for example, have tried to detect linear features that might be related to roadways or channel systems. They have furthermore investigated the potential to correlate settlement patterns with vegetation, hydrological, and geological features on the karst landscape, to isolate topographic features where sites are likely to be located, and to create detailed maps of vegetation and drainage pattern maps, in order to provide insight into the relationships among land cover, paleohydrology, water management, bajo systems, settlement pattern and subsistence strategies. Rothaus and De Morett (2001) focused on the search for anomalies; because of the multiple indicators (variables) involved, as well as uncertainty of the results, they used an approach based on visual detection rather than automated classification. They presumed that the use of landscape by prehistoric people associated with archaeological sites was much the same as the one involved in contemporary swidden agriculture. To be able to identify archaeological sites, subtle variations in spectral response had to be detected. The need to use multi-temporal and multi-scalar imagery to confirm potential anthropogenic anomalies and exclude transient environmental phenomena and data errors has been demonstrated.

While satellite remote sensing is one of the most discussed topics in archaeology, remarkably little has been published that details the steps researchers are using (Rothaus and De Morett 2001). Computer manipulation and processing of data is now essential to nearly all remote sensing, nearly as important as collecting the actual data, because noise, defects, and distortions in raw imagery obscures important information. Frequently, general purpose image processing, photogrammetry, or GIS software (which often includes elements of the first two) contains all the necessary tools; special-purpose software is sometimes also required (Kvamme 2005). The image analyses are mostly either visual or based on simple image enhancements. Most of the basic tools for archaeological prospection are based on inspection of aerial and satellite data and reconnaissance in the field. The potential of advanced processing methods have been realized only recently. Different authors propose the use of high-resolution optical and radar images, and pattern recognition techniques, to support archaeological prospection (e.g. Lira et al. 2005).

3 Data selection and processing

Our case study area is located in the central part of the Yucatan peninsula in Mexico (Fig. 1), and largely corresponds to the Calakmul Biosphere Reserve. The area has been practically depopulated during the last 30 years, and economic activities and exploitations have ceased. As a direct consequence, a tropical forest has been recovering in the whole protected area; only a few trails and roads remain. Visibility within dense vegetation is sometimes restricted to just a few meters, which makes terrestrial observations in most of the area practically impossible.

The archaeological database used in the study includes 63 sites with appropriate attributes – coordinates, type, and rank. In order to eliminate the influence of unknown lower rank sites and to use only the most important and characteristic sites only larger sites, i.e. major and medium centres (33 in total), were considered in most analyses. An additional limiting factor was the spectral representativeness of the locations. Sites that are close to settlements and roads or that have been cleared or cultivated (secondary vegetation) were not used in the analysis of satellite data, leaving 20 sites in the study set. Another problem is the chronology of the sites; it is known that Maya sites in the area are from different periods, but dating is too unclear to be considered in our analyses.

Most of the digital spatial data for our research was obtained from Mexican institutions CONABIO (Comisión Nacional para el Conocimiento y Uso de la Biodiversidad; G. García 2000) and INEGI (Instituto Nacional de Estadística, Geografía e Informática). Additionally, remote sensing data, i.e. aerial photographs and medium resolution satellite images, have been used. Although high-resolution satellite images (Ikonos, QuickBird) could provide better results in spatial sense, they were not used, primarily due to their prohibitively high cost, particularly in view of the scope of the study and the extent of the area surveyed.

A DEM has also been produced from two different hypsometrical data sources at the scale of 1:50,000 with 10 m height contour lines intervals. The accuracy of the produced DEM, with a cell size of 20 m, is a few meters, depending on the roughness of the actual terrain (Podobnikar and Oštir in press). A higher precision DEM, with a 6 m resolution, was produced for some areas with ortho-

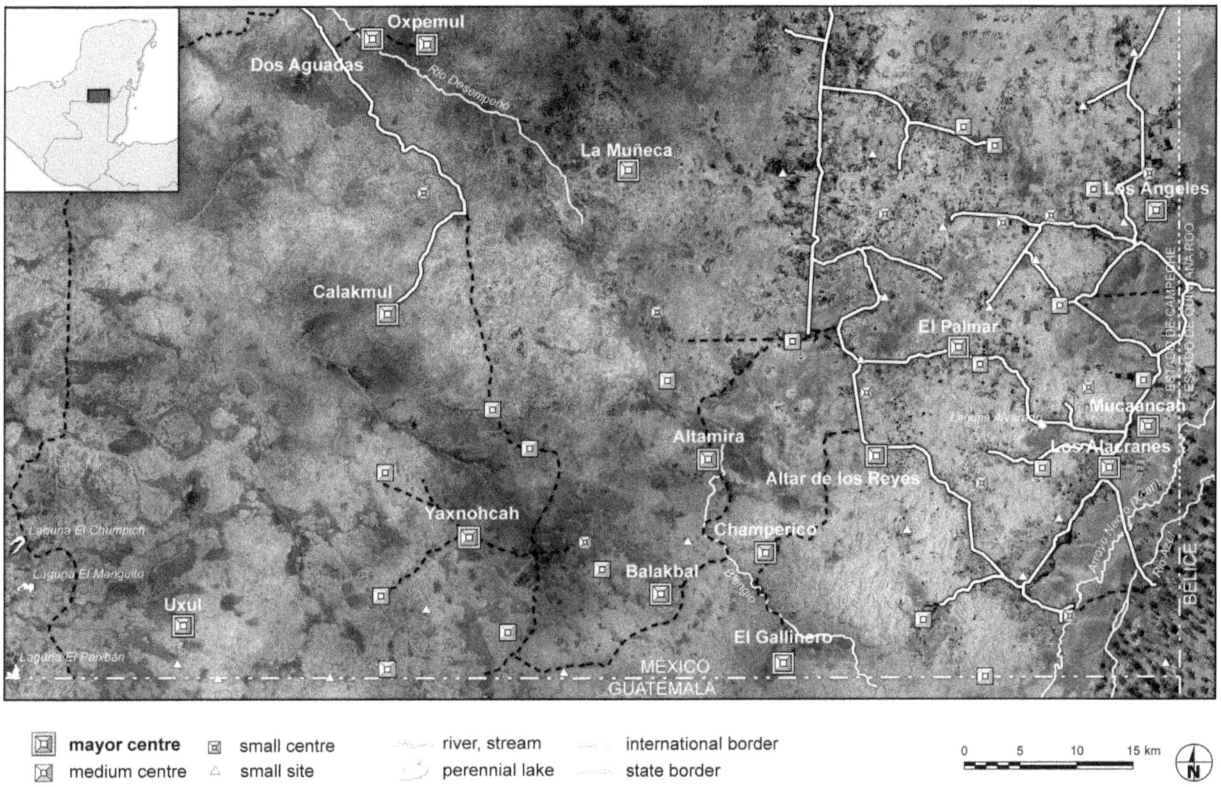

Figure 1 *Location of the study area. Archaeological sites are displayed over NDVI obtained from Landsat satellite image.*

photo processing (CONABIO 1995-96), but it represents the vegetation rather than the ground surface.

To detect archaeological sites in the central Yucatan Peninsula, both aerial photography and satellite remote sensing data have been used. Aerial photographs were taken in 1995 and 1996 to support the mapping of the Calakmul Biosphere Reserve. Within this project approximately 400 photographs have already been observed visually using a stereoscope, scanned and photogrametrically processed to provide high-resolution imagery and corresponding elevation models of the area. Another set of images will follow in the near future.

Besides aerial data, satellite images from different optical (SPOT and Landsat) and radar (ERS) sensors have also been used. First, all the images were georeferenced to the common projection (UTM on WGS84). Since the data used is of medium resolution, ranging from 10 m for the SPOT panchromatic to 30 m for Landsat, simple polynomial transformation has been applied. Nevertheless, for several reasons it was not easy to obtain accurate control points in the area. On one hand, the vegetation is very dense and there are few easily recognizable (human-made) structures present on both images and maps. It is also very difficult to relate features present on images with objects shown on maps, especially if maps are out-of-date. Additionally, the INEGI maps available at scale 1:50,000 are hardly appropriate for referencing purposes. Since the sensors used do not have complex geometrical distortions, first order polynomial transformation, obtained from between 50 and 100 control points per scene, resulted in positional accuracy in the range of one or two pixels, which is approximately 50 m.

Two radar images from the ERS (European Remote Sensing Satellite) have been interferometrically processed. The first was collected by the ERS 1 satellite on 1996-02-11 and the second by ERS 2 on 1996-04-22. The time difference was selected to observe differences in humidity and to relate them to anomalies. However, in the period between image acquisitions (more than two months) the vegetation changed significantly, reducing the correlation between images drastically and thus preventing the DEM production. The images were combined with coherence and several interesting features were observed visually, mostly related to contemporary human activities. Bajo areas were delimited, and the amount of water in aguadas can be observed in images acquired on both the February and April dates.

Although the application of radar in archaeology is well known (Lira et al. 2005; Sever and Irwin 2003; El Baz 1997; Adams et al. 1981), we have not been able to find a clear relation between radar response and site locations. This is partly due to environmental conditions that are extremely unfriendly to radar observation. Other authors have either observed areas with poorer vegetation (El Baz 1997) or used multifrequency or multipolarization radar (Sever and Irwin 2003). Some archaeological sites have been observed on a combination of ERS images, but the amount of "noise" was too high to perform a (semi)automatic detection. More promising results have been recently obtained with advanced filtering of fine beam Radarsat data (Lira et al. 2005).

Better results were obtained from SPOT and Landsat optical satellite data. Two panchromatic SPOT images, acquired on 2002-11-23 and 2001-03-31 covering different parts of the study area have been tested. The 10 m resolution of the first generations of SPOT systems is too coarse to detect structures directly, but enables observation of "unusual" features that can be related to archaeological sites. Simple and advanced edge detection was used, including filtering in the direction of the sun. Again, some of the sites were clearly visible on original or processed images, but the amount of non-relevant information was relatively high. Nonetheless, SPOT images have been successfully used as one of the mapping sources during the fieldwork. SPOT data has been also very extensively used for detecting environmental features whose locations and characteristics were relevant for management of fieldwork, e.g. communications, water and vegetation.

Despite the fact that Landsat has a lower spatial resolution than SPOT, its spectral characteristics are far superior. Seven spectral bands, ranging from visible, near and mid-infrared, to thermal infrared, with a resolution of 30 m, present a good base for environmental studies and detection of anomalies possibly related to archaeological features. A Landsat 7 Enhanced Thematic Mapper Plus (ETM+) image, acquired in on 2000-03-27, has been used. ETM+, in contrast to the previous versions of the sensor, includes a panchromatic band with the resolution of 15 m, enabling the sharpening of bands and thus a better delineation of features.

Before performing further site analyses, responses in the individual bands and different band ratios were studied. Bands 3, 4 and 1 were used to discriminate water and vegetation. Bands 3, 4, 5 and 7 are very sensitive to vegetation and are therefore most useful for detecting human impact on vegetation (Sabins 1997; Rothaus and De Morett 2001). Band rationing is a common technique used in digital image processing to enhance a contrast between selected and superfluous features; it is normally employed to identify vegetation concentrations. Band ratios have the effect of removing shadows and illumination variations caused by differences in ground slope, i.e. topographic impression (Sabins 1997; Richard and Jia 2006). In this manner, spectral signatures can be compared directly, regardless of different illumination conditions.

Vegetation has a very strong reflectance in the near infrared and a low reflectance in the red portion of the electromagnetic spectrum. The ratio of infrared to red response is therefore very sensitive to vegetation conditions (basically the presence of chlorophyll in leaves). A normalized differential vegetation index – NDVI, defined as (IR-R)/(IR+R) – has been shown to be correlated with green leaf biomass and the green leaf area index; it increases with an increase in either parameter. Other ratios related to vegetation or geology can also be connected with archaeological sites. Rothaus and De Morett (2001), for example, have used a combination of bands 4, 5 and 7 to compute the ratios 5/4, 7/5 and 7/4 and combined them into colour composite image (5/4 showing in red, 7/5 in green, and 7/4 in blue).

4 Environmental characteristics of the archaeological sites

Detection of archaeological sites is possible by observing contrasts between archaeological features and the natural background. When archaeological features possess physical properties different from those of the surrounding matrix, the distinction may be noticed (Kvamme 2005). Environmental characteristics of the sites have therefore been studied to find site indicators from Landsat ETM+ image data and the digital elevation model.

A buffer zone for each site has been produced. Different radii (50, 250, and 1000 m) were tested for the properties of the immediate vicinity and a larger influence area. Since the results followed a similar trend for all buffers, most of the analyses have been performed with a radius of 250 m. Indicators have been compared with the background, i.e. 60 points distributed randomly in the area. These random points could also contain archaeological sites, since the area has been surveyed only partly and not all the sites have been found. The proximity of bajo areas was also analysed to establish their spatial relation with the sites and to verify the influence of bajo vegetation on spectral response.

The digital values (reflectance) in individual bands of the Landsat ETM+ image, acquired on 2000-03-27, show differences between random points and archaeological sites, especially those of a higher rank (major and medium centres, Fig. 2a). However, the deviations are very small and only noticeable in bands 4, 5, and 7. Small differences can be enhanced by computing band ratios. Fig. 2b compares normalized differential vegetation index and band ratios 5/4, 7/5, and 7/4 of higher rank sites and random points. The archaeological sites have a slightly smaller vegetation index, meaning that the vegetation is less intensive or disturbed in growth (to be more specific – the amount of chlorophyll present in the leaves is smaller). A ratio of bands 4 and 5 often indicates density of canopy coverage. In vegetation sciences, the band ratio is normally 4/5, which renders denser canopy brighter. As the object of interest in this study is human activity, the ratio has been reversed (5/4) rendering sparser canopy brighter. The ratios 7/5 and 7/4, both of which indicate general vegetation vigour, are also useful (Rothaus and De Morett 2001). One can clearly see a substantial divergence of sites from random points in ratios 7/5 and 7/4 (the index for sites being almost 50% larger than for random points). The spectral characteristics of sites can be an indicator used in predictions. We anticipate that the use of multitemporal satellite data, covering different seasons and humidity conditions, could reveal changes in vegetation in more detail.

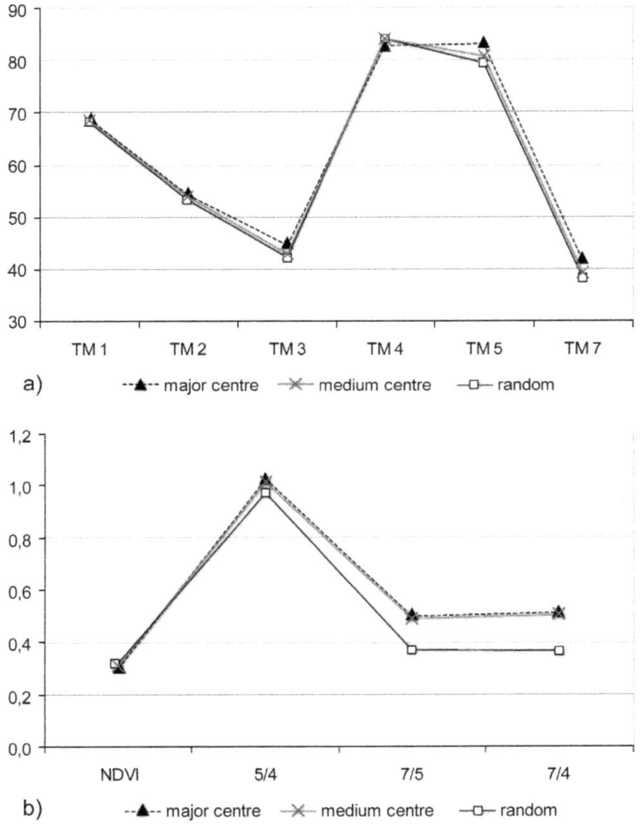

Figure 2 *Spectral signatures (a) and especially indices (b) show that larger sites differ from random points.*

A significant number of detected Maya sites are found close to bajo areas. Since the vegetation in bajos is characteristic (typically less intensive), these areas could influence the lower vegetation index found within the archaeological sites. The bajo areas have been therefore delineated by reclassifying the vegetation data provided by CONABIO. Fig. 3 shows higher response for bajos in all bands except band 4; this means sparser vegetation, enabling the reflectance of bare soil. Bajos also exhibit characteristic terrain – they are almost flat and well distinguished from the surroundings. Since none of the archaeological sites in our database lies within a bajo, we evaluated the number and influence of "bajo" random points on the statistical relevance of performed analyses. As only one random point was positioned in a bajo (meaning less than 2% of all random points) we can say that the results are not biased regarding satellite or terrain data.

In contrast, it is interesting to analyse the distance of sites from bajos and the extent of bajos in the surrounding area (a radius of 2000 m has been used, since the average distance between bajos and major sites is slightly more than 1000 m). For major centres, 17.8% of the surrounding area is covered by bajos. For medium, the figure is 13.4%, small, 12.2%, while for random points, it is 10.7%. In particular, larger sites are evidently closer to bajos than random points. An average distance for significant sites to a bajo is 880 m. This is only 2/5 of the value that one would expect for random points, which have an average distance of 2070 m (Podobnikar and Oštir in press). The analysis should be further improved by a better definition of a bajo (exclusion of smaller ones, clearer delineation), by use of elevation data and additional selection of (representative) sites.

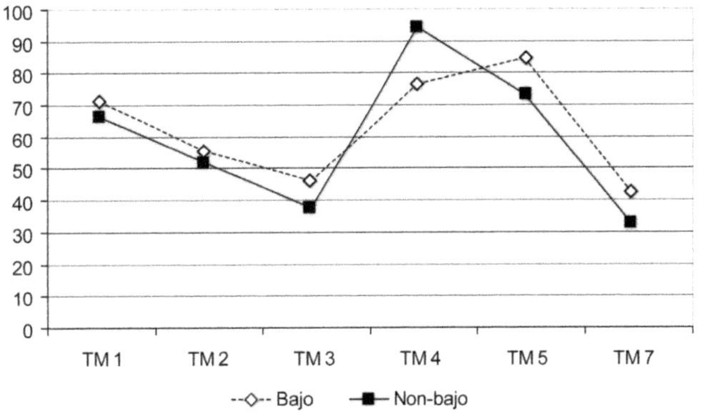

Figure 3 *Comparison of spectral properties for bajo and non-bajo areas.*

5 Conclusions

Geographical information systems have become one of the standard tools in archaeological research. In this study, we have used basic and advanced processing methods to determine some spatial properties of Maya sites in the central Yucatan Peninsula. This was the first step towards a model that could predict possible locations of unknown sites and thus simplify field surveys. The area is covered with dense vegetation and has been only partially prospected. The research performed so far suggests that up-to-date technology, especially GIS analyses and remote sensing image interpretation, supported with extensive knowledge of the natural environment and Maya culture, can contribute to a better understanding of their activities and give quantitative answers to settlement hypotheses.

To apply the GIS methods in archaeology – or any other discipline – good spatial datasets are required. In our project, an archaeological database of south-eastern Campeche area that includes 63 sites with appropriate attributes has been used. To eliminate "noise" in processing, only larger sites were considered for most of the analyses, i.e. major and medium centres (33 in total). Many additional spatial data obtained from different sources has been utilised, most importantly, a digital elevation model and remote sensing images. A DEM has been produced from hypsometrical data at scale 1:50,000 and has a final cell size of 20 m and vertical accuracy of several meters. The elevation data has been supplemented with satellite images from ERS, SPOT and, most significantly, Landsat, together with derived data, such as several indexes.

The study has been performed with several (partially independent) objectives in mind. First, we have tried to observe characteristics of archaeological sites with respect to relief and other peculiarities of the natural environment detected from satellite images. Next, the distances of the sites from bajos have been computed and compared to random points.

Detection of archaeological sites is possible by observing contrasts between areas with archaeological features and natural background. We have observed that environmental characteristics of major and medium centres differ from those of random points. The sites frequently occur in prominent areas, being characterized with light slopes, higher curvature and roughness of the terrain (high local differences in heights, meaning more agitated relief) and higher relative heights. It has also been statistically proven that spectral properties of the site areas, mostly indexes obtained from Landsat satellite images and related to vegetation, are different from their surroundings. Although the differences are generally small and significant only for certain parameters, one can say that the areas with a high concentration of archaeological site locations show disturbed or less intensive vegetation.

The results of the study are preliminary since the research is still ongoing. While some answers have been found through our work, many questions remain, and therefore the research will continue in several directions. We will use additional and better data sources, mostly acquired from processed aerial photographs and satellite images, i.e. high-resolution multispectral (or hyperspectral) satellite images, multitemporal Landsat data to cover seasonal changes, multipolarization or multifrequency radar images, and lidar for better elevation data of smaller areas.

References

ADAMS, R.E.W. 1981 Settlement Patterns of the Central Yucatan and Southern Campeche Regions, in *Lowland Maya Settlement Patterns* (W. Ashmore, ed.). Albuquerque, University of New Mexico Press: 211-257.

ADAMS, R.E.W., W.E. BROWN, and T.P. CULBERT 1981 Radar Mapping, Archeology, and Ancient Maya Land Use. *Science*, 213 (4515): 1457-1463.

EL-BAZ, Farouk 1997 Space Age Archaeology. *Scientific American*, 277 (2): 40-45.

FOLAN, W.J., J. MARCUS, S. PINCEMIN, M. DEL ROSARIO DOMÍNGUEZ CARRASCO, L. FLETCHER, and A.M. LÓPEZ 1995 Calakmul: New Data from an Ancient Maya Capital in Campeche, Mexico. *Latin American Antiquity*, 6 (4): 310-334.

HOLCOMB, D.W. 1998 Applications of Imaging Radar to Archaeological Research in *Manual of Remote Sensing, Volume 2, Principles and Applications of Imaging Radar* (F.M. Henderson and A.J. Lewis, eds.). New York, John Wiley and Sons: 769-776.

LIRA, J., P. LÓPEZ, and A. RODRIGUEZ 2005 Detection of Maya's archaeological sites using high resolution radar images. *International Journal of Remote Sensing*, 26 (6): 1245-1260.

KVAMME, Kenneth L. 2005 Terrestrial Remote Sensing in Archaeology in *Handbook Of Archaeological Methods* (H. Maschner and C. Chippindale, eds.). Lanham, AltaMira Press: 423-477.

PODOBNIKAR, T. and K. OŠTIR in press GIS and Remote Sensing Analysis, in *Reconocimiento arqueologico en el sureste de Campeche, Mexico: 1996-2005* (I. Šprajc, ed.), BAR International Series (Paris Monographs in American Archaeology). Oxford, BAR Publishing.

RICHARD, J.A., and X. JIA 2006 *Remote Sensing Digital Image Analysis: An Introduction*, Berlin, Springer.

ROTHAUS, R.M., and A.A. DE MORETT 2001 Landsat TM imagery in landscape archaeology: Detection and modelling in *Remote Sensing in Archaeology* (S. Campana and M. Forte, eds.). Firenze, All'Insegna del Giglio: 149-173.

RUPPERT, K., and J.H. DENISON 1943 *Archaeological Reconnaissance in Campeche, Quintana Roo, and Peten*. Washington, Carnegie Institution of Washington.

SABINS, F.F. 1997 *Remote Sensing – Principles and Interpretation*. New York, W.H. Freeman.

SATURNO, W.A., T.L. SEVER, D.E. IRWIN, and B.F. HOWELL 2006 Regional Scale Landscape Archaeology: 21st Century Remote Sensing Technology and the Ancient Maya in *The Manual of Remote Sensing, 3rd Edition (A Series): Volume 5: Remote Sensing of Human Settlements* (A.B. Rencz, M.K. Ridd and J.D. Hipple, eds.). Bethesda, American Society for Photogrammetry and Remote Sensing: 489-502.

SEVER, T.L., and D.E. IRWIN 2003 Landscape archaeology : Remote-sensing investigation of the ancient Maya in the Peten rainforest of northern Guatemala. *Ancient Mesoamerica*, 14: 113–122.

ŠPRAJC, I. 2002-2004 Maya sites and monuments in SE Campeche, Mexico. *Journal of Field Archaeology*, 29 (3-4): 385-407.

ŠPRAJC, I. (ed.) in press *Reconocimiento arqueologico en el sureste de Campeche, Mexico: 1996-2005*, BAR International Series (Paris Monographs in American Archaeology). Oxford, BAR Publishing.

Application of hyperspectral remote sensing to the Celtiberian city of Segeda

J. G. Rejas Ayuga,[1] F.Burillo Mozota,[2] R. López,[2] M. Farjas Abadía[3]

[1]National Institute of Aereospatial Technology (INTA), Spain
[2]Faculty of Humanities and Social Sciences of Teruel, University of Zaragoza (UZ), Spain
[3]E.T.S.I. in Topography, Geodesy and Cartography, Polytechnic University of Madrid (UPM), Spain
rejasaj@inte.es

1 Introduction

Non-invasive archaeological surveying, such as airborne remote sensing, is a research field with great potential, as evidenced by scientific publications (Scollar et al. 1990). The development of new image sensing techniques in the last decades, such as hyperspectral sensors, increases the possibilities of applying this technology to the detection and study of archaeological sites (Belvedere et al. 2001; Emmolo et al.; Farjas et al. 2003).

In this sense, it is of interest to analyse the relation between parameters such as temperature and reflectance (which can be obtained through the images) and the spatial distribution of particular mappable features. These relations are often displayed as micro- or macro-textures on the surface, which can be digitally detected and rendered (Mena, Malpica 2005).

With regard to the thermal band, the strategies for the classification and identification of patterns, in comparison with other zones of the spectrum, can vary significantly. Buried features that respond to a different temperature in relation to their surrounding could present an equal behaviour in the rest of the spectral bands when on surface, hence the importance of the thermal data as additional information about the separability of classes in a multivariate analysis.

Although remote sensing has been used since its origins in various fields of research, its application to particular fields requiring better spatial resolution and a larger number of narrower bands has not been possible until the implementation of hyperspectral sensors.

Techniques such as detection of thermal anomalies and classification based on spectral angle have proven to be very useful in distinguishing classes in the visible and infrared short wave (VIS-SWIR) as well as the thermal infrared (TIR) zone. They have been applied especially to geothermal exploration and survey (Belvedere et al. 2001; Goovaerts et al. 2004).

On the basis of this, the goals that we have established are the use, potentiation and application of data derived from airborne remote sensing (including active as well as passive sensors) for the development of methodologies capable of allowing the study of the Celtiberian city of Segeda.

This paper describes the process of hyperspectral data acquisition. We present the first results obtained from the processing of images and the analysis of the information. In addition, we outline the direction to be followed in the subsequent uses of a SAR (Synthetic Aperture Radar) sensor in a multi-source study.

2 The archaeological site of Segeda

The area of study consists of the archaeological site constituted by the Celtiberian city of Segeda (Zaragoza), located in the central area of the Iberian Peninsula, towards the North-East. The city is found on the Central Iberian Range, by the Jalón River (a stream of the Ebro River on its right margin). In its first phase, the indigenous city was built on the Hill of Mara (Zaragoza). Segeda was one of the first settlements to mint coins in Celtiberia, soon after 179 BCE (http://www.segeda.net; Burillo Mozota 2006).

Appian of Alexandria (*Iber*, 44-47) described Segeda as a 'great and powerful city of the Celtiberians known as Belli' and describes how Rome declared war on them for breaking the pacts that Gracchus had signed with the indigenous communities of the Middle Ebro Valley in 179 BCE. Livy (*per.* 47) points out that this war was the cause for the rescheduling of the election of the consuls from the Idus of March to the first of January. Moreover, Appian, Diodorus (31, 39-41) and Florus (1, 34, 3) mention that Nobilior, heading a Roman troop of 30,000 men, attacked a coalition of 25,000 Celtiberians in 153 BCE. This coalition had been put together by the two most important Celtiberian cities of the Iberian Range--Segeda and Numantia. The first encounter took place on the 23rd of August. The defeat suffered by the Roman Army caused this day, the day of Vulcan, to be considered a *dies nefastus*.

After these events, the city of Segeda was never more mentioned in the classical sources, although the coalition of Belli and Titti, headed by Segeda, was indeed mentioned in subsequent years. There is evidence that the city-state of Segeda survived, for it kept minting coins with legends containing the name of the city. In fact, it has been discovered that the citizens of Segeda erected a settlement *ex novo* in close proximity to the ruins of the previous one, on the archaeological site identified as Durón de Belmonte de Gracián. This second city was destroyed and abandoned on the first century CE, probably during the Sertorian Wars. At this point, the new political centre was transferred to Bilbilis Italica, on the Hill of Bámbola (Calatayud), at approximately 10 kilometres from the two cities of Segeda.

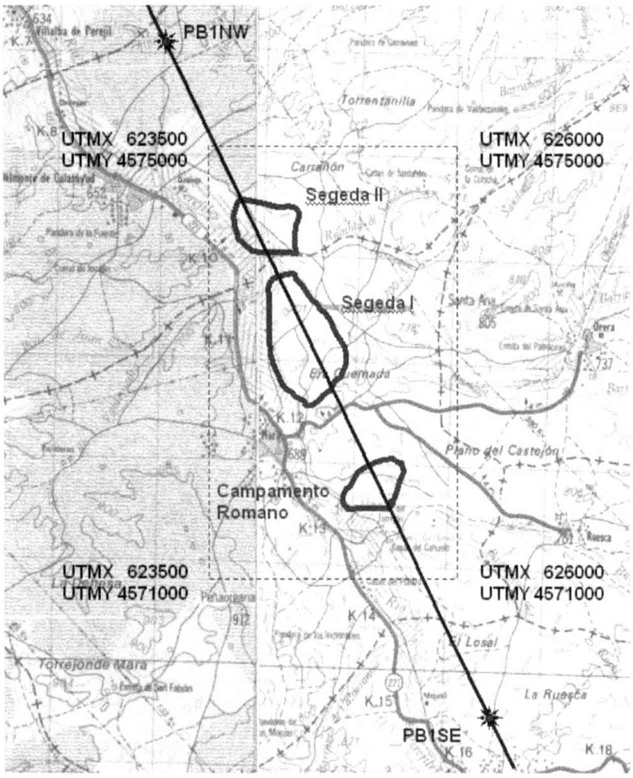

Figure 1 *Location of the Celtiberian (Segeda I) and Roman (Segeda II) cities. Flight parameters for AHS in 2005 and SAR in 2006.*

3 Data acquisition

The images of Segeda were obtained in July 2005 using a hyperspectral airborne sensor known as AHS (*Airborne Hyperspectral Scanner*) (http://www.argonst.com/). It consists of a mechanical broom sweeper known as *whiskbroom*. The AHS registers data in 12 bits in the atmospheric windows from the visible to the thermal infrared and divides the incident radiation into 5 individual spectrometers or 'ports'. It makes use of two black bodies as a source of reference, which can be controlled in temperatures ranging from –15° C to +25° C relative to the temperature of the sensor head.

The FOV (*Field of View*) and IFOV *(Instantaneous Field of View)* of the AHS are 89° and 2.5 mrad respectively. These parameters, added to the sweep speed (which oscillates between 6.25 and 35 Hz), specify its internal orientation as well as the spatial coverage and resolution at which data can be recorded. Thus, the AHS allows the obtainment of resolutions that range from 2.4 to 13.7 m for standard flight altitudes.

An inertial system POS/AV 410 from the company Applanix (http://www.applanix.com/) has been used as an auxiliary system. It consists of a GPS double-frequency receiver and an inertial IMU (*Inertial Measurement Unit*) platform. The integration of a system of these characteristics with remote sensing sensors allows the recording of geographical positioning data and angles of attitude of the ensor during the obtainment of high-precision images.

Parameters	Hyperspectral	SAR
Flight altitude	2100 m	1500 m
Terrain altitude	700 m	700 m
Spatial coverage	2000 m	2000 m
Spatial resolution	2.4 m	5 m
Azimut	320°	320 °

Alongside the airborne project, a field radiometry project has been carried out using the spectroradiometers GER 1500 and ASD, in search for the spectral characterisation of surfaces within the archaeological site.

Laboratory work has complemented the tasks carried out on the field. The ASD spectroradiometer has been applied to samples of minerals and material retrieved from the area of study. The principal objective of the radiometric readings is the atmospheric correction of the images and the obtainment of spectral signatures that will be subsequently used as a search pattern.

4 Data preprocessing

Before analysing the images and the associated auxiliary data it was necessary to carry out a set of tasks geared towards the correction of two types of distortion (radiometric and geometric), which take place during the obtainment of the data. Moreover, it was necessary to modify the parameters of the image to an absolute scale (in spatial and spectral terms), so as to make it compatible with results from other sources, sensors and dates.

The first step consisted of estimating the noise present in the hyperspectral image. A MNF (*Minimum Noise Fraction*) algorithm was applied. This allows the reduction of noise in the image through a transformation of its axes. The results suggested that we should discard channels 55 to 65 and channel 70, as they display a high signal/noise ratio.

With regard to the reflective bands (up to 2.5 μm), the original digital levels have been calibrated to radiances through calibration quotients obtained from the calibration 'integrator sphere' USS 400 (*Labsphere*) (http://www.labsphere.com/) covered with Spectraflect (*Labsphere*). A lineal interpolation has been applied to the thermal bands, using the recorded digital level and the temperature assigned to the reference black bodies (10 and 50° C), located before and after each scanned line.

In order to convert the radiance values in the sensor to reflectivity values on the terrain (a parameter that is indicative of the state of the materials and that can be compared with other spectra originating from libraries), we have adopted an empirical method, using the radiometric readings originating from the field as well as those obtained at the laboratory. In order to do this, the data obtained with GER 1500 and ASD were processed in Matlab, transforming the measures of radiance to values of reflectivity on the basis of radiance measures from the reference target (Spectralon). This generated the spectral signatures for each of the measured surfaces (Fig. 2).

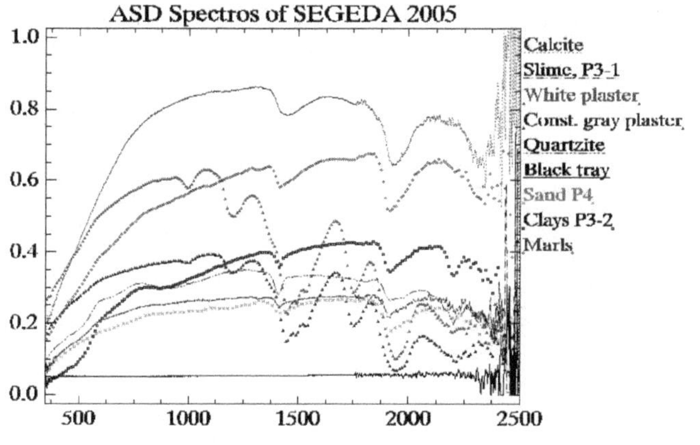

Figure 2 *Spectral signatures of Segeda.*

The algorithm '*Empirical Line*' (Smith, Milton 1999) was applied to port 1 of AHS through the spectral signatures of the 'bright' and 'dark' surfaces obtained from the area of study. This method allows us to achieve two goals in one step. On one hand we switch from radiance on the sensor to reflectance on the terrain. On the other hand, we correct the radiometric distortion produced by the atmosphere, which absorbs and disperses part of the radiation originating from the surface.

Another key set of tasks that have been approached involve the geometric correction and georeferencing of the hyperspectral images. A parametric georeferencing method was applied, known as direct georeferencing. This allows us to correct geometric distortions produced by movements of the platform taking place at the moment of obtaining the images and georeference each pixel of the image in a single automatic process, using positioning and orientation data measured through an inertial GPS/IMU system.

The software package POSPAC (http://www.applanix.com/) was used to calculate and process the data recorded by the inertial system. The trajectory has been calculated by solving a differential cinematic positioning in bi-frequence L1/L2 mode. The reference GPS stations that have been used are those of Yebes and La Rioja (ftp://ftp.geodesia.ign.es), of the National Geographical Institute (IGN), located at 200 and 215 kms respectively, and the station set up at the EUITTO of the Polytechnic University of Madrid, located at 295 kms from the area of study.

Once the flight trajectory was calculated, the images were geometrically corrected using a direct geocoding software called PARGE (http://www.rese.ch/products.html). This algorithm requires as input data the external orientation parameters (position and orientation) calculated from the inertial system as well as a digital elevation model (DEM). The DEM used in the georeferencing process was rasterised with the altimetry of the sheets of the IGN at a 25-metre resolution, resampled to 2.5 metres.

The hyperspectral images with a spatial resolution of 2.4 metres were merged with aerial orthophotos at a 0.5 m. resolution. This methodology produces images with a better spatial resolution than the originals whilst preserving the richness of the spectral information recorded by the instrument. In order to do this we used the algorithm *Imgefus*, implemented into the image handling software PCI. The resulting images were used in subsequent processes of classification and texture analysis.

5 Preliminary analyses of the hyperspectral images

The aim of the corrections carried out during the preprocessing stage is the obtainment of images with a better spatial resolution, perfectly georeferenced and converted into percentages of reflectivity and surface temperature for the VIS-SWIR channels, respectively. These transformations will allow us to detect thermal anomalies and apply a classificatory scheme based on the spectral angle. We will also be able to scale and identify the precise coordinates of the detected elements.

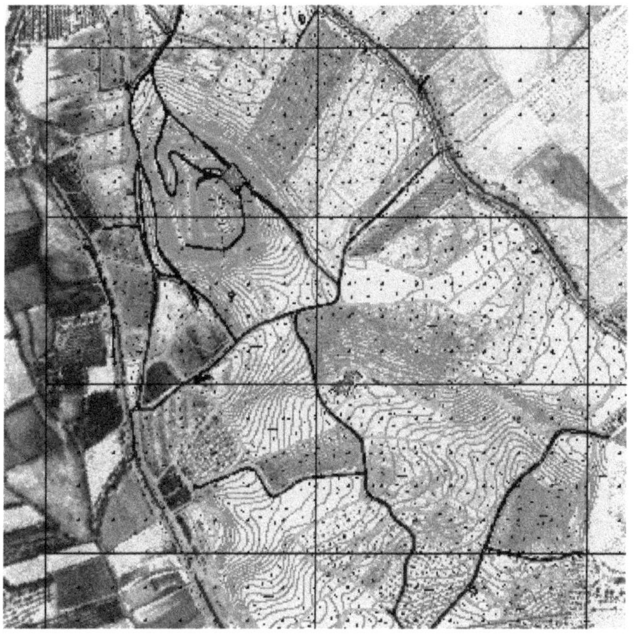

Figure 3 *AHS 78 channel with cartography.*

Two distinct techniques have been used. In the first stage, only the thermal bands of the AHS were used, discarding channels 60 to 65 and 70 of this port, as they display high levels of noise, as mentioned above. With this first set of tasks, we aim to identify pixels that respond differently in comparison to their surrounding. That is to say, we aim to calculate thermal anomalies.

In order to do this, we first applied a convolution to the entire image using a median filter. Next we produced a quotient of the image with the aim of thresholding the separability of spectral classes that display the channels 75 and 80 of the AHS (Rejas et al. 2005). The difference of the median filter corresponding to channels 80 and 76 has been divided into the sum of both channels. This function was corrected using the relation of the gain parameters of each of the channels.

The result of the transformation function is an image in which the detected pixels are highlighted as possible thermal anomalies in relation to their surrounding. It was observed that one of these anomalies displays a concentric circular shape with a maximum diameter of 150 m. It could belong to a buried stone structure or a wall.

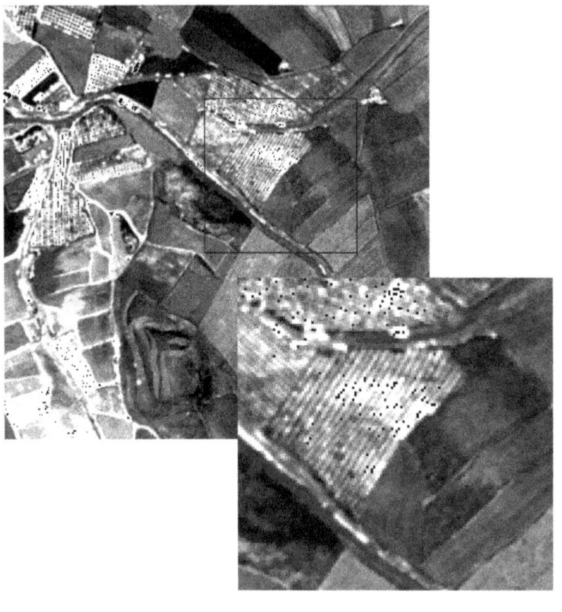

Figure 4 *Thermal anomaly in channel AHS 76 and detail of a zone close to the Hill.*

Next, as a result of the expectations created by the methodology and results obtained in (Farjas et al. 2003), we have applied a second technique of digital image classification based on spectral angle. In this stage of the analysis all AHS bands were used. The images of co-occurrence were calculated and several filters of high step were applied in an initial analysis of textures in order to achieve an approximation to the identification of borders on the surface that may appear on the images.

We then applied the *Spectral Angle Mapper* (SAM) algorithm of the image handling software package ENVI. This algorithm treats the spectrum as a vector in space with an equal dimensionality to the number of bands and determines the spectral similarity between both spectra—one adopted as a reference belonging to a endmember and the other one belonging to the pixel to be classified, in relation to the n-dimensional angle between them.

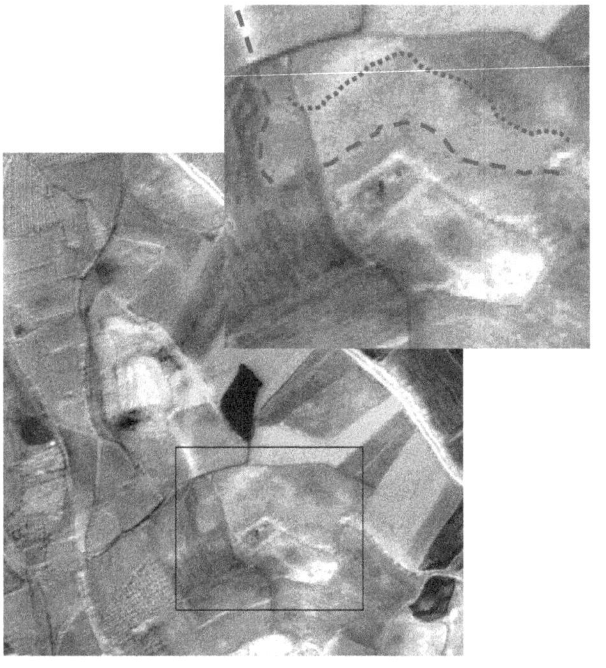

Figure 5 Alignments that could possibly indicate the remains of a wall.

In order to select the pure pixels present in the training areas used during the classification, we compared directly the spectra of the laboratory with those measured in situ using the apparent reflectance data of the sensor. The image pixel is assigned to the class with the lowest spectral angle, as the correlation with the reference spectrum will be higher the narrower the angle is.

The accuracy of the classification thus obtained has been tested using the error matrix method. We obtained an α parameter of 0.78. As a real value of the classes we have used the spectra measured over 15 check points. The margin of accuracy of the classification is over 79%. The percentage of non-classified pixels is 2% (a parameter that is highly dependable on the classes and the training applied).

SAM displays satisfactory results in the classification of this area. It distinguishes satisfactorily different groups of land covers, as is to be expected according to previous studies (Bor-Chen Kuo 2001) However our main interest consists in identifying linear limits that could be indicative of the presence of a Celtiberian wall. This result coincides with the borders that were detected when a Sobel filter was applied to the thermal channels.

6 Conclusions

This paper represents the first stages of the research project inquiring into the properties of hyperspectral data for the identification and classification of archaeological remains in the Celtiberian city of Segeda. The results demonstrate the great potential of the AHS sensor as an additional source of information for archaeological surveys. It has been confirmed that hyperspectral images are capable of distinguishing a large number of different materials (natural and manmade) on surface. Moreover, we have studied the relation between anomalies calculated through the thermal channels of the AHS and the identification of possible buried features. This improves archaeology's capabilities of locating and investigating features.

We believe that it is necessary to take the investigation further by integrating SAR data, which would allow us to contrast and increase the data obtained. The radiometric calibrations and the georeferencing processes that have been carried out are considered of critical importance to the validation of future results in a prolonged, multi-source study.

Acknowledgements

We are grateful to the Division of Remote Sensing of the Department of Earth Observation of INTA for the resources and data provided during the preparation of this work.

We are very grateful to Mr. Antonio Ruiz Verdú (CEDEX) for his ASD spectral signatures measurements.

This work is part of the I+D project: HUM2005-03369/HIST, financed by the Ministry of Education and Science and the FEDER funds.

References

BELVEDERE, O., BURGIO, A., CIRAOLO, G., LA LOGGIA G., MALTESE 2001. *Hyperespectral MIVIS data analyses for archaeological applications.* Fifth International Airborne Remote Sensing conference and Exhibition, San Francisco, 17-20 Sept. 2001.

BOR-CHEN KUO 2001. *Improved statistics estimation and feature extraction for hyperspectral data classification.* PhD Thesis and School of Electrical & Computer Engineering Technical Report TR-ECE 01-6, Purdue University December 2001 (88 pages).

EMMOLO D., FRANCO V., LO BRUTTO M., ORLANDO P. and VILLA B. *Hyperspectral techniques and gis for archaeological Investigation.* ISPRS Commission WG VII/4.

FARJAS M., REJAS J.G., GÓMEZ J. A., DE MIGUEL E., FERNÁNDEZ-RENAU A. 2003. A*irborne multispectral remote sensing application in archaeological areas.* Proceedings *The E-way into the four Dimensions of Cultural Heritage,* Vienna, Austria, April 2003.

MENA J.B. AND MALPICA J.A. 2005. *Color image segmentation based on three levels of texture statistical evaluation.* Applied Mathematics and Computation, 161 (2005) 1-17.

REJAS J.G., PRADO E., JIMÉNEZ M., FERNÁNDEZ-RENAU A., GÓMEZ J.A. AND DE MIGUEL E. 2005. Caracterización del sensor hiperespectral AHS para la georreferenciación directa de imágenes a partir de un sistema Inercial GPS/IMU. International congress 6th Geomatic week, Barcelona Febr. 2005.

SCOLLAR, I., TABBAGH A., HESSE, A., HERZOG I. 1990. Archaeological prospecting and remote sensing. Cambridge University Press, 1990.

GOOVAERTS P., JACQUEZ G., WARNER A.. AND CRABTREE B. 2004. *Detection of local anomalies in high resolution hyperspectral imagery using geostatistical filtering and local spatial statistics.* IEEE Transactions on Geoscience and Remote Sensing, 0-7803-8350—8/04/20.00 (c).

SMITH ,G.M. AND MILTON, E,J 1999, *The use of the empirical line method to calibrate remotely sensed data to reflectance.* International Journal of Remote Sensing, 20: 26553 – 2662.

BURILLO MOZOTA, F. 2006, "Segeda and Rome. The historical development of a Celtiberian city-state", in L. Abad; S. Ramallo & S. Keay (eds.), *Early Roman Town in Hispania Tarraconensis. Journal of Roman Archaeology. Supplementary Series Number 62*, Portsmouth Rhode Island, pp. 159-170.

http://www.segeda.net, website of the I+D research project of Segeda.

ftp://ftp.geodesia.ign.es, Data server of permanent GPS stations of the National Geographical Institute (IGN).

http://www.argonst.com/

http://www.labsphere.com/

http://www.rese.ch/products.html

http://www.applanix.com/

Using historical aerial photographs: the case of Pontecagnano and its territory (Salerno, Italy)

A. Rossi, A. Santoriello

Università degli Studi di Salerno - Dipartimento di Beni Culturali

Within this region historical sources place a geographical locality known as the *Ager Picentinus*, and the history of the area is centred on *Picentia.*[1]

There is little documentary evidence for the 'colony' of *Picentia* that has been contextualised in an historical process of extensive insight in which, the colonial phenomenon, dated only on a deductive basis to 268 BC, is no more than the result of an urban strategy that already occurred in the first half of the fourth century BC.[2]

The sources related that *Picentia* was destroyed after the Social War by Sulla; such destruction has also been found in excavations in the settlement area.[3] With regard to later phases, both written sources and archaeological evidence are rather fragmentary. This would suggest that the territory of *Picentia*, in various ways that are yet to be defined, was part of a larger territorial and administrative area, in which the colonies of *Paestum* and *Salernum* played central roles at different times.

In this context it is fundamental to focus on the events connected to the use and exploitation of the *ager publicus populi Romani* launched in the whole of Campania at the end of the third century BC, culminating in the numerous political/military and juridical changes that took place up until the formation of *municipia*, at the beginning of the first century BC.

Although the evolution of the agrarian landscape in the valley of Pontecagnano may seem very complex, the objective of this article is to provide an overview of the nature of the agricultural organisation and the forms of settlement and continuity in associated landscapes, in particular, those dating between the mid-Republican and the Imperial periods.

The territory

A general picture of agricultural landscape organisation was given in a study by M. Guy and S. Stefan for the Soprintendenza Archeologica di Salerno in 1990. This study employed both satellite images (Landsat TM, Spot Image) and multi-scale and multi-temporal aerial photographs, and led to the identification of two main systems of land subdivision: the first orientated north 31°-33° east, and the second north 45° east.

The first system north 31°-33° east has been dated by French researchers to the date of the foundation of *Picentia*. Its cadastral form is made up by parallel bands (canals, ditches, and general alignments) laid out at regular intervals of 210m following, according to the authors, the Greek model.

The second agrarian division north 45° east is instead more complex: this layout has been defined as generically earlier than the north 31°-33° east.

This research has benefited from new evidence made available from recent excavations carried out during works for the Salerno-Reggio Calabria motorway in the area of the ancient settlement of Pontecagnano, and it was seen to be a good moment to re-interpret the cartographic and aerial photographic documentation[4] of this area in order to identify forms and methods of settlement and the agricultural exploitation of this part of the territory.

A preliminary phase of this study consisted in recognising all the anomalies, such as recent land divisions, deep ploughing, land reclamation, infrastructure works, that have all contributed to significant changes to the agricultural landscape and caused the disappearance of numerous small properties.

During the second phase of work, the authors proceeded through different levels of logical analysis (characterisation, identification, classification, deduction) to naming all simple recognisable elements, which correspond on the ground to evidence for subsistence strategies directly and/or indirectly related to the ancient landscape.

[1] For a discussion on the *Ager Picentinus* cf. to Cerchiai 1995.
[2] For a recent re-interpretation of the sources and phases of *Picentia* see Giglio 2003, pp. 119-131.
[3] See Cinquantaquattro 1999, pp. 121-171.

[4] In order to integrate the new archaeological data into its ancient topographical context, the documents relating to the flights that carried out the aerial photographic campaign needed to be studied: RAF, scale 1:10000, on August 1943 (photos 4166-4170); WWS, scale 1:38000 ca., in July 1955 (photo 11337); IGM, scale 1:15000 ca. in 1956. On the basis of these documents, both analogical (reading the stereoscopic pairs, identification of marks etc.) and digital analyses were carried out with the aid of a WS Stereographic and a Vexcel 5000 scanner and the creation of a DTM (digital terrain model).

This work made it possible to confirm the existence of a first set of coherent alignments, orientated north 31°-34° east, from here onwards defined as cadastre "A" (Fig. 1). They are mostly seen on the north-eastern side of the plain, where it rises up to join the foot hills of the Picentini mountains. This set of alignments can be seen to have at least five main "segments" in a north-east direction with a recurring distance (with a higher percentage than others) of 210-213 m. Another frequently occurring distance is 420 m., i.e. the measurement of two narrow bands of 210 m. Four main stretches are identifiable on a north–west orientation; although they are structurally coherent they do not allow for reliable metric divisions to be identified.

On the whole, on the basis of the layout of the principal axes, the 'form' of the landscape can be defined – it was organised along elongated land parcels according to the isoclinic elements facing north-east.

In practice cadastre "A" generally confirms both the orientation and the distances already recognised by the French scholars.

With regard to the chronology, proposing absolute dates for landscape structures is always problematic and complex. Such difficulty is caused by their potential for change (of a legal and functional nature, etc.) and for decay that often prevents the survival of intermediate divisions and, sometimes, even of major boundaries. However, the excavation of two trenches (Tr. 14, Tr. 11) revealed the presence of two beaten earth roads on a north 31° east orientation; on the basis of the stratigraphic sequence and the material recovered, archaeologists have dated the successive layers from the Archaic period to the third quarter of the fourth century BC.

Analysis has also identified a second system orientated north 45° east (conventionally identified as cadastre "B", Fig. 1) already identified by Guy and Stefan. This system is in some cases legible for long stretches (984, 1357 m.), and has been identified all over the plain, particularly along the coast between the Picentino River and the Torrente Asa. Beyond, in the area of the so-called Lago Piccolo, land reclamation works and the chaotic land divisions have significantly affected the possibility of identifying reliable isoclinic orientations. However, in the area between the Picentino River and the Torrente Asa, 13 segments oriented according to longitude and seven according to latitude can be fairly reliably distinguished. It is also worth noting, how these are arranged generally, at regular intervals according to an average distance of 35-37 m., corresponding to an *actus quadratus,*[5] or corresponding to distances referable to fractions or multiples – for example, 11.80 m (40.04 *pedes*) or 70.71-71.72 m., 183 m. (239.12-242.54, 618 *pedes*). These seem to form part of a wider landscape subdivision, organised by elongated isomorphic allotments, following a particular form of geometric models typical of the *strigatio/scamnatio*.

A proposed date for cadastre "B" is from the sixth century BC. Once again archaeological data provide more precise dating elements: in a few of the excavation trenches of the motorway (5-7), roads and canals linked with necropoleis areas and aligned with cadastre "B" were excavated dating to the period between the first half of the fourth century BC and the end of the third century BC.

Contrary to the chronology proposed by the French scholars, cadastre "B", whilst dating to the Archaic period, certainly continues to be in use up until the beginning of the third century BC.

It can be generally hypothesized that the two landscape divisions, both adapted to geomorphological features of the land and to the available natural resources, had their origins in the sixth century BC and coexisted for a long period until the cadastre "B" system took over at the foundation of *Picentia*.

As mentioned above, cadastre "B" seems to resemble the *strigatio/scamnatio* scheme and, might even represent not only a reference system for the organisation of the landscape but also for the "urban" layout of *Picentia* itself. Archaeological investigations carried out within the area of the ancient settlement have, in fact, allowed a complex to be excavated that is structured around elongated blocks (in *strigas*) with a north-east/south-west orientation, which is coherent with the overall landscape structure.

Lastly, a third system was identified (cadastre "C", Fig. 1) which was like centuriation and orientated north 50° west. This system, which appears more articulated and widespread than the previous ones, is clearly recognisable thanks to the many fossilised alignments that have survived both in the modern boundaries of cadastral allotments and in current roads, both main roads and farm tracks.

Measurement criteria were applied to this last system to help to better and more precisely understand the overall development in plan, its functions and internal dynamics. It was hoped that this would avoid ending up with having only revealed a Roman cadastre by highlighting orthogonal topographical lines on an aerial photo or on a map.[6]

[5] It was decided to relate the *Actus* (= 120 feet) to a measurement of 35.52 m.

[6] For a discussion on the issues relating to metro-morphological analyses and on the fossilisation of cadastres refer to: F. Favory, 'Morphologie agraire isocline avec une limitation romaine. Acquis et problèmes', in G. Chouquer (ed.), *Les formes des paysages. 2. Archéologie des parcellaires, Actes du Colloque d'Orléans (mars 1996)*, Paris 1996, pp. 193-200, including bibliography; G. Chouquer, 'Dégradation et fossilisation de cadastres centuriés, in M. Clavel-Lévêque (ed.), *Cadastres et espace rural. Approches et réalités antiques,* Table ronde de Besançon (mai 1980), Paris 1983, pp. 137-157.

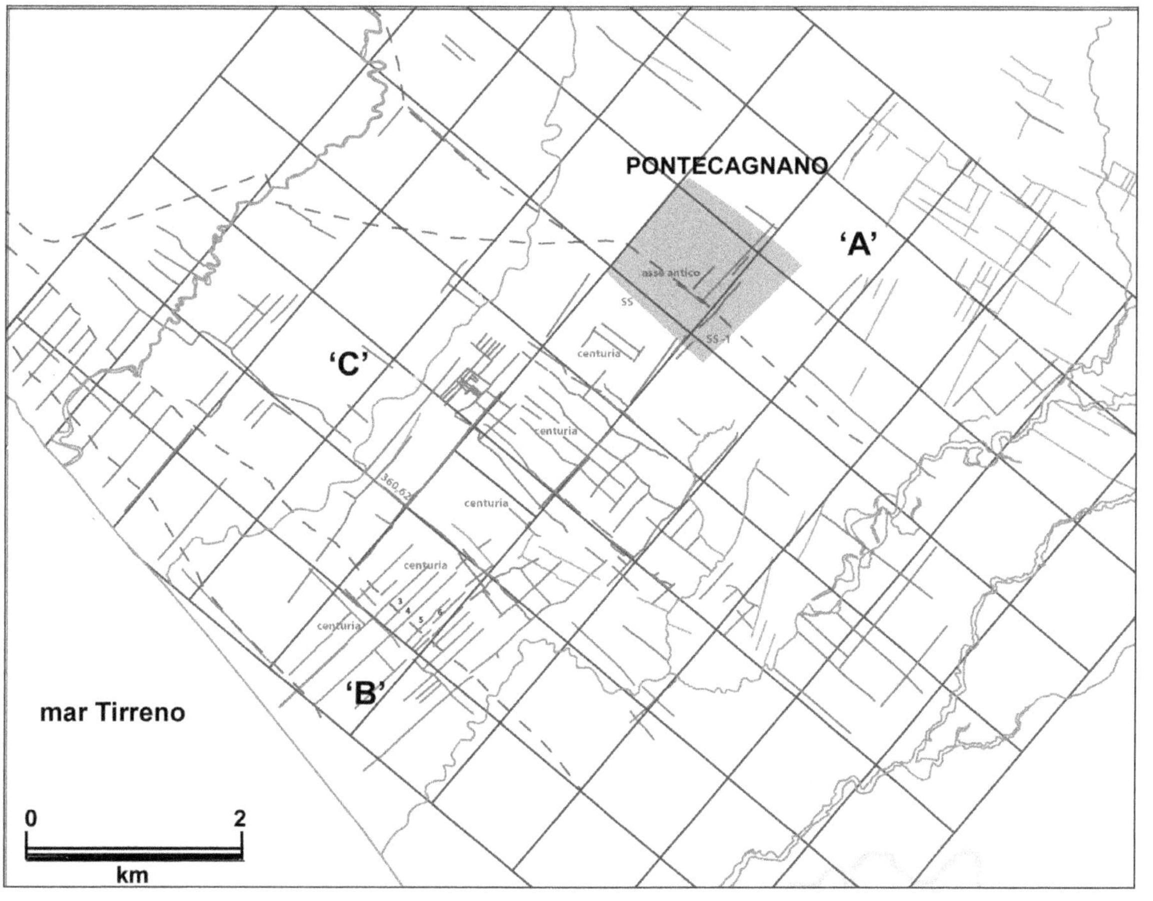

Figure 1 *Pontecagnano (in the Provincia di Salerno). Graphic reconstruction of cadastral systems identified on aerial photograph no. 11337, flight WWS, average scale of around 1:38000, July 1955.*

Such an approach produced immediate results allowing us to grasp the significance of a considerable agricultural planning of the landscape dated to somewhere between the second half of the second and the beginning of the first century BC.

Such planning seems moreover to have integrated and absorbed previous subdivisions and in particular that of the system oriented north 45° east, employing them in a new structure based on a centuriated *pertica.*[7]

The methodological approach used a theoretical reference system calibrated on centuries with a 20x20 *actus* module (Fig.1).

The theoretical value of a 20x20 *actus* corresponds to roughly 710 m. (a foot of 0.2957), although it must be noted that linear measurements carried out on the still visible alignments in the current landscape vary between 708 and 712 m.

In order to find a correspondence between the archaeological evidence gathered in the field and from aerial photo interpretation, an automatic grid based on the theoretical module of 710 m was used, superimposed on the 1:10000 and 1:5000 maps and on an aerial photograph (no. 11337 taken in 1955) on a scale calculated to be about 1:38000.

Metrological analyses of a more in depth nature were also carried out in order to calibrate the uniform image obtained from such superimposing various images and to verify correct proportions.

The area in which verifications were made is situated between the current settlement of Pontecagnano to the north, the Picentino River to the west, the Torrente Asa to the east and the coastline to the south where the continuation of the *limites* and alignments in general, are seemingly more evident.

On the available cartographic and photographic documents, measurements were made at regular intervals for each identified or hypothesised *centuria*. These correspond to the distances between vertical isocline elements and those between horizontal isocline elements. The results were not always uniform.

[7] In the *Gromatici Veteres* (p. 149, L. 10), the importance of the *consuetudo loci* and of the *natura loci* is often reiterated, when choosing the nature of the boundaries that shape the land cadastre.

On the aerial photo examined it was possible to observe a more or less constant variation of distances around 710 m.[8] Such a variation is even more noticeable on the cadastral alignments reproduced on aerial photogrammetric photographs; here it is important to note that some degree of imprecision in the measurements is due to the graphic documentation.[9] However, measurements carried out on the three different types of documentation and the verification of the data gathered, has allowed a more precise hypothetical reconstruction.

Generally, the values that occur most frequently are 712.17, 712.34, 711.49, 710.25, 708.56 m, which correspond to: 20.04, 20.05, 20.03, 19.99, 19.94 *actus*, in other words *centuria* of 20 *actus*, but also multiples of these.[10]

Such variation contrasts with the uniformity of the theoretical grid of 710m although it adapts well to a system of subdivisions of 20x20 *actus*.

Such values, associated to the overall photo interpretation of the intermediate alignments, reveal at the same time an internal, fairly complex, subdivision within the centuria. The definition and articulation of land divisions was partly thanks to the adoption of the partitions of previous systems. Such partitions, in particular those of the cadastre "B", appear to survive not only as active signs participating in the organisation of the new form of agrarian space, but also as fossil traces that were by then out of use (Fig.1).

This hypothesis appears even more certain if we take into consideration a sequence of measurements that constantly emerge, which are linkable, directly or indirectly, to those internal divisions (*rigores* and *limites intercisivi*) that are typical of the creation of a cadastre system, that could be defined as *per strigas et scamna in centuriis*. Such data is obtained from a recurrent series of identified alignments, more specifically those found near the area occupied by the ancient settlement of Pontecagnano, particularly in the area just beyond the railway tracks. In this stretch it was possible to identify a recurring measurement which varies from between 70.33 m. and 72.98 m. This value, comparable in particular with the isoclinic features on a north-west/south-east direction, coincides almost perfectly with regular divisions of two *actus quadrati* (one *iugerum*) and with distances in "standard" *pedes*.[11]

With regard to the chronology of this last cadastre we must proceed with caution both due to the limits imposed by the methodological approach but also due to the uncertainty, to date, of the archaeological data. However, on the basis of the evidence considered, the stratigraphical relationships that exist between the previous alignments (cadastres A and B) and, the analogies with the work carried out by A. Rossi at Salerno[12], it is possible to propose a date ranging from the third quarter of the second century and the first century. BC.

It is in this phase that the agricultural changes imposed by the Romans take an important step forward.

In order to more precisely contextualise cadastre "C" of Pontecagnano, it can be said that it is the third quarter of the second century BC that represents a crucial moment. It is during this phase that a crisis occurs in the production system based on small and medium production centres in favour of large estates and pasturelands, product of a more or less legal and widespread *occupatio* of public land by the wealthier classes and of a contemporary emigration of small landowners towards Rome and the centre/north of Italy.[13] These events forced the Senate to pass the *Lex Sempronia Agraria* and the subsequent reform attempts promoted by the Gracchus family. The special commission of *triumviri agris dandis adsignandis* in fact was commissioned to divide, according to the centuriation model, a vast area,[14] regulating its use for agricultural purposes and recreating the conditions for a "rebirth" of small and medium properties without damaging the by-then established interests of the *latifundium*.[15]

There is a lot of evidence, even archaeological, for both the activities of the commission and of the huge infrastructural works carried out at this time; one need only consider the Graccan legislation with its landscape definitions or the reference to the commission of Sant'Angelo in Formis[16] or those found in the Vallo di

[8] The figure 710 m., at times highlighted in a punctual manner, was obtained by calculating the mean value of the closest, most recurring distances (708-712 m.).

[9] For a discussion on geometrical issues and inaccuracies, cf.: M. Guy, 'Géométrie des parcellaires réguliers: le problème des dimensions', in G. Chouquer (ed.), Les formes des paysages. 2. Archéologie des parcellaires, Actes du Colloque d'Orléans (mars 1996), Paris 1996, pp. 180-192, and in particular pp. 180-182.

[10] Distances equal or immediately comparable to multiples of 20 *actus*, for example, 1424.15 equal to 40.09 *actus*, were systematically verified both among the horizontal isoclinic elements and the vertical ones. But also rates of 1200 *pedes* returned by linear distances of 354.97, 369.8, 368.33, in other words 10 *actus*.

[11] It was possible in fact to verify a series of distances in *pedes* - 120, 480, 600, 1200, 2400 – comparable to the pattern of the so-called "common" distances quoted in the work *Gromatici Veteres*, p. 11=L. 109, 211-218, 243, 254, 258.

[12] Rossi 1999, pp. 17-26.

[13] Gabba 1972, pp. 73-112.

[14] For a discussion on the controversial intervention carried out in the *Ager Campanus* cf. A. Russo, 'Nota minima sui Gracchi e l'*Ager Campanus*', in *Ager Campanus*, pp. 41-43; for a discussion on the Vallo di Diano cf. Franciosi 2002, 195-228; including annotated bibliography.

[15] For these aspects see A. Russo, 'Tiberio Gracco e la riforma graccana', in *La romanizzazione*, pp. 161-193; including bibliography.

[16] C.I.L. V, 1^2, 640

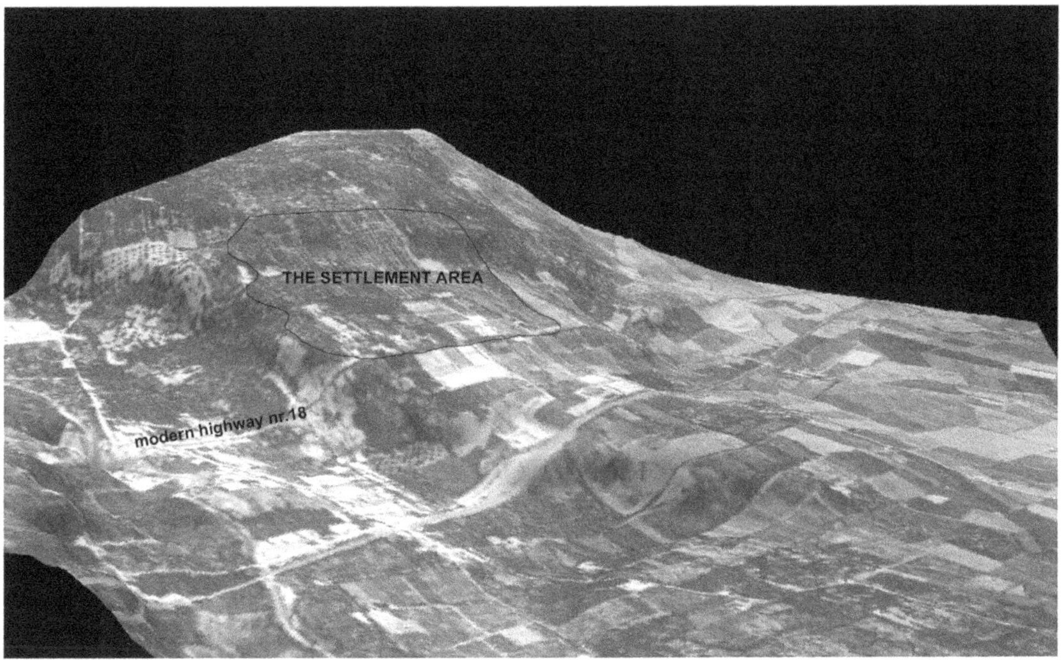

Figure 2 *Pontecagnano (in the Provincia di Salerno). DTM with draping of aerial photograph no. 11337, flight WWS, average scale of around 1:38000, July 1955. Note the plateau on which the ancient settlement once rose.*

Diano,[17] or to that exceptional document the *Lapis Pollae*, in which the construction of the consular road *Regio-Capuam* is praised along with the restitution of land to farmers by P. Popilius Laenas in 132 BC.[18]

The authors believe that it is within such dynamic processes, that the cadastral organisation seen in the Pontecagnano plain, should be placed. This should be considered in a single system integrated with the centuriation of *Salernum,*[19] the town on which it probably depended politically and administratively. Cadastre "C" shares its orientation and overall design, at the same time preserving as a peculiarity the internal divisions of previous layouts.

The subsequent destruction during the Sullan period of *Picentia* and its partial recovery during the Augustan period (in the first century BC) as shown by archaeological excavations, presumably did not bring about any major changes to the agricultural organisation of the area in light of the fact that, in the vast majority of cases, the lands assigned to the military took the form of previously defined divisions, such as ex-Latin colonies and small towns. It was easier to divide the land according to an established *modus.*[20]

The settlement area

The ancient settlement is located in an area surrounded on at least three of its sides by necropoleis, and was defined by the Fondazione Lerici during two geoarchaeological surveys.[21] The only archaeological remains are two pre-Roman sanctuaries and an area of the central part of the settlement where the remains of the city of *Picentia,* founded by the Romans in 268 BC, were found, and which currently form a small archaeological park.

Stereoscopic analysis of aerial photographs and photogrammetric reconstruction[22] have allowed us to identify the possible perimeter of the settlement. The

[17] V. Bracco, 'Un nuovo documento della centuriazione graccana. Il termine di Auletta', in RStorAnt, IX,1979, pp. 29-37 including bibliography. For the Gracchan name of Atena Lucana see G. Chouquer, F. Favory, Les arpenteurs romains, Paris 1992, p. 19.

[18] For a discussion on the different hypothesis for interpreting this document and on the *Regio-Capuam*, see A. Franciosi 2002, pp. 195-228.

[19] One might imagine an organisation based on the *pagus* as an administrative unit and within it, residential spaces (the *vici*), and other areas where the administrative and military functions (*oppida,* castra) took place. During this phase the settlement of *Picentia* might have corresponded to a settlement similar to a *vicus*. In general, with regard to these issues, cf. L. Capogrossi Colognesi, 'Pagi sanniti e *centuriatio* romana', in *Ager Campanus*, 2002, pp. 77-93.

[20] See: Sic. Flacco, p. 156, 1-3 Lach.= p.120, 7-9 Th.; Lib Col. 234, 9 Fundi; 236, 15 Praeneste; A. Rudorff, 'Gromatische Institutionen', in *Gromatici veteres*, II, Berlin 1852, pp. 372-373.

[21] For these aspects see A. Rossi, 'La topografia dell'abitato antico', in M. L. Nava - M. Osanna, (a cura di) *Lo spazio del rito. Santuari e culti in Italia meridionale tra indigeni e greci,* Atti delle giornate di sudio (Matera 27-29 giugno 2002), Siris, suppl.I, 2005, pp.195-196.

[22] The analysis of historical aerial photos was carried out with the aid of a WS Stereograph and of a Vexcel 5000 scanner and the creation of a DTM (Digital Terrain Model). For a discussion on the methodology adopted, cf. F.Piccarreta - G.Ceraudo, *Manuale di aerofotografia archeologica. Metodologia, tecniche e applicazioni*, Bari 2000.

micro survey revealed the presence of some particular contours of the land which, as morphological anomalies, highlight how the settlement's boundaries exploited the margins of the natural high-ground.

Evidence of an almost quadrangular outline was noted on aerial photos which follows the limits of a low plateau that had already been inhabited during the Archaic period[23] (Fig.2).

In particular it was noted how to the south the area is defined, with respect to the necropoleis, by a change in levels corresponding to the route of the modern provincial main road no.18: some large squared travertine blocks pertaining to a large masonry construction have often been found along the modern road.[24] To the west, the settlement area is defined by another, more marked change in level which coincides with the boundaries identified by the Lerici survey.[25]

It is however, difficult to ascertain the definite presence of a potential fortification. The latter, made up of a large ditch (25m) and an *agger*, was identified instead on the eastern side, during recent excavations (in 2005), carried out during the widening of the A3 motorway and directed by the Università di Salerno. The defensive circuit appears to date to the end of the sixth century BC and seems to define an orthogonal urban layout based on roads with a north-east/south-west orientation, defining urban blocks of about 45 m wide with a clear phase when the buildings were reconstructed, dating to the first half of the fourth century BC,[26] a phase which should be placed within a wider historical context.[27]

Subsequently, at the beginning of the third century BC, along the northern sector, the settlement is abandoned and is taken over by the suburban agricultural landscape of the Roman settlement.[28]

In conclusion, as has emerged in this study and as demonstrated elsewhere,[29] Roman *Picentia* does not seem to coincide perfectly with the boundaries of the Archaic and Classical city, and seems to be much smaller in size and limited to the central settlement area as defined by the Lerici survey.[30] After the foundation of the Roman colony in 268 BC, there was a reduction and a reorganisation of the urban area[31] accompanied by the definitive abandonment of the sanctuary areas and the workshop area.[32]

The new urban layout, although maintaining the same orientation as affected by the natural slopes of the land, seems to focus on the central area of the extensive settlement plateau, in the place where its most important road axis is preserved which, starting from the second century BC, may have become part of the *Regio-Capuam* consular road.

Another important stage in this territorial reorganisation occurred after the second Punic War,[33] when colonies of *cives Romanorum* were founded along the lower coast of the Tyrrhenian Sea to control the extensive territories taken from the Campani, as mentioned by Livy.[34] In this perspective, the colony of *Salernum* of 194 BC becomes the centre of the new territorial organisation, encompassing in its wider district, part of the territories of the Picenti and consigning their settlement to a minor role.

The new layout certainly involved a new agricultural organisation (cadastre "C") that was not directly linked to the single centuria allocated to colonists but more likely linked to the wider district.[35] The centurial system, in fact

---

[23] It is highly likely that the area reconstructed on the basis of aerial photo interpretation is to be identified with the Late Archaic and Classical layout. Although in the current state of research, this hypothesis cannot be backed up by conclusive evidence.

[24] See A. Lupia, 'I santuari di Pontecagnano: paesaggio, azioni rituali e offerte. La tipologia delle offerte nel santuario di Apollo', in M. L. Nava - M. Osanna, (eds.) Lo spazio del rito. Santuari e culti in Italia meridionale tra indigeni e greci, Atti delle giornate di studio (Matera 27-29 giugno 2002), Siris, suppl. I, 2005, footnote 56.

[25] On this side, the settlement does not appear to extend beyond the sanctuary area of Via Verdi, though in the current state of research it is difficult to trace a northern boundary: some indications were given by recent investigations carried out during works for the widening of the A3 motorway, which seem to confirm the peripheral nature on the other side of the motorway route. For the motorway excavations cf. Tocco Sciarelli 2003.

[26] Evidence for the settlement emerged during excavations carried out by a team led by Prof. G. Bailo Modesti of the Università 'L'Orientale' di Napoli and by Prof. L. Cerchiai of the Università di Salerno. For the fourth century BC phase cf. Giglio 2003, which contains a wide ranging discussion of sources in the context of the excavation data from the urban area.

[27] See, for example, the case of some Campanian cities: *Nuceria*, Pompeii, *Atella*, *Abella* e *Surrentum*, cf. Cerchiai 1995, pp.211-224.

[28] Even in the oriental sector (in the area between Via Cavalleggeri to the west and Via Palinuro to the east) the presence of a *fullonica* dating to the second century BC testifies to the peripheral nature of this area in comparison to the overall urban context. The presence of workshops found in this sector testified it to being peripheral already in the seventh century BC and through to the fifth century BC. (De Natale 1992, p.3).

[29] For a discussion on the boundaries of *Picentia* refer to the hypothesis formulated by Cinquantaquattro on the basis of the excavations (Cinquantaquattro 1999, pp. 121-122). Most recently, Giglio 2003.

[30] For a discussion on the extent of *Picentia* see Cinquantaquattro 1999, pp.153-154.

[31] Giglio 2003, pp. 120-121. For a reconstruction of the layout cf. Cinquantaquattro 1999, p. 153 and ff.

[32] Most recently Giglio 2003.

[33] See Strabo V, 4, 13, C 251.

[34] See Livy 34, 45, 1-2.

[35] For a discussion on the agricultural role of Roman colonisation, see G. G. Mason, 'The agrarian role of coloniae maritimae: 330-241 B.C.', in *Historia*, XLI, fasc. 1, 1992, pp. 75-87. The author places particular focus on the useless division proposed by Salmon (E. T. Salmon, 'The Coloniae Maritimae',

main seen on the coastline, appears to be an extension of the system already identified along the Salernitan coastal area,[36] which as seen above, by integrating an earlier layout had achieved an extensive and widespread system of organisation.

Following the eruption of AD 79, the *Picentia* settlement shrank further and appears to survive only along the route of the consular road.[37]

Acknowledgements

The authors wish to thank Prof. L. Cerchiai who allowed this fascinating research to be carried out, and the excavation team from the Università di Salerno and the Università di Napoli "L'orientale" who carried out the excavations for the enlargement of the A3 motorway (M. Viscione, C. Pellegrino, M. D'Andrea, D. Alfano, E. Caracciolo, P. Aurino, F. Basile, M. Mancusi and C. Schiano di Cola).

References

Ager Campanus G. Franciosi (a cura di), Ager Campanus, La storia dell'Ager Campanus. I problemi della limitatio e sua lettura attuale, Atti del Convegno Internazionale, Real Sito di S. Leucio (8-9 giugno 2001), Napoli 2002.

CERCHIAI I. 1995 *Campani*, Roma 1995.

CINQUANTAQUATTRO T. 1999 'Pontecagnano (Salerno). Saggi stratigrafici nell'abitato antico', in *BA* 28-30, 1999 (1994), pp. 121-171.

CINQUANTAQUATTRO T. 2001 *Pontecagnano. II. 6. L'agro picentino e la necropoli di località Caselle, AION ArchStAnt Quad.* 13, Napoli 2001.

De NATALE, S. 1992 'Pontecagnano II.2 la necropoli di S. Antonio: prop. ECI. 2. Tombe della Prima Età del Ferro, in *AION ArchStAnt, quad.8*, Napoli 1992.

FRANCIOSI, A. 2002 'La romanizzazione nel Vallo di Diano in età graccana', in *La romanizzazione*, pp. 195-228.

GABBA, E. 1972 'Urbanizzazione e rinnovamenti urbanistici nell'Italia centro-meridionale del I sec. a.C.', in *SCO,* XXI, 1972, pp. 73-112.

GIGLIO, M. 2003 'Picentia, fondazione romana?', in *AION ArchStAnt* 8, 2001 (2003), pp. 119-131.

GIGLIO, M. 2005 'Picentia: fenomenologia di una trasformazione', in G. Vitolo (a cura di), *Le città campane fra tarda antichità e alto medioevo*, Salerno 2005, pp. 379-392.

Gromatici Veteres F. BLUME, K. LACHMANN, Th. MOMMSEN, A. RUDORFF, *Die Schriften der römischen Feldmesser (Gromatici Veteres), I: Text; II: Erläuterungen*, Berlin 1848-1852.

LAGI De CARO, A. 1995 'Pontecagnano', in R. Cantilena (a cura di), *Caronte. Un obolo per l'aldilà*, 1995, pp. 345-346.

La romanizzazione G. FRANCIOSI (a cura di), *La romanizzazione della Campania antica,* 1, Napoli 2002.

ROSSI, A. 1999 'Alcune considerazioni sul territorio di *Salernum*', in *Apollo* XV, 1999, pp. 17-26.

RUSSO, A. 2002 'Nota minima sui Gracchi e l'*Ager Campanus*', in *Ager Campanus*, pp. 41-43.

TOCCO SCIARELLI, G. 2003 'L'attività archeologica della Soprintendenza delle province di Salerno, Avellino e Benevento nel 2002', in 'Ambiente e paesaggio nella Magna Grecia' *Atti del XLII Convegno di Studi sulla Magna Grecia*, Taranto 5-8 ottobre 2002, Taranto 2003, pp.631-638.

in *Athenaeum,* n.s., XLI, 1963, pp. 3-33) between a first phase of colonisation during the fourth century and the third century BC, characterised by a military-strategic role and, a second phase, during the second century BC, in which the *Coloniae Maritimae* play both an agricultural role whilst also controlling the territory.

[36] See Lagi De Caro 1995, pp. 345-346 and most recently, Rossi 1999, p. 24, where the territory of *Picentia* is hypothesized as pertaining to the colony of *Salernum* from the second century BC.

[37] For a discussion on Imperial and Late Antique *Picentia* see Giglio 2005.

Reality-based 3D modeling of the Angkorian temples using aerial images

Till Sonnemann, Martin Sauerbier, Fabio Remondino, Gerhard Schrotter

Institute of Geodesy and Photogrammetry, ETH Zurich, Switzerland - <msb><fabio><gerhard>@geod.baug.ethz.ch

Introduction

The documentation and visualization of large Cultural Heritage sites has become a very interesting and challenging research field in the recent years. Photogrammetry, the science of obtaining reliable measurements from images, is nowadays much more frequently used to generate visualization and animation products, documentation material for conservation, restoration and future planning as well as traditional maps of the investigated areas. This is mainly due to two facts. On the one hand, UNESCO is paying more attention to large cultural and natural sites: the 'Open Initiative' partnership was started with different space agencies to support and assist in the monitoring and documentation of World Heritage sites, natural hazards and for sustainable development using satellite data. On the other hand, a wide array of new technologies is nowadays available which greatly support the efficient generation, administration and analysis of such models. From the technological point of view data from high-resolution satellite sensors and digital aerial cameras are increasingly becoming available. Moreover, the automation of the photogrammetric processing chain has made great progress recently and different commercial software is nowadays available for non-expert users. In addition, Geographic Information System (GIS) technology provides many tools for data administration and analysis.

In this paper we report a study if aerial images are a useful source for a detailed 3D modeling of a historical site such as Angkor, Cambodia. A series of aerial images acquired in 1997 during a Japanese flight campaign over Cambodia covering the main temples of the old city of Angkor were chosen for this purpose. The images were oriented using additional points from a 1:5000 GIS model map. Afterwards, manual stereo measurements were used to create a digital terrain model (DTM) and an orthoimage. Four main temples and several other visible structures within the area were reconstructed in 3D. In the following sections, after a geographical and historical introduction to the Angkor site, the related works in the archaeological area will be summarized. Descriptions of our photogrammetric processing and modeling as well as conclusions end up this paper.

The UNESCO WH site of Angkor

Angkor is located in the central part of Cambodia, north of the Lake Boeng Tonle Sab, in the province of Siem Reap. The historic area spreads out over a largely forested and completely plain area of approximately 400 square kilometers, and only a few hills interrupt the plain. Angkor Archaeological Park contains the magnificent remains of the different capitals of the Khmer Empire, from the 9th to the 15th century. Many of the hundreds of ruins are still covered by vegetation and only a few, such as the famous temples of Angkor Wat and Bayon (at Angkor Thom) are under cultural heritage preservation. In 1992, the Angkor Archaeological Park was included into the UNESCO World Heritage List. Nowadays, UNESCO has set up a wide-ranging program in order to safeguard this symbolic site and its surroundings, nevertheless, many problems concerning its maintenance still remain.

Historical background

The early culture of Angkor, nowadays known as Cambodia, had socially and religiously developed under Indian influence and formed a society based on Hinduism. The rise of the Khmer empire began 802 AD, when they liberated themselves from Java to become the Kingdom of Kambuja. In its largest expansion, the Khmer ruled over an area that included the actual Cambodia and Thailand and even parts of Laos and Myanmar. Arts and architecture was influenced by Hinduism and later by Buddhism; this mixture of styles is visible at several temples, as some of the art also had been destroyed or redesigned with a new ruler. Today the only visible structures of the ancient Khmer empire are the sandstone built sanctuaries, because "[...] domestic dwellings have been built in [...] predominantly perishable materials" (Jacques, 1997) and not much remains today of the large cities which had spread around the temples but postholes.

The first Khmer capital surrounded the area's only larger hill with the central sanctuary Phnom Bakheng, built in 889-910, towering 60 m above the surroundings. Baphuon temple was built in the 11th century over a small mount north of Phnom Bakheng. Chinese historic documents state that the collapsed ruins were in the time of the Khmer covered with bronze plates. In the years 1113-1150 Angkor Wat was constructed, and until now it

Figure 1 *Map of central Angkor, showing coverage of aerial images. Ruins of Bayon temple (1181-1291).*

is one of the largest single religious structures in the world. As the translation of the name "the capital which is a temple" reveals, it contained a royal palace as well as the main sanctuary which was probably designated to Vishnu (Jacque, 1997). The structure gives an idea about Hindu cosmology with the five central towers standing for Mount Meru, the Hindu Olymp, the outer walls representing the mountains at the worlds end and the moat as the ocean beyond. The capital was relocated various times until Angkor Thom, "the great town", just north of the hill of Phnom Bakheng was founded 1181. The temple of Bayon was constructed in the years 1181-1219 a few kilometers north of Angkor Wat and close to Baphuon, in the centre of Angkor Thom as its main sanctuary; a magnificent piece of architecture in the style of Mahayana Buddhism that symbolized a Lotus-Blossom.

(Jacque, 1997). As impressive as the Khmer architecture is the water system with huge reservoirs and large city moats that surrounded the cities of Angkor Thom and Angkor Wat and canals which connected the occupied areas. They were probably not only for watering but used to transport the sandstones to the construction sites. Due to this well developed water system, several harvests per year were possible and this led to extreme wealth.

The last known stone inscription dates from 1327. It is assumed that due to the adoption of Theravada Buddhism in the 14th century, the ruler had lost the godlike position he had been in before. Therefore, no more large sanctuaries were built for his obeisance and the great architecture and the canal system was not maintained anymore. The rise of the Thai and their conquest of Angkor in 1431 is "the traditional (and probably inaccurate) date of what has been called the *abandonment of Angkor*" (Jacque, 1997), when the Khmer rulers had to move their capital to the southern part of their reign. There were several attempts to inhabit the area again, but droughts and floods and the advancing forests had probably destroyed the unguarded water system and within the centuries forests covered the temples. Although the ruins were still used as a sanctuaries by the local people, the Europeans the temples were only known due to tales from missionaries or explorers for several hundreds of years. French archaeologists visited the site in the 19th century. Since then various groups have excavated, reconstructed and preserved parts of the temples. A difficult task not only because of the vast area and the huge structures but also because of the political struggle Cambodia had been in the last century and which interrupted the work for several decades.

Related works in the study area

In the civil war the ruins were severely damaged, mostly due to art robbery. Nowadays the landscape has changed dramatically because of population growth and increasing tourism. This makes further protection of the historic sites necessary.

Several international teams have been or are working in cooperation with UNESCO, the local APSARA Authority and Cambodian universities on projects to safeguard the temples of Angkor, which include exchange projects and training of local experts and students. Besides the reconstruction work, a variety of remote sensing and photogrammetric methods were used to digitize the temples for cultural heritage purposes. A selection will be briefly introduced in the following (Tab.1):

- *Japanese Government team for safeguarding Angkor* (JSA, 1995-2005): within this project the northern library and other parts of Bayon were restored. A JICA flight campaign from 1997 produced detailed regional maps (1:5000). Keio University (Japan) used satellite data and kinematic GPS for a 3D digital reconstruction of Bayon and Angkor Wat (Watanabe et al., 1999). Sophia

University (Japan) laser-scanned several statues, and Tokyo University scanned the whole Bayon for a complete high-resolution reality based model (Ikeuchi et al, 2004, Kamakura et al., 2005). Moreover, a field campaign with a balloon flight delivered pictures of Bayon.

- *German APSARA Conservation Project* (GACP, 1995-2000, University of Applied Sciences, Cologne) works on the preservation of the APSARA relief of Angkor Wat (Leisen, 2000).

- The *Angkor Project Group* (APG, University of Heidelberg, Germany and Royal University of Fine Arts and Architecture, Phnom Penh, ongoing since 2005): concentrates on 3D-CAD modeling and terrain visualization based on 2D maps and digital images, focusing on the temples Bayon and Phnom Bakheng for reconstruction purposes.

- The *Banteay Srei Conservation Project* (BSCP, 2001-2005): in collaboration with the Swiss Agency of Development and Cooperation; the aim was to restore the Banteay Srei temple (Gerber & Sauer, 2003).

- *Greater Angkor Project* (GAP, 2001-2010, École française d'Extrême-Orient (EFEO), and University of Sydney): a variety of studies have been launched about the decline of Angkor which include mapping missions (RADAR data and aerial images were acquired in collaboration with JPL-NASA), on site investigation about the water system and the perished cities of the Khmer (Fletcher et al., 2003 and 2004) and detailed 3D-modeling of several sites.

- *Living with Heritage program* (LWH, inter-disciplinary program of the University of Sydney): the focus lies on present cultural heritage preservation.

Project	Internet Appearance
APSARA	http://www.autoriteapsara.org
JSA	http://angkor-jsa.org/
GAP	http://acl.arts.usyd.edu.au/angkor/gap/
Sophia U	http://angkorvat.jp
U Tokyo	http://www.cvl.iis.u-tokyo.ac.jp/research/bayon/
APG	http://scotty.iwr.uni-heidelberg.de/
GACP	http://www.gacp-angkor.de/
EFEO	http://angkor.wat.online.fr/

Table 1 *Research groups that work presently or have been working in Angkor and appear in the web [last visited September 2006]*

Photogrammetric modeling at ETH

The institute's first work in the Angkor area was the digital reconstruction of a smiling Buddha tower in Bayon (Visnovco "Topographic Mapping of the Angkor Archaeological Area". A Leica RC 20 aerial camera with a focal length of 153.18 mm was used for the project. Approximately 350 images were acquired from a flight altitude of about 780 meters above ground at an image scale of 1:5100. 60 mostly black and white copies were available for our purpose. The images cover an area of about 5.3 x 5.4 square kilometers, starting from the east end of the large water basins of the West Barray to the east of the Angkor Thom city moat. The most northern pictures include the northern moat of Angkor Thom, to the south they are bounded by the southern moat of Angkor Wat. The photos are arranged in seven strips with seven to ten photos each. Only eleven of these pictures are colored and those cover the main temples Bayon and Angkor Wat.

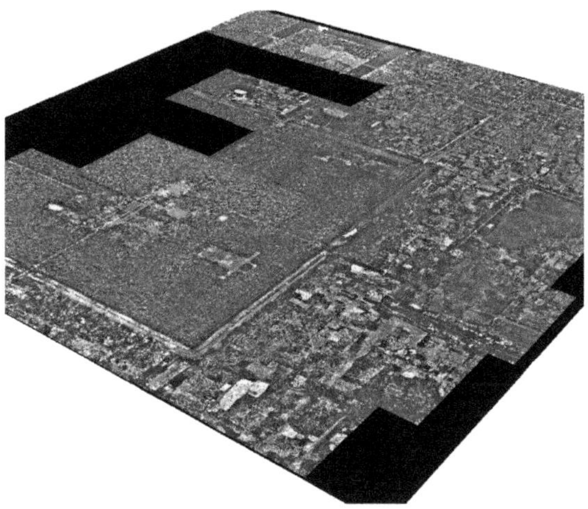

Figure 2 *The orthomosaic combines all aerial pictures that were included in the modeling. In the front lies the northwest corner of the Angkor Thom moat, in the far back is the temple of Angkor Wat. The blackened areas represent missing images in the available data set or images that could not be included due to orientation problems.*

All photos were digitized with a resolution of 15 microns using a photogrammetric scanner Vexcel Ultra Scan 5000. Tests with a 10-micron scan did not offer a higher quality of the digitized image. Only three of 22 original ground control points (GCP) from the flight campaign were located within the area of interest, therefore additional GCPs had to be found. Besides this, the image strips have a poor overlap of approximately 20%; hence most of the area, including the majority of the temples, is covered only by one strip. With this little overlap, the resulting orientation achieved by aerial triangulation would have been unstable. A solution for additional GCPs was found by using the digitized GIS map created in 1997. After bundle adjustment, a final global orientation accuracy of σ_0 = 7.3 μm was achieved. The average correction of control points had a RMS horizontal error of 1.66 m and a vertical RMS error of 1.31 m. Due to the strong vegetation inside the moat of Angkor Thom, problems occurred during aerial triangulation for several images and in the successive generation of a digital terrain model (DTM) and orthoimages. Therefore, two images were not included. The resulting DTM, textured with a single large grey scale orthoimage which is overlaid with colored orthoimages in the areas of Bayon and Angkor Wat, is shown in Fig. 2. The reconstruction of the main

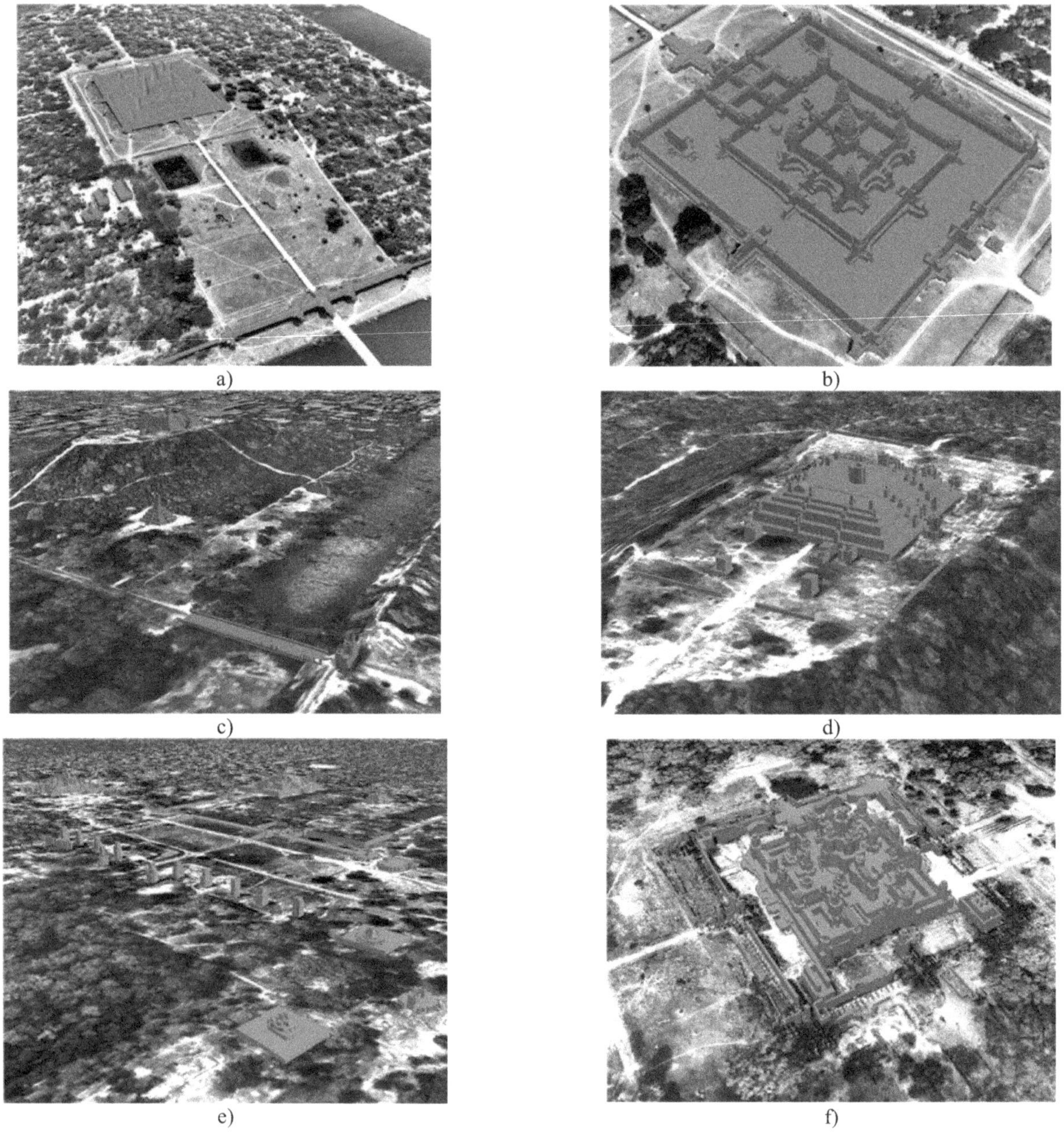

Figure 3 a) *Angkor Wat from bird view b) Angkor Wat c) Phnom Bakheng hill with surrounding temples d) Phnom Bakheng e) Angkor Thom with the temples of Bayon and Baphuon in the background f) Bayon temple.*

architectures of the Angkor area was afterwards performed by means of stereo-measurements on the oriented images.

Viewing and measuring in stereo allows a better reconstruction of the object of interest and increase the

reliability of the product. Mostly rooftops and inclined walls were outlined while, due to the low resolution images, the very detailed areas were simplified. To construct vertical walls of the temples, the measured horizontal outlined features were extruded from the top to the DTM base. Strong shadows that darken the north side of the buildings and the 20% overlap of the strips sometimes made a full model of the buildings impossible. Several attempts with different programs and additional ground control points extracted from the Japanese GIS map did not provide better results.

The texturing was accomplished with in house software. The cropped part of the aerial image together with the field of view and the TIN (Triangle Irregular Network) of facets provide the required information to start a GPU-based off-screen planar orthographic image mapping. The gained texture coordinates define the connection between the cropped image and the 3D model. The field of view is rendered using OpenGL and hardware acceleration. Preliminary results of the texturing of Angkor Wat show that with a small change in the view angle compared to the original image, and from distant view the texturing from aerial images is possible. A low resolution texturing is displayed on the institutes homepage.

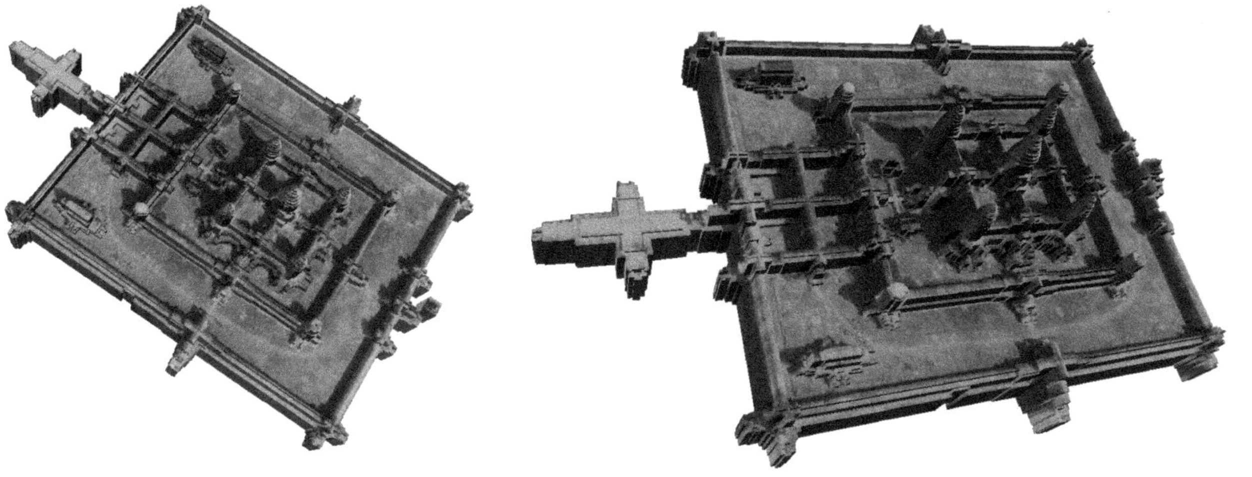

Figure 4 *Textured model of Angkor Wat.*

http://www.photogrammetry.ethz.ch/research/angkor3D.

For more detailed textures and for covering the side parts of the monuments, images from near surface photogrammetry would be of advantage.

A comparison with other data sources

The used aerial image data has a footprint of approximately 20 cm (pixel size of 15 μm). This value is not enough for an accurate and detailed digital reconstruction of non-rectangular or non-well preserved complex buildings. Very detailed structures had to be simplified in the modeling phase; neither entrances nor pillars with roofs can be displayed, as walls had been extracted from the rooftop to the ground. Certainly more accurate models could be generated with larger scale aerial images of using acquisition platforms like UAV or balloons and the use of additional terrestrial imagery. Nevertheless, the main structures could be successfully extracted with the available data, as it can be seen in Fig. 4. The question comes up if there is a possibility to use high-resolution satellite imagery for the modeling of similar sites. These image data are nowadays competing successfully with traditional aerial imagery, especially for the purposes of DTM generation or terrain study in such problematic areas. Nevertheless, regarding *Quickbird* or *IKONOS* data, as visible for example in Google Earth (Fig.5), even the high resolution of these images is not sufficient for architectural modeling, as it can be achieved with aerial imagery. Even with a set of stereo images, a detailed reconstruction of complex architectures as it is found in Angkor cannot be achieved at comparable accuracy. The Greater Angkor Project is currently preparing a super light plane flight campaign to take high resolution aerial images of the area, which could be an interesting comparison.

Further studies

At the Photogrammetry and Remote Sensing Group of ETH Zurich, the complete set of the aerial balloon images of Bayon is available. By including this data in the already processed one, parts of the 3D model could be improved significantly. Additionally, the complete set of aerial images was not used for modeling purposes yet. As there are various important archaeological sites within the surrounding area, this could be a promising approach to contribute to a better understanding of the temple allocation.

Conclusion

In this contribution we have presented the reality-based 3D modeling of some complex architectures of Angkor, Cambodia. The digital reconstruction, performed from a set of non-optimal aerial images, was overall successful, orthoimages of the complete area were derived and 3D models of several temples have been produced. The very few ground control points (GCP) in the original data, the strip overlap of 20%, the strong shadows in the images and the dense forest cover of the central part of Angkor Thom made the triangulation of all images and a full modeling of the structures an extremely difficult task. The very detailed and richly decorated architecture of the Khmer culture complicates the modeling with aerial photographs as well.

A complete series of colored images would have been of advantage for finding tie points for image orientation and would bring color into study and presentation tasks too.

Nevertheless, photogrammetric 3D-modeling is a useful and successful tool for archaeological landscape documentation of large cultural heritage sites, even if for simple visualization, preliminary and overall understanding of the area or documentation at smaller scale, satellite imagery could also be employed (Gruen et al., 2005). Together with terrestrial range- (Blais, 2004)

Figure 5 *Comparison of the aerial photos used for modeling (above) and satellite data (below, from Google Earth) showing Angkor Wat, and its southwestern corner, at the right the central part of Bayon. There is a visible difference in the resolution.*

or image-based (Remondino and El-Hakim, 2006) techniques, good results can be achieved and combined for cultural heritage site modeling.

The achieved results show that digital models resulting from photogrammetric processing are a more realistic and competitive edge versus CAD techniques and deliver useful information to safeguard and reconstruct an historic and natural landscape as a whole.

Acknowledgements

The team would like to thank JICA and the Kokusai Kogyo Co., Ltd for providing the aerial photos of the Angkor Area. Thanks to the GAP group of the University of Sydney for sending additional ground control points of the Angkor region which made the triangulation of the data possible. The first author would also like to thank Prof. Armin Gruen of ETH Zurich for giving him the opportunity to work on the project and the members of the Chair of Photogrammetry and Remote Sensing for their advice and constructive critics.

References

BLAIS, F. 2004: Review of 20 years of range sensor development. *Electronic Imaging*, 13(1): 232-240.

FLETCHER, R., EVANS D., TAPLEY I.J., AND A. K. MILNE 2004: Angkor: Extent, settlement pattern and ecology. Preliminary results of an AIRSAR survey in September 2000. *Bulletin of the Indo-Pacific Prehistory Association* 24: 130-8.

FLETCHER, R.J., BARBETTI, M., EVANS, D., THAN, H., SORITHY, I., CHAN, K., PENNY, D., POTTIER, C. AND T. SOMANEATH 2003: Redefining Angkor: Structure and environment in the largest, low density urban complex of the pre-industrial world. *UDAYA* 4: 107-21.

GERBER, M., AND A. SAUER 2003: *A high-precision VHR Laser range survey of the Banteay Srei temple complex (Greater Ankor, Cambodia).* Technical Report (3 parts), DEZA, Swiss Ministry of Foreign Affairs.

GRUEN, A., REMONDINO, F., AND L. ZHANG 2005: 3D Modeling and visualization of cultural heritage sites from high-resolution satellite imagery. *Proc. of the 25th Symposium of the European Association of Remote Sensing Laboratories (EARSeL),* pp. 611-620.

IKEUCHI, K., NAKAZAWA, A., HASEGAWA, K., AND T. OHISHI (2004): Representing Cultural Heritage in Digital Forms for VR Systems through Computer Vision Techniques. *GISIDEAS*, Sep. 16-18, 2004, Vietnam.

IKEUCHI, K., HASEGAWA, K., NAKAZAWA, A., TAKAMATSU, J. OISHI, T., MASUDA, T. 2006: Bayon Digital Archival Project. In: *Proceedings of the Tenth International Conference on Virtual Systems and Multimedia,* pages 334-343, November 2004.

JACQUE, C. 1997: *Angkor - Cities and Temples*, Thames and Hudson Ltd, London.

KAMAKURA, M., TAKESHI, O., TAKAMATSU, J. AND K. IKEUTCHI 2005: *Classification of Bayon*

Faces Using 3D models, Virtual Systems and MultiMedia, VSMM2005, Oct. 2005.

LEISEN, H. 2000: Problems of conservation and preservation of Apsaras in Angkor Wat. In: *Proc. of the Internat. Sympos. for the Conservation of Angkor Monument Sites, Tokyo October 21st, 2000,* JSA (ed.); pp 4-7.

REMONDINO, F., AND S. EL-HAKIM 2006: Image-based 3D modelling: a review. In: T*he Photogrammetric Record,* Vol.21 (115), September 2006, pp. 269-291.

VISNOVCOVA, J., ZHANG, L., AND A. GRUEN 2001: Generating a 3D Model of a Bayon Tower using non-metric Imagery. In: *Asian Journal of Geoinformatics,* Vol.2, No.1, September 2001, p.11-18.

WATANABE, N., USUDA, Y. AND S. KUBO 1999: Extracting Ruins in Angkor Region using Satellite Image Data. Conference Paper, *CIPA International Symposium 1999, October 3-6, 1999, Recife/ Olinda,* Brazil.

www.ingramcontent.com/pod-product-compliance
Ingram Content Group UK Ltd.
Pitfield, Milton Keynes, MK11 3LW, UK
UKHW061843041026
15591UKWH00015B/885

9 781841 719986